COMPLEX ECOLOGY
The Part–Whole Relation in Ecosystems

In Memoriam

GEORGE MASON VAN DYNE, 1932–1981

Early Systems Ecologist

Edited by
BERNARD C. PATTEN
University of Georgia

Coedited by
SVEN E. JØRGENSEN
University of Copenhagen

With a Foreword by
STANLEY I. AUERBACH
Oak Ridge National Laboratory

PRENTICE HALL PTR, Englewood Cliffs, New Jersey 07632

Library of Congress Cataloging-in-Publication Data

Complex ecology: the part–whole relation in ecosystems / edited by
    Bernard C. Patten : coedited by Sven E. Jørgensen : with a foreword
by Stanley I. Auerbach.
    p. cm.
    Includes bibliographical references and index.
    ISBN 0-13-161506-8
    1. Ecology—Simulation methods. 2. Biotic communities—Simulation
methods. 3. System theory. I. Patten, Bernard C.
II. Jørgensen, Sven Erik
QH541. 15.S5C65   1995
574.5'01'1—dc20   94-15451
                                                   CIP

Acquisitions editor: *Michael Hays*
Editorial/production supervision: *Raeia Maes*
Cover design: *Karen Marsilio*
Manufacturing manager: *Alexis R. Heydt*

© 1995 by Prentice Hall PTR
Prentice-Hall, Inc.
A Simon & Schuster Company
Englewood Cliffs, New Jersey 07632

Permission to publish the quotations of Dr. Van Dyne which appear at the beginning of each chapter has been courteously provided by Academic Press, Inc., Cambridge University Press, The Ohio Academy of Science, and University of Queensland Press.

All rights reserved. No part of this book may be reproduced, in any form or by any means, without permission in writing from the publisher.

Printed in the United States of America

10 9 8 7 6 5 4 3 2 1

ISBN 0-13-161506-8

PRENTICE-HALL INTERNATIONAL (UK) LIMITED, *London*
PRENTICE-HALL OF AUSTRALIA PTY. LIMITED, *Sydney*
PRENTICE-HALL CANADA INC., *Toronto*
PRENTICE-HALL HISPANOAMERICANA, S.A., *Mexico*
PRENTICE-HALL OF INDIA PRIVATE LIMITED, *New Delhi*
PRENTICE-HALL OF JAPAN, INC., *Tokyo*
SIMON & SCHUSTER ASIA PTE. LTD., *Singapore*
EDITORA PRENTICE-HALL DO BRASIL, LTDA., *Rio de Janeiro*

George Mason Van Dyne (1932-1981)
Early Systems Ecologist

Contents

Contributors ix

Editor's Preface: Why "Complex" Ecology? xiii

Coeditor's Preface: Complex Ecology in the 21st Century xvii

Dedication and Remembrance: Complex Person, Complex Ecologist—"Father of Systems Ecology?" xxi

Foreword: George M. Van Dyne—A Reminiscence xxvii

Ecosystems, Systems Ecology, and Systems Ecologists 1
 George M. Van Dyne

Part 1
MICROSCOPIC COMPLEXITY: DOWNWARD–INWARD IN THE ECOSYSTEM 29

1 Quantum Mechanics and Complex Ecology 34
 Sven E. Jørgensen

2   **Information Theory and Complex Ecology**   40
    Ramon Margalef

3   **The Far-from-Equilibrium Ecological Hinterlands**   51
    Lionel Johnson

4   **Cybernetic Theory of Complex Ecosystems**   104
    Milan Straškraba

5   **Entropy Control of Complex Ecological Processes**   130
    Peter Mauersberger

6   **Stochastic Theory of Complex Ecological Systems**   166
    Benzion S. Fleishman

Part 2
MACROSCOPIC COMPLEXITY: OUTWARD–UPWARD IN THE ECOSYSTEM   225

7   **Model Aggregation: Ecological Perspectives**   230
    William G. Cale

8   **Model Aggregation: Mathematical Perspectives**   242
    Nicholas K. Luckyanov

9   **Simulation Models of an Estuarine Macrophyte Ecosystem**   262
    W. Michael Kemp, Walter R. Boynton, and Albert J. Hermann

10 **Optimal Experimental Conditions to Validate Volatilization Models of Toxic Contaminants**   279
    Efraim Halfon

11 **Nonlinear Programming in Regulation of Deer Hunting Pressure**   295
    Gordon L. Swartzman and George M. Van Dyne

12 **Simulation Modeling in a Workshop Format**   311
    Gregor T. Auble, David B. Hamilton, James E. Roelle, and Austin K. Andrews

Part 3
COMPLEX ECOLOGICAL ORGANIZATION:
FEEDBACK AND STABILITY   335

13   The Mathematics of Community Stability   343
    Yuri M. Svirezhev and Dimitrii O. Logofet

14   Forest Ecosystem Stability: Revision of the
    Resistance–Resilience Model   372
    Jack B. Waide

15   Disturbance and Stress Effects on Ecological Systems   397
    David J. Rapport and Henry A. Regier

16   Societal Instability: Depressions and Wars as Consequences
    of Ineffective Feedback Control   415
    Kenneth E. F. Watt

17   The Nature and Significance of Feedback in
    Ecosystems   450
    Donald L. DeAngelis

18   Knowledge-based Large-scale Ecosystem Design   468
    Bernard P. Zeigler

19   Multicommodity Ecosystem Analysis: Dealing with the
    "Mixed Units" Problem in Flow and Compartmental
    Analysis   485
    Robert Costanza and Bruce Hannon

Part 4
COMPLEX ECOLOGICAL ORGANIZATION:
NETWORK TROPHIC DYNAMICS   509

20   Complex Food Webs   513
    Gary A. Polis

21   Ecosystem Trophic Foundations: Lindeman
    Exonerata   549
    Robert E. Ulanowicz

Part 5
COMPLEX ECOLOGICAL ORGANIZATION: EXTREMAL PRINCIPLES 561

22 Exergy and Ecological Systems Analysis 568
Sven E. Jørgensen

23 Exergy Principles and Exergical Systems in Ecological Modeling 585
Sven E. Jørgensen, Dimitrii O. Logofet, and Yuri M. Svirezhev

24 The Ecosystem as an Existential Computer 609
Michael Conrad

25 Information Theory and Ecological Networks 623
Hironori Hirata

26 Network Growth and Development: Ascendency 643
Robert E. Ulanowicz

Name Index 656

Subject Index 663

Contributors

Austin K. Andrews — National Biological Survey, National Ecology Research Center, 4512 McMurray Ave., Fort Collins, CO 80525-3400, USA

Gregor T. Auble — National Biological Survey, National Ecology Research Center, 4512 McMurray Ave., Fort Collins, CO 80525-3400, USA

Stanley I. Auerbach — Environmental Sciences Division, Oak Ridge National Laboratory, P. O. Box 2008, Oak Ridge, Tennessee 37831, USA

Walter R. Boynton — Chesapeake Biological Laboratory, University of Maryland, P. O. Box 38, Solomons, Maryland 20699, USA

William G. Cale — College of Natural Sciences and Mathematics, Indiana University of Pennsylvania, Indiana, Pennsylvania 15705, USA

Michael Conrad — Department of Computer Science and Biological Sciences, Wayne State University, Detroit, Michigan 48202, USA

Robert Costanza — Chesapeake Biological Laboratory, University of Maryland, P. O. Box 38, Solomons, Maryland 20688, USA

Donald L. DeAngelis — Environmental Sciences Division, Oak Ridge National Laboratory, P. O. Box 2008, Oak Ridge, Tennessee 37831, USA

Benzion S. Fleishman	Institute of Oceanology, Russian Academy of Sciences, 23 Krasikova Street, Moscow 117218, Russia
Efraim Halfon	National Water Research Institute, Canada Centre for Inland Waters, 867 Lakeshore, Burlington, Ontario L7R 4A6, Canada
David B. Hamilton	National Biological Survey, National Ecology Research Center, 4512 McMurray Ave., Fort Collins, CO 80525-3400, USA
Bruce Hannon	Geography Department, University of Illinois, Urbana, Illinois 61801, USA
Albert J. Hermann	Department of Oceanography, University of Washington, Seattle, Washington 98195, USA
Hironori Hirata	Department of Electrical and Electronics Engineering, Chiba University, 1-33 Yayoi-cho, Chiba-shi 260, Japan
Lionel Johnson	10201 Wildflower Place, Sydney, Victoria, British Columbia V8L 3R3, Canada
Sven E. Jørgensen	The Royal Danish School of Pharmacy, Department of Environmental Chemistry, Universitetsparken 2, DK2100 Copenhagen ø, Denmark
W. Michael Kemp	Center for Environmental and Estuarine Studies, University of Maryland, P. O. Box 775, Cambridge, Maryland 21613, USA
Dmitrii O. Logofet	Institute of Atmospheric Physics, Laboratory of Mathematical Ecology, Russian Academy of Sciences, 3 Pyzhevsky Lane, Moscow 109017, Russia
Nicholas K. Luckyanov	Department of Physiology and Biophysics, University of Miami, School of Medicine, P.O. Box 016430, Miami, Florida 33101, USA
Ramon Margalef	Department of Ecology, University of Barcelona, Diagonal 6465, Barcelona 28, Spain
Peter Mauersberger	Institute of Freshwater and Fish Ecology, Mueggelseedam 310, D-12587, Berlin, Germany
Bernard C. Patten	Institute of Ecology, University of Georgia, Athens, Georgia 30602, USA

Gary A. Polis	Department of General Biology, Vanderbilt University, P. O. Box 93B, Nashville, Tennessee 37235, USA
David J. Rapport	Institute for Research on Environment and Economy, University of Ottawa, 165 Waller Street, Ottawa, Ontario K1N 6N5, Canada
Henry A. Regier	Institute of Environmental Studies, University of Toronto, Toronto, Ontario M5S 1A4, Canada
James E. Roelle	National Biological Survey, National Ecology Research Center, 4512 McMurray Ave., Fort Collins, CO 80525-3400, USA
Milan Straškraba	Biomathematical Laboratory, Biological Research Center, Czech Academy of Sciences, Branisovska 31, Česke Budějovice, Czech Republic
Yuri M. Svirezhev	Potsdam Institüt für Klemafolgenforschung, Postfach 60 12 03, 14412 Potsdam, Germany
Gordon L. Swartzman	Center for Quantitative Science, HR-20 3737 15th Ave, N. E., # 304, University of Washington, Seattle, Washington 98195, USA
Robert E. Ulanowicz	Chesapeake Biological Laboratory, University of Maryland, P. O. Box 38, Solomons, Maryland 20699, USA
Jack B. Waide	FTN Associates, Ltd., 3 Innwood Circle, Suite 220, Little Rock, Arkansas 72211, USA
Kenneth E. F. Watt	Department of Zoology, Storer Hall, University of California at Davis, Davis, California 95616, USA
Bernard P. Zeigler	Artificial Intelligence Simulation Group, Department of Electrical and Computer Engineering, University of Arizona, Tucson, Arizona 85721, USA

Editor's Preface

Why "Complex" Ecology?

The Age of Ecology has arrived. Never in the history of the planet has there been a time of such deep, profound, and rapid change due to the activities of a single species. This is a dizzying time of sociopolitical upheaval and redefinition of modern, technological, postindustrial man within the new framework of a dominantly global economy and demonstrably holistic ecology. It is a time of "Gaia," and "Future Shock" as humanity undergoes transformation into the first holistic species ever to emerge from the crucible of biological evolution—a "hegemoniacal" species whose global reach now forces it to learn to take into account the ramifying consequences of its actions within the ubiquitous, invisible networks of nature. It is an adventuresome time filled with awesome potential for new dimensions and manifestations of the human experience. It is also a dangerous time with heavy penalties to be exacted, as population pushes against ever more impaired carrying capacity, for failing to master the new complexities humanity itself has engendered.

Human-induced complexification of the biosphere is the new challenge for science, societies, and individuals. If nature without humans was always complex in itself, it rarely mattered, for it was built on a plan where causality was local—limited to the proximate in space and time, because the consequences of individual lives were not powerful enough to propagate further. Evolution can arguably be stated to have occurred in response to immediate phenomena exerting the selection pressures science has come to recognize and forcing adaptations to these. There was never any need in evolutionary problem solving to take account of the spatial or temporal ramifications of local events. Indirect effects were small and diffused little outside the borders of constituted ecosystems, so evolution was unholistic and is conceived so by science. The human brain is constructed to deal with events that impinge directly on the sensory apparatus and relegates to an inchoate cognitive corner—intuition—concerns about systems, holism, and indirect effects.

Science is also so constructed and, because of this, ecology as science is at a crossroads today. Both it and general science are dominantly reductionistic, mechanistic, and analytical in a world increasingly shown to be whole, more than its clockworks, and irreducible, because to tease it apart is to lose the essence of its wholeness. Ecological science of the past provided the knowledge that made possible the new environmentalism. The "green revolution" had its factual roots in the descriptive ecology of most of this century—in names like Clements, Tansley, Leopold, Carlson, Lindeman, and the Odums. This was an ecology, however, that, while flirting with the themes of "systemness" in the organization of nature, found these largely intractable under its history-given empirical, experimental paradigm and, as exemplified in works like Peters's recent *A Critique for Ecology* (Cambridge University Press, 1991) is now turning away in denial. The complexity and wholeness commonly embraced as foundational themes in applied ecology—conservation, preservation, environmental ethics, ecological economics—are not the themes widely implemented in hard-core scientific ecology because we have not yet found a way to make workable science of philosophical holism.

This book, then, is about complexity in the natural world under growing anthropogenic influence and is motivated by an increasing need for ecology to develop substantive understanding of this in its own terms—not correlate, regress, or model it away as current methodologies tend to do. It holds that the ecology that has made possible awareness of essential holism in the human–nature relation has largely abandoned this ground in terms of basic science development out of either fear of complexity or its intractability. Scientific ecology is not presently laying down the capital for its future advancement as a science of whole systems, and, while denial of complexity may suit some purposes by keeping ecology empirical, it cannot in the end serve the broader needs of humanity.

There are real roadblocks in the way. Full-fledged, complex ecology, as conceived herein, is not the kind of science that lends itself readily to the immediate requirements of dissertation research, where how to do science is initially learned. Research of graduate students must be on well-defined problems (meaning disconnected from the world at large) and conducted over short periods of time. The requirements of complex ecology are not conducive to the quick production of products that ensure scientific success and advance careers or that later can qualify for the small-scale, short-term, piecemeal funding available. The systems calling is to a large extent incompatible with the compartmentalization of knowledge, fragmentation of disciplines, and normal prescriptions for success that are deeply ingrained in the scientific enterprise. These are some of the reasons for the widespread inattention to holism. But wholeness will out, and in the coming century the systems nature of nature and of the humans–nature relation in it are going to be expressed with a vengeance, and humanity will suffer for it, and all the more so because twentieth-century science failed to develop a holistic side soon enough. This is the why of "complex" ecology.

This book is dedicated to the memory of George Van Dyne, one of the first true systems ecologists, who struggled early against impossible odds to bring a semblance of multidisciplinarity and holism into his science. As an extension of the principles that drove Van Dyne, perhaps, *Complex Ecology* has the goal of bringing to explicit awareness and attention what he fully understood a generation ago: the real world that ecology would study is a world of incalculable complexity for which an adequate science has not yet been invented. This book will not invent it either; it is impossible to wave a wand and do so. If its authors will excuse this characterization, all things must come in their time, and in the transition of ecology from an empirical descriptive science to a powerful systems science of wholes, this is but an alchemic phase that we are presently in. The movement is just beginning, and the scientists who have contributed to this book are part of the same beginning that Van Dyne and others tried to initiate; only now the time is later and the inevitable consequences of failing to rise to the occasion are more crucially coming to pass. Reading the daily news or watching or listening to it on TV or radio verifies this all too readily.

This book does not address the systems ecology origins of the world's problems per se. It is concerned only with an exploration of some of the possible ingredients for a new science that can be called systems ecology, or complex ecology, a science that does not shy away from some of the most obvious facts of reality that are well recognized in nonscientific holism. It is intended to say to ecologists of the present generation, and more importantly to their students, and theirs, that the world is desperately complex and our science needs desperately to be made so also. Many of the chapters herein serve as apt illustration.

Part 1 explores ecological complexity from several microscopic perspectives provided by thermodynamics, quantum mechanics, information theory, and cybernetics. Part 2 examines macroscopic complexity, reviewing first the problem of model aggregation and then proceeding to a set of modeling applications. The last chapter in this section discusses modeling as a means to bring together diverse disciplinary expertise in the solution of complex applied environmental problems. Part 3 is concerned with the stability of ecological systems, its basis in feedback, the impacts of disturbance and stress, and the application of systems principles in the Biosphere II project in Arizona. Part 4 concerns trophic dynamics in the complex energy–matter flow webs of ecosystems. A complex food web is described, and a second chapter brings network perspectives to the original Lindeman trophic dynamics. The final section, Part 5, explores a range of extremal principles that has been introduced to explain the organization of ecosystems in optimization terms. "Exergy," "adaptability," and information flux culminating in "ascendency" are discussed.

One further note. This book was originally planned as a two-volume work but went to press as a single volume consisting of 26 chapters. Inevitably, the economics of modern-day publishing forced a size reduction, and the following chapters (original numbers) were removed:

Chapter (19). The Genotype–Phenotype–Envirotype Complex: Genetic and Ecological Inheritance in Holistic Evolution
F. John Odling–Smee (Brunel University of West London, UK) and Bernard C. Patten

Chapter (20). Indirect Effects in Complex Ecology
I. The Qualitative Theory
Bernard C. Patten, Tarzan Legovic (Rudjer Bošković Institute, Croatia), and Stuart J. Whipple (University of Georgia, USA)

Chapter (21). Indirect Effects in Complex Ecology
II. The Quantitative Theory
Bernard C. Patten and Masahiko Higashi (Kyoto University, Japan)

Chapter (25). Network Trophic Dynamics:
The Alpha and Omega of Food Web Organization
Bernard C. Patten, Masahiko Higashi, and Thomas P. Burns
(Science Applications International Corporation, USA)

(Epilogue) Hard Choices for Ecology
Bernard C. Patten

These chapters form the basis for a new book now in preparation.

This book was conceived shortly after Van Dyne's death in 1981; everyone connected with it has waited a long time to see it materialize. Thanks are due to the authors for letting their efforts be molded to the purpose of commemoration and for their forbearance through the many delays along the long way to publication. Those relatives and friends who sent photographs (Shirley, Julie, and Sallie, David Coleman, Freeman Smith, and Gerry Wright) are also thanked, as is Stan Auerbach for his remembrance in the foreword. I am especially grateful to my coeditor, Sven Jørgensen, who greatly assisted in management and planning; his chapters and the preface immediately following affirm his long-established commitment to complex ecology.

Bernard C. Patten

Coeditor's Preface

Complex Ecology in the 21st Century

COMPARISON WITH PHYSICS AND CHEMISTRY

The Danish humorist Storm Petersen said about 50 years ago, "It is difficult to make predictions ... particularly about the future." Anyhow, I shall attempt to make predictions on the development of systems ecology in the beginning of the next century. The natural basis for the predictions will be the trends in ecology and systems ecology that we can observe today.

We are in many respects in the same situation today as we were in physics and chemistry 80 to 100 years ago. In the second half of the nineteenth century and at the beginning of this century, physics and chemistry achieved many interesting analytical results that could not fit into the governing theories. To mention a few: How does radioactivity, discovered and examined by Madame Curie, occur? Some observations showed that light is waves, others that light is particles. Which were right? Why were elements emitting light of certain (predictive) frequencies only? The atomic weights of natural elements were a puzzle.

Developments in physics and chemistry worked hand in hand with those in technology. Many new technological "creations," which today we consider as an integral part of everyday life, appeared on the historical stage 80 to 120 years ago: films, airplanes, telephones, and electric lights, for example. Technological developments created a new scientific interest and reinforced the resources allocated to science, which led to new technological results, and so on. During the first couple of decades of this century the many observations of physics and chemistry were integrated in the form of holistic theories known as special and general relativity and quantum mechanics. This caused a further lift in technological development and even more societal interest in science.

We have a completely parallel situation in ecology today. During the last few decades, we have gathered numerous observations on ecosystem behavior, including reactions to the impacts of pollution. We in ecology are dealing with much more complex systems than those examined by physics and chemistry 100 years ago, but we also have better analytical tools, such as modern chemical instruments that give quick, accurate, and precise analytical measurements of even very small concentrations, and we have the explosive development in the use of the computer.

We also have our puzzles as physics and chemistry had: What causes sudden shifts in species' composition? What is the significance of species diversity for perturbations on ecosystems? Why are many ecosystem reactions the opposite of what is expected?

As the developments of physics and chemistry were reinforced by technological development, that of ecology is now being reinforced by developments in environmental technology and ecotechnology. Pollution and other environmental protection problems are linked to modern technological society and will probably never be solved completely. Therefore, it is expected that the coupling between ecology and environmental issues will continue to tighten, and the development of related sciences and technologies will accelerate even further in the next century. The resources for the continuous growth of ecology, complex ecology, systems ecology, and ecological modeling seem therefore guaranteed.

THE COMPLEXITY OF ECOSYSTEMS

It runs as a common thread throughout this volume and the entire production of George Van Dyne that we need much more than analytical results; we need to *synthesize*.

Ecosystems are very complex systems—so complex that we can never analyze all the components and relationships. Consequently, we cannot reach our goals by the analytical and reductionistic path of former science, but only through holism. We can never know everything, but we need not know everything. We need to know only the most critical relations to understand ecosystems and ecological reactions from a holistic point of view.

If we wished to describe the gas molecules in a room, we would not measure the velocity and direction of each molecule, but we would, by the use of thermodynamics, give the distribution pattern of these variables. Even if we could measure the velocity and direction of all molecules, this would be useless, because the pattern would change very rapidly, and it could not be computed on even the largest and fastest computer in the world.

The way forward in ecology is to analyze the important relations and to synthesize these to obtain a holistic picture, which will point toward the needed analytical results. In this way, it will be possible to approach asymptotically a better and better picture of nature; but we will never reach "the complete picture."

EXPECTED DEVELOPMENT OF SYSTEMS ECOLOGY

Let me now try to be audacious and make predictions about the development of systems ecology in the beginning of the next century. We can observe the distinct trends in this volume and in Van Dyne's work, and we can draw a parallel to chemistry and physics. However, because ecology is dealing with much more complex systems, we must also realize that the power of analytical methods is much weaker than in these physical sciences. In systems ecology, we will probably need to synthesize more and go directly to the complex system. We cannot draw conclusions from simple systems and apply them to complex ones as has often been done in physics and chemistry. We are dealing with irreducible systems (see Chapter 1), and we must therefore start at a certain relatively high level of complexity and use the computer and modeling as experimental tools. To interpret our observations, we must synthesize more than we analyze, but we need to do both. Also, due to the high complexity we shall not expect that one person can come up with one holistic theory that solves all problems. Present developments all contribute to a better understanding of ecosystem properties and behavior, but we need an integration of these approaches. Piece by piece, by the work of many systems ecologists, such synthesis will take place. We need sound scientific work that demonstrates that hierarchical organization, indirect effects, network theory, and extremal

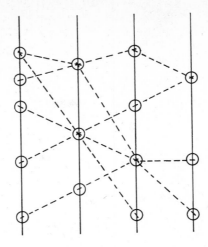

Figure 1

principles are just different sides of the same coin; they are complementary. Not surprisingly, complex systems need a complex of theories to expose all their many facets, and much experimental and theoretical work is needed to show that the various theories can be integrated. Figure 1 illustrates these ideas. Present theories are represented by solid lines. We need also some cross connections, as indicated by the broken lines, to get a fully valid theory.

Fractal theory, catastrophe theory, the theory of chaos, topology, and even to some extent a completely new mathematics may be needed to make cross connections. Let this be the challenge for the next (and much larger) generation of systems ecologists.

Probably, we need new methods to express ourselves and our observations. We are perhaps too much in the habit of thinking in numbers, instead of in patterns. An ecosystem situation is, for instance, much better illustrated by giving the distribution of species and various levels of certain components in different parts of the system than by a table with many numbers, which are not exact anyhow and which steadily change. In systems ecology, we may reformulate the old phrase "one picture is worth 1000 words" into "a picture (pattern) says more than 100 numbers." Still, we need to remember that it is impossible to know everything, and it is sufficient to know only the most crucial factors and relations, which may be better expressed by patterns or pictures than by numbers.

These new methods may change our computers and their languages. Today we use numbers and may translate them into patterns, but why not go directly to patterns? It should not be more difficult to translate the yes/no (0/1) code of computers into patterns rather than numbers. Geographic information systems are doing this now.

When we think how rapidly systems ecology and ecological modeling have grown during the past two decades, it should be expected that an integrated ecosystem theory based on pattern considerations and maybe some new mathematics is just around the corner. However, a much larger job is ahead of us than behind us. Although we shall become more and more systems ecologists, we should not expect "the integrated ecosystem theory" before the first decades of the next century, and maybe longer. It will not come as a revolution, but gradually from the contributions of many scientists. What we can expect to reach, let us say before 2020 (am I too optimistic?), will be a united theory that will integrate and explain many of our present observations—the holistic theory we all lack as a tool today. I do not dare to go beyond this point, but all sciences and also systems ecology will be exposed to new observations, new considerations, and new puzzles in the next 30 years. New challenges will therefore emerge and require new additions to ecosystem theory and systems ecology during this period.

Sven E. Jørgensen

Dedication and Remembrance

Complex Person, Complex Ecologist—"Father of Systems Ecology?"

Behold all flesh is as the grass,
and all the goodliness of man
is as the flower of grass;
for lo, the grass with'reth
and the flower thereof decayeth.
    I Peter 1:24

Blessed are the dead . . .
that they may rest from their labours
and that their works do follow after them.
    Revelation 14:13

George Van Dyne is the first systems ecologist in the history of the world to have died. He was a vigorous person with an uncommon zest for life and energy for achievement. Competitive by spirit, he would have scoffed at but enjoyed the notion that he could become the "Father of Systems Ecology" by being the first of his breed to succumb. This book, published in the fourteenth year after his passing, will serve to both honor and nominate him. Here was a complete man, as personally complex as the science he chose to pursue, who left his imprint on everything and everyone he touched. There was nothing neutral about him.

He was born in Pueblo, Colorado, on September 6, 1932. He died suddenly on his ranch in the foothills of the Front Range of the Rockies west of Fort Collins, Colorado, on Sunday, August 2, 1981, in the midst of a moment of activity with his family and animals. He was Professor of Range Science at Colorado State University at the time of his death, an internationally distinguished grassland and systems ecologist. He had authored, co-authored, or edited 9 books and monographs, 125 journal papers, and 25 special reports. His characteristically up to date "August 1981" *Curriculum Vitae* (August 1 was a Saturday; the c.v. had handwritten in ink alongside his birthdate, probably with GVD-trained secretarial efficiency on Monday morning, "Deceased August 2, 1981") listed 10 additional manuscripts submitted or in press and 25 others in preparation, which would be typical of the number of balls this man could keep in the air at one time. At least one of the 25 is now published, as Chapter 11 herein. The c.v. also documents 138 talks presented since 1968, 8 in Canada and 43 in other countries all over the globe—wherever the principles of systems ecology would grow grass. Van Dyne was an emissary and missionary for holistic ecology of complex ecosystems everywhere, an enthusiastic and infectious advocate of the notion that natural complexity wasn't so bad if science would just recognize it and stand up to it.

He did this in a full-out brawling way; his last c.v., for example, lists as his *"Major interests:* Systems Ecology; Soil–Plant–Animal Relationships; Modelling, Analysis, and Computer Applications to Biology Problems; Research Management and Planning; Nutrition of Range Plants and Animals." That he had no peer in the world in grassland ecology at the time of his death is only implicit in the list; the subject *per se* is not even entered. He served on numerous governmental panels in an advisory capacity and was a consultant to 28 agencies, foundations, academies, institutes, universities, and corporations. His major life achievement, buried in a list of other activities in his c.v., was as "Director and Principal Investigator, Grassland Biome Study, US/IBP (1967–1974)." In this, he organized literally hundreds of scientists, graduate students, technicians, and service people in a broad-scale coordinated frontal attack on the ecological complexities of an entire biome, the North American prairie.

The principal field site for this study was the Pawnee National Grassland northeast of Fort Collins, but 11 other sites were involved in the network. The Pawnee is now renamed, and grasslands became "steppes" in U.S. ecology in the 1980s; but the Grassland Biome program, the knowledge it generated about grasslands as complex ecosystems, and the precedents it established for future large-scale ecosystem studies (such as the present Long-Term Ecological Research (LTER) investigation of the U.S. National Science Foundation) remain as one of the indelible early achievements of the systems viewpoint in ecology, and singularly that of G. M. Van Dyne, although assuredly with the help and dedicated cooperation and efforts of many others.

Van Dyne established (got the approval and money, drew the plans, poured concrete, and drove nails) and directed the Natural Resource Ecology Laboratory at Colorado State University, which remains today one of the principal active centers for holistic ecology in the world. He brought $25 million in grants to his home university during the course of the International Biological Program (IBP) study. His efforts to guide and organize such a massive project burned him out emotionally for a time toward the end of his tenure as director, but never intellectually or spiritually. He went on to an active career in teaching and research in the Range Science department at Colorado State, never doubting for a moment that his approach was right; nature was tough and intractable, but with prodigious integrated effort would budge. The humans proved tougher, he had learned. This was a scientific visionary; doer of the impossible; husband, father, family man; professor to many students; gifted executive and organizer of ideas and people who left uncommon achievement in his wake.

Complex Ecology is dedicated to his memory. The volume contains two of his works, one previously unpublished in open literature. The introductory chapter, "Ecosystems, Systems Ecology, and Systems Ecologists," is a manifesto for systems ecology, vintage Van Dyne as good today as when it first appeared in the document gray literature of Oak Ridge National Laboratory in 1966 (ORNL-3957, UC-48, Biology and Medicine). Here he issues a call for a new ecology, as unrealized and even more urgent today than when he wrote the words. He recognizes in the ecosystem "unlimited size and complexity" of a kind and degree that demand a new science. He recognizes humans as manipulators and shapers of even global phenomena. His world is all connected together, and he sees the compartmentalization of science in learning, research, and applications as ultimately failing to elaborate a sound basis for sustainable man–nature coexistence on an uncompartmentalized planet. Human "induced instability of ecosystems" can propagate in the invisible networks of the biosphere and "is an important cause of economic, political, and social disturbances throughout the world." To combat the brain's fragmentation of reality and the translation of this into compartmental science and institutions, he calls for modeling, contrasting the reductive empiricism of experimental science with the abstractive holism of a needed systems science. He speaks of the partially known "green, pink, and brown" boxes (plants, animals, and environment) that it is the work of the systems ecologist to link into functioning wholes. He discourses on the importance of computers in this process, predicting developments now part of everyday life. The systems ecologist of tomorrow, he says in a typical turn of phrase, will have to be a "specialist in generalization." "Future leaders toward this goal must have the ability to organize concepts, things, and people." Multidisciplinary problems require interdisciplinary teams to address them, and the experiment with the latter that he conducted in the IBP still remains to be perfected. "The easy problems have been solved," he says; the difficult ones of unrestricted size and complexity remain.

George Van Dyne did not shrink from the intractability of scope or scale. He led a difficult scientific life by his own choices, and he led it hard, with fervor and commitment. He did not pursue a course of easy, discrete, traditional science to climb a ladder of facile academic success. He enjoyed the mixing of disciplines, exalted in the unknown, and reveled in the challenge of facing hard problems and leading others to face them too. He put every fiber of his physical and intellectual being into the early quest in ecology for true, substantive, systems knowledge, recognizing that the work was only beginning and that the demands of the next century would make his kind of complex ecology—systems ecology—mandatory. Some of these problems are taking form now; we know them in broad outline: overpopulation; overdevelopment; technological overshoot; growth economics; pollution and eutrophication; release of radioactivity, toxic substances, and the new products of a market-driven biotechnology into the environment; global change; greenhouse warming and sea level rise; lost biodiversity; AIDS.

Van Dyne was a pioneer and an inspiration, a person ahead of his time and without equal in his time. It was a privilege and inspiration to know him and be his friend. I hope this book may capture some of the flavor of the science of complexity that he foresaw and would have pursued more had he lived longer. That an ultimate systems approach to ecology will have to be done in something like his way in the future seems certain, and this durability of his central vision is why he should be recognized as the "Father of Systems Ecology."

The memorial service held for Van Dyne at Colorado State on August 6, 1981, included the following reading from the *Kansas Magazine* (1872) authored by John James Ingalls, U.S. Senator from Kansas during 1873–1891:

> Next in importance to the divine profusion of water, light, and air, those three great physical facts which render existence possible, may be reckoned the universal beneficence of grass.... Grass is the forgiveness of nature—her constant benediction. Fields trampled with battle, saturated with blood, torn with the ruts of cannon, grow green again with grass, and carnage is forgotten. Streets abandoned by traffic become grass-grown like rural lanes, and are obliterated. Forests decay, harvests perish, flowers vanish, but grass is immortal.... Its tenacious fibres hold the earth in its place, and prevent its soluble components from washing into the wasting sea. It invades the solitudes of deserts, climbs the inaccessible slopes and forbidding pinnacles of mountains, modifies climates, and determines the history, character, and destiny of nations. Unobtrusive and patient, it has immortal vigor and aggression. Banished from the thoroughfare and the field, it bides its time to return, and when vigilance is relaxed, or the dynasty has perished, it silently resumes the throne from which it has been expelled, but which it never abdicates. It bears no blazonry or bloom to charm the senses with fragrance or splendor, but its homely hue is more enchanting than the lily or the rose. It yields no fruit in earth or air, and yet should its harvest fail for a single year, famine would depopulate the world.... The primary form of food is grass. Grass feeds the ox: the ox nourishes man: man dies and goes to grass again; and so the tide of life, with everlasting repetition, in continuous circles, moves endlessly on and upward, and in more senses than one, all flesh is grass.

So long, George. You were really something.

Bernard C. Patten

George "Van Dynamo" in Action

Foreword

George M. Van Dyne—A Reminiscence[1]

Stanley I. Auerbach

Environmental Sciences Division, Oak Ridge National Laboratory[2]
Oak Ridge, TN 37831-6036, USA.

Each morning at Oak Ridge National Laboratory I park my car in a satellite lot recently created in what was about two acres of grass southwest of the present Environmental Sciences Building. Within the fescue-dominated margin of the lot stands a 10-foot strip of low fencing made of heavily galvanized steel. The strip is about 30 inches high, of which the first 24 inches is open mesh, with a solid rim comprising the top 6 inches. George Van Dyne installed that strip of fencing almost 25 years ago. Looking at it every morning triggers flashes of memory about George, who undoubtedly was one of the most unusual persons I have ever met. Professor Frank Pitelka, on an occasion a few years after the fence was installed, referred jokingly to George as "Van Dynamo." It may have been a joke, but it was an apt description. But to get back to the fence . . . .

George joined the Radiation Ecology Section of Oak Ridge National Laboratory in 1964, immediately after the completion of his doctoral studies in animal nutrition at the University of California at Davis. Why was George hired, and why did he choose to come to an ecological research unit that was focusing on problems associated with environmental radioactivity? First, he came to Oak Ridge because he was interested in ecology. Although his background and interests were tied to range science, he had already perceived that many of the kinds of scientific questions that were not being addressed were essentially ecological. Second, George came here because he was also interested in the applications of mathematics, especially in computer methods and approaches, and the National Laboratory had major computer resources. Another contributing factor was the presence on the staff of two other mathematical ecologists with budding ideas about systems, Jerry Olson and Bernard Patten. Here was an opportunity to form a unique combination in American ecology and create a new integrative approach.

A large tract of riverine floodplain had been recently cleared and planted in fescue grass in preparation for a large-scale field experiment simulating radioactive contamination effects on old-

---

[1]Publication No. 3139, Environmental Sciences Division.

[2]Operated by Martin Marietta Energy Systems, Inc., under contract DE-AC05-84OR21400 with the U.S. Department of Energy.

xxvii

field ecosystems. This experiment, which was being proposed to the Atomic Energy Commission, had as its central theme the long-term ecological effects of a nuclear war. It included the establishment of a large number of field enclosures for the study of mammals and other organisms. Being grassland, it was a natural draw for somebody with George's background and skills. He addressed himself to the planning and design with characteristic vigor. Because we were going to use enclosures to examine system-level effects, George was particularly concerned about the issue of vegetation sampling. He fretted about the effects of shade from the enclosure wall panels and how the shade would contribute to uneven growth responses of the vegetation. So he worked with the fabrication shops and designed a panel that would serve the enclosure purposes, yet at the same time, would allow light to enter and be rugged and relatively easy to fabricate. He had a section made and installed at a location where he could check on light transmission under conditions similar to those in the experimental area. He never bothered to remove it, nor did anyone else.

Our experimental radio-tracer work in ecosystems had stimulated an active interest in mathematical approaches, especially the then new field of systems analysis. He, Patten, and Olson raised the idea of establishing a new program in systems ecology at the University of Tennessee. It was an appropriate time because the National Laboratory and the university, with funding from the Ford Foundation, had established a special adjunct faculty program under which selected laboratory staff could teach part-time and direct research at the university. We began the necessary paper work and started negotiations with the university, and George and his colleagues began designing a three-quarter curriculum. Much had to be done. At the same time, planning and research activities demanded attention at the laboratory. Meanwhile, George was also studying computer operations at night and, on weekends, writing papers and engaging in intensive discussions with me and others on the need for large-scale research on ecosystems. Once he got an idea, George did not remain far from a typewriter. One day he asked my opinion about writing up some thoughts about systems analysis and ecology. Where could we get fast publication on something so new and specialized without a long review process? I recommended that he prepare a document for the laboratory's "gray literature" because this was speedy and we could distribute a large number of copies. Thus emerged the article that begins this book, "Ecosystems, Systems Ecology, and Systems Ecologists."[1] As such articles go, it was an instant best-seller in spite of being published in the often deprecated institutional documents, which were not peer reviewed. More than popularity was involved, however. The ideas expressed in that essay, which are widely implemented today, brought George to the attention of the admittedly small group of senior ecologists in the United States who were involved in developing ecosystem studies.

Why was this report so popular? The potential for use of systems analysis in ecological studies had already been recognized by such scientists as Holling[2] and Watt,[3] among others. E. P. Odum's textbook[4] had already stimulated growing interests in ecosystem ecology. At that time a number of ecologists were actively thinking about a new, large-scale approach to ecological research, the International Biological Program (IBP). Mathematical and numerical explorations of the properties of food webs had begun under several individual investigators and were encouraged by advances in population dynamics and in studies of two-species systems. The conceptualization of ecosystems, as presented in most ecology textbooks, was well developed. But few, if any, ecologists were using

---

[1] G. M. Van Dyne, 1966. Ecosystems, Systems Ecology, and Systems Ecologists. Oak Ridge National Laboratory. ORNL-3957. Oak Ridge, Tennessee.

[2] C. S. Holling, 1963. An experimental component analysis of population processes. *Mem. Entomol. Soc. Canada.* 32:22–32.

[3] K. E. F. Watt, 1965. An experimental graduate training program in biomathematics. *Bioscience* 15:777–780.

[4] E. P. Odum, 1959. *Fundamentals of Ecology,* 2d ed. W. B. Saunders, Philadelphia.

computers or were attempting to define ecosystem studies with models as prerequisites for analysis. George took all these ideas one major step forward by conceptualizing how the systems ecologist could hypothesize the fundamental cause-and-effect relationships in a system, then formulate these hypotheses into models, and ultimately test the output of the latter against the quantitative and qualitative behavior of an actual ecosystem. This was new thinking, a match to the new computers with which it would be possible for ecologists to organize, collect, and manipulate data on a scale that was earlier deemed unachievable. This was powerful stuff, and the new ecologists greeted it with great enthusiasm.

George's exemplar was the grassland ecosystems of the American West. The essay and George's subsequent talks and lectures attracted the attention of western ecologists, and after only a few years at Oak Ridge he was offered a position on the staff of Colorado State University. This was an opportunity he could not refuse; not only did it bring him back to his home area, but it offered him the opportunity to research and carry out his newly acquired ideas on his beloved grassland ecosystems. The timing was also highly propitious for involvement in the IBP. At Colorado State, George threw his energy into the planning and organizing of large-scale ecosystem research, both nationally, in support of the U.S. IBP efforts, and regionally, in the grassland biome. Within several months he and several collaborators had developed a prototype proposal for grassland research. This proposal called for coordinated research at several sites on the Great Plains ranging from Canada to Mexico. In October 1967, George held the first regional meeting at Fort Collins, Colorado. The report of this meeting (43 pages) was characteristic of George. It listed all the proposed participants and described the various proposed research sites. Research was outlined, including a variety of experiments, and program management was emphasized in 7 pages of detail! This grassland ecosystems report later served as a prototype for what became the U.S. Biome Program, or, as it was officially called, the Analysis of Ecosystems Program. George personally organized all aspects of the grassland project. Essentially, he created a whole new laboratory at CSU, the Natural Resource Ecology Laboratory (NREL), of which he was appointed the first director in 1970. He not only selected the site but also designed the building, obtained the funds, supervised the construction, and selected the equipment. No detail was too small or insignificant to escape his attention. A day represented 1440 minutes of opportunity to do work, and he drove himself and those around him to take advantage of every one of those minutes.

As leader of the Grassland Biome Project, Van Dyne simultaneously played several roles—manager, organizer, fund raiser, and researcher. As the latter, he pursued a number of issues that were associated with ecosystem dynamics but that also drew on his background in large-animal nutrition and physiology, as well as rangeland ecology. But his overriding dedication and commitment were to the development and use of models to guide and synthesize ecosystem research. Systems analysis and ecological modeling were the key elements of the U.S. Biome programs and were hardly known and certainly not used in the rest of the ecological research community here and abroad. Moreover, there was a certain amount of skepticism about this new "big ecology" of models, research teams, data banks, and the other panoplies of what came to be known as "Big Science." Van Dyne put a great deal of effort into stimulating interest in this approach. He spent a good part of the next three to five years carrying to others the message about grassland models and systems analysis as it applied to the study of ecosystem dynamics, proselytizing range scientists as well as others in all parts of the globe. One could hardly go to an international meeting during the 1969–1973 period without hearing of George's activities, whether the meeting was in Europe, Asia, Africa, or South America. He was remarkably effective, not only because of his drive and enthusiasm, but also because of his superb sense of organization of scientific details and his encyclopedic ability to recall the minutest details and data about any of a broad spectrum of scientific disciplines.

To say that he was somewhat dogmatic is a bit of an understatement. When he was manager of the program, his forcefulness more than occasionally resulted in controversy with other scientists, especially those who had similar traits. He firmly believed that his approach to ecosystem research in the grasslands was the appropriate model or paradigm for all the biome programs—an attitude that brought him into conflict with the other biome project directors, especially at program reviews. I was such a director for the IBP deciduous forest biome, and at one such review he took me to task for what he perceived as a program weakness, the lack of a biome-wide sampling and harvesting program to measure forest productivity by using his grassland approach. We got into quite an argument. To sample forests as extensively as he was sampling grasslands in the Great Plains would require resources far beyond anything heretofore conceived of in our program. Somewhat facetiously I noted that to cut, weigh, and grind trees from a number of sites would require large teams of tree cutters, dedicated railroad trains to carry the logs, and the use of several large timber mills. Aside from the enormous costs, I felt that I would probably not survive in my position after clear-cutting the first acreage. George was not convinced.

As a person with a fierce competitive spirit motivated by deep inner forces, he was a taskmaster who drove his colleagues, coworkers, and graduate students hard. But his demands on them barely equaled the effort and diligence he demanded of himself. Ultimately, pressures and demands for a more collegial approach led Van Dyne to step down from the directorship of the Biome Project and the Natural Resource Ecology Laboratory in 1974. As a faculty member of the Department of Range Science, he continued to work on grassland research, both in the United States and overseas. If anything, his research output and publications intensified during the following years.

At the time of his death, he had 35 manuscripts in press or in preparation. During a professional career of 25 years, he produced 125 papers in national and international scientific journals; he was the author, co-author, or editor of 9 books and monographs and 25 special reports. He delivered more than 150 presentations as an invited lecturer and consultant at 24 universities in the United States and at 20 universities in 15 foreign countries. His contributions were truly prodigious. All were accomplished with a special zeal and zest for life that were rarely matched among his contemporaries. On the day death struck, he was in full physical stride in the midst of outdoor activities with his family. He would have liked to have it said, in western metaphor, that he died "with his boots on" or that he died "in the saddle." Well, he did both, literally, on that summer day at his ranch in Colorado on August 2, 1981.

As I examine the chapter titles for this book and the names of their authors, I can't help feeling that George would be pleased, not necessarily because he would agree with all the ideas and concepts presented here, but because he would be pleased with the breadth and scope of systems ecology today. Complex ecology it is, and he helped make it that way. I believe the hidden, sensitive side of George would have reflected on the former students and colleagues, on the role that he played in the development of their ideas, and maybe on how this all began with that piece of fence that still stands in the East Tennessee sunshine.

REFERENCES

Jeffers, J. N. R. (ed). 1971. Mathematical Models in Ecology; Symposium (12th) of the British Ecological Society, Grange-over-Sands, Lancashire, March 1971, Blackwell, London. 398 pp.

"Systems for Stimulating the Development of Fundamental Research." 1978. Prepared by the Committee for Joint US/USSR Academy Study of Fundamental Science Policy. Commission on International Relations, National Academy of Sciences, National Research Council.

Van Dyne, G. M. 1966. Ecosystems, systems ecology, and systems ecologists. Oak Ridge National Laboratory. ORNL-3957. Oak Ridge, Tenn. 31 pp.

Van Dyne, G. M. 1971. Organization and management of an integrated ecological research program—with special emphasis on systems analysis, universities, and scientific cooperation. (In Jeffers, J.N.R., 1971 above).

Ecosystems, Systems Ecology, and Systems Ecologists

GEORGE M. VAN DYNE

Radiation Ecology Section, Health Physics Division
Oak Ridge National Laboratory, Oak Ridge, Tennessee
and
Associate Professor of Biology, University of Tennessee
Knoxville, Tennessee

ORNL-3957
June 1966

OAK RIDGE NATIONAL LABORATORY
Oak Ridge, Tennessee
operated by
UNION CARBIDE CORPORATION
for the
U. S. ATOMIC ENERGY COMMISSION

CONTENTS

Abstract 3

Acknowledgments 3

Introduction 3

Ecosystems 4
    Definitions 4
    The Ecosystem Concept: Unlimited Size and Complexity 4
    Trends in Ecological Research 5
    Ecosystem Components 5
    Dynamics 6
    Manipulation of Ecosystems 8

Systems Ecology 9
    Definitions 9
    Study of Ecosystems 10
    Conceptual Requirements 10
    Tools for Study of Ecosystems 11
    Operations Research and Systems Analysis Applications 13
    Importance of Digital Computers 14
    A Systems Approach 16

Systems Ecologists 16
    Definitions 16
    Systems Ecologists: Interdisciplinarians and Multidisciplinarians 18
    Old Problems—Old Techniques 19
    Old Problems—New Techniques 19
    New Problems—Old Techniques 19
    New Problems—New Techniques 19
    Some Pitfalls Facing Systems Ecologists 20
    Mathematical Training of Systems Ecologists 21
    Systems Ecology Research 21

Literature Cited 23

ABSTRACT

This paper defines and discusses ecosystems, systems ecology, and systems ecologists, in that order. Some properties of ecosystems and the ecosystem concept are given as a basis for defining the area of study called systems ecology. Problems, methods, tools, and approaches of systems ecology are considered in defining tasks, problems, and training of systems ecologists. The interdisciplinary nature of systems ecology research and the importance of computers in this research are considered. Examples of methods, concepts, and applications are drawn from a diverse body of ecological, natural resource management, and mathematical literature, which further illustrate the interdisciplinary nature of systems ecology. Advantages and limitations, with respect to total-ecosystem problems, of research by ecologists in universities, in state and federal experiment stations, and in national laboratories are compared. An example is given wherein, possibly under International Biological Program support, the skills and resources of these three groups of ecologists might be combined for integrated attack on nationally important ecosystem problems.

ACKNOWLEDGMENTS

In any organization in which there is frequent and free interchange it is difficult to clearly identify the origin of concepts and ideas. Much of the material herein is a product of cross-fertilization of ideas with S. I. Auerbach, J. S. Olson, and B. C. Patten. Generally, we agree in principle, although our approaches differ in practice. Several other colleagues in the Radiation Ecology Section also have provided constructive criticisms of this paper. Acknowledgment also is extended to J. W. Barrett, J. E. Cantlon, F. B. Golley, A. M. Schultz, E. G. Struxness, and K. E. F. Watt, who have criticized the manuscript. G. C. Battle is thanked for his editorial assistance. The author assumes responsibility for any misconceptions or errors in this paper and would appreciate comments and criticisms. Part of this paper is based on a lecture given to the Radioecology Institute, Oak Ridge Institute of Nuclear Studies, June 1965.

INTRODUCTION

The Radiation Ecology Section of the ORNL Health Physics Division has initiated a new program of studies in "Systems Ecology." Neither is this area of work in ecology clearly defined nor do all ecologists view it equally. As with any new field, systems ecology is beset with vociferous skeptics (largely those who have done well under the old conditions) but supported primarily by lukewarm champions (largely those who may do well under the new conditions). It is desirable to examine the subject more closely as a further basis for clarification of objectives of work in this area.

Often we feel that our own work and interests are of extra importance, but I do not propose systems ecology to be a new panacea, nor that we neglect more conventional approaches in ecology. However, I feel that a systems approach has much to offer in various phases of ecology and especially in renewable resource management. This leads to a brief review of some ecological concepts and to suggestions about the tools, training, and work of systems ecologists. It is in this context that this essay is offered.

The purposes of this paper are twofold:

(1) To discuss important properties of ecosystems and to consider application of recent techniques from systems analysis and other fields. This is done not only to help define systems ecology and to help prevent semantic confusion, but also to help define some of the needs, training, and perspective of systems ecologists.

(2) To provide an introduction to a selected part of the large and diverse literature, through 1965, which encompasses the interdisciplinary area of systems ecology.

ECOSYSTEMS

Definitions

In 1935 Tansley *(106)* introduced the term ecosystem, which he defined as the system resulting from the integration of all the living and nonliving factors of the environment. *Webster* now defines the term as a complex of ecological community and environment forming a functioning whole in nature. An ecosystem is a functional unit consisting of organisms (including man) and environmental variables of a specific area *(3)*. Macroclimate has an overriding impact on the other components, each of which is interrelated at least indirectly (Fig. 1). The term "eco" implies environment; the term "system" implies an interacting, interdependent complex.

Russian ecologists use the term ecosystem less frequently than the term biogeocoenosis, which Sukachev *(105)* defines as "any portion of the earth's surface containing a well-defined system of interacting living (vegetation, animals, microorganisms) and dead (lithosphere, atmosphere, hydrosphere) natural components...." Biogeocoenosis is derived from the Greek "bio" or life, "geo" or earth, and "koinos" or common. Sukachev and others feel that biogeocoenosis is the more accurate and descriptive term from an etymological viewpoint, but Sukachev recognizes that ecosystem is the older term and that the two terms are widely used as synonyms. For practical purposes and to avoid semantic argument, ecosystem and biogeocoenosis will be considered synonymous in this paper. Further discussion of ecosystem terminology is given by Schultz *(100)* and Maelzer *(73)*.

The Ecosystem Concept: Unlimited Size and Complexity

A system is an organization that functions in a particular way. The functions of an ecosystem include transformation, circulation, and accumulation of matter and flow of energy through the medium of living organisms and their activities and through natural physical processes. Some specific functional processes include photosynthesis, decomposition, herbivory, carnivory, parasitism, and symbiosis *(31)*. The ecosystem must be studied as a whole in order to understand energy transformations, the hydrologic cycle, or cycles of carbon, nitrogen, phosphorus, or other elements *(66, 68, 100)*.

The ecosystem is the fundamental unit of study in "pure" and "applied" ecology *(31, 81, 106)*. Directly or indirectly the ecosystem concept is useful in the management of renewable resources

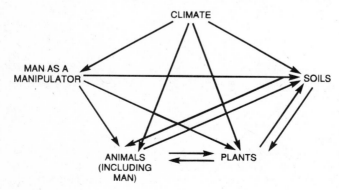

Figure 1. An ecosystem is an integrated complex of living and nonliving components. Each component is influenced by the others, with the possible exception of macroclimate. And now man is on the verge of exerting meaningful influence over macroclimate.

such as forests, ranges, watersheds, fisheries, wildlife, and agricultural crops and stock *(16, 26, 64, 72, 89, 91)*. Understanding the ecosystem concept is required in the disposal of radioactive wastes and in analyses of environmental pollution *(108)*. It has even found a place in medical studies of the digestive tract *(22)*.

The ecosystem as a unit is a complex level of organization. It contains both abiotic and biotic components. The order of increasing complexity is: cell < tissue < organ < organism < population < community < ecosystem. Although the ecosystem is the most complex level, the study of a given ecosystem is less complex in many instances than the study of the lower levels *(33)*.

We can consider steppe grasslands, deciduous forests, or oceans as examples of macroecosystems, as well as considering a given small plot or a spaceship and its contents as a microecosystem.

The term ecosystem is also used to describe a concept or approach of studying biotic–abiotic complexes. In applying the ecosystem concept there is no limit to size and complexity *(31, 34)*. We delineate boundaries of ecosystems chiefly for convenience of study, although some natural boundaries may occur (e.g., shore lines and air–water or soil–water interfaces for aquatic systems), and man often introduces distinct boundaries, such as fences and field edges. Most ecosystems are bounded in nature by gradual and indistinct boundaries.

In the sense that the term ecosystem implies a concept and not a unit of landscape or seascape, the emphasis is that the biologist must look beyond his particular biological entity (e.g., cells, tissues, organs, etc.) and must consider the interrelationships between these components and their environment. For example, in applying the ecosystem concept to the tissue level of organization (Fig. 6) the environment of a specific tissue would include the fluids, such as blood and lymph, that may bathe or pass through the tissue as well as surrounding tissues. Part of the environment, therefore, consists of other parts of the same organism as well as external components. Generally, the ecosystem concept is used in situations where at least several organisms are being considered.

Trends in Ecological Research

Although the concept of the ecosystem and many methods for studying ecosystems have been available for some time, only recently have many ecologists given more than lip service to the idea. Recently it has been suggested that the ecosystem is the rallying point for ecologists. There has been a gradual but distinct shift in emphasis in ecological studies and training from the description or inventory of ecosystems, or parts thereof, to the study of energy flow, nutrient cycles, and productivity of ecosystems. More workers are extending knowledge from the "anatomy" to the "physiology" of the environment. This requires different concepts, tools, and methods. The gradual change in emphasis from inventory to experimentation also requires more use of scientific methodology; this will be discussed below in the section "Systems Ecology."

Ecosystem Components

Jenny *(54–55)* discusses dependent and independent variables in ecosystems and shows their relationships by the following equations:

$$l, s, v, \quad \text{or} \quad a = f(L_0, P_x, t),$$

where the internal properties are l = ecosystem property, s = soil, v = vegetation, and a = animals. The external properties are L_0 = initial state of the system, P_x = external flux potentials, and t = the age of the system. External flux refers to the flux of nutrients, energy, etc., in from and out to adjoining systems and can be defined by

$$P_x = \left(\frac{P_{\text{out}} - P_{\text{in}}}{dX}\right) m,$$

where P_{out} and P_{in} are the flux out and in over a boundary thickness dX which has a permeation parameter m.

The controlling factors of the ecosystem are macroclimate, available organisms, and geological materials, where the last term includes parent material, relief, and ground water. Time is considered as a dimension in which the controlling factors operate, rather than as an environmental factor. The controlling factors are partially or entirely independent of each other. Each of the controlling factors is a composite of many separate elements, and each element is variable in time or space. Operationally, we may consider each controlling factor as a multiple-dimensioned matrix. Each change in a controlling agent in the ecosystem produces in time a corresponding change in the dependent elements of the ecosystem. In space and time there is a continuum of ecosystems.

Internal properties of ecosystems, such as rate of energy flow, might be considered as dependent factors which vary through time under the influence of a series of independent controlling factors. The dependent factors of the ecosystem are soil, the primary producers (vegetation), consumer organisms (herbivores and carnivores), decomposer organisms (bacteria, fungi, etc.), and microclimate. Each of these factors is dynamically dependent on the others (Fig. 1), and each is a product of the controlling agents operating through time.

Producers, consumers, and decomposers are not distributed at random in the abiotic part of an ecosystem. To maintain either dynamic equilibrium or ordered change in an ecosystem requires that a tremendous number of ordered interrelations exist among its dependent elements (82). To function properly ecosystems must process and store large amounts of information concerning past events, and they must possess homeostatic controls which enable them to utilize the stored information. This information may be expressed in amino acid and nucleotide sequences in genetic codes which have developed over evolutionary time, or it may be expressed in spatial or temporal patterns (20). For example, the changing patterns of plant populations and communities in secondary succession can be considered as expressions of genetically coded information. One species, population, or community is replaced by another with greater genetic potential for utilizing the resources of the changing environment.

Dynamics

Ecosystem changes may be caused by fluctuations in internal population interactions or by fluctuations of the controlling factors. Such changes may be cyclical or directional (14). Directional change from less complex to more complex communities may be considered as progression or succession; directional change from more complex to less complex communities may be considered as regression or retrogression; both are shown in Fig. 2.

Autogenic succession occurs when the controlling factors are stable and change is due to the effect of the system or some part of it on the microhabitat. Clements (15) formalized this process as migration, ecesis, competition, reaction, and stabilization. This type of primary succession produces changes which are usually gradual and continuous. Allogenic succession occurs when there is a change in the controlling factors. Most changes in the ecosystem are products of both allogenic and autogenic successions. Most macroecosystems can be said to be polygeneic and are the result of several climatic changes and erosion cycles. Purposeful alterations, such as disruption by man, in the controlling and controlled factors of the ecosystem may induce relatively permanent changes in the ecosystem.

Because ecosystems vary both temporally and spatially, and to prevent ambiguity, it is important to specify at least semiquantitative time and space scales. The importance of specifying

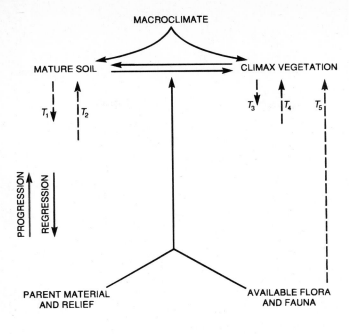

Figure 2. Ecosystems develop through time, under climatic control, from the original flora and fauna under a given set of relief and parent material conditions. A final dynamic equilibrium is reached in which there exist a mature soil and climax plant and animal populations.

a time scale is illustrated in Fig. 2, where the time for primary succession (see T_5 for progression in Fig. 2) is shown as much greater than the time requirement for man to disrupt the system and alter soil or vegetation (T_1 and T_3 in Fig. 2). In the process of retrogression, changes take place in the vegetation more rapidly than they do in the soil. Generally, the ecosystem will recover towards the stable state through a progressive process called secondary succession (T_2 and T_4). Again, the rate of progressive changes of soil properties is usually lower than that for vegetation. Recovery of the vegetation to the climax state may take an amount of time similar to that required for deterioration of the soil. Change in a given ecosystem component or property may be negligible in T_3 but considerable in T_5.

During progressive succession there is usually an increase in productivity, biomass, relative stability and regularity of populations, and diversity of species and life forms within the ecosystem (74). Finally, the ecosystem reaches a steady state or equilibrium, which is characterized by dynamic fluctuation rather than by directional change. This steady state of the ecosystem is referred to as climax (119). At climax the dependent factors are in balance with the controlling factors; the climax is an open steady state (101). A diversity of species and life forms occupies every available ecological niche at climax and, because there is a maximum number of links in the food web, the stability of the system is maximized (63). A maximum amount of the entering energy is used in maintenance of life. Fosberg (34) considers "that climax communities [are those] in which there is the greatest range and degree of exploitation or utilization of the available resources in the environment." There is no net output from an ecosystem in the climax state (86). Three states of ecosystems exist with regard to energy or nutrient balance: steady state or climax, positive balance or succession occurring, and negative balance or decadence and senescence (99).

There is continual interchange of matter and energy among contiguous ecosystems. This interchange or flux is an essential property of ecosystems. The fluxes in and out of an ecosystem may be difficult to measure accurately, but there is relatively less error in measuring flux in a macroecosystem than in a microecosystem, because usually the error in measurements is inversely proportional to the magnitude of the object, rate, or processes being measured. Also, the relative amount of relevant surface or area around an ecosystem decreases as its size increases; many of the measurement errors or biases occur at such interfaces because of subjective decisions in defining boundaries. Still, we may find it convenient to study microecosystems such as a sealed bottle containing nutrients, gases, organisms, and water. Essentially, this is the type of system we need to study in preparing for interplanetary travel. But even such discrete microcosms are not adiabatic with their environment, and ultimately they are dependent upon their environment for a continuing energy input.

When flux of some element in and out of a given ecosystem is negligible for a defined period of time we consider that ecosystem to be stable with regard to that element. The equilibrium is referred to as climax only if it is reached naturally. Other equilibria, or disclimaxes, can be maintained by man's intervention. Here is the essence of renewable resource management: maintaining disclimaxes at equilibrium for the benefit of man.

Manipulation of Ecosystems

Man is a vital part of most major ecosystems, and there is an increasing human awareness of man's part in them and his influence on them *(108)* (Fig. 3). Traces of his pesticides probably can be found in living organisms throughout the world. Humans are both parts of and manipulators of ecosystems. Induced instability of ecosystems is an important cause of economic, political, and

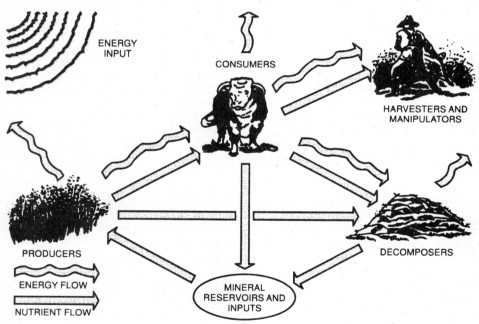

Figure 3. Man is both a spectator of and a participant in the functioning of ecosystems. He has manipulated ecosystems to maximize the flow of nutrients and energy to him from the producers and primary consumers. He has attempted to minimize the respiratory losses of energy from producers, consumers, and decomposers.

social disturbances throughout the world. In altering his environment in order to overcome its limitations to him, man learns that he often is faced with undesirable consequences of the environmental change *(13, 38, 39)*. In manipulating his environment (e.g., felling forests, burning grasslands or protecting them from fire, and draining marshes), seldom has he foreseen the full consequences of his action *(104)*.

Most ecosystems in our country were in climax states when civilized man began to affect them, but the economy of civilized man demanded that the ecosystems produce a removable product under his domination. In order to reach this goal he disrupted the climax ecosystems, perhaps by shortening food chains or by altering the diversity of life forms of primary producers. He has altered the rate of and amount of nutrients cycling through the system by such means as fertilization, both in aquatic and terrestrial systems (Fig. 3). In some instances the fertilization has been excessive and has led to undesirable side effects, such as algal blooms caused by excesses of organic wastes. In other instances man has altered the structure of ecosystems by simplifying them and diverting the flow of energy into his food products, such as in replacing a grassland and wild animals with a wheatfield. Eventually he has produced changes in some ecosystem properties, which in some instances has led to new quasi-stable levels. In other instances such changes have led to desertification, such as the result of centuries of overgrazing in the Middle East.

Man has also encountered difficulties when he attempts to return ecosystems to their native state or to preserve vegetation by the development of national parks or by control of predators *(104)*. In several instances ungulate populations have multiplied rapidly, outstripped the natural control by predators, exceeded the carrying capacity of their ranges, and severely damaged their habitat. Examples include the classical Kaibab mule deer problem *(94)* and the elk problem in Yellowstone National Park *(64)*. Man himself has had a direct and profound effect on some ecosystems he has attempted to maintain in a natural state, such as in Yosemite National Park *(39)*.

Exploitation of ecosystems is still occurring throughout the world, although the consequences are yet unknown. Systems ecology may contribute to a better understanding of ecosystems and ecosystem processes, which will help civilized man produce new and useful quasi-stable equilibria. We still need to know the long-term effects and profits of ecosystem manipulation, such as even further shortening of food chains, as human populations continue to increase exponentially and impose greater stresses on our world ecosystem. Knowledge about the entire ecosystem has become so important that ecologists can no longer be satisfied to be concerned with specific individuals, species, or populations in the ecosystem. In addition to plant ecologists, animal ecologists, microbial ecologists, etc., we must now train more and more young ecologists to confront the entire complexity of the ecosystem.

We are still in the process of developing technology and scientific knowledge that will enable us to better perceive the influences of our manipulations of ecosystems. Intelligent manipulation of ecosystems is increasing, and there is increased interest in scientifically defining the carrying capacity of ecosystems, as in the proposed International Biological Program, for example. Scientific ecology has clarified many cause-and-effect relationships in environmental change. An example of the value of basic ecological knowledge may be found in studies relating community stability, diversity, and biological control practices *(116)*.

SYSTEMS ECOLOGY

Definitions

The July 1964 issue of *BioScience* contained perspective articles by several noted ecologists, and the term systems ecology was used. E. P. Odum *(83)*, then president of the Ecological Society of America, used the term "systems ecology" as follows:

... the new ecology is thus a systems ecology—or to put it in other words, the new ecology deals with the structure and function of levels of organization beyond that of the individual and species.

We have been taught that ecology is the study of the relationships between organisms and their environment *(51)* and that ecology may be subdivided into autecology (of individuals or species), population ecology, and community ecology *(80)*. Systems ecology in a sense approximates communities ecology. The terms system and ecology both imply a holistic viewpoint. Just as ecology and ecosystems are considered by some people, systems ecology may be not so much an independent branch of study, but a point of view, a way of looking at things and explaining them, a concentration of pertinent concepts, facts, and data from various fields *(61)*.

Some workers consider the realm of systems ecology to be that of using mathematics to study ecological systems. Although application of mathematical techniques to study ecosystems is an important part of systems ecology, it is far from being all of it. Systems ecology can be broadly defined as the study of the development, dynamics, and disruption of ecosystems. I consider systems ecology to have two main phases—a theoretical and analytical phase and an experimental phase.

Earlier I stated that for studying function in ecology we need methods and concepts which are different from those for studying structure. Essentially, study of problems in systems ecology requires three groups of tools and processes: conceptual, mechanical, and mathematical.

Study of Ecosystems

The tools and processes required for systems ecology are different from those for conventional phases of ecology because of the complexity of the total ecosystem as compared with a segment of it. When we consider the totality of interactions of populations with one another and with their physical environs—i.e., ecosystem ecology—we face a new degree of complexity *(10)*. Other than some recent papers (e.g., ref. *41*) only a few reasonably adequate functional analyses of natural ecosystems exist *(80)*.

One of the major problems in systems ecology is that of analyzing and understanding interactions. Events in nature are seldom, if ever, caused by a single factor. They are due to multiple factors which are integrated by the organism or the ecosystem to produce an effect which we observe *(45)*. To further complicate the matter, various combinations of factors and their interactions may be interpreted and integrated by the ecosystem to produce the same end result.

Conceptual Requirements

A first conceptual requirement in systems ecology is clearer definition of problems. It is axiomatic that ambiguous use of terminology and an ambiguous statement of the problem lead to ambiguities of thought as well *(19)*. These statements apply to many fields, but, particularly here, clear definitions are required because of the type of people systems ecologists will be and the types of people with whom they will work (discussed further below). Furthermore, in using computers, which are essential tools for systems ecologists, it is necessary to formulate the problems precisely and to clearly delineate the factors involved.

A second conceptual requirement in systems ecology is more and better use of logic and scientific and statistical methods. Essentially, we can define scientific method as the pursuit of truth as determined by logic and experimentation. In scientific method we use the approach of systematic doubt to discover what the facts really are. Experimentation is one of several tools of scientific method used to eliminate untenable theories, that is, to test hypotheses *(32)*. Other experiments may be conducted to determine existing conditions, or to suggest hypotheses, etc. The conclusions from experiments may be criticized because the interpretation was faulty, or the original assumptions were faulty, or the experiment was poorly designed or badly executed *(88)*. Experimental design and statistical inference are aids in testing hypotheses.

Much past ecological research has not tested a hypothesis. There is a tendency for ecologists to pass over the primary phase of analysis. The lack of understanding of what is known already (inadequate knowledge of the literature, in part) is understandable because of the volume of material to be covered *(58)*. Glass *(40)* has clearly stated this dilemma—"the vastness of the scientific literature makes the search for general comprehension and perception of new relationships and possibilities every day more arduous." But inadequate examination of facts and data and inadequate formulation of hypotheses lead to uncritical selection of experiments testing poorly formulated hypotheses, and ecologists are often at fault here *(51)*. The experimental design is, essentially, the plan or strategy of the experiment to test clearly certain hypotheses *(32)*. Statistical methods are especially important in experimentation with ecosystems, because not all factors influencing the system can be controlled in the experiment without altering the system *(29)*. These uncontrollable factors lead to error or "noise" in our measurements, and inferences to be made from the results of experiments should be accompanied by probability statements *(32)*.

Eberhardt *(27)* has discussed many of the problems ecologists encounter in sampling, and has stressed the importance of statistical techniques in analysis of such problems. Methods of statistical inference are also useful in suggesting improvements in our mathematical models and in suggesting alterations in the design of future experiments. Some of the work initiated and developed by the late R. A. Fisher on partial correlation and regression is invaluable to us in evaluating independent and interaction effects in complex ecosystems where experimental control is neither possible nor desirable *(45)*.

The first two conceptual needs for systems ecology, mentioned above, lead naturally to the third, the approach of modeling (Fig. 4), with models which are mathematical abstractions of real world situations *(17, 107)*. In this process some real world situation is abstracted into a mathematical model or a mathematical system. Next, we apply mathematical argument to reach mathematical conclusions. The mathematical conclusions are then interpreted into their physical counterparts. In some instances we are able to proceed from the real world situation via experimentation to reach physical conclusions. In other instances, however, we cannot experiment with a situation that does not exist but may become real; examples are such situations as thermonuclear war and wide-scale environmental pollution *(50)*. In many cases we find it too costly to experiment; therefore mathematical modeling or mathematical experimentation may be especially useful.

Mathematical modeling is somewhat new to many conventional ecologists and, in part, is just as much an art as a science. To ensure that the model will be valid, the mathematical axioms must be translations of valid properties of the real world system. The application of mathematical argument gives rise to theorems which we hope can be interpreted to give new insight into our real world system. However, the value of these conclusions should, where possible, be verified by experimentation. We must then accept the conclusions or reject them and start over again. This procedure of modeling, interpretation, and verification is used in many engineering and scientific disciplines. The success of the procedure, however, depends on the existence of an adequate fund of basic knowledge about the system. This knowledge permits predictive calculations. Hollister *(50)* outlines some of the problems to be encountered in modeling ecological phenomena.

Tools for Study of Ecosystems

The above conceptual tools should provide a framework in which to attack the complex problems of systems ecology. To implement these methods in studying ecosystems we will need both physical and mathematical tools, including digital and analog computers and electrical, mechanical, and hydraulic simulation devices, and artificial populations *(44, 75, 85)*. The act of expressing and

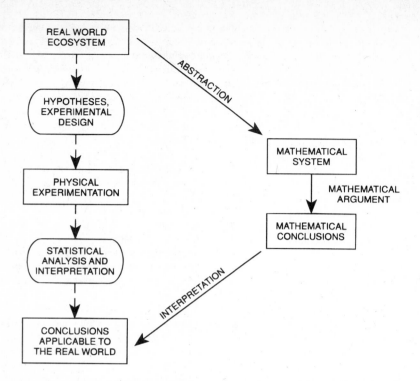

Figure 4. Two ways of experimenting with ecosystems. One involves the conventional process of formulating hypotheses, designing and conducting experiments, and analysis and interpretation of results. The second involves the abstraction of the system into a model, application of mathematical argument, and interpretation of mathematical conclusions.

testing biological problems with numerical, electrical, or hydraulic analogs often reveals some unsuspected relationships and leads to new approaches in investigation. In conducting experiments in systems ecology, more refined chemical analytical equipment will be needed, such as gas chromatographs, infrared gas analyzers, and recording spectrophotometers. Physical analytical equipment required includes micro-bomb calorimeters, biotelemetric equipment, and other electronic equipment useful for rapid, nondestructive sampling and measuring of plant and animal populations and parameters under field conditions.

The importance of these chemical and physical tools is apparent when one considers the amount and variety of apparatus required to construct and maintain even the simplest aquatic ecosystems or to transplant and manipulate naturally occurring ecosystems for detailed measurements (e.g., ref. *2*). A major reason for the scarcity of detailed studies of entire terrestrial ecosystems is that many ecologists are not trained to use many of the required, diversified tools. In other instances these tools may not be available to the ecologist. The systems ecologist cannot be an expert with each of these tools, but he must be aware of their applications and limitations in the study of components and processes in ecosystems. He will need to be conversant with the specialists in other disciplines who make increasing use of these modern and complex tools. For example, in the last 12 years there has been a sixfold increase in the use of large, expensive, and complex instruments in chemistry *(118)*.

One of the important tools is tracing with radioisotopes. This is valuable for identifying food chains, for determining the mass of nutrients in various compartments of an ecosystem, and for determining the time and extent of transfer of matter and energy among compartments within the ecosystem. Possibly we can use two, three, or more tags simultaneously in many ecological experiments if we select tracers which have appropriate physiological properties and types of radiation. Ionizing irradiation and selective poisons are other tools or treatments that can enable us to learn more about the function of ecosystem components without physically dismantling the entire ecosystem.

One interesting feature is that many of the new methods and instruments are simpler to use, although more expensive, because much laboratory skill in manipulation has been eliminated by instrumentation (40). The systems ecologist will still require the tools of conventional ecology, but he cannot rely on them alone. An important caution for the systems ecologist is that the problem should dictate the tools to be used; the opportunity to use a complex tool should not dictate the problem to be studied.

Operations Research and Systems Analysis Applications

Mathematical analysis will become increasingly important in providing advances in systems ecology. Large, fast digital computers have become available in the last 15 years and have allowed the development of special methods of analyzing and studying complex systems in industry and government. Most of these newer mathematical tools were developed in and are used primarily in two loosely defined and somewhat overlapping fields, operations research and systems analysis.

Operations research may be defined as the application of modern scientific techniques to problems involving the operation of a system looked upon as a whole (77). Included therein are any systematic, quantitative analyses aimed at improving efficiency in a situation where "efficient" is well understood (103).

Systems analysis is more difficult to define. Perhaps it can best be defined by opposites. The opposite of a systems approach is unsystematic or piecemeal consideration of problems; intuition may be taken as the opposite of analysis (46). Essentially systems analysis is any analysis to suggest a course of action arrived at by systematically examining the objectives, costs, effectiveness, and risks of alternative policies—and designing additional ones if those examined are found to be insufficient (93).

It is easily seen that operations research and systems analysis are both alike and different. They both contain elements from mathematical, statistical, and logical disciplines. In operations research, however, there usually is an unambiguous goal to be achieved, and the operations researcher is interested in optimization. The systems analyst faces a multiplicity of goals, a highly uncertain future, a frequent predominance of qualitative elements, and an exceedingly low probability of building an accurate and satisfactory model for his total problem (103). Because of the methods and techniques he can use effectively, the systems engineer has much to offer in study of ecosystems but he will need considerable guidance. In systems ecology he will be facing a collection and coupling of "green, pink, and brown" boxes (plants, animals, and physical environment) rather than the black boxes with which he is familiar (56). The interconnections between these boxes may be known only imperfectly, and the functional significance of the boxes will need to be established.

Some of the mathematical tools to be employed and examined in systems ecology include scientific decision-making procedures, theory of games, mathematical programming, theory of random processes, and methods of handling problems of inventory, allocation, and transportation (77).

Linear, nonlinear, and dynamic programming, which are especially important to the operations researcher, already show promise in ecology (4, 112, 115) and in management of renewable resources (12, 60, 69). Mathematical programming has already been used widely in agro-ecological problems, such as crop or yield prediction (97), in formulation of least-cost rations for livestock (110), and in farm management decisions (5). Game theory has been applied to decisions in cultivated-crop agriculture *(113)* and appears to have potential in dealing with wildland resources. Queuing theory and network flow appear to offer much in looking at problems of flow rates in ecosystems *(90)*. Margalef *(74)* has discussed and indicated some important applications of information theory in ecology. Cybernetics principles and techniques are also useful in studying biological systems *(37)*. Simulation is another important tool in operations research, although not limited to it. Mathematical simulation models have been used to study important resource problems, such as salmon population biology *(59, 95)*, and abstract systems *(36)*.

Importance of Digital Computers

Probably most systems ecology problems will be attacked first with deterministic models as first approximations *(70, 71)*. However, to increase their usefulness and their realism, stochastic elements will be involved in most models or an indeterministic point of view will be taken; for example, see Leslie *(65)*, Neyman and Scott *(79)*, and Jenkins and Halter *(53)*. This will require extensive use of digital computers, not only in simulation but also in analysis. Most stochastic models in ecology to date have been concerned with only one or two species rather than populations or ecosystems *(6)*. Stochastic simulation of biological models or processes has been a useful process in some problems *(109, 111)*. Many problems of modeling and analysis will require study and examination of the underlying statistical distributions *(28)*. In addition to the normal distribution, other distributions which will need examination and use in systems ecology problems include the Poisson, the exponential, and the log-normal. Monte Carlo methods will be especially valuable in developing, testing, and using stochastic or probabilistic models *(30, 67)*. Computers are essential in studying and using these statistical techniques in systems ecology. Other statistical aspects are discussed by Eberhardt *(29)*.

Compartment model methodology, implemented with both analog and digital computers, has proven its value in theoretical studies and is beginning to be put to use in analysis and extension of real data in medical *(9)* and ecological fields *(87, 90)*. Thus far, however, compartmental simultation models have been restricted to relatively simple ecological and agricultural situations, because most investigators have worked with analog systems of limited capacity *(1, 35)*, although simulation systems have been developed for and used with digital computers in the study of renewable resources (e.g., ref. *42*). Most systems ecologists will find it surprisingly easy to express many problems in the pseudoalgebraic languages, such as the many dialects of FORTRAN, used to communicate with digital computers.

An ecosystem might be depicted, as in Fig. 5, as composed of trophic levels represented as three-dimensional matrices. PRODUCER (i, j, k), CONSUMER (i, j, k), and DECOMPOSER (i, j, k) are matrices of species, individuals, and parts. The ranges of i, j, and k in each matrix are variable and depend upon the study. Matrices of transfer functions, depicted and simplified by the arrows in Fig. 5, are concerned with movement of matter or energy within individuals, between individuals, or between species. The latter two types of transfers may be between or within trophic levels. Also included in the figure is the fact that flux among contiguous ecosystems may be considered in matrix representation. Some of the transfer functions themselves may contain random noise and may be functions of a driving variable, such as macroclimate, acting on the system over time. Models or functions for macroclimatic influences may be constructed from actual data or may follow some prescribed hypothetical statistical distribution.

Consider the simplified case (Fig. 5) with only three parts per individual, three individuals per species, three species per trophic level, three trophic levels per ecosystem, and three ecosystems per problem. This leads to 3^5 microcompartments to be accounted for in addition to the many transfer functions interrelating the compartments. Many of the transfers will be zero, but this simplified model exceeds the capacity of most analog computers even if the problems of using various random function generators with an analog computer are bypassed. This example does not

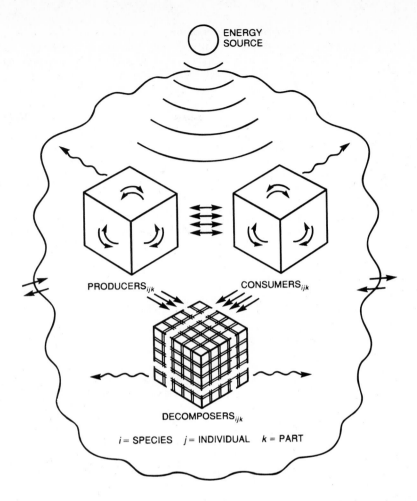

Figure 5. A matrix representation of ecosystems adapted to pseudoalgebraic computer languages. Each trophic level is represented as a three-dimensional matrix. The arrows, wavy for energy and straight for matter, represent matrices of transfer functions interconnecting parts within individuals, individuals of a species to each other and to other species, etc., on up to connections between contiguous ecosystems. The transfer functions may contain probabilistic components and may be probabilistic functions of external variables such as macroclimate.

indicate that analog computers will not play an important role in systems ecology, but only that they may be of limited value in many realistically complex situations. Their major role may be as teaching (and learning) tools and as components of hybrid (digital–analog) systems. The capabilities and versatility of digital computers in general are far greater than those of analog computers *(62)*.

Maximum use of most of the above mathematical tools and others by systems ecologists depends upon access to fast digital computers with large memory capacities *(115)*. Such access will be especially important in working with large complicated models where remote-console access to large central computers will be essential for efficient and rapid progress. Computer technology is approaching the point where the rate of debugging of programs is the limiting factor.

The role of computers in the future of systems ecology is too readily underestimated. Computers in tomorrow's technology will have larger and faster memories, remote consoles, and time-sharing systems. Some may accept hand-written notes and drawings, respond to human voices, and translate written words from one language into spoken words in another *(96)*. There will be vast networks of data stations and information banks, with information transmitted by laser channels over a global network. This network will be used not only by researchers but also by engineers, lawyers, medical men, and sociologists as well as government, industry, and the military. Computers could become tomorrow's reference library used by students in the university; they are already starting to revolutionize our present approaches to certain kinds of teaching. To utilize computers effectively in ecology we will have to state precisely what we know, what we do not know, and what we wish to know. Also, it will be necessary to assemble, analyze, identify, reduce, and store our ecological data and knowledge in a form retrievable by machine.

A Systems Approach

Systems ecology will call for an interdisciplinary team of systems ecologists, systems analysts and operations researchers (if they can be separated), conventional ecologists, mathematicians, computer technologists, and applied ecologists, including agriculturalists and natural resource managers of various disciplines. Systems ecologists studying ecosystems will devote at least as much time to delineating the problem as they will to solving it. This gives a hint as to the nature of the work of the systems ecologist.

The physical and mathematical tools to be used by this team are impressive, even though the list in the preceding sections is only partial. It serves to show that the systems ecologist will have to have more types of specialized training than did his predecessors. That different tools and methods may be needed to solve some of today's complex ecological problems is emphasized by the fact that many important contributors to advances in ecology in recent years may not be identified as ecologists *(92)*. This will be especially true of systems ecology in the future, even though ecologists must be generalists and systems ecologists also will, in part, have to be generalists. Still, there are probably few if any authentic generalists or truly great minds who are not firmly grounded in a specialty *(18)*. Most systems ecologists will serve their apprenticeship in basic fields. The conventional plant, animal, and aquatic ecologists will not be acceptable as systems ecologists, because they will lack the depth required in many specialties *(78)*.

This raises the difficult question of how to train a man to be a specialist in at least one field, to be able to converse well with specialists in several fields, and yet to have a holistic or systems viewpoint.

SYSTEMS ECOLOGISTS

Definitions

The systems ecologist of tomorrow may be defined as one type of scientist who is a specialist in generalization *(100)*. There are few if any systems ecologists today. Of today's biologists, perhaps some scientists in applied ecology can be considered systems ecologists (Fig. 6). In some applied ecology fields, such as forestry, it has been noted that the field is becoming so complex that more members of the profession will likely find it to their advantage to either specialize or become an exceptionally well-balanced generalist *(11)*. This trend is well established in many professional fields. In addition to the growing need for specialists in resource management, there also is a niche for the generalist *(120)*. Undergraduate programs have been developed to train such individuals.

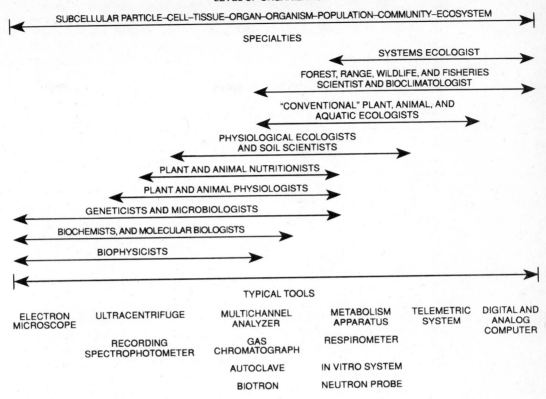

Figure 6. A schematic comparison of biologists, the level of organization of the media with which they work, and the tools they use. The double-ended arrows indicate general positions of the specialties. Tools especially, and to some extent the media, overlap widely for different specialties.

The applied ecologists, shown as the second group of specialists in Fig. 6, to some degree have in their training many elements of the training of the four groups of specialists listed below them. The applied ecologists are closer akin to the systems ecologists than are the conventional ecologists, because in their training and in their work they usually are more cognizant of the total ecosystem and its interrelations than are many conventional ecologists.

Consider, for example, a scientist responsible for the trout population in a Rocky Mountain forest. He realizes that the trout's well-being is inextricably related to the total environment. He must consider the impact of grazing, lumbering, mining, and road-building upon the response of the watershed to uncontrolled and fluctuating precipitation. He must consider also the inherent fertility of the watershed and its impact on populations of fish and fish foods. Superimposed upon this are other factors, such as insect control by wide-scale pesticide spraying, the problems of optimum rates and places of artificial stocking of streams, of seasons and levels of bag limits, of public relations, etc., ad infinitum. In contrast, for example, few plant ecologists thoroughly appreciate aquatic problems or communicate well with aquatic ecologists; few animal ecologists understand soil problems or communicate well with soil scientists.

The systems ecologist will require better mathematical, chemical, physical, and electronic training than either the applied or the conventional ecologist. Yet he must share their holistic way of thinking or approaching problems, and he must have a broad background in ecological subject matter. A lifetime may not be sufficient for any one person to prepare adequately to perform unassisted the synthesizing function, a major effort of the systems ecologist. This function requires the cooperation of specialists, and publication in each other's journals *(52)*. No individual will be able to direct or conduct research without consulting others to obtain a complete understanding of the processes within even most fairly simple ecosystems *(80)*. Future leaders toward this goal must have the ability to organize concepts, things, and people.

It has long been apparent to those in physical sciences that the ecologist, in the broad sense, must be an environmental specialist. For example, Jehn *(52)* suggested that an ecologist must simultaneously be a meteorologist, a soils physicist, a geologist, and a geographer. But because no one man can encompass all the required specialties, he must ally himself with these specialists *(61)*. Therefore, the greatest advances to be made in systems ecology will require the effort of an interdisciplinary team. How can this be done without losing the spontaneity and originality of the individual's personality?

Systems Ecologists: Interdisciplinarians and Multidisciplinarians

That interdisciplinary teams are required to solve many physicobiological problems of national importance is becoming more and more apparent *(43)*. For example, the understanding of pollution processes requires the cooperation "of [systems] ecologists, physiologists, biomathematicians, microclimatologists, geneticists, microbiologists, biochemists, chemists, morphologists, and taxonomists . . ." *(108)*. To work effectively as a member of an interdisciplinary team, the systems ecologist will need to establish a common vocabulary, an agreed-upon ideology, a set of reasonable goals, a common context for symbols, and ways for translating ideas into action *(57)*.

The systems ecologist is one of the types of interdisciplinary scientists who should be in great demand in the near future. It has been estimated *(98)*

> that about ten percent of our total national effort will be going into production, development, and research based on biophysics, biomedicine, bioengineering, and related computer projects by 1970. . . . then we must hurriedly prepare to train several thousand additional students to the Master's or Ph.D. level per year in this difficult field and we must anticipate at least a doubling of our teaching and research facilities in this interdiscipline once in each three years during this decade.

Systems ecologists can and should contribute heavily to these efforts. But the increasing importance of group effort and interdisciplinary teams in the study of major, man-created environmental problems creates new paradoxes. Large-scale, expensive research activity may decrease the flexibility and freedom which are intrinsic to research. Operation of an interdisciplinary team requires unique coordination and appreciation of contributions by different skills at many levels *(23)*. Interdisciplinary research must be reconciled with the continuing importance of distinctive contributions by highly talented and motivated individuals. Furthermore, it is historical fact that to date many major scientific achievements have been made by specialists—by scientists wearing blinders *(46)*.

Imagination and inventiveness, like the ability to work in an interdisciplinary team, are difficult to develop by training *(47)*. A successful systems ecologist will be one with the imagination to perceive an important problem before others do. He must have the inventiveness to devise and weigh alternatives for its solution. This emphasizes the multidisciplinary nature of systems ecologists.

In order to contribute effectively in the interdisciplinary team they must have sufficient depth in more than one specialty in order to make significant contributions to the solution of the problem. Thus, systems ecologists will need both breadth and depth of interests.

The role of the systems ecologist is complex and not well defined, nor is it easy to analyze. He may be viewed by many specialists of the interdisciplinary team as an amateur, and he may be viewed by his fellow ecologists with suspicion. Both views are justified until he proves his worth to all concerned. A major problem of the systems ecologist will be to convince an ecologist that a mathematical attack is useful and to convince the mathematician that his time and methods will be productive in ecology. Is there a natural course for this convincing to take? The interdisciplinary viewpoint does not rest solely on the biologist. A team composed of biologists with no mathematical and computer training and of engineers, mathematicians, and programmers with no biological training is doomed for failure *(62)*.

Consider next the major activities of the systems ecologist in terms of combinations of old and new problems attacked with old and new methods [*sensu* Bellman *(7)*].

Old Problems—Old Techniques

In attacking old problems with old techniques, systems ecologists will be looked upon by their colleagues as useful intermediaries between themselves and biometricians. Therein, systems ecologists are used by their colleagues to provide guidance in the use of rather fundamental mathematical principles and methods. Soon, however, the average ecologist will be better trained mathematically and will not require as much of this service from the systems ecologist. How soon this will be is not certain. These activities may well be considered as one useful part of the training of the young systems ecologist, but should not be the major activity of the senior systems ecologist.

Old Problems—New Techniques

Old problems can also be attacked with some of the new and challenging techniques, such as those from operations research discussed above. These efforts should have considerable feedback influence on some fields of applied mathematics and may serve to initiate further developments of new methods. The attack on old problems by the systems ecologist armed with new methods should be rewarding and interesting and should produce further insight into ecological systems in a relatively short period of time. This is so because much time and effort has already been expended in the development and application of these techniques in fields with direct ecological analogies—examples are the conduct of war with species competition, the management of a firm with dynamics of ecosystems, the manufacture of a product with problems of nutrient cycling and energy flow, and the planning of an economy with optimization and adaptation in ecosystems. Today's systems ecologists (if they exist) probably will work primarily in this and in the next capacity.

New Problems—Old Techniques

The systems ecologists will still find uses for old techniques to attack new problems. Here he will serve an important consulting and interpreting service to other ecologists. He will be able to use his insight into the problem and its analysis by drawing upon his accumulated experience. This will be an effort of significant importance for the systems ecologist in the near future.

New Problems—New Techniques

New problems attacked with new techniques will be the most challenging, most demanding, most rewarding—and most discouraging—efforts of the systems ecologist of tomorrow. A hint of the challenge and importance of these efforts is seen if we attempt to foresee the consequences of being able to model accurately, to simulate, and to predict behavior of macroecosystems in various fields of renewable resource management. Another challenge for systems ecologists is to work effectively in interdisciplinary teams which are now required and which soon will start to study in detail the movement and degradation of pollutants through complexes of ecosystems. Of similar importance tomorrow is the application of knowledge of ecosystem behavior to the development of support systems for exploration of space *(84).* These efforts will be very demanding, because they will require the closest cooperation of competent and interested specialists.

Such efforts offer ecologists all the tangible and intangible rewards for being first to make major contributions to a new area of study. However, these efforts can be most discouraging because their impact may not be prompt and success is not certain. Furthermore, significant achievements by the systems ecologist in this area may not be recognized immediately for their worth by many conventional and applied ecologists. Most ecologists are not trained to evaluate the nature of this work. Slobodkin *(102)* states that "the number of good quantitative ecologists is in the thirties or forties for the entire world...."

Some Pitfalls Facing Systems Ecologists

The availability of such powerful tools and equipment as gas chromatographs, telemetric devices, and computers will not make the solution of new problems trivial, nor will it make systems ecology research routine. Some of these powerful tools themselves are raising important problems.

A special problem exists with computers. In general, the larger and faster a computer is, the more economical it is, even for small problems, if there is sufficient work available. But, of necessity, operating procedures of large computer centers are rigid in order to maintain output. The "people problems" of getting small problems into large computers grow disproportionately with computer size *(98),* so remote access to and time sharing on these big computers will be essential. Without direct access to the computer, such as provided by remote consoles, many problems can be completed more rapidly with a hand calculator, although they may require several hours' work. Even though they could be run in a few seconds on a computer, the long delay or "turn-around" time in using a computer without direct access leads to inefficiency. Our research output is best and most efficient when we are able to progress at full speed, regardless of time of day or day of week, rather than to take days or weeks to complete a problem. Remote-console access to the large computers which will be required for many ecological problems will allow concentrated work and will in every sense give rapid results.

As compared with his predecessors, the systems ecologist will still have to acquire empirical data by means of experimental or literature research, but he will need a better grasp of the biological and physical interactions in the system he is studying, and he will have to apply more ingenuity and invention to formulating and analyzing his problems in order to make significant advances. The easy problems have been solved.

Another pitfall facing the systems ecologist of tomorrow, who may be an undergraduate today, is that often he has been given equations and their coefficients and has been asked to produce numerical or graphic solutions. The problem he faces when he leaves the "ivy-covered halls" is first to design experiments correctly and then to conduct them effectively before he even obtains experimental results. Then his problem will be to infer and derive the form of the equations and to determine analytically the magnitude of the coefficients. Needless to say, this will be a much different and more difficult task than that which he faces as a student. Perhaps it will be desirable to develop "co-op" training programs wherein the student may intersperse practical experience in his undergraduate program. Graduate students in systems ecology, of course, will have numerous opportunities to test the effectiveness of their training, especially in their research work.

Mathematical Training of Systems Ecologists

The training of systems ecologists in mathematics and computer sciences is an especially important part of their education. Watt's review (114) of the use of mathematics in population ecology gives numerous examples of mathematical methodologies and applications. Unfortunately, many ecologists receive little mathematical, statistical, or computer training, and this is only late in graduate school. An encouraging trend is the recent development of undergraduate biomathematical courses and curricula at several universities. For example, an undergraduate biologist at Colorado State University who has had college algebra can, in 12 quarter credits, complete courses in calculus and differential equations designed for biologists. Mathematical training for attacking four types of problems has been outlined for undergraduate students in biological, management, and social sciences (24).

Consider the four combinations generated by deterministic and stochastic phenomena, each with few or with many variables. Tools required in study of organized simplicity (i.e., deterministic × few variables) include the classical analytic geometry–calculus sequence, and difference and differential equations. Disorganized simplicity (i.e., stochastic × few variables) requires probability and statistics for analysis. Organized complexity requires linear algebra and many-variable advanced calculus. Study and use of complex stochastic models are needed for analysis of phenomena characterized as disorganized complexity. Computers are especially important in these last two areas, and computing practice in numerical analysis is equally important. For those systems ecologists who wish to specialize in computers in their undergraduate or graduate training, additional courses may be recommended (25).

Recently courses have been taught to ecologists which combine many systems-oriented mathematical approaches to ecological problems with practice in the use of digital computers (117) or use of both analog and digital computers (Systems Ecology, a yearlong graduate course at the University of Tennessee, which has been taught by B. C. Patten, J. S. Olson, and the author). Systems ecologists, however, should be trained in mathematics not so much for developing their skill in performing mass computations as for having the ingenuity of escaping them (47).

Systems Ecology Research

It is considered by some that existing ecological theory often has limitations in the rapid solution of many problems (76). Also, some consider that the reliance upon analogies from physics for the solution of ecological problems has distinct limitations (102). But the field essentially is a virgin area for tomorrow's better trained and better equipped systems ecologists. The dearth of quantitative ecologists has been mentioned above, and most ecologists have been isolated and not well supported. Hopefully, we will be able to develop centers wherein "critical masses" of systems ecologists will migrate and find suitable niches.

Perhaps a routine similar to the following will be of use to the systems ecologist (7). He will often have to guess at the fundamental cause-and-effect relationships in his system and may even have to guess about the basic variables. He will then test these hypotheses by comparing the quantitative and qualitative behavior of the real world with that predicted from his model. Fortunately, he will have varied and powerful tools available for the testing of complex (and realistic) as well as simple hypotheses. But there are an infinite number of hypotheses to test about any complex system, and most of these hypotheses will be wrong. Holling (48–49) shows by example how theory and experiment can be combined in a systems analysis of predation in a way to greatly reduce the number of hypotheses to be tested. The majority of the alternatives must be excluded by means which require, not computers, but only pencils, paper, and discriminating thought. The systems ecologist will have to develop the knack of feeding on negative information and use these negative hypotheses to guide his further experimentation and theory. In instances where experimentation or measurement of parameters is impossible without disruption of the systems, perturbation of the system followed by measurements may still give new insight for the definition of a model *(8)*.

A major hurdle to overcome in systems ecology research is lack of precedent in funding detailed and integrated research on complex ecological systems. This applies both to the theoretical and analytical phase and the experimental phase. In many respects no single organization has been working in depth on complex systems ecology problems, for such work may be beyond the role, objectives, or structure of existing organizations. Universities, national laboratories, and state and federal experiment stations each have some unique resources and capabilities for studying ecosystem problems. Some advantages and disadvantages of these three types of organizations with respect to "total systems" research, based on my experience in working in these organizations, are briefly outlined as follows.

In the past, ecological research conducted at universities generally has involved one investigator, or at most a few, on a part-time basis on problems of limited extent. Extensive and intensive interdepartmental cooperation in ecological research has been the exception rather than the rule. Many sources of funds are available to these researchers, but usually in amounts insufficient to attract permanent personnel and to support long-term ecosystem research. Although the economy of graduate student use has been exploited, often there is a lack of continuity in the conduct of long-term environmental researches. By the time the student gains competence and becomes capable of independent contributions to the project, he graduates and is lost to the project. Also, universities often lack controllable research areas or have conflicting needs for them.

Applied ecologists, in state fish and game departments or in state and federal agricultural experiment stations, for instance, often have controllable research areas on which to conduct long-term ecosystem research, but continuity again is impeded because of high turnover rate of personnel. Furthermore, their research funds often are restricted to only one or a few phases of the total ecosystem problem, and their funds have become more limited in recent years. Funds from many granting agencies may not be available to them for research, special training, or foreign study.

In the past ten years considerable ecological research has been conducted at several national laboratories. Ecologists in these laboratories have available many services, tools, and consultants which the university or experiment station scientists lack. Much of the ecological work in the national laboratories, however, has been concerned with specific needs of the funding agency. Most ecological researchers in these laboratories have come directly from liberal arts departments in universities, and these laboratory staffs are divided into subject matter groups for conduct of research. A total-system, interdisciplinary approach usually has not been implemented in their research. Although these ecologists are funded comparatively well, their costs are high due to the nature of their work. Although they may have long-term control over their research areas, the number of these laboratories is limited, and some important biomes, such as grasslands, are not within the boundaries of these laboratories.

I feel that perhaps research in systems ecology could encompass the advantages held by ecological researchers of the above three categories. This would require, however, some shifts concerning funding and conduct of research, and some shifts in administrative policy of the respective agencies or institutions. The exact nature and organization of such research is uncertain, although Dubos *(21)* has raised some interesting questions and has made some good suggestions about similar research in environmental biology.

The long-term impact on man of fundamental, total-ecosystem research should be recognized, and the framework should be developed for extensive and intensive intercooperation of these three groups of ecologists. Analytical and experimental research on total-ecosystem complexes should be initiated as soon as possible, if man is to benefit tomorrow, because most problems of environmental magnitude require many years of study before conclusions may be reached.

The proposed International Biological Program is, in several respects, a call for systems ecology research, both experimental and theoretical. This program could provide an incentive and a means for ecologists from universities, experiment stations, and national laboratories to work cooperatively and share funds, research areas, and talent. An example follows. Other such examples of needed research on total-system problems can be found in other parts of this continent and on other continents, in both terrestrial and aquatic ecosystems.

Consider the seminatural grassland ecosystems in the Great Plains. A more complete understanding of the structure and function of these ecosystems becomes increasingly essential as these lands are called upon to provide food, water, and recreation for tomorrow's growing populations. Several state and federal agricultural experiment stations hold sizable acreages of representative variations of grasslands, from the aspen parkland in Canada to the semidesert grasslands of Mexico. But at none of these stations is there a team equipped with suitable manpower or funds for intensive total-ecosystem research. Scientists at these experiment stations and nearby universities have accumulated considerable data and experience on and about these grassland ecosystems. There is no national laboratory in this vast area, and no university in the area can marshall many of the unique facilities, such as computing facilities and computer consultants, that are necessary for systems ecology research. Still, scientists at universities in the area can provide much necessary insight into these ecosystems and graduate students to help conduct ecosystem research in the area. With sufficient funds and planning it should be possible to combine the special skills and resources of all these groups and bring them to bear on problems of ultimate national importance at selected locations in the Great Plains.

LITERATURE CITED

(1) Arcus, P. L. 1963. An introduction to the use of simulation in the study of grazing management problems. Proc. New Zealand Soc. Animal Prod. 23:159–168.
(2) Armstrong, N. E., and H. T. Odum. 1964. Photoelectric ecosystem. Science 143:256–258.
(3) Bakuzis, E. V. 1959. Structural organization of forest ecosystems. Proc. Minn. Acad. Sci. 27:97–103.
(4) Barea, D. J. 1963. Analisis de ecosistemas en biologia, mediante programacion lineal. Archivos de Zootecnia 12:252–263.
(5) Barker, R. 1964. Use of linear programming in making farm management decisions. New York Agr. Exp. Sta. Bull. 993. 42 pp.
(6) Bartlett, M. S. 1960. Stochastic population models in ecology and epidemiology. Methuen and Co., Ltd. London. 90 pp.
(7) Bellman, R. 1961. Mathematical experimentation and biological research. Rand Corp. P-2300. 12 pp.
(8) Berman, M. 1963. A postulate to aid in model building. J. Theoret. Biol. 4:229–236.
(9) Berman, M. 1963. The formulation and testing of models. Ann. New York Acad. Sci. 108:182–194.
(10) Blair, W. F. 1964. The case for ecology. BioScience 14:17–19.
(11) Briegleb, P. A. 1965. The forester in a science-oriented society. J. Forestry 63:421–423.
(12) Broido, A., R. J. McConnen, and W. G. O'Regan. 1965. Some operations research applications in the conservation of wildland resources. Manage. Sci. 11:802–814.
(13) Caldwell, L. K. 1963. Environment: a new focus for public policy? Public Administration Review 23:132–139.
(14) Churchill, E. D., and H. C. Hanson. 1958. The concept of climax in arctic and alpine vegetation. Bot. Rev. 24:127–191.
(15) Clements, F. E. 1916. Plant succession: an analysis of the development of vegetation. Carnegie Inst. Wash. Publ. 242. 512 pp.
(16) Cole, L. C. 1958. The ecosphere. American Scientist 198:83–92.

(17) Coombs, C. H., H. Raiffa, and R. M. Thrall. 1954. Some views on mathematical models and measurement theory. *In:* Thrall, R. M., C. H. Coombs, and R. L. Davis (eds.). Decision Processes. John Wiley & Sons, Inc. New York. 332 pp.
(18) Dansereau, P. 1964. The future of ecology. BioScience 14:20–23.
(19) Davis, C. C. 1963. On questions of production and productivity in ecology. Arch. Hydrobiol. 59:145–161.
(20) Deevey, E. S. 1964. General and historical ecology. BioScience 14:33–35.
(21) Dubos, R. 1964. Environmental biology. BioScience 14:11–14.
(22) Dubos, R., and R. W. Schaedler. 1964. The digestive tract as an ecosystem. Amer. J. Med. Sci. 248:267–271.
(23) Duckworth, W. E. 1962. A guide to operational research. Metheun and Co., Ltd. London, England. 145 pp.
(24) Duren, W. L., Jr. (Chr.). 1964. Tentative recommendations for the undergraduate mathematics program of students in the biological, management and social sciences. Mathematical Association of America, Committee on the Undergraduate Program in Mathematics. 32 pp.
(25) Duren, W. L. (Chr.). 1964. Recommendations on the undergraduate mathematics program for work in computing. Mathematical Association of America, Committee on the Undergraduate Program in Mathematics. 29 pp.
(26) Dyksterhuis, E. J. 1958. Ecological principles in range evaluation. Bot. Rev. 24:253–272.
(27) Eberhardt, L. L. 1963. Problems in ecological sampling. Northwest Sci. 37:144–154.
(28) Eberhardt, L. L. 1965. Notes on ecological aspects of the aftermath of nuclear attack. pp. 13–25 *in:* Hollister, H., and L. L. Eberhardt. Problems in estimating the biological consequences of nuclear war. U. S. Atomic Energy Commission TAB-R-5.
(29) Eberhardt, L. L. 1965. Notes on the analysis of natural systems. pp. 27–40 *in:* Hollister, H., and L. L. Eberhardt. Problems in estimating the biological consequences of nuclear war. U.S. Atomic Energy Commission TAB-R-5.
(30) Elveback, L., J. P. Fox, and A. Varma. 1964. An extension of the Reed–Frost epidemic model for the study of competition between viral agents in the presence of interference. Amer. J. Hygiene 80:356–364.
(31) Evans, F. C. 1956. Ecosystem as the basic unit in ecology. Science 123:1127–1128.
(32) Feibleman, J. K. 1960. Testing hypotheses by experiment. Persp. Biol. and Med. 4:91–122.
(33) Feibleman, J. K. 1965. The integrative levels in nature. pp. 27–41 *in:* Kyle, B. (ed.). Focus on information and communication. ASLIB. London.
(34) Fosberg, F. R. 1965. The entropy concept in ecology. pp. 157–163 *in:* Symposium on Ecological Research in Humid Tropics Vegetation, Kuching, Sarawak, July 1963.
(35) Garfinkel, D., R. H. MacArthur, and R. Sack. 1964. Computer simulation and analysis of simple ecological systems. Ann. New York Acad. Sci. 115:943–951.
(36) Garfinkel, D., and R. Sack. 1964. Digital computer simulation of an ecological system, based on a modified mass action law. Ecology 45:502–507.
(37) George, F. H. 1965. Cybernetics and biology. Freeman and Co., San Francisco, California. 138 pp.
(38) George, J. L. 1964. Ecological considerations in chemical control: Implications to vertebrate wildlife. Bull. Entomol. Soc. Amer. 10:78–83.
(39) Gibbens, R. P., and H. F. Heady. 1964. The influence of modern man on the vegetation of Yosemite Valley. Calif. Agr. Exp. Sta. Manual 36. 44 pp.
(40) Glass, B. 1964. The critical state of the critical review article. Quart. Rev. Biol. 39:182–185.
(41) Golley, F. B. 1965. Structure and function of an old-field broomsedge community. Ecol. Monogr. 35:113–137.
(42) Gould, E. M., and W. G. O'Regan. 1965. Simulation, a step toward better forest planning. Harvard Forest Paper 13. 86 pp.

(43) Gross, P. M. (Chr.). 1962. Report of the committee on environmental health problems. Public Health Service Publ. 908. 288 pp.
(44) Harris, J. E. 1960. A review of the symposium: models and analogues in biology. Symp. Soc. Exp. Biol. 14:250–255.
(45) Hasler, A. D. 1964. Experimental limnology. BioScience 14:36–38.
(46) Hitch, C. 1955. An appreciation of systems analysis. Rand Corp. P-699. (Symposium on Problems and Methods in Military Operations Research, pp. 466–481.)
(47) Hoag, M. W. 1956. An introduction to systems analysis. Rand Corp. RM-1678. 21 pp.
(48) Holling, C. S. 1963. An experimental component analysis of population processes. Mem. Entomol. Soc. Canada. 32:22–32.
(49) Holling, C. S. 1965. The functional response of predators to prey density and its role in mimicry and population regulation. Mem. Entomol. Soc. Canada. 45:3–60.
(50) Hollister, H. 1965. Problems in estimating the biological consequences of nuclear war. pp. 1–11 *in:* Hollister, H., and L. L. Eberhardt. Problems in estimating the biological consequences of nuclear war. U.S. Atomic Energy Commission TAB-R-5.
(51) Hughes, R. D., and D. Walker. 1965. Education and training in ecology. Vestes 8:173–178.
(52) Jehn, K. H. 1950. The plant and animal environment: a frontier. Ecology 31:657–658.
(53) Jenkins, K. B., and A. N. Halter. 1963. A multi-stage stochastic replacement decision model. Ore. Agr. Exp. Sta. Tech. Bull. 67. 31 pp.
(54) Jenny, H. 1941. Factors of soil formation. McGraw-Hill Book Co. New York. 281 pp.
(55) Jenny, H. 1961. Derivation of the state factor equation. Soil Sci. Soc. Amer. Proc. 25:385–388.
(56) Jones, R. W. 1963. System theory and physiological processes: An engineer looks at physiology. Science 140:461–464.
(57) Kennedy, J. L. 1956. A display technique for planning. Rand Corp. P-965.
(58) Kramer, P. J. 1964. Strengthening the biological foundations of resource management. Trans. N. Amer. Wildlife and Natural Resources Conf. 29:58–68.
(59) Larkin, P. A., and A. S. Hourston. 1964. A model for simulation of the population biology of Pacific salmon. J. Fish Res. Bd. Canada 21:1245–1265.
(60) Leak, W. B. 1964. Estimating maximum allowable timber yields by linear programming. U.S. For. Serv. Res. Paper NE-17. 9 pp.
(61) Lebrun, J. 1964. Natural balances and scientific research. Impact of Sci. on Society 14:19–37.
(62) Ledley, R. S. 1965. Use of computers in biology and medicine. McGraw-Hill Book Co., Inc. New York. 965 pp.
(63) Leigh, E. G. 1965. On the relation between the productivity, biomass, diversity, and stability of a community. Proc. Nat. Acad. Sci. 53:777–783.
(64) Leopold, A. S., S. A. Cain, C. H. Cottam, I. N. Gabrielson, and T. L. Kimball. 1963. Wildlife management in the national parks. Amer. Forests 69:32–35, 61–63.
(65) Leslie, P. H. 1958. A stochastic model for studying the properties of certain biological systems by numerical methods. Biometrika 45:16–31.
(66) Lewis, J. K. 1959. The ecosystem in range management. Proc. Ann. Meet. Amer. Soc. of Range Manage. 12:23–26.
(67) Lloyd, M. 1962. Probability and stochastic processes in ecology. *In:* H. L. Lucas (ed.). The Cullowhee Conf. on Training in Biomath. Institute of Statistics, N. Car. St. U., Raleigh, N.C.
(68) Lotka, A. J. 1925. Elements of physical biology. Williams and Wilkins Co. Baltimore. 460 pp. (Also published by Dover Publications, Inc. New York. 1956.)
(69) Loucks, D. P. 1964. The development of an optimal program for sustained-yield management. J. Forestry 62:485–490.
(70) Lucas, H. L. 1960. Theory and mathematics in grassland problems. Proc. Intern. Grassland Cong. 8:732–736.

(71) Lucas, H. L. 1964. Stochastic elements in biological models; their sources and significance. pp. 355–383 *in:* Gurland, J. (ed.). Stochastic models in medicine and biology. U. Wisc. Press, Madison, Wisc. 393 pp.
(72) Lutz, H. J. 1963. Forest ecosystems: their maintenance, amelioration, and deterioration. J. Forestry 61:563–569.
(73) Maelzer, D. A. 1965. Environment, semantics, and system theory in ecology. J. Theoret. Biol. 8:395–402.
(74) Margalef, R. 1957. Information theory in ecology. Mem. Real Acad. Ciencias y Artes de Barcelona 23:373–449.
(75) Margalef, R. 1962. Modelos fiscos simplificados de poblaciones de organismos. Mem. Real Acad. Ciencias y Artes de Barcelona 24:83–146.
(76) Margalef, R. 1963. On certain unifying principles in ecology. The Amer. Nat. 97:357–374.
(77) Miller, I., and J. E. Freund. 1965. Probability and statistics for engineers. Prentice-Hall Inc., Englewood Cliffs, N. J. 432 pp.
(78) Miller, R. S. 1965. Summary report of the ecology study committee with recommendations for the future of ecology and the Ecological Society of America. Bull. Ecol. Soc. Amer. 46:61–82.
(79) Neyman, J., and E. L. Scott. 1959. Stochastic models of population dynamics. Science 130:303–308.
(80) O'Connor, F. B. 1964. Energy flow and population metabolism. Science Prog. 52:406–414.
(81) Odum, E. P. 1959. Fundamentals of ecology. W. B. Saunders Co. Philadelphia, Pa. 2nd Ed. 546 pp.
(82) Odum, E. P. 1963. Ecology. Holt, Rinehart, and Winston. New York. 152 pp.
(83) Odum, E. P. 1964. The new ecology. BioScience 14:14–16.
(84) Odum, H. T. 1963. Limits of remote ecosystems containing man. The Amer. Biol. Teacher 25:429–443.
(85) Odum, H. T. 1965. An electrical network model of the rain forest ecological system. U.S. Atomic Energy Commission PRNC 67.
(86) Odum, H. T., and R. C. Pinkerton. 1955. Time's speed regulator: the optimum efficiency for maximum output in physical and biological systems. Amer. Sci. 43:331–343.
(87) Olson, J. S. 1965. Equations for cesium transfer in a *Liriodendron* forest. Health Physics 11:1385–1392.
(88) Ostle, B. 1963. Statistics in research. Iowa St. Univ. Press. 2nd Ed. 585 pp.
(89) Ovington, J. D. 1960. The ecosystem concept as an aid to forest classification. Silva Fennica 105:73–76.
(90) Patten, B. C. 1964. The systems approach in radiation ecology. Oak Ridge National Laboratory Technical Memorandum 1008. 19 pp.
(91) Pechanec, J. F. 1964. Progress in research on native vegetation for resource management. Trans. N. Amer. Wildlife and Natural Resources Conf. 29:80–89.
(92) Platt, R. B., W. D. Billings, D. M. Gates, C. E. Olmsted, R. E. Shanks, and J. R. Tester. 1964. The importance of environment to life. BioScience 14:25–29.
(93) Quade, E. S. (ed.). 1964. Analysis for military decisions. Rand Corp. R-387. Rand McNally & Co., Chicago.
(94) Rasmussen, D. I. 1941. Biotic communities of Kaibab Plateau, Arizona. Ecol. Monogr. 11:229–275.
(95) Royce, W. F., D. E. Bevan, J. A. Crutchfield, G. J. Paulik, and R. L. Fletcher. 1963. Salmon gear limitation in northern Washington waters. U. Wash. Publ. in Fisheries (N.S.) 2:1–123.
(96) Sarnoff, D. 1964. The promise and challenge of the computer. Amer. Fed. Infor. Process. Soc. Conf. Proc. 26:3–10.
(97) Schaller, W. N., and G. W. Dean. 1965. Predicting regional crop production: an application of recursive programing. USDA Tech. Bull. 1329. 95 pp.

(98) Schmitt, O. H., and C. A. Caceres (eds.). 1964. Electronic and computer-assisted studies of bio-medical problems. C. C. Thomas. Springfield, Illinois. 314 pp.
(99) Schultz, A. M. 1961. Introduction to range management. U. California. Ditto notes. 116 pp.
(100) Schultz, A. M. 1965. The ecosystem as a conceptual tool in the management of natural resources. *In:* Parsons, J. J. (ed.). Symposium on quality and quantity in natural resource management (in press), manuscript 33 pp.
(101) Sears, P. B. 1963. The validity of ecological models. pp. 35–42. *in:* XVI International Congress of Zoology. Vol. 7. Science and Man Symposium—Nature, Man and Pesticides.
(102) Slobodkin, L. B. 1965. On the present incompleteness of mathematical ecology. Amer. Scientist 53:347–357.
(103) Specht, R. D. 1964. Systems analysis for the postattack environment: some reflections and suggestions. Rand Corp. RM-4030. 34 pp.
(104) Stone, E. C. 1965. Preserving vegetation in parks and wilderness. Science 150:1261–1267.
(105) Sukachev, V. N. 1960. Relationship of biogeocoenosis, ecosystem, and facies. Soviet Soil Science 1960:579–584.
(106) Tansley, A. G. 1935. The use and abuse of vegetational concepts and terms. Ecology 16:284–307.
(107) Thrall, R. M. 1964. Notes on mathematical models. U. Mich. Engin. Summer Conf.—Foundations and Tools for Operations Research and Management Sciences. Multilith. 97 pp.
(108) Tukey, J. W., *et al.* (Environmental Pollution Panel, President's Science Advisory Committee). 1965. Restoring the quality of our environment. Superintendent of Documents, U.S. Govt. Printing Office, Washington, D.C. 317 pp.
(109) Turner, F. B. 1965. Uptake of fallout radionuclides by mammals and a stochastic simulation of the process. pp. 800–820 *in:* Klement, A. W. (ed.). Radioactive fallout from nuclear weapons tests. U.S. Atomic Energy Commission Symp. Ser. 5. 953 pp.
(110) van de Panne, C., and W. Popp. 1963. Minimum-cost cattle feed under probabilistic protein constraints. Manage. Sci. 9:405–430.
(111) Van Dyne, G. M. 1965. Probabilistic estimates of range forage intake. Proc. West. Sect. Amer. Soc. Animal Sci. 16 (LXXVII): 1–6.
(112) Van Dyne, G. M. 1965. Application of some operations research techniques to food chain analysis problems. Health Physics 11:1511–1519.
(113) Walker, O. L., E. O. Heady, and J. T. Pesek. 1964. Application of game theoretic models to agricultural decision making. Agronomy J. 56:170–173.
(114) Watt, K. E. F. 1962. Use of mathematics in population ecology. Ann. Rev. Entom. 7:243–260.
(115) Watt, K. E. F. 1964. The use of mathematics and computers to determine optimal strategy and tactics for a given insect pest control problem. Canad. Entomol. 96:202–220.
(116) Watt, K. E. F. 1965. Community stability and the strategy of biological control. Canad. Entomol. 97:887–895.
(117) Watt, K. E. F. 1965. An experimental graduate training program in biomathematics. BioScience 15:777–780.
(118) Westheimer, F. H. (Chr.). 1965. Chemistry: opportunities and needs. Nat. Acad. Sci.—Nat. Res. Counc. Publ. 1292. Washington, D.C. 222 pp.
(119) Whittaker, R. H. 1953. A consideration of climax theory: the climax as a population and pattern. Ecol. Monogr. 23:41–78.
(120) Yambert, P. A. 1964. Is there a niche for the generalist? Trans. N. Amer. Wildlife and Natural Resources Conf. 29:352–372.

Part 1

MICROSCOPIC COMPLEXITY: DOWNWARD–INWARD INTO THE ECOSYSTEM

Science is strongest in its reductionist–mechanistic mode, gaining explanatory power to the extent that it can analyze the hierarchical world downward and inwardly to its component parts and processes. Although George Van Dyne was a holist in the sense that he took a broad, even global, view of his ecosystems, he was a close analyst of the myriad details within such systems at progressively more microscopic levels. He organized his multidisciplinary research programs and scientific teams accordingly to make use of the combined knowledge and expertise of numerous interactive disciplines and approaches. This section looks microscopically down the organizational hierarchy toward the component parts of complex ecological systems.

Chapter 1

In Chapter 1, coeditor Sven Jørgensen provides an "uncertainty principle" for complex ecology analogous to Heisenberg's well-known indeterminacy relation about the position and momentum of quanta in physics. Jørgensen argues that ecosystems are enormously complex. He puts this in numerical terms by computing an upper bound on the number of measurements available in a maximum research program compared to the number of observations needed to know an ecosystem. The sobering shortfall means that uncertainty in addressing ecological complexity is a given, a fact of scientific life. Put in modeling terms, models must always lack full information, and in fact modeling can be defined as a process of making homomorphic (many-to-one) mappings from reality into some knowledge frame. Systems of knowledge, belief systems, and the like, are always based on fragmentary information made to cohere within a frame of reference. Those who would study ecological complexity are faced with a dilemma. Ecosystems, because they operate as holistic units with everything at all levels of organization articulated by huge numbers of part–part and part–whole mutual dependency relationships, are irreducible. But, because of the enormity of the relationships involved, they cannot be known in every detail—they must be reduced in some decomposition process to know something of them. And to do this is to lose some of their essence. Just as the Heisenberg principle must be accepted by physicists, ecologists must accept similar limitations on ecological knowledge.

Chapter 2

Chapter 2, by Ramon Margalef, considers information theory aspects of the microscopic organization of ecosystems. The vision here is that the ecosystem is a "channel" for communication, and its component parts, such as its species, are the bearers of information with respect to this channel. The hierarchical character of complex ecological organization is acknowledged in discussion of the α (spatially local) and β (between patches) diversity of these alphabetlike bearers. Reinforcing Jørgensen's irreducibility property, Margalef writes "it is the total channel of the ecosystem that decides the issues of the uninterrupted game for survival." Complex ecologies become in the aggregate self-organizing systems; they have the power to hold their genomes constant, and as they do so they become autonomous. "The channel or the ecosystem represses rather than encourages change," just as "the use of language does not allow many changes in vocabulary. Novelty or invention is noxious to the working of any communication channel." Citing work on the importance of indirect effects, the author argues that this problem is manageable because connectivity is partial and temporally discontinuous within the changing ecosystem. But, on the other hand, since last states of processes become the initial states of their time continuations, where do the indirect effects go at discontinuities? They cannot just go away; they must be propagated into the next generation organization.

Chapter 3

Following on the quantum-mechanical and information-theoretic perspectives of the first two chapters, Lionel Johnson in Chapter 3 gives a biologist's attention to the thermodynamic foundations of complex ecology, improbably juxtaposing physics and ichthyology in a synthesis of complex ecological organization. At issue are concepts like the balance of nature and equilibrium in a world where change, not stasis, appears to be the rule. "The basic question to be examined . . . is whether or not equilibrium is a viable ecological concept . . . where change predominates." Reviewing various ecological views of equilibrium, Johnson concludes that "one state and one state only should be referred to without qualification as equilibrium: thermodynamic equilibrium, the state of maximum entropy in an isolated system." Biological systems are far from such equilibrium—the Prigogine property (systems evolve "toward a stationary state characterized by minimum entropy production compatible with the constraints imposed . . .")—but in stable configurations maintained by displacement-resisting homeostasis. The search for a core "homeostatic principle" requires simple exemplars, because normal biological complexity is obscuring, and this leads to the fish populations in Arctic lakes, which have been stable in an undisturbed state throughout postglacial time. The stability characteristics of these populations include both resistance, expressed as long-term invariant size and age distributions, and resilience, returning to the original state whenever displaced. In contrast, the small mammals of the surrounding tundra are the paradigm "pole of instability in the diversity–stability hypothesis."

The blind cave fishes of Zaire, cave fishes of the southern United States, marine and freshwater mollusks, giant tortoises of the Seychelles, white suckers in a Manitoba lake, tawny owls in Wytham Woods, United Kingdom; and, among plants, European virgin forests, the upland rain forest of Mauritius, and forests of northern lower Michigan; plus an extensive examination of the population structure of the Arctic char in lakes of the Canadian northwest—all these examples are reviewed to demonstrate the robustness of the conclusion that population structures emerge over long periods of time that achieve least attainable thermodynamic dissipation. "Least dissipation is the imposition of the maximum time delay, compatible with existence, on the energy flow through a species population, or ecodeme. Maximum time delay implies greatest biomass per unit energy input or least energy flux." Johnson's convergence to the Prigogine conclusion is stated more formally in the next chapter (by Straškraba) as an optimality principle, the minimization of excess entropy production. The alternative form, maximizing negentropy flux, where negative entropy corresponds to information, connects back to Margalef's ecosystem as a communication channel, where now communication may be interpreted as the process of maximizing information flow.

Johnson postulates that symmetric (reciprocal) interactions within ecodemes tend to minimize dissipation, whereas asymmetric interactions between ecodemes increase it. "This . . .," he writes in correspondence, "is a rather fundamental point, as the difference leads to hierarchy formation which is a conflict between forces tending to maintain symmetry and those tending to cause symmetry. Thus, hierarchy formation is a conflict between the two fundamentals of 'order': uniformity and series, or 'rank' and 'row.' The 'rank' in the hierarchy is then determined by the rate of dissipation, with the slowest at the head (having most control over resources)." Reviewing the thermodynamics of dissipative structures, the homeostatic steady states of biosystems far from equilibrium are referred to as "ergodynamic" equilibria around the optimal state of least dissipation as an attractor. For ecodemes, properties conferring least dissipation (and those exhibited by the Arctic char) include large size of individuals, greatest attainable mean age, low juvenile stocks, constant uniform size, and indeterminate age at death. "Homeokinesis" is the term given to extrapolation of least dissipation across hierarchical levels from individuals to populations to whole communities to achieve systemwide homeostasis. The evolution of hierarchies results, in complex ecological organization, involving (1) dissipative but coherent ecodemes, (2) asymmetric interdemic interactions, (3) movement toward ergodynamic equilibrium in ecological time, (4) acceleration of energy flux and resultant erosion of stability over evolutionary time, and (5) periodic resetting to new initial states by environmental discontinuities.

Chapter 4

In Chapter 4, Milan Straškraba continues the cybernetic development of goal-oriented attributes in complex ecological systems. He anchors his ideas to classical theoretical ecology by critically examining prey–predator models with respect to their suitability for representing ecosystems. The assumptions underlying stability studies of population models are really not very relevant for complex ecosystem theory, he concludes, which must take better account of the part–whole relation within ecosystems. A set of propositions is developed as "necessary conditions for realism" in ecological modeling: (1) Complex ecological systems are energy–matter open, far-from-equilibrium, dissipative structures with external driving; (2) their biotic components are actively interactive with their abiotic elements; (3) by their far-from-equilibrium status, their stability is more Lagrangian (bounded, or resistancelike) than like that of Liapunov (asymptotic, or resiliencelike around thermodynamic equilibrium as the attractor) and is determined collectively at all levels of biological organization; (4) linear dynamics occur under restricted environmental conditions, but nonlinear dynamics are more prevalent and allow adaptive system radiation over a broad range of environmental conditions; (5) adaptation, exhibited as near-optimal dynamic behavior, results from long-term, trial-and-error, stochastic interactions of components in which evolutionary mechanisms tend to damp input variability; and (6) adaptive and self-organizing though ecosystems may be, their characteristics at all levels of their hierarchical organization vary with inconstant conditions and disturbance. Straškraba sketches an approach to formulating cybernetic optimization models of complex ecological systems, with affinities to Mauersberger's and Fleishman's theories, which follow in the next chapters, but explicitly incorporating his six properties. Many of the themes of this chapter are recognizable from previous ones, and they also reappear in various connections throughout the remainder of the book.

Chapter 5

Information (negentropy) and entropy are well known as opposites, and in Chapter 5 Peter Mauersberger presents a mathematical theory of the entropy control of ecological complexity. The focus is on microscopic thermodynamics, but with macroscopic intent. Physical laws underlie and are necessary to biological ones and cannot be violated. All ecosystem processes, being irreversible, generate entropy; the second law of thermodynamics gives that local entropy production is always ≥ 0. Internal equilibration tendencies are opposed by external work that causes displacement. "Stationary states of ecosystems are nonequilibrium states far from the thermostatic equilibrium"—Johnson's "ergodynamic equilibria" (Chapter 3) restated in mathematical terms. Exergy, measuring displacement from equilibrium, is introduced; this will become one of several extremal (optimization) principles to be examined in Part 5. The application of thermodynamic formalism to aquatic ecosystems is demonstrated. Entropy production is always balanced by entropy export. The entropy balance equation is fundamental; it provides a means for unified treatment of physical, chemical, and biological processes, offering stability criteria, supporting the determination of bifurcation points, and giving an entry to macroscopic phenomenological, deterministic, stochastic, synergetic, and cybernetic aspects of complex ecological dynamics. Chaos is among the longer-term behaviors of such dynamics, and its existence imposes limits to predictability—another uncertainty principle perhaps. Examples of macroscopic descriptions of biological processes derived from thermodynamic considerations are presented. An optimality principle is stated; omitting the mathematics, "The development of the biocoenosis is controlled during the finite time interval ... in such a way that the time integral ... over the generalized excess entropy production ... becomes an extreme value (minimum) ... subject to the initial values ... and to the mass balance equations ... [L]ocally, but within a finite time interval, the deviation of the bioprocesses from a stable stationary state tends to a minimum." The equations of population dynamics are then interpreted in accordance with the entropy principle, illustrated with ecological examples, and the conclusion drawn that "the basic equations of theoretical ecology can and must be derived from the fundamental laws of physics, chemistry and biology, including the second law of thermodynamics."

Chapter 6

The mathematical approach is continued in Chapter 6, which represents a substantial formalization of complex ecological systems theory as developed in the former Soviet Union. The author, Benzion Fleishman, distinguishes between thermodynamics and "systemology," the latter a version of cybernetics appropriate to the quantitative analysis of complex systems. He considers stochastic versus deterministic objects; micro-, meso- and macro-states of processes; interacting versus noninteracting processes; system versus environment; elementary versus emergent processes; and for goal-directed systems tactical versus strategic goals. Four cybernetic properties of complex systems are formulated, which must come together in any realization of coherent behavior. These are "reliability," "noise immunity," "controllability," and "self-organizability"; they meet in different combinations. A distinction is made between Heisenberg's uncertainty principle in physics, which "establishes limits of measurement accuracy at the microlevel," and the situation in systemology, where "the mesolevel is considered the microlevel and ... principles of objective teleology ... allow for the identification of optimal model structures and behaviors at the macrolevel." In other words, systemology applies at the higher end of the complexity scale, whereas thermodynamics is appropriate to the lower end. And at the higher end, goals become relevant. Fleishman's treatment of optimality in complex system dynamics has "effectivity" as its central concept, "defined as the conditional probability of gaining goal A by the system A interacting with the environment B." Adaptation and preadaptation are formulated in a stability context, and the key characteristic of adaptive and preadaptive processes is regeneration. "Unlike simple systems with material (energetic) stability, it is more important for complex systems (biological among them) to be structurally and behaviorally stable in spite of variable material composition. Their specific property, therefore ... is the regeneration of dying elements and adaptation to changing environmental conditions." The reader will find this chapter mathematically challenging, but the rigor is not just arcane theory. Fleishman's systemology is motivated by practical concerns, and the second half of his chapter illustrates this through an application to fisheries management. He is very close to biologist Johnson (Chapter 3) in the objects of interest that motivate his theory.

Altogether, the chapters of Part 1 indicate broad concern in systems ecology at the present time with the development of an emerging general paradigm about natural complexity and the need to better formulate and understand in explicit terms the part–whole, microscopic–macroscopic relationship in ecosystems under such complexity.

1

Quantum Mechanics and Complex Ecology

SVEN ERIK JØRGENSEN

The Royal Danish School of Pharmacy
Department of Environmental Chemistry
Copenhagen, Denmark

In applying the ecosystem concept, there is no limit to size and complexity [Van Dyne, in *Chiasma* 8:59 (1970)].

INTRODUCTION

Quantum mechanics has taught us that there are limitations to the amount of knowledge we can obtain about nature from our observations. This is expressed in Heisenberg's uncertainty relation:

$$\Delta s \cdot \Delta p \geq \frac{1}{2} \bar{h}, \qquad (1.1)$$

where Δs is uncertainty in position, Δp is uncertainty in momentum, and h is Planck's constant = $6.626 \cdot 10^{-34} \cdot J \cdot s$; $\bar{h} = h/2\pi$. An attempt will be made in this paper to demonstrate that the application of quantum mechanics in physics has an analogy with what we experience today in ecology.

UNCERTAINTIES IN ECOLOGY

Some of the principles of quantum mechanics are (silently and slowly) being introduced and accepted in ecology, particularly in ecological modeling during the last fifteen years. An ecosystem is too complex to allow us to make the number of observations needed to set up a very detailed model—even if we still consider models with a complexity far from that of nature. The number of components (state variables) in an ecosystem is enormous.

If we estimate that the maximum amount of resources that can be devoted to one project corresponds to 10^8 measurements or determinations, the two extremes would be to apply these measurements to get one piece of information, but with a very high accuracy, or to attempt to squeeze as much information out of the data as possible for a very complex system. If we estimate the accuracy obtainable for one measurement to be 0.1 relatively, that is, a 10% standard deviation, we can in the first case obtain an accuracy of $0.1/\sqrt{10^8} = 10^{-5}$. In the latter case, the question is how many state variables can be considered in our model leaving us with a still fairly good picture of the system problem under consideration. If we have two dependent state variables and want to get a picture of their relation, we need at least three measurements. With two measurements, we cannot decide whether the function is linear or nonlinear. Correspondingly, if we have three dependent variables and want to get a picture of their variations and interactions, we need to describe the shape of a plane. Consequently, we need at least $3^2 = 9$ measurements. Finally, if we consider n state variables and cannot exclude any interrelationships, it would be necessary to have 3^{n-1} measurements. For $3^{n-1} < 10^8$, we can see that $n < 18$.

In accordance with these considerations, we might formulate a first approach to an approximate ecological uncertainty relation:

$$\frac{10^5 \cdot \Delta x}{\sqrt{3^{n-1}}} = 1, \tag{1.2}$$

where Δx is the relative accuracy (standard deviation) of one "situation" and n is the number of components in the model. Note that we have assumed dependent variables, for which a relationship is valid. Doubtless a model might often attempt relationships that can be omitted for the problem considered.

It is obvious that an increase in the number of dependent state variables will very rapidly require so many measurements that it becomes impossible to validate such a model due to a shortage of human resources.

The number of observations, 10^8, is to a certain extent chosen arbitrarily. Due to improved measuring techniques, we may in the future be able to raise the number of observations one or more orders of magnitude; but this does not change the fact that the number of observations needed is high and that we will always be forced to work with a number of observations that is far less than this number. This will prevent us from getting a complete and detailed picture. Even if we go to the limits given by quantum mechanics, the number of variables is still low compared to the number of components in an ecosystem. In accordance with (1.1), Δx would be on the order of 10^{-17}.

Another uncertainty relation may be used to give the upper limit of the number of observations:

$$\Delta t \cdot \Delta E \geq \frac{1}{2}\hbar, \tag{1.3}$$

where Δt is the uncertainty in time and ΔE is energy. If we use all the energy that the earth has received during its lifetime of 4.5 billion years, we get

$$\Delta E = 4.5 \cdot 10^9 \times 365.3 \times 24 \times 3600 \times 173.10^{15} \tag{1.4}$$
$$= 2.5 \times 10^{34} \text{ J}.$$

Δt would therefore be on the order of 10^{-67} sec. Consequently, an observation will take 10^{-67} sec, even if we use all the energy that has been available on earth. The number of observations since the universe was created $15 \cdot 10^9$ years ago is therefore

$$15 \cdot 10^9 \cdot 365.3 \times 24 \times 3600/10^{-67} = 4.7 \times 10^{84}. \tag{1.5}$$

This implies that we can replace 10^5 in equation (1.2) with 10^{59}, since

$$\frac{10^{-17}}{\sqrt{4.7 \times 10^{84}}} \cong 10^{-59}. \tag{1.6}$$

If we use $\Delta x = 1$ in equation (1.2), we get

$$\sqrt{3^{n-1}} = 10^{59},$$

or (1.7)

$$n \cong 237!$$

These results are completely in accordance with Niels Bohr, who expressed it as follows: "It is not possible to make an unambiguous picture (a model) of reality, as the uncertainty limits our knowledge." The uncertainty in the world of atoms is caused by the inevitable influence of the observer on the atomic particles, and in ecology it is caused by the enormous complexity and variability. No map (model) of the reality is correct. In cartography, there are many maps of the same piece of nature; one may be used by a pilot, another one by a car driver, a third one by a geologist, and so on. These various models reflect different viewpoints and purposes, and it is not possible to say which one is better than the other. They are all valid for different purposes. In other words, the theory of complementarity is indeed also valid in ecology.

This does not, on the other hand, imply that all models contribute equally to the understanding of nature. It is a general assumption that a model may be made more realistic by adding more and more connections, up to a point. The addition of new parameters after that point does not contribute further to improved simulations, but on the contrary, more parameters imply more uncertainty. Given a certain amount of data, the addition of new state variables or parameters beyond a certain model complexity does not add to our ability to model the system, but only adds unaccounted uncertainty.

These ideas are visualized in Fig. 1.1. The relationship between knowledge gained through a model and its complexity (measured as the number of state variables or connections) is shown for two levels of data quality and quantity. The question under discussion can be formulated in relation to this figure: How can we select the complexity and the structure of the model to assure that we are at the optimum for "knowledge gained," or for the best answer to the question posed to the model? Costanza and Sklar (1985) have examined articulation, accuracy, and effectiveness of 87 different mathematical models of wetlands. They use the following equation for the articulation index:

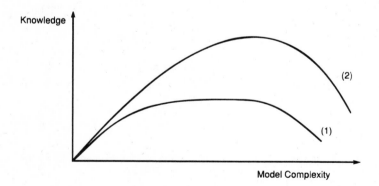

Figure 1.1 Knowledge versus model complexity measured, for example, by the number of state variables. Knowledge increases up to a certain level. Increased complexity beyond this level will not add to one's knowledge about the modeled system. At a high complexity level, one's knowledge might be decreased. Curve (2) corresponds to an available data set that is more comprehensive or has a better quality than that for (1).

$$A_i = \frac{N_i - 1}{k_i + (N_i - 1)} \times 100, \tag{1.8}$$

where A_i is the articulation index for mode i, N_i is the number of divisions in mode i, and k_i is a scaling factor for mode i. The number of divisions in each mode are the number of components or state variables for the component mode, N_c; the number of time steps for the time mode, N_t; and the number of spatial units for the space mode, N_s. The scaling factor was chosen to reflect the relative degree of difficulty of increasing the number of divisions in the mode and to give an idea of the maximum size of the most articulated existing models in each mode. The scaling factors selected for the components, time and space, respectively, were $k_c = 50$, $k_t = 1000$, and $k_s = 5000$.

Costanza and Sklar calculated the articulation index for both the model and the data, as it is relatively easy to run a simulation model with 10,000 time steps or more, whereas it is very difficult to collect supportive data at this frequency. This implies that articulation of the data is often the limiting factor. The average articulation index of the three modes was found for each of the 87 models examined.

An index of accuracy was calculated as the percentage of the total (historical) variation that was explained by the model, averaged over all three modes and stated as a fraction between 0 and 1. The average value was used to standardize the index across all three modes of articulation and to estimate model accuracy as a percentage of the total maximum accuracy possible.

Costanza and Sklar ranked the models by use of an index of effectiveness or explanatory power. This index was found as the coefficient of determination for each mode multiplied by the minimum of the data or model articulation index for that mode and then averaged over the three modes. The most effective model under this scheme is one that balances the costs of added articulation against the benefits of increased accuracy to best explain all the models of the system.

The results of this review are summarized in Figs. 1.2 and 1.3. Figure 1.2 indicates an interesting result as it supports the more philosophical statement shown in Fig. 1.1. Of course, all indexes must be considered relatively, and the authors state that the results should be considered as a hypothesis due to the small amount of supporting data.

The results from this model review support some interesting scientific perspectives. In the past, scientists have tended to narrow their questions to achieve higher accuracy. This leads to models with low complexity but high descriptive accuracy (Fig. 1.2). The results say much about little.

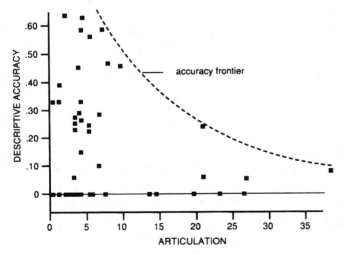

Figure 1.2 Plot of articulation index versus descriptive accuracy index for the models reviewed, showing the current accuracy frontier. (After Costanza and Sklar, 1985.)

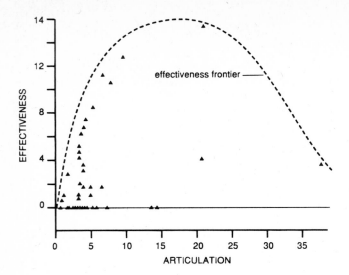

Figure 1.3 Plot of articulation index versus effectiveness index showing the current effectiveness frontier. (After Costanza and Sklar, 1985.)

Nature is, however, complex, and it becomes impossible to describe reactions of all species to the combinations of all possible impacts (forcing functions) by use of accurate answers to narrow questions. In physics it is impossible to know simultaneously the accurate location and velocity of a particle. This is in accordance with the uncertainty relations by Heisenberg and can be explained by Bohr's complementarity principle: the observer affects the object. In ecology we have a similar uncertainty relation: all components and processes cannot be described accurately at the same time. The product of the number of elements in the model and its descriptive accuracy has an upper limit, and the trade-off for the modeler is between knowing much about little and little about much. The complexity of nature can only be described by a statistically high number of elements, and scientists have therefore recently been looking at nature more comprehensively from the viewpoint of models and systems. This has enabled us to know a little more about much.

The implication of these considerations is that there is an optimum size or complexity of a model beyond which the benefits of additional knowledge are outweighed by the costs of lowered effectiveness (Fig. 1.3).

REDUCIBLE AND IRREDUCIBLE SYSTEMS

Another parallel in the development of ecological modeling may be drawn concerning the differences between reducible and irreducible systems (see Wolfram, 1984a, b). Physics has traditionally focused on reducible systems, for which a simple and superior description can be given. For such systems, we can reduce our observations to a few equations, which allow us to compute how the system develops. According to Wolfram, this may be an exception rather than a rule for natural systems, in particular for biological systems. For irreducible systems, we may know the rules of the systems, but we cannot survey the consequences of rules or make predictions unless we compute step by step how the system develops. There may be simple rules for evolution, for instance, but we can only decide on the development of biological evolution by observation of every step that it makes. The modeling of biological evolution could be carried out using a computer that contained all information about the living systems on earth, but this will never be possible to obtain.

For irreducible systems, the computer becomes an experimental tool that allows us to test consequences of rules or natural laws and compare them with other observations (Wolfram, 1984a, b). The great number of ecological models developed during the last two decades represents the use of experimental mathematics in ecology. They have all contributed to a better knowledge of ecosystems because they have all tested some rules (laws or equations), and the results carry us a smaller or larger step forward.

We can implement Darwin's theory in the application of experimental mathematics in the development of ecosystems. We can translate survival into thermodynamic terms (see, for instance, Jørgensen, 1982, and several of the following chapters in this book.) The concept of "fittest" is related to entire systems with all their complexity, variability, and stochastic nature. Therefore, in the future we will certainly reveal new principles along these lines and obtain new useful models of nature, but they will never give a complete picture.

DISCUSSION

Science develops useful models of nature. Models of atoms have given us a useful understanding of nuclear processes. The technical implications of these models are the best demonstration of their usefulness. Models of ecosystems are more recent, but in some cases they have also shown their usefulness in applications to a variety of environmental problems.

However, a complete, unambiguous picture of a complex ecological system does not exist today. Light may be considered as waves or particles, depending on the observations and the purpose of the light model. The same is valid for ecosystems; they may be described by the use of one model in one context or by other models in others. The validity of a model is dependent on the observations (which means the observer) and the purpose of the model.

The uses of models and holistic approaches in ecology in general have been criticized because they do not give a complete and unambiguous picture, but such a requirement can never be fulfilled because of inherent limitations in our ability to observe complex and irreducible systems such as ecosystems. Ecology is developing toward an acknowledgment of these limitations in our possibilities (see Jørgensen, 1992). Observations of single processes or biological details are naturally of importance; but when we put the detailed information together to form a picture of system behavior, we need to use models and accept their limitations caused by the overwhelming variability of nature—just as the physicists must accept Heisenberg's uncertainty principle and Bohr's complementarity theory. Modeling limitations are as valid in ecology as in physics, and the potential rewards of using them are equally great.

REFERENCES

COSTANZA, R., and SKLAR, F. H. 1985. Articulation, accuracy and effectiveness of mathematical models: A review of freshwater wetland applications. *Ecol. Mod.* 27:45–69.

JØRGENSEN, S. E. 1982. A holistic approach to ecological modelling by application of thermodynamics. In Mitsch, W. J., Ragade, R. K., Bosserman, R. W., and Dillon, J. A. (eds.), *Energetics and Systems*. Ann Arbor Science, Ann Arbor, Michigan, pp. 485–493.

JØRGENSEN, S. E. 1986. Structural dynamic model. *Ecol. Mod.* 31:1–9.

JØRGENSEN, S. E. 1992. *Integration of Ecosystem Theories: A Pattern*. Kluwer Academic, Dordrecht, The Netherlands. 385 pp.

WOLFRAM, S. 1984b. Computer software in science and mathematics. *Scient. American* 251:140–151.

WOLFRAM, S. 1984a. Cellular automata as models of complexity. *Nature* 311:419–424.

2

Information Theory and Complex Ecology

RAMON MARGALEF

Department of Ecology, University of Barcelona
Barcelona, Spain

The ecosystem is a fundamental unit of study in basic ecology [Van Dyne, in The Ecosystem Concept in Natural Resource Management, *Ch. X, p. 332 (1969), Academic Press].*

INTRODUCTION

I feel guilty of having introduced some confusion in relation to the concept of diversity in ecology, but I do not feel guilty about the conservatism and rigidity into which the whole subject has fallen. My approach to diversity is to use any suitable method, preferably borrowed from information theory, to quantify a concept that originated in the naturalistic contemplation of the biosphere, the richness or variety of complexity in nature. In the proper field of diversity, the goal is to find and apply a measure of how an ensemble or set is distributed over a number of subsets. Diversity is minimum (zero) if all the objects or subsets belong to the same class and maximum if each of them can be distinguished and put in a different class. Any quantitative and abridged measure of diversity may be relevant only in relation to the criterion of classification of sets into subsets that has been applied. It happens that diversities based on separate criteria of distribution (organisms distributed according to species, size, trophic level, speed; even "trophic species" if not misconstructed) show similar trends inside an ecosystem or a collection of comparable ecosystems. In my opinion, such regularities have more to do with thermodynamics than with statistics, and hence become suitable to base prediction on them. If, as is usual in ecology, the criterion of classification is the identification of individuals as members of different species, then diversity may have some direct meaning in relation to the processes of multiplication and all sorts of biological interactions that have led to the actual composition of the ecosystem and that maintain it as a combination of different species in certain proportions.

DIVERSITY IN SPACE AND TIME

The complex organization of an ecosystem is deployed in space, and diversity becomes necessarily a spectral measure. The spectra of diversity are continuous. The distinction between alpha diversity (over "small" areas) and beta diversity (comparison between patches that are separated and relatively large) has to be considered, at best, as a matter of convenience, and not as a way to bypass the need for diversity spectra. In the application of beta diversity to assess spatial heterogeneity, diversity is used as a variance; this is acceptable, but it does not seem to have further interest.

The significance of changes in diversity through time is more difficult to grasp than the differences over space, although this can be ascertained by working step by step as, for example, in Brillouin's expression. Stronger dynamics (high production/biomass or P/B ratio) are expected to drive down the values of diversity. An index of diversity that is very sensitive to dynamics is $k = \log S/\log N$, where S is number of species and N the number of individuals. The value of k is zero in a flow culture or chemostat, and $k = 1$ in an altogether dead system, such as in a museum display case. The measure k and its historical change provide an anticipation of which appears easier, to add a new individual or a new species to a system.

Interest in diversity grew when it was discovered that the abundances of different species living together exhibit a certain regularity. The species can be ordered in a way that abundances decrease or increase with regularity (that is, few common species, increasing numbers of rarer and rarer species). The regularity is in part an artifact resulting from manipulation (ordering) of the set, but the average of the quotients between any pair of numbers follows a certain tendency. The ratios are high in low-diversity systems and closer to 1 in high-diversity systems. The total number of species (never to be ascertained) is not the main determinant of diversity.

DIVERSITY MEASUREMENT

Besides the one mentioned, many quantitative expressions for diversity have been proposed. The more clever may relate the increment of the number of species to the increase in the number of individuals. The number of species represented in a collection of a fixed number of individuals caught at random may be the simplest index. In the limit, this reduces to the probability that two specimens extracted at random belong to the same species (Gini–Simpson index). The information measure of Shannon and Weaver has passed the test of many years of use. The Brillouin approach, from Boltzmann's entropy, is helpful if taken together with Shannon's expression. Both are complementary, in fact are the same, focusing on probability (Shannon) or on actual counts of quantified entities (Brillouin), which is more realistic in ecology. Gini (1912) and Simpson (1949) measure leads easily to imagining other indexes (the probability that the two specimens extracted at random belong to different species or else belong to different species that are not indifferent among them, as concerns biological interaction) and introduces the whole subject of compound probabilities. Usual diversity, restricted to a reckoning of separate individuals, has to be complemented by the considerations of finding together specific pairs, triplets, and so forth. Remember the usefulness of this approach in the study of language, where, after ascertaining the probabilities of the different letters, digrams, trigrams, and so on, one can prepare nonsense texts that imitate particular languages and even definite writers. Plain diversity gives, then, only an upper limit in a general procedure in which to place a more careful analysis of the associations between individuals of the same or different species, based either on simple relations of spatial proximity or on biological interaction (connectivity).

The fact that values of diversity (in Shannon's index) rarely exceed 5.3 bits/individual might suggest the existence of restrictions, perhaps very simple ones, in the usual processes leading to ecological organization. Other constraints may concern the number of actual functional links in relation to the number of possible links (connectivity).

In practice, diversity is computed on sets defined usually through two operations: (1) a physical process of taking "samples" and (2) the identification of the individuals that belong to one or a few taxonomic groups. Some examples are moths attracted to light, flowering plants counted in quadrats, and phytoplankton in small volumes of water sedimented and observed through an inverted microscope. The actual diversity of a whole ecosystem is unknown and probably unknowable and unexpressible. But there are reasons, or at least intuitions, to accept a positive correlation among the different diversities computed for different strata or different taxonomic groups in the whole ecosystem, although there are exceptions. This is the case of groups that expand and diversify under conditions that are not so favorable or even detrimental to the rest of the community. In marine phytoplankton, for instance, diversity of diatoms is not necessarily positively correlated with the diversity of dinoflagellates or other nondiatom phytoplankton.

ECOSYSTEM AS COMMUNICATION CHANNEL

Ecosystems belong to a special class of systems, those composed of replicable subsystems (species). We are free to consider the convenience of adopting other intermediate subsystems, like consortia or groups of coevolved species. Systems are not rigid but internally flexible, and they can adopt alternative states in a set of possible states. This is another contribution of the concept of diversity. The concept of information applies naturally to systems and to ecosystems. The ecosystem becomes a channel linking the past to the future, and its components appear as bearers of information with reference to this channel. This sort of self-reference may appear different from the behavior of a usual channel of information, with sender and receiver, codifiers and decodifiers. Once any eventual conceptual difficulty related to these differences is clarified, the common application of the concept of information should not offer more difficulties.

The shift from one to another state implies energy. One amount or parcel of energy cannot be used twice exactly in the same form. Perhaps the simplest way to state the essential is to remember that, to be recognized, energy has to interact with matter and leave some imprint. At the same time, energy changes or decays, shifting, for instance, in the case of photons, to a longer wavelength. In consequence, the loss of energy quality (usually expressed as an increase of entropy, which is not an entity, or at least no more an entity than phlogiston) is associated with an increase of information. All events leave traces, except perhaps a catastrophic erasure of a local area, and the general increase in entropy, usually expressed as a tendency toward a more uniform distribution of energy of lower quality, is matched by the increasing complexity of the traces left by the very decay of energy. Such relationships prompt relating or even equating information with negative entropy. In any description of the world, information appears necessarily associated with matter and energy, and the concept of system is necessary to provide the frame in which information is generated. The system *is* information, and its maintenance and accretion involve cybernetic behavior. But, at each level in the conceptual dissection of a system a particular level of information can be recognized. Diversity, as generally used in ecology, has to do with information at the level of assemblages (conventionally defined) of individuals belonging to different species. The elementary decision to make a step toward increasing redundancy or else to add new information is at the level of what is more likely, to add one individual or to add one species to the system. Under the influence of humans, it is easier to add one individual of an already existing species and even to complement this event with the extinction of some other species.

Interest in the consideration of information goes further than to pretend to justify or make more respectable the essentially naturalistic concept of diversity. In the case of ecosystems, as well in many other systems (social, economic, and so on), it appears that a correspondence may be established between the decay of energy and the increase of information. Besides, both related events may show some differences relative to space and time. If energy provides a bridge over space, information assures a persistence, or marks an evolution, that is expressed along time. Physical science can anticipate difficulties ahead when trying to deal with both simultaneously.

In an ecosystem, the major exchanges of energy, and presumably the largest increase in entropy, happen close to the site of the primary producers. In a more general way, differences are recognized also when the system is seen as made of cycles combined into larger cycles, with differences in the speed of turnover of the structures of different levels. Small component subsystems may last longer than the systems integrated by them (genes may last longer than species, the selfish gene idea, and species longer than ecosystems; words last longer than literary fads).

The major increase in information is apparent in the persistent structures closer to the large animals and trees and much more in species that are able to store information in individual memories and in structures or artifacts all around. The lack of total coincidence between entropy increase and information increase in time and space makes the functional structure of the ecosystem strongly anisotropic, and any simple structuralist definition (for example, of the ecological niche) may be inconsequential. Such facts find another expression in the principle usually named after St. Mark, with reference to the Gospel, that can be redefined in very broad terms, in the sense that the preexistence of a core of information acts as a center to accrete more information, which is paid for by the energy that decays or has been degraded in the past all around or in connected structures.

ANALOGY WITH LANGUAGE

Dealing with diversity, we are led to comparisons with information in the standard form of communication or information theory, often with the help of analogies borrowed from language. The analogy with language is reinforced because both ecosystems and languages (or discourses) belong to a class of systems made of parts (subsystems) that are self-replicating by their own power (biological organisms) or by an external agency (viruses, words, "second-order" individuals) (Cavalli-Sforza and Feldman, 1981). The instantaneous picture of an ecosystem comprises a number of species, each of them represented by a certain number of replicas or individuals. The proportions among the different species are seen as inherited from preceding states and are kept approximately constant by a number of feedback loops, negative in predator–prey relations, positive in competition. This system is projected into the future through a self-preserving mechanism that includes natural selection (positive feedback) and other possibilities for change. Thus, the ecosystem as a whole acts as a channel in which the relationship of the components, of each single individual to another, may be established. In fact, it is the total channel of the ecosystem that decides the issues of the uninterrupted gamble for survival. The channel or the ecosystem represses rather than encourages change. Genotypes can change faster than phenotypes, and usually do, and the evolution of organisms is fastest under domestication when they are disconnected from their native ecosystems.

The analogies between ecosystems and language can be taken a step further. In the same way as ecosystems and ecological succession repress potential evolution, the use of language does not allow many changes in vocabulary. Novelty or invention is noxious to the working of any communication channel. The nature of species and of words and the possible relations among them fix the average proportions of the components, as well as more complex conditions of proximity or relationship (syntax, style, connectivity, diversity). In nature, as in written texts, the "style" may be preserved at least for a while. In the case of language, besides (1) the replicating entities, or words, and (2) the way they connect, or style, there also exists a most important element: (3) what is said. The obvious analogy in ecosystems to "what is said" is the sequence of states that should offer a way to a deeper understanding of the works of nature.

The quantification or discontinuity in the components of the whole class of systems—words in languages, individuals in ecosystems—explains some significant properties of the same systems. Discontinuous components are always out of equilibrium. In ecosystems, we are forced to examine critically the usual approach to the analysis of interaction among species. The differential equations that neglect both space and the constraints of thermodynamics are not appropriate, and it is time to look elsewhere for improvements. Quantification of biomass is related to individual size and, in more dynamic terms, to the distribution of the average length of their allowed life span, two properties that may be functionally associated. They describe the unit of (creative) uncertainty over space and time, a window over which lack of equilibrium may be expected. Differential equations might be substituted by difference equations that are very sensitive to the initial values of the relevant variables. Appropriate selection of the values of the parameters assumed to be constant may lead to chaotic sequences, in which values eventually shift around some attractor (Kloeden and Mees, 1985). If we retain that parameters describing interaction among species are actually not constant, eventual chaotic behavior is expected to be at most only transitory. In the unrealistic case of regular cycles, information stops increasing soon; for a stochastic system, it may grow steadily; for a chaotic system, it is hard to make predictions.

DISTURBANCE AND STABILITY

The response of any ecosystem subjected to some disturbance or input of external energy can be described as a forced process, followed later by some wandering around in a field of low energy or indifference, in which prediction is difficult if not in the form of general statements about pattern. Stability has a tradition of confusion in ecology, implying some capacity of returning to previous states or else the property of slowing down change. A measure of stability can be associated with a function such as

$$\sum_{i=1}^{n} \left(\frac{N_i}{N}\right)\left(\frac{dN_i}{dt}\right) \to 0,$$

where N_i is number of individuals of species i and N is the total number of individuals. It is not excluded that a pattern describable by the persistence of local differences (between points A and B) should remain above a certain level:

$$\sum_{i=1}^{n} \left(\frac{N_i}{N}\right)\left(\frac{d[N_{iA} - N_{iB}]^2}{dt}\right) > \epsilon > 0.$$

This is relevant in relation to diversity spectra and information, as well as an expression of the transition from a forced and dynamic process to an indifferent pattern.

But such a pattern, on a topography of low energy, may be significant in terms of information. In fact, decisions important in evolution and in ecology may be grounded on such low-energy patterns, which are difficult to quantify. The comparison with language may be pertinent again. A statistical study of language concerning patterns of word formation and even of style, might not detect any difference between nonsense, things that make literary sense (a story or novel) but that might be impossible, and a "correct" description of reality.

MODELING

In is easy to criticize the usual approaches to modeling ecosystems, especially if criticism is extended to the success in explaining detailed patterns. The concerns for discontinuity and chaos have to be introduced in some form in the models of ecosystems. The classical equations of Lotka and Volterra can be improved by including space. This can be introduced in the aquatic case through consideration of sedimentation or vertical movements in general and also of turbulent diffusion. This approach applies naturally to the study of plankton and was proposed by Riley, Stommel, and Bumpus in a seminal paper in 1949. My work along the same lines (Margalef, 1978, 1985) has resulted in an expression that is perhaps too simple to be really useful but that reminds us of the double problems of space and thermodynamics, so poorly considered in the most common approaches. Powell and Richerson (1985) have been reaching conclusions along similar lines.

In general, production ($P = dB/dt$) can be related to a product of the available external energy doing work in the system, usually expressed as a decay of turbulent energy (A), by the covariance (C) in the distribution of the reactants that determine production (nutrients, light, cells, and so on). In a small system, inputs (I) from outside (in the form of nonrenewable resources) may be considered, as well as outputs (sedimentation and the like), but it suffices to enlarge the system and internalize the inputs and outputs to get an expression that applies to a system with perfectly renewable resources. An equivalent of the basic expression, $P = A \cdot C$, was proposed by von Smoluchowski in 1918 to describe chemical reactions in a space where a (colloidal) structure equivalent to stratification in lakes and oceans was growing. Problems posed by space do not disappear easily, since both A (turbulent diffusion) and C (covariance in the distribution of reactants) are strictly spectral quantities that scarcely have a meaning if not related to a definite size of a cell of measurement. So it must be.

The thermodynamic aspect is evident if we take derivatives with respect to time:

$$\frac{dB}{dt} = I + A \cdot C \qquad (I, \text{ inputs, eventually } = 0)$$

$$\underbrace{\left(\frac{d^2B}{dt^2}\right)}_{\text{deceleration of turnover}} = \underbrace{\left(C\frac{dA}{dt}\right)}_{\text{decay of energy}} + \underbrace{\left(A\frac{dC}{dt}\right)}_{\substack{\text{ecological segregation} \\ \text{yields information}}}$$

This expression is a reminder of the association between the decay of energy and the acquisition of information, and it describes the bare bones of successional change, with an outward expression in the form of the passage from process to pattern. This is not consistent with the acceptance of regular cycles, for example, of lynx–hare fame. Without totally excluding rhythms, development at the level of the ecosystem is more comparable to a sequence of "solitons," each of them started by some input from outside and going eventually through chaotic behavior, adding new layers to an eventual enrichment of the fine-grained pattern. Most observations fit this general model, which can be encapsulated in the general statement that no machine can turn twice in exactly the same way. *Mutatis mutandi* applies to economy, to language, and even to logic, as the successive manipulation of a concept often carries a shift in its sense, albeit unconsciously so.

Another word of caution concerning mathematical models. It is usual that the parameters adopted in a model are adjusted after accepting the hypothesis that the ecosystem is like a machine that turns indefinitely, keeping the same properties. But the real parameters, as far as they can be ascertained, lead to a model that shifts into new states naturally. The shift depends on the conversion of energy into information, including evolution or acceptance of new components (species).

DEVELOPMENT AND CHANGE

The historical development of systems is nonuniform. Long periods of slow change alternate with short periods of catastrophic or much more rapid change. If a disastrous input, in terms of energy, does not destroy organization, it can be used by the organization to increase itself. And any catastrophic event starts a process that eventually ends, adding further to the pattern. As far as disturbances and catastrophic events are concerned, the more important (more energetic) are less frequent, and the less important are more frequent. Anticipation or internalization of the most frequent disturbances (the circadian rhythm, for example) is an important job for evolution. The concept of filter may be useful. The biosphere reflects the suggested probabilistic distribution of the forcing inputs, both in ecological succession and in evolution.

I have been comparing ecosystems with channels that project information into the future. Probably it is appropriate, in very broad terms, to compare the biosphere to a set of channels placed side by side. The ecological segregation and the pattern formation in a domain of low energy should lead to a more "fibrous" structure, in which channels subdivide and segregate, becoming more independent. Inputs of (external) energy cause mixing and confusion between neighboring channels, with good opportunities for selection in hard environments and eventual acquisition of significantly new information. Ecosystems frequently have a "honeycomb" structure in their projection on a horizontal plane (Margalef, 1979), resulting from the spotlike combination of more mixed and more productive subsystems (often the result of random impacts) with more mature, more fine grained and often reticulated sections. It is easy to conceive such a pattern as the instantaneous section of a complex channel subjected to a spectrum of disturbance, as shown in Fig. 2.1. If organization is powerful enough to block destructive inputs from outside, it closes as well the door to the acquisition of new information. This is a principle introduced by Patten (1961) in an ecological context, and it can be reduced to the statement that information, if extensive enough, can block the way to the acquisition of new information at its level.

The alternative between catastrophic change (revolution) and slow historical development ("bureaucratization") may be paradigmatic of historical development in systems. Because regressive

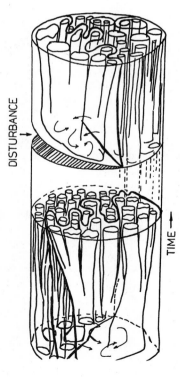

Figure 2.1 Highly idealized representation of the ecosystem channel. Disturbances, as inputs of external energy, induce mixing, loss of structure, increased turnover, and reset succession. Boundaries, eventually ergoclines, appear as high-tension interfaces and later slacken. Structure is reflected on space as patchiness and on time as "fibrosity." Changes at the biotic level are expressed in diversity, spectra of diversity, and connectivity. These descriptors help to characterize the "style" of the channel. A more complete insight into the "message" requires a deeper understanding of information, including genetic and cultural, and not only at the ecological level of species reaction and interaction.

changes are fast and undetermined, and progressive changes are slow and wander into a field of low energy, if a majority criterion is applied, all the biosphere seems caught in a process of succession. Only restricted places, disconnected and randomly distributed, experience resetting or regression, usually as the result of inputs of energy; but the major part of the biosphere is implied in a global succession that never reaches a final stage. Naturalists prefer or preferred succession over disturbance as a subject of study, and this is psychologically acceptable. Journalists prefer, of course, disasters to the more matter-of-course workings of succession or regular history. Revolutions happen from time to time; bureaucratization is overwhelmingly present.

The concept of asymmetries in time that manifest themselves as dicontinuities enhances the importance that has to be given to the functional anatomy of the environment. The environment can be dissected to locate the places where (external) energy does work relevant to the support of biological production. This machine structure, or working structure, has been dramatized by the proposal to designate as ergoclines (Legendre and Demers 1985) the places (usually surfaces or interfaces) where work is effective, as in marine and atmospheric fronts, thermoclines, boundaries of an emerging island or continent, and watersheds. At both sides of an ergocline, different ecological strategies may be selected, with perhaps another strategy in the very plane of the ergocline. The boundaries may be subjected to change or succession, from active interfaces (*limes convergens*, characterized by a high "surface tension," another useful analogy) to boundaries that are more passive, sinuous, or stretched (*limes divergens*), where the tension is lower, being distributed over a more extensive interface. If patchiness or honeycomb patterns have to be combined with more active boundaries, it is clear that the resulting patterns will be difficult to study and describe, but they are not wildly random.

DIVERSITY AND PATTERN

This conclusion emphasizes the need for a more detailed and careful application of all measurements involving information, among them diversity. Methods taken from picture analysis and pattern recognition have to be combined with the old naturalistic appreciation for diversity. Undeniably, complex spatial patterns are material in the decision processes (among them, natural selection) going on in the ecosystem. Important decisions can depend on complicated pattern matching. It can be argued that decision making by committee often proves to be no better than flipping a coin, and that perhaps ecosystems are less influenced by complicated internal relations than by energy flow and by forcing from the physical environment. But, consider briefly the generation of pattern by the colored scales in the wings of butterflies and its role in defense in relation to the visual organs and their analyzers in potential predators (a case in pattern matching). This example may be sobering in relation to any quick dismissal of the significance of complex structures or patterns in ecosystems.

Any plan to evaluate diversity more carefully immediately raises a number of supplementary problems. The use of the index k (log S/log N) lends itself to analyzing seasonal change and recognizing how in spring the number of individuals increases faster than the number of species, the trend reversing in fall, and how the different average life spans of the individuals of different species introduce irregular waves in any adopted measure. Diversity appears quantitatively intractable, if this has to be done in a simple way, and remains more a matter of feeling, as one would barely risk to propose an operational definition.

In any case, diversity sets an upper limit to possible interactions in the ecosystem, but the real level of interaction is much lower, and not only by reason of the limitation imposed by space. Thus is solved the problem of the relation between diversity and stability, fortunately resolved some time ago. Diversity in natural systems, even in artificial systems, has an upper limit that seems to be due not only to the logarithmic character of the usual index (Shannon's). In the distribution of individuals into breeding units or species, diversity usually stops close to 5.5 bits/individual. I proposed to split diversity into two components: (1) the number of species and (2) their inequal representation. This distinction has been adopted by many authors, who give the name of evenness (E) to an expression of the relative differences in representation of the different species. If H is Shannon's diversity and S is the number of species, $E = H/\log_2 S$, and $S^E = 2^H$. Personally, I do not believe that evenness adds anything really important to the concept of diversity (Fig. 2.2).

DIVERSITY AND CONNECTIVITY

The measure of diversity proposed by Gini and Simpson introduces joint probabilities, starting with pairs of elements. To develop the concept, association has to be based on spatial proximity (easier to work with) or on functional interaction (more interesting for the biologist). If p_i, p_j are the probabilities of species i, j, ($\Sigma p_i = 1$), and a_{ij} the probability of interaction between species i and species j (comprised between 0 and 1), the following expressions may be useful (Margalef and Gutiérrez, 1983):

$$\text{connectivity} = C = \frac{\sum_{i=1}^{n} a_{ij} p_i p_j}{\sum_{i=1}^{n} p_i^2} \quad \text{(1.6 in electronic nets)},$$

$$\begin{pmatrix}\text{relative connectivity}\\ \text{or connectance}\end{pmatrix} = C_{\text{rel}} = \frac{\sum_{i=1}^{n} a_{ij} p_i p_j}{\sum_{i=1}^{n} p_i p_j}.$$

We know that connectivity is partial. A system is neither rigid (fully connected: number of links proportional to the number of elements squared) nor disconnected (number of links inferior

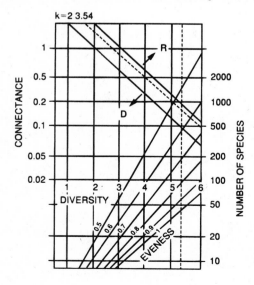

Figure 2.2 Relations between diversity ($H = -\sum_{i=1}^{n} p_i \log_2 p_i$), number of species ($S$), evenness ($E = H/\log_2 S$; $S^E = 2^H$), and connectance or relative-connectivity ($C = \sum_{i \neq j} a_{ij} p_i p_j / \sum_{i \neq j} p_i p_j) \cdot k = C \cdot 2^H$; k below 2 leads to decomposition of the system (D); $k = 4$ to rigidity and nonviability of the system (R). The interrupted line might represent a system with 100 species, evenness of 0.8, and diversity of 5.3.

to the number of elements less 1), and it would be interesting to know if the values of connectivity found in mechanical and electronic systems, and in language, are approached in their values in ecosystems. A number of authors have reviewed data concerning ecosystems and reach the conclusion that the product of the number of species times the relative connectivity (as the percent of the realized over the possible connections) falls between 2 and 12. I suspect that it is possible to define a closer range. If instead of S (number of species) we use S^E (number of species raised to evenness), the total number of links becomes proportional to $S^{(2-E)}$, and this explains the allometric relation between connectance and number of species (Cohen, 1988). Simulation of nets leads to the conclusion that $C_{rel} \cdot 2^H = C_{rel} \cdot S^E$ should remain between 2 and 4. The system is rigid at the value of 4, but is disjointed in the lower limit. Again, consideration of electronic nets suggests an intermediate value close to 3.5 and reduces the uncertainty of $S \cdot C_{rel} = 2$ to 12 to $S^E \cdot C_{rel} = 2$ to 4. Perhaps 3.5 (to be compared with the 1.6 value of connectivity; see formerly) is a limit to a trend, which is much more important than the upper limit to diversity.

CONCLUSION

We have to be diligent in exploring these and other constraints operative in systems made of replicable parts and improve the numerical analysis of the kind of data collected by ecologists. A practical question is to define the kind of interactions that have to be counted in relation to connectivity. In electronic nets we can count the wires. In ecosystems I would propose to count only direct interactions involving negative feedback. Patten (1982) has raised the problem of the meaning of indirect interactions; my impression is that this problem is manageable if one accepts that quantification and discontinuity can absorb much of the indirect effects. The next step should be to extend the study of interaction to ternary systems, which are crucial to understanding organization (two predator–prey binary systems in parallel, that is, two negative feedback loops, result in a disruptive or positive feedback loop expressed in competitive relations), and, because ternary systems offer a model of how potential connections are interrupted, to follow the development of hierarchy inside the systems. Not only ecosystems, but also neural nets, social systems, and the like, could be subjected to the same kind of analysis inside a common systems theory.

REFERENCES

CAVALLI-SFORZA, L. L., and FELDMAN, M. W. 1981. *Cultural Transmission and Evolution. A Quantitative Approach.* Princeton University Press, Princeton, New Jersey. 388 pp.

COHEN, J. E. 1988. Untangling "an entangled bank": recent facts and theories about community food webs. In A. Hastings (ed.), *Community Ecology.* Springer-Verlag, New York, pp. 72–91.

GINI, C. 1912. Variabilità e mutabilità. *Studi Econ.-Giuridice Fac. Giurisprudenza Univ. Cagliari* 3(2).

KLOEDEN, P. E., and MEES, A. I. 1985. Chaotic phenomena. *Bull. Math. Biol.* 47:697–738.

LEGENDRE, L., and DEMERS, S. 1985. Auxiliary energy, ergoclines and aquatic biological production. *Naturaliste Canadien (Rev. Ecol. Syst.)* 112:5–14.

MARGALEF, R. 1974. *Ecologia.* Ediciones Omega, Barcelona. 951 pp.

MARGALEF, R. 1978. Life-forms of phytoplankton as survival alternatives in an unstable environment. *Oceanologica Acta* 1:493–510.

MARGALEF, R. 1979. The organization of space. *Oikos* 33:152–159.

MARGALEF, R. 1985. From hydrodynamic processes to structure (information) and from information to process. In Ulanowicz R. E., and Platt T. (eds.), Ecosystem theory for Biological Oceanography. *Can. Bull. Fish Aquat. Sci.* 213:200–220.

MARGALEF, R. 1988. Ways to differentiation and diversity in ecosystems. In Velarde, M. G. (ed.), *Synergetics, Order and Chaos.* World Scientific, Singapore, pp. 441–447.

MARGALEF, R., and GUTIÉRREZ, E. 1983. How to introduce connectance in the frame of an expression for diversity. *Am. Nat.* 121:601–607.

PATTEN, B. C. 1961. Competitive exclusion. *Science* 134:1599–1601.

PATTEN, B. C. 1982. Indirect causality in ecosystems: its significance for environmental protection. In *Research on Fish and Wildlife Habitat.* Mason, W. T. and Iker, S. (eds.). U.S. Environmental Protection Agency, EPA-600/8-82-022. Washington, D.C., pp. 99–107.

POWELL, T., and RICHERSON, P. J. 1985. Temporal variation, spatial heterogeneity, and competition for resources in plankton systems: A theoretical model. *Am. Nat.* 125:431–464.

RILEY, G. A., STOMMEL, H., and BUMPUS, D. F. 1949. Quantitative ecology of the plankton of the Western-North Atlantic. *Bull. Bingham Ocean. Coll.* 12:1–169.

SIMPSON, E. H. 1949. Measurement of diversity. *Nature* 163:688.

VON SMOLUCHOWSKI, M. 1918. Versuch einer mathematischen Theorie der Koagulationskinetik kolloider Lösungen. *Zeitschrift f. physik. Chemie* 92:129–168.

3

The Far-from-Equilibrium Ecological Hinterlands

LIONEL JOHNSON

Freshwater Institute
Department of Fisheries and Oceans
Winnipeg, Manitoba, Canada

Knowledge about the entire ecosystem has become so important that ecologists can no longer be satisfied to be concerned with specific individual species or populations. In addition to plant ecologists, animal ecologists, microbial ecologists, etc., we must now train more and more young ecologists to confront the entire complexity of the ecosystem. They must be systems ecologists [Van Dyne, in *The Ecosystem Concept in Natural Resource Management*, Ch. X, p. 363 (1969), Academic Press].

OVERVIEW

Anomalous size and age structures of the dominant fish populations in undisturbed lakes of the Canadian Arctic stimulated a search for other populations with comparable configurations. These were found in diverse geographic areas, from dominant reptiles on a tropical island to cave fishes in Zaire. Similar configurations were also found in plants.

The characteristics in common were uniform size of individuals within a specific population (despite highly variable ages within each length group), great age, variable life span, and a dearth of juvenile replacement stock. All were essentially undisturbed populations, and all their ecosystems were highly autonomous. All these populations of indeterminate or indefinite growth pattern, in fact, exhibited the same structure as that of an undisturbed mammalian population. The conclusion is that dominant populations all exhibit characteristics of least energy flow or "least attainable dissipation." When disturbed, the populations increase their dissipation rates, but return to their original states when disturbing factors are removed.

From this starting point, it is hypothesized that each species population within an ecosystem tends to assume a state of least attainable dissipation and that interactions between species stimulate increased dissipation. Therefore, the greater the diversity of interactions, the greater the dissipation. Ecosystems are formed at the intersection of two trends, one tending to decelerate energy flow and the other to accelerate it. The acceleration trend is interpreted as a response to the physical principle of "least action" and the countervailing trend toward least dissipation as a response to "most action".

These trends operate in different time frames. That to least dissipation dominates system behavior in the short term, and that to accelerated energy flow dominates long term behavior. In the short term, the trend to deceleration must exceed that to acceleration if the system is to persist. Increased long term energy flux is seen in the general evolutionary tendencies toward increasing diversity and complexity.

These antagonistic processes are manifested in succession toward attainment of a "climax" state through decreased dissipation during ecological time and as "abscession" away from this state through the trend to increasing dissipation over evolutionary time. Interaction between these two trends gives rise to changes we recognize as evolution, but the trends also operate at other time scales, for example, metabolism, development, heredity, and ecology. This is a generalization of the concept of "heterochrony."

The difference in the time frames of these trends provides a "ratchet" mechanism imposing directionality on evolution. Two laws of "biodynamics" are developed, with correspondences to the laws of thermodynamics. These laws add to the laws of thermodynamics; they do not replace them.

INTRODUCTION

Since the fifth century B.C., when Herodotus first formulated the ideas that ultimately came to be expressed as the "balance of nature" (Egerton, 1973), the question of equilibrium has been in the forefront of biological thought. With the inception of ecology as a separate discipline in the late nineteenth century, the abstract notion of equilibrium assumed a pivotal role. From the lake as a "microcosm" (Forbes, 1887) to the vegetational "climax" (Clements, 1905, 1916, 1936; Whittaker, 1951, 1953; McIntosh, 1985), emphasis was placed on the balanced state.

Elton (1930) challenged this, maintaining that there was no such thing as a "balance of nature." Populations, he recognized, existed in a state of constant fluctuation. These and similar problems have rendered concepts like "stable" and "stability" virtually unusable unless strongly qualified. Subsequently, many ideas surrounding equilibrium and the stable climax have had to be discarded as inappropriate in a world where change, not stasis, appears to be the rule.

Nevertheless, the need for some homeostatic or stabilizing mechanism in the organic world seems inescapable if total chaos is to be avoided. In the world of the living there are undoubtedly elements of order not apparent in that of the nonliving. The basic question to be examined, therefore, is whether or not equilibrium is a viable ecological concept and, if so, what are the conditions necessary for its establishment and maintenance under conditions where change and fluctuation predominate. Before this can be attempted, the meaning and usage of the term "equilibrium" must be briefly considered.

In thermodynamics, equilibrium has the specialized connotation of a steady state attained in a isolated system at maximum entropy. This quite specific definition contrasts with usage in biology. Here, equilibrium has been employed in a number of different contexts, from the description of climax as a state of "dynamic equilibrium" (Tansley, 1935), to special meanings associated with the theory of island biogeography (MacArthur and Wilson, 1967) or that of "evolutionary equilibrium" (Hoffman, 1985). By contrast, the term "nonequilibrium" has been used to signify a mere absence of ecological equilibrium (Wiens, 1984). It has also been applied to systems of a special type in which a stationary state is maintained by energy input, although thermodynamic equilibrium is unattainable because of permeability of system boundaries to energy and entropy exchange (Prigogine, 1978).

The position adopted in this chapter is that all biological systems and processes are consistent with and dependent on thermodynamic principles; therefore, one state only should be referred to without qualification as equilibrium, and this is thermodynamic equilibrium. Biological systems have existed and maintained relatively constant states over long periods of time during which they have also been subject to change. Within any period of relative environmental constancy, populations fluctuate considerably at the local level. They exist far from thermodynamic equilibrium, depending for their existence on energy flows driving the processes of system maintenance. Therefore, they fit readily into the class of nonequilibrium systems.

Dynamic systems far from equilibrium may attain steady states in which free energy entering equals the rate of energy dissipation to heat. However, steady state does not in itself imply stability. An automobile engine may run at a steady state, given a certain depression of the accelerator, but the moment this pressure on the pedal is relaxed the speed of the engine slows. Stability implies prevention of change, and there are two types. One is that of a jet aircraft that employs feedback servomechanisms to correct deviations from straight-and-level flight. The deviations occur, even in the relatively constant environmental conditions that the aircraft normally experiences. The other type of stability is that of an armored tank, which by weight and other characteristics of design is impervious to even quite severe environmental fluctuations. The first type of stability, "resilience," necessitates constant energy input; the second, "resistance," depends on accumulated energy reserves. Biological stability depends on a combination of the two, in varying proportions according to environmental conditions.

Thus, a central feature of biological systems is their homeostasis, the maintenance of stable states far from equilibrium. As Lewontin (1957) states, "the central problem of evolution is the development of a theory of homeostasis which maintains the constant survival potential of the evolutionary unit." Slobodkin (1964) carries this further: "Homeostasis itself is the thing which evolutionary homeostasis is trying to hold constant." A homeostatic mechanism implies resistance to displacement or return to the initial state if temporarily displaced (as in the maintenance of blood pH). Waddington (1968a) has suggested that the term "homeorhesis," referring to stabilization of input and output flows, rather than the homeostasis of specific states, is a more appropriate term in the ecological context.

Thermodynamic systems may be divided into three broad categories: isolated, closed and open. An *isolated system* is one whose boundaries prevent exchange of either energy or matter with the external world; this is the basic reference system in thermodynamics. The isolated condition is essentially imaginary, since no system can be completely cut off from the world at large. An *open system* is one in which both energy and matter are exchanged. A *closed system* is a special case of an open system; in it, energy and entropy, but not matter, are exchanged across the system boundaries. A dynamic system exchanging energy and entropy with the external universe will reach a steady state if the system is closed and the energy input cycle and time delay imposed internally on the energy for the performance of work are both constant.

Biological systems have generally been considered open, although this categorization has, perhaps, been applied too broadly. Individual organisms are obviously open systems, since they depend for their existence on exchange of both energy and materials with the external physical world. However, openness in this broad sense is not a necessary condition for the existence of an ecosystem as it is with an organism; materials in an ecosystem may be internally recycled indefinitely. For example, certain permanently frozen lakes in Antarctica (Goldman, 1970) closely approach being closed and largely autonomous (see below); even in these circumstances, however, some gaseous exchange probably occurs. All other ecosystems in the biosphere are interconnected to a greater or lesser degree. However, at the limit, biospheric boundaries largely preclude exchange of significant quantities of matter. Thus, within these boundaries, the biosystem of the earth functions as a single closed unit.

An ecosystem is thus a subsystem of the whole. As defined by Tansley (1935), it is the functional combination of the "organism-complex" and the environmental factors to which it is exposed. To maintain its integrity, an ecosystem requires boundaries and, as Tansley maintains,

"The more relatively separate and autonomous the system, the more highly integrated it is and the greater the stability of its dynamic equilibrium."

An *autonomous ecosystem* may be defined as a sink for free energy in which all free energy entering as either sunlight or biomatter is retained until dissipated as heat. Free energy is energy capable of doing work; the rate at which work is done is a function of the rate at which energy is transformed from one form to another with the production of entropy.

As far as we know, with a few anthropogenic exceptions, the boundaries of the biosphere are completely impermeable to the exchange of biomatter. For long periods of geological time, the closed nature of the planetary system has limited biological fluctuation. Nevertheless, this closed condition is regularly violated by the entry of celestial objects and the escape of molecules attaining sufficient velocity to escape from Earth's gravity. These deviations probably have little significance most of the time. However, it has been suggested that, at infrequent but regular intervals, major fluctuations have been initiated by the incursion of large meteorites (Alvarez et al., 1980, 1984; Raup and Sepkoski, 1986). However, between these cataclysmic events the earth settles down within its boundaries, forming a single integrated system in which all parts contribute to the functioning of the whole.

If the isolated system functions as a reference system for physical thermodynamics, then the completely closed or autonomous ecosystem may similarly be regarded as the basic biological reference system. Such a system may be envisaged as a permanently frozen lake (or completely airtight greenhouse), allowing the input of light and the dissipation of heat through the transparent surface layer. In its simplest form the reference system would contain a simple food chain. Within its boundaries all materials would be recycled. Given a regular energy input cycle, the autonomous ecosystem should, if Tansley's axiom is correct, attain dynamic stability.

At each locality on the earth's surface, complementary flows of energy and matter, within partial boundaries, form unique ecosystems (Reiners, 1986). Individual ecosystems may vary greatly in their degree of autonomy, but all are nested within gradually widening constraints until ultimately bounded by the limits of the biosphere. The flows of energy and materials, regulated by the world's biota, drive the global biogeochemical cycles and induce changes (as in the increased oxygen content of the atmosphere and oceans) to which the system as a whole must then, itself, adapt (Lovelock, 1979).

Strong Inference

In his forward looking paper on developing the most appropriate approach to a complex scientific problem, Platt (1964) advocated the adoption of a policy of "strong inference":

> Strong inference, and the logical tree it generates, are to inductive reasoning what the syllogism is to deductive reasoning, in that it offers a regular method for reaching firm inductive conclusions one after the other as rapidly as possible.

Having developed the strong inference, Platt suggests that progress is rapid if the most likely solution is then selected and tested using the simplest system that has the necessary properties of interest. Similarly, Pagels (1982) states (of simple cosmic systems),

> These gifts of nature show how progress in understanding the universe is profoundly conditioned by the existence of a simple system in the environment. . . . Conceivably, the lack of such simple systems in other areas of science has inhibited progress.

This essay holds that such simple biological systems do exist, and they can serve as a base for development of a reference system. Comparable to a physical system, a biological reference system is also imaginary in that certain simplifying assumptions must be made, such as existence of a completely regular energy input regime and complete system autonomy. The study of simple biological systems allows strong inferences to be made and far reaching conclusions to be drawn.

Following Platt and Pagels, the study of ecosystems should rightly begin with the simplest system that can be found. A suitable starting point for such investigations should be autonomous ecosystems, functioning within a regular cycle of ambient energy, and they should have had sufficient time without disturbance to have reached a stationary state, if such a condition would develop.

Ecosystems approaching this ideal today can be found only in remoter parts of the earth. And ideas initiated by studies of previously undisturbed lakes in the Canadian Northwest Territories stimulated the concepts to be discussed below. Initially, these studies had a more prosaic objective than testing stability principles of biomechanics. But the attempt to understand fish production processes in these lakes ultimately necessitated exploration of more fundamental processes. The search for understanding led to more and more fundamental levels: what appeared to be a light at the end of each succeeding tunnel was merely an illuminated notice reading "exit only at the lower level."

It was evident that fish populations in virtually all the lakes examined had the appearance of great stability. The individuals were of large and uniform size and of great mean age, although actual ages of individuals within a size class were highly variable. Particularly noticeable was a virtual absence of juveniles in all samples.

These populations appeared to be stable in the sense of exhibiting relative constancy in size- and age-frequency distributions in the face of environmental fluctuations to which they were normally exposed. Further evidence of stability was shown in that, when intentionally displaced from their initial state and the disturbing factor then removed, they deviated and then returned to their original condition without evident oscillatory damping (Johnson, 1972, 1976, 1983). The evident stability of these fish populations was in striking contrast to the small mammal populations of the surrounding tundra.

These small mammal populations formed the pole of instability in the "diversity–stability hypothesis" which was popular at about the time our studies on arctic lakes began (MacArthur, 1955; Kolata, 1974). The diversity–stability hypothesis postulated that diverse systems are stable because of a large number of interconnecting energy pathways, whereas simple systems tend to oscillate. Although today somewhat discredited (Goodman, 1975), the diversity–stability hypothesis nevertheless accentuated the contrast between neighboring systems of the lakes and tundra. Since both systems were relatively simple, simplicity in itself could not explain the stability differences. The condition in arctic lakes stimulated a wider literature search for comparable systems. Examples of simple, apparently stable ecosystems were revealed at all latitudes. The full implication of the arctic lake ecosystems can be appreciated only after examination of a number of these interesting systems from various parts of the world. This preliminary overview establishes that the results obtained from arctic lakes are not merely of local significance but are truly of great generality. The results also help to eliminate certain conceptual difficulties arising from apparent misinterpretation of fish population data.

The study of naturally occurring biological phenomena can never be precise. It is therefore essential to try to identify salient features through the obscurity created by environmental "noise." Interpretation of such data frequently necessitates intuitive leaps across intellectual chasms that can be bridged only later by evidence from a wider range of observations, or perhaps by necessity once principal elements of the biological cryptogram have been put in place.

LAKES OF THE CANADIAN NORTHWEST TERRITORIES

Many lakes in the Canadian Northwest Territories exist in very isolated conditions far from human intervention and accessible only by air. These northern lakes are almost completely autonomous. They have poorly developed connections with other water bodies during the short summers and for remaining months of the year are, with rivers, ice covered and frozen solid. They have been free of human impact since their formation during retreat of the ice sheets in the closing stages of the Pleistocene. The ecosystems that developed in these conditions are relatively simple and in the northern most lakes consist of limited benthic and planktonic flora and fauna and a single species of fish, the Arctic char, *Salvelinus alpinus.*

The study of lakes in these remote regions of northern Canada began with a survey of the fishery potential of Great Bear Lake (Miller, 1947). This was followed in 1959 by an extensive survey of large lakes in the Barren Grounds (the vast area of mainland Canada north of the tree line) and in 1962 by a survey of the Arctic Archipelago (McPhail and Lindsey, 1970; Johnson, 1976). In 1961, a commercial fishery was started at Keller Lake, which provided an opportunity for more extensive observations (Johnson, 1972). These surveys were carried out by the Arctic Unit of the Fisheries Research Board of Canada. Between 1963 and 1965 a second, more detailed study of Great Bear Lake was conducted (Johnson, 1975a, b).

In 1974 a long-term study began on Arctic char, a species of great importance in the northern economy. Arctic char exists in both a freshwater form and an anadromous form. This study, centered on Nauyuk Lake (latitude 68° 22′ N, longitude 107° 40′ W) on the Kent Peninsula of the northern mainland, has continued to the present day. The program was primarily designed to investigate the biology of anadromous stocks of Arctic char (Johnson, 1980, 1989; Johnson and Burns, 1984). However, there occurred in the vicinity of Nauyuk Lake a number of small (<50 ha), highly autonomous lakes in which Arctic char was the sole fish species (Johnson 1994a, 1994b). These lakes and others with similar characteristics in various parts of the Arctic provide virtual laboratory conditions for the study of unexploited populations and their behavior under imposed change.

The lakes investigated range in latitude from just north of the tree line to Lake Hazen (latitude 81° 50′ N, longitude 71° 09′ W; surface area = 542 km^2) at the northern extremity of Ellesmere Island, the closest land to the north pole (Johnson, 1976, 1983) (Fig. 3.1). In size, the lakes range from Great Bear (31,000 km^2), which in surface area ranks ninth in world standing, to lakes of 50 ha or less. As might be expected, Lake Hazen at the northern end of the series has the smallest number of species. The single species of fish in this lake is Arctic char, which is the terminal species in the food chain. The plankton consists of one species of copepod, *Cyclops scutifer,* and two species of rotifer, *Keratella hiemalis* and *K. cochlearis* (McLaren, 1964). The benthic fauna is largely made up of several species of chironomid larvae that, when they pupate, form the major part of the diet of the char. Species richness increases toward the south, with dominance in the fish fauna being shared between Arctic char, lake char (lake trout), *Salvelinus namaycush,* and lake whitefish, *Coregonus clupeaformis.* The occurrence of Arctic char becomes less frequent with increasing distance from the coastline, dominance being assumed by lake trout and lake whitefish in varying proportions (Johnson, 1976, 1980). These primary species are supplemented by lake cisco, *C. artedii,* northern pike, *Esox lucius,* burbot, *Lota lota,* round whitefish, *Prosopium cylindraceum,* nine-spine stickleback, *Pungitius pungitius,* and the sculpins *Cottus cognatus* and *Myoxocephalus quadricornis (thompsonii).*

All the lakes are ice covered for eight to nine months of the year, with some of the larger lakes losing their ice completely only in relatively warm years. For the period of ice cover, the lakes are completely sequestered from the external world except for the small exchange of light and heat through a layer of ice up to 2 m thick. Northward of the Arctic Circle (66° 30′ N latitude) an increasing portion of the winter is spent in total darkness, with a correspondingly longer period of continuous daylight in summer. Temperature remains remarkably constant throughout the year. In large lakes, winter temperatures are between 1° and 3°C, and even in summer surface temperatures seldom rise much above 4°C (Johnson, 1966). In shallower lakes, surface temperatures may reach 12° to 14°C. Primary production is low. In a detailed study of Char Lake (latitude 74° 42′ N, longitude 94° 52′ W), Cornwallis Island, a value of 20 gC m^{-2} y^{-1} was obtained (Kalff and Welch, 1974; Welch and Kalff, 1974).

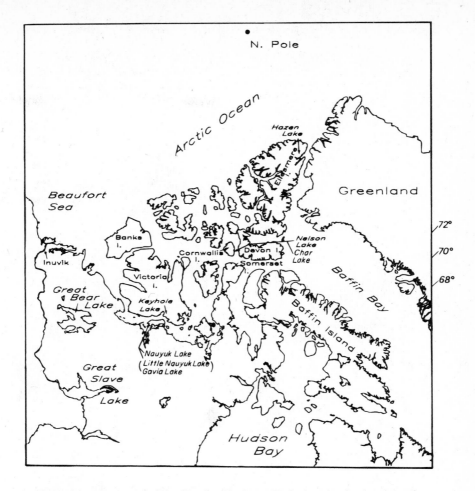

Figure 3.1 Sketch map of the Canadian Northwest Territories, showing locations of lakes referred to in the text.

In spring (June to July), when ice starts to melt around the edges of lakes, water birds arrive and begin nesting. The fishing birds, mostly loons, *Gavia* spp.; terns, *Sterna* spp.; and gulls, *Larus* spp., are virtually the only predators on the fish stocks. Since predation is confined mainly to smaller fish, above a certain size most fish are able to live out their physiological life span.

The structure of ecosystems in these lakes is relatively simple, with food chains being almost linear. Primary producing algae support an invertebrate population of benthic organisms, possibly mainly through a detrital and bacterial loop; then these invertebrates, largely chironomid larvae, form the staple food of the terminal fish species.

The biomass of fish in all lakes is high, but particularly so in those lakes where Arctic char is the only fish species. Fish biomass values range from 10 to 48 kg ha^{-1}, which is comparable with the total fish biomass in many multispecies lakes of temperate regions.

The outstanding features of all fish populations examined in these lakes are (1) individuals are generally old by temperate standards, large (up to 100 cm), and relatively uniform in size but of variable age; (2) their actual modal size varies considerably from lake to lake; and (3) the number of juvenile fish relative to adults is very small. These nearly universal characteristics indicate that the populations fluctuate very little over time. This apparent stability was evident also in the planktonic copepods of Char Lake (Rigler et al., 1974), where, over a three-year study, no measurable year-to-year fluctuation could be detected in the *Limnocalanus macrurus* population. In summary, Rigler (1975) stated that "there is no evidence to support the myth that simple systems are 'unstable.' " Stability seemed to be, as Tansley (1935) stated, a function of the autonomy of the system, not its simplicity.

In due course, repeated observations in a number of lakes, often at intervals of many years, confirmed that no measurable change in the fish populations had occurred. The maximum interval between observations was 23 years in the case of Lake Hazen (Johnson, 1976, 1983) (Fig. 3.2). The lakes, in fact, had all the appearances of existing in a "stable climax" state, the aquatic equivalent of the vegetational "climax" (Johnson, 1976, 1981). The major conceptual difference between these ecosystems and a forest system is that in lakes the dominant species are terminal in the food chain, whereas in the climax forest the dominants are primary producers.

Methods

The most effective sampling gear used in northern lakes is the gill net. This is invariably operated as a standard gang of multimeshed nets. It is recognized that fish samples generally exhibit a "bell-shaped" or "dome-shaped" catch curve (Ricker, 1975). The reason for this is often given as the "mesh selectivity" of specific mesh sizes for only fish of appropriate dimensions. For this reason, nets, particularly gill nets, are considered unreliable sampling gear in that samples taken are not representative of the population as a whole. Only the portion of a fish stock that is larger in size than the mode of the catch curve is fully vulnerable to the gear used. The inherent rationality of this explanation is unequivocal, and it has become an article of faith.

Conventional fishing theory assumes that the length of an individual fish is contingent on the population growth rate within a certain range of variation. The growth rate is a declining function of age. Since natural mortality affects all age classes (although acting most acutely on the younger ones), the number of individuals in each age class declines as size increases, generally following a negative exponential curve. Because the size of an individual is considered to be a relatively uniform function of time, the length–frequency distribution of the population is considered to follow a negative exponential curve similar to that of natural mortality.

While it is indisputable that mesh selectivity has an influence (a 10-cm fish is unlikely to be caught with a 10-cm mesh) and that length increases monotonically with time (although not necessarily at a uniform rate), the more restrictive aspects of fishing theory as described above were found inadequate to explain the structure of fish stocks of the Barren Grounds. When results from the wide range of mesh sizes used in sampling were combined, the lakes invariably yielded length–frequency distributions having a single dominant mode, not a series of modes corresponding to the various mesh sizes (Fig. 3.3). In certain cases, additional modes were exhibited at smaller and larger sizes than the dominant mode (Johnson 1976) (Figs. 3.4 and 3.5). It was also observed that identical series of nets used in a variety of lakes yielded modal lengths ranging from 10 cm in Gavia Lake (Johnson, 1994b) and Borup Fjord Lakes (Parker and Johnson, 1991) to 75 cm in Namaycush Lake (Johnson, 1976).

If a bell-shaped length–frequency distribution truly represents a population, and age at modal size is variable, the configuration obtained can result only from partial decoupling of age and growth. If this is the case, and the configuration is maintained over long periods, control of size in the single (dominant) fish species must be through internal self-regulation. This would represent an interesting departure from conventional theory, although in defense of the latter it must be stated that it is more likely to apply in exploited populations (where it is most frequently used) than in unexploited populations.

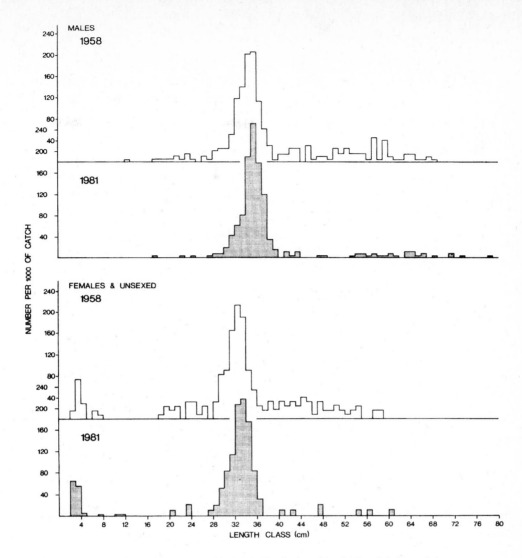

Figure 3.2 Lake Hazen. Length frequency distribution of Arctic char, *Salvelinus alpinus*, in 1958 and 1981. Frequencies represent catches from all mesh sizes combined (from Johnson, 1983).

In an attempt to resolve this conceptual problem and improve sampling of smaller fish, a new type of gill net was introduced to supplement that originally used (for details, see Johnson, 1983). The new gear provided a sequence of mesh sizes from 20 to 120 mm (stretched measure), following as closely as possible the ideal of a geometric increase in mesh size. There is every reason to believe that this gear samples effectively and repeatably down to a fish length of about 10 cm.

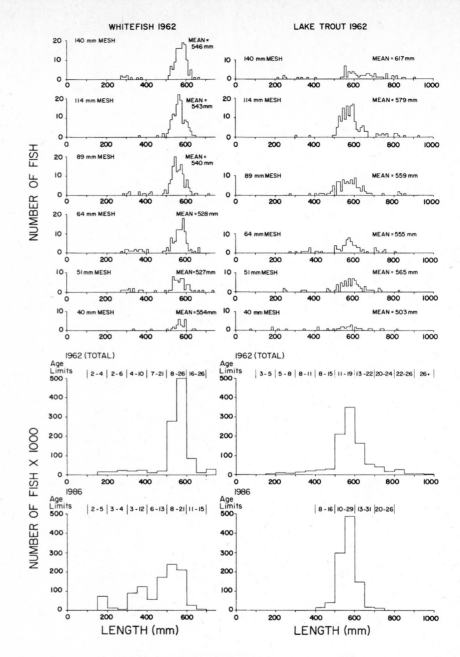

Figure 3.3 Keller Lake. Length frequency distribution of lake trout, *Salvelinus namaycush,* and lake whitefish, *Coregonus clupeaformis.* The combined catch from all nets and individual catches from various mesh sizes are shown (from Johnson, 1972). The lake was briefly resampled in 1986. Some small changes are apparent; the whitefish population is less uniform than in 1962, but the trout population is more uniform (Johnson, unpublished).

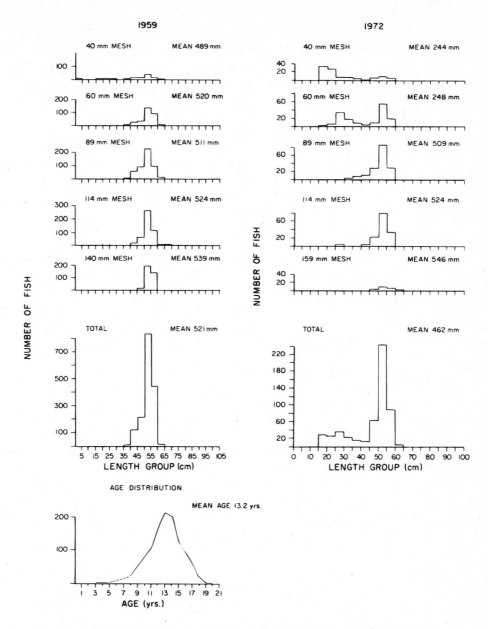

Figure 3.4 Lac La Martre. Length and age frequency distribution of lake whitefish, *Coregonus clupeaformis,* 1959 and 1972 (from Johnson, 1976).

Chap. 3   The Far-from-Equilibrium Ecological Hinterlands

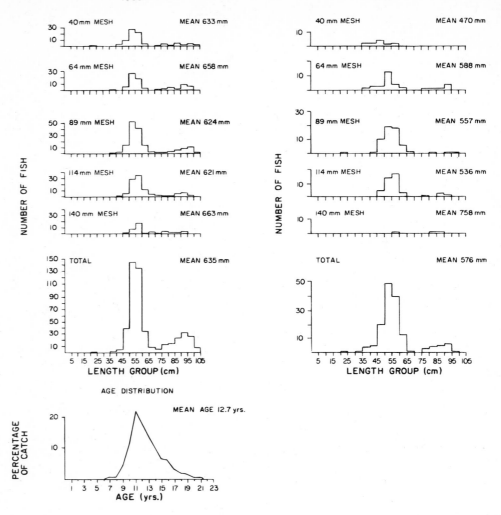

Figure 3.5 Lac La Martre. Length and age frequency distribution of lake trout, *Salvelinus namaycush*, by mesh size and combined catch for the years 1959 and 1972 (from Johnson, 1976).

Results

In Little Nauyuk Lake, the gear produced the characteristic bell-shaped length–frequency distribution, with a modal value of 22 cm, when used for the first time. This lake had not been previously fished or otherwise disturbed (Fig. 3.6). Interestingly, in neighboring Gavia Lake a length–frequency distribution was obtained very close to the accepted theoretical postulate, fish being most abundant in the 100- to 109-mm length class (the smallest size vulnerable to the gear) and declining in abundance as size increased. However, this apparently "normal" distribution subsequently appeared

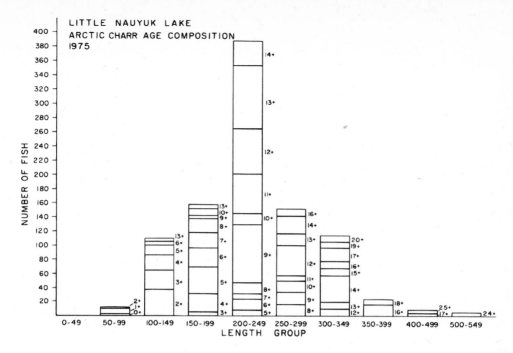

Figure 3.6 Little Nauyuk Lake, 1975. Length frequency distribution of Arctic char, *Salvelinus alpinus,* with ages allocated by length class. Apart from a small sample in 1974, this distribution represents that of an undisturbed lake (from Johnson, 1983).

to have resulted from unusual intraspecific effects. One very large cannibalistic char (64 cm) and several of medium size (40 to 50 cm) were found to be present. When these fish were removed early in the sampling program, the population structure rapidly assumed the more usual bell-shaped configuration (Fig. 3.7). The transition occurred in the short interval between 1975 and 1976, and the bell-shaped configuration subsequently remained virtually constant over the next five years despite considerable reduction of overall population density. The situation in this lake has not been duplicated.

Over the same period the char population of Little Nauyuk Lake was heavily exploited, similarly maintaining a constant configuration with a modal value of 22 cm (Fig. 3.8). These two series completely obviate the possibility that, at least for fish over 10 cm, mesh selection is responsible for the configuration of the catch curve. At the same time, the effectiveness of the gill net for all sizes of char above this length is fully established. The results from Gavia Lake indicate also that complete homogeneity in any series of biological data must never be expected.

These results have been examined in some detail because the concept of a population functioning as a coherent, dissipative unit, as developed below, depends on establishing that samples are representative of the whole population.

The low level and even apparent absence of juveniles observed in some populations stretches credibility; this is difficult to comprehend and even more difficult to explain. Nonetheless, as demonstrated later, this characteristic is not confined to fish, but is also observable in terrestrial plants and animals where sampling is not confounded by the indirectness of observation. Low juvenile populations have been observed sufficiently generally to indicate their reality. If a population is maintaining itself at a relatively constant density, potential recruits must exist somewhere. Jones

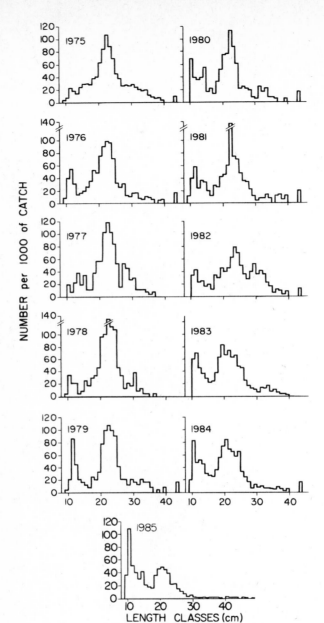

Figure 3.7 Little Nauyuk Lake. Length frequency distribution of Arctic char, *Salvelinus alpinus,* 1975–1985 (from Johnson, 1983, and previously unpublished material).

(1945) has noted, however, when referring to the same phenomenon in European virgin forests (where trees have a mean life span of 200 to 300 years), that the average number of recruits required annually is very small. The evidence in fish populations indicates that juveniles, when encountered, are found on the periphery of adult niche space, ready to move forward and join the adults when the opportunity occurs.

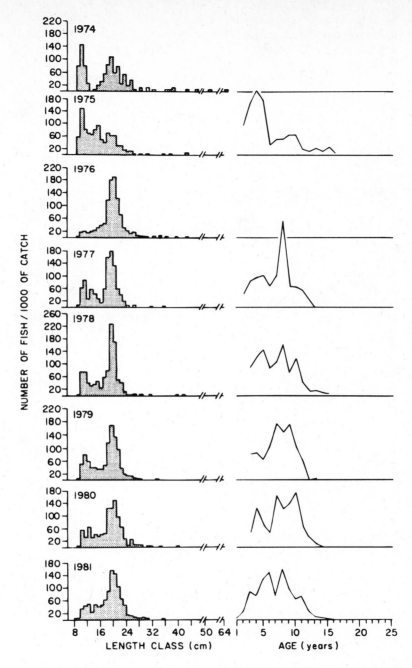

Figure 3.8 Gavia Lake. Length and age frequency distributions of Arctic char, *Salvelinus alpinus,* 1974–1981. Note the rapid change in length and age frequency distributions between 1975 and 1976, following removal of a number of large fish (from Johnson, 1983).

If the modal configuration is a true representation of the real population, then mechanisms must exist for its development and maintenance. Johnson (1976) postulated that there are at least two length modes, one of small, mostly younger fish maintained on the periphery of the adult niche space and the other of larger "adult" fish forming the "establishment" and occupying the prime habitat. Within each length class there is a relatively wide distribution of age. This is particularly noticeable within the modal length class, which includes fish of virtually all ages in the entire establishment. Segregation into the two groups is apparently effected by interactions. Juvenile fish are relegated to the niche periphery through suppression by adults, similarly to the suppression of seedlings and saplings in a climax forest. The juveniles are thus maintained in a "stack" from which individuals move forward into the establishment as size and opportunity allow. The transition occurs by rapid increase in growth rate over those lengths in the trough between the modes, then slowing as the modal size is approached. An appropriate analogy, with regard to size and age structure, is a professional sports team limited to a total number of players. The plausibility of this pattern has been corroborated by Sparholt (1985) and Riget et al. (1986) for Arctic char populations in Greenland.

It came as rather a surprise to find that, when subject to an experimental fishing regime, the modal configuration was maintained even as population density declined considerably. It was even more surprising to find that the anadromous stock of Arctic char at Nauyuk Lake, feeding in the relatively open conditions of the sea, maintained constant size structure over many years although simultaneously experiencing a great decline in abundance. This stock, which was fished by a local Inuit family, passed through a counting fence twice each year, enabling the whole seagoing population to be sampled without bias. When the population had been reduced to a very low level after five years of fishing, the annual sampling was suspended. When resumed in 1984, a change in configuration of the population had occurred; a new modal value, some 10 cm less than the original value (Fig. 3.9), was established (Johnson, 1989). Dempson and Green (1985), working on the coast of Labrador,

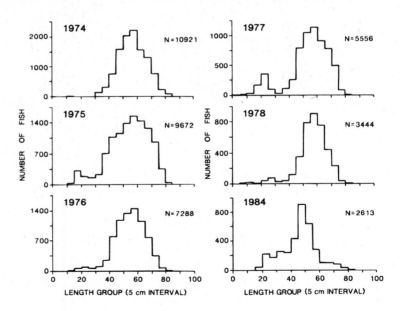

Figure 3.9 Nauyuk Lake. Length frequency distribution of the anadromous stock of Arctic char, *Salvelinus alpinus,* from 1974 to 1984. The histograms represent the total upstream run for each of the years indicated. The stock declined from 10,921 char in 1974 to 2613 in 1984 due to a local fishery (Johnson, 1980, and previously unpublished material).

similarly observed very small changes in stock configuration as the Arctic char population there declined during the course of a commercial fishery. It is evident that effective self-regulating mechanisms are operating well below the carrying capacity of the system. Density, it appears, is the factor controlling recruitment only at the steady state. Far from this climax, the population structure is actively maintained by a damping mechanism internal to the population. These mechanisms apparently allow the population to revert to its stationary state in an orderly, well-damped manner without oscillation when the perturbing factor is removed.

It is important to note that this population structure is identical to that of a mammalian population at "carrying capacity." The difference is that the mammalian population is of determinate growth pattern, whereas that of the fish population is of indeterminate type and therefore due to internal self-regulation. An indeterminate growth pattern is one for which growth continues throughout life.

Implications

It is evident from the structure of the Arctic char populations in these small lakes that the members function as a coherent group, not as a collection of individuals. This is evident in the maintenance of relatively constant size over long time periods, return to the original structure after severe disturbance, and retention of a constant size structure over severely declining population levels.

The actual structure of the population may be interpreted as one exhibiting "least attainable dissipation." This implies that the population proceeds toward a state in which, given the conditions prevailing and the genetic makeup of the species concerned, the largest biomass relative to energy intake is maintained. This is evident in gradual biomass increase following disturbance until an asymptote is approached. Least dissipation is indicated by absence of fluctuation, large size (and, concomitantly, relatively low specific metabolic rate), uniformity of size, great age, a low level of juveniles, and indeterminate age at death.

Such a distribution of energy accords with the general characteristics of all populations, which tend to increase toward the "carrying capacity" of the habitat concerned and then, within the limits imposed by environmental fluctuations, maintain relatively constant abundance. This condition is most evident in the dominant species in a food-chain hierarchy. Species lower in the hierarchy will maintain less biomass than those higher up, which are not subject to the constraints of shading, predation, or being grazed. In appropriate conditions a very wide variety of species, both plant and animal, can attain dominance. Therefore, as a primary inference, it may be stated that all species in an ecosystem tend to assume a state of least dissipation consistent with their position in the hierarchy and genetic makeup, within overall environmental constraints. Acceptance of this hypothesis provides the foundation for a chain of reasoning that has wide implications if it can be shown to be of general applicability.

THE ECOLOGICAL HINTERLAND

If the trend of arctic fish populations toward least dissipation is a general characteristic of all populations, then supporting evidence should be sought in undisturbed, autonomous systems at all latitudes. Preferably, the dominant species should be an organism with an indeterminate growth pattern. If the thesis is to be substantiated as a generalization, its manifestations should be evident in both terrestrial and aquatic ecosystems, with dominant species either plants or animals.

A number of examples, described below, have been assembled from the literature in which appropriate conditions appear to have been largely met. Most of these studies were made for other purposes; therefore, they frequently do not contain all the desirable information.

Ecosystems with Dominant Animal Populations

Blind cave fishes of Zaire. Among the world's simplest ecosystems are those in the caves of Zaire. The biology of these solution caves, occurring in a restricted region of limestone in the Congo Basin, was studied intensively by Heuts (1951). Uniquely, these ecosystems are comprised of a single species, the blind African cave fish *Caecobarbus geertsii*, which depends for its food on epigeal organisms carried down by floods during the rainy season. The rains last for only six months, with very little flow occurring during the remainder of the year; therefore, *Caecobarbus* subsists almost entirely without food for long periods. It is able to do this, according to Heuts, because of its very low rate of basal metabolism.

Heuts examined populations from six different caves which he referred to as populations A to F. The populations were entirely autonomous as there was no interconnection between caves. With the exception of population D, all were composed of uniformly sized fish of considerable age (Fig. 3.10), having modal lengths between 55 and 70 mm. Despite extensive searching, only specimens in their first or second year of life were collected from population D. Population D was also anomalous in that the adults were found exposed to light at the cave entrance, while juveniles were located farther underground. In all other populations, adults lived 100 to 300 m from the cave entrance in total darkness. In none of these populations were fish in a reproductive condition encountered.

Modal length in populations B and E was about 55 mm, with an age span in the modal-length class between 4 and 8 y. Populations A, D, and F were slightly larger; population F, the largest, had a modal length of 70 mm with an age spread within the modal-length class of 8 to 12 y. The largest fish in any population did not exceed 85 mm. Each population showed a different mean age. Heuts described population D as the only one having a normal age composition, "normal" implying a large number of young and a relatively small number of adults. The absence of juveniles in all other populations and the absence of sexually maturing individuals caused Heuts considerable consternation. He concluded:

> It seems very plausible in any case, that in such closed populations with strictly limited food resources, some kind of control mechanism of fertility or fecundity would exist, eventually correlated with the mean physiological and ecological features of each of the populations.

It is incontestable that reproduction in the populations other than D was taking place. It is, however, possible that behavioral segregation relegated the juvenile segment of the population to existence in some region where they were inaccessible to the sampling procedure. Such a possibility is indicated by the segregation observed in population D. As Heuts noted,

> Keeping in mind the special cave conditions, the closedness of the economical system, the food scarcity, and possibly highly specialized requirements of juvenile cave forms, the very

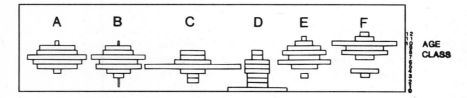

Figure 3.10 Age frequency distribution of six populations of the blind African cave fish, *Caecobarbus geertsii*, from the Congo Basin, Zaire (redrawn from Heuts, 1951).

peculiar aspects of the age distribution curves could be explained in either one of the following ways ... (1) intermittent reproduction at approximately 10 year intervals ..., or (2) [if] reproduction takes place every year, the juveniles are bound to live in special habitats ... inaccessible to man.

It is evident that in these highly autonomous conditions and with limited food supply, only a certain number of adults can survive. Since adults live for several years, the number of recruits required annually to maintain the relatively constant "establishment" population will be very small. For most efficient utilization of the restricted amount of energy available, the number of potential recruits would just exceed the number of replacements actually required.

Cave fishes of the southern United States. Comparable population structures to those of the Zaire cave fishes have been described by Poulson (1963) for North American cave fishes of the genus *Amblyopsis*. This genus shows a consistent reduction in eye development with increasing adaptation to a cave environment. Concomitant with cave adaptation, the genus as a whole also shows a gradual reduction in basal metabolic rate. The two most cave-adapted species are *A. spelaea,* the northern cavefish, and *A. rosae,* the Ozark cavefish. Both species live mainly on a diet of copepods, although larger fish (over 45 mm) also take isopods of the species *Asellus pellucidus,* amphipods of the species *Crangonyx gracilis,* as well as small salamanders and young of their own species. Poulson believed that the most significant factor in natural mortality before senescence was density-dependent cannibalism. Following hatching, the young live for some time in the gill cavities of adults; but after leaving this environment they virtually disappear for some time before reappearing as adults. Two optimal habitats of *A. spelaea* were completely censused over seven reproductive seasons, but reproducing females were found on only two occasions in one habitat and once in the other. "Furthermore," states Poulson, "out of a total of 127 females only 10 had eggs or young in their gill cavities." It is evident that reproduction has been reduced to an exceedingly low level.

Ambylopsis spelaea lives to about seven years of age and *A. rosae* to four years. Constancy of the age frequency distribution of the population was shown by repeated sampling over a seven-year period (Fig. 3.11). Length and age are closely correlated. Specimens of *A. rosae* of modal length (30 to 40 mm) were between the ages of three and five years; *A. speleae* of modal length (60 to 70 mm) were between four and six years old. Poulson comments, "It is nearly certain that the census figures for age groups 0 and 1 are erroneously low because census figures for age groups 2 and 3 two years later are higher." The recruitment mechanisms were unclear because of the difficulty of finding fish in the first years of life. Also, the results were almost identical to those obtained by Heuts in the Zaire caves.

The gradual loss of eyes in *Amblyopsis,* with concomitant reduction in basal metabolic needs, seems to be a specific case of evolution toward a state of reduced energy dissipation. In *Amblyopsis,* there is also a comparable trend toward an increase in parental investment in individual offspring and a consequent reduction in the number of offspring reared. This is achieved through hatching the eggs and rearing larval stages in the gill cavities, a technique that is doubtless very effective in the cave environment. Considerable reduction in metabolic expenditure can be achieved by reduction of nonfunctional sense organs. Eyes, it may be assumed, being very complex organs, have high energy requirements for their maintenance and operation. Therefore, in caves where they are no longer functional they can be reduced to a vestigial state with considerable decrease in energy dissipated per unit biomass.

Marine mollusks. Bottom deposits in the deep waters (895 to 1490 m) off the coast of Newfoundland in the Carson Canyon were sampled by Hutchings and Haedrich (1984) using trawls. Prominent in the samples were the bivalve mollusks *Yoldia thraciaeformis* and *Nuculana pernula*

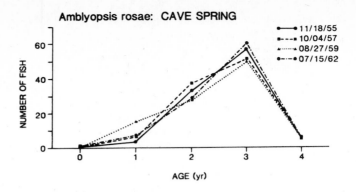

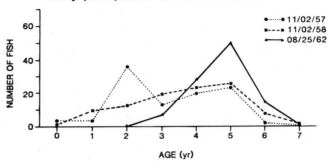

Figure 3.11 Age frequency distribution of two populations of the blind cave fish, *Amblyopsis,* from the southern United States. Above: *Amblyopsis rosae* from Cave Spring. Below: *Amblyopsis speleae* from Upper Twin Cave (redrawn from Poulson, 1963).

(Fig. 3.12). Relatively complete sampling was considered to have been achieved because of the large amounts of mud brought up by the trawl. Hutchings and Haedrich concluded:

> The size distribution of *Nuculana pernula* is characterized by an abundance of large individuals and a high degree of clustering around a single mode. Only 6 (3.3%) of 183 specimens have lengths under 20 mm and 62% of the population falls within a 2 mm size class (25–27 mm). The length frequency of *Yoldia thraciaeformis* is dominated by even larger individuals. Two (1.3%) of 149 specimens have lengths under 30 mm and 47% of the population falls within a 6 mm size class (35–41 mm).

The lower graph in Fig. 3.12 indicates that a "standard" growth curve can readily be developed from the same data.

Comparable size–frequency distributions for marine mollusks have been obtained by Radtke (1985) and Wieser et al. (1981). Radtke, working in the lagoon of Rose Atoll, a small isolated atoll in American Samoa, found that the giant clam, *Tridacna maxima,* maintained a bimodal length frequency over a number of years (Fig. 3.13). Sampling was done by divers. The small size of the lagoon (40 ha) and its relatively shallow water (maximum depth 16 m) and extreme clarity (visibility > 30 m) ensured comprehensive sampling of all areas. The maximum age of *Tridacna* was estimated to be about 18 y, and growth and size were, in this case, linearly related.

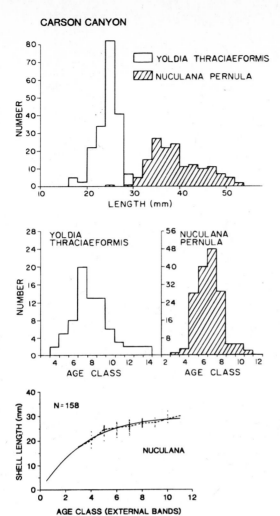

Figure 3.12 Length and age frequency distributions of the bivalve mollusks *Yoldia thraciaeformis* and *Nuculana pernula* from the deep waters of the northwest Atlantic Ocean. The lower diagram shows a "normal" growth curve obtained from the same data (redrawn from Hutchings and Haedrich, 1984).

Wieser and his coworkers, investigating the Bermudan tidal flats, found both bimodal and unimodal weight distributions in the cerithid snails *Batillaria minima* and *Cerithium lutosum* (Wieser et al., 1981). The absence of young specimens of *Batillaria* in one area caused Wieser to suggest that this species was being replaced by *Cerithium* under the stress of competition.

Freshwater mollusks. A dense population of the swan mussel *Anodonta grandis* was sampled by Green (1980) in Aeroplane Lake, near the town of Inuvik in the Canadian Northwest Territories. Sampling was carried out by divers using sieves. Not a single individual under 50 mm or under five years of age was obtained (Fig. 3.14) despite intensive sifting of the bottom deposits.

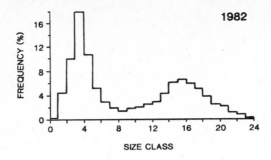

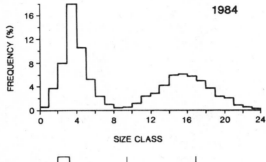

Figure 3.13 Length frequency distributions of the giant clam, *Tridacna maxima,* from Rose Atoll, American Samoa. Upper panel 1982; lower panel 1984. Size classes are presented in 10-mm increments (redrawn from Radtke, 1985).

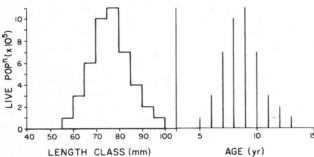

Figure 3.14 Length and age frequency distributions of the swan mussel, *Anodonta grandis,* from Aeroplane Lake, a small lake near Inuvik, Northwest Territories (redrawn from Green, 1980).

Giant tortoises of Aldabra Atoll.[1]   Aldabra, a dependency of the Republic of the Seychelles, is a coral atoll situated in the Indian Ocean 400 km north of Madagascar and 600 km east of the coast of Africa. It is composed of three main islands: Grande Terre, Malabar, and Ile Picard. The islands are uninhabited except for a small number of people operating a research station

---

[1] One of the cornerstones of science is its predictive capacity. Given the structure of arctic fish populations, it was necessary to test the generality of the inferences developed by examining structure in other autonomous ecosystems in other regions of the world, where the dominant species was an animal of indeterminate growth pattern. The most interesting candidate for examination was considered to be the giant tortoise population in Aldabra. This population seemed to have exactly the right characteristics of indeterminate growth, dominance, and isolation, in an otherwise totally different environment, that could provide the necessary data to substantiate the thesis. I was in the process of trying to convince authorities in the Canadian Department of Fisheries that a visit to Aldabra to investigate the tortoise population was in the best interests of fisheries science when I found that the necessary work had been done much more extensively than would have been possible in a single expedition.

funded by the British Royal Society and temporary workers on the coconut plantations. One of the most interesting features of the islands is their dense population of giant tortoises, *Testudo gigantea*. As far as is known, these are the only terrestrial ecosystems in which a reptilian population occupies the dominant role (Bourn and Coe, 1978). Indigenously, there were no mammals on the islands apart from bats, but goats, rats, dogs, cats, and mice have subsequently been introduced, giving rise to feral populations. During the nineteenth century the tortoise populations were severely depleted for their tortoiseshell, but in the last 70 years they have been strictly protected and now appear to have reached saturation levels. According to Stoddart and Wright (1967), "Aldabra is one of the least disturbed of all low-latitude islands, and for historical and environmental reasons, possesses an exceptionally rich and interesting fauna and flora." The climate is semiarid, with a wet season from December to April and a dry season from May to November. Total annual rainfall is around 1250 mm.

Aldabra is almost a mirror image of an arctic lake: it is tropical and terrestrial, and at the same time highly autonomous, being isolated by large expanses of ocean. The ecosystems are relatively simple in their structure compared with those of the wet tropics. The dominant species is a large vertebrate of indeterminate growth pattern.

The total tortoise population, estimated at some 150,000 individuals, is divided into five subpopulations, among which there appears to be very little interchange (Bourn and Coe, 1978). The low maintenance requirements of the tortoise allow it to develop unusually high densities, reaching about 27 individuals/ha on the island of Grande Terre (Gibson and Hamilton, 1983). Tortoises are long-lived, reaching an age of about 60 y, although older specimens have been recorded. In size, males may reach a carapace length of 100 cm, but females are somewhat smaller, reaching about 80 cm. The modal size class of tortoises at Cinq Cases on Grande Terre (carapace length, 35 to 45 cm) had an age spread of 10 to 20 y (Grubb, 1971), whereas on Malabar the modal size class (60 to 70 cm) was estimated to have ages between 12 and 22 y. Initial growth of young tortoises is quite rapid; a carapace length of 20 to 30 cm is attained by the time they are about five years old. Above this size they are virtually immune to predation (Swingland and Coe, 1978, 1979).

The various studies have all found that the length frequency distribution of the tortoise population follows a bell-shaped curve. Grubb (1971) provides data for the Cinq Cases population (Fig. 3.15), but he also indicates that he has reservations as to its true representativeness. However,

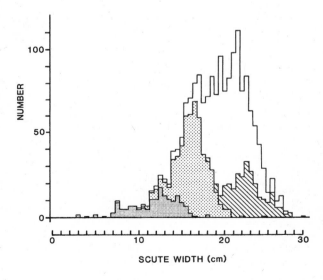

Figure 3.15 Length frequency distribution of the third vertebral scute (as a relative measure of overall size) from 1771 tortoises (*Testudo gigantea*) from Cinq Cases, Aldabra Atoll, Seychelles. Four categories are recognized: (1) up to 10 y old, accurately aged (heavy stipple); (2) 11 to 20 y old, roughly aged (light stipple); (3) adult males, unaged (oblique shading); and (4) females and unsexed individuals, unaged (blank) (redrawn from Grubb, 1971).

there is no obvious reason why he should question the validity of his data, except for the fact that it does not conform to accepted patterns. In the Anse Mais population of Grande Terre, Gaymer (1968) found that the population did not appear to be reproducing, because no individuals less than 350 mm were recorded (Fig. 3.16). At Takamaka, on Grande Terre, only 35 tortoises of carapace length less than 15 cm were observed, and most of these were found outside the experimental area. Gaymer (1968) stated,

> An estimated 500 tortoises live in this region, which indicates a low ratio of young to old in the population even assuming many small tortoises were missed. This seems rather unlikely, since marked small tortoises were re-found on several occasions during the 1964 visit.

Grubb (1971) stated that "because of the paucity of small tortoises in these samples a special effort was made to measure animals of about 20 cm carapace length or less".

Bourn and Coe (1978) suggested that the shape of the length frequency distribution indicates a declining population "in which pre-reproductive size classes formed a relatively small proportion." Swingland (1977) commented, "Giant tortoise hatchlings for the first 3 or 4 yr of life feed in areas where, because of the terrain, animals older than 4–5 yr cannot feed." Swingland also suggests that the population in the high-density areas is at the carrying capacity determined by available resources.

From the repeatability of the pattern and the absence of places for juveniles to hide, there seems little doubt that the samples collected provided a relatively true representation of the tortoise population. Each investigator produced the same general scenario of a low level of juveniles and a high density of large animals. There seems little possibility that large numbers of juveniles were living in some inaccessible region, and hence little chance that the distributions obtained were an artifact of sampling procedures.

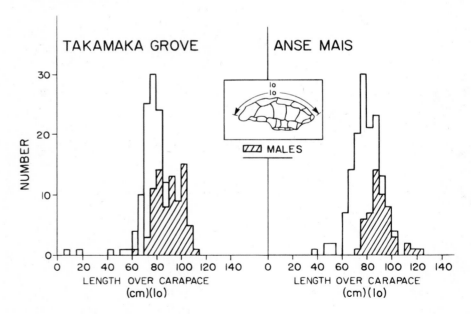

Figure 3.16 Carapace length frequency distributions of the giant land tortoise *Testudo gigantea* from two locations on Ile Grande Terre, Aldabra Atoll: Takamaka Grove and Anse Mais, in 1964–1965 (redrawn from Gaymer, 1968).

The tortoises provide, in addition, an interesting example of the manner in which a dominant species can stimulate to its own advantage energy flow at lower trophic levels. The most important food resource of the tortoise population is the closely cropped "tortoise turf" of the coastal grasslands. Production of this tortoise turf is an example of a "grazing lawn" in the sense of McNaughton (1984), where production of highly nutritious forage is stimulated by the grazing animal. Increased energy flow at a lower level is transformed into maintenance of increased biomass at the higher hierarchical level. This is the essential basis of agriculture.

The white sucker population in Heming Lake, northern Manitoba. One of the most extensive series of fisheries data from a lake under complete control of the investigator was collected at Heming Lake in northern Manitoba by G. H. Lawler (pers. comm.). Heming Lake (latitude 56° N, longitude 101° W) is a small lake within the boreal forest 500 km north of Winnipeg; it has a surface area of 386 ha. The original plan of the investigation was to test the hypothesis that the economically important parasites *Triaenophorus crassus* and *T. nodulosus* could be eliminated from their intermediate host, lake whitefish (*Coregonus clupeaformis*), by removing the final host, northern pike (*Esox lucius*), through heavy fishing (Lawler and Watson, 1958). The investigation started in 1952 and continued until 1976. Fishing pressure on all species was gradually increased until 1960, then relaxed. At the same time, all species in the lake were monitored at fixed locations using standard gangs of gill nets having a wide range of mesh sizes. The white sucker, *Catostomus commersoni*, population was not a target species in the main fishing operations.

In 1952, the essentially undisturbed sucker population showed a bimodal length frequency distribution, with a large mode at 35 cm and a smaller, poorly represented one at about 15 cm (Fig. 3.17). As fishing pressure increased, there was a gradual shift in the relative abundance of the two modes. By 1960, the large mode had virtually disappeared, but the small mode was very well represented. At this stage, the northern pike population had been sufficiently reduced for the purposes of the experiment to relax fishing pressure and allow the sucker population to revert to its former configuration. By 1962, the population was evidently in a transitional stage, forming a uniform

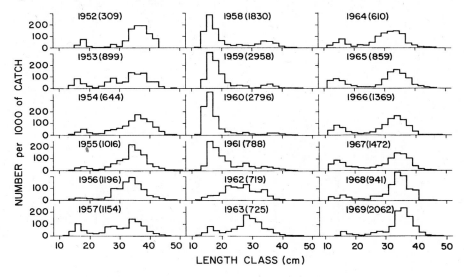

Figure 3.17 Length frequency distributions of white sucker, *Catostomus commersoni*, from Heming Lake, northern Manitoba, for the period 1952 to 1969 (G. H. Lawler and J. S. Campbell, pers. comm.).

group with a modal value of about 26 cm. This configuration lasted one year only. By the following year, the original bimodal configuration had been reestablished, and this state was maintained until this first phase of the operation was terminated.

There is a remarkable degree of cohesion maintained within the white sucker population. Number-at-length fluctuates in a highly damped manner without oscillatory damping. Modal values in the trough between the two standing modes (17 to 35 cm) appear not to occur, except as a transitional phenomenon.

Tawny owls in Wytham Woods, Oxford. Far from occurring in the ecological hinterlands, this example is from one of the world's leading ecological research reserves. A meticulous study on tawny owls, *Strix aluco,* was carried out by Southern (1970a, b) between 1947 and 1959 in Wytham Woods, a 520-ha reserve of woodlands maintained by Oxford University.

Being vertebrates of genetically determined size, tawny owls existing in a bimodal size frequency distribution in spring (and a unimodal distribution in fall) would be expected. The feature of significance in this case is not the population structure, but the manner in which the population functions coherently in the regulation of its own density. The damping of oscillations in population number is strictly comparable with those occurring in the white sucker population of Heming Lake.

The tawny owl study started in 1947 and continued through 1959. The study area appears to be well bounded, at least to the owls, by surrounding farmland. Total counts were made of most aspects of the owl's economy: number of eggs laid, number of hatchlings and fledglings, and the number of pairs occupying territories, as well as detailed studies on the abundance of potential food organisms and food actually consumed.

Fortuitously for the investigation, England experienced an exceptionally cold winter in 1946–1947, sufficiently severe to reduce considerably the resident tawny owl population of Wytham Woods. In the spring of 1947, at the start of the investigation, the population comprised 16 breeding pairs, which gradually and monotonically increased to 32 pairs by 1956. The population then remained at this asymptotic level until the investigation concluded in 1959 (Fig. 3.18). Population density was controlled through the need of each breeding pair to establish a territory.

From Fig. 3.18, it is apparent that the actual rate of increase in the number of breeding pairs was largely independent of the number of potential recruits. It was independent also of food resources. As with age and growth, the number of potential recruits and actual recruitment to the establishment

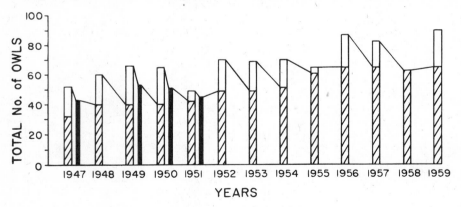

Figure 3.18 The population of tawny owls, *Strix aluco,* 1947–1959 in Wytham Woods, Oxford. Adults: shaded bars; potential recruits: blank. Solid bars show fall census figures in the years indicated. There was a gradual monotonic increase in the adult population following severe decline in the winter of 1947 (redrawn from Southern, 1970b).

population cannot be completely decoupled. Over the course of time, the number of territories within the area gradually increased until asymptotic density was reached. The fate of potential recruits unable to secure a territory is indicated by the fact that marked birds were occasionally found dead outside the study area, apparently having died of starvation. It appears that population regulation takes place at the most sensitive point in the life history, the point of recruitment to the establishment.

It is evident from the monotonic pattern of recovery that an efficient and effective damping mechanism existed. This implies a high degree of cohesion and communication among all members of the population.

Ecosystems with Plant Dominants

If the proposition that individual species function as coherent dissipative units tending toward a state of least dissipation is of general validity, then plants will exhibit similar distributional patterns. Such distributions will be most readily recognizable among the dominant tree species in a "climax" forest.

European virgin forests. In a review of European virgin forests, Jones (1945) concluded that the dominant tree species may exhibit a trunk-diameter distribution comparable with that which might be expected in a stand of uniform age. The age of these dominant tree species is frequently about 300 years. Jones stated:

> Many reports of even-aged stands are doubtless based on appearance only, or at most stem-diameter frequencies, which as will be seen later, can be misleading.... A far commoner form of uneven-aged forest has a regular canopy and has very much the appearance of an even-aged high forest. It is perhaps too much to say that all ages are present, but although old stems predominate, the dominant stand includes a wide range of ages. In an example of a Bosnian beech forest given by Cernak the age range was 200 years, though nearly 70% of the stems were between the limits of 170 and 200 years; the youngest stems were 90–100 years old. Regeneration in such forests often appears to be absent, and there is a striking deficiency of smaller stems, giving a diameter distribution which may approach that of an even-aged stand. This, however, does not necessarily mean that the forest is not reproducing itself and maintaining its present structure; the condition doubtless arises through the relatively small fraction of the life spent in growing from ground-level into a canopy when a gap is formed. Where the average length of life is 300 years only two or three gaps per hectare, each containing one or two young stems, would be sufficient to perpetuate the forest.

The upland rain forest of Mauritius. The island of Mauritius in the Indian Ocean, home of the now extinct dodo, *Raphus cucullatus,* has been subjected to the usual depredations of isolated islands through deforestation and the introduction of new species. In an attempt to record the characteristics of the original vegetation before it had all been destroyed or modified, Vaughan and Wiehe (1937, 1941) carried out extensive ecological investigations in various parts of the island. In plots in the upland rain forests, chosen for the absence of nonindigenous species and conformity with the perceived indigenous vegetation, detailed measurements were made of all species present.

The forest was characterized as being five layered, with three strata of trees, one of shrubs, and the ground flora. The uppermost stratum, composed of about 12 species, formed an open canopy of trees with umbrella-shaped crowns. As a rule, the main bole is rather short and straight without lateral branches until it has emerged from the closed second stratum. The third tree stratum forms an understory to the closed canopy. It is apparent that the forest is a highly organized structure.

The species from all three strata show a trunk-diameter frequency distribution approaching a negative exponential curve, although the right-hand tail is of great length (Fig. 3.19). Predominant in the upper story are the species *Mimusops maxima, Canarium mauritianum,* and *Calvaria major.* *Mimusops maxima* exhibited a well-defined bimodal distribution with modes at 7 and 37 cm, again with a long right-hand tail extending to values over 100 cm. *Canarium mauritianum* and *C. major* showed identical distributions, so their values were combined. The resulting curves were bimodal with peaks at 22 and 50 cm; no specimens under 12 cm could be found. Vaughan and Wiehe believed this absence of seedlings and saplings might be attributable to the gradual demise of certain species due to invading plants or animals. Absence of regeneration in the upper canopy species *Calvaria,* they suggested, might be due to extinction of the dodo. Seeds of this species are frequently associated with dodo remains, and it is possible that their passage through the gut of dodos assisted germination.

Such explanations can by no means be ruled out, but must be reconsidered in light of findings from other forests.

The forests of northern lower Michigan. Sakai and Sulak (1985) traced forest succession in northern lower Michigan over a period of four decades, from initial stages following recent clear-cutting to the attainment of mature forest. This is a most interesting example as it shows the conversion of a negative exponential distribution of stem-diameter frequency to a unimodal distribution after 40 years. It is of further interest in that three tree species, *Thuja occidentalis, Picea mariana,* and *Abies balsamea,* all function in concert (Fig. 3.20).

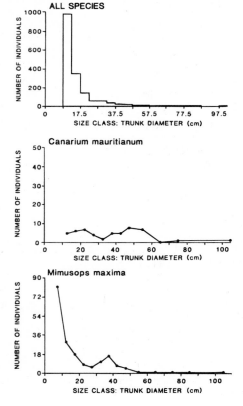

Figure 3.19 Trunk diameter distribution of trees in the upland rain forests of Mauritius. Upper panel: all species combined; middle panel: *Canarium mauritianum* and *Calvaria major*; lower panel: *Mimusops maxima*. No specimens under 12 cm in either *C. mauritianum* or *C. major* could be found in the plot examined (redrawn from Vaughan and Wiehe, 1941).

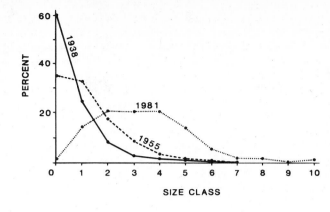

Figure 3.20 Changes in trunk-diameter frequencies in a plot of mixed conifers (white cedar, *Thuja occidentalis*; black spruce, *Picea mariana*; and balsam fir, *Abies balsamea*) in northern lowland Michigan. Size class 0: ≥1 m tall but <0.5 in. dbh (diameter at breast height); size class 1: ≥0.5 in. dbh but less than 1.5 in. dbh; and so on. 1938 N = 539; 1955 N = 609; 1981 N = 205 (redrawn from Sakai and Sulak, 1985).

STRONG INFERENCES

The primary conclusion from which all further inferences derive is that within an autonomous ecosystem individual species populations, or ecodemes, function coherently, proceeding toward a state of least dissipation. This may be taken as an expression of Lord Rayleigh's "principle of least energy dissipation." The prime emergent property of dominant populations in isolated systems is their inherent stability in the face of variable environmental conditions. Stability is implicit in the large size and great age of individuals, the uniformity of their size within a given ecosystem, great mean age, and the great spread of age within modal size groups. The initial inference is that within autonomous systems stability is attained through the capacity of species to move toward a state of least dissipation.

A second inference follows from the first. It is self-evident that the trend to least dissipation is not a universal characteristic of interactions within an ecosystem. However, despite interactions between individuals of different ecodemes, which lead to increased dissipation, the ecosystem as a whole, by mechanisms of cycling, storage delay, and internalization or closure, also attains a state of least dissipation. Arctic lakes or tropical rain forests attain greatest stability within the precinct of a closed system. Least dissipation implies the capacity to delay energy flow and to acquire and conserve resources (energy and materials) to the greatest extent possible. The longer the time delay imposed on energy flow, the greater the biomass and the greater the perturbation—the damping moment, or resistance stability, of the system. In that different species have different capacities to acquire and conserve resources, the system forms a hierarchy. At the head of the hierarchy is the dominant species, which is the one that can effect the best integration between acquisition and conservation under the prevailing conditions. Least dissipation is most readily observed in these dominant species since they are the ones least affected by grazing or predation. Least dissipation is manifested in the structural characteristics of large size, uniform size, great age, and indeterminate age at death.

The ability of an ecodeme to control resources is defined by Van Valen (1973) as its *fitness*. The greater the biomass of the ecodeme and the longer the mean life span of its component individuals, the greater the fitness, and the greater also the damping moment exerted on the system.

Symmetric and Asymmetric Interaction

A state of least dissipation is contingent on the uniformity of individuals in a population, which in turn implies symmetric interactions between them. Symmetric interaction denotes stasis, and "no-win, no-lose" encounters. It is evident that symmetric interaction and least dissipation are not universal characteristics of biological relationships in ecosystems. A major source of asymmetry is the consumption of one organism by another. This results in the expenditure (dissipation) of considerable energy above basal metabolic requirements. Although varying over wide limits, an ecological transfer efficiency of 10% to 15% is generally accepted as an average value (Slobodkin, 1964; Brylinsky and Mann, 1973). The *ecological efficiency* is the fraction of the energy in food that is converted into consumer biomass; the remaining 90% (less the unavailable portion) is dissipated in the performance of work essential to effect the transformation. Therefore, the greater the diversity of species in an ecosystem, the greater will be the number of asymmetric interactions and amount of energy dissipated, and the smaller the resultant standing biomass, relative to energy input.

Biological processes thus appear to exhibit two opposing trends, one toward least dissipation and the other toward accelerated dissipation. It is postulated that these have their origin in the tendency for energy to assume a uniform distribution (the state of maximum entropy) versus the requirement that this symmetry be disturbed if energy flow is to be maintained. Near thermodynamic equilibrium, the maintenance of this asymmetry necessitates that a certain amount of work be done and allows patterns to form in the energy flow. The more uniform the distribution, the less the work necessary and the less the dissipation. Within the limits of validity, increased flow leads to increased dissipation and a greater potential for work. Since energy passing through a system tends to take the path of least time, the biosystem evolves between two trends. One is toward deceleration, or retardation, of the energy flow and homogenization (uniform distribution); the other is toward the acceleration of energy flow and heterogeneity. According to Newton's laws, where there is acceleration there is a force. Biological systems may thus be thought to evolve within the field of overlap between two antagonistic "forces": accelerated energy flow promoted by direct asymmetric interrelations (such as feeding), and decelerated energy flow emergent from the mainly indirect symmetric properties of system organization (such as cycling, storage, and network closure), which confer least dissipation as the net trend in ecosystem development.

Ecosystem diversity tends to increase to the greatest extent possible commensurate with the characteristics of energy inputs and internal and external constraints. The greater the energy input, the greater the number of regular fluctuations, the fewer the erratic fluctuations (that is, less "noise"), the greater the number of identifiable signals, and the greater the diversity. Diversity decline during the final stages of succession (Whittaker, 1969; Odum, 1969) indicates the trend to reduced dissipation. At this time, the dominant species is a major constraint on diversity. Increase in diversity over evolutionary time indicates the ultimate supremacy of the trend to increased energy flow. There is thus a short-term route to stability through homogeneity and high biomass and a long-term route to stability through greatest possible diversification. The stability characteristics of a particular ecosystem are dependent on the point at which these two opposing trends settle down within the constraints of system autonomy and characteristics of the energy input.

Diverse systems notwithstanding, the greater the number of specialized primary producers or the greater the amount of energy available, the greater will be the total energy uptake. This may result in higher total biomass, but not necessarily higher system biomass per unit energy input. Thus, a diverse tropical system may have a higher biomass per unit area than one in the Arctic, but the energy input per unit biomass required to sustain the arctic system is very much smaller than in the tropical system.

A generalized statement can thus be formulated:

Symmetric interactions between like individuals tend to minimize the rate of dissipation; asymmetrical interactions tend to increase it.

Therefore, individual species populations, or ecodemes, functioning as dissipative units, assume a state of least dissipation compatible with existing constraints. Asymmetric interaction between ecodemes results in increased dissipation. Since these two antagonistic processes occur simultaneously, a system can exist and maintain its identity only in the vicinity of the null point where the two trends approach equality. This null point will be reached only in a closed system, where energy input and output are equal.

Energy entering the biosphere as light passes through at least one material cycle before being dissipated to the universe as heat (Morowitz, 1968). Passing through material cycles involves a time delay in energy flow, allowing work to be done. The rotations of the earth, functioning as a commutator, prevent processes from going to completion and preclude the assumption of a completely steady state. The commutator thus allows work to be done on a continual basis. Given time and a regular energy-input cycle, the earth system in its prebiotic state may be assumed to have attained a steady state as close to equilibrium as conditions allowed. The regularity of seasonal and diurnal changes in energy ambience created, in effect, a signal system unique at each point on Earth's surface. In these conditions, close to equilibrium, the energy flow through suitable materials gave rise to "dissipative structures" through work done. The rectilinear propagation of light energy was diverted into circular or cyclic pathways and time-delayed in its passage through materials.

Dissipative Structures

The first general statement in nonequilibrium thermodynamics was formulated by Onsager (1931) in his statement of reciprocal relations. These relations state in qualitative terms that

> When two or more irreversible transport processes (heat conduction, electrical conduction and diffusion) take place simultaneously in a thermodynamic system, the processes may interfere with each other.

In such circumstances, the system obeys the variational principle (an extension of Lord Rayleigh's principle of least dissipation of energy) in which *"the rate of increase of entropy* plays the role of a potential" (Onsager, 1931). This implies that the system inherently moves in the direction of a state in which the rate of dissipation is the least attainable within the prevailing constraints. The state of least dissipation functions as an "attractor" for the system. This role may be contrasted with that of "maximum entropy," which, in an equilibrium system, ensures entropy increase with time. Onsager's relations apply strictly only in the vicinity of thermodynamic equilibrium. Onsager's reciprocal relations mean that, close to thermodynamic equilibrium, if two energy transport processes interfere, the system tends to assume a state of least dissipation. This interference may lead to dissipative structures, patterns in energy flow maintained by the dissipation of energy (Prigogine and Wiame, 1946; Prigogine et al., 1972; Nicolis and Prigogine, 1977; Prigogine, 1978).

> Such flow structures exist only within certain boundary conditions and survive only on energy input dissipated in the maintenance of the flow pattern. (Neumann, 1978)

These structures exist far from equilibrium (Prigogine, 1978). Dissipative structures are maintained through work done imposing time delay on energy flow.

Within the range of validity, Prigogine and Stengers (1984) state that

> [the] system evolves toward a stationary state characterized by minimum entropy production compatible with the constraints imposed on the system.... When the boundary conditions prevent the system from going to equilibrium, it does the next best thing; it goes to a state of minimum entropy production—that is, to a state as close to equilibrium as "possible."

Gnaiger (1983, 1986) has referred to the dynamically active steady states tended to or reached in biological systems as *ergodynamic equilibria.*

According to the theorem of minimum entropy production, "if a system is perturbed the S [entropy] production will increase but the system reacts by coming back to the minimum value of the S production" (Prigogine, 1978). Dissipative structures, however, may exist far from equilibrium (Prigogine, 1978), and

> Far from equilibrium the system may still evolve to some steady state, but *in general this state can no longer be characterized in terms of some suitably chosen potential (such as entropy production for near-equilibrium states)* [emphasis added]. (Prigogine and Stengers, 1984)

From this point on, in the study of living systems, it seems necessary to part company with Prigogine and his co-workers, for it is apparent that a state of "least dissipation" does function as a potential within ecosystems. Prigogine and Stengers's statement may well apply to nonliving dissipative structures, but for living organisms it appears inapplicable.

Self-organization

Over evolutionary time dissipative systems become amplified by a process of self-organization. This results from slight ascendancy of the irreversible trend to continuously accelerating long-term energy flux, a process resisted by short-term ascendancy of cyclic dissipative processes. If the cyclic processes are to continue to exist in the face of overall acceleration of energy flow, they must evolve or be eliminated.

Self-organization may be compared with a sailboat "beating" against the wind, that is, making progress in the direction from which the wind is coming. In sailing theory, it is the pressure of the hull and keel (derived from wind in the sails) against lateral resistance of the water that produces the resultant motion to windward. Fluidity of the water allows motion at an angle of about 45° to the direction of the wind.

It appears that the nonequilibrium biological world, if not running exactly parallel with the conventional world of equilibrium thermodynamics, has significant correspondences. "Entropy production" functions as a potential comparable to "entropy" in the equilibrium system. An increase in "number of complexions" is equivalent to entropy increase; comparably, an increase in diversity is a second-law-like effect since it is brought about by an increase in the rate of entropy production. The ecological climax state, reached when energy input and output are equal, corresponds in this sense to thermodynamic equilibrium. Thus, biological systems appear to fall into a new and exciting pattern in which *dynamic* aspects of thermodynamics take on a full significance.

Time, for example, assumes a central position in biodynamics since life is an expression of the relative rates of many processes. Biological time is therefore "polychronic" in the sense of Hall (1983), with a multiplicity of activities taking place simultaneously. Many of these activities are antagonistic, tending to move the system in opposite directions. Monochronic, or sequential, time has little relevance to the biological world except in the evolutionary perspective.

These far-reaching inferences, arising out of the investigation of simple systems, appear to be parsimonious in relation to observed facts. Ultimately, biological systems appear dependent on the antagonism between homogeneity and heterogeneity, symmetry and asymmetry, and reversibility and irreversibility. Homogeneity, symmetry, and cyclic activity (reversibility) are all supported by an appropriate input of energy. Explicitly, the biosystem is contingent on five factors:

1. Individual ecodemes function as coherent dissipative units, tending to a state of least dissipation and uniform energy distribution through symmetric interaction and homeokinesis (see later).

2. Interaction between ecodemes (as units) is asymmetric, tending to increase the overall rate of dissipation.
3. Over ecological time systems proceed to ergodynamic equilibria provided (a) the tendency to least dissipation exceeds the trend to increasing dissipation, (b) the system is autonomous, and (c) the energy input cycle is regular.
4. Over evolutionary time the trend to acceleration of energy flow gradually erodes the stability attained during ecological time.
5. Abrupt environmental change of large magnitude reduces diversity, resetting the system to a new starting point.

The Ecodeme as a Dissipative Structure

An ecodeme is a species population; it may be defined as a group of like individuals interacting within its niche space. Within the niche boundaries, it is postulated, an ecodeme functions as a coherent dissipative unit tending to assume a state of least dissipation. Least dissipation is characterized by least energy expended per unit biomass maintained, within the constraints provided by the environment and the genetic makeup of the species.

A relative measure of dissipation rate, or least entropy production, is the production/biomass (P/B) ratio. The P/B ratio at steady state is a measure of the rate at which bioenergy must be added, relative to the biomass accumulated, to replace that lost through processes of repair and maintenance. It thus provides a relative measure of energy flow in terms of turnover time. Declining value of the P/B ratio, for example, over the course of succession of communities (Margalef, 1968; Odum, 1969), thus indicates a decrease in dissipation. Successional decline in P/B ratios will not be monotonic, since an ecosystem may pass through alternating phases of dominance and diversity (Harper, 1969). Decline in community P/B ratios will be most noticeable in that of dominant ecodemes.

In general, a dominant species exhibiting uniformity displays the characteristics of a *K-selected species* (MacArthur and Wilson, 1967; Southwood, 1977; Parry, 1981). These include low reproductive rate, long generation time, large size, and high survival rate, especially of reproductive stages. Species lower in the ecosystem hierarchy exhibit higher turnover rates, assuming characteristics of an *r-selected species*. These are high reproductive rate, short generation time, small size, high fecundity, and so on. All species have an intrinsic rate of reproduction (r) which is related to their (rate of) production (P). It is postulated that all species exhibit K-characteristics to the extent that, in order to exist, they must proceed toward a state of least dissipation. K therefore is equivalent to accumulated biomass (B). The ratio r/K then emerges as dimensionally equivalent to the P/B ratio and provides a relative measure, for the ecodeme, of the rate of dissipation. Similarly, the power/economy ratio of Gnaiger (1986) is dimensionally equivalent to the P/B ratio. Power depends on rate of energy flow, and economy (itself evidence of the trend to least dissipation) is measured in terms of accumulated biomass.

Least dissipation, at the ecodeme level, will be indicated by the following characteristics:

1. *Large individual size.* This is required to minimize metabolic demands relative to biomass. The total metabolic needs of an individual tend to increase over its growth period, but the specific metabolic rate declines as size increases (Kleiber, 1961). This is expressed by Odum (1956) as the inverse-size metabolic law. Least dissipation thus demands that the population be composed of individuals of as large a size as can be maintained. Size thus increases along the sequence from primary producer (active tissue) to terminal predator (Kerr, 1974). At each stage, dissipation is reduced; otherwise, the food chain would be unstable. Where the environment allows, metabolic rate in the adult is reduced to the lowest level compatible with existence, as previously observed for the cave faunas of Africa and the United States.

2. *Greatest attainable mean age.* It is self-evident that the longer the mean life span, the longer will be the turnover time of the population.

3. *Low level of juvenile replacement stock.* The longer the turnover time, the smaller will be the requirement for replacements in a steady-state ecosystem. Greatest efficiency of energy utilization will be attained if the high energy demands of reproduction and early growth stages are maintained at lowest levels compatible with survival.

4. *Constant uniform size.* Least continuous dissipation will be attained if large uniform individuals are maintained at constant density. Any departure from uniformity will tend to increase dissipation. Therefore, a trend to least dissipation necessitates the existence of a damping mechanism to maintain uniformity.

If individuals are removed from a uniform population, there will be loss of uniformity as juvenile replacements (recruits) move up to maintain numerical constancy. With curtailment of the life span by removal, turnover time will be reduced and growth and reproduction stimulated. Removal of individuals (for example, by fishing) should thus tend to push length frequency from a bell-shaped distributional pattern toward a curve having the functional form of a negative exponential. This effect was observed intraspecifically in Gavia Lake when the few large Arctic char were removed. Thus, the negative exponential curve represents the greatest possible rate of energy flow through a population, whereas the bell-shaped curve represents the effect of time-delayed energy flow. It is evident in Little Nauyuk and Gavia lakes (in the latter, once the large fish were removed) that there are powerful mechanisms maintaining uniformity of size structure in the face of severe disturbance. The functional mechanism for maintaining this uniformity, it is hypothesized, is "homeokinesis" (next section), the equipartitioning of energy available among the constituent individuals of a species. Homeokinesis is therefore an integral mechanism in establishing least dissipation.

In this context, a phenomenon of particular significance in undisturbed populations is the wide spread of ages within the modal size class. This spread in age is essential if uniformity and continuity are to be maintained. A partial decoupling of age and growth is implied, which will be most apparent in organisms of indeterminate growth pattern. If adult fish of virtually all ages accumulate in the modal size class of a population, then growth must be regulated by interactive processes. It is evident also that length and age cannot be completely decoupled. Medawar's first law of growth stipulates that size is a monotonically increasing function of age (Medawar, 1945). Therefore, the trend to uniformity must function within the constraints of Medawar's laws and the limited life span of the organism. This partial decoupling permits a good correlation between age and mean length, but necessitates a high degree of length variability within each age class. In vertebrates, such as mammals and birds, for which growth is of determinate pattern, size is largely under genetic control. Hence, beyond the age of maturity, there is little direct correspondence between age and size. This results in a size–age relationship identical in form to that in the populations previously discussed. Since it seems certain that determinate growth evolved from the indeterminate condition, it must be concluded that size uniformity is of great evolutionary significance.

5. *Indeterminate age at death.* Variation in age at death is essential to minimize cycling of the population. A set life span promotes strong fluctuation, as apparent in such inherently fluctuating organisms as 13- and 17-year cicadas, *Magicicada* spp., and Pacific salmon, *Oncorhynchus* spp.

To meet the above five conditions an ecodeme must function as a coherent unit, not merely as an uncoordinated collection of individuals. This implies that the interacting group must exhibit emergent properties not deducible from the study of individual behavior. Coherent behavior implies organization and communication between individuals and the loss of certain degrees of individual freedom. Although coherent behavior is apparently universal in aggregations of organisms from slime molds and bacteria to human societies, its relationship to population dynamics has been little studied. As pointed out by Reiners (1986), artificial barriers within ecosystem ecology have resulted in the omission of "for example, regulatory processes at the population level." Nevertheless, the existence of regulatory processes at the population level has been established in a few species such as the great tit, *Parus major,* and red grouse, *Lagopus lagopus scoticus* (Ebling and Stoddart, 1978). Its existence in the tawny owls of Wytham Woods, the white suckers of Heming Lake, and other species in simple ecosystems, seems incontrovertible.

Organization therefore seems to be a necessary, inherent property of all organisms. As Salthe (1985) states, "What is implied here, however, is that what appeared to many of us as radically new inventions in evolution, say social living, were in fact imminent in less differentiated systems of an earlier time."

Homeokinesis

If, as Prigogine and Wiame (1946) suggest, a single organism functions as a dissipative structure tending to least dissipation, then, for a group of interacting organisms to function as a dissipative unit, a mechanism inducing coherence and organization is necessary. Organization is possible only by achieving symmetric energy currents through symmetrical interaction. It also demands a goal (Pittendrigh, 1958; Mayr, 1976). The goal postulated here is the thermodynamic state of least dissipation. The physical basis for this interactive process of equipartitioning energy among symmetrical pathways has been described by Soodak and Iberall (1978) under the doctrine of homeokinesis. Homeokinesis equipartitions the available free energy among component "atomisms" (indivisible charged entities; see below) through an interactive process.

Homeokinesis is defined by Soodak and Iberall (1978) as "The achievement of homeostasis by means of a dynamic regulation scheme whereby the mean states of the internal variables are attained by the physical action of thermodynamic engines." However, "such complex atomisms do not equipartition the interaction energy per collisional cycle, but instead internally time-delay, process, and transform collisional inputs using many fluidlike dissipative mobile steps." Atomisms maintain nearly constant energy potentials with respect to time, in contradistinction to, for example, gas molecules in a closed container, which vary in their kinetic energy level at any instant in time, but equipartition the energy over a collisional cycle.

Essentially, homeokinesis ensures the maintenance of homogeneity. There is, of course, a limit to the difference between interacting entities over which homeokinesis will be effective. However, within this domain of effectiveness, the trend to homogeneity leads to a symmetrical state and no further change. Clearly, such a mechanism must be operating within the fish populations in northern lakes and, by extension, to all species.

Homeokinesis is graphically expressed in the Japanese proverb which holds that a nail that sticks up will soon be hammered down. However, within ecodemes, hierarchies at a secondary level may develop. Provided the degree of asymmetry among individuals is limited, this will, in turn, allow increased information to be coded, hence enabling a more flexible response to environmental variability to develop. Species may thus evolve toward a hierarchical organization, as in the wolf pack, or toward extreme uniformity, as in many fish schools. The former will allow individuality and experience to modify behavior; uniformity will require conformity through inherent behavior patterns. Homeokinesis is seen as the process regulating uniformity in the size of organisms in a particular ecodeme, thus inducing uniformity in size irrespective of age.

Action Principles

Underlying homeokinetic processes are two antagonistic "action" principles: *least action* and *most action*; the latter is equivalent to least dissipation.

The *principle of least action* was developed out of the *principle of least time* first enunciated by Pierre de Fermat in the seventeenth century. Fermat argued that the path of light through several transparent media of different refractive indexes corresponded to the path of least time, which is not necessarily the shortest distance between any two points under consideration. This principle was generalized by Maupertius and Lagrange, finally becoming "Hamilton's first principle of function" or, more preferably, the principle of least action, involving also a principle of stationary time (see below). "Action" is a function of energy and time. Davies (1980) states:

> The significant fact about the path of least action is that it coincides with the Newtonian path—the trajectory one would calculate from Newton's laws. . . . Complete chaos [in the universe] is thus averted because matter is lazy as well as undisciplined.

A time delay is necessary if work is to be performed; hence, energy passes through the system as rapidly as possible. However, in the biological world the instantaneous superiority of the trend to least dissipation induces the passing energy to perform more work if it is to increase its rate of passage. But again, no more work is done than is absolutely necessary (Saunders and Ho, 1981).

It is evident that there is a general trend toward increasing diversity over evolutionary time (Simpson, 1953, 1969; Raup and Sepkoski, 1982). There is also an increase in diversity from the poles to the wet tropics, which is associated with an increase in energy flow (Golley, 1972; Burger, 1981; Johnson, 1994a). It is therefore concluded that, with the increase in diversity over evolutionary time, there has been a concomitant increase in energy flux.

Such an increase in flow accounts for the directionality in evolution. The continuing trend to acceleration of energy flow stimulates increasing diversity and increasing complexity. Antagonism between the trends to least dissipation and increasing energy flux provides the basis for the homeostatic mechanism. Homeostasis will ensue if, in the short term, the trend to least dissipation is ascendant over the trend to accelerating energy flux at all stages before the climax.

Richard Feynman, in his lectures on mathematics (Feynman, 1963), discussed the principle of least action in terms of an interesting and intellectually stimulating byway of physics (actually, he called it "fun"), but for biologists it has a deeper significance. Action is defined as the difference between kinetic and potential energy at each point in time, integrated over the trajectory of the system. For example, a particle projected from A to B follows the path along which the integral of action has the least value, where

$$\text{action} = S = \int_{t_1}^{t_2} (KE - PE)\, dt.$$

Feynman comments that the term "least" does not imply either a maximum or a minimum. What is implied is that "the first-order change in the value of S, when you change the path, is zero." That is, a small change in path does not increase the elapsed time of the event. For this reason, Watson (1986) suggests that the correct designation should be the *principle of stationary time*. Feynman continues:

> [M]ost phenomena in mechanics and electrodynamics obey the principle of least action, but there is also a class of phenomena that does not. For example, if currents are made to go through a piece of material obeying Ohm's Law, the currents distribute themselves inside the piece so that the rate at which heat is generated is as little as possible; if everything is kept isothermal then it is possible to state that the rate at which entropy is generated is a minimum.

In such circumstances, a condition of least entropy production is that the currents be distributed as symmetrically as possible throughout the material.

Action has the dimensions [energy × time] and is measured, for example, in joule-seconds. Least action, in effect, is equivalent to "most dissipation" or "most entropy production," and "least dissipation" is equivalent to "most action."

Organisms may be regarded as entities composed of energy, matter and time; therefore, they are subject to action principles and their actions should be measured in joule-seconds. They exist in the second category of phenomena discussed by Feynman, those that generate least entropy. Least entropy production is equivalent to most action; that is, organisms accumulate energy to the greatest extent possible and conserve it as long as they can within the total ecosystem constraints. As a first approximation, the energy in biomass may be regarded as potential energy, and kinetic energy is the energy flux necessary to maintain such biomass. Part of the energy dissipated is for necessary basal metabolic purposes, but at steady state this may be considered constant and, for present purposes, disregarded.

At the most fundamental level, Planck's constant is the quantum of "action" as expressed in Heisenberg's uncertainty principle:

$$\Delta t \, \Delta E = \hbar ,$$

where t is time, E is energy, and $\hbar$ is Planck's constant. This relationship strongly suggests that organisms are amplifications of quantum phenomena.

Origin of the Biosystem

In view of the restrictions on the application of Onsager's reciprocity relations, it must be postulated that the biosystem originated when Earth's surface had cooled to a point close to thermodynamic equilibrium. At equilibrium, certain quantum effects occur spontaneously, involving the transfer of electrons. These quantum effects became amplified through a resonant response to environmental signals originating in intermittent solar energy inputs. Photons were captured, and their energy was gradually dissipated in work processes. In aquatic systems close to chemical equilibrium, cyclic reversible processes involving no change in entropy occur spontaneously. It is hypothesized that, under intermittent solar energy input, spontaneous cyclic processes occurring in the "primordial soup" were induced through resonance to prolong their activity cycles. Excited molecules temporarily stored energy, imposing a time delay on the energy flow. In accordance with Onsager's relations, the two opposing processes, increased time delay and a rapid return to the ground state, interfered with each other, inducing a state of least dissipation.

Least dissipation implies maximum time delay in energy flow through a system. In excited states, molecules of greater complexity could form, increasing time delays and reducing dissipation rates. Slow dissipation of energy as excited molecules returned to ground allowed the necessary work to be done to maintain complexity. In this way, dissipative structures were formed close to equilibrium. The most long lasting and robust processes were "selected" by a process of signal enhancement—repeated inscription of the representation of a weak signal on a receptive surface until the signal's characteristics, through repetition, become heavily outlined. These enhanced signals tied up available materials and became dominant. Energy was thereby delayed by storage and passage through material cycles. Cyclical energy transport and consequent time delays interferred with rectilinear energy transport as required by the principle of least action. Thus, biosystems assumed a state of least dissipation.

Initially, the complex molecules formed "atomisms" and a protoecosystem emerged. In this, the atomisms interacted symmetrically, maintaining uniformity through homeokinesis. Uniformly charged atomisms have the characteristics of indeterminacy required for the coding of information (Polanyi, 1968). The state of symmetry established initially was eventually broken by certain groups of atomisms responding to specific components in the signal medley to which they were exposed. The result was distortion of the initial symmetric state and formation of a hierarchy.

The energy accumulated through resonance in effect became trapped between two "attractors," most action and least action. Under specific constraints provided by the local environmental signal set, the biosystem developed within the intersection of the spheres of influence generated by these two attractors (Fig. 3.21). In the limit, the "most action attractor" may be regarded as a state like crystallization in which order is locked away for an indefinite period of time. At the opposite pole, the "least action attractor" can be considered as like combustion, or the instantaneous release of photons. Since both attractors are lethal, the living system can only viably exist at some point intermediate between them.

The symmetrically interacting atomisms, or protoorganisms, in the increasingly autonomous ecosystem, assumed a state of least dissipation corresponding to a *tabula rasa* of symbols equipotentially charged and therefore capable of coding information (Polanyi, 1968; Scheer, 1970). Identification of signals and information coding demand that work be done (Pierce, 1961), with a concomitant increase in entropy production. Signals exist as regular fluctuations in environment, and they are coded by the establishment of a pattern superimposed on the symbols. Coding thus results in distorting, but not breaking, the initial symmetry. Iberall et al. (1981) conclude that the system must be maintained in a state of *equipollence* (equal power) if it is to survive. The initial equipollent state attained may be regarded as a state of ergodynamic equilibrium (Gnaiger, 1983, 1986); it is a state of least work or least entropy production. Distortion of the initial symmetry implies change and more work done, but continued maintenance of the system demands a new ergodynamic equilibrium and a state of least dissipation appropriate to the new conditions.

At any instant in time, a given configuration of the ecosystem tends to a state of least dissipation. However, over longer periods, the principle of least action stimulates exploration of the total signal space and identification of the greatest number of signals possible. "Trapped" energy can only "escape" more rapidly and obey the principle of least action by an increase in internal work. This process is inhibited at each stage by the need to maintain a state of least dissipation. Thus, in the short term, the system assumes a configuration of "least work," while in the long term work done tends to increase. In this manner the second law of thermodynamics is confronted and overcome.

Stimulated by the principle of least action, the system is gradually driven farther and farther from thermodynamic equilibrium as internal work and entropy production increase. For this to be possible, a system, far from thermodynamic equilibrium but maintaining a state of ergodynamic equilibrium, must remain subject to Onsager's reciprocal relations. Thus, *provided the system remains*

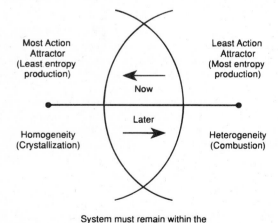

Figure 3.21 Diagram indicating interaction of the two fields of attraction within which the biosystem lives. "Succession" results from the tendency for the system to move toward the "crystallization" attractor in the short term. "Abscession" results from gradual movement toward "combustion" in the long term. The antagonism between the two forces in different time frames forms a ratchet mechanism that is responsible for evolution and its directionality.

close to ergodynamic equilibrium, it can evolve away from thermodynamic equilibrium. That is, it must remain sensitive to the small forces inducing most action. This is comparable to the assumption implicit in Newton's laws: that a system at rest is equivalent to one moving at constant velocity (Cohen, 1985). *Thermodynamic equilibrium is equivalent to a body at rest; ergodynamic equilibrium is equivalent to a body moving at constant velocity.*

By analogy, it is postulated that there is a correspondence between a system close to thermodynamic equilibrium and a nonequilibrium system that has reached a state of ergodynamic equilibrium. At ergodynamic equilibrium, a state of near symmetry is attained in which the internal forces inducing change are the least attainable. In these circumstances the delicate antagonisms postulated in Onsager's theorem can manifest themselves macroscopically as they become amplified over evolutionary time. Therefore, in the nonequilibrium world, a closed system experiencing a regular cycle of energy input and in a state of ergodynamic equilibrium corresponds to an isolated system in conventional thermodynamics. In these circumstances, entropy production, not maximum entropy, functions as the driving potential.

The energy trapped between the action principles thus proceeds to a state of ergodynamic equilibrium and least entropy production (least work) in any specific configuration of the ecosystem, but, being malleable, the system components may change. These changes allow the system to move in two directions: (1) toward the most-action attractor, manifested in more work performed (or energy dissipated) in support of greater diversity and complexity, or (2) toward the least-action attractor, resulting in reduced dissipation and increased biomass. Given the long-term evolutionary trend toward increased diversity and increased average complexity (Simpson 1953, 1969; Raup and Sepkoski, 1982, 1986: Sepkoski, et al., 1981), an event leading to an increase in diversity or complexity is more probable than one leading to a reduction in work. For a change in either direction to be viable, it must allow the system to settle again to a new ergodynamic equilibrium, even though in extreme cases drastic reordering of the species complement may be necessary.

Short-term behavior is thus dominated by the trend to reduce the power output of the system, but long-term behavior is dominated by the tendency for power to increase. If the ecosystem is to remain in ergodynamic equilibrium, the opposing trends must be nearly equal. This implies a dissipation rate approximately midway between zero and instantaneous dissipation, which is the region of maximum power and 50% efficiency (Odum and Pinkerton, 1955; Odum, 1956; Gnaiger, 1983, 1986). Ecosystems thus tend to develop maximum power.

Diversity

Ecosystem diversity tends to increase over evolutionary time to the greatest extent possible commensurate with characteristics of energy input and internal and external constraints. Diversity, reflecting species richness and equitability, measures the number of signals and their relative strength. High diversity represents a large number of signals of relatively uniform strength and a low signal-to-noise ratio. Low diversity indicates a highly pulsed energy input and high signal-to-noise ratio. Ecosystems in the wet tropics, with a generally uniform annual energy input and little variability, tend to be diverse. Arctic systems, with a highly pulsed energy input and a high noise level, tend to be simple. If a tropical ecosystem is exposed to a high degree of variability both within and between years, as in the coastal upwelling system off Peru, it is maintained in a simple state, but its production of biomass may be extremely high. Practically, this has allowed intensive exploitation of the anchoveta, *Engraulis ringens,* which for a time was one of the world's most productive fisheries (Schaefer, 1970).

Succession and Evolution

Operation of the antagonistic trends to least and most action can be readily seen in the process of succession, the development of a mature ecosystem following formation of new areas for colonization, such as on a glacial moraine, or in the formation of new land in a river delta. During early stages, there is a gradual increase in species diversity, with a concomitant increase in dissipation rate, which peaks in the penultimate stage. This is followed by diversity decline and an increase in biomass during terminal stages as the "climax" is approached (Clements, 1936; Whittaker, 1969; Odum, 1969). Conversely, diversity increases when the dominant species is removed (Connell, 1978). This indicates ascendancy of the trend to reduced dissipation over ecological time as ergodynamic equilibrium is approached, in agreement with the postulate that the production/biomass ratio (P/B) also declines with maturity (Margalef, 1968; Odum, 1969).

If an ecosystem is subjected to severe disturbance in which many species are eliminated, it may be reduced to a much simpler state, from which point diversity once more begins to increase. This is true for disturbances on both ecological and evolutionary time scales, although in the latter case elimination implies extinction. Successionally, the system approaches a steady state involving the species available, but this temporarily stable state is gradually eroded as new species evolve or invade from other regions over the course of time. Long-term stable climaxes are also prevented by geophysical and climatological changes and disturbance. Thus, ergodynamic-equilibrium seeking tendencies (to climax) in the short term are offset by longer-term forces of change. Even chemical reactions, which go to short-term conclusions under equilibrium theory, spontaneously pass to other states under the changing conditions of longer time. This does not deny chemical equilibrium theory and, by a parallel argument, neither should an end-state seeking property of ecological succession be denied. The operation of two forces, each being ascendant within a different time frame, functions as a ratchet mechanism slowly winching up the ecosystem to greater diversity, complexity and power. The pawl in the ratchet that prevents slipping back is irreversibility of the evolutionary process: the higher probability that any change results in an increase in entropy production.

If succession is the term summing up the processes leading to ergodynamic equilibrium, then the antagonistic trend leading to increasing dissipation and increasing power may be termed *abscession* (moving away). Equilibrated chemicals and mature communities both succeed toward stable end points in their shorter terms and then "absceed" away from these states over longer periods of time. Neither succession nor abscession by itself is the evolutionary process; it is out of asymmetric antagonism between the two of them that new species evolve.

Hierarchy Formation

At the most fundamental level, a hierarchy may be viewed as a dynamic conflict between the two basic characteristics of order: uniformity and sequence. Again, the pattern of symmetry opposed to asymmetry is evident. Superimposed on this notion is the requirement that in a hierarchy processes at "higher" levels proceed more slowly than those at "lower" levels (Waddington, 1968a, b; Simon, 1973; O'Neill, *et al.,* 1986). Waddington (1968a) states:

> Bastin [same symposium, vol. 2, pp. 252–265], considering the nature of the concept of a hierarchy of levels of organization, which plays such an important part in biology, argued that the only logical way in which it is possible to discriminate a number of activities into a hierarchy is by considering their reaction times, a higher level in the hierarchy always having a much longer reaction time than a level classified as lower. Again, Bohm [same symposium, vol. 2, pp. 18–60] discussed the concept of order as basic to both quantum mechanics and to fundamental biology, and argued that a hierarchical order or orders must eventually imply the existence of a "timeless order," thus emphasizing the importance of gross differences in reaction time.

Inherent differences between species in their rates of acquisition and dissipation of energy ensure the development of a hierarchy. The dominant species, exerting greatest control over available resources, impose a cascade of "top-down" control measures on the remainder of the system (Northcote, 1988). "Bottom-up" control is exercised, for example, by the primary producers controlling assimilation.

Each level higher in the hierarchy operates with an increasing reaction time, corresponding to a decreased reaction or dissipation rate. For a hierarchy to maintain itself, the various levels, or "holons" (Koestler, 1967), must be discrete. As Planck's theorem demands, energy distribution is discontinuous; there can be no intermediate energy values between adjacent levels. Within the simple autonomous reference ecosystem of the arctic lake, a hierarchy (the food chain) is established in which each succeeding level operates at a reduced rate of dissipation. There is a progression from rapidly turning over algae, through invertebrates (frequently via a detrital–bacterial "loop"), which may live two to three years, to the dominant and terminal fish species with extended life spans. In the forest, trees, by having protective devices against animal grazing, have likewise achieved very long turnover times, hence assume dominance. Subsidiary levels in the hierarchies of forest communities are formed by animals supported by the regular production of leaves, flowers, and consumable understory plants. High diversity results from the almost limitless ramification of these relationships. The greater the diversity is, the greater the work necessary to maintain the hierarchy. Thus, each holon within a hierarchy forms a coherent dissipative unit, each level constitutes a similar unit, and finally the entire hierarchy itself also comprises a distinct dissipative unit, all at different, nested levels of organization.

Generalization of Concepts

The mechanisms outlined above bring together several previously recognized generalizations.

Biological temperature. Temperature governs the rate of flow of energy within and between systems. The direction of flow is determined not by the total energy, but by the ratio of total energy to entropy.

The conceptual analysis of temperature has been developed from the tendency of one system to capture energy from another (Keenan et al., 1974). System A captures energy from system B if the entropy (S) of A relative to its total energy (E) is greater than the corresponding value of system B:

$$\left[\frac{\delta S_A}{\delta E_A}\right]_{\eta,\beta} > \left[\frac{\delta S_B}{\delta E_B}\right]_{\eta,\beta},$$

where δ represents a small change in the property of the value it precedes and η and β are the numbers of particles and constraints, respectively. Energy flows from system B to A when

$$\left[\frac{\delta E_A}{\delta S_A}\right]_{\eta,\beta} > \left[\frac{\delta E_B}{\delta S_B}\right]_{\eta,\beta}.$$

In thermodynamic terms, temperature (T) regulates the flow of energy from one system to another. Temperature is defined as the reciprocal of the capturing tendency and is usually measured in the arbitrarily assigned Kelvin scale (K):

$$T \equiv \left[\frac{\delta E}{\delta S}\right]_{\eta,\beta}.$$

If biological processes follow the same pattern, "biological temperature" can be evaluated by the ratio of total energy to entropy production. Again, as a first approximation if entropy production at the steady state (P) is substituted in this formula for entropy, and the total energy is taken as biomass energy (B), then *biological temperature* may be written as

$$BT \equiv \frac{\text{biomass}}{\text{production}}.$$

Thus, biological temperature can be expressed in terms of the components of action as the "biomass accumulation ratio," *B/P* (Whittaker and Woodwell, 1968). *B/P* is the reciprocal of the more frequently used production/biomass ratio (*P/B*) which, in the more actively conservative biological world, may be referred to as the *capturing tendency*.

MacArthur (in Leigh, 1965) concluded *that* the *P/B* ratio is dimensionally equivalent to the reciprocal of mean life span. Hence, capturing tendency can be expressed in terms of the reciprocal of turnover time (1/mean life span) (= reaction time). The faster the turnover rate (and the smaller the turnover time), the greater the capturing tendency. Hence, tropical systems, with high turnover rates, tend to capture energy from ecosystems at higher latitudes. This is against the expected direction based on biological temperature: from higher *BT* poleward to low *BT* nearer the equator, or from slowly turning over systems to those with more rapid turnover rates. This equatorward flow of energy is in the opposite direction to that dictated by physical temperature (from warmer to colder). The postulated geographic gradient in biological temperature is substantiated by the work of Mann and Brylinsky (1975), who showed that the *P/B* of freshwater systems decreases with increasing latitude. The faster turnover rates in tropical forests compared with temperate systems have been discussed by several authors (Whittaker, 1966; Golley, 1972; Reiners, 1972; Burger, 1981). It is evident also that the general increase in capturing tendency with decreasing latitude is amplified by increase in environmental temperature resulting from an increase in the basic metabolic rates of bioprocesses.

Despite faster turnover rates, the total biomass of tropical systems is much larger than that in arctic regions because much more total energy is assimilated. Qualitatively, energy flow from the Arctic toward the tropics can be seen in the annual movement of millions of birds that migrate poleward in spring to nest and rear their young, returning equatorward in the fall in greater numbers than on arrival. This results in energy transfer toward lower latitudes. Migration patterns evidently originate by seasonal inversion of the capturing tendency.

However, within an ecosystem, it is evident that energy moves in the opposite direction: from species populations with high rates of turnover to those with relatively long turnover times. This agrees with the conclusion that *P/B* declines along the food chain (Saunders et al., 1980). The capturing tendency within the ecosystem is thus reversed with respect to the movement between ecosystems: energy within the ecosystem flows from regions of low to high biological temperature. To effect this, internal work must be done to raise the temperature. With each succeeding level in the hierarchy the total energy cost accumulates, for it is not the simple energy content that must be counted, but the total energy required to produce the elaborated food supporting the higher levels (Odum, 1988).

There is thus an antagonism between the direction of energy flow within the ecosystem and that over the world system, and this represents the highest level of antagonsim between least and most action. It is the spatial equivalent of the antagonism between succession and abscession that operates over different time scales. There would appear to be an equivalence between space and time in biology as well as in physics.

Heterochrony. Waddington (1968b) points out that there are at least four significantly different biological time scales, those of metabolism, development, heredity, and evolution. To these can be added successional or ecological time. At each of these levels, as in a fractal, there is a self-similar interplay between the attractors, leading to increased or decreased entropy production. This interplay between action principles results in "heterochrony" (Russell, 1916; Gould, 1977) at all levels, a term that has previously been applied only to the evolutionary and developmental levels. *Heterochrony* has been defined by de Beer (1940; Gould, 1977) as

> a phyletic change in the onset or timing of development, so that the appearance or rate of development of a feature in a descendant ontogeny is either accelerated or retarded relative to the appearance or rate of development of the same feature in an ancestor's ontogeny.

Gould (1977), however, considers de Beer's definition too narrow, although he accepts it because of its general usage. Heterochrony, in its widest sense, implies change (increase or decrease) in the rate at which processes occur in all the above time scales and adaptation of the timing to the nature of the signal system. Ecologically, heterochrony implies adjustment to long- or short-term changes in local signal series. Short-term changes are effected by adjustment in growth and mortality rates at the ecodeme level and by changes in species complement at the ecosystem level. Massive heterochronic change is evident at geologic boundaries, such as between the Cretaceous and the Tertiary periods.

r- and K-Selection. As previously observed, a dominant species displays the characteristics of K-selection (MacArthur and Wilson, 1967; Southwood, 1977; Parry, 1981), whereas species lower in the dominance hierarchy exhibit characteristics of r-selection. All species have both a K-component, equivalent to the energy in their biomass (although the overall sense of K is described by "action," energy $\times$ time, rather better than instantaneous biomass energy), and an r-component, representing in broadest terms the energy flow as measured by replacement requirements. The ratio K/r, if measured in terms of biomass energy (B) and rate of replacement (analogous to P), can thus be taken as a further expression of biological temperature. As Gnaiger (1983, 1986) points out, the r/K ratio is a special case of his "power/economy ratio." Power depends on the rate of entropy production, and economy is measured in terms of accumulated biomass. Again, this is dimensionally equivalent to the P/B ratio. All three are thus relative measures of the capturing tendency.

Resonance. If organisms originated as a resonant response, they should exhibit characteristics of resonators. Resonance is essentially a process whereby energy dissipation is delayed by cyclic processes. If two oscillators are tuned to the same frequency and one is activated, the other will oscillate in response to output from the first. If the frequency of the driving oscillator is changed slightly, the driven oscillator will still respond, but less efficiently. Furthermore, if the driven oscillator is damped, it will respond to a wider range of driving frequencies. Within a certain response range, the more the driven oscillator is damped, the wider the range of frequencies to which it will respond.

Correspondingly, in an ecosystem, the dominant species is most highly damped and at the same time most responsive to the widest range of signals. In more usual ecological terms, dominants are species utilizing the widest range of resources (McNaughton and Wolf, 1970). This is readily seen in humans, the dominant, *Homo sapiens,* utilizing a limitless range of food and materials but still increasing the number of species exploited.

Competition can be interpreted as encroachment on the signal series of species utilizing neighboring "frequencies." Successful competition increases the damping moment, increasing the range of utilizable resources (that is, K-selection). Lower in the hierarchy, specialized species are finely tuned to the specific signals to which they respond (r-selection). Such species are closely adapted to the driving signal and therefore able to harvest the energy with considerable efficiency. Efficiency in this case is the ratio of energy in the signal to energy assimilated. During subsequent utilization of assimilated energy, the organism with the slower dissipation rate will be more efficient in terms of work done.

CONCLUSION

Naturally occurring, autonomous ecosystems from the High Arctic to the tropics, with dominant species ranging from fish, mollusks and reptiles, to birds and trees, show a remarkable degree of similarity in their structure. Species at the head of the ecosystem hierarchy proceed to a state of least dissipation, least entropy production, and least work. As dominant species characterize ecosystems, the latter as wholes exhibit the same trend to least dissipation. If this were not so, the system as such could not exist. Least dissipation is reflected in large size, uniformity in size, great age, and a low level of juvenile replacement stock. All these parameters are relative; there are no absolute values to provide a firm basis for comparison. The trend to least dissipation on which living systems are so dependent counteracts and overcomes the universal principle of least action. Least dissipation is equivalent to most action. Ecosystems, of necessity, function within the overlap between the domains of attraction of the least and most action attractors.

Within this area of overlap, ecosystems approach ergodynamic equilibria and a constant rate of energy dissipation, the classically recognized state of climax. The climax is a "minimax" condition (Lotka, 1925). It represents a state of least action as developed by the evolutionary process up to the moment under consideration and, at the same time, it represents a state of most action attainable by the current system configuration. Thus, any ecosystem (in a constant environment) assumes the appropriate state of least dissipation. Any viable changes thereafter, either of species complement or in the genetic makeup of a species, require a new minimax condition to be assumed, appropriate to the new configuration. The new minimax may involve either increased or decreased dissipation, although the former is slightly more probable. Given the greater probability of any change increasing the rate of dissipation, a ratchet mechanism is formed that imposes irreversibility, or directionality, on evolution.

These characteristics are most apparent in ecosystems at the extremes of conditions—in the ecological hinterlands. The more commonly studied realms of the living world are usually much more complex and difficult to decipher. Most normative systems, in addition to being more complex, are open and experience fluctuations due to environmental (input) variability and migrations. Furthermore, they have been disturbed by humans. They are not only far from thermodynamic equilibrium, but are also very far from ergodynamic equilibrium. They exist in a tension zone between successional "searching for equilibrium" in shorter time frames, to abscessional displacement away from equilibrium in the longer run. However, as soon as movement begins in one direction, new changes are superimposed, altering the ergodynamic equilibrium position. Ergodynamic equilibrium is thus like a mirage on the horizon to which the system is attracted. If change occurs in the basic conditions of existence, either biotic or physical, the mirage changes its position and the ecosystem tracks in the new direction in behavior reminiscent of cybernetic homing on a moving target. During geological epochs of relative environmental stasis, ergodynamic equilibria can be approached, and held for relatively long periods of time.

A great gulf therefore exists between upper level processes in and of ecosystems, and the day-to-day activities normally investigated in ecological studies. The discrepancy must somehow be bridged if a fair and satisfactory balance between the requirements of natural ecosystems and humans is ever to be achieved. Even where there has been no disturbance, many species do not appear to obey the rules. For example, various species of Pacific salmon (*Oncorhynchus* spp.) frequently accumulate great biomass and then die after a fixed, relatively short life span (Foerster, 1968; Hart, 1973). A fixed life span is inimical to the maintenance of a steady state and ultimately leads to population fluctuations. Nevertheless, these fluctuations exhibit a periodicity indicative of internal damping. Similarly, small mammals of the tundra at the pole of instability in the MacArthur diversity–stability hypothesis exhibit periodicity indicative of internal damping. Life styles such as those of salmon are difficult to interpret in detail, though they may be accounted for in a general way as opportunistic adaptation to signals of high intensity and short duration, ability to cross boundaries impermeable to most species, and the undoubted value of fluctuations to avoid predator entrainment.

Somewhat mystifying too is the apparent dearth and even complete absence of juveniles in many populations. This has been reported sufficiently frequently to indicate its reality, including in terrestrial systems not subject to significant sampling error. Undeniably, some mechanisms exist whereby juveniles grow to recruitment size. The solution to this problem can wait; it need not delay general conclusions. Suppression of juveniles by the establishment population, relegating them to the niche fringe of existence, occurs in many plant species, as recognized in the suppression of seedlings and saplings by forest dominants. Whatever the abundance and habitat of "missing" juveniles, it is evident that they do not form a significant part of the niche space of adults, which, in fact, comprises the major portion of the system.

The principal difference between ecosystems with trees and animals as dominant species is that in the forest system dominant plants are also the primary producers. Thus, there is no primary food chain comparable to that of the arctic lake reference system. This does not imply existence of additional principles; it merely indicates that in the right environment plants have a better capacity to acquire and accumulate energy than animals. Initially, plants were the dominant species in aquatic systems during much of the Precambrian, when stromatolites (symbiotic associations of algae and bacteria) covered benthic regions of the shallow seas (Golubic, 1976). Only when grazing animals evolved during the late Precambrian and Cambrian did systems begin to evolve more rapidly (Stanley, 1973a). Similarly, in forested regions of recent times, after removal of trees for human use, the system may be converted to grassland with the aid of a dominant grazing animal. This gives rise to food chains similar to, but even shorter than, those of small lakes. Such food chains, from grass to the dominant grazer, develop naturally in natural grasslands. The general principles remain the same; it is the attributes of the species which succeed that change.

Many minor problems resulting from innumerable apparent anomalies will inevitably occur in all realistic biological situations. This must not impede the search for general principles. Problems can be noted and put aside until a day arrives when their solution will be apparent. It is not always necessary to destroy old theories in the erection of new ones, as in the current trend to expunge Clements and climax from ecological thought, in favor of Gleasonian individualism and perpetual motion. Specifying different time frames in which two different processes operate (succession and abscission) solves most of the apparent problems and can lead to an integrated theory. A major difficulty is lack of time and a current shortage of available material free from disturbance whereby the state of "natural" ecosystems can be examined. Current changes put us in the position of gas molecules in an open container trying to develop the principles of statistical mechanics. Luckily, the container is complex and there are still a few compartments relatively unaffected by our own presence.

The importance of discrete boundaries to ensure that most of the interactions take place within the area of observation is well illustrated by the work of Wiens (1984) on communities of the open shrub steppe in New Mexico. This work provides a complete counterpoint to the systems described above and helps resolve discrepancies between the constancy (stability) of the arctic lake ecosystem and the fluctuating (unstable) character of the small mammal populations in the surrounding tundra. It also emphasizes the enormous difficulties encountered in attempting to establish general principles in such unbounded systems as the shrub steppe, northern tundra, or open ocean. It also underlines the difficulties of using bird populations for developing basic ecological theory, organisms on which so much effort in this direction has been expended. Wiens concludes:

> Bird communities pose additional problems. Many birds are extremely mobile, undertake short- or long-distance migrations, and demonstrate varying degrees of fidelity to previous breeding or wintering locations. These characteristics produce a substantial flux of individuals in local populations, and spread the sources of biological limitation of populations, and the

consequent determination of community structuring, over a large and undefined area. Because most local assemblages of birds contain mixtures of species that differ in migratory tendency and pathways and in longevity and fecundity, the *dynamics of any given local assemblage are likely to be driven by an amorphous complex of factors, the effects of which are likely to be different for almost every local assemblage.* This uncertainty decreases the probability that any "patterns" that seem apparent in such communities are real [emphasis added].

Nevertheless, on a broader basis, as Wiens points out, there is a pattern. Apparent near chaos at the local level becomes smoothed and reduced in a widened context as boundary constraints over the whole shrub-steppe biome become expressed. It is therefore postulated that the same biodynamic factors occurring in simple autonomous systems are also operating in the shrub-steppe, on the tundra, or within the ocean, but lack of local constraints precludes these effects from becoming recognizably manifested. Ergodynamic equilibrium is approached as boundaries expand into neighboring systems until ultimately the boundaries of the biosphere itself are reached.

In most ecosystems, frequent displacement of populations from ergodynamic equilibrium occurs, yet this does not result in a chaotic state. Populations are internally damped, as shown so clearly in the white sucker population of Heming Lake. Following disturbance, internal damping, a functional part of the trend to least dissipation, allows a well-damped return to the ground state. This process is implicit in the general acceptance of the fact that ecosystems recover if damaged (Cairns, 1980). Ecodemes and ecosystems should not be regarded as systems in ergodynamic equilibrium, but as *ergodynamic-equilibrium seeking systems.*

The central core of ecological energetics is generally considered to be the "second law of thermodynamics: that all systems tend toward maximum entropy and that open systems can be maintained by negentropy . . ." (Reiners, 1986). This statement needs considerable revision and modification if *dynamic* thermodynamic concepts are to be applied to biological systems. Biologically, maximum entropy is a nonallowable state since it implies a static condition in which living organisms cannot exist, nor could they exist if entropy increased with time (except in the sense that they exist through the gradual increase in entropy of the whole universe). The concept of maximum entropy is, in the first instance, important only as it applies to inanimate objects. In the living world, entropy production rates are of primary significance. Such time-dependent factors are not incorporated in the fundamental laws of thermodynamics; time appears only to give a general sense of direction, as in the statement "entropy increases with time." Thus, new laws are essential, but they must be additive and not in conflict with the original set.

The first law of biodynamics, it is suggested, should be a statement to the effect that, locally in time and space, free energy capable of being assimilated may be conserved in cyclic processes symmetrical with time. Near thermodynamic equilibrium, cyclic and rectilinear energy transport processes interfere with each other so that systems assume a least dissipation state that may be referred to as ergodynamic equilibrium. This law is a prerequisite of existence; it implies movement toward a state of maximum energy acquisition and least attainable entropy production commensurate with prevailing constraints. It may be described as the successional law imposing time delay on energy flow; it applies to any specific configuration of species under constant environmental conditions. It is effective in the short term. Energy is acquired and accumulated through a resonant response to signals in the environment. Cope's rule, which implies recognition of the trend for certain evolutionary lines to increase in size (Stanley, 1973b), with a concomitant reduction in specific metabolic rate (Kleiber, 1961), is the prime example of this law functioning in its evolutionary context.

The second law (which is supreme) acts in opposition to the first. It acts on changes that inevitably occur in any set configuration. It indicates that, far from equilibrium, living systems, although proceeding in the short term to least entropy production, tend to proceed irreversibly over the long term toward a state of increasing entropy production. This occurs under stimulus of the principle of least action. It may be described as the abscessional law.

The biosystem functions through the antagonism of these two laws, which, because of their operation in different time frames, function as a ratchet mechanism. The only way energy trapped between the two attractors can escape at a faster rate is for the power of the system to increase. This increased power must be dissipated through the identification and coding of an increasing number of signals and an increased complexity of mechanisms for acquisition and conservation of energy. For example, organs for mobility and weapons such as canine teeth serve both for acquisition and defense, but both require considerable energy output for their maintenance and operation. Conversely, species lose organs when they are no longer functional, as in the case of eyes of cave-dwelling animals. Despite such apparent cases of reversal, increasing diversity and increasing average complexity are the major characteristics of evolution. This general trend toward increasing dissipation under the stimulus of the principle of least action gives directionality to evolution. However, directionality, in the sense of increasing diversity, is not strictly monotonic. Environmental change or the ascendancy of a new species may temporarily reduce diversity, allowing an increase in biomass of species best adapted to the new conditions. At this point, diversity once more begins to increase. There is no reversal of the evolutionary process, merely termination of certain evolutionary lines and resetting of the starting position.

These two laws can operate only within the boundaries of a closed system and only close to ergodynamic equilibrium. Ultimately, the whole world must form a single functional unit because all subsystems are interconnected, not least by circulations of the atmosphere and nutrient and hydrologic cycles, but also by food webs and other causal networks that reticulate across all habitats, landscapes and seascapes, biomes and life zones. Hence the biosystem, gradually increasing its power over evolutionary time, has been able to move per force away from thermodynamic equilibrium. The exergical work done enables physical and chemical states to be maintained far from their equilibrium conditions, as postulated by Lovelock (1979) in the controversial *Gaia* hypothesis.

Although the rules of the nonequilibrial world are more complex and technically different from those of the equilibrium world, there are many similarities and correspondences. Living things, in many ways, exhibit an inverted dynamic to that of the nonliving world. It is an inverted world striving to achieve and maintain viability in the vicinity of ergodynamic equilibria. Work is done against the second-law imperative. The world is not chaotic; that is a property of mathematics. The real world is well ordered. Entropy production, not entropy, is the near-term ruler. Only when regular, life-sustaining flows of energy cease does the real world revert to the domain of equilibrium thermodynamics. Entropy, as always, in the long run, wins.

ACKNOWLEDGMENTS

The concept of this paper was originally developed for the Tromso Workshop of the International Society of Arctic Char Fanatics and published in abbreviated form in the Proceedings (*Proc. Third Workshop of I.S.A.C.F.,* Drottningholm, 1985). I thank those attending for their encouragement and interest.

I also thank Dr. Eric D. Schneider, Hawkwood Institute, Livingston, Montana, for his continued interest and stimulation, and for bringing to my attention the paper by Drs. Sakai and Sulak; Dr. Erich Gnaiger, University of Innsbruck, Austria, for much stimulating discussion; and Drs. J. Craig, J. W. Clayton and R. Stewart, and Mr. G. D. Koshinsky, of the Freshwater Institute, Winnipeg, for reading an earlier version of the manuscript and making many helpful comments. Lt. Henry Parker, R.N., stimulated the thoughts on heterochrony. I thank the editor of the *Canadian Journal of Fisheries and Aquatic Science* for permission to reproduce Fig. 3.14 (from Green, 1980) and Figs. 3.3 through 3.9 (from Johnson, 1976 and 1980); the editor of the *Journal of Marine Ecology, Progress Series,* for permission to reproduce Fig. 3.12 (Figs. 2, 3, 5, and 6 in Hutchings and Haedrich, 1984); the editor of the *American Midland Naturalist* for permission to reproduce Fig.

3.20 (Fig. 1 in Sakai and Sulak, 1985); the editor of *Philosophical Transactions of the Royal Society, London,* for permission to reproduce Fig. 3.15 (Fig. 32 in Grubb, 1971); the editor of the *Journal of Zoology, London,* for permission to reproduce Fig. 3.16 (Fig. 3 in Gaymer, 1968) and Fig. 3.18 (Fig. 8 in Southern, 1970b); the editor of the *Journal of Ecology* for permission to reproduce Fig. 3.19 (Fig. 1 in Vaughan and Wiehe, 1941); the editor of *Annales de la Societe Royale Zoologique de Belgique* for permission to reproduce Fig. 3.10 (Fig. 13 in Heuts, 1951); and Dr. R. Radke and the University of Hawaii for permission to reproduce Fig. 3.13 (Figs. 4 and 5 in Radtke, 1985). I am grateful also to Drs. G. H. Lawler and J. S. Campbell of the Department of Fisheries, Canada, for permission to use their data on Heming Lake, reproduced as Fig. 3.17; and Dr. J. A. Wiens, Department of Biology, Colorado State University, Fort Collins, for permission to quote from his paper on "Understanding the Non-Equilibrium World" (Wiens, 1984).

REFERENCES

ALVAREZ, W., ALVAREZ, W., ASARO, F., and MICHEL, H. V. 1980. Extra-terrestrial causes for the Cretaceous–Tertiary extinction. *Science* 208:1095–1108.

ALVAREZ, W., KAUFMANN, E. G., SYRLYK, F., ALVAREZ, L. W., ASARO, F., and MICHEL, H. V. 1984. Impact theory of mass extinctions and the invertebrate fossil record. *Science* 223:1135–1141.

BOURN, D., and COE, M. 1978. The size structure and distribution of the giant tortoise population on Aldabra. *Phil. Trans. Roy. Soc. Lond. (B)* 282:139–175.

BRYLINSKY, M., and MANN, K. H. 1973. An analysis of the factors governing productivity in lakes and reservoirs. *Limnol. Oceanog.* 18:4–14.

BURGER, C. 1981. Why are there so many kinds of flowering plants? *Bioscience* 31:572–591.

CAIRNS, J. (ed.). 1980. *The Recovery Process in Damaged Ecosystems.* Ann Arbor Science, Ann Arbor, Michigan. 167 pp.

CLEMENTS, F. E. 1905. *Research Methods in Ecology.* University of Nebraska Publishing Co., Lincoln. 334 pp.

CLEMENTS, F. E. 1916. *Plant Succession. An Analysis of the Development of Vegetation.* Carnegie Institute, Washington, Washington, D.C. Publ. No. 242. 512 pp.

CLEMENTS, F. E. 1936. The nature and structure of the climax. *J. Ecol.* 24:252–284.

COHEN, B. I. 1985. *The Birth of a New Physics.* W. W. Norton, New York. 258 pp.

CONNELL, J. H. 1978. Diversity in tropical rain forests and coral reefs. *Science* 199:1302–1310.

DAVIES, P. 1980. *Other Worlds: Space, Superspace, and the Quantum Universe.* Simon & Schuster, New York. 207 pp.

DE BEER, G. R. 1940. *Embryos and Ancestors.* Oxford, University Press, New York. 108 pp.

DEMPSON, B., and GREEN, J. M. 1985. Life history of anadromous arctic charr, *Salvelinus alpinus,* in the Fraser River, northern Labrador. *Can. J. Zool.* 63:315–324.

EBLING, F. J., and STODDART, D. M. (eds.). 1978. *Population Control by Social Behavior.* Symp. 23, Inst. Biol. London. 304 pp.

EGERTON, F. N. 1973. Changing concepts of the balance of nature. *Quart. Rev. Biol.* 48:322–350.

ELTON, C. 1930. *Animal Ecology and Evolution.* Oxford Clarendon Press, London. 96 pp.

FEYNMAN, R. P. 1963. *The Feynman Lectures on Physics,* Vol. 1. Leighton, R. B., and Sands, M. (eds.). Addison-Wesley, Reading, Massachusetts, pp. 19-1 to 19-4.

FOERSTER, R. E. 1968. The sockeye salmon, *Oncorhynchus nerka. Bull. Fish. Res. Board Can.* 162. Queen's Printer, Ottawa. 422 pp.

FORBES, S. A. 1887. The lake as a microcosm. *Bull. Sci. Assn. Peoria, Ill.* 1887:77–87.

GAYMER, R. 1968. The Indian Ocean giant tortoise, *Testudo gigantea,* on Aldabra. *J. Zool. Lond.* 154:341–363.

GIBSON, C. W. D., and HAMILTON, J. 1983. Feeding ecology and seasonal movements of giant tortoises on Aldabra Atoll. *Oecologia* 56:84–92.

GNAIGER, E. 1983. Heat dissipation and energetic efficiency in animal anoxobiosis: economy contra power. *J. Exp. Zool.* 228:471–490.

GNAIGER, E. 1986. Optimum efficiencies of energy transformation in anoxic metabolism: the strategies of power and economy. In Callow, P. (ed.), *Evolutionary Physiological Ecology.* Cambridge University Press, New York, pp. 7–36.

GOLDMAN, C. R. 1970. Antarctic freshwater systems. In Holdgate, M. R. (ed.), *Antarctic Ecology.* Academic Press, New York, pp. 609–627.

GOLLEY, F. B. 1972. Energy flux in ecosystems. In Wiens, J. A. (ed.), *Ecosystem Structure and Function.* Proc. 31st Ann. Colloquium Biol., Oregon State University Press, Corvallis, Oregon, pp. 69–90.

GOLUBIC, S. 1976. Organisms that build stromatolites. In Walter, M. R. (ed.), *Stromatolites.* Elsevier, New York, pp. 113–126.

GOODMAN, D. 1975. The theory of diversity–stability relationships in ecology. *Quart. Rev. Biol.* 50:237–266.

GOULD, S. J. 1977. *Ontogeny and Phylogeny.* Harvard University Belknap Press, Cambridge, Massachusetts. 501 pp.

GREEN, R. H. 1980. The role of the unionid clam population in the calcium budget of a small arctic lake. *Can. J. Fish. Aquat. Sci.* 37:219–224.

GRUBB, P. 1971. The growth, ecology, and population structure of giant tortoises on Aldabra. *Phil. Trans. Roy. Soc. Lond. (B)* 260:327–372.

HALL, E. T. 1983. *The Dance of Life: The Other Dimension of Time.* Doubleday, Garden City, New York. 250 pp.

HARPER, J. L. 1969. The role of production in vegetational diversity. In Woodwell, G. M., and Smith, F. H. (eds.), *Diversity and Stability in Ecological Systems.* Brookhaven Sympos. Biol. 22:48–62.

HART, J. L. 1973. *Pacific Fishes of Canada.* Bull. Fish. Res. Board. Can. 180. Queen's Printer, Ottawa.

HEUTS, M. J. 1951. Ecology, variation and adaptation of the blind African cave fish *Caecobarbus geertsii* Blgr. *Ann. Soc. Roy. Zool. de Belgique* 82:154–230.

HOFFMAN, A. 1985. Island biogeography and palaeobiology: in search for evolutionary equilibria. *Biol. Rev.* 60:455–471.

HUTCHINGS, J. A. and HAEDRICH, R. L. 1984. Growth and population structure in two species of bivalves (Nuculanidae) from the deep sea. *Mar. Ecol. Prog. Ser.* 17:135–142.

IBERALL, A., SOODAK, H. and ARENSBERG, C. 1981. Homeokinetic physics of societies—a new discipline: autonomous groups, cultures, polities. *In* Real, H., Ghista, D., and Rau G. (eds.), *Perspectives in Biomechanics,* Vol. 1, Part 1. Harwood Academic Press, New York, pp. 433–527.

JOHNSON, L. 1966. The temperature of maximum density of fresh water and its effect on circulation in Great Bear Lake. *J. Fish. Res. Board Can.* 23:969–973.

JOHNSON, L. 1972. Keller Lake: characteristics of a culturally unstressed community. *J. Fish. Res. Board Can.* 29:731–740.

JOHNSON, L. 1975a. Distribution of fish species in Great Bear Lake with reference to zooplankton, benthic invertebrates and ecological conditions. *J. Fish Res. Board Can.* 32:1989–2005.

JOHNSON, L. 1975b. Physical and chemical characteristics of Great Bear Lake. *J. Fish. Res. Board Can.* 32:1971–1987.

JOHNSON, L. 1976. Ecology of arctic populations of lake trout, *Salvelinus namaycush,* lake whitefish, *Coregonus clupeaformis,* arctic char, *S. alpinus,* and associated species in unexploited lakes of the Canadian Northwest Territories. *J. Fish. Res. Board Can.* 33:2459–2488.

JOHNSON, L. 1980. Arctic charr. In Balon, E. K. (ed.), *Charrs, Salmonid Fishes of the Genus Salvelinus.* Dr. J. W. Junk, The Hague, pp. 15–98.

JOHNSON, L. 1981. The thermodynamic origin of ecosystems. *Can. J. Fish. Aquat. Sci.* 38:571–590.

JOHNSON, L. 1983. Homeostatic characteristics of single species fish stocks in arctic lakes. *Can. J. Fish. Aquat. Sci.* 40:987–1024.

JOHNSON, L. 1988. The thermodynamic origin of ecosystems: a tale of broken symmetry. In Weber, B. H., Depew, D. J., and Smith, J. D. (eds.), *Entropy, Information and Evolution, New Perspectives on Physical and Biological Evolution.* M. I. T. Press, Cambridge, Massachusetts. 376 pp.

JOHNSON, L. 1989. The anadromous Arctic charr, *Salvelinus alpinus,* of Nauyuk Lake, Northwest Territories, Canada. *Physol. Ecol. Japan. Spec.* 1:201-227.

JOHNSON, L. 1994a. Pattern and process in ecological systems: a step toward the development of a general ecological theory. *Can. J. Fish. Aquat. Sci.* (January 1994).

JOHNSON, L. 1994b. Long-term experiments on the stability of two fish populations in previously unexploited Arctic lakes. *Can. J. Fish. Aquatic Sci.* (January 1994).

JOHNSON, L., and BURNS, B. (eds.). 1984. *Biology of the Arctic Charr.* University of Manitoba Press, Winnipeg. 584 pp.

JONES, E. W. 1945. The structure and reproduction of the forest of the north temperate zone. *New. Phytol.* 44:130–148.

KALFF, J., and WELCH, H. E. 1974. Phytoplankton production in Char Lake, a natural polar lake, and in Meretta Lake, a polluted polar lake, Cornwallis Island, Northwest Territories. *J. Fish. Res. Board Can.* 31:621–636.

KEENAN, J. H., HATSOPOULOS, G. N., and GYFTOPOULOS, E. P. 1974. *Thermodynamics.* Encycl. Brittanica. H. Benton, Chicago.

KERR, S. R. 1974. Theory of size distribution in ecological communities. *J. Fish. Res. Board. Can.* 31:1859–1862.

KLEIBER, M. 1961. *The Fire of Life.* Wiley, New York. 454 pp.

KOESTLER, A. 1967. *The Ghost in the Machine.* Hutchinson (Picador edn., 1975), London. 384 pp.

KOLATA, G. B. 1974. Theoretical ecology: beginnings of a predictive science. *Science* 183:400–401.

LAWLER, G. H., and WATSON, N. H. F. 1958. Limnological studies of Heming Lake, Manitoba, and two adjacent lakes. *J. Fish. Res. Board. Can.* 15:203–218.

LEIGH, E. G. 1965. On the relationship between productivity, biomass, diversity and stability of a community. *Proc. Nat. Acad. Sci. USA* 53:777–783.

LEWONTIN, R. C. 1957. The adaptations of populations to varying environments. *Cold Spring Harbor Symp. Quant. Biol.* 22:395–408.

LOTKA, A. J. 1925. *Elements of Physical Biology.* Williams and Wilkins, Baltimore. 460 pp. (Reprinted, 1956 as *Elements of Mathematical Biology*), Dover, New York. 465 pp.

LOVELOCK, J. E. 1979. *Gaia; A New Look at Life on Earth.* Oxford University Press, New York. 157 pp.

MACARTHUR, R. H. 1955. Fluctuations of animal populations and a measure of community stability. *Ecology* 36:533–536.

MACARTHUR, R. H., and WILSON, E. O. 1967. The theory of island biogeography. *Princeton Mon. Pop. Biol.,* Vol. 1. Princeton University Press, Princeton, New Jersey, pp. 1–203.

MANN, K. H., and BRYLINSKY, M. 1975. Estimating productivity of lakes and reservoirs. *In* Cameron, T. W. M., and Billingsley, L. W. (eds.), *Energy Flow, Its Biological Dimension.* Royal Society of Canada, Ottawa, pp. 221–226.

MARGALEF, R. 1968. *Perspectives in Ecological Theory.* Chicago University Press, Chicago. 111 pp.

MAYR, E. 1976. *Evolution and the Diversity of Life.* Harvard University Press, Cambridge, Massachusetts. 721 pp.

MCINTOSH, R. P. 1985. *The Background of Ecology: Concept and Theory.* Cambridge University Press, New York. 383 pp.

MCLAREN, I. A. 1964. Zooplankton of Lake Hazen, Ellesmere Island and a nearby pond with special reference to the copepod *Cyclops scutifer* Sars. *Can. J. Zool.* 42:613–629.

McNaughton, S. J. 1984. Grazing lawns: animals in herds, plant form and co-evolution. *Am. Nat.* 124:863–886.

McNaughton, S. J., and Wolf, L. L. 1970. Dominance and the niche in ecological systems. *Science* 1167:131–139.

McPhail, J. D., and Lindsey, C. C. 1970. Freshwater Fishes in Northwestern Canada and Alaska. *Bull. Fish. Res. Board Can. 173.* 381 pp.

Medawar, P. B. 1945. Size, shape, and age. In Clark, Le Gros, and Medawar, P. B. (eds.), *Essays on Growth and Form.* Oxford University Press, New York, pp. 157–187.

Miller, R. B. 1947. Great Bear Lake. *In* Northwest Canadian Fisheries Surveys 1944–45. *Bull. Fish. Res. Board Can.* 72:31–44.

Morowitz, H. J. 1968. *Energy Flow in Biology.* Academic Press, New York. 179 pp.

Neumann, E. 1978. Dissipative structures. Review of "Self-organization in non-equilibrium systems: from dissipative structure to order through fluctuations," by Nicolis, G. and Prigogine, I. Wiley–Interscience, New York. 492 pp. *Nature (London)* 271:785–786.

Nicolis, G., and Prigogine, I. 1977. *Self-organization in Non-equilibrium Systems: From Dissipative Structure to Order through Fluctuations.* Wiley–Interscience, New York. 492 pp.

Northcote, T. G. 1988. Fish in the structure of freshwater ecosystems: a "top-down" view. *Can. J. Fish. Aquat. Sci.* 45:361–379.

Odum, E. P. 1969. The strategy of ecosystem development. *Science* 164:262–270.

Odum, H. T. 1956. Efficiencies, size of organisms, and community structure. *Ecology* 37:592–597.

Odum, H. T. 1988. Self-organization, transformity and information. *Science* 242:1132–1139.

Odum, H. T., and Pinkerton, R. C. 1955. Time's speed regulator, the optimum efficiency for maximum power output in physical and biological systems. *Am. Sci.* 43:331–343.

O'Neill, R. V., De Angelis, D. L., Waide, J. B. and Allen, F. T. H. 1986. *A Hierarchical Concept of Ecosystems.* Princeton University Press, Princeton, New Jersey. 253 pp.

Onsager, L. 1931. Reciprocal relationships in irreversible processes. *Phys. Rev.* 37:405–426.

Pagels, H. R. 1982. *The Cosmic Code.* Bantam Books, New York. 333 pp.

Parker, H. H., and Johnson, L. 1991. Population structure, ecological segregation and reproduction in non-anadromous Arctic charr, *Salvelinus alpinus* (L.) in four unexploited lakes in the High Arctic. *J. Fish Biol.*:123–147.

Parry, G. D. 1981. The meaning of *r*- and *K*-selection. *Oecologia (Berlin)* 48:260–264.

Pierce, J. R. 1961. *Symbols, Signals and Noise: The Nature and Process of Communication.* Harper & Row, New York. 305 pp.

Pittendrigh, C. S. 1958. Adaptation, natural selection and behavior. In Roe, A., and Simpson, G. G. (eds.), *Behavior and Evolution.* Yale University Press, New Haven, Connecticut.

Platt, J. R. 1964. Strong inference. *Science* 146:347–353.

Polanyi, M. 1968. Life's irreducible structure. *Science* 160:1308–1312.

Poulson, T. L. 1963. Cave adaptation in Amblyopsid fishes. *Am. Midl. Nat.* 70:257–290.

Prigogine, I. 1978. Time, structure and fluctuations. *Science* 201:777–785.

Prigogine, I., and Stengers, I. 1984. *Order Out of Chaos: Man's New Dialogue with Nature.* Bantam Books, New York. 349 pp.

Prigogine, I., and Wiame, J. M. 1946. Biologie et thermodynamique des phenomenes irreversible. *Experientia* 2:451–453.

Prigogine, I., Nicolis, G., and Babloyantz, A. 1972. Thermodynamics of evolution. *Physics Today,* November 1972, pp. 23–28 and 38–44.

Radtke, R. 1985. Population dynamics of the giant clam *Tridacna maxima,* at Rose Atoll. *Rept. Hawaii Inst. Marine Biol.,* University of Hawaii, Honolulu.

Raup, D. M., and Sepkoski, J. J. 1982. Mass extinctions in the marine fossil record. *Science* 215:1501–1503.

RAUP, D. M., and SEPKOSKI, J. J. 1986. Periodic extinction of families and genera. *Science* 215:1501–1503.
REINERS, W. A. 1972. Structure and energetics of forests. *Ecol. Mon.* 42:71–94.
REINERS, W. A. 1986. Complementary models for ecosystems. *Am. Nat.* 127:59–73.
RICKER, W. E. 1975. Computation and interpretation of biological statistics of fish populations. Fish. Mar. Serv. Bull. 191, Dept. Env. Canada. 381 pp.
RIGET, F. F., NYGARD, K. H., and CHRISTENSEN, B. 1986. Population structure, ecological segregation and reproduction in a population of Arctic char (*Salvelinus alpinus*) from Lake Tasersuag, Greenland. *Can. J. Fish. Aquat. Sci.* 43:985–992.
RIGLER, F. H. 1975. The Char Lake project. In Cameron, T. W. M. and Billingsley, L. W. (eds.), *Energy Flow—Its Biological Dimension*. Royal Society, Ottawa, Canada, pp. 171–198.
RIGLER, F. H., MCCALLUM, M. E., and ROFF, J. C. 1974. Production of zooplankton in Char Lake. *J. Fish. Res. Board Can.* 31:637–646.
RUSSELL, E. S. 1916. *Form and Function: A Contribution to the History of Animal Morphology*. John Murray, London. Republished 1982. University of Chicago Press. 383 pp.
SAKAI, A. K., and SULAK, J. H. 1985. Four decades of secondary succession in two lowland permanent plots in northern Michigan. *Am. Midl. Nat.* 113:146–157.
SALTHE, S. N. 1985. *Evolving Hierarchical Systems*. Columbia University Press, New York. 343 pp.
SAUNDERS, H. L., CUMMINS, G. W., GAK, D. Z., PIECZYNSKA, E., STRASKRABOVA, V., and WETZEL, G. 1980. *In* Le Cren, E. D., and Lowe-McConnell, R. H. (eds.), *The Functioning of Freshwater Ecosystems*. Cambridge University Press, New York, pp. 341–392.
SAUNDERS, P. T., and HO, M. W. 1981. On the increase in complexity in evolution. II. The relative complexity and the principle of minimum increase. *J. Theor. Biol.* 90:515–530.
SCHAEFER, M. B. 1970. Men, birds and anchovies in the Peru Current. *Trans. Am. Fish. Soc.* 99:461–467.
SCHEER, B. T. 1970. A universal definition of work. *Bioscience* 26:505–506.
SEPKOSKI, J. J., BAMBACH, R. K., RAUP, D. M., and VALENTINE, J. W. 1981. Phanerozoic marine diversity and the fossil record. *Nature* 293:435–437.
SIMON, H. A. 1973. The organization of complex systems. *In* Pattee, H. H. (ed.), *Hierarchy Theory*. Braziller, New York, pp. 3–27.
SIMPSON, G. G. 1953. *The Major Features of Evolution*. Columbia University Press, New York. 434 pp.
SIMPSON, G. G. 1969. The first three billion years of community evolution. In Woodwell G. M., and Smith, F. H. (eds.), *Diversity and Stability in Ecosystems*. Brookhaven Symp. Biol. 22, pp. 162–177.
SLOBODKIN, L. B. 1964. The strategy of evolution. *Am. Sci.* 52:342–537.
SOODAK, H., and IBERALL, A. 1978. Homeokinesis: a physical science for complex systems. *Science* 201:579–582.
SOUTHERN, H. N. 1970a. Ecology at the crossroads. *J. Ecol.* 58:1–11.
SOUTHERN, H. N. 1970b. The natural control of a population of tawny owl (*Strix aluco*). *J. Zool. Lond.* 162:197–285.
SOUTHWOOD, T. R. E. 1977. Habitat: a template for ecological strategies? *J. An. Ecol.* 46:337–365.
SPARHOLT, H. 1985. The population, survival, growth, reproduction and food of arctic charr, *Salvelinus alpinus* (L.), in four unexploited lakes in Greenland. *J. Fish. Biol.* 26:313–330.
STANLEY, S. M. 1973a. An ecological theory for the sudden origin of multicellular life in the late Precambrian. *Proc. Nat. Acad. Sci. USA* 70:1486–1489.
STANLEY, S. M. 1973b. An explanation for Cope's Rule. *Evolution* 271:1–26.
STODDART, D. R., and WRIGHT, C. A. 1967. Ecology of Aldabra Atoll. *Nature (Lond.)* 213:1174–1177.

SWINGLAND, I. R. 1977. Reproductive effort and life history strategy of the Aldabran giant tortoise. *Nature (Lond.)* 269:402–404.

SWINGLAND, I. R., and COE, M. 1978. The natural regulation of giant tortoise populations on Aldabra Atoll. *J. Zool. Lond.* 186:285–309.

SWINGLAND, I. R., and COE, M. 1979. The natural regulation of giant tortoise populations on Aldabra Atoll: recruitment. *Phil. Trans. Roy. Soc. London (B)* 286:177–188.

TANSLEY, A. G. 1935. The use and abuse of vegetational concepts and terms. *Ecology* 16:284–307.

VAN VALEN, L. 1973. A new evolutionary law. *Evol. Theory.* 1:1–30.

VAUGHAN, R. E., and WIEHE, P. O. 1937. Studies on the vegetation of Mauritius. I. A preliminary survey of plant communities. *J. Ecol.* 25:289–343.

VAUGHAN, R. E., and WIEHE, P. O. 1941. Studies on the vegetation of Mauritius. III. Structure and development of the upland climax forest. *J. Ecol.* 29:127–160.

WADDINGTON, C. H. 1968a. The basic ideas of biology. In Waddington, C. H. (ed.), *Towards a Theoretical Biology, Vol. 1, Prolegomena.* Aldine, Chicago, pp. 1–32.

WADDINGTON, C. H. 1968b. Towards a theoretical biology. *Nature (Lond.)* 218:525–527.

WATSON, A. 1986. Physics—where the action is. *New. Sci.,* January 1986, pp. 42–44.

WELCH, H. E., and KALFF, J. 1974. Benthic photosynthesis and respiration in Char Lake. *J. Fish. Res. Board Can.* 31:609–620.

WHITTAKER, R. H. 1951. A criticism of the plant association and climax concepts. *Northwest Sci.* 25:17–31.

WHITTAKER, R. H. 1953. A consideration of climax theory: the climax as a population and pattern. *Ecol. Monogr.* 23:41–78.

WHITTAKER, R. H. 1966. Forest dimensions and production in the Great Smokey Mountains. *Ecology* 47:103–121.

WHITTAKER, R. H. 1969. Evolution of diversity in plant communities. In Woodwell, G. M., and Smith, F. H. (eds.), *Diversity and Stability in Ecological Systems.* Brookhaven Sympos. Biol. 22, Brookhaven National Laboratory, Upton, New York, pp. 178–196.

WHITTAKER, R. H., and WOODWELL, G. M. 1968. Dimension and production relations of trees and shrubs in the Brookhaven forest, New York. *J. Ecol.* 56:1–25.

WIENS, J. A. 1984. On understanding a non-equilibrium world: myth and reality in community patterns and processes. In Strong, D. G., Jr., Simberloff, D., Abele, L. G., and Thistle, A. B. (eds.), *Ecological Communities: Conceptual Issues and the Evidence.* Princeton University Press, Princeton, New Jersey, pp. 439–457.

WIESER, W., GRABNER, M., and KOCH, F. 1981. Distribution and migration of two Cerithid snails on a sand flat in Bermuda. *Marine Ecol.* 2:51–61.

4

Cybernetic Theory of Complex Ecosystems

MILAN STRAŠKRABA

Biomathematical Laboratory,
Biological Research Center,
Czechoslovak Academy of Sciences,
České Budějovice, Czechoslovakia

> The systems oriented studies rally around the ecosystem concept. The ecosystem concept was the higher order or level of concept required to provide the integration and interdisciplinarity. Effectively this was a "conceptual quantum jump".... The adoption of the ecosystem concept provided the theoretical and philosophical framework that was missing in the largely single-factor studies of classical ecology. Systems ecology is a new ecology.... [Van Dyne, in *Grasslands, Systems Analysis and Man*, Int. Biol. Program 19, Chap. 14, p. 885 (1980), Cambridge University Press].

INTRODUCTION

My goal in this chapter is to identify the basic features of ecosystems that have to be included in theoretical ecosystem models. Contemporary ecological theory (for example, May, 1975; Roughgarden et al., 1989; Wissel, 1990) is largely based on the Lotka–Volterra models of population dynamics (called hereafter the L–V equations or model). A number of studies have shown that the applicability of such models to natural situations is questionable (Brown, 1981; Hall, 1988).

Levin (1981) stressed that the L–V equations have had their value seriously undermined by overuse and by misguided attempts to parameterize them on the basis of data. Consequently, they should not be taken literally, but only as guides to theory and experimentation. The processes both Lotka (1925) and Volterra (1931) modeled were interactions between species populations where environmental influences are implicit. For ecosystem theory, we must account for a multiplicity of mutual interactions between different groups of organisms and between organisms and their environment.

There are basically two approaches: population models, which are general, and ecosystem models, which are mostly complex constructs lacking generality (Patten, 1976; Straškraba and

Gnauck, 1985; Jørgensen, 1983; Swartzman and Kaluzny, 1987; Wulff et al., 1989). What is needed for the development of ecological theory is a marriage of both approaches: simple models, which, however, will capture all the basic features and mechanisms of both populations and ecosystems.

To contribute to this goal I will first critically evaluate stability studies of theoretical models of population dynamics. Second, the assumptions underlying these stability studies will be compared with ecosystem features. Finally, a new concept of basic ecosystem features will be presented and theoretical ecosystem models consistent with these outlined.

STABILITY STUDIES OF PREY-PREDATOR MODELS

The literature on theoretical aspects of population dynamics is rather extensive (for example, reviews by May, 1979, for single populations, and Khanin et al., 1978; Svirezhev and Logofet, 1978; Nisbet and Gurney, 1982, for prey–predator interactions; May, 1984). My major goal is to use population models to derive general models that incorporate important ecosystem processes.

One way to categorize population dynamics models applicable for the development of ecosystem theory is by number of trophic levels:

1. Predator–prey population models (that is, two trophic levels)
2. Three and more trophic-level population models

First, I will examine predator–prey models to see how far they conform to what I call theoretical observations. This term comprises those characteristic features of reality that are of a qualitative rather than quantitative nature, derived either from generalized observations in nature or obtained for laboratory populations under strictly controlled and simplified conditions. For theoretical models it is not at all necessary that they match quantitatively some observed population trajectories. However, it is a question of subjective selection which aspects of reality viewed in this way are to be taken into consideration and how they are seen. One such important aspect is the stability versus diversity problem, which will be used as a point of examination.

The most commonly studied property of simple models is stability. Mathematical convenience rather than ecological usefulness seems to be the major driving force for the welter of stability analyses in population ecology (Ginzburg, 1980). Nevertheless, the stability problem is rather important from the ecological point of view. The difficulty lies in confronting the intuitive, many-sided meanings of stability by field ecologists and the usually much more restricted mathematical meaning by theoreticians.

The method most often used in L–V stability studies is Liapunov stability. *Liapunov stability* is a mathematical means for studying asymptotic behavior of differential equations with constant coefficients. A system is stable in the Liapunov sense if, for every initial time t_0 and for any region $X: |X - X(0)| < \epsilon$, there exists $\delta > 0$ such that if $[x(t_0) - \bar{x}] < \delta$, then $[x(t) - \bar{x}] < \epsilon$ for all t. Therefore, if we start from an arbitrary value $x(t_0)$ within δ units in state space of the steady-state value of the system $(\bar{x})$, the trajectory of the system will always remain within another region of ϵ units around the steady-state value. When the system trajectory leaves the region defined by ϵ, process stability (that is, in the sense of Liapunov) is lost. However, *Lagrange stability* can still be attained if some region exists within which the system trajectory is bounded (Thornton and Mulholland, 1974). This is a less stringent requirement, and it is more realistic.

Explanation for the discrepancy between Liapunov stability and real-world stability has been sought in several directions:

1. Stochasticity of both environmental conditions and population states
2. Dependence of stability inferences for linear systems on values of coefficients

3. Inclusion of nonlinearities in models
4. Covering several species of both prey and predators in complex food webs
5. Use of difference instead of differential equations
6. Structuring of models with respect to space, age, and social organization
7. Use of equations showing new qualities such as chaos and bifurcation

I will not deal here with the stochasticity issues, but will review some of the others.

Linear Systems and Linearization

Harwell et al. (1977) suggested that Liapunov stability analyzed only by means of system linearization gives useful inferences about a system only if it behaves linearly throughout its vector space. Goh (1975, 1980) showed that among all possible linear L–V systems, very few can be found that are not vulnerable to a continuous disturbance. Dubois (1979) concluded that inferences made from L–V models depend greatly on coefficient values. Therefore, strict restrictions on the utility of Liapunov methods are met, both structural and parametric.

Nonlinear Two-Species Models

Steele (1974) has used a simple model to investigate the effect of nonlinear feeding and threshold effects of food concentration on stability. Murdoch and Oaten (1975) and Aggrawala and Gopalsany (1980) showed that density dependence and time lags increase stability. Murdoch and Oaten (1975), Bazykin (1976), Schoener (1978), and Gruber (1976) studied the effect of satiation. In all instances, stability properties increased by the inclusion of nonlinearities, a finding that also applies to multispecies models.

Multispecies Models

The problem of many species has actually sharpened the debate about simple models. Kerner (1961), advocating the use of methods from statistical mechanics, studied two predators and one prey as early as 1960. The less successful predator gradually disappears in his model. Gardner and Ashby (1970) investigated the probability of stability in randomly created n-species models with differing degrees of connectance. They found a decrease in stability with increase of complexity (connectivity). Systematic investigations by May (1971, 1973a) on the connectance–stability relationship were extended by studies of De Angelis (1975), Robinson and Valentine (1979), Nunney (1980), and Abrams and Allison (1982). Nunney (1980) suggested that "the conclusions based upon linearized equations that do not derive from biologically realistic models are severely restrictive."

The above were mostly linear models, but the same conclusions have been obtained with some nonlinear multispecies models (Austin and Cook, 1974; Pimm, 1979). King and Pimm (1983) and Stenseth (1985) modified L–V models in such a way that stability increased with connectivity (instead of the usual pattern of decrease). The inclusion of some realistic nonlinear effects has also been shown to increase the Liapunov stability of simple multispecies models. Such effects include alternation of prey organisms and switching of predators (Murdoch and Oaten, 1975; Nunney, 1980; Townsend et al., 1986), self-competition of prey species (Koch, 1974), and competition between prey (Crow, 1977; Powell, 1980). Investigations aimed at elucidating Hutchinsons's paradox of planktonic algae (Greeney et al., 1973; Phillips, 1973; Peterson, 1975), concluded that the coexistence of these forms is possible in simplified models only when several nutrients are available.

Space, Age, and Social Structure

Classical investigations suggest that stability is achieved when a refuge exists for the prey species. Environmental heterogeneity and prey patchiness were shown to increase stability properties of both linear and nonlinear models (Maynard-Smith, 1974; Murdoch and Oaten, 1975; Nisbet and Gurney, 1978 and references therein; Nunney, 1980). Recently, interest in structural problems has increased in terms of thermodynamic considerations (for example, Segel and Levin, 1976; Dubois, 1979; Laplante, 1978).

Introduction of age structures into models of population dynamics (Sinko and Strefer, 1967; Smith and Wollkind, 1983) also changes considerably the predictions derived from L–V models. Stabilization effects of age-specific differences have been shown by Maynard-Smith and Slatkin (1973), whereas Smith and Wollkind (1983) observed destabilization as a result of the introduction of age structures into a model.

Socially structured populations were studied by Gurney and Nisbet (1979), who found that a dominance hierarchy is strongly beneficial to population stability.

Difference versus Differential Equations

Maynard-Smith (1978) and May (1973b) have expressed that discrete models are more realistic in many situations. They are, however, usually less stable than continuous models (May 1973b). Kindlmann and Rejmánek (1982) therefore attempted a quantitative comparison of both kinds of models. They obtained approximate solutions for the relations of stability to connectivity and the number of species, suggesting a common region where continuous models are always stable, but discrete ones are either stable or unstable. A methodology for assessing the response of predator–prey systems to external perturbations based on difference equations was presented by Silvert and Smith (1981).

Chaotic Behavior, Bifurcations, and Catastrophes

In hydrodynamics it has been shown that some simple deterministic systems of differential equations possess more complex stability behavior than either asymptotic or periodic attractor trajectories (Gleick, 1988). These remain bounded and their behavior deterministic; however, after a period of time they go into oscillations with no finite period. Moreover, starting the solution at any point arbitrarily close to the original conditions will sometimes result in a trajectory quite different from the given one (Ulanowicz, 1979).

The same chaotic behavior was found for certain simple, two-species models of density-dependent populations (May, 1975; May and Oster, 1976; Guckenheimer, 1977). Within specific ranges of parameter values, populations will not exhibit stable equilibrium points, but stable cycles or chaotic behavior instead. It seems difficult to distinguish between chaotic and periodic stochastically disturbed model populations even when powerful analytical methods, like spectral analysis, are used (Poole, 1977; Schaffer and Kot, 1986).

Chaotic behavior is more likely with several species (May and Oster, 1976) and with nonlinear models (Guckenheimer, 1977). Studies of host–parasite interactions (Beddington et al., 1975) and of competing species (Hassel and Comins, 1976) indicate that chaos will intercede at lower parameter values than in the single population cases.

Laboratory population behavior given in Nicholson (1954) suggested that chaos can also be found in reality. Later, more thorough investigations did not support this view. Thomas et al. (1980) studied 27 populations of *Drosophila* and found dynamic stability to occur. They claimed that chaos and limit cycles are mathematical artifacts rather than signs of reality. Similarly, Mueller and Ayala

(1981) have shown for 25 laboratory populations of *Drosophila melanogaster*, near their carrying capacity, that they either possess statistically significant equilibria or give results statistically indistinguishable from stability.

Another characteristic of simple-model behavior is bifurcation (May and Oster, 1976; May, 1977; Kempf, 1980). Levin (1978) shows that already a simple logistic model of the type

$$\dot{N} = rN(1 - N) \tag{4.1}$$

bifurcates. When the parameter r is varied, an infinite number of bifurcations appears.

From bifurcations there is only a step to catastrophes (Thom, 1975), with implications for ecology as summarized by Casti (1979), Kempf (1980), and Loehle (1989).

Trophic-level Models

Although, in the trophic-dynamic sense, prey and predators represent two trophic levels, I will designate as trophic-level models those that cover at least the most simplified trophic pyramid, having three or more levels. Particular emphasis will be on models that include trophic-level "zero": a substrate for autotrophic or heterotrophic organisms.

The first model belonging to this category is by Phillips (1973, 1978), who aimed at theoretical investigation of marine pelagic systems. His nutrient–algae–predator model appeared to be more stable than an otherwise comparable nutrient–prey system. Based on another controversial issue, Rosenzweig's (1979) "paradox of enrichment," the author has investigated exploitation in a three-trophic-level system (Rosenzweig, 1973). The model was extended further by allowing interspecific competition of predators (Wollkind, 1976). A similar investigation was done by Steele (1974), whose conclusions about the higher stability of a system with threshold effects were mentioned before.

Another problem was investigated by Austin and Cook (1974), who combined multispecies systems into trophic levels (up to six), but did not distinguish any regularities in stability concerning parallel and hierarchical species organizations. Nunney (1980) concluded that stability is increased by increased complexity as a result of distinguishing trophic levels. These conclusions were not supported by subsequent studies (Abrams and Allison, 1982; Smítalová, 1983). However, recognizing from the studies reviewed features like dependence of stability on parameter values, equation structure, and organizational characteristics modeled, not every kind of "trophic-level" distinction can be expected to result in increased stability. Positive or negative results in this respect are obtained only by chance or by the ability of an author to approach particular ecosystem structures. The only paper in which the problem has been studied with insight is that by Jeffries (1979); in his holistic ecosystem model the trophic hierarchy appears to be stabilizing.

One important feature was introduced into models of this type by Abrosov (1975) and Aponin and Bazykin (1977), who based their model on the principle of continuous cultivation. In contrast to most previous closed systems, they deal with an open system, with matter (and energy) flowing in and out. The inclusion of this characteristic was not evaluated from the viewpoint of stability, but was used for investigating management of an aquatic ecosystem. Systematic stability studies of the effect of including realistic features in the model by Sjöberg (1977) have suggested that the inclusion of feedback effects (recycling of nutrients) into a four-level ecosystem radically changed stability properties. According to the parameter values used, nonlinear (Michaelis–Menten) nutrient kinetics resulted in either an oscillatory or unstable system. A highly stabilizing effect was due to threshold feeding effects of zooplankton, as observed by Steele (1974). Levin et al. (1977), in a continuous system with a substrate, bacteria, bacteriophage, and bacteria preyed upon (infested by viruses), found that structural stability is possible only when the number of predators does not exceed the number of prey, or the latter the sum of resources and predators.

Inclusion of external driving (seasonality) into a two- and three-level model has the consequence of permanent coexistence of prey and predators for certain regions of parameter values

(Nisbet and Gurney, 1982; Britton, 1989). When evaluating prey–predator population studies, the asymptotic stability behavior of linear L–V equations with constant coefficients has often resulted in conclusions contradictory to natural observations. Moreover, it appears that stability results are highly sensitive to particular values of parameters and to the structure of the equations and mechanisms included. The simple inclusion of some nonlinearities does not make stability–complexity models closer to reality, as often claimed.

However, continuing investigation of modified two-species models suggests a number of more realistic mechanisms that can result in stable equilibria and coexistence of prey and predators. These mechanisms are threshold and nonlinear feeding, density dependence, time lags, and predator satiation. Similar observations were made for multispecies models, where, in addition, the importance of alternation of prey organisms, switching of predators, and self-competition among prey were recognized. Structures of a different kind, that is, space, age, and social structures, represent another kind of generally stabilizing effect.

This short, surely incomplete (see, for example, developments in Schoener, 1976, and Wangersky, 1978) and possibly biased overview suggests that there is little ground to suppose that L–V types of models are a useful basis for the development of theories of complex ecosystems.

CONFORMING LIAPUNOV STABILITY WITH ECOSYSTEM REALITY

The underlying assumptions of Liapunov stability important for evaluating its ecological usefulness are as follows:

1. For a linear system, Liapunov asymptotic stability is identical with global stability, but for nonlinear systems the necessary linearization results in only local stability being analyzable.
2. The coefficients of the system remain constant. Continuous perturbations due to coefficient changes are not analytically tractable.
3. By means of Liapunov stability analysis, only asymptotic values of the system are considered, and the only perturbations analyzable are those of the initial conditions, $x(t_0)$.
4. Only the behavior of the system is studied, not its stability to perturbations of structure.

Let us now examine the assumptions underlying stability analyses of simple ecological models and compare them with reality.

Assumption 1

As shown in the previous section, most students of theoretical models hold that nonlinearity is a realistic assumption about populations. The same seems to be generally recognized in both empirical (for example, Shugart, 1984, for terrestrial and Straškraba, 1980, for aquatic systems), as well as simulation studies (Jørgensen, 1983; Straškraba and Gnauck, 1985; Hanski, 1990) of ecosystems.

The only cautionary points against assuming ecosystem nonlinearity have been raised by Patten (1975, 1983). This author claimed that even if ecosystems were linear in nature, this property would never be observed because of the impossibility of performing a superposition experiment or observation on a complex system. All the relevant inputs to a given observable output could never be manipulated simultaneously. Therefore, even if a real ecosystem were truly linear, nonlinear behavior, what he (1983) termed "pseudononlinearity," would always be observed. Patten's points have been widely rejected, though not disproved, and because he raised them he has been misconstrued as an advocate of linear models of reality to the total exclusion of nonlinear formulations. As one of his supporting arguments, Patten stressed the popularity of linear relationships among ecologists. This cannot be considered anything more than a desire to grasp the easiest way of quantitative analysis being used for both linear and nonlinear (linearized by transformation) relationships. Moreover, the variance of ecological data is usually too high to allow recognition of any definite nonlinear form by statistical means unguided by theory. In the examples used by Patten (1975) to demonstrate the linearized ecosystem response, I can recognize only the smoothing out of input variability and nothing more.

The linearization hypothesis has been subjected to experimental testing by Dwyer and Perez (1983) on a marine microcosm. Time series analysis of the signals obtained resulted in a rejection of the hypothesis that the response of the microcosm is linear. The same can be deduced from an earlier experimental terrestrial microcosm study by Voris et al. (1980). Toxic doses of cadmium were added to a meadow microcosm, and nonlinearities of the global response were analyzed. The response was measured by harmonic analysis of the output signal. Functional complexity evaluated as the degree of response nonlinearity was positively correlated with a measure of stability. Also, the automatic recording of solar radiation inputs and changes in carbon dioxide as the pelagic system outputs in Slapy Reservoir, Czechoslovakia (Nesměrák and Straškraba, 1985), although analyzed by linearized spectral analytical methods, are evidently nonlinear (Straškraba, 1986). In none of these empirical observations, however, has Patten's (1983) pseudononlinearity property been ruled out.

The above studies were particularly concerned with the question of how far the short-term reaction of a specific ecosystem can be considered linear. Similar issues concern the nonlinearities of ecosystem component reactions and holistic ecosystem responses as recognized by comparative empirical studies of a number of ecosystems of a given type under conditions corresponding to a broad range of input conditions.

As to component reactions, in L–V equations the interaction coefficients a_{ij} as well as the "growth rate" coefficients b_{ij} are considered linear. Figure 4.1A and B shows a few examples of some observed forms of a_{ij} for algae interaction with their "prey" (nutrients) as well with their predator (filter feeders). The shape of what might be equated with b_{ij} is also given. In addition, the growth and interaction parameters depend on external variables like light, I, temperature, T, and others. Some typical shapes of such relationships are given in Figure 4.1C and D. It is difficult to imagine any reasonable linearization of such responses, at least for the whole operational range of the respective input variables. However, for partial coverage of the total range, some linearization will do well, as shown in Figure 4.1D. It is also true that superposition of curves for different species and the addition of stochastic noise, as well as neglect of the effect of additional variables, will spoil the figure. Perhaps nothing more than a simplistic relationship can be distinguishable statistically. Such relationships may be fairly useful for situations where nothing more is known, but their explanatory force is low and they cannot prove or disprove the linearity of component reactions.

An average shape of the reaction of an ecosystem of a particular type (say a forest or a lake) can be recognized from comparative studies of ecosystems located in different places with highly variable conditions. Figure 4.2 gives some observed shapes based on studies in the International Biological Program. The aquatic examples are based mostly on different lakes located along the axis, but several individual lakes have, in recent years, moved considerably (and in both directions) along the temperature–precipitation axis due to human impact. The reactions of individual lakes run in parallel with the average set. Only a few lakes have changed their average depth $\bar{z}$, but because the relationship is mediated by changes of mixed depth, a more easily manipulable variable, examples of considerable shift of one and the same lake or reservoir along the axis have been observed. The geographically based inputs for terrestrial vegetation, temperature, and precipitation will, in fact, cover more than one type of vegetation and span from desert to northern tundra. However, history teaches us that localities that once bore forests are deserts today.

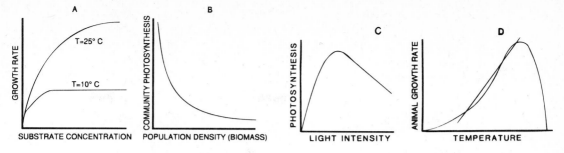

Figure 4.1 Examples of nonlinear interactions in ecosystems (corresponding to a_{ij} and b_{ij}) and relations of ecosystem processes to external driving variables. A. Interaction of two state variables for two values of an external driving variable. The figure represents growth rate of a plant (alga) related to substrate concentration, and each curve is described by a Michaelis–Menten type of equation (rectangular hyperbola): $y = y_{max} \cdot x /(a + x)$. In this case, $y_{max} = f(T)$, T being temperature. However, an identical approximation is often used for animal–plant interactions, for example, the feeding rate of an animal on plants. B. Example of nonlinear density dependence of growth rate. The curve was obtained in experiments with individual algal species, but the reaction of phytoplankton assemblages in lake investigations resulted in similar approximations (modified according to Straškraba and Gnauck, 1985). Moreover, similar relations were observed for some animal populations. C. Photosynthesis related to light intensity over the naturally occurring range of radiation. The degree of inhibition indicated by the figure represents an extreme example. In dependence on species and conditions (for example, spectral composition of light, intensity of turbulent mixing), inhibition may be much less evident or disappear. D. Animal growth rate related to temperature over the whole range of tolerance is typically nonmonotonic. As indicated, for a restricted range of temperature, linearization may provide a sufficient approximation.

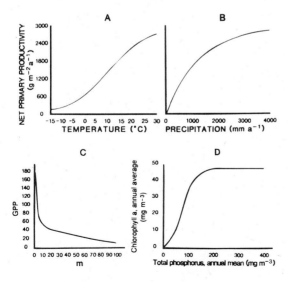

Figure 4.2 Nonlinear global ecosystem reactions to linear inputs as demonstrated by results of comparative studies during the International Biological Program. A, B. Terrestrial production (from Lieth, 1975). C, D. Aquatic production (from Straškraba, 1980a). A. Relationship of aboveground production of terrestrial vegetation to average annual temperature of the corresponding locality. B. The same related to annual precipitation. C. Average epilimnial gross primary production (GPP, kg m^{-3} yr^{-1}) in lakes of different depths. D. Dependence of average annual algal biomass (measured as chlorophyll a concentration) on the concentration of critical nutrient (phosphorus, mg m^{-3}) in lakes.

Assumption 2

To assume that populations are constant-parameter systems is to neglect any differences due to age, size, interaction, adaptation, genetic drift, and so on. As shown above, only age structure differentiation was introduced into the L–V models, mostly in the unsatisfactory form of discrete age categories. Parameters became age specific, but were otherwise constant. Parameter changes are typical phenomena in biological objects in general, and this is true for the ecosystem in particular, as one of the most complex biological objects. Examples are treated under proposition 6.

Assumption 3

For a system to behave asymptotically, a closed system with no inputs must be assumed. As shown by Patten (1979), the only equilibrium state of an (open) ecological system is the zero, dead state. For a constant steady state, the input has to be constant also, which is not the case for any real ecosystem. For the most part, solar radiation and temperature inputs are cyclic to some degree. Tropical ecosystems are mostly considered to live under more constant conditions, at least as far as solar radiation and temperature are concerned. However, recent studies have shown deterministic trends in environmental conditions in the tropics, for example, seasonal changes in precipitation. Even the deepest oceanic environment was found to be subject to significant changes in environmental variables. The stabilization of simple systems by periodicity of external disturbances and consequent periodicity of mortality resulting in alternation of density-dependent and density-independent growth was shown by Koch (1974b).

The possibility of evaluating only perturbations of the initial system conditions is a strong limitation. Perturbations of parameters and of inputs are of at least the same, if not much higher, importance for ecosystem studies (Wolawer, 1980; Recknagel, 1984).

Assumption 4

Structure is of the same importance as behavior, and a mathematical direction that has received attention only recently (Siljak, 1978) is to consider changing structure as a system perturbation. Structural problems are connected, for example, with the immigration of new species, management problems associated with cultivation of particular species, and the reduction of sensitive species due to indirect effects of human activity (for example, pollution problems). In addition to sensitivity of behavior to structure, the problem of sensitivity of structure to inputs is interesting from an ecological point of view, but unsolved by present methods. If some input (forcing) variables are changed, some species present in the system might disappear and, eventually, other species appear.

In spite of a few extensions to ecosystem problems, the dominant aspect under which the L–V equations were treated up to now was species interactions. Recent developments have demonstrated many features for introducing reality into the L–V equations. Several mathematically oriented authors have admitted that some conclusions obtained are relevant more to mathematics than ecology (for example, Ginzburg, 1980). For ecological theory, mere representation of realistic details and particularities does not seem to be an adequate goal.

The ecosystem point of view introduces new qualities into the problem. Mutual relations among plants, animals, and their abiotic environment, as they characterize a lake, woodland area, or pasture, have to be represented. We demonstrated that the assumptions on which simple models and their stability analysis were based do not apply. Therefore, L–V equations fail to provide an adequate theoretical basis for ecosystem studies.

The questions then arise as to what assumptions will be adequate for ecosystems and what features should be covered by theoretical ecosystem models.

THE NEW CONCEPT

The L–V equations represent the first ingenious mathematical simplification of the problem of organism interactions (Levin, 1981). They have played a positive role in the development of ecological theory. However, what both mathematicians and ecologists have attempted until now were mostly straightforward extensions of the theory to cope with more details and particulars. This was, perhaps, adequate for the original problem, although attempts to transfer the approach to much more complex interactions at the ecosystem level failed. Simplified theoretical representations of ecosystems have to be developed on a different basis, one that is adequate for the different setting. On purely intuitive mathematical grounds, this is perhaps not so easy, at least as seen from the near absence of adequate treatments. In this situation I feel justified, as a primarily empirical systems ecologist using bits of mathematics, to specify the characteristics that I think are most deeply rooted in the ecosystem concept.

In accord with Patten (1979), therefore, I present what I believe are basic features for a theory of complex ecosystems in the form of "propositions." This is close to Patten's "necessary conditions for realism in ecological models." Jeffries (1979) designed, via stability studies, a holistic ecosystem model that he believes, and not without reason in my opinion, to be consistent with the design and operation of a real ecosystem. A similar process was followed in Innis and Clark's (1978) investigation of the principles of ecosystem organization. Related lines of thought were developed in the papers of Ulanowicz (1979a,b; 1980).

The examples for substantiating the formulation of the propositions will be treated very loosely. It is difficult to give detailed references to obvious generalities that appear from many studies. Specific examples would degrade rather than stress the degree of generality that I feel the propositions possess. Some of the features treated here have already been represented in mathematical form in previous papers originating from empirical experiences with aquatic ecosystems and with complex ecosystem simulation models (Straškraba 1980a; Bakule and Straškraba, 1982, 1987). Different aspects of mathematical formalism have also appeared in papers mentioned in the present discussion. However, this is the first attempt at a systematic treatment. To facilitate communication with the more mathematically oriented scientific community, I will attempt to describe the basic features in a generalized mathematical form. By doing so, perhaps others will be stimulated to extend the mathematics further.

Proposition 1

Ecosystems are systems open with respect to energy, matter and information. Therefore, external driving and energy dissipation always exist, resulting in system states far from equilibrium.

From the point of view of thermodynamics, an ecosystem is, like any other complex biological system, far from equilibrium. The energy-dissipative character of such systems is beyond doubt (Jørgensen, 1983). The conclusion of Jeffries's (1979) stability investigation of holistic ecosystem models suggests, in fact, that the representation of energy dissipation is an important characteristic of ecosystem models.

Proposition 2

The biotic components of an ecosystem are in active interrelationship with some abiotic components.

Abiotic nature does not only provide external inputs to an ecosystem, but also an active part of it. Biotic–abiotic interaction is best evidenced during the colonization of bare substrates. During early evolution, for example, interaction of oxygen organisms is supposed to have played a major role. To satisfy this feature, a simple model can be considered an ecosystem model only if it includes at least one dynamic abiotic component.

Proposition 3

An ecosystem is always bounded. It's states cannot become negative and cannot exceed limits imposed by the laws of thermodynamics. It is, therefore, always Lagrange stable. Stability is connected with a number of phenomena at all levels of biological organization: nonlinearities and feedbacks at the physiological level; spatial, age, and social structures of populations; the "trophic" levels of complex multispecific food webs; hierarchies of trophic levels, and so on. Instabilities concern populations as ecosystem components.

I know of no example of a real ecosystem that, by endogenous processes, has simply faded away rather than changing into another state. So much for bifurcations and catastrophes at the ecosystem level. It is true that flourishing green was replaced by desert in the Sahara due to deforestation, but this desert is also a fully functioning ecosystem, an expression of a changed environment, and far less productive and favorable for humans. When populations of different species have disappeared and been replaced by others, ecosystem state has changed or, depending on definitions, another ecosystem has evolved from the former. It is the population as a subunit of the ecosystem that may become unstable, but many stabilizing mechanisms also act on the population and lower organizational levels.

Liapunov stability is not an adequate measure because ecosystems are far from equilibrium (Proposition 1), and the perturbational impulse to within δ remains near to equilibrium.

Proposition 4

Ecosystem components (even if linear—Patten's "pseudo nonlinearity") behave nonlinearly, but ecosystem dynamics are linear over a restricted range of conditions. Mechanisms resulting from the evolutionary self-design of ecosystems lead to the smoothing out of input variability. However, ecosystem behavior at large is empirically nonlinear due to the limiting operation of physical laws setting upper limits on existence. Nonlinearity assures ecosystem existence within a broad operating state space.

By the operational space of the ecosystem, we imply some range of conditions in which the ecosystem is able to exist. If we consider a given lake an ecosystem, its condition may range from shallow to deep, from almost no throughflow to rapid washout of organisms, from being surrounded by a woodland to just sand dunes, from low nutrient input to hypereutrophication, and so on.

The stabilizing mechanisms mentioned in Proposition 3 are also responsible for the smoothing out of input variability. Patten (1979) has enumerated a number of such mechanisms. According to Prigogine's thermodynamics, nonlinearity and dissipation are the physical forces of speciation and organization (Nicolis and Prigogine 1977; Feistel and Ebeling, 1982).

In mathematical models of different systems, linearity is usually connected with models having fixed or time-varying parameters and fixed state structures. This is not the case for ecosystems where, for several reasons, model parameters almost never are fixed, and variable state structure is the rule. The most obvious example of the latter is that component populations do not remain the same; different species appear during changed conditions—the essence of a variable structure, and usually nonlinear, system.

Proposition 5

An ecosystem is an adapted system. Adaptation results from long-term stochastic trial-and-error interactions of elements, leading to apparent optimality (or suboptimality) of the interacting elements' behavior.

In an early, ingenious study, Korostyshevsky et al. (1974) showed that the problem of instability of limited populations is solved when providing for genetic evolution. In stochastic population models, infinite coexistence of prey and predators is possible when mutation ability, heredity, natural selection, and capacity of the organism to extend the feasible range of environmental conditions are simultaneously included.

Coevolution, the simultaneous, balanced evolution of both prey and predators, plants and animals, honey and bees, is documented in a number of studies (Nitecki, 1983; Futuyma and Slatkin, 1983; Brown and Vincent, 1987). Similarly, the organism–environment couple is a result of such coevolution (Patten, 1982a, b). The characteristics of ecosystem elements cannot be considered arbitrary, but rather are finely tuned due to long-term interactions. On the gene level, the studies popularized by Dawkins (1976) suggest how balanced design of the whole chromosome results from purely stochastic encounters of "selfish" elements. Similarly, ecosystem optimality or adequate design is a macroscopic feature that describes the results of long-term history of stochastic encounters at the microscopic level. The winners of the evolutionary game are the components of the ecosystem, and the ecosystem is the coevolved result of the evolutionary game.

A popular question in some ecological circles (for example, discussions in the journal *Wiadomisci Ekologiczne*) is whether the behavior of the ecosystem is fully explained by the behavior of its elements, or are some higher "ecosystem" principles in operation (Wilson and Sober, 1989)? This, like the chicken versus egg first question, is irrelevant. The elements developed simultaneously in their mutual interactions and those with their environment. The behavior of the ecosystem as a whole is fully determined by its elements, but the latter are also fully determined by the behavior-constraining organization of the ecosystem. Neither egg nor chicken was first, and the same holds for the population and the ecosystem. Both come into simultaneous existence through mutual implication, and both proceed through gradual coevolutionary development. It is not the same question if we ask whether ecosystem behavior is due to the sum of reactions of species as we measure them in the laboratory in the absence of other organisms, taken out of context. The answer to this question is no. We have to consider the interactions among species. The successful series of investigations of animal foraging suggests that optimality theory (in the broad sense) provides us with clues as to which of many possible situations will happen.

From Propositions 3 and 5, model instability signifies modeling inadequacy, and therefore models cannot be used for postulates concerning real situations. Model structure and parameters cannot be arbitrary if real behavior of ecosystems is of concern. Only evolutionarily feasible structures and parameters are appropriate for ecosystem modeling. To determine these, macroscopic description by means of optimality formulations or game theory is relevant. We assume that optimal constructs will also possess maximum stability.

Suboptimality means that no singular points are reached, but only some broader region of satisfaction. There are two basic reasons that organisms will not be selected for some fixed, absolute optima during evolution:

1. Conditions of ecosystem existence continuously change over a relatively wide range, for example, due to changes in weather conditions between years or to population increase or decrease.
2. Survival of the organism is guaranteed only if many aspects of life, including some mutually opposing ones, are in some "optimal" state simultaneously.

Moreover, natural selection acts on a relative, not absolute, scale. Present species disappear only when displaced by a stronger competitor or when they become maladapted to a changed environment. It has been demonstrated by geneticists (Lande, 1982) that, because several genes are decisive for one character and portions of genetic material exchanged during reproduction bear several genes for unrelated characters, selection is not a straightforward process.

Proposition 6

An ecosystem is an adaptive, self-organizing system. Its features are not constant, but vary with disturbances and in different ranges of its operational space. Variable features concern both the behavior of individual elements and total structure.

Due to inconstancy of conditions, the parameters characterizing component organism reactions change. In some instances these changes have been shown to follow a common "ecosystem" reaction; in others we are aware only of changes at the individual species' level.

Empirical relations to organism size of such diverse ecological parameters as growth rates, respiration rates, home ranges, prey sizes, and many other characteristics have been recognized by numerous authors (see summaries by Peters, 1983, and Calder, 1988). The size relationships are valid both for individual species and for a span of sizes from bacteria to whales. Size-related structural trends have been recognized in the sea (for example, Sheldon et al., 1972, for plankton and Schwinghamer, 1981, for benthos). Similar concise trends related to major changes of environmental variables emerge from freshwater plankton studies (Watson and Kalff, 1981; Sprules et al., 1983).

The plasticity of ecological interactions is best demonstrated by the successful theoretical predictions obtained by optimum foraging theory for multiple prey–predator interactions. Interaction intensity varies highly according to absolute and relative quantities of possible prey species and their individual size and accessibility (Krebs and Davies, 1984; Werner and Hall, 1974). Competition between different predators modifies prey selection (Milinaki, 1982). Predators switch between food types due to image formation and learning (Krebs and Davies, 1984).

Short-term adaptations to changes in environmental variables are a well-established phenomenon. Phytoplankton communities adapt on a daily basis to changing light conditions by adjusting their chlorophyll *a* content (Platt, 1981). Temperature acclimation has been demonstrated in numerous studies of both plant and animal species (for example, Hochachka and Somero, 1973). For phytoplankton populations of a lake, it has been shown (Straškraba, 1976) that the shape of the photosynthesis–temperature dependency varies during a year in a systematic manner, related to seasonal ambient temperature variations.

Fairly rapid genetic changes due to organism interactions have been demonstrated in numerous studies (for example, King, 1980). One example is particularly well documented by using experimental arrangements compared to natural observations to demonstrate repeatability and predictability of short-term natural selection. Endler (1980) studied color patterns in different background substrates and under different predation regimes in the guppie (*Poecilia reticulata*) in its natural environment in Trinidad. Within a span of 10 generations, color patterns adapted to balancing predator avoidance on one-hand (leading to cryptic coloration) and sexual selection on the other (leading to conspicuousness). The direct impact of these adaptive changes on predator–prey relationships was not evaluated in this evolutionary study, but the inconstancy of parameters and adaptive events may be inferred.

When environmental conditions are altered by humans, the ecosystem is placed in another region within its operational space. Correspondingly, interactions within the ecosystem are modified. As an example, take the intensive use of insecticides for mosquito control. Strains of *Anopheles* resistant to the various chemicals have appeared and spread rapidly. During consequent applications, mosquitos were less affected, but cohabitants were reduced to a high degree, including some natural control agents of mosquitos. As a result of the unbalanced interactions, the pests became more abundant in treated than in untreated localities.

During a stronger alteration of conditions, the adaptation capabilities of some organisms present in the ecosystem are exhausted and such populations disappear. Other more suitable species take their place, and the structure of the ecosystem is changed. Examples include alterations of size and species composition of oceanic plankton with geographical position (Semina, 1972) and major changes in composition of freshwater phytoplankton along a trophic gradient (Watson and Kalff, 1981). Illustrated is the self-organizing, restructuring capability of ecosystems.

MATHEMATICAL FORMULATIONS OF THE NEW CONCEPT

The L–V models have the fairly simple configuration of a closed system with constant parameters and structure (Fig. 4.3), as described by

$$\dot{x}_1 = f_1(x_1, x_2, p), \qquad (4.2)$$
$$\dot{x}_2 = f_2(x_1, x_2, p),$$

with initial conditions $x(t = 0) = x_0$, time derivatives, $\dot{x}$, x_1 = prey, x_2 = predators, and p = constants representing interaction parameters. In matrix notation, providing for multiple interacting species,

$$\dot{x} = f(x, p) \qquad (4.3)$$

with initial conditions x_0, x = a vector of states, and p = a vector of parameters. In the following, the mentioning of x_0 will be omitted.

The open system characterized by Proposition 1 (Fig. 4.4), with external driving $z(t)$ and dissipation *diss*, is given by

$$\dot{x} = f(x, p, z(t), diss) \,. \qquad (4.4)$$

Here the external driving or input $z(t)$ may be a scalar variable or a vector. The time dependence of z is rather characteristic for natural ecosystems, particularly in respect to energy inputs (solar radiation, and also temperature, precipitation and other related meteorological inputs). The dissipative characteristic of the system is expressed by the parameter *diss*, which in fact is also a vector if we consider specific values for each state variable in x. The parameter *diss* can also be expressed as a fraction of input; this fraction will not be constant, but will depend on ecosystem function and structure.

Dynamic interrelations with the abiotic environment according to Proposition 2 are expressed in Fig. 4.5. Some component of the abiotic environment becomes a state variable of our system, x_a:

$$\dot{x} = f(x_a, x, p, z(t), diss) \,. \qquad (4.5)$$

By x_a, we typically represent the critical nutrient limiting ecosystem development, which might be nitrogen for land and sea, but phosphorus more typically in freshwater. According to the situation and aim of the study, other nutrients or interacting critical elements of the environment may become important. The reason for inclusion of x_a into the system is that critical concentrations are determined as much by input as by internal processes in the system. We cannot estimate system performance by just the amount of input.

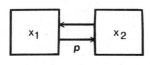

$\dot{x} = f(x, p)$
$x(t_0) = x_0$

x = state vector
p = constant rate parameters
x_0 = vector of initial conditions

Figure 4.3 Schematic representation of a Lotka–Volterra predator–prey model.

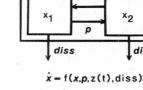

$\dot{x} = f(x, p, z(t), diss)$

$z(t)$ = external driving
$diss$ = dissipation of the system

Figure 4.4 An externally driven dissipative system viewed according to Proposition 1.

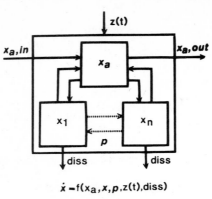
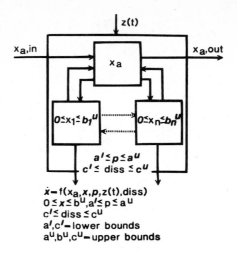

Figure 4.5 Dynamic interactions of the ecosystem with its abiotic environment, according to Proposition 2. The number of species is extended to n.

Figure 4.6 According to Proposition 3, the ecosystem is viewed as constrained by physical principles recognized in organism physiology and structure.

In Fig. 4.6, the bounded system is represented by

$$\dot{x} = f(x_a, x, p, z(t), diss) , \qquad (4.6)$$

where x cannot become negative and is bounded from above:

$$0 \leq x \leq b^u . \qquad (4.6a)$$

The expression of constraints for state variables in the form of (4.6a) is misleading. In fact, it is due to the curve form given by the equations and to the boundedness of parameters. If, in a computation, the value of x surpasses the limits given by (4.6a) the setting of the system is wrong, and only artifacts can be produced by using rules like

$$\text{if } x \leq 0 \quad \text{then } x = 0$$

or

$$\text{if } x \geq b^u \quad \text{then } x = b^u .$$

For parameters the inequality

$$a^l \leq p \leq a^u \qquad (4.6b)$$

holds true, where a^l may be both positive and negative for some processes. Also, there are physical limits to the dissipation parameters:

$$c^l \leq diss \leq c^u . \qquad (4.6c)$$

The boundedness expressed by inequalities (4.6a) through (4.6c) will be valid for subsequent equations, but will be omitted for simplicity.

Nonlinearity of ecosystem structure is stressed in Proposition 4 (Fig. 4.7). An appropriate state equation is

$$\dot{x} = f(x_a, x^k, p, z(t), diss), \quad (4.7)$$

with x^k, $k > 1$, representing the nonlinear character. Nonlinearity effects may not be apparent for specific values of $x_{a,in}$ and $z(t)$, which now characterize the system's position in the operational space. However, when the system moves within broad limits of $x_{a,in}$ or $z(t)$, the nonlinear effects become decisive.

The representation of system adaptability according to Proposition 5, as based on control theory, is given in Fig. 4.8, where system parameters are allowed to vary according to some control variable, u:

$$\dot{x} = f(x_a, x^k, p(u), u, z(t), diss). \quad (4.8)$$

Therefore, parameters are no longer fixed constants, but are variables, $p(u)$. Consider the case of temperature or light adaptation of the system. Here, parameters can be considered to vary as a function of the corresponding external time-dependent variable. Technical description will simply be in the form of parameter time dependence:

$$p(u) = p(z(t)) = p(t).$$

However, this shortcut obscures the origin of parameter variability, particularly if the effect is not straightforward, but mediated by the system performance. This will be, for example, the typical situation in the case of light adaptation, which is determined by incident light but also by other

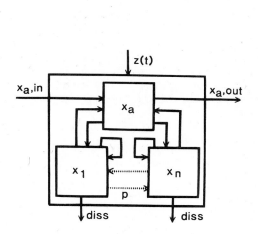

$\dot{x} = f(x_a, x_k, p, z(t), diss)$
x_k = vector of nonlinear states

Figure 4.7 The ecosystem viewed according to Proposition 4. In addition to boundedness, nonlinearities of state variables are produced, particularly by feedback mechanisms at the state-variable level. Additional feedback mechanisms are covered by the network structure.

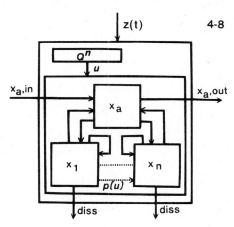

$\dot{x} = f(x_a, x_k, p(u), 3\ u, 3\ z(t), 3\ diss)$
u = control variable
Q^n = inherent control
$p(u)$ = variable (adaptive) parameters

Figure 4.8 Adaptation of the organisms is introduced according to Proposition 5. The parameters of the system are allowed to vary according to some rules.

variables, including the state x (for example, self-shading effects). An explicit formulation of $p(t)$ will be a complex task, evidently without any advantage. Therefore, we use in all instances the notation of variable parameters by means of $p(u)$, stressing the origin of variability.

In the preceding paragraph we have assumed that the function $p(u)$ is known. This is valid only to a very limited extent. Such functions have been empirically derived only for some particular situations. A general rule is needed, that is valid also for unknown situations. Adaptation is considered an evolutionary phenomenon, a result of some sort of selection between alternatives. Such a process is most generally described mathematically by means of a function Q, called for historical reasons the "goal function," a very misleading name. The use of Q does not distinguish between physical and living systems, with a rationale for existence, like economic systems, or without it. It does not imply any goal in the sense of aims or purposes; physical systems may seek optima without being "agents." Thus, the behavior of such systems is commonly best approximated by using a goal function. Biologists should not be deterred by the name. Mathematically, each goal function Q is associated with restrictions (constraints) for its application, which are inevitable for the formulation of the problem. This is another common source of misunderstanding: there is nothing like absolute optimality, only optimal selection among a given set of possibilities, characterized by the constraints. The specification of constraints may be broad and inexact or wrong, but in any case they belong to the problem. Another difficulty puzzling biologists is the existence of a common goal for a system as diversified as an ecosystem. Recent advances in cybernetics, particularly the hierarchical systems methodology (Salthe, 1985, 1989; O'Neil et al., 1986), allow a formulation of hierarchical or decentralized structures in which no common goal is formulated a priori, but results from specific goals of individual subsystems. Therefore, we consider the control theoretical and cybernetic formulation an adequate approximation of processes in an ecosystem. For interactions, a more adequate basis will be that using the theory of games, where time to reach a goal is not considered infinite, but provision is made for the steps performed by the individual game participants. However, development of the mathematical apparatus of game theory unfortunately does not allow one to reach applicable solutions in instances interesting from the viewpoint of an ecosystem, except for extremely simplified situations.

The last, proposition 6, considers the case of an ecosystem where not only function as characterized by the parameters p varies, but also structure, the representation of species in the ecosystem, and the existence or nonexistence of particular interrelations between these. Structure is not more constant (Fig. 4.9), but is similar to the parameters assumed to depend on some control variable, u:

$$\dot{x} = f(x_a, x^k(u), u, z(t), diss) . \qquad (4.9)$$

The problem of determining the function Q is not considered here in detail, only the successful empirical predictions based on the approach mentioned above.

DISCUSSION

What I have attempted to stress in this chapter is that the dominant directions for theoretical mathematical treatment of "ecosystems" are based on assumptions not in accord with basic ecosystem characteristics. Existing ecological models are basically aimed at representing two or more interacting species, not the complex interactions among many organisms and their complex environments. "Population–community" modeling on a small scale dominates "ecosystem" modeling on a large scale (Oksanen, 1988).

It is true that a mathematical model is just a crude approximation of reality. However, it is improbable that models not conforming to basic ecosystem features can be useful approximations.

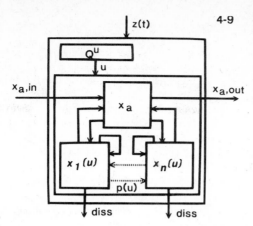

$\dot{x} = f(x_a, x_k(u), 4\ p(u), 4\ u, 4\ z(t), 4\ diss)$
$x_k(u) =$ variable structure

Figure 4.9 In addition to parameters, the structure of the ecosystem can also vary considerably (Proposition 6).

We are dealing here with theoretical models intended to represent holistic principles on which generalized theoretical, mathematical treatment of ecosystems can be based. As Levin (1981) has pointed out, theoretical models are not to be confused with realistic representation of specific hypotheses concerning observed behavior of particular systems. For this purpose, much more detailed and "realistic" representations like those in simulation studies are more adequate.

Stenseth (1984) points out the need for ecosystem theory and stresses that we may have to search for new mathematical methods or approaches to develop such theory. Evolutionary ecology, which has developed on the basis of modern biological thought, seems to be a good point of departure. However, it does not seem reasonable to assume that models of evolution concentrating on genetic mechanisms will be simply extendable to cover organisms and relationships in an ecosystem. Evolutionary models are based on evolutionary theory, and ecosystem models will likewise have to be based in ecosystem theory.

A particular characteristic of the theoretical approach suggested here is that it is based on phenomenological, holistic features of ecosystem function, not on specific processes within the ecosystem. The notion of optimality introduced into models of this kind might seem "Panglossian," but this is far from the case once we recognize that optimality is not just a leading principle, but also a useful methodology with a highly developed mathematical apparatus (cybernetics, control theory). Moreover, this is methodology aimed at dealing with complex systems.

In evolutionary biology, the notion of optimality was highly misunderstood for some time. Proponents of "optimality" used the term rather loosely and not as a means of rigorous mathematical approximation, and critics often did not grasp its mathematical meaning either (see Stearns, 1977; Gould and Lewontin, 1979). The assertion that "architectonic restrictions rather than optimality play an important role" indicates complete misunderstanding from the cybernetic point of view. Both goals and restrictions are necessary parts of optimization formulations. Stressing superiority of one model over another when both give the same qualitative predictions on the basis of "escapement from the metaphysical process of teleological speculation" (Ollason, 1980) does not seem to be a scientific but rather a religious or political argument. Scientifically, two models resulting in identical predictions are equivalent. Nevertheless, one may be superior for some purposes by specifying the more detailed mechanisms involved, whereas the other might be more favorable for its generality in other applications. This has to be decided on a scientific basis, either from the predictability viewpoint (Peters, 1983, 1991; Quinn and Dunham, 1983) or on the basis of the more general

Einsteinian definition of science as a logically uniform system of thought (Levin, 1981; Stenseth, 1984).

Due to widespread use and the evident predictive power of "optimum foraging models" in explaining certain rules of animal behavior, recent books with ecological orientation are beginning to clarify the issue (Krebs and Davies, 1984; McFarland, 1985; Cohen, 1987; Lundberg and Aström, 1990). Mathematically, optimality describes, in the broadest sense, the selection among different alternatives according to certain rules from some viewpoint. The same notion is used for description of systems reaching (or selecting) some defined state. Optimization methodology proved useful for describing a number of physical processes and chemical reactions (as in thermodynamics; see Lamprecht and Zotin, 1978) or if we have to deal with an organism, or if the issue is human economy or technical systems. In all instances we define a model (a mathematical *approximation of reality*) by describing in general terms what we observe. The following elements must be defined:

1. A function Q mathematically formalizing the selection criterion (what will be either "maximized or minimized")
2. Equations describing system behavior quantitatively and its reactions to values of parameters, u, among which selection is to be made
3. Restrictions characterizing the movement of the system and the space in which the search for parameter values is to be made.

Due to historical development of optimization methodology for economic and operations research applications, the technical terms used are misleading, particularly for biologists interested in process mechanisms and inexperienced with mathematics. The function Q, called a "goal function," is particularly perplexing because it suggests teleologically oriented objectives that are normally absent from mechanistic rationales. However, optimality notions as used in particle physics or chemical thermodynamics do not imply any kind of rational decision process on the part of particles or chemical substances. Nevertheless, self-organization of both particles and substances is taking place (Nicolis and Prigogine, 1977; Feistel and Ebeling, 1989). For organisms, it has been stressed that to use optimality methods does not imply that complex equations are being solved, for example, when flying or foraging (Krebs and Davies, 1984). In all such instances, only approximations to a certain degree of accuracy (Smith, 1987) of optional solutions of equations—a human representational system—are realized.

From the mathematical point of view, or more specifically the control-theoretic or cybernetic one, different optimality formulations are applicable to different problems, but also different approximations and methods may be applied to the same problem. Because all descriptions represent only an approximation of reality, it is a matter of comparison which one gives the best answer to the question being investigated. Papers by Maynard-Smith (1978), Hines (1987), and others have shown that game-theoretic formulations approximate what is going on during interactions between different organisms. This seems to be valid also for interactions within an ecosystem. The principal drawback of such methods, however, lies in limited mathematical tractability and restriction to static, few-species (player) interactions. Methods for dynamic multiplayer games are in development. Therefore, more general formulations using Bellman's, Pontryagin's, or other optimality principles, although less precise in expressing the actual state of affairs, are the only methods for more complex systems. Similarly, cybernetic methods of satisfaction or suboptimality might be more adequate to predict what is happening in an ecosystem, but such methods are only now in early stages of development.

When we see recently that optimality principles have found their place in evolutionary biology of populations, their role for higher levels of ecological organization is less appreciated. Stenseth (1984) argues that maximization of individual fitness over evolutionary time is based on genetic identity between different organs of an organism, but no such genetic identity to evolve "cooperation" by natural selection exists in a community or an ecosystem. Nevertheless, coevolution of different organisms is becoming generally recognized in evolutionary biology (see, for example, Futuyma and Slatkin, 1983; Nitecki, 1983; Stenseth, 1986; Brown and Vincent, 1987), suggesting that the information exchange between organisms is a force strong enough to act in natural selection. Once we recognize this force, there is no need to restrict it only to two-sided information exchange between organisms, for example, between prey and predators. On a more complicated basis, three levels will also be interacting, and, in fact, this has already been appreciated in behavioral ecology (for example, Milinaki's 1982 observations on intraspecific selection). Although I do not know of any direct observations demonstrating that four, five, or more species will be coevolving as do two, there seems to be no reason why it should not be so. Definitely, the rules to govern these kinds of interactions will be more general, compared to more specific forms of coevolution recognized among pairs of species. Therefore, methods to describe such ecosystem features will not necessarily be identical with those used in present models of coevolution.

Again, we have no a priori *scientific* grounds to accept or reject "Gaia"—the coevolved planetary biosphere—unless it appears that reality is or is not successfully predicted through such a paradigm.

The problem of approximating more complex situations by simpler methods, discussed above in connection with different notions of optimality, concerns also the issue of linearization. As stressed by Patten (1983), linear methods are often the only ones available, and many nonlinear problems have to be solved by linearization. Nonlinear methods are rapidly developing, but are far from reaching the degree of generality of linear mathematics. Numerical solutions of nonlinear problems lose not only mathematical ease and elegance, but they are also less general. Numerically, specific examples are solved rather than a generalized picture directly obtained. The use of linear methods for basically nonlinear problems is justified as a first approximation and to the extent that the general and elegant solutions obtained have predictive power for the problem in question. When this stage is reached, we have to leave the more secure route and resort to a more difficult but more scientifically appropriate way. We surely have not exhausted in ecology the possibilities of linear methodology, which have found such widespread use in engineering and technology. However, for many ecological questions of the day, linear approaches appear to be less and less adequate.

REFERENCES

ABRAMS, P. A., and ALLISON, T. D. 1982. Complexity, stability and functional responses. Am. Nat. 119:240–249.

ABROSOV, N. S. 1975. Theoretical investigations of the mechanisms of regulation of species structure of the community of autotrophic organisms. *Ekologyia (Moscow)* 6:5–15. (In Russian).

AGGRAWALA, B. D., and GOPALSANY, W. S. C. 1980. Recurrence and non-stationary coexistence in two-species competition. *Ecol. Mod.* 9:153–163.

APONIN, J. M., and BAZYKIN, A. D. 1977. Model of eutrophication in predator–prey systems. Int. Inst. App. Syst. Anal., Laxenburg, Austria, RM-77-16.

AUSTIN, M. P., and COOK, B. G. 1974. Ecosystem stability: a result from an abstract simulation. *J. Theor. Biol.* 45:435–458.

BAKULE, L., and STRAŠKRABA, M. 1982. On multi-objective optimization in aquatic ecosystems. *Ecol. Mod.* 17:75–82.

BAKULE, L., and STRAŠKRABA, M. 1987. On structural strategies in aquatic ecosystems. *Ecol. Mod.* 39:171–180.

BAZYKIN, A. D. 1976. Structural and dynamic stability of model predator–prey systems. Int. Inst. App. Syst. Anal., Laxenburg, Austria, RM-76-8.

BEDDINGTON, J. R., FREE, C. A., and LAWTON, J. M. 1975. Dynamic complexity in predator prey models framed in difference equations. *Nature* 225:58–60.
BRITTON, N. F. 1989. Aggregation and the competitive exclusion principle. *J. Theor. Biol.* 136:57–66.
BROWN, J. M. 1981. Two decades of homage to Santa Rosalia: toward a general theory of diversity. *Amer. Zool.* 27:877–888.
BROWN, J. S., and VINCENT, T. L. 1987. Predator–prey coevolution as an evolutionary game. In Cohen, Y. (ed.), *Applications of Control Theory in Ecology, Lecture Notes in Biomathematics*, 73. Springer, New York, pp. 83–100.
CALDER, W. A., III. 1988. *Size, Function and Life History.* Harvard University Press, Cambridge, Massachusetts. 448 pp.
CASTI, J. L. 1979. *Connectivity, Complexity and Catastrophe in Large Scale Systems.* Wiley, New York. 203 pp.
COHEN, Y. 1987. Applications of optimal impulse control to optimal foraging problems. In Cohen, Y. (ed.), *Applications of Control Theory in Ecology, Lecture Notes in Biomathematics*, 73. Springer, New York, pp. 39–56.
CONRAD, M. 1983. *Adaptability: The Significance of Variability from Molecule to Ecosystem.* Plenum Press, New York. 383 pp.
CROW, M. E. 1977. Increasing ecosystem complexity: how to lose your marbles and increase stability. *Abst. Bull. Ecol. Soc. Amer.,* June 1977, p. 12.
DAWKINS, R. 1976. *The Selfish Gene.* Oxford University Press, New York. 224 pp.
DE ANGELIS, D. L. 1975. Stability and connectance in food web models. *Ecology* 56:238–243.
DUBOIS, D. M. 1979. State of the art of predator–prey modelling. In Jørgensen, S. E. (ed.), *State of the Art in Ecological Modelling.* Int. Soc. Ecol. Modelling, Copenhagen, pp. 751–758.
DWYER, R. L., and PEREZ, K. T. 1983. An experimental examination of ecosystem linearization. *Am. Nat.* 127:305–323.
EBELING, W., and FEISTEL, R. 1982. *Physics of Self-organization and Evolution.* Akademie Verlag, Berlin. 451 pp. (In German).
ENDLER, J. A. 1980. Natural selection of color patterns in *Poecilia reticulata. Evolution* 34:76–91.
FEISTEL, R., and EBELING, W. 1989. *Evolution of Complex Systems.* Deutscher Verlag der Wissenschaften, Berlin. 248 pp.
FLEISHMAN, B. S. 1982. *Principles of Systemology.* Radio i Swjaz, Moscow. 368 pp. (In Russian).
FUTUYMA, D. J., and SLATKIN, M. (eds.) 1983. *Coevolution.* Sinauer, Sunderland, Massachussetts. 555 pp.
GARDNER, M. R., and ASHBY, W. R. 1970. Connectance of large dynamic (cybernetic) systems: critical values for stability. *Nature* 228:784.
GINZBURG, L. R. 1980. Ecological implications of natural selection. In Barigozzi, C. (ed.), *Vitto Volterra Symposium on Mathematical Models in Biology.* Springer, New York, pp. 171–183.
GLEICK, J. 1988. *Chaos. Making a New Science.* Penguin Books, New York. 352 pp.
GOH, B. S. 1975. Stability, vulnerability and persistence of complex ecosystems. *Ecol. Mod.* 1:105–116.
GOH, B. S. 1980. *Management and Analysis of Biological Populations.* Elsevier, Amsterdam. 288 pp.
GOULD, S. J., and LEWONTIN, R. C. 1979. The spandrels of San Marco and the Panglossian paradigm: or critique of the adaptionist programme. *Proc. Roy. Soc. London B* 205:581–598.
GREENEY, W. J., BELLA, D. A., and CURL, H. C. 1973. A theoretical approach to interspecific competition in phytoplankton communities. *Am. Nat.* 107:405–425.
GRUBER, B. 1976. The influence of saturation on predator–prey relations. *Theor. Pop. Biol.* 10:173–184.
GUCKENHEIMER, J. 1977. The dynamics of density dependent population models. *J. Math. Biol.* 4:101–104.

Gurney, W. S. C., and Nisbet, R. M. 1970. Ecological stability and social hierarchy. *Theor. Pop. Biol.* 16:48–80.

Hanski, I. 1990. Density dependence, regulation and variability in animal populations. *Phil. Trans. Roy. Soc. London B* 330:141–150.

Hall, C. A. S. 1988. An assessment of several of the historically most influential theoretical models used in ecology and of the data provided in their support. *Ecol. Mod.* 43:5–31.

Harwell, M. A., Cropper, W. P., and Ragsdale, H. L. 1977. Nutrient cycling and stability: a re-evaluation. *Ecology* 58:660–666.

Hassel, M. P., and Comins, H. W. 1976. Discrete time models for two species competition. *Theor. Pop. Biol.* 9:207–221.

Hines, W. G. S. 1987. Evolutionary stable strategies: a review of basic theory. *Theor. Popul. Biol.* 31:195–272.

Hochachka, P. W., and Somero, G. N. 1973. *Strategies of Biochemical Adaptation.* Saunders, Philadelphia. (Polish translation. Panstwowe Wydawnictwo Naukowe Warszaw). 455 pp.

Innis, G. S., and Clark, W. R. 1978. A self-organization approach to ecosystem modelling. In Innis, G. S. (ed.), *Grassland Simulation Model.* Springer, New York, pp. 779–787.

Jeffries, C. 1979. Stability of holistic ecosystem models. In Halfon, E. (ed.), *Theoretical Systems Ecology.* Academic Press, New York, pp. 489–504.

Jørgensen, S. E. 1983. *Application of Ecological Modelling in Environmental Management.* Developments in Environmental Modelling, Vol. 4A. Elsevier, Amsterdam. 735 pp.

Kempf, J. 1980. Multiple steady states and catastrophes in ecological models. *J. Intern. Soc. Ecol. Mod.* 2(3–4):55–79.

Kerner, E. H. 1961. On the Volterra–Lotka principle. *Bull. Math. Biophys.* 23:141–157.

Khanin, M. A., Dorfman, N. L., Bucharov, G. B., and Levadnyi, V. P. 1978. *Extremal Principles in Biology and Physiology.* Nauka, Moscow. 256 pp. (In Russian).

Kindlmann, P., and Rejmánek, M. 1982. Continuous or discrete models of multispecies systems: how much less stable are the latter ones? *J. Theor. Biol.* 94:989–993.

King, W. E. 1980. The genetic structure of zooplankton populations. In Kerfoot, W. (ed.), *Evolution and Ecology of Zooplankton Communities.* University Press of New England, Hanover, New Hampshire, pp. 325–328.

King, W., and Pimm S. L. 1983. Complexity, diversity, and stability: a reconciliation of theoretical and empirical results. *Am. Nat.* 122:229–239.

Koch, A. L. 1974a. Competitive coexistence of two predators utilizing the same prey under constant environmental conditions. *J. Theor. Biol.* 44:387–395.

Koch, A. L. 1974b. Coexistence resulting from an alternation of density dependent and density independent growth. *J. Theor. Biol.* 44:373–386.

Korostyshevsky, M. A., Schtabuoy, M. R., and Ratner, V. A. 1974. On some principles of evolution viewed as a stochastic process. *J. Theor. Biol.* 78:85–103.

Krebs, J. R., and Davies, N. B. 1984. *Behavioural Ecology. An Evolutionary Approach,* 2nd ed. Sinauer Assoc., Sunderland, Massachusetts. 493 pp.

Lamprecht, Y., and Zotin, S. J. (eds.) 1978. *Thermodynamics and Kinetics of Biological Processes.* Walter de Gruyter, Berlin. 428 pp.

Lande, R. 1982. A quantitative genetic theory of life history evolution. *Ecology* 63:607–613.

Laplante, J. P. 1978. Inhomogeneous steady-state distributions of species in predator–prey systems. A specific example. *J. Theor. Biol.* 81:29–95.

Levin, B. R., Stewart, F. M., and Chao, L. 1977. Resource limited growth, competition and predation: a model and experimental studies with bacteria and bacteriophage. *Am. Nat.* 111:3–24.

Levin, S. A. 1978. Pattern formation in ecological communities. In Steele, J. M. (ed.), *Spatial Pattern in Plankton Communities.* Plenum Press, New York, pp. 433–465.

Levin, S. A. 1981. The role of theoretical ecology in the description and understanding in heterogeneous environments. *Am. Zool.* 21:865–875.

Lieth, M. 1975. Modelling the primary productivity of the world. In Leith, H., and Whittaker, R. H. (eds.), *Primary Productivity in the Biosphere.* Springer-Verlag, New York, pp. 237–263.

Loehle, C. 1989. Catastrophe theory in ecology: a critical review and an example of the butterfly catastrophe. *Ecol. Mod.* 49:125–152.

Lotka, A. J. 1925. *Elements of Physical Biology.* Williams and Wilkins, Baltimore. 460 pp.

Lundberg, P., and Áström, M. 1990. Functional response of optimally foraging herbivores. *J. Theor. Biol.* 144:367–377.

May, R. 1984. *Exploitation of Marine Communities.* Dahlen-Konferenzen. Springer Verlag, Berlin. 336 pp.

May, R. M. 1971. Stability in multispecies community models. *Math. Biosci.* 12:59–79.

May, R. M. 1973a. *Stability and Complexity in Model Ecosystems.* Princeton University Press, Princeton, New Jersey. 235 pp.

May, R. M. 1973b. On relationships among various types of population models. *Am. Nat.* 107:46–57.

May, R. M. 1975. *Stability and Complexity in Model Ecosystems,* 2nd ed. Princeton University Press, Princeton, New Jersey. 265 pp.

May, R. M. 1977. Threshold and breakpoints in ecosystems with a multiplicity of stable states. *Nature* 269:471–477.

May, R. M. 1979. Simple models for single populations: an annotated bibliography. *Forschr. Zool.* 25(2/3):95–107.

May, R. M., and Oster, J. F. 1976. Bifurcations and dynamic complexity in simple ecological models. *Am. Nat.* 110:573–599.

Maynard-Smith, J. 1974. The theory of games and the evolution of animal conflict. *J. Theor. Biol.* 47:209–221.

Maynard-Smith, J. 1978. Optimization theory in evolution. *Ann. Rev. Ecol. Syst.* 9:31–56.

Maynard-Smith, J., and Slatkin, M. 1973. The stability of predator–prey systems. *Ecology* 54:384–391.

McFarland, D. 1985. *Animal Behaviour, Psychobiology, Ethology and Evolution.* Putman, London. 576 pp.

Milinaki, M. 1982. Optimal foraging: the influence of intraspecific competition on direct selection. *Behav. Ecol. Sociobiol.* 11:109–115.

Mueller, L. D., and Ayala, F. J. 1981. Dynamics of single species population growth: stability or chaos? *Ecology* 62:1148–1154.

Murdoch, W. W., and Oaten, A. 1975. Predation and population stability. *Adv. Ecol. Res.* 9:1–131.

Nesměrák, I., and Straškraba, M. 1985. Spectral analysis of the automatically recorded data from Slapy Reservoir, Czechoslovakia. *Int. Revue Hydrobiol.* 70:27–46.

Nicholson, A. J. 1954. An outline of the dynamics of animal populations. *Aust. J. Zool.* 2:9–65.

Nicolis, G., and Prigogine, L. 1977. *Self-Organization in Nonequilibrium Systems.* Wiley, New York. 491 pp.

Nisbet, R. M., and Gurney, W. S. C. 1978. Population dynamics in a periodically varying environment. *J. Theor. Biol.* 56:459–475.

Nisbet, R. M., and Gurney, W. S. C. 1982. *Modelling Fluctuating Populations.* Wiley, New York. 379 pp.

Nitecki, M. M. 1983. *Coevolution.* University of Chicago Press, Chicago. 275 pp.

Nunney, L. 1980. The stability of complex model ecosystems. *Am. Nat.* 115:639–649.

Oksanen, L. 1988. Ecosystem organization: mutualism and cybernetics or plain Darwinian struggle for existence? *Am. Nat.* 131:424–444.

Ollason, J. S. 1980. Learning to forage optimally? *Theor. Popul. Biol.* 18:47–56.

O'NEILL, R. V., DEANGELIS, D. L., WAIDE, J. B., and ALLEN, T. H. F. 1986. *A Hierarchical Concept of Ecosystems.* Princeton University Press, Princeton, New Jersey. 253 pp.

PATTEN, B. C. 1975. Ecosystem linearization: an evolutionary design problem. *Am. Nat.* 109:529–539.

PATTEN, B. C. 1976. *Systems Analysis and Simulation in Ecology,* Vol. IV. Academic Press, New York. 593 pp.

PATTEN, B. C. 1979. Necessary conditions for realism in ecological models. In Dame, R. F. (ed.), *Marsh–Estuarine Systems Simulation.* University of South Carolina Press, Columbia, pp. 237–247.

PATTEN, B. C. 1982a. Environs: relativistic elementary particles for ecology. *Am. Nat.* 119:179–219.

PATTEN, B. C. 1982b. Environs: coevolutionary units of ecosystems. In Mitsch, W. J., Ragade, R. K., Bosserman, R. W., and Dillon, J. A. (eds.), *Energetics and Systems.* Ann Arbor Science, Ann Arbor, Michigan, pp. 108–118.

PATTEN, B. C., 1983. Linearity enigmas in ecology. *Ecol. Mod.* 18:155–170.

PETERS, R. H. 1983. *The Ecological Implications of Body Size.* Cambridge University Press, Cambridge, Massachusetts. 329 pp.

PETERS, R. H. 1991. *A Critique for Ecology.* Cambridge University Press, Cambridge. 366 pages.

PETERSON, R. 1975. The paradox of the plankton: an equilibrium hypothesis. *Am. Nat.* 109:35–49.

PHILLIPS, O. M. 1973. The equilibrium and stability of simple marine biological systems. I. Primary nutrient consumers. *Am. Nat.* 107:73–93.

PHILLIPS, O. M. 1978. The equilibrium and stability of simple marine biological systems. III. Fluctuations and survival. *Am. Nat.* 112:745–757.

PIMM, S. L. 1979. Complexity and stability: another look at MacArthur's original hypothesis. *Oikos* 33:351–357.

PLATT, T. (ed.) 1981. Physiological basis of phytoplankton ecology. *Can. Bull. Fish. Aquat. Sci.* 210, Fisheries and Oceans, Ottawa, Canada. 346 pp.

POOLE, R. W. 1977. Periodic, pseudoperiodic, and chaotic population fluctuations. *Ecology* 58:210–213.

POWELL, R. A. 1980. Stability in a one-predator–three-prey community. *Am. Nat.* 115:567–579.

QUINN, J. F., and DUNHAM, S. E. 1983. On hypothesis testing in ecology and evolution. *Am. Nat.* 122:602–617.

RECKNAGEL, F. 1984. A comprehensive sensitivity analysis for an ecological simulation model. *Ecol. Mod.* 26:77–96.

ROBINSON, J. V., and VALENTINE, W. D. 1979. The concept of elasticity, invulnerability and invadability. *J. Theor. Biol.* 81:91–109.

ROSENZWEIG, M. L. 1973. Exploitation in three trophic levels. *Am. Nat.* 107:275–294.

ROSENZWEIG, M. L. 1979. Paradox of enrichment: destabilization of exploitation ecosystems in ecological time. *Science* 171:385–387.

ROUGHGARDEN, J., MAY, R. M., and LEVIN, S. A. 1989. *Perspectives in Ecological Theory.* Princeton University Press, Princeton, New Jersey. 325 pp.

SALTHE, S. N. 1985. *Evolving Hierarchical Systems. Their Structure and Representation.* Columbia University Press, New York. 343 pp.

SALTHE, S. N. 1989. Self-organization of/in hierarchically structured systems. *Syst. Res.* 6:199–208.

SCHAFFER, W. M., and KOT, M. 1986. Chaos in ecological systems: the coals that Newscastle forgot. *Trends in Ecol. and Evol.* 1:58–63.

SCHOENER, T. W. 1976. Alternatives to Lotka–Volterra competition: models of intermediate complexity. *Theor. Pop. Biol.* 10:309–333.

SCHOENER, T. W. 1978. Effects of density-restricted food encounters on some single-level competition models. *Theor. Popul. Biol.* 13:366–381.

SCHWINGHAMER, P. 1981. Characteristic size distributions of integral benthic communities. *Can. J. Fish. Aquat. Sci.* 38:1255–1263.

SEGEL, L. A., and LEVIN, S. A. 1976. Application of nonlinear stability theory for the study of the effects of diffusion on predator–prey interactions. In Piecirelli, R. A. (ed.), *Topics in Statistical Mechanics and Biophysics, Am. Inst. Physics Conf. Proc.* 27:123–152.

SEMEVSKI, F. N., and SEMENOV, C. U. 1982. *Mathematical Modelling of Ecological Processes.* Giodrometeoizdat, Leningrad. 280 pp. (In Russian).

SEMINA, M. J. 1972. The size of phytoplankton cells in the Pacific Ocean. *Int. Revue Hydrobiol.* 57:177–205.

SHELDON, R. W., PRAKASH, A., and SUTCLIFFE, W. H. 1972. The size distribution of particles in the ocean. *Limnol. Oceanogr.* 17:327–390.

SHUGART, H. H. 1984. *A Theory of Forest Dynamics. The Ecological Implications of Forest Succession Models.* Springer, New York. 278 pp.

SILJAK, D. D. 1978. *Large-scale Dynamic Systems.* North-Holland, New York. 416 pages.

SILVERT, W., and SMITH, W. R. 1981. The response of ecosystems to external perturbations. *Math. Biosci.* 55:279–306.

SINKO, J. W., and STREFER, W. 1967. A new model for age–size structure of a population. *Ecology* 78:910–918.

SJÖBERG, S. 1977. Are pelagic systems inherently unstable? A model study. *Ecol. Mod.* 3:17–37.

SMÍTALOVÁ, K. 1983. On the effect of the number of trophic levels on community stability. *Acta facultati rerum naturalium Universitatis comeniae Formatio et prof. naturae* 8:135–143.

SMITH, E. A. 1987. On fitness maximization, limited needs, and hunter-gatherer time allocation. *Ethol. Sociobiol.* 8:73–85.

SMITH, J. L., and WOLLKIND, D. J. 1983. Age structure in predator–prey systems: intraspecific carnivore interaction, passive diffusion, and the paradox of enrichment. *J. Math. Biol.* 17:275–288.

SPRULES, W. G., CASSELMAN, J. M., and SHUTER, B. J. 1983. Size distribution of pelagic particles in lakes. *Can. J. Fish. Aquat. Sci.* 40:1761–1769.

STEARNS, S. C. 1977. The evolution of life history traits: A critique of the theory and a review of the data. *Ann. Rev. Ecol. Syst.* 8:145–171.

STEELE, J. H. 1974. Stability of plankton ecosystems. In Usher, M. B., and Williamson, M. H. (eds.), *Ecological Stability.* Chapman and Hall, New York, pp. 179–194.

STENSETH, N. Ch. 1984. Why mathematical models in evolutionary ecology? In Cooley, J. M., and Golley, F. B. (eds.), *Trends in Ecological Research for the 1980s.* Plenum Press, New York, pp. 239–287.

STENSETH, N. 1985. The structure of food webs predicted from optimal food selection models: an alternative to Pimm's stability hypothesis. *Oikos* 44:361–364.

STENSETH, N. 1986. Darwinian evolution in ecosystems: a survey of some ideas and difficulties together with some possible solutions. In Casti, J. L., and Karlqvist, A. (eds.), *Complexity, Language, and Life: Mathematical Approaches,* Biomathematics Vol. 16. Springer, New York, pp. 105–145.

STRAŠKRABA, M. 1976. Development of an analytical phytoplankton model with parameters empirically related to dominant controlling variables. In Glaser, R., Unger, K., and Koch, M. (eds.), *Umwelbiophysik,* Arbeitstagung 19.10 bis 1.11.1973 Kühlungsborn. Akademie Verlag, Berlin, pp. 33–65.

STRAŠKRABA, M. 1980a. The effect of physical variables on freshwater production: analyses based on models. In LeCren, E. D., and Lowe McConnell, R. H. (eds.), *The Functioning of Freshwater Ecosystems.* Cambridge University Press, Cambridge, pp. 13–84.

STRAŠKRABA, M. 1980b. Cybernetic categories of ecosystem dynamics. *J. Int. Soc. Ecol. Mod.* 2:87–96.

STRAŠKRABA, M. 1986. Fitting dynamic water quality models for Slapy Reservoir. In Lerner, A. (ed.), *Monitoring to Detect Changes in Water Quality Series.* IAHS Publ. No. 157:327–336.

STRAŠKRABA, M., and GNAUCK, A. 1985. *Freshwater Ecosystems. Modelling and Simulation.* Elsevier, Amsterdam. 309 pp.

SVIREZHEV, Y. M., and LOGOFET, D. O. 1978. *Stability of Biological Communities.* Nauka, Moscow. 352 pp. (In Russian).

SWARTZMAN, G. L., and KALUZNY, S. P. 1987. *Ecological Simulation Primer.* Macmillan, New York. 370 pp.

THOM, R. 1975. *Structural Stability and Morphogenesis.* Benjamin, Reading, Massachusetts. 348 pp.

THOMAS, W. R., POMERANTZ, M. J., and GILPIN, M. E. 1980. Chaos, econometric growth and group selection for dynamical stability. *Ecology* 61:1312–1320.

THORNTON, K. W., and MULHOLLAND, R. J. 1974. Lagrange stability and ecological systems. *J. Theor. Biol.* 45:473–485.

TOWNSEND, C. R., WINFIELD, I. J., REWSON, G., and CRYER M. 1986. The response of young roach, *Rutilus rutilus,* to seasonal changes in abundance of microcrustacean prey: a field demonstration of switching. *Oikos* 46:372–378.

ULANOWICZ, R. E. 1979a. Prediction, chaos, and ecological perspectives. In Halfon, E. (ed.), *Theoretical Systems Ecology.* Academic Press, New York, pp. 107–117.

ULANOWICZ, R. E. 1979b. Complexity, stability and self-organization in natural communities. *Oecologia* 43:295–298.

ULANOWICZ, R. E. 1980. An hypothesis on the development of natural communities. *J. Theor. Biol.* 85:223–245.

VAN VALEN, L. M. 1983. How pervasive is coevolution? In Nitecki, M. H. (ed.), *Coevolution.* University of Chicago Press, Chicago, pp. 1–19.

VAN VORIS, P., O'NEIL, R. V., EMANUAL, W. R., and SHUGART, H. H. 1980. Functional complexity and ecosystem stability. *Ecology* 61:1352–1360.

VOLTERRA, V. 1931. *Lectures on the Mathematical Theory of the Struggle for Life.* Gauthier–Villars, Paris. 214 pp. (In French).

WANGERSKY, P. J. 1978. Lotka–Volterra population models. *Am. Rev. Ecol. Syst.* 9:189–218.

WATSON, S., and KALFF, J. 1981. Relationships between nannoplankton and lake trophic status. *Can. J. Fish. Aquat. Sci.* 38:960–967.

WERNER, E. E., and HALL, D. J. 1974. Optimal foraging and the size selection of prey by the bluegill sunfish (*Lepomis macrochirus*). *Ecology* 55:1216–1232.

WILSON, D. S., and SOBER, E. 1989. Reviving the superorganism. *J. Theor. Biol.* 136:337–356.

WISSEL, C. 1990. *Theoretische Ökologie.* Akademie-Verlag, Berlin. 299 pp.

WOLAWER, T. G. 1980. Effect of system structure, connectivity, and recipient control on the sensitivity characteristics of a model ecosystem. *Int. J. Systems Sci.* 11:291–308.

WOLLKIND, D. J. 1976. Exploitation in three trophic levels: an extension allowing intraspecies carnivore interaction. *Am. Nat.* 110:431–447.

WULFF, F., FIELD, J. G., and MANN, K. H. (eds.) 1989. *Network Analysis in Marine Ecology.* Springer Verlag, New York. 284 pp.

5

Entropy Control of Complex Ecological Processes

PETER MAUERSBERGER

Academy of Sciences
Institute for Geography and Geoecology, Department of Hydrology
Berlin

A system functions only if there is a driving force. [Van Dyne, in *The Ecosystem Concept in Natural Resource Management*, Ch. X, p. 329 (1969), Academic Press].

INTRODUCTION

Use of Fundamental Laws of Physics, Chemistry, and Biology in Ecosystem Analysis

The analysis and modeling of ecosystems should not be exclusively or predominantly based on observational data. At least, the correct type of model equations should be deduced from basic physical, chemical, and biological constraints, since we observe only incomplete sets of variables and selected processes that are controlled by several, simultaneously varying, independent variables. Furthermore, our measurements are confined to a limited time interval during which the ecosystem may not show the full variety of phenomena possible under the given conditions. This incomplete information about the ecosystem's behavior is insufficient for a deeper understanding of the functional mechanisms and for predicting the properties of the system under strongly varying conditions, the more so since self-organizing processes may change the situation drastically.

    To get a fuller understanding of processes in ecosystems, laboratory studies are performed under conditions more or less comparable with those in nature; but inevitably there are differences between experimental and natural conditions. Thus, a theory is required that covers a relatively wide range of situations. In order to gain deeper insight into the laws of ecosystems' behavior, the inductive approach to ecosystem analysis, which prefers knowledge from measurements as the basis of modeling, must be combined with the deductive approach, which deduces a priori knowledge about possible structures and the functioning of ecosystems from cybernetics and general laws of physics, chemistry, and biology.

Entropy Control of the Development of Ecosystems

Ecosystem development shows an irreversible and nonrepeatable history, a tendency expressed by the second law of thermodynamics (Mauersberger, 1991). Besides the balance equations of momentum, energy, and mass for all constituents of the ecosystem, the second law of thermodynamics must inevitably be taken into account. All processes within the ecosystem are of the irreversible type, thus producing entropy. The main effect of the intensive exchange of matter and energy between the ecosystem and its surroundings is not an accumulation of energy and matter, but export of the entropy generated inside the ecosystem. The ecosystem can maintain its state only by exporting the produced entropy with the help of flows of energy and matter across its boundary and through the system. Radiation, inflow, and outflow of dissolved or particulate material, as well as sedimentation and resuspension of matter and all biotic and abiotic constituents of the system, are involved in the exchange of matter, energy, and entropy. The structure, state, and further development of the ecosystem are regulated by the mutual effects of entropy-producing and entropy-reducing processes. The entropy principle controls the development of the ecosystem by determining which of the energetically possible processes actually occurs. While the entropy principle acts as a "director," the energy balance plays the equally important role of "bookkeeper."

THERMODYNAMIC BACKGROUND

Nonequilibrium Thermodynamics

A priori knowledge about internal control mechanisms, reaction on external impact, and strategies of ecosystems can be derived from generalized nonequilibrium thermodynamics (Serrin, 1986). The generalization consists of introducing biological variables and biological processes into the balance equations of mass, momentum, energy, and entropy of nonequilibrium thermodynamics (Mauersberger 1978, 1979). Because of the nonlinearity of the chemical and biological processes and of some of the physical processes inside the ecosystem, and because of the fact that ecosystems are far from being in thermostatic equilibrium, we will not refer to the local equilibrium hypothesis (which states that the relations established in thermostatics are locally valid even if the system as a whole is out of equilibrium). With the intention of going beyond the local equilibrium hypothesis, we give preference to a formalism that is as simple as possible: The variables specifying the state of local equilibrium, that is, internal energy, entropy, and specific volume, are supplemented by additional variables, ζ_i, which vanish or become nonessential in local equilibrium states. Equations expressing the evolution in time of these "deviation coordinates," ζ_i, have to be obtained from information about the system under consideration and its history. This generalization of the Gibbs relation bears some similarities to the theory of internal variables, as well as to extended irreversible thermodynamics (compare Casas-Vazques et al., 1984). The local entropy production can also be derived from the energy balance introducing the excess energy supply connected with irreversible processes (Mauersberger, 1990).

Generalized Gibbs Relation

It is postulated that the partial caloric equation of state of the ith constituent of the system contains the following variables: partial specific internal energy, u_i, per unit mass, partial specific entropy, s_i, per unit mass, specific volume, α_i, and the dimensionless deviation coordinate, ζ_i:

$$u_i = u_i(\alpha_i, s_i, \zeta_i) \, . \tag{5.1}$$

Instead of assuming this function to be known explicitly, for the generalized Gibbs relation we require

$$du_i = Tds_i - p_i d\alpha_i + e_i d\zeta_i, \tag{5.2}$$

where T denotes temperature and p_i the partial pressure of the ith constituent. The quantity e_i can be understood as supplementary energy or free enthalpy per unit mass connected with the nonequilibrium. Memory effects and influences of the history of the system must also be taken into account by e_i, in which case e_i follows from an integrodifferential equation. The nonequilibrium contribution $e_i d\zeta_i$ results from additional external work on the system against the tendency of returning to local equilibrium and, furthermore, it is coupled with changes in entropy by internal processes far from local equilibrium, such as fast chemical reactions, biological processes, structural changes in complex molecular arrangements, high-frequency phenomena, and so on. It must be assumed that e_i vanishes at local equilibrium.

Introducing specific free enthalpy, g_i, as a function of temperature, pressure, and deviation coordinate in

$$g_i(T, p_i, \zeta_i) = u_i - Ts_i + p_i\zeta_i, \tag{5.3}$$

we infer from eq. (5.2) that

$$dg_i = s_i\, dT + \alpha_i\, dp_i + e_i d\zeta_i \tag{5.4}$$

and

$$e_i = \frac{\partial g_i}{\partial \zeta_i}, \quad \frac{\partial e_i}{\partial T} = \frac{\partial^2 g_i}{\partial T \partial \zeta_i} = \frac{\partial s_i}{\partial \zeta_i}, \tag{5.5}$$

and so on. In general, all these variables (s_i, u_i, g_i, e_i, T, p_i, α_i, ζ_i) are functions of position $\vec{r}$ and time t.

With the exception of special but important situations, the increase of ζ_i is connected with increasing deviation from local equilibrium. But, nevertheless, ζ_i does not directly determine the distance of the nonequilibrium system from equilibrium, since that distance is defined by the value of entropy in comparison with entropy at local equilibrium.

Exergy

The difference between the entropy S of the ecosystem and the value S_{eq} at thermodynamic equilibrium is one of the factors that determines its exergy, Ex:

$$Ex = T(S_{eq} - S) = U + PV - TS - \sum_j X_j n_j. \tag{5.6}$$

U, V, and n_j denote the internal energy, volume, and mole numbers of the ecosystem. P, T, and X_j are properties of the external reservoirs with which the ecosystem is assumed to interact. Thus, exergy depends on the state of the surroundings and can be used to investigate ecosystems under the influence of changing external factors (Jørgensen and Mejer, 1981).

Thermodynamics of Ecosystems

In applying thermodynamics to the theory of ecosystems, it is assumed that:

- Biological phenomena, although not completely determined by the laws of macroscopic physics and chemistry, are not in contradiction to them.

- Physics succeeds in marking out the scope within which biological processes may occur, provided nonlinear irreversible processes are taken into account.
- Thermodynamic laws constitute constraints also on the self-adaptation and self-organization of ecosystems.
- Self-adaptation and self-organization enable living systems to develop structures and processes that are governed by typical biological laws in addition to physical and chemical laws.

Therefore, the thermodynamics of ecosystems results from nonlinear thermodynamics by two steps: First, biological state variables and processes are introduced into the balances of mass, momentum, and energy and into the Gibbs relation. Second, the system of basic equations becomes completed by adding suitable relations typical for biological processes.

Ecosystems consist of arbitrary but finite numbers of chemical constituents ($j = 1, 2, 3, \ldots$) and biological "components" ($k = 1, 2, 3, \ldots$), for instance, species or "groups" such as diatoms, blue-green algae, primary producers, populations of animals, and the like. The choice and definition of the "constituents" or "components" depend to a certain degree on the aims of modeling. The theory outlined here is independent of the special choice. From the basic theory, special models may be derived. A more detailed description takes into account chemicals stored inside the kth biological component and engaged in the internal storage of nutrients, energy, or internally (for example, by photosynthesis) produced organic substances (for example, hydrocarbons). These chemicals may be numbered $l = 1, 2, 3, \ldots$. The composition of the system changes, first, by the input and output taken into account by boundary conditions and the mass balance equations and, second, by internal chemical reactions ($r = 1, 2, 3, \ldots$) and biological processes ($n = 1, 2, 3, \ldots$), for instance, by photosynthesis, uptake of nutrients, and respiration.

Ecosystems are far from being in thermostatic equilibrium. The time evolution of the deviation coordinates, ζ_j, for the chemical constituents and of the deviation coordinates, ζ_{kl}, for the biological components results both from the extent, $d\xi_r$, of fast chemical reactions and from the extent, $d\xi_{kn}$, of biological processes. Therefore, we apply the following rate equations:

$$\frac{d\zeta_j}{dt} = \sum_r \nu_{jr}^* \mathcal{M}_j \frac{d\xi_r}{dt} - \sum_k \sum_n a_{jkn}^* \frac{d\xi_{kn}}{dt} + \ldots, \qquad (5.7)$$

$$\frac{d\zeta_{kl}}{dt} = \sum_n b_{kln}^* \frac{d\xi_{kn}}{dt} + \ldots, \qquad (5.8)$$

where dots indicate further effects far from local equilibrium, which we do not investigate here. The coefficients ν_{jr}^*, a_{jkn}^*, and b_{kln}^* are functions of the deviation coordinates and of the state variables, for instance, temperature and pressure. In shallow lakes, the dependence on pressure can be neglected.

Summing up the overall chemical and biological constituents of the ecosystem and introducing the quantities

$$V_r = -\sum_j \nu_{jr}^* N_j e_j \mathcal{M}_j \qquad (5.9)$$

and

$$V_{kn} = \sum_j a_{jkn}^* N_j e_j + \sum_l b_{kln}^* B_{kn} e_{kn}, \qquad (5.10)$$

we arrive at the following generalized Gibbs relation for the aquatic ecosystem:

$$\frac{du}{dt} = T\frac{ds}{dt} - p\frac{d\alpha}{dt} + \sum_j \mu_j \frac{dN_j}{dt} + \sum_k \mu_k \frac{dB_k}{dt} + \sum_k \sum_l \mu_{kl} \frac{dm_{kl}}{dt} + \sum_r V_r \frac{d\xi_r}{dt}$$

$$+ \sum_k \sum_n V_{kn} \frac{d\xi_{kn}}{dt},$$

(5.11)

where T, t, p = temperature, time, pressure $\left(p = \sum_i p_i\right)$

α, ρ = specific volume, mass density $\left(\rho = \sum_i \rho_i = 1/\alpha\right)$

u, s = specific internal energy per unit mass, specific entropy per unit mass of the mixture (water, chemicals, and biota)

M_j = molar mass of the jth chemical constituent

μ_j, μ_k = chemical potential per unit mass of the jth chemical constituent or kth biological component

μ_{kl} = chemical potential of the lth compound stored inside the kth biological component

N_j, B_k = mass fraction of the jth chemical or kth biological constituent ($N_j = \rho_j/\rho$, $B_k = \rho k/\rho$)

m_{kl} = mass fraction (in relation to biomass) of the lth storing compound inside the kth biological component ($m_{kl} = \rho_{kl}/\rho_k$)

ξ_r = extent of the rth chemical reaction

ξ_{kn} = extent of the nth biological process, kth biological component

V_r, V_{kn} = supplementary energies influenced by the history of the ecosystem and connected with the rth chemical reaction or nth biological process, which cause deviation from the equilibrium reference state

$\frac{d}{dt} = \frac{\partial}{\partial t} + \vec{v} \cdot \nabla$ = material time derivative.

The sum over i includes water, chemical constituents, and biological components.

THERMODYNAMICS OF AQUATIC ECOSYSTEMS

Aquatic Ecosystems

Aquatic ecosystems may serve as an example for the application of nonlinear thermodynamics of continuous systems to ecosystem analysis. Balances of mass, momentum, energy, and entropy are written as partial differential equations. For instance, the mass balances of the constituents and of the ecosystem itself are continuity equations [see eqs. (5.24) and (5.25)]. Allowance is made for the mass exchange between ecosystem components and due to other types of sources and sinks, such as fishing. Boundary conditions must take into account that ecosystems are open systems exchanging mass, momentum, and energy with their surroundings. Connected with these exchange processes, entropy produced by irreversible processes inside the ecosystem is exported, which is a condition for the further existence and development of the ecosystem.

Living and nonliving constituents of the ecosystem are regarded as being subject to external forces. Momentum may be transferred from one constituent to another even if the masses of these constituents remain constant. Aquatic organisms may move actively. Therefore, the momentum of any constituent need not be balanced by itself. The full system of balance equations is discussed by Mauersberger, 1978, 1985a. Here we confine ourselves to the balances of energy and entropy.

Energy-balance Equations

The kinetic energy of the motions $\vec{u}_i$ relative to the mean $\vec{v}$,

$$\vec{u}_i = \vec{v}_i - \vec{v}, \qquad \rho\vec{v} = \sum_i \rho_i \vec{v}_i \qquad (5.12)$$

is included in the specific kinetic energy ϵ per unit mass of the ecosystem:

$$\epsilon = \frac{1}{2}\vec{v}\cdot\vec{v} + \frac{1}{2}\sum_i N_i \vec{u}_i \cdot \vec{u}_i. \qquad (5.13)$$

From the balances of momentum of the constituents, there follows

$$\frac{d\epsilon}{dt} + q\epsilon = \rho\vec{v}\cdot\vec{K} + \sum_i \rho_i \vec{u}_i \cdot \vec{K}_j - S \cdot\cdot \nabla\vec{v}$$

$$- \sum_i S_i \cdot\cdot \nabla\vec{v}_i + \nabla\cdot\sum_i\left(S_i\cdot\vec{v}_i - \frac{1}{2}\vec{v}\cdot\vec{v}_i\rho_i\vec{u}_i\right) \qquad (5.14)$$

$$+ \sum_i\left(\vec{v}_i\cdot\rho_i\vec{q}_i + \frac{1}{2}\vec{v}_i\cdot\vec{v}_i q_i\right).$$

The time variation $d\epsilon/dt$ of the kinetic energy results from changes q of the total mass of the ecosystem, due to the work of external forces, $\vec{K} = \sum_i N_i \vec{K}_i$, and stress, $S = \sum_i S_i$, from partial forces, K_i, and stresses, S_i, acting on the ith constituent, to diffusive flow, $\rho_i \vec{u}_i$, from the supply of momentum to the ith constituent, $\rho_i \vec{q}_i$; and to creation or destruction q_i of the mass of the ith constituent.

The total energy of the ecosystem, consisting of internal energy, u, kinetic energy, ϵ, and electromagnetic energy, u_e, is increased by the work of external body forces, by stresses acting at the boundary, and by the inflow of kinetic, thermal, and electromagnetic energy (for instance, radiation) from outside the system. Therefore, the balance equation reads

$$\frac{d}{dt}\int_{V(t)}[u_e + \rho(u + \epsilon)]dV = \int_{V(t)}\sum_i \rho_i\vec{v}_i \cdot \vec{K}_i dV$$

$$+ \oint_{F_V(t)}\left[\sum_i \vec{v}_i\cdot S_i - \vec{W} - \vec{P}\right. \qquad (5.15)$$

$$\left.- \sum_i\left(\frac{1}{2}\vec{v}_i\cdot\vec{v}_i + h_i\right)\rho_i\vec{u}_i\right]\cdot \vec{df}.$$

$V(t)$ is a material volume and $F_V(t)$ its surface. $\vec{W}$ denotes the flow of nonmechanical energy (heat-flow density) and h_i the specific enthalpy of the ith constituent.

Since the propagation and absorption of electromagnetic energy (light energy and radiation) play a fundamental role in ecosystem dynamics, we have to take into account the balance of electromagnetic energy, u_e (see, for example, Phillips, 1962):

$$\frac{\partial u_e}{\partial t} + \text{div}(\vec{v}u_e + \vec{P}) = -\vec{j} \cdot \vec{E} - Q(u_e), \tag{5.16}$$

where $\vec{P}$ denotes the flux of electromagnetic energy, $\vec{j}$ the current density, $\vec{E}$ the electric field, $\vec{j} \cdot \vec{E}$ Ohm's heat, and Q the exchange of electromagnetic energy between the electromagnetic field (for example, radiation) and living or nonliving matter. It is reasonable to assume that

$$Q(u_e) = \sum_i \rho_i Q_i \gtreqless 0. \tag{5.17}$$

Entropy Balance Equation

Combining eqs. (5.14), (5.16), and the differential form of (5.15) by eliminating $d\epsilon/dt$ and du/dt, we derive the entropy balance:

$$\frac{\partial \rho s}{\partial t} + \text{div}\,\vec{S} = \sigma. \tag{5.18}$$

The local time variation of entropy, ρs, per unit volume results from the local entropy production, σ, and from the entropy flux density, $\vec{S}$, through the ecosystem and across its boundary. $\vec{S}$ is given by

$$\vec{S} = \rho s \vec{v} + \frac{\vec{W}}{T} - \sum_i \frac{(\mu_i - h_i)\vec{J}_i}{T} + \vec{S}'. \tag{5.19}$$

Convection of entropy, flow of nonmechanical energy (heat and radiation), diffusive material fluxes, and a nonequilibrium contribution, $\vec{S}'$, sum up to $\vec{S}$. The diffusive mass flow $\vec{J}_i$ is defined by $\vec{J}_i = \rho_i \vec{u}_i$.

The second law of thermodynamics states that σ is nonnegative:

$$\sigma \geq 0. \tag{5.20}$$

By the inequality (5.20), a trend in time is asserted for all processes in ecosystems. This is in accordance with the irreversible history of these systems. Equation (5.20) is one of the basic laws for the development of ecosystems. It is perhaps the most fundamental of all principles for predicting macroscopic events. The behavior of macrosystems such as ecosystems is not determined merely by their energy. Entropy must also be taken into account.

The local entropy production, σ, depends on all hydrophysical, hydrochemical, and hydrobiological processes in the ecosystem: heat flow, flow of matter, chemical reactions, biological processes, friction, extinction, transfer of mass and momentum between constituents, and so on. Writing only the first terms, we have

$$\sigma = \vec{W} \cdot \text{grad}\,\frac{1}{T} - \frac{1}{T}\sum_i \vec{J}_i \cdot \text{grad}\,\mu_i|_T \\ + \frac{1}{T}\sum_r A_r w_r + \frac{1}{T}\sum_k \sum_n A_{kn} R_{kn} + \ldots. \tag{5.21}$$

The rates of chemical reactions and R_{kn} of the biological processes are denoted by w_r and R_{kn}, respectively. A_r and A_{kn} are generalized affinities defined by

$$A_r = -\sum_j \nu_{jr}\mu_j M_j - V_r, \qquad (5.22)$$

and

$$A_{kn} = \mathfrak{S}_{kn} - \sum_j a_{jkn}\mu_j - \sum_l b_{kln}\mu_{kl} - V_{kn}. \qquad (5.23)$$

Here, ν_{jr} denotes stoichiometric coefficients, M_j molar masses, and $\mathfrak{S}_{kn}$ specific absorption of incident radiation by aquatic organisms of species k connected with the process R_{kn}. The coefficients a_{jkn} in the mass balance equations describe the consumption or production of chemical constituents by the kth biotic component during the nth biological process:

$$\rho\frac{dN_j}{dt} + \text{div}\,\vec{J}_j = \sum_r \nu_{jr}M_j w_r + \sum_k \sum_n a_{jkn}R_{kn} \pm \ldots, \qquad (5.24)$$

whereas the coefficients b_{kln} characterize the mass transfer between biological components of the ecosystem:

$$\rho\frac{d}{dt}(B_k m_{kl}) + \text{div}(m_{kl}\vec{J}_k) = \sum_n b_{kln}R_{kn} \pm \ldots . \qquad (5.25)$$

We observe that σ is a bilinear form,

$$\sigma = \sum_m J_m X_m, \qquad (5.26)$$

of quantities, J_m, which can be thought of as "flows," and of quantities, X_m, which act as generalized "forces" giving rise to these flows. For instance, we may identify

Flows:	$\vec{W}$	$\vec{J}_i$	w_r	R_{kn}	...
Forces:	$\text{grad}\,\frac{1}{T}$	$-\frac{1}{T}\text{grad}\,\mu_i\|_T$	A_r/T	A_{kn}/T	...

Regrouping of terms in equations (5.18), (5.19) and (5.21) is possible, leading to somewhat different definitions of the entropy flow density and of some forces and fluxes. But the chemical and biological terms in σ remain unaffected. Arbitrarily chosen partial sums of σ cannot be expected to be nonnegative, but the rates and affinities of biological processes are nonnegative by definition, since these processes are nonreversible.

The functional relations between flows and forces are assumed to be linear as far as transport phenomena are concerned (Onsager's theory). To set up a satisfactory description of chemical and biological processes, it is necessary to extend the theory to the nonlinear range. The relations between the rates, w_r, and the affinities, A_r, as well as those between the rates, R_{kn}, and the affinities, A_{kn}, are nonlinear functions.

Thus, the entropy balance equation and the second law of thermodynamics reduce a large variety of phenomena to a few equations that can serve as starting points for further investigations.

Affinities of Biological Processes

From the entropy balance equation, the affinities, the driving forces of the biological processes, follow as functions of the chemical potentials of the components taking part in these processes [eq. (5.23)]. To determine the dependence of the affinities on light intensity and state variables (temperature, nutrient concentrations, biomasses, and so on), the theory of dilute solutions (see Planck, 1913; Stumm and Morgan, 1970) is applied not only to the dissolved inorganic or organic materials in the water body, but also to the biota (Mauersberger, 1981). In such a way, we get

$$\mu_j = g_j(p, T) + \frac{R_0 T}{\mathcal{M}_j} \ln(x_j), \qquad (5.27)$$

$$\mu_k = g_k(p, T, m_{kl}) + \frac{R_0 T}{\mathcal{M}_k} \ln\left(\frac{\mathcal{M}}{\mathcal{M}_k} B_k\right), \qquad (5.28)$$

where g_j, g_k = specific free enthalpy of component j or k
$R_0, \mathcal{M}$ = universal gas constant, mean molar mass
$\mathcal{M}_j, \mathcal{M}_k$ = molar masses of the component j or k
x_j = molar fraction of the jth chemical component

and

$$g_k(p, T, m_{kl}) = g_k(p, T, O) + \sum_l \mu_{kl} m_{kl}. \qquad (5.29)$$

Introducing (5.27) and (5.28) into (5.23), the driving forces of the biological processes (primary production, uptake, respiration, grazing, and so forth) result as functions of the state variables. Making use of an optimal principle, which connects rates and affinities, it is possible to deduce at least the types of equations that describe how the rates of biological processes depend on temperature, light intensity, biomass, concentrations of nutrients and dissolved oxygen, and other parameters.

CONSEQUENCES FOR ECOSYSTEM DYNAMICS

Overview

The entropy balance plays a fundamental role in the theory of aquatic ecosystems. It allows for the unified treatment of physical, chemical, and biological processes and not only stimulates the derivation of rate–force relations, but also offers stability criteria, supports the determination of bifurcation points, and, last but not least, bridges the phenomenological, stochastic, and cybernetic approaches to ecosystem dynamics. Introducing the macroscopic description of biological processes into the mass-balance equations of the biological components of the ecosystem [for example, (5.25)], the basic equations of population dynamics result. Entropy controls ecological processes and the development of ecosystems (cf. Johnson, 1990).

The structure of the thermodynamic theory of ecosystems, which aims at developing a general method for the analysis and modeling of these systems, is shown in Fig. 5.1. Balance equations and additional assumptions, especially about the biological components, together with the generalized Gibbs relation, form the basis of the entropy balance out of which general as well as specific conclusions can be inferred. Compared with the application of irreversible thermodynamics to physics and chemistry, the thermodynamic theory of aquatic ecosystems is characterized not only by including water organisms, but also by combining thermodynamics with stochastic, cybernetic, and synergetic methods related to the description of biological phenomena.

Characterization of Ecosystems

Ecosystems are open systems. Their boundaries are permeable, permitting energy and matter to cross them. Effects of environmental constraints and influences on the system play an important role in the regulation and maintenance of the system's spatiotemporal as well as trophic organization

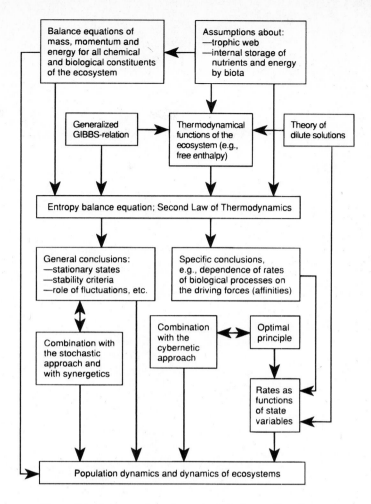

Figure 5.1 Structure of the thermodynamic theory of ecosystems.

and functioning. Ecosystems operate outside the realm of classical thermodynamics. Biological, chemical, and some of the physical processes are nonlinear. Stationary states of ecosystems are nonequilibrium states far from the thermostatic equilibrium. In the course of time, entropy does not tend to a maximum value, nor entropy production to a minimum. Entropy decreases when the order of organization and structure of the system increases. Entropy production is counterbalanced by export of entropy out of the system. Therefore, the entropy principle can contribute to the theory of ecosystems by offering:

- Conditions for the stability of stationary nonequilibrium states (*Fliessgleichgewichte*)
- Thresholds for changes in the ecosystem's regime, which are connected with bifurcation points of the basic system of differential equations

Resilience is a typical property of ecosystems (Holling, 1973, 1976). They are able to persist in a globally stable dynamic regime in spite of suddenly or (quasi-) periodically varying external influences. But if the fluctuations from outside or inside the system exceed a critical size so that they no longer can be compensated for by the present mechanisms of entropy export, then a qualitative change in the system's structure, functioning, and dynamics happens at this threshold. Under varying environmental constraints, the development of an ecosystem is characterized by a sequence of transitions to new regimes, which generates an adequate type of entropy export within a new type of organization and dynamics of the system.

Evolution Criterion, Stationary States, and Instabilities

If the concentrations at the boundaries of the ecosystem are kept fixed, the evolution of the system follows the evolution criterion of Glansdorff and Prigogine (1971):

$$\frac{\partial_X P}{\partial t} \equiv \int_V \sum_m J_m \frac{\partial X_m}{\partial t} dV \leq 0 , \qquad (5.30)$$

which states that the changes of the forces X_m always proceed in such a way as to lower the entropy production, P, of the ecosystem:

$$P = \int_V \sigma \, dV = \int_V \sum_m J_m X_m \, dV > 0 . \qquad (5.31)$$

In (5.30) and (5.31), the local entropy production, σ, of all physical, chemical, and biological processes is integrated over the volume, V, of the ecosystem.

Stationary states of the ecosystem are defined by time-independent values of state variables, forces, and flows (for example, s^*, u^*, ρ^*, ..., A_i^*/T^*, A_{kn}^*/T^*, ..., $\vec{W}^*$, $\vec{J}_i^*$, w_r^*, R_{kn}^*, ...). However, these variables, forces, and flows can depend on the space coordinates. Stationary states, denoted by X_m^*, J_m^*, are stable, if the excess entropy production $\delta_X P$ is positive definite:

$$\delta_X P = \int_V \delta_X \sigma \, dV = \int_V \sum_m \delta J_m \, \delta X_m \, dV > 0 . \qquad (5.32)$$

$\delta_X P$ is the part of the entropy production that arises from the "excess flows," $\delta J_m = J_m - J_m^*$, and "excess forces," $\delta X_m = X_m - X_m^*$, that is, from the deviations of the state, X_m, J_m, from the values, X_m^*, J_m^*, at the stable stationary state.

From the Gibbs relation (5.11), it can be inferred that, around the stationary state, entropy becomes a homogeneous function of second degree in the independent variables:

$$s - s^* = -\frac{1}{2} \sum_j \sum_{j'} \left(\frac{\partial \mu_j}{\partial N_{j'}}\right)^* \delta N_{j'} \, \delta N_j$$

$$-\frac{1}{2} \sum_k \sum_{k'} \left(\frac{\partial \mu_k}{\partial B_{k'}}\right)^* \delta B_{k'} \, \delta B_k - \frac{1}{2} \sum_r \sum_{r'} \left(\frac{\partial \zeta_r}{\partial V_{r'}}\right) \delta V_{r'} \, \delta V_r - , \dots . \qquad (5.33)$$

[For the sake of brevity, some additional terms have been suppressed in eq. (5.33)]. Therefore, the stability of a locally stationary state can be investigated using the Liapunov function

$$L = s - s^* - \frac{1}{T} |\vec{v} - \vec{v}^*|^2 < 0 , \qquad (5.34)$$

which requires that the necessary and sufficient thermodynamic stability conditions

$$\sum_m \sum_n a^*_{mn} \, \delta\eta_m \, \delta\eta_n > 0 \tag{5.35}$$

and, furthermore, the sufficient condition

$$\frac{dL}{dt} \geqq 0 \tag{5.36}$$

be fulfilled. In eq. (5.35), we used the abbreviations a^*_{mn} for $\partial\mu_j/\partial N_{j'}$, $\partial\mu_k/\partial B_{k'}$, $\partial\xi_r/\partial V_{r'}$, and $\partial\xi_{kn}/\partial V_{k'n'}$, and η_m for N_j, B_k, V_r, V_{kn}.

If the concentrations at the boundaries of the volume element under consideration are kept constant, then it follows that $\rho(dL/dt)$ is identical with the local excess entropy production $\delta^2\sigma$:

$$\rho\frac{dL}{dt} = \delta^2\sigma = \sum_m \delta J \, \delta X. \tag{5.37}$$

The vanishing of $\delta_X P$ or $\delta^2\sigma$, respectively, determines the thermodynamic threshold (or bifurcation point) that separates the stable regime ($\delta_X P > 0$ or $\delta^2\sigma > 0$) from the unstable regime ($\delta_X P < 0$ or $\delta^2\sigma < 0$). Beyond such a threshold another, perhaps radically different type of structure and functioning of the ecosystem comes into being.

For the management of water resources, the investigation of critical situations of aquatic ecosystems when their regime becomes unstable and a new type of structure and functioning originates, is at least as important as simulation of the annual course of part of the state variables, which is usually the aim of ecosystem modeling. Structural changes must be studied thoroughly before we can rely on forecasts with the help of models. Thermodynamic as well as mathematical aspects of evolutionary processes involving changes of structures are treated by Zotin, 1972; Rubin, 1976; Nicolis and Prigogine, 1977; Ebeling and Feistel, 1982; and others.

Combination of Deterministic, Stochastic, Synergetic, and Cybernetic Methods

The thermodynamic theory of aquatic ecosystems is based on the following set of macroscopic equations (Mauersberger, 1978, 1983):

- Mass balances of all constituents (continuity equations)
- Balance of momentum (Navier–Stokes equation)
- Balance of kinetic, internal, and electromagnetic energy
- Appropriate initial and boundary conditions
- Generalized Gibbs relation
- Entropy balance and second law of thermodynamics
- Rates of biological processes as functions of state variables

Derived from these basic equations, deterministic water-quality models are nonlinear initial boundary value problems of ordinary or partial differential equations (see, for example, Orlob, 1983; Straškraba and Gnauck, 1983). Within special intervals of parameters and variables, systems of nonlinear differential equations or nonlinear difference equations may allow for more than one solution. Solutions bifurcate at "bifurcation points." (Bifurcation theory is treated, for instance, by Ioos and Joseph, 1980; Salvadora, 1984.) The system of nonlinear differential or difference equations can be solved only with the help of computers. It is troublesome or even impossible to get, in all cases, an overview of the set of possible solutions by numerical integration. Therefore, attention is drawn to the fact that the vanishing of the excess entropy production, $\delta_x P = 0$, allows for the determination of bifurcation points.

To deal adequately with the qualitative changes of the structure and dynamics of ecosystems, in particular with bifurcations, the fluctuations of the state variables, forces, and flows must be taken into account. The behavior of nonlinear systems near bifurcations is one of the reasons for the combination of deterministic and stochastic methods and for the application of synergetics in ecosystem modeling. The very existence of many degrees of freedom in ecosystems automatically implies the appearance of fluctuations generated inside the system in addition to those introduced from outside. The expectation values of the stochastic approach to ecosystem dynamics correspond to the variables used in the deterministic description. But additional information is needed, especially about variances and further statistical characteristics that reflect the effects of fluctuations in nonlinear systems. This is especially the case if in the macroscopic description more than one physically acceptable (that is, real and bounded) solution of the nonlinear differential or difference equations exists and if the macroscopic equations do not justify a preference for one of them. Near the bifurcations, the statistical fluctuations can be amplified and finally may drive the average values to a new macroscopic state corresponding to another branch of the solution of the deterministic equations. Because of the significance of fluctuations for the development of ecosystems, stochastic methods must be applied in addition to deterministic equations (Mauersberger, 1988).

It is worth mentioning that methods of synergetics also offer powerful tools for dealing with fluctuations and for studying the stochastic dynamics in the transition zone near bifurcations (Haken, 1983). Synergetics focuses attention on situations in which the macroscopic behavior of the ecosystem undergoes drastic changes. Synergetics demonstrates how cooperation of subsystems brings about spatial, temporal, or functional structures on macroscopic scales. In cases when the system loses (linear) stability, a small number of state variables or only one state variable, the order parameters, dominates in the sense that it governs all other state variables ("slaving principle"). The complex problem of solving the full system of model equations is reduced to the solution of a small system of equations or of one differential equation, respectively.

Ecosystems are complicated structures. For management purposes, ecological models must be coupled to economic models. Often it is sufficient to analyze and predict some integral features of the system, for instance, the amount of seston or the total biomass of phytoplankton and the dominant species. To reduce the number of equations to be solved, the cybernetic approach to ecosystem modeling postulated optimization principles from which such integral features can be derived. A goal function, which the system is supposed to be seeking, is assumed. The model parameters or, in other cases, the species best suited to given conditions are determined by optimization procedures. While cybernetics contributes valuable knowledge, for instance, about the hierarchy of systems and subsystems and about the feedback mechanisms in complex systems, the formation and stability of structures of ecosystems, the feedback processes, and so on, depend strongly on physical, chemical, and biological constraints. Therefore, the thermodynamic method, which takes the basic physical, chemical, and biological laws fully into account, must be combined with the cybernetic approach. Comparing cybernetics with synergetics, we observe that both are concerned with control. Self-organization processes under changing control are studied by synergetics whereas in cybernetics procedures are devised for controlling a system so that it behaves in a desired manner.

Thus, we come to the conclusion that, to improve ecosystem modeling further, we have to succeed in combining deterministic and stochastic approaches as well as hydrothermodynamic, cybernetic, and synergetic methods.

Chaotic Behavior

In attempting to predict an ecosystem's behavior, we should be aware that predictability of even classical deterministic systems can, under certain circumstances, be limited to a short time. After this time, the initial and final states will be causally disconnected. When a control parameter is changed continuously, a nonlinear system may pass through a hierarchy of instabilities and different kinds of patterns: spatially homogeneous or inhomogeneous time-independent states, periodic motions and period-couplings, quasiperiodic motions (containing several independent frequencies), chaos, and various transitions between these behavioral states. Chaos is an irregular, but bounded, motion which is, however, the solution of deterministic equations and different from noise. While noise can be reduced by refinement of the measurements, this is impossible with respect to chaotic behavior (sometimes called intrinsic noise).

Chaotic regimes are a special type of solution of nonlinear difference or differential equations when the distance between two initially closely neighboring trajectories grows rapidly, for a short time even exponentially. If this is the case, then the future development of the ecosystem or subsystems can proceed along quite different routes, although the equations describing temporal evolution are entirely deterministic. It is reasonable to assume that such phenomena exist in ecosystems (Allen, 1990; Gutierrez and Almiral, 1989; Loehle, 1989). Therefore, the following questions arise: What are the limits of predictability of the development of a given ecosystem? Beyond which limits will further efforts to improve forecasts be fruitless? What role is played by intrinsic irregular variations of state variables, population densities, and energy fluxes? Answers are not yet quite clear, in spite of a growing number of investigations (for example, May 1974, 1977; Haken, 1981).

MACROSCOPIC DESCRIPTION OF BIOLOGICAL PROCESSES DERIVED FROM THERMODYNAMIC PRINCIPLES

Introduction

A great variety of formulas has been proposed and used in water-quality modeling to describe at the macroscopic level the dependence of biological processes on temperature, light intensity, pH, nutrient concentrations, and other external and internal controlling factors (see, for instance, Orlob, 1983; Straškraba and Gnauck, 1983). An excellent review of many of the early models has been given by Patten (1968). Examined with regard to their application for the simulation of observed data, these formulas are to a certain degree equally useful. But in view of the analysis of processes and the deeper understanding of ecosystems, this variety of formulas, revealing a lack of knowledge about the ways influencing factors interplay, is of limited value. The great number of more or less different mathematical formulations for each process results from the fact that it is difficult, or even impossible, to infer from field measurements how the rates of the biological processes depend on several simultaneously varying independent variables, the relative importance of which varies with time and space.

On the basis of thermodynamic theory, the mathematical formulations for the control of macroscopic biological processes (primary production, uptake, assimilation, respiration, and the like) by light intensity, nutrient concentrations, biomass, and temperature can be derived (Mauersberger, 1982d, 1984). Supposing that biological processes are neither completely determined by the laws of physics and chemistry nor in contradiction to them, additional equations typical for biological phenomena are necessary to bring the system of ecological equations to the full complement. From the entropy-balance equation it follows how forces that drive the physical, chemical, and biological processes depend on state variables, for example, on light intensity, temperature, nutrient concentrations, and biomass. But the nonlinear relations between biological process rates and driving forces (affinities) cannot be inferred from the entropy-balance equation. An additional optimization principle, which is to be understood in the sense of dynamic programming, has been proposed, from which at least the types of mathematical functions describing the dependences of rates on affinities can be derived.

Furthermore, information is needed about the influence of temperature and pH on enzymatically catalyzed reactions and about the dependence of photochemical reactions on the density of photons. Introducing the affinities as functions of state variables, the functional relationships between rates and state variables are determined. Continuously varying control variables can be introduced. This way of deriving the macroscopic description of biological processes from basic principles aims at ensuring against defects in the fundamental premises, and this is of great importance not only for any scientific discipline but also for management activities.

Optimization Principle

The state of the biocoenosis at any time is specified by the mass fractions, B_k, of the biomasses of the species $k = 1, 2, 3, \ldots, K$ as functions of space and time. The rates, Y_{kn}, or $R_{kn} = \rho_{ko} Y_{kn}$, respectively, of the biological processes, such as primary production, uptake, respiration, and grazing ($n = 1, 2, 3, \ldots$), are nonlinear functions of the driving forces, X_{kn}, of these processes. For notational simplicity, the sets B_k, Y_{kn}, and X_{kn} are denoted by B, Y, and X, and the dependence of these quantities on the local coordinates and on time is not written explicitly.

At any time, the local state of the biocoenosis can be characterized by its deviation from a stable stationary reference state (*Fliessgleichgewicht*). In thermodynamics of nonlinear irreversible physical and chemical processes, a stationary state Y_i^*, X_i^* is defined by

$$\sum_i Y_i^* (X_i - Y_i^*) = 0 . \tag{5.38}$$

This stationary state is stable if the excess entropy production is positive definite [see eqs. (5.36) and (5.37)]:

$$\sum_i (Y_i - Y_i^*)(X_i - X_i^*) \geq 0 . \tag{5.39}$$

Equations (5.37) and (5.38) are valid for small deviations from the state Y_i^*, X_i^*. Biological systems are able to widen out stationary states, for instance by adaptation. Therefore, we define the generalized excess entropy production caused by biological processes by

$$E = \sum_k B_k \sum_n \left\{ \int_{X_{kn}^*}^{X_{kn}} [Y_{kn}(X) - Y_{kn}^*] d\zeta_{kn} + \Lambda_{kn} \int_{X_{kn}^*}^{X_{kn}} Y_{kn}^* d\zeta_{kn} \right\} . \tag{5.40}$$

Condition (5.38) is taken into account in (5.40) with the help of Lagrangian multipliers, Λ_{kn}.

To determine the nonlinear relations between rates and affinities of biological processes, the following principle of dynamic programming is introduced (Mauersberger, 1982a,b; 1985): The development of the biocoenosis is controlled during the finite time interval of length τ by the affinities X in such a way that the time integral

$$I = \underset{X[\tau]}{\text{extr}} \int_{t-\tau}^{t} E(B(t'), X(t')) \, dt' \tag{5.41}$$

over the generalized excess entropy production, E, within $(t - \tau) \leq t' \leq t$ becomes an extreme value (minimum) subject to the initial values $B_0 \equiv B(t - \tau)$ and to the mass balance equations

$$\frac{dB_k(t)}{dt} = f_k(B(t), Y(X(t))) , \tag{5.42}$$

which connect B and X through the rates Y.

The optimization principle (5.41) postulates that locally, but within a finite time interval, the deviation of the bioprocesses from a stable stationary state tends to a minimum. Whereas in the classical calculus of variations the concept of a function yielding an extremal point in function space is used, in (5.41), as is usual in dynamic programming, multistage decision processes are considered yielding an extremum by a continuous policy (Bellman, 1957).

For extreme values inside the region $\{B, X\}$, where the X values vary freely, (5.41) for $j, k = 1, 2, 3, \ldots, K$ and $n = 1, 2, 3, \ldots$ leads to the equations

$$\frac{\partial E(B, X)}{\partial X_{kn}} + \sum_j \frac{\partial I}{\partial B_j} \frac{\partial f_j(B, Y(X))}{\partial X_{kn}} = 0, \qquad (5.43)$$

which allow for the derivation of the nonlinear functions $Y = Y(X)$.

Process Rates as Functions of Affinities

Cross processes, that is, the dependence of a rate on the affinities of other rates, are involved in the optimization principle (5.41), as well as in (5.43). Of course, for different species ($j \neq k$) the process rates, Y_{kn}, are only indirectly connected with the driving forces, X_{jm}, of other species through the state variables (light intensity, temperature, nutrient concentrations, biomass, and the like) upon which the affinities X_{kn}, X_{jm} depend. Therefore, it is assumed that

$$\frac{\partial Y_{kn}}{\partial X_{jm}} = \frac{\delta_{jk} \, \partial Y_{kn}}{\partial X_{km}}. \qquad (5.44)$$

If, furthermore, all the cross effects are supposed to be linear processes, eqs. (5.43) can be solved. Here we confine ourselves to the simplified solution when cross effects are neglected:

$$Y_{kn} = Y_{kn}^* \frac{1 - D_{kn} \exp[-c_k X_{kn} \operatorname{sign}(n)]}{1 - D_{kn} \exp[-c_k X_{kn}^* \operatorname{sign}(n)]}, \qquad (5.45)$$

where sign $(n) > 0$ for "producing" processes like primary production, but sign$(n) < 0$ for "reducing" processes like respiration. These solutions are valid only within the neighborhood of the stable stationary state Y^*, X^*. If more than one reference state and different branches of the solution exist, these branches are assumed to be of type (5.45) and to join steadily, but they must not converge with steadily varying tangents.

Furthermore, it is not necessary that all branches tend to zero if all driving forces X go to zero. It is only for the branch, the reference state Y^*, X^* which is nearest to $X = 0$, that $Y \to 0$ is certain for $X \to 0$. If, in ecosystem modeling, $Y(X)$ is assumed to be a steadily differentiable function for all X within $0 \leq X \leq \infty$, this is valid only under additional simplifying assumptions. The case of unsteadily varying tangents should be considered thoroughly.

In eq. (5.45), the coefficient c_k is determined by

$$\frac{1}{c_k} = \frac{\partial I}{\partial B_k}, \qquad (5.46)$$

that is, by the excess entropy production of the kth species in the time interval $(t - \tau) \ldots t$, thus taking into account the history of this population. The magnitude of c_k is small if the excess entropy production was high, and vice versa. Since, as a rule, the deviation from the stable stationary reference state is high during phases of rapid development, c_k is small during and immediately after such periods.

Defining the driving forces X_{kn} by

$$X_{kn} = \frac{\tilde{A}_{kn}}{R_0 T}, \qquad \tilde{A}_{kn} = A_{kn} \mathcal{M}_k, \tag{5.47}$$

for the rates of uptake, primary production, and other producing processes [sign(n) > 0], we infer from eq. (5.45) that

$$Y_{kn} = Y_{kn}^* \frac{1 - D_{kn} \exp(-c_k \tilde{A}_{kn}/(R_0 T))}{1 - D_{kn} \exp(-c_k \tilde{A}_{kn}^*/(R_0 T))}, \tag{5.48}$$

but for the rates of respiration and other reducing processes [sign(n) < 0],

$$Y_{kn} = Y_{kn}^* \frac{D_{kn} \exp(c_k \tilde{A}_{kn}/(R_0 T)) - 1}{D_{kn} \exp(c_k \tilde{A}_{kn}^*/(R_0 T)) - 1}. \tag{5.49}$$

Affinities as Functions of State Variables

For brevity, the abbreviation "process (k, n)" is used in the following for the nth biological process of the kth species. From the local entropy balance [or by introducing the mass balance eqs. (5.24) and (5.25) into the generalized Gibbs relation (5.11)], the affinity, A_{kn}, of the biological process (k, n), for instance (5.9), can be derived. Since the macroscopically defined biological processes are one-way processes, that is, taking place in one direction only, the condition

$$\tilde{A}_{kn} \geqq 0 \tag{5.50}$$

must be fulfilled.

The amount of energy internally stored in the kth species is increased by photosynthesis (light process) and respiration. This energy is used for the uptake of nutrients and assimilation. The change of internally stored energy and the mass transfer connected with the biological process (k, n) are treated as coupled simultaneous reactions. For the macroscopic description of the amount of internally stored energy, we use the quantity ϵ_k defined by

$$\epsilon_k = \exp\left(\frac{-\tilde{A}_{Ek}}{R_0 T}\right), \tag{5.51}$$

where $\tilde{A}_{Ek}$ is the affinity of the chemical process connected with changes of internally stored energy. Since $\epsilon_k \to 0$ when the internal energy storage becomes completely filled, ϵ_k may be called an "energy deficit."

From eqs. (5.23), (5.28), and (5.51), it follows that the affinity $\tilde{A}_{kn}$ can be written as

$$\tilde{A}_{kn} = R_0 T(f_{kn} - \ln Z_{kn}), \tag{5.52}$$

where, for convenience, the abbreviations

$$f_{kn} = -\frac{(\Delta_{kn}\tilde{G} + \tilde{V}_{kn})}{R_0 T} \tag{5.53}$$

and

$$Z_{kn} = \epsilon_k^{p_{kn}} \prod_1 x_i^{\nu_i} \tag{5.54}$$

have been introduced.

$\Delta_{kn}\tilde{G}$ is the standard free enthalpy of the mass transfer by the process (k, n). The molar fractions of the chemical and biological components participating in the process (k, n) are denoted by x_i. While Z_{kn} depends on the concentrations, f_{kn} first of all is a nonlinear function of temperature (and of pressure, if pressure varies intensively).

The affinity of photosynthesis, depending on the absorption of light, θ_k, may be written

$$\tilde{A}_k = \theta_k - \theta_k^{\min} \gtreqless 0, \qquad \theta_k = \vartheta_k \, \eta_k, \tag{5.55}$$

where

$$\theta_k^{\min} = \Delta_{kP}\tilde{G} + \tilde{V}_{kP} + R_0 T \ln Z_{kP}. \tag{5.56}$$

Thus, $\theta_k^{\min}$ is a function of temperature, pH, oxygen concentration, internally stored energy, and so on.

Process Rates as Functions of State Variables

Introducing eqs. (5.52) and (5.55) into (5.48) and (5.49), we obtain

$$R_{kn} = R_{kn}^{\max} \left[1 - D_{kn} \cdot Z_{kn}^{c_k} \cdot \exp(-c_k f_{kn}) \right] \tag{5.57}$$

for the rates, R_{kn}, of uptake and assimilation,

$$R_k = R_k^0 \left[D_{kR} \cdot Z_{kR}^{-c_k} \cdot \exp(c_k f_{kR}) - 1 \right] \tag{5.58}$$

for the rate, R_k, of respiration, and

$$P_k = P_k^{\max} \left[1 - D_{kP} \exp\left[\frac{-c_k(\theta_k - \theta_k^{\min})}{R_0 T} \right] \right] \tag{5.59}$$

$$= P_k^{\max} \left[1 - D_{kP} Z_{kP}^{c_k} \exp\left[-c_k \left(\frac{\theta_k}{R_0 T} + f_{kP} \right) \right] \right] \tag{5.60}$$

for the rate, P_k, of photosynthesis.

From eqs. (5.48), (5.49), and (5.57) through (5.60), it becomes evident that $R_{kn}^{\max}$, R_k^0, and $P_k^{\max}$ depend on c_k, that is, on the history of the population. These quantities increase during periods of rapid development. Furthermore, they may be functions of temperature. Very often it is assumed that

$$R_{kn}^{\max}, R_k^{(0)}, P_k^{\max} \sim \exp\left(\frac{-E}{R_0 T} \right), \tag{5.61}$$

where E is the activation energy of the special process.

The functions (5.57) through (5.60) describe the relationships between the rates of biological processes and the state variable in aquatic ecosystems. Coefficients that cannot be determined by the macroscopic (phenomenological) theory must be evaluated in accordance with microscopic theories, laboratory data, and field measurements.

The main influence of temperature on the rates (5.57) through (5.60) results from the functions f_{kn} and $\theta_k^{\min}$. Differentiating with respect to T and omitting henceforth the subscripts, we obtain from (5.57) for uptake, assimilation, and so forth,

$$\frac{\partial R}{\partial T} = (R^{\max} - R)c \frac{\partial f}{\partial T} + \cdots \tag{5.62}$$

and

$$\frac{\partial^2 R}{\partial T^2} = (R^{\max} - R)c \left(\frac{\partial^2 f}{\partial T^2} - c\left(\frac{\partial f}{\partial T}\right)^2 \right) + \cdots. \tag{5.63}$$

From (5.58), follows that for respiration

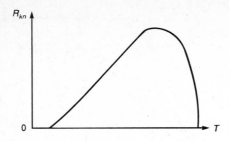

Figure 5.2 Relationship between process rate, R_{kn}, and temperature, T.

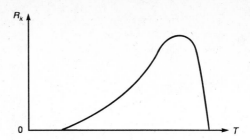

Figure 5.3 The rate R_k of respiration as a function of temperature T at constant values of oxygen concentration, biomass, and light intensity.

$$\frac{\partial R}{\partial T} = (R^{(0)} + R)c\frac{\partial f}{\partial T} + \cdots \quad (5.64)$$

and

$$\frac{\partial^2 R}{\partial T^2} = (R^{(0)} + R)c\left(\frac{\partial^2 f}{\partial T^2} + c\left(\frac{\partial f}{\partial T}\right)^2\right) + \cdots, \quad (5.65)$$

but photosynthesis, from (5.60),

$$\frac{\partial P}{\partial T} = c(P^{\max} - P)\left(\frac{\partial f}{\partial T} - \frac{\theta}{R_0 T^2}\right) + \cdots \quad (5.66)$$

and

$$\frac{\partial^2 P}{\partial T^2} = c(P^{\max} - P)\left(\frac{\partial^2 f}{\partial T^2} + \frac{2\theta}{R_0 T^3} - c\left(\frac{\partial f}{\partial T} - \frac{\theta}{R_0 T^2}\right)^2\right) + \cdots. \quad (5.67)$$

The dots indicate additional terms resulting from the time dependence of $R^{\max}$, $R^{(0)}$, and $P^{\max}$. Taking account of Van't Hoff's reaction isobar, we derive from (5.37) that

$$\frac{\partial f}{\partial T} = \frac{\Delta H}{R_0 T^2} - \frac{\partial [V/(R_0 T)]}{\partial T}. \quad (5.68)$$

The standard reaction enthalpy, ΔH, is $\Delta H < 0$ for respiration, but $\Delta H > 0$ for other processes, such as assimilation. Often it is possible to neglect $\partial \Delta H/\partial T$. This leads to

$$\frac{\partial^2 f}{\partial T^2} = -2\frac{\Delta H}{R_0 T^3} - \frac{\partial^2}{\partial T^2}\left(\frac{V}{R_0 T}\right). \quad (5.69)$$

Thus, relations between process rates and temperature have been derived. Figures 5.2 through 5.4

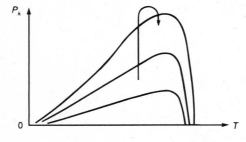

Figure 5.4 Type of dependence of the rate, P_k, of photosynthesis on temperature, T, at different intensities of irradiance. The arrow indicates the influence of increasing light intensity on this function.

illustrate these rates qualitatively as functions of temperature at constant values of oxygen and nutrient concentrations, pH, internal energy storage, and external light intensity. Influences of these other factors on rates of biological processes are discussed in the following two sections.

Uptake

Because of the ability of phytoplankton to store nutrients in excess of their requirements, their relative growth rate must not immediately respond to a change in the external nutrient concentrations. Also, it is known from chemostat studies that the specific growth rate depends on internal rather than external nutrient levels, whereas the specific uptake rate depends on both internal and external nutrient concentrations. Furthermore, energy generated by photosynthesis, as well as by respiration, and stored internally in the biomass is used for enzymatically controlled ion transport against concentration gradients from the external medium into the interior of the biomass. Therefore, the accumulation of surplus nutrients within the algae and the role of internally stored energy demand the separation of uptake rate, photosynthesis, and assimilation rate, if in the time scale used in the model the time lags between nutrient uptake, photosynthesis, and growth are of importance.

The rate, U_{kj}, of uptake of nutrient j (for example, nitrate) by species k depends on the molar fractions, x_j and x_{kj}, of the external or internal nutrient concentrations, respectively. We further use the mass fractions, N_{kj} of nutrient j stored internally in species k, and S_{jk} of the external nutrient concentration relative to the biomass of species k:

$$N_{kj} = \frac{\rho_{kj}}{\rho_k}, \qquad S_{jk} = \frac{\rho_j}{\rho_k} = \frac{N_j}{B_k}, \qquad (5.70)$$

$$N_j = \frac{\rho_j}{\rho}, \qquad B_k = \frac{\rho_k}{\rho}.$$

Equation (5.54) specializes into

$$Z_{Ukj} = \epsilon_k^{p_{U_{kj}}} \frac{x_{kj}}{x_j} = \epsilon_k^{p_{U_{kj}}} \frac{N_{kj}}{S_{jk}} \qquad (5.71)$$

and (5.57) becomes

$$U_{kj} = U_{kj}^{\max}(c_k)\left[1 - D_{U_{kj}}\left(\epsilon_k^{p_{U_{kj}}} \frac{N_{kj}}{S_{jk}} e^{-f U_{kj}}\right)^{c_k}\right] \geqq 0. \qquad (5.72)$$

Omitting the subscripts, eq. (5.72) reads

$$U = U^{\max}(c)\left[1 - D\left(\epsilon^p \frac{N}{S} e^{-f}\right)^c\right] \geqq 0 \qquad (5.73)$$

and, drawing special attention to particular independent variables, can be written in the following equivalent formulations:

$$\frac{U}{U^{\max}} = 1 - \left(\frac{N}{N^{\max}}\right)^c = 1 - \left(\frac{S^{\min}}{S}\right)^c = 1 - \left(\frac{\epsilon}{\epsilon^{\max}}\right)^{pc} \geqq 0. \qquad (5.74)$$

The uptake rate does not depend on the absolute value of the external nutrient concentration, but on the mass fraction of the external nutrient relative to the local value of the biomass. Under given conditions of temperature, amount of internally stored energy, external nutrient concentration, and biomass of the species, the internally stored nutrient can reach the maximum value

$$N^{max} = \frac{Se^{f(T)}}{\epsilon^p D^{1/c}}, \qquad (5.75)$$

which increases with increasing external nutrient concentration and with increasing internal energy available for uptake.

If only S is varied and all other variables are kept constant, then a minimum value

$$S^{min} = Ne^{-f(T)}\epsilon^p D^{1/c} \qquad (5.76)$$

exists, below which uptake of the nutrient is impossible. S^{min} increases with the increase of internal nutrient storage and of energy deficit, ϵ, which in turn must be smaller than ϵ^{max}:

$$\epsilon^{max} = \left(\frac{Se^{f(T)}}{ND^{1/c}}\right)^{1/p}. \qquad (5.77)$$

Depending on both internal and external variables, N^{max}, S^{min}, and ϵ^{max} characterize limits of uptake under momentary local conditions, but do not play the role of typical constants of the species.

The dependence of the uptake rate on independent variables as inferred from thermodynamic theory is shown in Figs. 5.5 through 5.7.

In water-quality modeling, the relation

$$U = \frac{U_{max}S}{K + S} \qquad (5.78)$$

is commonly used, denoting by S the external nutrient concentration but not the mass fraction ρ_j/ρ_k. Ignoring this very important difference, formulas (5.74) and (5.78) are qualitatively compared with each other in Fig. 5.8. Obviously, eq. (5.78) offers a good approximation. But (5.73) and (5.74) have the advantage of including the functional dependencies on other controlling factors like temperature, as well as internal storage of nutrients and energy.

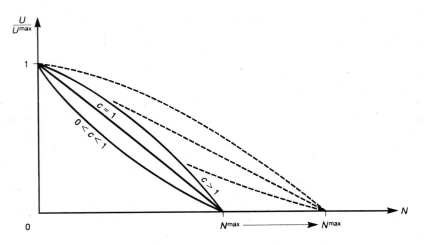

Figure 5.5 Uptake rate, U, as a function of the mass fraction, N, of internally stored nutrient relative to the biomass of the uptaking species. The arrow indicates the effect of increasing external nutrient concentration and increasing internal storage of energy available for uptake.

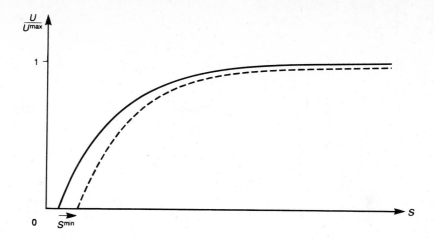

Figure 5.6 Uptake rate, U, as a function of the mass fraction, S, of external nutrient relative to the biomass of the uptaking species. The minimum value, S^{min}, increases if internal nutrient storage, N, and energy deficit, ϵ, are increasing or if (suboptimal) temperature, T, decreases.

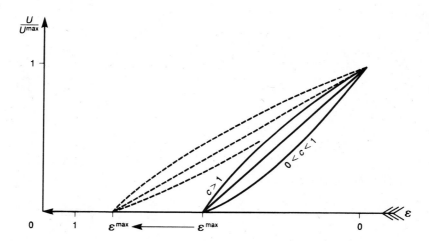

Figure 5.7 Uptake rate, U, as a function of the deficit, ϵ, of internally stored energy, the upper limit, ϵ^{max}, of which increases if (suboptimal) temperature, T, and external nutrient, S, increase, or internal nutrient storage, N, decreases.

Primary Production

Gross primary production can be modeled as one of the biological processes besides respiration, grazing, and mortality, if the time lags between uptake, photosynthesis, and assimilation are short compared with the time scale of modeling. If, furthermore, variations in the molar mass of the phytoplankton species are neglected, and if it can be assumed that approximately constant nutrient uptake and renewal of internally stored energy always ensure assimilation, then eq. (5.54) becomes

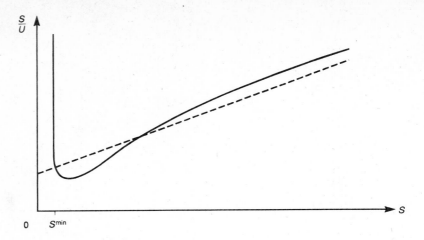

Figure 5.8 Comparison of the thermodynamically founded formula (5.74), with (5.78) shown by the broken line.

$$Z_{kP} \approx K_{kP} B_k \prod_j x_j^{\nu_j}, \tag{5.79}$$

with $K_{kP} \simeq$ constant. For a chemical constituent j produced by primary production (for example, oxygen), $\nu_j > 0$ by definition, but $\nu_j < 0$ for chemicals that are consumed during this process (HCO_3, NO_3, PO_4, and others). From eqs. (5.79) and (5.44), the following expression for the rate of gross primary production results if the number k of species is omitted:

$$P = P^{max}(c) \left[1 - D_P \left\{ BK_p \prod_j x_j^{\nu_j} \exp\left(-f_P - \frac{\theta}{R_0 T}\right) \right\}^c \right] \geqq 0. \tag{5.80}$$

This equation can be transformed by introducing the carrying capacity, B^{max}, and the limits, x_j^{min} and $[O_2]^{max}$, for the external nutrient j or the oxygen concentration $[O_2]$ in the water, respectively:

$$\frac{P}{P^{max}} = 1 - \left(\frac{B}{B^{max}}\right)^c = 1 - \left(\frac{x_j^{min}}{x_j}\right)^c = 1 - \left(\frac{[O_2]}{[O_2]^{max}}\right)^c \geqq 0. \tag{5.81}$$

The carrying capacity

$$B^{max} = \exp\left(f_P + \frac{\dfrac{\theta}{R_0 T}}{D_P^{1/c} K_P \prod_j x_j^{\nu_j}}\right) \tag{5.82}$$

is a function of light intensity (through θ), temperature, and the local concentrations of dissolved oxygen and other constituents taking part in primary production. B^{max} increases with increasing nutrient concentrations, but decreases with increasing concentration of oxygen in the water, which in turn is limited by $[O_2] \leqq [O_2]^{max}$. The molar mass, x_j, of a dissolved nutrient must exceed a minimum value

$$x_j^{\min} = D_P^{1/c} B K_P \prod_{\substack{m \\ (m \neq j)}} x_m^{v_m} \cdot \exp\left(-f_p - \frac{\theta}{R_0 T}\right), \tag{5.83}$$

which increases with biomass, dissolved oxygen, and decreasing other nutrients. Neither $B^{\max}$ and $[O_2]^{\max}$ nor $x_j^{\min}$ are constant attributes of the phytoplankton species. They characterize the local conditions for primary production, which depend on characteristics of the species, the history of the population, and the local state of the ecosystem, as can be seen from eqs. (5.82) and (5.83).

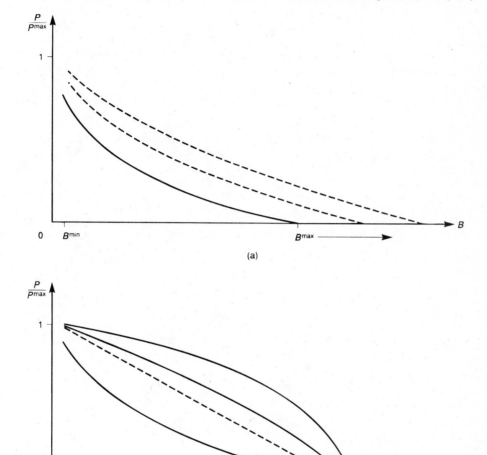

Figure 5.9 Rate of primary production, P, as a function of the biomass of the species under consideration: (a) for increasing nutrients, (suboptimal) temperature, and light intensity or decreasing oxygen concentration, but constant c_k; (b) for varying c_k, but constant T, x_j, $[O_2]$, and light intensity.

Figures 5.9 through 5.12 show the rate of gross primary production as a function of controlling factors as determined by equations (5.80) and (5.81), which have been inferred from thermodynamic principles.

Dominant Controlling Variables

One of the preconditions for progress in ecosystem modeling is the synthesis of methods of cybernetics and thermodynamics. In the cybernetic approach to ecosystem modeling, continuous control variables (for instance cell or colony volume) are used to characterize different species within functional groups of the biocoenosis (Straškraba et al., 1979). In this case, it is necessary to know the rates of biological processes as functions of the control variables. Applying especially the cell volume for the characterization of the dominating phytoplankton species, we have to know how photosynthesis, uptake, respiration, and the like, depend on cell volume. Radtke and Straškraba (1980) inferred these relations from data originating from measurements in lakes, reservoirs, and laboratories. They used linearized approximations, but stressed that in most instances the curves are nonlinear.

These nonlinear relationships can be derived by introducing continuous control variables into the thermodynamic optimization principle discussed previously (Mauersberger, 1985c). Instead of numbering the different phytoplankton species, $k = 1, 2, 3, \ldots$, one or more control variables, v, are introduced. The set of biomass fractions, B_k, of the phytoplankton species is replaced by the biomass fraction $B(v)$ of the dominant species. The sets of rates, Y_{kn}, and affinities, X_{kn}, are replaced by $Y_n(v)$ and $X_n(v)$. The state of phytoplankton at any time and place is specified by the mass fraction, $B(v(r, t))$, of the biomass of the dominating species. The rates of the processes, $n = 1, 2, 3, \ldots$ (photosynthesis, nutrient uptake, respiration, etc.) determine the further development of the phytoplankton. They are nonlinear functions of the affinities, $X_n(v)$, which act as driving forces for these processes.

The relationships between rates and affinities follow from the optimization principle:

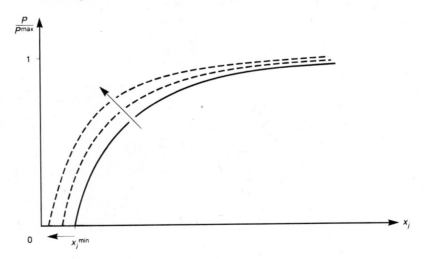

Figure 5.10 Rate of primary production, P, as a function of the (external) nutrient concentration, x_j (molar fraction). The arrow indicates the effect of decreasing biomass or increasing (suboptimal) temperature, light intensity, and concentrations of other nutrients (x_m; $m \neq j$).

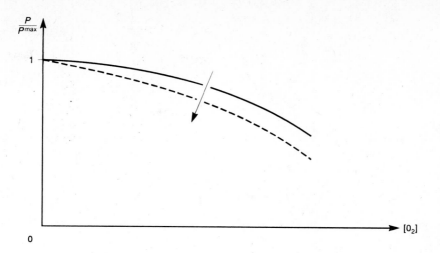

Figure 5.11 Rate of primary production, P, as a function of the oxygen concentration, $[O_2]$, in water. The arrow indicates increasing biomass.

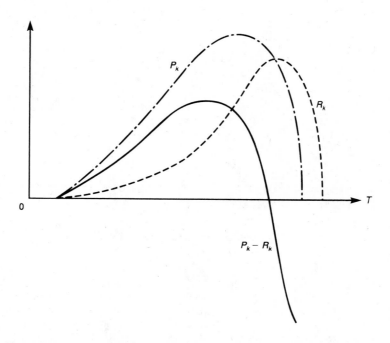

Figure 5.12 Temperature dependence of gross primary production, P_k, respiration, R_k, and net primary production rate, $(P_k - R_k)$, for constant light intensity, biomass, oxygen, and nutrient concentrations.

$$I = \underset{X[\tau]}{\text{extr}} \int_{t-\tau}^{t} E(B(v), X(v))dt', \qquad (5.84)$$

$$E = B(v) \sum_n \int_{X_n^*}^{X_n} [Y_n(X) + (\Lambda_n - 1)Y_n^*]dX_n, \qquad (5.85)$$

$$\frac{dB(v)}{dt} = f(B(v), Y(v)). \qquad (5.86)$$

This principle is to be understood in the sense of dynamic programming and corresponds to the postulate that the development of phytoplankton is controlled by driving forces, $X(v)$, during the finite time interval of length τ in such a way that the time integral (5.84) over the generalized excess entropy production E within $(t = \tau) \leq t' \leq t$ becomes an extreme value subject to the initial value $B(v(t - \tau))$ and to the mass balance equation (5.86), which connects B and X through Y. If more than one control variable is used, v should be interpreted as the set of these variables.

Inside the region $\{B, X\}$, where the X_n vary freely, it follows that from (5.84) for $m = 1, 2, 3, \ldots$,

$$\frac{\partial E(B, X)}{\partial X_m} + \frac{\partial I}{\partial B} \cdot \frac{\partial f(B, Y)}{\partial X_m} = 0. \qquad (5.87)$$

These differential equations permit the nonlinear functions $Y = Y(X)$ to be determined. The case where the extreme value lies on the boundary of the region $\{B, X\}$ is not investigated here.

The solution of (5.87) is

$$Y_m(X) = Y_m^* \frac{1 - D_m \exp(-cX_m \, \text{sign}(Y_m))}{1 - D_m^* \exp(-cX_m^* \, \text{sign}(Y_m))}, \qquad (5.88)$$

with

$$D_m = D_m^{(0)} - \sum_{\substack{r \\ (r \neq m)}} L_{mr} X_r, \; D_m^* = D_m^{(0)} - \sum_{\substack{r \\ (r \neq m)}} L_{mr} X_r^*. \qquad (5.89)$$

By the sum over r (with the exception of $r = m$), linear cross effects are taken into account. $D_m^{(0)}$ and the Onsager coefficients, L_{mr}, are independent of X. Introducing $X_m = X_m(v)$ into (5.88), the rates Y_m are derived as nonlinear functions of the control variable (Mauersberger and Straškraba, 1987).

BASIC EQUATIONS OF POPULATION DYNAMICS IN ACCORDANCE WITH THE ENTROPY PRINCIPLE

Laws of Population Dynamics in Aquatic Ecosystems

Suppose there is an overlap between generations so that the biomass of every species varies continuously. The growth of phyto- and zooplankton in aquatic ecosystems may then be described by differential equations that can be inferred from thermodynamic theory by introducing the rates of biological processes into the mass-balance equations. To simplify the presentation, we confine ourselves to the following formulation of mass-balance equations for the kth phytoplankton species:

$$\frac{dB_k}{dt} + \frac{1}{\rho} \text{div} \, \vec{J}_k = B_k P_k - B_k R_k - \sum_l g_{lk} Z_l G_l, \qquad (5.90)$$

taking into account primary production, respiration, and grazing by the lth zooplankton species, but neglecting changes in the internally stored nutrients and energy. Z_l is the mass fraction of the lth zooplankton species. The coefficients g_{lk} describe trophic interactions between phyto- and zooplankton.

Formulations of type (5.81) are applied for the rates of primary production P_k, respiration R_k, and grazing G_l:

$$P_k = P_k^{\max} \left[1 - \left(\frac{B_k}{B_k^{\max}} \right)^{c_k} \right], \tag{5.91}$$

$$R_k = R_k^{(0)} \left[\left(\frac{B_k}{B_k^{\min}} \right)^{c_k} - 1 \right], \tag{5.92}$$

$$G_l = G_l^{\max} \left[1 - \left(\frac{Z_l}{Z_l^{\max}} \right)^{c_l} \right]. \tag{5.93}$$

The upper limit, $Z_l^{\max}$ of Z_l, is proportional to the mass fractions of the prey populations. The dependence of G_l on temperature and biomasses Z_l, B_k corresponds with P_k as a function of T, B_k, and x_j (Figs. 5.9, 5.10, and 5.12). This result is in accordance with observation (for example, Horn, 1981).

From eqs. (5.90) through (5.93), it follows for $k = 1, 2, 3, \ldots$, that

$$\frac{dB_k}{dt} + \frac{1}{\rho} \operatorname{div} \vec{J}_k = a_k(c_k) \cdot B_k - b_k(c_k, T, \theta, x_j, [O_2]) \cdot B_k^{1+c_k} - \sum_l G_l^{\max} Z_l + \sum_l d_{kl} \frac{Z_l^{1+c_l}}{B_k^{q_k+c_l}}. \tag{5.94}$$

dB_k/dt is the material derivative. Equations (5.94) are partial differential equations governing phytoplankton dynamics. Differing from the Lotka–Volterra model, they are generalized "logistic" laws similar to those used by Bertalanffy (1942) and many other authors. The coefficients and the noninteger exponents depend on the local state variables and on the history of both the phytoplankton and the zooplankton populations. Analogous equations can be derived from zooplankton dynamics (see, for example, Mauersberger, 1982c).

Further equations (balances of chemical constituents, momentum, energy, and so on) describe chemical reactions, sorption and desorption, physical processes like currents, turbulent transport and mixing, extinction of light, and so on. Boundary conditions reflect constraints acting from the surrounding world on the fluid motions and on components of the aquatic ecosystem. By offering the full set of equations for the multispecies community of an aquatic ecosystem, the thermodynamic theory shows how biological, chemical, and physical interactions affect population and community dynamics. All the coefficients or parameters of the theory are physically, chemically, or biologically defined quantities.

Inasmuch as there is no overlap between successive generations, so that population growth occurs in discrete steps, difference equations (with time as a discrete variable) are the appropriate tool. The full system of differential or difference equations can only be solved with the help of computers since the phenomena, and therefore the equations of population dynamics, are nonlinear and extremely complex. In the following sections we concentrate our attention on two simple examples that are solvable without a computer.

Growth of Phytoplankton

In investigating phytoplankton growth, let us agree on the following restrictions: Hydrodynamic effects are neglected, and only primary production, respiration, and mortality (described as a linear process proportional to phytoplankton biomass) are taken into account. The carrying capacity $B_k^{\max}$, defined by eq. (5.82), is assumed to be independent of time. Then (5.90) becomes

$$\frac{dB_k}{dt} = P_k B_k - R_k B_k - K_k B_k \qquad (K_k \approx \text{constant}) . \qquad (5.95)$$

Introducing (5.91) and (5.92), defining $y(t)$ by

$$y(t) = \frac{B_k(t)}{B_k^{\max}} \qquad (0 < y \leqq 1) , \qquad (5.96)$$

and using the abbreviations

$$a = P_k^{\max} + R_k^{(0)} - K_k \gtreqless 0, \qquad (5.97)$$

$$b = P_k^{\max} + R_k^{(0)} \left(\frac{B_k^{\max}}{B_k^{\min}}\right)^{c_k} > 0 , \qquad (5.98)$$

we obtain from (5.95) the differential equation

$$\frac{dy(t)}{dt} = ay(t) - by(t)^{1+c} \qquad (c \equiv c_k). \qquad (5.99)$$

The solution of (5.99) or (5.95), respectively, is

$$B_k(t) = \frac{B_k(\infty)}{\left\{1 - \left[1 - \left(\frac{B_k(\infty)}{B_k(0)}\right)^{c_k}\right] \exp(-ac_k t)\right\}^{1/c_k}} , \qquad (5.100)$$

where $B_k(0)$ denotes the initial value, but $B_k(\infty)$ is the asymptotic value of $B_k(t)$ for $t \to \infty$, which is given by

$$B_k(\infty) = \begin{cases} (a/b)^{1/c_k}, & \text{if } a \geqq 0, \\ 0, & \text{if } a \leqq 0. \end{cases} \qquad (5.101)$$

In the initial phase, that is, if $\exp(-ac_k t) \approx 1 - ac_k t$, it follows from (5.100) that phytoplankton growth is of hyperbolic type if and only if $B_k(0) < B_k(\infty)$:

$$B_k(t) \approx \frac{B_k(0)}{\{1 - [1 - B_k(0)/B_k(\infty))^{c_k}] ac_k t\}^{1/c_k}} . \qquad (5.102)$$

Figure 5.13 illustrates the solution (5.100). It should be emphasized that this result is inferred (under simplifying assumptions) from basic thermodynamic equations.

Multiple Stationary States

Locally stationary states of the kth phytoplankton species, defined by $\partial B_k/\partial t = 0$, offer examples for structural instabilities under changing external conditions. Taking into account, as in eq. (5.85), losses (for instance by mortality) proportional to B_k, from (5.90) we obtain

$$\frac{\partial B_k}{\partial t} = B_k(P_k - R_k - K_k) - \sum_l g_{lk} Z_l G_l - Q_k , \qquad (5.103)$$

where Q_k describes the hydrodynamically determined inflow/outflow of phytoplankton biomass:

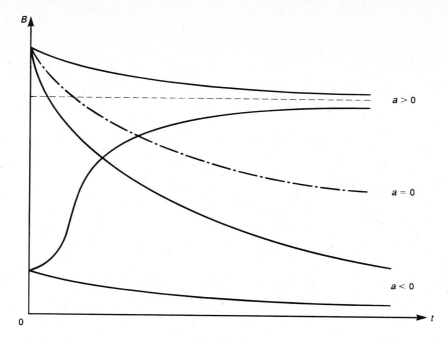

Figure 5.13 Phytoplankton growth according to eq. (5.100). The parameter a is defined by eq. (5.97). If $a > 0$ and $B_k(0) < B_k(\infty)$, there is hyperbolic growth during the initial phase.

$$Q_k = \vec{v} \cdot \nabla B_k + \frac{1}{\rho} \operatorname{div} \vec{J}_k . \tag{5.104}$$

The condition $\partial B_k/\partial t = 0$ will be discussed under the simplifying assumptions that grazing can be approximated by

$$\sum_l g_{lk} Z_l G_l \approx ZG = ZG^{\max} [1 - K_Z Z^c B_k^{-m}] \tag{5.105}$$

and that temperature, light intensity, and nutrient concentrations are kept constant, so that

$$\frac{\partial B_k^{\max}}{\partial t} = 0 . \tag{5.106}$$

Introducing (5.91), (5.92), (5.96), (5.97), and (5.98) into (5.103), we finally obtain

$$\frac{\partial y(t)}{\partial t} = -b\left[y^{1+c} - \frac{a}{b} y + p - r y^{-m} \right] \tag{5.107}$$

with

$$p = \frac{Q_k + ZG^{\max}}{b B_k^{\max}} \gtreqless 0, \tag{5.108}$$

$$r = \frac{K_z G^{\max} Z^{1+c_z}}{b(B_k^{\max})^m} > 0, \quad \text{when } Z > 0 . \tag{5.109}$$

Since a, b, p, and r do not depend explicitly on y, it is possible to introduce a potential function, $W(y)$:

Chap. 5   Entropy Control of Complex Ecological Processes

$$\frac{\partial y}{\partial t} = -\frac{\partial W(y)}{\partial y}, \qquad (5.110)$$

$$W(y) = \frac{y^{2+c}}{2+c} - \frac{ay^2}{(2b) + py - rf(y)}, \qquad (5.111)$$

$$f(y) = \begin{cases} \ln(y), & \text{if } m = 1, \\ \dfrac{y^{1-m}}{1-m}, & \text{if } m \neq 1. \end{cases} \qquad (5.112)$$

In such exceptional cases, phytoplankton dynamics can be investigated with the help of methods from catastrophe theory (Thom, 1972). Stationary states, $y(t) = \bar{y}$ = constant, are determined by $\partial W/\partial y = 0$. They are stable if $\partial^2 W/\partial y^2 > 0$ for $y = \bar{y}$. Structural instabilities result if $\partial W/\partial y = 0$ and $\partial^2 W/\partial y^2 = 0$ for $y = y_0$.

Figure 5.14 gives an incomplete glimpse into the situation. If the losses by mortality are less than primary production ($a > 0$), then:

- Two stable stationary states separated by an unstable stationary state are possible if $p_2 < p < p_3$; that is, phytoplankton biomass becomes reduced to a sufficient extent by grazing and hydrodynamic processes ($Q_k > 0$).
- Only one stable state exists if grazing is accompanied by convective inflow of biomass ($Q_k < 0$; $p_1 < p < p_2$).
- Stationary states are impossible if production and inflow of biomass intensively go across grazing and mortality ($p < p_1$) or if outflow and grazing exceed all bounds ($p > p_3$), respectively.

Assuming that temperature, light intensity, and nutrient concentrations are time independent, it has been shown in the preceding discussion that instabilities are induced by convection, diffusion, and grazing. It should not be overlooked that the thresholds, p_j, depend, through a, b, and r, on

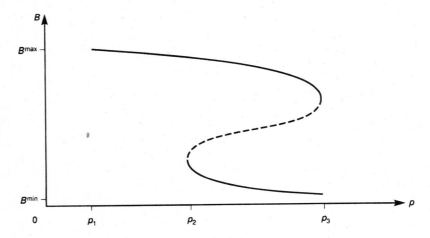

Figure 5.14 Multiple stationary states of the phytoplankton biomass in dependence of the parameter p defined by eq. (5.108). The broken line denotes the unstable stationary states situated between two stable stationary states if $p_2 < p < p_3$.

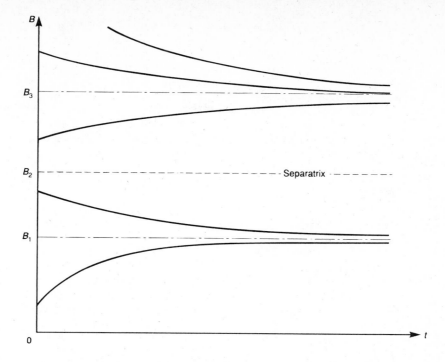

Figure 5.15 Time-dependent solutions, $B(t)$, are influenced by stationary states B_1, B_3, which act as attractors, and by an unstable stationary state B_2, which separates two regions of possible solutions of the differential equation for phytoplankton development. Simplifying the problem, B_j are assumed time independent.

temperature, light, and nutrients and, furthermore, through c_k and c_z on the history of the phytoplankton and the zooplankton populations. Especially, in the limit $r \to 0$,

$$p_1 \to \frac{a}{b-1} \quad \text{and} \quad p_2 \to p_3 \to c\left[\frac{a}{(1+c)b}\right]^{1+1/c}. \tag{5.113}$$

Consequently, these thresholds vary with spatial or temporal changes of physical and chemical state variables.

The existence of multiple stationary states is of importance also for the development of the biomass $B_k(t)$ in time. Unstable stationary states serve to separate regions of possible solutions, $B_k(t)$, of the differential equations of population dynamics. Within such regions, which are selected by initial conditions, the solutions $B_k(t)$ show a tendency to approach a stable stationary state acting as an "attractor." Figure 5.15 demonstrates this situation under time-independent controlling factors.

Closing Remarks

Only when the interactions of more than one species of organisms are considered in both time and space can ecological situations be investigated and understood. Patchiness, species diversity, cyclic changes in the community, and the more or less pronounced invariance of functions exhibited by the ecosystem against superimposed environmental fluctuations, as well as other ecological phenomena, can be dealt with. But this is not the purpose of the present chapter, which aims at making evident that the basic equations of theoretical ecology can and must be derived from fundamental laws of physics, chemistry, and biology, including the second law of thermodynamics. Of course, it is possible to develop a model solely on the basis of observations and to check afterward if all fundamental constraints are fulfilled or not. But the deductive approach allows for more effective in situ observations and measurements in laboratories, the results of which can serve for validation of the theory and for the determination of values of macroscopic model parameters.

Advances in ecological research require further development of theory in combination with measurements and modeling. There is a real challenge to improve scale-specific versions of theory in order to understand phenomena emergent at higher levels of biological organization (Agren & Bosatta, 1990; Roughgarden et al., 1989; Ulanowicz, 1986, 1989; Mauersberger, 1991).

REFERENCES

ÅGREN, G. I., and BOSATTA, E. 1990. Theory and model or art and technology in ecology. *Ecol. Mod.* 50:213–220.

ALLEN, J. C. 1990. Factors contributing to chaos in population feedback systems. *Ecol. Mod.* 51:281–298.

BELLMAN, R. 1957. *Dynamic Programming.* Princeton University Press, Princeton, New Jersey. 339 pp.

BERTALANFFY, L. von 1942. *Theoretical Biology.* Borntraeger, Berlin (in German). 125 pp.

CASAS-VAZQUES, J., JOU, D., and LEBON, G. (eds.) 1984. *Recent Developments in Nonequilibrium Thermodynamics.* Proceedings, Workshop Barcelona, Spain, 1983. Lecture Notes in Physics, Vol. 199. Springer-Verlag, New York. 485 pp.

EBELING, W., and FEISTEL, R. 1982. *Physics of Self-organization and Evolution.* Akademie-Verlag, Berlin (in German). 451 pp.

GLANSDORFF, F., and PRIGOGINE, I. 1971. *Thermodynamic Theory of Structure, Stability and Fluctuations.* Wiley, New York. 306 pp.

GUTIERREZ, E., and ALMIRAL, H. 1989. Temporal properties of some biological systems and their fractal attractors. *Bull. Math. Biol.* 51:785–806.

HAKEN, H. (ed.) 1981. *Chaos and Order in Nature.* Springer Series in Synergetics, Vol. 11. Springer-Verlag, New York. 275 pp.

HAKEN, H. 1983. *Advanced Synergetics.* Springer-Verlag, New York. 356 pp.

HOLLING, C. S. 1973. Resilience and stability of ecological systems. *Ann. Rev. Ecol. Syst.* 4:1–23.

HOLLING, C. S. 1976. Resilience and stability in ecosystems. In Jantseh, E., and Waddington, C. H. (eds.), *Evolution and Consciousness: Human Systems in Transition.* Addison-Wesley, Reading, Massachusetts, pp. 73–92.

HORN, W. 1981. Phytoplankton losses due to zooplankton grazing in a drinking water reservoir. *Int. Revue Hydrobiol.* 66:787–810.

IOOS, G., and JOSEPH, D. D. 1980. *Elementary Stability and Bifurcation Theory.* Springer-Verlag, New York. 286 pp.

JOHNSON, L. 1990. The thermodynamics of ecosystems. In Hutzinger, O. (ed.), *The Handbook of Environmental Chemistry, Vol. 1.: The Natural Environment and Biogeochemical Cycles.* Springer-Verlag, New York, pp. 1–47.

LOEHLE, C. 1989. Catastrophe theory in ecology: A critical review and an example of the butterfly catastrophe. *Ecol. Mod.* 49:125–152.

JØRGENSEN, S. E., and MEJER, H. F. 1981. Application of exergy in ecological models. In Dubois, D. (ed.), *Progress in Ecological Modelling.* Editions CEBEDOC, Liege, Belgium. pp. 311–347.

MAUERSBERGER, P. 1978. On the theoretical basis of modelling the quality of surface and subsurface waters. *Proc. Baden-Symposium 1978.* IAHS-AISH Publ. No. 125:14–23.

MAUERSBERGER, P. 1979. On the role of entropy in water quality modeling. *Ecol. Mod.* 7:191–199.

MAUERSBERGER, P. 1981. Entropy and free enthalpy in the aquatic ecosystem. *Acta Hydrophysica* 26:67–90 (in German).

MAUERSBERGER, P. 1982a. Determination of the nonlinear relations between rates and affinities of producing and reducing processes in the aquatic ecosystem. *Acta Hydrophysica* 27:125–130 (in German).

MAUERSBERGER, P. 1982b. Optimization principle in the theory of aquatic ecosystems. *Z. Meteorol.* 32:272–277 (in German).

MAUERSBERGER, P. 1982c. Logistic growth laws for phyto- and zooplankton. *Ecol. Mod.* 17:57–63.

MAUERSBERGER, P. 1982d. Rates of primary production, respiration and grazing in accordance with the balances of energy and entropy. *Ecol. Mod.* 17:1–10.

MAUERSBERGER, P. 1983. General principles in deterministic water quality modeling. In Orlob, G. T. (ed.), *Mathematical Modeling of Water Quality: Streams, Lakes, and Reservoirs.* Wiley, New York, pp. 42–115.

MAUERSBERGER, P. 1984. Thermodynamic theory of the control of processes in aquatic ecosystems by temperature and light intensity. *Gerlands Beitr. Geophysik* 93:314–322.

MAUERSBERGER, P. 1985a. Local entropy production in aquatic ecosystems. *Acta Hydrophysica* 29:235–258.

MAUERSBERGER, P. 1985b. Optimal control of biological processes in aquatic ecosystems. *Gerlands Beitr. Geophysik* 94:141–147.

MAUERSBERGER, P. 1985c. Dominant controlling variables in the theory of biological processes in aquatic ecosystems. *Gerlands Beitr. Geophysik* 94:161–165.

MAUERSBERGER, P. 1988. Generalized Gibbs equation in the theory of aquatic ecosystems. *Acta Hydrophysica.* 32:27–36.

MAUERSBERGER, P. 1990. Fundamentals of the thermodynamic theory of aquatic ecosystems. *Acta Hydrophysica* 34:29–43.

MAUERSBERGER, P. 1991. The role of fundamental laws of physics, chemistry and biology in limnological research. *Acta Hydrophysica* 35:21–31.

MAUERSBERGER, P., and STRAŠKRABA, M. 1987. Two approaches to generalized ecosystem modelling: thermodynamic and cybernetic. *Ecol. Mod.* 39:161–169.

MAY, R. M. 1974. Biological populations with nonoverlapping generations. Stable points, stable cycles, and chaos. *Science* 186:645–647.

MAY, R. M. 1977. Thresholds and breakpoints in ecosystems with a multiplicity of stable states. *Nature* 269:471–477.

NICOLIS, G., and PRIGOGINE, I. 1977. *Self-organization in Nonequilibrium Systems.* Wiley, New York. 491 pp.

ORLOB, G. T. (ed.) 1983. *Mathematical Modeling of Water Quality: Streams, Lakes, and Reservoirs.* International Series on Applied Systems Analysis, Vol. 12. Wiley, New York. 518 pp.

PATTEN, B. C. 1968. Mathematical models of plankton production. *Int. Revue Ges. Hydrobiol.* 53:357–408.

PHILLIPS, M. 1962. Classical electrodynamics. In Flügge, S. (ed.), *Handbuch d. Physik—Encyclopedia of Physics,* Vol. 4. Springer, New York, pp. 1–108.

PLANCK, M. 1913. *Lectures about Thermodynamics.* 4th Ed. Veit, Leipzig (in German). 23 pp.

RADTKE, E., and STRAŠKRABA, M. 1980. Self-optimization in a phytoplankton model. *Ecol. Mod.* 9:247–268.

ROUGHGARDEN, J., MAY, R., and LEVIN, S. (eds.) 1989. *Perspectives in Ecological Theory.* Princeton University Press, Princeton, New Jersey. 394 pp.

RUBIN, A. B. 1976. *Thermodynamics of Biological Processes.* Moscow University Press, Moscow (in Russian). 238 pp.

SALVADORA, L. (ed.) 1984. *Bifurcation Theory and Applications.* Lecture Notes in Mathematics, Vol. 1057. Springer-Verlag, New York. 233 pp.

SERRIN, J. (ed.) 1986. *New Perspectives in Thermodynamics.* Springer-Verlag, New York. 260 pp.

STRAŠKRABA, M. 1979. Natural control mechanisms in models of aquatic ecosystems. *Ecol. Mod.* 6:305–321.

STRAŠKRABA, M., and GNAUCK, A. 1983. *Aquatic Ecosystems—Modeling and Simulation.* Fischer-Verlag, Jena (in German): English edition: *Freshwater Ecosystems; Modelling and Simulation.* Elsevier, Amsterdam and Oxford University Press, New York, 1985. 309 pp.

STUMM, W., and MORGAN, J. J. 1970. *Aquatic Chemistry.* Wiley, New York. 583 pp.

THOM, R. 1972. *Stabilité Structurelle et Morphogenèse.* Benjamin, New York. 362 pp.

TRUESDELL, C. 1957. Sulle basi della termomeccanica. *Rend. Lincei* (8) 22:33–38, 158–166.

TRUESDELL, C., and TOUPIN, R. A. 1960. The classical field theories. In Flügge, S. (Ed.), *Handbuch d. Physik—Encyclopedia of Physics,* Vol. III/1. Springer-Verlag, New York, pp. 226–793.

ULANOWICZ, R. E. 1986. *Growth and Development: Ecosystems Phenomenology.* Springer-Verlag, New York. 203 pp.

ULANOWICZ, R. E. 1988. On the importance of higher-level models in ecology. *Ecol. Mod.* 43:45–56.

ZOTIN, A. I. 1972. *Thermodynamics Aspects of Developmental Biology.* Karger, Basel, Switzerland. 159 pp.

GLOSSARY OF NOTATIONS

Dimensions: E: energy; K: temperature; L: length; M: mass; T: time.

A_{kn} [$E\,M^{-1}$]:	Affinity of the nth biological process, species number k [see eq. (5.23)]
$\tilde{A}_{kn}$ [$E\,\text{mol}^{-1}$]:	$\tilde{A}_{kn} = A_{kn}\,\mathcal{M}_k$
A_r [$E\,\text{mol}^{-1}$]:	Affinity of the rth chemical reaction
B_k [$M\,M^{-1}$]:	Mass fraction of the kth species, $B_k = \rho_k/\rho$
c_k, c [1]:	See eq. (5.46)
f_{kn}, f [1]:	See eq. (5.53)
N_j [$M\,M^{-1}$]:	Mass fraction of the jth chemical constituent, $N_j = \rho_j/\rho$
$\mathcal{M}_j, \mathcal{M}_k$ [$M\,\text{mol}^{-1}$]:	Molar masses of the chemical or biological components, respectively
P_k, P [T^{-1}]:	Rate of primary production or of photosynthesis, respectively; entropy production in the section "Evolution Criterion, Stationary States, and Instabilities"
R_k, R [T^{-1}]:	Rate of respiration by the kth species
$R_{kn} = \rho_k Y_{kn}$ [$M\,L^{-3}\,T^{-1}$]:	Rate of the nth biological process, species k [see eq. (5.24)]
R_0 [$E\,\text{mol}^{-1}\,K^{-1}$]:	Universal gas constant
s [$E\,M^{-1}\,K^{-1}$]:	Specific entropy of the total system
T [K]:	Temperature
t [T]:	Time
u [$E\,M^{-1}$]:	Specific internal energy of the total system
U_{kj}, U [T^{-1}]:	Uptake rate, species k, nutrient j
$\vec{v}$ [$L\,T^{-1}$]:	Mean velocity [see eq. (5.19)]
V_r [$E\,\text{mol}^{-1}$]:	Supplementary energy connected with the rth chemical reaction far from thermostatic equilibrium [see eq. (5.11)]
V_{kn} [$E\,M^{-1}$]:	Specific supplementary energy connected with the nth biological process, species k
$\tilde{V}_{kn}$ [$E\,\text{mol}^{-1}$]:	$\tilde{V}_{kn} = V_{kn}\,\mathcal{M}_k$
w_r [$\text{mol}\,L^{-3}\,T^{-1}$]:	Rate of the rth chemical reaction

x_j [mol mol^{-1}]: Molar fraction of the jth constituent
X_{kn} [1]: Driving force of the nth biological process, species k [see eq. (5.47)]
Y_{kn} [T^{-1}]: Rate of the nth biological process, species k
Z_l, Z [1]: Mass fraction of the lth zooplankton species
α [$L^3 M^{-1}$]: Specific volume, $\alpha = 1/\rho$
ϵ_k [1]: Energy deficit [see eq. (5.51)]
ϑ_{kn} [$E\,M^{-1}$]: Specific absorption of incident radiation by the kth species connected with the nth biological process
θ_{kn} [E mol^{-1}]: $\theta_{kn} = \epsilon_{kn}\,\mathcal{M}_k$
μ_j, μ_k, μ_k [$E\,M^{-1}$]: Chemical potentials [see eq. (5.27)]
ν_{jr} [1]: Stoichiometric coefficients [see eq. (5.24)]
ρ [$M\,L^{-3}$]: Mass density of the ecosystem
ρ_l [$M\,L^{-3}$]: Mass density of the ith constituent
σ [$E\,L^{-3}\,T^{-1}\,K^{-1}$]: Local entropy production [see eq. (5.21)]
ξ_r [mol M^{-1}]: Extent of the rth chemical reaction
ξ_{kn} [$M\,M^{-1}$]: Extent of the nth biological process, species k

6

Stochastic Theory of Complex Ecological Systems

BENZION S. FLEISHMAN

Institute of Oceanology
Russian Academy of Sciences
Moscow

A systems approach requires (i) combining, condensing, and synthesizing a great amount of information concerning components of the system; (ii) examining in detail the structure of the system; (iii) translating this knowledge of system components, functions and structures into models of the system; and (iv) using the models to derive new insights into the operation, management and utilization of the system. [Van Dyne, in *Grasslands, Systems Analysis and Man*, Int. Biol. Program 19, Chap. 14, p. 883 (1980), Cambridge University Press]

The original and persistent problems of ecology arose from the frustrating attempts to study these enormously complex systems with tools and languages designed for much simpler ones. [Van Dyne, in *The Ecosystem Concept in Natural Resource Management*, Ch. X, p. 334 (1969), Academic Press]

INTRODUCTION: OVERVIEW

This chapter introduces a theory of complex "systemology" into ecology. Differences between the paradigms of physics and systemology do not rule out analogy in considering and comparing their models. For example, the systemological "minimax" as developed in this chapter may be taken to represent another step toward the elimination of indeterminacy by physical ergodicity. Both physics and systemology identify their models at the same mesolevel. This chapter presents an analytical review of models of more and more active and detailed complex systems. A model of the initial state structuring of complex system organization and behavior is considered for the first time. The necessity for complex systems to regenerate is established. The complex system model is detailed at macro-, meso-, and microlevels. The parametric interrelation of these models is presented. An effect corresponding to the known "overcatch" effect of trade populations appears in the models. All the models are of stochastic character. A broad ecological interpretation of the obtained results is presented. For the first time, the logarithmic law of growth is derived from the hyperbolic law reducing the reliability of complex (specifically, biological) systems.

PHYSICS AND SYSTEMOLOGY

Apart from inner stimuli, the progress of theoretical biology is connected with the development of physical and cybernetic theories of the nature of life. This chapter deals mainly with theoretical results of ecology (above-organism biology) obtained on the basis of systemological ideas. "Systemology" represents a version of cybernetics specifically concerned with the broad range of interrelations with physics.

The thermodynamic approach as well as the wider theory of nonequilibrium states contributes much to the understanding of molecular biology processes (Prigogine, 1980). Investigations of this kind are being continued in the field of ecology (Mauersberger, 1982). Early cybernetic ideas (Wiener, 1948; Ashby, 1958) have enriched biology, and now they are becoming fruitful in ecology (Straškraba and Gnauck, 1983; Bakule and Straškraba, 1984).

Until recently, any important achievement in natural sciences was invariably based on fundamentals of physics. Only recently have the fundamental works by Kotel'nikov (1956) and Shannon (1948) initiated a productive field of cybernetics, that is, *systemology,* which can also be called the theory of potential effectivity of complex systems (Fleishman, 1971). Previously, the differences between physical and systemological paradigms have been emphasized (Fleishman, 1982a). Here, I shall endeavor to find elements of continuity between physics and systemology as a basis for better understanding of systemology applications to complex ecology.

PROCESSES AND SYSTEMS

Processes, Monoprocesses, and Polyprocesses

The definitions of a process and a system are based on the following philosophical categories: substances of material and energy, reflection and information, relationship and interaction, preference and goal, and striving and activity. The forms of existence of these categories are space and time, discreteness and continuity, and stochasticity and determinacy.

Continuous-deterministic considerations are a particular limit case of discrete-stochastic considerations; that is why the latter are used below. Let us introduce the notation for a discrete set $\mathbf{X} = \{X\}$ of elements X in the case that they are unordered, and $\mathbf{X} = (X)$ in the ordered case.

A *stochastic object,* $\mathcal{X} = (\mathbf{X}, \mathbf{P})$, is an ensemble, $\mathbf{X} = (X)$, of objects, X, and a distribution, $\mathbf{P} = (P(X))$, of probabilities, $P(X) \geq 0$, of object realization $\left(\sum_{X \in \mathbf{X}} P(X) = 1 \right)$. A *deterministic object* can be considered as a degenerate case of the stochastic object, $\mathcal{X} = X$, when the corresponding probability $P(X) = 1$. For *a stochastic vector,* $\vec{\mathcal{X}} = (\mathcal{X}_i) = (\mathbf{X}, \mathbf{P})$, with stochastic components, $\mathcal{X}_i = (\mathbf{X}_i, \mathbf{P}_i)$ $(i = \overline{1, N})$, we have $\mathbf{X} = (X)$, $\mathbf{P} = (P(X))$, $\mathbf{X}_i = (X_i)$, $\mathbf{P}_i = (P_i(X_i))$, $\mathbf{X} = (\mathbf{X}_i)$, and $X = (X_i)$.

We divide the index set $I = (1, 2, \ldots, N)$ into two subsets $I = E \cup \overline{E}$, where $E = (i_1, \ldots, i_k)$ is an arbitrary subset, and $\overline{E} = (i_{k+1}, \ldots, i_{N-k})$ is its additional subset. Now, we form two corresponding stochastic subvectors from $\vec{\mathcal{X}}$-vector components, $\vec{\mathcal{X}} = (\vec{\mathcal{X}}_E, \vec{\mathcal{X}}_{\overline{E}})$, where $\vec{\mathcal{X}}_E = (\mathcal{X}_{i_1}, \ldots, \mathcal{X}_{i_k})$ and $\vec{\mathcal{X}}_{\overline{E}} = (\mathcal{X}_{i_{k+1}}, \ldots, \mathcal{X}_{i_{N-k}})$. Then, due to coordination of distributions, we get

$$\sum_{X_i \in \mathbf{X}_i, i \in \overline{E}} P(X) = P(X_E) \quad \text{and} \quad \sum_{X_i \in \mathbf{X}_i, i \in E} P(X) = P(X_{\overline{E}}) . \tag{6.1}$$

We call the stochastic vectors $\vec{\mathcal{X}}_E$ and $\vec{\mathcal{X}}_{\bar{E}}$ *independent* of each other if for all $X \in \mathbf{X}$ we get

$$P(X) = P(X_E) \cdot P(X_{\bar{E}}). \tag{6.2}$$

If relation (6.2) is not fulfilled even for one $X \in \mathbf{X}$, then the stochastic vectors $\vec{\mathcal{X}}_E$ and $\vec{\mathcal{X}}_{\bar{E}}$ are called *dependent*. This definition covers $n > 2$ stochastic vectors. If no subset $E \subseteq I$ is allowed to divide the stochastic vector $\vec{\mathcal{X}}$ into two independent subvectors, $\vec{\mathcal{X}}_E$ and $\vec{\mathcal{X}}_{\bar{E}}$, then it is called *integrity*.

Let us consider the four-dimensional space–time $R_4 = (y)$ of points $y = (x, t)$ with discrete spatial, $x = (x_1, x_2, x_3) \in R_3$, and temporal, t, coordinates ($t = 0, 1, 2, \ldots$), and with the discreteness interval taken as unity. We consider the point domain $Y = \{(x, t)\} \in R_4$ with its boundary subset $Y' = \{(x', t')\} \subset Y$ for each value of t, we consider projections of the domain Y and its boundary Y' onto space R_3. We get the corresponding domain $G_t = (x_t)$ and its boundary subset $\Gamma_t = (x'_t) \subset G_t$. For our further aims, it is important to introduce the sequences $G = (G_t)$ and $\Gamma = (\Gamma_t)$, $t = 0, 1, 2, \ldots$.

Now, consider the N-dimensional phase space $R_N = (\vec{X})$ of phase vectors $\vec{X} = (X_i)$, $i = \overline{1, N}$. We correlate a *phase vector* $\vec{X}_{x,t}$ with each point $y = (x, t) \in G$. Then the domain $G_t = (x_t)$ and its boundary subdomain $\Gamma_t = (x'_t)$ will correspond to the domain $X_{G_t} = (\vec{X}_{x,t})$ and its boundary subdomain $X_{\Gamma_t} = (\vec{X}_{x',t})$; the sequences $G = (G_t)$ and $\Gamma = (\Gamma_t)$ will correspond to the sequences $X_G = (X_{G_t})$ and $X_\Gamma = (X_{\Gamma_t})$. The sequence X_G will be referred to below as the *trajectory* in phase space, and the sequence X_Γ is its *boundary values*. Note that, at the level of phase vectors, the latter is a subset of the former.

A *process* is an integral stochastic trajectory, $\mathcal{X} = \mathcal{X}_G = (\mathbf{X}_G, \mathbf{P}_G)$.

A *process microstate* is a stochastic domain of values $\mathcal{X}_t = \mathcal{X}_{G_t} = (\mathbf{X}_{G_t}, \mathbf{P}_{G_t})$ at an instant t. A microstate is called the $\mathcal{X}_0 = \mathcal{X}_{G_0}$ *initial value of a process*.

Boundary values of a process are stochastic boundary values, $\mathcal{X}_\Gamma = (\mathbf{X}_\Gamma, \mathbf{P}_\Gamma)$.

A *process macrostate* is an s-dimensional functional, $\Phi = \Phi(\mathcal{X}) = (u_j)$, $j = \overline{1, s}$, found in the process. Its components, $u_j = u_j(\mathbf{P}_G)$, called *macroparameters*, depend on the distributions $\mathbf{P}_G$. For example, for thermodynamic processes, the entropy macroparameter has the form $u = H = C \log P$, where C is constant and P is the probability of a typical microstate.

In the deterministic case of degenerate distribution, the P_G corresponding process is called *deterministic*. In this case only, the macroparameters $u_j = u_j(X_G)$ become functionals of trajectories themselves. It follows from (6.1) that with a process $\mathcal{X} = \mathcal{X}_G$ given, it is possible to determine its initial, $\mathcal{X}_0$, and boundary, $\mathcal{X}_\Gamma$, values; however, the inverse statement will be untrue, as the ensembles of phase vectors of the boundary values are a subset of the first one.

Theoretical physics is concentrated on the solution of the inverse problem, that is, with the initial, $\mathcal{X}_0$, and boundary, $\mathcal{X}_\Gamma$, values of the process given. It is required to regenerate the process $\mathcal{X}$ using "laws of nature" in the form of D (differential equations) with respect to space x and time t for phase vectors, taking into account restrictions imposed on macroparameters (conservation laws being most important among these). Therefore, the aim of theoretical physics is to obtain the dependence

$$\mathcal{X} = F_D(\mathcal{X}_0, \mathcal{X}_\Gamma), \tag{6.3}$$

that is, an explicit expression of the corresponding distributions

$$\mathbf{P}_G = F_D(\mathbf{P}_{G_0}, \mathbf{P}_\Gamma) \tag{6.4}$$

and, in the deterministic case, the expression

$$\mathbf{X}_G = F_D(\mathbf{X}_{G_0}, \mathbf{X}_\Gamma). \tag{6.5}$$

It is not an exaggeration to say that during the last three centuries human genius has been challenged by such problems, and amazing progress has been made in the field (Einstein, 1936).

Let us divide the process $\mathfrak{X} = \mathfrak{X}_G = (\mathbf{X}_G, \mathbf{P}_G)$ into two subprocesses $\mathfrak{X} = \mathfrak{X}_G = (\mathfrak{X}_G^{(0,t-1)}, \mathfrak{X}_G^{(t,\infty)})$ taking place before and after an arbitrary instant of time, t, where $\mathfrak{X}_G^{(0,t-1)} = (\mathbf{X}_G^{(0,t-1)}, \mathbf{P}_G^{(0,t-1)})$ and $\mathfrak{X}_G^{(t,\infty)} = (\mathbf{X}_G^{(t,\infty)}, \mathbf{P}_G^{(t,\infty)})$. If there is a limit, $\mathbf{P}_G^{(\infty,\infty)} = \lim_{t \to \infty} \mathbf{P}_G^{(t,\infty)}$, independent of the initial distribution $\mathbf{P}_{G_0}$, then the process is called *ergodic*. This means that for any infinitesimal $\epsilon > 0$ there is such a time value, $t = t_\epsilon$, from which onward the distance (for example, Euclidean distance) between distributions of the process is $\rho(\mathbf{P}_G^{(t,\infty)}, \mathbf{P}_G^{(\infty,\infty)}) \leq \epsilon$. The parts of the process $\mathfrak{X}_G^{(0,t_\epsilon-1)}$ and $\mathfrak{X}_G^{(t_\epsilon,\infty)}$ are called *transitional* and *stationary* processes, respectively, and t_ϵ is the *time of the transitional process*.

The above definition of ergodicity is used to derive the ergodicity of microstates of the process. In this case, an *ergodic* process is one with limit $\mathbf{P}_{G_\infty} = \lim_{t \to \infty} \mathbf{P}_{G_t}$, of instantaneous distribution $\mathbf{P}_{G_t}$ independent of the initial distribution $\mathbf{P}_{G_0}$. The stationary and transitional processes and time t_ϵ of the latter are determined in the same way. So, over its stationary portion $\mathfrak{X}_G^{(t_\epsilon,\infty)}$, an ergodic process is independent of its initial value; that is, according to (6.3),

$$\mathfrak{X}_G^{(t_\epsilon,\infty)} \simeq F_D(\mathfrak{X}_F). \tag{6.6}$$

If the domain degenerates into a point, then, generally, quantum-mechanical considerations are carried on there, and in the deterministic case, statistical ones. In the deterministic case of the nondegenerate domain G, we consider dynamics of continuous media: gas, liquid, and solid. For the deterministic case, when a domain covers the whole space R_3 (not necessarily with the Euclidean metric), theories of physical fields are developed (Einstein, 1936).

The characteristic time, t_c, of a process is the time required for a significant change to occur in distribution of the process microstate. One of the possible mathematical determinations of the value t_c is as follows. Consider the maximum of a certain (Euclidean, for example) distance, $\max_{0 \leq t,t' \leq T} \rho(\mathbf{P}_{G_t}, \mathbf{P}_{G_{t'}}) = \varphi(T)$, between two instantaneous distributions of microstates at the instants t and t' over the time interval $[0, T]$. It is clear that the function $c = \varphi(T)$ reduces monotonically with increasing T. That is why there exists an inverse function, $t_c = \varphi^{-1}(c)$, and this leads to the value t_c.

For each process, all other processes are divided into three classes according to their characteristic times. Those possessing similar characteristic times are called similarly mutable processes; those with significantly shorter or longer characteristic times are called noises or constants, respectively.

Two dependent (independent) processes, $\mathfrak{X}_1$ and $\mathfrak{X}_2$, respectively, are called below *interacting* (*noninteracting*). In the former case, the compound process $\mathfrak{X} = (\mathfrak{X}_1, \mathfrak{X}_2)$ is called their *interaction*. This definition covers $n > 2$ processes. Two interacting processes $\mathfrak{X}_1 = (\mathbf{X}_1, \mathbf{P}_1)$ and $\mathfrak{X}_2 = (\mathbf{X}_2, \mathbf{P}_2)$, the interaction of which $\mathfrak{X} = (\mathfrak{X}_1, \mathfrak{X}_2)$ is known (unknown), are called a *monoprocess* (*polyprocess*). This definition covers $n > 2$ processes. This means that for a polyprocess the distributions $\mathbf{P}_1$ and $\mathbf{P}_2$ are known, but its $\mathbf{P}$ is unknown and in principle cannot be regenerated from $\mathbf{P}_1$ and $\mathbf{P}_2$, although the ensemble $\mathbf{X} = (\mathbf{X}_1, \mathbf{X}_2)$ can be regenerated from $\mathbf{X}_1$ and $\mathbf{X}_2$.

All processes, with rare exceptions (gravitational ones, for example), interact with others under natural (nonlaboratory) conditions. From Galileo on, physicists have discovered laws concerning macroparameters of substantial processes (among them, the conservation laws of matter and energy have proved to be most important). The difficulty of discovery derived from the necessity of abstracting processes from side effects in order to obtain a "pure" law (such as, in neglecting friction in order to state the law of inertia). The Newtonian laws of mechanics, for example, allowed for the fairly accurate description of planetary routes (a gravitational process). However, up to now, physics has been incapable of describing polyprocesses (such as weather) with good accuracy. Also, by its nature, physics has not been able to derive methods for describing informational processes.

Complex and Pseudocomplex Systems

Let us consider the ensemble $A = \{A_e\}$ consisting of at least two *substantial* objects, A_e, marked with elements e of an abstract set $U = \{e\}$. Let the element e be related to the elementary process $\mathcal{X}_e$, and the pair of objects $(A_e, A_{e'})$ be related to paired interaction of elementary interaction processes, $\mathcal{X}_{ee'} = (\mathcal{X}_e, \mathcal{X}_{e'})$. Then, for all objects of the ensemble A, their *joint interaction* is $\mathcal{X}_U = (\mathcal{X}_e, e \in U)$.

For some pairs of elements (e, e'), the corresponding processes $\mathcal{X}_e$ and $\mathcal{X}_{e'}$ are interacting; we consider only these pairs of elements (e, e') and call them interprocess *relationships*. A set of relationships is denoted by $V = \{(e, e')\}$.

The *structure*, $|A|$, of an ensemble, $A = \{A_e\}$, is a pair of sets, $|A| = (U, V)$, of elements U and their relationships V. So, by definition, a structure coincides with the mathematical notion of a "graph."

Environment B is the joint process including all elementary processes $\mathcal{X}_e$ of substantial objects, $A_e \notin A$, beyond the ensemble A.

A *system*, $A = (|A|, \overline{A})$, is a pair $(|A|, \overline{A})$, where $|A| = (U, V)$ is structure, and $\overline{A} = \mathcal{X}_U$ is the joint process of its elements (if emergent—next paragraph); in this case the process is called the *behavior* of the system.

By definition, an *emergent* process is a joint process, $\mathcal{X}_U$, of system elements (arising in a system only), which differs *in quality* from its constituent elementary processes, $\mathcal{X}_e$. Unlike them, an emergent process can be not only substantial (material), but also energetical or even informational. That is why, for example, behavioristic interaction of a system and its environment, although a joint process of *all* elementary processes by definition, is still an individual interaction for each system.

Interaction of a system A and its environment B is, first, an external behavioristic interaction between system behavior, $\overline{A}$, and that of B as processes, which amounts to specification of system behavioral boundary conditions by environment; as to the system, it prescribes interaction law peculiarities, $D = D(A)$ [see (6.3)]:

$$\overline{A} = \mathcal{X}_G = F_{D(A)}[\mathcal{X}_{G_0}, \mathcal{X}_\Gamma(B)]. \tag{6.7}$$

Second, in the course of interaction the environment B has an internal structural effect on elements and relationships of the system:

$$|A| = |A(B)|. \tag{6.8}$$

The *system microstate*, $A(t) = (|A(t)|, \overline{A}(t))$, at the instant t is system structure $|A(t)|$ at that instant together with the microstate of its behavior, $\overline{A}(t) = \mathcal{X}_{G_t}$. The initial state of the system is $A(0)$; the transitional and stationary states of the system are corresponding states of its behavior.

The *system macrostate*, A, is connected with its interaction with environment B, and is determined by functionals of (u, v), which specify resource exchange, $u = \Phi(A, B) = (u_j)$, for each resource, $v = (v_k)$. Here, *macroparameters* can be of temporal, substantial, or informational character.

For systems $A = (|A|, \overline{A})$, there are two characteristic times—structural, $t_{|A|}$, and behavioral, $t_{\overline{A}} = t_c$—of their behavior as a process. *Structural characteristic time*, $t_{|A|}$, is defined as the time elapsing between two structural changes (exclusion or addition of at least one element or relationship) of the system, construed as a graph, $|A| = \Gamma = (U, V)$. Structural mutation of a system is called its *evolution*, and the time $t_{|A|}$ is called the characteristic time of its evolution. For example, the characteristic time of evolution of biological systems is much greater than that of social systems; as to subclasses of biological systems, evolutionary time varies widely. The individual morphological structure connected with species formation has a great characteristic time of evolution. This significantly exceeds the characteristic time of the evolution of population age structure. The latter is of the same order as that of the characteristic time of evolution of community species composition, with short environmental characteristic times. When these times are of the same order as the characteristic time of evolution of species formation, the latter can coincide with the characteristic time of evolution of community species composition.

If the behavior of a system is a polyprocess (or monoprocess), then this system is *pseudocomplex* (*pseudosimple*) or indeterminate (determinate). Since definition of polyprocesses is connected with lack of knowledge about the full description of the process, the notion "pseudocomplex" is connected with ignorance. In general, in physics, complexity is synonymous with ignorance.

Physics manages to describe simple systems quite accurately; as to complex systems in the broad sense, their description cannot be achieved because, by definition, there is no full description of their polyprocess behavior (examples such as of weather and the majority of geophysical systems illustrate the statement). Here, physical descriptions are replaced by statistical ones, with a variety of multidimensional statistical methods that do not expose but rather bypass the inner mechanisms of the phenomena.

The *goal $\underline{A}$* of the system, A, is a certain preferable system macrostate. The macrostate allowing best achievement of a goal is called *optimal*. Since the system macrostate is determined by the (u, v) exchange, the goal can be presented as a certain optimal (useful) (u, v) exchange, $[\underline{A} = (u, v)]$.

The *decision act* of a system is to choose goal-including alternatives controlled by probabilities of subjective interpretation. If system behavior has (or lacks) decision acts, then the system is *complex* (*simple*) or *purposeful* (*spontaneous*). Thus, in systemology, complexity is an inner property of systems that possess decision acts with probabilities of subjective character. In spontaneous systems, on the other hand, decision probabilities are of objective character.

Experimental methods have been developed for the study of purposeful human (Acoff and Emery, 1972) and animal (Hinde, 1970) behavior. The same methods are used to study the behavior of automata simulating the purposeful behavior of animals. These methods allow the separation of the act of decision from the behavior of the mentioned systems and the estimation of corresponding subjective probabilities. That is why the behavioral expediency of individuals and simulating automata are beyond doubt. However, there is doubt as to the purposefulness of behavior in above-the-organism biological systems, that is, populations, communities, ecosystems, and the biosphere.

To my mind, the definition of purposefulness with the decision act as its criterion, and the experimental methods of separating this act, cannot help but separate it also from the behavior of at least the first two above-organism biological systems. Besides, undeniable facts prove that individual goals are subordinate to population goals, and the hypothesis of population goal subordination to community goals also seems quite correct. These problems are closely connected with goal hierarchy and an observed system hierarchy. Let us start with the former.

Generally, system goals form a multilevel hierarchy (a tree). At the highest level is the *strategic* goal; *tactical* goals occupy lower levels. Tactical goals at a lower level are subordinate to goals of higher levels. So, the whole hierarchy is subordinate to the strategic goal. Down the hierarchy, the diversity of tactical goals increases, together with the ability of the system to preset its concrete forms, the sequence of their fulfillment, and the time when one tactical goal is to be replaced by another. As to the strategic goal, the system is not free to preset one for itself. Who can do it then?

This question can be answered only after considering the existing hierarchy of systems, biological systems among them (Fig. 6.1). Within the hierarchy, each system belongs to a higher system as an element and includes (as elements) a subsystem of a lower hierarchical level. Then the answer is simple. The strategic goal of a system is preset by the higher system.

The strategic goal of purposeful systems (including biological systems) is as follows: to survive and contribute to the survival of the higher system.

The systems possessing a degenerate deterministic behavior are *deterministic*; in the general case, they are called *stochastic*. It follows from the above definitions that purposeful systems can

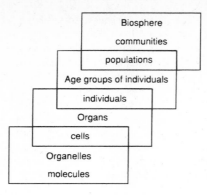

Figure 6.1 Hierarchy of biological systems, with alternating metabolic and structural levels.

be deterministic, and spontaneous systems stochastic, because regardless of the unpredictability of a purposeful system as to tactical goal change, with such a goal in view the system can behave quite determinately just because of its purposefulness. At the same time, spontaneous systems (physical systems included) can demonstrate quite chaotic and even extremely unpredictable, lottery-type behavior.

Among examples of simple deterministic physical systems are Newtonian mechanical systems comprising elements in the form of material points with gravitational interactions, as well as systems in statistical physics in which elements have the form of interacting molecules.

Wiener (1948) took living organisms and control devices like servomechanisms as prototypes of cybernetic systems. Their behavior was characterized mainly by a large number of feedbacks and homeostasis. However, these qualities were not enough to distinguish them from nonliving, natural, pseudocomplex systems. An attempt to count the reception of information (as well as control) among the distinctive features of cybernetic systems, with information interpreted as diversity, has added nothing, since in nonliving nature diversity exists together with the simplest forms of its reflection. That is why the criterion of presence of the decision act is the only thing in behavior that allows us to distinguish purposeful cybernetic systems as living, and automata simulating their behavior as distinct from spontaneous physical systems. This causes the appearance of a narrower class of complex systems, because complex pseudosystems (polyprocess, nonliving systems) remain outside of this class. Note that Acoff and Emery (1972) deal with a still narrower subclass of purposeful systems caused by the variability of tactical goals.

Establishing the broadest possible class of nonphysical, purposeful systems was not only a solution of the classification problem. It had a far more important end in view, one connected with the existence of specific, nonphysical laws of systemology. Compared with the difficulty of the statement of physics laws in systemology, the problem was aggravated by the fact that, apart from abstracting from side circumstances, one also had to abstract from all physical substances in order to get "pure" systemological properties (see below).

That is why, when dealing with the class of pseudocomplex systems, neither Wiener and Ashby nor their followers in the field of the theory of automatic control could discover the laws of systemology. For the same reasons, pioneers in cybernetics saw in information theory a theory of information "measurement" (as if it were a certain physical substance), instead of a theory of potential noise immunity. Only in the fundamental works by Kotel'nikov (1956) and Shannon (1948), of the same importance for systemology as the works by Galileo for physics, did the theory of the potential effectivity of complex systems start to exist (Fleishman, 1971). In these works, the following cybernetic properties were studied as "pure" ones: reliability (R property), noise immunity (I property), controllability (C property), and self-organizability (L property). This approach led to the establishment of limit laws of systemology which play the same role as the conservation laws in physics.

The modern state of physics was formed after continuous deterministic considerations were replaced by discrete stochastic ones and phenomenological considerations by model ones. These modern physical ideas were the basis of systemology, which dealt with discrete systems characterized by significantly more indeterminate behavior as compared with the indeterminacy of behavior of physical systems. This was reflected in the necessity for subjective interpretation of the probabilities of the decision act, while the probabilities of spontaneous behavior of physical systems had the usual frequency interpretation.

During the initial period of cybernetics development, an organism served as a prototype of cybernetic systems; later, in the period of its systemological development, the prototype more and more tended to above-organism ecological biosystems, that is, populations, communities, ecosystems, and the biosphere.

Although all systemological structures are of model character, there are still some with one-level models (no connection between micro- and macrolevels), and this makes them close to phenomenological laws of physics (to the Newtonian laws of mechanics, for example). Among systemological models of primary R, I, and C properties, such one-level models are those describing I and C properties of informational and control interactions of purposeful systems (like phenomenological laws of substantial interaction of physical bodies). However, the model of the R property (reliability) and its relative survivability model (RC property) are two-level models already.

Let us discuss the physical bases of the systemologically important minimax principle used in model considerations of ergodic processes to eliminate indeterminacy of their initial values.

The Minimax Principle

We consider the macroparameter case of (u, v) exchange between a system, $A = (|A|, \overline{A})$, and its environment B. With eqs. (6.7) and (6.8) taken into account, the resource, v, received by the system as the functional $v = \Phi(A, B)$ has the form

$$v = \Phi \{(|A(B)|, F_{D(A)} [\mathfrak{X}_{G_0}, \mathfrak{X}_\Gamma(B)]), B\} = \tilde{I}(\mathfrak{X}_{G_0}, A, B). \tag{6.9}$$

Indeterminacy of the initial value, $\mathfrak{X}_{G_0}$, makes theoretical physics pass to the consideration of ergodic systems in a stationary state. Let $A = (|A|, A)$ be a system with a constant structure, $|A|$ = constant, functioning in a constant environment, B = constant, and having ergodic behavior $\overline{A}(t)$. This case in connected to a real case with characteristic times $t_{\overline{A}} \ll t_{|A|} \ll t_B$. Then, for the system stationary state, with eq. (6.6) taken into account, (6.9) gives the following expression for the system-independent macroparameter, v, of the initial value, $\mathfrak{X}_{G_0}$:

$$v = I(A, B). \tag{6.10}$$

However, indeterminacy of environment B is caused by unknown environmental behavior. We can know only the class, $\mathfrak{B}_b$, of environments with certain restrictions imposed on their members and presented in the form of fixed values of macroparameters, b, like integral characteristics of "positive" (enhancing) or "negative" (suppressing) environmental effects on the system. Such indeterminacy is typical for systemology and for complex ecology in particular. With the functional $v = I(A, B)$ assumed to be a useful resource for the system, indeterminacy of the concrete effect of the environment B from the class $\mathfrak{B}_b$ can be eliminated by means of consideration of the "worst" variant, $B_{opt}(A, b)$, in the class; that is,

$$I(A, B_{opt}(A, b)) = \min_{B \in \mathfrak{B}_b} I(A, B). \tag{6.11}$$

But, behavioral indeterminacy of systems, especially in the case of purposeful behavior, remains the worst variant in passing from physical to systemological problems. At this stage, it seems natural to use the optimum principle as a consequence of the purposefulness of system A. This principle of systemology has been adopted from biology (Rashevsky, 1960) and successfully used in ecology (Fleishman, 1971; Semevskiy and Semenov, 1982).

Indeed, let system A belong to the class $\mathcal{A}_a$, where integral parameters a are fixed at values good for obtaining the useful resource v. Then, we should search for such an optimal $\mathcal{A}_a$ class system, $A_{opt}(a, b)$, that satisfies

$$f(a, b) = I(A_{opt}(a, b), B_{opt}(A_{opt}(a, b), b)) = \max_{A \in \mathcal{A}_a} \min_{B \in \mathcal{B}_b} I(A, B) . \tag{6.12}$$

Note, with another sequence of extrema, that is, with

$$g(a, b) = I(A_{opt}(a, B_{opt}(a, b)), B_{opt}(a, b)) = \min_{B \in \mathcal{B}_b} \max_{A \in \mathcal{A}_a} I(A, B) , \tag{6.13}$$

in the general case we would obtain different values of $g(a, b) \neq f(a, b)$ (absence of the saddle point in game-theory terms).

It is important to stress that max min and min max can be interpreted in terms of the fundamental properties of system behavior, "stimulus–response" and "response–stimulus" (the latter being more effective).

Possibility of Identifying Physical and Systemological Models and the Original System at the Mesolevel Only

Let us consider a most important problem of identification (experimental verification) of the theory of physics and systemology in the investigation of a natural original system. For physics, a particular case might be astronomic objects, and for systemology, artificial objects (Fleishman, 1982a). In the case in question, it is possible to identify the theory and experience by measurements or direct observations only at the space–time level of an experimentalist. In contrast to the micro- and macrolevels, this level is called a mesolevel.

In physics, the mesolevel is considered the macrolevel (with determined macroparameters), and the widely known Heisenberg's principle of indeterminacy establishes limits of measurement accuracy at the microlevel. In systemology, the mesolevel is considered the microlevel (except for microspatial systems), and the widely known principles of objective teleology (Fleishman, 1977a, 1982a) allow for the identification of optimal model structures and behaviors as well as of the original system at this level, while macroparameters are determined at the macrolevel.

So, although identification is carried out at the mesolevel in both physics and systemology, in fact, for the former it is at the macrolevel and for the latter at the microlevel. These statements directly concern ecology as above-organism macrobiology.

LAWS OF POTENTIAL EFFECTIVITY

In order of increasing activity, the primary properties of complex systems can be considered as follows: reliability (R property), noise immunity (I property), controllability (C property), and self-organizability (L property), as well as combinations of primary properties among which survivability (RC property) is best studied.

For each X and XY property we consider corresponding X and XY sections, A_X, A_{XY}, and B_X, B_{XY}, for systems A and environments B. These are subsystems of the system A with which parts (subenvironments) of the environment B interact. In the same way, the strategic goal, $\underline{A}$, of the system crowns the hierarchy, with tactical goals, $\underline{A} = (\underline{A}_X, \underline{A}_{XY})$, being subordinate to it.

The main notion of the theory of potential effectivity is *effectivity*, $P = P(\underline{A}|A, B)$, which can be defined as the conditional probability of gaining goal $\underline{A}$ by system A interacting with environment B. *Potential effectivity* is the maximal possible value of P within given limits. The total effectivity, P, of the system, A, and effectivities, P_X and P_{XY}, of its separate sections, A_X and A_{XY}, are interrelated. The main task is to estimate P on the basis of P_X and P_{XY} (a decomposition problem, corresponding to the problem of process trajectory description with the use of initial and boundary conditions in physics). Since effectivity P is the main goal functional in the theory, it is important to find its relationship with the macroparameter v considered above as the goal functional.

Models and Laws

Reliability: Model of R section, A_R. The goal $\underline{A}_R$ of the system A_R coincides with the strategic goal, which is survival. A_R is a two-level model with $t_{|A_R|} \sim t_{\overline{A}_R} \ll t_{B_R}$. At the instant t, the microstate of the structure $|A_R| = (U, \cdot)$ consists of $N_A(t)$ elements $U = \{e_\gamma\}$ ($\gamma = \overline{1, N_A(t)}$. As a result of interaction with the environment B_R, each of them perishes (or survives) with a probability $p_\gamma (1 - p_\gamma)$, ($P_{G_i} = (p_\gamma, 1 - p_\gamma)$). The behavior $\overline{A}$ consists in finding its microstate (alive and dead elements) with a delay assumed to be equal to unity. Then a living system can be replenished with an arbitrary number of elements (or it can regenerate them) without being affected by the indifferent environment. The system is considered to be dead from the moment when more than a $(1 - \theta)$th part of its elements is dead.

The system gains lifetime, t, by "paying" the environment with lives of its elements e_γ. In other words, here the (u_R, v_R) exchange has the form of (N_A, t) exchange. An optimal (N_A^0, t) exchange is an exchange creating the situation where the minimal number of compensating elements is sufficient to ensure an infinitely large value of t [reliability, $P_R(t)$, is infinitely close to unity].

The R law (Fleishman, 1964) states that the optimal (N_A^0, t) exchange takes place when

$$N_A^0 = C_R \log t, \qquad (6.14)$$

where $C_R = C_R((p_\gamma))$ is the fundamental constant depending on conditions of element and system death. Thus, for unlimited existence (immortality) of the system, at least logarithmic growth of element number, $N^0 = N^0(t)$, in time t is necessary but not sufficient. That is why finite systems (with $N_A(t) = $ constant) are mortal. Ecological consequences of the R law will be discussed in a later section.

The theory of reliability, originating in the 1950s, met the requirement of ensuring trouble-free operation of complex radioelectronic systems. By now, it has turned into the fundamental theory of trouble-free functioning of complex systems (incorporating failing elements) of *any physical nature*. Reliability of finite systems has been studied, and a variety of methods for their lifetime prolongation has been worked out; among these are element duplication and effective trouble-search algorithms.

Noise immunity: Model of I section, A_I. A_I is a one-level model with $t_{\overline{A}_I} \ll t_{|A_I|} \ll t_{B_I}$. The system tactical goal, A_I, is to receive a certain number, M, of important signals coming from the environment, B_I. These signals are received in the form of coded words, symbol by symbol, one symbol per a discrete instant of time, $t = 1, 2, \ldots$. This process is disturbed by environmental noise, B_I, prescribed by the Markovian $(a \times a)$ matrix, $[p_\alpha^\beta]$, of probabilities, p_α^β, of distortion of the alphabet, $\alpha = \overline{1, a}$, used to code the signal (p_α^β is the conditional probability of receipt of the symbol β, provided the symbol α was transmitted). In its structure, $|A_I|$, the system comprises $M = a^{tH_I}$ ($0 \leq H_I \leq 1$) ways of coding "t-long" words, which comprise all signals likely to come from the environment B_I. The procedure (algorithm) of distinguishing (decoding, identifying) one of M transmitted signals (which one is unknown) from its noise-distorted "t-long" code word corresponds to behavior, $\overline{A}_I$, of the system A_I.

Noise immunity, $P_I(t)$, is the probability of correct identification (decoding), M, of "t-long" signals. The greater the word length, t (identification period), the greater the number of signals, M, that can be distinguished by a system possessing noise immunity infinitesimally close to unity. System A_I can be said to take from the environment, B_I, a certain number, M, of useful signals that are paid for with a certain lifetime period, t, of A_I; that is, in this case the (u_I, v_I) exchange has the form of (t, M) exchange. An optimal (t, M^0) exchange ensures that, with a quite high value of t, almost reliable identification of the maximal number, $M = a^{tH_I}$, of signals takes place. Evidently, this value, $M = a^{tH_I}$ [or the value $H_I = H_I(A_I, B_I) = H_I((|A_I|, \overline{A}_I), [p_\alpha^\beta])$, which is the same], depends on the noise level and on methods of signal coding and decoding.

The *I law* (Shannon, 1948) states that the optimal (t, M^0) exchange takes place when

$$M^0 = M(t) = \exp(C_I t), \qquad (6.15)$$

where

$$C_I = C_I([p_\alpha^\beta]) = \max_{|A_I|} \max_{\overline{A}_I} H_I((|A_I|, \overline{A}_I), [p_\alpha^\beta]) \qquad (6.16)$$

is the fundamental $[p_\alpha^\beta]$ matrix-dependent constant (communication channel capacity in information-theory terms). The *I* law was formulated with an important assumption about the independence of noise processes at different instants.

Information theory originated from necessities of communication techniques. After the general case, $M =$ constant (Kotel'nikov, 1956), and the asymptotic case, $M = \exp(t H_I)$, $t \to \infty$ (Shannon, 1948), were considered, a fundamental theory of potential noise immunity in the reception of *signals of arbitrary physical nature* evolved. The main result was to establish the optimal algorithms of transmission ($|A_I|$) and faultless reception ($\overline{A}_I$) of the maximal possible number, $M_0 = e^{tC_I}$, of signals, given the noise level (*I* law). In addition, the theory comprises methods of optimal information coding, thus ensuring the maximal compression of information and elimination of its excessive parts.

Since the potential noise immunity is unattainable, information theory offers many practical algorithms of noise-proof transmission and reception of information. Information theory has contributed to the development of the theory of pattern identification (although the latter is not as complete). The theory of detection of one signal against the background of noise is a particular case of information theory. The latter is profoundly connected with mathematical statistics and, through this, with decision theory and game theory.

Controllability: Model of C section, A_C. A_C is a one-level model with $t_{\overline{A}_C} \sim t_{\overline{B}_C} \ll t_{|A_C|} \sim t_{|B_C|}$. We consider a conflict between systems A_C and B_C in a symmetrical situation of competition for important resources, the quantity of which, $K =$ constant, is limited. If the resource is vitally important, the goals of the competing systems, A_C and B_C, are strategic; otherwise, they are tactical.

The conflict is a succession of separate contests at discrete instants of time, $t = 1, 2, \ldots$. Each system, A_C and B_C, has a set of alternative actions, $X = \{x\}$ and $Y = \{y\}$, respectively. For each pair of actions, (x, y), a gain function is determined indicating the quantity of the resource, $M(x, y)$, acquired by system A_C [system B_C loses $-M(x, y)$]. The system structures are not determined (the systems are structurally amorphous). Their behaviors, $\overline{A}_C$ and $\overline{B}_C$, consist of choice (without informing each other) of actions x and y in each separate contest, formed as acts of decision, that is, in accordance with distributions of probabilities, $p_A(x)$ and $p_B(y)$. The conflict is not time limited, and it does not cease until the whole amount (quantity K) of resources is captured by one of the systems (and the other is completely ruined, or even dead if the resource was vitally important). The value $|M(x, y)| \ll K$ is supposed to be small. Then, the instantaneous average gain of system

A_C is equal to

$$H_C = H_C(p_A, p_B) = \sum_{x \in X, y \in Y} M(x, y)\, p_A(x)\, p_B(x),$$

while that of system B_C is $-H_C$. The fixed value of the resource, K, can be gained by A_C or B_C during the random time t (or average time, $\bar{t}$). Note that, depending on the sign of the value, $H_C > 0$ or $H_C < 0$, the gaining system is A_C or B_C, respectively, with the probability p_C (controllability) close to unity. It can be said that, for a certain quantity of the resource, system A_C or B_C pays a certain average period of its lifetime; that is, the (u_C, v_C) exchange is the $(\bar{t}, K)$ exchange. The optimal $(\bar{t}, K^0)$ exchange ensures that one of the systems can with near certainty gain the maximal quantity of the resource, K^0, during a fixed average competition time, $\bar{t}$. Using results from dynamic programming (Bellman, 1957), we can get the C law (Fleishman, 1966a).

The C law states that the optimal $(\bar{t}, K^0)$ exchange takes place when $K^0 = K(\bar{t}) = C_C\bar{t}$, where

$$C_C = C_C([M(x, y)]) = \max_{p_A} \min_{p_B} H_C(p_A, p_B) = \min_{p_B} \max_{p_A} H_C(p_A, p_B) \qquad (6.17)$$

is the gain function-dependent fundamental constant (*value* of the game in game-theory terms).

The C law is based on the assumption that actions of systems are independent in separate conflicts. Note that the model A_I is more sophisticated than A_C, since the former undertakes optimization of both structure, $|A_I|$, and, behavior, $\bar{A}_I$, while the latter deals only with optimization of behaviors, $\bar{A}_C$ and $\bar{B}_C$. In model A_C, the minimax principle (widely applied in systemology) was first used.

Control theory was created to meet the requirements of industrial automation as a theory of automatic management of control systems and controlled objects under asymmetric conditions. It had a great effect on early cybernetic ideas of feedback and homeostasis. However, its role was always limited to transformation of physical equations describing either object movement or a physicochemical process by the addition of control variables.

In parallel with the theory of automata and control, game theory was developed for economic problems and conflict situations (von Neumann and Morgenstern, 1944). Game theory is free of physical description of the controlled object and deals with the "pure" control phenomenon by expressing the gain of competing sides directly through their control effects (*without a description of the physical nature* of the controlled object). Game theory is symmetrical with regard to the sides in conflict.

The main result of game theory consists of establishing optimal control effects of the sides, meaning that a deviation from these affects the side in question. It is important to note that, despite determinacy of the task, optimal control effects are of probabilistic character, in the general case. With one side indifferent, the theory of optimal decisions connected with mathematical statistics appears to be a particular case of game theory. The theory has also been generalized for the case of more than two sides, which can also make groups. A theory of nonconflict games has only recently been developed.

Survivability: Model of RC section, A_{RC}. A_{RC} is a two-level model with $t_{\bar{A}_{RC}} \sim t_{\bar{B}_{RC}} \ll t_{|A_{RC}|} \sim t_{|B_{RC}|}$. The structure, $|A_{RC}| = (U^A, \cdot)$, of the system consist of $U^A = \{e_{A\gamma}, e_{R\gamma'}\}$ vitally important and protective A and R elements ($\gamma = \overline{1, N_A}$, $\gamma' = \overline{1, N_R}$), respectively.

Consider a conflict situation where the structure, $|B_{RC}| = (U^B, \cdot)$, of the system B_{RC} has a set of $U^B = \{e_{C\delta}\}$ ($\delta = \overline{1, N_C}$) elements eliminating A and R elements. In their turn, the latter can affect activities of C elements by eliminating or neutralizing them. Prior to interaction, each system, A_{RC} and B_{RC}, has the limited numbers, N_A, N_R, and N_C, respectively, of A, R, and C elements. At discrete instants $t = 1, 2, \ldots$, separate contests between systems A_{RC} and B_{RC} take place; in their course, system B_{RC} distributes separate portions of its C elements to eliminate certain quantities of A and

R elements of system A_{RC} (this is behavior of system B_{RC}). Behavior of system A_{RC} against B_{RC} is determined at each instant t by the probability $p = p(N_R(t))$ of elimination (or neutralization) of one C element by $N_R(t)$ R elements available in system A_{RC} at the instant t. Unlike the model A_R, here the phenomenon of regeneration is not taken into account due to the short duration of interaction of systems A_{RC} and B_{RC} compared with the recoil period for *preadaptation* of their structures, $|A_{RC}|$ and $|B_{RC}|$, to the next conflict. Specifically, a certain fixed quantity, E_R, of an active protective substratum (energy, for example) is evenly distributed by system A_{RC} among R elements, thus determining their activity.

The conflict between systems A_{RC} and B_{RC} is considered to be over after B_{RC} has lost all its C elements. Then, if A_{RC} still contains a certain θ_Ath part of the initial quantity, N_A, of its A elements, it is considered to be alive; if this part is smaller, the system is considered dead. So, the strategic goal of a system is to survive. The goal of system B_{RC}—eliminating system A_{RC}—can be considered as tactical.

System A_{RC} tends to exchange its A and R elements for the greatest possible number of C elements of system B_{RC}, since exhaustion of the latter with simultaneous survival of the θ_Ath part of its own N_A A elements keeps it alive. So here we have $(\vec{N}, N_C)$ exchange, where the vector $|A_{RC}(t)| = \vec{N} = (N_A, N_R)$ determines the initial structural state of system A_{RC}. The optimal $(\vec{N}, N_C^0)$ exchange is such an $(\vec{N}, N_C^0)$ exchange, which leads to the maximal value, $N_C = N_C^0$, of C elements with the optimal number, $N_R^0 = N_R(N_A)$, of R elements corresponding to fixed values of N_A and E_R, when a probability $1 - p_{RC}$ of death of the system is close to unity ($p_{RC} = p_{RC}(N_A)$ — survivability of the system).

The *RC law* (Fleishman, 1966b) states that the optimal $[(N_A, N_R^0), N_C^0]$ exchange takes place when $N_C = H_{RC}(A_{RC}, B_{RC})N_A$ and $N_C^0 = C_{RC} N_A$, where

$$C_{RC} = \max_{A_{RC}} \min_{B_{RC}} H_{RC}(A_{RC}, B_{RC}) = (1 - \theta_A) \exp(cE_R) \quad (c > 0), \qquad (6.18)$$

is the fundamental constant depending on the quality of R elements, and the optimal value, $N_R = N_R^0$, is

$$N_C^0 = (1 - \theta_A)cE_R N_A . \qquad (6.19)$$

The *RC* law is formulated with an important assumption about independence of the death of A and R elements at different instants of the conflict between systems A_{RC} and B_{RC}. As follows from the *RC* law, with $N_C < N_C^0$ number of C elements of system B_{RC} available, the optimal system, A_{RC}^0, survives with a probability close to unity. Ecological consequences of the *RC* law are taken up later.

It is important (for biological applications in particular) that the models A_C and A_{RC} can be used not only in conflict situations. They are also helpful in indifferent situations when, due to behavioristic indeterminacy of the environment, B, system A patterns its behavior on the "worst" possible environmental behavior to ensure its effectiveness.

Self-organizability: Models of L sections, A_L. Self-organizability of a system is its ability to improve its properties by means of adaptation or, possibly, preadaptation to varying environmental conditions (L property) or self-perfection, when insulated from the environment. The theory of self-organizability is extremely complicated. That is why only some fragments of it have been developed as yet. At first, they were connected only with the progressive structural evolution of more and more complex mathematical models. First studies were by J. von Neumann, who dealt with self-reproducing cellular automata that, having reached a certain critical state of complexity, started to reproduce automata more complex than themselves. Then, with the use of mutable classical automata, a set of prognostic computer experiments (evolutionary modeling) was carried out (Fogel et al., 1966). Finally, in a paper by Ivakhnenko (1971) and a series of other papers by the same author, the "method of group argument account" (MGAA) was developed with the use of mutable polynomials. For MGAA, existence of the best prognostic polynomial of optimal complexity has been discovered. The most important feature of MGAA is naturally incorporated in its algorithms, a rational sorting out of polynomial variants. For a theoretical estimate of prognosis accuracy dependence on the number of involved details, the results of information theory have been used.

The self-organization effect is based not only on the mutation mechanism; development of its fundamentals is also connected with the theory of optimal pattern classification and recognition and with adaptation theory. Theoretical analysis of the adaptation effect is based on theoretical and game principles and allows study of adaptation and preadaptation of regenerative systems under conditions of variable environmental pressure. Regenerative systems are the subject of a later section and are not discussed in detail here. The spontaneous effect of self-organization is studied by means of synergetics (Haken, 1983). Synergetics is based on the physical ideas and means of the theory of nonequilibrium states (Prigogine, 1980).

Asymptotic Expressions for Goal Functionals P_X and v_X and Their Relationship

Since the goal $\underline{A}_X$ of the model of X section A_X is expressed through (u_X, v_X) exchange, for the corresponding effectivity we get $P_X = P(A_X|A_X, B_X) = P[(u_X, v_X)|A_X, B_X]$. At great values of u_X [and corresponding great values of $v_X = v_X(u_X, A_X, B_X)$], for existing models of X sections A_X, there exists a limit law of potential effectivity (Fleishman, 1971):

$$P_X = P((u_x, v_x)|A_x, B_x) \xrightarrow{u_X \to \infty} \begin{cases} 1, & \text{when } v_X < v_X^0, \\ 0, & \text{when } v_X > v_X^0, \end{cases} \qquad (6.20)$$

where

$$v_X^0 = v(u_X) = \max_{A_x \in \mathcal{A}_X} \min_{B_x \in \mathcal{B}_X} v_x(u_X, A_X, B_X) \qquad (6.21)$$

is the fundamental value involved in the optimal (u_X, v_X^0) exchange, and $\mathcal{A}_X$ and $\mathcal{B}_X$ are broad classes of X sections, A_X and B_X. If the class $\mathcal{B}_X$ degenerates into one X section, B_X, then (6.21) contains only one maximum. There arises a special question about replacement of extrema in (6.21). The theory of potential effectivity gives explicit analytical expressions, P_X, for a number of X sections.

Below, we shall use the lower Boolean estimate of probability, $P_t = P(X)$, of joint arrival, $X = (X_1, \ldots, X_s, \ldots, X_t)$ at time t, of the events X_s,

$$P_t \geq 1 - P(\overline{X}_1, \ldots, \overline{X}_{t_1}) - \sum_{s=t_1+1}^{t} P(\overline{X}_s), \qquad (6.22)$$

where $\overline{X}$ is an addition to the event X, as well as the expression of distribution $P_E = (P_E(X))$ truncated by the event $E \subseteq X$. Let $P(E) = \sum_{X \in E} P(X)$. Then, by definition,

$$P_E(X) = \begin{cases} p(X)/p(E), & \text{when } X \in E, \\ 0, & \text{when } X \notin E, \end{cases} \qquad (6.23)$$

and, if $\tilde{E} \subseteq E$, the notation will be $P_E(\tilde{E}) = \sum_{X \in \tilde{E}} P_E(X)$.

Let us start with some special Boolean estimates. If, for $\epsilon > 0$ and $\eta = t_1 P(\overline{X}_{t_1})/\epsilon$, it is required that

$$P(\overline{X}_s) \leq P(\overline{X}_{t_1}) \left(\frac{t_1}{s}\right)^{1+\eta}. \tag{6.24}$$

Then, following (6.22) and using the integral estimate of the sum, we obtain

$$P_t \geq 1 - P(\overline{X}_1, \ldots, \overline{X}_{t_1}) - \epsilon\left(1 - \left(\frac{t_1}{s}\right)^\eta\right). \tag{6.25}$$

Now consider the case of pseudoindependent events (Fleishman, 1982a, b), assuming events X_s to be binary and random, $X_s = \begin{Bmatrix} 0, & 1 \\ 1 - p_s, & p_s \end{Bmatrix}$, and independent before truncation. Let $E = E_0 \cup E_1$, where

$$E_0 = \left\{X: \sum_{s=1}^{t} X_s = 0\right\} \text{ and } E_1 = \left\{X: \sum_{s=1}^{t} X_s = 1\right\}.$$

Then

$$P(E_0) = \prod_{s=1}^{t}(1 - p_s), \quad P(E_1) = P(E_0)\sum_{s=1}^{t}\frac{p_s}{1 - p_s},$$

and, assuming $E_1 = \tilde{E}$, we get from (6.23)

$$P_E(E_1) = \left(1 + \left(\sum_{s=1}^{t}(p_s^{-1} - 1)^{-1}\right)^{-1}\right)^{-1}. \tag{6.26}$$

Let us consider an asymptotic estimate as $t \to \infty$ of the left part of relation (6.22); assuming $t_1 = 0$,

$$P_t \geq 1 - \frac{t}{f_1}, \tag{6.27}$$

where a hypothetical limit is

$$f_1^{-1} = \lim_{t \to \infty} \frac{1}{t}\sum_{s=1}^{t} P(\overline{X}_s). \tag{6.28}$$

Similarly, for relation (6.23), with $t_1 \to \infty$,

$$P_E(\tilde{E}) = (1 + (tf_2)^{-1})^{-1} \simeq 1 - (tf_2)^{-1}, \tag{6.29}$$

with a hypothetical limit

$$f_2^{-1} = \lim_{t \to \infty} t\left[\frac{1 - P(\tilde{E})}{P(E)}\right]. \tag{6.30}$$

Note that, in the case of pseudoindependent events, this limit exists.

Let us interpret the probability $P_t = P_R(t)$ through reliability (that is, the probability of section A_R living up to instant t) and the probability $P(\overline{X}_s) = 1 - \mu(s)$ $[P(X_s) = \mu(s)]$ through *viability (nonviability)*, that is, the instantaneous probability of survival or (death) of section A_R. Then, with $t_1 = 1$ and assuming $\mu(1) = 1 - P_R(1) = \varepsilon > 0$, under the condition

$$\mu(s) \leq \varepsilon \left(\frac{1}{s}\right)^2 \tag{6.31}$$

we get from (6.24) and (6.25) the estimates

$$P_R(t) \geq 1 - \varepsilon \left[2 - \left(\frac{1}{t}\right)\right], \qquad P_R(\infty) \geq 1 - 2\varepsilon. \tag{6.32}$$

Let us interpret probabilities p_σ and $1 - p_\sigma$ as probabilities of death and survival of the system resulting from the σth cause and let us assume that a system can perish only once. Then, $P_E(E_1)$ can be taken as the probability of system death, $\mu(s)$, resulting from joint action of s causes. In particular, assuming that $s = 2$, $\mu(2) = \mu$, $p_1 = x$, and $p_2 = y$, we get

$$\mu = \frac{x + y - 2xy}{1 - xy}. \tag{6.33}$$

Let us consider a two-level model, A_R, of reliability when the instantaneous system state, $A_R(s)$, contains $N_A(s)$ elements. Then, at the instant s, with $N_A(s) \to \infty$ as $s \to \infty$, its viability has asymptotic expressions:

$$1 - \mu(s) \begin{cases} = 1 - \exp[-f_2 N_A(s)] & (a) \\ \geq 1 - \dfrac{1}{f_2 N_A(s)} & (b) \end{cases} . \tag{6.34}$$

The first is for the case of independent death of separate elements according to the Bernoullian scheme (Fleishman, 1971); the second describes the truncated distribution $1 - \mu(s) = P_E(\tilde{E}_1)$, where f_2 is determined by the relation (6.29) and (6.30) in which $t = s$ has been replaced by $N_A(s)$.

Using the condition (6.31) and the following from relation (6.32), as well as relations (6.34), we get

$$P_R(t) \geq 1 - \varepsilon(2 - t^{-1}) \quad \text{and} \quad P_R(\infty) \geq 1 - 2\varepsilon, \quad \text{when } N_A(s) \begin{cases} = C_R \ln s & (a) \\ \geq C'_R s^2 & (b) \end{cases}, \tag{6.35}$$

and when $N_A(s) = \text{constant } [\mu(s) = \mu = \text{constant}]$,

$$P_R(t) \begin{cases} = (1 - \mu)^t = \exp\left(\dfrac{-t}{f_R}\right) \\ \geq 1 - t\mu = \dfrac{1 - t}{f_R} \end{cases} \quad \text{and} \quad f_R = \begin{cases} [\ln(1 - \mu)^{-1}]^{-1} & (a) \\ \mu^{-1} & (b) \end{cases}, \tag{6.36}$$

with $(N_A(t), t)$ exchange, where $C_R = 2/f_2$ and $C'_R = 1/f_2 \, \varepsilon$ are the fundamental constants [see eq. (6.14)] for (a) (Fleishman, 1971) and (b), respectively. Note the peculiar feature of the R section with increasing t; that is, $P_R(t)$ always decreases, tending to constant $P_R(\infty)$ $(0 \leq P_R(\infty) \leq 1)$. Now, we proceed to other X sections $(X \neq R)$.

With peculiarities of one-level (t, v_X) exchanges taken into account (Fleishman, 1971), for $X = I, C$ we get, similarly,

$$P_X(t) \simeq \begin{cases} 1 - \exp(-t_{fX}) \\ \exp(-t_{fX}) \end{cases} \text{when} \quad \begin{matrix} v_X < v_X^0 \\ v_X > v_X^0 \end{matrix}, \tag{6.37}$$

where

$$f_X = f(H_X, C_X), \quad v_X = \varphi_X(tH_X), \quad v_X^0 = \varphi_X(tC_X),$$

and constants H_X and C_X are determined by (6.15), (6.16), and (6.17).

For survivability, with $[(N_A, N_R(N_A)), N_C]$ exchange between the systems A_{RC} and B_{RC}, there is (provided $N_A \to \infty$) an asymptotic relation (Fleishman, 1971):

$$P_{RC}(N_A) = \begin{cases} 1 - \exp(-N_A f_{RC}) \\ \exp(-N_A f_{RC}) \end{cases} \text{with} \quad \begin{matrix} N_C < N_C^0 \\ N_C > \tilde{N}_C^0 > N_C^0 \end{matrix}, \tag{6.38}$$

where

$$f_{RC} = f_{RC}(H_{RC}, C_{RC}), \quad N_C = \varphi(N_A H_{RC}), \quad N_C^0 = \varphi(N_A C_{RC}),$$

and the constants H_{RC} and C_{RC} are determined by (6.18) and (6.19). Here, everywhere the functions f_X and f_{RC} are expressed through the k function, $f(H, C) = k(\psi(H), \psi(C))$, and functions $y = \varphi(x)$ and $y = \psi(x)$ monotonically increase with increasing arguments. Besides, the k function monotonically increases with the increasing difference modulus of its arguments:

$$k(\theta, p) = \theta \ln\left(\frac{\theta}{p}\right) + (1 - \theta) \ln\left(\frac{1 - \theta}{1 - p}\right) \xrightarrow{|\theta - P| \to 0} \frac{(\theta - p)^2}{2\theta(1 - \theta)} \tag{6.39}$$

and

$$\varphi_I(x) = \exp(x), \quad \varphi_C(x) = \varphi_{RC}(x) = x. \tag{6.40}$$

It follows from (6.37) to (6.40) that, with fixed v_X and v_{RC} and increasing v_X^0 and v_{RC}^0, respective effectivities, P_X and P_{RC}, also grow. This proves their validity as goal functionals.

Decomposition of the System A

Additive Decomposition: Criterion of Equidurability.
Let $\bar{R}$ be the set of all X properties of a system A except for $X = R$. Then, using the Boolean formula, we get for effectivity P of the system $A = \{A_X\}$:

$$P \geq P_R + P_{\bar{R}} - 1, \quad P_{\bar{R}} \geq 1 - \sum_{X \in \bar{R}}(1 - P_X). \tag{6.41}$$

Since, in the last estimate, the order of deviation from unity is determined by the highest order of an item (with a fixed number n of X properties), then it seems natural to choose all items of the same order (the principle of section equidurability):

$$1 - P_x \simeq (1 - P_{\bar{R}})/n. \tag{6.42}$$

This and expressions (6.37), with $n = $ constant and $t \to \infty$, allow an estimate for $P_{\bar{R}}$ in the form

$$P_{\bar{R}} = P_{\bar{R}}(t) \begin{cases} \simeq 1 - \exp(-f_{\bar{R}}t) & \text{(a)} \\ \geq 1 - (f_{\bar{R}}t)^{-1} & \text{(b)} \end{cases}, \tag{6.43}$$

where $f_{\bar{R}} = f_X = $ constant and the estimate (b) is given for the case when all corresponding X sections are considered at truncated distributions.

Multiplicative decomposition of the system $A = (A_R, A_{\bar{R}})$ and its effectivity macroparameter, f. Let us consider the system effectivity $P = P(t)$ during the final functioning period $(0, t)$. The system consists of two subsystems: a reliability system, A_R, and an active system, $A_{\bar{R}}$. Correspondingly, the system goal $\underline{A} = (\underline{A}_R, \underline{A}_{\bar{R}})$ splits into two subgoals: a general goal, $\underline{A}_R$, and a tactical one, $\underline{A}_{\bar{R}}$. That is why the total effectivity, $P(t) = P(\underline{A}|A,B)$, as the joint probability of gaining the goals, is equal to the product [see expressions (6.36) and (6.43)]:

$$P(t) = P_R(t) \cdot P_{\bar{R}}(t) \begin{cases} = \exp(-f_R^{-1}t) \cdot (1 - \exp(-f_{\bar{R}}t)) & (a) \\ \geq (1 - f_R^{-1}t)(1 - (f_{\bar{R}}t)^{-1}) & (b) \end{cases}, \quad (6.44)$$

$$P_{max} = \max_t P(t) = P(t_{opt}) = P(f) \quad \text{and} \quad t_{opt} = t(f),$$

where macroparameter effectivity (Fleishman, 1971),

$$f = f_R \cdot f_{\bar{R}}, \quad (6.45)$$

and for the pragmatically significant asymptotic case (as $f \to \infty$),

$$1 - P(f) \simeq \frac{t(f)}{f_{\bar{R}}} \simeq \begin{cases} \dfrac{\ln f}{f} & (a) \\ \dfrac{1}{\sqrt{f}} & (b) \end{cases}, \quad (6.46)$$

which demonstrates an asymptotically worse convergence to unity of the effectivity in case (b) as compared to (a).

Primary optimal structuring of the system, A. Up to now we have considered system A with a preset structure, $|A|$. Now let us consider an amorphous two-level formation to represent a developing (still nonregenerative) system, $A = \{e_\gamma\}$, consisting of primary elements, e_γ ($\gamma = \overline{1, N}$), and their relationships. Accidental relationships between elements form relatively short-term structures, so we shall assume one order of their characteristic times to be $t_{\bar{A}} \sim t_{|A|}$. To reflect an "amorphous chaos" instead of the "trajectory" presentation of the system through time t, we shall consider the system to have been preassigned statistically, with the distribution $\mathbf{P} = (P(X))$, $X = (X_1, \ldots, X_n)$, of states of its enlarged secondary elements (relationships included), $E_i = \{e_{\gamma i}\}$, $A = \{E_i\}$, which, according to a certain indication F, can be in the states $F(E_i) = X_i \in \mathbf{X}_i$. This predetermination takes place in the time period $(0, t)$. Having introduced the macrolevel of goal functionals of (u, v) exchange, we pass to a three-level consideration when predetermination of the goal $\underline{A} = (u, v^0)$ in the form of the outlet (u, v^0) exchange leads, generally speaking, to a certain set of realizing microstates, $E_0 \subset \mathbf{X} = (X_i)$. Here, effectivity of system A has the form

$$P = \sum_{X \in E_0} P(X) = P(E_0). \quad (6.47)$$

Now consider two structurings, $A = \{e_\gamma\} = \{E_i^R\} = \{E_j^{\bar{R}}\}$, of the same system, A (initially amorphous), consisting of primary elements with intersecting secondary elements, $i = \overline{1, n_R}$, $j = \overline{1, n_{\bar{R}}}$. Each structuring leads to its own subsystem of system A. The first will be associated below with the reliability subsystem, A_R, and the second with the active subsystem, $A_{\bar{R}}$. As a result, we get the decomposition $A = (A_R, A_{\bar{R}})$.

Let us find the dependence between dimensions n_R and $n_{\bar{R}}$. It is natural to structure the reliability subsystem, A_R, after the active one, $A_{\bar{R}}$, keeping in mind the usefulness of doubling active elements to ensure more reliable fulfillment of their functions. That is why

$$n_{\bar{R}} = \sum_{i=1}^{n_R} n_{\bar{R}i} = \alpha n_R, \qquad (6.48)$$

where $\alpha = \dfrac{1}{n_R} \sum_{i=1}^{n_R} n_{\bar{R}i}$ is the average number of active elements in a reliability element assumed to be fixed.

If goals of the subsystems, $A_R = E_0^R$ and $A_{\bar{R}} = E_0^{\bar{R}}$, are given, it is possible to determine the goal of system $A = (A_R, A_{\bar{R}})$ and projections of the set E_0^R onto sets E_i^R in the form E_{0i}^R, $i = \overline{1, n_R}$, as well as a certain "permissible" set of states $E^{\bar{R}} \leq \mathbf{X}^{\bar{R}}$ such that $E_0^{\bar{R}} \leq E^{\bar{R}}$. Assume that subsystem A_R consists of n_R elements comprising all elements working in parallel. Then, to ensure reliability of system A, it is possible to put into operation *all* reliability elements. This leads to the Boolean estimate (6.27), where, assuming $t = n_R$, $P_t = P_R$, $X_i = E_i^R$, and $f_1 = f_R$, we get the estimate (6.46b) for the probability $P_R(n_R)$. Similarly, taking into account (6.48), and assuming $t = n_R = \alpha n_{\bar{R}}$, $P_{\bar{R}}(t) = P_{\bar{R}}(n_{\bar{R}})$, and $f_{\bar{R}} = \alpha f'_{\bar{R}}$ in (6.43b), we get an estimate for the probability $P = P(n_R)$. After that, effectivity of the system, $P = P(n_R)$, as a function of a number of reliability elements turns into the product $P(n_R) = P_R(n_R) \cdot P_{\bar{R}}(n_R)$, and when $n_R = t(f)$, this leads to the maximum, $P_{\max} = \max_{n_R} P(n_R) = P(f)$, where the macroparameter $f = \alpha f_R f'_{\bar{R}}$, and f_R and $f'_{\bar{R}}$ are determined from (6.28) and (6.30). At the same time, in the asymptotic case, as $f \to \infty$, we get for $P(f)$ and $n_{R_{opt}} = \tau(f)$ the asymptotic estimates (6.46b).

Ecological Consequences of the Reliability Limit Law

Now, as indicated previously, we take up some of the ecological consequences. A successful primary structuring of system $A(f \gg 1)$ can lead to a reliability section, $A_R = \{E_i^R\}$, with an evolving structure, $|A|$, containing at every instant $n_R = m_s$ exclusive elements, E_i^R, each comprising n_{is} primary elements, so that the total number of primary elements is

$$\sum_{i=1}^{m_s} \epsilon \, E_i^R = m_s \, n_{is}. \qquad (6.49)$$

Here, dynamic stability $(t_{|A|} \gg t_{\bar{A}})$ is ensured by m_s *structural elements* (SE) E_i^R and $L \leq m_s(m_s - 1)/2$, and their interrelationships at the structural level, and n_s primary homogeneous interchangeable and renewed *metabolic elements* (ME) at the metabolic level. In particular, when $m_s = 1$, structural degeneration takes place and system A turns into a metabolic system consisting of $n_s = n_{1s}$ primary elements (see previously the R section model of A_R).

Death of system A at the metabolic level is connected with the death of at least one of its SE E_i^R. An SE is considered dead if at the instant $s + 1$ the number of its MEs becomes less than $n_{is}^{cr} = \theta n_{is}$ ($0 < \theta_{is} < 1$). Death of system A at the structural level is connected with lack of integrity. The system is considered to have integrity at an instant s if any division into two demands the elimination of not less than $l_s^{cr} = \theta_s l_s$ ($0 < \theta_s < 1 < l_s < m_s - 1$) relationships between the corresponding SEs. Assume, at the instant s, that the MEs incorporated in the SE E_i^R survive (or perish) independently with probability $p_{is}(1 - p_{is})$, and that SEs keep (or lose) their interrelationships independently with probability $d_s(1 - d_s)$. Then, asymptotically, with $n_{is} \to \infty$ and $\theta_{is} < p_{is}$, the probability, P_{is}, of SE E_i^R survivability has the form

$$P_{is} \sim 1 - \exp(-n_{is} k_{is}), \qquad (6.50)$$

and with $l_s \to \infty$ and $\theta_s < d_s$, the probability, D_s, of its nonseparation has the form

$$D_s \sim 1 - \exp(-l_s k_s), \qquad (6.51)$$

where $k_{is} = k(\theta_{is}, p_{is})$ and $k_s = k(\theta_s, d_s)$ (Fleishman, 1971).

At $n_{is} = \Gamma n_s / k_{is} m_s$, relations (6.49) and (6.50), where $\Gamma = \lim_{s \to \infty} \Gamma_s$, and $\Gamma_s = m_s / \sum_{i=1}^{m_s} k_{is}^{-1}$, allow expression of the maximal metabolic viability (Fleishman, 1971):

$$1 - \mu_1(s) \sim \left[1 - \exp\left(\frac{-\Gamma}{m_s}\right)\right]^{m_s} \sim \begin{cases} \exp(-cm_s^{|\delta|}) \\ 1 - cm_s^{-\delta} \\ 1 - \exp(-(1+\delta)m_s^{\zeta}) \end{cases} \qquad (6.52)$$

$$\text{with } n_s = \frac{\ln c^{-1}}{\Gamma} m_s + \frac{1+\delta}{\Gamma} m_s \times \begin{cases} \ln m_s, & \delta < 0 \\ \ln m_s \\ m_s^{\zeta} \end{cases} \delta > 0,$$

where $c, \zeta > 0$, and δ are arbitrary constants.

Relation (6.51) and the generalization of results by Gilbert (1959) allow structural viability to be obtained:

$$1 - \mu_2(s) \sim 1 - m_s \exp(-l_s k), \qquad (6.53)$$

where $k = \lim_{s \to \infty} ks$.

With (6.52) and (6.53) taken into account, the maximal value of the total viability, $1 - \mu(s)$, of the system can be achieved at $\mu_1(s) \sim \mu_2(s)$, or $l_s = \Gamma n_s / km_s$ (it is quite the same), and has the form

$$1 - \mu(s) \sim 1 - \mu_1(s) - \mu_2(s) \sim 1 - 2\mu_1(s) \sim 1 - \mu_1(s). \qquad (6.54)$$

Now, we use requirements (6.31) of the hyperbolic reliability law to find the dependence of the numbers n_s and m_s of metabolic (ME) and structural (SE) elements, respectively, on the age s of the system A. It follows from (6.51), (6.52), and (6.54) that

$$m_s \sim \begin{cases} (c/\epsilon)^{1/\delta} s^{c_1} \\ c_2 (\ln s)^{1/\zeta} \end{cases}, \quad n_s \sim \begin{cases} c_3 s^{c_1} \ln s \\ c_4 (\ln s)^{(1+\zeta)/\zeta} \end{cases}, \qquad (6.55)$$

where

$$c_1 = 2/\delta, \quad c_2 = [2/(1+\delta)]^{1/\zeta}, \quad c_3 = \frac{1+\delta}{\Gamma \delta} 2 (c/\epsilon)^{1/\delta}, \quad c_4 = \frac{1+\delta}{\Gamma} c_2^{1+\zeta},$$

and $\epsilon, \zeta,$ and $\delta > 0$, and for a degenerate metabolic case $m_s = 1$, the relation is (6.25).

Now we may biologically interpret the results obtained (Fleishman, 1991). As is found empirically, with increasing age the viability of biological systems increases with hyperbolically decreasing reliability (accordingly, the number of individuals of a certain age belonging to a nonreplenished generation also decreases (Pianka, 1966; see Fig. 6.2), and with logarithmically growing dimensions (Bakman's law; Bakman, 1943). However, connection between these facts, particularly the quantitative connection, was not known. The developed theory fills the gap. Specifically, among its consequences are Bakman's logarithmic law and relations (6.14) and (6.55) at $\zeta \gg 1$, when the metabolic effect (ME) prevails over the structural effect (SE). The first is to be

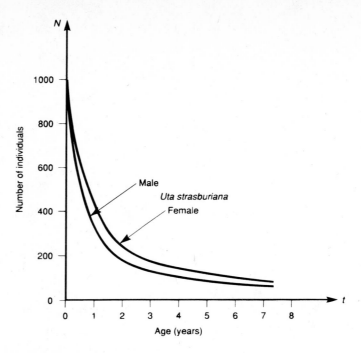

Figure 6.2 Number, $N = 1000\, P_R(t)$, of "t-years" old male and female lizards, *Uta strasburiana* (Tinkle, 1967). $P_R(t)$ is the probability of reaching age t by an individual (reliability).

found at the level of organelle molecules, organ cells, and individuals of an age group or population; the second, at the level of cell organelles, individuals' organs, age groups of individuals belonging to a population, and communities of the biosphere. So the law of alteration of the mentioned levels in the hierarchy of biological systems (Fig. 6.1) takes place.

Note that the general form of hyperbolic conditions (6.31) of nonviability decrease allows further analysis of this effect with the aim of expressing nonviability, $\mu(s)$, through different factors leading an individual to death (see later, at the end of the section on theoretical parameter identification).

Secondary Optimization on Integral Macroparameters and Ecological Consequences of the Survivability Limit Law

Let there be a limited resource K of material or energy. It is necessary to distribute the resources, $K_R + K_{\bar{R}} = K$, between two subsystems, A_R and $A_{\bar{R}}$, to support their *parameter* effectivities, f_R, and $f_{\bar{R}}$. This is to be done in such a way that effectivity, $P_{\max} = P(f)$, of the system, $A = (A_R, A_{\bar{R}})$, which is a monotonically increasing function of the effectivity parameter, $f = \alpha f_R f'_{\bar{R}}$, could become the maximum, $P_{\max\,\max} = P(f_{\max})$, where $f_{\max} = \alpha\, f_R^0 \cdot f_{\bar{R}}^{'0} = \mathscr{F}(K)$ is the corresponding maximal value of f. Since K_R and $K_{\bar{R}}$ monotonically increase with f_R and $f'_{\bar{R}}$, they can be expressed accurately enough through parabolas

$$K_R = k_R f_R^{b_R} \quad \text{and} \quad K_{\bar{R}} = k_{\bar{R}} f_{\bar{R}}^{'b_{\bar{R}}}, \tag{6.56}$$

with positive coefficients. When solving this extremum problem, the optimal values

$$f_R^0 = [(b_R/k_R)/(b_R + b_{\bar{R}})]^{1/b_R} K^{1/b_R} \quad \text{and} \quad f_{\bar{R}}^{\prime 0} = [(b_{\bar{R}}/k_{\bar{R}})/(b_R + b_{\bar{R}})]^{1/b_{\bar{R}}} K^{1/b_{\bar{R}}} \tag{6.57}$$

are obtained, leading to the maximal value

$$f_{\max} = \mathscr{F}(K) = \alpha f_R^0 \cdot f_{\bar{R}}^0. \tag{6.58}$$

These relations still remain to be ecologically interpreted. The secondary optimization and profound ecological interpretation are connected with the survivability model, A_{RC} (Fleishman, 1971).

Preadaptation of system A_{RC} to the environmental system B_{RC} takes place prior to their interaction and consists of the following. Let the system have a limited quantity, E, of vital creative substrate (matter or energy) that it can evenly distribute among N_A of its A elements so that each share is $e = E/N_A$. Let each A element reproduce from the primary substratum secondary substrata, and the protecting one among them in the amount of $e_R = f(e) = f(E/N_A)$. Then its total amount is

$$E_R = N_A f E/N_A = g_E(N_A). \tag{6.59}$$

The general superadditive law of system effectivity relative to the elements ("strength is in unity") makes us believe that decrease of the function $g_E(x)$ with increasing x reflects the effect of primary substrate splitting, which affects the reproduction of secondary substrata. Thus,

$$g_E(x) > 0, \quad g_E'(x) < 0, \quad \text{and} \quad g_E(x) \xrightarrow[x \to \infty]{} 0. \tag{6.60}$$

Having put (6.59) into (6.18), we get

$$N_C^0 = N_E(N_A) = (1 - \theta_A) N_A \exp[c\, g_E(N_A)]. \tag{6.61}$$

It follows from the conditions of (6.60) that as $N_A \to \infty$, $g_E(N_A) \to 0$, and thus (6.61) has the asymptote:

$$N_C^0 = N_E(N_A) \xrightarrow[N_A \to \infty]{} (1 - \theta_A) N_A. \tag{6.62}$$

So, the function $N_E^0 = N_E(N_A)$ can have peculiarities only at finite values N_A about the initial value $N_A = 1$.

Fleishman (1971) presents an analysis of these peculiarities for the general case, as well as for the simplest hyperbolic predetermination,

$$g_E(x) = \frac{\beta}{x + \gamma}, \tag{6.63}$$

where

$$\beta = \frac{BE^\delta}{\delta - 1} \quad \text{and} \quad \gamma = \frac{2 - \delta}{\delta - 1}, \tag{6.63'}$$

and for a parabolic predetermination,

$$f(E) = BE^\delta, \tag{6.64}$$

where $B > 0$ and $1 < \delta \leq 2$, which allows approximating a broad class of real functions with reasonable accuracy. In this case, relation (6.60) gives an explicit expression:

$$N_C^0 = N_E(N_A) = (1 - \theta_A) N_A \exp\left[\frac{c\beta E^\delta}{(\delta - 1)\left(N_A + \frac{2 - \delta}{\delta - 1}\right)}\right]. \tag{6.65}$$

Here, with $E < E_{cr}$ (E_{cr} is the critical value), N_C^0 monotonically increases with increasing N_A and with $E > E_{cr}$; at first, it reaches the maximum, $N_C^{max} = N_E(N_A^0)$, and then the minimum, $N_C^{min} = N_E(N_A^1)$. The variables E_{cr} and N_A^0 have the form

$$E_{cr} = \left[\frac{4(2-\delta)}{cB}\right]^{1/\delta} \quad \text{and} \quad N_A^0 = c\beta \left[\frac{(1-\sqrt{1-(E_{cr}/E)^\delta})}{2}\right]^2. \quad (6.66)$$

Figure 6.3 presents a family of curves, $\log_{10} N_C$, at different values, $E \geq E_{cr}$, depending on $\log_{10} N_A$, and at fixed values $\theta_A = 0.5$, $c = B = 1$, and $\delta = 1.01$ at which $E_{cr} \simeq 4$.

The results can be interpreted as follows. The value of vital primary substratum determines the ways in which system A_{RC} can ensure its maximal survivability. If $E \leq E_{cr}$, then there is only one way to enlarge survivability, that is, by increasing the number of elements N_A, which become more and more defenseless. If $E > E_{cr}$, there appears another way. It consists of structuring a comparatively small number of well-protected A elements (see Fig. 6.3). Both means of survivability

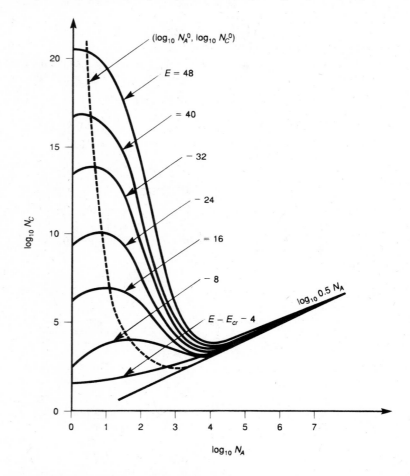

Figure 6.3 The two ways of survivability. Dependence of $\log_{10} N_C$ on $\log_{10} N_A$ at different values of E, and $c = B = 1$, $\theta = 0.5$, and $\delta = 1.01$; $\log_{10} N^0_A$ is the optimal value leading to the maximal value of $\log_{10}(N^0{}_C)$. The trajectory of $(\log_{10} N^0{}_A, \log_{10} N^0{}_C)$ is depicted by the broken curve.

of a population to ensure its stability; the other, the formation of modest-sized populations of well-protected multicellular individuals (for example, warm-blooded animals with white blood corpuscles and other defense mechanisms). Note that the limitless growth of the number of system elements to ensure as high a system reliability as desired follows from the reliability limit R law.

Let us consider animals at three hierarchical levels: cell, individual, and a species-forming set of individuals. Let n be the number of cells in an individual and N be the number of individuals in the species. Then n can be considered a complexity characteristic of the species' individual, and the product Nn characterizes the distribution of the species' cells, which is a manifestation of its individuals' stability. Von Forster (1962) stated an experimentally determined dependency,

$$\log_{10} Nn = 15 + 0.52 \log_{10} n, \qquad (6.67)$$

for known species, from unicellular forms through human beings (Fig. 6.4). So, the growth of species complexity (growth, n) is accompanied by the growth of its stability (growth, Nn). Von Forster opposes this to the fact that growth of the atomic weight (complexity) of chemicals is accompanied by reducing spatial distribution (stability). For example, hydrogen, with an atomic weight equal to unity, is the most widely distributed gas in the universe.

To obtain a theoretical dependence corresponding to the experimental relation (6.67), it is natural to use the model of survivability. Note that animals do possess this active property and to a greater extent than plants. The number N_A^0 can be interpreted through the number $N = N_A^0$ of the species' individuals; the number E can be interpreted through the species' biomass. Assume that

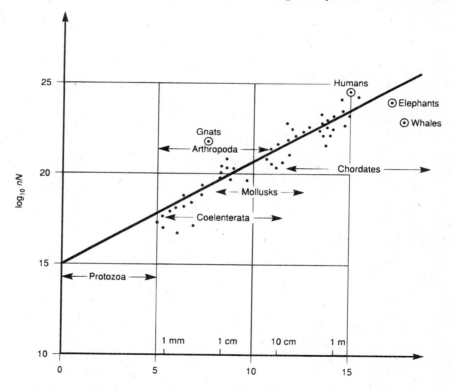

Figure 6.4 The diagram "complexity (n)–stability (Nn)" (von Forster, 1962). Nn is the distribution of cells of different animal species (one individual has n cells).

species' individuals; the number E can be interpreted through the species' biomass. Assume that R and A elements are located in one species member forming its active defense part. The possibility of such an interpretation is based on the proportionality of N_R^0 and N_A^0 numbers of these elements in the optimal case [see eq. (6.19)]. Then, assume that the individual-forming A and R elements consist of primary elements (cells) amounting to the number n, with the biomass of a cell equal to b. The biomass of the species is then

$$E = nNb. \qquad (6.68)$$

Now consider the ecologically interesting case when $E \gg E_{cr}$ ($E_{cr} / E \ll 1$). It follows from (6.66) that

$$N \simeq c\beta \left[\frac{(E_{cr}/E)^\delta}{4} \right]^2. \qquad (6.69)$$

Having put (6.63') for β and (6.66) for E_{cr} into (6.69), we get after simple transformations

$$N \simeq \frac{C_1}{E^\delta}, \qquad (6.70)$$

where $C_1 = (2 - \delta)^2/cB(\delta - 1)$. Putting (6.68) for E into (6.70), and multiplying both sides of the obtained expression by n, after simple logarithmic transformations we get

$$\log_{10} nN \simeq (1 + \delta)^{-1} \log_{10} C_2 + (1 + \delta)^{-1} \log_{10} n, \qquad (6.71)$$

where $C_2 = C_1/b^\delta$. With the value of δ close to unity (see an example in Fig. 6.3), we get from (6.71)

$$\log nN \simeq C + 0.5 \log_{10} n, \qquad (6.72)$$

where $C = 0.5 \log_{10} C_2$.

The sense of the constant C is as follows. For the unicellular ($n = 1$) case, parameter C can be interpreted as the parameter of critical "density." With $C = 14$ or 15, it is approximately a single-celled animal per square meter of earth's surface. With the value $C \simeq 15$ put into the theoretical relation (6.79), the latter becomes rather close to the empirical relation (6.67).

So, a consequence of the reliability R law is the necessity to widen complex (specifically biological) systems; from the survivability RC law, it is necessary to pass from unicellular to two-level (cell-incorporating individuals) multicellular organisms. Thus, based on the single strategic goal of survival, even the primary R and RC laws explain the necessity of at least a two-level structural organization and *regeneration* of complex (specifically biological) systems.

Note that the physical explanation of cell division as a result of the metabolic demand of a certain relationship between their volume and surface can have an alternative consisting of the idea of almost "flat" or "spongy" creatures. The next section is devoted to the theory of adaptation of regenerative systems.

ADAPTATION OF REGENERATIVE SYSTEMS

Adaptation and Preadaptation: The Adaptive Cycle

R and RC sections reflect the stability of complex systems as related to their death from damage done to their structure, either by the environment or by another system.

Another important aspect of the stability of complex systems is that of structural deformation under certain outside influences. This occurs as a result of "pulse" effects of the environment passing to a new stationary state during which the system must adapt its structure to this new state or "elasticity" or "resilience" (Holling, 1978). However, to avoid physical analogies, the term "adaptivity" will be used and denoted symbolically as the fundamental L property (self-organizability).

Unlike simple systems with material (energetic) stability, it is more important for complex systems (biological among them) to be structurally and behaviorally stable in spite of variable material composition. Their specific property, therefore (theoretical basis above), is the regeneration of dying elements and adaptation to changing environmental conditions. Both aspects were considered for the first time in classical models of cellular automata by von Neumann (1966) and in models of collective behavior of stochastic automata (Tsetlin, 1968).

Let T_A be the time of adaptation of a complex system, A_L, to a new environment, B_L, which is in a stationary state during the period T_B. If the system adjusts by its reactions and structure to environmental stimuli during the period $T_A < T_B$, it is said *to adapt* to the surroundings (stimulus–response scheme). Now, if the system, to the detriment of its adaptation to environment B, during period T_B adapts its structure to environment B' following B in time, it is said to be *preadapted* to environment B' (response–stimulus scheme). The survivability model above is an example of a model allowing for preadaptation. Undoubtedly, for preadaptation, the system needs a predictor of future states. The simplest predictor is one used to predict a periodic sequence of stationary environments with period h. It should possess only a long-term individual or "genetic" memory covering the time period $T > h$. The most complex predictor is the intellectual one of humans. Now, let us pass to mathematical formalization of the introduced notions (Fleishman, 1977a,b).

A class of systems $\mathcal{A}$ is fixed, as well as classes of environments $\mathcal{B}$, which are arranged in relation to the system so that for each pair of classes of environments, $\mathcal{B}_1$ and $\mathcal{B}_2$, a relationship, $\mathcal{B}_2 \leq \mathcal{B}_1$, is given (meaning that $\mathcal{B}_2$ is "worse" than or equivalent to $\mathcal{B}_1$). Within the class $\mathcal{A}$, the system $A = A[B(\mathcal{B})]$ is considered *optimally preadapted* to the poorest environment $B(\mathcal{B})$ of the class $\mathcal{B}$ if

$$\min_{B \in \mathcal{B}} \max_{A \in \mathcal{A}} v(u, A, B) = v\{u, A[B(\mathcal{B})], B(\mathcal{B})\} . \tag{6.73}$$

Within the class $\mathcal{A}$, the system $A = A(\mathcal{B})$ is *optimally adapted* to the environment class $\mathcal{B}$ if

$$\max_{A \in \mathcal{A}} \min_{B \in \mathcal{B}} v(u, A, B) = v\{u, A(\mathcal{B}), B[A(\mathcal{B}), \mathcal{B}]\}, \tag{6.74}$$

where, for system A, $B(A, \mathcal{B})$ is the poorest member of the environmental class $\mathcal{B}$.

In this way, the mechanism response–stimulus or stimulus–response corresponds to the minimax or maximin, respectively, of the goal functional v. Let the adaptive *operative characteristic* (OC), $m(\mathcal{B}_1, \mathcal{B}_2) = v\{u, A(\mathcal{B}_2), B[A(\mathcal{B}_2), \mathcal{B}_1]\}$, be determined as the value of the goal functional v after adaptation of the system to the environmental class $\mathcal{B}_2$, where in fact the class of environments $\mathcal{B}_1$ takes place. If $\mathcal{B}_2 \leq \mathcal{B}_1$, then for OC the following sequence of inequalities of the "adaptive cycle" should be fulfilled:

$$m(\mathcal{B}_1, \mathcal{B}_2) \geq m(\mathcal{B}_1, \mathcal{B}_2) \geq m(\mathcal{B}_2, \mathcal{B}_2) \geq m(\mathcal{B}_2, \mathcal{B}_1) . \tag{6.75}$$

The preadaptive OC and inequalities of the preadaptive cycle are determined in the same way.

To each environment B, a scalar value $w = f(B)$ is set, which is referred to below as pressure on system A. A class of environments, $\mathcal{B}_w = \{B: f(B) = w = \text{constant}\}$, is defined, with a fixed value for pressure; $w = $ constant. The set of environments can be ordered, $\mathcal{B}_{w_2} \leq \mathcal{B}_{w_1}$, if $w_1 \leq w_2$. For OC environmental pressures on the system, $m(w_1, w_2) = m(\mathcal{B}_{w_1}, \mathcal{B}_{w_2})$ is also determined. Then the sequence of inequalities (6.75) of the adaptive cycle for the environmental pressures on the system, at $w_1 \leq w_2$, has the form

$$m(w_1, w_2) \geq m(w_1, w_2) \geq m(w_2, w_2) \geq m(w_2, w_1). \tag{6.76}$$

General Regenerative System Model and Its Particular Cases

This section follows Fleishman (1982c, 1983, 1984). The general model of an a-component regenerative system, $A_{a\vec{b}} = \{A_\alpha\}$, comprises a components, $A_\alpha = \{e_{\alpha\beta\gamma}\}$, which in turn consist of $N_{\alpha\beta}$ elements, $e_{\alpha\beta\gamma}$ ($\alpha = \overline{1, a}$, $\beta = \overline{1, b_\alpha}$, $\gamma = \overline{1, N_{\alpha\beta}}$, $\vec{B} = (b_\alpha)$). The elements $e_{\alpha 1\gamma}$ originate; elements $e_{\alpha\beta\gamma}$ pass into elements $e_{\alpha\beta+1\gamma}$ ($\beta = \overline{1, \beta_\alpha - 1}$); all elements $e_{\alpha\beta\gamma}$ can disappear or lose connections with the elements $e_{\alpha'\beta'\gamma'}$ with probabilities $\lambda_{\alpha 1}$, $\lambda_{\alpha\beta}$, $\mu_{\alpha\beta}$, and $\mu_{\alpha\beta}^{\alpha'\beta'}$, respectively. In addition, movements of elements $e_{\alpha\beta\gamma}$ from the cell x into the cell x' of a discrete space, $R_3 = \{x\}$, are possible, and these are regulated by a Markovian transition matrix, $\mathcal{P} = [p_x^{x'}]$, which, for transition probabilities $p_x^{x'} \geq 0$, $\sum_{x'} p_x^{x'} = 1$ holds true. Functioning of the system $A_{a\vec{b}}$ takes place in a two-component environment, $B = \{B^+, B^-\}$, the components of which, $B^+ = \{e^+\}$ ("positive") and $B^- = \{e^-\}$ ("negative") for the system $A_{a\vec{b}}$, consist of N^+ useful and N^- detrimental elements, which are renewed by the system with probabilities p^+ and p^-, respectively. In general, this study deals with a dynamic picture of system $A_{a\vec{b}}$'s interaction with environment B when all mentioned probabilities depend on discrete time, $t = 1, 2, \ldots$, and on the position of cell x within the space $R_3 = \{x\}$.

In the theory of potential effectivity, a number of models can be considered particular cases of the general model, $A_{a\vec{b}}$, of the regenerative system. These include the nonadaptive regenerative model of reliability, $A_R = A_{11}$, nonregenerative models that are the adaptive integrity model, $A_{RI} = A_{11}$ (Fleishman, 1980), the preadaptive survivability model, $A_{RC} = A_{21}$, and, finally, the adaptive regenerative models, A_{a1} and A_{1h}.

The system is considered dead if, as a result of its interaction with the environment, either a certain part (θth) of the initial number of its elements ceases to function (in model A_R) or a part of its vitally important ("working") elements becomes inoperative (in model A_{RC}). The same is true for systems that lose their integrity (in model A_{RI}).

Let us consider the general model of a regenerative system $A_{a\vec{b}}$. The goal, $\underline{A_{a\vec{b}}} = (\Delta\lambda, E_{a\vec{b}}\nu)$, of the system is a change, $\Delta\lambda = (\Delta\lambda_{\alpha\beta})$, of the regenerative possibilities of elements (provided $\overline{\lambda} = \sum_{\alpha\beta} \lambda_{\alpha\beta} \Big/ \sum_\alpha b_\alpha = $ constant), which maximizes the mathematical expectation $E_{a\vec{b}}\nu$ of the total number $\nu_{a\vec{b}}$ of its elements. In the general case $a, b_\alpha \geq 2$, the dependence (6.33) is considered:

$$\mu_{\alpha\beta} = \mu(x_{\alpha\beta}, y_{\alpha\beta}) = \frac{x_{\alpha\beta} + y_{\alpha\beta} - 2x_{\alpha\beta}y_{\alpha\beta}}{1 - x_{\alpha\beta}y_{\alpha\beta}}, \qquad (6.77)$$

where $x_{\alpha\beta}$ and $y_{\alpha\beta}$ are probabilities of elements' deaths caused by both internal and external factors, respectively. The system $A_{a\vec{b}} = (x, \lambda)$ with a fixed structure, $x = (x_{\alpha\beta})$, adapts its behavior, $\lambda = (\lambda_{\alpha\beta})$, to the pressure $w = \overline{y} = \sum_{\alpha\beta} y_{\alpha\beta} \Big/ \sum_\alpha b_\alpha$ of the environment, $B = (\overline{\lambda}, \overline{y})$. When the behavior of both the environment and system $A_{a\vec{b}}$, as well as the structure of the latter, is periodic with period h, then parameters of the environment and the system are considered periodically dependent on the index β (that is, they are time varying).

System $A_{a1} = (\vec{x}, \vec{y})$ in Uniform Space

Consider, following Fleishman (1977a,b), a dynamic model of a regenerative system for $a \geq 2$ and $b_\alpha = 1$, when the conditional probabilities $\lambda_{\alpha 1} = \lambda_\alpha(t)$, $x_{\alpha 1} = x_\alpha(t)$, and $y_{\alpha 1} = y_\alpha(t)$ depend on time t, but not on the position of elements within the N-cell space, R_3. In this case, $E\nu_{a1}(t) = N a\overline{p}(t)$, where $\bar{p}(t) = \sum_{\alpha=1}^{a} p_\alpha(t)/a$, and $p_\alpha(t)$, the absolute probability of the presence of element $e_{\alpha 1\gamma}$ in a cell of space R_3, is independent of the space cell; that is, degeneration of the Markovian matrix $\mathcal{P} = [\vec{p}(t)]$ takes place. The asymptotic stationary case as $t \to \infty$ is investigated, when $\lambda_\alpha(t) \to \lambda_\alpha$, $x_\alpha(t) \to x_\alpha$, $y_\alpha(t) \to y_\alpha$, and $Ev_{a1}(t) \to Ev_{a1}(\vec{p}(t) \to \vec{p})$. As the goal function, a value is taken,

$$M_{\vec{x}}(\vec{\lambda}, \vec{y}) \triangleq \frac{\bar{p}}{\lambda}(1 - a\bar{p}) = (\bar{\lambda}a)^{-1} \sum_{\alpha=1}^{a} \frac{\lambda_\alpha}{\mu}(x_\alpha, y_\alpha), \qquad (6.78)$$

which increases with increase of $\bar{p}$; its expression follows from the equation determining

$$p_\alpha(t+1) = p_\alpha(t)(1 - \mu_\alpha(t)) + p_0(t)\lambda_\alpha(t). \qquad (6.78')$$

Here, with $\bar{y}_1 = 1 - \bar{z}_1 = w_1$ and $\bar{y}_2 = 1 - \bar{z}_2 = w_2$, OC (the adaptive operative characteristic) has the form

$$m(\bar{z}_1, \bar{z}_2) = 1 + v(\bar{z}_2) + \delta(\bar{z}_1, \bar{z}_2)\sqrt{v'(\bar{z}_1)v'(\bar{z}_2)}(\bar{z}_1 - \bar{z}_2), \qquad (6.79)$$

with $m(\bar{z}_1, \bar{z}_2) \triangleq M_{\vec{x}}\{\vec{\lambda}(\lambda, \bar{y}_2), \vec{y}[\vec{\lambda}(\lambda, \bar{y}_2), \bar{y}_1]\}$ and $\gamma(\bar{z}_1, \bar{z}_2) = \sqrt{v'(\bar{z}_2)v_0'(\bar{z}_1)/v'(\bar{z}_1)v_0'(\bar{z}_2)}$, where the function $v(\bar{z})$ represents the solution of the transcendental equation

$$\bar{z} = \overline{[(1+X)^{-1} + v^{-1}]^{-1}} \qquad (6.80)$$

in relation to v. In eq. (6.80) $X_\alpha = x_\alpha/(1 - x_\alpha)$ $(\alpha = \overline{1, a})$ and $v_0(\bar{z}) = [\bar{z}^{-1} - (1 + \overline{X})^{-1}]^{-1}$; the conditional differences of the adaptation cycle OC do not depend on $\vec{\lambda}$ and are simply expressed by means of parameters, $\gamma = \gamma(\bar{z}_1, \bar{z}_2)$ and $\delta = [v(\bar{z}_1) - v(\bar{z}_2)]/(\bar{z}_1 - \bar{z}_2)\sqrt{v'(\bar{z}_1)v'(\bar{z}_2)}$ (Fleishman, 1977b, 1982a).

Analytical solution of all problems of the adaptation cycle depends on the explicit solution of eq. (6.80), which, for the case $a = 2$, can be solved when it becomes a quadratic equation. The explicit solution can also be obtained for the asymptotic case $a \to \infty$, with a "linear" structure $\vec{x}$, when

$$(1 + X_\alpha)^{-1} = (1 + X_1)^{-1} + \frac{\Delta(\alpha - 1)}{a - 1}, \qquad \Delta = (1 + X_1)^{-1} - (1 + X_a)^{-1} \qquad (6.81)$$

and

$$v(\bar{z}) = v_\Delta(\bar{z}) = [g_\Delta(\bar{z})^{-1} + g_\Delta(1 + \overline{X})^{-1}]^{-1}, \qquad g_\Delta(z) = \frac{[\exp(\Delta z) - 1]}{\Delta}.$$

The Periodic System A_{1h} in Nonhomogeneous Space

Consider, after Fleishman (1982c, 1983, 1984), a model of the regenerative system $A_{a\vec{b}}$ for the case $a = 1$ and $b_\alpha = t$, when parameter β signifies the age of element $e_{1\beta\gamma}$. Unlike in model A_{a1}, it is assumed that in one cell x a few elements can originate and stray for a long period of time; that is, the considered cell is n times "magnified" as compared to a model A_{a1} cell. The probabilities $\vec{\lambda}(t) = (\lambda_x(t))$, $\vec{x}(t) = (x_x(t))$, and $y(t) = (y_x(t))$ depend not only on time t, but also on cell position, x, in the space R_3. Stochastic migration of elements takes place in the model. The asymptotic case for $t \to \infty$ of the periodic system, $A_{1t} \to A_{1h}$ with period h, is investigated, when $Ev_{1t} \to Ev_{1h}$ and all probabilities appear for integer values of $t = \sigma + lh$ to be periodic functions $\varphi(t)$ of the argument $\sigma = \overline{0, h-1}$, $\varphi(t) = \varphi(\sigma)$ with period h.

It is possible to show that, for the matrices $\mathcal{P}(\sigma) = [p^x(\sigma)]$ or $\mathcal{P}(\sigma) \neq [p^x(\sigma)]$, which have real nonrepeating main values, if $x_x(\sigma) = x(\sigma)$ and $\mu_x(\sigma) \to 0$, the explicit expression Ev_{1h} has the lower estimate

$$Ev_{1h} = n \sum_{\sigma=0}^{h-1} \overline{\lambda}(\sigma) A(\sigma)^{h-\sigma} (E - A(\sigma)^h)^{-1} \vec{e} \geq Ev_{1h} = \frac{nN}{h} \sum_{\sigma=1}^{h-1} \frac{\overline{\lambda}(\sigma)}{\overline{\mu}(\sigma)}. \quad (6.82)$$

Here, $A(\sigma) = D(1 - \vec{\mu}(\sigma))$ is the $N \times N$ diagonal matrix with vector $1 - \vec{\mu}(\sigma)$ along the diagonal, and $\vec{e}$ is the singular column vector, $E = D(\vec{e})$, $\overline{\lambda}(\sigma) = N^{-1} \sum_{x=1}^{N} \lambda_x(\sigma)$,

$$\overline{\mu}(\sigma) = \begin{cases} \tilde{x}(\sigma) + y(\sigma) & \sigma \in \{\sigma: p_x^{x'}(\sigma) \neq p^{x'}(\sigma)\} \\ \overline{x}(\sigma) + \overline{y}(\sigma) & \sigma \in \{\sigma: p_x^{x'}(\sigma) = p^{x'}(\sigma)\} \end{cases},$$

where $\tilde{x}(\sigma) = x(\sigma)$, $y(\sigma) = \min_x y_x(\sigma)$, $\theta = 2(1 - p_x^{x'}(\sigma)) \leq x(\sigma) \leq 0$ $(\tilde{\mu}(\sigma))$, $\tilde{\mu}(\sigma) = \max_x \mu_x(\sigma)$, $\overline{x}(\sigma) = \sum_{x=1}^{N} p^x(\sigma) x_x(\sigma)$, and $y(\sigma) = \sum_{x=1}^{N} p^x(\sigma) y_x(\sigma)$, where $x_x(\sigma)$ and $y_x(\sigma)$ are probabilities of elements' deaths due to internal and external causes, respectively. Comparison of the goal functional (6.78) of system A_{a1} and the lower estimate (6.82) of A_{1h} shows that, with the latter assumed to be the goal functional of A_{1h}, all solutions of the adaptation cycle extremum problems (with appropriate interpretation of their restrictions) coincide for both systems. Models A_{a1} and A_{1h} possess an ecological interpretation: the former as a model of an a-species community and the latter as a model of an h-age migratory population.

The Nonregenerative System A_{11} with Multigoal Structural Adaptation

Fleishman (1974, 1980) considered the model $A_{Rl} = (|A_{Rl}|, \overline{A}_{Rl})$ of a system interacting with the environment for such a short period of time that the effect of regeneration can be neglected. The system adapts by its structure, $|A_{Rl}| = \Gamma_{N\mu_{11}^{11}}$ (a stochastic graph), to the environment, $B = (N^+, \mu_{11}^{11})$, which offers a limited number, N^+, of useful elements, and damages, with probability μ_{11}^{11}, the information connections between the system's elements. Tactical goals of the system include the preservation of its integrity (connectedness of its stochastic graph) in both the indifferent A_{Rl}^1 and conflict A_{Rl}^2 situations, and also successful procurement of useful elements A_{Rl}^3. For each of these goals, an optimal behavior, A_{Rl}^i, and structure, $|A_{Rl}|^i$, are formulated, determined by their optimum numbers of elements, $N_{opt} = N_{opt}^i$ ($i = 1, 2, 3$). In this way, multigoal structural optimization of system A_{Rl} takes place with due respect to internal and external conditions. The system A_{Rl} represents a degenerate nonregenerative case of system $A_{a\vec{b}} = A_{11}$ with parameters $a = b_\alpha = 1$ and $w = \mu_{11}^{11}$, and can be interpreted, for example, as the model of an N-individual fish school, which, in the absence of a leader, is characterized by schooling behavior, informational relationships, and imitation-type control.

Estimates of Time Taken to Reach Steady State and General Remarks

Let $\eta(t) = |1 - Ev(t)/Ev|$ be the relative difference (at instant t) of the value of $Ev(t)$ from its stationary value, Ev. Denote $\eta_{a1}(1)$, $\eta_{a1}(t)$, and $\eta_{1h}(t)$ for the systems A_{a1} and A_{1h}, respectively. Using (6.78) and (6.82), it is possible to show (Fleishman, 1983, 1984) that if with $t \to \infty$ the time-dependent probabilities $\lambda_{\alpha 1}(t)$ and $\mu_{\alpha 1}(t)$ have the limits λ_α and μ_α, respectively, then at any $\epsilon > 0$, for all $t > t_\epsilon$, $\eta(t) < \epsilon$ will be true, where

$$t_\epsilon = \begin{cases} \left[1 + \ln\left[\dfrac{\eta_{a1}(1)}{\epsilon}\right]\right]/\ln\left(\dfrac{1}{k_{a1}}\right) & \text{for } A_{a1} \\ \ln\left(\dfrac{1}{\epsilon}\right)/\ln\left(\dfrac{1}{k_{1h}}\right) & \text{for } A_{1h} \end{cases} \quad (6.83)$$

and

$$k_{a1} = 1 - \min_{\alpha, t} \mu_\alpha(t) + a \max_{\alpha, t} \lambda_\alpha(t) + \max_\alpha |\lambda_\alpha(t) - \lambda_\alpha| + \max_\alpha |\mu_\alpha(t) - \mu_\alpha| < 1 \quad (6.83')$$

$$k_{1h} = \min_\sigma \overline{\mu}(\sigma).$$

The developed theory allows us to obtain probable estimates of deviation of considered stochastic values, ν, from the theoretical expected value, $E\nu$, even if estimates of dispersion are known instead of dispersion values. For example, if the stochastic value $\nu < C + E\nu$ is known from biology (where it is evident that $C = N - E\nu_{a1}$ and $C = \eta N' - E\nu_{1h}$), then, using the estimate of dispersion $D\nu < CE\nu$ and the Tchebyshev inequality, we get

$$P = Q(\nu \leq E\nu + K\sqrt{C\,E\nu}) \geq 1 - K^{-2} \quad (K > 1). \quad (6.84)$$

The considered aspects of stability can be found in different components and elements of system $A_{a\vec{v}}$. Model A_R reflects only the general tendency of regenerative systems to unrestricted regeneration. Other models are more detailed. Model A_{Rt} illustrates the multigoal behavior of elements within a certain aggregation (microlevel). Models A_{RC} and A_{1h} give details of behavior of components of the system $A_{a\vec{v}}$ (mesolevel); in the former this is obtained by incorporating defense elements into the working elements in the form of their defense portions (Fleishman, 1971). Finally, model A_{a1} refers to the regenerative system itself with undetailed components (macrolevel); the individuality of its components is determined by their structure, x. The negative competitive interaction of the components is represented *only* by the restricting relationship $\sum_{\alpha=1}^{a} \lambda_\alpha = a\overline{\lambda} = $ constant, connected with the limited number of useful elements of the environment. Evidently, the model does not allow for relations of the "prey–predator" type.

REGENERATIVE SYSTEM CONTROL ALLOWING FOR ADAPTATION

Based on the previously developed stochastic theory of community adaptation to impulse environmental effects (Fleishman, 1977b, 1984), a general problem of community control will be formulated and solved. The solution provides specific interpretations for free trade and the cultivation of useful and suppression of harmful communities. To be considered are parametrically interrelated macro-, meso-, and micromodels of community, population, and cohorts. The cohort model describes general conditions for survival of juvenile stages. The macro- and mesomodels, in the asymptotic case of increasing pressure on the community, show the effect of decreasing withdrawal after its certain rise (corresponding to the effect of overcatching). Methods to identify theoretical parameters by using experimental data will be discussed.

Deterministic Theory of Fisheries

The mathematical theory of community control appeared in its specific version as a mathematical theory of fisheries in 1918 after the pioneering work published by Baranov (1918). Today this theory embodies a stable system of concepts and a number of mathematical models. The basic concepts of fishery theory are stock, B, and recruitment, R, of the exploited population, A, living in environment G, with trade conditions F, and catch C, of the trade system H.

The basic model is the dynamic one showing changes in the stock:

$$\frac{dB}{dt} = \Phi(B), \qquad (6.85)$$

where the function $\Phi(B)$ covers the specific conditions of the commercial situation. One distinguishes piecemeal, C_N, weighted, C_w, and balanced (equilibrium), C_E, catches (with $\Phi(B)$ equal to zero).

Based on eq. (6.85) and the well-known formula of L. von Bertalanffy, the works of Beverton and Holt (1957) and Ricker (1975) reveal the expression of the catch, $C_w = g(F)$, while Gurman (1981) provides algorithms for yielding an optimal catch, $C_w = g(F)$, throughout community succession as a function of fishing effort, F. Pella (1969) determines the equilibrium catch, $C_E = h(F)$, as a function of fishing efforts F in the stationary (climax) state of the fishery.

The equations obtained, $C_w = f(F)$ and $C_E = h(F)$, can be justified by their single maxima reflecting the observed overcatch effect. However, the functions $f(\cdot)$, $g(\cdot)$, and $h(\cdot)$ are determined by the type of function $\Phi(\cdot)$. In this way, $h(F) = \Phi[\Psi(F)]$, where $\Psi(F) = \Omega^{-1}(F)$ and $\Omega(B) = \Phi(B)/B$. Therefore, the adequacy of these equations is connected to the reality of the function $\Phi(B)$. The latter was justly criticized by Semevskiy and Semenov (1982) because it disregards the population's adaptation to environmental pressure. When properly considered in this chapter, the population's adaptation revealed model (6.85) to be closer to reality. Nevertheless, no author has yet examined adaptation of a fishery to fishing efforts and environmental pressure within the framework of model (6.85). Moreover, being a deterministic model, this does not reflect the stochastic pattern of the commercial activity.

The theory of community control developed later tries to cover the two gaps in the deterministic theory mentioned above. This theory constructs for the stationary climax state a stochastic model of optimal control relating community adaptation to local effects of the controlling system and the environment. Such local effects may differently influence the stock, B. In any case, its dependence, $B = b(F, R)$, on the integral parameters of pressure, F, and the environmental resource, R, restricting recruitment may be derived only by solving extremum problems with limitations due to parameters F and R. This dependence is important in both theory and practice. The main results of this chapter are drawn from such dependence.

General Formulation of the Problem

Community control may be considered as a certain interaction between three systems of constant structure, the community A, the controlling system H, and the environment G. Here the states of the systems are described only by their behaviors, which are reduced to adaptation $\overline{A}$ of community A to effects H and G of the controlling system and the environment, respectively. Such control can be presented in terms of (u, v) exchanges (Fleishman, 1970). Let us consider (W, B) exchange between the community and environment and (F, C) exchange between the community and the controlling system. In the latter, B and C are the stock and removal from it (withdrawal), and W and F are certain integral characteristics of the pressure rendered on the community from the environment and the controlling system, respectively. The recruitment, R, of the stock may depend only on the environment, $[R = g(G)]$, or on the controlling system, $[R = g(G, H)]$.

In the first case, the control is called free trade, and withdrawal from the stock, the catch (Fig. 6.5b). In the second case, the control is called cultivation of useful (Fig. 6.5c) or suppression of the harmful community (Fig. 6.5d), while withdrawal is called the cropping or suppression effect, respectively. In particular, when there is no control, the community naturally adapts to the changing environmental pressure from W to F (Fig. 6.5a).

The effect of natural adaptation of the community was studied by means of the so-called adaptation cycle based on the dependence of the stock, $B = \beta(A, G)$, and on the state of community A and that in the environment, G and G' (Fleishman, 1977b, 1984), where

$$\beta(A, G) = B, \qquad \beta(A, G') = B' \in \mathfrak{B}. \tag{6.86}$$

With general formulation of the problem as controlling in the case of community states $A \in \mathfrak{A}$ and effects of the controlling system $H \in \mathfrak{H}$, we establish the following goal functions:

$$\beta(A, H) = B \in \mathfrak{B} \quad \text{and} \quad \gamma(H, A) = C \in \mathfrak{C}, \tag{6.87}$$

where $\mathfrak{B}$ and $\mathfrak{C}$ are regulated sets. In this way we can fix the critical subset of stocks, $\mathfrak{B}_{cr} \subset \mathfrak{B}$, so that if, due to the effect of the controlling system, $B \in \mathfrak{B}$, the community is regarded as perished.

Later we introduce a set of pairs, $\{(F, R)\} = \mathfrak{F} \times \mathfrak{R}$, monoparameter subsets, $\mathfrak{H}_F \subset \mathfrak{H}$, and $\mathfrak{A}_R \subset \mathfrak{A}$, of the effects of the controlling system and the states of the community with fixed pressure and replenishment, as well as two-parameter subsets with regard to ordering, and "preferable" subsets, $\mathfrak{C}_{FR} \subset \mathfrak{C}$ and $\mathfrak{B}_{FR} \subset \mathfrak{B}$, of respective values of the purpose functions.

It is necessary to find such maximum two-parameter subsets $\mathfrak{H}_{FR} \subset \mathfrak{H}_F$ and $\mathfrak{A}_{FR} \subset \mathfrak{A}_R$ of the effects of the controlling systems and community states for which the following relationships would prove valid:

$$\begin{aligned} \{C: C = \gamma(H, A) \; \forall \; H \in \mathfrak{H}_{FR} \quad \text{and} \quad A \in \mathfrak{A}_{FR}\} \subseteq \mathfrak{C}_{FR} \\ \{B: B = \beta(A, H) \; \forall \; H \in \mathfrak{H}_{FR} \quad \text{and} \quad A \in \mathfrak{A}_{FR}\} \subseteq \mathfrak{B}_{FR} \end{aligned} \tag{6.88}$$

With the given recruitment, R, the pressure F of the controlling system is regarded as permissible if $F \in \mathfrak{F}(R) \subset \mathfrak{F}$, where $\mathfrak{F}(R) = \{F: \mathfrak{B}_{FR} \cap \overline{\mathfrak{B}}_{cr} \neq \emptyset\}$; that is, the subset of permissible pressures corresponds to the subset $\mathfrak{H}_{FR}^*$, where $F \in \mathfrak{F}(R)$ shows the permissible effects of the controlling system.

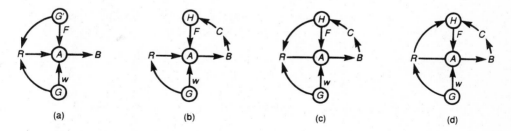

(a)   (b)   (c)   (d)

Figure 6.5 Scheme of interaction in community A, its resource B, recruitment R, environment G, and controlling system H. In the absence of a controlling system H: (a) Natural adaptation of the community to changing pressure of the environment from W to F and recruitment, $R = f(G)$, which depends on the environment. In the presence of the controlling system rendering its pressure F upon the community: (b) Free trade when $R = f(G)$ and cases when recruitment, $R = f(G, H)$, depends on the environment and the controlling system related to (c) cultivation of a useful community, and (d) suppression of a harmful community.

In general, there are two different ways to solve this problem. First, system H affects the community, which then adapts to this effect and to replenishment (the stimulus–response hypothesis). In accordance with this, we shall first determine a subset of "preferable" effects,

$$\mathfrak{H}_{FR}(A) = \{H: \gamma(H, A) = C \in \mathfrak{C}_{FR}\},$$

and then the subset sought:

$$\mathfrak{A}_{FR} = \{A: \beta(A, H) = B \in \mathfrak{B}_{FR}, H \in \mathfrak{H}_{FR}(A)\},$$

and (6.89)

$$\mathfrak{H}_{FR} = \{H: \gamma(H, A) = C \in \mathfrak{C}_{FR}, A \in \mathfrak{A}_{FR}\}.$$

The second solution is obtained if first the community adapts to our impact and replenishment and then we influence and replenish it (the response–stimulus hypothesis). In accordance with this, we first determine the subset of "preferable" community states,

$$\mathfrak{A}'_{FR}(H) = \{A: \beta(A, H) = B \in \mathfrak{B}_{FR}\},$$

and then the sought subsets:

$$\mathfrak{H}'_{FR} = \{H: \gamma(H, A) = C \in \mathfrak{C}_{FR}, A \in \mathfrak{A}'_{FR}(H)\},$$

$$\mathfrak{A}'_{FR} = \{A: \beta(A, H) = B \in \mathfrak{B}_{FR}, H \in \mathfrak{H}'_{FR}\}.$$

(6.90)

After that, we make a choice between the two hypotheses, which, generally speaking, results in different solutions, $(\mathfrak{A}_{FR}, \mathfrak{H}_{FR})$ and $(\mathfrak{A}'_{FR}, \mathfrak{H}'_{FR})$; these solutions are then compared with the original (the objective teleology of Fleishman, 1977a, 1982a).

Let us consider a particular case of the general problem of control without withdrawal C when pressure F is not exercised by us but by another environmental state, $G' = H$ (Fig. 6.5a). Here we assume that $\gamma(H, A) = \gamma(G', A) = \beta(A, G')$. Instead of the preferable subset $\mathfrak{C}_{FR}$, we shall analyze the subset $\mathfrak{B}'_{FR} \subset \mathfrak{B}$, which is "preferable" for the environment G' in the following sense. To make the estimates more reliable, since the environmental effects, G', on the community are uncertain, we choose among them such effects that, with the pressure F fixed, will result in the subset $\mathfrak{B}'_{FR}$ that will be most "unfavorable" for the community.

In this case it is necessary to find such maximum two-parameter subsets, $\mathfrak{C}_{FR} \subset \mathfrak{C}$ and $\mathfrak{A}_{FR} \subset \mathfrak{A}$, for which the controlling relations (6.88) turn into the controlling relations:

$$\{B': B' = \beta(A', G') \,\forall\, G' \in \mathfrak{C}_{FR} \text{ and } A' \in \tilde{\mathfrak{A}}_{FR}\} \subseteq \mathfrak{B}'_{FR},$$

$$\{B: B = \beta(A, G') \,\forall\, G' \in \mathfrak{C}_{FR} \text{ and } A \in \mathfrak{A}_{FR}\} \subseteq \mathfrak{B}_{FR}.$$

(6.88′)

Accordingly, the solutions determined by relations (6.89) and (6.90) turn into solutions

$$\mathfrak{A}_{FR} = \{A: \beta(A, G') = B \in \mathfrak{B}_{FR}, G' \in \mathfrak{C}_{FR}(A)\},$$

(6.89′)

where

$$\mathfrak{C}_{FR}(A') = \{G': \beta(A', G') = B' \in \mathfrak{B}'_{FR}\} \quad \text{and} \quad \mathfrak{C}_{FR} = \{G': \beta(A', G') = B' \in \mathfrak{B}'_{FR}, A' \in \tilde{\mathfrak{A}}_{FR}\}$$

in the case of the stimulus–response hypothesis, and

$$\mathfrak{C}'_{FR} = \{G': \beta(A', G') = B' \in \mathfrak{B}'_{FR}, A' \in \tilde{\mathfrak{A}}'_{FR}(G')\},$$

(6.90′)

where

$$\tilde{\mathfrak{A}}'_{FR}(G') = \{A: \beta(A, G') = B \in \mathfrak{B}_{FR}\} \quad \text{and} \quad \tilde{\mathfrak{A}}'_{FR} = \{A: \beta(A, G') = B \in \mathfrak{B}_{FR}, G' \in \mathfrak{C}'_{FR}\}$$

in the case of the response–stimulus hypothesis.

If it is possible to choose such subsets $\mathfrak{C}_{FR}$ and $\mathfrak{B}'_{FR}$ with which the respective subsets $\mathfrak{A}_{FR}$ and $\tilde{\mathfrak{A}}_{FR}$ or $\mathfrak{A}'_{FR}$ and $\tilde{\mathfrak{A}}'$ coincide, then we can say that there is some *correspondence* between the models of control in case of adaptation and natural adaptation in the stimulus–response hypothesis and that of the response–stimulus, respectively. When the subsets $\mathfrak{B}_{FR} = \{B_{FR}^{\max}\}$, $\mathfrak{B}'_{FR} = \{B_{FR}^{\min}\}$ and $\mathfrak{C}_{FR} = \{C_{FR}^{\max}\}$ degenerate into maximum and minimum elements of sets $\mathfrak{B}$ and $\mathfrak{C}$, the relations (6.89), (6.90), (6.89'), and (6.90') degenerate into relations of the trinary two-matrix game:

$$\max_{H \in \mathfrak{H}_F} \gamma(H, A) = \gamma[H_F(A), A]$$

$$\max_{A \in \mathfrak{A}_R} \beta[A, H_F(A)] = \beta[A_{FR}, H_F(A_{FR})] = B_{FR}^{\max} \qquad (6.89'')$$

$$\gamma[H_F(A_{FR}), A_{FR}] = C_{FR}^{\max}$$

in the case of the stimulus–response hypothesis, and

$$\max_{A \in \mathfrak{A}_R} \beta(A, H) = \beta[A'_R(H), H]$$

$$\max_{H \in \mathfrak{H}_F} \gamma[H, A'_R(H)] = \gamma[H_{FR}, A'_R(H_{FR})] = \tilde{C}_{FR}^{\max} \qquad (6.90'')$$

$$\beta[A'_R(H_{FR}), H_{FR}] = \tilde{B}_{FR}^{\max}$$

in the case of the response–stimulus hypothesis, and the binary matrix game with generally noncoinciding expressions,

$$\max_{A \in \mathfrak{A}_R} \min_{G \in \mathfrak{C}_F} \beta(A, G) = \beta[A''_{FR}, G_F(A''_{FR})] = b(F, R) \qquad (6.91)$$

$$\neq \min_{G \in \mathfrak{C}_F} \max_{A \in \mathfrak{A}_R} \beta(A, G) = \beta[A''_R(G_{FR}), G_{FR}] = \tilde{b}(F, R),$$

in the respective cases of the stimulus–response and response–stimulus hypotheses. If they are equal, the corresponding solution is called a saddle point. If the solutions A_{FR} and A''_{FR} or $A'_R(H_{FR})$ and $A''_R(G_{FR})$ coincide, we can speak about some correspondence between the main models of control under adaptation and natural adaptation.

The explicit solutions (6.89) and (6.90) of eq. (6.88) may be of excessive nature for a discrete case only when the sets $\mathfrak{H}_F$ and $\mathfrak{A}$ are not great or in a continuous case employing the apparatus of mathematical analysis. In this case, the main functions $C = \gamma(H, A)$ and $B = \beta(A, H)$ are considered as genuine continuously differentiable functions with real multidimensional arguments. Then, for a degenerative case, the extremals in the relations (6.89''), (6.90''), and (6.91), with the analytical restrictions $\mathfrak{A}_R$ and $\mathfrak{H}_F$ on the arguments, will be found by solving problems of conventional extremals. This approach is employed later. Herein some critical value of B_{cr} of the unit B plays the part of a critical set, $\mathfrak{B}_{cr}$.

Starting from the solution obtained in the degenerative case, it is possible to derive a solution for the general case by using the continuity of functions in $\beta(A, G)$ and $\gamma(H, A)$ in the following way. Let the extremum values with some of their "permissible" proximities,

$$B_{FR}^{\max} = \beta(A_{\text{opt}}, G_{\text{opt}}) \in \mathfrak{B}_{FR}, \quad B_{FR}^{\min} = \beta(A'_{\text{opt}}, G_{\text{opt}}) \in \mathfrak{B}'_{FR}$$

$$B_{FR}^{\max} = \beta(A_{\text{opt}}, G_{\text{opt}}) \in \mathfrak{B}_{FR}, \quad C_{FR}^{\max} = \gamma(H_{\text{opt}}, A'_{\text{opt}}) \in \mathfrak{C}_{FR},$$

Chap. 6   Stochastic Theory of Complex Ecological Systems

be derived by optimal values with their respective proximities $A_{opt} \in \mathfrak{A}_{FR}$, $G_{opt} \in \mathfrak{E}_{FR}$, $G'_{opt} \in \mathfrak{E}'_{FR}$, $A'_{opt} \in \mathfrak{A}'_{FR}$, and $H_{opt} \in \mathfrak{H}_{FR}$. Then the latter proximities may be considered as solutions of the problems in their more realistic and general form.

Basic Relations

We have presented above the general formulation of the problem of optimal control and a scheme for its solution with limitations such as subsets $\mathfrak{A}_R$, $\mathfrak{H}_F$, and $\mathfrak{E}_F$, as well as the goal functions $C = \gamma(H, A)$, $B' = \beta(A, H)$, and $B = \beta(H, G)$. The further task is to obtain real expressions for limitations and goal functions in examining the communities from the actual class of regenerative systems (Fleishman, 1984). As already mentioned, all studies involve stationary (climax) states if community behavior can be divided into time intervals during which the community adapts to the transition (succession) regime from one stationary pressure of the controlling system to another. The values of these time intervals were determined by Fleishman (1984).

The community A is specified at three space–time levels: the macrolevel, by using the model $A = A_{a1}$ of populations in the community ($a \geq 2$); the mesolevel, by using the model A_{1h} of cohorts in the population; and the microlevel, by using the model A_{11} of fish schools in the cohorts (Fleishman, 1980, 1984). These models, indicated by MAM, MEM, and MIM, have the following specific space–time scales: (10^3 km, year), (10^2 km, month), and (10 km, day), respectively. In this way, the model MAM is a nondegenerate ($a \geq 2$) and the model MEM a degenerate ($a = 1$) model of the community. Later, depending on the three types of control (free trade, cultivation, and suppression), the models will have the designations MAM, MEM; MAM⁺MEM⁺; and MAM⁻, MEM⁻. The integral parameters of these models are marked accordingly.

In the stochastic theory of the natural adaptation of community A to impulse pressure of the environment G in a stationary case, we can see the following explicit task of the community, $A = (|A|, \overline{A})$, and the environment, $G = (|G|, \overline{G})$, as systems specified by the structure, behaviors, and limitations $\mathfrak{A}_R$ and $\mathfrak{E}_F$ for MAM:

$$A_{a1} = (\vec{x}, \vec{\lambda}), \qquad G = (\cdot, \vec{y}), \qquad \mathfrak{A}_R = \{\vec{\lambda}: \sum_{\alpha=1}^{a} \lambda_\alpha = a\overline{\lambda} = R\}, \tag{6.92}$$

$$\mathfrak{E}_F = \{\vec{y}: \sum_{\alpha=1}^{a} y_\alpha = a\overline{y} = F\},$$

where $\vec{x} = (x_\alpha)$, $\vec{y} = (y_\alpha)$, and $\vec{\lambda} = (\lambda_\alpha)$. Similarly, we have the expression for a periodic population, A_{1h} with period h, and composed of σ groups of individuals (of all cohorts whose individuals were born in the same season, σ, respective of their age), as well as for a periodic model of the environment G with period h, and the limitations $\mathfrak{A}_{R\alpha}^\alpha$ and $\mathfrak{E}_{F\alpha}^\alpha$ for MEM:

$$A_{1h} = (\vec{x}_{\alpha 1}, \vec{\lambda}_\alpha), \qquad G_\alpha = (\cdot, \vec{y}_\alpha), \qquad \mathfrak{A}_{R\alpha}^\alpha = \{\vec{\lambda}_\alpha: \sum_{\sigma=0}^{h-1} \lambda_{\alpha\sigma} = h\overline{\lambda}_\alpha = R_\alpha\}, \tag{6.93}$$

$$\mathfrak{E}_{F\alpha}^\alpha = \{\vec{y}_\alpha: \sum_{\sigma=0}^{h-1} y_{\alpha\sigma} = h\overline{y}_\alpha = F_\alpha\},$$

where λ_α, $\lambda_{\alpha\sigma}$; x_α, $x_{\alpha\sigma}$; and y_α, $y_{\alpha\sigma}$ are the stationary values of instantaneous probabilities of births and deaths of individuals in the σ group of the α population due to some external and internal causes.

To obtain stationary goal functions, the following relations are used in the transition regime (Fleishman, 1982a). The instantaneous mathematical expectation, $N_\alpha(t) = N_{p_\alpha}(t)$, of the resource quantity of the α population, where N is the number of space elements, $p_\alpha(t)$ ($\alpha = \overline{1, a}$), is the probability of the element's being occupied by an α individual, and $p_0(t)$ the probability of its vacancy, are all interrelated with one another through the probabilities $p_\alpha(t)$ by the following relations in the model MAM:

$$p_\alpha(t+1) = [1 - \mu_\alpha(t)] p_\alpha(t) + \lambda_\alpha(t) p_0(t). \tag{6.94}$$

The instantaneous mathematical expectation $N_\alpha(t)$ of the resource quantity of the α population is expressed as the total expectation of the resource quantities of s cohorts, $N_{\alpha s}(t)$, (its replenishment) by the following relation in the model MEM:

$$N_\alpha(t) \stackrel{\Delta}{=} \sum_{s=0}^{t} N_{\alpha s}(t) = \sum_{s=0}^{t} L_\alpha(s) [1 - F_{\alpha s}(t)], \tag{6.95}$$

where

$$L_\alpha(s) = n\lambda_{\alpha s} \tag{6.96a}$$

and

$$1 - F_{\alpha s}(t) \begin{cases} = \prod_{\tau=s}^{t} [1 - \mu_{\alpha s}(\tau)] \\ > 1 - \sum_{\tau=s}^{t} \mu_{\alpha s}(\tau) \end{cases} \tag{6.96b}$$

show the probability of survival of individuals of the s cohort until the moment t in the case of their independent death (6.96a) and in the general case (6.96b). In these expressions, $\lambda_{\alpha s}$ is the probability of an individual's birth in the s cohort, n is the number of space elements in the model MEM, and μ is the total probability of death from external and internal causes. The latter looks as follows (Fleishman, 1977b, 1982a):

$$\mu = \xi + \eta = x + y + o(xy), \tag{6.97}$$

whereupon

$$\xi = \frac{x(1-y)}{(1-xy)} = x + o(x^2), \quad x = \frac{\xi}{1-\eta},$$
$$\eta = \frac{y(1-x)}{(1-xy)} = y + o(y^2), \quad y = \frac{\eta}{1-\xi}. \tag{6.98}$$

In the stationary case, with $t \to \infty$, by using the relations (6.94), (6.95), and (6.96), we can express the instantaneous mathematical expectations, $N_\alpha = N p_\alpha$ and $N_{\alpha\sigma} = \sum_{k=0}^{\infty} N_{\alpha\sigma+kh}$, of resource quantities in the α population of the community and its σ group as $\mu_{\alpha\sigma} \to 0$, respectively (Fleishman, 1977b, 1984)

$$\frac{N_\alpha}{N} = p_\alpha = \frac{\lambda_\alpha p_0}{\mu_\alpha} = \left(\frac{\lambda_\alpha}{\mu_\alpha}\right)\left(1 + \sum_{\alpha=1}^{a} \frac{\lambda_\alpha}{\mu_\alpha}\right)^{-1} \tag{6.99}$$

and

$$\frac{N_{\alpha\sigma}}{N'} = \frac{\lambda_{\alpha\sigma}}{\mu_{\alpha\sigma} + o(1)}, \tag{6.100}$$

where $N' = n/h$.

From relations (6.99) and (6.100), expressions for instantaneous normalized resources were obtained for annual cases,

$$B \triangleq \sum_{\alpha=1}^{a} \frac{N_\alpha}{N} = \sum_{\alpha=1}^{a} P_\alpha = \left[1 + \left(\sum_{\alpha=1}^{a} \frac{\lambda_\alpha}{\mu_\alpha}\right)^{-1}\right]^{-1} = \beta(A, G) = \beta(\vec{\lambda}, \vec{y}), \qquad (6.101)$$

whence $[1 - \beta(\vec{\lambda}, \vec{y})^{-1}]^{-1} = \sum_{\alpha=1}^{a} \lambda_\alpha/\mu_\alpha$, and seasonal cases,

$$B^\alpha \triangleq \sum_{\sigma=0}^{h-1} \frac{N_{\alpha\sigma}}{N'} = \sum_{\sigma=0}^{h-1} \frac{\lambda_{\alpha\sigma}}{\mu_{\alpha\sigma}} + o(1) = \beta^\alpha(A, G) = \beta^\alpha(\vec{\lambda}_\alpha, \vec{y}_\alpha). \qquad (6.102)$$

The same expressions, (6.101) and (6.102), are found for the goal functions $\beta(A, H)$ and $\beta^\alpha(A, H)$. However, the general formulation of the optimal control problem needs linear conditions that are more general than (6.92) and (6.93):

$$\mathfrak{A}_R = \left\{\vec{\lambda}: \sum_\alpha r_\alpha \lambda_\alpha = R\right\}, \qquad \mathfrak{H}_F = \left\{\vec{y}: \sum_\alpha f_\alpha y_\alpha = F\right\} \qquad (6.103)$$

$$\mathfrak{A}_R^\alpha = \left\{\vec{\lambda}_\alpha: \sum_\sigma r_{\alpha\sigma} \lambda_{\alpha\sigma} = R_\alpha\right\}, \qquad \mathfrak{H}_{F_\alpha}^\alpha = \left\{\vec{y}_\alpha: \sum_\sigma f_{\alpha\sigma} y_{\alpha\sigma} = F_\alpha\right\}. \qquad (6.104)$$

Their essence, as well as the interpretation of the positive coefficients r and f, are presented below.

Let us define the mathematical expectations of instantaneous quantities of withdrawals from the α population of the community and its s cohort in the models MAM and MEM, respectively, as $Y_\alpha(t) \triangleq \tilde{N}_\alpha(t+1) - N_\alpha(t+1)$ and $Y_{\alpha s}(t) \triangleq \tilde{N}_{\alpha s}(t) - N_{\alpha s}(t)$, where $\tilde{N}_\alpha(t+1) = N\tilde{p}_\alpha(t+1)$, $\tilde{p}_\alpha(t+1) = [1 - x_\alpha(t)]p_\alpha(t) + \lambda_\alpha(t)p_o(t)$ and $\tilde{N}_{\alpha s}(t+1) = N_{\alpha s}(t)[1 - x_{\alpha s}(t)]/[1 - \mu_{\alpha s}(t)]$. Using relations (6.94) and (6.95), it is possible to demonstrate that in the stationary case with $t \to \infty$ and y_α and $y_{\alpha\sigma} \to 0$, the stationary values of Y_α and $Y_{\alpha\sigma} = \sum_{k=0}^{\infty} Y_{\sigma+kh}$ look as follows for periodic functions $\lambda_{\alpha s}(s)$, $x_{\alpha s}(s)$, and $y_{\alpha s}(s)$ with period h in terms of the argument s:

$$Y_\alpha = N_\alpha y_\alpha + o(y_\alpha^2) \quad \text{and} \quad Y_{\alpha\sigma} = N_{\alpha\sigma} y_{\alpha\sigma} + o(y_{\alpha\sigma}^2). \qquad (6.105)$$

From relations (6.99), (6.100), and (6.105), we derive expressions for instantaneous normalized withdrawals per year,

$$C \triangleq \sum_\alpha \frac{c_\alpha Y_\alpha}{N} = \sum_\alpha c_\alpha p_\alpha y_\alpha = (1-B) \sum_\alpha c_\alpha \frac{\lambda_\alpha}{\mu_\alpha} y_\alpha = \gamma(H, A) = \gamma(\vec{y}, \vec{\lambda}), \qquad (6.106)$$

and per season,

$$C^\alpha \triangleq \sum_\sigma \frac{c_{\alpha\sigma} Y_{\alpha\sigma}}{N'} = \sum_\sigma c_{\alpha\sigma} \frac{\lambda_{\alpha\sigma}}{\mu_{\alpha\sigma}} y_{\alpha\sigma} = \gamma^\alpha(H, A) = \gamma^\alpha(\vec{y}_\alpha, \vec{\lambda}_\alpha). \qquad (6.107)$$

Their essence and the interpretation of the positive coefficients C are presented below.

Using the relations (6.96), (6.101), (6.102), (6.106), and (6.107), the main estimates of the withdrawals by the values of the resources are

$$(1-B)\left(\frac{\underline{d}_1 R - \bar{d}_2 B}{1-B}\right) \leq C \leq (1-B)\left(\frac{\bar{d}_1 R - \underline{d}_2 B}{1-B}\right)$$

and (6.108)

$$\underline{d}_{1\alpha} R_\alpha - \bar{d}_{2\alpha} B^\alpha \leq C_\alpha \leq \bar{d}_{1\alpha} R_\alpha - \underline{d}_{2\alpha} B^2,$$

where

$$\underline{d}_1 = \min\frac{C_\alpha}{r_\alpha} \leq \bar{d}_1 = \max\frac{C_\alpha}{r_\alpha}, \qquad \underline{d}_2 = \min(c_\alpha x_\alpha) \leq \bar{d}_2 = \max(c_\alpha x_\alpha),$$

and (6.109)

$$\underline{d}_{1\alpha} = \min\frac{C_{\alpha\sigma}}{r_{\alpha\sigma}} \leq \bar{d}_{1\alpha} = \max\frac{C_{\alpha\sigma}}{r_{\alpha\sigma}}, \qquad \underline{d}_{2\alpha} = \min(c_{\alpha\sigma} x_{\alpha\sigma}) \leq \bar{d}_{2\alpha} = \max(c_{\alpha\sigma} x_{\alpha\sigma}).$$

Since, with growing values of the resources B and B^α, the lower and upper estimates of the withdrawals C and C^α prove to be their monotonically diminishing functions, with sufficient proximity of these estimates the minimum (or maximum) of resources $B_{\min}$ (or $B_{\max}$) will correspond to the maximum (minimum) of withdrawals $C_{\max}$ ($C_{\min}$). That is, in the case in question there is correspondence between the basic models.

Let us now find the explicit expression for $B_{\max\min} = \max_{\vec{x}} \min_{\vec{y}} B = b(F, R)$ and $B_{\min\max} = \min_{\vec{y}} \max_{\vec{x}} B = \bar{b}(F, R)$ by using relations (6.91) with the explicit expressions $\beta(\vec{\lambda}, \vec{y})$ and $\beta^\alpha(\vec{\lambda}_\alpha, \vec{y}_\alpha)$ defined by relations (6.101) and (6.102) with the conditions (6.103) and (6.104) and the additive dependence $\mu = x + y$ (6.97). The latter is accepted with disregarded values $o(y^2)$ in other relations also discussed.

Employing Lagrangian multipliers, the extremum expressions $b(F, R)$ and $\bar{b}(F, R)$, and the optimal values of $\vec{\lambda}$ and $\vec{y}$ leading to them, can be found. The choice between the hypothesis stimulus–response and response–stimulus, as in the similar case $r_\alpha = f_\alpha = 1$ considered in Fleishman (1977a, b), will result in a degenerate community in the case of the response–stimulus hypothesis. Therefore, we accept the stimulus–response hypothesis. For it, the value

$$B_{\max\min} = b(F, R) = \left[1 + \left(\frac{d_4 R}{d_3 + F}\right)^{-1}\right]^{-1} \qquad (6.110)$$

is obtained when

$$\lambda_\alpha = \lambda_\alpha^0 = \frac{d_4^{-1} R f_\alpha}{r_\alpha^2} \quad \text{and} \quad y_\alpha = y_\alpha^0 = \frac{d_4^{-1}(d_3 + F)}{r_\alpha} - x_\alpha, \qquad (6.111)$$

where $d_3 = \sum_\alpha f_\alpha x_\alpha$ and $d_4 = \sum_\alpha f_\alpha/r_\alpha$. Similarly, the value

$$B_{\max\min}^\alpha = b^\alpha(F_\alpha, R_\alpha) = \frac{d_{4\alpha} R}{d_{3\alpha} + F_\alpha} \qquad (6.112)$$

is derived when

$$\lambda_{\alpha\sigma} = \lambda_{\alpha\sigma}^0 = \frac{d_{4\alpha}^{-1} R_\alpha f_{\alpha\sigma}}{r_{\alpha\sigma}^2} \quad \text{and} \quad y_{\alpha\sigma} = y_{\alpha\sigma}^0 = \frac{d_{4\alpha}^{-1}(d_{3\alpha} + F_\alpha)}{r_{\alpha\sigma}} - x_{\alpha\sigma}, \qquad (6.113)$$

where $d_{3\alpha} = \sum_\sigma f_{\alpha\sigma} x_{\alpha\sigma}$ and $d_{4\alpha} = \sum_\sigma f_{\alpha\sigma}/r_{\alpha\sigma}$.

The accurate solution of the extremum problem in question, when the nonlinear dependence $\mu = \xi + \eta$ of x and y and $c_\alpha = c_{\alpha\sigma} = f_\alpha = f_{\alpha\sigma} = r_\alpha = r_{\alpha\sigma} = 1$, is presented in Fleishman (1977b, 1984). The adapted value λ^0 is still dependent not only on R but on F as well. It follows from the above that in the first approximation the community and population adapt only to the environmental resources R allocated for their replenishment, and this is ecologically justified. Having substituted (6.110) and (6.112) into the corresponding estimates (6.108), after some simple transformations we obtain

$$\underline{D}\,\underline{c}(F, R) \leq C \leq \overline{D}\,\overline{c}(F, R) \tag{6.114}$$

$$\underline{D}_\alpha\,\underline{c}^\alpha(F_\alpha, R_\alpha) \leq C^\alpha \leq \overline{D}_\alpha\,\overline{c}^\alpha(F_\alpha, R_\alpha)$$

where

$$\underline{c}(F, R) = b(F, R)(F + d_3 - \overline{d}_5) \leq \overline{c}(F, R) = b(F, R)(F + d_3 - \underline{d}_5),$$

$$\underline{c}^\alpha(F_\alpha, R_\alpha) = b^\alpha(F_\alpha, R_\alpha)(F_\alpha + d_{3\alpha} - \overline{d}_{4\alpha}) \leq \overline{c}^\alpha(F_\alpha, R_\alpha) = b^\alpha(F_\alpha, R_\alpha)(F_\alpha + d_{3\alpha} - \underline{d}_{4\alpha})$$

$$\underline{D} = \frac{d_1}{d_4} \leq \overline{D} = \frac{\overline{d}_1}{d_4}, \quad \underline{d}_5 = \frac{d_2 d_4}{d_1} \leq \overline{d}_5 = \frac{\overline{d}_2 d_4}{d_1} \tag{6.115}$$

$$\underline{D}_\alpha = \frac{d_{1\alpha}}{d_{4\alpha}} \leq \overline{D}_\alpha = \frac{\overline{d}_{1\alpha}}{d_{4\alpha}}, \quad \underline{d}_{5\alpha} = \frac{d_{2\alpha} d_{4\alpha}}{d_{1\alpha}} \leq \overline{d}_{5\alpha} = \frac{\overline{d}_{2\alpha} d_{4\alpha}}{d_{1\alpha}}.$$

Let us now proceed to consideration of control of the community in the MIM model. To this end, we can study the space element in the MAM or MEM models with stationary probability $p = N_\alpha/N$ or $p = N_{\alpha\sigma}/N'$ of having the α individual determined by relations (6.94) or (6.100) (this probability is proportional to the average density, ρ, of the α individual in the range under discussion).

Assume that the region of withdrawal composed of N^* elements is divided into a number l of equal sections. Each section is exploited by a single type of harvesting equipment. Then,

$$p \stackrel{\Delta}{=} \frac{N_\alpha}{N} \sim \frac{N_\alpha^*}{N^*} = \frac{1}{N^*}\sum_{i=1}^{l} N_{\alpha i}^* = \frac{1}{l}\sum_{i=1}^{l} \frac{N_{\alpha i}^* l}{N^*} = \frac{1}{l}\sum_{i=1}^{l} p_i = \text{constant}, \tag{6.116}$$

where N_α^* is the total number of α individuals in the withdrawal region, $N_{\alpha i}^*$ is the unknown number of α individuals in the i section, and $p_i = N_{\alpha i}^* l/N^*$ is its corresponding frequency. Let all l types of equipment be allocated with some fixed quantity of the resource, K (say fuel, for example). For each kind, i, the resource amount, k_i, is such that

$$K = \sum_{i=1}^{l} k_i. \tag{6.117}$$

Assume that $D(k)$ is the conventional probability of harvesting, with a particular kind of gear, an α individual whose value is equal to k. This probability increases monotonically with increasing k. Then, the mathematical expectation of catch by all l equipment types will be:

$$Y \stackrel{\Delta}{=} \sum_{i=1}^{l} Y_i = N^* \sum_{i=1}^{l} p_i D(k_i) = Y(\vec{p}, \vec{k}). \tag{6.118}$$

Here, $Y_i = N_{\alpha i}^* D(k_i)$ is the expectation of withdrawal by the ith type of equipment that captures the quantity k_i of the resource. It is required to maximize Y in terms of its variable $\vec{k} = (k_i)$ with regard to its most unfavorable minimal entity by using the unknown vector $\vec{p} = (p_i)$. This meaning of $Y = Y_{\text{max/min}}$ in this case will be the local mean optimal withdrawal minimally guaranteed (say, the catch).

Theorem I. The mean minimally guaranteed withdrawal with limitations (6.116) and (6.117) will occur with any values of $\vec{p} = (p_i)$ and $\vec{k} = (K/l)$. It will be equivalent to

$$Y = Y_{\text{max min}} = \max_{\vec{k}\ (6.117)} \min_{\vec{p}\ (6.116)} Y(\vec{p}, \vec{k}) = N^* l p D \frac{K}{l}. \qquad (6.119)$$

Proof. From eq. (6.118), we have

$$\min_{\vec{p}} Y(\vec{p}, \vec{k}) = Y(\vec{p}_0, \vec{k}) = N^* l p D(k_{i_0}) \leq Y(\vec{p}, \vec{k}), \qquad (6.120)$$

where $\vec{p}_0 = \underbrace{(0, \ldots, 0, \overset{i_0}{lp}, 0, \ldots, 0)}_{l}$ and i_0 is found from the relation $k_{i_0} = \min_{1 \leq i \leq l} k_i$. Later, bearing in mind that $D(k)$ grows monotonically with augmentation of k, and in accordance with relation (6.117), $k_{i_0} \leq K/l$, we obtain from relation (6.120) the following:

$$Y_{\text{max min}} = \max_{\vec{k}\ (6.117)} \min_{\vec{p}\ (6.116)} Y(\vec{p}, \vec{k}) = \max_{\vec{k}\ (6.117)} N^* l p D(k_{i_0}) = N^* l p \frac{K}{l},$$

which was required to be proved.

Thus, using mean values calculated in the MAM and MEM models, or individuals' densities with the optimal strategy of their withdrawal under local conditions of the MIM model, the actual withdrawal is not less than that predicted by the MAM and MEM models. Fleishman (1974, 1977b, 1980) has provided an information theory of fish-school formation which permits evaluating the size, m, of schools based on external and internal factors. The m school is an elementary ecological unit in the MIM model.

Optimization of Control by Integral Parameters

All further computations are made for the previously discussed case when the main models correspond to one another. To this end, it is necessary that $\underline{D} \sim \overline{D}$ and $\underline{D}_\alpha \sim \overline{D}_\alpha$. To have this, it is sufficient that $\underline{d}_1 \sim \overline{d}_1$ and $\underline{d}_{1\alpha} \sim \overline{d}_{1\alpha}$ (see relation (6.115)). The latter relations are more accurate the closer are the dependencies $c_\alpha/r_\alpha \sim d_1$ and $c_{\alpha\sigma}/r_{\alpha\sigma} \sim d_{1\alpha}$ to the constants d_1 and $d_{1\alpha}$.

Ecologically, the constancy of these relations is justified if we interpret the coefficient c_α by the price of biomass of the α individual. The biomass should be proportional to the α individual's requirements in the environmental resources specified by the coefficient r_α. So, later we shall assume that $\underline{D} \sim \overline{D} \sim D \sim d_1/d_4$ and $\underline{D}_\alpha \sim \overline{D}_\alpha \sim D_\alpha \sim d_{1\alpha}/d_{4\alpha}$. Then, to make the main models in the MAM correspond, it is sufficient to consider the asymptotic case $F \to \infty$ ($F \gg d_3$) [see relations (6.110) and (6.114)]. Here we have

$$B_{\text{max min}} = b(F, R) \sim \left[1 + \left(\frac{d_4 R}{F}\right)^{-1}\right]^{-1} = \frac{d_4 R}{d_4 R + F}, \qquad (6.121)$$

$$C_{\text{min max}}/D = c(F, R) = b(F, R) F \sim \left[1 + \left(\frac{d_4 R}{F}\right)^{-1}\right]^{-1} F = \frac{d_4 R F}{d_4 R + F}.$$

To make the main models in the MAM correspond at final values of F_α, it is necessary to require the constants $\underline{d}_{5\alpha} \sim \bar{d}_{5\alpha} \sim d_{5\alpha}$ to be close [see relations (6.112) and (6.114)]. The more accurate the latter relation is made, the closer the products $c_{\alpha\sigma}x_{\alpha\sigma} = d_{5\alpha}$ are to each other. The constancy of these products is ecologically justified if we interpret the coefficient $c_{\alpha\sigma}$ through the α individual's biomass, which is proportional to its average life span (the inversely proportional instantaneous probability of the individual's death, $x_{\alpha\sigma}$). Here we have [see relations 6.112 and 6.114]

$$B^\alpha_{\text{max min}} = b^\alpha(F_\alpha, R_\alpha) \sim \frac{d_{4\alpha}R_\alpha}{F_\alpha + d_{3\alpha}}$$

and

$$\frac{C^\alpha_{\text{min max}}}{D_\alpha} = c^\alpha(F_\alpha, R_\alpha) = b^\alpha(F_\alpha, R_\alpha)(F_\alpha + d_{3\alpha} - d_{5\alpha}) = \frac{d_{4\alpha}R_\alpha(F_\alpha + d_{3\alpha} - d_{5\alpha})}{F_\alpha + d_{3\alpha}}. \quad (6.122)$$

Let us now consider the particular interpretations of relations (6.121) and (6.122). We begin with free trade, when withdrawals (C and C^α) are interpreted through catches, and the pressures (F and F_α) through harvesting efforts. The ecological justification of max-min estimates of the resources has been discussed in Fleishman (1977b, 1982a). Equally justified are the min-max catch estimates. Indeed, by varying the values $\vec{y} = (y_\alpha)$ and $\vec{y}_\alpha = (y_{\alpha\sigma})$, the commercial system becomes interested in maximizing the catches C and C^α. At the same time the community, while adapting to trade effects and environmental resources by means of the replenishment specified by values $\vec{\lambda} = (\lambda_\alpha)$ and $\vec{\lambda}_\alpha = (\lambda_{\alpha\sigma})$, is "interested" in maximizing its reserves, B and B', which minimize the catches C and C'. When analyzing relations (6.121) and (6.122), we can see that with growth of fishing efforts, F and F_α, to preserve a constant quantity of the resource the summary resources of the environment, R and R_α, must grow linearly, together with the growth of F and F_α, respectively. The values of catches thereby linearly increase with growth of harvesting efforts, F and F_α. If we consider the stable total environmental resource R and R_α = constants, hyperbolic decreases in the resource reserve will be accompanied by logistic growth of catches with increasing F and F_α.

Of greatest theoretical and practical interest is the case of deteriorating environmental conditions accompanied by increasing trade efforts. Formally, this case can be described by the following conditions: $d_4R = F^{-u}$ and $d_{4\alpha}R_\alpha = F'^{-u}$, where $F'_\alpha = [(F_\alpha - d_{5\alpha})/d_{5\alpha}]^{1/(1+u)}$ $[F_\alpha = d_{5\alpha}(F'^{1+u}_\alpha + 1)]$ and where $u > 0$ specifies the degree of environmental deterioration; this value can be infinitesimally close to zero.

In this case, relations (6.121) and (6.122) pass into relations

$$B_{\text{max min}} = b(F) \sim (1 + F^{1+u})^{-1} \leq \frac{1}{F^{1+u}} \quad (6.123)$$

$$\frac{C_{\text{min max}}}{D} = c(F) = b(F)F \sim (1 + F^{1+u})^{-1} F = (F^{-1} + F^u)^{-1} < \frac{1}{F^u}, \quad (6.124)$$

and

$$B^\alpha_{\text{max min}} = b^\alpha(F_\alpha) = b(F'_\alpha) \sim (1 + F'^{1+u}_\alpha)^{-1} \leq \frac{1}{F'^{1+u}_\alpha} \quad (6.125)$$

$$\frac{C^\alpha_{\text{min max}}}{D_\alpha} = C^\alpha(F_\alpha) = C(F'_\alpha) = b(F'_\alpha)F' \sim (1 + F'^{1+u}_\alpha)^{-1} F'_\alpha = (F'^{-1}_\alpha + F'^u_\alpha)^{-1} < \frac{1}{F'^u_\alpha}. \quad (6.126)$$

Relations (6.123) to (6.126) permit us to find the maximum values, $C_{\text{min max}}$ and $C^\alpha_{\text{min max}}$, through F and the variables F_α and $F'_\alpha = (F_\alpha/d_{5\alpha} - 1)^{1/(1+u)}$ linked by a monotonic dependence. The maximum corresponds to the overcatch effect when the catch increases to the maximum with the increase in F or F_α and then declines.

Theorem 2. In the asymptotic case, $F \to \infty$ and $R = F^{-u}/d_4$, where $u > 0$, the maximum of the value $C_{\min \max}/D$,

$$\max_F \frac{C_{\min \max}}{D} = c(F_0) = [u^{1/(1+u)} + u^{-u/(1+u)}]^{-1}, \quad (6.127)$$

is obtained when

$$F = F_0 = u^{-1/(1+u)} \quad (6.128)$$

in which

$$B_{\max \min} = b(F_0) = \frac{u}{1+u}. \quad (6.129)$$

Similarly, with $F'_\alpha = (F_\alpha/d_{5\alpha}^{-1})^{1/(1+u)} \to \infty$ and $R_\alpha = F'^{-u}/d_{5\alpha}$, where $u > 0$, the maximum value of $C^\alpha_{\min \max}/D_\alpha$,

$$\max_{F_\alpha} \frac{C^\alpha_{\min \max}}{D_\alpha} \sim c(F'_{\alpha 0}) = (u^{1/(1+u)} + u^{-u/(1+u)})^{-1}, \quad (6.130)$$

is derived for the value

$$F'_\alpha = F'_{\alpha 0} = u^{-1/(1+u)} \quad (6.131)$$

with which

$$B^\alpha_{\max \min} = b(F'_{\alpha 0}) = \frac{u}{1+u}. \quad (6.132)$$

Figure 6.6 shows the dependence of (6.124), $C_{\min \max}/D = c(F)$, on the value of $\log F$ with different values of the parameter u. The same dependence occurs in the case of $C^\alpha_{\min \max}/D = c(F'_\alpha)$, determined by the value of $\log F'_\alpha$ on which it depends [see relation (6.126)].

It should be noted that all these equations, beginning with (6.121) and (6.122), are asymptotic as F and $F'_\alpha \to \infty$. Therefore, the outcomes of Theorem 2 are valid only when F_0 and $F'_{\alpha 0} \to \infty$, that is, when $u \to 0$. For this case, relations (6.127) to (6.132) pass into relations

$$\max_F \frac{C_{\min \max}}{D} = c(F_0) = 1 - u \ln\left(\frac{1}{u}\right) + o(u)$$

$$B_{\max \min} = b(F_0) = u + o(u^2) \quad (6.133)$$

$$F_0 = \frac{1}{u} - \ln\left(\frac{1}{u}\right) + o(u \ln u)$$

and

$$\max_{F'} \frac{C^\alpha_{\min \max}}{D_\alpha} = c(F'_{\alpha 0}) = 1 - u \ln\left(\frac{1}{u}\right) + o(u)$$

$$B^\alpha_{\max \min} = b(F'_{\alpha 0}) = u + o(u^2) \quad (6.134)$$

$$F'_{\alpha 0} = \frac{1}{u} - \ln\left(\frac{1}{u}\right) + o(u \ln u).$$

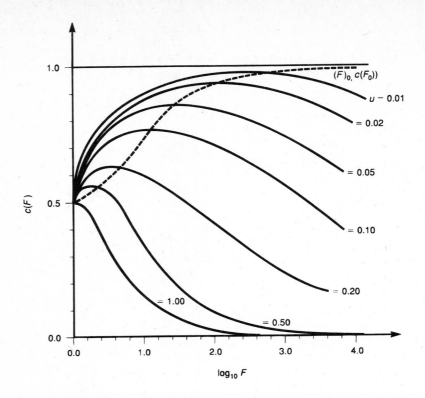

Figure 6.6 The overcatch effect. Dependence of standardized (normalized) annual catch, $c(F) = C/D$, on the logarithm of harvesting efforts, $\log F$, with the environmental resource quantity $R = d_4 F^{-u}$, where C is the annual catch, D and d_4 are constants depending on specific features of the exploited resource, and $U > 0$ is the parameter of environmental deterioration (depletion of environmental resources) with the growth of harvesting effort. F_0 is the optimal effort resulting in the maximum value, $c(F_0)$, of the function $c(F)$. The trajectory of $(F_0, c(F_0))$ is shown by the broken curve.

The optimal quantities of the environmental resources, R_0 and $R_{\alpha 0}$, are expressed through optimal values of the parameters F_0 and $F'_{\alpha 0}$ in accordance with $R_0 = F_0^{-u}/d_4$ and $R_{\alpha 0} = F'^u_{\alpha 0}/d_{4\alpha}$.

Let us now examine the relative losses $\eta \overset{\Delta}{=} (C_{\min \max} - C)/C_{\min \max} = (C(F_0) - C(F))/C(F_0)$ in the catch due to nonoptimal setting of fishing effort determined by the relative differential $\xi \overset{\Delta}{=} (F_0 - F)/F_0$. Since in our case $u \sim 0$, the parameters $F_{\alpha 0}$ and $F'_{\alpha 0}$ are approximately linearly related and hence we have for the relative remainders $\xi_\alpha \overset{\Delta}{=} (F_{\alpha 0} - F_\alpha)/F_{\alpha 0} \sim (F'_{\alpha 0} - F'_\alpha)/F'_{\alpha 0}$. Using relations (6.123), (6.126), (6.133), and (6.134), the connection between η and ξ is found to be as follows:

$$\eta = 1.5u\xi^2 + o(\xi^3). \qquad (6.135)$$

A similar dependence also exists for the value

$$\eta_\alpha = 1.5u\xi_\alpha^2 + o(\xi_\alpha^3). \qquad (6.136)$$

The asymptotic case of small values of $u > 0$ corresponds to a slow, almost imperceptible "deterioration" of the environment, which appears when the value $C_{\min \max}$ gains its maximum against F, and $C^\alpha_{\min \max}$ against F_α. This corresponds to the known effect of overcatch observed in commercial fishing practice (Ricker, 1975). It results in increasing catch with effort, growing to some critical value, F_0 or $F_{\alpha 0}$, where the catch reaches its maximum, following which, with further increase of $F > F_0$ and $F_\alpha > F_{\alpha 0}$, it decreases.

In the case we have considered, the critical values of B_{cr} and B^α_{cr} determine the critical value of the parameters $u = B_{cr}$ and $u = B^\alpha_{cr}$, which in turn, according to relations (6.133) and (6.134), specify the optimal permissible trade efforts, $1/B_{cr} - \ln(1/B_{cr}) \leq F_0$ and $1/B^\alpha_{cr} - \ln(1/B^\alpha_{cr}) \leq F'_0$; the latter are the permissible trade effect, $\vec{y}^0 = (y^0_\alpha)$ and $\vec{y}^0_\alpha = (y^0_{\alpha a})$ [see relations (6.111) and (6.113) with $F = F_0$ and $F_0 = F'_0$]. Numerical computations can be conveniently made by using the set of curves in Fig. 6.6.

Let us now examine relations (6.121) and (6.122) in the case of cultivating a useful community or suppressing a harmful one. The concept of a "harmful community" means as a rule, the degenerate case $a = 1$. This may be taken to correspond, for example, to weeds and pests of agricultural crops, nonvaluable species spoiling valuable ones, parasites suppressing cultured species, and such biological damage as produced by overgrowth of technical and engineering objects or transport vehicles (as in the case of marine fouling organisms). In the latter case, we can speak of the entire community as being harmful.

The maximization of yields C in cultivated systems is as justified as that of resource B by communities. In cases of suppression, the suppressing system is "interested" only in minimization of resource B. When the main models correspond to one another, the latter case is equivalent to the maximization of withdrawal.

In both these cases, as mentioned before, the controlling system H, unlike in the free trade case, together with the environment G, through its resources, affects recruitment, $R = f(H, G)$, of the resource. In a first approximation of the dependence, $f(\cdot, \cdot)$ may be regarded as additive. Then

$$R = R_G \pm R_H, \quad R_\alpha = R^\alpha_G \pm R^\alpha_H, \qquad (6.137)$$

where R_G, R^α_G and R_H, R^α_H are the environmental resources. If the controlling system is expressed in terms of the MAM and MEM models, respectively, then summation corresponds to cultivation and subtraction to suppression of the community. Both cultivation and suppression require efforts F and F_α for withdrawals to be made from community's resources—from yield in the first case and from the destroyed part of the resource in the second.

Let the controlling system possess certain limited means, K and K_α, to accomplish each of the measures mentioned. It is necessary to distribute these means rationally between the resources required for replenishment and for harvesting efforts in the useful community case, or for destruction of replenishment and part of the resource in the case of a harmful community. Then the following limitations will be observed:

$$k_1 R_H + k_2 F = K = \text{constant}$$

and $\qquad (6.138)$

$$k_{1\alpha} R^\alpha_H + k_{2\alpha} F_\alpha = K_\alpha = \text{constant},$$

where k_1 and $k_{1\alpha}$ denote the cost of growing or destroying one unit of replenishment, and k_2 and $k_{2\alpha}$ are the cost of efforts for harvesting one unit of yield or destroying one unit of resources for cultivation and suppression of the community, respectively.

It is necessary to find an optimal distribution of means that will result in maximum yield, max $C_{\min \max}$, in case of cultivation, and in minimization of the resource, min $B_{\max \min}$ in case of suppression of the community. If the critical value of the resource is B_{cr}, then in the first case it is required that min $B_{\max \min} \geq B_{cr}$, and in the second that min $B_{\max \min} < B_{cr}$. Let us solve the extremum problems for $B_{\text{max min}} = b(F, R)$ and $C_{\text{min max}}/D = c(F, R)$ by using the variables F and R under the conditions specified in (6.137) and (6.138).

Employing relations (6.121), (6.137), and (6.138), we obtain the F-optimized expression for the MAM model:

$$B^{\pm}_{\text{max min}} = b^{\pm}(F) = \left(\frac{1 + vF}{u^{\pm} \mp F}\right)^{-1} \tag{6.139}$$

$$\frac{C^{\pm}_{\text{min max}}}{D} = C^{\pm}(F) = b^{\pm}(F)F = \left(\frac{1/F + v}{u^{\pm} \mp F}\right)^{-1},$$

where $u^{\pm} = (k_1 B_G \pm K)/k_2$, $v = k_1/d_4 k_2$, and $D < F < u^{\pm}$.

Using relations (6.131), (6.137), and (6.138), we obtain for the $F^{\pm}_{\alpha}$-optimized expressions in the MEM model:

$$B^{\alpha\pm}_{\text{max min}} = b^{\pm}_{\alpha}(F^{\pm}_{\alpha}) = \frac{1 \mp F^{\pm}_{\alpha}}{F^{\pm}_{\alpha}} \tag{6.140}$$

$$\frac{C^{\alpha\pm}_{\text{min max}}}{D_{\alpha}d_{4\alpha}R^{\pm}_{\alpha}} = C^{\pm}_{\alpha}(F^{\pm}_{\alpha}) = b^{\pm}_{\alpha}(F^{\pm}_{\alpha})(F^{\pm}_{\alpha} - w^{\pm}) = \frac{(1 \mp F^{\pm}_{\alpha})(F^{\pm}_{\alpha} - w^{\pm})}{F^{\pm}_{\alpha}},$$

where

$$R^{\pm}_{\alpha} = R^{\alpha}_G \pm (K_{\alpha} \pm d_{3\alpha} k_{2\alpha})/k_{1\alpha}, \quad w^{\pm} = d_{5\alpha} k_{2\alpha}/k_{1\alpha} R^{\pm}_{\alpha},$$

$$F^{\pm}_{\alpha} = \frac{(F_{\alpha} + d_{3\alpha})k_{2\alpha}}{k_{1\alpha} R^{\pm}_{\alpha}}, \quad \text{and} \quad w^{\pm} < F^{\pm}_{\alpha}, \quad F^{+}_{\alpha} < 1.$$

Here, the upper indexes $+$ and $-$ relate to the cases of cultivation and suppression, respectively. Relations (6.139) and (6.140) permit us to find the maximum values $C^{+}_{\text{min max}}$ and $C^{\alpha+}_{\text{min max}}$ by using the variables F^+ and F^+_{α} (the latter is monotonically connected with the variable F_{α}) in the case of cultivation.

Theorem 3. The maximum value of $C^{+}_{\text{min max}}/D$,

$$\max_{F^+} \frac{C^{+}_{\text{min max}}}{D} = c^{+}(F^+_0) = b^{+}(F^+_0)F^+_0 = [\sqrt{u^+}/(1 + \sqrt{v})]^2, \tag{6.141}$$

is obtained for the value

$$F^+ = F^+_0 = \frac{u^+}{1 + \sqrt{v}}, \tag{6.142}$$

with which

$$B^{+}_{\text{max min}} = b^{+}(F^+_0) = \frac{1}{1 + \sqrt{v}}. \tag{6.143}$$

The maximum value of $C^{\alpha+}_{\text{min max}}$,

$$\max_{F^+_{\alpha}} \frac{C^{\alpha+}_{\text{min max}}}{D_{\alpha}d_{4\alpha}R^+_{\alpha}} = c^+_{\alpha}(F^+_{\alpha 0}) = b^+_{\alpha}(F^+_{\alpha 0})(F^+_{\alpha 0} - w^+) = (1 - \sqrt{w^+})^2, \tag{6.144}$$

is obtained for the value

$$F_\alpha^+ = F_{\alpha 0}^+ = \sqrt{w^+}, \qquad (6.145)$$

with which

$$B_{\text{max min}}^{\alpha+} = b_\alpha^+(F_{\alpha 0}^+) = \frac{1 - \sqrt{w^+}}{\sqrt{w^+}}. \qquad (6.146)$$

Let us estimate for the MAM model the relative yield losses, $\eta \triangleq [c^+(F_0^+) - C^+(F^+)]/c^+(F_0^+)$, as a function of relative deviations, $\xi^+ \triangleq (F_0^+ - F^+)/F_0^+$, of harvesting efforts from their optimal value. In a similar way, we shall have for the MEM model

$$\eta_\alpha^+ \triangleq c_\alpha^+(F_{\alpha 0}^+) - \frac{C_\alpha^+(F_\alpha^+)}{C_\alpha^+(F_{\alpha 0}^+)} \quad \text{and} \quad \xi_\alpha^+ \triangleq \frac{F_{\alpha 0}^+ - F_\alpha^+}{F_{\alpha 0}^+}.$$

It can be shown that

$$\eta^+ = \left(\frac{1}{\sqrt{v}}\right)\xi^{+2} + o(\xi^{+3}) \quad \text{and} \quad \eta_\alpha^+ = \frac{\sqrt{w^+}}{(1 - \sqrt{w^+})^2}\xi_\alpha^{+2} + o(\xi_\alpha^{+3}). \qquad (6.147)$$

In the particular case under discussion, the critical values of the resources, B_{cr}^+ and $B_{cr}^{\alpha+}$, determine the critical values of the parameters $v = 1/B_{cr}^+ - 1$ and $w^+ = 1/(1 + B_{cr}^{\alpha+})$ [see relations (6.141) and (6.143)]. In their turn, the latter determine the maximum permissible efforts for harvesting the yield:

$$F_0^+ < u^+ B_{cr}^+ \quad \text{and} \quad F_{\alpha 0}^+ < \left(\frac{B_{cr}^{\alpha+}}{1 + B_{cr}^{\alpha+}}\right)^2.$$

Let us now consider the case of the suppressed community. From relations (6.139) and (6.140), we have for the MAM and MEM models:

$$B_{\text{max min}}^- = b^-(F^-) = \left(\frac{1 + vF^-}{u^- + F^-}\right)^{-1} \geq \frac{1}{1 + v} \qquad (6.148)$$

$$B_{\text{max min}}^{\alpha-} = b_\alpha^-(F_\alpha^-) = 1 + \frac{1}{F_\alpha^-} \geq 1,$$

which are monotonically diminishing with increasing values, F^- and F_α^-, of resource dependence. On the other hand, due to the limitations in (6.138), we have $F \leq K/k_2$ and $F_\alpha^- \leq K_\alpha/k_{2\alpha}$. Therefore, if the critical values of the resource, B_{cr}^- and $B_{cr}^{\alpha-}$, of the suppressed community are given, then because of relation (6.148) the following inequalities must be satisfied:

$$\frac{u^-(1 - B_{cr}^-)}{(1 + v)B_{cr}^- - 1} \leq F^- \leq \frac{K}{k_2} \qquad (6.149)$$

$$\frac{1}{1 - B_{cr}^{\alpha-}} \leq F_\alpha^- \leq \frac{K_\alpha}{k_{2\alpha}}.$$

As follows from relation (6.149), the problem of suppressing the community may be solved only when the conditions described below are satisfied:

$$K \geq \frac{k_2 u^-(1 - B_{cr}^-)}{(1 + v)B_{cr}^- - 1} \quad \text{and} \quad K_\alpha \geq \frac{k_{2\alpha}}{1 - B_{cr}^{\alpha-}}. \qquad (6.150)$$

Parametric Relations among the MAM, MEM, and MIM Models

So far, explicit relations among the MAM, MEM, and MIM models have not been discussed, since each was the object of independent study and optimization. Let us now consider a new problem for optimizing these models "from top to bottom," that is, from MAM to MEM when the MAM model imposes limitations on the variables in MEM and the MIM model supplies model MEM with initial values.

It follows from relations (6.101), (6.102), and (6.105) that the expectation of the annual resources, N_α, and withdrawals, Y_α, in the α population has a double macro- and meso-concept:

$$N_\alpha = \sum_\sigma N_{\alpha\sigma} = N' \sum_\sigma \frac{\lambda_{\alpha\sigma}}{\mu_{\alpha\sigma}} = Np_\alpha \qquad \sum_\sigma \frac{\lambda_{\alpha\sigma}}{\mu_{\alpha\sigma}} = \frac{Np_\alpha}{N'}$$
$$\text{or} \qquad (6.151)$$
$$Y_\alpha = \sum_\sigma Y_{\alpha\sigma} = N_\alpha \sum_\sigma y_{\alpha\sigma} = N_\alpha y_\alpha \qquad \sum_\sigma y_{\alpha\sigma} = y_\alpha .$$

Inasmuch as the optimal values of the parameters N_α and Y_α, or of p_α and y_α (which is the same), are derived in the macromodel, relation (6.151) shows limitations for the variables $y_{\alpha\sigma}$ and $\lambda_{\alpha\sigma}$ of the mesomodel. The pressure on the community is specified here by its duration, $T_\alpha = \sum_\sigma T_{\alpha\sigma}$, in the year. The duration of the pressure, $T_{\alpha\sigma}$, in separate seasons shows the alternating variables.

The variables $y_{\alpha\sigma}$ and $\lambda_{\alpha\sigma}$ are imposed here with limitations, the essence of which, as well as the interpretation of coefficients, is given below:

$$\sum_\sigma t_{\alpha\sigma} y_{\alpha\sigma} = T_\alpha = \text{constant} \qquad (6.152)$$

and

$$\sum_\sigma g_{\alpha\sigma} \lambda_{\alpha\sigma}^s = G_\alpha = \text{constant,} \qquad (6.153)$$

where $s > 1$. It should be noted that the latter limitation is not related to external resources of the environment, but to internal and physiological resources of the α individuals that are exploited for reproduction and to protect recruitment.

Let us now find a conventional extremum of the value T_α when we fix the variables $\vec{\lambda}_\alpha = (\lambda_{\alpha\sigma})$ in terms of the variables $\vec{y}_\alpha = (y_{\alpha\sigma})(\sigma = \overline{0, h-1})$ with limitations (6.151). By using Lagrangian multipliers, the minimum value of T_α,

$$T_{\alpha\,\text{min}} = T(\vec{\lambda}_\alpha) = \sum_\sigma \sqrt{\lambda_{\alpha\sigma}(t_{\alpha\sigma} + \theta)} \cdot \sum_\sigma t_{\alpha\sigma} \sqrt{\lambda_{\alpha\sigma}/(t_{\alpha\sigma} + \theta)} - \sum_\sigma t_{\alpha\sigma} x_{\alpha\sigma} , \qquad (6.154)$$

is obtained when

$$y_{\alpha\sigma} = y_{\alpha\sigma}^0 = y(\vec{\lambda}_\alpha) = \frac{\sqrt{\lambda_{\alpha\sigma}/(t_{\alpha\sigma} + \theta)} \left(y_\alpha + \sum_\sigma x_{\alpha\sigma} \right)}{\sum_\sigma \sqrt{\lambda_{\alpha\sigma}/(t_{\alpha\sigma} + \theta)}} - x_{\alpha\sigma} , \qquad (6.155)$$

where the nonnegative parameter θ is found from the transcendental equation:

$$\sum_\sigma \sqrt{\lambda_{\alpha\sigma}(t_{\alpha\sigma} + \theta)} \cdot \sum_\sigma \sqrt{\lambda_{\alpha\sigma}/(t_{\alpha\sigma} + \theta)} = p_\alpha \left(y_\alpha + \sum_\sigma x_{\alpha\sigma} \right). \qquad (6.156)$$

Inasmuch as the right parts of relations (6.154) and (6.156) are monotonic functions of the parameter θ, we obtain the estimates

$$\left(\sum_\sigma \sqrt{t_{\alpha\sigma} \lambda_{\alpha\sigma}}\right)^2 \leq T_{\alpha\,\min} \leq \sum_\sigma \sqrt{\lambda_{\alpha\sigma}} \cdot \sum_\sigma \sqrt{t_{\alpha\sigma} \lambda_{\alpha\sigma}}$$

$$\left(\sum_\sigma \sqrt{\lambda_{\alpha\sigma}}\right)^2 \leq p_\alpha\left(y_\alpha + \sum_\sigma x_{\alpha\sigma}\right) \leq \sum_\sigma \sqrt{\lambda_{\alpha\sigma}/t_{\alpha\sigma}} \cdot \sum_\sigma \sqrt{t_{\alpha\sigma} \lambda_{\alpha\sigma}}.$$

(6.157)

In this way, we have derived the upper estimate of the value $T_{\alpha\,\min}$, while the upper estimate of the value $\sum_\sigma \sqrt{\lambda_{\alpha\sigma}} \leq \sqrt{p_\alpha\left(y_\alpha + \sum_\sigma X_{\alpha\sigma}\right)}$ can be used as a limitation in deriving the conventional extremum of the value G_α in terms of the variables $\vec{\lambda}_\alpha = (\lambda_{\alpha\sigma})$ ($\sigma = \overline{0, h-1}$).

Employing Lagrangian multipliers, it is possible to show that the minimum of G_α is

$$G_{\alpha\,\min} = \left[p_\alpha\left(y_\alpha + \sum_\sigma x_{\alpha\sigma}\right)\right]^s \cdot \left[\sum_\sigma g_{\alpha\sigma}^{-1/(2s-1)}\right]^{-(2s-1)},$$

(6.158)

and can be found when $\lambda_{\alpha\sigma} = \lambda_{\alpha\sigma}^0$, where

$$\lambda_{\alpha\sigma}^0 = \left[p_\alpha\left(y_\alpha + \sum_\sigma x_{\alpha\sigma}\right)\right] \left[\sum_\sigma g_{\alpha\sigma}^{-1/(2s-1)}\right]^{-2} g_{\alpha\sigma}^{-2/(2s-1)}.$$

(6.159)

The final form of $T_{\alpha\,\min} = T(\vec{\lambda}_{\alpha\sigma})$ will appear as $T_{\alpha\,\min}^0 = T(\vec{\lambda}_\alpha^0)$, and will be found when $y_{\alpha\sigma}^0 = y_{\alpha\sigma}^{00} = y_{\alpha\sigma}^0(\vec{\lambda}_\alpha^0)$ [see relations (6.154) and (6.155)]. The minimums so obtained are justified because, in order to save the efforts to suppress the community, we are interested in decreasing the duration of pressure (say, because of fuel expenses) at some fixed withdrawal (y_α = constant) by employing the optimal strategy $T_\alpha^0 = (T_{\alpha\sigma}^0)$, which takes into account the specific properties of different seasons. On the other hand, the α population "is interested" in saving various means, for example, of power supply so as to produce progeny at the fixed mathemetical expectation of its resource; to this end, it will optimally distribute this pressure between its σ groups $[L_\alpha^0 = (L_{\alpha\sigma}^0)]$.

The maximum permissible expected value of the numbers of juvenile cohorts, $N(t) = N(1)P_R(t)$ as $t \to \infty$ (when the juvenile stage is over and the initial ages are reached proceeding up to the adult phase), coincides with the initial expectation, $L_{\alpha\sigma}$, of the number of recruitments in the σ groups of the α population. Therefore, in accordance with relation (6.25), we have

$$L_{\alpha\sigma} = n\lambda_{\alpha\sigma} = N(\infty) \geq N(1)\left[P_R(t_1) - \frac{t_1\mu(t_1)}{\eta}\right]$$

(6.160)

(the mesolevel is "infinite" for the macrolevel). Later, we present experimental evaluation methods for the parameters of relation (6.160).

Ecological Interpretation of Constants, Parameters, and Limitations

The pressure exerted by the controlling system on the community (for instance, expansion of trade) can be characterized in the MAM model by the total quantity, F, of the same or similar withdrawal equipment (say, vessels) distributed in terms of F_α ($\alpha = \overline{1, a}$), the equipment needed for exploiting the α population. In the symmetric regime of the search, each of the F_α tools, in its turn, is divided into l_α independently searching groups, each having $u_\alpha = F_\alpha/l_\alpha$ tools. The interaction of the equipment in every group is such that any occasional finding of an accumulation of α individuals, at least by one tool of the group, the encounter probability being p_α and the detection probability D_α, or any discovery of this accumulation, the probability being $1 - (1 - D_\alpha)^{u\alpha}$, will be known to all tools of the group, which will then start their joint withdrawal. It is assumed that the number of tools in the group is sufficient for withdrawing the whole accumulation. For this to occur, it is necessary that $w_\alpha u_\alpha = m_\alpha$, where w_α is the number of α individuals that can be withdrawn by one tool, and m_α is the expected values of the number of α individuals in the accumulation. Let the mathematical expectations of the number of independent subregions, which a group of tools can thus investigate within the withdrawal regions, be equal to k_α. Then, with the natural assumption on the small probability of D_α, with regard to relation (6.151), the value Y_α will have a double concept:

$$Y_\alpha = N p_\alpha y_\alpha = k_\alpha l_\alpha [1 - (1 - D_\alpha)^{u\alpha}] p_\alpha w_\alpha \sim k_\alpha F_\alpha p_\alpha m_\alpha.$$

Hence, $F_\alpha = f_\alpha y_\alpha$, where

$$f_\alpha = \frac{N}{k_\alpha D_\alpha m_\alpha}, \tag{6.161}$$

and the initial limitation on the pressure will be:

$$\sum_\alpha F_\alpha = \sum_\alpha f_\alpha y_\alpha = F = \text{constant}.$$

Let us consider the environmental resources that chiefly limit recruitment of the α population and serve as the source of competition between α individuals. In the case of free trade, the food-spawning areas, and so on, comprise such limiting resources. Let us assume the dependence of recruitment, $L_\alpha = N\lambda_\alpha$, on the quantity of resources, R_α. If, in first approximation, this dependence is linear, $L_\alpha = (N/r_\alpha) R_\alpha$, where

$$r_\alpha = \frac{N R_\alpha}{L_\alpha} \tag{6.162}$$

is a certain ecological constant, then limitations on the total resources, $\sum_\alpha R_\alpha = R = \text{constant}$, will be expressed as $\sum_\alpha r_\alpha \lambda_\alpha = R$.

In the MEM model the pressure of the controlling system on the community is studied in terms of how seasonal conditions influence the efficiency of harvesting equipment in their contact with α individuals. This efficiency is specified by the value, $v_{\alpha\sigma}$, of the mean quantity of α individuals withdrawn per one time unit by one unit of equipment. Therefore, the mathematical expectation of seasonal withdrawals has a double concept: $Y_{\alpha\sigma} = N p_\alpha y_{\alpha\sigma} = F_\alpha v_{\alpha\sigma} T_{\alpha\sigma}$. Hence, $T_{\alpha\sigma} = t_{\alpha\sigma} y_{\alpha\sigma}$, where

$$t_{\alpha\sigma} = \frac{N p_\alpha}{F_\alpha v_{\alpha\sigma}} \tag{6.163}$$

and, inasmuch as the optimal values of the parameters $p_\alpha = p_\alpha^0$ and $F_\alpha = F_\alpha^0$ are calculated in the MAM model, the value $t_{\alpha\sigma}$ is constant here. In this way, $T_\alpha = \sum_\alpha T_{\alpha\sigma} = \sum_\alpha t_{\alpha\sigma} y_{\alpha\sigma}$.

Unlike the MAM model, in the MEM model we examine the dependence of recruitment of the σ group in the α population, not on the external environmental conditions but on the internal resources of the producers (for instance, the power resources related to reproduction and protection of progeny). By using concepts of general physiology (for instance, see Pianka, 1966), we can present the dependence of the expectation of recruitment, $L_{\alpha\sigma} = n\lambda_{\alpha\sigma}$, of the σ group on the reproductive resource, $R_{\alpha\sigma}$, as a convex parabola, $L_{\alpha\sigma} = R_{\alpha\sigma}^{1/s}/g_{\alpha\sigma}^{1/s}$, where

$$g_{\alpha\sigma} = \frac{R_{\alpha\sigma}}{L_{\alpha\sigma}^s} \tag{6.164}$$

and $S > 1$ are some physiological constants. It is the resource $R_\alpha = \sum_\sigma R_{\alpha\sigma}$ common for all σ groups of α populations, divided into the constant n^s. This resource looks as follows:

$$\sum_\sigma g_{\alpha\sigma}\lambda^s_{\alpha s} = \frac{R_\alpha}{n^s} = G_\alpha.$$

Let us now examine ecologically the instantaneous probability, $\mu(t)$, of a juvenile individual's death in the MIM model. This value is of primary importance in formulating the maximum permissible conditions for the juvenile cohort's survival. The juvenile stage in an animal's life is peculiar because of mortality due to various reasons. The survival probability, $1 - \mu(t)$, increases with time, t, catch volume, $\Delta V(t)$, of the distribution range, $V(t)$, and the food ration, $\mathcal{R}(t)$. By a cohort we mean a nonsupplemented generation of individuals of the same age that, generally speaking, is aggregated into m schools in which the value of m is variable as a function of the conditions.

Divide the juvenile cohort distribution range volume of $V(t)$ at the instant t into $A(t) = V(t)/\Delta V$ elements, each of volume ΔV. These elements may contain (or not) food particles and individuals with the probabilities $p_f(t)(1 - p_f(t) = q_f(t))$ and $p(t)(1 - p(t) = q(t))$ independent of one another. The food and individuals' densities are $\rho_f(t) = p_f(t)/\Delta V$ and $\rho(t) = p(t)/\Delta V$, respectively.

When eating, individuals and predators consume the volumes

$$\Delta V(t) \le \Delta V_{cr}(t) \quad \text{and} \quad \Delta V_p(t) \le \Delta V_{p\,cr}(t), \tag{6.165}$$

which do not exceed critical values to which quantities of captured elements correspond:

$$n(t) = \frac{\Delta V(t)}{\Delta V} \le n_{cr}(t) = \frac{\Delta V_{cr}(t)}{\Delta V}$$

$$n_p(t) = \frac{\Delta V_p(t)}{\Delta V} \le n_{p\,cr}(t) = \frac{\Delta V_{p\,cr}(t)}{\Delta V}. \tag{6.166}$$

The food particles and individuals are taken in this process in the quantities

$$m_f \le m_f(t) = \mathcal{R}(t)\Delta Q_f \quad \text{and} \quad m' < m = \frac{\mathcal{R}_p}{\Delta Q}, \tag{6.167}$$

which do not exceed the rations $m_f(t)$ and m corresponding to the minimum calorie rations, $\mathcal{R}(t)$ and $\mathcal{R}_p$, where ΔQ_f and ΔQ represent the caloric content of one food particle and individual, respectively. When juvenile individuals receive no rations, they die. Unlike them, predators, generally speaking, do not die but supplement their ration elsewhere.

In connection with the above, individuals' death from malnutrition may occur if:

1. The captured elements, $n_{cr}(t)$, contain less than $m_f(t)$ of food particles (Bernoulli scheme).
2. To obtain the ration, $m_f(t)$, it is necessary to catch a quantity of elements exceeding the value $n_{cr}(t)$ (Pascal scheme).

The instantaneous probability, $x_f(t)$, of an individual's death in these two cases has one and the same value

$$x_f(t) = \sum_{m=0}^{m_f(t)-1} B(m) \equiv \sum_{n=n_{cr}(t)}^{\infty} P(n), \tag{6.168}$$

where

$$B(m) = C^m_{n_{cr}(t)} p_f^m(t) q_f(t)^{n_{cr}(t)-m} \quad \text{and} \quad P(n) = C^{m_f(t)-1}_{n-1} p_f(t)^{m_f(t)} q_f(t)^{n-m_f(t)},$$

whereupon individuals withdraw $\eta_f(t)$ food particles and do not get overfed [$\eta_f(t) \le m_f(t)$]. Therefore, the expected value of $\eta_f(t)$ appears as

$$E\eta_f(t) = \sum_{m=0}^{m_f(t)-1} mB(m) + m_f(t)[1 - x_f(t)]. \quad (6.169)$$

In particular, with $m_f(t) = 1$, from eqs. (6.168) and (6.169) we have

$$x_f(t) = p_f(t) \quad \text{and} \quad E\eta_f(t) = 1 - x_f(t) = 1 - p_f(t) = q_f(t). \quad (6.170)$$

In the asymptotic case, $n_{cr}(t) \to \infty$, when $m_f(t) = \theta(t)n_{cr}(t) \to \infty$ ($0 \le \theta(t) \le \theta < 1$) with the conditions (Fleishman, 1971, 1982a)

$$\theta(t) < p_f(t), \quad (6.171)$$

we have [see relations (6.38) and (6.39)]

$$x_f(t) \sim \exp[-n_{cr}(t) k(\theta(t), p_f(t))], \quad (6.172)$$

where $k(\theta, p) = \theta \ln(\theta/p) + (1 - \theta) \ln[(1 - \theta)/(1 - p)]$, and, with $\theta \le p$, $k(\theta, p) \simeq (\theta - p)^2/2p(1 - p)$. If $p_f(t) = p_f =$ constant, and

$$\theta(t) \le \theta < p_f, \quad (6.173)$$

we obtain the upper estimate

$$x_f(t) \le \exp[-n_{cr}(t)k(\theta, p_f)], \quad (6.174)$$

showing $x_f(t)$ to be exponentially bounded and tending to zero with increasing $n_{cr}(t)$.

At the instant t, the cohort has the distribution range volume of $V(t)$ and consists of $v(t)$ m schools, which we identify with α individuals or simply with individuals. Assume the cohort is grazing in a field, $v_f(t)$, of aggregated food particles, which we shall simply call food particles. The cohort's range is visited by $v_p(t)$ predators. Assume that within the discrete time interval from t to $t + 1$ there are noncrossing subintervals of the cohort's active nourishment and of predators' visits to its range. Calculating the utilization of individuals and food particles, we obtain a system of finite-difference equations:

$$v(t + 1) = v(t) - \sum_{i=1}^{v_p(t)} \eta_i(t) - \sum_{j=1}^{v(t)} \epsilon_j(t), \quad v_f(t + 1) = v_f(t) - \sum_{k=1}^{v(t)} \eta_k(t). \quad (6.175)$$

Here, $\eta_i(t)$ and $v_p(t)$ and $\eta_k(t)$ and $v(t)$ are random and independent values of individuals and food particles, respectively, equally distributed with $i = \overline{1, v_p(t)}$ and $k = \overline{1, v(t)}$, and $\epsilon_j(t) = \begin{pmatrix} 0 & 1 \\ 1-x_f(t) & x_f(t) \end{pmatrix}$ and $v(t)$ are independent random values.

Let us examine the mathematical expectations:

$$Ev(t) = A(t)p(t),$$
$$Ev_f(t) = A(t)p_f(t), \quad (6.176)$$
$$E\eta(t) = n_p(t)p(t),$$

where $p(t)$ and $p_f(t)$ are the respective probabilities. Let us also find the instantaneous probability, $x_f(t)$, of an individual's death from a certain mean number of predators, $Ev_f(t)$:

$$x_f(t) \triangleq \frac{Ev_f(t)E\eta(t)}{Ev(t)} \equiv \frac{Ev_f(t)n_f(t)}{A(t)}. \qquad (6.177)$$

It should be noted that the predator also has a certain critical ration of m individuals with the critical number of the captured elements, n_{cr}. Therefore, by using the identical concepts of (6.177), we can express the probabilities $x_f(t)$ as:

$$x_f(t) = \begin{cases} \dfrac{Ev_f(t)m}{Ev(t)} & t \leq t'_{cr} \\ \dfrac{Ev_f(t)n_{f\,cr}}{A(t)} & t > t'_{cr} \end{cases} \text{with} \qquad (6.178)$$

where t'_{cr} is the critical time until which, due to the high concentration of individuals, the predator will get its ration m by catching the number of elements $n_f(t) \leq n_{fcr}$, which will be increasing with time $t < t'_{cr}$. Then, getting at $t = t'_{cr}$ the value $n_p(t'_{cr}) = n_{pcr}$ and exceeding it, the predator will catch a certain stable number of elements, n_{pcr}, but will more and more undereat its ration, m, and get fewer and fewer individuals, $m' < m$, due to their decreasing concentration. Relations (6.177) show that with a constant mean number of predators, $Ev_p(t) = Ev_p = $ constant, due to decrease in the value $Ev(t)$ (because of increasing t) the value $x_p(t)$ grows until the moment t'_{cr}; then, with increase in t, this value decreases. This is related to the augmentation of $A(t)$. Such behavior of $x_p(t)$ is due to the ever-increasing decrease of individuals that, with increasing t, more and more dominates over their removal by predators.

Taking the mathematical expectation from both parts of each relation composing (6.175) and employing the lemma for the expected value of a random number of random summands, as well as the values from (6.176) and (6.177), we shall assume that the range $(A(t+1) \sim A(t))$ is growing slowly with increasing time, t. Then, by dividing this value into the value $A(t) > 0$, we get from eq. (6.175) a finite-difference relation for the respective probabilities with the condition $x_p + x_f \leq 1$:

$$\begin{aligned} p(t+1) &= p(t) - [x_p(t) + x_f(t)]p(t) \\ p_f(t+1) &= p_f(t) - p(t)E\eta_f \end{aligned} \qquad (6.179)$$

In the absence of predators, when $x_p(t) = 0$ and with the minimum critical ration $m_f(t) = 1$, by using relations (6.170) we obtain a particular case of system (6.179),

$$\begin{aligned} p(t+1) &= p(t) - p_f(t)p(t) \\ p_f(t+1) &= p_f(t) - p(t)[1 - p_f(t)] \end{aligned} \qquad (6.180)$$

which has an explicit analytical solution.

In general, system (6.180) has no analytical solution. It is natural to suggest that the food concentration is practically constant, that is, when $p(t) = 0$ $(p_f(t))$ relations (6.179) will lead to $p_f(t) = p_f = $ constant. In this case increase in $x_f(t)$ is related only to growth of the critical ration, $m_f(t)$. Then, we finally have from eqs. (6.179):

$$p(t+1) = p(1) \prod_{\tau=1}^{t} [1 - x_p(\tau) - x_f(\tau)]. \qquad (6.181)$$

Expression (6.181) shows the probability of an individual's survival until the moment $t + 1$. It is clear that this does not increase with increased t.

A basic question is that of the limiting value of $p(\infty) = \lim_{t \to \infty} p(t)$ or, more exactly, of the conditions under which $p(\infty) > 0$. It is obvious that to reach a limit it is necessary for $x_p(t)$ and $x_f(t)$ to tend to zero with increasing t. However, the adequacy of the conditions depends on the order of the tendency of these variables to approach zero [see eqs. (6.31) and (6.32)]. Here we examine the case of time-sequential impact on the individual from predators and from malnutrition. If both of these factors acted simultaneously, then the probability, $\mu(t)$, of an individual's death due to both factors would look as follows:

$$\mu(t) = \mu[x_p(t), x_f(t)], \quad (6.182)$$

where, according to relation (6.97), $\mu(x, y) = (x + y - 2xy)/(1 - xy)$. But, inasmuch as in this case $x_p(t)$ and $x_f(t)$ tend to zero with increase of t, both cases asymptotically coincide. Finally, the order of tendency to zero in the value $\mu(t)$ is determined by the order of the maximum:

$$\mu(t) \sim \max\,[x_p(t), x_f(t)]. \quad (6.183)$$

It should be emphasized again that the theory developed here concerns only those communities that are in the state of climax, experiencing controlling impacts lasting for many years. Therefore, when speaking about plant communities, this theory may be related only to perennial agricultural crops or to weeds. The theory does not consider the most productive agricultural crops harvested in successional states.

Identification of Theoretical Parameters by Empirical Data

The values encountered in community-controlling models can be divided into explicit and latent. Among the latter we consider the expected values of resource quantities, N_α, $N_{\alpha\sigma}$, and m; the explicit values include expectations of withdrawals, Y_α, $Y_{\alpha\sigma}$, and Y_i. The last three values are practically not estimated in our calculations at all. It is common knowledge that simplified statistical interpretation of explicit values as samplings from latent ones (general populations) is not acceptable (see Ricker, 1975). It is impossible to estimate latent values and find their links with explicit values if we do not use any prior theoretical hypothesis. The problem of identifying certain theoretical values by empirical data means to identify (estimate) them by empirical analogues. For latent values, this is possible only in those rare cases when we organize a continuous experiment to determine the biological parameters of the community.

Let us begin by identifying the parameter in the MAM model. There are obviously such estimates as

$$P_\alpha \simeq P_\alpha^* = \begin{cases} S_\alpha/S, & \text{for sedentary } \alpha \text{ individuals} \\ \rho_\alpha \Delta S, & \text{for motile } \alpha \text{ individuals,} \end{cases} \quad (6.184)$$

where S and S_α are the range areas occupied by the community and of the α population, respectively; ρ_α and ΔS are the density of α individuals and its mean specific "life space." The latter can be identified by a continuous account of some characteristic section of the area S_α, where α individuals in quantity n_α ($\Delta S = \sum_\alpha \rho_\alpha S_\alpha/n_\alpha$) are found. Then we estimate the values

$$N_\alpha \sim N_\alpha^* = \frac{S_\alpha}{\Delta S}\,(\alpha = \overline{0, a}) \quad \text{and} \quad N \sim N^* = \sum_{\alpha=0}^{a} N_\alpha^*,$$

where S_0 is the "free area" ($S_0 = N_0^* \Delta S = N^* \Delta S p_0^*$). Using the mean life span, $\overline{\tau}_\alpha$, of the α individual in years, we assess the probability

$$x_\alpha \sim x_\alpha^* = \frac{1}{\overline{\tau}_\alpha}. \quad (6.185)$$

To identify the birth probability, λ_α, of the α individual is difficult due to the factors mentioned above. This probability can be obtained by averaging in terms of σ groups the lower estimates of the mesoparameters, $\lambda_{\alpha\sigma}$, in accordance with expression (6.160). However, the right side of this expression, in its turn, requires experimental assessment of the parameters t_1, $P_R(t_1)$, and $\mu(t_1)$ in terms of the respective empirical curves of the juvenile cohorts' survival (see Pianka, 1966). Such lower estimates should be accompanied with more formal and approximate estimates, $\lambda_\alpha \sim \lambda_\alpha^* = L_\alpha^*/N_0^*$, where L_α^* shows the quantity of young α individuals.

Finally, the parameter y_α may be evaluated as $y_\alpha \sim y_\alpha^* = Y_\alpha^*/N_\alpha^*$, where Y_α^* is the actual annual harvest of α individuals. The accuracy of the estimate so obtained is verified by formula (6.99). Similar estimates are made for the mesoparameters $\lambda_{\alpha\sigma}$, $X_{\alpha\sigma}$, and $y_{\alpha\sigma}$. Further identification of the theoretical parameters and their values in the MAM model is carried out by using relations (6.111).

The important constants d_3 and d_4 depend on the parameters f_α and r_α, which are hard to identify theoretically [see relations (6.161) and (6.162)]. Therefore, it is essential to specify the constants d_3 and d_4 directly through the experimentally defined parameters λ_α^*, x_α^*, and y_α^*. To this end, let us consider two values of the pressure, F and F' ($F < F'$). With the help of relations (6.111) we obtain

$$\frac{y_\alpha' + x_\alpha}{y_\alpha + x_\alpha} = \frac{F' + d_3}{F + d_3}. \tag{6.186}$$

Hence,

$$d_3 = (F' - F)\left[(y_\alpha' + x_\alpha)/(y_\alpha + x_\alpha) - 1\right]^{-1} - F$$

and

$$d_4 = \frac{(d_3 + F)^2}{d_3 R} \sum_\alpha \frac{\lambda_\alpha x_\alpha}{(x_\alpha + y_\alpha)^2}.$$

With the hypothesis of intense pressure, when $y_\alpha \gg x_\alpha$ and $F_\alpha \gg d_3$, from relation (6.186) we obtain an important equation that does not depend on x_α and d_3:

$$y_\alpha' \sim \frac{F'}{F} y_\alpha. \tag{6.187}$$

From (6.111), with changing environmental conditions, we come to R and R' through

$$\lambda_\alpha' \sim \frac{R'}{R} \lambda_\alpha. \tag{6.188}$$

Specifying, as described above, the empirical values $y_\alpha \sim y_\alpha^*$ and $\lambda_\alpha \sim \lambda_\alpha^*$ with the values F and R, by using formulas (6.187) and (6.188) we can predict the values y_α' and λ_α' for the intense pressure case, and by using relations (6.99) we can find the value p_α'. This enables us to predict the values B' and C' with the help of formulas (6.101) and (6.106), as well as to correlate them with the theoretical values estimated through formula (6.110) and assessments (6.120).
In particular, from relations (6.184) and (6.187), we have

$$\frac{S_\alpha}{S_0} \frac{S_\alpha'}{S_0'} \sim \frac{y_\alpha'}{y_\alpha} \sim \frac{F'}{F}. \tag{6.189}$$

Let us now examine a special experimental method for "biomodeling" of commercial harvesting under conditions of controlled and uncontrolled ecological experimentation. In the uncontrolled part of the community, the expected value of prey numbers, N_α, and of predators, $M_\beta = Mq_\beta$, is equal to Mq_β, composing the q_β part of their total number, $\left(M\sum_\beta q_\beta = 1\right)$. Then the mathematical expectation for the total catch of predators when they pursue α individuals is $Y_\alpha = \sum_\beta q_\beta \mathcal{R}_{\alpha\beta} = M\mathcal{R}_\alpha$, where $\mathcal{R}_{\alpha\beta}$ and $\mathcal{R}_\alpha$ are the expected values of the respective quantities of α individuals eaten per year (annual ration) by each predator separately and all of them together. It is possible to consider formally that there is one "summary" predator whose expected numbers are M and whose ration is $\mathcal{R}_\alpha$. Under natural conditions this predator seeks the α individuals employing some part of its population, whose expected numbers, M_α, are proportional to those of the α individuals, $M_\alpha = rr_\alpha^{-1}N_\alpha$, where r and r_α are constants, the first of which determines the predator's activity. In this case, $N_\alpha y_\alpha = Y_\alpha = M_\alpha \mathcal{R}_\alpha = rN_\alpha r_\alpha^{-1}\mathcal{R}_\alpha$. Hence,

$$r_\alpha y_\alpha = r\mathcal{R}_\alpha = F_\alpha$$

and

$$\sum_\alpha r_\alpha y_\alpha = r\sum_\alpha \mathcal{R}_\alpha = r\mathcal{R} = F, \qquad (6.190)$$

where the annual ration, $\mathcal{R}_\alpha$, and the total ration, $\mathcal{R}$, may be interpreted as values proportional to the trade conditions, F and F_α. Therefore, this situation may be treated in terms of the previously developed theory of commercial exploitation.

Now, let us fix in the controlled part of the community the quantities $N_\alpha^* = $ constant of the prey, and $M^* = $ constant of the predator with its two kinds of activity, lower and higher levels of its trade efforts, $F' = c'\mathcal{R}' > F = c\mathcal{R}$. This situation may be characterized by a pair of values, c and $\mathcal{R}_\alpha$, of which the first determines numbers of the predator and the second specifies its ration. Let $c' > c$ and $\mathcal{R}'_\alpha > \mathcal{R}_\alpha$. Then, by using relation (6.189) we can obtain the basic equation permitting us to predict parameters of the uncontrolled portion of the community by those of its controlled part in the case of intense exploitation, $F'_\alpha/F_\alpha \sim F'/F$:

$$\gamma_\alpha = \frac{S_\alpha}{S_0}\frac{S'_\alpha}{S'_0} \sim \frac{y'_\alpha}{y_\alpha} \sim \frac{F'_\alpha}{F_\alpha} \sim \frac{F'}{F} = \frac{c'}{c}\cdot\frac{\mathcal{R}'_\alpha}{\mathcal{R}_\alpha}. \qquad (6.191)$$

To illustrate a "biomodel" of trade, consider Menge (1975), who studied competition in a New England rocky intertidal community between a mussel, *Mytilus edulis* ($\alpha = 1$), a barnacle, *Balanus balanoides* ($\alpha = 2$), and involving a gastropod predator, *Thais lapillus*. Under field experimental conditions, $M^* = 5$ predators were introduced into controlled sections of the community containing $N_1^* = 80$ *Mytilus edulis*. There were two sets of tests: s_1 included nine tests a month and a half long; s_2 consisted of six tests of one month each. The predators consumed prey under different environmental conditions that affected their activity, which decreased as a function of desiccation and the animals' resistance to wave action.

Table 6.1 illustrates results expressed on a monthly basis. The total quantities of *M. edulis* eaten, Y_1 and Y'_1, correspond to the respective probabilities $\bar{y}_1 = Y_1/sN_1^*$ and $\bar{y}'_1 = Y'_1/sN_1^*$. Table 6.1 shows that values of the relation $(\mathcal{R}'_1/\mathcal{R}) \le 2$ were observed as a function of the predator's activity and different wave strengths. Means cannot be employed here since the number of observed sections under different conditions is not known. Inasmuch as *Mytilus edulis* lives for ten years and *Balanus balanoides* only five years, we can correlate the variable x_α with the data in Table 6.1 by using the values y_α, taking the time unit equal to one month. From equation (6.185), $x_1 \sim x_1^* = 0.008$ and $x_2 = x_2^* = 0.016$, and we can see that $\bar{y}_\alpha \gg x_\alpha$; in other words, the prevailing conditions are those of intensive trade.

Let us now consider the data presented by Menge (op. cit.) for the uncontrolled experimental part of the community. The author examined vacant sections, S_0 and S'_0, and those occupied by

TABLE 6.1 Quantities Y_1 and Y'_1 of *Mytilus edulis* consumed per square meter per month by *Thais*, and associated predation probabilities y_1 and y'_1

		$s = s_1 = 9$		$s = s_2 = 6$		
Desiccation ↓ Wave strength →		Strong	Weak	Strong	Weak	
High	Y_1	24	96	75	128	$\tilde{y}_1$
		0.05	0.2	0.15	0.26	
Low	Y'_1	44	106	43	183	
		0.09	0.2	0.08	0.37	
$\tilde{y}'_1/\tilde{y}_1 = Y'_1/Y_1 = \mathcal{R}'_1/\mathcal{R}_1$		2	1	0.7	1.5	$\tilde{y}'_1$

Data from Menge, 1975.

Mytilus edulis, S_1 and S'_1, and *Balanus balanoides*, S_2 and S'_2, and recorded densities, ρ and ρ', (individuals/m^2) of the predator *Thais* on unprotected and protected parts of the community. Results are summarized in Table 6.2. From the data in Tables 6.1 and 6.2, we can conclude that relation (6.191) in the case of *Mytilus edulis* considered is quite satisfactory since $\gamma_1 = 9.8 \geq (c'/c)(\mathcal{R}'_1/\mathcal{R}_1) = 5(\mathcal{R}'_1/\mathcal{R}_1)$.

The identification of parameters for deterministic models of exploited populations is well described in Ricker (1975). No similar identification for the MEM model has been made so far.

Let us now consider the identification of parameters in the MIM model. Numerous examples of hyperbolic diminution of instantaneous death and survival probabilities to time t have been summarized by Pianka (1966). The present theory not only corresponds to reality, but it also gives the critical value (equal to unity) of the degree of the hyperbola representing the instantaneous probability, $\mu(t)$, of an individual's death. Unfortunately, most empirical data do not consider this value, but that of the probability of an individual's survival until the given instant $P_R(t)$. Therefore, it is important to have indirect ways to estimate the probability $\mu(t)$ from other experimental data. From relations (6.166) and (6.167) and condition (6.173), we have

$$n_{cr}(t) = \frac{\Delta V_{cr}(t)}{\Delta V} > \frac{m(t)}{p_f} = \frac{\mathcal{R}(t)}{p_f \Delta Q_f} \qquad (6.192)$$

and

$$\theta(t) = \frac{m(t)}{n_{cr}(t)} = \frac{\mathcal{R}(t)}{\Delta V_{cr}(t)} \cdot \frac{\Delta V}{\Delta Q_f} < \theta < p_f. \qquad (6.193)$$

Using von Bertalanffy's formula for juvenile individuals (small t), we obtain the following value for one individual's weight:

$$Q(t) = Q_\infty [1 - \exp(-kt)]^3 \sim Q_\infty k^3 t^3, \qquad (6.194)$$

TABLE 6.2 Percent of predators on vacant (S_0 and S'_0) and occupied (*Mytilus*, S_1 and S'_1; *Balanus*, S_2 and S'_2) plots in unprotected (passive predator) and protected (active predator) parts of the community

Predator	Section, α	Vacant ($\alpha = 0$)	*Mytilus* ($\alpha = 1$)	*Balanus* ($\alpha = 2$)	Density of predators per m^2	
Passive	$S_\alpha(\%)$	5	85	6	53	ρ
active	$S'_\alpha(\%)$	15	26	2	251	ρ'
$\gamma_\alpha = (S_\alpha/S_0)/(S'_\alpha/S'_0)$			9.8	9.0	$\rho'/\rho = c'/c = 5$	

where k is a constant and Q_∞ is the ultimate weight of the individual. Assuming the individual's ration $\mathcal{R}(t)$ is proportional to its weight, we have

$$\mathcal{R}(t) = c_1 Q(t) \approx c_1 Q_\infty k^3 t^3. \tag{6.195}$$

The critical prey quantity, $\Delta V_{cr}(t)$, captured by the individual will then be

$$\Delta V_{cr}(t) = \pi r_m^2(t)\, w(t)\, t_{cr}, \tag{6.196}$$

where the parameters $r_m = r_m(t)$ show definite (reliable) discovery of the food object by the m school and $w = w(t)$ is the speed of the m school. These values were specified by Fleishman (1980), but in this case they depend on the time t.

Substituting relations (9.195) and (9.196) into (9.174), we obtain

$$x_f(t) \leq \exp\left[\frac{-\pi r_m^2(t) w(t) t_{cr} k(\theta, p_f)}{\Delta V}\right] \leq \exp(-c_2 t^3), \tag{6.197}$$

provided that

$$r_m^2(t) w(t) > c t^3, \tag{6.198}$$

where $c = c_1 Q_\infty k^3 \Delta V / \theta \pi t_{cr} \Delta Q_f$ and $c_2 = c \pi t_{cr} k(\theta, p_f)/\Delta V$.

Let us further estimate $x_p(t)$ for the values $t > t'_{cr}$. From relation (9.178), with the fixed expected values of predators' numbers, $Ev_p(t) = Ev_p = $ constant, by using expression (6.166) we have

$$x_p(t) = \frac{Ev_p\, \Delta V_{p\,cr}}{A(t)\, \Delta V}. \tag{6.199}$$

In the segment $1 \leq t < t'_{cr}$ the value $x_p(t)$ increases inversely proportional to the decreasing value of the expectation of the number, $Ev(t)$, of the cohort. Therefore, in the segment in question, $x_p(t) \gg x_f(t)$, and it is worthwhile to compare values beginning with the moment $t_1 = t'_{cr}$. That is why, finally, conditions (6.114), (6.162), and (6.183) will look as follows. If

$$x(t) = \max_{t > t'_{cr}} (x_p(t), x_f(t)) \tag{6.200}$$

$$= \max_{* \, t > t'_{cr}} \left\{\frac{Ev_p\, \Delta V_{p\,cr}}{A(t)\, \Delta V}, \exp(-c_2 t^3)\right\} \leq x_p(t'_{cr})\left(\frac{t'_{cr}}{t}\right)^{1+\eta},$$

then [see relation (6.160)]

$$N(\infty) \geq \frac{N(t'_{cr}) - N(1) x(t'_{cr}) t'_{cr}}{\eta}. \tag{6.201}$$

The choice of $x_p(t)$ or of $x_f(t)$ as the limiting one in relation (6.200) depends on the order of the exponential growth of the number of elements in the range $A(t)$. The relations $x_p(t) \gtrless x_f(t)$ take place as a function of relations

$$A(t) \lessgtr c_3 \exp(c_3 t^3), \tag{6.202}$$

where $c_3 = Ev_p \Delta V_{p\,cr}/\Delta V$. In the last case, it is necessary to require that $A(t) > C_4 t^{1+\eta}$, where $C_4 = Ev_p \Delta V_{p\,cr}/x(t'_{cr})$.

REFERENCES

ACOFF, R. L., and EMERY, F. E. 1972. *On Purposeful Systems.* Aldine Atherton, Chicago and New York. 280 pp.

ASHBY, W. R. 1958. *An Introduction to Cybernetics.* Wiley, New York. 320 pp.

BAKULE, L., and STRAŠKRABA, M. 1984. On optimality in multispecies ecosystems. *Ecol. Mod.* 26:33–39.

BAKMAN, G. 1943. *Wachstum und Organische. Zeit.* Leipzig. 195 pp.

BARANOV, F. I. 1918. To the problem of biological fundamentals of fisheries. *Izv. otdela rybolovstva i nauchn. prom. issledovaniy.* 1/2. Petrograd, pp. 84–128 (in Russian).

BELLMAN, R. 1957. *Dynamic Programming.* Princeton University Press, Princeton, New Jersey. 200 pp.

BEVERTON, R. J. H., and HOLT, S. T. 1957. *Dynamics of Exploited Fish Populations.* Fish Invest. 2, Ser. 11, No. 19. 533 pp.

EINSTEIN, A. 1936. Physik und realität. *Franklin Inst. Journal* 221:313–347.

FLEISHMAN, B. S. 1964. Statistical sequential analysis and automatic self-control. In *Uchenye zapiski po statistike,* v. 8. Nauka, Moscow, pp. 51–77 (in Russian).

FLEISHMAN, B. S. 1966a. Statistical limits of effectivity of complex systems. In *Prikladnye zadachi tehnicheskoi kibernetiki.* Sovetskoe Radio Moscow, pp. 13–42 (in Russian).

FLEISHMAN, B. S. 1966b. On survivability of complex systems. *Izv. Akad. Nauk SSSR. Tech. Kibern.* 5:28–38 (in Russian).

FLEISHMAN, B. S. 1970. The state of theory of potential efficiency of complex systems. *Cybernetica* 133(4):199–211.

FLEISHMAN, B. S. 1971. *Elements of the Theory of Potential Efficiency of Complex Systems.* Sovetskoe Radio, Moscow. 224 pp. (in Russian).

FLEISHMAN, B. S. 1974. On the optimization of the model "Swarm and aggregated food." *Tez dokl. Vsesoiuzn. seminara "Mat. modelirovanie morskikh ekosistem."* Naukova dumka, Kiev. pp. 42–44 (in Russian).

FLEISHMAN, B. S. 1977a. Control over the biosystems and their objective teleology. *IFAC—Symposium on Control Mechanisms in Bio- and Ecosystems,* September 11–16, 1977. *Biocommunication, Ecosystems* 5:76–80.

FLEISHMAN, B. S. 1977b. Stochastic models of community. In *Oceanology. Biology of the Oceans,* Vol. 2. *Biological Productivity of the Oceans.* Nauka, Moscow, pp. 76–88 (in Russian).

FLEISHMAN, B. S. 1980. Information theory of collective behaviour applied to fish schools. *Int. Soc. Ecol. Mod. J.* 2:34–41.

FLEISHMAN, B. S. 1982a. *Fundamentals of Systemology.* Radio i svyaz'. 368 pp. (in Russian).

FLEISHMAN, B. S. 1982b. Stochastic theory of ecological interactions. *Ecol. Mod.* 17:64–73.

FLEISHMAN, B. S. 1982c. Prediction of community state on the basis of its adaptation and succession cycle. *Acta Hydrophysica,* Berlin, Bd. 2M, H. 314, pp. 167–185 (in Russian).

FLEISHMAN, B. S. 1983. On adaptation theory of regenerative systems. *Fourth Formator Symposium on Mathematical Methods for the Analysis of Large-Scale Systems,* May 18–21, 1982. Publ. Hause, Czechoslovak Acad. Sci., Prague, pp. 207–222.

FLEISHMAN, B. S. 1984. Contribution to the theory of adaptation with application to ecology. *Ecol. Mod.* 26:21–31.

FLEISHMAN, B. S. 1987. Stochastic theory of community control. *Ecol. Mod.* 39:121–159.

FLEISHMAN, B. S. 1991. Hyperbolic law of reliability and its logarithmic effects in ecology. *Ecol. Mod.* 55:75–88.

FOGEL, L. I., OWENS, A. J., and WALSH, M. J. 1966. *Artificial Intelligence through Simulated Evolution.* Wiley, New York. 164 pp.

GILBERT, E. N. 1959. Random graphs. *Ann. Math. Stat.* 30(4):1141–1144.

GURMAN, V. N. (ed.). 1981. *Models of Environmental Resources Control.* Nauka, Moscow. 264 pp. (in Russian).
HAKEN, H. 1983. *Advanced Synergetics. Instability Hierarchies of Self-organizing Systems and Devices.* Springer-Verlag, New York. 282 pp.
HINDE, R. A. 1970. *Animal Behaviour. A Synthesis of Ethology and Comparative Psychology,* 2nd ed. McGraw-Hill, New York. 590 pp.
HOLLING, C. S. 1978. *Adaptive Environmental Assessment and Management.* Wiley, New York. 377 pp.
IVAKHNENKO, A. G. 1971. *Heuristic Self-organizing Systems in Technical Cybernetics.* Technika, Kiev. 372 pp. (in Russian).
MAUERSBERGER, P. 1982. Rates of primary production, respiration and grazing in accordance with the balances of energy and entropy. *Ecol. Mod.* 17:1–10.
MENGE, B. A. 1975. Ecological implications of patterns of rocky intertidal community structure and behavior along an environmental gradient. In *The Ecology of Fouling Communities.* Duke University Marine Laboratory, Beaufort, North Carolina. pp. 155–180.
PELLA, J., and TOMLINSON, P. 1969. A generalized stock production model. *Amer. Topical Tuna Comm. Bull.* 3:421–496.
PIANKA, E. R. 1966. *Evolutionary Ecology.* Harper & Row, New York. 356 pp.
PRIGOGINE, I. 1980. *From Being to Becoming. Time and Complexity in the Physical Sciences.* W. H. Freeman, New York. 218 pp.
RASHEVSKY, N. 1960. *Mathematical Biophysics,* Vol. 1. Dover, New York. 488 pp.
RICKER, W. E. 1975. Computation and Interpretation of Biological Statistics of Fish Populations. Bulletin 191, Department of Environment, Fisheries and Marine Service, Ottawa. 266 pp.
SEMEVSKIY, F. N., and SEMENOV, S. M. 1982. *Mathematical Modelling of Ecological Processes.* Gidrometeoizdat, Leningrad. 280 pp. (in Russian).
SHANNON, C. 1948. A Mathematical Theory of Communication. *Bell. System Tech. J.* 623–655.
STRAŠKRABA, M. and GNAUCK, A. 1983. *Aquatische Ökosysteme. Modellierung und Simulation.* VEB Gustav Fischer Verlag, Jena. 279 pp.
TINKLE, D. W. 1967. The Life and Demography of the Side-blotched Lizard *Uta strasburiana.* Misc. Publ. Mus. Zool. Mich., No. 132. 182 pp.
TSETLIN, M. L. 1968. *Investigations on Theory of Automata and Modelling in Biological Sciences.* Nauka, Moscow. 316 pp. (in Russian).
VON FORSTER, H. 1962. Bio-logik. In Bernard, E. E. and Kabe, M. R. (eds.), *Biological Prototypes and Synthetic Systems.* Plenum Press, New York, pp. 7–23.
VON NEUMANN, J. D., and MORGENSTERN, O. 1944. *Theory of Games and Economic Behavior.* Princeton University Press, Princeton, New Jersey. 625 pp.
VON NEUMANN, J. D. 1966. *Theory of Self-reproducing Automata.* University of Illinois Press, Urbana. 382 pp.
WIENER, N. 1948. *Cybernetics. Control and Communication in the Animal and the Machine.* Wiley, New York; Herman et Cie, Paris. 192 pp.

Part 2

MACROSCOPIC COMPLEXITY: OUTWARD–UPWARD TOWARD THE ECOSYSTEM

Science is not at its best in the consideration of wholes. "Divide and analyze" is its dictum, and the holistic aspects of its phenomena are frequently only weakly inferrable from a knowledge of parts. Surprises always occur, for example, as "side effects" or "emergent properties." George Van Dyne well understood Sven Jørgensen's uncertainty principle (Chapter 1); he wrote "synthesis" objectives and subprojects into his research proposals to serve as goals beyond the mechanism–reductionism of normal science. Pragmatist though he was in grantsmanship and organizing the research of himself and others, he would not abandon his sense of the complete system or the effort to know wholes just because, like Swiss cheese, most of his commandable knowledge about them was holes. Van Dyne's life's work is a testament to an article of faith that meaningful knowledge of ecological complexity does not depend on knowing everything. With uncommon clairvoyance and a stoic tolerance for frustration and setback, he looked through the newly developing macroscope of his time for his world view, using computers and modeling to conform information and data about parts into inferred properties of wholes. Part 2 glances along the hierarchical axis toward the higher levels of complex ecological organization, on the edge where irreducibility and the absolute need to reduce meet in unresolvable conflict.

Chapter 7

Chapter 7 by William Cale is the first of two contributions on model aggregation—what happens when lower-level properties must be foregone in a synthesis of higher-level behavior. Models are abstractions, homomorphisms, shadows of real systems consisting of parts and relations that are aggregates of real parts and relations. A brief history of the aggregation problem in ecological modeling and the study of its consequences is given. Incorrect conceptualization, model formalism, and measurement error all contribute, with lumping of categories, to the fictionalization of reality. The study of aggregation consequences in simulation modeling proceeds by constructing a reference system—a known universe, as it were, that can serve as a surrogate for truth. The system can be modified in a variety of aggregated configurations to explore the consequences. "This process avoids the epistemological dilemma of directly claiming anything about how an aggregate model behaves compared to how an ecosystem behaves, but leaves intact the opportunity to extend understanding of aggregation from the confined reality of a reference system to the broader context of modeling ecosystems." Three types of error measures for assessment are bias, relative error (normalized bias), and total error (square of integral bias). Three aggregation theorems from an earlier paper are reviewed to show "that problems associated with aggregation permeate the science"; they are not just peculiar to modeling. Variability is the greatest source of aggregation error, and all ecological data are distributed. An investigation of parameter distributions in a simple population model demonstrates that aggregation errors are smallest when parameters have stronger, rather than weaker, central tendencies in the distributions of their values. Explorations with several specific model systems lead to a number of illustrative results, and, finally, some guidelines are given about when and when not to aggregate in model construction. The author concludes that, "While there is some reason for optimism when dealing with quite simple networks or single trophic levels, there are no proven guidelines for handling more complex experimental or theoretical problems." The modeling of complex ecology remains an enigma but, Cale says, "we are learning."

Chapter 8

In Chapter 8, Nickolas Luckyanov presents a mathematically oriented account of aggregation, beginning with a formal definition of the problem. Following Cale's prescription, he establishes a reference (mathematical) system upon which "aggregative transformations" can be investigated. The idea is to find functions in the variables of the aggregated system that behave like the functions in the original unaggregated variables. The latter functions are then called "representable." A set of theorems is stated and proved defining conditions for representable functions. In the general case, aggregating transformations are functions of the whole set of initial variables, but in ecological modeling where original categories (for example, trophic levels) may already be lumped, aggregated variables must be defined in terms of nonintersecting subsets of initial variables. Models are then said to be "separable," and conditions are developed for this property. Concepts are illustrated by a two-level predator–prey model. The aggregation of linear systems is then considered and a distinction made between aggregating them in real ("R-aggregatable") and imaginary domains. Several theorems are proved giving conditions for R-aggregation. Separability of linear models is then taken up, and an algorithm developed for "approximate linear separability." The author concludes with the observation that, while some progress in solving model aggregation problems has been made, more work is needed. The problems are by no means as intractable as they might at first seem.

Chapter 9

In Chapter 9, Michael Kemp, Walter Boynton, and Albert Hermann describe simulation modeling in the design of research on the estuarine macrophyte communities of Chesapeake Bay, USA. These communities provide habitat and feeding areas for fish and waterfowl, they serve as filters to trap and bind suspended sediments, and they function as storages and sinks for nutrients. Pollution conditions have led to their decline, and this chapter describes combined simulation and empirical approaches to the investigation of causes and their mitigation. Three questions are considered: what are the responsible factors, what is the role of macrophytes in the larger estuarine ecosystem, and what management actions would be best for restoration? Mechanistic relations are determined from controlled experiments, interpreted for natural conditions and extrapolated to the broader estuarine and related ecosystems. Research methods are organized hierarchically, from bench-scale bioassays with laboratory microcosms to field mesocosms and then descriptive field studies. Two approaches to simulation modeling are used: (1) a downward hierarchical decomposition of macrophyte ecosystems into six subsystem models to balance the needs for mechanistic detail against conceptual and computational limitations, and (2) a management-motivated upward extension to the entire estuarine ecosystem, including the socioeconomic sphere. The subsystems of the first approach, in which carbon and nitrogen are transferred between 45 state variables, are autotrophs, epibiota, water and plankton, benthos, mobile invertebrates, and nekton. Each of these is described in detail. The management model traces human impacts on macrophytes to fishery and other resources. Five submodels, comprising 15 state variables, are producers; sediments; water with dissolved nutrients, herbicides, and suspended sediments; herbivorous invertebrates; and fish. The authors conclude with an evaluation of their approach. Ecological modeling can usefully complement experimental research by identifying poorly understood relationships for clarification, helping to explain field and laboratory results, sharpening vague ideas into specific testable hypotheses, and evaluating management alternatives in terms of possible immediate and long-term socioeconomic consequences.

Chapter 10

In Chapter 10, Efraim Halfon is concerned with the modeling of toxic substances in the freshwater environment. There are some 60,000 modern chemicals in use today, and incomplete knowledge of relationships between chemical structure and environmental behavior makes prediction of environmental fates and effects, two aspects of ecotoxicology, very difficult. Most fate processes follow first-order kinetics, however, making linear models appropriate for their description. Modeling falls into two categories, formulating dynamics and concentrating on statistical studies focusing on equilibrium conditions. Factors usually incorporated include rates of loading (input), processes of degradation, and transport; and field testing is needed to calibrate and validate the models. Halfon reviews the relationship between physical and chemical properties of substances and their environmental behavior. He points out that secondary derivatives also with toxic potential may derive from degradation processes and that models generally do not take such by-products into consideration. Equilibrium and dynamic models are reviewed. The former are most useful in indicating which environmental compartment or sector would be the main recipient of a contaminant under constant conditions. Dynamic models treat nonconstant behavior and can yield information about residence time. Aggregated models can give useful, but less accurate, predictions. After reviewing model uncertainty, the formulation and testing of models for volatilization processes are described in some detail to illustrate the typical range of considerations that enter into complex environmental modeling. Modeling for ecotoxicology is not fundamentally different from other ecological modeling—both are complex.

Chapter 11

George Van Dyne was a systems ideologue, to be sure, but many of his actual writings were technical expositions, unlofty in tone or content, on specific, fine-grained issues or topics. Grassland processes, animal nutrition, design of equipment, methods for field or laboratory sampling, mathematical or computational analyses, modeling, resource- and range-management protocols, organizing research teams, teaching—these were the component parts of his brand of down-to-earth holism. A philosopher of science George certainly was, but not out of the armchair or of the eloquent literary variety. He wore his sleeves rolled up; he crammed his organized mind and dirtied his boots and hands with minute details; he did things in field and laboratory, and his science followed his doing. He was a hunter and sportsman in the Colorado (USA) montane back country, and Chapter 11 describes a holistic, but pragmatic, approach to deer population management that came out of this recreation he enjoyed.

In this chapter, Gordon Swartzman presents a previously unpublished paper that he and Van Dyne wrote together in 1973. The subject is one in natural resources management, one of the main themes throughout all George's work. He saw resource systems as complexly interwoven with their surrounding ecology, and he argued for ecosystem approaches to their study and management. Swartzman points out that many of the things for which Van Dyne stood have come to pass; "there has been increased awareness that management of natural resources, especially over wide areas, requires multiple objectives and necessitates consideration of the noncontrollable aspects of the resource." This paper considers how to distribute deer-hunting pressure unevenly over the entire state of Colorado as suits local conditions and argues with characteristic attention to holism and a systems view that hunters as well as the hunted should be studied. An optimization framework in nonlinear programming is presented for setting bag limits, adjusting hunting-season duration, and limiting the number of licenses issued in various management units in order to maximize the statewide harvest without decimating local deer populations or interfering with human activity in developed areas. From a brief background in linear programming, reformulation to introduce nonlinear inequality constraints and stochastic coefficients is described to define an objective function whose maximization corresponds to the optimum statewide deer kill. The function includes, by week and region, data on numbers of hunters, bag limits, success rates, and a decision variable to hunt or not. The details of its formulation amount to a type of study of what might be termed "resource systems analysis." Results for the linear deterministic versus nonlinear stochastic approach are presented, compared, and evaluated. Beginning with 8402 original animals, the deterministic model yields 8435 as the postharvest population size, and the stochastic approach yields 9180 for a single-year result. Successive iterations are recommended to generate longer-term predictions. Noting the limitations of their demonstration model, the authors conclude by observing that, while such problems may be analytically difficult, combined efforts of resource managers and ecological systems analysts can yield quantitative decision making in the sphere of renewable resources management.

Chapter 12

Chapter 12, also motivated by the pragmatic needs of resource decision making, implements one of the kinds of activities Van Dyne championed—modeling-guided workshops. One of his students, Carl Walters of the University of British Columbia, was instrumental in developing with C. S. Holling the workshop method called *adaptive environmental assessment and management* (AEA). In this chapter, Gregor Auble, David Hamilton, James Roelle, and Austin Andrews describe the use of workshop simulation modeling, according to the AEA prescription, for assessing environmental impacts of human development on fish and wildlife resources. Between 1978 and 1986, thirteen different natural resource systems were evaluated. The authors note that integrative benefits accrue to participants even in the absence of real predictive capability of the resulting simulation model. It is the power to organize different perspectives (of developers, conservationists, managers, and so on) into a coherent view of a subject that is of greatest value. The modeling process facilitates the reconciliation of objectives, identification of assumptions, and clear multiperspective conceptualization of the resource use problem; it emphasizes interactions between components of the resource system; it provides a logical framework for information synthesis and the juxtaposition of socioeconomic and ecological concerns; and it allows computer gaming to explore the possible consequences of management decisions.

In workshop planning, a preworkshop scoping meeting is held to describe the subject system and associated problems, identify questions for which answers will be sought, consider alternative modeling approaches, plan logistics, and select participants (to include subject matter specialists, policy decision makers, and managers). A workshop generally lasts five days. Its activities include bounding the model; formulating a "looking outward" interaction matrix; developing (programming and debugging by staff) submodels; coupling the submodels; and interactively gaming with the resultant whole system model. Applications to 13 projects are briefly described, summarizing purposes, time and space scales, and the basic factors considered. Two projects are described in further detail. One project evaluated complex land-use patterns (for recreation, hunting, logging, and energy development) affecting elk populations in the Jackson Hole, Wyoming (USA), region over a 50-year time horizon. The other assessed the consequences of alternative flood control schemes at the Malheur Lake National Wildlife Refuge in Oregon (USA). An evaluation of the workshop simulation modeling method is offered. Benefits include improved communication among participants; better systems understanding of causal relationships, interactions, and the dynamic character of the resource system; and improved ability to see different perspectives on diverse issues. Limitations are mainly shortness of time and deficiencies in the produced models, which of necessity cannot be very refined. Despite its utility and successes, the approach was terminated for administrative reasons by the sponsoring agency. The authors conclude, nevertheless, by reiterating the need to bring systems ecology perspectives and methods to bear on complex resource management issues. They observe that this is taking place rather broadly in the resource fields through the introduction of computer technologies such as geographic information systems and expert systems, to which simulation modeling may eventually become coupled. In its applied orientation, this chapter differs from many others in the book, which are theoretical; but its pragmatism clearly reflects the Van Dyne philosophy that, no matter how complex the science of complex systems becomes, it should always pertain to solving real problems in the real world.

7

Model Aggregation: Ecological Perspectives

WILLIAM G. CALE

College of Natural Sciences and Mathematics
Indiana University of Pennsylvania
Indiana, Pennsylvania

> Models provide a paradigm or a structure into which to embed our theories, data and experience . . . they provide a conceptual framework of logical interrelationships. Models are integrating mechanisms because they allow us to take data from field, laboratory, and armchair and couple it in a unique way [Van Dyne, in *Ohio J. Sci.* 78, 191 (1978)].

INTRODUCTION

With clarity and simplicity, George Van Dyne expressed the essence of what two decades later have become vital interrelated topics in the study of theoretical ecology: hierarchical organization and aggregation. Implicit in the study of natural systems is the concept that integrated processes acting together produce an entity, the ecosystem, that can be investigated as a unit. Van Dyne was especially interested in this perspective and in developing methods to simulate ecosystem behavior using computers. We can easily infer from the above quotation that his belief was that understanding (and therefore modeling) useful ecosystem properties does not require an understanding of all the complexity of the processes that are part of ecosystems. This is aggregation, the lumping together of lower-level parts and processes to reduce complexity, while retaining (we hope) the integrity of prescribed aspects of ecosystem behavior.

In this chapter we will explore the theory of aggregation, especially as it relates to simulation models. However, problems of aggregation extend in obvious ways to experimental and theoretical ecology; therefore, a review of those topics and how they relate to modeling will also be considered. To begin, a brief historical perspective is provided.

Models are abstractions of reality, representing only limited aspects of the behavior of real systems (Patten, 1971). Thus, the modeler makes a priori decisions about what portions of a system are to be treated explicitly and what aspects are to be subsumed into aggregate model variables. This decision process establishes what is called a homomorphism, the many-to-one correspondence between model components and system components. Models are not miniature versions of real systems; they are simplifications of real systems and as such contain parts that are aggregates.

Beginning in the late 1960s and extending through most of the 1970s, ecologists devoted a great deal of effort to developing large-scale simulation models of natural systems. At least one product of that effort, obvious today, was the absolute realization of the impossibility of gathering concurrent time series of data representative of the species living in a place. It was not so much due to any technical limitation, although that problem remains very real (for example, in the study of below ground biological processes), but rather to the sheer complexity of ecosystems. Numbers of individuals within species, numbers of species, subtle trophic interactions, spatial heterogeneity, temporal dynamics, and so on, forced ecological modelers to limit the conceptual framework of their models and, concomitantly, to defend a simplified outlook. By the end of the 1970s, it had become obvious that ecologists should investigate the consequences of a problem closely analogous to one that had been of interest to economists (Ijiri, 1971 and Chipman, 1975, 1976 provide introductions to the economic literature) for over 30 years: the aggregation of variables in a macrosystem (ecosystem) to form a microsystem (model).

The first paper in the open ecological literature dealing with aggregation was published by Zeigler in 1976. Zeigler has studied aggregation from a systems theoretic perspective and has contributed extensively (e.g., Zeigler, 1971, 1976, 1979, 1980, and this volume) to our general understanding of aggregation. Webster (1975) made brief mention of aggregation in an appendix to his dissertation, pointing out the impossibility of retaining a correct transient response in an aggregate model variable described by constant coefficients. But it was not until 1979 that the results from a series of studies undertaken by theoretical ecologists on the topic of aggregation were published.

O'Neill and Rust (1979) used linear and nonlinear mass-balance models to examine the consequences of lumping two or more interacting components into a single variable. Cale and Odell (1979, 1980) conducted similar analyses and proved several results specific to linear systems. In a very important paper, O'Neill (1979) demonstrated that the functional form describing an aggregate may be completely unrelated to the general description appropriate to any single member of the composite. And Cohen (1979) began an exploration of the aggregation problem in population ecology and proved it impossible to compute an average growth-rate parameter for age-structured populations subject to stochastic variation in their vital rates. Thus, in a review paper written by O'Neill and Gardner (1979), aggregation was listed as one of the principal sources of uncertainty (along with measurement error, incorrect model formalism, and natural variability) in ecological model output. By the close of the decade, it was well understood by a rather small group of theoreticians that aggregation introduces error into models and that this error is (for practical purposes) unavoidable and unpredictable.

Continuing research into aggregation has sought strategies to minimize introduced error (Gardner et al., 1982; Iwasa et al., 1987, 1989; Rastetter et al., 1992), examined the relationship of aggregation to other aspects of ecological theory (Cale et al., 1983a; Allen et al., 1984; Cohen, 1985; Cale and O'Neill, 1988; Gard, 1988; Alymkulov, 1991), explored the problem in different modeling contexts (Gardner et al., 1980; Innis and Rexstad, 1983; de Caprariis, 1984; Sugihara et al., 1984), further developed the mathematical theory of ecological aggregation (Schaffer, 1981;

Luckyanov et al., 1983; Cale et al., 1983b), and examined ways in which the theory can be applied to experimental ecology (Gardner et al., 1982; Cale and McKown, 1986; Bartell et al., 1988; Rastetter et al., 1992). A summary of the important results from these and other studies along with a brief presentation of new results on the aggregation of organisms into species forms the remainder of this chapter.

DEVELOPING ECOLOGICAL AGGREGATION THEORY

Defining the Problem

Developing theory that is to apply at the ecosystem level presents a logical problem that sets ecology apart from the other sciences. Testing ecosystem theory at the appropriate temporal and spatial scales is usually not possible. Ecologists are not at liberty to manipulate the world's ecosystems at will. Even if that were possible, human life is not nearly long enough to examine the dynamics that scientists believe may occur over hundreds of years. Thus, experimentalists may exploit natural events (such as succession following glacial retreat) or design manageable projects examining parts of ecosystems. Theoreticians must evaluate their ideas against necessarily incomplete data sets or postulate an assumed truth against which theoretical constructs are tested. This latter approach has been especially fruitful in a wide variety of theoretical endeavors, including population biology, population genetics, and ecosystem modeling.

The aggregation problem, then, can be investigated by constructing a reference system. This reference system is completely specified as to structure, mathematical formalism, and parameter values. Since this system can be solved either analytically or by numerical methods, it can be treated as a surrogate for truth and used as a bench mark against which we can compare other systems. In particular, we may condense the reference system and test the dynamics of this dimensionally smaller (aggregate) model against the original. This process avoids the epistemological dilemma of directly claiming anything about how an aggregate model behaves compared to how an ecosystem behaves, but leaves intact the opportunity to extend understanding of aggregation from the confined reality of a reference system to the broader context of modeling ecosystems.

Imagine that the reference system, S, contains p distinct state variables. Each of the p state variables is perfectly described by a generalized first-order, nonlinear differential equation of the type

$$\dot{Y}_i = I_i(t, Y) + f_i(Y) - g_i(Y) - h_i(Y), \qquad i = 1, 2, \ldots, p \qquad (7.1)$$

in which

I_i = a linear or nonlinear environmental input function

f_i, g_i = functions describing interactions between state variables (they may be any linear or nonlinear function of time and Y)

h_i = a linear or nonlinear function of time and Y describing environmental loss.

Equation (7.1) is developed to describe mass or energy transfers and its solution is known, at least to an arbitrary degree of numerical accuracy.

A model of S may contain only m equations, $m < p$. Of the m model equations, n will be aggregates ($n \leq m$). Each of these n aggregate equations will represent two or more state variables in the original reference system, S. The modeler establishes this homomorphism, and one of the goals of aggregation research is to understand which types of correspondences optimally preserve the system characteristic of interest. Functions in the model are developed such that, at equilibrium, mass-balance characteristics (that is, input–output relationships) of the model are identical to those in S. Therefore, by setting the initial conditions in the model to correspond to the initial conditions in S, differences between the model and S will arise only during transient response. These differences may be measured by any of several error statistics.

Three measures of error that seem to be most meaningful for ecological investigation are (1) bias, (2) relative error, and (3) total error. Let $Y_A(t)$ be an aggregate model state variable representing r variables in S. Bias is then defined as

$$B(t) = Y_A(t) - \sum Y_i(t) , \qquad (7.2)$$

where the sum is taken over the appropriate r variables in S. Bias simply measures the difference at any time t between the state of the aggregate model variable and the summed states of the reference variables it represents. Relative error normalizes bias:

$$E_R(t) = \frac{B(t)}{Y_i(t)} . \qquad (7.3)$$

Multiplication by 100 expresses relative error as the percentage deviation at time t of the aggregate state from the true (reference) state. Total error is the integrated squared bias:

$$E_T = \int B(t)^2 \, dt. \qquad (7.4)$$

Squaring eliminates possible cancelling effects since bias can change sign within a single simulation (Cale and Odell, 1979).

The choice of an error statistic depends on the modeling objective. If we are modeling a system variable in which threshold effects may be important, as in the case of dissolved oxygen in a stream or lake, then bias must be minimized. Frequently, we are interested in closely approximating system state, but some error can be tolerated. It would be satisfactory when modeling most populations to know that model projections were never more than a fixed percentage in error. Aggregations constructed to bound relative error within such a percentage achieve the objective. Finally, the goal may be to estimate a long-term average system state. In this case the model must be nearly correct most of the time, but occasional deviations will not greatly affect the result. Total error should be computed. In all cases, error becomes zero when $Y_A(t) = \overset{r}{\Sigma} Y_i(t)$ for all t.

Theoretical Results

The general theory establishing zero error conditions for aggregation in generalized, nonlinear mass-balance models was presented by Cale et al. (1983b; see also Iwasa et al., 1987, 1989). Their theorems and corollaries are given next, together with explanations. Interested readers are referred to the 1983b paper for proofs.

Theorem 1. In any system composed entirely of state variables characterized by linear environmental loss functions and state independent inputs, the error due to aggregation to a single state equation is zero whenever the original loss functions are identical.

Corollary 1.1. Given a system as described by Theorem 1, the state of an aggregate model of m equations, $p > m > 1$, will be the same as the state of the original system for all time.

Corollary 1.2. The individual state variables of an aggregate model as described in Corollary 1.1 may have error when their state is compared to the summed states of the system variables they represent.

The early investigations (Zeigler, 1976; O'Neill and Rust, 1979; Cale and Odell, 1979) all discovered what appeared to be a robust property: the nature of the loss function was critical to aggregation. Theorem 1 proves that nonlinear interactions between components are not a source of aggregation error whenever inputs do not depend on state and *all* loss functions are both linear and identical. Generally, the aggregate model is not a condensation of the original system to one state variable, and Corollary 1.1 proves that as long as the conditions of the theorem are not violated, any aggregation will mimic total system state without error. However, even in the case under discussion, direct comparisons between model variables and the system variable(s) that they represent will yield error (Corollary 1.2). This includes even those model variables that are not aggregates.

Theorem 2. In any system where each state variable remains, for all time, in constant proportion to every other state variable, it is possible to construct a single-state-variable model that has zero error.

Corollary 2.1. A single-state-variable mass-balance aggregation model for the system described by Theorem 2 may have error.

Corollary 2.2. The conditions of Theorems 1 and 2 define the only circumstances for which it is possible to aggregate a system of p equations to one of m equations with a zero error result.

There is a well-known criterion in economic theory, the Lange–Hicks condition (Simon and Ando, 1961), that states that two or more variables that move together may be aggregated. For zero error, it is not only required that variables move together, but also that they remain in constant proportion to each other (Theorem 2). It is very important to note that Theorem 2 proves only the existence of a function, Y^*, which could mimic total system state with zero error. Corollary 2.1 demonstrates that Y^* is not necessarily the model we would derive from equilibrium mass-balance considerations. Indeed, we can infer from the proof of the corollary that it is vanishingly unlikely that Y^* would match an ecological mass-balance model.

Corollary 2.2 is the most important result for theoretical ecology. It establishes that the conditions set forth in the first two theorems are the only conditions that permit zero error aggregation. These conditions are not likely to be encountered in any natural system. Thus, aggregation error becomes a reality of the science of ecology. The theoretical results extend beyond modeling to all aspects of basic and applied ecology. Some of these aspects will be discussed in the next section.

Theorem 3. For each aggregation model of S, there exists at least one p-tuple of initial conditions in S for which the aggregation model simulates S with minimum error.

The final theorem states an intuitive reality. Its significance is that there exist states in the system from which a model simulation can begin and yield results with minimum introduced aggregation error. It is presently not known how to derive or estimate these states.

AGGREGATION AS A PART OF ECOLOGY

Field Ecology and Transmutations

It is important that students of ecology come to understand that problems associated with aggregation permeate the science. While much of our knowledge about aggregation was developed in the context of ecosystem modeling, the phenomenon itself appears in many other familiar research situations:

determining process rates, deriving empirical (or process) relationships, estimating population fitness and biotic potential, and drawing inferences (such as system stability characteristics) from classical competition theory. The most common source of aggregation error in classical ecological investigation is variability within the system. This problem is encountered in virtually all field studies and is generally ignored because the investigator is unaware that a problem exists. Gardner et al. (1982) provide the following example.

Suppose an investigator is interested in determining the growth rate (or death rate or respiration rate) of a population of organisms. These organisms exist in the field in a variety of local habitats and are distributed across the range of tolerance of the species. The investigator is aware of possible local effects and designs an experimental protocol (using, for example, stratified random sampling methods) that accounts for actual circumstances. Measurements from this experiment are used to calculate population growth as an average, accordingly weighted (perhaps by frequency of occurrence of individuals within each location or by size of the sampling areas). Such a procedure is familiar not only in field studies but also in laboratory settings.

An aggregation error has been introduced by the above procedure. Suppose it was experimentally determined that each local group within the population exhibited one of five growth rates (0.1, 0.3, 0.5, 0.7, and 0.9) and that all rates were equally weighted. It seems obvious to claim an average growth rate of 0.5 for the population. This is wrong. If we consider exponential increase, then the claim of a single value expressing average growth is equivalent to

$$N_0 e^{0.5t} = N_1 e^{0.1t} + N_2 e^{0.3t} + N_3 e^{0.5t} + N_4 e^{0.7t} + N_5 e^{0.9t} \tag{7.5}$$

in which $N_0 = \sum_{i=1}^{5} N_i$ at $t = 0$. A well-known result from calculus proves the impossibility of representing a sum of exponentials by a single exponential term. Equation (7.5) is an equality only when $t = 0$. It is impossible to pool different rates that describe one process into a single aggregate rate without introducing error. The generality of this conclusion is a consequence of Theorem 1.

The illustration has obvious extensions. We cannot characterize lake primary productivity by averaging across each phytoplankton population. Poikilothermic populations living in thermally stratified environments have no average respiration rate. Differential predation pressure must be explicitly evaluated, since it is not possible to speak about average loss. Quite simply, the dynamics of aggregates are different from the sum of the dynamics of the components that comprise them.

Descriptions that are adequate at one level of organization usually do not suffice to describe behavior at higher (aggregate) levels of organizations. A change occurs as we move from one hierarchical level to another. This phenomenon has been aptly described (O'Neill, 1979) as a "transmutation across hierarchical levels." That the exponential model [eq. (7.5)] fails for the population, while successfully describing specific classes within the population, suggests that a completely different description is needed at the population level. O'Neill (1979) develops several examples that are elegant in their simplicity and compelling due to their connection to common ecological experience.

Thus, we are asked to consider a population composed of organisms that respond to an environmental variable such as temperature according to a threshold (step) function. For each organism, a critical value, T_c, exists that partitions that organism's response into two distinct regimes. In principle, the T_c of a given individual may be experimentally determined. The point O'Neill makes is that genetic variability will result in a distribution of T_c values across the population. While the appropriate descriptor at the organism level was a step function, O'Neill shows that genetic variability results in an arctangent function (when T_c approaches a normal distribution) or a linear function (when T_c is uniform) when working with the whole population.

Discussions about the implications of hierarchical organization in ecological research are beginning to appear in the literature (Webster, 1979; Allen and Starr, 1982; Sugihara, 1983; Allen et al., 1984; O'Neill et al., 1986). Continuing investigations may lead to methods for conceptualizing ecosystems that are distinctly different from the organism–population–community paradigm that has been so powerful for so long. Suggestive of discovering new ways to characterize ecosystems, Luckyanov et al. (1983) found that the product of population sizes was the appropriate transformation in an aggregate model of n prey species originally described by Volterra equations. As in all the sciences, understanding that our present approach is not completely satisfactory to address the questions being asked becomes the impetus for change (Kuhn, 1970).

It is apparent that one problem of fundamental importance in ecology is to identify the levels of organization in ecosystems that may be nearly perfectly described mathematically. If we must reduce our frame of reference to subatomic levels before a consistent description is feasible, then ecology becomes lost as a science. However, we know that this is not the case, not only from experience in ecology but also from understanding something of the behavior of systems in general. The water molecule is worthwhile to investigate as an entity (or level of organization) because it has properties unknowable from an analysis of hydrogen and oxygen. The analogy is easily extended to ecosystems.

We can build a convincing case that individual organisms are entities that can be very accurately described mathematically. Physiological ecology has demonstrated that a knowledge of appropriate morphological features together with attributes such as age and sex is sufficient to allow extremely accurate prediction of such processes as respiration, ingestion, assimilation, reproductive potential, and life expectancy. All the characteristics we would like to describe for the population are predictable for single organisms. Yet, in ecosystem modeling research, it is the population that is treated as an entity. Given that we know this aggregation will not be perfect and the transmutation function will probably not be known, it becomes necessary to evaluate the consequences of aggregating organisms into populations.

A preliminary analysis is given in Table 7.1. Two well-known models, the first described by density-independent losses and the second by density-dependent (logistic) kinetics, are assumed to be adequate for predicting the dynamic behavior of individuals. Twelve cases were studied. In each situation a population of 500 individuals was created by drawing 500 parameter sets from the distributions shown in Table 7.1. A single population equation of the same form as that describing

TABLE 7.1 Parameter descriptions for two model types

Model type	Case	Input		Parameters		Maximum E_R
		Distribution	Description	Distribution	Description	
$\dot{X} = I - a \cdot X$	1	C	$I = 10$	U	$0 < a < 2$	0.524
	2	C	$I = 10$	N	$a = 1 \pm 0.3$	0.103
	3	C	$I = 10$	T	$0 < a < 2$	0.260
	4	N	$I = 10 \pm 3$	N	$a = 1 \pm 0.3$	0.073
	5	T	$5 < I < 15$	T	$0 < a < 2$	0.274
$\dot{X} = I - a \cdot X^2$	1	C	$I = 10$	U	$0 < a < 2$	0.288
	2	C	$I = 10$	N	$a = 1 \pm 0.3$	0.040
	3	C	$I = 10$	T	$0 < a < 2$	0.107
	4	N	$I = 10 \pm 3$	N	$a = 1 \pm 0.3$	0.012
	5	T	$5 < I < 15$	T	$0 < a < 2$	0.108
	6	N	$I = 10 \pm 3$	N	$a = 3 \pm 1$	0.023
	7	T	$5 < I < 15$	U	$0 < a < 2$	0.298

Parameters sets for 500 individuals in each of 12 cases are drawn from constant (C), uniform (U), triangular (T), and normal (N) distributions as indicated. The summed solution to 500 state equations representing the population is compared through time to the solution of a single aggregate equation to compute maximum relative error, E_R.

individuals in the collection was parameterized from mean values of the underlying distributions. Forced response simulations were run to equilibrium, and the summed state of 500 individuals was compared to the state of the aggregate population equation at each time step. Maximum relative error is reported in Table 7.1.

The results are encouraging. In each model, errors are smallest when parameters are characterized by distributions with a clear central tendency. The uniform distribution, with no central tendency, caused maximum relative errors to exceed 28%. Fortunately, we would not expect to encounter the uniform distribution often in a natural population. Aggregations where parameters were characterized by normal distributions, a situation we would expect to encounter frequently, performed with less than 8% error. Comparisons between models for the same parameter distribution show that the logistic formulation introduces less aggregation error than the linear model. A tentative conclusion is that the population may be an acceptable level of organization for mathematical description, especially if density-dependent losses apply. However, as suggested by the earlier theoretical presentation and further discussed in the next section, aggregations of populations must be handled with care.

Implications of Aggregation for Present Ecological Theory

In addition to issues that aggregation raises in the context of field or laboratory investigation and simulation modeling, there are also significant implications for theoretical ecology.

An important example of inherent aggregation error was recently provided by Cohen (1985). He proved that it is impossible to develop a single metric describing fitness for an entire population when that population contains individuals or groups that differ in fitness. The study is quite significant, since the concept of population fitness is central to much of modern theory in population genetics, and Cohen has demonstrated that we cannot develop a consistent statistic to describe it. It is ironic that it took 11 years to finally get this important contribution to science through the peer review process and into print. His critics were not eager to acknowledge the result.

Ecological competition has been a source of intense debate over the past several years. The main issue seems to be one of methodology, concerned with experimental design and hypothesis testing. A recent volume by Salt (1984) explores this subject in some detail. There is another aspect to this question, however, regarding the theory of resource competition.

Cale and O'Neill (1988) analyzed whether two models of the same system, one an exploitative resource model and the other a mass-balance flow model, are consistent with each other. That is, do the two models reach identical equilibrium states, exhibit the same stability characteristics, and have the same dynamic response behavior in the neighborhood of equilibrium? Competition models do not explicitly consider a shared resource; rather, they use competition coefficients to express the negative impact of each population on the others. Mass-balance models, on the other hand, describe not only the consuming populations, but also the resource. Competition models are therefore dimensionally smaller than mass-balance models. For example, in a two-consumer, one-resource system the former model has two equations, the latter three.

It is known that these two representations are not exact as long as the competition coefficients are constants (Abrams, 1980), but matching total dynamics is seldom an issue; competition models are typically analyzed in the neighborhood of equilibrium. Will aggregation in the competition model (the subsuming of resource state variables into constant coefficients) affect the ability of that model to capture system dynamics close to equilibrium?

It was demonstrated analytically and illustrated by simulation that the two representations are inconsistent whenever resources are scarce. Only when resources are abundant (and the phenomenon of interest, resource competition, is therefore unimportant) are the models consistent. Furthermore, stable competition models can be made unstable by simple alterations in time scale. Mass-balance models retain their stability properties under all transformations of time. These studies demonstrate that the two bodies of theory, when applied to the same system, do not yield the same result. The authors point out that they do not claim that either model is right or wrong, only that aggregation error should be an important consideration in theoretical ecology.

USING AGGREGATION THEORY

While we know that aggregation introduces error in simulation studies, field research, and theoretical development, our understanding of the magnitude of this error and the conditions that tend to minimize (or exacerbate) it is still limited. Several studies have addressed this question and guidelines have emerged that are in accord with the theory. Since it is not practical to consider carrying out ecological research under zero error conditions, these guidelines are of the type that attempt to ensure that introduced error is small.

Gardner et al. (1982) studied four fundamental types of ecological interactions and their aggregate representation, as follows:

1. Parallel case, in which two populations are causally independent. In an energy flow model, two populations of primary producers would be causally independent. Aggregate: two populations are aggregated into one.
2. Two resource–one common consumer (triangle I) case, in which a consuming population such as an herbivore uses two plant resources simultaneously. Aggregate: two resources are aggregated; consumer remains separate.
3. One resource–two consumer (triangle II) case, in which a single primary producer is consumed simultaneously by two herbivores. Aggregate: resource remains separate; two consumers aggregated.
4. Series case, in which two populations are linked through material or energy flow. Aggregate: two populations are aggregated into one.

The authors identified 40 distinct and ecologically meaningful mathematical relationships that could be used to describe these interactions. Then an aggregation model was developed for each case and its transient response compared to the disaggregated case. The purpose was to identify robust conditions for aggregation, situations that allow aggregation regardless of the underlying type of ecological interaction and the mathematics that describes it.

In the parallel case, no more than 10% relative error resulted from aggregation, regardless of the mathematical process formulation, whenever output rates differed by no more than a factor of 3. Similarly, in both triangle cases, if output rates were within a factor of 2, then relative error was bounded under 10%. Interestingly, the rate of loss is more important than the process description (density dependent or density independent) that characterizes it. Likewise, since interaction mechanisms (for example, linear donor control, linear recipient control, cross products, or Michaelis–Menten) may be preserved when aggregating two components in the same trophic level, aggregation error can be controlled. Extending these results beyond two component aggregations in small networks has not been attempted.

Gardner et al. (1982) further concluded that series aggregations should be avoided. Loss of the interaction term results in aggregate dynamics that are completely different from those of the original system. The sole circumstance where series aggregation introduces small error occurs when both loss rates are nearly identical and density independent and the interaction term is linear, donor-controlled.

Sugihara et al. (1984) considered the question of multispecies fisheries management in the context of aggregation theory. They concluded that relaxing the strict requirements of Theorem 2 to the more general Lange–Hicks criterion should be acceptable. But they correctly point out that we must be sensitive to our modeling objectives when aggregating, since scale (especially temporal scale) is extremely relevant. As an example, it was suggested that, to model a property such as yield (generally best described by processes on a seasonal or annual scale), we should not focus attention on "microlevel fast processes," in other words, on aggregate processes at the same level of resolution as the phenomenon of interest.

A similar conclusion was reached by de Capraiis (1984), who evaluated modeling approaches for lake ecosystems. In oligotrophic and eutrophic lakes, de Capraiis argues that fluctuations in abiotic variables are slow compared to biotic components. It is then demonstrated that low-resolution (highly aggregated) models will serve quite well in these systems for predicting abiotic state (treat the biota as constant since their rapid fluctuations do not affect slow-moving abiotic variables) or biotic conditions (treat abiotic conditions as constant since they move so slowly compared to the biota). However, in mesotrophic lakes it is necessary to use high-resolution process models because the interaction between biotic and abiotic variables will be significant.

CONCLUDING REMARKS

Among the difficulties encountered when studying complex natural systems is that of avoiding or minimizing the introduction of error due to aggregation. Although there is some reason for optimism when dealing with quite simple networks or single trophic levels, there are no proven guidelines for handling more complex experimental or theoretical problems. Indeed, such guidelines may never be developed since the analysis of all possible ecological interactions and mathematical descriptions does not seem feasible. At our present level of understanding, ecologists can be aware of the aggregation problem and avoid those situations known to introduce large errors.

Aggregation error results from our efforts to simplify or reduce complexity. That seems to beg the question of whether we know how to simplify. I think the answer is that we are learning. Part of that learning involves our slowly developing a willingness (perhaps a tolerance) to think about the organization of ecosystems in nontraditional terms. The small group of ecologists working on hierarchical approaches to ecosystem ecology represents one effort at redirecting how we conceptualize and study natural systems. Aggregation is a problem that will be solved, if in fact a solution is possible, by discovering ways to partition ecosystems into entities that can be characterized mathematically and that have properties meaningful to questions of interest to people.

ACKNOWLEDGMENTS

The author gratefully acknowledges support provided by the National Science Foundation, most recently through grant BSR-8211836 to the University of Texas at Dallas.

REFERENCES

ABRAMS, P. A. 1980. Are competition coefficients constant? Inductive versus deductive approaches. *Am. Nat.* 116:730–735.

ALLEN, T. F. H., and STARR, T. B. 1982. *Hierarchy: Perspectives for Ecological Complexity.* Univ. Chicago Press, Chicago. 310 pp.

ALLEN, T. F. H., O'NEILL, R. V., and HOEKSTRA, T. W. 1984. Interlevel Relations in Ecological Research and Management: Some Working Principles from Hierarchy Theory. USDA Forest Service General Technical Report RM-110, Rocky Mountain Forest and Range Experiment Station, Fort Collins, Colorado. 11 pp.

ALYMKULOV, E. D.-A. 1991. Aggregation in models of ecosystems dynamics and stability problems. *Ecol. Mod.* 58:383–386.

BARTELL, S. M., CALE, W. G., O'NEILL, R. V., and GARDNER, R. H. 1988. Aggregation error, research objectives and relevant community structure. *Ecol. Mod.* 41:157–168.

CALE, W. G., and MCKOWN, M. P. 1986. A cost analysis technique for research management and design. *Env. Manag.* 10(1):89–96.

CALE, W. G., and ODELL, P. L. 1979. Concerning aggregation in ecosystem modeling. In Halfon, E. (ed.), *Theoretical Systems Ecology*. Academic Press, New York, pp. 55–77.

CALE, W. G., and ODELL, P. L. 1980. Behavior of aggregate state variables in ecosystem models. *Math. Biosci.* 49:121–137.

CALE, W. G., and O'NEILL, R. V. 1988. Aggregation and consistency problems in theoretical models of exploitative resource competition. *Ecol. Mod.* 40:97–109.

CALE, W. G., O'NEILL, R. V., and SHUGART, H. H. 1983a. Development and application of desirable ecological models. *Ecol. Mod.* 18:171–186.

CALE, W. G., O'NEILL, R. V., and GARDNER, R. H. 1983b. Aggregation error in nonlinear ecological models. *J. Theor. Biol.* 100:539–550.

CHIPMAN, J. S. 1975. Optimal aggregation in large scale econometric models. *Sankhya, Series C* 37:121–159.

CHIPMAN, J. S. 1976. Estimation and aggregation in econometrics: an application of the theory of generalized inverses. In Nashed, M. Z. (ed.), *Generalized Inverses and Applications*. Academic Press, New York, pp. 549–769.

COHEN, J. E. 1979. Comparative statics and stochastic dynamics of age-structured populations. *Theor. Pop. Biol.* 16(2):159–171.

COHEN, J. E. 1985. Can fitness be aggregated? *Am. Nat.* 125(5):716–729.

DE CAPRARIIS, P. 1984. A note on the sufficiency of low resolution ecosystem models. *Ecol. Mod.* 21:199–207.

GARD, T. C. 1988. Aggregation in stochastic ecosystem models. *Ecol. Mod.* 44:153.

GARDNER, R. H., MANKIN, J. B., and EMANUEL, W. R. 1980. A comparison of three carbon models. *Ecol. Mod.* 8:313–332.

GARDNER, R. H., CALE, W. G., and O'NEILL, R. V. 1982. Robust analysis of aggregation error. *Ecology* 63(6):1771–1779.

IJIRI, Y. 1971. Fundamental queries in aggregation theory. *J. Am. Stat. Assoc.* 66(336):766–782.

INNIS, G., and REXSTAD, E. 1983. Simulation model simplification techniques. *Simulation,* July, pp. 7–15.

IWASA, Y., ANDREASEN, V., and LEVIN, S. 1987. Aggregation in model ecosystems. I. Perfect aggregation. *Ecol. Mod.* 37:287–302.

IWASA, Y., LEVIN, S., and ANDREASEN, V. 1989. Aggregation in model ecosystems. II. Approximate aggregation. *IMA J. Math. Applied Med. Biol.* 6:1–23.

KUHN, T. S. 1970. *The Structure of Scientific Revolutions*. University of Chicago Press, Chicago. 210 pp.

LUCKYANOV, N. K., SVIREZHEV, Y. M., and VORONKOVA, O. V., 1983. Aggregation of variables in simulation models of water ecosystems. *Ecol. Mod.* 18:235–240.

O'NEILL, R. V. 1979. Transmutations across hierarchical levels. In Innis, G. S., and O'Neill, R. V. (eds.), *Systems Analysis of Ecosystems*. International Cooperative Publishing House, Fairland, Maryland, pp. 59–78.

O'NEILL, R. V., and GARDNER, R. H. 1979. Sources of uncertainty in ecological models. In Zeigler, B. P., Elzas, M. S., Klir, G. J., and Oren, T. I. (eds.), *Methodology in Systems Modelling and Simulation*. North-Holland, Amsterdam, pp. 447–463.

O'NEILL, R. V., and RUST, B. 1979. Aggregation error in ecological models. *Ecol. Mod.* 7:91–105.

O'Neill, R. V., and DeAngelis, D. L., Waide, J. B., and Allen, T. F. H. 1986. *A Hierarchical Concept of the Ecosystem.* Princeton University Press, Princeton, New Jersey. 253 pp.

Patten, B. C. 1971. A primer for ecological modeling and simulation with analog and digital computers. In Patten, B. C. (ed.), *Systems Analysis and Simulation in Ecology,* Vol. I. Academic Press, New York, pp. 3–121.

Rastetter, E. B., King, A. W., Cosby, B. J., Hornberger, G. M., O'Neill, R. V., and Hobbie, J. E. 1992. Aggregating fine-scale ecological knowledge to model coarser-scale attributes of ecosystems. *Ecol. Applications* 2(1):55–70.

Salt, G. W. (ed.). 1984. *Ecology and Evolutionary Biology, A Round Table on Research.* University of Chicago Press, Chicago. 130 pp.

Schaffer, W. M. 1981. Ecological abstraction: the consequences of reduced dimensionality in ecological models. *Ecol. Monogr.* 51(4):383–401.

Simon, H. A., and Ando, A. 1961. Aggregation of variables in dynamic systems. *Econometrica* 29(2):111–138.

Sugihara, G. 1983. Peeling apart nature. *Nature* 304:94.

Sugihara, G., et al. 1984. Ecosystem dynamics—group report. In May, R. M. (ed.), *Exploitation of Marine Communities.* Springer-Verlag, New York, pp. 131–153.

Webster, J. R. 1975. Analysis of Potassium and Calcium Dynamics in Stream Ecosystems on Three Southern Appalachian Watersheds of Contrasting Vegetation. Ph.D. dissertation, University of Georgia, Athens. 232 pp.

Webster, J. R. 1979. Hierarchical organization of ecosystems. In Halfon, E. (ed.), *Theoretical Systems Ecology.* Academic Press, New York, pp. 119–129.

Zeigler, B. P. 1971. A note on system modelling, aggregation and reductionism. *Int. J. Biomed. Computation* 2:277–280.

Zeigler, B. P. 1976. The aggregation problem. In Patten, B. C. (ed.), *Systems Analysis and Simulation in Ecology,* Vol. IV. Academic Press, New York, pp. 299–311.

Zeigler, B. P. 1979. Multilevel multiformalism modeling: an ecosystem example. In Halfon, E. (ed.), *Theoretical Systems Ecology.* Academic Press, New York, pp. 17–54.

Zeigler, B. P. 1980. Simplification of biochemical reaction systems. In Segel, L. A. (ed.), *Mathematical Models in Molecular and Cellular Biology.* Cambridge University Press, Cambridge, Massachusetts, pp. 112–145.

8
Model Aggregation: Mathematical Perspectives

NICHOLAS K. LUCKYANOV

Rosenteil School of Marine and Atmospheric Sciences
University of Miami
Miami, Florida

> The functional ecologist concentrated on the analysis of mechanisms underlying the action of ecological systems. . . . It soon became apparent, however, that if the models were based on reasonably complex assumptions the models became completely intractable mathematically. Therefore the model builder quickly learned to develop models based on a small number of simple assumptions and to live with the fact that his models did not correspond very closely to the real world. While the models were thereby holistic and integrative, they suffered the major drawback that they were unrealistic. [Van Dyne, in *The Ecosystem Concept in Natural Resource Management*, Ch. X, p. 333–4 (1969), Academic Press].

INTRODUCTION

The present increased interest in simulation models in ecology can be explained by the fact that, so far, ecosystems have been relatively little studied and only certain facts are known to researchers. The development of models provides an opportunity to combine these facts into a more or less harmonious system. Presently, the scope of our knowledge, as well as the gaps in it, is becoming outlined. This, in turn, will stimulate new research. However, due to the still existing absence of mathematically grounded methods for modeling ecological systems, the forecasting value and generality of results give rise to a certain element of doubt.

One of the most difficult problems in modeling ecological systems is that of dimension. Even such a comparatively simple object as an oligotrophic water body may account for more than 80 types of zooplankton, 350 types of benthos, about 50 types of fish, and other organisms. Obviously, the inclusion in a model of all the taxonomic units would drastically complicate it and make it more difficult to understand and interpret. In addition, requirements on the amount and accuracy of experimental information would be stricter, and the time needed for programming, debugging, and computation would be prolonged. To decrease model dimensions, it is necessary to perform variable aggregation, that is, to unite some variables of the system that are related to one another by certain properties into some sort of categories, each of which is a new variable with properties defined by the aggregation laws.

The problem of aggregation of the components of complex systems is of topical interest in various fields of knowledge dealing with mathematical modeling. The most profound investigations on the theory of aggregation were undertaken in economic systems modeling (Matin, 1985). The fact is that large dimension and a great number of details in economic models are as bad, from the point of view of decision making, as the lack of information about the object under consideration. On the other hand, decision makers are interested in the maximum amount of information about the controlled object presented in the simplest form available. For example, this information can be presented as some combination of integral parameters. This problem leads to the problem of aggregate search. The same problem arises in describing the kinetics of petrochemical reactions, where systems of equations of very large dimension are obtained (Zhorov et al., 1967). Statistical physics (Landau and Lifshitz, 1980) is another example of the development of an aggregated physical description of interacting particle ensembles.

In the present paper, notions of aggregation and separability are introduced and examples of their application to ecological models are given. The main properties of aggregated and separated systems are considered. Then, an algorithm for approximate separability of linear systems, applicable to a wide range of models, is constructed and its properties studied.

THE GENERAL AGGREGATION PROBLEM

Problem Statement

Let us consider the following system of differential equations:

$$\dot{x}_i = f_i(X), \qquad i = \overline{1, N}. \tag{8.1}$$

We can change the variables in system (8.1) as follows:

$$y_j = \varphi_j(X), \qquad j = \overline{1, M}, \quad M \le N. \tag{8.2}$$

We then get a new system:

$$\dot{y}_j = F_j(X) = \sum_{i=1}^{N} \frac{\partial \varphi_i(X)}{\partial x_i} f_i(X). \tag{8.3}$$

Definition 1. Let us call the substitution of variables (8.2) an *aggregative transformation* for system (8.1), and the new system of differential equations (8.3) an *aggregated system*, if (1) $M < N$ and (2) $\Phi_j(Y) = F_j(X)$, that is, if we can find $\Phi_j(Y)$ functions such that functions $F_j(X)$ may be represented as functions of the variables Y only.

We shall consider the problem of finding the aggregation conditions for ecosystem models. For this we will solve the problem of representability of many variable functions, $F(X)$, as a function of only those variables, Y, that are related to X as in eqs. (8.2).

Representability of Functions in Aggregated Variables

Let us consider a multivariable function $F(X)$, which can be differentiated with respect to all N variables, and let us make the change of variables (8.2) in such a way that every function $\varphi_i(X)$ can be differentiated with respect to all its variables. Let us introduce the matrices:

$$M1 = \begin{bmatrix} \dfrac{\partial \varphi_1}{\partial x_1} & \dfrac{\partial \varphi_2}{\partial x_1} & \cdots & \dfrac{\partial \varphi_M}{\partial x_1} \\ \cdots & \cdots & \cdots & \cdots \\ \dfrac{\partial \varphi_1}{\partial x_N} & \dfrac{\partial \varphi_2}{\partial x_N} & \cdots & \dfrac{\partial \varphi_M}{\partial x_N} \end{bmatrix}, \quad M2 = \begin{bmatrix} \dfrac{\partial \varphi_1}{\partial x_1} & \dfrac{\partial \varphi_2}{\partial x_1} & \cdots & \dfrac{\partial \varphi_M}{\partial x_1} & \dfrac{\partial F}{\partial x_1} \\ \cdots & \cdots & \cdots & \cdots & \cdots \\ \dfrac{\partial \varphi_1}{\partial x_N} & \dfrac{\partial \varphi_2}{\partial x_N} & \cdots & \dfrac{\partial \varphi_M}{\partial x_N} & \dfrac{\partial F}{\partial x_N} \end{bmatrix}.$$

Definition 2. Let us call the function $F(X)$ *representable in the aggregated variables* defined by equations (8.2) if we can find a function $\Phi(Y)$ of M variables such that the following equation holds:

$$\Phi(Y) = \Phi(\varphi_j(X)) = F(X). \tag{8.4}$$

We can prove the following theorem (Luckyanov, 1981; Luckyanov et al., 1983).

Theorem 1. A function $F(X)$ is representable in the aggregated variables (8.2) if and only if the matrices $M1$ and $M2$ have the same rank.

Necessity. Let us assume that relation (8.4) is true. Differentiating it with respect to all its variables x_i, we get an algebraic system of N linear equations for each of M derivatives, $\partial \Phi(Y)/\partial y_i$:

$$\frac{\partial F(X)}{\partial x_i} = \sum_{j=1}^{M} \frac{\partial \Phi(Y)}{\partial y_j} \frac{\partial \varphi_j(X)}{\partial x_i}, \quad i = \overline{1, N}; \quad j = \overline{1, M}. \tag{8.5}$$

System (8.5) should have a nontrivial solution as it follows from the existence of a function $\Phi(Y)$ by assumption. Thus, the ranks of matrices $M1$ and $M2$ are equal.

Sufficiency. Matrices $M1$ and $M2$ have equal ranks. This means that there exist functions $\psi_j(X)$ such that the following system is true:

$$\frac{\partial F(X)}{\partial x_i} = \sum_{j=1}^{M} \psi_j(X) \frac{\partial \varphi_j(X)}{\partial x_i}, \quad i = \overline{1, N}; \quad j = \overline{1, M}. \tag{8.6}$$

Equations (8.6) mean that a relation exists between the functions $F(X)$ and $\varphi_j(X)$ such that, when $F(X)$ has been differentiated with respect to each of its variables, the function coefficient at each derivative for all arguments of each function $\varphi_j(X)$ does not depend on the number of variables i. By multiplying the right and left sides of system (8.6) by dx_i, summarizing, and then changing the order of summing on the right side, we get

$$dF(X) = \sum_{j=1}^{M} \psi_j(X) \, d\varphi_j(X).$$

This relation means that there exists a function Φ of M variables such that putting values $\psi_j(X)$

into it gives the function $F(X)$. Functions $\psi_j(X)$ are derivatives obtained as a result of differentiating the function $\Phi(Y)$ with respect to y_j:

$$\psi_j(X) = \partial\Phi(X)/\partial y_j .$$

Thus, the theorem is proved.

Corollary 1. Any differentiable function of aggregated variables of the function $F(X)$ is an aggregating variable.

Proof. Let us take a set of M differentiable functions $\eta_j(Y)$. Putting their values into the matrices $M1$ and $M2$, we can see that such an action corresponds to the multiplication of matrices $M1$ and $M2$ columns by derivatives $\partial\eta_j(y_j)/\partial y_j$. This does not change the matrix ranks for derivatives other than zero. This means, according to Theorem 1, that functions $\eta_j(y_j)$ are also aggregating variables.

MODEL SEPARABILITY

In the previous section, a general case was considered when aggregating transformations (8.2) are functions of the whole set of initial variables. However, in dealing with ecological models, we often encounter a somewhat different type of aggregation for which each set of variables of one trophic or energetic level is aggregated, and variables are constructed of nonoverlapping subsets of initial variables. In this case, the problem of aggregation of variables of a function of many variables becomes a problem of dividing the variables into nonoverlapping subsets. Let us introduce the following definition.

Definition 3. We shall say that the set of variables X of an n-dimensional function $F(X)$ is *m-separable* if we can find m ($m < n$) functions $\{\varphi_k(\mathring{X}_k)\}$, $k = \overline{1, m}$, dependent on variables of nonoverlapping subsets $\mathring{X}_k$ ($\mathring{X}_k \subset X$), such that the function $F(X)$ is representable in the following aggregated variables:

$$Y_k = \varphi_k(\mathring{X}_k), \quad k = \overline{1, m}; \quad \mathring{X}_i \cap \mathring{X}_j = \emptyset, i \neq j. \tag{8.7}$$

Let us prove the criterion of m-separability (Voronkova and Luckyanov, 1985).

Theorem 2. A set of variables X of an n-dimensional function $F(X)$ that can be differentiated by each of its variables is m-separable if and only if there are m functions ($m < n$) $\varphi_k(\mathring{X}_k)$ ($k = \overline{1, m}$; $\mathring{X}_K \subset X$; $\mathring{X}_i \cap \mathring{X}_j = \emptyset, i \neq j$) differentiated with respect to all its variables, such that the following equations are true:

$$\frac{\partial\varphi_k(\mathring{X}_k)}{\partial x_i}\frac{\partial F(X)}{\partial x_j} = \frac{\partial\varphi_k(\mathring{X}_k)}{\partial x_j}\frac{\partial F(X)}{\partial x_i}, \quad x_i, x_j \in \mathring{X}_k . \tag{8.8}$$

Necessity. Let us number the function variables so that the numbers of variables belonging to set $\mathring{X}_1$ are less than variable numbers belonging to set $\mathring{X}_2$, which, in their turn, are less than variable numbers belonging to set $\mathring{X}_3$, and so on. Consequently, this numbering of variables makes the expanded Jacobian matrix (matrix $M2$) look like a set of vertical columns:

$$M2 = \begin{bmatrix} \frac{\partial \varphi_1}{\partial x_1} & 0 & 0 & \cdots & 0 & 0 & \frac{\partial F}{\partial x_1} \\ \frac{\partial \varphi_2}{\partial x_2} & 0 & 0 & \cdots & 0 & 0 & \frac{\partial F}{\partial x_2} \\ \hdashline \frac{\partial \varphi_1}{\partial x_{k_1}} & 0 & 0 & \cdots & 0 & 0 & \frac{\partial F}{\partial x_{k_1}} \\ 0 & \frac{\partial \varphi_2}{\partial x_{k_1+1}} & 0 & \cdots & 0 & 0 & \frac{\partial F}{\partial x_{k_1+1}} \\ \hdashline 0 & \frac{\partial \varphi_2}{\partial x_{k_1+k_2}} & 0 & \cdots & 0 & 0 & \frac{\partial F}{\partial x_{k_1+k_2}} \\ 0 & & \frac{\partial \varphi_3}{\partial x_{k_1+k_2+1}} & \cdots & 0 & 0 & \frac{\partial F}{\partial x_{k_1+k_2+1}} \\ \hdashline 0 & 0 & 0 & \cdots & \frac{\partial \varphi_{m-1}}{\partial x_{m-1 \atop \sum_{i=1} k_i}} & 0 & \frac{\partial F}{\partial x_{m-1 \atop \sum_{i=1} k_i}} \\ 0 & 0 & 0 & \cdots & 0 & \frac{\partial \varphi_m}{\partial x_{m-1 \atop \sum_{i=1} k_i+1}} & \frac{\partial F}{\partial x_{m-1 \atop \sum_{i=1} k_i+1}} \\ \hdashline \end{bmatrix}, \quad (8.9)$$

where k_i is the number of variables that constitute the set $\mathring{X}_i$.

By definition, each function $\varphi_i(\mathring{X})$ is a function of its variables. So, the rank of the Jacobian matrix, corresponding to matrix (8.9), is equal to m. Then, according to Theorem 1, the rank of matrix (8.9) should also be equal to m if the transformation (8.7) is an aggregating one.

Let us consider all minors of matrix (8.9) of the order $m + 1$. We shall not consider such minors that are equal to zero, because their columns are equal to zero. Then, other minors should be of the following structure: each should have at least one row with a derivative of each of the functions from the set $\{\varphi_k(\mathring{X}_k)\}$. Because the number of such functions is equal to $m + 1$, and the number of rows in minors is $m + 1$, then obviously each minor should have two rows with different derivatives of one and the same function.

Let us assume that such a "double" function in a certain minor would be the function $\varphi_k(\mathring{X}_k)$. Then, let us assume that the derivatives of the function $\varphi_k(\mathring{X}_k)$ included in the given minor are calculated by the variables $x_i, x_j \in \mathring{X}_k$.

The majority, that is, $m - 1$, of columns of the given minor consists of only one term other than zero. Thus, expanding the determinant corresponding to the given minor in these columns and equating it to zero, we obtain that the rank of the given minor is determined by the value of the determinant of the second order:

$$\begin{vmatrix} \dfrac{\partial \varphi_k}{\partial x_i} & \dfrac{\partial F}{\partial x_i} \\ \dfrac{\partial \varphi_k}{\partial x_\lambda} & \dfrac{\partial F}{\partial x_\lambda} \end{vmatrix}.$$

Expanding this determinant and equating it to zero, we get the relation (8.8).

Varying the functions $\varphi_k(\mathring{X}_k)$, as well as the variables by which the functions are differentiated, we may show that relation (8.8) is true for each of m functions $\varphi_k(\mathring{X}_k)$ when differentiated by each of its variables. Thus, necessity is proved.

Sufficiency. It is easy to show, using relations (8.8), that the ranks of matrix $M2$ and matrix $M1$ corresponding to it are equal. To do this, as we did in proving necessity, it is necessary to consider all order $m + 1$ minors of matrix (8.9), which are not equal to zero. Expanding determinants, corresponding to minors, in "scarcely filled columns," we get that the rank of each is equal to m. From this it follows, according to Theorem 1, that the n-dimensional function $F(X)$ will be representable in new aggregated variables, determined by system (8.7). This proves the sufficiency of conditions (8.8).

Corollary 2. If two separated subsets $\mathring{X}_1$ and $\mathring{X}_2$ of the set of variables of the function $F(X)$ have at least one common element, then the total subset $\mathring{X} = \mathring{X}_1 + \mathring{X}_2$ will also be separable.

Proof. If we suppose that corresponding derivatives are not equal to zero, eqs. (8.8) for the subset $\mathring{X}_1$ can then be written

$$\frac{\partial F(X)/\partial x_j}{\partial F(X)/\partial x_i} = \psi_{ji}(\mathring{X}_1), \qquad x_i, x_j \in \mathring{X}_1.$$

Let us write down eqs. (8.8) in the same way for the set $\mathring{X}_2$ and assume that x_j will be common to both sets. If we divide all the relations for the subset $\mathring{X}_2$ with derivatives $\partial F(X)/\partial x_j$ by all the relations for the set $\mathring{X}_2$ with derivatives $\partial F(X)/\partial x_j$, we get a group of the following relations:

$$\frac{\partial F(X)/\partial x_k}{\partial F(X)/\partial x_i} = \frac{\psi_{ji}(\mathring{X}_1)}{\eta_{jk}(\mathring{X}_2)} = \xi_{ki}(\mathring{X}_1 + \mathring{X}_2),$$

where $\eta_{jk}(\mathring{X}_2)$ is a function on the right side of the relations for the subset $\mathring{X}_2$.

Let us assume that the subset $\mathring{X}_1$ is M-dimensional and the subset $\mathring{X}_2$ is L-dimensional. Then we get $(M - 1)(L - 1)$ last equations. According to Theorem 2, for each of the sets $\mathring{X}_1$ and $\mathring{X}_2$, we have $M(M - 1)/2$ and $L(L - 1)/2$ equalities of type (8.8) correspondingly. Joining these equations with the latter relations, we get $(M + L - 1)(M + L - 2)/2$ conditions for the total subset $\mathring{X}_1 + \mathring{X}_2$ with dimension $M + L - 1$. From this it follows that the function $F(X)$ is such that a problem of finding a function $M + L - 1$ of variables $\mathring{X}_1 + \mathring{X}_2$ that would satisfy the terms of Theorem 2 may be set.

The criterion of separability (8.8) solves some problems of mathematical ecology that bear no relation to the aggregation problem. For instance, let us consider the problem of trophic function appearance (Luckyanov et al., 1983). Let us take a model of a two-level predator–prey type system having the following form:

$$\begin{aligned} \dot{x}_i &= a_i x_i - b_i f_i(x_i) x_3 + \psi_i(x_i) \\ \dot{x}_3 &= -a_3 x_3 + x_3 \sum_{i=1}^{2} c_i f_i(x_i), \quad i = 1, 2, \end{aligned} \qquad (8.10)$$

where x_1, x_2, x_3 are population numbers of prey and predators accordingly. Here, a_3, a_i are coefficients of natural increase of prey and predator death rates, $f_i(x_i)$ is a trophic function of the ith prey, and $\psi_i(x_i)$ is a self-inhibition function.

Let us assume that in constructing model (8.10) the prey population was artificially divided into two (generally unequal) parts; for example, in system (8.10) the equations for prey describe the same population. Then the aggregating function has to be linear; for example,

$$y = \varphi(x_1, x_2) = K_1 x_1 + K_2 x_2 .$$

Substituting this into (8.10), we obtain the following equation for the aggregating variable, y:

$$y_1 = \sum_{i=1}^{2} K_i [a_i x_i - b f_i(x_i) x_3].$$

Then, making use of the separability condition (8.8) for the right part of the last equation, we get

$$a_1 - b_1 f'_1(x_1) x_3 = a_2 - b_2 f'_2(x_2) x_3 .$$

Assuming values of trophic coefficients in the equations and describing different parts of the same population as being the same, we come to the conclusion that

$$f_1(x_1) = d x_1 + g; \qquad f_2(x_2) = d x_2 + g, \qquad (8.11)$$

where d and g are constant coefficients. Thus, the condition of linearity of a within-species aggregating function gives us the trophic functions (8.11). Generally, such a trophic function may be represented as a piecewise linear function:

$$f(x) = \begin{cases} 0, & 0 \leq x < \alpha_1 \\ -A + \dfrac{A}{\alpha_1} x, & \alpha_1 \leq x < \alpha_2 \\ -A + \dfrac{A}{\alpha_1} \alpha_2, & \alpha_2 \leq x \end{cases} .$$

To estimate the properties of a piecewise linear trophic function, a set of computer runs was conducted to compare the properties of models with a piecewise linear and rational-fractional trophic function. In a known model of an oligotrophic lake, where a rational-fractional trophic function was used, this was replaced by an approximating piecewise linear function. The results showed that if the original trophic function was accurately approximated by a piecewise function all the results obtained from the original model could also be obtained by the model with the piecewise trophic function.

AGGREGATION OF LINEAR SYSTEMS

Linear models are widely used in ecological modeling, (O'Neill, 1979; Grant et al., 1988; Woolhouse and Harmsen, 1991), mathematical economics (Ven, 1976, Leontief and Duchin, 1986), pharmacology (Byrd et al., 1982) and other fields. Thus, methods of linear system aggregation are of great practical value. Further elaboration of aggregation methods for linear systems should stimulate a search for solution of the general aggregation problem.

Consider a linear system

$$Y = AX, \qquad (8.12)$$

where X and Y are vectors of the same dimension n, and A is a square $n \times n$ matrix. Let us change the variables in system (8.12) as follows:

$$\overline{X} = BX; \qquad \overline{Y} = BY, \qquad (8.13)$$

where B is an $n \times m$ matrix.

Definition 4. We shall call the substituted variables (8.13) a *linear aggregative transformation* for the system (8.12) if (1) $m < n$ and (2) the transformed variable $\overline{Y}$ can be represented as a function of only the transformed variable $\overline{X}$. The task is to find an aggregating matrix, B, and to study the properties of the aggregated system corresponding to those of matrix A.

General Solution of the Problem

Let us formulate the general solution in the form of a theorem (Luckyanov, 1984).

Theorem 3. For each square matrix, A, representing the linear system (8.12), we can construct a matrix polynominal $P_m(A')$ of power m ($m < n$) such that the $m \times n$ matrix, B, made up of several eigenvectors of $P_m(A')$ will be a linear aggregative transformation, transforming system (8.12) of order n to the system of order m ($m < n$).

Proof. Substituting Y from (8.12) for $\overline{Y}$ in (8.13), we get an expression relating $\overline{Y}$ to X:

$$\overline{Y} = BAX. \qquad (8.14)$$

Obviously, the question here is whether it is possible to represent the new variables $\overline{Y}$ as functions of the new variables $\overline{X}$. The problem of this type for multivariate functions has been solved in Theorem 1. As follows from (8.13), in the linear case for each of the multidimensional functions $\overline{y}_i(X)$, the Jacobian matrix corresponds to matrix B^T, where the symbol ^{T} denotes transposition. The extended matrix for each function $\overline{y}_i(X)$ is formed by adjoining the column $(B(I)A)^T$. Such a form of the additional term is determined by the fact that, from (8.14),

$$\overline{y}_i = B(I)AX,$$

where $B(I)$ is the ith row of matrix B. Generally, from the equality of the ranks of the Jacobian matrix and the extended one, it follows that the additional column is a linear combination of the columns of the Jacobian matrix. As a result, we get m conditions:

$$\sum_{i=1}^{m} \alpha_{ij} B^T(I) = A^T B(J), \qquad J = \overline{1, m}. \qquad (8.15)$$

Solving system (8.15) with respect to each of the columns $B^T(K)$, we get the equations

$$\left(\sum_{i=1}^{m} \gamma_i (A^T)^i \right) B^T(K) = \lambda B^T(K), \qquad K = \overline{1, m}. \qquad (8.16)$$

Equations (8.16) prove the theorem.

Comment 1. Equations (8.16) give an algorithm for the determination of aggregative linear transformations for the linear system (8.12). In fact, $\sum_{i=1}^{m} \gamma_i (A^T)^T$ with coefficients γ_i, determined by a_{ij}, constitutes the aggregative matrix.

Comment 2. It may seem that Theorem 3 contradicts an intuitive impression that not every linear system can be aggregated. In reality, the theorem by no means contradicts this hypothesis, but, on the contrary, proves it. The fact is that, when an intuitive aggregation is made, real transformations B are sought for real matrices A, while the majority of eigenvectors of the polynominal matrix (8.16) are imaginary, even for real matrices A.

To distinguish between linear systems, aggregated in real and imaginary domains, let us introduce the following definition.

Definition 5. Let us denote as *R-aggregatable* such a linear system for which at least one real linear aggregating transformation B exists. We can prove the following theorem.

Theorem 4. A real matrix A of linear system (8.12) is R-aggregatable if it has at least one real eigenvalue.

Proof. Let us aggregate system (8.12) to the first order. According to Theorem 3, aggregating transformations for such systems are the eigenvectors of the matrix A, and since one of the eigenvalues λ_i is real, the eigenvector corresponding to this λ_i has to be real, too. Thus, the considered system is R-aggregatable.

Definition 6. Let us call the transformed system

$$\overline{Y} = C\overline{X}$$

fully aggregatable if the matrix C is nonsingular.

Theorem 5. The linear system (8.12) is fully R-aggregatable to a system of the order l if matrix A is real, nonsingular, and has l real eigenvalues.

Proof. As is known from Gantmacher (1977), the eigenvalues $\overline{\lambda}_i$ of a matrix polynominal $g(A)$ are related to eigenvalues λ_i of matrix A by

$$\overline{\lambda}_i = g(\lambda_i).$$

From this follows that, if matrix A has l real eigenvalues, then for the matrix polynominal $g(A)$ we can get l real eigenvalues, too. Real eigenvectors, corresponding to these eigenvalues, will constitute the matrix B of the real aggregating transformation with linear independent lines. Since the rank of a matrix, which is the result of multiplication of the square matrix by the nonsingular one, does not change (Gantmacher, 1977), the real matrix C of order $\{m \times m\}$ is also nonsingular.

Obviously, when constructing aggregating transformations, the properties of matrix polynominals are of great importance. Especially interesting is the problem of interaction between matrix eigenvectors and the matrix polynominal constructed from it. The following theorem gives an answer to this problem (Luckyanov, 1984).

Theorem 6. For any subset, Ω, of m eigenvectors of the matrix A $(n \times n)$ $(n > m)$, corresponding to m simple eigenvalues of matrix A, a matrix polynominal may be constructed such that any linear combination of these vectors will be the eigenvector of this matrix polynominal.

Proof. As is known from Gantmacher (1977), any linear combination of eigenvectors of a matrix is itself an eigenvector of the matrix if and only if these vectors correspond to one and the same eigenvalue. On the other hand, each of the eigenvectors, $\{b_i\}$, of the matrix A, is at the same time an eigenvector of any polynominal $P(A)$ and, what is more, the eigenvalue λ, corresponding to it, is a polynominal $P(\lambda_i)$, where λ_i is the eigenvalue of matrix A corresponding to b_i. The polynominal $P(A)$ satisfies the terms of the theorem if, for all the eigenvalues λ_i corresponding to the vectors included in Ω, it would have the same value:

$$\lambda = P(\lambda_i), \quad i = 1, 2, \ldots, m.$$

Writing this polynominal in a usual form, we get a system of m linear algebraic equations with respect to P_j coefficients of the polynominal P:

$$\sum_{j=1}^{m} P_j \lambda_i^j = \lambda, \quad i = 1, 2, \ldots, m. \tag{8.17}$$

We can see from (8.17) that the Van der Mond determinant is the determinant of the system. Thus, it is always different from zero if none of the $\{\lambda_i\}$ are equal to zero and they all differ from one another ($\lambda_i \neq \lambda_k$, $i \neq k$). Thus, if all the above-mentioned conditions are observed, system (8.17) has only one solution, which gives the coefficients of the unknown matrix polynominal.

SEPARABILITY OF LINEAR MODELS

The problem of linear separability is a subproblem of the problem of linear aggregation considered in the previous section. Thus, the general statement of the problem will be the same. In the given linear system (8.12), we make a change of variables (8.13). The specific feature of the separability problem is that the aggregation of components is such that $S = n - m + 1$ variables of the system (8.10) become mapped into one component while others remain unchanged. Suppose that the aggregated variable in (8.13) is the last one. Then, for the matrix B we get

$$B = \begin{bmatrix} & & i & j & k & & \\ 1 & 0 & \cdots\cdots\cdots\cdots\cdots\cdots\cdots\cdots\cdots\cdots\cdots\cdots\cdots\cdots & 0 \\ 0 & 1 & 0 & \cdots\cdots\cdots\cdots\cdots\cdots\cdots\cdots\cdots\cdots\cdots & 0 \\ \cdots\cdots\cdots\cdots\cdots\cdots\cdots\cdots\cdots\cdots\cdots\cdots\cdots\cdots\cdots\cdots\cdots\cdots\cdots\cdots \\ 0 & \cdots\cdots\cdots\cdots\cdots\cdots\cdots\cdots\cdots\cdots\cdots\cdots & 0 & 1 \\ 0 & \cdots\cdots\cdots\cdots & \gamma_i^0 & \gamma_j^0 & \gamma_k^0 & \cdots\cdots\cdots & 0 \end{bmatrix} \tag{8.18}$$

Here index j over the column indicates the column number and entries γ_j are constants.

Separability Condition for Linear Systems

According to Theorem 2, a subset $\mathring{X}$ of the variables of a multidimensional function $F(X)$, $\mathring{X} \subset X$, is separable if and only if there is a function $\varphi(\mathring{X})$ such that the equations

$$\frac{\partial F(X)}{\partial x_i} \frac{\partial \varphi(\mathring{X})}{\partial x_j} = \frac{\partial F(X)}{\partial x_j} \frac{\partial \varphi(\mathring{X})}{\partial x_i} \tag{8.19}$$

are true for all $x_i, x_j \in \mathring{X}$.

Obviously, when aggregating (8.13) with matrix (8.18), each variable Y_l of the transformed system

Chap. 8   Model Aggregation: Mathematical Perspectives

$$\overline{Y} = BAX \tag{8.20}$$

should be a multidimensional function with a separable subset of variables $\mathring{X}$, with aggregated variables x_i, x_j, x_k. The separating function $\varphi(\mathring{X})$, according to (8.13) and (8.18) will be as follows:

$$\overline{X}(m) = B(m, *)X, \tag{8.21}$$

where $B(m, *)$ is the last row of matrix B. Substituting $\overline{Y}_l$ from (8.20), which has the following form,

$$\overline{X}(m) = B^{-1}A^{-1}\overline{Y}_l,$$

and $\overline{X}(m)$ from (8.21) into criterion (8.19), we get the criterion of separability of variables for linear systems:

$$B(l, *) [B(m, j) A(*, i) - B(m, i)A(*, j)] = 0, \quad l = \overline{1, m}, \tag{8.22}$$

where $A(*, i)$ is the ith column of matrix A, $i, j \in \Omega$; the set Ω includes the numbers of aggregated variables $i, j, \ldots, k$.

As follows from (8.22), a rather limited class of linear systems can be exactly separated. However, problems of this type quite often have to be solved for arbitrary systems as well. Thus, the problem arises of approximate separability of a group of variables for linear systems of type (8.12). Substituting in (8.22) the corresponding rows from matrix (8.18), we note that condition (8.22) divides matrix A into three parts:

(1) Matrix A columns whose numbers do not belong to the set Ω. Condition (8.22) does not concern these columns.

(2) Intersection of columns with numbers belonging to the set Ω and rows with numbers that do not belong to Ω. For these "subcolumns," condition (8.22) is satisfied only when these subcolumns are mutually proportional. This result is acquired when substituting rows $B(l, *)$, with $l = \overline{1, m-1}$, into (8.22).

(3) Substitution of the last matrix B row from (8.18) into (8.22) gives the whole-scale condition (8.22) for the intersection of columns and rows of matrix A with the numbers of set Ω. Now let us introduce a submatrix W of dimension $s \times s$, comprised of those matrix A members that lie on the intersection of columns and rows of matrix A with numbers from the set Ω. Matrix W describes the interaction of aggregated variables:

$$W = \begin{bmatrix} a_{ii} \; a_{ij} \; \ldots \; a_{ik} \\ a_{ji} \; a_{jj} \; \ldots \; a_{jk} \\ \ldots \ldots \ldots \end{bmatrix}.$$

Introducing the vector γ of dimension s with coordinates $\{\gamma_i\}$ from matrix (8.18), and assuming that all $\{\gamma_i\}$ differ from zero, we may present relations (8.22) with the aid of matrix W in the following form:

$$\frac{\gamma W(*, 1)}{\gamma_i} = \frac{\gamma W(*, 2)}{\gamma_j} = \cdots = \frac{\gamma W(*, m)}{\gamma_k}. \tag{8.23}$$

Assuming that the relations in (8.23) are equal to a constant, λ, we get the following linear system:

$$\begin{aligned} \gamma_i a_{ii} + \gamma_j a_{ji} + \cdots + \gamma_k a_{ki} &= \lambda \gamma_i \\ \gamma_i a_{ij} + \gamma_j a_{jj} + \cdots + \gamma_k a_{kj} &= \lambda \gamma_j \,. \\ \gamma_k a_{ik} + \gamma_j a_{jk} + \cdots + \gamma_k a_{kk} &= \lambda \gamma_k \end{aligned} \tag{8.24}$$

From (8.24) it follows that, if corresponding variables of system (8.12) are linearly separable and the separating matrix is as in (8.18), then the separability coefficients $\{\gamma_i\}$ from (8.18) should constitute an eigenvector of the submatrix W.

ALGORITHM FOR APPROXIMATE LINEAR SEPARABILITY

Obviously, if we take into account the whole condition (8.22), we come to a rather nonuniform and complex problem. Besides, it is easy to see that the condition of mutual proportionality of complete columns of matrix A with numbers belonging to Ω also satisfies condition (8.22). Thus, later we shall use a simplified (sufficient) condition of separability of a group of variables, that is, the condition of proportionality of matrix A columns with numbers from the set Ω.

Algorithm of Matrix Approximation (Alymkulov and Luckyanov, 1986)

Consider system (8.12) with an arbitrary real square matrix A. To aggregate it with the aid of transformations (8.13) and (8.18), it is necessary to construct a matrix $\tilde{A}$ that would be most similar to the original matrix A and at the same time would satisfy the simplified terms of (8.22). Let us try to construct matrix $\tilde{A}$ in the following form: columns with numbers not belonging to the set Ω are equal to corresponding columns of matrix A, whereas columns with numbers from the set Ω should be proportional to each other. Thus, the matrix should be as follows:

$$\tilde{A} = \begin{bmatrix} & & & i & & & j & & & k & & \\ a_{11} & a_{12} & \cdots & \alpha_i\beta_1 & a_{i\,1+1} & \cdots & \alpha_j\beta_1 & a_{1\,j+1} & \cdots & \alpha_k\beta_1 & a_{1\,k+1} & \cdots & a_{1n} \\ a_{21} & a_{22} & \cdots & \alpha_i\beta_2 & a_{2\,i+1} & \cdots & \alpha_j\beta_2 & a_{2\,j+1} & \cdots & \alpha_k\beta_2 & a_{2\,k+1} & \cdots & a_{2n} \\ & & & & & \vdots & & & & & & & \\ a_{n1} & a_{n2} & \cdots & \alpha_i\beta_n & a_{n\,i+1} & \cdots & \alpha_j\beta_n & a_{n\,j+1} & \cdots & \alpha_k\beta_w & a_{w\,k+1} & \cdots & a_{nn} \end{bmatrix}, \quad (8.25)$$

where elements a_{pq} are equal to elements of matrix A, respectively, α_j is a proportionality coefficient for the jth column, and β_l is the lth element of the "base" column. The values of the coefficients $\{\alpha_i\}$ and $\{\beta_l\}$ are to be determined. To estimate the proximity between matrices A and $\tilde{A}$, let us introduce a square functional, J, of distance between vectors Y determined by (12) and $\tilde{Y}$ of the form

$$\tilde{Y} = \tilde{A}X. \qquad (8.26)$$

Now the functional J will have the following form:

$$J = \Phi[(\tilde{Y} - Y)^2].$$

Substituting in the last equation functions Y and $\tilde{Y}$, determined by eqs. (8.12) and (8.26), we get

$$J = \Phi[X^T(\tilde{A}^T - A^T)(\tilde{A} - A)X], \qquad (8.27)$$

where ^{T} again means transposition. If we suppose that vector X is a known time function, then the functional (8.27) can be written as

$$J = \int_{T_1}^{T_2} X^T(\tilde{A}^T - A^T)(\tilde{A} - A)X\,dt, \qquad (8.28)$$

where T_1 and T_2 are time limits of a vector function, $X(t)$.

Let us set the task of finding parameters $\{\alpha_i\}$ and $\{\beta_l\}$ such that the functional (8.28) can be minimized. Substituting in (8.28) the matrix $\tilde{A}$ determined by (8.25), we get the following expression for J:

$$J = \sum_{r=1}^{n} \sum_{p,q \in \Omega} (\alpha_p \beta_r - a_{rp})(\alpha_q \beta_r - a_{rq}) X_{pq}, \tag{8.29}$$

where X_{pq} is determined by the following equation:

$$X_{pq} = \int_{T_1}^{T_2} X_p(t) X_q(t) \, dt. \tag{8.30}$$

As follows from (8.29), the functional J depends only on those elements of matrix A that are on the columns with numbers from the set Ω. Let us designate such a "submatrix" of matrix A by V; let us also introduce vectors α and β and matrix X with the corresponding parameters $\{\alpha_i\}$, $\{\beta_l\}$, and X_{pq}, determined by (8.30). Now the functional (8.28) can be presented as

$$J = \int_{T_1}^{T_2} X^T (\alpha \beta^T - V^T)(\beta \alpha^T - V) X \, dt. \tag{8.31}$$

Differentiating J from (8.31) with respect to one of the unknown parameters α_i, we get

$$\frac{\partial J}{\partial \alpha_i} = \overline{V} \int_{T_1}^{T_2} X^T [K1(i) \alpha^T + \alpha K1^T(i)] X \, dt - \int_{T_1}^{T_2} X^T [K1(i) \beta^T V + V^T \beta K1^T(i)] X \, dt,$$

where $V = \beta^T \beta$ and $K1(i)$ is an s-dimensional vector $(K1(i)j = \delta_{ij}$; δ_{ij} is K, Kronecker's delta).

Equating each derivative $\partial J/\partial \alpha_i$ to zero, we get a general expression for the whole set of derivatives of the functional J from (8.31) with respect to each α_i:

$$\overline{V} X \alpha = X V^T \beta.$$

Assuming now that matrix X is nonsingular and $\overline{V} \neq 0$, we get $\overline{V} \alpha = V^T \beta$.

$$\alpha = \frac{V^T \beta}{\overline{V}} \tag{8.32}$$

Let us differentiate J from (8.31) with respect to one of the unknown parameters, β_j, and introduce vector $K2(i)$ of dimension $w \times s$, which is similar to $K1(i)$. Now, reasoning in the way in which we arrived at eq. (8.32) and considering that $\alpha^T X \alpha \neq 0$, we obtain an expression for vector β:

$$\beta = \frac{V X \alpha}{\alpha^T X \alpha}. \tag{8.33}$$

Comment 3. Equations (8.32) and (8.33) enable us to construct an iterative procedure for finding parameter values $\{\alpha_p\}$ and $\{\beta_i\}$ minimizing the functional (8.28). To do this, given a certain set of values $\{\alpha_p^0\}$, with the aid of (8.33) we can find the parameter values $\{\beta_r^1\}$. Later, having acquired this, we can find, using (8.32), the values of $\{\alpha_p^1\}$ and again substitute them in (8.33) to get new values $\{\beta_r^2\}$, and so on.

We can present this process in an operational form if we substitute (8.32) into (8.33):

$$\beta = \frac{\beta^T \beta}{\beta^T V X V^T \beta} V X V^T \beta. \tag{8.34}$$

Thus, the problem of existence and uniqueness of the functional (8.28) minimum arises, as well as the problem of convergence of the iterative procedure, proposed in Comment 3.

Example 1

In a paper by Slatyer (1977), a Markov model of forest succession was described. The process of transition from a phase of a forest development consisting of gray birch (*Betula*), black gum (*Nyssa*), red maple (*Acer*), and beech (*Fagus*) was described by a 4 × 4 matrix representing transition probabilities (rows to columns) from one state to another during 50 years:

$$\begin{bmatrix} 0.05 & 0.01 & 0.00 & 0.00 \\ 0.36 & 0.57 & 0.14 & 0.01 \\ 0.50 & 0.25 & 0.55 & 0.03 \\ 0.09 & 0.17 & 0.31 & 0.96 \end{bmatrix}. \tag{A1}$$

Here, the first row and column positions denote gray birch; the second, black gum; the third, red maple; and the fourth, beech. A calculation of system behavior was also made, with initial conditions given by the vector $v_0 = (100, 0, 0, 0)^T$.

Suppose for some reason it is necessary to combine the data on black gum and red maple, which represent intermediate stages on the landscape. The question of interest is whether the given area will remain "intermediate," and we thus come to the problem of separating the second and third columns of matrix (A1).

By means of the algorithm discussed in Comment 3, the following approximation matrix (A1) can be obtained:

$$A = \begin{pmatrix} 0.005 & 0.004 & 0.004 & 0.00 \\ 0.36 & 0.31 & 0.33 & 0.01 \\ 0.50 & 0.41 & 0.435 & 0.03 \\ 0.09 & 0.24 & 0.258 & 0.96 \end{pmatrix}$$

According to eq. (8.14), the transformation and the new matrix have the following form:

$$B = \begin{pmatrix} 1 & 0 & 0 & 0 \\ 0 & 0.96 & 1.03 & 0 \\ 0 & 0 & 0 & 1 \end{pmatrix}; \quad C = \begin{pmatrix} 0.05 & 0.01 & 0.00 \\ 0.86 & 0.74 & 0.04 \\ 0.09 & 0.25 & 0.96 \end{pmatrix}. \tag{A2}$$

Table 8.1 shows the system trajectories generated by matrices (A1) and C [contracted according to the separability law (A2)]. As seen from the table, the values obtained by means of matrix C are rather close to those obtained by means of matrix A and aggregation.

Analysis of Iterative Procedure (Alymkulov and Luckyanov, 1986)

First, let us consider some features of the operator VXV^T. It is easy to show that this operator is nonnegative and symmetric. In fact,

$$VXV^T = (VX^{1/2})(VX^{1/2})^T$$

and

$$(u_0(VX^{1/2})(VX^{1/2})^T, u) = ((VX^{1/2})^T u, (VX^{1/2})^T u) \geq 0.$$

TABLE 8.1 Two-hundred-year trajectories of four tree species in two cases: (a) unaggregated, matrix A1; and (b) aggregated, matrix C.

	Age of Forest (years)				
	0	50	100	150	200
			(a)		
Gray birch	100	5	1	0	0
Black gum	0	36	29	23	18
Red maple	0	50	39	30	35
Beech	0	9	31	47	58
			(b)		
Gray birch	100	5	1	0	0
Intermediate component	0	86	68	52	41
Beech	0	9	31	48	59

Data represent percentages of the total population at each time.

This means that, in a certain orthonormalized basis, the matrix has a diagonal form with nonnegative elements along the diagonal. Obviously, the eigenvectors $\{e_i\}$ of the operator are invariant (that is, immovable at iterations). Thus, the eigenvectors determine the stationary points of the iterative procedure and, subsequently, extrema of the functional (8.28).

Let us consider now the question of the uniqueness of the minimum. To achieve this, we can reshuffle the λ_i so that

$$\lambda_1 > \lambda_2 \geq \ldots \geq \lambda_w. \tag{8.35}$$

Let β_0 be the initial approximation and $\beta_0 = \sum_i c_i e_i$, where c_i are coefficients of expansion in basis. From (8.32) and (8.33), we get

$$\beta_1 = \frac{\sum_i c_i^2 \sum_i \lambda_i c_i e_i}{\sum_i \lambda_i c_i^2},$$

$$\beta_2 = \frac{\sum_i c_i^2 \sum_i \lambda_i^2 c_i^2}{\sum_i \lambda_i c_i^2 \sum_i \lambda_i^3 c_i^2} \sum_i \lambda_i^2 c_i e_i,$$

and, in general,

$$\beta_w = \frac{\left[\left(\sum_i c_i^2\right)\left(\sum_i \lambda_i^2 c_i^2\right) \ldots \left(\sum_i \lambda_i^{2n-2} c_i^2\right)\right]}{\left(\sum_i \lambda_i c_i^2\right)\left(\sum_i \lambda_i^3 c_i^2\right) \ldots \left(\sum_i \lambda_i^{2n-1} c_i^2\right)} \sum_i \lambda_i^2 c_i e_i. \tag{8.36}$$

Let us assume that $c_1 \neq 0$, in which case the denominator of (8.33) is other than zero.

Theorem 7. If condition (8.35) is satisfied and $c_1 \neq 0$, then the functional minimum $J(\beta, \alpha)$ is unique and is reached on the vector l_1, corresponding to the greatest eigenvalue λ_1; that is,

$$J(\mathbf{e}_i) > J(\mathbf{e}_1), \quad \text{for } i > 1.$$

Before proving Theorem 7, let us note that a minimum of the square function $J(\beta, \alpha)$ does exist.

Proof. We are interested in $\min J(\alpha, \beta)$. Let us fix β and start searching for $\min J(\beta, \alpha)$. It exists also at the point of the minimum $\alpha = V^T\beta/\beta^T\beta$. However, $\partial J/\partial \alpha = 0$ at one point $\alpha = V^T\beta/\beta^T\beta$; that is, this is the point of the minimum and it is unique.

Let us take as an initial approximation

$$\beta_0 = \mathbf{e}_i + \mathbf{e}_1.$$

Then

$$\beta_1 = \lambda_{i1} \mathbf{e}_i + \lambda_1 \mathbf{e}_1,$$

$$\beta_{-n} = \mathbf{e}_i + \left(\frac{\lambda_i}{\lambda_1}\right)^n \mathbf{e}_1, \tag{8.37}$$

$$\beta_n = \mathbf{e}_1 + \left(\frac{\lambda_1}{\lambda_i}\right)^n \mathbf{e}.$$

Let us show that $J(\beta_{-n}, \alpha_{-n}) > J(\beta_{-n+1}, \alpha_{-n+1})$. Now, we fix β_{-n}. With the aid of (8.32) we search for the corresponding α_{-n}. $\min_\beta J(\beta, \alpha_{-n})$ gives $\partial J/\partial \beta = 0$, whence

$$\beta_{-n+1} = \frac{VX\alpha_{-n}}{\alpha^T_{-n} V\alpha_{-n}}.$$

Each iterative step improves:

$$J(\beta_{-n}, \alpha_{-n}) \geq J(\beta_{-n+1}, \alpha_{-n}) \geq J(\beta_{-n+1}, \alpha_{-n+1}).$$

As follows from (8.36), the first inequality is strict, $[J(\beta_{-n}) \neq J(\beta_{-n+1})]$, because $\beta_{-n} \neq \beta_{-n+1}$. Thus, $J(\beta_{-n}, \alpha_{-n}) > J(\beta_{-n+1}, \alpha_{-n+1})$ and, in general,

$$J(\beta_{-n}, \alpha_{-n}) > J(\beta_{-n+1}, \alpha_{-n+1}) > \ldots > J(\beta_0, \alpha_0) > J(\beta_1, \alpha_1) > \ldots > J(\beta_n, \alpha_n).$$

For $n \to \infty$, $J(\beta_{-n}, \alpha_{-n}) \to J(\mathbf{e}_i)$ and $J(\beta_n, \alpha_n) \to J(\mathbf{e}_1)$. In the extreme we get $J(\mathbf{e}_i) > J(\mathbf{e}_1)$, $i > 1$.

Let us consider the question of convergence of norms. As follows from (8.36),

$$\|\beta_k\| = \frac{\left(\sum_i c_i^2\right) \ldots \left(\sum_i \lambda_i^{2n-2} c_i^2\right)}{\left(\sum_i \lambda_i c_i^2\right) \ldots \left(\sum_i \lambda_i^{2n-1} c_i^2\right)} \left(\sum_i \lambda_i^{2n} c_i^2\right)^{1/2}. \tag{8.38}$$

Let us assume that $h_k = \|\beta_k\|/\|\beta_{k-1}\|$ and analyze the convergence $\prod_{k=1}^{\infty} h_k$. For this it is sufficient to show the convergence of $\sum_{k=1}^{\infty} |h_k - 1|$, where

$$h_k - 1 = \frac{\sum_{i=2}^{n} \lambda_i^{2k-2} c_i^2 \left(\sum_{i=2}^{n} \lambda_i^{2k} c_i^2\right)^{1/2} - \sum_{i=2}^{n} \lambda_i^{2k-1} c_i^2 \left(\sum_{i=1}^{n} \lambda_i^{2k+1} c_i^2\right)^{1/2}}{\sum_{i=1}^{n} \lambda_i^{2k-1} c_i^2 \left(\sum_{i=1}^{n} \lambda_i^{2k-2} c_i^2\right)^{1/2}}.$$

Let us divide the numerator and denominator by $\lambda_1^{3k-2} c_1^2$ to get:

$$h_k - 1 = \frac{\left[1 + \sum_{i=2}^{n}\left(\frac{\lambda_i}{\lambda_1}\right)^{2k-2}\left(\frac{c_i}{c_1}\right)^2\right]\left[1 + \sum_{i=2}^{n}\left(\frac{\lambda_i}{\lambda_1}\right)^{2k}\left(\frac{c_i}{c_1}\right)^2\right]^{1/2} - \left[1 + \sum_{i=2}^{n}\left(\frac{\lambda_i}{\lambda_1}\right)^{2k-1}\left(\frac{c_i}{c_1}\right)^2\right]\left[1 + 2\sum_{i=2}^{n}\left(\frac{\lambda_i}{\lambda_1}\right)^{2k+1}\left(\frac{c_i}{c_1}\right)^2\right]^{1/2}}{\left[1 + \sum_{i=2}^{n}\left(\frac{\lambda_i}{\lambda_1}\right)^{2k-1}\left(\frac{c_i}{c_1}\right)^2\right]\left[1 + \sum_{i=2}^{n}\left(\frac{\lambda_i}{\lambda_1}\right)^{2k-2}\left(\frac{c_i}{c_1}\right)^2\right]^{1/2}}.$$

We can expand the numerator into a Taylor series:

$$1 + \sum_{i=2}^{n}\left(\frac{\lambda_i}{\lambda_1}\right)^{2k-2}\left(\frac{c_i}{c_1}\right)^2 + \frac{1}{2}\sum_{i=2}^{n}\left(\frac{\lambda_i}{\lambda_1}\right)^{2k}\left(\frac{c_i}{c_1}\right)^2 - 1 - \sum_{i=2}^{n}\left(\frac{\lambda_i}{\lambda_1}\right)^{2k-1}\left(\frac{c_i}{c_1}\right)^2 \quad (8.39)$$

$$- \frac{1}{2}\sum_{i=2}^{n}\left(\frac{\lambda_i}{\lambda_1}\right)^{2k+1}\left(\frac{c_i}{c_1}\right)^2 + o(\cdot).$$

The denominator has the order of 1 and does not affect the convergence process. For any i, each sum in (8.39) is a geometric progression, and the whole numerator is a finite sum of geometric progressions; that is, the series $\sum_{k=1}^{\infty}|h_k - 1|$ does converge.

Thus, the sequence $\{\|\beta_k\|\}$ converges, and note that it converges to $\mathbf{e}_1$. Let us show this. The vector $\overline{\beta}$ is collinear to the vector $\mathbf{e}_1 + \sum_{i=2}^{n}\left(\frac{\lambda_i}{\lambda_1}\right)^k\left(\frac{c_i}{c_1}\right)\mathbf{e}_i$, and

$$\left\|\mathbf{e}_1 + \sum_{i=2}^{n}\left(\frac{\lambda_i}{\lambda_1}\right)^k\left(\frac{c_i}{c_1}\right)\mathbf{e}_i\right\| \underset{k\to\infty}{\to} \|\mathbf{e}_1\|.$$

Thus, $\beta_k/\|\beta_k\| \to \mathbf{e}_1/\|\mathbf{e}_1\|$, and as the basis is orthonormalized, $\beta_k/\|\beta_k\| \to \bar{\mathbf{e}}_1$. Thus, the sequences $\{\|\beta_k\|\}$ and $(\beta_k/\|\beta_k\|)$ converge; consequently, the sequence $\{\beta_k\}$ converges to the vector collinear to $\bar{\mathbf{e}}_1$.

The minimum of the function $J(\beta, \alpha)$ does exist if, by expanding $\beta_0 = \sum_i c_i\bar{\mathbf{e}}_i$ ($c_1 \neq 0$) (this was shown above). The uniqueness of the minimum gives the condition $\lambda_1 > \lambda_2 \geq \ldots \geq \lambda_n$. If λ_1 has a multiplicity greater than 1, then the minimum will be any vector of the subspace, stretched on the eigenvectors, corresponding to λ_1. The procedure will converge to the projection of the initial approximation on this space. The matrix of the aggregated system is easily derived from the equation $B\tilde{A} = CB$ if the structures of matrices B and $\tilde{A}$ are known. The matrix has the following form:

$$C = \left[\begin{array}{c|c} & \beta_1 \\ & \vdots \\ & \beta_k \\ \hline \alpha V & \lambda \end{array}\right], \quad (8.40)$$

where λ is the eigenvalue of the matrix W, which describes the interaction of separable variables in the initial matrix A.

Example 2

A simulation model for a large oligotrophic lake was described by Voinov et al. (1981). The main processes were described by the state variables phytoplankton, zooplankton, benthos, fishes, bacterioplankton, detritus, and nutrients, expressed as concentrations or densities per unit water volume.

By an aggregation algorithm this seven-component model was aggregated to a six-component one, combining bacterioplankton and detritus. These components were chosen for aggregation because methods of identifying and counting bacterioplankton are poorly developed, the role of bacterioplankton in lake ecosystems is still in question, and detritus and bacterioplankton usually are not discriminated in empirical studies (Loucks et al., 1979; Riley and Stefan, 1988).

A nonlinear system of differential equations was linearized around a steady state. Analysis of the model results parameterized for Onega Lake (Voronkova, 1984) showed that in winter months (November to April) the concentration of components in the lake does not change, or changes only insignificantly. Thus, the procedure of substituting the original model with a linearized one is justified, at least for these months. For a typical winter temperature of $+3°C$, the seven-component linearized model is as follows:

$$\dot{x}_1 = -0.4407x_1 - 1.5908x_2 + 3.3368x_7$$
$$\dot{x}_2 = 0.5519x_1 - 2.3497x_2 - 0.5032x_3 - 0.629x_4 + 2.3812x_5$$
$$\dot{x}_3 = 0.0065x_2 - 5x_3$$
$$\dot{x}_4 = 0.0014x_2 - 0.0023x_3 - 0.0003x_4 + 0.0005x_5 \quad (A3)$$
$$\dot{x}_5 = -1.4101x_2 - 0.0019x_3 - 0.0098x_4 - 1.9427x_5 + 0.2452x_6$$
$$\dot{x}_6 = 0.15x_1 + 0.9684x_2 + 0.386x_3 + 0.3899x_4 - 1.0144x_5$$
$$\quad - 0.253x_6 + 0.4561x_7$$
$$\dot{x}_7 = -1.2626x_1 - 0.145x_5 + 0.8368x_6 - 3.474x_7$$

For system (A3),

$$B = \begin{bmatrix} 1 & 0 & 0 & 0 & 0 & 0 & 0 \\ 0 & 1 & 0 & 0 & 0 & 0 & 0 \\ 0 & 0 & 1 & 0 & 0 & 0 & 0 \\ 0 & 0 & 0 & 1 & 0 & 0 & 0 \\ 0 & 0 & 0 & 0 & 0 & 0 & 1 \\ 0 & 0 & 0 & 0 & -0.58 & 0.84 & 0 \end{bmatrix}. \quad (A4)$$

Then, matrix C for the system expressed in terms of separated variables is

$$C = \begin{bmatrix} -0.44 & -1.59 & 0 & 0 & 3.34 & 0 \\ 0.55 & -2.35 & -0.50 & -0.63 & 0 & 0.1 \\ 0 & 0.007 & -5.00 & 0 & 0 & 0 \\ 0 & 0.001 & -0.002 & 0 & 0 & 0 \\ -1.26 & 0 & 0 & 0 & 0.46 & 0.8 \\ 0.126 & 1.631 & 0.325 & 0.333 & 0.36 & 0.156 \end{bmatrix}. \quad (A5)$$

Thus, system (A3) is substituted by system (A5), which, in matrix form, can be written as

$$\dot{\overline{X}} = C\overline{X},$$

where $\overline{X} = (x_1, x_2, x_3, x_4, x_5, x_6)^T$. Here, components 1 to 4 correspond to the same components in (A3), while x_5 corresponds to original x_7, and x_6 is an aggregate of x_5 and x_6 in (A3).

Comparison of the results between models (A3) and (A6) shows that they are fairly close, well within the limits of precision of the method.

CONCLUSION

The task of decreasing the dimension of mathematical models of ecosystems is one of the most important in simulation modeling theory. Until quite recently, this field has been poorly studied. Now, a certain step forward has been made in aggregation theory, as well as in the construction of variable aggregation methods for real systems. However, the aggregation problem is far from being solved in a general way. It is our hope that the results presented above will stimulate further research in this field. Model aggregation is crucial to all future modeling of complex systems.

REFERENCES

ALYMKULOV, E. D., and LUCKYANOV, N. K. 1986. Method of approximative aggregation in linear systems. *J. Comp. Math. Phys.* 26:1316–1324 (In Russian).

BYRD, R. A., YOUNG, J. F., Kimmel, C. A., Morris, M. D., and Holson, J. F. 1982. Computer simulation of mirex pharmokinetics in the rat. *Toxicol. Appl. Pharmacol.* 66:182–192.

GANTMACHER, F. R. 1977. *The Theory of Matrices.* Chelsea Publishing, New York. Vol. 1, 374 pp.; Vol. 2, 277 pp.

GRANT, W. E., MATIS, J. H., and MILLER, W. 1988. Forecasting commercial harvest of marine shrimp using a Markov chain model. *Ecol. Mod.* 43:183–195.

LANDAU, L. D., and LIFSHITZ, E. M. 1980. *Statistical Physics.* Pergamon Press, New York. 387 pp.

LEONTIEF, W. W., and DUCHIN, F. 1986. *The Future Impact of Automation on Workers.* Oxford University Press, New York. 170 pp.

LOUCKS, O. L., (many coauthors). 1979. In Breck, J. E., Prentki, R. T., and Loucks, O. L. (eds.), *Aquatic Plants, Lake Management, and Ecosystem Consequences of Lake Harvesting.* Proc. Conf. February 14–16, 1979. Madison, Wisconsin. 435 pp.

LUCKYANOV, N. K. 1981. Aggregation in simulation models of ecological systems. *Izv. AN SSSR, Tekh. Kibernet.* 5:30–35. (In Russian).

LUCKYANOV, N. K. 1984. Linear aggregation and separability of models in ecology. *Ecol. Mod.* 21:1–12.

LUCKYANOV, N. K., SVIREZHEV, Y. M., and VORONKOVA, O. V. 1983. Aggregation of variables in simulation models of water ecosystems. *Ecol. Mod.* 18:235–240.

MATIN, A. V. 1985. *Decomposition and Aggregation in Optimization of Economic Models.* Sovetskoe Radio, Nauka, Moscow (in Russian). 69 pp.

O'NEILL, R. V. 1979. A review of linear compartmental analysis in ecosystem science. In Matis, J. H., Patten, B. C., and White, G. C. (eds.), *Compartmental Analysis of Ecosystem Models.* International Coop. Publ. House, Fairland, Maryland. 368 pp.

RILEY, M. J., and STEFAN, H. G. 1988. MINLAKE: a dynamic lake water quality simulation model. *Ecol. Mod.* 43:155–182.

SLATYER, R. O. (ed.). 1977. Dynamic Changes in Terrestrial Ecosystems: Patterns of Change, Techniques for Study and Application to Management. MAB Technical Notes, 4. UNESCO, Paris. 30 pp.

VEN, V. L. 1976. Linear model aggregation. *Izv. AN SSSR, Tekh. Kibernet.* 2:3–11 (in Russian).

VOINOV, A. A., VORONKOVA, O. V., LUCKYANOV, N. K., and SVIREZHEV, Y. M. 1981. Modelling of lake ecosystems. In *Problems of Ecological Monitoring and Ecosystems Modelling.* Leningrad Hydrometeoizdat 4:204–234 (in Russian).

VORONKOVA, O. V. 1984. Ecological-geographic review of Onega Lake. In *Mathematical Models of Water Ecosystems.* Computer Center, USSR Academy of Science, Moscow (in Russian), pp. 31–74.

VORONKOVA, O. V. and LUCKYANOV, N. K. 1985. Linear models separability. In *Problems of Ecological Monitoring and Ecosystems Modelling.* Leningrad *Hydrometeoizdat* 7:209–218 (in Russian).

WOOLHOUSE, M. E., and HARMSEN, R., 1991. Population dynamics of *Aphis pomi:* a transition matrix approach. *Ecol. Mod.* 55:103–112.

ZHOROV, Y. N., PANCHENKOV G. M., et al. 1967. *Mathematical Description and Optimization of Processes of Oil Refining and Oil Chemistry.* Leningrad, Khimia (in Russian).

9

Simulation Models of an Estuarine Macrophyte Ecosystem[1,2]

W. MICHAEL KEMP
Horn Point Environmental Laboratories, Cambridge, Maryland

WALTER R. BOYNTON
Chesapeake Biological Laboratory, Solomons, Maryland

ALBERT J. HERMANN
Department of Oceanography, University of Washington, Seattle, Washington

A model is a synthesis of microhypotheses, i.e., a synthesis of information about structure and function of the system [Van Dyne, in *Ohio J. Sci.* 78, 197 (1978)].

Final simulation models are complex in appearance, but they are built up from simple mathematical statements and statistical distributions which represent the functions, interrelationships, and values attributed to the real world ecosystem. Numerical simulation is important; it is the only known technique which is capable of representing the complexities of real ecosystems [Van Dyne, in *Proc. XI Int. Grassland Conf.*, p. A142 (1970), U. Queensland Press].

INTRODUCTION

Estuaries are complex and dynamic ecological systems that interact with human societies in many ways. These coastal ecosystems provide a bountiful source of fisheries production and diverse commercial and recreational opportunities. Natural biogeochemical processes within these systems

---

[1] University of Maryland Center for Environmental and Estuarine Studies Contribution No. HPEL-94-2499.

[2] An earlier version of this paper appeared in a special NOAA report (Marine Ecological Modeling. K. Turgeon (ed.), National Oceanic and Atmospheric Administration, Environmental Data and Information Service, Washington, D.C., 1983).

are also capable of transforming many wastes emanating from human activities into useful components of regional and global cycles. Excessive waste inputs can, however, overwhelm the natural cycles. In such cases, alternative uses of estuarine resources may conflict with one another. In general, environmental research in estuarine systems has contributed significantly to both the basic understanding of ecological processes and the resolution of these natural resource conflicts.

In recent decades, many estuaries have undergone substantial modifications in species composition and ecological relations. One such change, which has occurred in Chesapeake Bay since the mid-1960s, is the drastic decline of the submersed vascular plants that once dominated the estuary's littoral region. Coincident with this loss of aquatic plants, there have been significant changes in water quality (Stevenson and Confer, 1978), as well as declines and shifts in various fisheries in the Bay (Boynton et al., 1979).

An extensive research effort, involving field and laboratory experiments, combined with ecological modeling and quantitative resource assessment, was undertaken to investigate this problem for upper Chesapeake Bay (Kemp et al., 1980). These studies documented the importance of macrophyte communities as habitats and feeding areas for diverse fish and waterfowl, as "filters" which trap and bind suspended sediments, and as storages and sinks for plant nutrients (Ward et al., 1984; Kemp et al., 1984). Research results have also suggested that continuing increases in sediment and, especially, nutrient loading from the bay's watershed have led to serious deterioration of light conditions needed for macrophyte growth; other anthropogenic wastes, such as agricultural herbicides, may have added additional (but lesser) stresses leading to the macrophyte decline (Kemp et al., 1983a).

In this chapter, we describe the simulation modeling framework that was used to organize, focus, and elaborate a broad empirical research program investigating the loss of macrophyte communities in upper Chesapeake Bay. We provide selected results of modeling studies to illustrate the wide range of conditions the models were capable of simulating, the ability of models to provide insight into complex ecological relations, and the utility of model hindcasts and forecasts in developing resource management strategies.

RESEARCH ORGANIZATION AND DESIGN

This research program addressed three broad questions. What factors were responsible for the macrophyte decline? To what extent do such macrophyte communities influence the overall estuarine ecosystem and which ecological interactions are most critical for maintenance of the macrophyte populations? What resource management options are most likely to succeed in restoration of the macrophyte communities? We established at the outset the following philosophical objectives for our research design (Levins, 1966): to use "controlled" experiments for discerning mechanistic relations among ecological factors, to interpret these experimental results in relation to "realistic" conditions actually occurring in nature, and to extend these results into a "general" context relevant for the whole estuary and other related ecosystems (Kemp et al., 1980). To meet these broad objectives, we employed a hierarchical array of research methods, combining bench-scale bioassays with laboratory microcosms, outdoor mesocosms, and descriptive field studies.

A variety of conceptual and simulation models were utilized to integrate this research program. It was reasoned that models could facilitate the coupling of experimental findings on relationships of causality or influence (Patten, 1985) with the inherently holistic perspective of descriptive in situ observations (Odum, 1984). Furthermore, simulation models were used to confer generality upon specific results at either end of the controllability–realism spectrum (Kemp et al., 1980). This was done by constructing, calibrating, and verifying models with data from a variety of systems. Thus, for example, simulation models were used to examine the ecological effects of various changes in water quality conditions characteristic of different regions of Chesapeake Bay. These models were also used to interpolate and extrapolate the experimental results over a wide range of environmental conditions observed (past or present) in nature.

Two distinctly different strategies for simulation modeling were central in this research program. One strategy was directed primarily toward understanding the dynamic behavior of the macrophyte ecosystems, including carbon and nutrient flux and cycling, resource competition, and trophic interactions. A hierarchical perception was used to decompose the complex macrophyte ecosystem into a cluster of simplified subsystem models. This allowed sufficient ecological detail to be maintained against conceptual and computational limitations. The second modeling approach emphasized the role of these plant communities in a larger context of the entire estuarine system, including socioeconomic considerations. In this case, an aggregated version of the overall macrophyte ecosystem model was developed, emphasizing interactions with human systems. This model was placed into a sequence of cascading connections of influence, relating human uses of the estuary for waste disposal to macrophyte ecosystem dynamics, to human uses of the estuary as a source of fisheries harvest and recreational activities.

MACROPHYTE ECOSYSTEM MODEL

Ecosystem Modeling Framework

An initial step in developing a simulation model of the macrophyte ecosystem involved identifying both an appropriate level of aggregation and the essential state variables. Population time constants (Goodall 1974; Schaffer, 1981), as well as life histories, trophic relations, and habitats (Boling et al., 1975), were considered in defining aggregated biological state variables. Other simplifications include consideration of only one plant nutrient as potentially rate limiting.

In an effort to retain the essential ecological features of this system while maintaining computational tractability, the macrophyte system was decomposed into six subsystem models. Similar approaches have been used by previous investigators (Goodall, 1974; Overton, 1975; McIntire and Colby, 1978). These subsystem models contain between six and ten state variables, a total similar to the size of previously reported macrophyte simulation models (Titus et al., 1975; Belyaev et al., 1977; Ferguson and Adams, 1979; Short, 1980; Weber et al., 1981; Verhagen and Nienhuis, 1983; Van Montfrans et al., 1984).

Subsystems were defined so as to maximize internal interactions and minimize connection with external variables and feedbacks from other subsystem models (Simon, 1973). The resulting subsystems are (1) the autotroph model, which considers competition for light and nutrients among major photosynthetic groups of organisms; (2) the epibiota model, which describes the habitat on macrophyte leaf surfaces; (3) the water/plankton model, which includes suspended and dissolved substances; (4) the benthos model, which includes organisms, sediments, and biogeochemical processes; (5) the mobile invertebrates model, which simulates populations moving among other subsystems; and (6) the nekton model, which includes higher trophic levels supported by production from other subsystems. There are 45 state variables contained in all six subsystem models; however, 8 of these reappear in more than one subsystem. This redundancy of variables means that the state spaces overlap, and it further ensures consistency in the overall behavior of the macrophyte ecosystem model and its subsystem simulations. The number of common variables in subsystem models decreases away from the autotroph model, suggesting a reduction in the number of direct interactions among variables at higher trophic levels.

These models were designed to represent a unit area of water and sediment in an estuarine macrophyte ecosystem, with spatial averaging implied. Both carbon (C) and nitrogen (N) are modeled in this scheme, where N is conserved within the model during all transactions, while C is transformed (with CO_2 making the difference) as needed according to prescribed C:N ratios for all biological state variables. Flows of both C and N are crucial to the behavior of this ecosystem. However, to include both with completely conserved materials would require nearly twice the number of variables. Several previous modeling studies have explicitly considered both C and N (e.g., Walsh, 1975 a,b; Kremer and Nixon, 1978; Hopkinson and Day, 1977; Najarian and Taft, 1981). However, most ecosystem models have been confined to tracing the flows of either carbon (energy) or nutrients, but not both (Najarian and Harleman, 1977; Wetzel and Wiegert, 1983).

The mathematical structure of this model uses nonlinear, first-order, differential equations simulated by finite difference techniques. There is one equation for each state variable, and each term in an equation represents an interaction between variables. The time step for rectangular integration was set at 2 hours. In the following section, we describe the structure and report selected simulation results for one macrophyte subsystem, the autotroph model.

Autotroph Model Structure

A major objective in developing the autotroph subsystem model was to examine the consequence of different environmental conditions on the competitive balance among the primary producers in a macrophyte-dominated community. The general structure of this model is depicted in Fig. 9.1, where phytoplankton, epiflora, macrophytes, and benthic microalgae all compete for limited availabilities of light and nutrients. Competition for light occurs through direct shading, while nutrient competition involves two separate sources of nitrogen (water column and sediment pore waters), which undergo periodic depletion of supplies. Only the rooted vascular plants have direct access to both nutrient sources. The original versions of this model included only seven state variables connected to numerous external factors (Kemp et al., 1983b); however, the current model includes both inorganic and organic detritus, as well as sessile (colonial) epifauna, as part of the total epiphytic community inhabiting macrophyte leaf surfaces.

The nature of mathematical formulations used in this and related models can be illustrated with the primary production term in the macrophyte growth equation:

$$P = [C/N][ATTEN][LKIN][TEMP][NKIN][LAI]. \tag{9.1}$$

Here, production (P) is a multiplicative function of six auxiliary variables: [C/N], the nitrogen-to-carbon conversion; [ATTEN], the light attenuation relation; [LKIN], the photosynthesis–irradiance function; [TEMP], the temperature kinetics; [NKIN], the nitrogen uptake relation; and [LAI], an index of leaf area representing the ability to absorb photons. Light attenuation follows a simple Beer–Lambert relation, with various materials contributing to the effect (e.g., Parsons et al., 1977):

$$I_z = I_0 e^{-kz}, \tag{9.2}$$

where I_z and I_0 are light levels at depth z and at the water surface, respectively. The attenuation coefficient k is taken as the sum of individual k's for seston, phytoplankton, epiphytic material, and vascular plant leaves, where each k is a linear function of the amount of material per square meter, with the overall intercept attributable to dissolved substances and the water itself. The photosynthesis–irradiance relation is approximated by a rectangular hyperbola (Parsons et al., 1977):

$$P = P_m \frac{I_z}{I_k + I_z}, \tag{9.3}$$

where P_m is the light-saturated (maximum) photosynthesis, and I_k is the light level at the intersection of P_m and the initial slope. Data for all the light relations were obtained from experiments in our laboratory using the macrophyte species *Potamogeton perfoliatus* (Goldsborough, 1983). The temperature (T) function used is a simple Arrhenius relation,

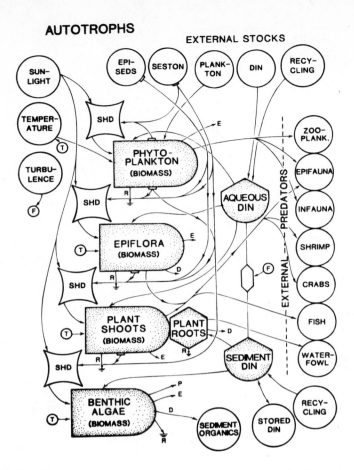

Figure 9.1 Conceptual diagram of autotroph ecosystem model depicting interactions among four autotrophic groups (phytoplankton, epiflora, macrophyte plants, and benthic algae) that compete for limited availabilities of light and dissolved inorganic nitrogen (DIN). Sunlight reaching each autotroph is reduced by shading (SHD) associated with the autotrophs themselves, along with seston and epiphytic sediments (Epi-Seds) on macrophyte leaves. External forcing functions are represented by circles; interactions indicated by lines with arrows; state variables are represented by shaded symbols. Symbols here are based on Odum (1971) (adapted from Kemp et al., 1983b).

$$\text{TEMP} = e^{-(K_t/T)} . \tag{9.4}$$

Values of K_t were obtained from the literature for related macrophyte species (Titus and Adams, 1979; Barko and Smart, 1981), and these were calibrated for *P. perfoliatus* using field data (Kemp et al., 1984). A higher-order equation (Johnson et al., 1974), which accounts for stress at high temperature via protein denaturation, was used in related models (see "Management Model" section).

Little information was available concerning the appropriate algebraic expression for describing macrophyte nitrogen uptake (V) from two sources (water column and sediment pore water). Most published experimental studies have not addressed the question of appropriate rate kinetics for simultaneous uptake from two sources (Iizumi and Hattori, 1982; Thursby and Harlin, 1982; Short and McRoy, 1984). The formulation used here was analogous to the Michaelis–Menten relation, with a single, overall maximum uptake rate [$V_m = f(P_m)$], but differing half-saturation constants for each of the two uptake routes:

$$V = V_m \frac{N_a + k^*N_b}{K_s + (N_a + k^*N_b)}, \tag{9.5}$$

where N_a and N_b are water-column and pore-water nitrogen concentrations (primarily NH_4^+), K is the half-saturation constant for uptake of N_a, and (K_s/k^*) is the half-saturation for N_b. The need to calibrate this kinetic expression and the absence of information in the literature motivated us to conduct appropriate experiments with *P. perfoliatus* (Kemp et al., 1981, 1984). Similar expressions were used to describe light, nutrient, and temperature interactions in primary production of other autotrophic groups in this model.

Autotroph Model Results

The dynamic behavior of the macrophyte ecosystem was simulated for three different physical environments (Figs. 9.2 and 9.3): an open embayment characterized by rapid exchange with external estuarine waters, a protected cove with more restricted tidal flushing, and experimentally fertilized

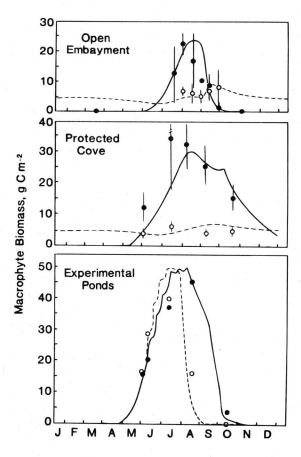

Figure 9.2 Annual distributions of macrophyte biomass in an open embayment and a protected cove in upper Chesapeake Bay and the experimental ponds receiving bay water. Lines represent output from the autotroph ecosystem model, while circles and intersecting vertical bars are means ± standard errors for field data. In the upper and middle panels, solid lines and filled circles are for shoot (leaf and stem) biomass, and dashed lines and open circles are for root (plus rhizome) biomass. In the bottom panel, solid lines and filled circles are for ponds receiving low fertilization rates, and dashed lines and open circles are for ponds receiving high fertilization rates.

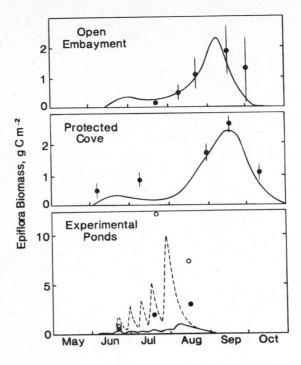

Figure 9.3 Temporal distributions of epifloral biomass on macrophyte leaves from May through October as estimated from the autotroph model and field data. See Fig. 9.2 for further explanation.

ponds with limited exchange and no tidal mixing. The model was originally calibrated with data from the open embayment; it was then verified for the protected cove situation, changing only the external forcing functions. Finally, the model was used to predict the outcome of fertilization experiments conducted in pond "mesocosms" contemporaneously with model simulations.

In general, the correspondence between model output and empirical data for standing stocks of autotrophic groups was very good. Simple linear regressions between model and data means for macrophyte shoot biomass were highly significant in all three cases, with the model explaining 89%, 62%, and 95%, respectively, for the three conditions. Model output correlated well with data means for epiflora also (Fig. 9.3), with r^2 values of 0.99 and 0.89 for the two field conditions (only three data points were available for pond experiments). Where an estimate of the sampling variance was available for empirical measurements, the model trace generally fell within one (always within two) standard errors about the mean.

Qualitative differences for autotroph abundances among the three sites were also captured in the simulation. Peak biomasses for macrophytes and their associated epiflora were slightly higher (20% to 30%) for the protected cove site compared to the open embayment (Figs. 9.2 and 9.3). Standing stocks reached levels two to five times greater in the experimental ponds than in the embayment. The length of the macrophyte growing season (defined by the presence of shoot biomass) increased from about 4 months in the open embayment to 7 or 8 months at the cove and pond sites (Fig. 9.2). Sensitivity analyses indicated that these differences in maximum abundance and growing season duration were largely the result of differences in flushing and mixing rates. Higher rates of water exchange between the model ecosystem and the external estuary allowed higher phytoplankton, seston, and nutrient concentrations, which stimulated epiflora and phytoplankton production, but inhibited macrophyte growth. With increased mixing rates, particulate matter tended to remain in suspension more readily, leading to greater attenuation of light available for macrophyte growth. These and related results are discussed elsewhere in greater detail (Kemp et al., 1981).

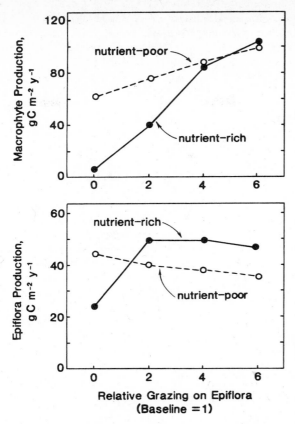

Figure 9.4 The autotroph model's simulated effects of changing rates of herbivorous grazing on epiflora in terms of production by macrophytes (upper panel) and epiflora (lower) under nutrient-poor and nutrient-rich conditions.

Simulation studies with this model revealed that the four autotrophic groups would coexist under unperturbed conditions of nutrient and seston concentrations. Temporal separation of peak growth periods and spatial separation of habitats effectively ameliorated the competition for light and nutrients among these autotrophs. Nutrient enrichment of estuarine waters, however, led to enhanced growth of phytoplankton and epiflora at the expense of the macrophytes, which were inhibited by increased light attenuation. Herbivorous grazing by invertebrates on epiflora allowed coexistence of epiflora and their macrophyte hosts even under extreme nutrient enrichment (Fig. 9.4). Although increased grazing pressure had relatively little effect ($\pm 30\%$ to 50%) on either epiflora or macrophyte production with low nutrient inputs, under nutrient-rich conditions grazing led to substantial increases in production of both epiflora ($2\times$) and macrophytes ($10\times$). The potential importance of grazing on epiflora in the competition between epiflora and their hosts has been discussed previously (Van Montfrans et al., 1984); however, potential interacting effects of grazing and nutrient enrichment have not been considered. In this situation, grazers function much like "keystone predators" by removing the advantage of one competitor over another (Paine, 1980).

MACROPHYTE MANAGEMENT MODELING

Management Modeling Framework

Parallel to the macrophyte ecosystem modeling effort, a system of resource management models was developed for focusing on the multiple interactions between socioeconomic systems and the estuarine macrophyte ecosystems. The objective of this management modeling was to assist in utilizing scientific information toward the protection and management of submersed vascular plants in Chesapeake Bay. These models were intended to be used both in assessing factors potentially contributing to the decline in macrophyte abundance and in estimating the consequences of this decline in terms of changes in fish production, sedimentation, and nutrient fluxes.

The management modeling framework consisted of a network of interconnected models that traced the influence of human activities on macrophyte populations, which in turn affect fisheries and other resources valued by society (Boynton et al., 1981; Kemp et al., 1983b). In this scheme, a hydrologic-chemical runoff model (Holtan and Yaramanglou, 1979) combined meterological conditions with agricultural practices to calculate delivery of nutrients, sediments, and herbicides from fields to the estuary. The transport, flushing, and transformations of these substances in the estuary were estimated using simple, steady-state, box models (Officer, 1980). A macrophyte management model was used to simulate the behavior of submersed vascular plants, other autotrophs, nutrients, sediments, invertebrates, and fish in relation to these inputs from adjacent land and water (Kemp et al., 1981). The benefits and costs of alternative agricultural or waste-treatment practices that influence fisheries and other resource values via changes in water quality and macrophyte growth were assessed using resource evaluation models (Kahn and Kemp, 1985; Boynton et al., 1981).

Connections between submodels in this management modeling framework are generally unidirectional, with feedback occurring only indirectly through the management decision process. For example, materials enter the estuary from the watershed, and the estuary, per se, has little direct influence on watershed activities. In this scheme the modeler serves as the interface between connected submodels, and piecewise simulations can be performed with no loss of information since feedbacks are weak. In other words, the output information from simulations in one submodel is used by the modeler to define input conditions for the next submodel in the sequence.

Macrophyte Management Model Structure

At the focal point of this resource management framework is the macrophyte management model. This model emphasizes interactions between macrophyte ecosystems and human systems (Fig. 9.5). Specifically, water quality effects on macrophyte production and abundance are included along with the habitat and food-chain factors by which macrophytes influence fish production. The structure of this model aggregates much of the complexity that had been emphasized in the macrophyte ecosystem submodels (Kemp et al., 1983b). Sensitivity analyses performed for the ecosystem submodels provided some guidance on appropriate strategies of aggregation, wherein crucial variables and pathways were perserved, and less sensitive factors were either omitted or combined.

The general structure of the macrophyte management model is comprised of 15 state variables organized into five groups: (1) the three major primary producers or photosynthetic groups, all competing for limited light and nutrient resources; (2) the sediments and their associated chemistry; (3) the water with its dissolved nutrients (DIN) and herbicides (HCD), as well as suspended particulate matter (SPM); (4) the herbivorous invertebrate secondary producers at the lower end of the food chain; and (5) the fish, which are generally tertiary producers at the top of the ecological food chain. These state variables are driven by 11 seasonally varying, external forcing functions. This model also includes new state variables (viz., dissolved and adsorbed herbicides) not occurring in the ecological submodels, but included here because of their potential importance in resource management. The differential equations defining this model are essentially similar to those used in the autotroph model. However, these equations tend to be less mechanistic and more linear. These forms are consistent with the concept of increasing linearity of systems with increasing degree of aggregation (e.g., Patten, 1975; Odum, 1983).

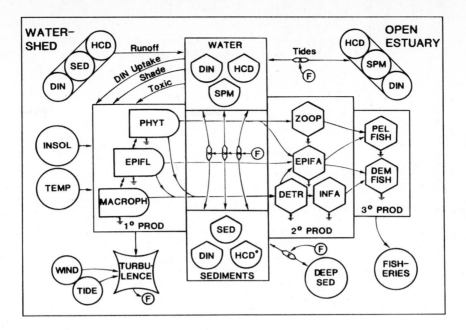

Figure 9.5 Conceptual diagram of macrophyte management model depicting interactions among three major primary producer (1° PROD) groups (PHYT, phytoplankton; EPIFL, epiflora; MACROPH, macrophytes) competing for limited availabilities of sunlight (INSOL) and dissolved inorganic nitrogen (DIN). Autotroph production, which is influenced by shading effects (including suspended particulate matter, SPM), DIN concentrations, and toxicity of herbicides dissolved in water or sorbed to sediments (HCD, HCD*), supports secondary production (2° PROD), including zooplankton (ZOOP), epifauna (EPIFA), infauna (INFA), and detritus (DETR); and tertiary production (3° PROD), including pelagic (PEL) and demersal (DEM) fish. See Fig. 9.1 for explanation of symbols (adapted from Kemp et al., 1983b).

Macrophyte Management Model Results

The general behavior of the macrophyte management model correlated well with field data from the open embayment site (Fig. 9.6). As with the autotroph model, macrophyte and epiflora biomasses were reproduced reasonably well in simulation. Similarly, simulations of fish and invertebrate population abundances corresponded well with measurements. For this model, we have made no attempt at verification with a second data set, largely because data were unavailable for many key variables. By changing nutrient, sediment, and herbicide loading rates in the model, simulations were obtained corresponding to both pre-1970 and post-1980 water quality conditions (Fig. 9.6). In the later simulation, macrophyte biomass was decreased by about 40%, with a reduction (about 20%) in summer abundance of demersal fishes. Such a small difference in fish biomass would be difficult to detect empirically; however, simulation studies revealed that total loss in macrophytes would result in >50% decrease in demersal fish. Model sensitivity analyses indicated that most of this loss of fish abundance was attributable to reductions in food and refuge from predators, rather than simple loss of habitat where fish produced elsewhere congregate locally. This distinction would be difficult to discern without the aid of the simulation model.

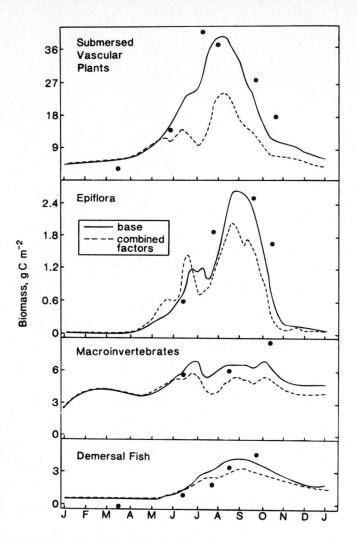

Figure 9.6 Annual distributions of biomasses for submersed vascular plants (shoots and roots), epiflora, macroinvertebrates, and demersal fish. Solid lines and circles represent output from macrophyte management model and data means, respectively, for an open embayment in upper Chesapeake Bay (1970s); dashed lines indicate model output for conditions in the 1980s at the same site.

Multiple simulation experiments with this model allowed consideration of the relative effects of herbicide, sediment, and nutrient loading on macrophyte production (Fig. 9.7). Here, growth of submersed plants exhibits little response to changes in herbicide loading from the watershed. Rapid dilution, degradation, and sorption to sediments by the major herbicides in this region, combined with low toxicity of degradation products and a degree of resistance exhibited by these macrophytes, all contribute to this minimal effect (Kemp et al., 1983a). Sediment inputs produce a more dramatic effect on macrophyte production, following, essentially, an exponential relation. However, much of the total estuarine sediment loading is derived from natural processes, such as shore erosion, which

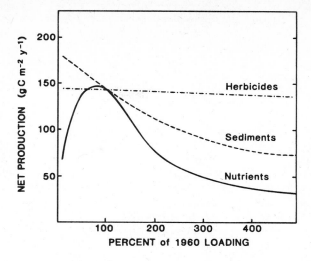

Figure 9.7 The macrophyte management model's simulated effects of changing watershed loading rates for herbicides, sediments, and nutrients on annual net production of macrophytes. Loading rates are adjusted relative to values estimated for 1960 (adapted from Boynton et al., 1981, and Kemp et al., 1983b).

are difficult to manage (Kemp et al., 1984). Nutrient (and in particular nitrogen) loading at low levels causes an enhanced macrophyte growth, whereas reduced production results from inputs greater than estimated 1960 rates. As emphasized above in relation to the autotroph model, decreased vascular plant production at high nitrogen levels results from enhanced growth of planktonic and epiphytic algae, which effectively reduce light available to the macrophytes (Twilley et al., 1985). In combination, historical changes in these water quality variables between 1930 and the present appear sufficient to account for the observed decline in Chesapeake Bay macrophytes (Kemp et al., 1983a).

RETROSPECTIVE CONSIDERATIONS

The roles of ecological modeling in scientific research and natural resource management have been much debated over the last 20 years (Jeffers, 1973; Mar, 1974; Cooper, 1975; Watt, 1975; Wiegert, 1975). Some of the often mentioned utilities of modeling in environmental research include organizing research objectives and methods, identifying missing information or poorly understood relationships, formulating and formalizing scientific hypotheses, interpolating and extrapolating from a given data base, and testing sensitivities of model variables in relation to their real-world counterparts. In her review, Pielou (1981) concluded that many of these points, while conceptually valid, are often overstated. For the research program described in this paper, an attempt was made to utilize models toward most of these objectives. In the following section, we take a retrospective view of this modeling effort (emphasizing the two models presented here) to identify examples of how these models may have enhanced the overall research effort.

Understanding Ecological Interactions

In the conceptualization stages of developing the autotroph model, we were forced to recognize that one very basic aspect of macrophyte physiology, nutrient uptake kinetics, had not been described for whole plants in the scientific literature. Although experimental evidence suggested that, for a given plant, root and shoot uptake of nitrogen and phosphorus were both important and generally interdependent (e.g., Thursby and Harlin, 1982), a kinetic formulation for whole-plant uptake was lacking. Thus, the demands of model description (to express relationships in precise and explicit terms) identified for us a suite of empirical experiments of fundamental importance for understanding macrophyte ecology and physiology (Kemp et al., 1994). Once this kinetic relationship was defined, the simulation model was able to calculate the relative importance of root versus shoot uptake of nitrogen under different environmental conditions.

In considering competitive relations among autotrophic groups, model descriptions included numerous interactions, ranging from nutrient exchange to allelopathy to light attenuation. Experiments with the autotroph model demonstrated the crucial significance of light attenuation in determining the outcome of the competition under various simulated conditions. Modeling experiments also illustrated the relative importance of phytoplankton and epiflora in attenuating light from macrophytes under different degrees of fertilization. Sensitivity analyses with the autotroph model helped to explain the marked differences observed for growth cycles of macrophyte abundance in "exposed" versus "protected" sites. Apparently, reductions in flushing rate associated with protected coves allowed macrophytes to compete more effectively with microalgae for water-column nutrients. This mechanism contributed to extended growing seasons in coves, consistent with historical in situ observations (Orth and Moore, 1983).

Although several recent papers have suggested that variations in grazing pressure on epiflora might strongly affect competition between epiflora and their host macrophytes, empirical evidence had appeared inconsistent (Kemp et al., 1994). Simulation experiments led to the formulation of a testable hypothesis involving interacting effects of grazing and nutrient enrichment. Simulation suggested that under oligotrophic conditions grazing had little effect on epiflora or macrophyte production, but that in eutrophic situations increased grazing resulted in enhanced production of both macrophytes and epiflora.

These examples illustrate how ecological modeling can complement empirical research by identifying poorly understood relationships worthy of empirical study, by aiding in the explanation of field and laboratory results, and by clarifying and formalizing loosely defined ideas into testable scientific hypotheses.

Resource Management Applications

In this study, numerous experiments were conducted to determine the potential importance of different factors (such as sediment loading, herbicide runoff, and nutrient enrichment) as sources of stress for macrophyte growth in the Bay (Kemp et al., 1983a). All these experiments examined one factor in isolation from the others, and the controlled nature of these studies makes simple extrapolation of results to conditions in nature tenuous. The models presented in this paper, however, provided a vehicle for integrating empirical results and extrapolating to nature. The autotroph model allowed results from mesocosm fertilization studies to be extended to actual estuarine situations. This was done by numerically supplementing mesocosm results (reproduced by the model) with natural processes such as tidal exchange and wind mixing, which had been omitted from the empirical experiments because of logistical limitations.

The macrophyte management model was used to compare the relative importance of herbicides, suspended sediments, and nutrients as stressors for macrophyte growth under various scenarios. For conditions at one field site in the early 1980s, nutrient enrichment and sediment loading were shown to have many times greater impact than herbicide inputs on macrophyte abundance, and nutrient enrichment of bay waters was estimated as the single most important factor contributing to the macrophyte decline. These results were influential in the formulation of state and federal agency policies to reduce eutrophication of Chesapeake Bay. The modeling studies also supported a government policy of encouraging minimum-tillage agriculture in the estuary's watershed, since this approach tends to reduce nutrient and sediment losses from farmlands at the expense of increased application of herbicides for weed control (Kemp et al., 1994).

The macrophyte management model also demonstrated the nature of an interaction between macrophytes and fish at opposite ends of the trophic web. Because of the complexity of relationships, including feedback effects (grazing on epiflora, nutrient regeneration, and the like), such connections would be otherwise difficult to establish. Although field data had shown far greater abundances of fish in vegetated as compared to bare habitats, sensitivity analyses with the model allowed the relative importance of different characteristics of the macrophyte habitat (for example, increased food versus refuge from predators) to be partitioned. Model results were combined with resource economic analyses to estimate the shadow-priced value of macrophytes for commercial and recreational fisheries in Chesapeake Bay to be in excess of $1 million dollars per year (Kahn and Kemp, 1985). This habitat role of macrophyte beds, as clarified by modeling analyses, has been recognized by government agencies and incorporated into current fisheries management policies. Thus, we conclude that ecological modeling of submersed macrophyte communities in Chesapeake Bay has contributed both to basic understanding of estuarine ecology and to judicious management of the bay's natural resources.

ACKNOWLEDGMENTS

Numerous individuals have contributed directly and indirectly to the research summarized here. Much of the analysis and computer programming were done by S. Bollinger, while additional computation and interaction were provided by T. Schueler, R. Walker, R. Thomann, and K. Lezon. Assistance in the relevant data collection and experimentation was provided by S. Bunker, J. Cunningham, K. Kaumeyer, L. Lubbers, D. Marbury, M. Meteyer, J. Metz, K. Staver, and R. Twilley. We are indebted to R. Costanza, T. Dolan, D. Flemer, J. Kremer, H. Odum, J. Smullen, S. Taylor, R. Walker, R. Wetzel, M. Yaramanglou, J. Zucchetto, and many others for stimulating interactions on the conceptual aspects of this work. This research was supported by grants from the U.S. Environmental Protection Agency Nos. R805932010 and X003248010 and from the Maryland Department of Natural Resources No. C18-80-430(82). Financial support for digital computation was provided by the University of Maryland Computer Science Center. Typing was done by J. Gilliard and drafting by F. Younger and D. Kennedy.

REFERENCES

BARKO, J. W., and SMART, R. M. 1981. Comparative influences of light and temperature on the growth and metabolism of selected submerged freshwater macrophytes. *Ecol. Monogr.* 51:219–235.

BELYAEV, V. I., KHAILOV, K. M., and OKHOTNIKOV, I. N. 1977. Mathematical simulation of a marine coastal ecosystem containing macrophytes. *Aquat. Bot.* 3:315–328.

BOLING, R. H., GOODMAN, E. D., VAN SICKLE, J. A., ZIMMER, J. O., CUMMING, K. W., PETERSON, R. C., and REICE, S. R. 1975. Toward a model of detritus processing in a woodland stream. *Ecology* 56:141–151.

BOYNTON, W. R., KEMP, W. M., and STEVENSON, J. C. 1979. A research design for understanding and managing complex environmental systems. In Johnson, R. R. and McCormick, J. F. (eds.), *Strategies for Protection and Management of Floodplain Wetlands and other Riparian Ecosystems.* GTR-WO-12. U.S. Dept. Agric., Forest Service, Washington, D.C., pp. 168–177.

BOYNTON, W. R., KEMP, W. M., HERMANN, A. J., KAHN, J. R., SCHUELER, T. R., BOLLINGER, S., LONERGAN, S. C., STEVENSON, J. C., TWILLEY, R., STAVER, K., and ZUCCHETTO, J. 1981. An analysis of the energetic and surplus economic values associated with the decline of submerged macrophytic communities in Chesapeake Bay. In Mitsch, W. J., Bosserman, R. W., and Klopatek, J. (eds.), *Energy and Ecological Modeling.* Elsevier, Amsterdam, pp. 441–454.

COOPER, C. F. 1975. Ecosystem models and environmental policy. *Simulation* 27:133–138.

FERGUSON, R. L. AND ADAMS, S. M. 1979. A mathematical model of trophic dynamics in estuarine seagrass communities. In Dame R. F. (ed.), *Marsh–Estuarine Systems Simulation.* University of South Carolina Press, Columbia, South Carolina, pp. 41–70.

GOODALL, D. W. 1974. Problems of scale and detail in ecological modeling. *J. Environ. Mgmt.* 2:149–157.

GOLDSBOROUGH, W. J. 1983. Response of the submersed vascular plant, *Potamogeton perfoliatus* L., to variable light regimes in an estuarine environment. Ms. thesis, University of Maryland, College Park, Maryland.

HOLTAN, H. N., and YARAMANGLOU, M. 1979. Procedures Manual for Sediment, Phosphorus and Nitrogen Transport Computations with USDAHL. Maryland Agric. Exp. Sta., Univ. of Maryland, College Park, Maryland. 34 pp.

HOPKINSON, C. S., and DAY, J. W. 1977. A model of the Barataria Bay salt marsh ecosystem. In Hall, C. A. S., and Day, J. W. (eds.), *Ecosystem Modeling in Theory and Practice.* John Wiley, New York. pp. 236–265.

IIZUMI, H., and HATTORI, A. 1982. Growth and organic production of eelgrass (*Zostera marina*) in temperate waters of the Pacific coast of Japan. III. The kinetics of nitrogen uptake. *Aquatic Botany* 12:245–256.

JEFFERS, J. N. R. 1973. Systems modelling and analysis in resource management. *J. Environ. Mgmt.* 1:13–28.

JOHNSON, F. H., EYRING, H., and STOVER, B. J. 1974. *The Theory of Rate Processes in Biology and Medicine.* Wiley, New York. 703 pp.

KAHN, J. R., and KEMP, W. M. 1985. Economic losses associated with degradation of an ecosystem: The case of submerged aquatic vegetation in Chesapeake Bay. *J. Econ. Environ. Mgmt.* 12:246–263.

KEMP, W. M., LEWIS, M. R., CUNNINGHAM, J. J., STEVENSON, J. C., and BOYNTON, W. R. 1980. Microcosms, macrophytes, and hierarchies: Environmental research in the Chesapeake Bay. In Giesy, J. P. (ed.), *Microcosms in Ecological Research.* Symp. Ser. 52 (CONF-781101) NTIS, Springfield, Virginia, pp. 911–936.

KEMP, W. M., BOYNTON, W. R., STEVENSON, J. C., and MEANS, J. C. (eds.). 1981. Submerged Aquatic Vegetation in Chesapeake Bay. Univ. Maryland Center Environ. Estuar. Stud. Ref. No. 80-168, Cambridge, Maryland. pp. 785.

KEMP, W. M., BOYNTON, W. R., STEVENSON, J. C., TWILLEY, R. R., and MEANS, J. C. 1983a. The decline of submerged vascular plants in upper Chesapeake Bay: Summary of results concerning possible causes. *Mar. Technol. Soc. J.* 17:78–89.

KEMP, W. M., HERMANN, A. J., and BOYNTON, W. R. 1983b. A simulation modeling framework for ecological research in complex systems. The case of submerged vegetation in upper Chesapeake Bay. In Turgeon, K. W. (ed.), *Marine Ecosystem Modeling.* U.S. Department of Commerce, Washington, D.C., pp. 131–158.

KEMP, W. M., BOYNTON, W. R., TWILLEY, R. R., STEVENSON, J. C., and WARD, L. G. 1984. Influences of submersed vascular plants on ecological processes in upper Chesapeake Bay. In Kennedy, V. S. (ed.), *The Estuary as a Filter.* Academic Press, New York, pp. 367–394.

KEMP, W. M., BOYNTON, W. R., and HERMANN, A. J. 1994. Ecosystem modeling and energy analysis of submerged aquatic vegetation in Chesapeake Bay. In C. A. S. Hall (ed), *Maximum Power.* University of Colorado Press, Niwot (in press).

KREMER, J. N., and NIXON, S. W. 1978. *A Coastal Marine Ecosystem: Simulation and Analysis.* Springer-Verlag, New York, 217 pp.

LEVINS, R. 1966. The strategy of model building in population biology. *Amer. Scient.* 54:421–431.

Mar, B. W. 1974. Problems encountered in multi-disciplinary resources and environmental simulation models development. *J. Environ. Mgmt.* 2:83–100.

McIntire, C. D., and Colby, J. A. 1978. A hierarchical model of lotic ecosystems. *Ecol. Monogr.* 48:167–190.

Najarian, T. O., and Harleman, D. R. F. 1977. A real time model of nitrogen-cycle dynamics in an estuarine system. *Prog. Water Technol.* 84:323–345.

Najarian, T. O., and Taft, J. L. 1981. Nitrogen-cycle model for aquatic systems: Analysis. *J. Env. Eng. Div.* (Am. Soc. Civil Eng.) 107 (EE6):1141–1156.

Odum, H. T. 1971. *Environment, Power and Society.* Wiley, New York, 331 pp.

Odum, H. T. 1983. *Systems Ecology: An Introduction.* Wiley, New York, 644 pp.

Odum, E. P. 1984. The mesocosm. *BioScience* 34:558–562.

Officer, C. B. 1980. Box models revisited. In Hamilton, P., and McDonald, K. B. (eds.), *Estuarine and Wetland Processes.* Plenum, New York, pp. 65–114.

Orth, R. J., and Moore, K. A. 1983. Chesapeake Bay: An unprecedented decline in submerged aquatic vegetation. *Science* 222:51–53.

Overton, W. S. 1975. The ecosystem modeling approach in the coniferous forest biome. In Patten, B. C. (ed.), *Systems Analysis and Simulation in Ecology,* Vol. 3. Academic Press, New York, pp. 117–138.

Paine, R. T. 1980. Food webs: linkage, interaction strength and community infrastructure. *J. Anim. Ecol.* 49:667–685.

Parsons, T. R., Takahashi, M., and Hargrave, B. 1977. *Biological Oceanographic Processes,* 2nd ed. Pergamon, New York, 332 pp.

Patten, B. C. 1975. Ecosystem linearization: an evolutionary design problem. *Am. Nat.* 109:529–537.

Patten, B. C. 1985. Energy cycling, length of food chains and direct versus indirect effects in ecosystems. In Ulanowicz, R. E., and Platt, T. (eds.). Ecosystem Theory for Biological Oceanography. *Can. Bull. Fish. Aquat. Sci.* 213:119–138.

Pielou, E. C. 1981. The usefulness of ecological models: A stock-taking. *Quart. Rev. Biol.* 56:17–31.

Schaffer, W. M. 1981. Ecological abstraction: The consequences of reduced dimensionality in ecological models. *Ecol. Monogr.* 51:383–401.

Short, F. T. 1980. A simulation model of the seagrass production system. In McRoy, C. P., and Phillips, R. (eds.) *Handbook of Seagrass Biology: An Ecosystem Perspective.* Garland STPM Press, New York, pp. 277–295.

Short, F. T., and McRoy, C. P. 1984. Nitrogen uptake by leaves and roots of the seagrass, *Zostera marina* L. *Botanica Marina* 27:547–555.

Simon, H. A. 1973. The organization of complex systems. In Pattee, H. H. (ed.), *Hierarchy Theory. The Challenge of Complex Systems.* Braziller, New York, pp. 1–27.

Stevenson, J. C., and Confer, N. M. (eds.). 1978. Summary of Available Information on Chesapeake Bay Submerged Vegetation. FWS/OBS-78/66, Fish and Wildlife Service, U.S. Department of Interior, Washington, D.C., 335 pp.

Thursby, G. B., and Harlin, M. M. 1982. Leaf–root interaction in the uptake of ammonium by *Zostera marina. Marine Biology* 72:109–112.

Titus, J. E., and Adams, M. S. 1979. Coexistence and the comparative light relations of the submerged macrophytes, *Myriophyllum spicatum* and *Vallisneria americana. Oecologia* 40:273–286.

Titus, J., Goldstein, R. A., Adams, M. S., Mankin, J. B., O'Neill, R. V., Weiler, P. R., Shugart, H. H., and Booth, R. S. 1975. A production model for *Myriophyllum spicatum L. Ecology* 56:1129–1138.

Twilley, R. R., Kemp, W. M., Staver, K. W., Stevenson, J. C., and Boynton, W. R. 1985. Nutrient enrichment of estuarine submersed plant communities. I. Algal growth and effects on production of plants and associated communities. *Mar. Ecol. Progr. Ser.* 23:178–191.

VAN MONTFRANS, J., WETZEL, R. L., and ORTH, R. J. 1984. Epiphyte–grazer relationships in seagrass meadows: consequences for seagrass growth and production. *Estuaries* 7(4A):189–309.

VERHAGEN, J. H. G., and NIENHUIS, P. H. 1983. A simulation model of production, seasonal changes in biomass and distribution of eelgrass (*Zostera marina*) in Lake Grevelingen. *Mar. Ecol. Progr. Ser.* 10:187–195.

WALSH, J. J. 1975a. A spatial simulation model of the Peruvian upwelling ecosystem. *Deep-Sea Res.* 22:201–236.

WALSH, J. J. 1975b. Utility of systems models: a consideration of some possible feedback loops of the Peruvian upwelling ecosystem. In Cronin, L. E. (ed.) *Estuarine Research.* Academic Press, New York, pp. 617–633.

WARD, L. G., KEMP, W. M., and BOYNTON, W. R. 1984. The influence of waves and seagrass communities on suspended particulates in an estuarine embayment. *Mar. Geol.* 59:85–103.

WATT, K. E. F. 1975. Critique and comparison of biome ecosystem modeling. In Patten, B. C. (ed.), *Systems Analysis and Simulation in Ecology,* Vol. 3. Academic Press, New York, pp. 139–152.

WEBER, J. A., TENHUREN, J. D., WESTRIN, S. S., YOCUM, C. S., and GATES, D. M. 1981. An analytical model of photosynthetic response of aquatic plants to inorganic carbon and pH. *Ecology* 62:697–705.

WETZEL, R. L., and WIEGERT, R. G. 1983. Ecosystem simulation models: Tools for the investigation and analysis of nitrogen dynamics in coastal and marine ecosystems. In Carpenter, E. J., and Capone, D. G. (eds.), *Nitrogen in the Marine Environment.* Academic Press, New York, pp. 869–892.

WIEGERT, R. G. 1975. Simulation models of ecosystems. *Ann. Rev. Ecol. Systematics* 6:311–338.

10

Optimal Experimental Conditions to Validate Volatilization Models of Toxic Contaminants

EFRAIM HALFON

National Water Research Institute
Canada Centre for Inland Waters
Burlington, Ontario, Canada

A question that arises in conducting and synthesizing systems-oriented studies, is how much detail to include in the system. A facetious definition I heard once indicated that "in instances where you have from one to three variables you have a science, when you have from four to seven variables in a study you have an art, and when you have more than seven variables you have a system"! [Van Dyne, in *Grasslands, Systems Analysis and Man*, Int. Biol. Program 19, Chap. 14, p. 889 (1980), Cambridge U. Press].

INTRODUCTION

Knowledge of the fate of toxic contaminants in the environment is important because of the large number of chemicals (about 60,000) presently in use. A variety of volatile compounds are routinely found in water and air (see Kaiser et al., 1983; Lane et. al., 1992; Sweet and Vermette, 1992). Mathematical models have been used (Neely and Blau, 1977; Lassiter et al., 1979; Burns et al., 1982; McCall et al., 1983; Halfon, 1984a, b, 1986; Thomann, 1989; Halfon and Oliver, 1990; Mackay and Paterson, 1991) to predict contaminant fates, but incomplete knowledge of the relationship between chemical structure and environmental behavior (Kaiser, 1984) makes predictions uncertain. Here, basic principles of fate models are briefly reviewed and limitations of model formulations for a specific process, the volatilization of a toxic contaminant from a water body to the atmosphere, are analyzed. The concluding section includes recommendations about appropriate experimental conditions to assess the validity of two alternative models.

Past Endeavors

Prediction of the fate of toxic contaminants in the environment was made possible in the 1970s by the realization that most environmental fate processes follow first-order kinetics; therefore, linear models [see eq. (10.1)] are an appropriate representation of contaminant behavior (Lassiter et al., 1979). In the 1960s, the main purpose of scientific studies was only to monitor environmental concentrations of pollutants, usually several years after contamination of the environment. For example, Lake Ontario was subject to contaminants as early as 1909 with the establishment of industries on the shores of the Niagara River. Contaminant loading rates peaked in 1960–1963 (Durham and Oliver, 1983). Reduction in these rates preceded widespread recognition of a pollution problem in Lake Ontario that occurred in the 1970s. The historical pollution trends are recorded in the bottom sediments of Lake Ontario.

Early investigations of the fate of toxic contaminants led to two distinct but related lines of research: studies on model ecosystems focusing on the dynamics of the fate of the contaminants and statistical studies focusing on equilibrium conditions. This dichotomy still exists in modeling strategies: some mathematical models include equilibrium conditions [for example, the fugacity approach of Mackay and Paterson (1982, 1991) and the model EXAMS (Burns et al., 1982)], while others are strictly dynamic [for example, the models PEST (Park et al., 1980) and TOXFATE (Halfon, 1986; Halfon and Oliver, 1990)].

Mathematical models predict concentrations of contaminants in different compartments, or state variables, according to (1) loading rates into the system (inputs); (2) degradation (photolysis, hydrolysis, oxidation, biodegradation, and so on) rates (parameters or submodels); and (3) transport rates (parameters or submodels) between compartments (volatilization to the atmosphere, wet and dry deposition, adsorption on soil particles and sediments in the aquatic environment, and resuspension from bottom sediments and/or currents). Halfon (1984b) has reviewed the data base necessary to develop and verify ecosystem fate models. Figure 10.1 shows compartments and environmental processes that affect the behavior of contaminants. Figure 10.2 shows compartments often used in modeling exercises and chemical properties usually associated with contaminant partition in these compartments. At present, many factors (for example, microclimate, wind, spatial variability, and uncertainty in the model structure and parameter values) preclude the use of laboratory data and mathematical models alone to set environmental standards. Field testing is still necessary, and mathematical models can help in integrating this information and possibly in directing data collection efforts.

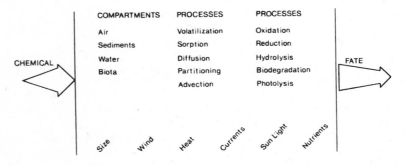

Figure 10.1 State variables, environmental factors, and transfer and degradation processes present in fate models.

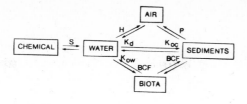

S = saturated water solubility; P = vapor pressure;
K_{ow} = octanol-water partition coefficient ($\frac{\text{conc in octanol}}{\text{conc in water}}$);
BCF = bioconcentration factor ($\frac{\text{conc in organism}}{\text{conc in medium}}$);
H = Henry's law constant ($\frac{\text{conc in air}}{\text{conc in water}}$);
K_d = soil sorption coefficient ($\frac{\text{conc in soil}}{\text{conc in water}}$);
K_{oc} = soil sorption coefficient expressed on an organic carbon basis.

Figure 10.2 Physical properties used in equilibrium models.

Physical and Chemical Properties Related to Environmental Behavior

Transfer and degradation parameters constrain the behavior predicted by a mathematical model; transfer parameters are based on the physicochemical properties of the contaminant and the environment. The measurement of water solubility, octanol–water partition coefficient (K_{ow}), and soil adsorption coefficient (K_{oc}) have been standardized. Nevertheless, the water pH has a large influence if the chemical is ionizable. Vapor pressure is difficult to measure, especially for contaminants of very low volatility. To reduce the impact of lack of experimental data, quantitative structure activity relationships (QSAR) have been developed (Kaiser, 1984): physical and chemical properties of the contaminants, measured in the laboratory or estimated from the molecular structure, have been used to predict the environmental behavior of the contaminant itself. Metcalf et al. (1971) and Santschi (1985), among others, built model ecosystems or mesocosms that were used to follow the movement and degradation of contaminants under controlled conditions.

A large number of statistical models, usually linear regressions, has been published relating properties such as the octanol–water partition coefficient (K_{ow}) to bioconcentrations in fish, adsorption on suspended sediments, and so on. The main problem, still debated, is whether this information is sufficient to make extrapolations from the laboratory to the field and whether mathematical models developed from laboratory data are valid under field conditions. Linear regression models can be used if appropriate observations are not available; errors and uncertainty must be taken into consideration when predictions are made based on these models. For example, the boiling point of a chemical can be used to compute its vapor pressure and, if the solubility is known, to predict the Henry's law constant, or the ratio of equilibrium concentrations of a chemical in the water and in the air. This ratio is always used in models to compute the volatility of a chemical. Halfon (1985a) analyzed the appropriate procedures for computing the statistical linear regression models when both variables are subject to measurement error and natural variability, which is the case in most ecotoxicological models.

Toxic contaminants in the environment may be degraded or converted into other chemicals that may also be toxic. Most mathematical models do not simulate the fate of these by-products and, therefore, degraded forms are considered lost from the system. These degradation processes can be of a variety of classes, including hydrolysis, oxidation, reduction, substitution, elimination, isomerization, and ion exchange. Most of these environmental reactions are bimolecular (for example, in hydrolysis, water is a reagent). Nevertheless, when one of the reactants is present in excess, the process becomes a pseudo-first-order process and can be included in the model as a linear formulation.

EQUILIBRIUM AND DYNAMIC MODELS

The basic framework of fate models is usually based on linear ordinary differential equations; nevertheless, any process formulation might be nonlinear. Within this framework, two approaches have been used. In the dynamic approach, the changes in concentration over time are considered important. In the equilibrium approach, the main assumption is that enough time has lapsed for the contaminant to reach equilibrium in the environment. This second approach has been used by Mackay and Paterson (1982; 1991), Burns et al. (1982), and McCall et al. (1983). The last model has a structure of the following kind:

$$C_{sed} \rightarrow C_w \rightarrow C_a \rightarrow C_{sw} \rightarrow C_s$$
$$\downarrow$$
$$C_f$$

where the C's represent concentrations of the contaminant in the water (C_w), sediment (C_{sed}), fish (C_f), air (C_a), soil water (C_{sw}), and soil (C_s). EXAMS was developed at the U.S. Environmental Protection Agency in Athens, Georgia, by Burns et al. (1982) to integrate information on the chemical properties of a contaminant with the physical characteristics of an aquatic environment and produce acceptable fate predictions. EXAMS is used for regulatory purposes and to estimate the average behavior of new chemicals (prior to marketing) in a lake, river, or pond.

The importance of equilibrium models lies in their ability to indicate which compartment would be the main recipient of the contaminant if conditions remain constant through time. This approach is useful for regulatory agencies to assess the possible hazard level of a new contaminant. Predictions, on average, agree with monitoring data, but do not agree well for specific sites.

Dynamic models can describe site-specific nonequilibrium conditions and can address the question of residence time. One of the earliest of these models was developed by Neely and Blau (1977). This model describes the behavior of a chemical, chlorpyrifos, in a fish pond; transfer among the different compartments depends on six parameters that describe volatilization, hydrolysis, uptake by fish, excretion by fish, adsorption to soil, and desorption from soil. The model has a linear formulation, and the equation describing concentrations in water is as follows:

$$v \frac{dC_w}{dt} = -k_1 A C_w - k_2 V C_w - k_3 F C_w + k_4 F C_f - k_5 S C_s + k_6 S C_s, \qquad (10.1)$$

where v is the volume (m^3) of the pond, A is its area (m^2), F is fish biomass (g), S is the weight of soil (g), and k_1 to k_6 are parameters that quantify the different transport and removal processes. Most dynamic fate models, with more or fewer state variables and parameters, have a similar structure.

Halfon and Maguire (1981) developed stochastic fate models to describe the variability in the data that is often observed when multiple measurements are taken of the same compartment. Their model describes the fate of fenitrothion in three compartments: the surface microlayer, the main water body, and the bottom sediment of a pond. Model parameters were estimated by fitting a stochastic model to the whole range of observations; thus, parameter values were quantified as ranges rather than as best-fit values. When used for prediction, the model computes the time needed for fenitrothion to be removed from a pond, after an aerial spraying, with a 95% probability. The 95% removal time was estimated to be 20% higher than the time predicted by the deterministic model, thus showing that pollution persisted longer than previously expected. In the same paper, Halfon and Maguire also studied the problem of model aggregation, or the predictions made by an aggregate model with two or one spatial compartments, rather than by the original three. They concluded that aggregated models give good predictions, but with less accuracy: models with more parameters are site specific, whereas the more aggregated models are more general, but more theoretical.

MODEL UNCERTAINTY IN ECOTOXICOLOGY

Two major problems in developing fate models are choice of state variables and quantification of parameter values. Choice of appropriate state variables is a common problem in systems ecology. Halfon (1983a, b; 1985b) has analyzed the choice process using system methods. He concluded that no best model structure exists, but that an appropriate choice may be made by using an appropriate decision-making tool.

Choice and quantification of parameters is an even more difficult problem. In a simple linear formulation, the parameters are a composite of several processes. For example, in Neely and Blau's (1977) model, the parameter k_1 describes volatilization; k_1 can also be represented as a submodel describing the behavior of a contaminant at the air–water interface, as in the two-layer model of Whitman (1923). Later in this chapter the volatilization submodel is described in detail, but the important point is that any parameter can be expanded into a submodel and given a more accurate description. Even so, no unique formulation has been universally accepted for any particular process; field and laboratory experiments are performed under specified conditions. Generalization of a set of data to a model that is generally applicable, for example, to all aquatic environments, is difficult. Quantification of parameters is therefore arbitrary in most fate models, even when empirical regression analyses are used; this area can benefit from interaction between modelers and ecotoxicologists.

VOLATILIZATION SUBMODELS

The transfer rate of a toxic contaminant from water (parameter k_1 in Neely and Blau's model) to air depends on its vapor pressure, molecular weight, and water solubility, and on environmental conditions at the air–water interface (wind speed and temperature) and the relative ratio of the water volume and surface area. Liss (1973) developed a model framework based on Whitman's (1923) boundary layer formulation, with two microlayers at the water interface controlling the movement of a chemical from water to air (Fig. 10.3). The first layer, on the water side, has a resistance (k_{liq}) to the movement of a chemical across the interface. The second layer, on the air side, has a resistance (k_{gas}). The total air–water interface resistance is the harmonic mean or

$$kv = \frac{1}{1/k_{gas} + 1/k_{liq}}, \qquad (10.2)$$

where kv is the volatilization rate (cm h^{-1}) constant of a chemical from water. A physical rather than mathematical description of kv is presented later in eq (10.28). Liss (1973) performed experiments in the ocean to estimate the transfer rate of oxygen and carbon monoxide from ocean water to the

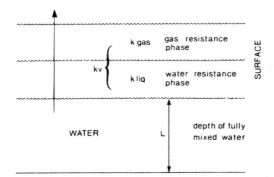

Figure 10.3 Two-layer air–water interface.

atmosphere. This two-layer volatilization model is generally accepted and has been incorporated in various fate models such as EXAMS and PEST. Even though the general concept was accepted, EXAMS and PEST (from here on called models A and B, respectively) have two different formulations to compute the parameters k_{gas} and k_{liq} and therefore kv. In model A, molecular oxygen is used as a reference chemical; in model B, benzene is used.

The formulation of a correct volatilization model is important to resolve the question, much debated in the Great Lakes community (Oliver, 1984), of whether some water bodies are a sink for some toxic contaminants or only a transition zone. For this purpose the outputs of the two models A and B were compared under the same conditions of wind and temperature. From the analysis of the results, a set of experimental conditions were derived that should produce data necessary to discriminate between the two models and to verify which model is more accurate. The two models have been simulated at wind speeds ranging from 1 to 15 m/s and at temperatures of 1° to 25°C for two chemicals, hexachlorobenzene (HCB) and the more volatile 1,2,4-trichlorobenzene (1,2,4-TCB). Both are often found near industrial discharges.

Model Structures

The basic structure of the two volatilization models is the two-layer boundary system. The flux of a chemical across the air–water interface is proportional to the concentration difference C in water and in air. The parameter kv has dimensions [LT^{-1}], usually centimeters per hour.

If the concentrations in air and water are in equilibrium, then

$$C_a = H * C_w, \qquad (10.3)$$

where H, the dimensionless Henry's law constant, is

$$\frac{C_a \text{ (equilibrium concentration in gas phase [g/cm}^3 \text{ air])}}{C_w \text{ (equilibrium concentration of nonionized dissolved chemical in liquid phase [g/cm}^3 \text{ water])}}.$$

Alternatively, the Henry's law constant can also have dimensions of $m^3 - $ atm/mol if the following definition is used:

$$H = \frac{\text{vapor pressure [atm]}}{\text{water solubility [mole m}^{-3}\text{]}}.$$

Liss (1973) estimated that the diffusion coefficient for oxygen is about 104 times greater in air than for the gas dissolved in water. Therefore, diffusive processes in water rather than in air likely control its transfer across the air–water interface. This assumption is valid for most contaminants present in water; nevertheless, Mackay (personal comm., December 1986) thinks that the movement of PAHs (partially aromatic hydrocarbons) from water to the atmosphere is probably controlled by processes in air rather than in water. Oxygen has a Henry's law constant of about 30 ($m^3 - $ atm/mol).

According to Fick's law, the flux of a chemical across the water boundary, assuming a higher concentration in water than in air, is

$$F = D \frac{dC}{dZ}, \qquad (10.4)$$

where D is the diffusion constant of the chemical [$L^2 T^{-1}$], C is the concentration of the compound, and Z is the infinitesimal layer thickness. Likewise, the flux across the aqueous interface is

$$F = \frac{D}{(RT)(dC/dZ)}, \qquad (10.5)$$

where D is the diffusion constant of the chemical in air, R is the gas constant [8.206 10^{-5} $m^3 - $ atm/(mol K)], and T is the temperature (K).

Model A. In model A the gas resistance of the compound of interest is proportional to the relative transfer of water vapor; thus

$$k_{gas} = \frac{wf \, H \, \sqrt{18/mwt}}{RT}, \qquad (10.6)$$

where wf (cm h^{-1}) is a function of wind speed (m sec^{-1}),

$$wf = 0.1857 + 11.36 U^* \quad \text{(Liss, 1973)}, \qquad (10.7)$$

and 18 is the molecular weight of water, mwt is the molecular weight of the chemical of interest (284.89 for HCB and 181.45 for 1,2,4-TCB), and U^* is the wind speed at the air–water interface or friction velocity. Equation (10.6) states that the higher the molecular weight of the contaminant in relation to water is, the slower the flux across the interface.

The empirical eq. (10.7) (Liss, 1973) for wind speed correction is valid for a friction velocity U^* in the range of 1.6 to 8.2 m/s, and it accounts for 98.3% of the variance of the exchange constant k_{gas}. For windspeed (U_{10} [cm/s]) measured at a reference height (Z_{10} [cm]) of 10 meters, the friction velocity can be computed as

$$U^* = U_{10} \, \text{sqrt} \, \sqrt{C_D}, \qquad (10.8)$$

where the shear stress C_D (g cm^{-2}), for velocities $1 < U_{10} < 15$ m/s, is

$$C_D = 5 \times 10^{-5} \times \sqrt{U_{10}} \quad \text{(Wu, 1969)}. \qquad (10.9)$$

In model A, k_{liq} is computed using oxygen as reference or

$$k_{liq} = k_{O_2} \sqrt{(32/mwt)}, \qquad (10.10)$$

where 32 is the molecular weight of oxygen. The parameter k_{O_2}, or transfer rate of molecular oxygen at the interface, is a function of temperature (Kramer, 1974) as

$$k_{O_2} = k_{O_2} \, 1.024^{(t-20)}, \qquad (10.11)$$

where t is the temperature in degrees Celsius. The parameter k_{O_2} is computed according to calculations made by Banks (1975), who estimated that at wind velocities under 5.5 m/s the reaeration coefficient is proportional to the square root of the wind velocity, and, for higher wind speeds, the oxygen constant is proportional to the square of the wind speed, or

$$k1 = 4.19 \, 10^{-6} \sqrt{(U_{10})}, \quad \text{for } u < 5.5 \text{ m/s}, \qquad (10.12)$$

$$k1 = 3.27 \, 10^{-7} \, (U_{10})^2, \quad \text{for } u \geq 5.5 \text{ m/s}, \qquad (10.13)$$

and

$$k_{O_2} = k1 \, (\text{m/s}) * 3600 \, (\text{s/h}) * 100 \, [\text{cm m}^{-1}]. \qquad (10.14)$$

Model B. In model B, the main assumption is that the reference chemical is benzene rather than oxygen; furthermore, the flux of a chemical across the interface is assumed inversely proportional to the molal volume (V_{tom}) of the compound of interest, rather than inversely proportional to the molecular weight as in model A. The molal volume (V_{tom}) of water (V_{H_2O}) is 14.4 (cm^3/g mole); of benzene, 96; of hexachlorobenzene, 246.6; and of trichlorobenzene, 169.8 (volumes computed with data from Perry, 1963). Note that the molecular weight of a compound is known perfectly, whereas the molal volumes must be measured; therefore, the data used in model B might be uncertain. In model B, k_{gas} is computed as

$$k_{gas} = \frac{wf\,H\,(V_{H_2O}/V_{tom})^{0.6}}{RT}, \tag{10.15}$$

where wf was defined in eq. (10.7), H is the Henry's law constant, V_{tom} is the molal volume of the chemical of interest, in this case HCB and 1,2,4-TCB, T is the temperature (K), and R is the gas constant. The exponent 0.6 in equations (10.15) and (10.16) was used by Park et al. (1980) according to findings by Othmer and Thaker (1966). The computation of k_{gas} in eq. (10.15) is similar to the computation of k_{gas} in eq. (10.6); the main difference is that different chemical properties are used.

Model B includes two equations for the computation of the correction factor k_{ben} which has the same function as k_{O_2} is eq. (10.10); for wind velocities of <3 m/s

$$k_{ben} = \frac{(1.287\ \text{wind velocity}/3) + 2.5}{100} \times 1.016^{(t-20)} \left(\frac{96}{V_{tom}}\right)^{0.6}, \tag{10.16}$$

where the molal volume of benzene is 96 and the temperature correction factor is based on reaeration studies of Streeter et al. (1936). This equation was empirically derived by Park et al. (1980), who computed the relation between k_{ben} and wind velocity as 1.287/3 wind velocity + 2.5. The factor 100 in eq. (10.16) is used to convert the units from meters per hour to centimeters per hour. The exponent 0.6 has also been used in eq. (10.15) according to results from Othmer and Thaker (1966). For wind velocities over 3 m/s, the model incorporates the experimental findings of Cohen et al. (1978) as

$$k_{liq} = k_{ben}\,1.016^{(t-20)} \left(\frac{96}{V_{tom}}\right)^{0.6}, \tag{10.17}$$

where

$$k_{ben} = (11.4\,\text{Re}^{0.195} - 4.1), \tag{10.18}$$

and Re is the Reynolds number. In model B, the resistance in the liquid layer depends not on the relative molecular weights of the contaminant of interest and oxygen, but on wind-induced turbulence at the air–water interface. Model B's main assumption is therefore that the water viscosity at the air–water interface has much less importance than wind-induced turbulence; thus, model B predicts higher volatilization rates than model A.

Wu (1969) used the roughness Reynolds number, Re, to characterize the turbulent flow over a roughened free air–water interface. Cohen et al. (1978), using a wind tunnel and an aquarium with a stirrer, showed that a correction factor, k_{ben}, for the liquid resistance is a well-behaved function of Re at wind speeds below 9 m/s. This function is

$$k_{ben} = 11.4\,\text{Re}^{0.195} - 4.1 \quad \text{at stirrer speed of 540 rpm}. \tag{10.19}$$

The Reynolds number is defined as

$$\text{Re} = \frac{Z_0 U^*}{a}, \tag{10.20}$$

where Z_0 is the effective roughness height and a is the air kinematic viscosity with a value of 0.17 (cm^2 s^{-1}) at 20°C (Sabersky and Acosta, 1964). Using the logarithmic wind velocity profile near the interface,

$$U = \frac{U^*}{K} \ln \frac{Z}{Z_0}, \qquad (10.21)$$

where k is the Von Karman constant taken to be 0.4; from eq. (10.8),

$$\text{sqrt } \sqrt{C_D} = \frac{U^*}{U} \qquad (10.22)$$

or

$$\frac{U}{U^*} = (C_D)^{-0.5} = \frac{1}{K} \ln \frac{Z}{Z_0} \qquad (10.23)$$

or

$$\frac{K}{\sqrt{C_D}} = \ln \frac{Z}{Z_0} \qquad (10.24)$$

or, by taking the antilogarithms of both sides and rearranging,

$$Z_0 = Z \exp \frac{-K}{\text{sqrt}\sqrt{C_D}}. \qquad (10.25)$$

By substituting eqs. (10.8) and (10.25) in eq. (10.20), we obtain the formula

$$Re = \frac{Z_{10} U_{10} \sqrt{C_D}}{a \exp (K/\sqrt{C_D})}, \qquad (10.26)$$

which can be used in the empirical eq. (10.19) (Cohen et al., 1978).

Model B computes values of k_{liq} that are very close to those observed by Cohen et al., for example, 2.9 (cm/h) (predicted and observed) for benzene at 20°C and 3.34 observed versus 3.39 predicted for toluene at 20°C.

Computation of Volatilization Rates from a Water Body

The volatilization rate kv [eq. (10.2)], computed either from model A or B, can be used to calculate the relative removal rate, kv^c [h^{-1}], of a contaminant from a given body of water by

$$kv^c = \frac{kv \quad [\text{cm h}^{-1}]}{L \quad [\text{cm}^3 \text{ cm}^{-2}]},$$

where L is the average depth, or the ratio of the fully mixed water volume and the surface area. The parameter kv^c can be used in the model

$$-d[C]/dt = kv^c [C] \qquad (10.27)$$

to compute the relative changes in water concentration due to volatilization. The constant kv^c can be expressed in term of the mass-transfer rates of the substance across liquid- and gas-phase boundary layers. The general expression for kv^c is

$$kv^c = \frac{1}{L} \left[\frac{1}{k_{liq}} + \frac{RT}{H \, k_{gas}} \right]^{-1}. \qquad (10.28)$$

In Lake St. Clair, a shallow lake, L is about 439, and in Lake Ontario, 8333 (cm^2 cm^{-3}); in a laboratory cylindrical microcosm with a radius of 200 cm and 100 cm deep, this ratio is 100.

TABLE 10.1 Prediction of the model parameters k_{gas}, k_{liq}, and kv for the two contaminants HCB and 1,2,4-TCB at a wind speed of 7 m/s and at a temperature of 5°C

	Model A		Model B	
	HCB	1,2,4-TCB	HCB	1,2,4-TCB
k_{gas}	17.6	184.1	13.0	135.2
k_{liq}	1.3	1.7	4.3	5.3
kv	1.2	1.6	3.2	5.1
		Half-life (days)		
Microcosm	2.4	1.8	0.9	0.6
Lake St. Clair	11	8	4	2.5
Lake Ontario	201	150	75	47

Units are in cm/h for kv, k_{liq}, and k_{gas}. The half-lives of the two chemicals are computed using eq. (29).

The half-life of a contaminant in a given water body under different wind and temperature conditions can be computed as

$$\text{half-life units [days, d]} = \frac{0.693}{[\ln 2]} \frac{L/}{[\text{cm}^3 \text{ cm}^{-2}]} \frac{kv}{[\text{cm h}^{-1}]} \frac{24}{[\text{h d}^{-1}]}. \quad (10.29)$$

Table 10.1 shows the values computed by the two models for the two chemicals HCB and 1,2,4-TCB, using a wind speed of 7 m/s at a temperature of 5°C; this table also shows the contaminants' half-lives predicted by models A and B in Lake St. Clair and Lake Ontario and in the microcosm assuming constant environmental conditions. In this example the lakes and the microcosm are assumed fully mixed; nevertheless, this situation might occur only seldom in nature, and half-life estimates might be too high. Likewise, from knowing the experimental half-life of a contaminant, it is difficult to estimate the volatilization rate, since rarely is a lake fully mixed and thermal stratification might change the value of L [eq. (10.28)] daily in an unpredictable and nonmeasurable manner.

DETERMINATION OF EXPERIMENTAL CONDITIONS TO DISCRIMINATE THE TWO MODELS

The difference in model formulation and assumption is important from a practical point of view to assess the fate of, among others, HCB and 1,2,4-TCB in the Great Lakes. For this purpose, field and laboratory experiments must be performed at different wind speeds and at different temperatures, with the results compared with the theoretical predictions. To find the optimal environmental conditions to discriminate the predictions made by models A and B, simulations were performed for HCB and 1,2,4-TCB at wind velocities of 1 to 15 m/s and temperatures of 1° to 25°C. 1,2,4-TCB is more volatile, with a Henry's law constant of $3.34 \cdot 10^{-2}$ (m^3 − atm/mol), than HCB, with a Henry's law constant of $4 \cdot 10^{-4}$ (m^3 − atm/mol).

For each simulation run, the three parameters kv, k_{gas}, and k_{liq} were computed for all combinations of wind and temperature; here the hypothesis is that experiments to differentiate between the two models should be conducted at environmental conditions where the predictions of the two models are most different. The conclusions follow.

1. k_{gas}. Results show that experiments to differentiate between the two formulations should be performed at high wind speeds and low temperature conditions (Fig. 10.4). This result might be counterintuitive, given the fact that volatilization rates are directly proportional to wind speed and temperature; however, since the purpose is to differentiate between the two models, which have temperature formulations that differ mostly at low temperatures, experiments performed at lower temperatures are better even if absolute volatilization rates might be slower than at high temperatures. For example, at a temperature of 5°C and a wind speed of 10 m/s, model A predicts a k_{gas} for HCB of 35.5 cm/h and model B, 18.9 cm/h, a difference that should be measurable experimentally. Note that a high k_{gas} means less resistance in the gas phase.

2. k_{liq}. To differentiate the two model formulations, experiments should be conducted at wind speeds of 6 to 11 m/s and at temperatures higher than 15°C; indeed, higher than 20°C would be even better (Fig. 10.5). At wind speeds higher than 11 m/s, simulations of the two models tend to converge. Estimated values of k_{liq} at a wind speed of 10 m/s at a temperature of 25°C are 4.35 cm/h for model A and 8.52 cm/h for model B; thus, in model A the water phase has a higher resistance than in model B. Cohen et al. (1978) state that three regions might be considered to characterize the relation between k_{liq} and wind velocity. At wind velocities below 3 m/s, the flow is aerodynamically smooth (Wu, 1969), and k_{liq} should be in the range of 1 to 3 cm/h. At these low wind speeds, it is difficult to measure k_{liq}. At velocities higher than 10 m/s, wave breaking might commence and field measurements might be difficult; k_{liq} might be as high as 70 cm/h. In the intermediate region of 3 to 10 m/s, k_{liq} might range in value from 3.5 to 30 cm/h. Table 10.1 shows that model B predicts a k_{liq} of 4.3 and 5.3 cm/h for HCB and 1,2,4-TCB, while model A predicts a much lower 1.3 and 1.7, respectively. Model A might therefore underestimate volatilization rates.

3. kv. Experiments should be conducted at wind speeds of 9 to 10 m/s at temperatures of 10° to 20°C (Fig. 10.6). These environmental conditions are a compromise between the need for high wind speeds and low temperatures to estimate k_{gas} and the relatively higher temperatures needed and wind speeds not higher than 11 m/s to estimate k_{liq}. At a wind speed of 7 m/s and a temperature of 10°C, model A predicts a global volatilization rate of 2.7 m/h, while model B predicts 4.9 m/h, or almost double the volatilization rate.

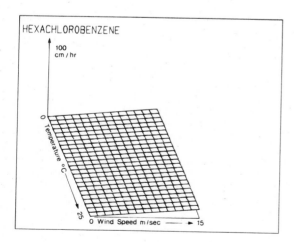 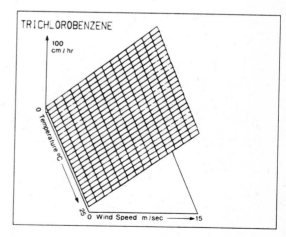

Figure 10.4 Difference between k_{gas} values (cm/h) predicted by model A and model B at wind speeds of 1 to 15 m/s and temperatures of 0° to 25°C. $k_{gas(A)}$ values are always higher (less resistance) than $k_{gas(B)}$ values.

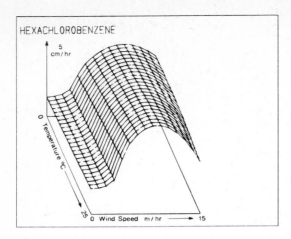
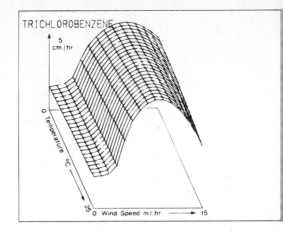

Figure 10.5 Difference between k_{liq} values (cm/h) predicted by model A and model B at wind speeds of 1 to 15 m/s and temperatures of 0° to 25°C. $k_{liq(B)}$ values are always higher (less resistance) than $k_{liq(A)}$ values.

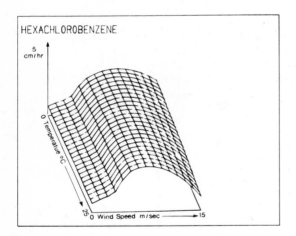
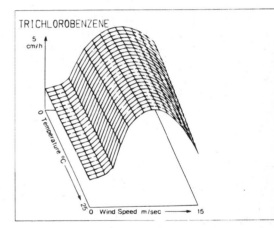

Figure 10.6 Difference between kv values (cm/h) predicted by model A and model B at wind speeds of 1 to 15 m/s and temperatures of 0° to 25°C. kv_B values are always higher (less resistance) than kv_A values.

4. Choice of chemical. From the analysis of Fig. 10.6, we can note that the differences between the two models are higher for a volatile compound such as 1,2,4-TCB than for HCB. At a windspeed of 10 m/s and a temperature of 10°C, the difference for the predicted kv between model A and model B is 4.2 cm/h for 1,2,4-TCB and 2.2 cm/h for HCB. The difference is much higher than for the volatile 1,2,4-TCB. The same high differences can be seen for the other two parameters, k_{gas} and k_{liq}. Therefore, the most volatile compound should be chosen for the experiments. However, at parity of vapor pressure and molecular weight, volatile substances are less soluble than less volatile chemicals (Mackay et al. 1979), as can be seen from the definition of Henry's law constant:

$$H = \frac{\text{vapor pressure [atm]}}{\text{solubility [mole/m}^3\text{]}} .$$

For chemicals of low-solubility experiments, the correct results might be more difficult to obtain. The recommendation is therefore to use the most volatile chemicals at the indicated temperature and wind conditions.

5. The limiting factor, k_{gas} or k_{liq}. In modeling terms, the question can be rephrased as follows: To which parameter is the mathematical model of volatilization most sensitive? Liss (1973) stated that the transfer of oxygen or carbon dioxide largely depends on resistance in the liquid phase and that experiments in a wind tunnel show that the exchange constant increases approximately as the square of the wind velocity. Sutherland (1978, as cited by Park et al., 1980) states that for environmental contaminants the usual ratio between k_{gas} and k_{liq} is about 50 to 250, thus confirming the predominance of k_{liq} as a limiting factor. Cohen et al. (1978) confirm the importance of the water resistance, and they also state that in the literature they cannot find a simple relation between $k1$ and wind speed. Their experimental results show a linear relation between the Reynolds number, nonlinearly dependent on wind speed, and $k1$. Table 10.1 also shows that kv depends more on a correct value of k_{liq} than of k_{gas} for HCB and 1,2,4-TCB.

DISCUSSION

In this chapter, some principles for modeling the fate of toxic contaminants in the environment have been reviewed; from a theoretical point of view, fate models are not different from other ecological models. The problems of the uncertainty of model structure and incomplete quantification of parameter values are not new and need more theoretical analysis (Halfon, 1983a,b; 1985b). Some problems of model structure, however, can benefit from the judicious interfacing of theoretical models and field work; furthermore, this detailed analysis of volatilization models has suggested specific measurements, that, if performed, can improve the model structure. The problem of model adequacy is not only theoretical, but it can be approached from a practical point of view with specific results in mind. For example, both models A and B might not be completely adequate, but by a detailed analysis of theoretical behavior a third model, model C, might be derived and validated with field measurements. The following comments have general applicability:

1. Model A has been widely used in the literature and is generally accepted to model the volatilization process of toxic contaminants from aquatic ecosystems, and it computes volatilization rates of the right order of magnitude. Nevertheless, this model could be improved.

2. Both models A and B include empirically derived equations, eqs. (10.7) and (10.19), respectively. These formulations reduce the theoretical and general validity of the models, especially because no replicate experiments were done by other investigators. Nevertheless, from a practical point of view these models are acceptable. Model B incorporates some physical parameters, such as the Reynolds number, that might be important in a mechanistically accurate volatilization model, but it also includes parameters such as the exponent 0.6 [eq. (10.17)] and the intercept −4.1 [eq. (10.18)] that, by their uncertainty, lower the confidence in the model. It is recommended that empirical relationships in the models should be eliminated and should be replaced by formulations based on the physics of the system.

3. Experiments should be performed at wind-speed and temperature conditions where differences between the two models are largest to provide a better model discrimination, and not at environmental conditions where volatilization rates are maximum. From a practical point of view, the main problem is the extrapolation of kv values obtained in the environment from values obtained in the laboratory. In fact, in the laboratory the parameter L [eq. (10.28)] is much smaller than in any open environment; thus, wall effects might play an important role. In the field the value of L

might be continuously changing. Table 10.1 shows that model B predicts a volatilization rate about three times higher than predicted by model A and consequently shorter half-lives. Unfortunately, this factor of 3 in volatilization rates might not be detected with field experiments (Maguire, National Water Resources Institute, personal communication, April 1986) given the uncertainty of the value of L [eq. (10.28)] under experimental conditions. A solution to this dilemma would be to perform volatilization experiments in well-stratified lakes or uniformly mixed ponds so that a precise value of L can be determined.

4. Even if we develop a perfect model C, we are still confronted with the practical use of the model. In fact, if we use a model for a large area like Lake Ontario, temperature and wind conditions might have a large spatial variability with no lakewide averages. Thus, prediction of volatilization rates in the field will always be uncertain, but within a restricted range of values, given the knowledge of water temperature (from hydrodynamic models) and wind conditions (from climatic models).

5. The final problem regards the volatilization of chlorobenzenes from the Great Lakes. Is Lake Ontario a sink or a transition zone (Oliver, 1984) for some volatile toxic contaminants? A simulation analysis using the TOXFATE model (Halfon, 1986) showed that about 50% of HCB and up to 80% to 90% of 1,2,4-TCB might be volatilized to the atmosphere. TOXFATE uses a formulation as in model A. If a formulation of model B had been used, volatilization rates predicted by the model would have been even greater. An analysis of results presented in this paper shows that (1) the volatilization rate kv is more dependent on k_{liq}, the resistance in the water phase, than on k_{gas} (Mackay, personal communication, June 1986, thinks that PAHs might be an exception); (2) model B compares well with laboratory data obtained by Cohen et al. (1978); and (3) model A is more conservative than B; thus field volatilization rates might be even higher than predicted by the model TOXFATE. This analysis leads to the confirmation of Oliver's (1984) hypothesis of high volatilization rates in Lake Ontario.

ACKNOWLEDGMENTS

Many thanks to Farrell Boyce and Dr. Mark Donelan for their comments about the physical structure of the air–water interface, to Dr. R. J. Maguire for his comments on the chemical aspects of the volatilization process, and to Drs. G. Auble, E. Ongley, and J. Roelle for their comments on the manuscript, which increased readability and understanding.

REFERENCES

BANKS, R. B. 1975. Some features of wind action on shallow lakes. *J. Environ. Eng. Div., Proc. ASCE* 101(EES):813–827.

BLAU, G. E., NEELY, W. B., and BRANSON, D. R. 1975. Ecokinetics: a study of the fate and distribution of chemicals in laboratory ecosystems. *Am. Inst. Chem. Eng. J.* 21:854–861.

BURNS, L. A., CLINE, S. M., and LASSITER, R. R. 1982. *Exposure Analysis Modeling System (EXAMS): User Manual and System Documentation*. U.S. Environmental Protection Agency, Environmental Research Laboratory, Athens, Georgia. 443 pp.

COHEN, Y., COCCHIO, W., and MACKAY, D. 1978. Laboratory study of liquid-phase controlled volatilization rates in presence of wind waves. *Env. Sci. Technol.* 12:553–558.

DURHAM, R. W., and OLIVER, B. G. 1983. History of Lake Ontario contamination from the Niagara River by sediment radiodating and chlorinated hydrocarbon analysis. *J. Great Lakes Res.* 9:160–168.

HALFON, E. 1983a. Is there a best model structure? I. Modeling the fate of a toxic substance in a lake. *Ecol. Mod.* 20:135–152.

HALFON, E. 1983b. Is there a best model structure? II. Comparing the model structures of different fate models. *Ecol. Mod.* 20:153–163.

HALFON, E. 1984a. Error analysis and simulation of Mirex behaviour in Lake Ontario. *Ecol. Mod.* 22:213–252.

HALFON, E. 1984b. Predicting the environmental fate of toxic contaminants in large lakes: Data requirements for mathematical models. In Kaiser, K. L. E (ed.), *QSAR in Environmental Toxicology*. D. Reidel, Boston, pp. 137–151.

HALFON, E. 1985a. The regression method in ecotoxicology. A better formulation using the geometric mean functional regression. *Env. Sci. Technol.* 19:747–749.

HALFON, E. 1985b. Is there a best model structure? III. Testing the goodness of fit. *Ecol. Mod.* 27:15–23.

HALFON, E. 1986. Modelling the fate of Mirex and Lindane off the Niagara River mouth. *Ecol. Mod.* 33:13–33.

HALFON, E., and MAGUIRE, R. J. 1981. Distribution and transformation of fenitrothion sprayed on a pond: modeling under uncertainty. In Beck, M. B., and van Straten, G. (eds.), *Uncertainty and Forecasting of Water Quality*. Springer-Verlag, New York, pp. 117–128.

HALFON, E. H., and OLIVER, B. G. 1990. Simulation and data analysis of four chlorobenzenes in a large system, Lake Ontario, with TOXFATE, a contaminant fate model, In Jorgensen, S. E. (ed.), *Ecological Modelling in the 1990's,* Elsevier, New York, pp. 197–214.

KAISER, K. 1984. *QSAR in Environmental Toxicology*. D. Reidel, Boston, 406 pp.

KAISER, K. L. E., COMBA, M. E., and HUNEAULT, H. 1983. Volatile halo carbons in the Niagara River and in Lake Ontario. *J. Great Lakes Res.* 9:212–223.

KRAMER, G. R. 1974. Predicting reaeration coefficients for a polluted estuary. *J. Environ. Eng. Div., Proc. ASCE* 100(EE1):77–92.

LANE, D. A., JOHNSON, N. D., HANLEY, M. J., SCHROEDER, W. H., and ORD, D. T. 1992. Gas- and particle phase concentrations of alpha-hexachlorocyclohethane, gamma-hexachlorocyclohexane, and hexachlorobenzene in Ontario air. *Env. Sci. Technol.* 26:126–132.

LASSITER, R. R., BAUGHMAN, G. L., and BURNS, L. A. 1979. Fate of toxic organic substances in the aquatic environment. In Jorgensen, S. E. (ed.), *State of the Art in Ecological Modelling,* International Society of Ecological Modelling, Copenhagen, Denmark, pp. 219–246.

LISS, P. S. 1973. Processes of gas exchange across an air–water interface. *Deep-Sea Res.* 20:221–238.

MACKAY, D., and PATERSON, S. 1982. Fugacity revisited. *Environ. Sci. Technol.* 654A–660A.

MACKAY, D., and PATERSON, S. 1991. Evaluating the multimedia fate of organic chemicals: a level III fugacity model. *Env. Sci. Technol.* 25:427–436.

MACKAY, D., SHIU, W. Y., and SUTHERLAND, R. P. 1979. Determination of air–water Henry's law constant for hydrophobic pollutants. *Environ. Sci. Technol.* 13:333–337.

MCCALL, P. J., LASKOWSKI, D. A., SWANN, R. L., and DISHBURGER, H. J. 1983. Estimation of environmental partitioning of organic chemicals in model ecosystems. *Residue Rev.* 85:231–244.

METCALF, R., SINGHA, G. K., and KAPPOR, I. P. 1971. Model ecosystem for the evaluation of pesticide biodegradability and ecological magnification. *Environ. Sci. Technol.* 5:709–713.

NEELY, W. B., and BLAU, G. E. 1977. The use of laboratory data to predict the distribution of chlorpyrifos in a fish pond. In Kahn, M. A. Q. (ed.), *Pesticides in Aquatic Environments*. Plenum Press, New York, pp. 145–163.

OLIVER, B. G. 1984. Distribution and pathways of some chlorinated benzenes in the Niagara River and Lake Ontario. *Water Poll. Res. J. Canada* 19:47–58.

OTHMER, D. F., and THAKER, M. S. 1966. Correlating diffusion coefficients in liquid. *Ind. Eng. Chem.* 45:589–600.

PARK, R. A., et al. 1980. Modeling Transport and Behavior of Pesticides and Other Toxic Organic Materials in Aquatic Environments. Report No. 7, Center for Ecological Modelling, Rensselaer Polytechnic Institute. Troy, New York. 165 pp.

PERRY, R. H. 1963. *Chemical Engineers' Handbook*, 4th ed. McGraw-Hill, New York, various pages.

SABERSKY, R. H., and ACOSTA, A. J. 1964. *Fluid Flow: A First Course in Fluid Mechanics.* Macmillan Co., New York. 393 pp.

SANTSCHI, P. H. 1985. The MERL mesocosm approach for studying sediment–water interactions in ecotoxicology. *Environ. Technol. Letters* 6:335–350.

STREETER, H. W., WRIGHT, C. T., and KEHR, R. W. 1936. Measures of natural oxidation in polluted streams. III. An experimental study of atmospheric reaeration under stream-flow conditions. *Sewage Works J.* 8:282–316.

SUTHERLAND, R. P. 1978 (as cited by Park et al., 1980). Determination of Henry's Law Constants. M.A. thesis, University of Toronto, Canada.

SWEET, C. W., and VERMETTE, S. 1992. Toxic volatile organic compounds in urban air in Illinois. *Env. Sci. Technol.* 26:165–172.

THOMANN, R. V. 1989. Bioaccumulation model of organic chemical distribution in aquatic food chains. *Env. Sci. Technol.* 23:699–707.

WHITMAN, W. G. 1923. The two-film theory of gas absorption. *Chem. Metal. Eng.* 29:146–148.

WU, J. J. 1969. Wind stress and surface roughness at air–sea interface. *J. Geophys. Res.* 74:444–455.

11

Nonlinear Programming in Regulation of Deer Hunting Pressure

GORDON L. SWARTZMAN

Center for Quantitative Science
University of Washington
Seattle, Washington

GEORGE M. VAN DYNE

Formerly of Natural Resource Ecology Laboratory
Colorado State University
Fort Collins, Colorado

Science is now in the early stages of a major revolution caused by the increase in computational and mathematical power afforded by computers. It is evident that the magnitude of this increase is so large that it is rather difficult to comprehend and appreciate. [Van Dyne, in *The Ecosystem Concept in Natural Resource Management*, Ch. X, p. 340 (1969), Academic Press].

Systems ecology can be considered to be the growing science of studying the development, dynamics and disruption of ecosystems. It uses a large number of mathematical and analytical tools from systems analysis and operations research [Van Dyne, in *Grasslands, Systems Analysis and Man*, Int. Biol. Program 19, Chap. 14, p. 883 (1980), Cambridge University Press].

PREFACE BY G. L. SWARTZMAN

Since this paper was written in 1973, there has been an increased awareness that management of natural resources, especially over wide areas, requires multiple objectives and necessitates consideration of noncontrollable aspects of the resource. Tools like linear programming are now commonplace in timber and range management. Stochastic models have made their appearance in fisheries, and risk analysis is recognized as an important part of impact assessment by government agencies and industries concerned with such problems.

In this paper, George Van Dyne and I argued for a new approach to herd management, that is, hunter management. Although it is patently obvious that the issuing of permits to hunters controls the general level of hunting pressure, it is not clear how this pressure can be distributed over a state in such a way as to sufficiently harvest, yet not decimate, herds in each management region. This paper also recognizes that the key to regulating hunting pressure lies as much in characterizing the population of hunters as in assessment of the resource. This is only beginning to be realized in resource management. In fisheries, for example, chronic overfishing of many stocks has resulted in the enforcement of area closures and trip limits (analogous to hunters' bag limits). These impositions result in changes in the pattern of fishing that depend on fleet characteristics (fishing power, experience, flexibility to change fishing gear, and so on). The management challenge of the 1980s appeared to be representing the behavior of resource harvesters or manipulators as they respond to regulation. In some ways it is difficult to believe that this realization was so long in coming.

INTRODUCTION

This paper examines the possible application of nonlinear programming to the problem of trying to best allocate hunting pressure throughout a state. A general framework is presented for setting bag limits, lengthening or shortening the hunting season, and limiting numbers of licenses in various game management units. The objectives are to have as large a total kill as possible throughout the state without decimating deer populations in any region, while spreading hunting pressure away from areas of high human population densities. Extensions of the problem (for example, effects of weather) are also discussed. An example problem dealing with a single hunting management region is solved in both deterministic and stochastic formulations, and some of the implications of these solutions are discussed.

Game managers in many states are confronted with the problem of hunter distribution. Hunting is often concentrated near areas of high human population density, while deer herds frequently predominate in areas away from major metropolitan districts. This imbalance of human and animal populations may result in high kills in areas of high human populations and low kills in other areas. This may lead to overpopulation and large losses of deer because of winter starvation in areas of low human populations.

Game managers and game commissions wish to have as large a total deer kill as possible throughout a state without having so large a kill in any region that the kill might be detrimental to the herds. They would also like to balance the hunting load throughout the state. Liberalizing the bag limit has been used as a way to bring heavier hunter pressure into previously underhunted areas. For example, Kimball (1955) reported that underhunted management units in Colorado have bag limits of "hunters' choice (either sex), 2 deer, 3 deer, or multiple license." These differential bag limits have been used effectively to shift pressure away from overhunted units (Hay et al., 1961).

Altering the hunting season has also been used to manipulate hunting pressure. Preseason hunts have been effective in bringing additional hunters into previously underhunted areas. Postseason hunts often have not been effective because of a decline in hunter interest, bad weather, and bucks (males) being in rut (Hunter and Yeager, 1949). Another alternative is to limit the number of permits in overhunted areas. Excessive pressures on herds near large cities in Idaho were reduced by drawing a limited number of permits for those areas (Rasmussen and Doman, 1947).

Out-of-state hunters are encouraged in some states. Besides bringing extra revenue into the state, it has been reported (Hay et al., 1961) that nonresidents hunt less accessible areas and hunted more days than resident hunters. Klein (1959) reported that accessibility had a major influence on distributing hunting pressure. Building access roads into underhunted areas has been effective in increasing hunter pressure in these areas, even though several studies have shown that the majority of hunters stay very close to roads (Leopold et al., 1951; Stenlund et al., 1952). Another possible source of control is to open the season on weekends in areas of low hunting pressure and in the middle of the week only in areas of high pressure. Murphy (1965) reported that opening the Missouri season early in the week resulted in well-distributed hunting pressure throughout the season.

The game manager has data and information available on a variety of factors useful in decision making. Some of this information is imprecise, although still useful. Although it is evident that state game managers have at their disposal several methods by which they can manipulate hunting pressures, political decisions may override biological considerations in setting seasons and bag limits. For purposes of this paper, we will consider only some of the biological factors in management decision making. The problem, then, is how managers can use information on herd size and age structure, range condition survey, and hunting effectiveness studies in such a way as to provide optimum hunting for the state. Our purpose in this paper is to view hunting management as a nonlinear programming problem to provide an optimization framework for addressing this question and to illustrate, with simple examples, how this framework might be used by game managers.

A BRIEF CONSIDERATION OF MATHEMATICAL PROGRAMMING

Linear programming is the most commonly used operations research technique and can be used to introduce the general formulation of our type of problem. Davis (1967) has applied linear programming to a game management problem, and the reader may wish to refer to that paper for another interpretation of linear functions. Briefly, in linear programming we seek the maximum (or minimum) of a linear objective function subject to linear constraint inequalities:

$$\text{Maximize} \sum_j c_j x_j$$
$$\text{subject to} \sum_j a_{ij} x_j \leq b_i ,$$

where $x_j \geq 0$ for $i = 1, 2, \ldots, m$. The c_j, a_{ij}, and b_i represent constants in the formulation and are derived from data and experience, as will be shown for the case considered in this paper. The x_j, which appear in both the objective function and constraint inequalities, are the unknowns whose values are found in the linear programming solution.

In linear programming both the objective function and constraint inequalities have a linear mathematical form. Furthermore, the c_j, a_{ij}, and b_i are presumed to be known with certainty. In our formulation of the game management problem, however, nonlinear inequalities and stochastic coefficients must be used to reflect uncertainty about those parameters and nonlinear causal relationships between variables.

DEFINITION OF VARIABLES AND OBJECTIVE FUNCTION

Consider the problem of deer hunting in, say, one of the western states of the United States. The species in question is the mule deer (*Odocoileus hemionus*). Let subscript i denote a hunting management region in the state. In many states these regions are chosen to represent deer herds and are generally separate, large watershed basins or groups of smaller watersheds. Let j denote the week of the season. In this treatment, we consider the season in terms of weeks, with $j = 1$ being the week starting October 1 in our example. This is early for the start of a hunting season and would correspond to having a "preseason" hunt. Let k denote the type of hunter. We distinguish three hunter types: (1) in-state urban hunters, (2) in-state rural or local hunters, and (3) out-of-state hunters. These hunters differ in their skill and familiarity with the herds, but not in the weapons used for hunting. Let l be the age class or sex class of deer: (1) fawns, (2) yearling males, (3) yearling females, (4) adult males, and (5) adult females.

We will consider three types of control exercised by the game manager in this formulation: control over the bag limit, the season, or the number of permits issued in each management region. Let B_{il} denote the bag limit on deer type l in hunting region i. Let Q_{ij} denote whether hunting is allowed in region i on week j. If Q_{ij} equals zero, hunting is not allowed during that week, while if Q_{ij} equals 1, it is allowed. Let M_{ik} denote the maximum number of permits issued to hunters of type k in hunting region i.

The two main objectives of the game manager are to maximize the total kill over the season for the whole state, subject to limitations so as not to decimate herds in any area, and to distribute hunting pressure proportionately to deer populations. In the mathematical programming formulation, the first of these objectives will be handled through the objective function, while the second will be included in the constraints to be discussed later.

Maximizing the deer kill throughout the state may be expressed as maximizing the objective function

$$Q = \sum_i \sum_j \sum_k H_{ijk} D_{ijk} Q_{ij} \sum_l S_{ijkl} . \tag{11.1}$$

Here H_{ijk} denotes the total number of hunters of type k in region i on the jth week of the season. This will be considered to be a function of the bag limit in region i. D_{ijk} denotes the average number of days hunted per week in area i in week j by a hunter of type k. Q_{ij}, which is controlled by the game manager, has a 0 or 1 value, depending on whether hunting is or is not allowed in area i on week j. S_{ijkl} denotes the average success by a hunter of type k in area i on week j with deer of type l. Note that H_{ijk}, D_{ijk}, and S_{ijkl} must be inferred from information obtained from previous hunting seasons. Some of the variables may also be expressed as functions of bag limits set on the number of deer to be killed in various areas.

In summary, the total kill throughout the state may be seen as the sum of the kill success per hunter per day over all areas (S), times the number of days hunted in a particular week (D), times the number of hunters involved in that week (H), times a decision variable (Q) determining whether hunting is allowed that week.

CONSTRAINTS ON THE SOLUTION

Constraints are imposed on the game manager in his or her work. Some of these are included as constraints in the mathematical programming formulation of the game management problem. The first constraint relates to the distribution of deer kill throughout the state. Let N_{il} denote the number of deer (estimated) of type l in area i at the start of the hunting season. In the following formulation we consider percentage of kill rather than the total number of animals killed. Total kill could be entered into the constraints as well, but to do this would complicate the problem by requiring simultaneous consideration of forage quantity and quality, among other variables. The reader may wish to refer to Davis (1967), who considered some aspects of herd–forage relationships.

For the present problem, we wish that at least a minimum percentage of the deer are killed so that the herd will not become so large as to destroy its range and eventually starve in a severe winter. C_{il} denotes this minimum percentage kill for region i for deer type l. Also, no more than a maximum percentage of the deer should be killed so that deer herd sizes may be maintained. We denote this maximum percentage of kill in area i of deer type l by K_{il}. These maximum and minimum kill percentages are area specific and can be adjusted from year to year as population estimates and range quality in each area change. For example, if the deer herd in a given area is clearly overgrazing, the minimum percentage of kill would be large, while if the herd were clearly reduced relative to the area's carrying capacity, the maximum percentage of kill would be very low (or zero). Notice that, although the intention of these constraints is to assure herd and forage protection, they are expressed in terms of the kill (rather than the surviving animals) because that is what the manager best controls. Parameters C_{il} and K_{il} are based on the manager's assessment of the herd and its forage. We can then express these limitations as a pair of inequalities,

$$C_{il}N_{il} \le \sum_k \sum_j Q_{ij} H_{ijk} D_{ijk} S_{ijkl} \le K_{il}N_{il}, \qquad (11.2)$$

for each i and l.

It might be possible to limit the number of hunters in an area by issuing only a limited number of permits in that area, as expressed by

$$H_{ijk} \le M_{ik} \qquad (11.3)$$

for each i, j, and k. In cases where permits are not restricted, M_{ik} is assumed to be arbitrarily large.

We also constrain the problem so that no more deer per hunter are killed than the individual seasonal bag limit as expressed by

$$\sum_j \sum_k D_{ijk} S_{ijkl} Q_{ij} \le B_{il} \qquad (11.4)$$

for each i and l. Notice that this inequality constrains the average kill per hunter throughout the season to be less than or equal to the bag limit. This inequality does not mean that a particular hunter cannot kill more than his bag limit, while others kill less; however, it is assumed that hunters are law abiding and that this constraint is realistic.

To complete this formulation as a mathematical programming problem, we must show how the decision variables Q_{ij} and B_{il} affect the objective function. Following a discussion of this, we consider parameter estimation, where the difficulty lies in obtaining parameters, the feasibility of solutions, and possible extensions of the problem.

DECISION VARIABLES IN THE OBJECTIVE FUNCTION

Notice that the decision variable B_{il}, the bag limit for deer type l in area i, does not appear in the objective function (11.1). However, bag limits in an area do influence the number of hunters in that area. This relationship is important because it is a key to distributing hunting pressure throughout a state, along with limiting the number of licenses.

Let us assume that the number of hunters that will hunt in a given area is directly, but nonlinearly, affected by the bag limit for that area (Fig. 11.1). Here we are concerned with the total bag limit or the sum over all age classes and sex classes of deer:

$$B_i = \sum_l l_l B_{il}.$$

The l_l are weightings of the importance to hunters of hunting deer in a particular age class and sex class. The relationship could be different for different types of hunters (k). For example, numbers of local hunters in a given region are probably less likely to be influenced by bag limits than are numbers of out-of-state hunters. However, in this example, no consideration is made of this fact.

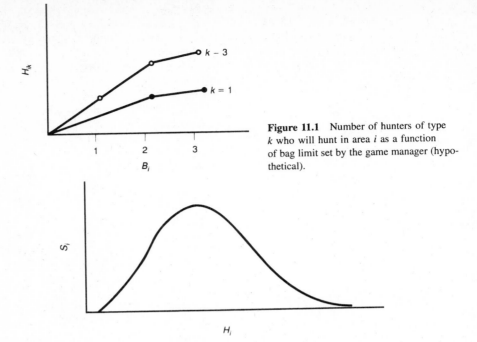

Figure 11.1 Number of hunters of type k who will hunt in area i as a function of bag limit set by the game manager (hypothetical).

Figure 11.2 Kill success per hunter per day as a function of hunting pressure (hypothetical).

The number of hunters in an area affects hunter success in that area. At low hunting pressure an increasing number of hunters *increases* individual hunter success at first because of the increased probability of hunter–deer contacts, but further increases in hunting pressure reduce success per hunter due to interference, as shown in Fig. 11.2. Obviously, kill success per hunter is not only a function of the number of hunters. Kill success may also be related to size of the deer population in a given area. Thus, the function shown in Fig. 11.2 could be changed to include another independent variable, deer density, and perhaps an interaction between number of hunters and number of deer as affecting kill success per hunter per day. We would expect that, as the number of deer increased for a constant number of hunters, there would be increasing hunter success. There should be an interaction between the number of hunters and deer density because a large number of hunters in an area with a high deer density should cause the deer to move and become more susceptible to hunting.

We have now related both hunter success and number of hunters (variables in the objective function) to bag limits. These nonlinear relations result in nonlinearities in both the objective function and constraints. Eberhardt (1960) has discussed factors affecting hunter success, and the type of information he presents could be used to further modify or supplement the preceding relationship.

ESTIMATION OF PARAMETERS

The objective function (11.1) can be rewritten to express the relationships in Figs. 11.1 and 11.2 as

$$Q = \sum_i \sum_j \sum_k f(B_i) D_{ijk} Q_{ij} \sum_l g(f(B_i)) \,. \tag{11.5}$$

Here, $f(B_i)$ denotes the functional relationship in Fig. 11.1 for number of hunters hunting, and

$g(f(B_i))$ denotes another functional effect of bag limits on hunter success through number of hunters as an intermediate variable (that is, bag limit controls number of hunters, which controls hunter success, as in Fig. 11.2). The assumption in this equation is that hunter success, S_{ijkl}, depends only on bag limit, B_i, and not on time of year or hunter type.

Estimates of coefficients like D_{ijk}, the average number of hunting days per week per hunter, and N_{il}, the deer population sizes, are obtained from game managers. Their information sources include licenses issued, check stations set up in each management area, aerial surveys of deer populations, field observation tagging procedures, and interviews with hunters after the season. As these data are presently obtained in most states, they provide only approximate estimates for the parameters in this problem (Maguire and Severinghaus, 1954; Powell, 1958). However, new techniques such as satellite monitoring and electronic surveillance of herd size, as well as computerization of records, may increase the accuracy, precision, and usefulness of these estimates. Also, improved interviewing techniques have proved helpful (James et al., 1964). Initial solution of the optimization problem formulated, even with inaccurate parameter estimates, might well be valuable in focusing attention on data collection needs.

Questions arise about constraints (11.2), which set minimum and maximum kill limits in each area. How can the game manager ensure that these limits are met? Clearly, the limitation cannot be exactly met in practice. At best, we can hope that the limits will be met with some probability and, as such, the constraint would become stochastic. Solution techniques, as yet, are not well developed or programmed for all forms of nonlinear programming problems with stochastic coefficients. Also, the curves in Figs. 11.1 and 11.2, only represent average values, whereas, in reality, the effect could be better represented as an average from some probability distribution.

There are three independent decision variables, Q_{ij}, B_{il}, and M_{ik}, that can be assigned a range of values. A solution that is optimal with respect to all three of these variables cannot feasibly be obtained with a single optimization run using nonlinear programming techniques. This is mainly due to the size of the problem. With 100 management areas, four age-sex classes, three hunter types, and ten possible weeks in the hunting season, we have 1000 Q_{ij}'s, 400 bag limits, and potentially 300 M_{ik}'s. This is too large a problem for a solution using conventional nonlinear programming solution techniques.

Another problem is that of continuity of seasons. The game manager traditionally sets seasons in blocks of time. That is, seasons usually encompass consecutive weeks. However, an optimal solution to the problem as outlined in this paper could result in seasons of nonconsecutive weeks. Generally, this would be undesirable due to the enforcement problem and inventory of hunter success at checking stations. A formulation having only consecutive-week seasons might be made by replacing Q_{ij} by two new variables, one denoting the first week of the season and the other the length of the season. As an alternative, a nonlinear programming solution could be obtained for the optimal bag limit for a particular length of season (of consecutive weeks), with numbers of permits issued being limited in particular regions. The nonlinear programming solutions could then be compared with those for other possible seasons with the same license limitations and a final decision then made on the best of these. The problem could be solved, as it is stated above, using a constrained nonlinear programming algorithm (Fiacco and McCormick, 1964; Himmelblau, 1972).

The problem as it is presently structured is a dynamic problem; that is, time (weeks j) is included as a variable in the problem. Furthermore, the Q_{ij} variable represents a weekly decision on whether or not to allow hunting. Problems with decisions at fixed stages fall into the realm of dynamic programming. Therefore, the problem of which season to set for optimal deer harvest under a given permit limitation and bag limit combination could be structured and solved using a dynamic programming algorithm. Dynamic programming was used by Romesburg (1972) to optimize white-tailed deer harvest over a 10- to 15-year horizon, but that problem did not have nearly as many decision alternatives. To our knowledge, dynamic programming has never been used to solve a problem of the size of the one here formulated.

HUNTING ACCESS AND WEATHER CONSIDERATIONS

The number of hunters of a particular type on a specific week in a given area does not depend solely on the bag limit in that area; it also depends on which week it is. Several studies (for example, Kimball, 1955; Taber and Dasmann, 1957) have shown that hunting pressure is heaviest at the beginning of the main season (excluding early bow-hunting seasons) and decreases toward the end of the season. This change in the number of hunters also depends on hunter success during the season. For example, if hunters are successful at the beginning of the season, their numbers drop toward the end of the season. If hunters are unsuccessful due to a severe storm early in the season, the number of hunters might later increase if and when weather conditions improved. Hunter success also depends on whether hunting had been allowed during the previous week.

Also, hunter success depends on access road conditions in a particular area (denoted by R_i and quantified in some fashion) and is probably heavily influenced by weather conditions. Let us denote the weather in area i on week j by W_{ij}. Weather conditions also have a large effect on hunter density in a particular area; dry weather results in hunters being active, but the noise produced by their walking on a dry forest floor may alarm deer and result in a low kill (Fobes, 1945; Maine Department of Inland Fisheries and Game, 1963).

Taking all these effects into consideration, we can rewrite the objective function (11.5) as follows, where f' and g' are generalized functions we do not describe explicitly:

$$Q = \sum_i \sum_j \sum_k f'\left(B_i, j, \sum_{k'=1}^{k-1} \sum_l S_{ijk'l}, W_{ij}, R_i\right) D_{ijk} Q_{ij} \sum_l g'(H_{ijk}, W_{ij}, Q_{i(j-1)}). \quad (11.6)$$

Here, deer kill is a function (f') of bag limits B_i, the week j, previous hunting success $\sum_{k'=1}^{k-1} \sum_l S_{ijk'l}$, weather W_{ij}, access road conditions in a given area R_i, number of days hunted per hunter D_{ijk}, the season Q_{ij}, and hunter success, which is additionally a function (g') of the number of hunters H_{ijk}, the weather W_{ij}, the type of deer l, and whether hunting was allowed during the previous week $Q_{i(j-1)}$.

SIMPLIFIED NUMERICAL EXAMPLE OF A DETERMINISTIC AND STOCHASTIC PROBLEM

Although the previous discussion has developed a real-life problem through the formulation stage, it has not provided any discussion of insights pertaining to a solution. Furthermore, with the present data base such a solution would only be a mathematical exercise. Programming techniques are most valuable at present for the insights that can be gained from the solution and from "exercising the model" to ascertain effects of parameter changes from which management alternatives may then proceed. Therefore, we formulate a simplified problem in a single game management unit where a data base does exist and analysis can be carried out. Whereas the above formulations represent the future and goals of mathematical programming use in natural resource management, the following numerical problem and subsequent discussion represent work constrained by the present state of the art and data.

The game management problem we have chosen is that of optimizing kill in a single herd area while ensuring with some probability that the herd survives into the next year. This problem handles the fact that overwintering death rates are not deterministic; it makes them stochastic with normal distributions. The normal distribution assumption is made for convenience, although other distributions could be used. The kill is "optimal" in the sense of the mathematical programming formulation, not in the sense of the maximum growth rate at the inflection point of a population growth curve. We restrict consideration to a single hunting management region where we wish, as before, to maximize the kill, assuming that we know the deer populations by age class and sex class for the current year. We also assume that we can control the kill rate of each age- and sex-class group (which is possible to some extent by the methods discussed earlier). Also, it is assumed that management region areas are sufficiently large "natural herd units" so that immigration and emigration are negligible.

We define numbers in the age-classes and sex-classes as follows: N_1 = fawns, N_2 = yearling males, N_3 = yearling females, N_4 = adult males, and N_5 = adult females. If X_j's denote kill rates for a deer of age- and sex-class j (kill per unit deer), then the objective function becomes:

$$\text{maximize total kill } Q = \sum_j X_j N_j. \tag{11.7}$$

The following constraints are considered:

1. We wish the size of next year's herd to be at least as large as that of this year's because the assumption is made here that the range is not being used to capacity. A stable deer population is assured if the reproduction of animals surviving the hunt is at least as great as the combined hunt kill and natural winter mortality. In the more general problem of the previous section, we handled this constraint by maintaining some percentage of the present herd size in a region at the end of the season. Later this constraint is made more realistic by including death rates as a stochastic variable. The deterministic form of this constraint is expressed by the inequality

$$r_3(N_3 - X_3N_3 - d_3N_3) + r_5(N_5 - X_5N_5 - d_5N_5) \geq N_1X_1 + N_2X_2 + N_3X_3 + N_4X_4 + N_5X_5$$
$$+ d_1(N_1 - N_1X_1) + d_2(N_2 - N_2X_2) + d_3(N_3 - N_3X_3) + d_4(N_4 - N_4X_4) + d_5(N_5 - N_5X_5). \tag{11.8}$$

Here, r_3 is the reproduction rate for yearling females and r_5 that for adult females. The d_i's represent natural winter death rates per individual, respectively, for fawns, yearling males, yearling females, adult males, and adult females. For purposes of simplification, death rates are assumed equal during the winter for yearlings and adults of the same sex.

2. We wish the posthunt ratio of breeding females to breeding males to be 8:1 at most. This is to ensure that sufficient males are left alive to mate with the remaining breeding females. This constraint is expressed by

$$N_3 + N_5 - N_3X_3 - N_5X_5 \leq 8(N_2 + N_4 - N_2X_2 - N_4X_4). \tag{11.9}$$

3. We wish to replenish the yearling population for the next year so as not to have a population overloaded with older adults. Thus, fawns surviving into the next year must be at least as great as the number of yearlings in the present population. This is expressed by the inequality

$$N_1 - N_1X_1 - d_1(N_1 - N_1X_1) \geq N_2 + N_3. \tag{11.10}$$

4. We also wish that no more than 95% of remaining age classes is killed. This is a check to assure that a severe winter does not decimate an age class and is expressed by

$$X_i \leq 0.95, \quad \text{for } i = 2, 3, \ldots, 5. \tag{11.11}$$

5. Additional constraints, for purposes of realism and analytical necessity, are that the harvest must be nonnegative, or

$$X_i \geq 0, \quad \text{for } i = 1, 2, \ldots, 5. \tag{11.12}$$

This problem, as expressed in eqs. (11.7) through (11.12), is a linear programming problem with twelve constraints and five decision variables (these are the X_i's, the kill fractions in each age

TABLE 11.1 Parameters for stochastic and linear programming game management formulations

				Death rate	
Index	Animal group	Initial number N_i	Reproduction rate r_i	Mean d_i	Variance $\sigma^2_{d_i}$
1	Fawns	3278	0.00	0.420	0.0284
2	Yearling males	556	0.00	0.066	0.0000
3	Yearling females	858	0.94	0.152	0.0129
4	Adult males	1098	0.00	0.066	0.0000
5	Adult females	2612	1.65	0.152	0.0129
	Total	8402	—	—	—

class considered). The example data in Table 11.1 are adapted from a situation described by Taber and Dasmann (1957, 1958). We used initial numbers based on their data as well as their reproductive rates. But their death rates were estimated on the assumption of a steady-state population without harvest by hunting. Because we assume hunting in our example, we have reduced the estimates for mean death rate. We have also estimated a variance for death rates of fawns, yearling females, and adult females.

STOCHASTIC PARAMETERS IN THE LINEAR PROGRAMMING PROBLEM

Fawn and female mortality is increased more by severe winters than for other age classes and sex classes. Therefore, we made year-to-year death rates (d_1, d_3, and d_5) normally distributed stochastic (random) variables with respective variances of 0.0284, 0.0129, and 0.0129 and covariances of zero. The problem then becomes a linear programming problem with some stochastic parameters.

If any of the coefficients a_{ij} in a constraint

$$\sum_{j=1}^{n} a_{ij} x_j \leq b_i$$

are stochastic, then these can be said to have some mean $\bar{a}_{ij}$ and variance σ^2_{ij}. If the assumption can be made that the stochastic a_{ij}'s come from a normal probability distribution, with these distributions being independent of each other if there is more than one stochastic a_{ij}, then this stochastic constraint can be transformed into an equivalent nonlinear constraint. The constraint equation has coefficients depending on the $\bar{a}_{ij}$, σ^2_{ij}, and on the probability (call it P) that we wish the constraint to be satisfied. Since we have random parameters, we can never say with certainty that a given constraint will be satisfied. The form of the stochastic constraint transformed into its equivalent nonlinear deterministic constraint is given by (Charnes and Cooper, 1962; Bracken and McCormick, 1968):

$$\sum_{j=1}^{n} \bar{a}_{ij} x_j + k_1 \left(\sum_{j=1}^{n} \sigma^2_{ij} x_j^2 \right)^{1/2} \leq b_i, \qquad (11.13)$$

where $\sigma^2_{ij} = 0$ for all nonstochastic coefficients, and k_1 is a number chosen from the normal probability distribution that depends on the probability P. If we want to assure with probability q that our constraint is met, then

$$P\left[\frac{x-\mu}{\sigma}\right] = q. \tag{11.14}$$

We can use a table of normal deviates, pick the value of q, and then find z to satisfy the expression; thus

$$\frac{x-\mu}{\sigma} = z,$$

where x is the value we seek, μ is our estimate of the mean, and σ is an estimate of the standard deviation. This is a common approach in a statistical theory as outlined by Steel and Torrie (1960) and Sengupta (1972).

Only constraints (11.8) and (11.10) contain d_1, d_3, or d_5 (the stochastic death rates) and thus are the only constraints affected by stochasticization of the model. Let us rewrite these two constraints using the theory described above and the equations for the linear model [eqs. (11.8) to (11.12)]. If we choose $P = 0.6$ (60% probability of satisfying each of these constraints), then $k_1 = 0.842$ and $k_2 = 0.25$. We can expand the general formulation for the constraint for maintaining herd size [eq. (11.8)] as follows:

$$N_1(d_1-1)X_1 + N_2(d_2-1)X_2 + N_3(d_3-r_3-1)X_3 + N_4(d_4-1)X_4 + N_5(d_5-r_5-1)X_5 \\ \geq (d_2N_2 - r_3N_3 + d_4N_4 - r_5N_5) + (d_1N_1 + d_3N_3 + r_3d_3N_3 + d_5N_5 + r_5d_5N_5). \tag{11.15}$$

In our analyses we had arbitrary estimates of variances for d_1, d_3, and d_5. To account for this variance (σ^2_i), the right side is changed according to the method described in eq. (11.14) by adding an expression, $k_2\sigma_c$, where k_2 is a tabulated value related to the probability level selected (60% in our case). The σ^2_c is a common variance obtained here from using all terms in the right side containing stochastic variables. Thus,

$$\sigma_c = 809 = [(0.169 \cdot 3278)^2 + (0.114 \cdot 858)^2 + (0.94 \cdot 0.114 \cdot 858)^2 \\ + (0.114 \cdot 2612)^2 + (1.65 \cdot 0.114 \cdot 2612)^2]^{1/2}. \tag{11.16}$$

The left side is changed according to the theory outlined in eq. (11.13).

A similar procedure is used for the constraint related to replenishing yearlings. Both the linear and nonlinear programming problems had a linear objective function with five variables. The linear programming formulation has twelve linear inequality constraints, and the nonlinear programming formulation has ten linear inequality constraints and two nonlinear ones (Table 11.2). Both problems were solved with the computer code FLEXIPLEX (Himmelblau, 1972).

RESULTS OF THE LINEAR DETERMINISTIC AND NONLINEAR STOCHASTIC FORMULATIONS

Results of these analyses are presented in Table 11.2 for various age–sex groups and summarized in Fig. 11.3 for the total herd. Starting with a total herd size, before hunting, of 8402 (see Table 11.1), in the linear deterministic case some 30% of the animals were harvested, as compared to 25% in the situation where winter death loss rates were assumed to be stochastic and resulted in the nonlinear programming formulation of the optimization problem. In both instances about one-fourth of the fawns and none of the adult females were harvested, with more in the stochastic example, but the total number of males (yearling + adult) harvested differed by less than 5%. The major difference in the harvest was that in the linear deterministic case some 369 yearling females were taken, compared to only 32 in the nonlinear stochastic situation. The small harvests allowed for fawns and females probably are not unrealistic and perhaps could cover illegal or accidental hunting of these groups in a bucks (males) only situation.

TABLE 11.2 Results of analyses of the linear deterministic problem and the nonlinear stochastic problem

Age–Sex Group	Harvest percent	Late fall population	Winter death loss	Spring population	Reproduction	Final population
Deterministic						
Fawns	26	2406	1010	1396	—	4045
Yearling males	95	28	2	26	—	698
Yearling females	43	489	74	415	390	698
Adult males	67	362	24	338	—	364
Adult females	0	2612	397	2215	3655	2630
Stochastic						
Fawns	26	2436	1023	1413	—	4379
Yearling males	42	320	21	299	—	706
Yearling females	4	826	55	771	724	707
Adult males	90	110	7	103	—	402
Adult females	0	2612	397	2215	3655	2986

Numbers are rounded to nearest animal.

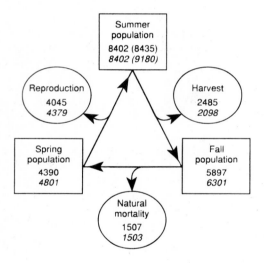

Figure 11.3 Summary of dynamics of total herd size with optimal harvesting with either deterministic or stochastic (italics) estimates of winter death rates for fawns, yearling females, and adult females. Summer populations are given for year n and year $n + 1$ in parentheses.

Harvesting a high proportion of yearling females is highly counterintuitive. We normally expect does (females) not to be harvested or, if they are, that there will be little difference between shooting a yearling versus a mature doe. In contrast, the predicted percentage of adult females to be harvested was near zero, which agrees with common experience. Low doe harvests here suggest that doe seasons could only be defended under conditions of excessive deer population sizes. This phenomenon may be clarified by looking at yearling to adult female ratios. In the original herd composition there was 33% as many yearling females as adult females. The solution for the deterministically formulated problem gave about 27% as many yearlings as adults, and the solution for the stochastically formulated problem gave 24% as many yearling as adult females. The lower percentage of yearlings in the stochastic problem illustrates a general phenomenon of "stochasticizing" a problem formulation. In the stochastic problem, the overall final herd size was the largest of the three instances (9180 individuals). The number of fawns needed to maintain the population was 48% in both final herds and 39% in the initial herd. With uncertainty in the data and information used in formulating the problem, larger populations and larger replacement populations must be maintained. Reproductive females represented about 40% of the initial and final herds. Slightly more adult females, proportionately, were in the final herd in the stochastic case due to the low reproductive rates for yearling does as compared with mature does: 0.94 and 1.65, respectively.

Examination of the solution vector, that is, the optimal percent kill of each age–sex class and of the constraints [eqs. (11.8) through (11.12), with parameter values from Table 11.1] shows only the constraint of killing less than 5% of the adult females as superfluous. In actuality, the algorithm attempted to drive the adult female harvest to a negative number. The constraints for replenishing yearlings, for maintaining male : female ratios, and for maintaining herd size were limiting in both linear and nonlinear formulations.

The objective function contains an estimate of the initial number, N_i, of each age and sex group in the herd. In practice, these estimates may contain error. There is an approach in the literature to handling uncertainty in the c_j's of the general problem formulation, assuming normally distributed errors (Sengupta, 1972). We modify the objective function as follows:

$$Q = \sum_{j=1}^{n} \bar{c}_j x_j - k \left(\sum_{j=1}^{n} \sum_{i=1}^{m} x_j \sigma_{ij}^2 x_i \right)^{1/2}, \qquad (11.17)$$

where $\bar{c}_j$ are estimates of the means and σ_{ij}^2 are estimates of the variances.

Assuming covariances among initial population estimates to be zero and assuming variances $\sigma_{N_j}^2$ we could reformulate our objective function into a nonlinear form to account for variability in initial population estimates:

$$Q = \sum_{j=1}^{5} x_j N_j - k_3 \left(\sum_{j=1}^{5} x_j^2 \sigma_{N_j}^2 \right)^{1/2}, \qquad (11.18)$$

where k_3 would be selected as was k_1 (see the preceding text and Table 11.2). This would offer no particular problem, but examination of eq. (11.15) and Table 11.2 shows that the N_j are found in three of the constraints. Thus, the variance in estimates of N_j would have to be considered there also. Similarly, we should examine the effect of variance in the estimates of the reproductive rates for yearling and adult does, as these terms are found in the first constraint (see Table 11.2). Alternatively, we can vary the initial population sizes and examine the influence on the harvest and final population size, as is done next for adult males.

The linear formulation in the long run is relatively insensitive to initial conditions. For example, increasing the adult male population by about 5% had the following impact on harvest: increased herd size gave 2536 animals harvested or 30% of the herd and about the same percent for normal herd size (see Table 11.2). Increased initial herd size resulted in a slightly larger final herd size, but herd composition by age–sex class was almost identical. The major difference due to changing the initial population estimate for adult males, a difficult group to census, was in the harvests. With increased herd size there were 1329 males harvested as compared to 1264 with normal herd size, and the percentage of adult males harvested was greatly increased in the former case.

Many simplifying assumptions were made in the preceding formulation. For example, both mortality and natality rates might vary according to the relationship between population size and the range conditions. Not only would mean values for natality and mortality vary, but also variances could be expected to change. For example, if a large amount of forage were available per unit of population, then mortality rates would be expected to be small and relatively constant as compared with a situation where population size fully saturated the available range.

We have assumed that covariances among the death rate estimates are zero and that variances of the estimate of male death rates are zero. In practice, this may not be the case, which would lead to a more complicated formulation of our nonlinear problem. In fact, there probably is a nonzero correlation between death and reproductive rates, which would further complicate problem formulation. We have, however, illustrated the general methods to account for stochastic terms in the constraints and objective function of a linear programming problem.

In the analyses presented, total population size for the original conditions was 8402 animals. Under the deterministic solution, resultant population size is 8435 and under stochastic analysis, 9180. However, the only constraint in the nonlinear programming formulation that affects population size directly is that the number of yearlings in any given year must be equal to or greater than the number of yearlings in the preceding year. However, the analysis shown above considered only a single year. The resultant population vector after the predicted harvest could be used as initial conditions for another year's analysis. Successive calculations year by year could give a prediction of population size and age- and sex-class distribution over a long period of time. It would be interesting to determine if a steady state size and distribution of population would be reached by such a procedure. Furthermore, we could then formulate the problem as a dynamic programming problem, rather than as a sequence of year-by-year nonlinear programming problems, and compare results. A dynamic programming formulation, however, is beyond the scope of this chapter.

POTENTIAL APPLICATIONS

The above-described approaches are largely of an exploratory nature. Extensions of the problem can be added to the framework presented. The model could be modified to include several game species interdependently. Stocking and breeding practices could also give game managers new control of animal populations, as already is done with fishes. Big game represent an extremely valuable publicly owned resource. Even small increases in efficiency of use can realize major financial and other values. Operations research and systems analysis procedures are of considerable utility in industries where only economic return is considered. Therefore, it is reasonable to expect that nonlinear and dynamic programming techniques can also play a major role in future game management.

Our example considered only a limited number of variables in a single area. In real management situations, there are many variables and many areas in a state. For example, Colorado has 101 game management units in four regional areas. Big game hunting is big business, yet game managers have been slow to introduce analytical techniques into their decision making. Instead, most decisions seem to be made qualitatively by regional managers, often without knowledge of conditions in, or impacts of their decisions on, other regions. Considerable amounts of information are being accumulated in many states from relatively extensive and expensive management studies and surveys of hunting pressures, hunting success, game herd age and sex structure, range condition, and trend. Yet it appears this valuable information is not being fully utilized in the largely qualitative decision-making process now in effect.

We have explained and used relatively few variables in our example formulation. Even so, it is already possible to see that formulations of many natural resource management optimization problems call for current state-of-the-art techniques from the operations research field. These examples give some insight into the type of work that should be done in future applications of operations research techniques to game and fish management. We hope that more operations research specialists are attracted to work in this area. Although the problems are difficult analytically, the combined efforts of experienced resource managers, ecological systems analysts, and operations researchers can lead to quantifying the decision-making process.

ACKNOWLEDGMENTS

We thank E. Prenzlow of the Colorado Department of Game, Fish, and Parks and R. Hansen and H. Steinhoff of Colorado State University for their constructive comments. We also thank E. Kautz and Z. Abramsky for help with the illustrative examples.

This paper reports on work supported in part by the U.S. National Science Foundation, Grant BMS73-02027 A02 to the Grassland Biome, U.S. International Biological Program, for "Analysis of Structure, Function, and Utilization of Grassland Ecosystems."

REFERENCES

BRACKEN, J., and MCCORMICK, G. P. 1968. *Selected Applications of Nonlinear Programming.* Wiley, New York. 110 pp.

CHARNES, A., and COOPER, W. W. 1962. Chance constraints and normal deviates. *J. Am. Stat. Assoc.* 57(297):134–148.

DAVIS, L. S. 1967. Dynamic programming for deer management planning. *J. Wildl. Mgmt.* 31(4):667–679.

EBERHARDT, L. 1960. Estimation of Vital Characteristics of Michigan Deer Herds. Michigan Dept. Cons., Game Div. Rept. No. 2282. 192 pp.

FIACCO, A. V., and MCCORMICK, G. P. 1964. Computational algorithm for the sequential unconstrained minimization technique for nonlinear programming. *Mgmt. Sci.* 10(4):601–617.

FOBES, C. B. 1945. Weather and the kill of white-tailed deer in Maine. *J. Wildl. Mgmt.* 9(4):76–78.

HAY, K. G., HUNTER, G. M., and ROBBINS, L. 1961. Big Game Management in Colorado 1949–1958: A Ten Year Survey of Applied Big Game Management. Colorado Dept. Game and Fish Tech. Bull. 8. 112 pp.

HIMMELBLAU, D. W. 1972. *Applied Nonlinear Programming.* McGraw-Hill, New York. 498 pp.

HUNTER, G. N., and YEAGER, L. E. 1949. Big-game management in Colorado. *J. Wildl. Mgmt.* 13(4):392–411.

JAMES, G. A., JOHNSON, F. M., and BARICK, F. B. 1964. Relations between hunter access and deer kill in North Carolina. *Trans. N. Am. Wildl. Nat. Resources Conf.* 29:454–463.

KIMBALL, T. L. 1955. The application of controlled hunting to the take of big game. *Proc. Intern. Assoc. Game Conser. Comm.* 45:95–102.

KLEIN, D. R. 1959. Evaluation of Hunter Harvest. Alaska Dept. Fish and Game, Job Completion Rept., Fed. Aid Project W-3-R-13, pp. 20–24.

LEOPOLD, A. S., RILEY, T., MCCAIN, R., and TEVIS, L., Jr. 1951. The Jawbone Deer Herd. Calif. Game Fish Dept. Game Bull. 4. 139 pp.

MAGUIRE, H. F., and SEVERINGHAUS, C. W. 1954. Wariness as an influence on age composition of white-tailed deer killed by hunters. *J. New York Fish and Game* 1(1):98–109.

Maine Department of Inland Fisheries and Game. 1963. Deer season—1963. *Game Div. Leafl. Ser.* 21:1–21.

MURPHY, D. A. 1965. Effects of various opening days on deer harvest and hunting pressure. *Proc. Southeastern Assoc. Game Fish Comm.* 19:141–146.

POWELL, L. E. 1958. Hunter report cards versus questionnaires. *Proc. Western Assoc. State Game Fish Comm. Conf.* 38:235–238.

RASMUSSEN, D. I., and DOMAN, E. R. 1947. Planning of management for programs for western big-game herds. *Trans. N. Am. Wildl. Conf.* 12:204–210.

ROMESBURG, C. H. 1972. *Prescriptive Decision Models for Managing the Harvest of White-tail Deer Herds.* University of Pittsburgh, Pittsburgh, Pennsylvania. 283 pp.

SENGUPTA, J. K. 1972. *Stochastic Programming—Methods and Applications.* North Holland, Amsterdam. 313 pp.

STEEL, R. G. D., and TORRIE, J. H. 1960. *Principles and Procedures of Statistics—With Special Reference to the Biological Sciences.* McGraw-Hill, New York. 481 pp.

STENLUND, M. H., MORSE, M. A., BURCALOW, D. W., ZORICHAK, J. L., and NELSEN, B. A. 1952. White-tailed deer bag checks, Gegoka Management Unit, Superior National Forest. *J. Wildl. Mgmt.* 16(1):58–63.

TABER, R. D., and DASMANN, R. F. 1957. The dynamics of three natural populations of the deer *Odocoileus hemionus columbianus. Ecology* 38(2):233–246.

TABER, R. D., and DASMANN, R. F. 1958. The Black-Tailed Deer of the Chaparral; Its Life History and Management in the North Coast Range of California. California Game and Fish Dept. Game Bull. 8. 163 pp.

12

Simulation Modeling in a Workshop Format

GREGOR T. AUBLE, DAVID B. HAMILTON, JAMES E. ROELLE, AND AUSTIN K. ANDREWS

U.S. National Biological Survey
National Ecology Research Center
Fort Collins, Colorado

> What are the claims that these model builders and model users are making about models contributing to better renewable resource management? Complex ecosystems must be simplified for purposes of management, and models serve a role in conceptual simplification [Van Dyne, in *Ohio J. Sci.* 78, 191 (1978)].

INTRODUCTION

In this chapter, we describe and evaluate our experience using simulation modeling in a workshop format. This experience was accumulated as part of an experiment to test workshop modeling as a tool for the U.S. Fish and Wildlife Service (the Service). The basic mission of the Service is to conserve, protect, and enhance fish and wildlife and their habitats for the continuing benefit of the people. In an effort to better fulfill this mission, a program was initiated in the mid-1970s to develop improved methods for assessing the impacts of development activities on fish and wildlife resources. As part of this program, the Service decided to explore the utility of a collection of systems analysis and group dynamics procedures (known as Adaptive Environmental Assessment and Management) being developed by C. S. Holling, C. Walters, and their associates at the University of British Columbia and the International Institute for Applied Systems Analysis (IIASA, Laxenburg, Austria). Workshop simulation modeling was the central and most clearly specified aspect of the approach.

Initial training in workshop modeling techniques was provided by the University of British Columbia. Between 1978 and 1986, we conducted simulation modeling workshops concerning 13 different natural resource systems; these applications constitute the experience evaluated in this chapter. During the course of these applications, we interacted with three other groups using this type of simulation modeling approach: the University of British Columbia, IIASA, and Environmental and Social Systems Analysts Ltd. (ESSA), a consulting firm in Vancouver, British Columbia.

The benchmark description of the workshop modeling process is a book entitled *Adaptive Environmental Assessment and Management* (Holling, 1978). Several general evaluations of the process have been conducted (for example, International Institute for Applied Systems Analysis, 1979; Environmental and Social Systems Analysts, 1982). Furthermore, a variety of specific case studies has been described (for example, Baskerville, 1979; Clark, Jones, and Holling, 1979; Walters, Hilborn, and Peterman, 1975; Wood, 1979), including several Service projects that form the basis for this chapter (Andrews et al., 1982; Hirsch, Andrews, and Roelle, 1979; Roelle and Auble, 1983).

PROCESS

Rationale

The workshop simulation modeling process involves a variety of interested parties in the design and construction of a simulation model of a natural resource system. A principal benefit of simulation modeling is the increased understanding of system structure and behavior that the modeler gains from the modeling process. This benefit usually accrues even in the absence of real predictive power in the resulting model. Modelers have often been frustrated, however, by the fact that this benefit has not extended to potential users of simulation models (Watt, 1977). We believe this is largely because those users have not actively participated in the modeling process. A workshop setting provides an opportunity for potential users to participate in the modeling process and to acquire a better understanding of the natural resource system. More specifically, the following aspects of workshop simulation modeling are worth emphasizing.

1. Construction of a specific simulation model provides a common objective for a variety of agencies and groups with different, and often competing, interests and responsibilities.
2. The modeling process focuses attention on interrelationships and connections between components of the resource system. These connections often represent the points at which agencies and other interested groups must interface in their attempts to deal with the complexities of the resource issue.
3. Model building requires a clear statement of assumptions being made, sometimes subconsciously, in day-to-day management and decision making. A statement of assumptions and their implications provides other workshop participants with a better understanding of how management decisions are made and represents a set of clearly posed questions or hypotheses for analysis.
4. The simulation model provides a logical framework for synthesizing existing information.
5. The attempt to quantify environmental, economic, and social processes associated with a resource issue quickly and objectively identifies gaps in data or flaws in the conceptual understanding of the system. Such information can be used to establish research priorities.
6. With sufficient refinement, models may provide a reasonable representation of the consequences of various management or decision alternatives.

Activities

The workshop modeling process may involve from one to several workshops of three to five days' duration. If the primary objective is to initiate or improve interaction among various agencies or interest groups, a single workshop may suffice. In cases requiring detailed analysis, a series of workshops may be conducted over a period of several months to several years. The relatively short time frame of each workshop forces participants to focus on important issues. The time between workshops is used for data collection and analysis.

Workshops are convened by a group of facilitators, or workshop staff, trained in systems analysis, computer modeling, policy analysis, and group dynamics. The staff has four primary functions: to moderate the workshops; to assist participants in formulating and constructing a computerized simulation model of the resource system under consideration; to assist participants in the interpretation of model output; and to aid in the integration of workshop results into relevant management activities and research. Although workshop participants provide subject matter expertise, some knowledge of hydrology, water chemistry, soils, vegetation dynamics, wildlife population dynamics and habitat requirements, and economics is required by the staff. Our group has formal training in wetland ecology, fishery science, general ecology, wildlife management, computer simulation modeling, statistics, and economics. Through nine years of workshops, the group has developed some familiarity with the fields of hydrology, water chemistry, and soils.

The stated purpose of an initial workshop is to have a simulation model running at the end of one week. The remainder of this section describes the activities that typically occur before and during an initial workshop for a project.

Preworkshop scoping meeting. Prior to an initial workshop, a scoping meeting is held to familiarize the staff with the specific resource problem and to plan the workshop. The scoping meeting is attended by the workshop staff and a client or sponsor. Specific activities at the scoping meeting include a general description of the natural resource system, specification of questions to be addressed by the analysis; conceptualization of possible model structures, planning the logistics of the workshop, and selection of participants.

The participants are critical to the success of the workshop process. It is important to have a cross section of people, representing not only multiple disciplines, but also multiple levels within organizations, including:

1. Subject matter specialists who have the best understanding of how various parts of the system work
2. Decision makers who formulate policies based on their best projections of what those policies will mean and who have the best understanding of associated political constraints
3. Managers who implement management actions to produce desired changes in the system and have the best understanding of implementation constraints

Experience has shown that workshops work best with 25 to 30 participants. It is difficult to get enough subject matter specialists and all agencies and interest groups represented with fewer people; however, it is difficult to keep more than 30 people occupied productively during the workshop.

Beginning the workshop. An initial workshop generally lasts five days with participants arriving on Sunday and leaving on Friday (Table 12.1). A mixer is usually scheduled on Sunday evening for participants to meet each other informally and for the facilitators to meet the participants and identify their expectations or reservations about the week ahead. This informal interaction sets the tone for the workshop.

Monday morning starts with welcoming comments by the client or workshop host. Participants are asked to introduce themselves and state their expectations for the workshop. This provides an opportunity to voice personal views or agency positions early in the workshop, thereby minimizing these types of position statements later in the week. Following introductions, the workshop staff

TABLE 12.1 Typical workshop agenda

Time	Activity
Sunday P.M.	Mixer
Monday A.M.	Introductions and overview
Monday P.M.	Bounding exercise
	(Staff meeting: Submodel definition)
Tuesday A.M.	Looking outward matrix exercise
Tuesday P.M.	Subgroup meetings
	(Staff meeting: Variable naming)
Wednesday	Subgroup meetings
	(Staff: Submodel programming and debugging)
Thursday A.M.	Subgroup reports
	(Staff: Submodel programming and debugging)
Thursday P.M.	Scenario development
	(Staff: Submodel programming and debugging)
	(Staff meeting: Submodel linking and model debugging)
Friday A.M.	Interactive gaming and discussion

explains what a simulation model is (since most participants have little or no experience with models) and describes the agenda for the week.

Bounding the model. Bounding is the process of deciding what will and will not be included in the workshop simulation model in terms of basic components, geographic area, and time frame. The bounding process at an initial workshop stresses simplification for better understanding. Later workshops may add additional detail or complexity, depending on management and decision-making needs.

Bounding is approached through a group discussion of actions (activities that management can undertake to manipulate the system) and indicators (performance measures used to evaluate the response of the system). At this point in the workshop, the objectives and relative merits of actions and indicators are *not* discussed; such discussions generally involve substantial conflict because of the differing responsibilities and viewpoints of agencies and groups represented. Identification of actions and indicators initiates model construction by specifying an important subset of the input and output variables. The focus in the bounding exercise is on input variables that might be managed. Additional input variables (for example, precipitation) are added later as necessary to simulate the desired output variables.

Discussion subsequently turns to the spatial and temporal resolution necessary to represent the components and processes implied by the set of actions and indicators. Spatial resolution is defined both in terms of the total geographic area represented in the model and also the subdivisions required to represent the dynamics of the processes involved. Temporal resolution is defined both in terms of a basic time step of model calculations (difference equations) and the number of iterations needed to cover the time horizon of interest.

On Monday evening, the workshop staff partitions the resource system into three to six subsystems, or submodels, based on the results of the bounding exercise. Criteria used in selecting the submodels include minimizing information transfers between submodels (each subgroup considers a relatively isolated part of the whole system), efficiently allocating people (each subgroup represents the concerns and expertise of a group of participants), and distributing work equally among workshop facilitators (because limited time is available for coding and debugging).

Looking outward matrix exercise. Workshop participants next define the interactions between submodels by constructing a *looking outward matrix,* which is a modification of an interaction matrix (see, for example, Leopold et al., 1971). Submodels are arrayed as both the row and column headings of a matrix. For each column of the matrix (that is, for each submodel), participants are asked what information they need from other submodels in order to predict how their submodel would behave under various management alternatives. The elements in a given column thus contain the inputs a submodel needs from other submodels. The elements in a given row specify the outputs a submodel must produce for other submodels.

Construction of the looking outward matrix achieves several objectives. First, the matrix elements, along with the actions and indicators, define in some detail the structure of the overall model and the submodels. Second, the process "prunes" the model and subsequent analysis. Experts generally want to include much more detail than is needed for an analysis. This process focuses on information from other areas that the experts need in order to make reasonable predictions about their areas of expertise, rather than allowing them to list all the information that could be provided. Finally, the process of constructing the matrix facilitates interdisciplinary communication and identifies interactions that may cause unexpected effects as the biological, physical, and economic aspects of the system interact.

Submodel development. Once the overall model structure is defined, participants meet in smaller subgroups (one for each submodel) to model the internal dynamics of each component. A workshop staff member is assigned to each group to facilitate discussions and write computer code. The basic charge to each subgroup is the following: given a set of actions (bounding exercise) to be represented in the model and a set of inputs available from other submodels (looking outward matrix), describe the mechanisms and processes that occur in your submodel to produce the set of indicator variables (bounding exercise) that you must represent, as well as the outputs from your submodel that are required by other submodels (looking outward matrix). Subgroup meetings generally require one and a half days during the workshop. It should be emphasized that the participants decide the form and parameters of the difference equations constituting the actual model. The staff facilitates this process by leading discussion and by asking that relationships be expressed as "rules for change." Facilitators often comment on the logical consistency or practicality of a particular formulation and suggest suitable approaches. However, it is the participants' model and their decisions are implemented by the facilitators.

Because each staff member programs a submodel independently, the workshop staff meets Tuesday evening and agrees on names and units of measure for all variables used by more than one submodel. The list is inevitably altered as subgroup meetings continue the following day. This Tuesday evening staff meeting also resolves problems with model conceptualization that often arise as subgroups discuss how to model their component (for example, a monthly time step is most appropriate for the hydrology submodel, but the fish model requires an estimate for the 24-hour minimum dissolved oxygen concentration). Following subgroup meetings, the workshop staff begins writing computer code.

Subgroup reports and scenario development. On Thursday, participants present overviews of each submodel. These presentations provide a better understanding of the subject area of each submodel and an appreciation of strengths and weaknesses. Participants also combine actions identified in the bounding exercise into management or development scenarios to be simulated. For this exercise, participants are often grouped by agencies or philosophical viewpoints so that scenarios represent extreme differences. During these sessions, the workshop staff continues the process of programming and debugging.

Submodel linking and model debugging. On Thursday evening, the workshop staff meets to link the various submodels and test scenarios developed by participants. Despite efforts throughout the week, problems arise when the submodels are linked together. These problems include misunderstandings about variable names or units of measure, programming or conceptual errors that don't show up until actual output from other submodels is used or until actual scenarios are run, and scenarios that are not specified in sufficient detail.

Interactive gaming and discussion. The purpose of this session is threefold: to build some level of credibility in the model, to identify problems with the model, and to provide additional interdisciplinary communication. These objectives are accomplished by a process of progressive model *in*validation. The session begins by choosing and running a scenario that establishes some credibility in the model. Additional scenarios are then run and participants discuss model output and suggest model changes. To the extent possible, the workshop staff implements those changes in front of the group and reruns the scenarios. While model results that match expectations are reassuring, they do not always provide the best learning opportunity. Therefore, extreme scenarios are also run in an attempt to invalidate the model. Such extreme scenarios push the model beyond its limits and produce unexpected or unreasonable behavior. Subsequent discussions concerning why the model behaved as it did can be very rewarding. Problems stemming from programming errors or simplifying assumptions are briefly noted. Discussions that identify conceptual gaps in understanding of the resource system are more interesting and can result in the development of a list of priority research needs. The most worthwhile and interesting discussions occur when unexpected model behavior results from complex physical, biological, and economic interactions. These results may ultimately be judged correct, but are initially unexpected because participants tend to evaluate the resource issue from restricted disciplinary or agency viewpoints (for example, the mixed stock harvest situation described in a later section). The workshop closes with a summary of discussion points and a listing of priority research or information needs, which are subsequently incorporated into a workshop report prepared by the facilitators.

Software and Hardware

Simulation models are programmed in FORTRAN with the aid of several utility software packages. SIMCON (Hilborn, 1973) is a general simulation controller for difference equations that facilitates data input, stores variable values at the end of each model iteration, and provides both printed and plotted output. This package allows the workshop staff to focus on developing code for the simulation model (rather than input–output) during the short time frame of a workshop. Several additional utilities allow rapid interactive testing and debugging.

To date, we have developed these workshop simulation models on a mainframe computer accessed with graphics terminals. We have used this hardware configuration because the models usually have large output storage requirements, building a complete model in five days requires fast response, interactive gaming requires fast turnaround to keep the discussion moving quickly, and rapid graphics output is required to display multiple variables changing through time.

We have also periodically evaluated microcomputers for workshop model development and gaming. Available FORTRAN compilers, debugging software, and simulation control packages are as good as mainframe packages. Furthermore, microcomputers with several megabytes of internal memory (or virtual memory capability) and large hard disks provide sufficient storage. However, in 1985 when the Service terminated its evaluation of the workshop simulation approach, most of these models were complex enough to require at least 5 minutes of execution time on a microcomputer. While this might be acceptable in some settings, it was not fast enough for the interactive gaming and discussions that are so important in a workshop. Recent advances in both hardware and software have made microcomputers a viable platform for workshop simulation.

APPLICATIONS

Overview

During the period 1978–1986, we used the approach described above on a total of 13 projects. While no two of these projects were exactly alike in terms of model components, five dealt largely with water management issues, two with environmental contaminants, one with the dynamics of a managed wildlife population, two with agricultural–environmental interactions, and three with industrial–environmental interactions (Table 12.2). Objectives for the projects were highly variable, although many involved assessment of some type of environmental impact. Other objectives included design of mitigation plans, synthesis of research results, and identification of key research and monitoring needs. Improving communication among interest groups was in many cases a primary, but often unstated, objective.

The spatial scale of these models was most often on the order of a watershed or river basin. In many cases, the basic spatial unit was further subdivided into logical pieces, such as river reaches or herd ranges. On a few occasions, processes involving spatial scales broader than that used for the basic model were also represented. For example, part of the Sacramento–San Joaquin model considered ducks in the entire Pacific flyway, but focused on the importance of land-use practices in the Central Valley of California and their impacts on habitat for wintering waterfowl.

Basic time steps in the models ranged from a day to a year, depending largely on the interval over which significant changes in variables of interest might be expected to occur and thus how often information had to be exchanged between submodels. In some cases, the basic time step was subdivided into smaller increments in order to represent adequately the dynamics of a more rapid process. For example, the acidic deposition model operated on a basic time step of one week, but changes in plankton populations were calculated at much more frequent intervals.

The models produced in these projects also varied considerably in their size and complexity, having as few as three and as many as six submodels. The smallest of the models contained about 200 variables (or 2500 considering all elements of arrays as separate variables) and about 500 lines of executable FORTRAN code (exclusive of the general iteration and input–output functions handled by the simulation control package). The largest model contained about 700 variables (44,000 array elements) and 2500 lines of executable code. Individual submodels ranged from 25 to 150 variables (75 to 25,000 array elements, respectively) and from 50 to 500 lines of executable code. Size and complexity were influenced by many factors, the most important of which were the complexity of the problem being addressed and the amount of information and understanding available. There was also a tendency for size to increase as additional time was devoted to model development. Some projects involved only a single workshop and the resulting models were usually small. Other projects involved up to three workshops with numerous other smaller meetings. Models resulting from these efforts tended to be the largest.

Jackson Hole Application

Several aspects of this project are described here to serve as examples of the results of individual workshop exercises and to illustrate the kinds of computations developed in a workshop setting. Calculations of the elk population dynamics submodel are described in some detail. Many submodels have been simpler (for example, representations of wildlife in terms of habitat suitability indexes that depend on physical or vegetation variables, as in the Malheur model described later). However, some submodels (for example, plume dynamics in the Drilling Fluids model and several water routing and allocation submodels) have been considerably more complex. Further details of this particular project are described in Roelle and Auble (1983) and Andrews et al. (1981b).

TABLE 12.2 Characteristics of projects and models

Project	Purpose	Spatial scale	Time scale	Basic factors considered
Water management				
Truckee–Carson (Andrews et al., 1980)	Assess effects of alternative water management strategies to aid in design of USGS river quality assessment	Two connected river basins with five reservoirs, fourteen river reaches, and associated agricultural lands and wildlife habitat	Seasonally for 70 years	Available supply of water; competing demands; allocation of water according to priority system; water routing; impacts on water quality, fish, and wildlife
Sacramento–San Joaquin	Assess effects of alternative water management strategies on fish and wildlife	Two connected river basins with ten reservoirs, nine river reaches, and associated agricultural lands and wildlife habitat	Monthly for 40 years	Available supply of water; competing demands; allocation of water according to priority system; water routing; land-use changes; impacts on fish and waterfowl habitat
Susitna	Design mitigation plan for proposed hydroelectric facility	River basin with two reservoirs, two river reaches, and associated wildlife habitat	Monthly for 40 years	Reservoir operations; instream flows; impacts on fish; vegetation succession and habitat manipulations; impacts on moose, bear, and other wildlife
Platte River (Nebraska Natural Resources Commission, 1985; Roelle et al., 1983)	Assess effects of alternative water development projects	River basin with several proposed reservoirs, eight river reaches, and associated drainage areas	Monthly for 50 years	Reservoir operations and water routing; groundwater exchanges; agricultural and municipal–industrial water use; agricultural production; instream flows; impacts on riparian vegetation and wildlife; economic values
Malheur Lake (Hamilton et al., 1986)	Assess impacts of proposed drainage canal on fish and wildlife	Enclosed basin with 1-ft elevation zones	Monthly for 40 years	Inflows, outflows, and lake levels; cover-type changes; impacts on habitats of colonial nesting birds and waterfowl
Environmental contaminants				
Drilling Fluids (Auble et al., 1983, 1984a)	Synthesize results of research program concerning fate and effects of drilling muds and cuttings in marine environment	Wedge-shaped discharge plume with representative 1-m plots at selected distances downcurrent	Daily for 20 years	Dispersion of muds and cuttings in water column; deposition and sediment interactions; impacts on benthic communities
Acidic Deposition (Andrews et al., 1982; U.S. Fish and Wildlife Service, 1982)	Identify additional needed research	Small watershed with lake, inflows, outflows, and associated forests	Weekly for 20 years	Chemistry of inflows, lake, and outflows; dynamics of plankton and benthos; dynamics of fish populations

Population management			
Jackson Hole (Andrews et al., 1981b; Roelle and Auble, 1983)	Assess effects of alternative management strategies on elk populations	Large valley and surroundings uplands with five elk herds, five summer ranges, and three winter ranges	Seasonally for 50 years	Elk movements, reproduction, natural mortality, and harvest
Agricultural–environmental interactions			
Integrated Pest Management (Johnson et al., 1983)	Assess environmental and economic consequences of alternative pest management strategies	Hypothetical watershed with 12 types of agricultural fields	Daily within selected seasons for 5 years	Agricultural production decisions; crop production; pest population dynamics; transport of agricultural chemicals; impacts on adjacent streams and avian populations
North Dakota Wetlands (Andrews et al., 1981a)	Identify preservation and protection strategies for wetlands in North Dakota	Hypothetical watershed with associated cover types and ownership patterns	Annually for 50 years	Hydrologic condition of wetlands; land-use changes in response to various protection programs; response of waterfowl to habitat conditions
Industrial–environmental interactions			
Beluga Coal (McNamee et al., 1982)	Assess impacts of proposed coal development on human population and fish and wildlife resources	Four drainage basins	Annually for 60 years	Facilities development and operation; habitat destruction or alteration; changes in human population; changes in streamflows and water quality; impacts on moose and fish populations
Mobile Bay (Hamilton et al., 1982)	Assess impacts from development in Mobile Bay on fish and wildlife	Mobile Bay and two surrounding counties with bay subdivided into seven areas	Annually for 30 years	Human population growth and movement; industrial development; changes in air and water quality; impacts on fish resources
Merritt Island (Hamilton et al., 1985a)	Identify needs for monitoring impacts of Kennedy Space Center operations on Merritt Island National Wildlife Refuge	Hypothetical management unit including uplands, impoundment, and estuary	Monthly for 50 years	Surface and groundwater hydrology; operation and management of mosquito control impoundments; upland habitat manipulations; inputs of contaminants from Kennedy Space Center operations; impacts on sea grass beds

Land ownership in the area around Jackson, Wyoming, is a complex of public (including Grand Teton National Park, John D. Rockefeller, Jr. Memorial Parkway, Yellowstone National Park, Bridger–Teton National Forest, and the National Elk Refuge) and private holdings. Mobile wildlife species, particularly elk, utilize these lands in complex patterns to satisfy their seasonal habitat requirements and are managed by several agencies with different, sometimes conflicting, objectives. These agencies, as well as many private citizens, are concerned that the combination of a complex management situation and increasing pressure for development of private lands may have serious implications for some of the wildlife resources that contribute significantly to the attractiveness of the Jackson area.

The objectives of the simulation modeling workshop were to promote communication among interests and agencies concerned with resource development and management in the Jackson area; to identify potential conflicts within and among management agency objectives and policies, particularly with regard to elk; and to identify important interrelationships between wildlife management strategies and attendant issues such as population growth, recreation, logging, and energy development. Participants included representatives from the Wyoming Game and Fish Department, National Elk Refuge, Grand Teton National Park, Bridger–Teton National Forest, Teton County Planning Commission, Jackson city government, outfitting and ranching industries, Jackson Hole Chamber of Commerce, and the local environmental community.

Model structure. The overall geographic boundary for the model was the total area utilized annually by elk that winter in Jackson Hole. This area (Fig. 12.1) was subdivided into five elk summer areas (southern Yellowstone National Park, Grand Teton National Park, the Teton

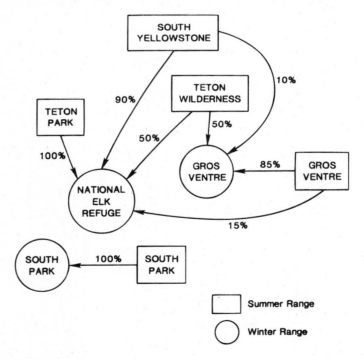

Figure 12.1 Diagrammatic representation of spatial units reflecting herd segments and migration patterns.

Wilderness portion of Bridger–Teton National Forest, the Gros Ventre River drainage, and South Park), three elk winter areas (National Elk Refuge, the Gros Ventre River drainage, and South Park), and three areas of concentrated human activity (Jackson, the Gros Ventre River drainage, and South Park). The time horizon was 50 years.

The time step was annual with certain calculations partitioned into three seasons defined by major elk movements and human recreational activities. The fall season (September–November) included most of the elk hunting season and the time when elk move from summer range to winter range. The winter season (December–April) represented the primary period of winter recreation (downhill and cross-country skiing and snowmobiling) and the time when elk are on their winter ranges. The summer season (May–August) included the period of summer tourism and the time of elk migration to and use of their summer ranges.

The actions and indicators identified at the workshop (depicted on the main diagonal of the matrix in Fig. 12.2) were grouped into three submodels: human population, elk, and other wildlife. The interactions among submodels that were identified in the looking outward matrix exercise are depicted as off-diagonal elements in Fig. 12.2.

Elk submodel calculations. The difference equations updating the elk population were structured as a series of distinct seasonal calculations. The fall calculations pertained primarily to migration and harvest. Hunting mortality was divided into two components, one occurring before the fall migration, when elk are separated into their summer segments, and one occurring after the migration, when elk are mixed on the winter ranges. In both cases, the concept of a target or desired postharvest population was used to compute the number of elk available for harvest. Elk in excess of these targets, which were set for both summer and winter herds, were considered to be a harvestable surplus. The targets were used in the model to reflect the idea that management agencies have at least general goals for the number of elk in each herd.

A premigration harvest was simulated on each summer herd segment except the one utilizing Grand Teton National Park. South Yellowstone elk are not actually harvested until after they leave the boundaries of Yellowstone National Park; however, in the model, this harvest occurs before the migration. A harvest is conducted in Teton Park, but it is confined to the southern and eastern portions. The assumption made was that the Teton Park harvest is directed at the mixed population of animals wintering on or around the National Elk Refuge and therefore, in the model, constitutes a postmigration harvest. Postmigration harvests were also allowed on the Gros Ventre and South Park winter ranges.

The number of hunters needed for each harvest was estimated by dividing the desired harvest by the hunter success ratio for the previous year. In each case the number of hunters could be limited by factors such as agency policies or the physical capacity of the harvest area. Realized harvest for each area was then calculated on the basis of the number of hunters present and a new success ratio computed from a randomly generated snowfall factor. Total harvest was distributed among four sex and age cohorts (calves, yearling bulls, cows, and adult bulls) according to average proportions observed in each herd segment in recent years. Elk were assumed to be completely mixed on the winter ranges; therefore, postmigration harvests were assumed to occur in proportion to the number of elk contributed by each of the summer segments.

Postharvest survival on the winter feeding grounds was assumed to be a function of three factors: a mortality rate determined by winter food supply, an additional mortality rate determined by food supply the previous summer, and a base mortality rate occurring even with adequate summer and winter food. The mortality rate due to winter food supply was based on food availability per elk per day. The total food requirement for each wintering herd was calculated using the number of elk arriving on the winter range, an average number of days spent there, and an average daily forage requirement per elk. Total usable natural forage was estimated by multiplying the number of acres of winter range by the average amount of forage available per acre. The number of acres

	THIS SUBMODEL NEEDS INFORMATION CONCERNING		
FROM THIS SUBMODEL	**Population Growth**	**Elk**	**Other Wildlife**
Population Growth	Actions Regulate human population growth Regulate housing density Regulate access to public lands Enlarge existing ski areas Regulate snowmobile use Regulate oil and gas drilling Regulate logging Indicators Human population size Level of recreation activities Level of energy development Level of logging Acres in various land use categories	Acres of private land developed Acres logged Acres in oil and gas development Miles of road constructed Snowmobile use	Acres of private land developed
Elk	Number of elk Number of hunters Number of elk harvested Deviation of elk populations from targets	Actions Regulate supplemental elk food Regulate elk harvest Regulate natural forage production Indicators Number of elk Number of elk harvested Cost of supplemental elk food	Number of elk wintering in South Park
Other Wildlife	Number of bald eagles Deviation of bald eagle populations from targets Number of mule deer Deviation of mule deer populations from targets	- No information required -	Actions Create nondevelopment buffer zone Regulate recreational activity in winter habitats Indicators Number of bald eagle nests Number of wintering bald eagles Number of mule deer

Figure 12.2 Actions and indicators (main diagonal), and results of the looking outward matrix exercise (off-diagonal) for the Jackson Hole model.

of winter range was modified by factors such as development of private lands, snowmobiling and cross-country skiing activity (Fig. 12.2), and a randomly generated snowfall factor. The requirement for supplemental food was then calculated by subtracting the amount of natural forage available from the total food requirement for the herd. The amount of supplemental food available each year was limited to a fixed amount (reflecting fixed agency budgets) plus any unused food carried over in storage from previous years. The total amount of food available was then converted to an amount available per animal per day, which in turn was used to calculate a winter mortality rate. Total winter mortality was computed using this mortality rate based on winter food supply, a similar mortality rate based on food supply the previous summer (see discussion of summer dynamics later), and a base mortality rate for each wintering herd.

At the beginning of each summer season, elk surviving the fall and winter were assumed to move back to the summer range from which they came. Calf production per cow was computed from a pregnancy rate and a brucellosis infection rate, both of which were determined by the density of elk in the previous winter. The mortality rate due to summer food supply was calculated by dividing the summer period into two equal parts. The total amount of food required for each summer herd segment in the first half of the season was calculated from the number of days in the period, the number of elk present, and an average amount of food required per elk per day. The amount of food available was computed from the area of summer range, as modified by factors such as oil and gas production, road construction, and logging (Fig. 12.2), and the average amount of forage available per unit area. If the available summer forage exceeded the requirement of the herd, animals were assumed to stay in that geographic area. If the forage requirement exceeded the amount available, animals not finding sufficient forage were assumed to disperse in fixed proportions to other summering areas. Following this dispersal, similar calculations were performed for the second half of the summer season, and a mortality rate was calculated based on the forage available per elk per day. As mentioned above, this mortality rate was assumed to be manifested during the subsequent winter.

Results. The spatial pattern and assumptions of the elk submodel produce some interesting constraints on the behavior of various herd segments (which are managed by a complex of federal and state agencies). A fixed food supply on the winter range results in density-dependent mortality that controls the number of elk that survive the winter period and therefore limits the total number of elk that return to the summer ranges. The distribution of this total number of elk among the summer herd segments depends on their relative net rates of increase (the sum of all production and mortality factors). Herd segments having higher net rates of increase return proportionately more elk to their summer ranges and therefore constitute an increasing proportion of the winter populations until something happens to equalize the net rates of increase for all summer herd segments sharing a winter range.

This situation has important implications because of the extent to which various herd segments are managed by different agencies with different objectives. Some of the ways in which net rates of increase could become balanced (for example, food limitation or disease on the summer range, reductions in harvest of the summer herd segments during migration) create severe difficulties for the agencies involved.

Malheur Lake Application

A second application involved use of simulation modeling workshops to analyze the potential consequences of flood-control alternatives at Malheur Lake. In this example, we provide less detail concerning model structure and somewhat more information about the benefits of the modeling process.

Malheur Lake is the largest freshwater marsh in the western contiguous United States and is one of the main management units of the Malheur National Wildlife Refuge in southeastern Oregon. The marsh provides excellent waterfowl production habitat, as well as vital migration habitat for birds in the Pacific flyway. Water shortages have typically been a problem in this semiarid area. However, record snowfalls and cool summers have recently caused Malheur Lake to rise to its highest level in recorded history. This has resulted in the loss of approximately 57,000 acres of important wildlife habitat through deep inundation of marsh communities, as well as extensive flooding of local ranches, roads, and railroad lines.

In response to this situation, the U.S. Army Corps of Engineers conducted a feasibility study on a set of flood-control alternatives for Malheur Lake. Under the Fish and Wildlife Coordination Act (16 U.S.C. 661–666; 48 Stat. 401, as amended), the Portland, Oregon, Ecological Services Field Office of the Service had responsibility for preparing a Coordination Act Report documenting potential impacts of the alternatives. Input to the report was also required from the Malheur National Wildlife Refuge (a different division within the Service) because of potential impacts on the refuge. State water resource and fish and wildlife agencies were interested in the alternatives because of potential out-of-basin diversions and associated potential impacts of downstream flooding and loss of a cold-water fishery. Local landowners were interested because of potential flood-protection benefits. A series of modeling workshops was conducted to facilitate discussions among these groups and to develop a simulation model that the U.S. Fish and Wildlife Service could use to analyze proposed alternatives (Hamilton et al., 1986).

Model structure. The simulation model consisted of three submodels: hydrology, vegetation, and wildlife. The primary actions identified in the bounding exercise were several drainage canals from Malheur Lake that varied in capacity, intake elevation, and operating schedule. The major indicators produced by the model were lake elevation, canal flow into an adjacent river basin, area of various cover types, and indexes of abundance or habitat suitability for a number of wildlife species. The model considered a time horizon of 50 years, the planned life of the proposed projects. Hydrologic indicators were calculated monthly, while indicators of vegetation and wildlife were calculated annually. Spatially, the model represented the Malheur–Harney Lakes Basin. The area was divided into two subbasins (Malheur Lake and Harney–Mud Lakes) for hydrologic calculations and into 15 management units for cover calculations.

The hydrology submodel calculated changes in lake volume, elevation, and surface area based on simple water budgets for Malheur and Harney–Mud lakes (Fig. 12.3). Monthly surface inflows (corresponding to data from U.S. Geological Survey gaging stations), precipitation, and evaporation for the period of record (1928–1984) were read from an input file. Because analysis of water projects is highly dependent on assumptions concerning future precipitation and runoff, a set of parameters allowed selected portions of the hydrologic record to be repeated or eliminated so that alternative future runoff scenarios could be generated.

Surface inflows were adjusted based on irrigation diversions and storage in proposed upstream reservoir projects. Groundwater inflows were constant in normal runoff years and were calculated as a proportion of surface inflows when annual runoff exceeded the 25-year return interval flow. A running average of surface inflows over several years and the associated 25-year return interval flow were used to introduce a time lag in groundwater response. Proposed drainage canals were the only surface outflows from the lake. Canal flow was calculated based on lake elevation, canal intake structure elevation, canal capacity, and canal operating criteria (for example, in some scenarios no discharge was allowed when the receiving basin was flooding or when irrigators were diverting from the receiving basin).

Each new lake volume was converted to a corresponding elevation and estimate of surface area based on tables provided by the U.S. Geological Survey. The structure of the submodel was

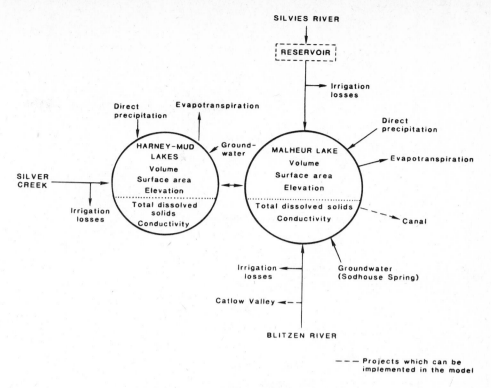

Figure 12.3 Malheur Lake hydrology submodel.

thus similar to the HEC-5 model (Chu and Bowers, 1977) used by the Corps of Engineers for their hydrologic analysis of alternatives.

The vegetation submodel calculated changes in the areal extent of five wetland and three upland cover types based on assumptions concerning natural succession and the hydrologic regime calculated by the hydrology submodel. To convert monthly lake elevations to depths of inundation for various cover types at various relative elevations, the total area of each spatial unit was divided into 20 one-foot elevational zones. The transitions between cover types considered by the submodel are shown in Fig. 12.4. A change matrix specified the conditions under which each cover type was converted to other cover types and the fractional rates of change occurring in a year in which the conditions were met. For example, 25% of the wet meadow in a zone was converted to emergent vegetation if the maximum water depth in the zone during the previous year was between 1 and 4 feet and the zone was not dry at any time during the growing season. Although the vegetation submodel was relatively simple in concept, the net vegetation changes over time caused by a hydrologic modification influencing 21 possible transitions among seven cover types in 20 elevation zones in two lake basins were not always intuitively obvious.

The wildlife submodel calculated measures of wildlife populations or habitat suitability for seven avian species or species groups based on hydrologic and vegetative conditions. Population models for each species would have been desirable; however, most of the important species are migratory and, therefore, many important processes (for example, winter mortality) occur in areas far removed from the Malheur–Harney Basin. As an example of the wildlife calculations, the number

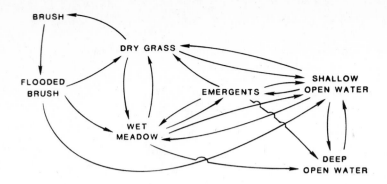

Figure 12.4 Transition pathways among cover types in the Malheur Lake model.

of breeding pairs of Canada geese was calculated from a regression ($r = 0.84$) using the number of acres of the emergent cover type inundated with between 1 and 6 feet of water.

Results. The simulation model was very useful for illustrating the consequences of proposed flood-control alternatives, as well as for exploring the implications of general assumptions being made in the impact assessment. For example, the analysis of alternatives in general, and model behavior in particular, was critically dependent on assumptions concerning future runoff. Under an assumption of relatively moderate future hydrologic conditions, there were only small differences in simulated behavior between the no action and canal alternatives; under an assumption that extreme flooding will reoccur during the lifetime of a project, some canals had a pronounced effect on peak flood levels and thus the extent to which wetland vegetation would be affected by deep inundation. While the importance of runoff assumptions would appear obvious, we have seen many cases in which differences in basic assumptions (sometimes unstated) among agencies have resulted in different impact predictions. Another example of the model's utility concerns the vegetation and wildlife impacts of flood-control alternatives. In several instances, model indicators differed from the initial expectations of workshop participants. This was often because participants were assuming a relatively fixed period of time required for different vegetation types to recover based on "average" conditions implied by a particular hydrologic alternative. However, recovery times varied substantially depending on particular hydrologic sequences and, thus, alternatives needed to be evaluated in terms of the likelihood that specific sequences might occur, rather than in terms of "average" expected conditions.

In a more general sense, the series of workshops provided a nonadversarial forum for affected agencies and interest groups to discuss flood-control alternatives. Although the workshops did not develop a consensus on the best alternative, they did significantly narrow the range of alternatives for detailed evaluation.

EVALUATION

As is the case with any self-evaluation, there is a strong possibility that our perspective is biased. Nonetheless, we have tried to judge workshop simulation modeling objectively in terms of its success in providing the intended benefits of communication, understanding of the system, prediction, and decision making. In addition, we have tried to identify the characteristics of suitable applications of a workshop modeling approach.

Producing Intended Benefits

Communication. The collaborative construction of a simulation model in a workshop setting definitely results in enhanced communication among participants. The group dynamics of the meeting itself are almost always "successful." Participants interact meaningfully about the real issues, a diverse set of people works toward the common objective of model construction, and participants generally feel that the workshop was a reasonable investment of their time and effort. Utilization of model construction as a neutral focus for the meeting almost always works in the sense that participants are able to engage in structured discussion about the natural resource system and its management without being locked into traditional stances or adversarial relationships.

Understanding of system. In our experience, workshop modeling increases shared understanding of the natural resource system by participants. Two general types of new understanding are involved. The first is the result of focused communication. The modeling provides a framework and opportunity for participants to communicate their knowledge of respective pieces or parts of the system. The second involves the kind of "systems understanding" that is generated by attempting to couple pieces of a system together and by exploring the net results of a large number of individual assumptions and relationships. This area is where workshop simulation modeling has real advantages over alternative formats (for example, a meeting consisting of presented papers). In some sense, workshop modeling, as we have conducted it, can be viewed as an attempt to involve a larger set of participants directly in the process of model construction so as to allow them to gain the understanding that comes from actually trying to develop a model, regardless of the quality of the resulting model. Success in cultivating this kind of learning by doing is difficult to quantify or document. Nonetheless, we can give several examples of the type of "systems understanding" gained by participants.

Clarification of Coupling Variables. The looking outward matrix, which identifies variables "passed" between submodels, focuses attention on the variables that couple different processes and, in some cases, disciplines. The difficulties encountered in this exercise are often nothing more than tedious (but straightforward) problems of aggregation, partitioning, or dimensional conversion. Sometimes, however, much more fundamental difficulties are encountered. Perhaps the best example of this concerns the nature of exposure to contaminants. In several workshops (for example, Drilling Fluids and Acidic Deposition), understanding of biological effects has been predominantly structured as responses to fixed-length, fixed-concentration exposures that are often used in laboratory toxicity tests. In some cases, the types of environmental or field exposures predicted by fate calculations are qualitatively and dramatically different (for example, concentration changes of orders of magnitude within short time scales). This problem is not "solved" by a modeling effort, but the modeling effort does force investigators of both fate and effects to recognize the similarities and differences between laboratory and field exposures.

System-level Constraints. The natural resource systems modeled in these workshops often are subject to multiple competing uses or demands. The structure of a simulation model can reveal fundamental constraints that limit the extent to which these demands can be simultaneously satisfied. Such constraints are often apparent in large-scale hydrologic systems where the total amount of water and the drainage pattern make it impossible to meet all competing demands for water. A more subtle type of behavioral constraint occurs when different subsets of a population are subject to differential mortality during one phase of their life cycle and are subject to a common, density-dependent mortality during a phase of the life cycle in which they are mixed together. In this situation, the spatial structure of the system imposes ratio constraints on the behavior of individual components. This type of behavior was evident in the Jackson Hole elk submodel (described previously) in which elk share a common winter range with a common density-dependent survival and then separate to relatively distinct summer ranges where they are subject to differential hunting pressure. We have also encountered this type of constraint with multiple runs of an anadromous fish species that are subject both to run-specific freshwater mortality or recruitment and to a common, mixed-stock ocean harvest.

Clarification of Causal Relationships. The explicit relationships required to develop a set of coupled difference equations are often well beyond the current level of understanding. Thus, modeling activities have generally done as much to highlight poorly known relationships as to document known relationships. For example, we have often had difficulty representing interactions between animals and their habitats. In particular, calculations of suitable food production, even when adjusted by factors reflecting availability, have almost always overestimated population numbers so severely that no reasonable, quantitative relationships could be identified between population size and the food component of habitat. Furthermore, we have found it very difficult to develop reasonable, quantitative relationships expressing the feedback effects of animals on their food supply. We have also had difficulty representing interactions between species. We have generally had little information available, for example, concerning the functional and numerical responses of predators to changes in prey abundance. This has been the case in a number of workshops both for natural predators and for human hunting and fishing pressure.

The point is not that these relationships are unknown in general, but rather that they are known in some systems and thus are expected to be of importance in the particular system being modeled. The modeling activity stresses the importance of understanding these types of relationships for a particular system, relative to the need to accumulate additional information that is primarily descriptive.

Dynamic Paradigm. One of the objectives of the simulation modeling activity is to refine the underlying conceptual models that are used in the management of a natural resource system. In this sense, a simulation model represents a way of thinking about the system as fundamentally dynamic and interacting. This perspective has been new and difficult for some participants to the extent that it contrasts, or collides, with other entrenched approaches to environmental assessment. Perhaps the most frequently encountered alternative approach might be called "classify and list." While this is a necessary precursor to a dynamic analysis, we have often found it difficult to move beyond continued refinement of lists describing the different types of impacts, cover types, or species to a consideration of relationships determining change over time. Consideration of how the system will change, in some cases, is limited to what might be called a "one-step projection" approach. In this approach, the analysis is limited to comparing the difference (for example, with and without project) between essentially static system states, rather than comparing alternative trajectories over time.

Prediction. The simulation models developed in these collaborative workshops have generally produced qualitatively reasonable behavior. However, our overall judgment is that they are very limited in terms of quantitative prediction. A major cause of this limited predictive power is the fact that the model itself is not the ultimate objective of the workshop modeling process. It is used as a focal point, but the ultimate objective is really the understanding that comes from modeling, rather than the production of a reliable "black-box" predictor.

In particular, the workshop process often proceeds too quickly in the sense that participants do not fully understand the general idea of simulation modeling or the objective of a particular step in the construction of a simulation model until after they have done it and seen the result. Long explanations severely stall the momentum of the workshop without significantly increasing the level of understanding. However, the alternative of learning by doing can result in a model that is poorly structured. We have often found it difficult to establish workable spatial and temporal scales appropriate for all the processes of interest. The looking outward matrix exercise, which establishes the basic connecting variables between submodels, is central to the model structure; however, it is often not given enough attention in the workshop because it is a confusing and tedious exercise.

Another limitation of the models constructed in a single workshop is that they are generally not refined due to the time scale of their construction. Much additional effort to estimate parameters and to verify and validate these models would generally be necessary to produce any real confidence in their quantitative predictive capability. Finally, the predictive ability of these models is often limited by real unknowns concerning the system. We consider that the modeling activity is successful to the extent that we identify these unknowns and, perhaps, estimate their relative importance. However, this is a less than completely satisfying result for those responsible for management decisions concerning the natural resource system.

The best predictive ability has been in the area of robust conclusions that depend on the logical structure of the system, rather than on the specific values of parameters or the exact form of relationships between variables. The system-level constraints described earlier in this section are examples of this type of result.

Decision making. We have not found the simulation modeling workshop to be an effective forum for decision making. The process used in the workshop is fundamentally not a decision analysis framework. In particular, the process explicitly ignores the relative weighting of multiple output variables that is necessary for optimization. This is done in order to focus attention on resolving how the system operates, rather than on the more difficult resolution of conflicting value systems.

Despite this design emphasis away from decision making, the expectation was that the communication and common understanding developed at a modeling workshop would form the foundation for better decisions regarding the natural resource system. We believe this has been the case. However, this is extremely difficult to document because the improvement consists of a large number of small changes in the perceptions of individuals, rather than in a formal group decision.

Several factors have limited our ability to extend the results of modeling workshops to explicit changes in the management of the systems being modeled. One is the lack of well-defined, "bottom-line" results from the modeling activity other than the model itself, which is admittedly not a fully reliable predictive tool. Another difficulty stems from the nature of decision-making authority over the types of natural resource systems being modeled. Our idealized "decision maker" category of workshop participant is a severe oversimplification. Much more frequently, we have found that decision-making authority is spread very diffusely among and within a number of institutions and often localized to a large number of small decisions each made with relatively restricted mandates, criteria, and options.

Suitability

We have found workshop simulation modeling to be valuable, especially in the areas of increased communication and understanding of a dynamic natural resource system. It is not, however, a panacea for addressing environmental problems. There are very real drawbacks in terms of cost (both in relatively highly trained staff and in participants' time) and the lack of defined products other than the model and its description. Some of the characteristics of what seem to us to be suitable applications of the approach are described next.

1. Consideration of the natural resource system, and thus the model, can be structured around a relatively simple organizing principle, with complexity entering into the details of how that principle operates. Examples include the movement of water in a hydrologic system, a migration pattern of animals, and the plume dynamics of discharged materials. In particular, we have had good success with hydrologically oriented workshops, perhaps because hydrologists tend to be generally familar with simulation modeling. Workshops structured around such central, organizing principles have been more successful than those in which the concerns span a number of fundamentally different processes connected only by geography or institutional responsibilities.

2. Communication and the development of a common perspective are important objectives. These are the most reliably produced benefits from the workshop modeling process and are often important objectives in a situation involving many actors (institutions, interests, and types of expertise). A modeling workshop is less appropriate in situations with one primary actor that has sufficient knowledge and power or when resources could obviously be better allocated to addressing a well-defined and critical research need.

3. The management issues have an important dynamic component. Examples of such processes include population dynamics, succession, or a sequence of management actions. Situations involving primarily static, geographic issues (for example, facility-siting decisions) are less suitable for the type of modeling workshops we have conducted.

4. The participants and client for the workshop are perceptive and knowledgeable. We have generally had the pleasure of working with excellent workshop participants and clients. Expertise, personalities, and representation of all critical interests are essential. The client for the workshop plays a critical role in identifying good participants and ultimately has the most important role in making a meaningful translation of the workshop results back into institutional decisions or policies. Ideally, the client must be perceptive concerning expectations and objectives for the workshop, the larger institutional framework, the people involved, and the natural resource system itself. Often the specific client, issues, and objectives are not clear. Part of the objective of the workshop is their clarification. However, individuals are needed who are able to define and redefine these unclear aspects before, during, and after the workshop.

CONCLUSION

In 1985, the Service terminated its formal project to apply, refine, and evaluate the workshop simulation approach. Staff members were assigned a mix of other duties, allowing the Service to retain the basic capability for application to suitable future projects of special concern (for example, the Malheur application), while allocating some of the agency's resources to other priority activities. Although there was general recognition of the validity and benefits of the approach, there were also some very real limitations, especially in the context of a mission-oriented agency such as the Service. These included a substantial amount of fuzziness or inability to communicate clearly exactly what the process was (in part due to confusion between the specific techniques of constructing a simulation model and the more philosophical goals of developing flexible management approaches to dynamic, interacting systems) and the lack of specific, well-defined results in terms of existing institutional and decision structures. Another factor was some degree of contrast between an emphasis on neutral facilitation of workshops and representation of the fish and wildlife resource concerns that constitute the agency's mission. This is not a problem in principle; however, it is a factor in terms of allocating limited agency resources.

We believe that the type of simulation modeling workshop we have described will continue to be useful in situations that have the characteristics previously specified (for example, many actors or interests; need for communication and common understanding; and complex, dynamic systems with relatively simple organizing principles). Other situations require other approaches. In several cases, we have utilized similar techniques and procedures for situations that did not seem especially suitable for a large-scale, five-day simulation modeling workshop. Sometimes quantification and the full specification of equations is not appropriate, and many benefits can be obtained from going no farther than conceptual modeling (for example, Auble et al., 1985; Roelle et al., 1984a). However, we have found it difficult in conceptual modeling to attain the clarity that is forced by an attempt to quantify relationships. We have also conducted smaller workshops involving a restricted set of interests (five to ten participants), time scale (two to three days), and scope of problem. To date, the most successful of these have involved water management problems on national wildlife refuges (for example, Auble et al., 1984b; Hamilton et al., 1985b; Roelle et al., 1984b; Roelle, Hamilton, and Asherin, 1990; Roelle and Hamilton, 1990). These applications involve a relatively restricted set of options and reasonably well defined, although temporally complex, objectives (for example, habitat conditions for various species).

In conclusion, we offer some brief speculation on future directions in the general attempt to bring the potential of systems ecology to bear on the real world of natural resource management. Two trends seem to be especially important. The first is the increasing penetration of appropriate elements of systems analysis and simulation modeling into the general disciplines of natural resource science. In the future, we see such techniques more frequently applied by managers and scientists as an integral part of normal management, analysis, and assessment, and perhaps less frequently conducted by a distinct group of modeling specialists. The second is increased coupling between simulation models and other computerized technologies (for example, geographic information systems and expert systems) in natural resource management. For example, although many analyses incorporate a large amount of spatial pattern and detail, there has been a tendency to concentrate either on a detailed representation of the dynamics of a spatially simple system with a simulation model or on a cartographically detailed representation of an essentially static spatial pattern with a geographic information system.

In many management problems, a relatively simple representation of how a simple spatial pattern changes over time would be more useful. Recent examples of this type of model include Costanza, Sklar, and White, 1990, and Walters, Gunderson, and Holling, 1992. Simulation models might achieve greater representation capability and more effective user interface with the incorporation of expert systems. Possible applications include expert systems to help estimate model parameters, interpret output, or recommend management actions, or to represent portions of a natural resource issue that are characterized by inexact reasoning (that is, good judgment), rather than numeric calculation. As an example of this last application, many of the decision processes incorporated in our past simulation models (for example, some water allocation decisions) might have been better represented as expert systems. Likewise, expert systems could attain greater problem-solving capability by making use of the dynamic behavior offered by simulation models. For example, an expert system to address certain waterfowl management issues on a marsh might use a simulation model to predict water level or water quality changes in the marsh throughout the year.

ACKNOWLEDGMENTS

This chapter was prepared by employees of the U.S. Fish and Wildlife Service as part of their official duties and, therefore, may not be copyrighted. R. Ellison and R. Johnson were integral members of the group conducting these modeling workshops and made essential contributions to the work described. C. S. Holling and C. Walters were, in large part, responsible for the initial development of this workshop modeling approach and its subsequent transfer to the Service. We gratefully acknowledge the contributions of additional individuals who have served as facilitators in one or more of the modeling workshops: P. Bunnell, R. Everitt, A. Farmer, W. Gazey, J. Heasley, D. Marmorek, P. McNamee, T. Shoemaker, J. Sisler, L. Smith, N. Sonntag, M. Staley, and T. Waddle. Finally, we wish to thank the literally hundreds of participants who have attended these workshops and who have contributed so freely of their time, creativity, and patience.

REFERENCES

ANDREWS, A. K., ELLISON, R. A., HAMILTON, D. B., and ROELLE, J. E. 1980. Application of the Adaptive Environmental Assessment Methodology to the Truckee–Carson River Quality Assessment. Unpublished. Available from U.S. Fish and Wildlife Service, Fort Collins, Colorado. 64 pp.

ANDREWS, A. K., AUBLE, G. T., ELLISON, R. A., HAMILTON, D. B., and ROELLE, J. E. 1981a. Results of a Modeling Workshop Concerning Preservation and Protection of Wetlands in North Dakota. Unpublished. Available from U.S. Fish and Wildlife Service, Fort Collins, Colorado. 61 pp.

ANDREWS, A. K., ELLISON, R. A., HAMILTON, D. B., ROELLE, J. E., and MCNAMEE, P. J. 1981b. Results of a Modeling Workshop Concerning Resource Development and Management in Jackson Hole, Wyoming. Unpublished. Available from U.S. Fish and Wildlife Service, Fort Collins, Colorado. 74 pp.

ANDREWS, A. K., AUBLE, G. T., ELLISON, R. A., HAMILTON, D. B., ROELLE, J. E., MARMOREK, D. R., and LOUCKS, O. L. 1982. Impacts of acid precipitation on watershed ecosystems: an application of the adaptive environmental assessment process. In Mitsch, W. J., Bosserman, R. W., and Klopatek, J. M. (eds.), *Developments in Environmental Modeling, 1: Energy and Ecological Modeling.* Elsevier Scientific, New York, pp. 393–400.

AUBLE, G. T., HAMILTON, D. B., and ROELLE, J. E. 1984b. Refuge Management Analyses: Levee Alternatives at Clarence Cannon National Wildlife Refuge. W/AEAG-84/06, U.S. Fish and Wildlife Service, Fort Collins, Colorado. 40 pp.

AUBLE, G. T., ANDREWS, A. K., ELLISON, R. A., HAMILTON, D. B., JOHNSON, R. L., ROELLE, J. E., and MARMOREK, D. R. 1983. Results of an Adaptive Environmental Assessment Modeling Workshop Concerning Potential Impacts of Drilling Muds and Cuttings on the Marine Environment. PB-83-114165. NTIS, Springfield, Virginia. 64 pp.

AUBLE, G. T., ANDREWS, A. K., HAMILTON, D. B., ROELLE, J. E., and SHOEMAKER, T. G. 1984a. A Workshop Model Simulating Fate and Effect of Drilling Muds and Cuttings on Benthic Communities. WELUT-85/02. U.S. Fish and Wildlife Service, Fort Collins, Colorado. 189 pp.

AUBLE, G. T., ANDREWS, A. K., HAMILTON, D. B., and ROELLE, J. E. 1985. Fish and Wildlife Mitigation Options for Port Development in Tampa Bay: Results of a Workshop. NCET Open File Report 85-2. U.S. Fish and Wildlife Service, Slidell, Louisiana. 36 pp.

BASKERVILLE, G. 1979. Implementation of adaptive approach in provincial and federal forestry agencies. In Adaptive Environmental Assessment and Management: Current Progress and Prospects for the Approach. CP-79-9. International Institute for Applied Systems Analysis, Laxenburg, Austria. 45 pp.

CHU, C. S., and BOWERS, C. E. 1977. Computer Programs in Water Resources. WRRC Bull. 97. University of Minnesota, Minneapolis, Minnesota.

CLARK, W. C., JONES, D. D., and HOLLING, C. S. 1979. Lessons for ecological policy design: a case study of ecosystem management. *Ecol. Mod.* 7:1–53.

COSTANZA, R., SKLAR F. H., and WHITE, M. L. 1990. Modeling coastal landscape dynamics. BioScience *40:*91–107.

Environmental and Social Systems Analysts. 1982. Review and Evaluation of Adaptive Environmental Assessment and Management. Environment Canada, Vancouver, British Columbia. 116 pp.

HAMILTON, D. B., ROELLE, J. E., and ELLISON, R. A. 1985b. Refuge Management Analyses: Water Management Alternatives at Squaw Creek National Wildlife Refuge. WELUT-85/W12. U.S. Fish and Wildlife Service, Fort Collins, Colorado. 31 pp.

HAMILTON, D. B., ANDREWS, A. K., AUBLE, G. T., ELLISON, R. A., JOHNSON, R. L., ROELLE, J. E., and STALEY, M. J. 1982. Results of a Modeling Workshop Concerning Economic and Environmental Trends and Concomitant Resource Management Issues in the Mobile Bay Area. Unpublished. Available from U.S. Fish and Wildlife Service, Fort Collins, Colorado. 84 pp.

HAMILTON, D. B., ANDREWS, A. K., AUBLE, G. T., ELLISON, R. A., FARMER, A. H., and ROELLE, J. E. 1985a. Environmental Systems and Management Activities on the Kennedy Space Center, Merritt Island, Florida: Results of a Modeling Workshop. WELUT-85/W05. U.S. Fish and Wildlife Service, Fort Collins, Colorado. 130 pp.

HAMILTON, D. B., ANDREWS, A. K., AUBLE, G. T., ELLISON, R. A., and ROELLE, J. E. 1986. Effects of Flood Control Alternatives on Fish and Wildlife Resources of the Malheur–Harney Lakes Basin. NEC-86/20. U.S. Fish and Wildlife Service, Fort Collins, Colorado. 85 pp.

HILBORN, R. 1973. A control system for FORTRAN simulation programming. *Simulation* 20:172–175.

HIRSCH, A., ANDREWS, A. K., and ROELLE, J. E. 1979. Implementing adaptive environmental assessment in an operating agency. In Adaptive Environmental Assessment and Management: Current Progress and Prospects for the Approach. CP-79-9. International Institute for Applied Systems Analysis, Laxenburg, Austria. 15 pp.

HOLLING, C. S. (ed.) 1978. Adaptive Environmental Assessment and Management. Wiley, New York. 377 pp.

International Institute for Applied Systems Analysis 1979. *Adaptive Environmental Assessment and Management: Current Progress and Prospects for the Approach.* CP-79-9. International Institute for Applied Systems Analysis, Laxenburg, Austria, 19 pp.

JOHNSON, R. L., ANDREWS, A. K., AUBLE, G. T., ELLISON, R. A., HAMILTON, D. B., ROELLE, J. E., and McNAMEE, P. J. 1983. Evaluating Environmental and Economic Consequences of Alternative Pest Management Strategies: Results of Modeling Workshops. Unpublished. Available from U.S. Fish and Wildlife Service, Fort Collins, Colorado. 38 pp.

LEOPOLD, L. B., CLARK, F. E., HANSHAW, B. B., and BALSLEY, J. R. 1971. A Procedure for Evaluating Environmental Impact. Geol. Surv. Circ. 645. U.S. Government Printing Office, Washington, D.C. 13 pp.

McNAMEE, P. J., ANDREWS, A. K., AUBLE, G. T., ELLISON, R. A., JOHNSON, R. L., HAMILTON, D. B., and ROELLE, J. E. 1982. Results of a Modeling Workshop Concerning Development of the Beluga Coal Resource in Alaska. Unpublished. Available from U.S. Fish and Wildlife Service, Fort Collins, Colorado. 60 pp.

Nebraska Natural Resources Commission. 1985. Platte River Forum for the Future. Nebraska Natural Resources Commission, Lincoln, Nebraska. 32 pp.

ROELLE, J. E., and AUBLE, G. T. 1983. Resource development and management in Jackson Hole, Wyoming. In Laurenroth, W. K., Skogerboe, G. V., and Flug, M. (eds.), *Developments in Environmental Modelling, 5: Analysis of Ecological Systems, State-of-the-Art in Ecological Modeling.* Elsevier Scientific, New York, pp. 303–311.

ROELLE, J. E., and HAMILTON, D. B. 1990. Suwannee River Sill and Fire Management Alternatives at the Okefenokee National Wildlife Refuge. Unpublished. Available from U.S. Fish and Wildlife Service, Fort Collins, Colorado. 39 pp.

ROELLE, J. E., AUBLE, G. T., and HAMILTON, D. B. 1984b. Refuge Management Analyses: Research Needs for Muscatatuck National Wildlife Refuge. W/AEAG-84/W05. U.S. Fish and Wildlife Service, Fort Collins, Colorado. 30 pp.

ROELLE, J. E., HAMILTON, D. B., and ASHERIN, D. A. 1990. Management Alternatives for the Gregory Landing Division, Mark Twain National Wildlife Refuge. Unpublished. Available from U.S. Fish and Wildlife Service, Fort Collins, Colorado. 20 pp.

ROELLE, J. E., ANDREWS, A. K., AUBLE, G. T., HAMILTON, D. B., and JOHNSON, R. L. 1983. Platte River Forum for the Future. Workshop Model Documentation. Unpublished. Available from U.S. Fish and Wildlife Service, Fort Collins, Colorado. 251 pp.

ROELLE, J. E., ANDREWS, A. K., AUBLE, G. T., HAMILTON, D. B., JOHNSON, R. L., and SISLER, J. F. 1984a. A Framework for Assessing the Impacts of Acidic Deposition on Forest and Aquatic Resources: Results of a Workshop for the National Acid Precipitation Assessment Program. FWS/OBS-84/16. U.S. Fish and Wildlife Service, Fort Collins, Colorado. 80 pp.

U.S. Fish and Wildlife Service. 1982. Effects of Acid Precipitation on Aquatic Resources: Results of Modeling Workshops. FWS/OBS-80/40.12. U.S. Fish and Wildlife Service, Fort Collins, Colorado. 129 pp.

WALTERS, C., GUNDERSON, L. and HOLLING, C. S. 1992. Experimental policies for water management in the Everglades. *Ecol. Applic.* 2:189–202.

WALTERS, C. J., HILBORN, R., and PETERMAN, R. 1975. Computer simulation of barren-ground caribou dynamics. *Ecol. Mod.* 1:303–315.

WATT, K. E. F. 1977. Why won't anyone believe us? *Simulation* 28:1–3.

WOOD, F. A. 1979. Experience in implementing adaptive management and assessment. In Adaptive Environmental Assessment and Management: Current Progress and Prospects for the Approach. CP-79-9. International Institute for Applied Systems Analysis, Laxenburg, Austria. 4 pp.

Part 3

COMPLEX ECOLOGICAL ORGANIZATION: FEEDBACK AND STABILITY

George Van Dyne was everyperson's ecologist. He had distinct ideas about the nature of ecosystems and the kinds of information that would be necessary to understand them as complex hierarchical organizations, and he was a willing and able advocate of particular approaches. But he was also open to everything and everyone. He welcomed scientific diversity; it was a match to natural diversity as well as the span of his own imagination. His programs were organized by consensus in a process in which he exerted strong leadership, but not enough to compromise democratic give-and-take between qualified equals. The seven chapters of this section reflect the kind of scientific ecumenism that Van Dyne espoused. He was interested in the salient properties inherent in the design and function of ecosystems, such as stability, disturbance and stress, and feedback, as these chapters discuss. Inadequate feedback is identified as the source of undesirable properties in Chapter 16, and in Chapter 18 the building of an actual large-scale synthetic ecosystem embodying some of the principles of complex ecological organization is discussed.

Chapter 13

Chapter 13 continues the mathematical approach to complex ecology, providing a review of community stability theory. Its authors, Yuri Svirezhev and Dimitrii Logofet, are as yet unrecognized world authorities on this subject, having previously published the definitive scholarly work (*Stability of Biological Communities,* English edition, 1983, Mir, Moscow) that in its mathematical rigor and geographic point of origin has not received the amount of attention accorded other treatments. Consider its chapters:

1. Stability of isolated populations
2. Discrete population models
3. Predator–prey type models
4. n-species communities
5. Trophic chains and communities with vertical structure

6. Ecological niche overlap and stability of horizontally structured communities
7. Extreme properties of ecological systems
8. Stability of spatially distributed ecosystems
9. Stability and complexity of ecosystem models

Svirezhev and Logofet should be required reading for all ecologists with interests in formal theoretical foundations for complex ecology. The more recent work of Logofet should also be consulted (*Matrices and Graphs, Stability Problems in Mathematical Ecology,* 1992, CRC Press).

In Chapter 13 the authors continue some of the themes advanced in their book. They begin by noting "that the stability of a biological population, community, or ecosystem is an intrinsic ability to resist perturbations coming in abundance from the environment. . . ." However, although the intuitive notion is clear, it is difficult to provide a precise and unambiguous "stable" definition. The diversity–stability hypothesis and some of its ramifications are noted. Echoing some of the connections between thermodynamics and stability made in Chapters 3 through 6, Lagrange and Liapunov concepts of stability in mathematics are reviewed and compared with ecological concepts. While Liapunov stability is stronger, and the Lagrange concept nearer to ecological meanings, "there exists a whole class of ecological models where both notions turn out to be equivalent." Chaotic regimes and strange attractors are pointed up as recent issues in population-dynamics modeling that initially cast doubt on the predictive ability of population models. But these characteristics, say the authors, have turned out to be ecologically interpretable. Technical details follow. The Liapunov concept depends upon properties of the community matrix, and matrix-stability properties do match ecological meanings. The qualitative stability of model ecosystems reduces to sign stability of the community matrix. A theorem giving a matrix sign-stability criterion is proved. "Only those structures satisfy the criterion which are either predation communities . . . [or their unions] linked one to another by amensal or commensal relations forming no direct cycles in the total community." Thus, sign stability of community matrices "admits the complete description of the class of sign-stable patterns in terms of the types of intra- and interspecies relations." The criteria for sign stability turn out to be useful in applications to vertically structured communities such as trophic chains. Conditions for stable coexistence in models of horizontally (for example, competitively) structured communities are also considered. The topics of trophic chains and horizontally developed communities are extensively developed. In concluding by calling attention to several topics beyond the scope of the chapter, the authors refer readers to their book for further details. This is good advice for those who admire the demonstration of rigor in ecology.

Chapter 14

Chapter 14 continues the stability theme with an application to forest ecosystems. Jack Waide, an author of the resistance–resilience model of ecosystem stability, points out that while these concepts have been useful in organizing research, there have been published criticisms. The purpose of his chapter is to provide a revised theoretical interpretation of resistance and resilience. Five factors have contributed to confusion about ecosystem stability: (1) improper coupling of the system to its physicochemical environment; (2) inadequate attention to scale, and specifically the built-in bias from human life span that rapidly turning over systems (like plankton) are unstable while slower ones (forests) are stable; (3) confusion caused by the complexity–stability debate; (4) underappreciation of the system concept itself in efforts to derive ecosystem-level properties solely from those of constituent populations; and (5) misinterpreting criticisms of succession, climax, and balance of nature as criticisms of stability. This chapter explores these issues, treating "ecosystems as hierarchical biogeochemical systems . . . [whose] resistance and resilience represent scaling variables which reflect the space–time [evolutionary] coupling of entire biotic assemblages to specific physicochemical environments ... [to] regulate the exchange, transformation, and storage of energy and essential elements by ecosystems."

In a review of mathematical stability concepts, a distinction is made between stability of ecosystems and stability of differential equations. But a differential equations model of the ecosystem is formulated to provide definitions and a framework for the investigation of ecological stability, which concerns response to perturbation. A progressively stronger set of stability definitions is introduced, reflecting the mathematical approach of the previous and several earlier chapters. It is observed that while formal definitions may be overly restrictive for ecological application, they provide a solid foundation when applied in an ecologically relevant manner. Thus, for example, the concept of ecosystems as "conditionally stable attractors in environmentally determined basins of attraction" is both conceptually useful and mathematically tractable. Restrictions on the application of mathematical stability definitions are noted: (1) ecosystem stability is not equivalent to constancy of function over time; (2) it is not necessarily related to species persistence, constancy, or composition over space or time; (3) ecosystem stability is not usefully tested by severe or persistent disturbances that completely degrade system structure and its coupling to environment; (4) normal environmental fluctuations or occurrences (for example wildfire) are not disturbances at the scale of the ecosystem; (5) stability concerns not the behavior of a specific trajectory within a state space, but the configuration of the space itself; and (6) stability is a scale-dependent property—species populations may be locally extinguished by disturbances in ecosystems that will recover.

The ecosystem as a hierarchical biogeochemical system is next described. Far-from-equilibrium thermodynamics and hierarchy concepts are the centerpieces of a conception "of the evolutionary emergence of ... [eco]systems as open, dissipative structures in spatially and temporally variable environments ... at various scales. ... [providing] the rationale for scale-dependent, macroscopic analyses of stability ..." The part–whole relationship is that "at a specific spatiotemporal scale, the processes of energy dissipation and biogeochemical cycling represent functional constraints on the behavior and evolution of species. But, it is the species themselves which 'carry out' these functional processes and which are responsible for the scale-dependent evolutionary coupling of the total biotic structure (i.e., the ecosystem) to the physicochemical environment." In its partness and wholeness, the ecosystem represents a "dual hierarchy ... insensitive to considerable variation in species composition. Thus, distinct species assemblages appear to realize comparable levels of structure and function, and a single basin of attraction defined in macroscopic structural and functional terms may encompass several distinct basins defined in relation to the persistence of species assemblages." In developing the concept of the ecosystem as a conditionally stable system, Waide explores thermodynamic, hierarchical, and natural-history perspectives, and then he reviews evidence from experimental ecosystem research. Finally, the concepts of resistance and resilience are revised and brought into conformity with the prior considerations. Following review of the original model for these concepts, revised interpretations are presented and supported by modern data. Drawing the contrast between species- and process-oriented approaches to complex ecology, Waide concludes that, "Determining relationships between the ecosystem viewed as a hierarchy of functional processes versus a hierarchy of species assemblages remains a major challenge for empirical and theoretical ecologists."

Chapter 15

In Chapter 15, David Rapport and Henry Regier consider natural disturbances and ecosystem responses to them, which may be rejuvenating or pathological. "Ecosystems thrive on disturbance," they write, but intermediate disturbance is best. Too much or too little is damaging. Human intervention is everywhere. If this is well informed, healthy ecosystems can result. Too often the process is uninformed (or uncaring). Degradative sequences tend to run in opposite directions to normal maturation sequences of unstressed ecosystems, which involve "progressively greater internal integration, less temporal variability in key components, greater adaptive or rheostatic capabilities with respect to the usual small to moderate external influences but lesser adaptive capacity to major unusual external influences." Diagnostic versus nondiagnostic stress symptoms are distinguished; both relate to the degree of ecosystem stress, but only the former provides guidance on appropriate therapeutic treatment. Processes by which pathology develops are complex, especially when multiple harmful stresses are involved. Controlled experiments on whole intact ecosystems are extremely valuable for determining stress and recovery characteristics. A table of stress symptoms is given, and it is observed that the "ecosystem distress syndrome" contains a simplification: despite differing degenerative pathways the degraded end states exhibit some broad and recognizable similarities in terms of modified function. This is von Bertalanffy's "equifinality" principle. In the end state of stressed ecosystems primary production decreases, decomposition and nutrient recycling are impaired, species diversity is reduced, parasitism and other negative interactions increase while mutualism and other positive interactions decrease, and both component and system stability decrease making disintegration more likely. Intense stress may extinguish all life.

Processes of ecosystem stress adaptation are not well explored, but some critical properties contributing resistance and resilience have been identified. Open ecosystems are more resistant; exchange of biota and products is possible permitting renewal and depuration. So are larger ecosystems, which have greater capacity to withstand local disturbances. Type, intensity, and multiplicity of stress factors are all important determinants of the stress response; even timing can make a difference. Catastrophic disturbance will lead to catastrophic change. Temporal and spatial dimensions of the stress response are reviewed in the context of buffering and compensating mechanisms. The authors argue for a holistic approach to the study of degradative and rehabilitative processes. Human presence, in absence of committed husbandry, always means stress, and the human factor in the total stress equation around the globe, "epicoenology," needs expanded attention. "As one considers the broad range of stresses, and the increasing number of ecosystems which are damaged by one or a combination of stresses, a picture emerges of the spread of a blight across a landscape leaving only small pockets unaffected. . . . The patchwork may superficially resemble the work of natural disturbances . . . [but s]tressed ecosystems tend to cluster near regions of human settlement . . . or downstream, atmospherically and hydrologically, from such centers." Disturbance (less chronic) and stress (more chronic) are distinguished as degraded states of ecosystems. A classification of states (natural; cultural–productive; urbanized; and stressed–degraded by harvest; by improper culture or harvest; and by pollution, restructuring or other abuses) and transformations between these is offered. Several examples of degradation and rehabilitation processes are given, and with a touch of cynicism the observation made that, "Throughout the whole [human] development process, lip-service will have been paid to safeguarding productive functions of the ecosystem. The usual outcome has been that productivity and other highly-valued aspects of the ecosystem have been sacrificed." Man is an abuser of nature, and the science and practice of proper ecodevelopment are poorly developed. Restoration back to the "natural" category of states "may have occurred in the distant past with collapse of ancient civilizations."

Chapter 16

Chapter 16 by Kenneth Watt takes the themes of *Complex Ecology* to the sociopolitical sphere, without which this book would be incomplete, for it is through instabilities in this domain—human threats to resistance and resilience of world ecosystems—that impetus for further scientific development of the complex man–nature relationship will come. In 1989 correspondence with the editor, Watt voiced his own version of the Van Dyne systems worldview in no uncertain terms: "Clearly, science in general and ecology also are in danger of being washed out to sea in a tidal wave of fragmentation and reductionism. . . . We [in the United States] are starting to pay the price for a dummed-out educational system. . . ." His chapter summarizes macrolevel findings of a two-decade long project involving efforts of about a hundred people to model society. The project was sparked in the late 1960s by ominous trends noted in overpopulation, pollution, resource depletion and degradation, and rapid atmospheric change with implications for climate change. Many teams at the University of California–Davis began modeling and simulation projects to explore the implications of these trends. The motivation for systems ecologists to become involved was that long runs of social data were available, unlike in ecology, and parallels between human society and ecological systems are close. Both are feedback systems (Chapter 17) in which disturbance and stress (Chapter 15) are causes of unstable (Chapters 13 and 14) behavior expressed in wars and depressions as unintended consequences of normal world functioning. In addition, long-wave phenomena in the world economy are analogous to well-known population cycles in ecology.

Watt reviews stability theory as it applies to society. The confusion in ecological stability concepts pointed up by Waide (Chapter 14) is reiterated here. Seven different ecological stability principles are identified as having become confused in the ecological literature, referred to as *omnivory, fine-grained landscape, fast homeostasis, rapid resource exploitation, flattened pyramid, connectance,* and *migration* principles. Two broad categories of complex system designs for adjustment to potentially destabilizing disturbances are (1) spreading risk (examples are given) and (2) facilitating homeostasis—by three methods: shortening time delays in negative feedback control loops, enhancing resource flux, and lowering and broadening hierarchical levels of organization (for example, ecological pyramids, administrative layers). With the seven stability principles in mind the author asks of his research, "What is the significance for human society of inadequate omnivory, excessive delay in feedback mechanisms, curtailment of resource supplies, multileveled hierarchical administrative structures and their vulnerability to catastrophic instability in the face of massive perturbations, and excessive connectance. . . ?" The time frame for his model-generated answers is decades or longer.

Laying a background in long-wave theory, and noting fifty-year cycles in several social phenomena, Watt takes up the topic of war as system destabilization. He notes that major wars have "an alarmingly cyclical character. The great powers have all gone to war with each other about every fifty years throughout modern times . . . wars [and their indirect manifestations like retiring war debt and the demographic impacts of war dead] are the key phenomena in setting the timing mechanism for long waves." Depressions, which terminate long-wave cycles, set the stage for subsequent wars. The model is described, including sector analyses, parameterization and other considerations for its design and development. It consists of nineteen sectors for each major nation, linked in a global version to those of other nations by flows of interest and inflation rates, money, goods, and people. Details are given for the energy, demographic, and agricultural subsystems. Then, depression and war cycles are discussed as consequences of lagged feedback control, with two examples given in illustration. Lessons for ecologists include: (1) complex models require complex programs to produce them, and both become unwieldy; (2) little relationship exists between model complexity and predictive ability; (3) diverse models require access to diverse literature and specialists; (4) model complexity demands state-of-the-art hardware and software; and (5) the seven stability principles are worth keeping in mind. The chapter ends with a gracious tribute to George Van Dyne by one who also deserves accolade himself as an early founder and "father" of systems ecology. Watt's unique manner of facing complex ecology through the views afforded by long-term data series is well known and well displayed in the chapter.

Chapter 17

In Chapter 17, Donald DeAngelis discusses feedback. He notes that feedback is intrinsic in the design of man-made systems. He cites the application of feedback models in biology and the human sphere, noting that concepts of cybernetics and system theory are twentieth-century developments that "may provide a synthesis of the mechanistic and organic views and correct the oversimplifications of both." A feedback system is separated from others by its ability to respond to a changing environment with changed behavior. He asks if ecosystems are cybernetic and do their feedbacks represent merely fortuitous phenomena or coherent organization. He discusses the population as a regulated feedback system; density-dependence represents tight regulation and density-independence loose regulation by environment. Homeostasis (maintained stasis) and homeorhesis (directed change) are distinguished and the role of feedback in both are discussed. Positive feedback, mutual reinforcement through a closed loop of two or more system variables, is introduced, and an example of it cited from the precybernetics literature of Frederick E. Clements: ". . . the habitat causes the plant to function and grow, and the plant then reacts upon the habitat, changing one or more of its factors in decisive or appreciable degree. The two procedures are mutually complementary. . . ."

The author comments on the early contribution of Van Dyne to organizing views of ecosystems and ways to model them around cybernetic and general systems theory themes. He reviews a debate in the literature over whether ecosystems are cybernetic or not, in the process pointing up ambiguity in the cybernetic concept permitting it to be applied to perhaps too wide a range of systems. He then proceeds to an examination of feedback in two ecological processes where cybernetic design has previously been inferred—succession (facilitation, tolerance, and inhibition models) and nutrient cycling (consumer regulation). He concludes that the existence of feedback in ecosystems is insufficient evidence for their being considered cybernetic systems. In ecosystems, like "markets" and unlike "firms" in economics, the "individual unit acts for itself. From the coaction of like units emerges a structure that affects and constrains all of them." The ecosystem "is not itself an agent, though it constrains the actions of individual agents whose activities create it." It may possess "epiphenomena" but not, like its members, self-interested "strategies." And, local competitions and struggles for existence may give rise to higher-level favorable dependencies, such as mutualisms. Nevertheless, the author concludes that "the holistic appeal of the cybernetic view of the ecosystem [may have] to give way to less majestic, but more appropriate descriptions of nature . . . that have greater explanatory power."

Chapter 18

If complex ecology and complex models are defeating, by exhaustion of resources and people, then other approaches to tractability must be discovered. The previous chapters in this section have dealt with theoretical but academic aspects of ecological complexity. This one, by Bernard Zeigler, describes the theory behind the design and construction of a synthetic, self-contained, bioregenerative macrocosm to explore issues that the direct study of natural ecosystems will not allow. Historical precedents for technology following science are numerous, but the question in this chapter is whether systems ecology is mature enough to provide a foundation for "eco-engineering." Zeigler, a theorist without peer in the field of computer simulation modeling, describes in the terminology and writing style of computer scientists and engineers an approach to the design of the controversial closed environment, Biosphere II, which has been operating for the past several years north of Tucson, Arizona.

He begins by noting that "Systems ecology rises to the status of a mature science when it serves as the basis for the design of 'artificial' ecosystems. . . ." The Biosphere II project, which has created an engineered replica of the earth's biosphere "I" in a 2.5-acre enclosure under glass, is such a program. Among its other features, the new biosphere is to be a faster simulator of earth's essential ecological processes. Five biome modules are incorporated into the design: jungle, savanna, marsh, desert, and ocean; eight people have lived within the enclosure. The overriding objective, in addition to a set of specific technical ones, is to prove the feasibility of indefinite persistence. Why is a modeling theorist and information scientist involved? " . . . because simulation models are to be called upon to predict the future consequences of management and control actions." Zeigler outlines the architecture of the control and management system. It has five layers: interaction, control, information, knowledge, and executive; the knowledge layer is described in detail as an example. Concepts of knowledge representation are reviewed, and terminology and definitions are provided. "Structure" and "behavior" are distinguished; structure may be "decomposed" into components and components reconstituted to systems by "coupling." A "taxonomic relation" defines what is decomposed and coupled. The decomposition, taxonomic, and coupling relations are combined to form a representation scheme, termed the "system entity structure," whose "entities" are conceptual components of reality, models for which reside in a "model base" of the system. Concepts that artificial intelligence and simulation modeling bring to the design of self-sufficient environments are summarized. These are called "knowledge-based" because they depend on representing expert knowledge in computer-usable form.

The author concludes, "In a volume dedicated to a visionary who was responsible in large part for the advent of systems ecology, this challenge of biospherics to leap to the next plateau is especially significant." Van Dyne, who early recognized computer simulation as one of the critical elements of systems ecology, would have relished this discourse from one visionary to another, and the concrete reality, and even the knotty problems, of Biosphere II. Readers wishing background in artificial intelligence and computer science may want to refer to Zeigler's book, *Multifacetted Modelling and Discrete Event Simulation* (Academic Press) or to the less formal summary of his approach in Chapter 2 of J. Sampson's *Biological Information Processing* (Wiley).

Chapter 19

Causality, directly or indirectly manifested, is a complex phenomenon involving combined energy–matter and signal flows of many different kinds and qualities in complex ecological systems. It is never fully implemented in modeling, only approximated. For example, in network models usually only a single conservative variable like energy or a nutrient is transferred between compartments; multicommodity flows are generally not modeled or are modeled only in a limited fashion due to the technical difficulties involved. In Chapter 19, Robert Costanza and Bruce Hannon consider some of the problems in approaching the ideal. They review basic mathematical forms of compartmental flow analysis, indicate the importance of what they call "net input intensity factors," discuss the need to deal with multiple commodities and joint products formed from them in realistic ecosystem models, develop a general mathematical framework for a multicommodity approach, and illustrate this with examples.

A fundamental issue of interconnections based on dissimilar components is "*commensurability*—the ability to measure the components in the same addable (commensurable) units. Both economics and ecology must deal with this 'apples and oranges' issue, although we feel its importance in ecology has not yet been fully appreciated." Input–output flow analysis requires only that total output of each commodity in a system equal total input, without the need to add different quantities; compartmental methods, on the other hand, require that inputs and outputs be added, and since these are usually qualitatively different in the multicommodity case, they are noncommensurable. To solve this problem, a set of intensity factors analogous to prices in economics must be calculated. The authors describe how to do this in a mathematical flow accounting procedure based on their unique formulation of input–output analysis for ecological applications. Hannon, of course, introduced input–output analysis into ecology in a seminal paper (*J. Theor. Biol.* 41:535) that has spawned several variations on a theme such as "environ" theory, network approaches to trophic dynamics, and "ascendancy" theory. A multiproduct ecosystem flow accounting diagram is given that aids interpretation of variables involved in the mathematical formulation of joint products. The formulation is applied to a single-commodity compartment model to clarify a number of points, and then problems in application to the incommensurable, multiple-commodity inputs and outputs of the global biosphere are explored. A table of commensurable flows in "embodied solar energy units" is derived by multiplying each commodity by a corresponding energy intensity coefficient. Comparisons of the relative importance of incommensurable quantities become possible, and it is observed, for instance, that fresh water represents 33% of direct inputs to manufactured goods versus fossil fuel at 29%; regarding the importance of hydrologic versus carbon cycles, "Water is three and carbon two orders of magnitude larger than manufactured goods in terms of total inputs. . . . The urban economy is shown to be very dependent on inputs from soil, atmosphere, and deep geology (for fossil fuels)." The authors conclude that "Analysis of ecosystems must include several different commodities measured in physical (noncommensurable) units. Flow [input–output] analysis can be used to derive a set of unique intensity factors that . . . allow physical flows to be put in commensurable units and the 'mixed units' problem to be solved . . . [to permit] more realistic and comprehensive analysis of ecosystem interdependence."

13

The Mathematics of Community Stability

YURI M. SVIREZHEV

Potsdam Institute for Climate Change Study
Potsdam

DIMITRII O. LOGOFET

Institute of Atmospheric Physics
Russian Academy of Sciences
Moscow

For more than 40 years theoretical population ecologists have used the early model formats of Volterra, Lotka and Gause, so one might ask "Are simulation models new?" A quick review of the types of models reviewed in several chapters of the present volume will show that a conceptual quantum jump has been made. The models now include dynamics of the entire system and follow flow of energy, water, carbon, numbers and other currencies. Thus, a start has been made in thinking of the complexity of ecosystems in an organized, quantitative and synthetic way [Van Dyne, in *Grasslands, Systems Analysis and Man*, Int. Biol. Program 19, Chap. 14, p. 890 (1980), Cambridge U. Press].

INTRODUCTION AND OVERVIEW

The idea that the stability of a biological population, community, or ecosystem is an intrinsic ability to resist perturbations coming in abundance from the environment (including those due to humans) is confirmed both by the course of development of biological science and the history of human activities affecting nature. Although the notion of stability of a biological system seems obvious and intuitively clear, it is a difficult problem to provide the concept with a precise and unambiguous definition. So far, this heavily overworked term has found no established ("stable") definition that covers all cases. Several approaches exist to the identification and analysis of an ecological system's stability, of which the following two are fundamentally distinct.

The first approach attempts to connect stability with other system characteristics that can be readily and directly measured. For example, many ecologists consider it almost an axiom that communities that are more complex in structure and richer in number of species are necessarily more stable (MacArthur, 1955; Elton, 1958; Pielou, 1969). Naturally, this motivates the use of indexes of species diversity (in particular, information entropy or some of its analogues) as measures of community stability (see review by Goodman, 1975). The use of diversity to measure stability results in a number of theoretical paradoxes (Logofet and Svirezhev, 1977), and, in addition, empirical data available are quite insufficient to establish the approach as well grounded (Svirezhev and Logofet, 1978).

The second approach relies on the investigation of stability properties of mathematical models describing the temporal dynamics of the system under study. There is a highly advanced mathematical theory of stability, with numerous applications in science and engineering, for which stability always has a rigorous definition. If, therefore, we have a descriptively adequate mathematical model of a system, then the stability problem is reduced to the task of choosing and analyzing that formal notion of stability that best represents meaningful ideas of the stability of the modeled system's dynamic behavior.

Granted, here we make a substitution as we replace the object itself by a mathematical model whose adequacy is merely postulated at this stage. But such an approach is quite justifiable since stability analysis of the model enables us to propose certain hypotheses concerning the laws of functioning of the modeled object, whose presence or absence in reality makes it possible to judge also the model's adequacy (Logofet, 1981a; Logofet and Alexandrov, 1983 a, b). Also, the requirement that the model's behavior should possess a certain stability provides an effective criterion for the selection of appropriate models and restricts arbitrariness in options for model parameters (Logofet and Svirezhev, 1980).

Both of the identified approaches have their benefits and limitations. The prognostic ability of the "model" approach in applications is one of the most evident of its advantages (Logofet, 1981a). In this chapter a review is given of certain stability concepts, and respective methods of analysis, for application to mathematical models describing the dynamics of multispecies communities or ecosystems. In the next section we investigate to what extent the notions of Liapunov and Lagrange stability correspond to "ecological" ideas of stability and illustrate some formal relations between these two concepts. While Liapunov stability is, in general, a stronger requirement than the Lagrange property, which is closer to the intuitive ecological concept, there exists a whole class of ecological models for which both notions turn out to be equivalent. In the following section on Volterra models we present other considerations demonstrating the advantage of Liapunov stability analysis applied to ecological models. Being dependent on stability properties of the community matrix, the Liapunov concept puts forward the entire hierarchy of various matrix stability properties that, although used earlier in other fields such as mathematical economics, also make sense ecologically. Next, the notion of the qualitative stability of model ecosystems is reduced to that of the sign stability of the community matrix by means of the Liapunov stability of a feasible equilibrium state. Recently established, the criterion of sign stability of an arbitrary $n \times n$ matrix admits a complete description of the class of sign-stable patterns in terms of the types of intra- and interspecific relations. This criterion turns out to be useful not only in practice, but also in such theoretical applications as the theory of "vertically structured" communities (section on trophic chains). In the penultimate section we consider conditions of stable coexistence in models of "horizontally structured" (for example, by competition) communities. And, finally, speculations concerning perspectives in mathematical research on community stability are presented in the concluding section.

ECOLOGICAL AND MATHEMATICAL DEFINITIONS OF STABILITY

Analyzing a variety of propositions concerning which ecological systems should be considered stable, three levels of applicability of the "stability" concept may be recognized (Svirezhev and Logofet, 1978). First, a whole geographic region or landscape may be, for a certain relatively long period, invariant in time. Such broad regions consist of biotic communities, which are higher level aggregates of multispecific populations occupying a common territory and showing a certain pattern of trophic interactions and metabolism. At this second level, a community may be considered stable if the number of its member species remains constant over sufficiently long intervals of time. It is this ecological definition that appears closest to various mathematical definitions of stability. The third group of requirements refers to individual populations rather than to communities. A community is believed to be stable when numbers of component populations do not undergo sharp fluctuations. This definition is closest to physical (or thermodynamic) notions of system stability.

In thermodynamics a system is considered stable when large fluctuations that can take it far from equilibrium, or even destroy it, are unlikely. Evidently, general thermodynamic concepts (for instance, the stability principle associated with the second law of thermodynamics) should be applicable to biological systems. By these concepts, any closed system with an energy flow (be it the entire biosphere or only a small lake) most likely develops toward a steady state involving certain feedback mechanisms. When this state is attained, the transfer of energy usually proceeds in one direction and at a constant rate; this condition corresponds to the thermodynamic stability principle. The predator–prey system of ecology is a classic example of such a flow-through system. Unfortunately, thermodynamic stability gives no effective method of judging the stability of a concrete biological community or ecosystem.

Let us pass from a real ecological object to its mathematical model formalized as a system of equations for n components:

$$\frac{dN_i}{dt} = F_i(N_1, \ldots, N_n) \qquad i = 1, \ldots, n. \tag{13.1}$$

Liapunov stability, which is the most fundamental and best developed concept in mathematical stability theory, means a certain closeness between a perturbed trajectory, $\mathbf{N}(t) = [N_1(t), \ldots, N_n(t)]$ and some unperturbed (and often equilibrial) solution $\mathbf{N}^*(t)$, provided the perturbations of an initial state are sufficiently small.

At the same time, the ecological sense of stability corresponds more closely to the requirement that $\mathbf{N}(t)$ must remain within a bounded domain lying in the interior, P_+^n, of the positive orthant, P, of the phase space for model (13.1), provided the perturbations of $\mathbf{N}(0)$ are also bounded. This property, which below is called *ecological stability,* is a particular case of Lagrange stability, and obviously a weaker requirement compared to the Liapunov concept.

Conditions for the Lagrange stability of some population models have been the subject of several publications (Thornton and Mulholland, 1974; Brauer, 1979; Logofet and Svirezhev, 1985; Adzhabyan and Logofet, 1992a, b), but there is still a lack of sufficiently general, well-developed methods. Recent research in various branches of population dynamics theory has shown that ecologically stable behavior can be realized not only in the form of regular or damped oscillations around an equilibrium state, but also in a variety of other ways. In particular, certain population models are characterized by the existence of several nontrivial equilibria rather than a single one (see review by Levin, 1979). Depending on the magnitude of perturbations, the system can pass from the attractive domain of one equilibrium state into that of another or (with small permanent random perturbations) can make cyclical passages among different stable states while spending most of the time in their close vicinities (Freidlin and Svetlosanov, 1976; Sidorin, 1981). Such behavior has served as a reason for introducing the notion of "resilience" into considerations of ecological stability (Holling, 1973, 1976; Grümm, 1976). It can be shown (Logofet, 1981b) that resilience is a stronger property of a dynamical system than Lagrange stability, but weaker than Liapunov stability.

The Lagrange stability notion also covers the behavior of some models from the topological catastrophe theory of Thom (1969), which in fact is a theory concerning singularities in smooth mapping (Arnold, Varchenko, and Guseyn-Zadeh, 1982). It turns out that an ecosystem's dynamic behavior, which is classified as a "catastrophe" in the topological sense (Jones, 1975), may not be a catastrophe in the ecological sense (Logofet and Svirezhev, 1979; Logofet, 1981b; Svirezhev,

1987), but rather corresponds simply to some reorganization of the ecosystem while preserving its general stability.

Closely related to ecological stability, the problem of the existence of "chaotic" or "pseudostochastic" regimes and "strange attractors" is also among recent issues in population dynamics models. These regimes were discovered in systems of deterministic equations, both difference and differential (see reviews by Yatsalo, 1984, and Shaffer, 1985). As a result, the prognostic capability of population models is called into question (May, 1975). However, recent studies have shown that chaos and strange attractors often arise from natural conditions that are ecologically interpretable (Svirezhev and Voinov, 1982; Logofet and Svirezhev, 1983; Alekseev and Kornilovsky, 1985; Shaffer and Kot, 1985; Svirezhev and Logofet, 1985).

Thus, the above-mentioned "nonclassical" concepts of ecological stability extend the class of models that provide an adequate description of dynamic behavior in a real system. But reasons still exist to support the Liapunov approach to stability analysis of ecological models, in addition to the fact that this usually serves as an initial step in studies of nonclassical concepts resulting in the elimination of "trivial" stability cases. These reasons are considered in the next section.

LIAPUNOV AND ECOLOGICAL STABILITY IN VOLTERRA MODELS

If, in eq. (13.1), a natural increase or decrease in population sizes is described by linear terms and a self-limiting counterplay between species by second-order terms, with no explicit relation to t, then this brings us to *Volterra-type models:*

$$\frac{dN_i}{dt} = \epsilon_i N_i - \sum_{j=1}^{n} \gamma_{ij} N_i N_j, \quad i = 1, \ldots, n, \quad (13.2)$$

historically the first comprehensively investigated systems of population dynamics equations (Volterra, 1931).

System (13.2) was derived from the "encounter principle" and "equivalence hypothesis," which has often caused criticism of the equations as "artificial." Nevertheless, eq. (13.2) can be derived from quite natural prerequisites. By accounting for balance relations and conservation laws (Svirezhev and Logofet, 1978), the scope of their applicability in practical situations increases.

In studying systems of type (13.2), Volterra considered two special classes, conservative and dissipative systems. System (13.2) is said to be *conservative* (in the Volterra sense) if there exists a set of positive numbers $\alpha_1, \ldots, \alpha_n$ such that

$$F(N_1, \ldots, N_n) = \sum_{i,j=1}^{n} \alpha_i \gamma_{ij} N_i N_j \equiv 0.$$

If the quadratic form $F(N_1, \ldots, N_n)$ is positive definite, the system is called *dissipative*.

If the quantity $V(\mathbf{N}) = \sum_{i=1}^{n} \alpha_i V_i$ is interpreted as some measure of total community biomass, then the sense of the above-given definitions is that species interactions do not affect the course of their total biomass in the conservative case, but do retard biomass increase in the dissipative case. In the presence of a feasible equilibrium state, $\mathbf{N}^* > \mathbf{0}$, various results were proved concerning the properties of these systems (for example, the existence of a motion integral in the conservative case and the convergence of trajectories to $\mathbf{N}^*$). Such results have been confirmed in the framework of Liapunov methods (Svirezhev and Logofet, 1983), the equilibrium $\mathbf{N}^*$ turning out to be locally nonasymptotically stable in a conservative system and globally (in the entire P_+^n domain) asymptotically stable in a dissipative system. But if one or several components of the equilibrium solution become negative, the corresponding species disappear from the community. Thus, the condition of existence and Liapunov stability of a feasible equilibrium **N*** is not only a sufficient condition, but also a necessary one for ecological stability in both conservative and dissipative systems.

While the equivalence between Liapunov and Lagrangian stabilities in these systems can be explained from the specific form of the equations, a recently proposed general method for estimating a stability domain in the model phase space depends on determination of a Liapunov stability domain in the parameter space of a modified model (Logofet and Svirezhev, 1985; Svirezhev and Logofet, 1985). This points out a fundamental relationship between Liapunov and Lagrange notions of stability.

Finally, some structural properties of real ecological systems appear closely correlated with the Liapunov stability of corresponding models. For instance, in work by Pimm (1980a) where stability properties in a statistical ensemble of food web community matrices were compared by parameters (like the number of trophic levels, omnivorous links, and connectedness), real webs differed from purely random structures by those features that confer stability in a community matrix.

These arguments are sufficient for developing the Liapunov stability concept as applied to mathematical models for major types of intra- and interspecific relational structures in biological communities. Subsequent sections of this chapter present such findings.

THE HIERARCHY AMONG STABILITY SUBSETS OF COMMUNITY MATRICES

Stability of the Community Matrix

In the theory of multispecies systems, an important part is played by the community matrix, which, for a general model of form (13.1), is defined as

$$A = \left[\frac{\partial F_i}{\partial N_j} \bigg|_{\mathbf{N}^*} \right], \qquad (13.3)$$

where **N*** > **0** denotes a feasible equilibrium state. The fundamental link between the community matrix and the Liapunov stability of an equilibrium in model (13.2) is that the stability of matrix (13.3) (in the sense that all Re $\lambda_i(A) < 0$) is sufficient (and in the linear case, necessary and sufficient) for Liapunov stability of the equilibrium. If there exists $\lambda_j(A)$ such that Re $\lambda_j > 0$, then the equilibrium is unstable. Therefore, the most general method of analyzing Liapunov stability in a model system is to linearize the equations in the vicinity of the equilibrium **N*** (or any other pertinent "reference" solution) and to examine properties of matrix (13.3). In some cases, it may be possible to construct a proper Liapunov function, thus achieving the possibility of judging not only the stability itself, but also the domain of stability, both in phase and parameter space (Goel, Maitra, and Montroll, 1971; Pykh, 1983; Svirezhev and Logofet, 1978).

For example, in eq. (13.2)-type Volterra systems a community matrix has the form

$$A = - \operatorname{diag} \{N_1^*, \ldots, N_n^*\} \Gamma, \qquad \Gamma = [\gamma_{ij}], \qquad (13.4)$$

and in this case it lacks the property of being either conservative or dissipative. The stability analysis reduces to investigating the set of matrix A's eigenvalues. Note that stability thus depends both on the pattern of species interactions (that is, on matrix Γ) and on characteristics inherent in the individual species themselves (that is, beyond any interactions represented by the coefficients ϵ_i, which, given matrix Γ, are in one-to-one correspondence with **N***).

D-stability

If, however, we require stability to be a property of matrix Γ alone and to be preserved under any feasible values of ϵ_i, then the notion of *D-stability* arises. This is well-known in mathematical economics (Arrow and McManus, 1958). Matrix A is called *D-stable* if the product DA preserves stability for any diagonal matrix D from the set, D_n^+, of all diagonal matrices with positive entries on the principal diagonal. D-stability is thus a property only of the pattern of species interactions, independent of their rates of natural increase or decrease.

D-stable matrices play an important role in applied research on stability and are the subject of intensive study in matrix theory. For instance, quite a number of sufficient conditions of D-stability are known (Johnson, 1974; Datta, 1978; Hartfiel, 1980). But most of these turn out to be only particular implications of the general hierarchy of stability notions described next.

Total Stability

An important subclass of D-stable matrices is constituted by *totally stable matrices* (Quirk and Ruppert, 1965), those matrices any of whose principal submatrices (that is, submatrices symmetric with respect to the principal diagonal) are D-stable. It follows immediately that any principal submatrix of a totally stable matrix is also totally stable. Thus, in community terms, total stability is apparently interpreted in the following way: the community that remains after any subset of species is excluded from the interaction structure retains its D-stability.

Any totally stable matrix is, clearly, also D-stable. An important property of the set of totally stable matrices is that the set is open, and any inner point of the set of D-stable matrices represents a matrix that is also totally stable (Hartfiel, 1980). Hence, the subset of totally stable matrices is the topological interior of the set of D-stable matrices.

Volterra Dissipativeness

The notion of a dissipative system of equations can also be applied to matrices (Svirezhev and Logofet, 1983). Matrix A is called *dissipative* (in the Volterra sense) if there exists a positive diagonal matrix $\mathscr{A}$ such that the quadratic form $<\mathscr{A}\mathbf{x}, \mathbf{x}> = F(x_1, \ldots, x_n)$ is negative definite. In a slightly different but equivalent form, matrix dissipativeness has also been defined under the terms *diagonal stability* (Barker, Berman, and Plemmons, 1978) and *positive D-dissipativeness* (Pykh, 1983).

Does Dissipativeness Ensue from Total Stability?

Any dissipative matrix can be shown to be totally stable, but the question as to whether the sets of dissipative matrices and of totally stable ones are coincident in the general case of $n \geq 3$ was considered an unsolved point in matrix theory (Barker et al., 1978). The following two theorems give constructive criteria for determining whether an $n \times n$ matrix of general form is dissipative and totally stable.

Theorem 1. A matrix

$$A = -\begin{bmatrix} 1 & a & b \\ c & 1 & d \\ e & f & 1 \end{bmatrix} \quad (13.5)$$

is dissipative if and only if all the principal minors of $(-A)$ are positive and the following condition holds:

$$d(-A) + E_2(-A) > \inf_{\substack{x, y > 0, \\ x \in I_2(a, c)}} \left\{ Q(x, y) = \frac{x(a^2 - abf) + (c^2 - cde)}{x} \right.$$

$$\left. + \frac{y(d^2 - bed) + (b^2 - aef)}{y} + \frac{xy(b^2 - abd) + (e^2 - cef)}{xy} \right\}, \quad (13.6)$$

where $d(-A)$ designates the determinant and $E_2(-A)$ the sum of principal second-order minors of matrix $(-A)$, and $I_2(a, c)$ is the interval where $a^2x^2 + (2ac - 4)x + c^2 < 0$.

Theorem 2. Matrix A is totally stable if and only if all the principal minors of $-A$ are positive and the following condition holds:

$$d(-A) - E_2(-A) \equiv ade + bcf - 2$$

$$< 2\left[\sqrt{(1 - ac)(1 - be)} + \sqrt{(1 - ac)(1 - df)} \right. \quad (13.7)$$

$$\left. + \sqrt{(1 - be)(1 - df)} \right].$$

The search for an example of a totally stable, but nondissipative, matrix is now reduced to identifying a matrix $-A$ with positive principal minors meeting condition (13.7), but violating (13.6). These properties are inherent, for example, in matrices A_1 and A_2:

$$A_1 = -\begin{bmatrix} 1 & 1 & 2 \\ 0.95 & 1 & 1.92 \\ 0.49 & 0.495 & 1 \end{bmatrix}, \quad A_2 = -\begin{bmatrix} -1 & -4 & -5 \\ -0.2 & -1 & -0.5 \\ -0.25 & -2 & -1 \end{bmatrix}.$$

Matrix A_1 corresponds to a community of three competing species with definite relations between intra- and interspecies competition coefficients, while the structure of matrix A_2 reflects an omnivorous situation such as occurs in some trophic webs (Pimm and Lawton, 1978).

As with matrices A_1 and A_2, any matrix within certain vicinities of them in matrix space possesses the same properties, so the set of totally stable matrices is topologically wider than that of dissipative ones. Thus, the relations between the notions in question constitute, in fact, the proper hierarchy:

| Dissipativeness | → | Total stability | → | D-stability | → | Stability | . (13.8)

Within the subset of normal matrices (that is, when $AA^T = A^TA$, A^T the transpose of A) all the relations in (13.8) are equivalent (Svirezhev and Logofet, 1978). Normal matrices include, for example, symmetric ones (as in competition communities; see later) and skew-symmetric matrices as in classical Volterra prey–predator systems). For a normal community matrix, therefore, the global stability of a feasible equilibrium N^* (with any positive initial values) follows from its local stability.

Dissipativeness and Diagonal Dominance

In stability studies of multispecies models, either in the linear approximation or by means of the Liapunov function, the problem is rather commonly reduced to the question of whether or not the community matrix is dissipative (Goh, 1977; Takeuchi, Adachi, and Tokumara, 1978; Pykh, 1983;

Svirezhev and Logofet, 1978). Because of the uniqueness and global stability of the feasible equilibrium state, a dissipative Volterra system can be recognized as a multispecies analogue to the logistic law of single population growth, where the dynamics are stabilized by intraspecific feedback. That is, self-regulation or self-limitation by population density occurs. This seems to be the circumstance by which we can justify the search for and interpretation of sufficient conditions for dissipativeness in terms such that intraspecific regulation predominates over interactions among different species. Such formulations are particularly widespread in studies of competition community models (May, 1973, 1974; Yodzis, 1978; Svirezhev and Logofet, 1978), of "connective" stability of the community matrix (Šiljak, 1974, 1975; Svirezhev and Logofet, 1983), and of stochastic analogues of ordinary differential equations for population dynamics (Ladde and Šiljak, 1975). The respective formal property of the community matrix reduces to the notion of the "positive dominating diagonal" (Nikaido, 1968) or that of "diagonal quasi-dominance" (Quirk and Saposnik, 1968).

Matrix A of size $n \times n$ is called *diagonally quasi-dominant* (formally $A \in q^D$) if there exists a set of positive numbers $\pi_1, \pi_2, \ldots, \pi_n$ such that

$$\pi_i a_{ii} > \sum_{j \neq i} \pi_j |a_{ij}|, \qquad i = 1, \ldots, n. \tag{13.9}$$

It is known (Moylan, 1977) that, if a matrix A is quasi-dominant, then $(-A)$ is dissipative, so quasi-dominant matrices, or strictly speaking, their negations, $-A$, constitute a subclass of dissipative matrices. In this subclass, global stability of the equilibrium is provided by intraspecific regulation predominating over interspecific interactions.

M-matrix as a Special Case of Diagonal Dominance

If, for instance, we consider the case where the contrast between intraspecific regulation and species interaction is expressed in its extreme form, we will have the following sign structure of the community matrix A:

$$S = \begin{bmatrix} - & \oplus & \oplus & \cdots & \oplus \\ \oplus & - & \oplus & \cdots & \oplus \\ \vdots & \vdots & \vdots & & \vdots \\ \oplus & \oplus & \oplus & \cdots & - \end{bmatrix}, \tag{13.10}$$

where the symbol $\oplus$ stands for either $+$ or 0. This structure corresponds to a community of n mutualistic species, each undergoing a density self-regulation. If matrix $(-A)$ is quasi-dominant here, then it represents the *M-matrix* (Nikaido, 1968), which has been intensively studied in mathematical economics (Leontief, 1966).

If the dynamics of a community with structure (13.10) is described by a Volterra system (13.2) or by some of its generalizations (Grossberg, 1977; Pykh, 1983), then, by a criterion for its being an *M*-matrix (Nikaido, 1968), we can show that the existence of a feasible equilibrium $\mathcal{N}^*$ in the case of all $\varepsilon_i > 0$ is equivalent to the fact that matrix Γ [with a (13.10) negative sign structure] is an *M*-matrix. Therefore, the matrix of a corresponding linear approximation to the model is quasi-dominant and hence dissipative. This conclusion implies the global pattern of $\mathcal{N}^*$'s stability, which is also typical for generalized Volterra systems with dissipative community matrices (Pykh, 1983). This general result renders the stability of many particular models of mutualism readily explainable (Kostitzin, 1934; Boucher, 1985).

Quasi-recessive Matrices

As noted above, for a quasi-dominant interaction matrix A the matrix $-A$ is dissipative, so diagram (13.8) can be supplemented by the following relation:

$$\boxed{\text{Quasi-dominance}} \longrightarrow \boxed{\text{Dissipativeness}} \quad . \tag{13.11}$$

To clarify whether the relation is reversible, we consider a class of matrices possessing the property that in a certain sense, is opposite to the diagonal dominance.

Some definitions follow. Matrix A will be called *diagonally recessive in columns (in rows)* if in any of its jth columns (or ith rows) there is an entry a_{ij} such that $|a_{ij}| \geq |a_{ii}|$. Matrix A will be called *diagonally recessive* if it has diagonal recessiveness in its columns or rows. Finally, matrix A will be called *diagonally quasi-recessive* (formally, $A \in qR$) if there exists a diagonal matrix $K = \text{diag }\{K_1, K_2, \ldots, K_n\}$ with all $K_i > 0$ (formally, $K \in D_n^+$) such that matrix AK is diagonally recessive.

Class qR turns out to be closed with respect to transposition and multiplication by any diagonal matrix $D \in D_n^+$, and it can be formally proved that $qR \cap qD = \varnothing$; that is, the classes of diagonally quasi-dominant and quasi-recessive matrices do not intersect.

An example of a diagonally recessive matrix can be given by the matrix (Logofet, 1989)

$$A = \begin{bmatrix} 1 & 7/6 & 1/2 \\ 3/4 & 1 & 3/2 \\ 3/2 & 1/2 & 1 \end{bmatrix}, \tag{13.12}$$

which is recessive both in columns and rows. Note also that all principal minors of (13.12) are positive, and the matrix $-A$ meets the conditions of Theorem 1 on dissipativeness of a 3×3 matrix. Since the recessiveness conditions for matrix (13.12) turn into strict inequalities, there exists a vicinity of matrix A consisting entirely of recessive matrices. Intersection of the corresponding vicinity of $-A$ with a vicinity of $-A$ in the (open) set of dissipative matrices yields an open subset of the matrices that is both dissipative and diagonally recessive.

This example indicates that the set of dissipative matrices is far from being exhausted by the diagonally quasi-dominant matrices alone, so the hierarchy (13.8) is to be complemented by the irreversible relation (13.11). The equivalence of all the properties in hierarchy (13.8) for a normal matrix can no longer be extended to relation (13.11). The ecological implication from this relation is that the global pattern of equilibrium stability, inherent in systems with dissipative community matrices, can be provided not only by strong intraspecific regulation (quasi-dominance), but also, when self-regulation is weak, by means of interspecific relations (diagonal quasi-recessiveness).

Further interpreting example (13.12) as a competition matrix for a three-species community, we may state that a feasible equilibrium will be stable in this community despite the fact that the competition effects of certain species on the others (for example, of the second on the first, of the third on the second, and so on) are stronger than intraspecific competition. While this competition pattern may seem fairly exotic, due to the nonsymmetry of its matrix (13.12), another example of a matrix with the same absolute values of its entries, and hence with the same recessive property, is

$$A = \begin{bmatrix} -1 & -7/6 & -1/2 \\ 3/4 & -1 & -3/2 \\ 3/2 & 1/2 & -1 \end{bmatrix}. \tag{13.13}$$

This matrix can be associated with a community of more natural structure, that is, a three-species, prey–predator chain where species 1 is the prey of both species 2 and 3. A special kind of nonsymmetry in matrix entries can now be attributed to natural biomass-balance relations in prey–predator interactions. Matrix A (13.12) is also dissipative by Theorem 1.

Within the Volterra class of population models, a community similar to (13.13) can be described by the following system of equations:

$$\dot{N}_1 = N_1(\epsilon_1 - \gamma_{11}N_1 - \gamma_{12}N_2 - \gamma_{13}N_3),$$
$$\dot{N}_2 = N_2(\epsilon_2 - \gamma_{21}N_1 - \gamma_{22}N_2 - \gamma_{23}N_3), \qquad (13.14)$$
$$\dot{N}_3 = N_3(\epsilon_3 - \gamma_{31}N_1 - \gamma_{32}N_2 - \gamma_{33}N_3).$$

The interaction coefficients γ_{ij} are given by the matrix

$$\Gamma = -\begin{bmatrix} -1 & -7/6 & -1/2 \\ 3/4 & -1 & -3/2 \\ 3/8 & 1/8 & -1/4 \end{bmatrix}. \qquad (13.15)$$

If $\epsilon = [8/3, 7/4, -1/4]^T$, then the feasible equilibrium has the form

$$\mathbf{N}^* = \Gamma^{-1}\epsilon = [1, 1, 4]^T$$

and, in accordance with formula (13.4),

$$-\mathrm{diag}\{1, 1, 4\}\,\Gamma = A,$$

where A is given in (13.13). Since the system is dissipative, the equilibrium $\mathbf{N}^*$ is globally stable in the interior of the positive orthant, P_+^3, of the phase space for model (13.14).

Note that, in this example, we had to choose a chain that is "paradoxical" (in the sense that $N_3^* > N_2^*$; see Svirezhev and Logofet, 1983, Chapter V) in order to derive the unnatural relation $|a_{31}| > |a_{13}|$, which is apparently related to the recessiveness of matrix (13.13), from the quite feasible inequality $|\gamma_{31}| < |\gamma_{13}|$, of "predation nonsymmetry" in the interaction matrix Γ (13.15).

The belief that the stability of ecosystems should be necessarily provided by self-regulation dominating over interspecific relations (Goh, 1979) has already been called into question. For instance, Jeffries (1979) noted that, even from the standpoint of qualitative (or "sign") stability (see next section), it is quite unnecessary for self-regulation to be at each vertex of the community graph (that is, in each species) and to dominate over other interactions. But, it is sufficient for it to be present at some "critical" vertices only, at some nonzero level of intensity (see the "color text" in the next section). Within the set of dissipative quasi-recessive matrices, however, the stability is provided with no self-regulation dominance at any vertex at all.

From all the above speculations, we may see that whether a particular community matrix falls into a particular set in the hierarchy of stability properties depends on both the qualitative pattern and quantitative relations among matrix entries. During the previous two decades, however, an approach has also been developing that relies only on the qualitative structure of matrices and, hence, of systems with which they are associated. This approach is presented in the next section.

QUALITATIVE STABILITY OF A MODEL COMMUNITY AND SIGN STABILITY OF ITS MATRIX

Definition of Sign Stability

To identify a property from the hierarchy described above for a particular community matrix, we should definitely know the signs and magnitudes of the matrix elements. However, the qualitative structure of intra- and interspecific relations in system (13.1), that is, the pattern of self- and mutual influences among species, can be represented by the sign structure of matrix (13.3) alone,

$$S = [\text{sign } a_{ij}], \quad \text{where sign } a_{ij} = \left\{\begin{array}{c}+\\0\\-\end{array}\right\}, \tag{13.16}$$

or by an associated graph. Therefore, the idea arose logically in theoretical ecology to identify a class of model systems whose stability can be determined only by the pattern of this sign structure or that of its corresponding graph (May, 1973; Levins, 1974). For instance, in a conservative system like (13.2), stability reduces to the existence of a feasible equilibrium $\mathbf{N^*} > \mathbf{0}$, thus being guaranteed (under certain technical assumptions) only by nonsingularity of matrix Γ. The latter turns out to be equivalent to an ability of the trophic graph to be subdivided into several nonoverlapping prey–predator pairs (Yorke and Anderson, 1973; Svirezhev and Logofet, 1978).

In more general cases, the theory of qualitative stability in model ecosystems (May, 1973; Logofet, 1978) and its formalization through the notion of sign stability do not impose any a priori restraints on the system of model equations (13.1), except for the existence of a feasible equilibrium. Then, the sign stability of matrix A (Quirk and Ruppert, 1965) that denotes stability of A together with that of any matrix B of the same sign structure (that is, $[\text{sign } b_{ij}] = [\text{sign } a_{ij}]$) implies that stability is preserved whatever the quantitative variations in a community's intra- and interspecific links, provided that their qualitative structure (that is, the sign pattern) is not changed.

In more formal terms, matrix sign stability can be defined through the notion of their "sign equivalence" (Logofet and Ul'janov, 1982), which turns out to be an equivalence relation in the general sense of set theory (Van der Waerden, 1937), thus generating a partition of the set of all real $n \times n$ matrices into equivalence classes. In the linear space of real $n \times n$ matrices, each of these classes represents a convex (unpointed) cone that, in fact, is a coordinate "half-plane" with a certain part of the n^2 matrix elements being zero and the rest having definite signs.

Matrix A is called *sign stable* if any matrix B is stable when it is sign equivalent to A.

Signed Directed Graphs

Sign stability is clearly a property of the whole equivalence class of matrices that can be well represented by its sign structure S of the form (3.16) or by the *signed directed graph* (SDG), in which there is a directed link from a vertex j to vertex i when $a_{ij} \neq 0$, the link being assigned the sign of a_{ij}, that is, the sign of species j's influence on species i. For example, the SDG that is shown in Fig. 13.1a is associated with matrix

$$A_1 = \begin{bmatrix} -a & b & c \\ 0 & 0 & -d \\ 0 & e & 0 \end{bmatrix},$$

and all the matrices equivalent to

$$A_2 = \begin{bmatrix} 0 & -1 & 0 & 0 & 0 \\ 1 & 0 & -1 & 0 & 0 \\ 0 & 1 & -1 & -1 & 0 \\ 0 & 0 & 1 & 0 & -1 \\ 0 & 0 & 0 & 1 & 0 \end{bmatrix}$$

are represented by the SDG shown in Fig. 13.1b.

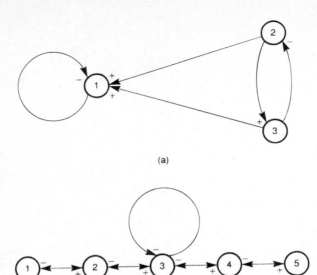

Figure 13.1 Examples of signed, directed graphs. (a) Self-limited species 1 is connected by commensal links with the prey (2)–predator (3) pair. (b) Species 5 feeds on species 4, feeds on species 3, and so on; species 3 is self-limited.

Necessary Conditions for Sign Stability

Relevant conditions for sign stability of a matrix were first obtained in mathematical economics (Quirk and Ruppert, 1965) and then reformulated in ecological terms (May, 1973; Logofet, 1978). However, the problem of giving a complete description of the set of sign-stable patterns still remained unsolved since no conditions were found that would be both necessary and sufficient (that is, a criterion) for sign stability of an arbitrary matrix. Subsequently, the sufficient conditions (Jeffries, 1974) and later the necessary and sufficient ones (Jeffries, Klee, and Van den Driessche, 1977; Logofet and Ulianov, 1982; Logofet and Ul'janov, 1982) were obtained, and this made it possible to describe the complete set of sign-stable patterns in terms of pairwise species interaction types (Logofet and Ulianov, 1982; Svirezhev and Logofet, 1983).

According to Quirk and Ruppert (1965), sign-stability conditions for an $n \times n$ matrix $A = [a_{ij}]$ are the following:

(i) products $a_{ij}a_{ji} \leq 0$ for any pair of indexes $i \neq j$;
(ii) $a_{ii} \leq 0$ for any i, and $a_{kk} < 0$ for some k;
(iii) for any set of three or more distinct indexes $i_1 \neq i_2 \neq \ldots \neq i_m$ ($m \geq 3$), the product $a_{i_1 i_2} a_{i_2 i_3} \ldots a_{i_m i_1} = 0$; and
(iv) there exists a nonzero term in the expansion of det A.

According to condition (i), a qualitatively stable community should contain no relations of competition ($--$) or of mutualism ($++$). Condition (ii) states that there must be no self-increasing species and at least one species must be self-limiting. Condition (iii) denotes the absence of closed directed loops (cycles) having more than two links in the corresponding SDG. Finally, condition (iv) is equivalent to the existence of a partition of the SDG into a number of cycles of length 2 or 1 such that their total length equals n (Logofet, 1978; Logofet and Ulianov, 1982).

Conditions (i) and (ii) were proved to be necessary by the Rouse–Hurvitz criterion combined with the admissibility of any large variations in magnitudes of nonzero a_{ij}'s (Quirk and Ruppert,

1965; Svirezhev and Logofet, 1978). The proof of the necessity of condition (iii) relies on the above ideas of sign-equivalence classes of matrices (Logofet and Ulianov, 1982). Condition (iv) is clearly necessary.

Note that both SDGs in Fig. 13.1 satisfy all the conditions (i) to (iv). However, neither matrix A_1 nor A_2 is stable, each having a pair of purely imaginary eigenvalues (Jeffries, 1974; Logofet, 1978). Thus, the set of conditions (i) to (iv) is only necessary but insufficient for sign stability.

Sign-stability Criterion for a Nondecomposable Matrix

It can be proved (Quirk and Ruppert, 1965; Svirezhev and Logofet, 1978) that if a matrix A is nondecomposable (that is, its graph is strongly connected in the sense that there exists a directed path of finite length between any pair of vertices) and meets condition (iii) pertaining to the absence of cycles, then A has a symmetric 0 pattern. A matrix A is said to have a *symmetric 0 pattern* if, for any pair of indexes $i \neq j$, the condition $a_{ij} = 0$ implies $a_{ji} = 0$. It follows that nondecomposable matrices that satisfy the necessary conditions of sign stability are those strongly connected structures that consist of links of prey–predator type $(+ -)$ only, the *predation matrices* or *graphs* (Jeffries, 1974). Since the eigenvalue set of a decomposable matrix A is the union of the eigenvalue sets for all its indecomposable diagonal blocks, the verification of sign-stability conditions for an arbitrary $n \times n$ matrix is to be reduced to the examination of whether each predation subgraph in the corresponding SDG satisfies conditions (i) to (iv). To unify the terminology, a connected component consisting of one vertex may be termed a *trivial predation subgraph*.

In such an SDG, all the predation subgraphs can be obtained by merely canceling all the amensal $(0-)$ and commensal $(0+)$ links. For example, the SDG in Fig. 13.1a contains two predation subgraphs, $\{1\}$ and $\{2–3\}$, while the SDG in Fig. 13.1b consists entirely of a single predation graph.

The formulation of the sign-stability criterion for a predation submatrix is concerned with the color test (Jeffries, 1974; Logofet, 1978). A predation graph is said to *pass the color test* if each of its vertices may be colored black or white such that:

(a) each self-limiting vertex is black, ;

(b) there exist white vertices and each white vertex is connected by (a pair of) predation links to at least one other white vertex, ○——○; and
(c) each black vertex connected by a predation link to one white vertex is connected to at least one other white vertex, ○——●——○.

For a predation matrix to be sign stable, it is necessary and sufficient that conditions (i), (iii), and (iv) be satisfied as well as condition:

(ii)′ the predation graph must fail the color test.

The scheme for the sufficiency proof was proposed by Jeffries (1974) and developed in detail by Svirezhev and Logofet (1978), while the necessity of all the same conditions was proved by Logofet and Ul'janov (Logofet and Ul'janov, 1982; Logofet and Ulianov, 1982).

From this criterion it becomes clear why, for example, matrices A_1 and A_2 associated with the SDGs in Fig. 13.1 are not sign stable. While the predation subgraph $\{1\}$ in Fig. 13.1a fails the color test, the subgraph $\{2–3\}$ passes it. The predation graph in Fig. 13.1b also passes the color test, but if the self-regulation loop is transferred into any other vertex, the test will fail.

When substituted for condition (ii), condition (ii)' indicates that not only the presence of self-limiting species but also their special allocation within community structures is of critical importance for sign stability.

Sign-stability Criterion for an Arbitrary $n \times n$ Matrix

According to all the above results, for an arbitrary $n \times n$ matrix we have the following:

Theorem 3 (Logofet and Ul'janov, 1982). An $n \times n$ matrix A is sign stable if and only if conditions (i), (ii)', (iii), and (iv) hold for any predation subgraph in the SDG of matrix A.

When interpreting this theorem in ecological terms, we obtain a complete description of the class of sign-stable patterns (Logofet and Ulianov, 1982). These are (1) either predation patterns, that is, the patterns of only a prey–predator relation type (resource–consumer, host–parasite, and so on) with a special location of self-limiting species (so that the color test is violated), or (2) any combinations of the predation communities described above that are linked one to another by amensal or commensal links forming no direct cycles in the total community graph.

As a consequence of Theorem 3, the following theorem on sign stability of a matrix with all negative diagonal entries can now be easily obtained, whereas formerly it required some special considerations (Quirk and Ruppert, 1965).

Theorem 4. An $n \times n$ matrix A with all $a_{ii} < 0$ is sign stable if and only if conditions (ii) and (iii) hold. In other words, there must be no competition or mutualism relations in the community, and the SDG must contain no cycles longer than 2 (Svirezhev and Logofet, 1983).

Sign Stability in the Hierarchy of Stability Subsets of Matrices

By means of Theorem 4, a relation can be established between sign stability and other previously considered properties. The following theorem is true (Logofet, 1986).

Theorem 5. A sign-stable matrix $A = [a_{ij}]$ with all $a_{ii} < 0$ is dissipative.

The sign-stability criterion thus provides further development of the hierarchy (13.8) and (13.11) of the matrix-stability subsets up to the diagram shown in Fig. 13.2. Here, the sets that are open in the linear space of $n \times n$ matrices are conventionally depicted as curvilinear contours, while those that are not open are shown as rectangular ones. The upper vertical half-plane corresponds to matrices with a negative principal diagonal, the horizontal half-plane to those with a nonpositive diagonal (and a nonzero trace), and the lower vertical half-plane to matrices having some positive entries in the principal diagonal.

Application of the Sign-stability Criterion

Note that the class of sign-stable communities has appeared to be rather narrow. In particular, those models cannot enter it that take into account a closed turnover of nutrients or in which a predator feeds on more than one trophic level, situations that are fairly common in ecological modeling.

However, the lack of sign stability does not yet mean that the system cannot be stable in principle. The sign stability just requires the maximum possible region of stability in the parameter space of the model. When violated, sign stability indicates that the stable dynamic behavior of a

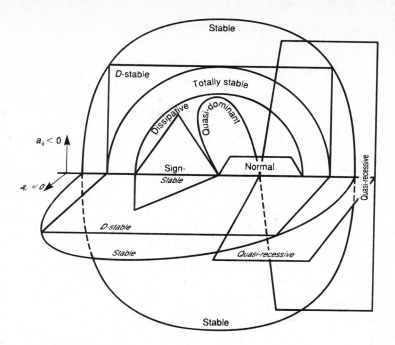

Figure 13.2 Relations among stability subsets of $n \times n$ matrices, $n \geq 3$.

system, subject to variations in the magnitude of intra- and interspecific links, is more vulnerable to perturbations. Thus, the sign-stability criterion represents a convenient tool for preliminary, qualitative analysis of the trophic and other structures in an ecosystem from the viewpoint of stability in a corresponding dynamic model because, to be applied, it requires nothing but a depiction of the sign pattern by its SDG. In particular, the criterion demonstrates that the presence of species with density self-regulation and their special location within the community structure are of principal importance for stability. In some cases, the criterion would be able to indicate those links whose presence or absence could affect the stability of the whole system.

As an example of its application, we may cite the analysis of structures that were aggregated (to different degrees) from a complete scheme of biological interactions known for a cotton agrobiocoenosis (Logofet and Junusov, 1979). This example has enabled us to explain, from the population-dynamics point of view, a number of empirically observed phenomena. These include a stabilizing effect of competition among insect pests, a destabilizing effect of the cotton-wilt disease agents, and a destabilizing effect of competition between parasites and predators of pests. The latter must be especially accounted for in planning particular programs of biological pest control.

Another example of application (Logofet and Svirezhev, 1984) concerns a case study on the development of a simulation dynamic model for a desert community in Tigrovaya Balka Reserve, Tajikestan SSR (Usmanov et al., 1981). Although providing only some heuristic considerations rather than a rigorous method, the sign-stability theory prompted a way of aggregating some ecosystem components by combining some vertices of the graph that occupy similar trophic positions and generate closed loops into one vertex. Excluding such loops and hence diminishing the number of points where the sign-stability criterion is violated in a particular graph, we were able to indicate those components and links that were of principal importance for model stability and needed special care in monitoring and modeling.

STABILITY OF TROPHIC CHAINS WITH A PERMANENT INFLOW OF EXTERNAL RESOURCE

Equations for Simple Trophic Chains

In the previous sections, no a priori restrictions were imposed on the structure of systems under consideration, but rather the restrictions resulted from a corresponding analysis. At the same time, when the question concerns a biological community or ecosystem, at least two types of characteristic structures can be distinguished: vertically structured and horizontally structured communities (Svirezhev and Logofet, 1978). By a *horizontal structure* we mean interactions of species within a single trophic level; *vertical structure* implies interactions between trophic levels only.

Any real community is described by a fairly complicated web of trophic interactions. However, by means of aggregation into trophic levels or by distinguishing the dominant paths of energy transfer, the structure can often be represented by a simple trophic chain by which biomass is transferred from one trophic level to the next (Svirezhev and Logofet, 1978). So natural and simple, these structures have been the subject of some studies in the framework of mathematical models (Ehman, 1966; Goel, Maitra, and Montroll, 1971). Although Volterra models with an "inert component" of a biogeocoenosis showed a possibility for any finite number of trophic levels to exist, this possibility appears to be in contradiction with ecological reality. The formalism given next is founded on the notion of a generalized "resource" R inflowing to the system and feeding the first trophic level at a certain rate Q. In a real ecosystem, this may be either energy or some chemical elements of vital importance (for example, carbon, nitrogen, or phosphorus). Depending on the particular meaning of R, a trophic chain may be open (acyclic) or partially closed by a resource cycle.

With certain simplifying assumptions, natural considerations of matter and energy balances lead to the following systems of equations for biomass dynamics, N_i, designating the biomass of the ith trophic level ($N_0 = R$), or of the ith "species" to preserve the terminology (Svirezhev and Logofet, 1978):

(a) *Open chain*

$$\frac{dR}{dt} = Q - V_0(R)N_1 ,$$

$$\frac{dN_i}{dt} = -m_i N_i + k_i V_{i-1}(N_{i-1})N_i - V_i(N_i)N_{i+1} , \qquad i = 1, 2, \cdots, n-1 , \qquad (13.17)$$

$$\frac{dN_n}{dt} = -m_n N_n + k_n V_{n-1}(N_{n-1})N_n .$$

(b) *Partially closed chain*

$$\frac{dR}{dt} = Q - V_0(R)N_1 + \sum_{i=1}^{n} a_i m_i N_i ,$$

$$\frac{dN_i}{dt} = -m_i N_i + k_i V_{i-1}(N_{i-1})N_i - V_i(N_i)N_{i+1} , \qquad i = 1, 2, \cdots, n-1 , \qquad (13.18)$$

$$\frac{dN_n}{dt} = -m_n N_n + k_n V_{n-1}(N_{n-1})N_n .$$

Here, $V_i(N_i)$ denotes per capita rate of uptake of the ith species biomass by the $(i + 1)$th species; k_{i+1} is the efficiency of processing the biomass of the ith species into the $(i + 1)$th species' biomass ($0 < k_{i+1} < 1$); m_i is mortality at the ith trophic level; a_i is that portion of dead biomass of the ith species that is returned into the "resource" level due to the activity of decomposers ($0 \leq a_i < 1$).

Trophic Chain of Length q

If system (13.17) or (13.18) has a stable equilibrium state in which only the first q components are nonzero, then this state is naturally identified with a *trophic chain of length q*, and the question arises as to what should be the rate (Q) of external resource inflow in order that just this equilibrium be realized in the system [given the trophic functions $V_i(N_i)$ and the parameters k_i, m_i, and a_i]. In other words, what are the conditions for a stable trophic chain of length q to exist?

When neither of the species has an abundance of its trophic resource, the Volterra description of trophic functions is pertinent: $V_i(N_i) = \alpha_i N_i$ ($i = 0, 1, \ldots, n$), the equations (13.17) and (13.18) reducing into equations of an "almost Volterra" type. In these equations, the expressions for biomass values at the equilibrium, $\mathbf{N}_q^* = [N_0^*, N_1^*, \ldots, N_q^*, 0, \ldots, 0]$, can be obtained in explicit form. Moreover, it can be proved that if $\mathbf{N}_q^* > \mathbf{0}$, then all the preceding $N_i^* > 0$ ($i = 0, 1, \ldots, q-1$), so the notion of trophic-chain length introduced above proves to be correct (Svirezhev and Logofet, 1978).

The feasibility condition $N_q^* > 0$ brings about a restriction on Q from below,

$$Q > Q^*(q) \tag{13.19}$$

in the case of an open chain, and a restriction

$$Q > \tilde{Q}^*(q) \tag{13.20}$$

in the case of a partially closed chain, where $Q^*(q)$ and $\tilde{Q}^*(q)$ are certain functions of model parameters (see Appendix A). It can be shown also that

$$Q^*(q) > \tilde{Q}^*(q) ; \tag{13.21}$$

that is, in a partially closed chain the rate Q may be less than that in an open chain of the same length.

The stability problem for equilibria of the form $\mathbf{N}_q^* = [N_0^*, N_1^*, \ldots, N_q^*, 0, \ldots, 0]$ reduces to that of stability conditions for a matrix of system (13.17) or (13.18) when linearized at this equilibrium point. The sign-stability criterion appears useful in searching for these conditions. When system (3.17) is linearized at the point $\mathbf{N}_q^*$, its matrix has, in fact, the form

$$F = \begin{bmatrix} A_q & 0 \\ 0 & D_{n-q} \end{bmatrix}, \tag{13.22}$$

where A_q is a Jacobian $(q + 1) \times (q + 1)$ matrix and D_{n-q} is a diagonal $(n - q) \times (n - q)$ matrix (see Appendix B). If D_{n-q} is stable, then the stability of F (13.22) depends on the properties of matrix A_q alone. From the corresponding SDG in Fig. 13.3 it can readily be seen that all the conditions (i) to (iv) of the sign-stability criterion are satisfied for matrix A_q, the color test failing under any actual set of negative or zero diagonal entries $-h_i$, $i = 1, 2, \ldots, q$. Hence, A_q is stable under any feasible values of model parameters, whereas the stability condition for D_{n-q} yields a restriction on Q from above:

$$Q < Q^*(q + 1) \tag{13.23}$$

(see Appendix B). From (13.19) and (13.23), it follows that an open chain of length q exists steadily if and only if the rate of external resource inflow is bounded from above and below,

$$Q^*(q) < Q < Q^*(q + 1), \tag{13.24}$$

by certain functions of model parameters specified in Appendix A.

Figure 13.3 SDG associated with matrix $\tilde{A}_q$. Broken arrows correspond to $h_i > 0$; they vanish when $h_i = 0$.

Stability in a Partially Closed Chain

When the chain is partially closed [system (13.18)], its "closing" terms $C_i = a_i m_i$ ($i = 1, \ldots, n$) modify the linearization matrix, similar to (13.22), into the form

$$\tilde{F} = \begin{bmatrix} \tilde{A}_q & C_{(q+1) \times (n-q)} \\ 0 & D_{n-q} \end{bmatrix}. \tag{13.25}$$

Although the sign-stability criterion is no longer satisfied with the submatrix $\tilde{A}_q$, other methods (see Appendix C) can provide the stability conditions needed. In the important case for which closure is realized only at the primary producer level, a necessary and sufficient condition for the chain of length q to exist appears to be similar to restriction (13.24):

$$\tilde{Q}^*(q) < Q < \tilde{Q}^*(q+1), \tag{13.26}$$

where function $\tilde{Q}^*(q)$ is specified in Appendix C. In more general cases, conditions (13.26) appear to be only necessary but insufficient for the stable existence of a partially closed trophic chain.

The main conclusion, however, retains its general sense that the whole range of possible rates of external resource inflow is divided by the points $Q^*(1), Q^*(2), \ldots, Q^*(q), \ldots$ [respectively, by $\tilde{Q}^*(q)$, $q = 1, 2, \ldots$, for a partially closed chain] into consecutive intervals, each interval featuring the stable existence of a trophic chain of a definite length only. In other words, energy inflowing to the system acquires a certain quantification in which, as follows from the analysis of sequences $\{Q^*(q)\}$ and $\{\tilde{Q}^*(q)\}$ (Svirezhev and Logofet, 1978), the higher the number of a next trophic level, the larger is the "quantum" of energy that must enter to provide fixation of this level in the ecosystem.

Some Generalizations

The main conclusions of this theory still prevail if we reject the Volterra description of trophic functions $V_i(N_i)$, the analogy being more complete for an open trophic chain (Svirezhev and Logofet, 1978). Interesting effects occur when the trophic functions have a saturation level. With trophic functions of Holling's type I form (Holling, 1965), a trophic chain is always unstable; with types II and III, a chain loses its stability only when equilibrium values N_i^* are sufficiently large (that is, when the trophic levels are fairly abundant). The faster the saturation occurs, the lower are the values of N_i^* that cause degradation of the chain, the stability disappearing when saturation occurs in at least one trophic level. Competition within a trophic level stabilizes the system and promotes conservation of the trophic chain.

Application Examples

A practical verification of the mathematical theory of trophic-chain stability was based on Odum's (1957) data on yields and energy flows in the five-level ecosystem of Silver Springs in Florida, USA. The actual rate of energy inflow to this ecosystem was found to fall in the interval $Q^*(4) < Q < Q^*(5)$ (Svirezhev and Logofet, 1978). That is, the primary production rate and other ecosystem parameters were such that they provided for the theoretical existence of only four trophic levels, while the fifth level was not able to become realized in this ecosystem. The empirical data confirmed the conclusion completely.

The mathematical theory of trophic chains also gives a method for estimating the energy inflow rate that is necessary for the fifth level to exist, as well as a method for determining the percentage decrease in the production rate of the first level that would cause loss of stability, and hence elimination of the highest trophic levels. An approach is also given to model investigation of other issues in the theory of trophic structures. In particular, these results permitted a theoretical estimation of the minimal protected area necessary for persistence of a predator at the top of a trophic chain (Logofet and Svirezhev, 1984). A decrease in the area below this estimation would result in loss of stability at the highest trophic level and its elimination from the community, rather than in a corresponding decrease in biomass of each trophic level. Principal nonlinearity of ecological interactions takes place here.

As noted by Pimm and Kitching (1987) in their study of water-filled cavities in trees, the variability in the yearly energy input (leaf fall) is a critical factor in shortening the food chains, which appear to have different numbers of trophic levels in different types of forests with similar (average) values of energy input. These observations are explainable by our theory since a variation in the Q flow can easily shift it into the interval where a shorter chain steadily exists. As the interval bounds are themselves determined by the ecological parameters of a chain under consideration, the chains in different forests may well have different numbers of trophic levels under the same value of energy input Q. Yet the theory certainly needs further development, particularly for the case of variable rather than constant values of Q.

ECOLOGICAL NICHE OVERLAP AND STABILITY OF HORIZONTALLY STRUCTURED COMMUNITIES

Competitive Exclusion Principle and Dynamic Equations for a Competition Community

This section is devoted to stability in models of interactions within a single trophic level or, in other words, of *horizontally structured communities*. The most meaningful findings in this field were made for communities of several species competing for common resources. These models do not explicitly consider the dynamics of the resources, being based on the concepts of ecological niche and niche overlap of competing species (MacArthur, 1968; May and MacArthur, 1972). Experimentally established by Gause (1934), the competitive exclusion principle, according to which two species with similar ecological requirements cannot coexist in a single habitat, seems to reveal a theoretical generalization of a certain limit to niche overlap in a stable community (Svirezhev and Logofet, 1978). But, in the deterministic systems of equations for the dynamics of a competition community (May and MacArthur, 1972; May, 1974),

$$\frac{dN_i}{dt} = r_i N_i - \frac{r_i}{K_i} \sum_{j=1}^{n} \alpha_{ij} N_i N_j, \quad i = 1, \ldots, n, \qquad (13.27)$$

which were deduced from considerations of niche overlap (r_i denotes an intrinsic rate of natural increase, K_i a carrying capacity for the ith species, and α_{ij} a measure of overlap between the ith and jth niches), the authors failed to show any limitations on niche overlap (expressed as density of "species packing"). To achieve this, they had to introduce stochastic versions of eqs. (13.27).

Dissipativeness of a Competition Matrix

If the competition coefficients

$$\alpha_{ij} = \int f_i(x) f_j(x)\, dx, \qquad (13.28)$$

where f_i and f_j are "utilization functions" of ith and jth species defined on the "resource spectrum," and the integration is over the whole spectrum (May and MacArthur, 1972), and if there exists in system (13.27) an equilibrium state $\mathbf{N}^* > \mathbf{0}$, then the corresponding community matrix is dissipative in the sense previously described (Svirezhev and Logofet, 1978). It follows that $\mathbf{N}^*$ is globally stable in P_+^n and Liapunov stability is equivalent to ecological stability in system (13.27). In other words, stable coexistence in system (13.27) reduces to the existence of a positive solution to the linear system

$$A\mathbf{N}^* = \mathbf{K}, \qquad \mathbf{K} = [K_1, \ldots, K_n]^T > \mathbf{0}. \qquad (13.29)$$

Stability Measure for Competition Structures

It has been shown (Svirezhev and Logofet, 1983) by a geometric interpretation for positiveness of the solution to (13.29) that it is just at this point that limitations on niche overlap appear (even in the deterministic case) to provide stable coexistence in model (13.27).

If A is a nonsingular, non-negative $n \times n$ matrix, then a linear transformation defined by A contracts the orthant P^n onto an n-hedral angle AP^n with generatrices $A\mathbf{e}_j$ ($\mathbf{e}_j = [0, \ldots, 0; 1, 0, \ldots, 0]^T$, $j = 1, \ldots, n$. The existence of a positive solution to system (13.29) is then equivalent to the fact that vector $\mathbf{K}$ belongs to the interior of AP^n. As an evident measure of how many such vectors $\mathbf{K}$ there are in AP^n, we may propose the ratios of n volumes of the cones AP^n and P^n or of $(n-1)$ volumes of their sections to, for example, the unit sphere S^n (Fig. 13.4a), or the standard simplex Σ^n (Fig. 13.4b). In the latter case, this measure can be calculated fairly easily (Svirezhev and Logofet, 1978):

$$\mu_\Sigma(A) = \frac{\sigma^n(A)}{\Sigma^n} = \frac{|\det(A)|}{\|A\mathbf{e}_1\|_\Sigma \cdots \|A\mathbf{e}_n\|_\Sigma}. \qquad (13.30)$$

It varies within the range $0 \leq \mu_\Sigma \leq 1$ and may serve as a measure of the stability of a competition

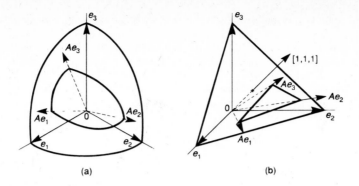

Figure 13.4 The definition of an "equilibrium" measure for a competition matrix A ($n = 3$). (a) In the Euclidean metric; (b) for the norm $\|x\|_\Sigma = |x_1| + |x_2| + |x_3|$ [the direction of $\mathbf{K} = [1, 1, 1]^T$ goes beyond the simplex $\sigma^3(A)$].

structure given by a dissipative matrix A of competition coefficients. It now becomes clear how the equilibrium $\mathbf{N}^* > \mathbf{0}$ can disappear in system (13.27) with a fixed set of K_i values, a phenomenon that has been observed in some models (Svirezhev and Logofet, 1978).

For example, if the utilization functions f_i are given as the densities of normal distributions with the same variance w and with equal distances d between centers of neighboring niches (May and MacArthur, 1972), then the competition coefficients (13.28) have the form

$$\alpha_{ij} = a^{|i-j|^2}, \qquad a = \exp\left(\frac{-d^2}{4w^2}\right). \tag{13.31}$$

When species packing becomes more dense, that is, when $a \to 1$, the simplex $\sigma^n(A)$ contracts so much that the vector $\mathbf{K}$ falls outside its contour (Fig. 13.4b) and, as a result, system (13.27) loses its stability and competitive exclusion of some species takes place.

Partly Positive Equilibria

Similar considerations can indicate which subsets of an existing set of n species constitute a stable composition in a given environment (with a given matrix A and vector $\mathbf{K}$). The problem reduces to the existence of a partly positive equilibrium state $\mathbf{N}^\circ \geq \mathbf{0}$, where only some of n components are positive and the rest are zero (Svirezhev and Logofet, 1978). The positive coordinates must constitute a solution to the "truncated" system of equations,

$$A'\mathbf{N}' = \mathbf{K}', \qquad \mathbf{K}' = [K_{i_1}, \ldots, K_{i_{n-m}}]^T, \tag{13.32}$$

where matrix A' is derived from matrix A by canceling certain m of its rows and the corresponding columns ($m = 1, \ldots, n - 1$). This means that the linear operator A is contracted to the subspace R^{n-m}, into which the vector $\mathbf{K}$ is also projected. In general, projection of the n-hedral angle AP^n is wider than the image of the positive orthant P^{n-m} with the transformation A': $P_m(AP^n) \supset A'P^{n-m}$. But for a positive solution to (13.32) to exist, it is necessary and sufficient that the projection of $\mathbf{K}$ fall within the $(n - m)$-hedral angle:

$$\mathbf{K}' = p_m(\mathbf{K}) \in A'P^{n-m}. \tag{13.33}$$

In different examples, both the validity and violation of condition (13.33) turned out to be possible (Svirezhev and Logofet, 1978). In particular, a situation was observed where the composition that resulted from elimination of one species was not stable, but degraded further by losing some other species as well, until a positive solution to the truncated system (13.32) developed. This means that, in such a model, invasion or other successful introduction of the species that were lost cannot proceed sequentially—those species can become established in the community only in parallel.

Another Representation of Competition Coefficients

Since, on the one hand, an accurate determination of the utilization functions $f_i(x)$ is difficult to realize in practice and on the other, different methods give estimates of niche overlap that are far from being equal (Harner and Whitmore, 1977; Abrams, 1980; Petraitis, 1981), it is interesting to study dynamic models with competition coefficients of a more general form than (13.28) or (13.31). Such a study was undertaken (Logofet, 1975; Logofet and Svirezhev, 1977) under the assumption that a measure of overlap between the ith and jth niches is a decreasing function, $\alpha(z)$, of a generalized "distance" between these niches in the resource space:

$$\alpha_{ij} = \alpha(|i - j|), \quad \alpha(0) = 1, \qquad (13.34)$$

the "distance" being represented by the difference between species' numbers i and j.

Equation (13.34) still gives a symmetric matrix of competition coefficients, $A = [\alpha_{ij}]$, and gives rise to certain properties of the model (13.27), but these constraints are still insufficient to provide the existence and, at least, local stability of the equilibrium $\mathbf{N}^* > \mathbf{0}$ (since matrix A is normal, the latter also means global stability, as previously developed). Sufficient conditions for stability are established by the following theorem:

Theorem 6 (Logofet, 1975). If a monotone, decreasing function $\alpha(z)$ of an integer variable z meets the condition of strict convexity,

$$\alpha(z) < \frac{1}{2}[\alpha(z - 1) + \alpha(z + 1)], \qquad (13.35)$$

then matrix $A = [\alpha_{ij}]$ with entries (13.34) is positive definite. Thus, when the competition coefficients (13.34) decrease in a convex manner, that is, the decrease proceeds sufficiently smoothly, matrix A is positive definite and the community is dissipative.

In particular, it was just this pattern of decrease that was revealed in some particular cases of competition matrices considered by May (1974) and Svirezhev and Logofet (1978), as well as in matrix (13.31), with values of a sufficiently far from 1. Positive definiteness of all these matrices illustrates Theorem 6. Note that the above-mentioned disappearance of the positive equilibrium state in the system with a competition matrix (13.31) with values of a close to 1 is also accompanied by violation of the convexity conditions (13.35).

A general inference from all these results is that the limits for niche overlap in a stable competition community should not only be determined by the stability conditions of the competition matrix, but should also be an implication from conditions for the existence of positive or partly positive equilibria. Here, dependence of the limits to niche overlap on values of environmental carrying capacities for n competing species naturally arises. In the dissipative case, it is the positiveness, or feasibility conditions, that determine the stable-coexistence combinations of species from the original set of n species, thus imposing theoretical conditions for the successful introduction or invasion of species into a community with an incomplete set of competitors.

CONCLUSIONS

The problem of formalizing stability properties, which are intuitively clear for the systems of mathematical ecology (biological populations, communities, ecosystems and the like), reduces to the elaboration of formal analogues to the notion of "ecological" stability in the framework of mathematical models for the dynamics of multicomponent systems. As we have seen, whereas the notion of Lagrangian stability is somewhat nearer to ecological ideas of stability than the concept of Liapunov stability of a model equilibrium, the latter admits the development of a rather orderly theory of stability for the main structural types of model communities. Briefly, the main features of this theory are the following:

1. A strict hierarchy of properties of a stable community $n \times n$ matrix was established: D-stability $\Leftarrow$ total stability $\Leftarrow$ diagonal stability (dissipativeness) $\Leftarrow$ diagonal quasidominance. A new class of matrices, quasi-recessive matrices, has been defined and studied, and shown in principle that community dynamics can be stabilized not only by intraspecific self-regulation, but also as a result of prevailing interactions among different species.
2. A constructive theorem on the criterion of sign stability of a matrix has been proved. The criterion is formulated in terms of a signed directed graph (SDG) representing a qualitative structure of intra- and interspecific relations in the community. Only those structures satisfy the criterion that are either predation $(+-)$ communities, that is, have only prey–predator relations with special locations for self-limiting species, or the unions of predation communities linked one to another by amensal $(0-)$ or commensal $(0+)$ relations forming no direct cycles in the total community. The position of sign-stable matrices was determined within the hierarchy of stable community matrix properties, and a picture obtained for the topology of $n \times n$ matrix stability subsets. The sign-stability criterion appears to be a convenient tool for preliminary qualitative analysis of a multispecific assemblage from the standpoint of stability in its dynamic model.
3. On the basis of the sign-stability criterion and feasibility conditions, a theory has been developed for stable communities with a vertical structure, for example, trophic chains of a definite length and either open or partially closed to a resource component. Given the mortality and trophic function parameters in all the trophic levels, the stable number of trophic levels (the length of the chain) is strongly determined by the rate of external resource inflow to the ecosystem.
4. In models of a community of n competing species, the limits to ecological niche overlaps that should exist according to the competitive exclusion principle arise from the feasibility and stability conditions. In particular, a convex pattern of decrease in competition coefficients with increasing distance between species in the resource space turns out to be sufficient in many cases. A relative portion of the surface that is cut off the standard simplex in P^n by a cone of positive directions of the community matrix is a convenient quantitative measure of community stability, which admits comparison of different competition patterns. Given a structure of the resource space and ecological niches, and given the rates of natural increase of competing species, a stable species composition and the conditions for successful introduction of new species can be determined from the analysis of feasibility and stability conditions for the corresponding equilibria.

Such problems as the analysis of qualitative structure in species interactions in biocoenoses, estimation of external resource inflow rates necessary to maintain trophic chains of certain lengths and forecast the stable length of chains under given rates of resource inflow, estimation of minimal areas necessary for protected species at the tops of trophic chains, and others can be mentioned as examples of potential practical applications of ecological stability theory.

The above-described results from the development of the Liapunov concept in systems ecology are concentrated about "lumped" models of multispecific communities, models that consider neither the age nor spatial distribution of a population. Thus, they certainly do not exhaust the variety of possible applications of even the Liapunov concept alone. Stability issues in models of age-specific dynamics, and questions of the stabilizing or destabilizing effects of migration and random diffusion processes in models of spatially distributed communities remain beyond the scope of this chapter. Some results concerning these problems can be found in corresponding chapters of the books by Svirezhev and Logofet (1983) and Logofet (1992).

It is clear that stability concepts, other than those of Liapunov or Lagrange, which are so numerous in mathematics, can also find application in systems ecology. When, however, we are concerned with population or community models, not all the mathematical stability concepts are appropriate. For instance, if a feasible equilibrium $\mathbf{N}^*$ should be stable in the mean-square sense, that is, in the sense that the integral of all deviations

$$\int_{t=t_0}^{T} \| \mathbf{N}(t) - \mathbf{N}^* \| \, dt$$

is small enough, then some (or even all) of $N_i(t)$ could be close (or even equal) to zero within sufficiently short intervals of time. Actually, this should be interpreted as the extinction of some (or all) of the populations and could hardly correspond to any realistic ecological concept of stability.

Liapunov stability, as the most mathematically developed concept, gives scope for the broad application of matrix and graph theories to ecological stability analysis, the limits to applicability in any particular case being determined by the real "ecological" content of the problem under study. At the same time, investigation of community-matrix stability is far from being the only appropriate direction for matrix theory applications in mathematical systems ecology. Another important pathway lies in the study of compartment models and balance schemes or, in more general terms, in the analysis of ecological networks. To find out the points of contiguity and fundamental relations among these several areas is the task of future theoretical research in complex ecology.

APPENDIX A: FEASIBILITY CONDITIONS FOR A CHAIN OF LENGTH Q

If we define the following quantities,

$$H_{2s-1} = g_1 g_3 \ldots g_{2s-1}, \qquad H_{2s} = g_2 g_4 \ldots g_{2s}, \qquad g_i = k_i \alpha_{i-1}/\alpha_i \, ;$$

$$f_{2s-1} = \frac{\mu_1}{H_1} + \frac{\mu_3}{H_3} + \ldots + \frac{\mu_{2s-1}}{H_{2s-1}}, \qquad f_{2s} = \frac{\mu_2}{H_2} + \frac{\mu_4}{H_4} + \ldots + \frac{\mu_{2s}}{H_{2s}}, \qquad (A.1)$$

$$s = 1, 2, \ldots; \qquad \mu_i = m_i/\alpha_i, \qquad i = 1, \ldots, n,$$

where parameters m_i and k_i are taken from the model equations (13.17) or (13.18), and α_i from the Volterra form of trophic functions, $V_i(N_i) = \alpha_i N_i$ ($i = 0, 1, \ldots, n$), then the condition $N_q^* > 0$ can be algebraically shown to result in the following restriction on Q:

$$Q > Q^*(q) = \alpha_0 f_{q-1} f_q \, . \tag{A.2}$$

For a partially closed system (13.18), the feasibility condition will have the form

$$Q > \tilde{Q}^*(q) = Q^*(q) - \begin{cases} [(f_{2s-1}\varphi_s + f_{2s}\psi_s) - \sigma_{2s}], & \text{if } q = 2S, \\ [(f_{2s}\psi_{s+1} + f_{2s+1}\varphi_s) - \sigma_{2s+1}], & \text{if } q = 2S + 1, \end{cases} \tag{A.3}$$

where

$$\varphi_S = \sum_{j=1}^{S} a_{2j} m_{2j} H_{2j-1}, \qquad \psi_S = \sum_{j=1}^{S} a_{2j-1} m_{2j-1} H_{2j-2}, \qquad \sigma_q = \sum_{j=1}^{q} a_j m_j f_j H_{j-1}, \qquad H_0 = 1 \, . \tag{A.4}$$

Note that the expressions in brackets in (A.3) are always positive; that is, in a partially closed chain the rate Q may be less than that in an open chain of the same length.

APPENDIX B: STABILITY CONDITIONS FOR AN OPEN CHAIN OF LENGTH Q

When system (13.17) is linearized at the point $\mathbf{N}_q^*$, its matrix has the following form:

$$F = \begin{bmatrix} A_q & 0 \\ 0 & D_{n-q} \end{bmatrix}, \qquad \text{(B.1)}$$

where A_q is a Jacobian $(q + 1) \times (q + 1)$ matrix of the form

$$A_q = [a_{ij}] = \begin{bmatrix} -b_0 & -d_0 & & & & & \\ b_1 & -h_1 & -d_1 & \cdots & & & 0 \\ & & \vdots & & & & \\ & & & & \vdots & & \\ 0 & & \cdots & b_{q-1} & -h_{q-1} & -d_{q-1} \\ & & & & b_q & -h_q \end{bmatrix}, \qquad \text{(B.2)}$$

with the entries $b_0 = \alpha_0 N_1^*$, $b_i = k_i \alpha_{i-1} N_i^*$, $d_0 = \alpha_0 N_0^*$, $d_i = \alpha_i N_i^*$, $h_i = 0$, $i = 1, \ldots, q$, and

$$D_{n-k} = \text{diag}\{-m_{q+1} + k_{q+1}\alpha_q N_q^*, -m_{q+2}, \ldots, -m_n\}.$$

If the eigenvalues of A_q are denoted by μ_i, then those of F will be equal to

$$\lambda_i = \begin{cases} \mu_i, & i = 0, 1, \ldots, q, \\ k_{q+1}\alpha_q N_q^* - m_q, & i = q + 1, \\ -m_i, & i = q + 2, \ldots, n, \end{cases}$$

that is, the stability of F depends exclusively on properties of the submatrix A_q and the first diagonal term of D_{n-k}. As follows from the sign-stability criterion, matrix A_q is stable for any feasible values of the model parameters (see Fig. 13.3 for the corresponding SDG). Hence, the only condition for matrix F's stability follows from the requirement that $\lambda_{q+1} < 0$ and reduces to the condition that

$$N_q^* < m_{q+1}/\alpha_q k_{q+1},$$

which, in turn, yields

$$Q < Q^*(q + 1), \qquad \text{(B.3)}$$

where Q^* is a function defined in (A.2). From (A.2) and (B.3) it follows that the equilibrium $\mathbf{N}_q^*$ is feasible and stable in the linear approximation of system (13.17) if and only if

$$Q^*(q) < Q < Q^*(q + 1). \qquad \text{(B.4)}$$

APPENDIX C: STABILITY CONDITIONS FOR A PARTIALLY CLOSED CHAIN OF LENGTH Q

For a partially closed chain [system (13.18)], the first row of the linearization matrix, which is similar to (B.1), is modified by the "closing" terms $C_i = a_i m_i$ ($i = 1, \ldots, n$) into the form

$$\tilde{F} = F + C_{1 \times (n+1)} = \begin{bmatrix} \tilde{A}_q & C_{(q+1) \times (n-q)} \\ 0 & D_{n-q} \end{bmatrix}. \qquad \text{(C.1)}$$

Now the sign-stability criterion can no longer be satisfied with submatrix $\tilde{A}_q$, but its characteristic polynomial meets certain recurrent relations, which give a possibility of finding, for a particular value of q, the condition that $\mathbf{N}_q^*$ is stable. In the particular case where $C_1 > 0$, $C_i = 0$, $i = 2, \ldots, n$, the feasibility and stability conditions are proved (Svirezhev and Logofet, 1978) to be similar to the restriction (B.4):

$$\tilde{Q}^*(q) < Q < \tilde{Q}^*(q + 1), \tag{C.2}$$

where

$$\tilde{Q}^*(q) = Q^*(q) - \begin{cases} a_1 m_1 f_q, & \text{for an even } q, \\ a_1 m_1 f_{q-1}, & \text{for an odd } q. \end{cases} \tag{C.3}$$

In a more general case, condition (C.2) with the sequence $\tilde{Q}^*(q)$, $q = 1, 2, \ldots$ defined in (A.3) is only necessary but insufficient for a feasible $\mathbf{N}_q^*$ to be stable.

REFERENCES

ABRAMS, P. 1980. Some comments on measuring niche overlap. *Ecology*, 61:44–49.

ADZHABYAN, N. A., and LOGOFET, D. O. 1992a. Population size dynamics in trophic chains. In Izrael, Y. A. (ed.), *Problems of Ecological Monitoring and Ecosystem Modelling*, Vol. 14. Gidrometeoizdat, St. Petersburg, pp. 135–153 (in Russian).

ADZHABYAN, N. A., and LOGOFET, D. O. 1992b. Extension of the stability concept in a mathematical theory of trophic chains. *Zhurnal Obschei Biologii (Journal of General Biology)* 53:81–87 (in Russian).

ALEKSEEV, V. V., and KORNILOVSKY, A. N. 1985. Ecosystem stochasticity model. *Ecol. Mod.* 28:217–229.

ARNOLD, V. I., VARCHENKO, A. N., and GUSEYN-ZADEH, S. M. 1982. *Singularities of Differentiable Mappings*. Nauka, Moscow. 304 pp. (in Russian).

ARROW, K. J. and MCMANUS, M. 1958. A note on dynamical stability. *Econometrica* 26:448–454.

BARKER, G. P., BERMAN, A., and PLEMMONS, R. J. 1978. Positive diagonal solutions to the Lyapunov equations. *Linear Multilinear Algebra* 5:249–256.

BOUCHER, D. H. 1985. Lotka–Volterra models of mutualism and positive density-dependence. *Ecol. Mod.* 27:251–270.

BRAUER, F. 1979. Boundedness of solutions of predator–prey systems. *Theor. Pop. Biol.* 15:268–273.

DATTA, B. N. 1978. Stability and D-stability. *Linear Algebra Appl.* 21:135–141.

EHMAN, T. I. 1966. On some mathematical models of biogeocenoses. In *Problems of Cybernetics*, Issue 16. Nauka, Moscow, pp. 191–202 (in Russian).

ELTON, C. S. 1958. *The Ecology of Invasions of Animals and Plants*. Methuen, London, 181 pp.

FREIDLIN, M. I., and SVETLOSANOV, V. A. 1976. The effects of small random disturbances on steady states of ecological systems. *Zhurnal Obschey Biologii (J. General Biol.)* 37:715–719 (in Russian).

GAUSE, G. F. 1934. *The Struggle for Existence*. Williams and Wilkins, Baltimore, 163 pp.

GOEL, N. S., MAITRA, S. C., and MONTROLL, E. W. 1971. On the Volterra and other nonlinear models of interacting populations. *Rev. Mod. Phys.* 43:231–276.

GOH, B. S. 1977. Stability in many-species systems. *Amer. Nat.* 111:135–143.

GOH, B. S. 1979. Robust stability concepts for ecosystem models. In Halfon, E. (ed.), *Theoretical Systems Ecology. Advances and Case Studies*. Academic Press, New York, pp. 467–487.

GOODMAN, D. 1975. The theory of diversity–stability relationships in ecology. *Quart. Rev. Biol.* 50:237–266.

GROSSBERG, S. 1977. Pattern formation by the global limits of a nonlinear competitive interaction in n dimensions. *J. Math. Biol.* 4:237–256.

GRÜMM, H. R. 1976. Definitions of Resilience. IIASA Research Report RR-76-5. International Institute for Applied Systems Analysis, Laxenburg, Austria. 20 pp.

HARNER, E. J., and WHITMORE, R. C. 1977. Multivariate measures of niche overlap using discriminant analysis. *Theor. Pop. Biol.* 12:21–36.

HARTFIEL, D. J. 1980. Concerning the interior of D-stable matrices. *Linear Algebra Appl.* 30:201–207.

HOLLING, C. S. 1965. The functional response of predator to prey density and its role in mimicry and population regulation. *Mem. Entomol. Soc. Canada* 45:1–60.

HOLLING, C. S. 1973. Resilience and stability of ecological systems. *Ann. Rev. Ecol. Syst.* 4:1–23.

HOLLING, C. S. 1976. Resilience and stability of ecosystems. *Evol. Consciousness Hum. Syst. Transit.* Addison-Wesley, Reading, Massachusetts, pp. 73–92.

JEFFRIES, C. 1974. Qualitative stability and digraphs in model ecosystems. *Ecology* 55:1415–1419.

JEFFRIES, C. 1979. Stability of holistic ecosystem models. In Halfon, E. (ed.), *Theoretical Systems Ecology. Advances and Case Studies.* Academic Press, New York, pp. 487–504.

JEFFRIES, C., KLEE, V., and VAN DEN DRIESSCHE, P. 1977. When is a matrix sign stable? *Can. J. Math.* 29:315–326.

JOHNSON, C. R. 1974. Sufficient conditions for D-stability. *J. Econom. Theory* 9:53–62.

JONES, D. D. 1975. The Application of Catastrophe Theory to Ecological Systems. IIASA Research Report RR-75-15. International Institute for Applied Systems Analysis, Laxenburg, Austria. 57 pp.

KOSTITZIN, V. A. 1934. *Simbiose, Parasitisme et Evolution (Etude mathématique).* Hermann, Paris. 47 pp.

LADDE, G. S., and SILJAK, D. D. 1975. Stochastic stability and instability of model ecosystems. In Proc. IFAC 6th World Congr. Boston–Cambridge, 1975, Part 3. Pittsburg, Pennsylvania, pp. 55.4/1–55.4/7.

LEONTIEF, W. W. 1966. *Input–output Economics.* Oxford University Press, New York. 561 pp.

LEVIN, S. 1979. Multiple equilibria in ecological models. In Trudy mezhdunarodnogo simposiuma po problemam matematicheskogo modelirovaniya protsessov vzaimodejstviya chelovecheskoy aktivnosti i okruzhayuschey sredy (*Proc. Intern. Symp. on Problems of Mathematical Modelling in Processes of Interaction between Human Activity and Environment*), Telavi, USSR, September 1978, Vol. 2. Computer Center, USSR Acad. Sci., Moscow, pp. 61–75 (in English).

LEVINS, R. 1974. Problems of signed digraphs in ecological theory. In Levin, S. A. (ed.), *Ecosystem Analysis and Applications.* SIAM Institute for Mathematics and Society, Research Applications Conference on Ecosystems, July 1–5, 1974, Alta, Utah. Society for Industrial and Applied Mathematics, Philadelphia, Pennsylvania, pp. 264–277.

LOGOFET, D. O. 1975. On the stability of a class of matrices arising in the mathematical theory of biological associations. *Soviet Math. Dokl.* 16:523–527.

LOGOFET, D. O. 1978. Towards qualitative stability of ecosystems. *Zhurnal Obshchey Biologii (J. Gen. Biol.)* 39:817–822 (in Russian).

LOGOFET, D. O. 1981a. What is mathematical ecology? In Svirezhev, Y. M., and Pasekov, V. P., (eds.) *Mathematical Models in Ecology and Genetics.* Nauka, Moscow, pp. 8–17 (in Russian).

LOGOFET, D. O. 1981b. Towards the problem of ecological system stability. In *Proc. Int. Symp. on Problems of Mathematical Modelling in Processes of Interaction between Human Activity and Environment* (Telavi, USSR, September 1978), Vol. 2. Computer Center, USSR Acad. Sci., Moscow, pp. 61–75 (in Russian).

LOGOFET, D. O. 1987. On the Hierarchy of Subsets of Stable Matrices. *Soviet Math. Dokl.* 34(2):247–250.

LOGOFET, D. O. 1989. Do there exist diagonally stable matrices without dominating diagonal? *Soviet Math. Dokl.* 38:113–115.

LOGOFET, D. O. 1992. *Matrices and Graphs, Stability Problems in Mathematical Ecology.* CRC Press, Boca Raton, 308 pp.

LOGOFET, D. O., and ALEXANDROV, G. A. 1983a. Modelling of a matter cycle in a mesotrophic bog ecosystem: I. Linear analysis of carbon environs. *Ecol. Mod.* 21:247–258.

LOGOFET, D. O., and ALEXANDROV, G. A. 1983b. Modelling of a matter cycle in a mesotrophic bog ecosystem: II. Dynamic model and ecological succession. *Ecol. Mod.* 21:259–276.

LOGOFET, D. O., and JUNUSOV, M. K. 1979. The issues of qualitative stability and regularization in dynamic models of the cotton agrobiocoenosis. In Svirezhev, Y. M. (ed.), *Control and Optimization in Ecological Systems.* Issues of Cybernetics, Vol. 52. USSR Acad. Sci., Moscow, pp. 62–74 (in Russian).

LOGOFET, D. O., and SVIREZHEV, Y. M. 1977. Stability in models of interacting populations. In *Problems of Cybernetics,* Issue 32. Nauka, Moscow, pp. 187–202 (in Russian).

LOGOFET, D. O., and SVIREZHEV, Y. M. 1979. Ecological modelling, catastrophe theory and some new concepts of stability. In Samet, P. A. (ed.), *EuroIFIP 79.* North-Holland, Amsterdam, pp. 287–392.

LOGOFET, D. O., and SVIREZHEV, Y. M. 1980. The model for human population dynamics as a part of the global biosphere model: some aspects of modelling in a dialogue regime. *Ecol. Mod.* 9:269–280.

LOGOFET, D. O., and SVIREZHEV, Y. M. 1983. Stability concepts for biological systems. In *Problems of Ecological Monitoring and Ecosystem Modelling,* Vol. 6. Gidrometeoizdat, Leningrad, pp. 159–171. (in Russian).

LOGOFET, D. O., and SVIREZHEV, Y. M. 1984. Modelling of population and ecosystem dynamics under reserve conditions. In *Conservation, Science and Society.* Contrib. First Intern. Biosphere Reserve Congress, Minsk, Byelorussia, USSR, September 26–October 2, 1983, Vol. 2. UNESCO-UNEP, pp. 331–339.

LOGOFET, D. O., and SVIREZHEV, Y. M. 1985. Ecological stability and the Lagrangian concept. A new glance at the problem. In *Problems of Ecological Monitoring and Ecosystem Modelling,* Vol. 7. Gidrometeoizdat, Leningrad, pp. 253–258 (in Russian).

LOGOFET, D. O., and ULIJANOV, N. B. 1982. Necessary and sufficient conditions for sign stability of matrices. *Soviet Math. Dokl.* 25:676–680.

LOGOFET, D. O., and ULIJANOV, N. B. 1982. Sign stability in model ecosystems: a complete class of sign-stable patterns. *Ecol. Mod.* 16:173–189.

MACARTHUR, R. H. 1955. Fluctuations of animal populations and a measure of community stability. *Ecology* 36:533–536.

MACARTHUR, R. H. 1968. The theory of niche. In Lewontin, R. (ed.), *Population Biology and Evolution.* Syracuse University Press, Syracuse, New York, pp. 159–176.

MAY, R. M. 1973. *Stability and Complexity in Model Ecosystems.* Princeton University Press, Princeton, New Jersey. 265 pp.

MAY, R. M. 1974. On the theory of niche overlap. *Theor. Pop. Biol.* 5:297–332.

MAY, R. M. 1975. Biological populations obeying difference equations: stable points, stable cycles and chaos. *J. Theor. Biol* 51:511–524.

MAY, R. M., and MACARTHUR, R. H. 1972. Niche overlap as a function of environmental variability. *Proc. Nat. Acad. Sci. USA* 69:1109–1113.

MOYLAN, P. J. 1977. Matrices with positive principal minors. *Linear Algebra Appl.* 17:53–58.

NIKAIDO, H. 1968. *Convex Structures and Economic Theory.* Academic Press, New York. 405 pp.

ODUM, H. T. 1957. Trophic structure and productivity of Silver Springs, Florida. *Ecol. Mon.* 27:55–112.

PETRAITIS, P. S. 1981. Algebraic and graphical relationships among niche breadth measures. *Ecology* 62:545–548.

PIELOU, E. C. 1969. *An Introduction to Mathematical Ecology.* Wiley-Interscience, New York. 286 pp.

PIMM, S. L. 1980a. Properties of food webs. *Ecology* 61:219–225.

PIMM, S. L. 1980b. Food web design and the effects of species deletion. *Oikos* 35:139–149.

PIMM, S. L., and KITCHING, R. L. 1987. The determinants of food chain lengths. *Oikos* 50:302–307.

PIMM, S. L. and LAWTON, J. H. 1978. On feeding on more than one trophic level. *Nature* 275:542–544.

PYKH, Y. A. 1983. *Equilibrium and Stability in Models of Population Dynamics.* Nauka, Moscow. 198 pp. (in Russian).

QUIRK, J. P., and RUPPERT, R. 1965. Qualitative economics and the stability of equilibrium. *Rev. Econ. Studies* 32:311–326.

QUIRK, J., and SAPOSNIK, R. 1968. *Introduction to General Equilibrium and Welfare Economics.* McGraw-Hill, New York. 221 pp.

SHAFFER, W. M. 1985. Order and chaos in ecological systems. *Ecology* 66:93–106.

SHAFFER, W. M., and KOT, M. 1985. Do strange attractors govern ecological systems? *Bioscience* 35:342–350.

SIDORIN, A. P. 1981. Behaviour of a population with several steady states in a random environment. In Svirezhev, Y. M., and Pasekov, V. P. (eds.), *Mathematical Models in Ecology and Genetics.* Nauka, Moscow, pp. 71–74 (in Russian).

ŠILJAK, D. D. 1974. Connective stability of complex ecosystems. *Nature* 249:280.

ŠILJAK, D. D. 1975. When is a complex system stable? *Math. Biosci.,* 25:25–50.

SVIREZHEV, Y. M., 1987. *Nonlinear Waves, Dissipative Structures and Catastrophes in Ecology.* Nauka, Moscow. 368 pp. (in Russian).

SVIREZHEV, Y. M., and LOGOFET, D. O. 1978. *Stability of Biological Communities,* Nauka, Moscow. 352 pp. (in Russian). English translation 1983: Mir, Moscow. 319 pp.

SVIREZHEV, Y. M., and LOGOFET, D. O. 1985. Complicated dynamics in simple models of ecological systems. *Mathematical Research,* Vol. 23, *Lotka-Volterra Approach to Cooperation and Competition in Dynamic Systems.* Akademie-Verlag, Berlin, pp. 13–22.

SVIREZHEV, Y. M., and VOINOV, A. A. 1982. A shallow lake as a dynamic system. In *Proc. Intern. Sci. Workshop Ecosystem Dynamics in Freshwater Wetlands and Shallow Water Bodies.* USSR, Minsk-Pinsk-Tskhaltoubo, July 12–26, 1981, Vol. 2. Center of Intern. Projects GKNT, Moscow, pp. 156–173.

TAKEUCHI, Y., ADACHI, N., and TOKUMARA, H. 1978. Global stability of ecosystems of the generalized Volterra type. *Math. Biosci.* 42:119–136.

THOM, R. 1969. Topological models in biology. *Topology* 8:315–335.

THORNTON, K. W., and MULHOLLAND, R. J. 1974. Lagrange stability and ecological systems. *J. Theor. Biol.* 45:473–485.

USMANOV, Z. D., SAPOZHNIKOV, G. N., ISMAILOV, M. A., CHERENKOV, S. N., BLAGOVESCHENSKAYA, S. T., and YAKOVLEV, E. P. 1981. Modelling of desert community dynamics in Tigrovaya Balka Reserve. *Dokl. Tadjik SSR Acad. Sci.* 34:629–632 (in Russian).

VAN DER WAERDEN, B. L., 1937. *Moderne Algebra,* Vol. 1. Springer, Berlin. 272 pp. (in German).

YATSALO, B. I. 1984. *Complicated Dynamic Behaviours in Models of Closed Ecological Systems.* Physics-mathematics candidate dissertation. Moscow State University, Moscow. 112 pp. (in Russian).

VOLTERRA, V. 1931. *Leçons sur la Theorie Mathématique de la Lutte pour la Vie.* Gauthiers-Villars, Paris. 214 pp.

YODZIS, P. 1978. *Competition for Space and the Structure of Ecological Communities* (Lecture Notes in Biomathematics, Vol. 25). Springer, New York. 191 pp.

YORKE, J. A., and ANDERSON, W. N. 1973. Predator–prey patterns. *Proc. Nat. Acad. Sci. USA* 70:2069–2071.

14

Ecosystem Stability: Revision of the Resistance–Resilience Model

JACK B. WAIDE

USDA Forest Service
Southern Forest Experiment Station
Forest Hydrology Laboratory
Oxford, Mississippi

We need to manage and to use our natural resources more wisely and yet more intensively in the future. To do this we need to incorporate more of our experience, our data, and our theory into the decision-making process. We can use simulation models in this synthesis effort to advantage. We can perform management experiments with ecosystem level models, generate output from those experiments, and condense and interpret this output in a manner useful to the management agency personnel. The result will be better resource management decisions based on scientifically and technically defendable information which will have greater internal consistency and which will produce better results under many conditions [Van Dyne, in *Ohio J. Sci.* 78, 190 (1978)].

INTRODUCTION

The resistance–resilience model of ecosystem relative stability or response to disturbance (Webster, Waide, and Patten 1975) has provided part of the conceptual foundation for ecosystem research at the Coweeta Hydrologic Station in the mountains of North Carolina, USA (Monk et al., 1977). It has also provided a point of departure for analyzing responses of stream ecosystems to experimental disturbances (Webster et al., 1985) and long-term forest responses to intensive management (Waide and Swank, 1976; Swank and Waide, 1980). However, both methodological and conceptual criticisms of this model have been published. Also, recent advances in ecosystem science affect the theoretical basis of the model. In response to both factors, I present here a revised theoretical interpretation of the resistance–resilience model.

The general topic of stability has received considerable attention in the ecological literature. Part of this emphasis derives from earlier natural history approaches to ecosystem analysis and management (for example, Lutz, 1963). A more recent stimulus of extensive interest in ecological stability is the complexity–stability hypothesis (Odum, 1953; MacArthur, 1955; May, 1973; Goodman, 1975; Pimm, 1984). Although considerable research has been conducted on this topic, theories of ecosystem stability remain in a confused and unsatisfactory state. Pimm (1984) suggested several reasons for this, particularly that investigators have employed inconsistent measures or meanings of the basic concepts under study. The most revealing aspect of Pimm's analysis is that it excluded from consideration those macroscopic variables that should be the primary focus of ecosystem stability discussions.

To motivate the discussions that follow, I suggest that several major factors have contributed to current confusion over ecosystem stability. First, many theorists have adopted a naive view of ecosystem coupling to a specific physicochemical environment (see also Patten 1978; O'Neill et al., 1986). Thus, ecosystem dynamics and properties are often analyzed and compared abstracted from the environmental constraints to which they are intimately coupled. Such an approach—best typified by numerous temperate–tropical comparisons that motivated early interest in stability—leads to inappropriate analyses of ecosystem dynamics and stability. One major point I wish to make is that, in the context of the physicochemical environment to which they are coupled, natural ecosystems are (conditionally) stable systems. Thus, the whole issue of ecosystem stability focusing on whether they are stable or not has received improper emphasis.

A second cause of current confusion over ecosystem stability relates to Egerton's (1973) perceptive review of the "balance of nature concept" and its influence on ecological theory. As a direct consequence of this "oldest ecological theory," ecologists have tended to view the natural world through the human spatiotemporal perspective (see also Allen and Starr, 1982; O'Neill et al., 1986). Thus, high-turnover systems such as planktonic ecosystems are frequently considered unstable, whereas ecosystems such as forests, which exhibit less rapid dynamics, are viewed as stable. If detailed studies of natural ecosystems over the past several decades have taught us one major lesson, it is that ecosystems are scale-dependent systems and that perceptions of change referenced to the human life span provide an inappropriate basis for judging ecosystem dynamics and stability. Reliance on such approaches leads to the improper linking of stability with constancy and to the failure to appreciate the significance of ecosystem dynamics in space and time.

Third, the complexity–stability debate has itself caused difficulties in developing useful macroscopic theories of ecosystem stability. This continuing area of research has in some sense fostered the view that ecosystems can somehow be randomly assembled from constituent populations and the resultant stability properties quantified. Such approaches ignore the fundamental structuring imposed on biological populations by ecosystem constraints of matter and energy processing. Also, legitimate concerns in applied ecology, such as the behavior of ecological systems simplified by humans or the conservation of genetic diversity, have become intertwined with the complexity–stability hypothesis (Goodman, 1975). This has created problems for analyzing ecosystem persistence in spatially and temporally variable environments and for examining ecosystem responses to anthropogenic disturbance.

A fourth difficulty with discussions of ecosystem stability emerges from the failure of ecologists to define relationships between ecosystem properties and characteristics of constituent species populations (Schindler et al., 1980; O'Neill and Waide, 1981; O'Neill et al., 1986). The main points are that ecosystem characteristics and dynamics cannot be predicted from those of constituent species populations and that ecologists have devised no generally acceptable means of decomposing ecosystems into subsystems with functional properties that commute with measures of intact system dynamics (see also Rosen, 1972). Thus, ecosystem stability is not to be judged solely in relation to species persistence in space or time or to constancy of species composition.

Finally, as in many areas of ecosystem science, discussions of stability have been clouded by the lingering shadow of arguments over older, deterministic theories of succession and climax. Thus, many purported criticisms of ecosystem stability theory (for example, Botkin, 1980) have actually been criticisms of classical but inappropriate models of succession and climax, as well as of the underlying balance of nature concept (see also Richards and Charley, 1983/1984).

The major intent of this chapter is to clarify discussions of ecosystem stability. Based on a functional, integrative approach to the analysis of ecosystem dynamics and stability, I relate observable macroscopic properties of natural ecosystems to their stability in the context of a theoretical view of ecosystems as hierarchical biogeochemical systems. Moreover, I reevaluate the relative stability model of Webster, Waide, and Patten (1975) in relation to this theoretical view of ecosystems and suggest that resistance and resilience represent scaling variables that reflect the space–time coupling of entire biotic assemblages to specific physicochemical environments. The basic thesis underlying my arguments is that macroscopic structural and functional attributes of ecosystems reflect the evolutionary coupling of constituent biotic populations to physicochemical environmental variables that regulate the exchange, transformation, and storage of energy and essential elements by ecosystems. These macroscopic properties reflect both the persistence of natural ecosystems in spatially and temporally variable environments and the extent and rate of ecosystem response to exogenous disturbance. Arguments presented here represent the refinement and extension of discussions in Waide and Jager (1981).

BRIEF REVIEW OF MATHEMATICAL STABILITY CONCEPTS

This chapter focuses on the stability of ecosystems, rather than on differential equations. However, as a point of departure for what follows, this section provides a brief review of mathematical stability definitions. These ideas are covered in greater detail elsewhere (see, for example, Lewontin, 1970; Rosen, 1970; May, 1973; Waide and Webster, 1976; Pimm, 1982).

Consider the behavior of a general system whose state at time t is characterized by the vector $x(t)$ of n time-dependent state variables and whose time dynamics are described by the vector-valued dynamic equations (defined on an appropriate state space X, a subspace of Euclidean n-space R^n)

$$\dot{x}(t) = F(x, p, t), \quad x(t_0) = x_0.$$

Here the dot notation denotes differentiation with respect to time, p is a vector of parameters (process rates, inputs), and x_0 represents initial state at time t_0.

For this system, we define a singular point x' as a point in state space at which time derivatives vanish; that is, x' is a singular point if $F(x', p) = 0$. The concept of stability is concerned with the configuration of the n-dimensional state space X as defined by the F. Are there one or several singular points x' within this space, and how do solutions to F behave in the neighborhood of each singular point? If the equations F admit only one singular point, we are concerned with the global stability of the system; if they admit two or more singular points, we are interested in the local stability or behavior of the system in a small region of state space around each.

Intuitively, the concept of stability relates to whether a system whose nominal or undisturbed dynamics as described by the functions F remain essentially unaltered following some imposed disturbance or whether the system has been changed substantially by that disturbance. Here, disturbance may relate abstractly to some temporary change in the functional form of F or to a temporary displacement or "bump" in the value of x or p. Table 14.1 provides a series of increasingly restrictive formal definitions of what it means for a system to be essentially unaltered, that is, stable. The term trajectory refers to the time change in system state, from some initial state x_0 to some final state $x(t_1)$, which may or may not be a singular point.

The least restrictive definition, boundedness, requires that two trajectories that begin arbitrarily close to one another never diverge farther than some upper bound M. More restrictively, a system is weakly stable if it always remains near a singular point (in a geometric sense, within an ϵ

TABLE 14.1 Brief review of mathematical stability definitions

Type of stability	Definition
Bounded	Let x, $\hat{x}$ be two system trajectories. If for any $\delta > 0$ there exists $M > 0$ such that $|x_0 - \hat{x}_0| < \delta$ implies that $|x(t) - \hat{x}(t)| \leq M$, all $t \geq t_0$.
Weakly stable (stable in sense of Liapunov)	If for any $\epsilon > 0$, there exists $\delta > 0$ such that $|x_0 - x'| < \delta$ implies that $|x(t) - x'| < \epsilon$, all $t > t_0$.
Asymptotically stable	If the system is weakly stable, and if there exists $\alpha > 0$ such that $|x_0 - x'| < \alpha$ implies $x(t) \to x'$ as $t \to \infty$.
Conditionally (asymptotically) stable	If the system converges to x' [$x(t) \to x'$] only from within a bounded region of state space. *Attractor* refers to the conditionally stable singular point or trajectory; *basin (domain) of attraction* refers to the bounded region of state space within which the system is stable.

The system defined by equations F (see text) is said to possess the type of stability specified in the left-hand column if it satisfies the analogous definition specified in the right-hand column.

neighborhood) once it is "bumped" away from it and asymptotically stable if it uniquely returns to the singular point as time progresses. The final definition in Table 14.1 has the greatest relevance to questions of ecosystem stability: a conditionally stable system is one that converges to a particular singular point, not from anywhere in state space X, but only from a bounded region of X around x'. The singular point to which the system converges is termed an *attractor*, and the region of state space within which convergence to x' occurs is termed the *domain* or *basin of attraction*.

Although these definitions capture the essence of mathematical stability theory, they are overly restrictive for ecological application. In many cases the system under study, rather than converging over time to a singular point x', may exhibit oscillatory behavior. If the oscillations have regular period and amplitude, as determined uniquely by the value of p rather than by some arbitrary value of x_0, then the system is exhibiting limit-cycle behavior. The stability properties of limit cycles may be defined analogously to those of singular points (Table 14.1). Further generalizations of these ideas are also possible, adding additional aspects of ecological realism. For example, recent advances in the definition of attractors within dynamical systems theory (reviewed by Stewart, 1985) may provide a particularly useful starting point for ecosystem stability considerations.

Although these are extremely abstract concepts that apply to the study of many specific kinds of systems, they represent more than mathematical window-dressing for ecologists. When applied in an ecologically relevant manner, these ideas and their formal generalizations provide a solid foundation for theoretical investigations of ecosystem dynamics and stability. Based on these mathematical definitions, I develop the argument that ecosystems may be usefully viewed abstractly as conditionally stable attractors in environmentally determined basins of attraction. Holling (1973, and subsequent elaborations) has reached the same conclusion in a population-theoretic context and has employed the concepts of conditional stability and limit cycle in his definition of ecosystem resilience (which differs from the Webster, Waide, and Patten, 1975, usage of this term).

Central to these ideas are the abstract notions of state space and basin of attraction. Although these ideas may initially appear vague to many ecologists, the ecological literature contains concrete examples of state-space diagrams. Whittaker's (1975, Fig. 4.1) diagram of the boundaries of terrestrial biomes in relation to mean annual temperature and precipitation may be viewed as an illustration of ecosystem organization, at extremely broad scales of space and time, depicted as basins of attraction within an environmentally defined state space. So, too, may Holdridge's (1967) life-zone classification of vegetation. Similarly, Whittaker's (1956) diagram of the distribution of vegetation in the Great Smokey Mountains in relation to topography and elevation and a comparable diagram for vegetation distribution within the Coweeta Basin are examples of ecosystem organization depicted as basins of attraction within environmentally defined state spaces at finer scales of space and time. Thus, these theoretical notions have definite ecological utility.

APPLICABILITY OF STABILITY DEFINITIONS IN ECOSYSTEM ANALYSIS

The extensive literature on ecosystem stability is characterized by vague and inconsistent uses of basic terms (Pimm, 1984) and by ecologically naive applications of the mathematical definitions just reviewed. As a consequence, some ecologists have concluded that classical stability concepts have limited ecological utility. In contrast, a more useful view is that these mathematical ideas must be applied in an ecologically meaningful fashion if continuing research is to produce ecological insights of broad generality, rather than numerical results of mathematical interest only. In this spirit, I briefly elaborate several restrictions on the application of mathematical stability definitions in ecosystem analysis. These restrictions represent constraints on the ecosystem attributes to be considered and on the evidence to be examined in judging the stability of natural ecosystems. Such restrictions are necessary since many ecologists appear not to appreciate the essential fact that stability characteristics of ecosystems differ meaningfully from those of populations, food webs, and communities.

First, ecosystem stability is not equivalent to apparent constancy of function over time. The tendency to equate stability with constancy results from the influence of the balance of nature concept on ecological thinking. Constancy may imply dynamic stability (that is, ability to recover following disturbance), but it may also imply lack of disturbance, the occurrence of which is essential to the testing of stability. Thus, for example, frequent statements in the literature that forests are more stable than grasslands or plankton are without meaningful ecological content. Such statements reflect human perception of rates of change in the natural world in relation to our spatiotemperal perspective and life span. They imply nothing about the functional organization of the three ecosystem types, each of which must be judged conditionally stable in relation to a particular physicochemical environment.

Second, ecosystem stability is not necessarily or uniquely related to species persistence or to constancy of species composition over space or time. To the extent that species diversity represents a functional redundancy at the level of the ecosystem (Whittaker and Woodwell, 1972; O'Neill et al., 1986), macroscopic ecosystem properties are insensitive to some range of variation in species composition. To this same extent, species turnover at various scales of space and time becomes an important mechanism for maintaining ecosystem integrity in response to environmental variation at those same scales. This restriction does not imply, however, that analyses of ecosystem dynamics are unrelated to species characteristics, as many ecosystem theorists have been quick to assume. To borrow Hutchinson's (1965) apt metaphor, species represent the current cast of actors in the contemporary ecological theater; at least some of the biotic components of ecosystems must respond following disturbance or the entire ecosystem will collapse. But, in general, species turnover in space and time appears to be one mechanism preserving ecosystem functional integrity, so ecosystem stability should not be judged solely on constancy of species composition. The challenge remains to determine experimentally when species turnover reflects function-preserving redundancy and when it reflects directional change to a functionally different ecosystem.

A direct corollary of this second restriction is that the stability of ecosystems has little to do with supposed tests of the complexity–stability hypothesis. Ecosystem trophic complexity may be related to the functional redundancy of component species, as well as to the spectrum of frequency signals impinging on a given ecosystem in a given physicochemical environment. It may also promote the functional constancy of the system at certain space–time scales (McNaughton, 1985).

But, in relation to the macroscopic view of functional integrity developed here, no universal relationship exists between ecosystem stability and trophic complexity.

A third important restriction relates to confusion in the use of the term disturbance. Ecosystem stability is not usefully tested by persistent, severe disturbances that completely degrade system structure, that destroy the coupling between an ecosystem and a particular physicochemical environment, or that represent permanent directional changes in that environment. Ecosystems are recognized here as being only conditionally stable within certain bounded regions of environmentally defined state space. While it is important to determine ecological and environmental factors that define the bounds of ecosystem organization viewed abstractly as basins of attraction, it is not particularly useful to discuss, under the heading of stability, the types of disturbance enumerated above. This point is particularly relevant to the consideration of ecosystem responses to anthropogenic insults, especially if such insults cause permanent changes in environments beyond the ability of ecosystems to respond and if they do not parallel impacts that are part of the evolutionary history of natural ecosystems.

Other confusions concerning the nature of disturbances in the context of stability theory also exist. At the scale at which they occur, oscillations in climatic variables do not represent an ecosystem disturbance any more than does the occurrence of wildfire in woodland, savanna, or grassland ecosystems subjected to burning on a statistically regular basis. Such oscillations form part of the signal structure of the physicochemical environment impinging on a given ecosystem, and they have become incorporated into the biological structure of the ecosystem through the evolution of life history characteristics and rate processes attuned to these oscillations (see also O'Neill et al., 1986). Thus, at a certain spatiotemporal scale, ecosystem dynamics have become coupled to these abiotic oscillations, which should not be viewed as disturbances at that scale. McNaughton's (1985) detailed analysis of the coupling of vertebrate grazing and primary production to space–time patterns of rainfall in tropical savannas is an excellent case in point.

Another confusion common in the literature is that ecosystem stability is concerned with the configuration of an environmentally defined state space, not with the behavior of any specific trajectory (that is, time course of system behavior) within that space. For example, consider the ecosystem present at a specific point on the earth's surface (for example, oak–hickory forest on a mid-elevation north-facing slope at Coweeta). Following long-term climatic change associated with glaciation, the ecosystem present at that geographic location would change dramatically (for example, to spruce-dominated boreal forest). Whereas many authors would interpret this as reflecting ecosystem instability (for example, Davis, 1976), it probably reflects just the opposite. If the new ecosystem state at this point now lies (in an abstract geometric sense) within the bounded region of state space defined by the climate newly impinging on that location, then this shift reflects conditional stability. In other words, the configuration of the abstract state space has not changed, but its physical realization on the landscape has been altered due to temporal dynamics in macroclimate. In reference to this example, Davis (1976) has identified life history characteristics that have allowed tree species to invade successfully under closed canopies as being important in the long-term response of eastern North American forests to glaciation. In the context developed here, such adaptations become important mechanisms that preserve ecosystem functional integrity at the spatial scale of the eastern deciduous forest and at the temporal scale of glaciation events.

A final, significant restriction on the use of classical stability concepts in ecosystem analysis is implicit in most of the points above: stability is a scale-dependent property of ecosystems. Definition of the focal ecosystem of interest is not arbitrary, but it requires delineation of the space–time scale over which relevant ecosystem dynamics are to be observed and of the processes that regulate observable dynamics at that scale. Thus, dynamic behaviors that appear stable at one space–time scale of observation may not be at another. In exactly the same way, disturbances must be defined in a scale-dependent fashion (see also Gerritsen and Patten, 1985). At a relatively fine spatiotemporal scale of observation, low-frequency environmental oscillations appear as factors disturbing ecosystem integrity. At a broader space–time scale of focus, these same physicochemical frequency signals have become incorporated into ecosystem structure through similar life history adaptations of constituent species to common environmental oscillations. At this broader space–time scale, the ecosystem is stable in respect to these oscillations, which are not disturbances but rather frequency signals to which ecosystem dynamics are coupled.

THEORETICAL CONTEXT FOR MACROSCOPIC APPROACHES TO ECOSYSTEM STABILITY

The above restrictions on ecological stability analyses are required because we presently lack a comprehensive synthetic theory of the ecosystem. Such a theory should provide the basis not only for analyzing ecosystem dynamics and processes regulating them across scales of space and time, but also for viewing ecosystems abstractly as conditionally stable attractors within environmentally defined basins of attraction. Because they are essential to arguments presented here, key elements of an emerging synthetic theory of the ecosystem are briefly summarized in this section (see Schindler et al., 1980; Waide et al., 1980; O'Neill and Waide, 1981; Allen, O'Neill, and Hoekstra, 1984a; O'Neill et al., 1986 for further details). This theoretical approach rests upon conceptualization of the ecosystem in macroscopic structural and functional terms as a hierarchical biogeochemical system.

Based on the formalism provided by irreversible or nonequilibrium thermodynamics (Prigogine, 1978, 1980), the ecosystem like any persistent biological structure (Morowitz, 1978; Mercer, 1981), may be viewed as a configuration of matter and energy that persists in far-from-equilibrium states by dissipating biologically elaborated, free-energy gradients, thereby forming organic structures out of inorganic elements mobilized from the physicochemical environment. Thus, the ecosystem may be viewed as a functional, biogeochemical system (Fig. 14.1); energy in solar radiation is converted into chemical bond energy by autotrophic photosynthesis. This chemical bond energy represents a free-energy gradient that is dissipated in order to build persistent, high-energy organic structures out of low-energy inorganic compounds acquired from the surrounding geochemical matrix. Upon death, organic structures are decomposed, with energy being dissipated as heat and contained elements returned to the geochemical matrix in a state of low chemical potential.

Hence, it is appropriate to view biogeochemical element cycles both as a necessary consequence of energy dissipation at the level of the ecosystem (Morowitz, 1966, 1974) and as allowing or facilitating a certain level of energy dissipation governed by physicochemical constraints on biological element mobilization and recycling (Webster, Waide, and Patten, 1975). Thus, the ecosystem may be conceptualized macroscopically as an open, dissipative structure (Blackburn, 1973) and as a persistent organic configuration maintained in far-from-equilibrium states by coupled levels of energy dissipation and biogeochemical cycling. This thermodynamic perspective provides the theoretical basis for relating the stability of natural ecosystems to macroscopic structural and functional characteristics.

The second component of the theoretical viewpoint adopted here, which provides the basis for scale-dependent analyses of ecosystem dynamics, derives from concepts of hierarchical organization (Simon, 1962; Weiss, 1971; Pattee, 1973) as applied in an ecological context (Overton, 1975; Webster, 1979; Schindler et al., 1980; Waide et al., 1980; O'Neill and Waide, 1981; Allen and Starr, 1982; Allen, O'Neill, and Hoekstra, 1984a; O'Neill et al., 1986). Central to hierarchical approaches is a concern with time constants associated with levels of system organization; definition of system structure is contingent on the design of sampling strategies or filters that reveal frequencies at which relevant system behaviors occur. Hierarchical systems are structured into a series of vertical levels, with higher levels corresponding to lower frequencies of behavior (that is, slower time constants) and lower organizational levels to higher frequencies. Further structuring of hierarchical systems

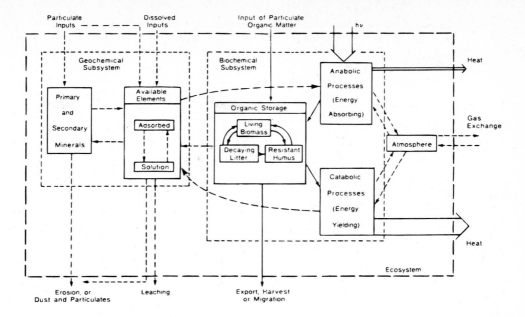

Figure 14.1 Conceptual model of the ecosystem as a biogeochemical system. Boxes in the diagram represent important storage pools of elements within ecosystems or processes involving the synthesis and degradation of high-energy organic structures out of low-energy inorganic compounds. Double arrows depict the flow of energy through the system, solid single arrows represent transfer of elements bound in organic forms, and dashed arrows symbolize transfers of inorganic elements in solid, solution, or gaseous forms. The biochemical subsystem is seen to depend on constant energy input and dissipation and on continual exchanges of elements with the atmosphere and the geochemical subsystem. The entire ecosystem is thus represented thermodynamically as an open, dissipative structure, exchanging both energy and matter with the surrounding biosphere (diagram modified and adapted from Bormann and Likens, 1979; Schindler et al., 1980).

is based on strengths of interactions among components within a given frequency- or rate-defined level; strongly interacting components form subsystems at a given hierarchical level and are functionally delimited from other groups of strongly interacting components within that same level at points where interaction strengths are reduced in magnitude. In such hierarchical systems, interactions between subsystems at a given level define the dominant time constant at the next higher level, so subsystems at one level become individual components at the next higher level. Thus, each level is effectively and functionally isolated from other levels and is at once a constraint on lower levels and constrained by higher levels. Each level may thus be viewed as a persistent level of system organization and as a structure with dynamics that are essentially independent of detailed, higher-frequency behaviors at lower organizational levels.

What is suggested here, and discussed more completely elsewhere (for example, see O'Neill et al., 1986, for a thorough but slightly different treatment of these ideas), is that the structuring of ecosystems as hierarchical or scale-dependent systems is a natural consequence of the evolutionary emergence of these systems as open, dissipative structures in spatially and temporally variable environments. Natural physicochemical environments may be characterized by oscillations of relevant variables at various scales of space and time (for example, frequency components in time series of precipitation and temperature, hydrodynamic processes in aquatic environments, and statistical recurrence intervals for wildfire or major windstorm). Such oscillations represent frequency signals that become incorporated into the structure of the ecosystem through the evolution of common life history adaptations to similar frequency signals by constituent species populations. Through this process of evolutionary incorporation (Allen and Iltis, 1980), ecosystem dynamics become coupled to environmental oscillations at ever-wider scales of space and time. As a consequence, high-frequency dynamics reflect ecosystem behaviors at relatively fine scales of space and time, whereas low-frequency dynamics occur over broader space–time scales. Thus, ecological sampling schemes essentially involve the design of scale-dependent filters or windows through which we recognize ecosystem organization and dynamics at certain scales of space and time (see also Allen, Sadowski, and Woodhead, 1984b).

The theoretical conceptualization of ecosystems as hierarchical biogeochemical systems provides the rationale for scale-dependent, macroscopic analyses of stability in relation to observable structural and functional characteristics of ecosystems. This approach also provides a basis for resolving conflicting approaches to stability by ecologists working at population and ecosystem levels of organization. That is, the ecosystem must be recognized as a dual hierarchy (O'Neill et al., 1986), both of functional processes and of species populations. The functional or process view of the ecosystem, the central concern in this analysis, focuses attention on macroscopic structural and functional characteristics of ecosystems and their coupling to physicochemical environments at definable spatiotemporal scales. The species-centered view of the ecosystem, on the other hand, is concerned with the dynamics, persistence, and interactions of constituent biotic populations. In relation to the physicochemical environment defined at a specific spatiotemporal scale, the processes of energy dissipation and biogeochemical cycling represent functional constraints on the behavior and evolution of species. But it is the species themselves that "carry out" these functional processes and that are responsible for the scale-dependent evolutionary coupling of the total biotic structure (that is, the ecosystem) to the physicochemical environment (Waide, 1982).

This view of the ecosystem as a dual hierarchy may provide resolution to otherwise disparate ecological views of stability. Ecosystem ecologists argue in a macroscopic structural and functional sense for viewing ecosystems as conditionally stable systems in environmentally defined basins of attraction. Population-oriented ecologists focus instead on the persistence of species assemblages and on multiple stable states of ecosystem organization (Sutherland, 1974; May, 1977; Holling, 1981). Such approaches may not be inconsistent. Viewed in terms of a hierarchy of processes, ecosystem structure and function are insensitive to considerable variation in species composition. Thus, distinct species assemblages appear to realize comparable levels of structure and function, and a single basin of attraction defined in macroscopic structural and functional terms may encompass several distinct basins defined in relation to the persistence of species assemblages. Much of the purported evidence in support of multiple stable states has been discounted in part because previous studies did not analyze the dynamics of species assemblages over appropriate scales of space and time (Connell and Sousa, 1983). However, this does not invalidate the suggested relation between ecosystem- and population-oriented definitions of stability in terms of a dual hierarchy. This approach essentially represents a functional view of evolutionary convergence at the level of the ecosystem. Investigating relationships between the dynamics and stability of the ecosystem viewed as a hierarchy of functional processes, as opposed to a hierarchy of species assemblages, represents a major challenge for contemporary ecology and one that is currently receiving insufficient attention.

THE ECOSYSTEM: A CONDITIONALLY STABLE SYSTEM

Previous sections of this chapter have provided the theoretical basis for a macroscopic approach to ecosystem stability. Based on these ideas, I summarize in this section explicit arguments derived from irreversible thermodynamics, hierarchy theory, natural history observations, and results of experimental ecosystem research to support my contention that ecosystems may be viewed abstractly as *conditionally stable attractors* within an environmentally defined state space. These arguments provide the basis for the revised resistance–resilience model presented in the next section.

Thermodynamic Perspective

From the perspective of nonequilibrium thermodynamics, the biological structures that persist in far-from-equilibrium states are only metastable (Morowitz, 1978; Prigogine, 1980). In such dissipative structures, organizational integrity is maintained by a certain level of energy dissipation (in fact, energy dissipation or entropy production represents a Liapunov function for dissipative structures that guarantees their local stability; Prigogine, 1980). Higher or lower levels of energy dissipation (entropy production) result in functionally different organizational states.

This thermodynamic view of metastable states is equivalent to the view of the ecosystem as a conditionally stable system. From the thermodynamic perspective, the ecosystem is a persistent structure maintained in nonequilibrium states by coupled levels of energy dissipation and biogeochemical cycling (Fig. 14.1). Energy fixation and subsequent dissipation result from metabolic activity by the existing organic structure and provide for the elaboration of new structure. This metabolic function represents an autocatalytic capacity at the level of the ecosystem, which may result in unbounded and hence unstable growth. It is specifically the coupling of energy dissipation to biogeochemical cycling and consequent kinetic limitations on biotic mobilization and recycling of essential elements that place bounds on autocatalytic growth processes (Webster, Waide, and Patten, 1975). These kinetic limitations on element cycling result from the space–time coupling of a specific biotic structure (that is, ecosystem) to a given physicochemical environment. Factors that break this scale-dependent coupling destroy the balance between energy dissipation and biogeochemical cycling and cause the ecosystem to collapse or expand to functionally different organizational states. Hence, the ecosystem is a conditionally stable (or metastable) attractor within an environmentally defined state space.

Hierarchical Perspective

Implicit in the hierarchical approach to biological systems (Simon, 1962; Weiss, 1971; Pattee, 1973) is the concept of stability. This is made explicit in Bronowski's (1972) discussion of stratified stability. Each level in the hierarchy of systems is viewed as a stable structure independent of structural details and high-frequency behaviors associated with lower organizational levels. Each level is thus viewed as a stable level of system organization and evolution that provides the template for further evolutionary advance to higher organizational levels. Thus, hierarchical organization and stratified stability are inherent in the evolution of complex biological systems.

The view of ecosystems as hierarchical or scale-dependent systems is equivalent to this concept of stratified stability (see also O'Neill et al., 1986). Because macroscopic characteristics of ecosystems have become evolutionarily coupled to environmental oscillations at a certain scale of space and time, the dynamics and functional integrity of the ecosystem are governed by processes operating at that scale, independently of processes operating at other scales. Factors that destroy this space–time coupling of ecosystem structure and function to specific frequency ranges of environmental signals result in a functionally different system. Thus, stratified stability is another way of stating that conditional stability is a scale-dependent attribute of ecosystems.

Natural History Perspective

Ecosystem analysis, in its rush to become more rigorous, must not ignore the critical importance of natural history research. I suggest that natural history observations provide additional evidence for the present macroscopic focus on ecosystem (conditional) stability. Early natural history observations identified persistent space–time patterns or regularities in the distribution of organic structures over the earth, particularly in relation to similar patterns in physicochemical environments (see O'Neill et al., 1986, for a discussion of the relation of these observations to the emergence of a scale-dependent view of ecosystems).

In relation to these observations, the physicochemical environment may be viewed as being textured at various scales of space and time by circulation patterns of the atmosphere and oceans and by consequent scale-dependent discontinuities in the distribution of solar radiation, moisture, and essential elements. By placing specific physicochemical constraints on levels of energy dissipation and biogeochemical cycling achieved by the organic structures we term ecosystems, this texturing defines (abstractly) the scale-dependent limits of basins of attraction within an environmentally defined state space. Thus, it is appropriate to view ecosystems as conditionally stable systems in relation to textured physicochemical environments and to elucidate relationships between macroscopic ecosystem attributes and physicochemical constraints on biotic processes at various scales of space and time.

Evidence from Experimental Ecosystem Research

Finally, results of intensive research on natural ecosystems are beginning to provide explicit experimental evidence for the macroscopic view of ecosystems as conditionally stable systems. Such evidence is of two main types: (1) research on space–time dynamics of ecosystem structure and function in relation to oscillations in relevant environmental variables, and (2) data on mechanisms and rates of ecosystem recovery following experimental manipulation.

The first category includes, among others, studies by McNaughton (1985) on coupled dynamics of vertebrate grazing and grass production in relation to space–time patterns of rainfall in African savanna ecosystems, research by Sprugel and Bormann (1981) on dynamics of productivity and biogeochemistry in northeastern wave-regenerated balsam-fir forests, and considerable research on dynamics of coniferous forests in response to space–time patterns in the occurrence of wildfire and defoliation (Holling, 1981). In each case the persistence and functional integrity of the ecosystem is maintained by the coupling of certain biotic processes to the space–time pattern of oscillation in relevant environmental variables. If the space–time pattern of triggering environmental oscillations is altered, as in the case of wildfire suppression, then the ecosystem undergoes change to a functionally new organizational state, revealing the conditional stability of the previous state. It is exactly this type of experimental evidence that forms the basis for dynamic biological oceanography and limnology (Legendre and Demers, 1984), which views the scale-dependent organization of aquatic ecosystems as being coupled to hydrodynamic processes across scales of space and time.

The second category of evidence includes macroscopic studies of ecosystem response to experimental manipulation, such as well-known forest clear-cutting experiments at Hubbard Brook, New Hampshire (Bormann and Likens, 1979) and Coweeta (Swank and Crossley, 1988). In both cases, the state disruption of forest removal did not destroy the fundamental coupling of the forest ecosystem to the physicochemical environment, and a variety of biological processes were responsible for rapid recovery as measured by macroscopic structural and functional variables (Swank and Waide, 1980).

In the case of Hubbard Brook, results of watershed manipulation experiments, together with results of forest simulation models and studies on surrounding forested stands, were formalized as the "shifting mosaic steady-state model" (Bormann and Likens, 1979). This model accounts for the persistence of northern hardwood forests in relation to regional environmental oscillations and for forest recovery from experimental disturbance. This model focuses on recovery and persistence in macroscopic structural and functional terms and views the northern hardwood forest as a mosaic of developmental phases that, at a certain space–time scale, achieves a conditionally stable steady state.

TABLE 14.2 Estimated turnover rates and time constants for various macroscopic measures of forest ecosystem recovery from clear-cutting at Coweeta[a]

Ecosystem state measure	Watershed	Turnover rate[b] (τ^{-1}, year^{-1})	Time constant[b] (τ, year)
Streamflow	7	0.17	6
	13 (first cut)	0.085	12
	13 (second cut)	0.11	9
	28	0.17	6
	37	0.14	7
Solute Export:	7		
NO$_3$–N		0.20	5
K		0.17	6
Ca		0.17	6
Stream (NO$_3$–N)	Several[c]	0.14	7
Leaf area index	13	0.12	8
Leaf biomass	13	0.091	11
Stand density	13, 28	0.062	16
Basal area	13, 28	0.050	20

[a]Data derived from a variety of watershed clear-cutting experiments performed at Coweeta.
[b]Time constant defined as the time required for 0.632 of the total response to occur; the turnover rate is the inverse of the time constant.
[c]Based on stream chemistry from Watersheds 3, 7, 10, 13, 19, 22, 28, 37, 40, and 41.

In the case of Coweeta, numerous watershed-scale clear-cutting experiments have been conducted. From these experiments, data on a variety of structural and functional measures of ecosystem state are available for assessing rate and extent of ecosystem recovery (for example, streamflow, stream nutrient concentrations, nutrient export, leaf area and biomass, tree density and basal area). Analyses of postcutting dynamics based on this extensive data base reveal common patterns and rates of ecosystem recovery as measured by these structural and functional variables (Table 14.2). These results provide strong evidence both for the macroscopic recovery of southern Appalachian forests following clear-cutting and for the high predictability of ecosystem response to experimental disturbance when the fundamental ecosystem–environment coupling has not been disrupted.

REVISIONS TO THE RESISTANCE–RESILIENCE MODEL OF ECOSYSTEM RELATIVE STABILITY

Arguments presented above provide the theoretical basis for scale-dependent, macroscopic approaches to ecosystem stability and for viewing the ecosystem abstractly as a conditionally stable system coupled to a textured physicochemical environment at a certain scale of space and time. Thus, they also provide the stimulus for analyses of relative rather than absolute stability and for examining the comparative responses of ecosystems to experimental manipulation. Similar arguments provided justification for the original resistance–resilience model of relative stability (Webster, Waide, and Patten, 1975). However, the theoretical basis for these earlier arguments was not fully elaborated, and the model itself was based on a somewhat inappropriate mathematical formalism. This has generated conceptual and methodological criticisms of the resistance–resilience model, as well as confusions in model interpretation and application. To alleviate these problems, I first reexamine the original model and then reinterpret the concepts of resistance and resilience in relation to the theoretical arguments developed above.

Original Resistance-Resilience Model of Relative Stability

In defining resistance and resilience, Webster, Waide, and Patten (1975) restricted attention to ecosystem dynamics within definable basins of attraction. They were primarily concerned with the comparative or relative stability of ecosystems and with ecosystem characteristics associated with different levels of relative stability. Questions of relative stability concern the nature of ecosystem response to disturbance or displacement from nominal levels of structure and function.

The concepts of resistance and resilience were specifically defined in relation to the time course of change in some macroscopic measure of ecosystem state (for example, biomass, total nitrogen export) in response to a specific disturbance. Resistance was related (inversely) to the maximum extent of state displacement following disturbance, whereas resilience was related to the rate of recovery to the nominal state. Webster, Waide, and Patten (1975) suggested that these definitions of resistance and resilience as two components of ecosystem relative stability identified alternative aspects of ecosystem persistence in spatially and temporally variable environments. Resistance to displacement results from the elaboration and maintenance of ecosystem structure, whereas resilience following displacement is related to ecosystem metabolism or turnover rate. In the closed biogeochemical cycles of the biosphere (Hutchinson, 1948), the observable structural and functional properties of ecosystems were seen as being determined by a realized balance between factors favoring resistance and resilience. Thus, characteristics of biogeochemical element cycles were of central concern in analyzing components of ecosystem relative stability.

Webster, Waide, and Patten (1975) constructed and parameterized a hypothetical model of ecosystem biogeochemistry for eight idealized ecosystem types. Model responses to comparable disturbances were fit to the equation for a second-order damped oscillator. This allowed the concepts of resistance and resilience to be quantified (Child and Shugart, 1972; Waide et al., 1974), the eight ecosystem types to be ranked in terms of relative resistance and resilience, and specific ecosystem attributes associated with resistance and resilience to be identified. Resistance was shown to be related to large amounts of element storage in abiotic or biotic compartments and to long turnover times of these storage pools; resilience was related to rapid turnover and recycling rates. Thus, resistance and resilience were seen as being inverse concepts. Certain environments favor ecosystems maintaining high organic storage that turns over slowly, attributes that contribute to ecosystem persistence by enhancing resistance to displacement. In contrast, ecosystems in other environments maintain low organic storage but turn over rapidly and thus persist by responding rapidly to environmental oscillation (Webster, Waide, and Patten, 1975).

These conclusions were generally supported by results of other studies (Marks, 1974; O'Neill et al., 1975; Vitousek et al., 1979; Waring and Franklin, 1979; Watson and Loucks, 1979; Swank and Waide, 1980; Van Voris et al., 1980). However, experimental tests of these ideas using laboratory microcosms (Dwyer et al., 1978; Leffler, 1978) provided only partial support, and other investigators criticized both the conceptual basis of the resistance–resilience model (Botkin, 1980) and the analytical methods employed (Harwell, Cropper, and Ragsdale, 1977).

Revised Interpretation of Resistance and Resilience

Although the resistance–resilience model has provided a useful approach to characterizing ecosystem responses to exogenous disturbance, the model is not consistent with the theoretical approach to ecosystem analysis presented earlier. Together with specific criticisms of the model and other practical concerns (Waide and Jager, 1981), this suggests the need for a revised interpretation of resistance and resilience. To motivate this revision, I suggest that the original focus on resistance and resilience as two different aspects of ecosystem response to disturbance obscures more basic principles of ecosystem organization. Environments impinging on ecosystems and defining limits of basins of attraction are characterized not only by certain scale-dependent mean values, but also by oscillations about these means.

Over evolutionary time, the biota of every ecosystem have evolved in response to statistically regular oscillations of certain types, magnitudes, and periodicites. Thus, the physicochemical environment impinging on a given ecosystem exhibits a certain frequency or signal structure. The ecosystem coupled to this environment similarly exhibits a certain biotic turnover structure, that is, a complement of sizes, longevities, and metabolic or turnover rates among its constituent biota. In other words, over evolutionary time, the frequency structure of the environment has become incorporated into the biological turnover structure of the ecosystem through the life history evolution of constituent biota. This is an ecosystem, rather than a species-level property, and is a specific means of ecosystem–environment coupling.

In this context, consider briefly a temperate deciduous forest and a planktonic ecosystem characteristic of the open ocean as idealized end points along a continuum of ecosystem–environment frequency couplings. In the planktonic ecosystem, minimal biotic structure is present, and essential elements are not effectively retained within the planktonic assemblage, which is thus constantly losing elements to lower depths via settling of organisms and particulate matter (Pomeroy, 1970; 1975). Continuing productivity depends on hydrodynamic mixing processes for regeneration of essential elements. Thus, the physicochemical environment impinging on this ecosystem is characterized by frequent fluctuations in element availability and by hydrodynamic turbulence. A biotic structure characterized by small, short-lived autotrophs having high turnover rates represents a clear coupling to this environment. Such an ecosystem–environment coupling, characterized by rapid biotic responses to recurring oscillations in element availability, would be referred to as highly resilient in the context of Webster, Waide, and Patten (1975).

Forests, by contrast, are characterized by massive structures that retain large pools of elements in living and decaying organic tissues. These elements are released gradually as organic tissues decompose, with the released elements being taken up by root–mycorrhizal associations and free-living microbes so that essential elements are effectively retained and recirculated within the ecosystem. This large biotic structure, the result of a high rate of net production maintained over long generation times in an environment that oscillates at a much lower frequency than in the planktonic system, effectively moderates its own environment and buffers itself against oscillations in climatic conditions and in moisture and element availability. Due to the inverse relation between mass and metabolic rate inherent in biological organization, element pools in this ecosystem turn over slowly. Following Webster, Waide, and Patten (1975), this ecosystem would be referred to as highly resistant.

Although it may seem appealing to discuss differences between planktonic and forest ecosystems in the context of the original resistance–resilience model, I suggest that it is trivial to do so, in that what is involved is a more fundamental difference between ecosystems and associated physicochemical environments. That is, plankton and forest represent different biotic turnover structures coupled to environments that differ substantially in the level of productivity and organic mass they will support, as well as in the frequencies at which they oscillate (that is, in frequency structures). It is simply not useful to compare these systems in the context of "extent of displacement" and "rate of recovery" following some disturbance. These are two functionally different ecosystems undergoing dynamics at different space–time scales, and it is not reasonable to expect them to respond similarly to a given disturbance.

To make this point more clearly and to elucidate further relationships between revised concepts of resistance and resilience and the theoretical treatment of the ecosystem as a hierarchical biogeochemical system, I suggest that three macroscopic variables are sufficient to characterize the essential properties of ecosystems and their space–time couplings to specific physicochemical environments. Each variable actually has a dual character: a property of the physicochemical environment (Fig. 14.2a) realized by an associated property of the ecosystem (Fig. 14.2b).

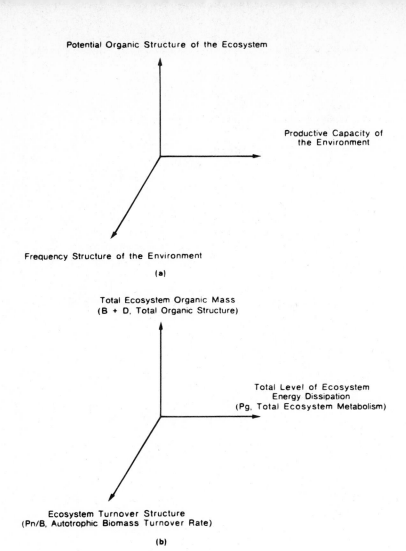

Figure 14.2 State-space diagram for representing ecosystem organization and properties in terms of the three macroscopic variables defined in the text. Each variable has a dual character: (a) an abstract property of the environment as realized by (b) an associated property of the ecosystem.

The first variable may be termed the productive capacity of the environment. Each physico-chemical environment, based on characteristic climatic or hydrodynamic conditions and element availability, will support a certain level of biotic productivity. Because it is impossible to measure this capacity in the absence of an ecological system to realize it, the gross autotrophic productivity of the ecosystem (Pg) represents a useful index of this capacity. From a thermodynamic perspective, Pg represents the total level of ecosystem metabolism, that is, the total level of energy fixation–dissipation implicitly coupled to a certain level of biogeochemical cycling.

The second macroscopic variable is related to the frequency structure of the environment, as incorporated into and measured by the biotic turnover structure of the ecosystem. Depending on the space–time scale of analysis, various measures of this variable are possible (note that measures of the other two variables must be referenced to this same scale). For this analysis, the biomass turnover rate Pn/B—the ratio of net autotrophic productivity to autotrophic biomass—is taken as a useful index of this variable, appropriate at the scale of the local ecosystem. For example, this is suggested as a measure of the dominant frequency of oscillation exhibited by a local plankton patch in response to small-scale hydrodynamic controls over biotic growth processes and element availability or by a local forest stand in response to relatively short-term oscillations (years to decades) in climatic variables or occurrences of wildfire or major windstorms.

The third macroscopic variable is the total ecosystem structure which can be maintained in a given environment. This is measured by the actual organic matter standing crop maintained by an ecosystem in a given environment, quantified as the sum of autotrophic biomass (B) and total nonliving organic matter (D).

These three macroscopic variables provide the basis for representing ecosystem organization graphically in relation to the theoretical concepts developed herein. That is, the ecosystem may be represented, abstractly in relation to the three-dimensional state space of Fig. 14.2, as a persistent organic structure maintained in far-from-equilibrium states ($B + D$) by coupled levels of energy dissipation and biogeochemical cycling (Pg) at a certain scale of space and time (Pn/B). Reichle, O'Neill, and Harris (1975, Fig. 2, p. 31) earlier utilized a similar graphical scheme to illustrate ecosystem organization in a different theoretical context.

To apply these concepts to ecosystems, representative data for a variety of terrestrial, wetland, freshwater, and marine ecosystems have been composited (Table 14.3). These values were derived primarily from original summaries in Rodin and Bazilevich (1967), Whittaker and Likens (1973a, b), and Waide and Jager (1981), plus additional supporting sources. Where direct comparisons are possible, these data appear similar (order-of-magnitude agreement or better) to the more extensive ecosystem inventories of Olson, Watts, and Allison (1983).

Relationships among these three macroscopic variables are graphed in Fig. 14.3. Each stick and ball in this figure corresponds to a given ecosystem type, interpreted as a locus of ecosystem organization or as a localized basin of attraction in an environmentally defined state space. Contrasts in organization among ecosystem types are apparent in the full three-dimensional diagram (Fig. 14.3a) and are further clarified by projections of points onto each of the three two-dimensional axes (Fig. 3b–d).

In these diagrams, the points do not vary continuously along each axis, but instead tend to appear as clusters of points separated by approximately order-of-magnitude changes in the value of the appropriate macroscopic variable. This is particularly true along the organic structure ($B + D$) and biomass turnover rate (Pn/B) axes, suggesting nearly quantum jumps in ecosystem structure due to increases in energy dissipation coupled to environments at increased scales of space and time. Each of these clusters represents ecosystems grouped according to the physiognomy of dominant autotrophs. Thus, localized at the back right of Fig. 14.3a and characterized by large organic structure and high metabolism but slow turnover is a series of closed forests. At successively lower values of ecosystem metabolism (Pg) and organic structure ($B + D$), but higher rates of biomass turnover (Pn/B), are three additional ecosystem clusters: terrestrial ecosystems dominated by open–canopy woody species or grasses; shallow or nearshore freshwater, wetland, or marine ecosystems dominated by emergent vascular plants; and open-water ecosystems dominated by planktonic autotrophs. Within these clusters other functional relationships are also apparent. For example, within each cluster a consistent, positive relationship exists between ecosystem metabolism and biomass turnover rate, a surrogate measure for the turnover rate of essential elements in that environment.

TABLE 14.3 Summary data on macroscopic structural and functional properties of a variety of terrestrial, wetland, freshwater, and marine ecosystem types.

Ecosystem type	Symbol[a]	Pg	Pn	B	$B+D$	Pn/B
Tropical rain forest	TRF	9.0	2.7	48	68	0.056
Tropical seasonal forest	TSF	5.0	2.0	38	58	0.053
Temperate evergreen forest	TEF	3.25	1.3	35	62	0.037
Temperate deciduous forest	TDF	2.9	1.2	34	58	0.035
Boreal forest	BF	2.0	0.8	22	56	0.035
Swamp forest[b]	SF	2.8	1.1	17.5	35	0.063
Woodland–shrubland	WS	2.0	0.8	5.8	20	0.138
Mangrove swamp	MS	6.0	4.0	6.0	28	0.670
Savanna	Sa	1.5	0.9	4.0	12	0.225
Grassland	Gr	1.2	0.7	1.8	20	0.400
Desert–shrub	DS	0.3	0.2	0.65	12	0.277
Extreme desert	De	0.005	0.003	0.02	0.02	0.150
Tundra	Tu	0.26	0.16	1.0	35	0.160
Bog	Bg	0.30	0.18	0.86	34	0.210
Palustrine wetland[c]	PW	2.1	1.3	1.0	1.7	1.30
Salt–tidal marsh	SM	2.5	1.5	3.3	3.5	0.454
Lacustrine, littoral	Lt	1.4	1.1	0.46	1.8	2.37
Estuarine, littoral	EL	6.3	3.6	0.75	1.6	4.76
Marine reef–algal bed	RAB	3.8	2.3	1.5	1.6	1.53
Streams, creeks, springs	SC	0.64	0.34	0.097	0.81	3.46
Open estuary	Es	0.58	0.34	0.005	0.12	64.4
Lacustrine, limnetic	Lm	0.33	0.25	0.004	0.073	63.0
Coastal upwelling	Up	2.7	1.6	0.011	0.040	146.
Continental shelf	Sh	0.40	0.24	0.005	0.030	48.0
Open ocean	Oc	0.21	0.12	0.003	0.023	41.7

Relationships among these macroscopic ecosystem properties are graphed in Fig. 14.3. The variables Pg, Pn, B, and D, respectively, refer to estimated values of gross and net primary productivity (kg dry matter m^{-2} y^{-1}), total autotrophic biomass (kg m^{-2}), and total nonliving organic matter (that is, total detritus, kg m^{-2}). The autotrophic biomass turnover rate, Pn/B, is in units of y^{-1}. Most of the data shown here were taken or calculated form the earlier summaries in Whittaker and Likens (1973a, b), Rodin and Bazilevich (1967), and Waide and Jager (1981). Supporting data were compiled from Reiners (1973), Wetzel and Rich (1973), Woodwell et al. (1978), Whittaker (1975), and Schlesinger (1977). In general, data on terrestrial ecosystems represent composites of values in the Whittaker–Likens and Rodin–Bazilevich summaries; data on wetland, freshwater, and marine ecosystems were largely composited from the Whittaker–Likens and Waide–Jager summaries. In cases where productivity estimates were available only in terms of Pn, values of Pg were estimated using the following Pn/Pg ratios: for TRF, 0.3; for other ecosystems containing woody vegetation, 0.4; and for all other ecosystems, 0.6. These estimated ratios were taken from the sources cited above, plus McNaughton and Wolf (1979).

[a]These symbols are used to identify points plotted in Fig. 14.3.
[b]Combination of values shown in Waide and Jager (1981) for bog forests and wooded swamps.
[c]Combination of values shown in Waide and Jager (1981) for fens, meadow marshes, and lotic marshes.

Of greatest relevance to discussions here is the strong inverse relationship ($r^2 = 0.94$) between total organic structure and the biomass turnover rate (Fig. 14.3d) (excluding the single point for extreme desert, De, which appears as an outlier on all axes). These two variables were shown by Webster, Waide, and Patten (1975) to be strongly correlated with resistance and resilience, respectively. Thus, this diagram represents the postulated inverse relationship between resistance and resilience as originally defined. However, Fig. 14.3 and the underlying theoretical arguments show this to involve more general relationships among fundamental, macroscopic properties of ecosystems viewed as hierarchical biogeochemical systems. Figure 14.3d depicts more than simple differences among ecosystems in terms of resistance to and resilience following disturbance. This figure represents fundamental differences among functionally distinct ecosystems maintaining structure and undergoing dynamics at different scales of space and time. It is trivial to refer to these differences in terms of resistance and resilience. Rather, inasmuch as mass and metabolic or turnover rate are inversely related at all levels of biological organization (Conrad, 1973), resistance and resilience

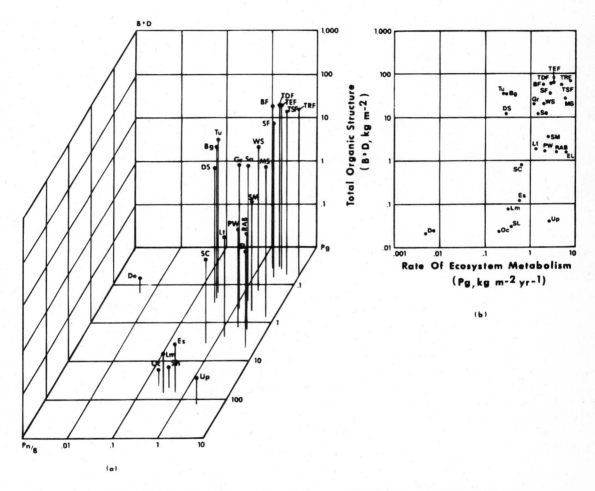

Figure 14.3 Graphical representation of structural and functional properties of various ecosystems in terms of three macroscopic variables: the total organic structure of the ecosystem ($B + D$, kg m^{-2}), the rate of ecosystem metabolism (Pg, kg m^{-2} y^{-1}), and (next page) the autotrophic biomass turnover rate (Pn/B, y^{-1}). Graphs shown here are based on data and symbols presented in Table 14.3. In part (a), the full three-dimensional relationship among these variables is graphed. In parts (b) to (d), projections of the individual points onto the three two-dimensional axes are presented. Part (d) depicts the inverse relationship between the apparent resistance and resilience as originally defined by Webster, Waide, and Patten (1975).

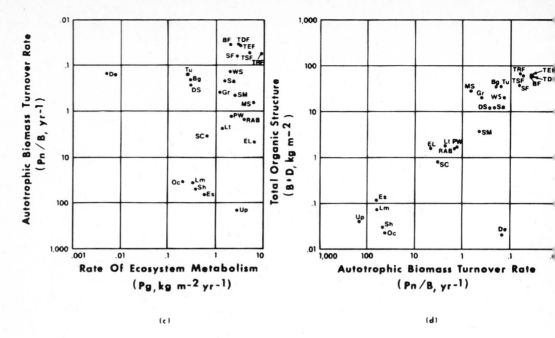

Figure 14.3 Continued.

are reinterpreted theoretically here as scaling variables that reflect the space–time coupling of the biotic structures we term ecosystems to specific physicochemical environments.

Hence, the terms resistance and resilience should not be applied to analyses of postdisturbance dynamics by functionally distinct ecosystems. Such dynamics are shown here to result from fundamental scale-dependent properties and rate constants inherent in the organization of natural ecosystems. However, the concepts of resistance and resilience still have utility, particularly in a management context, where their application is restricted to postdisturbance dynamics within a given basin of attraction, that is, when applied to different local realizations of a given ecosystem type, which exhibit comparable levels of organic structure and energy dissipation–biogeochemical cycling at a common scale of space and time (for example, to the recovery of different mixed hardwood stands following management intervention). Even here, the concept of resistance requires slight revision. What is important is not the absolute magnitude of the change in state following disturbance, but whether the change is or is not sufficiently great to push the ecosystem into a new basin of attraction, thereby causing it to assume a new and functionally distinct organizational state. Thus, resistance should be measured in relation to the size of the relevant basin of attraction, that is, the magnitude of state change or disturbance that can be tolerated and still allow functional integrity to be preserved, but beyond which recovery is impossible. The interpretation of resilience remains unchanged: the rate of recovery to the previous macroscopic state, for disturbances that do not push the ecosystem into new basins of attraction. Note that the nominal macroscopic state can be defined abstractly in relation to a singular point, a limit cycle, or other relevant trajectory.

Finally, what factors regulate ecosystem resistance and resilience as redefined here? Again, biogeochemical considerations appear critical, and the earlier results of Webster, Waide, and Patten (1975) generally apply. Hence, resistance is enhanced by the presence of large stores of essential elements, partly in abiotic pools, but especially in slowly decaying organic forms. Where such pools are absent or are largely depleted by a given disturbance or where biological populations are prevented from utilizing these reserves (for example, due to toxicants), the ecosystem may be pushed into a new basin of attraction. Functional changes in planktonic ecosystems of the lower Great Lakes, presumably due to eutrophication-induced silica depletion (Schelske et al. 1983), serve as an unusual example of this point. Similarly, resilience is related to the rate at which essential elements are recycled and made available for uptake by regenerating biota following disturbance. Where rates of element mobilization and uptake are severely limited by climatic or physicochemical constraints, where available elements are physically immobilized or lost from the ecosystem in solution or gaseous form, or where the disturbance inhibits rates of recycling and uptake, then the rate of ecosystem recovery will be reduced.

CONCLUSIONS

Global patterns of circulation in the atmosphere and oceans, as related to continental position and topography, produce scale-dependent discontinuities in the distribution of solar energy, moisture, and essential elements. These discontinuities represent a texturing of the earth's environment in space and time, which places scale-dependent physicochemical constraints on the exchange, transformation, and storage of energy and essential elements by living organisms. Over evolutionary time, this space–time texture has become incorporated into biotic structures through similar life history adaptations of species populations to shared environmental constraints and frequency signals.

Thus, coupled to a structured physicochemical environment is a textured biosphere. From this mosaic of living forms we abstract—at certain scales of space and time—persistent units of structure or pattern for study as ecosystems. The ecosystem is thus viewed as a scale-dependent macroscopic unit of study and as a persistent structure maintained in organized states far from thermal equilibrium by coupled processes of energy dissipation and biogeochemical cycling. The organization and behavior of the ecosystem depend on functional processes operating at the scale of observation. Factors that interrupt the space–time coupling of the ecosystem to a specific physicochemical environment decouple the processes of energy dissipation and element cycling from the existing structure and force the ecosystem into a new organizational state. The ecosystem may thus be viewed abstractly (in terms of its dynamic behaviors) as a conditionally stable attractor or (as a unit of organization in a textured biosphere) as a basin of attraction in an environmentally defined state space.

Viewed in terms of the inverse relation between mass and metabolism inherent in biological organization, the concepts of resistance and resilience are reinterpreted in the present theoretical context as scaling variables that reflect the space–time coupling of specific biotic assemblages to physicochemical environments. These terms should not be applied to the response of distinctly different ecosystems to exogenous disturbances. Such postdisturbance responses of functionally different ecosystems reflect dynamics inherent in scale-dependent ecosystem organization.

However, the concepts of resistance and resilience may be usefully applied to disturbance responses within a given basin of attraction, that is, to the disturbance responses of different local realizations of a given ecosystem type exhibiting comparable levels of structure and function at similar spatiotemporal scales. In this restricted context, and particularly in relation to ecosystem management, it becomes important to determine factors that prevent the ecosystem from assuming a new organizational state following disturbance (ecosystem resistance defined in relation to the size of the basin of attraction) and that regulate its rate of recovery to the undisturbed or nominal state (ecosystem resilience). These two components of ecosystem relative stability are related to characteristics of biogeochemical element cycles, particularly the presence of large element storages in living and slowly decaying organic pools (resistance) and the rate at which elements are mobilized and recycled by biota (resilience).

Finally, the macroscopic approach adopted here clarifies much of the existing confusion concerning ecosystem stability and helps to resolve conflicting approaches to stability among population and ecosystem theorists. Implicit in the present theoretical approach to ecosystems is the concept of conditional stability, suggesting that excessive attention has been given to this topic in the past. However, even though macroscopic requirements of matter and energy processing impose constraints on behaviors of species components of ecosystems, sufficient behavioral flexibility appears to result from individualistic adaptations of species populations to result in substantial functional redundancy in ecosystem composition. Thus, functional integrity at a macroscopic level may appear as multiple stable states in reference to the persistence of species assemblages. Such a functional view of evolutionary convergence at the level of the ecosystem may be especially important in two circumstances: in climatically similar regions on different continents (due to different evolutionary histories of distinct biotas) and following successive disturbances of the same ecosystem (compositional differences due to differential sampling of the total environmental frequency structure by the available species pool). However, hard evidence in support of such a view of ecosystem functional convergence is presently lacking. Determining relationships between the ecosystem viewed as a hierarchy of functional processes versus as a hierarchy of species assemblages remains a major challenge for empirical and theoretical ecologists.

REFERENCES

ALLEN, T. F. H., and ILTIS, H. H. 1980. Overconnected collapse to higher levels: urban and agricultural origins, a case study. In Banathy, B. H. (ed.), *Systems Science and Science, Proceedings of the Twenty-fourth Annual North American Meeting of the Society for General Systems Research,* Systems Science Institute, Louisville, Kentucky, pp. 96–103.

ALLEN, T. F. H., and STARR, T. B., 1982. *Hierarchy: Perspectives for Ecological Complexity.* University of Chicago Press, Chicago. 310 pp.

ALLEN, T. F. H., O'NEILL, R. V., and HOEKSTRA, T. W. 1984a. *Interlevel Relations in Ecological Research and Management: Some Working Principles from Hierarchy Theory.* Gen. Tech. Rpt. RM-110. USDA Forest Service, Rocky Mountain Forest and Range Experiment Station, Fort Collins, Colorado. 11 pp.

ALLEN, T. F. H., SADOWSKI, D. A., and WOODHEAD, N. 1984b. Data transformation as a scaling operation in ordination of plankton. *Vegetatio* 56:147–160.

BLACKBURN, T. R. 1973. Information and the ecology of scholars. *Science* 181:1141–1146.

BORMANN, F. H., and LIKENS, G. E. 1979. *Pattern and Process in a Forested Ecosystem.* Springer-Verlag, New York. 253 pp.

BOTKIN, D. B. 1980. A grandfather clock down the staircase: stability and disturbance in natural ecosystems. In Waring, R. H. (ed.), *Forests: Fresh Perspectives from Ecosystem Analysis.* Oregon State University Press, Corvallis, pp. 1–10.

BRONOWSKI, J. 1972. New concepts in the evolution of complexity, Part II. *Amer. Scholar* 42:110–122.

CHILD, G. I., and SHUGART, H. H., JR. 1972. Frequency response analysis of magnesium cycling in a tropical forest ecosystem. In Patten, B. C. (ed.), *Systems Analysis and Simulation in Ecology,* Vol. 2. Academic Press, New York, pp. 103–135.

CONNELL, J. H., and SOUSA, W. P. 1983. On the evidence needed to judge ecological stability or persistence. *Am. Nat.* 121:789–824.

CONRAD, M. 1973. Thermodynamic extremal principles in evolution. *Biophysik* 9:191–196.

DAVIS, M. B. 1976. Pleistocene biography of temperate deciduous forests. *Geoscience Man* 13:13–26.

DWYER, R. L., NIXON, S. W., OVIATT, C. A., PEREZ, K. T., and SMAYDA, T. J. 1978. Frequency response of a marine ecosystem subjected to time-varying inputs. In Thorp, J. H., and Gibbons,

J. W. (eds.), Energy and Environmental Stress in Aquatic Systems. DOE Symp. Series (CONF-771114), National Technical Information Services, Springfield, Virginia, pp. 19–38.

EGERTON, F. N. 1973. Changing concepts of the balance of nature. *Quart. Rev. Biol.* 48:322–350.

GERRITSEN, J., and PATTEN, B. C. 1985. System theory formulation of ecological disturbance. *Ecol. Mod.* 29:383–397.

GOODMAN, D. 1975. The theory of diversity–stability relationships in ecology. *Quart. Rev. Biol.* 50:237–266.

HARWELL, M. A., CROPPER, W. P., and RAGSDALE, H. L. 1977. Nutrient cycling and stability: a reevaluation. *Ecology* 58:660–666.

HOLDRIDGE, L. 1967. *Life Zone Ecology.* Tropical Science Center, San Jose, Costa Rica. 206 pp.

HOLLING, C. S. 1973. Resilience and stability of ecological systems. *Ann. Rev. Ecol. Syst.* 4:1–24.

HOLLING, C. S. 1981. Forest insects, forest fire, and resilience. In Mooney, H. A., Bonnickson, T. M., Christensen, N. L., Lotan, J. E., and Reiners, W. E. (technical coordinators), *Proc. Conf. Fire Regimes and Ecosystem Properties,* Gen. Tech. Rep. WO-26. USDA Forest Service, Washington, D.C., pp. 445–464.

HUTCHINSON, G. E. 1948. Circular casual systems in ecology. *Ann. New York Acad. Sci.* 50:221–246.

HUTCHINSON, G. E. 1965. *The Ecological Theater and the Evolutionary Play.* Yale University, Press, New Haven, Connecticut. 139 pp.

LEFFLER, J. W. 1978. Ecosystem response to stress in aquatic microcosms. In Thorp, J. H., and Gibbons, J. W., (eds.), Energy and Environmental Stress in Aquatic Systems. U.S. Dept. of Energy Symp. Ser. (CONF-771114), National Technical Information Services, Springfield, Virginia, pp. 102–119.

LEGENDRE, L., and DEMERS, S. 1984. Towards dynamic biological oceanography and limnology. *Can. J. Fish. Aquatic Sci.* 41:2–19.

LEWONTIN, R. C. 1970. The meaning of stability. *Brookhaven Symposia Biol.* 22:13–24.

LUTZ, H. J. 1963. Forest ecosystems: their maintenance, amelioration, and deterioration. *J. Forestry* 61:563–569.

MACARTHUR, R. H. 1955. Fluctuations of animal populations and a measure of community stability. *Ecology* 36:533–536.

MARKS, P. L. 1974. The role of pin cherry (*Prunus pennsylvanica* L.) in the maintenance of stability in northern hardwood ecosystems. *Ecol. Mon.* 44:73–88.

MAY, R. M. 1973. *Stability and Complexity in Model Ecosystems.* Princeton University Press, Princeton, New Jersey. 235 pp.

MAY, R. M. 1977. Thresholds and breakpoints in ecosystems with a multiplicity of stable states. *Nature* 269:471–477.

MCNAUGHTON, S. J. 1985. The ecology of a grazing ecosystem: the Serengeti. *Ecol. Mon.* 55:259–294.

MCNAUGHTON, S. J., and WOLF, L. L. 1979. *General Ecology,* 2nd ed. Holt, Reinhart and Winston, New York. 702 pp.

MERCER, E. H. 1981. *The Foundations of Biological Theory.* Wiley, New York. 232 pp.

MONK, C. D., CROSSLEY, D. A., JR., TODD, R. L., SWANK, W. T., WAIDE, J. B., and WEBSTER, J. R. 1977. An overview of nutrient cycling research at Coweeta Hydrologic Laboratory. In Correll, D. L. (ed.), Watershed Research in Eastern North America. Chesapeake Bay Center for Environmental Studies, Smithsonian Institution, Edgewater, Maryland, pp. 35–50.

MOROWITZ, H. J. 1966. Physical background of cycles in biological systems. *J. Theor. Biol.* 13:60–62.

MOROWITZ, H. J. 1974. The derivation of ecological relationships from physical and chemical principles. *Proc. Nat. Acad. Sci. USA* 71:2335–2336.

MOROWITZ, H. J. 1978. *Foundations of Bioenergetics.* Academic Press, New York. 344 pp.

ODUM, E. P. 1953. *Fundamentals of Ecology.* W.B. Saunders, Philadelphia. 384 pp.

OLSON, J. S., WATTS, J. A., and ALLISON, L. J. 1983. Carbon in live vegetation of major world ecosystems. ORNL-5862. Environmental Sciences Division, Oak Ridge National Laboratory, Oak Ridge, Tennessee. 164 pp.

O'NEILL, R. V., and WAIDE, J. B. 1981. Ecosystem theory and the unexpected: implications for environmental toxicology. In Cornaby, B. W. (ed.), *Management of Toxic Substances in Our Ecosystems: Taming the Medusa*. Ann Arbor Science Publishers, Ann Arbor, Michigan, pp. 43–73.

O'NEILL, R. V., HARRIS, W. F., AUSMUS, B. S., and REICHLE, D. E. 1975. A theoretical basis for ecosystem analysis with particular reference to element cycling. In Howell, F. G., Gentry, J. B., and Smith, M. H. (eds.), Mineral Cycling in Southeastern Ecosystems. Energy Research and Development Administration (ERDA) Symp. Series (CONF-740513), National Technical Information Services, Springfield, Virginia, pp. 28–40.

O'NEILL, R. V., DEANGELIS, D. L., WAIDE, J. B., and ALLEN, T. H. F. 1986. *A Hierarchical Concept of Ecosystems*. Princeton University Press, Princeton, New Jersey. 253 pp.

OVERTON, W. S. 1975. The ecosystem modeling approach in the Coniferous Forest Biome. In Patten, B. C. (ed.), *Systems Analysis and Simulation in Ecology*, Vol. 3. Academic Press, New York, pp. 117–138.

PATTEE, H. H. (ed.) 1973. *Hierarchy Theory: The Challenge of Complex Systems*. George Braziller, New York. 156 pp.

PATTEN, B. C. 1978. Systems approach to the concept of environment. *Ohio J. Sci.* 78:206–222.

PIMM, S. L. 1982. *Food Webs*. Chapman and Hall, London. 219 pp.

PIMM, S. L. 1984. The complexity and stability of ecosystems. *Nature* 307:321–326.

POMEROY, L. R. 1970. The strategy of mineral cycling. *Ann. Rev. Ecol. Syst.* 1:171–190.

POMEROY, L. R. 1975. Mineral cycling in marine ecosystems. In Howell, F. G., Gentry, J. B., and Smith, M. H. (eds.), Mineral Cycling in Southeastern Ecosystems. Energy Research and Development Administration (ERDA) Symp. Ser. (CONF-740513), National Technical Information Services, Springfield, Virginia, pp. 209–223.

PRIGOGINE, I. 1978. Time, structure, and fluctuations. *Science* 201:777–785.

PRIGOGINE, I. 1980. *From Being to Becoming: Time and Complexity in the Physical Sciences*. W. H. Freeman, San Francisco. 272 pp.

REICHLE, D. E., O'NEILL, R. V., and HARRIS, W. F. 1975. Principles of energy and material exchange in ecosystems. In van Dobben, W. H., and Lowe-McConnell, R. H. (eds.), *Unifying Concepts in Ecology*, Dr. W. Junk, The Hague, pp. 27–43.

REINERS, W. A. 1973. Terrestrial detritus and the carbon cycle. In Woodwell, G. M., and Pecan, E. V. (eds.), Carbon and the Biosphere. Energy Research and Development Administration (ERDA) Symp. Series (CONF-720510), National Technical Information Services, Springfield, Virginia, pp. 303–327.

RICHARDS, B. N., and CHARLEY, J. L. 1983/1984. Mineral cycling processes and system stability in the Eucalypt forest. *Forest Ecol. Management* 7:31–47.

RODIN, L. E., and BAZILEVICH, N. I. 1967. *Production and Mineral Cycling in Terrestrial Vegetation*. Oliver and Boyd, Edinburgh. 288 pp.

ROSEN, R. 1970. *Dynamical System Theory in Biology*, Vol. 1: *Stability Theory and Its Applications*. Wiley, New York. 302 pp.

ROSEN, R. 1972. On the decomposition of a dynamical system into noninteracting subsystems. *Bull. Math. Biol.* 34:337–341.

SCHELSKE, C. L., STOERMER, E. F., CONLEY, D. J., ROBBINS, J. A., and GLOVER, R. M. 1983. Early eutrophication in the Lower Great Lakes: new evidence from biogenic silica in sediments. *Science* 222:320–322.

SCHINDLER, J. E., WAIDE, J. B., WALDRON, M. C., HAINS, J. J., SCHREINER, S. P., FREEMAN, M. L., BENZ, S. L., PETTIGREW, D. R., SCHISSEL, L. A., and CLARK, P. J. 1980. A microcosm approach to the study of biogeochemical systems: 1. Theoretical rationale. In Giesy, J. P. (ed.),

Microcosms in Ecological Research. U.S. Department of Energy Symp. Ser. (CONF-781101), National Technical Information Services, Springfield, Virginia, pp. 192–203.

SCHLESINGER, W. H. 1977. Carbon balance in terrestrial detritus. *Ann. Rev. Ecol. Syst.* 8:51–82.

SIMON, H. A. 1962. The architecture of complexity. *Proc. Amer. Philos. Soc.* 106:467–482.

SPRUGEL, D. G., and BORMANN, F. H. 1981. Natural disturbance and the steady state in high-altitude balsam fir forests. *Science* 211:390–393.

STEWART, I. 1985. Dynamical systems: attraction in a new idea. *Nature* 317:573–574.

SUTHERLAND, J. P. 1974. Multiple stable points in natural communities. *Amer. Nat.* 108:859–873.

SWANK, W. T., and WAIDE, J. B. 1980. Interpretation of nutrient cycling research in a management context: evaluating potential effects of alternative management strategies on site productivity. In Waring, R. W. (ed.), *Forests: Fresh Perspectives from Ecosystem Analysis.* Oregon State University Press, Corvallis, Oregon, pp. 137–158.

SWANK, W. T., and CROSSLEY, D. A. (eds.). 1988. *Forest Hydrology and Ecology at Coweeta.* Springer-Verlag, New York. 469 pp.

VAN VORIS, P., O'NEILL, R. V., EMMANUEL, W. R., and SHUGART, H. H. 1980. Functional complexity and ecosystem stability. *Ecology* 61:1352–1360.

VITOUSEK, P. M., GOSZ, J. R., GRIER, C. C., MELILLO, J. M., REINERS, W. A., and TODD, R. L. 1978. Nitrate losses from disturbed ecosystems. *Science* 204:469–474.

WAIDE, J. B. 1982. Comments on: using qualitative analysis to understand perturbations to marine ecosystems in the field and laboratory. In Archibald, P. A. (ed.), Environmental Biology State-of-the-Art Seminar, EPA-600/9-82-007. Office of Exploratory Research, Environmental Protection Agency, Washington, D.C., pp. 119–122.

WAIDE, J. B., and JAGER, Y. 1981. Methods for investigating the stability of wetland ecosystems. Report to the Institute of Water Resources, U.S. Army Corps of Engineers, Ft. Belvoir.

WAIDE, J. B., and SWANK, W. T. 1976. Nutrient cycling and the stability of ecosystems: implications for forest management in the southeastern U.S. *Soc. Am. Foresters Proc.* 1975:404–424.

WAIDE, J. B., and WEBSTER, J. R. 1976. Engineering systems analysis: applicability to ecosystems. In Patten, B. C. (ed.), *Systems Analysis and Simulation in Ecology,* Vol. 4. Academic Press, New York, pp. 329–371.

WAIDE, J. B., KREBS, J. E., CLARKSON, S. P., and SETZLER, E. M. 1974. A linear systems analysis of the calcium cycle in a forested watershed ecosystem. In Rosen, R., and Snell, F. M. (eds.), *Progress in Theoretical Biology,* Vol. 3. Academic Press, New York, pp. 261–345.

WAIDE, J. B., SCHINDLER, J. E., WALDRAN, M. C., HAINS, J. J., SCHREINER, S. P., FREEDMAN, M. L., BENZ, S. L., PETTIGREW, D. R., SCHISSEL, L. A., and CLARK, P. J. 1980. A microcosm approach to the study of biogeochemical systems: responses of aquatic laboratory microcosms to physical, chemical, and biological perturbations. In Giesy, J. P. (ed.), Microcosms in Ecological Research. U.S. Dept. of Energy Symp. Ser. (CONF-781101), National Technical Information Services, Springfield, Virginia, pp. 204–223.

WARING, R. H., and FRANKLIN, J. F. 1979. Evergreen coniferous forests of the Pacific Northwest. *Science* 204:1380–1386.

WATSON, V., and LOUCKS, O. L. 1979. An analysis of turnover times in a lake ecosystem and some implications for system properties. In Halfon, E. (ed.), *Theoretical Systems Ecology: Advances and Case Studies.* Academic Press, New York, pp. 355–383.

WEBSTER, J. R. 1979. Hierarchical organization and ecosystems. In Halfon, E. (ed.), *Theoretical Systems Ecology: Advances and Case Studies.* Academic Press, New York, pp. 119–131.

WEBSTER, J. R., WAIDE, J. B., and PATTEN, B. C. 1975. Nutrient recycling and the stability of ecosystems. In Howell, F. G., Gentry, J. B., and Smith, M. H. (eds.), Mineral Cycling in Southeastern Ecosystems. Energy Research and Development Administration (ERDA) Symp. Ser. (CONF-740513), National Technical Information Services, Springfield, Virginia, pp. 1–27.

Webster, J. R., Gurtz, M. E., Hains, J. J., Meyer, J. L., Swank, W. T., Waide, J. B., and Wallace, J. B. 1985. Stability of stream ecosystems. In Barnes, J. R., and Minshall, G. W. (eds.), *Stream Ecology.* Plenum, New York, pp. 375–395.

Weiss, P. A. 1971. The basic concept of hierarchic systems. In Weiss, P. A. (ed.), *Hierarchically Organized Systems in Theory and Practice.* Hafner, New York, pp. 1–44.

Wetzel, R. G., and Rich, P. H. 1973. Carbon in freshwater ecosystems. In Woodwell, G. M., and Pecan, R. V. (eds.), Carbon and the Biosphere. Energy Research and Development Administration (ERDA) Symp. Ser. (CONF-720510), National Technical Information Services, Springfield, Virginia, pp. 241–263.

Whittaker, R. H. 1956. Vegetation of the Great Smoky Mountains. *Ecol. Mon.* 26:1–80.

Whittaker, R. H. 1975. *Communities and Ecosystems,* 2nd ed. Macmillan, New York. 385 pp.

Whittaker, R. H., and Likens, G. E. 1973a. Carbon in the biota. In Woodwell, G. M., and Pecan, E. V. (eds.), Carbon and the Biosphere. Energy Research and Development Administration (ERDA) Symp. Ser. (CONF-720510), National Technical Information Services, Springfield, Virginia, pp. 281–302.

Whittaker, R. H., and Likens, G. E. (eds.) 1973b. The primary production of the biosphere. *Human Ecol.* 1:299–369.

Whittaker, R. H., and Woodwell, G. M. 1972. The evolution of natural communities. In Wiens, J. A. (ed.), *Ecosystem Structure and Function.* Oregon State University Press, Corvallis, Oregon, pp. 137–160.

Woodwell, G. M., Whittaker, R. H., Reiners, W. A., Likens, G. E., Delwiche, C. C., and Botkin, D. B. 1978. The biota and the world carbon budget. *Science* 199:141–146.

15

Disturbance and Stress Effects on Ecological Systems

DAVID J. RAPPORT

Institute for Research on Environment and Economy
University of Ottawa and Statistics Canada
Ottawa, Ontario

HENRY A. REGIER

Institute for Environmental Studies
University of Toronto
Toronto, Ontario

> Man is a vital part of most ecosystems, and there is an increasing human awareness of man's part in them and his influence on them. . . . Humans are both parts of, and manipulators of ecosystems. Induced instability of ecosystems is an important cause of economic, political, and social disturbance throughout the world. . . . In manipulating his environment . . . seldom has [man] foreseen the full consequences of his action [Van Dyne, in *Chiasma* 8:59–60 (1970)].

INTRODUCTION AND SUMMARY

Natural disturbance may promote rejuvenation of the natural ecosystem, informed husbandry may produce a desirable cultured ecosystem, and harmful stress leads, by definition, to undesirable ecosystem pathology. The process by which pathology develops is complex especially with interactive and synergistic consequences of multiple harmful stresses. The ecosystem distress syndrome provides a simplification: despite differing degenerative pathways, the degraded end states usually, but not always, exhibit some broad similarities. Once an ecosystem transforms due to moderately intense stress, it may develop into a stress-adapted but crippled state in which some lost functions reappear in a different guise. If the stresses become extremely intense, then all life is extinguished locally, which is the ultimate end state of a degraded ecosystem.

The process of ecosystem adaptation and accommodation to stress is not well explored scientifically. However, some critical properties conferring resistance have been identified, and it appears that certain vulnerable structural features (for example, soil integrity and wetland vegetation) may prove to be "Achilles heels" in a degeneration process.

The temporal and spatial development of ecosystem pathology mirrors the complexity of the multiple stresses, each acting with its own peculiar spatial and temporal character. A holistic approach to ecosystem transformations—both degradative and rehabilitative—is useful for identifying and describing various transitional sequences between a natural state, some degraded state due to conventional development and urbanization, and a more fully productive and sustainable state due to ecodevelopment of a natural state or redevelopment of a degraded state. As human pressures intensify and in the absence of sufficient commitments to stewardship and husbandry, stress effects will predictably intensify and spread further across the ecosystems of the biosphere. The study of *epicoenology,* concerning the vast effects of human stresses on ecosystems, thus appears to be an important field for expanded research.

STRESS AND DISTURBANCE

Ecosystems thrive on disturbance. Many owe their continuity and vigor to periodic disruptions of structure and process. Vogal (1980) provides an exhaustive review of these rejuvenation processes in systems ranging from streams to deserts. The classic example is the role of fire in maintaining many forest and grassland ecosystems that can scarcely do without fire. If the tall grass prairie is not periodically burned or grazed, productivity and species diversity declines. In the drier regions, the absence of such disturbance favors takeover of the grassland by woody species with the result that tall grass prairie transforms to shrub or forestland (Risser, 1987). Many wetlands fringing rivers, lakes, and seas persist because of occasional floods and storms of some intensity (Prince and D'Itri, 1985).

In maintaining ecosystems that persist because of disturbances, it appears that the type of disturbance is important and also the frequency and intensity of disturbances (Sousa 1984). Studies by Paine (1979) on the fortunes of the sea palm community of the exposed coast of the state of Washington tell the story. Too little or too much wave action leads to demise of this community; a certain frequency and intensity of disturbance maintains it.

The terms disturbance and stress are used interchangeably by some ecologists. We distinguish between them: one kind of disruption, a *disturbance,* helps to revitalize the ecosystem (Godron and Forman, 1983); the other, a *stress,* debilitates it, may cripple it, and if sufficiently intense may extinguish it. Selye (1973) distinguished eustress and distress to differentiate various kinds of challenges to the health of organisms. *Eustress* stimulates normal vigorous behavior, evokes adaptive responses, and thus strengthens the well-being of the organism. *Distress* may be followed by types of responses that may protect against demise but do not lead to enhanced vitality of the system. Something that is a eustress at moderate levels of intensity may be a distress at intense levels; similarly, an intense natural disturbance may cripple an ecosystem occasionally. Intense stressful disruptions lead eventually to extreme pathology and to the absence of life in the locale.

Humans intervene deliberately in natural or degraded ecosystems to modify them for the direct benefit of humans. If such interventions are well informed, then the result may be termed sustainable development or ecodevelopment (Friend and Rapport, 1991). Through the cultural practices of informed ecosystem husbandry, humans continually intervene to sustain ecosystemic productivity for the desired resources, to prevent natural ecological recovery to a pristine state, and also to forestall harmful ecological degradation. The latter condition is seldom achieved, especially not on urban lands. The practices of uninformed "husbandry" become harmful stresses, as defined above.

It may be noted that the terms *disruption, disturbance,* and *stress* refer to some external influence, intervention, force, input, and the like, onto or into the system, that is, some modification of the natural system's usual external driving variables. The cause or agency of an external stress may be termed a *stressor,* that is, someone or something that exerts a stress. The immediate effect on a system of some externally caused disruption or the longer-term response of the system to such a disruption is sometimes termed a stress, which would be inconsistent with our use of the term here. For this latter phenomenon, we here use the terms *response, symptom, effect,* and *strain,* with the system then being in a *stressed* state.

MATURATION AND REJUVENATION

In the late 1960s, a number of ecologists noted that not all the symptoms exhibited by an ecosystem when subjected by humans to a particular stress were distinctively diagnostic of that stress; a number of stresses evoked some similar responses or symptoms (Woodwell, 1967, 1970; Regier, Applegate, and Ryder, 1969; Regier and Cowell, 1972). With respect to Lake Erie, this realization helped to explain why there was so much dissent among the experts as to what was "killing" Lake Erie (Regier and Hartman, 1973). Nondiagnostic symptoms were used by researchers specializing in particular stresses to attribute the overall effects resulting from a combination of causes to one particular cause (in which they had a vested interest). There was conflict between experts over toxic loading, siltation, eutrophication, overfishing, and even radionuclide loading. One-stress experts used nondiagnostic symptoms to support their claim for research support and personal recognition. Incidentally, there was no significant risk that the Lake Erie ecosystem would be transformed into a nonliving state, that is, that it would "die" utterly. By analogy with organismic systems, Lake Erie had become obese, intoxicated, flaccid, incontinent, and flatulent.

Features of the general stress syndrome of nondiagnostic symptoms, which occurred commonly with a wide variety of stresses by humans, were summarized and analyzed by Rapport, Regier, and Hutchinson (1985). That some of the detailed effects of many human stresses on ecosystems were different from those associated with the normal rejuvenating disturbances of nature was clarified further for freshwater ecosystems by Steedman and Regier (1987). Human stresses often cause crippling or pathological rejuvenescence from which full recovery through natural secondary succession may be very slow or incomplete, and, hence, deliberate rehabilitative and mitigative interventions by humans may be necessary to effect satisfactory recovery (Francis *et al.*, 1979; Rapport, 1989a, b; 1990a, b).

Steedman and Regier (1987) (also see Odum, 1985) showed that the usual ecosystemic degradation sequence with intensifying stresses was, to some extent, in the opposite direction of the usual maturation sequence of unstressed systems in general (Bertalanffy, 1975; also see Davidson, 1983). Primary ecological succession and secondary succession following some noncrippling disruption or stress are special cases of Bertalanffy's general inference on system "maturation." The latter involves progressively greater internal integration, less temporal variability in key components, and improved adaptive or rheostatic capabilities with respect to the usual small to moderate external influences, but lesser adaptive capacity to major unusual external influences.

To determine whether a modified ecosystem is in a state of normal disturbance or of pathological degradation, we need to examine more than a haphazard assortment of stress symptoms or system characteristics. Nondiagnostic symptoms may not be of much help. Current and historic information on types and intensities of natural disturbances and human stresses will be needed. The ecosystem should be examined to determine whether key abiotic or biotic structures have been lost or permanently deformed or crippled. As our scientific understanding develops further, a complicated practice of ecosystem diagnosis and therapy will develop (Francis *et al.*, 1979; Rapport, Regier, and Thorpe, 1981; Rapport, 1992a, b, c).

STRESSED ECOSYSTEMS

The behavior of pathological ecosystems is becoming better understood as case histories accumulate and as carefully controlled stress interventions are performed on whole ecosystems (for example, Bormann and Likens, 1979a, b; Schindler et al., 1985; Schindler, 1988).

Table 15.1 provides a comparison of some trends in stressed ecosystems as identified by Odum (1985) and by Rapport, Regier, and Hutchinson (1985) with the features characteristic of stressed landscapes as summarized by Godron and Forman (1983). For convenience, we have used Odum's major headings: energetics, nutrient cycling, community structure, and general system-level trends.

Some similarities are apparent in properties of stressed ecosystems and landscapes. Stresses act to disrupt energy processing, with the result that primary productivity in relation to community respiration becomes unbalanced. Odum inferred that community respiration increases as the damage caused by stress is repaired, "diverting energy from growth and production to maintenance." At the same time, it is often the case that primary productivity declines with stress. The result is a net decrease in landscape productivity with stress. Mitigative cultural practices may involve purposeful fertilization of or external subsidy to infertile soil as in the cultivated landscape (Godron and Forman, 1983).

Stress interrupts the process of nutrient recycling by a variety of mechanisms. Clear-cutting and other physical disturbances to vegetated ecosystems increase lateral transport across the surface

TABLE 15.1 Major symptoms of stress in ecosystems and landscapes

System property	Ecosystem (Odum, 1985)	Ecosystem (Rapport et al., 1985)	Landscape (Godron and Forman, 1983)
Energetics			
Community respiration	−	?	?
Primary production	?	−	?
Net landscape production	?	?	−
Nutrient flow			
Horizontal transport	+	+	+
Community structure			
Species diversity	−	−	−
r-selected species	+	+	?
Short-lived species	+	+	?
Smaller biota	+	+	?
Food chain length	−	?	?
Exotic species	?	+	?
System features			
Openness	+	+	+
Succession reversal	*	*	?
Metastability	?	?	−
Negative interaction	+	?	?
Disease incidence	?	+	?
Mutualism	−	?	?
Population self-regulation	?	−	?
Resource use efficiency	−	?	?
Boundary distinctiveness	?	?	+
Boundary linearity	?	?	+

Symbols: −, a decrease; +, an increase; *, yes; and ?, not specified in source reference.

of water and materials dissolved in it or mixed with it. Air pollution, by rendering the decomposer functions less efficient, also leads to increased horizontal flow of nutrients and reduced vertical flow. The consequence is a system in which nutrients are exported to neighboring systems. But one ecosystem's loss is another's gain; in many cases the aquatic systems then gain nutrients lost by the terrestrial and to an excess so that eutrophication (that is, overfeeding) results. (A hypereutrophic, methane-producing pond is like an obese, flatulent organism.)

The most visible symptoms of stressed ecosystems are in the community structure. Stressed systems have a more monotonous appearance. Species diversity is reduced, and what species remain are often the smaller, less valued forms, making the landscape visually less interesting. Opportunistic species and r-strategists, especially in the form of exotic species, thrive under many types of stress. Stressed environments come to be dominated by rapidly reproducing, rapidly colonizing, nonself-regulatory species of the smaller varieties; these are normally present in nonstressed environments, but in low numbers (Rapport, Regier, and Hutchinson, 1985).

Concerning ecosystem features of stressed environments, the three accounts in Table 15.1 complement one another. One feature that crops up both in Odum (1985) and in Rapport, Regier, and Hutchinson (1985) is that ecosystems may appear at first glance to revert to earlier stages of development. But this is misleading. The full suite of effects of one or more human stresses on an ecosystem includes some effects that are not at all typical of natural disturbances or natural resets of the ecosystem's self-control capabilities.

By analogy (or homology?), it may be noted that, when adult human organisms become diseased or are injured, they usually discontinue some activities typical of adults and may revert to some activities more typical of infants. They may stay in bed longer, eat simpler foods, become incontinent, be more passive and dependent on others, and so on. But such pathological phenomena are not usually interpreted as indicative of rejuvenation; hence we may refer to them as *pathological rejuvenation* if we want to use the term rejuvenation at all in this context.

The other "systems" features of Table 15.1 cover a wide and interesting terrain. Odum infers that parasitism and other negative interactions increase, while mutualism and other positive interactions decrease with increasing intensity of harmful stress. Here we need to discriminate since the effects of some disease vectors on ecologically dominant species may serve as the periodic disturbances that naturally rejuvenate an ecosystem. But Rapport, Regier, and Hutchinson (1985) point to examples of increases in disease incidence that occur commonly in stressed ecosystems. Karr (1991) used the incidence within a sample of fish of disease tumors, fin damage, and other anomalies as inversely related to the "biotic integrity" of a stream ecosystem. Steedman (1991) found that infestation of fish by certain external parasites increased with the degree of degradation of streams subject to a mix of stresses in urban watersheds.

Reduced stability (or increased temporal variability) in components or in the total system is also characteristic of stressed environments. Note the increasing fluctuations of some fish populations before their collapse by 1960 in Lake Erie (Regier and Hartman, 1973; Regier and Baskerville, 1986). Godron and Forman (1983) point to a reduction in metastability; that is, systems in stressed locales are more readily transposed from their current structural configuration to a different kind of form than are unstressed systems.

ECOSYSTEM DISINTEGRATION

Bertalanffy's principle of equifinality applies to a system that is "maturing," that is, progressing within a benign environment in which external disturbances and stresses do not incapacitate internal self-organizing processes that lead toward a "mature state." His principle states that a variety of pathways may lead to much the same kind of state that is defined not as a unique set of numerical and categorical manifestations, but rather in a more functional, generic way. Living entities such as organisms and ecosystems do not fully repeat previous states, but do possess and exercise rheostatic capabilities toward a state of "integrity" or "health." A full life is more a convoluted path than a stationary point, apparently at all levels of organization.

An equifinality principle appears also to apply to the degradation sequence under intensifying human stresses, and perhaps also under abnormally high levels of natural disturbance. The end state is the *general stress syndrome* with a variety of features common to the effects of most stresses. But each stress also entrains its particular symptoms. The paths by which different stresses cause a mature ecosystem to become transformed into one that exhibits the general stress syndrome are not coincident. For example, in an aquatic ecosystem, the initial responses to increasing eutrophication are often quite different from those of low-intensity conventional fishing or of low levels of turbidity due to erosion of clay soils, but the end states under severe stress share similarities. The sequences and the end states of ecosystem degradation are not as well explicated as are the sequences and end states of ecosystem recovery (Odum, 1985; Bonsdorff, 1985).

With respect to disintegration, we may compare findings on the acidification of a lake in the Canadian Boreal Shield (Schindler *et al.,* 1985; Schindler, 1988) and findings on the impacts of air pollution stress on forest ecosystems (Smith, 1981).

In a study by Schindler and his colleagues, a small lake was manipulated by adding sulfuric acid over a period of 8 years to bring the pH down from 6.8 to 5.0. Numerous biological and chemical variables were monitored, and the observations were compared with those made for a similar lake that remained relatively unstressed. Table 15.2 summarizes the findings. The first signs of change were in the composition of the phytoplankton; some sensitive species were replaced by more tolerant forms. By the third year, when the pH had declined to 5.9, some dominant species declined to very low numbers, and some rare species disappeared. In the fourth year, thick visible mats of a filamentous alga of the genus *Mougeotia,* not previously noticed, formed in the littoral areas. Sensitive species of fish (such as sculpin) declined, and crayfish appeared to be suffering physiologically. As the pH declined further, a high incidence of a microsporozoan parasite appeared in the crayfish, further weakening that population. Replacements continued in both the algae and diatom communities, with acidophilic species replacing the former dominants. As the pH dropped to nearly 5.0, lake trout changed their spawning locations perhaps because shoreline areas, now covered with filamentous algae, were no longer suitable. Surprisingly, even after 8 years of lower pH, primary productivity and nutrient recycling were maintained close to the reference levels.

TABLE 15.2 Stages in acidification of a small boreal shield lake

Year	Acidity (pH)	Recorded impacts, highlights
1976	6.8	None
1977	6.1	Some species of phytoplankton replaced by others.
1978	5.9	Declines in opossum shrimp (*Mysis relicta*) and fathead minnow (*Pimephales promelas*).
1979	5.6	Filamentous algae (*Mougeotia*) form highly visible thick mats in littoral areas; loss of sensitive species.
1980	5.6	Crayfish (*Orconectes*) infected with microsporozoan parasite, egg masses infected with fungi; failure in recruitment of lake trout (*Salvelinus namaycush*).
1981	5.0	Recruitment failures in several more fish species; crayfish decline to only a few percent of initial population.
1982	5.1	Spawning location of lake trout (*Salvelinus namaycush*) changed as preferred nearshore habitat became unsuitable because of filamentous algae cover; no young of the year in fish species.
1983	5.1	Disappearance of crayfish, mayfly (*Hexagenia*), and other taxa that were previously abundant; no successful recruitment in any fish species.

Nutrient recycling was apparently upheld by microchemical buffering that occurred at the sediment–water interface.

The signs of air pollution on forest ecosystems have a characteristic temporal sequence as summarized by Smith (1981) and further developed by Bormann (1982, 1985); see Table 15.3. The first signs of stress detected were alterations to some aspect of the life cycle of sensitive species or individuals. These changes might involve reduced photosynthesis for some sensitive plants, or changes in reproductive capacity, or changes in the predisposition to insect or fungus attacks. With increased pollution, stressed populations of sensitive species decline to low numbers or become extinct. With further stress, the size distribution of plants is reduced and the forest becomes increasingly dominated by the small, scattered shrubs and herbs, including a number of invading species not previously present. As the process proceeds, further primary productivity is reduced. Bormann (1982) suggests that this reduction is due to the impaired capacity of the forest to substitute tolerant for intolerant species as stress proceeds. In the final stages, some impairment in nutrient cycling is also evident.

In both cases, it is the behavior and ultimately the reproductive failures in sensitive species that offer early signs of stress. Reductions in productivity and nutrient cycling apparently did not occur in the acidified lake. Based on Schindler's work and that of Harvey (1982) and others, we note that acidification of freshwaters by humans results in relatively few symptoms typical of the general stress syndrome (Rapport, Regier, and Hutchinson, 1985).

Few studies have been undertaken of interactions among stresses, whether synergistic or antagonistic, where two or more stresses are operating. Figure 15.1 is based on the informed judgment of an interdisciplinary group of experts with respect to the forces affecting Southern Green Bay, Lake Michigan. Tracing particular effects to stresses, singly or in combination, can become hopelessly complex in such situations. However, a simple though not fully general generalization emerges: whether triggered by stresses acting simply or jointly, many symptoms of the general distress syndrome appear.

TABLE 15.3 Pattern of decline of forest ecosystems under a pollution stress

Level of anthropogenic pollution	Severity of impact on ecosystem	Expected ecological effects
Insignificant	Pristine state persists	None
Low level	Relatively unaffected	Ecosystem serves as a sink for pollutants
Levels inimical to sensitive species	If continued, impairment of competitive ability of sensitive species	Reduced photosynthesis; changes in reproductive capacity; changes in vulnerability to insect or fungus attacks; deleterious effects on nutrient cycling
Increased pollution stress	Resistant species substitute for sensitive ones	Sensitive species lost from the system, mostly due to secondary effects like fungal or insect attacks; failure in pollination due to deleterious effects on honeybees or other sensitive animals
Severe levels of pollution	Large-scale changes in original system	Large plants, shrubs, and trees of all species die off; toxic concentrations of accumulated pollutants limit many species; ability of system to regulate biogeochemical cycles severely diminished
Very heavy pollution	Completely degraded ecosystem	Small taxa highly tolerant of harmful substances persist in an unorganized mess

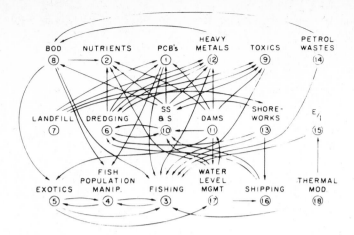

Figure 15.1 Digraph of interactions among stresses acting on Southern Green Bay, Wisconsin (from Harris et al., 1982). The symbol SS & S means suspended solids and sediments; E/I means entrainment and impingement of biota in water intakes.

ADAPTATION

An ecosystem may respond adaptively to stress due to a wide variety of mechanisms that enable the system to resist the stress or to recover from the stress when it is relaxed, or an ecosystem may accommodate to the stress by altering its structure and behavior. With respect to resistance, open ecosystems can be expected to be more resistant than relatively closed systems. The more open the system is, the greater the opportunity for repairing damage to species and populations through immigration and also for dissipating toxic materials, if only through flushing them to other systems and, in the process, diluting the direct local effects. The larger the system is, the greater the probability of structural and functional redundancy and the less likely it is that ecosystem-wide damage will follow local damage. In this, of course, it is not merely a question of the area, but often more of the edges and interfaces that constitute critical habitat. Thus, the destruction of a macrophyte bed that fringes only a small part of an otherwise steep basin of an aquatic ecosystem is likely of more importance than a similar area lost in a shallow, vegetated estuary. The function of shoreline vegetation and wetlands is increasingly appreciated as a modulator of nutrients, taking up quantities in a highly reactive state and subsequently releasing them gradually in a more refractory state. This slows the rate of eutrophication of the recipient water body. Vegetation may have a similar function in relative detoxification of contaminated water (Prince and D'Itri, 1985; Harris et al., 1988).

If application of a stress is short-lived and natural resistance of the ecosystem is high, the damaged area can be expected to bounce back. The speed of the recovery will likely be a function of the factors conferring the resistance, for example, the openness of the system, the ease with which new colonizers can reach the damaged area, the self-cleansing capabilities through sedimentation, and so on (Cairns, 1986, 1990). Speed of recovery will also depend on the temperature of the local climate (Regier, Holmes, and Pauly, 1991). Wetlands fringing rivers and lakes tend to have these properties with respect to moderate changes in water levels (Patterson and Whillans, 1985). The composition of the soft-bottom macrobenthic communities of the Swedish Coast of the Baltic Sea recovery has tracked the reductions in effluents from pulp mills (Cedarwall, 1986).

The type of stress also has relevance. Some stress-induced transformations are of a more permanent character than others. Physical alterations of habitat structure, particularly the permanent destruction of breeding and spawning grounds, tend to have long-term consequences. Appropriate new physical features may form only slowly under the natural abiotic processes.

Timing of the stress intervention with respect to the metabolism and seasonal state of the recipient ecosystem may affect both its severity and the resistance of the ecosystem. In a grassland system, similar fires have different consequences in late spring or early spring. In late spring, a burn has the effect of maximizing forage yields and enhancing the growth of warm-season perennials, while early spring burns reduce yield but increase species diversity (Risser, 1987).

Fisheries managers have for centuries operated on a very general inference that harvesting of fish during the spawning season is more destructive than at other times. With many species, behavior important to successful spawning is apparently severely disrupted by conventional fishery practices (Regier, Applegate, and Ryder, 1969).

The evolutionary history of the ecosystem also has relevance to its ability to cope with stress. Margalef (1975), Fisher (1977), Holling (1985, 1986), and others contend that those ecosystems that have evolved in relatively tranquil environments readily collapse under abnormal stress. Part of the explanation for this might lie in the concept of *preadaptation*, a phenomenon that arises entirely fortuitously. Essentially, a community or ecosystem that occurs in an environment with frequent disturbances of one sort or another (for example, fire in the boreal forest, drought in the prairies, seasonal nutrient flux in estuaries, or floods in wetlands) may be preadapted to a stress that is novel to the ecosystem, but that in some way mimics a disturbance with respect to which the ecosystem has evolved. In general, the terrestrial ecosystems that are periodically denuded by fires or drought have plants that protect their reproductive organs (intercalary meristems) underground. Such adaptations serve well when the same systems are periodically denuded by novel stresses such as SO_x fumigation (Rapport, Regier, and Hutchinson, 1985).

Some mechanisms of resistance or coping with stress, as described above, are only temporarily efficacious. If a stress is intense enough or prolonged enough, even at low levels, ecosystems eventually may transform drastically. The resulting transformation has been variously termed degradation, ecosystem breakdown, collapse, and devolution, depending on the predilections of the observer (see McIntosh, 1980).

Following catastrophic change, a secondary process of adaptation may occur whereby the ecosystem becomes what we term *stress adapted*. Stress-adapted ecosystems may occur where natural systems are driven into new regions of stabilized degradation in which the continuing external influences of humans become the overriding determinant of system properties. These may include badly urbanized areas, grossly polluted industrial zones, severely polluted estuaries, and the like. Case histories are insufficient to generalize on how colonization by stress-adapted biota occurs in such stressed ecosystems once sufficient time has passed. We surmise that some of the pathological symptoms might eventually become less intense, even though the levels of stress are not relaxed. For example, alternative methods of nutrient recycling and primary productivity may proliferate in such stress-adapted systems. How such a stress-adapted ecosystem will differ from cases less severely stressed and now readily observable has not been researched intensively from the present perspective.

TEMPORAL DEVELOPMENT OF ECOSYSTEM PATHOLOGY

Is it more characteristic of ecosystems under stress that symptoms develop gradually and monotonically, or do symptoms suddenly appear after some threshold is crossed? To the extent that ecosystems possess mechanisms whereby stress for a time is neutralized or resisted, it may be that the appearance of severe symptoms is a sudden consequence of the exhaustion of a stress-buffering or stress-compensating capability (Holling, 1986).

The acidification process in buffered lakes provides an example. A particular drainage basin may possess several kinds of buffering capacity, each of which operates predominantly in a different pH range. When a particular buffering type of capacity is exhausted, the lake shifts rapidly to a more acidic condition. Consequently, with progressive acidification, the temporal sequence of the pH of lake water should approximate a downward series of steps. The edge of a step should correspond to the exhaustion of a particular type of buffering capacity. In a rather poorly buffered small lake, Schindler *et al.* (1985) showed that pH dropped progressively and symptoms developed gradually, beginning with physiological and population effects on sensitive species. But this may not be typical of the gradual acidification of a landscape, since the small lake was subjected to a comparatively large loading over a short interval of time, and any stabilizing mechanisms that may have operated had the loading been more gradual may have been overwhelmed. The temporal development of symptoms may be related to the nature of the intensification through time of a stress. A stress of low intensity that increases only gradually may be accommodated to a higher level of intensity without drastic transformation than a stress of high intensity applied almost instantaneously.

Different indicators reflect the different features of the development of pathology. If the focus is on the efficiency of nutrient processing or the capacity for net primary production, then development of pathology may be seen as a more gradual process. If, instead, symptoms are in terms of disappearance of some sensitive species, then the symptoms may appear suddenly. Much may depend on whether the stresses strongly affect key organizing centers of the ecosystem (Steedman and Regier, 1987). For example, if the stress quickly removes a large amount of the wetland vegetation in a small lake, then the ramifications on the entire lake system are likely to be sudden and severe, because such crucial components of the ecosystem may serve as central organizing features of the entire system (Regier *et al.*, 1988).

SPATIAL DEVELOPMENT OF ECOSYSTEM PATHOLOGY

The one-dimensional catenary sequence of effects downstream from an outfall of a sanitary sewer has been a focus of study for many decades (Wilhm, 1975). Here, distance from the source provides a measure of the time from first impact on the moving water mass. A saprogen system of indicator organisms was developed and a variety of useful generalizations was applied in practice to manage such outfalls. Immediately below an outfall, there occurs a linear sequence of intensifying degradation which reaches a low point some distance downstream. This depends on the ratio of the loading of putrescible materials to the original oxygen concentration, plus oxygen loading rates into the water, the stream's turbulence, temperature, and so on. Downstream from the low point, the river progressively exhibits recovery from the stress effects.

Patrick (1967) reported changes in the communities in two-dimensional estuaries under pollution stress. In this and many similar studies (for example, Leppakoski, 1975; Bonsdorff, 1985), the spatial distribution of pathology mirrors the stress gradient established in the aquatic systems under point source pollution. Similar work has been done on terrestrial systems (Gorham and Gordon, 1960; Woodwell, 1967, 1970; Whitby *et al.*, 1976). The most extreme manifestations of pathology occur near the source of stress; in the case of chemical stress, this reflects the intensity or concentration of toxics or organics in the ecosystem. Species richness, species diversity, and the composition of species (especially the proportion of pollutant-tolerant to intolerant forms) generally reflect well the stress gradients.

For the three-dimensional dynamic processes of the atmosphere and large water bodies through which pathology in such ecosystems develops spatially, fewer generalizations are available. How do the symptoms of stress spread spatially through large ecosystems and landscapes? To what extent do ecosystems export pathology to neighboring systems, that is, infect others? The simplest case is when the stress applied to the system emanates from a single point source. The subsequent spatial development of symptoms would likely be governed by both the stress gradient and the dynamic interactions among the water, soil, and atmospheric parts of the system. Thus, in a lake or coastal area, a pollutant or nutrient stress may spread initially with the shoreline currents from the mouth of a polluted stream or from a point source. Hydrodynamically, a lake is something like an intermittent one-sided river, with the shore of the lake resembling the shore of a large river, ecologically (Steedman and Regier, 1987). The pollutant may be stirred up by the currents, partly inactivated by particles that absorb some chemicals, and eventually be distributed throughout the whole three-dimensional ecosystem by complicated hydrological and atmospheric currents, by migrating organisms, which may be perceived as "biotic currents," and by bioaccumulation within food webs through biochemical processes (Mackay, 1991).

As in the temporal development of pathology, much likely depends on the spatial configuration of the stresses applied to the ecosystem or landscape. The spatial development of symptoms mirrors in some respect the spatial and temporal development of stresses. For example, in the Lower Great Lakes, the initial symptoms of pathology were along the shore in areas of human settlement and concentrated activity. With the growth of coastal settlement, indicators of pathology were apparent farther along in the inshore regions. As stresses intensified, effects became apparent offshore and eventually in the whole basin (Regier and Baskerville, 1986; Regier, 1986). A three-dimensional representation would be very complex; it would reflect both patterns of human settlement and ecosystemic processes that spread impacts into the various dimensions of the ecosystem. Overall, the various stresses have their most immediate impacts in the streams and in the nearshore water and then spread outward into the offshore and deeper water (Whillans, 1979). In large water bodies, the nearshore is an ecological extension of rivers and estuaries, both in freshwater and marine systems (Steedman and Regier, 1987). Oceans also possess rivers "with two sides" as in the case of the Gulf Stream and Kuroshio Current; such phenomena are relatively rare and intermittent in large freshwater bodies, for example, in the Bodensee. In any case, water bodies are invariably complicated, structurally and dynamically.

STRESS EPICOENOLOGY

An epidemic is a disease outbreak acting on (*epi*) people (*demos*). An epidemic may result from disease organisms, environmental factors, or psychosocial causes. Consider the effects of human stresses on ecosystems over space and time from a perspective like that of epidemiology, or what we may term *epicoenology*. Here, *coen* derives from the Greek term for common, *koinos,* and this implies community; biogeocoenosis is generally used by ecologists in Eastern Europe to denote ecosystem.

Many critics of the currently dominant form of civilization in the world refer to it as pathogenic, not only as it relates to other human civilizations, but also to other aspects of our biosphere. The most general issue in the study of epicoenology would be to comprehend this dominant strain of humanity as a disease vector or scourge on the biosphere. We recognize this, but not in the extreme form popularized by Sale (1985).

Some kinds of epidemics are spread by direct contact between infected and susceptible humans; some proliferate due to sputa, excreta, or other leavings of infected people; some are transmitted by intermediate species or hosts; and so on. The role of humans in epicoenology may bear more comprehensive study. The willful or unintended introduction by humans of many undesirable exotic species is an example of such a role (Bird and Rapport, 1986). In a somewhat more abstract way, the infection of one human culture with destructive practices toward nature by the dominant culture may be mentioned: here aspects of human culture are the vectors of pathological state.

As we consider the broad range of stresses and the increasing number of ecosystems that are damaged by one or some combination of stresses, a picture emerges of the spread of a blight across the landscape, leaving only small pockets unaffected. These may be naturally resistant to the stresses or may persist in areas that fortuitously remained out of reach of stress regimes. The patchwork may superficially resemble the work of natural disturbances as in the appearance of a forest on a regional scale after spotty clear-cut logging operations, or after natural blowdowns (Loucks, 1983). There are usually some essential differences. Stressed ecosystems tend to cluster near regions of human settlement (see Bird and Rapport, 1986; Rapport et al., 1993). Stresses as compared to natural perturbations often cause deeper damage to an ecosystem, rendering it incapable of quick recovery through the normal processes of secondary succession. Stresses are generally not one-shot impacts, but tend to persist and intensify.

If stresses have damaged sensitive structures, such as the marshlands and shoals of aquatic systems that provide special and scarce breeding and feeding grounds for fish and waterfowl, then their abatement in itself may not lead to rapid recolonization by desired biota. Damage to these critical organizing centers (Steedman and Regier, 1987, 1990; Harris et al., 1988) may permanently cripple the ecosystem, unless humans introduce artificial features as surrogates of those that were lost. In this, the mitigative technology is rapidly developing, and the transplantation of entire marshlands is becoming feasible, as in the Campbell River estuary, British Columbia (Bird and Rapport, 1986, p. 97).

Thus, the difference between disturbance and stress in this context has implications for the behavior of the system in the recovery phase. Stresses often persist over long time periods in the same region and are not randomly scattered over space and time. Disturbances tend to leave healthy systems distributed in proximity to the disturbed areas. Intensely stressed ecosystems are often enmeshed in a moderately stressed region near centers of human activity, or downstream—hydrologically, atmospherically, and biotically—from such centers. From the perspective of the natural ecosystem, such human centers are disorganizing in their impact and are often located in or near areas that were once centers of natural self-organization, such as the mouths of rivers. Neighboring ecosystems may also be damaged, reducing the prospect for automatic recovery facilitated by nearby healthy systems. And, finally, the effect of stress may often be to damage critical elements or processes of the ecosystem, precluding automatic reestablishment of the system following relaxation of the stresses.

An example relates to pulp mills and other industries in estuaries and bays of the Baltic Sea, particularly along the Swedish and Finnish coasts of the Gulf of Bothnia. In these cases, the distribution of stressed ecosystems superficially resembles a random pattern of disturbance. With recognition that damage to the estuaries is a result of municipal and industrial pollution, such loadings have been reduced in many instances, and recovery to more desirable conditions is occurring in local areas (Cedarwall, 1986). However, the picture for the Bothnian Sea as a whole is more that of intensifying stress from nutrient and toxic loadings, as well as overharvesting. Thus, areas of isolated stress are showing signs of coalescing into wider regions of continuous pathology even while severe stresses in some locales are abating and partial local recovery is apparent (Rapport, 1989b). This resembles cases in the northeastern United States and in the lower Great Lakes basin.

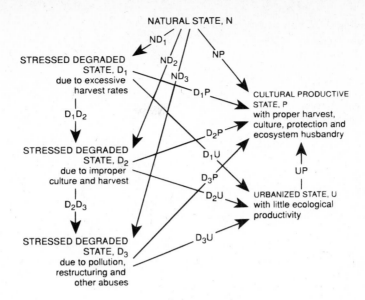

Figure 15.2 Classification of ecosystem states and some types of transformations between states (shown by arrows).

REHABILITATIVE THERAPY OF DEGRADED ECOSYSTEMS

In the Great Lakes, major efforts have been underway, starting about 1955, to reverse the effects on various human abuses on the lakes (NRC/RSC, 1985). The term *remediation* is generally used with respect to efforts to reverse or combat pollution; in practice, it means a partial reduction of the loadings of some pollutants. *Rehabilitation* is used with respect to efforts to improve the association of species and the productivity of the most desirable species. Rehabilitation usually involves quite thorough remediation plus mitigative, restorative, or enhancement efforts to foster ecosystem recovery (Francis *et al.*, 1979).

Figure 15.2 provides an overview of regional degradation and rehabilitation. Through the abuses that regularly accompany conventional exploitative economic development, natural ecosystems become degraded unnecessarily; see the three ND arrows in the figure. Often the development sequence has progressed from excessive harvesting through inappropriate crop culture to industrial pollution; see the two DD arrows. Some degraded parts of a region have become urbanized usually with intensely harmful practices; see the three DU arrows.

Throughout the whole development process, lip service will have been paid to safeguarding productive functions of the ecosystem. The usual outcome has been that productivity and other highly valued features of the ecosystem have been sacrificed. Resolve is growing that much of the productive capability must be rehabilitated; see the three DP and the one UP arrows. The science and practice of how this might be achieved is evolving (Hartig and Zarull, 1992; Regier, 1992), and it must be achieved locally and regionally in the face of intensifying global threats, such as climate change.

Pressures will intensify to transform some natural states into properly productive states; see the one NP arrow. Again, the science and practice of such ecodevelopment is inadequately understood. Not shown in the figure are any arrows pointing back to the natural state, N. Such restoration may have occurred over centuries since the distant past with collapse of some ancient civilizations. A return of an ice age to cover the entire Great Lakes Basin with glaciers would provide an eventual opportunity for a new beginning, although some locales where thermonuclear electric generating plants once stood might then be unnaturally radioactive.

TOWARD ECOSYSTEM INTEGRITY

In recent decades the people of the various jurisdictions of the Great Lakes have committed themselves progressively, thoroughly, and comprehensively to a goal of environmental health or *ecosystem integrity* (Edwards and Regier, 1990; Regier, 1992). In practice, this appears to us to imply three major strategies, as follows.

Representative near-pristine, dynamic natural areas must be preserved in perpetuity to the extent of something over 10% of the area of each of the region's "ecological elements," such as the watersheds. All such areal preserves must be interlinked in a dynamic network of corridors, themselves in a near natural state. Many of these areas and corridors are sensitively used for recreational purposes. Stewardship skills are evolving (Lerner, 1993).

Areas that are currently degraded to a moderate degree must be rehabilitated to a more desirable state. Most urban locales fall in this class, as do many intensely used farms and rural locales. Here, desirable species, both indigenous and exotic, that are tolerant of intense but careful human interaction will thrive. Self-organizing, healthy ecosystemic behavior will be fostered to help contain the economic costs of the necessary husbandry (Kauffman *et al.,* 1992).

Many locales around the edges of the Great Lakes were degraded to an abject state of an ecological slum. These generally require major corrective interventions to help the local ecosystem out of its crippled state. The necessary remediation is gradually being expanded into a more comprehensive form of ecosystem therapy (Hartig and Zarull, 1992).

In the Greater Toronto Area, within the Great Lakes Basin, all levels of government currently share a broad commitment to these three interactive strategies toward ecosystem integrity (see, for example, RCFTW, 1992). A kind of redemocratization is occurring (Edwards and Regier, 1990; Francis, 1990; Lerner, 1993) because the existing conventional agencies in the various levels of government find it difficult to collaborate effectively in the usual legalistic, bureaucratic context. Hence, new forms of governance are evolving, with actor systems, stakeholder groups, co-management, and partnerships. The redemocratization is providing opportunities for the professional expertise of the therapists of ecosystem integrity.

Science particularly relevant to epicoenology and ecosystem therapy in the Great Lakes context is also evolving, in theory (Kay, 1991) and in practice (Edwards and Regier, 1990; Regier, 1992).

Policies and codes of ethics—at the levels of agencies, corporations, professions, and individuals—are being developed to be consistent, at least in part, with a commitment to ecosystem integrity. The evolution of these ethical commitments is being fostered through relevant philosophical study (Westra, 1989).

ACKNOWLEDGMENT

The Donner Canadian Foundation and the Social Sciences and Humanities Research Council of Canada provided funding for the second author's contribution to this work.

REFERENCES

BERTALANFFY, L. VON. 1975. *Perspectives on General Systems Theory.* Braziller, New York. 183 pp.
BIRD, P. M., and RAPPORT, D. J. 1986. State of the Environment Report for Canada. Minister of Supply and Services Canada, Ottawa. EN 21-54/1986E. 263 pp.

BONSDORFF, E. 1985. Recovery Potential of the Fauna of Brackish-water Softbottoms and Marine Intertidal Rockpools. Abo Akademi, Finland (Academic Dissertation).

BORMANN, F. H. 1982. Air pollution stress and energy policy. In Reidel, C. H. (ed.), *New England Prospects: Critical Choice in a Time of Change.* University Press of New England, Hanover, New Hampshire, pp. 85–140.

BORMANN, F. H. 1985. Air pollution and forests: an ecosystem perspective. *Biosci.* 35:434–441.

BORMANN, F. H., and LIKENS, G. E. 1979a. *Pattern and Process in a Forested Ecosystem.* Springer-Verlag, New York. 253 pp.

BORMANN, F. H., and LIKENS, G. E. 1979b. Catastrophic disturbance and the steady state in northern hardwood forests. *Amer. Scientist* 67(6):660–669.

CAIRNS, J. 1986. Freshwater. In Sadar, M. H., et al. (eds.), Proc. Workshop on Cumulative Environmental Impacts: a Binational Perspective. Canadian Environmental Assessment Research Council and U.S. National Research Council. Ottawa, Ministry of Supply and Services Canada. EN 106-2/1985, pp. 39–43.

CAIRNS, J. 1990. Prediction, validation, monitoring and mitigation of anthropogenic effects upon natural systems. *Environ. Auditor* 2(1):19–25.

CEDARWALL, M. 1986. State of the Swedish coastal zone of the Gulf of Bothnia. In Publication of the Water Research Institute, Third Gulf of Bothnia Symposium, Pori, Finland. Water Research Inst., Nat. Board of Waters, Helsinki, Finland, pp. 122–128.

DAVIDSON, M. 1983. *Uncommon Sense: The Life and Thought of Ludwig von Bertalanffy (1901–1970), Father of General Systems Theory.* J. P. Tarcher, Los Angeles, California. 247 pp.

EDWARDS, C. J., and REGIER, H. A. (eds.). 1990. An Ecosystem Approach to the Integrity of the Great Lakes in Turbulent Times. Great Lakes Fishery Commission, Spec. Pub. 90-4. 299 pp.

FISHER, N. S. 1977. On the differential sensitivity of estuarine and open-ocean diatoms to exotic chemical stress. *Amer. Nat.* 111:871–895.

FRANCIS, G. R. 1990. Flexible governance. In Edwards, C. J., and Regier, H. A. (eds.), An Ecosystem Approach to the Integrity of the Great Lakes in Turbulent Times. Great Lakes Fishery Commission Spec. Pub. 90-4, pp. 195–207.

FRANCIS, G. R., MAGNUSON, J. J., REGIER, H. A., and TALHELM, D. R. 1979. Rehabilitating Great Lakes Ecosystems. Ann Arbor, Michigan. Great Lakes Fishing Comm. Tech. Rep. No. 37. 99 pp.

FRIEND, A. M., and RAPPORT, D. J. 1991. Evolution of macroinformation systems for sustainable development. *Ecol. Econ.* 3:59–76.

GODRON, M., and FORMAN, R. T. T. 1983. Landscape modification and changing ecological characteristics. In Mooney, H. A., and Godron, M. (eds.), *Disturbance and Ecosystems.* Springer-Verlag, Berlin and New York, pp. 12–28.

GORHAM, E., and GORDON, A. G. 1960. Some effects of smelter pollution northeast of Falconbridge, Ontario. *Can. J. Bot.* 38:307–312.

HARRIS, H. J., HARRIS, V. A., REGIER, H. A., and RAPPORT, D. J. 1988. Importance of the nearshore area for sustainable redevelopment in the Great Lakes with observations on the Baltic Sea. *Ambio* 17:112–120.

HARRIS, H. J., TALHELM, D. R., MAGNUSON, J. J., and FORBES, F. M. 1982. Green Bay in the Future—A Rehabilitative Prospectus. Great Lakes Fish. Comm. Tech. Ref. 38, Ann Arbor, Michigan. 59 pp.

HARTIG, J. H., and ZARULL, M. A. (eds.). 1992. *Under RAPs: Toward Grassroots Ecological Democracy in the Great Lakes Basin.* University of Michigan Press, Ann Arbor, Michigan. 289 pp.

HARVEY, H. H. 1982. Population responses of fishes in acidified waters. In Johnson, R. E. (ed.), Acid Rain/Fisheries, Proceedings of the International Symposium on Acidic Precipitation in Northeastern North America, Ithaca, New York, August 2–5, 1984. American Fisheries Society, Bethesda, Maryland, pp. 227–242.

HOLLING, C. S. (ed.). 1978. *Adaptive Environmental Assessment and Management.* Wiley, New York. 377 pp.

HOLLING, C. S. 1985. Resilience of ecosystems, local surprise and global change. In Malone, T. F., and Roderer, J. G. (eds.), *Global Change.* Cambridge University Press, New York, pp. 228–269.

HOLLING, C. S. 1986. The resilience of terrestrial ecosystems: local surprise and global change. In Clark, W. C., and Munn, R. E. (eds.). *Sustainable Development of the Biosphere.* Cambridge University Press, New York, pp. 292–317.

KARR, J. R. 1991. Biological integrity: a long-neglected aspect of water resource management. *Ecol. Applications* 1:66–84.

KAUFFMAN, J., RENNICK, P., REGIER, H. A., HOLMES, J. A., and WICHERT, G. A. 1992. Metro Waterfront Environmental Study. Waterfront Planning Section, Metropolitan Planning Department, Toronto, Ontario. 88 pp. (reproduced).

KAY, J. J. 1991. A nonequilibrium thermodynamic framework for discussing ecosystem integrity. *Environ. Management,* 15(4):483–495.

LEPPAKOSKI, E. 1975. Assessment of degree of pollution on the basis of macrozoobenthos in marine brackish-water environments. *Abo Akademi, Acta Academise Aboensis,* Ser. B., Vol. 35, No. 3. 90 pp.

LERNER, S. 1993. Environmental Stewardship. Department of Geography, University of Waterloo Publication Series, No. 39, Waterloo, Ontario. 454 pp.

LOUCKS, O. L. 1983. New light on the changing forest. In Flader, S. L. (ed.), *The Great Lakes Forest: An Environmental and Social History.* University of Minnesota Press, Minneapolis, Minnesota, pp. 17–32.

MACKAY, D. 1991. *Multimedia Environmental Models: The Fugacity Approach.* Lewis Publishers, Chelsea, Michigan. 260 pp.

MARGALEF, R. 1975. Human impact on transportation and diversity in ecosystems. How far is extrapolation valid? In Proceedings of the 1st International Congress of Ecology, Structure, Functioning and Management of Ecosystems. The Hague, Sept. 8–14, Centre for Agricultural Publishing and Documentation, Wageningen, Netherlands, pp 237–241.

MCINTOSH, R. P. 1980. The relationship between succession and the recovery process in ecosystems. In Cairns, J. (ed.), *The Recovery Process in Damaged Ecosystems.* Ann Arbor Science, Ann Arbor, Michigan, pp. 11–62.

NRC/RSC. 1985. The Great Lakes Water Quality Agreement: An Evolving Instrument for Ecosystem Management. Report of a Non-Governmental Review under Auspices of U.S. National Research Council and Royal Society of Canada, National Academy Press, Washington, D.C. 224 pp.

ODUM, E. P. 1969. The strategy of ecosystem development. *Science* 164:262–270.

ODUM, E. P. 1985. Trends expected in stressed ecosystems. *Bioscience* 35:419–422.

PAINE, R. T. 1979. Disaster, catastrophe and local persistence of the sea palm *Postelsia palmaeformis. Science* 205:685–687.

PATRICK, R. 1967. Diatom communities in estuaries, *Am. Assoc. Adv. Sci. Publ.* 83:311–315.

PATTERSON, N., and WHILLANS, T. H. 1985. Human interference with natural water levels in the context of other cultural stresses in Great Lakes wetlands. In Prince, H. H., and D'Itri, F. M. (eds.), *Coastal Wetlands.* Lewis Publishers Chelsea, Michigan, pp. 209–251.

PRINCE, H. H., and D'ITRI, F. M. (eds.). 1985. *Coastal Wetlands.* Lewis Publishers, Chelsea, Michigan, 286 pp.

RCFTW. 1992. Regeneration, Toronto's Waterfront and the Sustainable City: Final Report. David Crombie, Commissioner; Royal Commission on the Future of the Toronto Waterfront, Toronto, Canada. 530 pp.

RAPPORT, D. J. 1989a. What constitutes ecosystem health? *Perspectives Biol. Medicine* 33:120–132.

RAPPORT, D. J. 1989b. Symptoms of pathology in the Gulf of Bothnia (Baltic Sea). *Biol. J. Linnean Soc.* 37:33–49.

Rapport, D. J. 1990a. Criteria for ecological indicators. *Environ. Monitoring Assessment* 15:273–275.

Rapport, D. J. 1990b. Challenges in the detection and diagnosis of pathological change in aquatic ecosystems. *J. Great Lakes Research* 16(4):609–618.

Rapport, D. J. 1992a. Defining the practice of clinical ecology. In Costanza, R., Norton, G., and Haskell, B. (eds.), *Ecosystem Health: New Goals for Environmental Management.* Island Press, Washington, D.C. 269 pp.

Rapport, D. J. 1992b. Evolution of indicators of ecosystem health. In *Proc. Intern. Symp. Ecological Indicators.* Academic Press, New York, pp. 121–133.

Rapport, D. J. 1992c. Evaluating ecosystem health. *J. Aquatic Ecosystem Health* 1:15–24.

Rapport, D. J., Smith, J. C., Schultz, C., Haas, G., Friend, A., and Dunn, G. 1993. Baseline Environmental Analysis of the Ontario Hydro "Balance of Power" Plan. Torrie Smith Associates, Ottawa.

Rapport, D. J., Regier, H. A., and Hutchinson, T. C. 1985. Ecosystem behavior under stress. *Amer. Natur.* 125:617–640.

Rapport, D. J., Regier, H. A., and Thorpe, C. 1981. Diagnosis, prognosis and treatment of ecosystems under stress. In Barrett, G. W., and Rosenberg, R. (eds.), *Stress Effects on Natural Ecosystems.* Wiley, New York, pp. 269–280.

Regier, H. A. 1986. Progress with remediation, rehabilitation and the ecosystem approach. *Alternatives* 13(3):46–54.

Regier, H. A. 1990. Great Lakes Charter: institutional framework for ecosystem integrity. In Bankert, L. S., and Flint, R. W. (eds.), Contemporary and Emerging Issues in the Great Lakes. Great Lakes Monograph No. 3, State University of New York at Buffalo, pp. 1–6.

Regier, H. A. 1992. Ecosystem integrity in the Great Lakes Basin: an historical sketch of ideas and actions. *J. Aquatic Ecosystem Health* 1:25–37.

Regier, H. A., and Baskerville, G. L. 1986. Sustainable redevelopment of regional ecosystems degraded by exploitive development. In Clark, W. C., and Munn, R. E. (eds.), *Sustainable Development of the Biosphere.* IIASA Symposium, Laxenburg, Austria, August 1984. Cambridge University Press, Cambridge, England, pp. 75–101.

Regier, H. A., and Cowell, E. B. 1972. Applications of ecosystems theory, succession, diversity, stability, stress and conservation. *Biol. Conservation* 4:83–88.

Regier, H. A., and Hartman, W. L. 1973. Lake Erie's fish community: 150 years of cultural stresses. *Science* 180:1248–1255.

Regier, H. A., Applegate, V. C., and Ryder, R. A. 1969. Ecology and management of the walleye in Western Lake Erie. *Great Lakes Fish. Comm. Tech. Rep. 15.* 101 pp.

Regier, H. A., Holmes, J. A., and Pauly, D. 1991. Influence of temperature changes on aquatic ecosystems: an interpretation of empirical data. *Trans. Amer. Fish. Soc.* 119:374–389.

Regier, H. A., Tuunainen, P., Russek, Z., and Persson, L. E. 1988. Rehabilitative redevelopment of the fish and fisheries of the Baltic Sea and the Great Lakes. *Ambio* 17:121–130.

Risser, P. G. 1987. Landscape ecology: state-of-the-art. In Turner, M. G. (ed.), *Landscape Heterogeneity and Disturbance.* Springer-Verlag, New York, pp. 3–14.

Sale, K. 1985. *Dwellers in the Land, a Bioregional Vision.* Sierra Club Books, San Francisco, California. 217 pp.

Schindler, D. W. 1988. Effects of acid rain on freshwater ecosystems. *Science* 239:149–159.

Schindler, D. W., Mills, K. H., Malley, D. F., Findlay, D. L., Shearer, J. A., Davies, I. J., Turner, M. A., Linsey, G. A., and Cruikshank, D. R. 1985. Long-term ecosystem stress: the effects of years of experimental acidification on a small lake. *Science* 226:1395–1401.

Selye, H. 1973. The evolution of the stress concept. *Amer. Scientist* 61:692–699.

Smith, W. H. 1981. *Air Pollution and Forests: Interactions between Air Contaminants and Forest Ecosystems.* Springer-Verlag, New York. 519 pp.

SOUSA, W. P. 1984. Role of disturbance in natural communities. *Amer. Rev. Ecol.*, Sep. 15:535–591.

STEEDMAN, R. J. 1991. Occurrence and environmental correlates of black spot disease in stream fishes near Toronto, Ontario. *Trans. Amer. Fish. Soc.* 120:494–499.

STEEDMAN, R. J., and REGIER, H. A. 1987. Ecosystem science for the Great Lakes: perspectives on degradative and rehabilitative transformations. *Can. Jour. Fish. Aquat. Sci.* 44 (suppl. 2):95–103.

STEEDMAN, R. J., and REGIER, H. A. 1990. Ecological bases for an understanding of ecosystem integrity in the Great Lakes Basin. In Edwards, C. J., and Regier, H. A. (eds.). An Ecosystem Approach to the Integrity of the Great Lakes in Turbulent Times. Great Lakes Fishery Commission Spec. Pub. 90-4, pp. 257–270.

VOGAL, R. J. 1980. The ecological factors that produce perturbation-dependent ecosystems. In Cairns, J. (ed.), *The Recovery Process in Damaged Ecosystems.* Ann Arbor Science, Ann Arbor, Michigan, pp. 63–74.

WESTRA, L. 1989. Respect, dignity and integrity: an environmental proposal for ethics. *Epistemologia* 12:215–230.

WHILLANS, T. H. 1979. Historic transformations of fish communities in three Great Lakes bays. *J. Great Lakes Res.* 5:195–215.

WHITBY, L. M. J., STOKES, P. M., HUTCHINSON, T. C., and MYSLIK, G. 1976. Ecological consequence of acidic and heavy metal discharges from the Sudbury smelters. *Can. Mineral.* 14:47–57.

WILHM, J. L. 1975. Biological indicators of pollution. Whitton, B. A. (ed.). *River Ecology.* University of California Press, Berkeley, California, pp. 375–402.

WOODWELL, G. M. 1967. Radiation and the patterns of nature. *Science* 156:461–470.

WOODWELL, G. M. 1970. Effects of pollution on the structure and physiology of ecosystems. *Science* 168:429–433.

16

Societal Instability: Depressions and Wars as Consequences of Ineffective Feedback Control

KENNETH E. F. WATT

Department of Zoology, University of California
Davis, California

Ecology: The Sobering Science. Man is a vital part of most major ecosystems, and awareness is increasing of his part in them and his influence on them. Humans are both parts of and manipulators of ecosystems, and induced instability of ecosystems is an important cause of economic, political, and social disturbances throughout the world. In altering his environment in order to overcome its limitations to him, man often is faced with undesirable consequences. Man's activities affect his environment in a way he cannot predict [Van Dyne, in The Ecosystem Concept in Natural Resource Management, Ch. X, p. 362–3 (1969), Academic Press].

HISTORY OF THIS PROJECT

This chapter, digested from a book by the author (Watt, 1992), summarizes findings of most interest to an ecological audience from a two-decade project to model society at the University of California at Davis. Output has been extremely voluminous. Only macrolevel system properties of nation states and the international system will be discussed here.

About a hundred people have contributed to this project as staff, graduate students, postdoctoral fellows, faculty colleagues, outside advisors, or consultants. At least 20 people contributed ideas, theories, methods, or other help without which the project would not have arrived at its present state. In the interest of brevity, the issue of authorship of particular ideas will not be mentioned again until a final section, in which the source for particular elements in the work will be specified.

There were several original motives for this work. By 1968, many people had noticed independently some ominous trends: overpopulation, pollution, resource depletion and degradation, and gradually more rapid changes in the physical and chemical composition of the atmosphere, with implications for climate change. Many teams began modeling and simulation projects to explore the implications of these trends.

A second concern was the possible existence of other prospective threatening developments not yet recognized. To illustrate, interactions between different subsystems of human society might generate new types of problems that no one had yet noticed. Several recent reviews cover the results and methods of most such modeling projects: the book *Groping in the Dark* (Meadows, Richardson, and Bruckman, 1982) and two excellent and comprehensive surveys by Hughes (1980, 1985).

Systems ecologists had two special reasons for becoming involved with this effort to model society. Tests of ecological theories using ecological data are often bedeviled by both inadequate sampling volume and veiled sources of measurement bias, leading to data afflicted by both high sampling variance and inaccuracy. Both of these hinder the process of strong inferential hypothesis testing. Social data are often based on the entire population of entities, thus facilitating very rigorous statistical testing of hypotheses, with unequivocal results. Also, ecology has few very long runs of data, whereas some social time series extend from the present back to 1720 or earlier.

Thus, social data provide ecologists with a new means of testing ecological theories. Furthermore, there is an astonishing array of correspondences between human society and ecological systems. Odum (1983) devotes 107 pages of his masterwork to correspondences between human and ecological systems. Analogies between different system levels is a major theme of the huge treatise on system theory by Miller (1978). Energy storages and flows are critical for understanding human society. More generally, human society can be understood only in terms of conceptual models that recognize the significance of storages as well as flows. Storages have already been recognized as central elements in modern ecosystem theory (for example, Odum, 1983; O'Neill, 1976). Recognition of the critical significance of storages is a prerequisite for modeling such phenomena as economic decline brought on by excessive overhang of inventory (a storage, not a flow) over a market. Other examples of storages in modern societies are the size of the federal government's public debt and the amount of oil remaining in the ground at prices people are prepared to pay for it in competition with other energy sources.

Population growth rates, birth and death rates, and age structure are necessary ingredients in comprehensive, realistic models of human society. Human society shows far more evidence of population density-dependent effects than nonecological scholars have noticed, in part because they haven't been looking for them. A wide variety of social phenomena, including war and international trade, can be understood within an ecological context. In short, ecologists have several reasons for being interested in social systems, and many ecologists have pursued this interest (for example, Gever et al., 1986; Hall, Cleveland, and Kaufmann, 1986; Odum, 1983; Watt, 1974, 1982; Watt et al., 1975, 1977). The new journal *Ecological Economics* represents a culmination of this trend.

Since about 1978, three new motives for modeling social systems have become apparent. The stability of complex living systems gradually became of more concern to ecologists, resulting in a large, diverse, if controversial and confusing, literature. Possible applications of ecological stability theory to the stability of complex social systems became of interest. This idea is linked to two others. At about the same time, as the world economy slowed down, there was a resurgence of interest in the possible existence of long-wave oscillations in the world economy (typically referred to as "Kondratieff waves"), waves of roughly half a century in length being terminated by international depressions (Freeman, 1983). A diverse international literature on this theme from many countries and disciplines has resulted. Freeman (1983) includes summary statements from 17 individuals or teams. Systems ecologists and other systems scientists were bound to wonder if these long waves were fundamentally similar to other well-known long waves, such as insect or 10-year furbearer cycles. This in turn suggests that simulation models could expose the mechanisms causing long waves and recommend strategies for managing them.

Finally, statistical analyses of historical data suggest that wars are not unfortunate accidents, but rather unintended effects of the normal functioning of the present world system (Craig and

Watt, 1985; Watt, Craig, and Auburn, 1988). Wars and depressions are alternate consequences of the same systemic forces. Indeed, Boulding (1962) concluded that wars and depressions are alternate manifestations of stress in social systems, a conclusion supported by the research reported here.

Current thinking on various issues raised in this brief introduction will now be reviewed as background to a discussion of methods and findings.

ECOLOGICAL STABILITY THEORY AND THE STABILITY OF SOCIETY

Controversy exists as to whether the theorems of ecological stability theory have any correspondence to observed phenomena in nature. Goodman (1975) and McNaughton (1977) exemplify the opposite sides. This controversy is unfortunate because ecological stability theory seems to have great potential for rationalizing social phenomena. Accordingly, Watt and Craig (1986) attempted to deal with the confusion on this matter and also show how ecological stability theory could be applied to society. It appeared that seven different ecological stability principles had become confused in the literature. One means of distinguishing the principles was to classify them on the basis of possible types of response by living systems to potential sources of instability. The scheme is presented in Table 16.1. Living systems could respond to potential sources of instability by adjustment or by avoidance. Two basic strategies of adjustment are possible: spreading the risk or expediting or facilitating homeostasis. Two different categories of risk spreading are possible: increasing the number of intersubstitutable resources on which the system is dependent and increasing the diversity (mosaic nature) of the environment.

The first of these was the topic of MacArthur's (1955) note. It is also the basis for the insurance business and the old adage, "Don't put all your eggs in one basket." For human societies,

TABLE 16.1 Classification of ecological stability theories

Type of response to prospective source of instability	Strategy for dealing with prospective source of instability	Number and identification of ecological stability principle, with representative authors
Adjustment	Spread risk • With respect to resources	1. The "omnivory principle" (MacArthur, 1955)
	• Through selection or production of a fine-grained landscape	2. The "fine-grained landscape principle" (Morris, 1963; Cronon, 1983; Harris, 1984)
	Expedite or facilitate homeostasis • Minimize feedback delay	3. The "fast homeostasis principle" (Hutchinson, 1948; Wangersky and Cunningham, 1956, 1957a, b)
	• By using resources faster	4. The "rapid resource exploitation principle" (O'Neill, 1976)
	• By flattening hierarchically organized systems	5. The "flattened pyramid principle" (Force, 1974; Beer, 1975; Butzer, 1980)
Avoidance	Minimize impact through diminishing connectance or reaction strength or number of system components, or by buffering, diapause, hibernation, and the like	6. The "connectance principle" (Gardner and Ashby, 1970; May, 1974; Lawlor, 1978)
	Flee or emigrate from source of potential instability	7. The "migration principle"

this principle means that it is potentially destabilizing to become critically dependent on a particular resource being depleted. That was the situation for Imperial Spain with respect to oak for shipbuilding in 1590 (Braudel, 1972) and England with respect to mineral resources in 1790 (Watt and Craig, 1986), and it is the situation with respect to crude oil for the United States now.

Morris (1963) discovered that spruce budworm populations were more unstable during an outbreak (would build to much greater densities) where the environment was in large, contiguous even-aged stands than where the forest was broken up into smaller stands and the landscape was more diverse. Cronon (1983) has shown that the North American Indians of New England understood the value of maintaining the landscape as a fine-grained, patchy mosaic through the use of controlled burning. This increases the diversity of kinds of foods available, and if one type of patch succumbs to a disaster (for example, a plant disease), there will be other types still able to provide food. Stability of the human population is maintained by avoiding a high degree of instability in the resource base on which it depends, as occurs, for example, if vast tracts are planted to corn, wheat, or potatoes, any of which could succumb to a plant pathogen outbreak.

The other broad category of adjustment to potential destabilizing perturbations involves expediting or facilitating homeostasis. Three strategies are available.

The first of these involves shortening the time delays in negative feedback control loops. It has been pointed out by Hutchinson (1948), Wangersky and Cunningham (1956, 1957a, b) and May (1974), among others, that the degree of instability increases with increased time delays in density-dependent negative feedback population control loops. Indeed, it has been shown using simulation that the well-known, 10-year wildlife cycle of the boreal forest is accounted for by time delays of the specific length that have been measured in the snowshoe hare–vegetation interactions by Keith and his students (see, for example, Cary and Keith, 1979; Pease, Vowles, and Keith, 1979; Vaughn and Keith, 1981). From May (1981), we would expect to find that 52-year waves in human society originate in homeostatic control mechanisms with time lags of roughly a quarter this duration, or about 13 years.

O'Neill (1978) has shown by simulation using key parameters from diverse ecosystems that ecosystem stability increases with rate of resource flux. An implication is that curtailment of resource flux would foster instability in human society and other living systems.

A third theorem on homeostasis and stability has been independently discovered in insect ecology by Force (1974), management science and operations research by Beer (1975), and Egyptology by Butzer (1982). In all living systems, too many hierarchical levels and too narrow a base relative to the height, that is, too high and steep-sided a pyramid structure, foster excessive dynamic rigidity in response to small perturbations, setting the stage for catastrophic instability in the face of major perturbations. Too many hierarchical levels increase the noise/signal ratio for vertical transmission of feedback control messages. This destabilizing mechanism operates, for example, where the large number of administrative layers in government or corporate structures cut off top managers from contact with realistic information or warnings about the system being managed.

Two other theorems are concerned with avoiding sources of instability. A series of writers in many fields, including Gardner and Ashby (1970), May (1974), and Lawlor (1978), have dealt with connectance between system elements. Systems are more likely to be unstable if they have too many species (system components), excessive connectance (too many linkages between each system component and other system components), or too much interaction strength. An illustration is the destabilization of North American society during the 1967 Arab–Israeli war by increased interaction strength of the connection to Middle East oil reserves due to increasingly rapid depletion of domestic reserves after 1970.

Finally, the impact of a potential source of destabilization may be eliminated by emigration, as when birds or people migrate.

Part of the confusion in the ecological literature may be due to a failure to distinguish between these seven ideas. Also, some seem to be saying opposite things to others of the seven, unless terms of reference are specified precisely. Thus, the first ("omnivory") theorem seems to equate increasing complexity with increasing system stability, whereas the connectance theorem seems to equate increasing complexity with increasing system instability. This confusion should evaporate with more precise definition of the phenomenon being modeled.

The research reported in this chapter seeks sources of instability in human society with these seven principles in mind. What is the significance for human society of inadequate omnivory, excessive delay in feedback mechanisms, curtailment of resource supplies, multileveled hierarchical administrative structures and their vulnerability to catastrophic instability in the face of massive perturbations, and excessive connectance, as of the U.S. agricultural sector to foreign markets, and the international trade in oil (as when the United States exports agricultural commodities to raise the foreign exchange to pay for oil imports)?

The focus of this research is not on instability associated with high-frequency, short-wavelength fluctuations, but rather with low-frequency oscillations with wavelengths of decades or longer.

LONG WAVES

There is a rich, long-established literature suggesting the existence of long waves in world society. Four points about these waves seem of paramount importance.

In very high quality data sets, such as the U.S. wholesale price index (Shirk, 1985), there are clear nadirs at 1789, 1843, 1896, and 1932. There are clear zeniths at 1814, 1864, and 1920. Thus the average internadir wavelength was 48 years, and the average interzenith wavelength was 53 years. This type of observation generated the notion of a roughly half-century wave in societies.

Whatever this phenomenon, it is remarkably synchronized across developed nations. For those data sets and nations where the time series are sufficiently complete or accurate to allow for determination of nadirs and zeniths, they coincide to within months of each other with respect to nations and phenomena (Rostow, 1979).

Furthermore, these long waves are now being discovered in types of data where they had hitherto not been noticed: computer analysis of the language in political speeches and speeches by monarchs also reveals a 50-year cycle (for example, see Namenwirth, 1973).

These waves are by no means regular. The nadirs were 54, 53, and 37 years apart. The wave-making mechanism is far less like a "ticking clock" than the mechanism that generates, for example, the 10-year lynx cycle.

We might expect that a phenomenon as important as these waves and the international depressions that terminate each wave would already be well understood. Surprisingly, every explanation put forth to explain these waves, or depressions, has run into criticism. Thus, even though the monetary contraction theory of depressions is believed by some to be the key element, it has been shown by Temin (1976) to be only a proximate, not an ultimate cause. There is still plenty of research to be done on a phenomenon of great importance.

WAR AS SYSTEM DESTABILIZATION

Recently, an even more compelling reason for looking into long waves in society has developed. It now appears that they are somehow connected to wars. Dyer (1985) notes that if we define a "world war" as a war in which all the great powers of the time are involved, there have been six, not two in modern times. His list includes the Thirty Years' War of 1618–1648, the War of the Spanish Succession, 1702–1714, the Seven Years' War of 1756–1763, the Revolutionary and Napoleonic Wars of 1791–1814, and the two world wars of 1914–1918, and 1939–1945. He notes that, apart from the long nineteenth-century gap, the list has an "alarmingly cyclical character. The great powers have all gone to war with each other about every 50 years throughout modern times." He further notes that even the "long peace" is deceptive because practically every great power fought one or more other wars between 1854 and 1870. His hypothesis is that peace treaties are accurate descriptions of power relationships in the world when they are signed, but after about 50 years, the relationships have altered. By then, one or more frustrated nations or nations with slipping power initiate war to formalize the new power structure. This paper suggests that wars are the key phenomena in setting the timing mechanism for long waves. However, it also suggests that two other war-related phenomena are important components of the wave-generating mechanism: retirement of war debt and demographic loops initiated by massive numbers of war deaths.

From the comparison of statistics on the number of war deaths worldwide with measures of economic health worldwide, it was discovered that the peak years for war deaths almost exactly coincide with the economic peaks, and the nadir years for the number of deaths in war occurred in 1838, 1888, and 1928—7, 8, and 4 years prior to economic nadirs (Craig and Watt, 1985).

This suggests that the state of the economy affects the probability of war or that the occurrence of war affects the subsequent state of the economy. Or wars and the state of the economy are linked in a circular, causal feedback loop (see for example, Goldstein, 1988).

It is argued hereinafter that the key to understanding this association is a little-known technical subdiscipline within macroeconomics, described in *Techniques of Treasury Debt Management* (Gaines, 1962), and the interaction of economic and demographic loops.

A few scholars have understood for a long time that the debts incurred by government during wartime and the gradually increasing interest associated with those debts involved enormous perturbations to society (Hirst, 1915). By 1932, it was understood that intergovernmental war debts on a massive scale created extremely serious problems. For example, if Germany was to retire its massive debt to other countries after 1920, it was necessary for Germany to have a massive net export surplus in order to raise the foreign currency required. But all other nations wanted to have export surpluses in order to stimulate their domestic economies. Clearly, it was not possible to have both (Moulton and Pasvolsky, 1932). By the time of World War II, many economists realized that wars set the stage for subsequent economic downturns (for example, Dickinson, 1940; Knight, 1940; Thorp, 1941). Clark (1931) and Gaines (1962) suggested that the retirement of war debt is one of the key causes for destabilization of the world system. Money comes into circulation at an unusually high rate when the citizenry receives money for its victory or liberty bonds. Consequently, within a few years overinvestment, particularly in energy production, in many nations causes simultaneous supersaturation of markets. Thus, wars set the stage for subsequent depressions.

The causality also goes in the other direction: depressions set the stage for subsequent wars. Rapid buildup in the armaments industry is used as a means of counteracting economic stagnation or depression, but simultaneous arms races in many nations create an international climate of suspicion, which increases the probability of war. This pattern was particularly clear in the 1910s and the 1930s. The association between rapid military buildup and short-term economic prosperity was rarely clearer than in Germany from 1932 to 1938: the unemployment rate dropped from 30.1% to 2.1%, coincident with an increase in crude steel output from 5.8 million tons to 22.7 million tons (Mitchell, 1980).

THE FIVE SOURCES OF INSTABILITY IN THE GLOBAL SYSTEM

This section reviews data illustrating the phenomena described by the model presented subsequently.

Exhaustive statistical analysis of historical data, conducted since 1968, suggests that only five phenomena are required to account for almost all the instability in the model reported here, given the focus on long- as opposed to short-wave fluctuations. The first of these is the status of the energy resource.

Figure 16.1 depicts 7-year moving averages of the United States wholesale price index plotted against year number. This graph is typical of many depicting the long-run change in some social variable. The year-to-year variance appears to be composed of one component attributable to variation around a trend and another component attributable to the trend itself. Why is the trend rising?

Four kinds of arguments support the proposition that this rising trend is due to the increasing cost of obtaining energy. Study of the correlation between highly aggregated price indexes and their

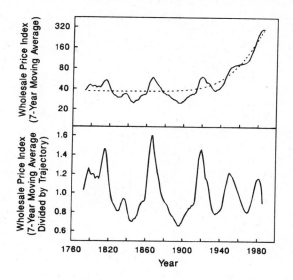

Figure 16.1 The distinction between the trend or trajectory for a variable measuring historical change in a system and the wavelike fluctuations about the trajectory. The example is the U.S. wholesale commodity price index (from splicing data series E-52[1] and E-23 in the *Historical Statistics of the United States,* updated with more recent statistics from Tables 750 and 755 in the 1989 edition of the *Statistical Abstracts of the United States* (109th ed.) and still more recent data in newspapers). The effect of high-frequency fluctuations has been eliminated by taking 7-year moving averages. In the top panel, the continuous line represents the moving average, and the dashed line is the trajectory, computed from $37 + \exp(-6.71 + 10.0594Y)$, where Y represents year number, less 1779. In the bottom panel, the value of the moving average for each year has been divided by the corresponding value for the trajectory so as to expose the wavelike pattern after removal of the effect of the trajectory. Note that the wavelength and amplitude of oscillation are dampening. Also note that every peak in the ratio plotted in the bottom panel comes either just after the end of a war (as in 1816, 1866, 1920, and 1950) or after an energy supply crunch, as in 1982. This makes the point that one systemic consequence of a war is an energy supply shortage, which results in excessive diversion of capital investment to the energy sector after the war. Energy and federal government public debt are two different means of expressing one of the fundamental systems phenomena critical for determining the long-run fate of nations: patterns of capital allocation over sectors of the economy.

---

[1] Throughout this paper, designations such as E-52, M-76, and the like, refer to data series maintained by the U.S. Department of Commerce, Bureau of the Census. Readers wishing to check or reproduce the author's work may find these data series in the *Historical Statistics of the United States* (HSUS) and updated annually in the *Statistical Abstracts of the United States* (SAUS), for which exact citations appear in the reference list.

subcomponents reveals that the energy component is the system driver: it has risen both more rapidly and more steadily than any other component.

The timing of changes in a variety of other social trends coincides with increases in the cost of energy discovery and production in the United States. Figure 16.2 illustrates and supports this contention. The four variables plotted against cumulative U.S. oil production are the ratio of a wage index to the consumer price index, that is, a measure of consumer purchasing power, the wage index (annual wage for contract construction workers), the consumer price index, and the constant-dollar cost to discover a marginal barrel of crude oil in the United States. The sharp downward break in wages, upward break in prices, and upward break in oil discovery costs all began at a cumulative oil production of about 100 billion barrels, half the ultimately recoverable total U.S. oil resource. This suggests that depletion of oil reserves and, more generally, of fossil fuel reserves is implicated in the changes in wages and prices.

The third argument comes from studying previous energy crises throughout history. The ratio of wages to price of firewood in England began a rapid decrease in 1540 when forests began to be depleted. The ratio of wages to firewood price began increasing in 1640, at which time coal shipments from Newcastle had increased from 50,000 tons, in 1570, to 480,000 tons (Watt, Craig, and Auburn, 1988). Thus, the ratio of wages to energy costs is an expression of the status of the energy resource.

The fourth argument combines theory with data. The independent variable in Fig. 16.2 is the cumulative oil production up to and including each year, not the production of energy each year. In the model described hereinafter, the state variable used to express the status of the energy resource is the cumulative domestic production of all types of fossil energy. Why this measure? What is the relation between this measure and three types of social variables: measures of wages, measures of prices, and measures of debt? What determines how much a person gets paid?

A contract construction worker hammers nails and saws wood, tasks that have remained quite similar for a long time, yet wages have continued to increase until recently. Also, if an individual migrates from a small city to a large city or from a small country to a large country, the wage increases. In each case, the person and the task may not change at all, but the wage increases. The capital stock with which a person works determines the wage, for a given level of competence and training. This capital stock is best measured by the current critical limiting resource for society. That was the availability of firewood in sixteenth-century England; in the United States today, it is the cumulative U.S. production of fossil fuel (which was used to manufacture the capital stock). Furthermore, it is the domestic production only that increases the capital stock because capital must leave a country to pay for imported energy.

The argument as to why cumulative fossil energy production is the current primary determinant of all prices in society seems clear. The amount of energy already used determines the extent to which readily accessible energy reserves have already been exploited. This determines the difficulty of exploiting those reserves that are still left (on the widely accepted assumption that the shallowest, most accessible, and most high-grade mineral resources are, on average, exploited first). The difficulty of exploiting the most accessible of the remaining reserves each year determines the cost of production; this cost is then passed all through society because energy costs enter into all other costs.

Consider the relation between energy and debt. What determines how much a borrower is prepared to borrow, how much a lender is prepared to lend, and the interest rates? Clearly, all these are related to the coupled notions of growth in resource use and the risk component of interest rates. Money is borrowed only when lender and borrower share the belief that the loan can be invested so as to repay principal and interest and, in addition, earn a profit for the borrower. This will be true only in a situation where high growth in return from the investment is likely, and that, in turn, typically depends on the availability of energy cheap enough to allow for that growth.

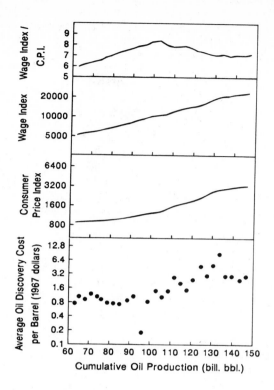

Figure 16.2 Demonstration of the coinciding changes in the cost of discovering fossil energy and consumer purchasing power (average wages divided by the consumer price index). Panel 1: consumer purchasing power; panel 2: index of average national wage; panel 3: consumer price index (1967 = 1000); panel 4: cost to discover one additional barrel of oil in the United States, in 1967 dollars. The independent variable for all four panels is the cumulative U.S. production of domestic crude oil, in billions of barrels. Average wage from D-745, Cumulative Price Index from E-135, both updated using recent data from *Statistical Abstracts of the United States.* Average constant-dollar cost to discover one new barrel of oil from drilling cost (purchasing power of the dollar)/barrels of new discoveries. Volume of new discoveries from reserves − reserves previous year + production this year. Oil data from Tables 776, 1252, and 1254, *Statistical Abstracts of the United States* for 1984 and corresponding tables for subsequent years. The consumer purchasing power in the top panel is inversely related to the oil discovery cost, typically with a 1-year lag. Thus, the energy discovery cost has ripple effects that gradually permeate all subsystems within society (to and including musical concerts, for example, the size of bands and orchestras being determined by the relative costs of energy and labor). The most convincing feature of this graph is that consumer purchasing power dropped precipitously after the surge in energy discovery costs in 1982 and then rose again after energy discovery costs dropped in the last few years. This supports the argument that there is an important, but overlooked causal connection between energy discovery cost and the performance of the economy.

To illustrate the effect of resource depletion on the risk component of the interest rate, the economics of exploratory drilling for oil and gas in the United States is instructive. From 1970 to 1982, the cost of all such drilling rose from 2.6 to 39.4 billion current dollars, yet proven reserves of oil dropped from 39 to 27.9 billion barrels (Tables 1252 and 1254, *Statistical Abstracts of the United States*, 1984). This large increase in the risk/reward ratio must have a big impact on the free market in venture capital and hence the society-wide risk component of interest rates.

Four other phenomena were discovered to be important in determining the dynamic behavior of the world system. The first of these is illustrated by the first three panels of Fig. 16.3. For each of these, the effect of the cumulative fossil energy production has been removed from the variable by a method that will be explained in a later section on modeling methodology. These panels depict, respectively, the birthrate of women 15 to 44 years old, the size of the population 20 to 24 years old, and the average wage of contract construction workers. This last is being used as an index for the wage level throughout society. Note that the depressions of 1843, 1896, and 1933 all appear as downward oscillations in the birthrate line. Declines in the birthrate line appear as declines in the line for number of 20 to 24-year-olds, 20 to 24 years later. Furthermore, years of unusually low numbers of 20 to 24-year-olds coincide with years of unusually high numbers of births, and vice versa. That is, the supply-to-demand ratio for numbers of 20 to 24-year-olds (in the labor force) determines the economic prospects for young people and this, in turn, determines their birth intentions. This homeostatic control mechanism (sizes of populations 22 years after birth determine birth rates) has a 22-year delay. One major source of instability in modern societies is thus identified: the long delay in this loop. This loop affects literally everything else going on in modern societies. To begin with, unusually high and low numbers of 20 to 24-year-olds depress and elevate the

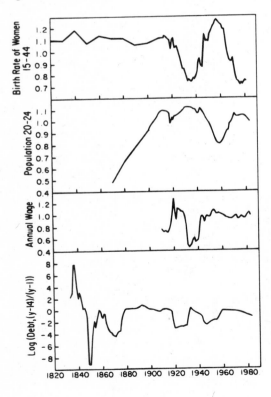

Figure 16.3 Oscillatory patterns observed in time series for U.S. birthrates, population size for 20 to 24-year-olds, annual wages of contract construction workers, and rate of change in the federal government public debt over a 13-year period. For the top three panels, the plotted data are the ratios of observed values to values computed from functions of the cumulative U.S. fossil fuel production, as measured in quadrillions of BTUs. (Data series: panel 1, B-6; panel 2: A-124; panel 3: D-745; panel 4: Y-493; all *Historical Statistics of the United States*, 1975).

average wage in society, all other things being equal, as indicated by the third panel of Fig. 16.3. This, in turn, affects the purchasing power of consumers for all quantities, and that, in turn, affects demand–supply trends and therefore prices. Thus, for example, the age distribution of society affects fuel prices, particularly because fuel consumption per person per year in vehicles peaks sharply as a function of age in males (Watt et al., 1975).

The effect of supply–demand for 20 to 24-year-olds on wages is not particularly clear in Fig. 16.3 because there is an interaction with another of the important phenomena driving society, depicted in the last panel of Fig. 16.3. The variable plotted is the logarithm of the ratio of the U.S. federal government debt 14 years ago to the debt the previous year. The higher this ratio is, the larger the amount of money that moved from the federal government to the private sector over the 13-year period ending last year and, hence, the higher the degree of supersaturation of all markets we would expect to find this year. Very high values of this ratio coincide with depressions; very low values coincide with periods of rapid economic growth. When this ratio is high, wages are low, and when it is low, wages are high (compare the third and fourth panels of Fig. 16.3). These two panels demonstrate that the excessively rapid retirement of war debt by government is a key determinant of depressions since wages are a key index of the state of economic health of a society. Rapid retirement of federal government public debt, as in the United States in the 1920s, occurs years after the rapid increase of this debt. A second source of long time lags, and hence instability in society, has been identified.

It is important to clarify precisely what is meant by "rapid retirement of federal government public debt." It is not the total magnitude of the debt incurred in war that is important or the amount by which the debt decreased per unit of time, but the amount by which the debt is reduced as a proportion of the debt outstanding at the beginning of a 13-year interval. To illustrate, the total debt (almost entirely due to the Civil War) was $2.756 billion in 1866 and had been reduced to $2.299 billion by 1879. The ratio of the former to the latter was 1.20. Then from 1879 to 1892 the debt was further reduced from $2.299 billion to $968 million, with a ratio of 2.37. The large amount of money introduced into the economy by this rapid retirement of federal government public debt produced an investment boom, as revealed in statistics on, for example, the number of new urban dwelling units started between 1879 and 1892. Markets became supersaturated and prices collapsed, triggering the depression of the mid-1890s. To illustrate, the building cost index dropped 15% from 1886 to 1897.

The huge Civil War debt had been a continuing potential threat to the economy since the end of the war. However, the timing of the postwar depression was set after 1879, when the rate of decrease in the debt was made to increase sharply.

The second panel of Fig. 16.3 reveals another linkage between war and the rest of the system. Note the extraordinarily low number of 20 to 24-year-olds in the United States in 1870, relative to the long-run trajectory for this variable. During the U.S. Civil War, 622,000 U.S. soldiers died (Dyer, 1985); there were only 1.9 million 10 to 14-year-old males in the country in 1860. This figure exposes the magnitude of the demographic impact in the only war where U.S. casualties were as high a proportion of total population as the war deaths for many nations in subsequent wars. It will be seen that a war with this level of demographic impact initiates harmonic oscillation in the demographic system. The population shortfall in young adult males leads to decreased competition in the labor force and improved economic prospects for the survivors. This leads to a surge in their birthrates and, ultimately, to overshooting of the long-run trajectory for 20 to 24-year-olds, followed by undershooting, then overshooting, repeatedly. This oscillatory pattern, with a 40-year wavelength, is clear in Fig. 16.3. The alternating undershooting and overshooting of the population trajectories lead to overshooting and undershooting of the wage trajectories. These, in turn, produce economic good times and bad times through their effect on consumer purchasing power and sales. Thus, serious wars, with massive war deaths, initiate long-term demographic cycles that in turn reinforce long-run economic cycles.

Another phenomenon that this research found to be an important determinant of social dynamics is the life cycle of international trade balances between nations. Consider a nation, E, that exports any commodity to an importing nation, I. At the outset of the long-run trade life cycle, E gradually increases its capacity to produce the commodity. Through this period, exports from E to I gradually increase. Finally, exports become so large that I can no longer afford the foreign exchange deficit on the transfer of money for the commodity and begins to increase its own production of the commodity, thus reducing the market for exports from E. As more and more nations do this, the volume of exports of the commodity from E declines from its peak. Furthermore, it may take some time before the producers in E accept that this loss of market is occurring, so they may continue to produce for an export market that no longer exists, long after it has ceased to exist. This is currently true of farmers in the United States, for example. A third source of instability due to long lags in modern society has been identified. It was noted earlier that major depressions tend to be international phenomena: depressions occur within months of each other in all developed nations. Why? What is the nature of the force synchronizing the economies of different nations? Interest and inflation rates are transmitted across international boundaries. Summarizing a major study by Darby et al. (1983), the main mechanism for international transmission of inflation is via the balance-of-payments effects of goods and assets substitutability. Thus, inflation in any major nation quickly becomes converted into inflation in all countries: the excess money supply in the inflating country translates into excess purchasing power spilling over into all countries. The foreign central banks are then reluctant to bear the cost of reducing the impact of this inflation rate in their own countries, so their money supply growth rates bend upward also.

The preceding argument is intended to justify using a simple model of U.S. society as a surrogate for a global model.

STRUCTURE OF THE SYSTEMS MODEL

The utility of highly complex models is being challenged in the technical literature on forecasting and planning (for example, Ascher, 1978; Armstrong, 1978; McClean, 1977; Makridakis et al., 1984). Complex models have three faults. They are difficult to understand, resistant to attempts at statistical falsification, and sensitive to violation of assumptions underlying the model in the data-generating process. Therefore, they are not readily amenable to the scientific method of a strong inferential hypothesis test and in a sense, therefore, are not the products of scientific activity in the traditional sense. Consequently, the program at Davis gradually evolved in the direction of extreme model simplicity, in which effort is focused primarily on the attempt to identify and incorporate into the model a handful of variables that can be demonstrated to be central and critical for the dynamic behavior of the system under consideration.

To this end, in building a model of society to account for wars and depressions, instead of attempting to model all sectors and phenomena, only those phenomena were included that appear to be necessary to explain the dynamic behavior of the society. Also, rather than modeling all sectors, agriculture was used as a surrogate for all production and consumption sectors. There are four reasons for this. First, agriculture is perhaps the only sector on which all the data are in the public domain. Variables are available, or can be computed from available variables, that reflect the true state of the system, rather than being primarily the result of arbitrary judgments about how to prepare the books on financial results. To illustrate, annual financial statements on, for example, an aircraft manufacturer determine overhead using assumptions about future sales. At the end of the first year of manufacturing a new model when, say, 30 copies of the model have been built, should all overhead costs of research, development, and tooling up for manufacturing be prorated over those 30 copies or some larger number? Clearly, unless overhead is prorated over some larger number of copies, the net profit figures will be deflated, which will discourage investors. The solution is to assume that the ultimate number of copies of this model to be manufactured will be some number, X, computed so as to yield a profit on the entire model that will be satisfactory to present and potential investors. The problem is that the true value of X cannot be known in advance. Thus, for many sectors of the society, there are formidable problems in interpreting financial statements. These problems are less severe for agriculture, where much of the overhead and gross profit follow trajectories that are much more predictable than for heavy industry.

Agriculture is a much more important sector than it appears at first because it affects many other parts of society, from the corporations that supply the machinery and chemical inputs to the corporations that transport inputs to the farm and outputs away from the farm and ultimately to retail outlets. Thus, statistics on agriculture, in a narrow sense, are surrogate statistics for a large segment of society.

Agriculture has been an important leading indicator of changes that would occur later in the rest of society. Thus, by 1927 the agricultural sector showed clear evidence of a decline that would not appear in the rest of society until late in 1929.

The agricultural sector epitomizes phenomena occurring throughout society. As fossil energy reserves are depleted, the return to energy expenditure throughout society follows the law of diminishing returns, whereas the energy input costs to all activities grow as an exponential or supraexponential function of cumulative production. Thus, simply multiplying the statistics on the agricultural sector by a gradually changing factor captures essential features characteristic of all sectors.

With agriculture, as with the rest of the model, the system measures chosen were in many cases different from those conventionally used. However, the attempt was to select measures that seem most revealing of the system state and least vulnerable to ambiguous or equivocal interpretations. Thus, in the case of agriculture, the performance measure selected was the realized net income of farm operators from farming, divided by the total interest paid on farm debt, including real estate, and nonreal-estate interest. This index was selected because the denominator, the interest, absolutely must be paid each year if the farm operator is to avoid foreclosure. Thus, this denominator seems like an appropriate measure, of unequivocal meaning, against which to evaluate net income. Figure 16.4 shows the time course of this variable. Its peak in this century was about 40, after World War

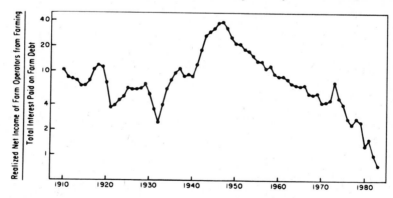

Figure 16.4 Trend in one of the key measures of system performance: the net return to the U.S. agricultural sector each year, expressed as a proportion of the annual total interest payment on farm debt associated with land, equipment, and supplies. (Data from Table 1133, *Statistical Abstracts of the United States*, 1984; and corresponding tables in previous editions and in *Historical Statistics of the United States*, and *Economic Indicators of the Farm Sector, Production and Efficiency Statistics,* 1981, U.S. Department of Agriculture (USDA), and *Income and Balance Sheet Statistics,* 1981, USDA.)

II, and it now stands at 0.75. The graph reveals the stimulating effect of two world wars and the depressing effects of recession in 1920 and depression in 1930–1934. The graph suggests that some type of long-wave generating mechanism is at work and supports the notion that there are feedback controls with long time delays at work in modern societies. Figure 16.5 illustrates how causal pathways flow into and out of the farm sector and how this is typical of the way events occur in sequence in each sector of society. The top panel reveals how the federal government public debt increased sharply during World War I and then was gradually decreased throughout the 1920s. This decrease translated into a high rate of increase in money for the private sector, and that, in turn, made possible a big increase in mortgage debt (panel 2) and nonmortgage debt in the farm sector. This borrowed money in the farm sector made possible, among other innovations, a transition from horse-drawn to tractor-drawn machinery (panel 3). That, in turn, elevated the index of total farm productivity above its long-run trend line (panel 4). Elevation of farm commodity supply relative to demand depressed unit prices for food, and this depressed farm income (panel 5). That, in turn, elevated the ratio of farm mortgage debt to farm income.

Corresponding statistics on all other sectors show the same type of pattern. Thus, for example, because of excessive building during the 1920s, the dwelling unit vacancy rate was seven times as high in 1930 as in 1920, evidence of market supersaturation (Wickens, 1941).

All the ideas mentioned to this point are assembled into a system flow chart in Fig. 16.6. In this chart, plus signs on an arrow imply a positive relation; negative signs imply an inverse relation. Two parallel lines perpendicular to an arrow imply a time lag. Each arrow represents an equation, or a term in an equation, that was fitted to a historical data series using either or both of the Matyas or Simplex algorithms (Schwefel, 1981).

This flow chart consists of four subsystems: a farm subsystem (upper left), government budget subsystem (upper right), demography subsystem (lower left), and energy subsystem (lower

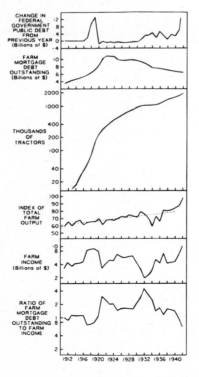

Figure 16.5 Graphic demonstration of the way causal pathways flow between key state variables in the systems model of Fig. 16.6. [Data sources: Y-493, K-361, K-184, K-414 (1947–1949 = 100), K-284, K-361 and K-284 in *Historical Statistics of the United States*, 1975.]

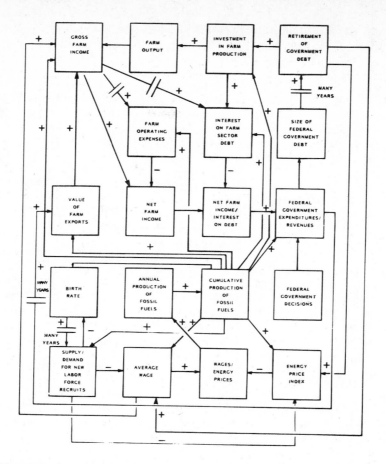

Figure 16.6 The model for a typical nation. The agricultural sector is used as a surrogate for all production and consumption sectors because the quality of the data base for agriculture is excellent and complete. The global model consists of one system of this form for each major nation, with the models being linked by flows of interest and inflation rates, money, goods, and people.

right). The long lags that generate the instability in this system originate at four points: from the time that federal government budgets are increased sharply (1) to the time that values of farm exports increase sharply, and (2) to the time that government debt is retired; and from the time that birthrates change until (3) the time those changes are reflected in the population aged 20 to 24 years, and (4) the time that annual production of fossil fuels has an immediate response to a change in wages relative to energy prices, but also a delayed response, with a lag of 9 years.

In all cases, the form of the equations and the parameter values were derived from actual data using time series that extend back from 1983 to 1910 or earlier, if data were available.

The equations used in the model were each selected after an exhaustive set of hypothesis tests to determine which independent variables were most important in accounting for long-wave fluctuations in the dependent variable and which mathematical structure best expressed the form of the relationship.

Chap. 16   Societal Instability

OBTAINING STRUCTURE AND PARAMETER VALUES FOR EACH EQUATION

A principal concern in the program reported here has been development of a formal protocol for determining the structure and parameter values for each equation in the model. In this section, a step-by-step explanation of this process is presented for five reasons. First, sufficient detail is given so that other workers can apply the method to different data sets. Then the reason for each step is explained in terms of ecological theory. The process is illustrated using a data set, the number of births in France, that illustrates several of the important phenomena typically encountered. The results from this data set are used, first, to give a statistical justification for the particular analytic process and, second, to explain the implications for ecological phenomena generally.

The French population data are from Mitchell (1980); the data on cumulative U.S. fossil fuel production were computed using a cumulative sum of time series M-76 (*Historical Statistics of the United States*, updated using Table 950, *Statistical Abstracts of the United States*, 1984).

Step 1. For each variable to be included in the equation, assemble the longest possible data series that can be found. The reason for this emphasis on the length of the data series is that slow change in the trend or trajectory may be veiled by wide-amplitude oscillations or fluctuations about the trend if only a short run of data is inspected. Also, an oscillatory pattern may be initiated or sustained by perturbations that only occur rarely. In terms of ecological theory, this remark about the need for a long data series implies that the variation in ecological variables must be separated into two categories: that due to fluctuation in the carrying capacity, brought on by change in the resource base (an ecosystem phenomenon), and change in the pattern of fluctuation about the carrying capacity generated by population phenomena.

Step 2. Determine the shape of the trajectory produced by change in the status of the current critical limiting resource. The purpose of this step is to distinguish variance attributable to change in the trajectory (carrying-capacity change) from variance due to oscillation about the trajectory. In some cases, as with the wholesale price index plotted in Fig. 16.1, the shape of the trajectory is fairly obvious from inspection of a graph. In other cases, as with the trend since 1800 in the number of births in France (top panel of Fig. 16.7), the long-run trend is obscured by recent violent fluctuations about the trend. In such cases, the shape of the trend is exposed by the use of very long moving averages. In this instance, the fine line in the top panel is a 71-year moving average. This step exposes the effect of the current most critical limiting resource on the carrying capacity. In general, as that resource becomes depleted, all price indexes gradually trend upward (as in Fig. 16.1), and all indexes of population size and economic activity trend gradually down (as in the top panel of Fig. 16.7).

Step 3. Determine the appropriate form of the equation describing the relation between the dependent variable and the status of the current critical limiting resource. The purpose of this step is to make possible a correction for the effect of changing carrying capacity on the dependent variable, so as to isolate and expose the effect of population-level phenomena on fluctuation about the trajectory determined by carrying-capacity changes. The variable used for the independent variable is the cumulative production of fossil fuel in the United States. This variable was obviously not causally connected to a decline in French birthrates from 1880 to 1900. Rather, for the period up to 1973, it is being used as a surrogate variable for synchronous, time-dependent, long-run changes in world society, including France, that gradually elevate prices, wages, the standard of living, and the level of application of technology; depress birthrates; and, more generally, drive the demographic transition. Cumulative U.S. fossil fuel production as the independent variable for a variety of other variables throughout this chapter is used in the same sense: it is a surrogate for various time-dependent changes, of which the status of the fuel resource is only one part of a complex of interrelated processes.

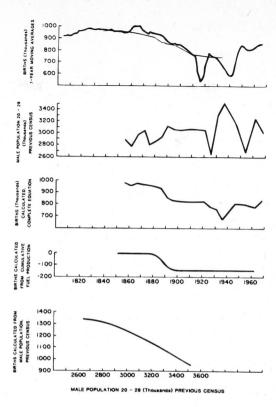

Figure 16.7 Number of births each year in France: an example of the issues raised by the process of fitting equations to data for a global model. Top panel: number of births in France each year, in thousands, 7-year moving average (thick line) and 71-year moving average (thin line). Second panel: male population in France, 20 to 29 years of age, in thousands; plotted data are for year of previous census, not current year. Third panel: number of births calculated from male population at previous census and cumulative production of fossil energy in the United States, in quadrillion BTUs. Fourth panel: component of calculated births computed from energy alone. Fifth panel: component of calculated births computed from male population at previous census alone. French demographic data from Mitchell (1980).

After 1973, however, the causal connection of the U.S. fossil fuel resource to the international system becomes more clear-cut: this variable has by then become the critical determinant of the world energy market. That is, no other country after 1973 could charge so much for its exported energy that the principal world energy consumer, the United States, would switch to its own domestic resources. This theory argues, in effect, that prices for Saudi Arabian crude oil are not determined by the cost of production of that crude oil, which is very low, but by the cost of producing U.S. crude oil.

The appropriate equation expressing the relation between U.S. cumulative energy production and, for example, the number of French births is determined by plotting the latter against the former and then picking the equation on the basis of the shape of the graph. It has been discovered that just nine different equations account for the relation between long-run trends in most social variables and the status of the energy resource. Seven of these are the most commonly encountered equations in any statistics or curve-fitting text, and many are found in almost all ecology texts, such as the exponential and logistic equations. Two additional equations have proved to be useful. One is the reverse logistic,

$$Y = a - \frac{a}{1.0 + b \exp(-cX)},$$

which describes, for example, the declining number of French births as a function of cumulative U.S. energy production. The other is a modified exponential growth form, in which a constant in the exponent is replaced by a declining linear function of the independent variable:

Chap. 16   Societal Instability

$$Y = a + \exp[bX + \exp(c + dX)].$$

Step 4. Once the appropriate form of this function has been selected, the best-fitting parameter values are selected by use of an iterative, nonlinear parameter estimation algorithm (Lootsma, 1972; Schwefel, 1981). This process is accomplished using a combination of Matyas and Simplex algorithms. The process is time consuming, and not straightforward because the algorithm may become trapped in an $n + 1$-dimensional local minimum region (one dimension for the residual sum of squares and n for the n parameters to be estimated). If that happens, the process is restarted at the same location using either or both of two devices. One is to switch methods (Simplex to Matyas, or vice versa). The other is to enlarge the volume within which the search is being conducted in case the algorithm is trapped in a steep-sided $n + 1$-dimensional canyon.

Step 5. Having obtained the parameter values that produce the lowest residual variance, the effect of changing carrying capacity on the dependent variable is eliminated by dividing observed Y values by Y values calculated from the equation fitted in step 4 or by subtracting calculated from observed values. Whether division or subtraction is used is based on guesswork about the final form of the equation.

Step 6. The ratio or difference just obtained is then plotted against values of the variable or variables hypothesized to account for deviations of the dependent variable about its trajectory. The object is to determine the form of the term in the final equation for births that accounts for deviations around the trend line or trajectory. In the case of births in France, a reasonable hypothesis is that births are most strongly related to the size of the French male population 20 to 29 years of age. Why was that variable selected? From common sense, we would expect that variable, which has undergone violent oscillations due to the trench warfare of World War I, to be the most likely explanation for the violent fluctuations about their trajectory observed in the number of French births. Inspection of the second panel in Fig. 16.7 supports that guess. An ecologist would expect that the number of births would be very low when population size was unusually high and that the number of births would be unusually high when population size was unusually low. As expected, the number of births in France had peaks in the 1920s and around 1950, when the population of males 20 to 29 years of age was unusually low relative to its long-run trajectory.

Hereinafter, the source of the raw data used to compute each statistical series that has not previously been mentioned will be identified by a reference in brackets. In the interests of brevity, where possible, data series will be identified by the letter-and-number code used by the U.S. Bureau of the Census (*Historical Statistics of the United States*, 1975).

Step 7. Use nonlinear estimation to obtain the values of the parameters in the final form of the equation. In this instance, that form was

$$B = [\exp(a - bP) - \exp(c - dP)] - \frac{e}{1.0 + f\exp(-gC)},$$

in which B represents the number of births each year, in thousands (Mitchell, 1980), P represents the male French population 20 to 29 years of age each year, in thousands (Mitchell, 1980), and C represents cumulative U.S. production of fossil fuel in quadrillion BTUs (M-76).

The fitted parameter values were

a	11.9	b	0.00110	c	12.35	d	0.00132
e	145	f	75.5	g	0.0772		

The equation, with these parameters, accounted for all but 10.9% of the variance in the number of births in France. The values calculated from the equation are depicted in the third panel of Fig. 16.7. The behavior of the two terms in the equation is depicted in the fourth and fifth panels of the Fig. 16.7.

Step 8. Observed values are divided by values calculated from the equation, and the ratio is plotted against both year number and each of the independent variables, in turn, to determine the source of major discrepancies between observed and calculated values.

Step 9. More terms and variables are added on the basis of the results from the previous step until no further significant reduction in the residual variance can be achieved.

Eight central issues for ecological modelers are exposed by the results from applying this procedure to the French birth data.

1. In this instance, as in every other case considered in this paper, the dependent variable is regulated by the interaction between an ecosystem phenomenon, which regulates the behavior of the long-run trajectory, and a population phenomenon, which regulates fluctuation about the trajectory. The residual variance is much higher if either of these is excluded from the equation. To illustrate, ignoring the gradual change in the trajectory, the first term in the equation accounts for only 74% of the variance in the number of births, year to year. Thus, an equation intended to account for a high proportion of the time-to-time variance in a process of this type must include separate terms for ecosystem-level and population-level regulating processes.

2. Rare events are very important in originating and sustaining oscillatory behavior. In this case, the major historical source of oscillations about the long-run trajectory in French births was the trench warfare of World War I; many men were locked into the trenches for long periods, separated from mates. The number of births in 1916 was depressed about 27% below its trajectory, and the number of men 20 to 29 years of age was by 1920 depressed far below the trajectory for that population group. The unusually small number of births in 1916 showed up as a depressed number of 20 to 29-year-olds in the period 1936–1945, and this resulted in elevated birth numbers at that time (the inverse relation between number of parents and number of births is expressed by the bottom panel of Fig. 16.7).

A rare event, the massive war deaths of the U.S. Civil War, in the same fashion initiated an oscillatory demographic–economic process in the United States (second panel of Fig. 16.3). This phenomenon of initiation and maintenance of oscillation has been reviewed by May (1981). For the differential equation form of the logistic growth equation with a time lag of duration T, the response of the system to a perturbation depends on the size of the product rT, where r is the intrinsic rate of natural growth. For human populations of Europe and North America of the last two centuries, r is about 0.032 per year (as measured from growth rates during periods of unconstrained or exponential growth). If rT is smaller than $1/e$, there is a monotonically damped stable point (in this case, trajectory); if rT is greater than $1/e$ but less than $\pi/2$, there is an oscillatory approach to the trajectory; and if rT exceeds $\pi/2$, the population exhibits stable limit cycles. It will be shown subsequently that there are two time lags in the demographic system, one 13 years long and one about 24 years long. For the larger of these, rT is 0.032(24), or 0.77, which exceeds $1/e$; consequently, perturbations, such as massive war deaths, initiate oscillatory approaches to the trajectory. However, if perturbations occur two or three times a century, the oscillations will be sustained. This would happen if, for example, wars initiate a process that after a lag results in another type of perturbation, such as a depression. In short, a system appearing to be chaotic or subject to catastrophe may result from the interaction between three different types of mechanisms: (1) a demographic or population subsystem with a long time lag, and hence a tendency to long-wavelength; slowly dampening oscillations about a trajectory; and two other mechanisms that perturb the system two or more times a century; (2) wars; and (3) depressions following wars, with a lag originating in the federal government war debt retirement process. This is reminiscent of ecological phenomena in which lag-producing processes endogenous to ecosystems interact with infrequent perturbations, such as Vesuvian-type volcanic eruptions (see, for example, Watt, 1973).

3. The reason why the equation accounts for only about 89% of the variance in French births is that this simple model cannot account for the birth depression brought on by the trench warfare around 1916. This model cannot account for war because it deals only with phenomena at the national level, and war is an international phenomenon. Thus, to account completely for national-level phenomena, we need a hierarchically organized model with at least two levels, one for the national and one for the international level. Furthermore, the two levels need to be coupled in a simulation program so that at each iteration certain variables at one level can affect processes at the other level.

4. Births in France are typical of many processes encountered in ecological or global modeling, in that an effect at one time is the result of a cause at a previous time. Here, births in one census year are more closely correlated with male population in the previous census year. This implies not only the 9-month lag for fetal development, but a several-year perceptual lag. The significance of the lags throughout human systems is that long lags imply an increased probability of sustained oscillations.

5. The equation that best explains variance in French births is structurally different from that best describing variance in U.S. births. The reason is that births in much of Europe are now a decreasing function of cumulative energy use, whereas in North America they are still an increasing function of energy use. Thus, different nations are at different stages in a long-run, multicentury developmental process, and one form of equation will not suffice for the corresponding phenomenon in every nation.

6. It is sometimes asked whether this type of research is simply curve fitting, or is it dealing with fundamental laws of nature. One means of gaining more confidence that basic principles are being dealt with is to conduct comparative studies of many countries and many time periods. If, for example, the same mechanism produces three consecutive world depressions, then there is more reason to believe that a fundamental principle is being revealed. Similarly, if the same type of effect shows up in many countries simultaneously for the same apparent reason, this provides more confidence that it is not simply a statistical artifact that is involved.

The status of U.S. fossil energy reserves is important because the world energy market became altered permanently, coincident with the time that half the ultimate recoverable U.S. oil resource was depleted. This is the basis for the notion that it is the status of the U.S. resource that sets the market for energy in each country, not the cost of production in that country.

7. Wars are stochastic, not deterministic, phenomena. The probability of war is increased in proportion to the severity of a previous depression or recession, which in turn is a long-lagged response to the previous war. Also, the probability of war is affected by the nature of governmental decisions in several countries. Thus, a realistic simulation model that can mimic the world system that produces wars and depressions must contain a stochastic element, and it must allow for different governmental decisions.

8. One reason why complex living systems may have a chaotic, surprising, catastrophe-prone appearance is the long lag separating causes and effects. Depressions and wars, particularly, exemplify surprising catastrophes that may be separated from their causes in both time and space. This notion will be illustrated by the analyses presented subsequently.

THE ENERGY SUBSYSTEM

The meaning and implications of the equations in each of the subsystems will be discussed in this and subsequent sections.

Annual U.S. production of fossil fuels is determined by the ratio of wages to the wholesale energy price. Therefore, the operation cycle for the energy sector for each year begins with computation of the current values for wages and the wholesale fuel price index.

The following equation accounts for all but 0.00219 of the year-to-year variance in the wage index from 1910 to 1983 (of total = 1.000):

$$W = a \exp[bC \exp(c + dC)] + e + fX1 + gX2, \qquad (16.1)$$

where W = average annual wage of contract construction workers, in current U.S. dollars (D-745),
C = cumulative U.S. production of fossil fuel, up to and including this year, in quads (C will be defined thus hereinafter) (M-76),
$X1$ = ln (ratio of value of the U.S. federal government public debt 14 years ago to the value last year); this logarithm is a measure of the likelihood of market supersaturation and hence of depression (Y-493),
$X2$ = population 20 to 24 years old (A-124).

Note that the three independent variables correspond to the three arrows pointing toward the box labeled "Average Wage" in the lower-left quadrant of Fig. 16.6.

The following are the estimated parameter values for eq. (16.1):

a	7.43×10^2		e	1.96×10^3
b	1.62×10^{-3}		f	-2.20×10^2
c	4.25×10^{-3}		g	-2.63×10^{-1}
d	-8.84×10^{-5}			

The significance of eq. (16.1) is that wages are rising exponentially with respect to cumulative U.S. fossil fuel production, but the exponential coefficient itself is a decreasing function of cumulative fuel production. One interpretation is that, as cumulative fuel production increases, the market for laborers supersaturates because the main demand for workers in the society shifts from primary and secondary to tertiary and quaternary industries (that is high technology and service activities). From inspection of the coefficients, downward deviations from the trend line for wages are produced by either market supersaturation or a surplus of 20 to 24-year-olds relative to demand.

The implications of the very low residual variance are far-reaching. Wages are a good measure of the general state of a society. The fact that this simple, three-independent variable, seven-parameter model can account for about 99.8% of the year-to-year variance in wages suggests that this equation is likely a realistic, and close to complete, "explanation" for depressions. They arise because government has been retiring its war debt too rapidly, thus introducing money into the private sector fast enough to produce excess productive capacity in all markets simultaneously and because the labor market becomes glutted by the excess number of young people seeking new jobs.

The following equation accounted for all but 0.0116 of the variance in the wholesale fuel price index from 1910 to 1983:

$$P = \exp(aC) + bX1 + cX2 + (dX2)^2 + eX3 + (fX3)^2, \qquad (16.2)$$

where P = wholesale price index, fuels (E-29),
$X1$ = ln (ratio of the value of the U.S. federal government public debt 14 years ago to the value last year),
$X2$ = wholesale price index, fuels, 3 years ago,
$X3$ = U.S. total population, 20 to 24 years of age.

Parameter estimates for eq. (16.2) are as follows:

a	-2.70×10^{-3}		d	-1.92×10^{-3}
b	-5.93		e	-1.96×10^2
c	0.290		f	-1.49×10^6

Equation (16.2) means that wholesale fuel prices are increasing as an exponential function of cumulative fuel production because, as more fossil fuel resources are exploited, the remaining reserves become progressively more costly to exploit (the remaining reserves are at greater depths and in progressively more inaccessible locations, such as the deep outer continental shelf). Deviations about that trajectory are almost entirely accounted for by the rate of retirement of federal government public debt, the size of the fuel price index 3 years previously, and the number of 20 to 24-year-olds. Very high supply relative to demand for 20 to 24-year-olds depresses fuel prices. The reason is that fuel use is a very steeply peaked function of age (in males only); 28-year-old males drive twice as many miles per year as 17-year-olds or 67-year-olds (Watt et al., 1975). Therefore, the fuel market is sensitive to large year-to-year changes in the abundance of 20 to 24-year-olds. Unusually large numbers depress wages of that group; decreased wages, in turn, decrease the amount they can pay for fuel. Thus, surprisingly, there is a statistical connection between population growth rates and fuel prices 20 to 24 years later. This is another mechanism that introduces long time delays, and hence wide-amplitude instability, into society.

Fuel prices are also depressed when all markets are supersaturated (when depressions or recessions have been brought on by excessive rate of retirement of federal government public debt). They are also depressed when fuel prices 3 years back were unusually high.

The following equation accounted for all but 0.0309 of the year-to-year variance in the domestic production of fossil fuel:

$$F = a + bX1 + (cX1)^2 + dX2 + (eX2)^2 + fX3 + (gX3)^2, \qquad (16.3)$$

where F = domestic production of fossil fuel, in quads (M-76),
$X1$ = average annual wage of contract construction workers, divided by the wholesale price index for fuels; thus, $X1$ is an index of the purchasing power of a typical worker in society with respect to fuels and energy (D-745, E-29),
$X2$ = the same as $X1$, but with a lag of 9 years,
$X3$ = domestic production of fuel 3 years ago.

Parameter values for eq. (16.3) are as follows:

a	4.29		e	3.39×10^{-3}
b	0.436		f	0.454
c	-1.87×10^{-3}		g	5.73×10^{-4}
d	-0.142			

There are several noteworthy points about this equation. First, the domestic production of fuel is sensitive to consumer purchasing power for fuel 9 years ago because part of the response of consumers to increased fuel prices occurs after a long lag. The reason for this, in turn, is that depreciation is typically a larger proportion of the expense associated with any fuel-using system than fuel cost. Or the decrease in cost that would result from trading in an old car to get a new, more fuel-efficient car is less than the increase in depreciation would be. Accordingly, part of the response to increased fuel prices is revealed only after people begin trading in their capital items that use fuel. This argument implies even longer lags for very major items, such as power-generating plants and steel mills. Failure to take account of this very long lagged response of energy production and consumption is why most energy models have consistently overestimated future demand trajectories.

The last term in this equation is required because there is a high degree of inertia in the energy production system.

THE DEMOGRAPHIC SUBSYSTEM

The following equation accounts for all but 0.00206 of the year-to-year variance in the population of 20 to 24-year-olds since 1909:

$$Y = a + bX1 + cC, \qquad (16.4)$$

where Y = total U.S. population 20 to 24 years old, in thousands, and
$X1$ = total number of U.S. live births, 20 to 24 years previous, in thousands (B-1).

Parameter values are as follows:

$$a \quad -2.94 \times 10^3$$
$$b \quad -0.870$$
$$c \quad -2.31$$

This equation means that the population of 20 to 24-year-olds is not a fixed proportion of the number of live births 20 to 24 years previously, but rather a rising proportion, with the rate of increase depending on cumulative fossil fuel production. Here, the last-named variable is a surrogate for the standard of living in the United States and the world and captures the effect of these standards on life expectancy from birth and the volume of international migration.

The following equation accounts for all but 0.055 of the year-to-year variance in the number of live births in the United States:

$$Y = a + bC + cX1 + (dX1)^2 + eX2 + (fX2)^2, \qquad (16.5)$$

where Y = U.S. total number of live births, in millions,
C = cumulative U.S. fossil fuel production in the previous year,
$X1$ = U.S. total population 20 to 24 years old in thousands, the previous year,
$X2$ = U.S. total number of live births, per year (Y–13).

The values for parameters are as follows:

$$\begin{array}{llll} a & -15.6 & d & -1.80 \times 10^{-8} \\ b & -0.00242 & e & -3.98 \\ c & -0.000903 & f & -0.558 \end{array}$$

This equation implies that there is a gradually rising trend in births with the increased level of affluence associated with higher cumulative fossil fuel production, that is, with the capital stock of society. However, births each year are depressed if the number of 20 to 24-year-olds is excessive, given the level of affluence of the society. Furthermore, there is a 13-year rhythm running through the birthrate data: years of unusually low births are followed by years of unusually large numbers of births 13 years later, and vice versa. The high residual variance reflects the relatively large impact on total births of war-related, short-term perturbations. Birthrates are unusually low when many service personnel are abroad and unusually high a year after their return from war.

THE AGRICULTURE SUBSYSTEM

Agricultural output (K-414) is very strongly determined by the U.S. cumulative production of fossil fuel. This simply reflects the fact that agricultural yields are strongly influenced by all forms of agricultural technology inputs. All but 0.0301 of the year-to-year variance in the index of farm output is accounted for by the following equation:

$$Y = a + bC, \qquad (16.6)$$

where $a = 26.7$,
$b = 0.0307$.

Equation (16.6) represents that component of the variance in farm output accounted for by the arrow in Fig. 16.6 pointing toward "Investment in Farm Production" and originating at "Cumulative Production of Fossil Fuels."

One central equation in this model is that for realized gross farm income. This equation is basically a surrogate for a large family of equations that would be used to account for gross national income. The following equation contains the four independent variables that would occur in any equations to account for year-to-year variance in income and accounts for all but 0.00604 of year-to-year variance in gross farm income:

$$I = a \exp[bC \exp(c + dC)] + e + fX1 + gX2 + (eX2)^2, \qquad (16.7)$$

where the parameters are

C = cumulative U.S. production of fossil fuel,

$X1$ = average annual wage of contract construction workers,

$X2$ = value of crude food exports (U-215).

Proceeding in this fashion, we can develop a complete set of dynamic equations for use in simulating the behavior of any socioeconomic–environmental phenomenon. But how do we know that these equations have any real application to the empirical world, rather than being simply products of a curve-fitting exercise? One approach to validation is to divide data into blocks for different time periods, get fitted parameter values for one time period, and then conduct simulations to determine if the equations and fitted parameters for one period mimic events in other periods. This process can be made very comprehensive by testing separate parts of each equation separately to ensure that the qualitative effects of each term are the same for different centuries.

To illustrate, suppose we test the hypothesis that the rate of decline in federal government public debt is, in fact, the critical timing variable for depressions. The data were separated into two blocks: a recent block for the period 1910 to 1984, inclusive, and an old block for 1846 to 1909. The equation

$$Y = \exp(a + bX1) + c + dX2 + (eX2)^2$$

was fitted, where

Y = U.S. wholesale price index,

$X1$ = the year number,

$X2$ = 7 + ln (federal government public debt, value in Y-14/value in Y-1).

The fitted parameters were $a = 158$, $b = 0.0823$, $c = 13.5$, $d = 20.3$, and $e = -2.42$. This equation yielded a deep depression in the 1930s, with 1933 being the lowest year. In fact, 1932 was the lowest year. When this same equation was applied to the period 1846 to 1909, it produced simulated depressions when they in fact occurred, in the 1840s and 1890s. The actual nadir year was 1896; the simulated nadir year was 1893.

DEPRESSIONS AS CONSEQUENCES OF LAGGED FEEDBACK CONTROL

Two of the system measures in the flow chart (Fig. 16.6) express the impact of all the key forces determining the dynamics of modern societies: the wage index (the box in row 5, column 2) and the ratio of net farm income to the total interest on farm debt [the box (3, 3)]. That is, these two variables may be regarded as surrogates for variables expressing phenomena occurring simultaneously in all economic sectors.

Experiments were conducted with three means of demonstrating the implications of sets of equations describing the dynamic behavior of complex systems. (1) The most commonly used procedure was to construct a simulation model, which was then used to generate sets of scenarios, each of which demonstrates the consequences of producing a simulated history resulting from different parameter values in equations, or switch setting for policy variables. (2) A less frequently used procedure was to use dynamic programming, or a variation on that method, in which the computer is programmed to select the optimal policy at each time step so as to produce some desired end state for the system, or trajectory, of certain state variables. (3) The third approach, which is discussed hereinafter, was to expose the implications of the equations by constructing graphs showing the long-run impacts of each component of an equation, separately or in groups of independent variables.

In the present case, where time series are available for many of the variables extending over several centuries, detailed analysis of past history by "tearing the equations apart" (the third approach) seems to provide the simplest and most transparent means of explaining the dynamics of the system. Furthermore, the first two procedures have not been particularly revealing: the results are obvious corollaries of the structure of the equation set. Thus, for example, the second approach reveals that the way to optimize the standard of living over centuries is to retire war debt more slowly than it has traditionally been retired (so as to prevent market supersaturation after wars) or, better yet, not to have wars at all. Specifically, computer output suggested that wars result in temporary boosts in standard of living in the few years prior to wars, during wars, and for several years after wars, for citizens of nations where there was no war-related destruction. However, the decreased standard of living in other years, relative to the actual historical trajectories for wages, was not offset by these temporary boosts. However, this result could have been predicted by noting that the estimated value for the parameter f in eq. (16.1) was negative.

The key to understanding this model is eq. (16.1) for wages of contract construction workers. Figure 16.8 depicts the extent to which this equation, in fact, accounts for the timing and severity of economic fluctuations since 1909.

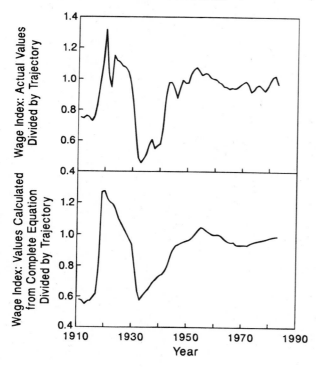

Figure 16.8 Graphic depiction of goodness of fit of the equation for the average annual wage of contract construction workers. The top panel is the time trend for the observed wage, divided by the value of the wage for each year calculated from an exponential function of cumulative U.S. fossil energy production. The bottom panel depicts the trend in the value for annual wage calculated from eq. (16.1), divided by the value computed from the same exponential function. Thus, we are comparing the observed values for the pattern of fluctuation around the trend line with the calculated values for the deviations around the trend line. Equation (16.1) clearly accounts for the timing and magnitude of changes in wages and hence is useful for dissecting the mechanisms producing major economic changes.

In the top panel, the observed value for wages in each year is divided by the value computed from an exponential function of cumulative fuel production. In the bottom panel, the estimate for average wage computed from eq. (16.1) is divided by the same exponential function. This is a very harsh, strong inferential hypothesis test. In effect, only the extent to which the equation can account for deviations around the trend is being considered, after the vast majority of the year-to-year variance (attributable to the trend) has already been removed. It will be seen that the equation accounts for all the major qualitative features of the deviation of wages around their trend line since 1909. Observed and calculated data show a vigorous increase, associated with World War I, a collapse from 1921 to the nadir of the Great Depression, a sharp increase to about 1953, then a shallow decrease to about 1968, and a recent period of little change.

What was the relative importance of the key system drivers in producing this pattern?

Figure 16.9 isolates the component of the variation attributable to cumulative energy production alone (top panel), population of 20 to 24-year-olds alone (second panel), and federal government public debt alone (third panel). Clearly, from the top panel, energy is implicated in the recent downturn in wages. That is, cheap energy is no longer being substituted for relatively more expensive labor. Now, increasingly, relatively cheaper labor will be substituted for relatively and progressively more expensive energy. Competition both within and among work forces in different nations is now sharply decreasing the rate of increase in wages relative to the rate of increase in energy prices, and hence all prices.

The second and third panels of Fig. 16.9 expose the relative importance of the size of the population 20 to 24 years old and the rate of change in federal government public debt over a 13-year period on the portion of the variance in wages not energy related. This isolation of factor effects was achieved by considering separate terms of eq. (16.1) one at a time. Evidently, the timing and severity of the Great Depression were due to simultaneous glutting of both labor and all other markets, both of which became progressively more serious after 1920. However, the population effect was of secondary importance to the effect induced by excessive retirement of war debt. We

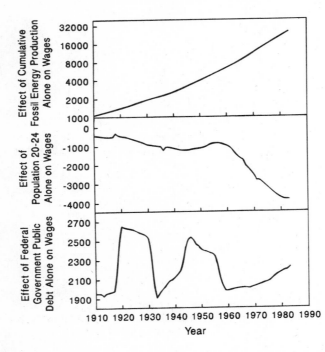

Figure 16.9 This graph isolates the separate effects on wages of energy-related phenomena (top panel), population-related phenomena (second panel), and federal government public debt (third panel). The Great Depression was caused primarily by excessively rapid retirement of the federal government public debt, secondarily by unusually numerous 20 to 24-year-olds, relative to their trend line. Recent "stagflation" is due to those two factors interacting with a third: phenomena related to energy depletion.

have documented every step in the market saturation argument for other sectors of society: this glutting process can be detected in everything from increasing vacancy rates in apartments (for example, Wickens, 1941) to increasing scrappage rates for cars, intensifying through the 1920s and early 1930s.

Figure 16.9 further reveals that the recovery from the Great Depression was aided by the clearing of the labor and other markets. To explain this phenomenon more comprehensively, it is necessary to trace a causal sequence flowing from birthrates to the size of the 20 to 24-year-old age cohort 20 to 24 years later, to the effect on wages, and hence on consumer purchasing power.

Perhaps a verbal explanation of the sequence will be more comprehensible than a discussion tied to the equation set. Because of excessively rapid retirement of the Civil War debt, excessively rapid rate of population growth after the Civil War deaths, and simultaneous phenomena occurring in other countries, there was a world depression in the 1890s. Birthrates were depressed during this depression, which produced small year classes of people 20 to 24 years later and set the stage for the prosperity of the early 1920s. However, the economic boom immediately following the 1890s depression elevated the number of live births, leading to the glutting of the labor market in the late 1920s and early 1930s. This depressed wages and contributed to the Great Depression. The low birthrates around 1933 produced a small age cohort of 20 to 24-year-olds in the mid-1950s, which decreased competition, elevated wages, and resulted in high birthrates. Thus, a picture emerges of a cyclical phenomenon that has a tendency to go into relaxation oscillations (note the increasing amplitude of oscillation in the top two panels of Fig. 16.3 for birthrate and population). The reason for the weak feedback control is that the time lag is so long: about 22 years from conception to the time the conceived child is attempting to make permanent entry into the labor force.

The phenomenon just described interacts and resonates with another linked cycle-generating mechanism produced by the oscillating loops involved with war and the repayment of war debt. The world depression of the 1890s resulted from excessively rapid repayment of war debt. The same was true of the 1930s depression. Only the depression of the 1840s had a different cause in the United States: the excessively rapid retirement of public debt during the late 1830s resulting from the vast sales by the federal government of public lands (Smith and Cole, 1935; Gaines, 1962).

The key role of federal government public debt retirement as the tripping mechanism for depressions is exposed by the bottom panel of Fig. 16.3. A high ratio of debt 14 years ago to debt last year implies a rapid reduction of that debt over the interval. Rapid retirement, in turn, is associated with depressions. Peaks in the bottom panel of that figure are invariably associated with depressions or wars.

Note, however, that the recent world economic "stagflation" is of a historically novel character: energy-related phenomena are playing a central role for the first time (the decline recently in the line of the top panel in Fig. 16.9).

Another means of exposing the system of forces producing economic downturns is dissection of the equation for gross realized farm income. Figure 16.10 compares observed and calculated deviations from the trend in this variable. Figure 16.11 shows the relative roles of energy-related phenomena and values of exports in producing the observed trend in gross realized farm income. Increased energy intensity had a positive effect on agriculture until just after World War II and has had an increasingly negative impact since, possibly due to its effect on market supersaturation. The curves show why agriculture is presently in difficulty: these two factors, together with depressed wages, are causing depressed gross realized farm income at present.

Another cause of recent stagflation is exposed by comparing the trends in gross farm income, farm overhead, and farm interest payments. Recently, the rate of increase of farm overhead with respect to cumulative energy production is dropping less rapidly than the rate for gross realized farm income, and the rate is declining still less rapidly for interest on farm debt. In other words, agriculture has become caught in a systems trap in which depleting energy reserves are driving costs and interest rates up faster than the return from production can be increased, given the system

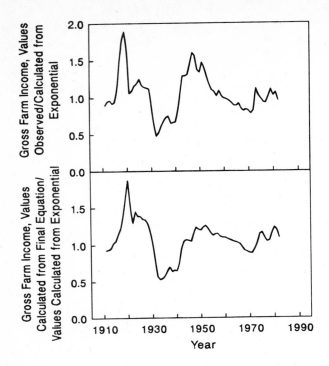

Figure 16.10 Pattern of deviations about the trend line in gross realized farm income in the observed data set (top panel) compared with those in the values calculated from eq. (16.7) (bottom panel). The model accounts for 98.8% of the variance in the entire data set and 99.5% of the year-to-year variance in nonwar-affected years.

of forces suppressing growth in consumer purchasing power. This situation is a metaphor for many sectors in society (consider the parallels to the airlines, for example). The way out of this trap is to decrease dependency on fossil fuel, insofar as possible, by shifting to inputs for which costs are less strongly affected by the depletion of fossil fuel reserves.

WARS AS EFFECTS OF LONG-LAGGED FEEDBACK CONTROL

To understand the system of forces operating to produce wars, we must imagine a world system consisting of several interacting nations, each of which is changing in accord with the system depicted in the flowchart of Fig. 16.6. To simulate such a system, a two-level hierarchically organized simulation model is required, with the fundamental structure of the type first explained by Mesarovic and Pestel (1974). In each time step, the state of several nations or groups of nations is updated, and the relationships between them are simulated. A war begins when any one of the nations in the world system is in an economic downturn, to which the government responds (1) by directing the resulting dissatisfaction and hostility of the electorate outward toward other nations, and (2) by utilizing rapid buildup in armaments as a means of dealing with underutilization of both labor capacity and capacity in heavy industrial markets (steel, shipbuilding, aircraft, tanks, and the like). Either of these responses increases the probability of war by alarming surrounding nations concerning the likelihood of a lightning attack. Two examples, well-documented, illustrate how economic stress leads into war.

The period 1905 to 1914 was characterized by an extraordinary number of days lost to industrial disputes in France, Germany, and the United Kingdom (Mitchell, 1980). While the data are spotty and incomplete, the available data series and anecdotal evidence suggest that economic growth was even more erratic during this period in Czarist Russia (see, for example, Mitchell, 1980;

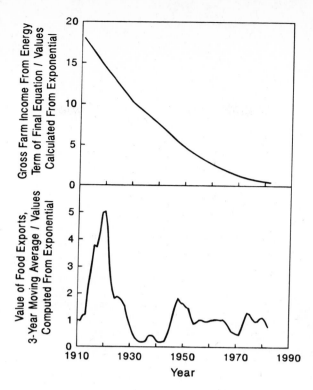

Figure 16.11 Isolation of the effects on gross realized farm income of energy-related phenomena (top panel) and food exports (bottom panel). Simulations of this type are conducted by suppressing the effect of all independent variables except one in each run, using a complicated fitted equation. To illustrate, suppose we wish to hold a parabolic term constant; we use 1, 0, and 0 for the three fitted regression coefficients. This procedure allows us to explore the separate effects of each independent variable on a dependent variable, one variable at a time.

Rostow, 1978). The available evidence suggests that the proximate cause for World War I was general mobilization in Russia as a means of dealing with internal dissension on a mass scale, rather than the assassination of the Archduke of Austria, which occurred much earlier in the summer of 1914. As Fay (1930) explains, an "apparently well-informed Russian . . . asserts that . . . general mobilization was strongly urged as a salutary measure against . . . internal industrial and revolutionary danger, rather than as a necessary military precaution against German attack." All major Russian cities were plagued with an extensive workingmen's strike at the time of the mobilization.

Hitler's rise in Germany was associated with a vast armaments buildup that resulted in a great improvement in the German economy. An unemployment rate of 30.1% in 1932 had dropped to 2.1% in 1938 (Mitchell, 1980).

Thus, we can postulate a long-run dynamic world system in which cycles come about through two linked mechanisms. A previous war (or simultaneous set of wars) creates economic stress in one or more nations subsequently either through economic stress resulting from punitive payments of war debt to other countries (as with Germany from 1920 to 1933) or through retirement of war debt leading to market supersaturation (as with the United States at the same time). One or more of the key nations in the world system then attempts to deal with these war-caused problems by an armaments buildup and by directing the hostility of the citizenry outward. This, in turn, sharply increases the probability of a subsequent war, which will be followed by a depression, which sets the stage for the next war.

Linked with this war debt mechanism is a demographic mechanism: massive war deaths lead to increased postwar wages for the survivors, higher birthrates, higher populations of 20 to 24-year-olds 20 to 24 years later, and resulting economic hard times. Governments try to drag their nations out of the hard times with weapons systems buildups, which increase the likelihood of the next war. This war-to-war wavelength can be as long as 49 years (1865 to 1914) or as brief as 20 years (1919 to 1939). In the latter case, a punitive and unrealistic postwar settlement (the Versailles Treaty) intensified the typical postwar competitive stresses between nations and shortened the time to the next war.

LESSONS FOR ECOLOGISTS

Experience gained in developing this model suggests several lessons for ecologists. All the statements that follow are only opinions, but they are opinions based on experience over almost two decades, during much of which at least one dedicated microcomputer was working 24 hours a day, 365 days a year.

There are tremendous institutional pressures on modeling teams to build highly complex models. Each client for model output is primarily concerned with the specifics of particular state variables of importance to them. A driving force promoting model complexity is to ensure that all prospective model clients, interest groups, or political actors find items in the output of concern to them. Another driving force fostering model complexity is the notion that the global properties of the system are generated by details of the dynamic interaction between several system subcomponents.

As a result, econometric, energy demand, transportation demand, and ecological models with very large numbers of equations and state variables have been produced. Experience has shown that these models are very costly and labor intensive to maintain and update and quickly become incomprehensible as their complexity increases, and certainly no client, audience, interest group, or policy actor will take the time to comprehend the documentation for such models. Indeed, the documentation is too voluminous to be publishable, and this inhibits dissemination of the model and its results. The largest of the Davis models is described in several thousand pages of text, with the volume describing the data-base management system itself being 552 pages long. The largest of the U.S. national energy models fill many volumes, which are collectively probably understood by no one. Indeed, no single person ever completely grasped the Davis model: different parts were developed by different teams, as was the case with all very large models. The volume of documentation associated with the IBP biome studies models was also enormous.

Worse, a consensus is developing that there isn't a monotonic, increasing, statistically discernible relationship between the number of variables or equations and the predictive utility of a model (for example, Armstrong, 1978; Ascher, 1978; Ascher and Overholt, 1983; Makridakis et al., 1984; McClean, 1977). Rather, it appears that the predictive utility of a model is most strongly influenced by the degree of understanding of the key linkages between system subcomponents built into the model. The highest predictive utility is being discovered in intermediate-sized, not extremely large, models (Makridakis et al., 1984). Odum (1983) also advocates the use of small models.

The response to this experience in the program described here has been to turn in another direction. An enormous amount of effort is expended, not on elaborating a model, but rather on exhaustive statistical analysis to identify a handful of state variables that interact to account for almost all the variance year to year in the dynamic behavior of the system. These variables are then incorporated into the simplest possible systems model, which uses the minimum possible equation set and individual equations of the absolute minimum complexity required to account for most of the variance in the behavior of the "dependent" variable for each equation.

Experience has shown that the model structure most likely to account for the behavior of complex systems will utilize one or a very few state variables from each of a number of different factor clusters, rather than a large number of variables from one, or a very few, factor clusters. Traditional educational programs train the researcher to view the world from within particular factor clusters (for example, demography, economics, resource geology, or political decision making). Models built on that approach are likely to be rather imperfect maps of the real world. The reason is a high degree of intervariable correlation within each of the factor clusters; once one factor has been incorporated from a factor cluster into a model, we have already accounted for most of the variance attributable to that factor cluster.

Because of the very wide variety of types of variables and phenomena that must be built into a model of a complex system, it has been found necessary to pool insights from an extremely diverse literature and to depend on personal conversations with an extraordinary diversity of specialists. The work reported here was the outcome of an evolutionary process of research over two decades. During this period, about a hundred people made contributions, without which the work would not have arrived at the present state. People both within and outside the group who had a large impact on the process of evolution toward the final product include John Brewer (conceptual models, computational methods), Paul Craig (conceptual models, the war–economy link, stability theory), Sherman Stein (conceptual models, mathematical and statistical foundations), Peter Hunter (the demography–economy–resources linkages), Jill Auburn (conceptual models, software systems development), Nance Mosman (urban systems, software systems development and advice), Jeff Young and John Mitchiner (conceptual models), George David (the significance of the federal government public debt, money and banking, war debt, market saturation, economic theory), Hilary Stinton (government statistics, conceptual models), Meredith Pierce (statistics, data sources), John Flory (urban systems), Robert Boyd (the world energy market), Leonard Myrup (meteorology, climatology), Reid Bryson and Herbert Lamb (historical climatology), Ted Foin (population age structure and its implications), Norm Glass (population age structure, land use), Raul Berrios Loyala (the notion of applying mathematical and statistical analysis to the "structure of history"), C. S. Holling (the idea of linking stability theory to historical analysis), Earl Cook (the role of resources in determining historical trends), Herman Speith (the importance of historical perspective, which led us to Oswald Spengler, Ortega y Gasset, and, finally, Fernand Braudel), R. J. Smeed (the link between transportation systems and land use), David Deamer (the role of rare events, such as volcanic eruptions, as reinforcers and sustainers of oscillatory behavior in complex systems), and Robert May (stability theory applied to complex phenomena). That list was organized by subject, rather than by importance of the contribution. M. King Hubbert and Richard Bellman were very important in shaping the thinking of this program at a very early stage in the 1960s.

No single person or team could live long enough to develop the breadth of knowledge required to deal with social and environmental problems of such high complexity. Networking to a large, diverse group of scholars has become mandatory.

This research has led to a view of how complex systems operate in terms of the stability theory that Robert May has brought into ecology (see, for example, May, 1981). For complex socioeconomic–environmental systems, it appears that the product of the intrinsic rate of natural increase, r, and the length of critical time delays in homeostatic mechanisms, T, is long enough so that return to a trajectory after a major perturbation is too great to allow for monotonic damping, but not long enough to pass the threshold that permits true limit cycles. Rather, the product rT produces a susceptibility to oscillatory damping to the trajectory in response to a major perturbation. However, modern postindustrial societies seem to be characterized by two classes of linked perturbations, wars and depressions, which occur sufficiently frequently to reperturb systems before the oscillations from the last perturbation have subsided. The result is a kind of quasi-limit cycling, in which the amplitude of oscillation diminishes somewhat as the elapsed time since the last major perturbation lengthens.

The work at Davis has been critically dependent on state-of-the-art, hardware–software systems because of the astronomical computer requirement for curve-fitting historical data series (the computer requirement for simulation is minuscule by comparison). Batch processing on mainframes proved to be very inefficient for this purpose. The operation is now tied to intensely interactive custom software designed by Jill Auburn operating in a Turbo Pascal (of Borland International) environment on an IBM PC AT with an 80287 arithmetic coprocessor for added speed. The curve-fitting algorithms allow alternation between the Simplex method and the artificial intelligence routines due to Matyas. The books by Schwefel (1981) and Lootsma (1972) have been the most informative sources of insight on this topic.

In this author's opinion, one of the most useful products to appear from ecology has been the seven stability principles referred to earlier. They are an important part of the theoretical background for this work. Their usefulness is underlined by their rediscovery, independently, in many other fields. It would be a shame if ecology turned its back on these principles, not because of their failure to map onto reality, but because of conflicts or confusion arising in the literature resulting from semantic imprecision.

Specifically, this paper points to three classes of potentially destabilizing forces for modern nation states, which can be interpreted in terms of the ecological stability principles.

1. Post-1973 lowered economic growth rates in industrial societies resulted from inadequate omnivory with respect to energy-using and generating systems. Rather than spreading the risk of societal failure over many alternative systems, the fundamental fabric of society became critically dependent on liquid fuel for vehicle propulsion, thus violating principle 1. The energy crisis that resulted when U.S. oil discovery costs began to escalate sharply was dealt with by massive increases in oil imports from remote sources, a variant of the migration principle. This increased our connectance to other nation states, which then became sources of supply instability. It would have been better to impose a high retail gasoline tax (as in Europe), which would have produced a fast homeostatic response in the form of increased system efficiency and some switching to alternative systems. Our society has become too committed to using high rates of resource flux to maintain stability, a strategy with poor intermediate-term prospects as a stabilizing mechanism when the requisite resources are being depleted rapidly (U.S. crude oil reserves are already more than half gone).

2. A second class of perturbations result from a high degree of connectance between different nation states with inadequate homeostatic control over perturbations traveling down the connectors. Wars occur after an economic perturbation in one nation is inadequately damped by homeostatic responses from other nations, such as increased loans, economic assistance, investments, or purchases. Once wars occur, they initiate a chain reaction of destabilizations, tripped by the retirement of war debt, which finally exhausts the homeostatic capacity of a society to absorb increased supply by upward adjustment in demand.

3. A third class of social destabilizing mechanisms arises because the homeostatic suppression of elevated birthrates by elevated unemployment rates operates only after a 22-year lag, when the supernumarary age class attempts permanent entry into the labor force. A useful homeostatic role of government would be education of the population concerning the deleterious future effects of high birthrates on unemployment rates, wages and prices, and so on. Related consequences of high birthrates include elevated crime rates, police costs, and educational taxes. A particularly compelling argument for birth control is that unusually large numbers of 20 to 24-year-old males tend to be followed by massive war deaths, which clear the market for that demographic group.

Retrospectively, it appears that the single step that contributed most to progress in the program at Davis was the realization that systems are affected both by changes in trajectories and by fluctuations about trajectories, that the two must be represented by separate terms in equations, and that trajectories were under control of the current most critical limiting resource.

It has been a great pleasure to contribute to this project in honor of the memory of George Van Dyne: a fine fellow, a prodigious worker, a towering intellect, and one of the great pioneers and inspirational leaders of systems ecology.

REFERENCES

ARMSTRONG, J. S. 1978. *Long Range Forecasting. From Crystal Ball to Computer.* Wiley-Interscience, New York. 612 pp.

ASCHER, W. 1978. *Forecasting. An Appraisal for Policy-makers and Planners.* Johns Hopkins University Press, Baltimore. 239 pp.

Ascher, W., and Overholt, W. H. 1983. *Strategic Planning and Forecasting. Political Risk and Economic Opportunity.* Wiley-Interscience, New York. 311 pp.

Beer, S. 1975. *Platform for Change.* Wiley, New York. 457 pp.

Boulding, K. E. 1962. *Conflict and Defense—A General* Theory. Harper & Row, New York. 349 pp.

Braudel, F. 1972, 1973. The Mediterranean and the Mediterranean World in the Age of Phillip II. Vols. 1 and 2. Harper & Row, New York. 1375 pp., continuous numbering.

Butzer, K. W. 1980. Civilizations: organisms or systems? *Am. Scient.* 68:517–523.

Cary, J. R., and Keith, L. B. 1979. Reproductive changes in the 10-year cycle of snowshoe hares. *Can. J. Zool.* 57:375–390.

Clark, J. M. 1931. *The Costs of the World War to the American People.* Yale University Press, New Haven, Connecticut. 316 pp.

Craig, P. P., and Watt, K. E. F. 1985. The Kondratieff cycle and war. *Futurist* 19 (April):25–27.

Cronon, W. 1983. *Changes in the Land. Indians, Colonists, and the Ecology of New England.* Hill and Wang, New York. 241 pp.

Darby, M. R., Lothian, J. R., Gandolfi, A. E., Schwartz, A. J., and Stockman, A. C. 1983. *The International Transmission of Inflation.* University of Chicago Press, Chicago. 727 pp.

Dickinson, F. G. 1940. An aftercost of the world war to the United States. *Amer. Econ. Rev.* 30 (suppl.):326–339.

Dyer, G. 1985. *War.* Crown Publishers, New York. 272 pp.

Fay, S. B. 1929. *The Origins of the World War,* Vol. II. Macmillan, New York. 577 pp.

Force, D. C. 1974. Ecology of insect host–parasitoid communities. *Science* 184:624–632.

Freeman, C. (ed.) 1983. *Long Waves in the World Economy.* Butterworths, London. 245 pp.

Gaines, T. C. 1962. *Techniques of Treasury Debt Management.* Free Press of Glencoe, New York. 317 pp.

Gardner, M. R., and Ashby, W. R. 1970. Connectance of large dynamical (cybernetic) systems: critical values for stability. *Nature* 288:784.

Gever, J., Kaufmann, R., Skole, D., and Vorosmarty, C. 1986. *Beyond Oil. The Threat to Food and Fuel in the Coming Decades.* Ballinger, Cambridge, Massachussetts. 304 pp.

Goldstein, J. S. 1988. *Long Cycles. Prosperity and War in the Modern Age.* Yale University Press, New Haven, Connecticut, 433 pp.

Goodman, D. 1975. The theory of diversity–stability relationships in ecology. *Quart. Rev. Biol.* 50:237–266.

Hall, C. A. S., Cleveland, C. J., and Kaufmann, R. K. 1986. *Energy and Resource Quality. The Ecology of the Economic Process.* Wiley, New York. 560 pp.

Harris, L. D. 1984. *The Fragmented Forest. Island Biogeography Theory and the Preservation of Biotic Diversity.* University of Chicago Press, Chicago. 211 pp.

Hirst, F. W. 1915. *The Political Economy of War.* J. M. Dent and Sons, London. 327 pp.

Historical Statistics of the United States. Bicentennial Edition. Colonial Times to 1970. 1975. U.S. Bureau of the Census, U.S. Dept. of Commerce, Washington, D.C. 1232 pp.

Hughes, B. B. 1980. *World Modeling. The Mesarovic–Pestel World Model in the Context of Its Contemporaries.* Lexington Books, Lexington, Massachusetts. 227 pp.

Hughes, B. B. 1985. *World Futures. A Critical Analysis of Alternatives.* Johns Hopkins University Press, Baltimore. 243 pp.

Hutchinson, G. E. 1948. Circular causal systems in ecology. *Ann. N.Y. Acad. Sci.* 50:221–246.

Knight, B. 1940. Postwar costs of a new war. *Am. Econ. Rev.* 30(suppl.):340–350.

Lawlor, L. R. 1978. A comment on randomly constructed model ecosystems. *Amer. Nat.* 112:445–447.

Lootsma, F. A. (ed.) 1972. *Numerical Methods for Non-linear Optimization.* Academic Press, New York. 439 pp.

MACARTHUR, R. H. 1955. Fluctuations of animal populations, and a measure of community stability. *Ecology* 36:533–536.

MAKRIDAKIS, S., ANDERSEN, A., CARBONE, R., FILDES, R., HIBON, M., LEWANDOWSKI, R., NEWTON, J., PARZEN, E. and WINKLER, R. 1984. *The Forecasting Accuracy of Major Time Series Methods.* Wiley, New York. 301 pp.

MAY, R. M. 1974. *Stability and Complexity in Model Ecosystems,* 2nd ed. Princeton University Press, Princeton, New Jersey. 265 pp.

MAY, R. M. (ed.) 1981. *Theoretical Ecology.* Sinauer Associates, Sunderland, Massachusetts. 489 pp.

MCCLEAN, M. 1977. Getting the problem right—a role for structural modeling. In Linstone, H. A., and Simmonds, W. H. C. (eds.), *Futures Research—New Directions.* Addison-Wesley, Reading, Massachusetts, pp. 144–157.

MCNAUGHTON, S. J. 1977. Diversity and stability of ecological communities: a comment on the role of empiricism in ecology. *Amer. Nat.* 111:515–525.

MEADOWS, D., RICHARDSON, J., and BRUCKMAN, G. 1982. *Groping in the Dark. The First Decade of Global Modelling.* Wiley, New York. 311 pp.

MESAROVIC, M., and PESTEL, E. 1974. *Mankind at the Turning Point.* Reader's Digest Press, New York. 210 pp.

MILLER, J. G. 1978. *Living Systems Theory.* McGraw-Hill, New York. 1102 pp.

MITCHELL, B. R. 1980. *European Historical Statistics 1750–1975,* 2nd rev. ed. Facts on File, New York. 868 pp.

MORRIS, R. F. (ed.). 1963. The Dynamics of Epidemic Spruce Budworm populations. *Memoirs Ent. Soc. Canada,* No. 31. 332 pp.

MOULTON, H. G., and PASVOLSKY, L. 1932. *War Debts and World Prosperity.* For the Brookings Institution by the Century Company, New York. 498 pp.

NAMENWIRTH, J. Z. 1973. The wheels of time and the interdependence of value change. *J. Interdisciplinary Hist.* 3:649–683.

ODUM, H. T. 1983. *Systems Ecology. An Introduction.* Wiley, New York. 644 pp.

O'NEILL, R. V. 1976. Ecosystem persistence and heterotrophic regulation. *Ecology* 57:1244–1253.

PEASE, J. L., VOWLES, R. H., and KEITH, L. B. 1979. Interaction of snowshoe hares and woody vegetation. *J. Wildlife Management* 43:43–60.

ROSTOW, W. W. 1979. *The World Economy: History and Prospect.* University of Texas Press, Austin, Texas. 833 pp.

SCHWEFEL, H.-P. 1981. *Numerical Optimization of Computer Models.* Wiley, New York. 389 pp.

SHIRK, G. 1985. The producer price index, 1720–1984. *Cycles* 36(1):16–20.

SMITH, W. B., and COLE, A. H. 1935. *Fluctuations in American Business 1790–1860.* Harvard University Press, Cambridge, Massachusetts. 195 pp.

Statistical Abstracts of the United States 1985, 105th ed. 1984. Bureau of the Census, U.S. Dept. of Commerce, Washington, D.C. 991 pp.

TEMIN, P. 1976. *Did Monetary Forces Cause the Great Depression?* W. W. Norton, New York. 201 pp.

THORP, W. L. 1941. Postwar depressions. *Amer. Econ. Rev.* 30(suppl.):352–361.

VAUGHN, M. R., and KEITH, L. B. 1981. Demographic response of experimental snowshoe hare populations to overwinter food shortage. *J. Wildlife Management* 45:354–380.

WANGERSKY, P. J., and CUNNINGHAM, W. J. 1956. On time lags in equations of growth. *Proc. Nat. Acad. Sci. USA* 42:699–702.

WANGERSKY, P. J., and CUNNINGHAM, W. J. 1957a. Time lag in prey predator population models. *Ecology* 38:136–139.

WANGERSKY, P. J., and CUNNINGHAM, W. J. 1957b. Time lag in population models. *Cold Spring Harbor Symp. Quant. Biol.* 22:329–337.

WATT, K. E. F. 1973. *Principles of Environmental Science.* McGraw-Hill, New York. 319 pp.

WATT, K. E. F. 1974. *The Titanic Effect. Planning for the Unthinkable*. Sinauer Associates, Sunderland, Massachusetts. 268 pp.

WATT, K. E. F. 1982. *Understanding the Environment*. Allyn and Bacon, Boston. 431 pp.

WATT, K. E. F. 1992. *Taming the Future: A Revolutionary Breakthrough in Scientific Forecasting*. The Contextured Web Press, Davis, California. 232 pp.

WATT, K. E. F., and CRAIG, P. P. 1986. System stability principles. *Systems Res.* 3:191–201.

WATT, K. E. F., CRAIG, P. P., and AUBURN, J. S. 1988. World economic modeling. In Ehrlich, P., and Holdren, J. (eds.), *The Cassandra Conference*. Texas A & M University Press, College Station, Texas. 330 pp.

WATT, K. E. F., YOUNG, J. W., MITCHENER, J. L., and BREWER, J. W. 1975. A simulation of the use of energy and land at the national level. *Simulation* 24:129–153.

WATT, K. E. F., MOLLOY, L. F., VARSHNEY, C. K., WEEKS, D., and WIROSARDJONO, S. 1977. *The Unsteady State. Environmental Problems, Growth, and Culture*. University Press of Hawaii, Honolulu. 287 pp.

WICKENS, D. L. 1941. *Residential Real Estate*. National Bureau of Economic Research, New York. 305 pp.

17

The Nature and Significance of Feedback in Ecosystems

DONALD L. DEANGELIS

Environmental Sciences Division
Oak Ridge National Laboratory
Oak Ridge, Tennessee

In the sense that the term ecosystem implies a concept and not a unit of landscape, the emphasis is that the biologist must look beyond his particular biological entity (e. g., cells, tissues, organisms, etc.) and must consider the interrelationships between these components and their environment [Van Dyne, in Chiasma 8:59 (1970)].

Fragmental studies, no matter how challenging, are still not holistic [Van Dyne, in The Ecosystem Concept in Natural Resource Management, Ch. X, p. 334 (1969), Academic Press].

INTRODUCTION

The idea of feedback is intrinsic to the design of complex automatic devices. Engineers use feedback to help steer ships, guide missiles, and control all types of production processes. Information from the "output" or performance of the system, say the current direction of a ship, is used to modify its future behavior when necessary, in this case by a steering correction.

Understanding the role of feedback has also become important in formulating conceptual and analytic models of natural systems. Models of genetic systems (Jacob and Monod, 1961), the nervous system (Ashby, 1960), the whole organism (Cannon, 1932; Wiener, 1948), and even political systems (Easton, 1965) have been developed. These systems can be described from the viewpoint that internal feedbacks regulate the dynamics.

There is nothing new in attempting to make sense of the world by organizing thought around comprehensible systems such as human-designed machines. Precisely working mechanical clocks captured the imagination of scientists and philosophers in the seventeenth century and, reinforced by Newton's mathematical laws of moving bodies, led to the view of the universe as a giant clockwork mechanism. This mechanistic model fell out of fashion during the nineteenth century, as it proved to be incapable of accommodating what was being learned about evolutionary changes in the world. A metaphor oriented around organismic growth and evolution replaced the clockwork mechanism. However, most variations of the metaphor contained some degree of vitalism or entelechy, which was incompatible with what was known about physical science.

The intellectual developments of the twentieth century, which include cybernetics, general systems theory, and nonequilibrium thermodynamics, have given hope that a more encompassing model may provide a synthesis of the mechanistic and organic views and correct the oversimplifications of both. These new perspectives focus on organization as opposed to either simplistic materialism or entelechy. Feedback is a basic component of this organization. The concept of feedback adds a new and more subtle dimension to mechanical systems. Unlike a clockwork mechanism, which, after winding up simply goes through autonomous, predetermined dynamics until it runs down, a feedback cybernetic system can react to a changing environment by continuously adjusting its behavior and even seeking its own source of energy and materials. Cybernetics has thus proved to be a key idea in interpreting the physiological control mechanisms in individual organisms.

In this chapter I will discuss the importance of feedbacks in some ecological systems, with an emphasis on what they imply about the nature of the system as a whole. Do the feedbacks allow us to conclude that an ecosystem is a type of cybernetic system? Alternatively, are the feedbacks merely fortuitous phenomena that do not reflect any sort of coherent organization? Finally, do the observed feedbacks in the ecosystem suggest some identifiable system type other than a cybernetic system? I want to stress at the outset that these are as much ideological questions as they are scientific questions. Nonetheless, resolution of questions such as these ultimately contributes to scientific understanding.

A central theme in ecology is the feedback loop between a species population and its environment. Consider a single population. By producing offspring, the population, in effect, sends messages to the environment. The environment sends return messages to the population in the form of the number of offspring that have survived (Belis, 1970) (Fig. 17.1). There are different possibilities

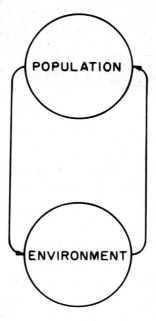

Figure 17.1 Feedback relationship between a species population and its environment.

for how tight the coupling is between these messages. At one extreme, the two may be tightly coupled; the population has a well-defined equilibrium level, and any deviation from it (an excess or deficit in the population level) is followed fairly regularly by a decrease or increase in survival of offspring.

This system can be called a *closed-loop* feedback system because there are reciprocal effects. An increase in population above some level affects the environment. Assume first that it diminishes some vital resources in the environment or stimulates the increase of predators or parasites. The changed environment, in turn, affects the population negatively by decreasing the chances of survival of individuals. The feedback loop in this case is called *negative feedback*. The system is self-regulating or *homeostatic,* because it regulates the population close to some equilibrium value or set point. It can be called cybernetic in the sense that cybernetics is the science of self-regulating systems.

How effectively regulation works depends on the degree of coupling between the messages. Suppose the coupling is very loose so that the population can increase to large sizes before the environment exerts a noticeable feedback effect on survival. In this case, regulation may be very ineffective. The time lags in the system caused by the length of time a message takes to have an effect can result in large population fluctuations. Another cause of deviation of the system from homeostasis is the prevalence of stochastic fluctuations in the environment. Suppose that the population is affected by occasional random environmental occurrences, entirely independent of population size, that drastically reduce it now and then, thus preventing the population from normally reaching those sizes where feedback effects from the environment influence it. In this situation, the regulation of the population is *open loop* because the message of the environment to the population does not depend on the message from the population to the environment. This system would generally not be called cybernetic.

Examples of both apparently tight regulation of population levels by the environment (density dependence) and loose regulation (density independence) have been reported in natural systems. The question of whether density-dependent or density-independent regulation is more fundamental has been the subject of controversy among ecologists (for example, see McLaren, 1971; DeAngelis and Waterhouse, 1987).

The system in Fig. 17.1 has been discussed from the viewpoint of an equilibrium level or set point. However, homeostasis is only one possible aspect of cybernetic system behavior. Orderly directed change, or *homeorhesis,* is another. One way in which homeorhesis can occur is when increasing population density improves the quality of the environment (from the point of view of the population) rather than diminishing it, so an increase in population feeds back to increased survival. Environmental quality and population can increase simultaneously. This mutual reinforcement, through a closed loop of two or more system variables, is called a *positive-feedback* loop. This kind of feedback occurs in many natural populations. Consumers of seeds and fruits disperse those resources and hence enhance the quality of their own environment. Limited grazing by many species increases the yield from the grazed resource, at least up to a point. For example, the feeding by *Cebus* monkeys on the terminal branches of *Gustavia* trees increases branching and hence the future crop of terminal buds (Oppenheimer and Lang, 1969). Positive feedback is part of many human-made systems and can be called cybernetic if the interaction steers the variables toward a goal.

The concepts discussed above with reference to a single population can be extended to the vastly more complex concept of the ecosystem. The ecosystem has been defined as a "unit of biological organization (that is, a 'community') in a given area interacting with the physical environment . . ." (Odum, 1969). The degree to which the term "unit of biological organization" is appropriate depends in part on the degree to which its component parts cohere. This means we must know the extent to which the components are linked by a network of feedbacks that either regulates the system at a fixed set point or drives it in a systematic manner toward a fixed goal.

The concept of feedback has been present in ecological theory since its early days, even before the word "feedback" entered the written language, which, according to Judson (1980), was in 1920. The idea was implicit in Clements's (1916) discussion of plant succession: "In short, the habitat causes the plant to function and grow, and the plant then reacts upon the habitat, changing one or more of its factors in decisive or appreciable degree. The two procedures are mutually complementary...." In a different aspect of ecology, food web theory, the Lotka–Volterra equations (Volterra, 1931) are classic equations of both ecological food web dynamics and feedback oscillation theory (see Minorsky, 1962). However, the key step toward preparing the way for the current systems ecology view was the coupling of the concepts of energy flow and nutrient cycling with the trophic-level concept to form an overall view of the ecosystem (Lindeman, 1942).

The application of general systems theory and cybernetics to ecosystems was extended by H. T. Odum (1960), E. P. Odum, (1969), Van Dyne (1966, 1969a, 1969b), Patten (1964), Olson (1965), Watt (1968), and others. The efforts of these ecologists owed much to the refinement of feedback systems analysis for the study of complex industrial and economic systems by Forrester (1961, 1971). Forrester provided a systematic methodology for analyzing complex problems that started with identifying the appropriate system variables and parameters and then tracing the cause-and-effect loops linking the variables. There is no better description of how this approach can be applied to ecosystems than that provided by Van Dyne (1969a). Van Dyne describes the method as one of logically organizing data in three steps: first into word models, then into picture models, and finally into mathematical models.

The work of these pioneers in systems ecology stands as a great achievement, and their methods of analysis will play a pivotal role in further improving our understanding of ecosystems. However, questions remain concerning the interpretation of cybernetic and general systems models as applied to ecosystems. The proposal of a "cybernetic paradigm" for ecosystems has drawn criticism, partly for reasons that parallel the criticisms of the density-dependent view of population regulation (Fig. 17.1); that is, the feedback couplings are not strong and pervasive enough to constitute a coherent cybernetic system. In addition, Engelberg and Boyarsky (1979) have argued that the feedbacks in ecological systems are merely brute flows of energy, not the transmittals of information of true cybernetic regulation. Furthermore, they noted that ecosystems do not appear to have the equivalents of goals or set points, such as even the simplest servomechanisms (for example, thermostats) have.

Responses to the Engelberg and Boyarsky (1979) article have stressed the myriad relationships among the components of any reasonably complex ecosystem (Jordan, 1981; Knight and Swaney, 1981; McNaughton and Coughenour, 1981; Patten and Odum, 1981). These include intricate mutualistic patterns, interspecific pheromone messages, and nutrient-recycling mechanisms. Beyond this, the responses pointed out that both homeostatic feedbacks and orderly, consistent patterns of development occur, although the time scales of these phenomena might be long.

One obvious fact revealed by that debate is that "cybernetic" means many things to many people. Even the self-organizing patterns of wind currents are classed as cybernetic by Knight and Swaney (1981). The weakness of this view is that it allows nearly any physical or chemical phenomenon to be interpreted as cybernetic, which diminishes the usefulness of the term in categorizing systems. A precise, unambiguous, and totally satisfactory definition of cybernetic may not be possible. However, a succinct one that may be fairly acceptable is given by Corning (1983), according to whom a cybernetic system is "a dynamic set of processes organized and internally directed toward certain goals or end states." He further defines key elements of cybernetic systems as including "(a) the setting of goals (and perhaps subgoals); (b) the implementation of actions designed to coordinate the behavior of the system and its parts toward goal attainment; and furtherance of these activities, processes of (c) communication, and (d) control."

Certainly, Engelberg and Boyarsky (1979) overlooked the huge numbers of feedbacks, both conspicuous and subtle, in ecosystems. These can exercise communication and control, satisfying (c) and (d) above. It remains, however, to be seen how well ecosystems can fulfill elements (a) and (b) of Corning's (1983) definition. We can accept the ecosystem as a highly complex feedback system but still question whether these attributes commonly ascribed to cybernetic systems hold for ecosystems. Can properties of homeostasis and goal direction be unambiguously assigned to ecosystems? If these properties do exist, are they readily explained in terms of individual species characteristics and interspecific interactions that would be favored by natural selection, or are higher-level principles necessary? Does the ecosystem optimize production, nutrient recycling, or other objective functions, as is often asserted? If such optimization does exist, it would be strong evidence for emergent ecosystem-level principles.

It would be impossible in a brief space to survey all the areas of ecology where feedbacks are important. Here only two of the many ecosystem processes that have been interpreted through the "cybernetic paradigm" will be considered. However, these occupy central places in ecosystem theory. The degree to which empirical and theoretical work supports such a paradigm will be examined with respect to ecological succession, sometimes said to exemplify goal direction or homeorhesis, and consumer regulation of primary production, possibly representing a homeostatic mechanism.

FEEDBACK IN ECOLOGICAL SUCCESSION

Succession is the pattern by which species in a given spatial area replace others and are, in turn, replaced through time. Two general types are generally distinguished. Primary succession refers to the successional process when it starts on a new, abiotic substrate. Secondary succession begins when an existing ecosystem is disturbed (for example, when a forest stand is clear-cut). Connell and Slatyer (1977) distinguish three types of successional models: facilitation, tolerance, and inhibition, each of which can be described in terms of feedback.

Facilitation Model

The concept of facilitation succession, or "relay floristics," goes back to Clements (1916) and has been echoed in standard textbooks on ecology up to the present time. There are three main ideas in this model. (1) Of the species that arrive on a newly opened space, only a few especially adapted ones can colonize it successfully. (2) These early colonists modify the environment so that later successional species can survive, but the environment becomes no more favorable, or in fact becomes less favorable, for the early colonists. (3) The new colonists replace the old ones. Several waves of new colonists may replace each other sequentially.

A feedback diagram of this situation is shown in Fig. 17.2. The compartments of this model are a series of species types, $X_1, X_2, \ldots, X_n$, and a measure of environmental quality, Y. Note that the feedbacks have been drawn from Y to show that they control the rates of increase, a_{ii}, of each vegetation type. The succession $X_1 \to X_2 \to \ldots \to X_n$ is from early pioneer species to late successional species. Soil depth may be the indicator of environmental quality, for example. The reciprocal effects of vegetation and soil depth can be seen in the signs of the interactions. The vegetation has a positive effect on the soil in all cases. However, the effect of the soil on the vegetation may vary. At first, the effect is negative for a particular vegetation type. Then, when Y reaches a high enough level, it promotes the growth of that type. For still higher values of Y, the effect decreases and may become negative again (Fig. 17.2b). For each succeeding vegetation type, this progression is shifted further toward higher values of Y.

An important point in this model is that the feedback loops $(X_1, Y), (X_2, Y), \ldots, (X_n, Y)$ all go through phases where they are positive-feedback loops. It is easy to infer from this model that goal direction, in a cybernetic sense, is involved in the succession process. Gutierrez and Fey (1980) have abstracted this view of succession to a series of feedback loops, primarily positive, that involve the basic biomass-development and nutrient-flow processes only (Fig. 17.3). There is no explicit

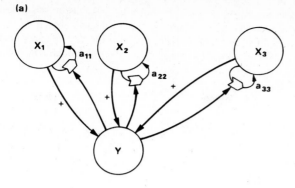

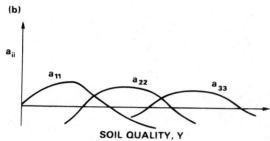

Figure 17.2 Feedback conceptualization of the facilitation–succession model. The variables $X_1, X_2, \ldots, X_n$ represent the population levels of n species, Y is a measure of the state of the environment, and the a_{ii}'s are population self-regulation effects that are influenced by Y.

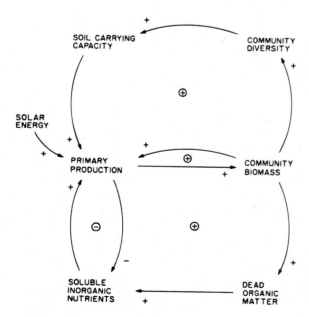

Figure 17.3 Conceptualization of principal feedback loops involved in secondary succession. (From Gutierrez and Fey, 1980. Reprinted by permission.)

Chap. 17 The Nature and Significance of Feedback in Ecosystems representation of species here. The authors attribute a fairly deterministic cast to this process-level succession, saying "secondary succession is the action of an ecosystem that returns the system to climax.... While humanlike volition cannot be attributed to a grassland, the restoration forces operate like a control system or servomechanism designed to approach and maintain the climax equilibrium...."

Tolerance Model

In the tolerance model, which was formulated first by MacArthur and Connell (1966), there are three main assumptions. (1) Any species that arrives on a plot is able to establish itself. (2) Early occupants modify the environment so that it is less favorable for them, but this modification has little effect on later arrivers. The effects on the environment in this case are not changes in soil quality, but reductions in light and nutrient availability through crowding. (3) Early colonists could survive the changes in the environment, but they are overshadowed by later arrivals. In feedback terms, each species type has a negative effect on the environment, degrading the environment for the next species. Thus, the feedback loops in this model are negative. The environmental feedback strongly affects early successional species and less strongly affects the later species, which, therefore, take over.

Inhibition Model

In the inhibition model, early arrivers affect later arrivers by negatively affecting soil and other environmental characteristics, and no species has an intrinsic advantage. As long as early species persist, they prevent later ones from entering. As in the tolerance model, the predominant feedback loops are negative. Eventually, later successional species can enter because the early ones have short life spans; that is, the a_{ii}'s of the early colonizers turn negative. An example of an inhibition system in nature is the *Calluna* heath. *Calluna* roots produce a substance that inhibits the mycorrhizal fungi of *Picea* (Miles, 1979), thus preventing trees from moving in.

In both the tolerance and inhibition models, the life cycles of individual plants are stressed, whereas in the facilitation model, the individual species are seen as component parts of an ecosystem-level process. A detailed elaboration of the individual life cycle viewpoint is the model of Grime (1979), who describes succession in terms of individual plant strategies. In particular, Grime distinguishes three types of vegetation strategies:

1. Ruderals: tolerant to low stress and high disturbance rates
2. Competitors: tolerant to low stress and low disturbance rates
3. Stress tolerators: tolerant to high stress and low disturbance rates

Ruderals are natural pioneer species because they move into a disturbed area quickly. However, they are eventually outgrown by competitors that move in later. Like ruderals, competitors cannot tolerate stresses such as light depletion, so as crowding occurs they do less well. The stress tolerators, which move in last, take over.

During succession, the vegetation affects the environment in various ways. Light is reduced and minerals are depleted. The feedback loops are again primarily negative, as shown in Fig. 17.4. Disturbances periodically reset various spatial areas to the ruderal-dominated state. (If disturbances were frequent enough, only ruderals would persist.) Therefore, succession in this model is an interplay among three main species strategies, each of which, under normal circumstances, is successful enough in its own way to survive evolutionarily.

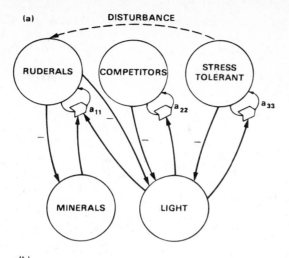

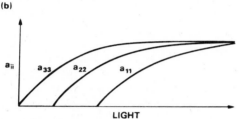

Figure 17.4 Feedback conceptualization of successional process as a competition among three plant strategies (adapted from Grime, 1979). Minerals and light affect the a_{ii}'s of the populations.

All the succession models discussed above involve feedbacks. However, the different models represent very different interpretations of what is going on. The facilitation model is suggestive of system-level processes, or of a kind of goal direction of the ecosystem as a whole, at a level above that of individual species. The tolerance and inhibition models suggest that the course of succession is set primarily by individual species strategies and the competitive interactions among these species.

Which model or models are best supported by empirical evidence? As Turner (1983) has reported, field experiments suggest that inhibition-dominated, or at least tolerance-dominated, succession is by far the most common (Standing, 1976; Connell and Slatyer, 1977; Sutherland and Karlson, 1977; Lubchenko and Menge, 1978; Sousa, 1979; Dean and Hurd, 1980; Day and Osman, 1981; Dean, 1981; Schoener and Schoener, 1981). Botkin (1981) showed that detailed computer models of secondary forest succession also supported tolerance or inhibition succession. In all but one of the few studies that indicate facilitation, the facilitation is not obligate (Turner, 1983).

This widespread presence of some level of inhibition implies that succession may not proceed with the uniformity that facilitation models would predict. Succession can be halted or even retrogress if inhibitory negative feedbacks gain the upper hand. As an example, Olson (1958) noted that succession on Lake Michigan sand dunes, rather than proceeding at one stage from black oak cover to a mesophytic forest, could be arrested by a buildup of soil acidity.

The above evidence concerning the models of ecological succession lends little support to the idea of systems-level homeorhesis or goal-seeking behavior. Attempts such as that of Gutierrez and Fey (1980) to abstract the process of succession to a few ecosystem-level feedback loops seem especially risky. This point has been made before, with great force, by Drury and Nisbet (1973), who showed numerous exceptions to many of the generalizations that have been made concerning succession. These considerations should not discourage the search for ecosystem-level generalizations with regard to succession. However, such generalizations will not be simple, and they may require attention to the game-theory aspects of the species involved in succession.

CONSUMER REGULATION AND NUTRIENT CYCLING

Another aspect of the cybernetic concept of the ecosystem is the homeostatic function of consumers of vegetation. The dynamics of two-species models of systems of this type have been studied now for many years. The models have stimulated the interest of systems ecologists because the consumer seems to fill the role of a control element, stabilizing systems (at least in some of the models) and recycling essential nutrients. A few notable and representative discussions of this cybernetic view are summarized next.

Mattson and Addy (1975) list several ways in which phytophagous insects in a forest seem to act like cybernetic regulators: (1) they affect plants at their vital areas of growth and, hence, can have a disproportionate influence; (2) consumers are almost always present where there are plants; (3) the plant–consumer relation is coevolved and strong; (4) the insects can respond to host changes; and (5) the plants can respond to the insects. The authors go on to say that the insect grazers have much more potential than decomposers to alter abiotic fluxes and, furthermore, that they tend to ensure consistent and optimal output of plant production for a particular site.

Hayward and Phillipson (1979) discuss the influence of small mammals as parts of ecosystems, supporting the ideas of Petrusewicz and Grodzinski (1975). The latter authors hypothesized that an ecosystem that included only producers and decomposers would be a poor one and have a low level of complexity. The consumer would have the function of providing a storage "shunt" for the system. Decomposers might not be able to respond quickly to excess production in favorable environmental times. Because nutrients may be locked up for long periods of time, this could have an adverse effect on production. A consumer would be able to respond to this storage quickly and short-circuit it (Fig. 17.5).

O'Neill (1976) has also hypothesized a favorable effect of consumers. Heterotrophs in a system can operate to control the rate of energy production by autotrophs. Relatively small changes in heterotroph biomass could be enough to respond to perturbations to the system and to restore

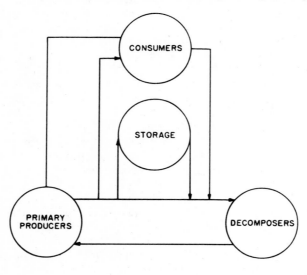

Figure 17.5 Consumers acting as a storage shunt in transfer of materials from primary producers to decomposers. (Adapted from Hayward and Phillipson, 1979. Reprinted by permission from Chapman and Hall.)

equilibrium. Gutierrez and Fey (1980) simply claim, without further comment, that consumers, through their diversity, provide structural stability.

Lovelock (1979) may have gone further than any other exponent of the cybernetic view in proposing that many properties of the entire earth are maintained at nearly constant levels by feedbacks originating within the biosphere. For example, the earth's temperature and concentrations of atmospheric gases appear to be relatively stable at levels far from what they would be in the absence of biota, and at levels, in fact, that are, according to Lovelock, optimal for living things. Other mechanisms are the biological methylization of sulfur in seawater, which allows the sulfur to be transported back to land via the atmosphere and helps maintain living organisms there (see also Margulis, 1981).

The views of these authors may differ from one another in detail, but all subscribe to the idea that there are meaningful ecosystem-level set points for such characteristics as production and diversity and that a network of feedbacks operates to optimize one or more of these characteristics. These attempts to find such ecosystem-level principles are certainly worthwhile. However, a rigorous pursuit of this goal means that certain questions, discussed next, should be considered.

The first question relates to the evolutionary perspective. Why would an ecosystem have evolved to optimize production, stability, diversity, or any other index? Rather than a whole ecosystem, consider first a single species. Optimization is an easy concept to imagine for a species that may be, or may have been, in competition with other species (see, for example, Schoener, 1971). The competing species that most efficiently utilizes the available resources and best avoids predators will, in principle, replace the others. Thus, it is reasonable to assume that species have evolved in ways that optimize their capabilities. However, even for the case of a single species, it is difficult to verify that competitive mechanisms exist for this optimization to take place. The idea of optimization becomes all the more difficult to justify when applied to whole ecosystems, for which it is very hard to prove the existence of mechanisms such as competition.

A second question concerns how we would go about demonstrating unambiguously that the ecosystem is optimizing something; production, stability, diversity, recycling, or anything else. Optimality theory is in general not testable. Any measurement can be expressed as the optimal solution to some problem, but there may be no way of knowing if the problem is relevant or not (Staddon and Hinson, 1983). Only constrained optimization is testable. Therefore, any statement that an ecosystem is in some way optimal must include a complete list of constraints under which this optimization has been obtained. A list of such constraints may be extremely difficult to provide.

I will focus a bit more on this second question before returning to the first. Difficulties arise even in defining basic ideas. Consider the concept of "stability." Consumers are said to help maintain stability, but how exactly is this meant? There is some lack of consistency in the literature. For example, as cited earlier, Hayward and Phillipson (1979) suggested that consumers, by acting as a shunt around the inactive organic matter storage compartment, may avoid delays in nutrient circulation and thus help stabilize the system. On the other hand, O'Neill and Reichle (1980) stated that a large "pool of inactive organic matter" plays a stabilizing role by buffering the ecosystem against "short-term fluctuations in environmental conditions." The conclusions of Hayward and Phillipson and O'Neill and Reichle, although contradictory, may both be right in some sense. In fact, Watson and Loucks (1979) have suggested that ecosystems "balance [the] two mutually exclusive aspects of stability" by having both large, slow pools and small, fast pools of nutrients. However, it should be apparent that, by the same type of logic, almost any observed characteristic of an ecosystem could be interpreted as some kind of stabilization strategy.

The same problem afflicts Lovelock's (1979) arguments that biotic feedback, acting as a cybernetic mechanism on the biosphere level, has stabilized a global environment favorable to life. The given fact that life occurs on earth, even if its strikes us as improbable, cannot, by itself, be used to argue that evolution has necessarily led to mechanisms on a global scale for the stabilization of conditions suitable for life. The mere fact that human beings exist on earth makes it inevitable that they will observe that conditions have remained favorable for life on earth for the past 2 billion years or more. However, we cannot say whether this is a matter of homeostatic behavior or simply chance. For example, there is strong evidence that the earth originally had a reducing atmosphere, which changed, after the evolution of photosynthesizing organisms, to an oxidizing atmosphere. Lovelock interpreted this as a favorable change brought about by life to facilitate further evolution. Suppose, however, that the earth's atmosphere had changed from a reducing one to an oxidizing one and that all life on earth had become extinct as a result. This would have been an obvious case for which life was not able to impose homeostatic regulation on its environment. However, in this case there would, by necessity, be no sentient beings around to observe this lack of homeostasis. For all we know, this extinction of life may have happened on countless other planets; the earth could be an anomaly. These conjectures are meant to point out a basic flaw in Lovelock's hypothesis, which is a case of a more general sort of erroneous logic discussed under the "anthropic cosmological principle" by Barrow and Tipler (1986).

The above comments are directed at the property of ecological stability, but any attempt to verify empirically that feedback networks act to optimize other characteristics (for example, productivity or diversity) will confront similar problems of ambiguity. It is more profitable at this point to return to the first question and see if there are a priori reasons to expect an ecosystem to behave as a coherent system and optimize certain properties.

There are at least three ways by which an ecosystem can be imagined to have evolved toward a kind of optimization. (1) Behavior of species populations may somehow result in optimization of the whole system through mutual adjustments of their interactions. (2) Ecosystems may compete with each other as wholes, forcing optimization on each other in the same way that competing corporations or national economies do (for example, see Dunbar, 1960). (3) There may be emergent laws, not obvious from consideration of lower levels such as populations, that drive the systems in such a direction.

A description of the first proposed cause of optimization, with respect to the stabilization of primary production, is given by O'Neill and Reichle (1980): "The mechanisms by which ecosystems establish a persistent energy base translate into competitive interactions among primary producer populations. A given number of populations, with limited light, nutrients, and water, will interact and tend to pack the niche space to support the maximum primary production that can be sustained by available resources. The ecosystem strategy is simply the result of the population processes." What is most important in the paper by O'Neill and Reichle is that they do not merely propose optimization at the ecosystem level, but they also outline the reasons for expecting it.

The species-oriented view can make very specific predictions regarding the stability of particular consumer–resource systems. As an example, Rosenzweig (1971) showed that many consumer systems have a state-plane appearance similar to that shown in Fig. 17.6, where $X = 0$ is the zero isocline of the resource and $Y = 0$ is the zero isocline of the consumer. If the consumer isocline is to the right of the peak of the resource isocline, the system is stable, as the system perhaps "should" be if the "good of the system" is so defined. However, if the consumer isocline is to the left of the peak, the system is unstable. Schaffer and Rosenzweig (1978) showed that the self-interest of individual consumers feeding at a higher rate will shift the consumer isocline to the left, that is, toward instability over evolutionary time scales. Only if the prey is able to evolve as rapidly as the predator, to escape predation and thus shift the predator isocline back to the right, can the system remain stable.

Schaffer and Rosenzweig's (1978) theoretical model is partly testable. We can observe consumer–resource systems in which the consumer is likely to evolve much faster than its resource (insects versus trees) and those for which the rates are equal. Schaffer and Rosenzweig have found that all cases of sizable consumer–resource instabilities fall into the first class. We might add to this the studies of Bakker (1983) regarding predator–prey dynamics between carnivores and ungulates. Bakker showed that ungulate adaptations for escape have evolved faster than the predators'

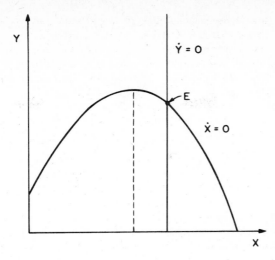

Figure 17.6 State-plane representation of consumer–resource interaction. Point E is the equilibrium point. If E is to the right of peak P, the system is stable; otherwise, it is unstable.

abilities to pursue them. It appears in nature that carnivores and ungulates do not generate severe cycles. The implications of this work are that the view of consumer–resource interactions based on natural selection at the individual level appears able to explain the behavior observed in nature. Thus, it appears that behavior at the level of species dynamics and interactions can explain some apparent tendencies toward optimization, such as maximum possible primary production, but can also account for observed nonoptimal behavior, such as strong population oscillations in ecosystems.

If all important cases of strong feedback interactions within an ecosystem can be explained in terms of selection on the level of the individual, there would be no justification for ascribing any higher-level goals to the ecosystem. Thus, the ecosystem would not fit Corning's (1983) definition of a cybernetic system.

Still, there remains the question of nutrient cycling. Consumers seem to help recycle nutrients faster than might be possible in their absence (see, for example, Sterner 1986). Dyer, DeAngelis, and Post (1986) have shown that nutrient recycling caused by moderate levels of consumption could increase biomass production above the level when there is no consumption. This hints at the possibility of cybernetic autotroph–consumer systems that operate to optimize production.

It is not inconceivable for such interactions to have evolved toward some optimal point through the second mechanism listed earlier, that of ecosystems competing with each other as wholes. Wilson (1980) described how a hypothetical system involving earthworms and plants could evolve toward greater efficiency in this way, even though at the level of individual selection there might not be any pressure to do so. Suppose, for example, there are two plant genotypes, one (type A) that produces detritus that favors earthworm production and another (type B) that does not. The presence of type A will stimulate earthworms and, indirectly, improve the quality of the soil. This cannot, in itself, lead to increased representation of type A over type B in the community, because the improved soil quality is assumed to help types A and B equally. However, as Wilson suggests, plants are generally spread nonuniformly on the landscape, so there may be patches of predominately type A genotypes and patches of predominately type B. The former patches will be more productive, and the effect of greater productivity will have the net effect of helping type A more than it does type B. In effect, the two patch or microecosystem types compete, and type A has the advantage over the long term. A similar argument could be made for the evolution toward earthworm genotypes that are more effective in improving the soil.

Such a scenario seems plausible, and a similar type of scenario could apply to consumer–autotroph interactions as well. Competition among spatially partially isolated ecosystems would tend to select for properties such as greater production and efficiency. These properties could serve as goals of ecosystem operation and might suffice to allow the system to be called cybernetic.

Despite the above arguments, broad-brush generalizations that consumers always aid in recycling do not seem warranted. For example, Jordan (1982) studied the recycling of nutrients in the Amazon rain forest, where conservation of nutrients is at a premium. Rather than "encouraging" herbivores to participate in rapid recycling of nutrients, according to Jordan, tree strategies in the Amazon rain forest may be even more intensely occupied by discouraging herbivores through production of large amounts of phenol toxins than tree strategies in ecosystems where nutrients are not in such short supply. Another relevant study is that of Vitousek (1982), who gathered information on woody plants in a wide variety of low-nutrient sites. Looking at such sites from an ecosystem perspective, we might expect the trees to foster mechanisms that would encourage decomposers to mineralize nitrogen quickly to maintain nitrogen availability for the woody plants. Instead, the litter produced by the woody plants has such a high carbon:nitrogen ratio that decomposition is slow, and the recirculation of nitrogen is so very slow that deterioration of the vegetation can set in. Thus, there are reasons to doubt that higher-level selection has been present in these systems.

The third way in which an ecosystem might be imagined to evolve toward some sort of optimization is through emergent laws that are not reducible to lower-level explanations such as population dynamics. This sort of evolution might imply some purposiveness of the ecosystem as a whole, which appears to be an attribute of the global biosphere as interpreted by Lovelock (1979) in his Gaia hypothesis. Most evolutionary theorists reject this view (for example, Dawkins, 1982, pp. 234–236). However, other suggestions for emergent laws deserve careful attention, such as Ulanowicz's (1986) view that the quantitative expression for system organization, based on feedbacks, that he calls "ascendency" necessarily increases during ecosystem development. Ulanowicz views ascendency, as well as material and energy feedbacks in ecosystems, as nonreductionistic concepts or agencies. I will not consider the important concept of ascendency any further here, except to say that I am in friendly disagreement that this concept, along with feedbacks, is nonreductionistic. I believe that the optimization of ascendency, if this is universal, must somehow emerge from behaviors at lower levels.

The above points are raised not as objections to an ecosystem-level point of view, but as indications of the kinds of situations that any coherent theory must be expected to explain most convincingly. It should be noted that it is not difficult to explain such observations from a viewpoint oriented around the selfish behavior of individual species operating within environmental constraints.

DISCUSSION AND CONCLUSIONS

The point of this paper has been to look at some ways in which feedback enters our ideas about the organization of ecosystems. The prevalence of feedbacks in ecosystems has often led to the belief that ecosystems behave as cybernetic systems. I believe it is illusory to draw such conclusions from the mere existence of feedbacks. Some comprehension of the holistic characteristics of ecosystems can be gained only with detailed knowledge of the effects of the feedbacks at the species level. I have briefly considered ecological succession and the effect of consumers in systems as examples of cases where cybernetic design has previously been inferred. Both conceptual and mathematical models of succession have taken a turn away from the system-level cybernetic view in recent years toward models more oriented around individual species strategies (for example, Shugart, 1984). Papers like that of Schaffer and Rosenzweig (1978) describe a parallel species strategy approach to interactions between organisms of different trophic levels.

Is the ecosystem cybernetic in nature? That is, are there "organic" interdependencies among species that cause the whole ecosystem to act in a goal-directed manner, either homeostatic or homeorhetic? Unless it can be shown that Wilson's (1980) proposed mechanism for selection among interacting systems of species is very common, I would suggest that the evidence at present does not support an affirmative answer to these questions. Since the subject of this paper is feedbacks, I have not examined other lines of evidence relating to the coherence of ecosystems. However, other types of observations, such as the spatial patterns of individual species, also tend to support the "principle of species individuality," at least among plant species (see, for example, Whittaker, 1970).

Yet, it would be an equally erroneous inference to say that the ecosystem is nothing more than a set of species that happen to be in spatial proximity. Corning (1983) offers a middle-ground conception of the ecosystem. He distinguishes between cybernetic systems and ecosystems in the same way that economists differentiate between firms and markets. For a definition of "market," Corning quotes from Waltz (1979): "The market arises out of the activities of separate units—persons and firms—whose aims and efforts are directed not toward creating an order but rather toward fulfilling their own internally defined interests by whatever means they can muster. The individual unit acts for itself. From the coaction of like units emerges a structure that affects and constrains all of them." Waltz is careful to note that the market is not itself an agent, although it constrains the actions of the individual agents whose activities create it. Hayek (1973) has described the market in a similar way. Some of his remarks are worth quoting here because they dismiss the idea of higher-level purpose: "Most important . . . is the relation of a spontaneous order to the conception of purpose. . . . Since such an order has not been created by an outside agency, the order as such also can have no purpose, although its existence may be very serviceable to the individuals which move within such order" (quote from Barrow and Tipler, 1986). It is reasonable to think of an ecosystem similarly to the way Waltz and Hayek think of the market, where species play the same roles in the ecosystem as persons and firms do in the market.

This view of the ecosystem as arising from the selfish interactions of species populations has been emphasized by some ecosystem theorists (for example, O'Neill and Reichle, 1980). However, it is important that the consequences of this view to the cybernetic interpretation of the ecosystem be fully elucidated. While an ecosystem is rich in feedbacks, both of energy and information, the species are the active mediators in all these feedbacks, and their behavior is conditioned by the many kinds of adaptation resulting from natural selection. It is doubtful that principles relevant to the macroscale of the ecosystem can be derived without reference to the microscale advantages and disadvantages to the individual organisms that compose species populations. This is taken for granted in economics. For example, no appreciable flow of money into various investment sectors, whatever their utility to the society as a whole, can be expected unless individual investors anticipate profits from the investment.

These considerations in no way diminish the importance of synergistic effects in ecosystems. Natural selection at the organism or species level does not prohibit, and in fact encourages, the formation of various dependencies such as mutualisms. As Corning (1983) remarks, synergistic phenomena do not necessarily require cybernetic systems with "overarching, systemic goals" and are commonplace in economics as analyzed by such theoretical approaches as "positive sum games" and "coalition theory." However, some of the system-level phenomena observed in ecosystems, such as slow or fast turnover pools of nutrients, are more appropriately classed as "epiphenomena," rather than ecosystem strategies. These epiphenomena result from the various abilities or inabilities of species to "make a living" in various niches that involve the processing of these nutrients in some way.

The fact that organisms are active, self-interested processors of energy and nutrients leads to apparent paradoxes not expected from straightforward analysis of energy and nutrient flows. For example, Rosenzweig (1971) showed theoretically that something very unexpected could happen when a predator–prey system was "enriched" by the addition of extra resources for the prey. The system could become highly unstable and go into oscillations. It is likely that this sort of behavior is not merely theoretical, but can occur in real systems. Such phenomena would be difficult to explain by theories that ignore individual species.

For the above reasons, I believe that a fruitful approach for ecologists seeking generalizations at the ecosystem level may be to give more consideration to the interactions between the macroscale and the microscale. A recent example of an analysis that focuses on the effects of species optimization on the total system is that of Abrams (1982). Abrams considered models of a three-species food chain (resource–consumer–predator) in which the consumer can vary its time spent foraging in a way that optimizes its fitness. Fitness here depends on the amount of resource obtained and susceptibility to predation, both of which increase as foraging time increases. Abrams showed that the degree of vulnerability to predation of the consumer can drastically affect its feeding on the resource and, therefore, the whole dynamics of the ecosystem. The way in which factors such as those explored by Abrams can affect the overall energy and nutrient flows needs to be explored for whole ecosystems in which each species is maximizing its fitness.

A step in this direction is the food web study carried out by Post and Pimm (1983). In randomly assembling food webs based on differential equation models of individual species, they found that the number of species that could stably be accommodated approaches a fairly well defined and consistent limit. No doubt regularities of this kind will emerge from further studies, but as parts of far more complex patterns than previous inferences from the cybernetic point of view have suggested.

The approach suggested here may not have the holistic appeal of the cybernetic view of the ecosystem. However, just as the clockwork paradigm eventually had to give way to less majestic, but more appropriate descriptions of nature, the cybernetic view needs to be replaced by conceptions that are perhaps less tidy, but that have greater explanatory power.

ACKNOWLEDGMENTS

I greatly appreciate the comments of R. V. O'Neill, H. H. Shugart, Jr., K. E. F. Watt, and A. M. Shapiro on early drafts of this paper. This research was sponsored by the National Science Foundation's Ecosystem Studies Program under Interagency Agreement No. BSR-80-21024, with the U.S. Department of Energy under Contract No. DE-ACO5-84OR21400 with Martin Marietta Energy Systems, Inc. Publication No. #2375, Environmental Sciences Division, Oak Ridge National Laboratory.

REFERENCES

ABRAMS, P. A. 1982. Functional responses of optimal foragers. *Amer. Nat.* 120:382–390.
ASHBY, W. R. 1960. *Design for a Brain.* Chapman and Hall, London. 286 pp.
BAKKER, R. T. 1983. The deer flees and the wolf pursues: incongruities in predator–prey coevolution. In Futuyma, D. J., and Slatkin, M. (eds.), *Coevolution.* Sinauer Associates, Sunderland, Massachusetts, pp. 350–382.
BARROW, J. D., and TIPLER, F. J. 1986. *The Anthropic Cosmological Principle.* Oxford University Press, New York. 706 pp.
BELIS, M. 1970. The cybernetics of evolving systems. In Rose, J. (ed.), *Progress of Cybernetics,* Vol. 3. Gordon and Breach, New York, pp. 1121–1130.
BOTKIN, D. B. 1981. Causality and succession. In West, D. C., Shugart, H. H., and Botkin, D. B. (eds.), *Forest Succession: Concepts and Application.* Springer-Verlag, New York, pp. 36–55.
CANNON, W. B. 1932. *The Wisdom of the Body.* W. W. Norton, New York. 333 pp.

Clements, F. E. 1916. *Plant Succession: An Analysis of the Development of Vegetation.* Carnegie Inst. Publ. 242, Washington, D.C., pp. 1–512.

Connell, J. H., and Slayter, R. O. 1977. Mechanisms of succession in natural communities and their role in community stability and organization. *Amer. Nat.* 111:1119–1144.

Corning, P. A. 1983. *The Synergism Hypothesis: A Theory of Progressive Evolution.* McGraw-Hill, New York. 492 pp.

Dawkins, R. 1982. *The Extended Phenotype.* W. H. Freeman, San Francisco. 307 pp.

Day, R. W., and Osman, R. W. 1981. Predation by *Patiria miniata* (Asteroidea) on bryozoans: prey diversity may depend on the mechanisms of succession. *Oecologia* 51:300–309.

Dean, T. A. 1981. Structural aspects of sessile invertebrates as organizing forces in an estuarine fouling community. *J. Exp. Mar. Biol. Ecol.* 53:163–180.

Dean, T. A., and Hurd, L. E. 1980. Development in an estuarine fouling community: the influence of early colonists on later arrivals. *Oecologia* 46:295–301.

DeAngelis, D. L., and Waterhouse, J. C. 1987. Equilibrium and nonequilibrium concepts in ecological models. *Ecol. Mon.* 57:1–21.

Drury, W. H., Nisbet, I. C. T. 1973. Succession. *J. Arnold Arboretum* 54:331–368.

Dunbar, M. J. 1960. The evolution of stability in marine environments: natural selection at the level of ecosystems. *Amer. Nat.* 94:129–136.

Dyer, M. I., DeAngelis, D. L., and Post, W. M. 1986. A model of herbivore feedback on plant productivity. *Math. Biosc.* 79:171–184.

Easton, D. 1965. *A Systems Analysis of Political Life.* Wiley, New York. 833 pp.

Engelberg, J., and Boyarsky, L. L. 1979. The noncybernetic nature of ecosystems. *Amer. Nat.* 114:317–324.

Forrester, J. W. 1961. *Industrial Dynamics.* MIT Press, Cambridge, Massachusetts. 464 pp.

Forrester, J. W. 1971. Counter-intuitive nature of social systems. *Technology Review* 73:53-60.

Grime, J. P. 1979. *Plant Strategies and Vegetation Processes.* Wiley, New York. 222 pp.

Gutierrez, L. T., and Fey, W. R. 1980. *Ecosystem Succession.* MIT Press, Cambridge, Massachusetts. 231 pp.

Hayek, F. A. 1973. *Law, Legislation, and Liberty. Vol. 1: Rules and Order.* University of Chicago Press, Chicago. 763 pp.

Hayward, G. C., and Phillipson, J. 1979. Community structure and functional role of small mammals in ecosystems. In Stoddard, D. M. (ed.), *Ecology of Small Mammals.* Chapman and Hall, London, pp. 135–211.

Jacob, F., and Monod, J. 1961. On the regulation of gene activity. *Cold Spring Harbor Symp. Quant. Biol.* 26:193–211.

Jordan, C. F. 1981. Do ecosystems exist? *Amer. Nat.* 118:284–287.

Jordan, C. F. 1982. Amazon rain forests. *Amer. Scient.* 70:394–401.

Judson, H. F. 1980. *The Search for Solutions.* Holt, Rhinehart, and Winston, New York. 211 pp.

Knight, R. L., and Swaney, D. P. 1981. In defense of ecosystems. *Amer. Nat.* 117:991–992.

Lindeman, R. L. 1942. The trophic-dynamic aspect of ecology. *Ecology* 23:399–418.

Lovelock, J. E. 1979. *Gaia. A New Look at Life on Earth.* Oxford University Press, New York. 157 pp.

Lubchenko, J., and Menge, B. A. 1978. Community development and persistence in a low rocky intertidal zone. *Ecol. Mon.* 48:76–94.

MacArthur, R. H., and Connell, J. H. 1966. *The Biology of Populations.* Wiley, New York. 200 pp.

Margulis, L. 1981. *Symbiosis in Cell Evolution.* W. H. Freeman, San Francisco. 419 pp.

Mattson, W. J., and Addy, N. D. 1975. Phytophagous insects as regulators of forest primary production. *Science* 190:515–522.

McLaren, I. A. (ed.). 1971. *Natural Regulation of Animal Populations.* Atherton Press, New York. 195 pp.

McNaughton, S. J., and Coughenour, M. B. 1981. The cybernetic nature of ecosystems. *Amer. Nat.* 117:985–990.

Miles, L. 1979. *Vegetation Dynamics.* Chapman and Hall, London. 80 pp.

Minorsky, N. 1962. Nonlinear Oscillations. Van Nostrand Reinhold, Princeton, New Jersey. 714 pp.

Odum, H. T. 1960. Ecological potential and analog circuits for the ecosystem. *Amer. Scient.* 48:1–8.

Odum, E. P. 1969. The strategy of ecosystem development. *Science* 164:262–270.

Olson, J. S. 1958. Rates of succession and soil changes on southern Lake Michigan sand dunes. *Bot. Gaz.* 119:125–170.

Olson, J. S. 1965. Equations for cesium transfer in a *Liriodendron* forest. *Health Physics* 11:1385–1392.

O'Neill, R. V. 1976. Ecosystem persistence and heterotrophic regulation. *Ecology* 57:1244–1253.

O'Neill, R. V., and Reichle, D. E. 1980. Dimensions of ecosystem theory. In *Forests: Fresh Perspectives from Ecosystem Analysis.* Fortieth Annual Biology Colloquium. Oregon State University Press, Corvallis, Oregon, pp. 11–26.

Oppenheimer, J. R., and Lang, G. E. 1969. *Cebus* monkeys: effect on branching of *Gustavia* trees. Science 165:187-188.

Patten, B. C. 1964. The Systems Approach in Radiation Ecology. ORNL/TM-1008. Oak Ridge National Laboratory, Oak Ridge, Tennessee. 19 pp.

Patten, B. C., and Odum, E. P. 1981. The cybernetic nature of ecosystems. *Amer. Nat.* 117:985–990.

Petrusewicz, K., and W. Grodzinski. 1975. The role of herbivore consumers in various ecosystems. In Reichle, D. E., and Goodall, D. M. (eds.), *Productivity of World Ecosystems.* National Academy of Sciences, Washington, D.C., pp. 64–70.

Post, W. M., and Pimm, S. L. 1983. Community assembly and food web stability. *Math. Biosc.* 64:169–192.

Rosenzweig, M. L. 1971. Paradox of enrichment: destabilization of exploitation ecosystems in ecological time. *Science* 171:385–387.

Rosenzweig, M. L. 1973. Evolution of the predator isocline. *Evolution* 27:84–94.

Schaffer, W. M., and Rosenzweig, M. L. (1978). Homage to the red queen. 1. Coevolution of predators and their victims. *Theor. Pop. Biol.* 14:135–157.

Schoener, T. W. 1971. Theory of feeding strategies. *Ann. Rev. Ecol. Syst.* 2:369–404.

Schoener, A., and Schoener, T. W. 1981. The dynamics of the species–area relation in marine fouling systems. 1. Biological correlates of changes in species–area slope. *Amer. Nat.* 118:339–360.

Shugart, H. H. 1984. *A Theory of Forest Dynamics.* Springer-Verlag, New York. 278 pp.

Sousa, W. P. 1979. Experimental investigations of disturbance and ecological succession on a rocky intertidal algal community. *Ecol. Mon.* 49:227–254.

Staddon, J. E. R., and J. M. Hinson. 1983. Optimization: a result of a mechanism? *Science* 221:976.

Standing, J. D. 1976. Fouling community structure: effects of the hydroid, *Obelia dichotoma,* on larval recruitment. In Mackie, G. O. (ed.), *Coelenterate Ecology and Behavior.* Plenum Press, New York, pp. 155–164.

Sterner, R. W. 1986. Herbivores' direct and indirect effects on algal populations. *Science* 231:605–607.

Sutherland, J. P., and Karlson, R. H. 1977. Development and stability of the fouling community at Beaufort, North Carolina. *Ecol. Mon.* 47:425–446.

Turner, T. 1983. Facilitation as a successional mechanism in a rocky intertidal community. *Amer. Nat.* 121:729.

Ulanowicz, R. E. 1986. *Growth and Development, Ecosystems Phenomenology.* Springer-Verlag, New York. 203 pp.

Van Dyne, G. M. 1966. Ecosystems, Systems Ecology, and Systems Ecologists. ORNL 3957. Oak Ridge National Laboratory, Oak Ridge, Tennessee. 40 pp. (Reprinted as the Introduction to this volume.)

Van Dyne, G. M. 1969a. Grasslands Management, Research, and Training Viewed in a Systems Context. Range Science Department, Science Series No. 3, Colorado State University, Fort Collins, Colorado. 50 pp.

Van Dyne, G. M. (ed.). 1969b. *The Ecosystem Concept in Natural Resource Management.* Academic Press, New York. 383 pp.

Vitousek, P. 1982. Nutrient cycling and nutrient use efficiency. *Amer. Nat.* 119:553–572.

Volterra, V. 1931. *Leçons sur la theorie Mathematique de la Lutte pour la Vie.* Gauthier-Villars, Paris. 214 pp.

Waltz, K. N. 1979. *Theory of International Politics.* Addison-Wesley, Reading, Massachusetts. 250 pp.

Watson, V., and Loucks, O. L. 1979. An analysis of turnover times in a lake ecosystem. In Halfon, E. (ed.), *Theoretical Systems Ecology.* Academic Press, New York, pp. 356–384.

Watt, K. E. F. 1968. *Ecology and Resource Management.* McGraw-Hill, New York. 450 pp.

Whittaker, R. H. 1970. *Communities and Ecosystems.* Macmillan, New York. 158 pp.

Wiener, N. 1948. *Cybernetics.* Wiley, New York, and Herman et Cie, Paris. 194 pp.

Wilson, D. S. 1980. *The Natural Selection of Populations and Communities.* Benjamin/Cummings, Menlo Park, California. 186 pp.

18

Knowledge-based Large-scale Ecosystem Design

BERNARD P. ZEIGLER

Artificial Intelligence Simulation Group
Department of Electrical and Computer Engineering
The University of Arizona
Tucson, Arizona

A Macrohypothesis of Interconnected Microhypotheses. Let us consider the lower set of hypotheses noted above as microhypotheses. These microhypotheses are concerned with the mechanisms by which matter and energy move through an ecosystem over time. If we couple these microhypotheses together we have a macrohypothesis, i.e., a model of the system. When we can express the microhypotheses in mathematical functions, and when we couple these expressions, the result is a simulation model [Van Dyne, in *Ohio J. Sci.* 78, 193–4 (1978)].

> In the high Sonoran desert north of Tucson, amid blooming cacti, rattlesnakes and Gila monsters, a remarkable building is taking shape. Covering 1.3 hectares (3.15 acres) and sheltered under a gleaming, 26-meter-high (85-ft.) cathedral-like lattice-work roof of steel tubing and glass, Biosphere II is both an architectural wonder and a scientific tour de force. . . . [E]ight people will be sealed inside for two years, getting nothing from the outside but information, electricity and sunshine. Along with 3,800 plants and dozens of species of invertebrates, mammals and other living organisms, they will form the largest self-sustaining ecosystem ever built.
>
> The human inhabitants of this mini-world—four men and four women, all single— . . . include a physician, a botanist, a marine biologist and experts on engineering and agriculture. . . .
>
> The $60 million experiment, financed by a group of venture capitalists . . . has two basic purposes. One is to test ideas for building outposts on other planets, where long stays would be common and resupply impossible. But Biosphere II is more than just the prototype of a space colony. It is a means of learning more about how the earth—"Biosphere I," in project jargon—sustains itself through the recycling of water, air and nutrients. . . .
>
> Scientists have been developing the physical plant of Biosphere II . . . using techniques that have enabled modern zoos and botanical gardens to put diverse habitats together in relatively narrow confines. At the same time, they have searched the world for representative flora and fauna that can re-create five different miniature biomes, or ecosystems: rain forest, savanna,

desert, ocean and marsh.

The results are spectacular. The structure, built on a slope, is dominated by a soaring Amazonian rain forest, lush with 300 species of plants. At its periphery, tree ferns and bromeliads flank a stream that leads to a mountainside flood-plain forest and an open vista of tropical savanna. There, plants from Africa, Australia and South America bask in a less humid atmosphere, where bees and hummingbirds help pollinate plants and a colony of termites aids in the decomposition of dying material.

A transition zone of thorn scrub from Madagascar and Mexico leads onto a Baja California desert biome. The stream, meanwhile, meanders to the saltwater marsh (transported in sections from the Florida Everglades) that gives onto the 10.6-meter-deep (35-ft.) ocean with its own coral reef and waves that can rise as high as 1.2 meters (4 ft.). Mangroves in the marsh are host to frogs, turtles and crabs, and the ocean includes 1,000 species of plants and animals.

The wilderness biomes, stretching along the horizontal axis of the T-shaped structure, will be nice places to visit, but the eight Biosphereans will not live there. Their home is in the stem of the T, where they will grow their food in a 0.2-hectare (0.5-acre) area. . . . [T]he farm is already producing crops. Rice grows in flooded paddies that are shared with tilapias—African fish—which eat algae and water ferns and in turn fertilize the water with their waste. Papaya and bananas are ripening in moist heat and late-summer sun. Sorghum, amaranth, dill, oregano, soybeans, corn, tomatoes, onions and other crops are all growing in compost without pesticides and with only natural predators, such as spiders, wasps, lacewings and lady-bugs, to keep voracious insects away.

The absence of pesticides and the emphasis on natural fertilizer are designed not only to keep the experiment as untainted as possible, but also to protect the health of the human consumers; because all the air and water in Biosphere II is continually recycled and regenerated, it is important that no poisons be introduced into the system anywhere. . . .

Isolated as they are, the Biosphereans . . . will have computer and voice communications with the outside world and their own Mission Control. In case of emergency, someone can be removed through an air lock without interfering with the functioning of the closed environment. . . .

[T]he Biosphereans will spend about four hours a day doing scientific work and four hours on food production. Eggs will be collected from the Biospherean chickens, milk from the Biospherean goats, fish from the rice paddies or the ocean, meat from a plentiful supply of Vietnamese potbellied pigs. . . .

It has been a formidable task to organize these details—assembling plant strains, microbes, insects, and putting them together with bats, bush babies, lizards, tortoises and other life forms. No one expects all the species to survive; . . . between a quarter and a third will become extinct during the two-year period. But that is part of the experiment as well. Scientists do not necessarily know which plants and animals are best suited to self-contained habitats, and trial and error is the only way to be sure. . . . If it works, it will . . . be hailed as one of the most concrete contributions ever made to understanding the workings of Biosphere I.

Copyright 1990 The Time Inc. Magazine Company. Reprinted by permission.

INTRODUCTION

Systems ecology rises to the status of a mature science when its methods serve as the design basis for a completely artificial, human-engineered ecosystem. Laboratory microcosms and mesocosms have an established tradition in ecological research, but a materially closed, balanced, full-scale "macrocosm" with human occupants, capable of indefinite, stable existence in a viable state, represents a new level of complex ecological engineering not, until recently, attempted. The concept of a closed synthetic ecosystem goes far beyond the usual technologies for managing natural landscape units for human purposes, such as in the stocking of fish in a lake or harvesting forests for maximum sustained yields of timber products. The goal of its construction would be nothing less than the creation and maintenance of a self-sustaining life support system meeting certain specified objectives, such as stability, persistence, and materialistic independence from the outside environment.

Modern materials technology has certainly made it possible to fabricate large enough structures to enclose such a macroecosystem and seal it off from its surroundings. And electronic and computer advances of the past few decades have provided the information-processing capability necessary to "wire" such a system for observation and control. But further developments are needed in what might be termed "knowledge-based design" to permit full coupling of physical (including biological) and informational elements. Modeling and simulation would have to be used intensively in concert with a theory from information science, of great potential but little known to ecologists, *knowledge representation*. The engineering of realistically scaled synthetic ecosystems is a prime example of the need for *multifaceted modeling* methodology, as expounded by Zeigler (1984, 1990), to deal coherently with the complexities involved.

The purpose of this chapter is to explain how knowledge representation and multifaceted modeling can be combined and applied in the design, observation, modeling, and control of human-made ecosystems. In this, a modeling theorist–information scientist will be trying to communicate with ecologists through at least two sets of not very overlapping paradigms and jargons. Therefore, an effort will be made to explain the artificial intelligence concepts used, although it would not be appropriate to reproduce expositions of these topics readily available elsewhere. Consequently, readers of this book unfamiliar with this background may wish to consult some of the references given before proceeding further. To assist with unfamiliar terminology, a glossary of terms has been assembled and appears at the end of the chapter.

LARGE-SCALE SELF-SUSTAINING ECOSYSTEMS

A *biosphere* is a stable, complex, evolving system containing life, composed of various ecosystems operating in a synergetic equilibrium, essentially closed to material input or output, and open to energy and information exchanges. The earth ecosystem is the only such biosphere currently known. However, projects to establish other examples are being undertaken by space and other agencies in the United States (NASA) and the Soviet Union.

Biosphere II is a unique, nongovernmentally sponsored project to create an analog of the earth ecosystem ("Biosphere I") in an encapsulated structure spanning 2.5 acres located 20 miles north of Tucson, Arizona. This project is intended to provide a model environment that can greatly enhance understanding of the present biosphere. Knowledge gained experimenting with Biosphere II could be invaluable in solving many of earth's ecosystemic problems. Principles established could help in the building of support systems for projected space stations, in colonization of the Antarctic, and in human expeditions below the earth's surface, in radiation impregnable habitats.

Biosphere II is contained within a glass enclosure and an underground seal that isolate it from gaseous and other material exchange with the surrounding environment. The system, while materially closed, is open with respect to energy (the sun being the major source) and information (for external management and control). Enclosed within the glass-paneled structure are several modules, each replicating a "biome" (self-sufficient ecosystem). These include rainforest, savanna, marsh, desert, and ocean. In addition, there is a module for human habitation and one for intensive agriculture. Eight people, called "biosphereans," are to live and work in the biosphere. They are to be supported by an external staff, who will monitor conditions and provide assistance via electronic communication.

The immediate objective of the Arizona project is proof of concept: to demonstrate that a quasi-self-contained ecosystem, the first of this scale, can remain "alive" indefinitely. Secondarily, the environment should be able to support the existence of its human inhabitants. Successful conclusion of this first effort should enable future applications of the self-contained life-support system concept to focus on improving the quality of life provided for the biosphereans.

Biosphere II can be viewed as one instance of a class of artificial environments with self-sustaining properties. It is distinguished from others, such as space shuttles, by its total reliance on bioregenerative processes. It therefore presents unique design challenges that must rely on ecological considerations for their solution. The control system of a self-sustaining bioregenerative environment, such as Biosphere II, must meet the following requirements:

1. Control the mechanical systems that regulate the exchange of energy, gases, moisture, and organisms between the biome modules. Maintenance of gas balance is a major control problem. Buffers for carbon dioxide and other gases are many orders of magnitude smaller in relation to the scale of the plant than for earth.
2. Provide management functions for scheduling agricultural activities such as seeding, irrigation, harvesting, and pest control.
3. Provide extensive data collection and analysis capabilities. An important goal of instrumentation is to make it possible to understand why a deviant (desirable or undesirable) situation arose and to replicate the conditions leading up to it.
4. Recognize deviant behavior patterns and aid in determining corrective actions. Major problems are expected in the control of plant and animal diseases due to the absence of species that would normally serve as natural inhibitors.
5. Over the long term, the system should become more knowledgeable about the behavior of the ambient environment and employ this knowledge to diminish dependence on external support for guidance.

Environmental control systems of this kind make use of simulation models in combination with *expert systems,* computer software intentionally built to embody significant components of human expertise. The simulation models are called on to predict future consequences of management and control actions. The expert systems use such predictions to do such things as provide early warning of impending danger states, help develop fixes and "work-arounds" in crises situations, and plan planting schedules taking microclimatic constraints into account.

ARCHITECTURE OF CONTROL–MANAGEMENT SYSTEMS

A possible architecture of a bioregenerative control–management system is portrayed in Fig. 18.1. There are five layers: interaction, control, information, knowledge, and executive. Each represents distinct functions to be described. Orthogonal to this functional hierarchy is an adaptive component, capable of monitoring all layers, with the objective of improving system performance and increasing its autonomy (Zeigler, 1992).

Interaction layer. This layer contains the components that interact directly with the biosphere proper. Included are sensors ranging from direct measurement devices for temperature, humidity, and the like, to sophisticated data processors such as gas content analyzers. Actuators put into effect control system commands, for example, to move window louvers or turn on water sprinklers.

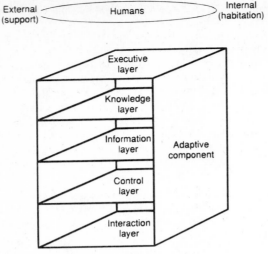

Figure 18.1 Architecture of control–management system for a synthetic bioregenerative environment.

Control layer. This layer contains the systems to control the lower-level actuators on the basis of sensor-supplied data and control policies determined at higher levels. Such control systems include conventional automatic controllers, but eventually may also include decision elements based on numerical and symbolic models operated fast enough for real-time response.

Information layer. This layer contains the data management systems that keep track of data captured at lower levels and organizes the data into a complete history of the physical and operational state of the biosphere. This history includes time-stamped records of both environmental conditions and concurrent control and management events facilitating later analysis by the adaptive component. Such behavioral series may also be compressed in the form of summary models.

Knowledge layer. This component contains the knowledge base for the biosphere's structure and behavior. As explained later, structural knowledge is represented in a "system entity structure" that encodes decomposition, taxonomic, and coupling relationships involving the physical, biological, environmental, and informational components. Behavioral knowledge is represented in simulation models and expert systems. Simulation models stored in a model base can regenerate and predict the dynamic behavior of heat flows, crop growth, gas budgets, and so on. Expert systems embody the expertise for management functions, such as scheduling agricultural activities and diagnosing malfunctions. They also provide policy settings for lower-level controllers. Expert systems can employ knowledge in the system entity structure directly (inferencing) or to access models in the model base and data (in the information layer).

Executive layer: This layer is responsible for deploying and coordinating the expert systems in the knowledge layer. The executive subsystem also contains the human-interactive interface, which supports knowledge engineering (transferral of expertise and other knowledge), information queries (concerning past, present, or expected behavior), and supervisory commands, including overrides of executive and lower-level decisions.

REPRESENTATION SCHEMES AND KNOWLEDGE

This section concerns the problem of representing knowledge to support the operation of the knowledge layer in the foregoing architecture. Concepts of knowledge representation are reviewed, first, and in subsequent sections the particular form of knowledge representation developed for the multifaceted modeling methodology (Zeigler, 1984, 1990) is discussed.

A *representation scheme* (Fig. 18.2) is a means of representing reality in computerized form. Each *representation* adhering to the pattern laid down by the scheme consists of four kinds of features or slots. *Operations* are procedures that can create, modify, and destroy representations or their components. *Questions* are procedures that can be used to interrogate the representation to get answers. Operations and questions are internal features of the representation in that they are meaningful independent of outside reality. However, a representation is not useful as something mapped onto itself, but only in reference to something else. Thus, the third feature of a representation is the designation of *what* it purports to represent. Finally, there must be a means of putting into *correspondence* the features of the representation with the reality it claims to represent.

Classical differential equation models fit this concept of representation scheme. Operations include the writing of equations and the assignment of values to initial conditions and parameters. Questions that can be addressed to the model concern the dynamic behavior of its variables; these can be answered to the extent that suitable analysis or simulation methods are available.

More generally, we can consider *formalisms* developed to express dynamical models as representation schemes, whether they be classical differential equations or the numerous, more recently developed vehicles of expression, such as discrete event specifications, automata, or Petri nets. Knowledge representation schemes originating with artificial intelligence (AI) research are also included. Such schemes will be considered later.

Having a representation of an entity is obviously not enough to claim knowledge about it. The representation becomes "knowledge" only when it is accurate, that is, *knowledge* is defined as valid representation. We turn to "homomorphism," a concept from mathematics.

A *homomorphism* is a correspondence between states of a pair of objects that is preserved under all relevant operations. A homomorphic relationship can be depicted by commutative diagrams such as that in Fig. 18.3. [See Zeigler, 1984, for a detailed discussion of homomorphism in the context of systems theory, modeling, and simulation.] In connection with representation schemes, the correspondence in question should associate states of the real-world entity with those of its representation such that, when a question is asked about the entity, the same answer is produced from the representation that would be obtained from a corresponding direct observation on the entity. The states of the entity and its representation will correspond if, for each real-world action that changes the entity state, there is a corresponding operation that updates the representation state accordingly.

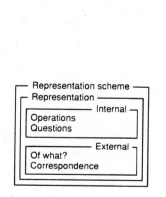

Figure 18.2 Basic elements of a knowledge-representation scheme.

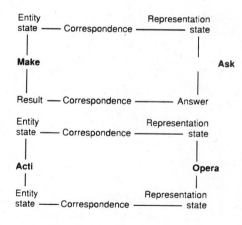

Figure 18.3 Homomorphism criterion for validity.

Thus, we say that a system has knowledge about an entity if it possesses a valid representation of the entity; that is, a homomorphism exists between the representation and the entity itself. It follows that valid dynamical models constitute a form of knowledge just as do valid AI schemes. A system can learn (gain knowledge) about an object by generating and validating representations or models of it. Some fundamental systems theory is applicable. For example, the *epistemological hierarchy* of Klir (1985) provides "levels of system specification" at which knowledge may be acquired, as well as conditions under which validation is possible (Zeigler, 1984).

The following dimensions of representation provide insight into how AI representation schemes go beyond classical model formalisms (Bobrow, 1985).

Inference. AI schemes have built-in mechanisms that can add facts to the representation state without further input from the real world.

Access. They have mechanisms to link units and structures to provide access to appropriate facts.

Matching. Representations may be compared for equality, similarity, and dissimilarity; action can be guided by such comparisons.

Meta-knowledge. A system may have knowledge about its own structure and operation that can be used in its acquisition and utilization of knowledge.

Something is said to be *knowable* by a system if it has a representation scheme to generate representations for it. In this sense, only a small (but important) fraction of reality is knowable with classical modeling formalisms. For example, a Newtonian model of the earth–sun system represents an orbital trajectory, but it cannot represent other immediately relevant knowledge. Indeed, it has no way of expressing what are the real-world entities involved or their parameters, such as masses, appearing in the equations. Questions can be asked about an orbit, but not about whether it is that of the earth around the sun. Nonmechanical properties of the entities not appearing in the equations cannot be known from them. Nor can the model make inferences or access its representations and compare them. It does not possess any meta-knowledge about them. Predictions obtained by running a dynamic model forward in time are an important form of inferencing. The point here, however, is that such a model is not capable of making such predictions without outside control.

The *frame hierarchy* concepts of artificial intelligence (Minsky, 1977) extend the scope of computer-representable reality. A frame contains knowledge about a particular object and is linked to other frames in a taxonomic relationship. To represent biome modules in Biosphere II, we might start by having a frame for Biome Module (Fig. 18.4). Each slot in this frame would represent an attribute of Biome Module. Associated with a slot may be descriptive information, such as the range of possible values of the attribute and its current value.

The Biome Module is a *class object* because there are several instances of it (Rainforest Module, Savanna Module, and so on). In the case of such a class object, the value in a slot represents a default value inherited by its instances. Also, associated with a slot are procedures such as how to compute the attribute value and what other procedures to trigger if the value is recomputed. As an instance of Biome Module, the frame for Jungle Module would include slots inherited from the latter. However, these would be modified to reflect the particular nature of Jungle Module as a Biome Module, rather than, say, the Savanna Module. In addition, Jungle Module would have slots that are not representative of biome modules in general, for example, having to do with species that only exist under high temperature and humidity conditions.

The organization of frames might form the hierarchy shown in Fig. 18.5. The Jungle Module, Marsh Module, and other biome modules would be linked to Biome Module via the *isa* link. The Intensive Agriculture Module and Habitation Module might similarly be taken as instances of the generic class Special Module, and both Special Module and Biome Module might be considered

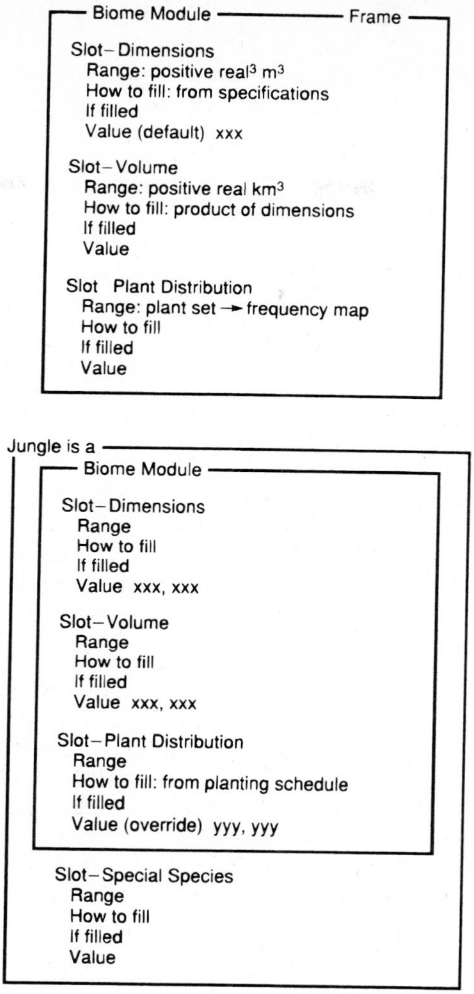

```
┌─ Biome Module ──────── Frame ─┐
│  Slot—Dimensions              │
│    Range: positive real³ m³   │
│    How to fill: from specifications │
│    If filled                  │
│    Value (default) xxx        │
│  Slot—Volume                  │
│    Range: positive real km³   │
│    How to fill: product of dimensions │
│    If filled                  │
│    Value                      │
│  Slot  Plant Distribution     │
│    Range: plant set → frequency map │
│    How to fill                │
│    If filled                  │
│    Value                      │
└───────────────────────────────┘

Jungle is a ─┐
  ┌─ Biome Module ─────────────┐
  │  Slot—Dimensions           │
  │    Range                   │
  │    How to fill             │
  │    If filled               │
  │    Value  xxx, xxx         │
  │  Slot—Volume               │
  │    Range                   │
  │    How to fill             │
  │    If filled               │
  │    Value  xxx, xxx         │
  │  Slot—Plant Distribution   │
  │    Range                   │
  │    How to fill: from planting schedule │
  │    If filled               │
  │    Value (override)  yyy, yyy │
  │  Slot—Special Species      │
  │    Range                   │
  │    How to fill             │
  │    If filled               │
  │    Value                   │
  └────────────────────────────┘
```

Figure 18.4 Concept of a frame: a Biome Module frame and its specialization.

as specialized classes of an even more generic class, Biosphere Module. The *ako* (a kind of) link represents this specialization relationship. The *ako* link differs from the *isa* link in that it relates classes to more general objects, rather than instances to classes.

In the hierarchy, inheritance of slots flows downward along the *ako* and *isa* links. Such inheritance and the triggering of procedures constitute forms of inference automatically performed by the system. For example, if a new slot is added to Biosphere Module, it will automatically be available in all subordinate frames. When a new value for a module parameter is obtained, the associated *IF FILLED* slot could trigger updating of the values of related parameters.

The frame formalism is said to be a fundamentally *declarative* ("knowing that") form of knowledge representation, since attribute values can be presented in propositional form as *facts* amenable to general-purpose logical inferencing. However, the *procedural attachment* realized in the *IF FILLED* and *HOW TO FILL* type of fields adds a dimension of *procedural* ("knowing how") representation capability (Intellicorp, 1984).

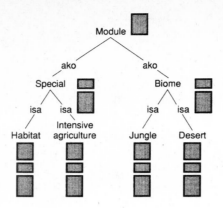

Figure 18.5 Semantic net.

SYSTEM-THEORETIC REPRESENTATION

Modern computer simulation offers a form of knowledge representation closely related to the frame hierarchy. Here, the worldviews of discrete-event simulation have been found to be highly compatible with the representation schemes of artificial intelligence (O'Keefe, 1986). Object-oriented programming [LOOPS (Bobrow and Stefik, 1983), FLAVORS (Weinreb, Moon, and Stallman, 1983), TI Scheme (Texas Instruments, 1986)] can be viewed as providing a computational basis for the frame hierarchy by allowing the programmer to associate methods (procedures) with objects that are inherited just as other slots are. Such methods can perform operations on the global object state (the ensemble of its slots) and invoke each other in a manner picturesquely called "message passing." It is not surprising, therefore, that languages are being developed to express both the dynamic knowledge of discrete-event formalisms and the declarative knowledge of AI-frame paradigms (see Klahr, 1986, and other articles in the same volume). These developments have been unified in a more fundamental paradigm that draws its inspiration from the system theory view of the world (reviewed by Pichler, 1985) and stems from the system-theoretic representation of simulation models for multifaceted modeling methodology (Zeigler, 1986, 1990).

System theory distinguishes between system *structure* (the inner constitution of a system) and *behavior* (its outer manifestation). Regarding structure, the theory provides the concepts of *decomposition,* how a system may be broken down into component systems, and *coupling,* how these components may be combined to reconstitute the original system. Thus, decomposition and coupling should be fundamental relations in a knowledge-representation scheme. System theory, however, has not focused on a third fundamental kind of relation, *taxonomic,* which concerns the admissible variants of a component and their specializations, exhibited, for example, by the generalization hierarchy of frames.

Regarding system behavior, it is useful to distinguish between *causal* and *empirical* representations. *Empirical representation* refers to actual records of data (time histories of variable values) gathered from a real system or model. *Causal representations* are integrated into units called *models,* which can be interpreted by suitable simulators to generate data in empirical form.

The System Entity Structure–Model Base

As a step toward a complete knowledge-representation scheme, the framework illustrated in Fig. 18.6 has been proposed. Decomposition, taxonomic, and coupling relationships are combined in a representation scheme called the *system entity structure* (Zeigler, 1984). This is a declarative scheme related to frame-theoretic and object-based representations. The *model base* contains models that

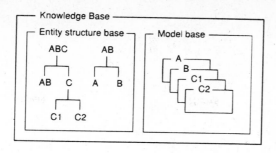

Figure 18.6 Knowledge representation to unify structural and behavioral knowledge.

are procedural in character, expressed in the classical and AI-derived formalisms mentioned earlier. This scheme is not complete since it does not deal with the fine-grained causal relations from which models are synthesized or with empirical raw data. The first should be the focus of major research, while the second is essentially available in database technology (Standridge, 1986). The *entities* of the entity structure refer to conceptual components of reality for which models may reside in the model base. Also associated with entities (as with other object types to be discussed) are slots for attribute-knowledge representation. An entity may have several *aspects*, defined as denoting a decomposition and therefore having several subentities. An entity may also have several *specializations*, defined as representing a classification of possible variants of the entity.

Figure 18.7 illustrates how knowledge about a synthetic biosphere might be represented in the system entity structure. We see that the biosphere is decomposed into modules, representing the collection of modules. The decomposition is formally represented by an aspect called module-decomposition. A module is a generic type for which at least one model exists in the model base. Slots associated with the entities contain information concerning the variables and parameters appearing in the associated models. A module has slots for describing the various physical attributes, such as dimensions, volumes, gas concentrations, and contained plants and animals and their distribution. Models for the various modules can be treated as components to be coupled together to compose a module for the overall system. The *coupling* of such models is associated with the aspect that subsumes the corresponding entities. For example, coupling of the modules, associated with the module-decomposition aspect, might mediate the flow of gases or microbes between the modules as a function of the connecting portal size.

Figure 18.7 also indicates the presence of another decomposition for the biosphere, which is encoded in the biota aspect. Using the biota aspect, a model of Biosphere II is built up from component models for Animals, Humans, and Plants by satisfying coupling constraints associated

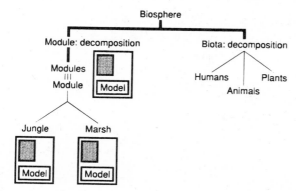

Figure 18.7 System entity-structure representation of a biosphere (partial).

with this aspect. Whereas the entity–aspect relation conveys decomposition knowledge, the entity–specialization relation represents taxonomic knowledge. In Fig. 18.8, the entity Plant has two specializations (classification schemes), Crop class and Season type, each of which consists of several entities. Specializations can be thought of as partitions; the intersection of two partitions forms a finer partition whose blocks are intersections of the originals.

In pruning, when one entity from a specialization is selected, it inherits the substructure (slots, aspects, and remaining specializations) of its parent. Thus, as in Fig. 18.8, if Soybean is chosen from Crop class forming Soybean Plant, it inherits the Season-type specialization, as well as all the attributes of Plant. Choosing Long Season from the latter specialization yields, finally, Long–season Soybean Plant, with all the substructure of Long Season, Soybean, and Plant.

It may happen that not all combinations of specialization choices represent actual possibilities. In this case, constraints can be added to the specializations to express allowable combinations. For example, if Soybean always has a Short Growing Season, the alternative combination would be excluded by the constraints.

The *multiple entity* ES represents the set of currently existing model components of type E. ES always has an aspect depicted as in Fig. 18.9a with three vertical lines, which is its *multiple*

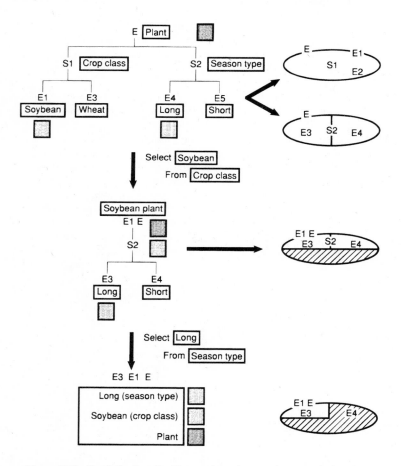

Figure 18.8 Specializations for Plant, and pruning to select a particular plant.

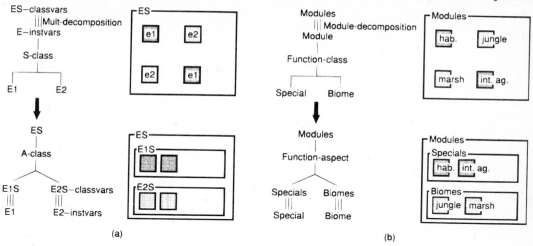

Figure 18.9 (a) Multiple entity concept and a transformation that removes the specialization. (b) Application to biosphere entity.

decomposition into the individual entities E. Class variables, carrying aggregation and distribution information, are associated with ES, whereas instance variables, belonging to each instance of E, are associated with E. When E has a specialization S, the model component denoted by ES naturally may contain components of the types represented by S. However, a *transformation* of the entity structure yields the form shown in Fig. 18.9a, in which ES has an aspect A class, consisting of the multiple entities of the specialization S class. The corresponding model of ES is obtained by coupling together the models E1S and E2S.

Figure 18.9b shows how these ideas apply to the modules in a biosphere. With the multiple entity, Modules, we associate such class variables as Number, the number of individual Modules, and also such aggregate variables as Average Volume. In general, for any attribute belonging to each instance, there are a host of aggregation and distribution attributes that can be associated with the multiple entity. Also illustrated is the interpretation of the transformation that results in expressing Modules as a decomposition of Special Modules and Biome Modules.

Accessing Knowledge in the System Entity Structure–Model Base

Earlier, it was indicated that the utility of a knowledge-representation scheme depends on the operations, queries, or transformations it accommodates. In this regard, the knowledge-based framework just introduced is intended to be *generative* in nature, that is, a compact representation scheme that can be unfolded to generate the family of all possible models synthesizable from components in the model base. The user, whether human or artificial, is a goal-directed agent that can interrogate the knowledge base and synthesize a model using *pruning* operations that ultimately reduce the structure to a *composition tree*. This contains all the information needed to synthesize a model in hierarchical fashion from components in the model base.

The generative capability of the entity structure is illustrated by the ability to attach entities and aspects to more than one place, thus generating new decompositions from existing ones. The entity-structure axiom of *uniformity* assures that any occurrence of an entity, aspect, or specialization carries with it the same substructure. Thus, if the Biota aspect is attached to the Module entities, a decomposition is obtained for each Module relating to Biota decomposition. This enables construction of finer models, in which each Module contains Human, Animal, and Plant subcomponents.

THE SYSTEM ENTITY STRUCTURE–MODEL BASE IN THE BIOREGENERATIVE CONTEXT

This section describes the role of the system entity structure–model base in supporting the objectives of bioregenerative system research and development.

Bioregenerative Environmental Design

The entity structure–model base knowledge-representation scheme serves as a means of organizing possible configurations of a system to be designed. Just as in the design of conventional artifacts, the design of a synthetic bioregenerative environment must select combinations of components (biome ecosystems, plants and animals, planting and harvesting schedules) that mutually support one another in achieving system goals. Rozenblit (1985) describes how pruning of the entity structure for a system-design domain serves as a basis for the generation of families of design models that can be simulated and evaluated relative to design objectives.

An important consideration in design of a bioregenerative environment is that, unlike a well-understood artifactual domain, many conditions encountered and questions raised are novel and knowledge is lacking. In this "ill-defined systems" context, it is especially important to employ extensive cross-checking of computations, for example, the predictions of models at various levels of aggregation. As previously indicated, the organization of these models is facilitated by the entity structure.

Behavior Observation Archiving

The system entity-structure representation serves as a framework upon which to organize the vast amounts of data to be collected for observation and analysis of system behavior. The data can be sorted into slots organized, first, according to entities representing conceptual components of the real system and, then, into the various levels of aggregation represented in the entity structure.

Biosphere Model Construction and Validation

Data collected and stored in an archiving system compatible with the entity structure facilitate the calibration and validation of models. Since the data are organized by entity and level of aggregation, they directly relate to models in the model base that are organized according to the same scheme. Since real-system data and model-generated data can be directly compared, a basis exists for automatic model calibration and validation.

Deviant Behavior Diagnosis and Correction

Novel bioregenerative systems are expected to encounter many new kinds of deviant behaviors, such as gas imbalances and plant epidemics, that must be dealt with in a timely manner. Simulations to discover corrective actions and predictions of their effects will play a major role in such failure management. Availability of models at various levels of aggregation will once again be crucial in such procedures. Coarse models, with fast simulation characteristics, may be employed to develop initial responses in crisis situations. More accurate, dissaggregated models may then be employed to validate and refine the initial responses or, indeed, to abort them in favor of better considered responses. Organization of these models is facilitated by the system entity structure.

SUMMARY

In a volume dedicated to a visionary responsible in large part for the advent of systems ecology, the challenge of creating life-supporting, self-sustaining, bioregenerative systems is especially significant. This chapter has reviewed the central ideas of knowledge representation and multifaceted modeling methodology in relation to this challenge. The central tenet is that the systems theory-based knowledge representation encodable in the systems entity structure–model base is crucial to successful design, implementation, and control of a bioregenerative system. If this hypothesis is confirmed in the successful bringing into existence of new artificial biospheres, it will surely be a candidate for supporting our attempts to live in harmony on earth.

ACKNOWLEDGMENT

This chapter is based in part on the author's consulting work with the Environmental Research Laboratory of the University of Arizona, whose assistance is gratefully acknowledged.

REFERENCES

ADELSBERGER, H. H., POACH, U. W., SHANNON, R. B., and WILLIAMS, G. N. 1986. Rule based object oriented simulation systems. In Luker, P. A., and Adelsberger, H. H. (eds.), *Intelligent Simulation Environments.* Simulation Series, Vol. 17. Society for Computer Simulation, San Diego, California, pp. 107–112.

BELOGUS, D. 1985. Multifacetted Modelling and Simulation: A Software Engineering Implementation. Doctoral dissertation. Weizmann Institute of Science, Rehovot, Israel. 197 pp.

BOBROW, D. G. (ed.). 1985. *Qualitative Reasoning about Physical Systems.* MIT Press, Cambridge, Massachusetts. 495 pp.

BOBROW, D. G., and STEFIK, M. J. 1983. The LOOPS Manual. Xerox Corporation, Palo Alto, California. 10 pp.

FRANTA, W. R. 1977. *A Process View of Simulation.* North-Holland, Amsterdam. 244 pp.

FUTO, I. 1985. Combined discrete/continuous modeling and problem solving. In Birtwistle, G. (ed.), *AI, Graphics and Simulation.* Society for Computer Simulation, San Diego, California, pp. 117–124.

HAYES, N. 1989. Biosphere II—a prototype for the future. *IEEE Computer 22* (May 5):11.

HOGEWEG, P., and HESPER, B. 1986. Knowledge seeking in variable structure models. In Elzas, M. S., Oren, T. I., and Zeigler, B. P. (eds.), *Modelling and Simulation in the Artificial Intelligence Era.* North-Holland, Amsterdam, pp. 227–244.

INTELLICORP 1984. The Knowledge Engineering Environment. Intellicorp, Menlo Park, California. 35 pp.

KERCHOFFS, E. J. H., and VANSTEENKISTE, G. C. 1986. The impact of advanced information processing on simulation—an illustrative review. *Simulation* 46:17–26.

KLAHR, P. 1986. Expressibility in ROSS, an object-oriented simulation system. In Vansteenkiste, G. C., Kerchoffs, E. J. H., and Zeigler, B. P. (eds.), *Artificial Intelligence in Simulation.* Society for Computer Simulation, San Diego, California, pp. 136–139.

KLIR, G. J. 1985. *Architecture of Systems Problem Solving*. Plenum Press, New York. 539 pp.

MCDERMOTT, J. 1982. R1: a rule-based configurer of computer systems. *Artificial Intelligence* 19:39–88.

MINSKY, M. 1977. Frame-system theory. In Johnson-Laird, K., and Wason, A. (eds.), *Thinking*. Cambridge University Press, New York, pp. 355–376.

O'KEEFE, R. 1986. Simulation and expert systems—a taxonomy and some examples. *Simulation* 46(1):10–16.

OREN, T. I. 1986a. Knowledge bases for an advanced simulation environment. In Luker, P. A., and Adelsberger, H. H. (eds.), *Intelligent Simulation Environments*. Simulation Series, Vol. 17. Society for Computer Simulation, San Diego, California, pp. 16–22.

OREN, T. I. 1986b. Implications of machine learning in simulation. In Elzas, M. S., Oren, T. I., and Zeigler, B. P. (eds.), *Modelling and Simulation Methodology in the Artificial Intelligence Era*. North-Holland, Amsterdam, pp. 41–57.

PICHLER, F. 1985. Dynamical systems concepts. In Trappl, R. (ed.), *Source-book in Cybernetics and Systems Research*. Hemisphere Publishing Co., Vienna, pp. 161–164.

RAJOGOPALAN, R. 1986. The role of qualitative reasoning in simulation. In Vansteenkiste, G. C., Kerchoffs, E. J. H., and Zeigler, B. P. (eds.), *Artificial Intelligence in Simulation*. Society for Computer Simulation, San Diego, California, pp. 9–26.

REDDY, Y. V., FOX, M. S., and HUSAIN, N. 1985. Automating the Analysis of Simulations in KBS. Proc. Society for Computer Simulation Multiconference, Society for Computer Simulation, San Diego, California, pp. 34–40.

REDDY, Y. V., FOX, M. S., HUSAIN, N., and MCROBERTS, M. 1986. The knowledge-based simulation system. *IEEE Software* March:26–37.

ROBERTSON, P. 1986. A rule based expert simulation environment. In Luker, P. A., and Adelsberger, H. H. (eds.), *Intelligent Simulation Environments*. Simulation Series, Vol. 17. Society for Computer Simulation, San Diego, California, pp. 9–22.

ROZENBLIT, J. W. 1985. A Conceptual Basis for Model-based System Design. Doctoral dissertation. Wayne State University, Detroit, Michigan. 132 pp.

ROZENBLIT, J. W., and ZEIGLER, B. P. 1985. Concepts for knowledge-based system design environments. In *Proceedings Winter Simulation Conference*, San Francisco. Society for Computer Simulation, San Diego, pp. 326–331.

STANDRIDGE, C. R. 1986. Simulation data bases. In Singh, M. (ed.), *Encyclopedia of Systems and Control*. Pergamon Press, New York, pp. 893–894.

TEXAS INSTRUMENTS, Inc. 1986. TI Scheme Language Reference Manual. Texas Instruments, Dallas Texas. 280 pp.

Time. 1990. Noah's ark—the sequel. *Time Magazine,* September 24, 1990, pp. 72–73.

WEINREB, D., MOON, D., and STALLMAN, R. 1983. LISP Machine Manual. MIT Press, Cambridge, Massachusetts.

ZEIGLER, B. P. 1984. *Multifacetted Modelling and Discrete Event Simulation*. Academic Press, New York. 372 pp.

ZEIGLER, B. P. 1985. System-theoretic representation of simulation models. *IEEE Trans.* 16(1):19–34.

ZEIGLER, B. P. 1986. Toward a simulation methodology for variable structure modelling. In Elzas, M. S., Oren, T. I., and Zeigler, B. P. (eds.), *Modelling and Simulation Methodology in the Artificial Intelligence Era*. North-Holland, Amsterdam, pp. 195–210.

ZEIGLER, B. P. 1987. Knowledge representation from Minsky to Newton and beyond. *Appl. Artificial Intelligence* 1:87–107.

ZEIGLER, B. P. 1990. *Object-oriented Simulation with Hierarchical Modular Models*. Academic Press, New York. 395 pp.

ZEIGLER, B. P. 1992. Endomorphic modelling concepts for high autonomy architectures. *Appl. Artificial Intelligence* 6(1):19–44.

ZEIGLER, B. P., and REYNOLDS, R. 1985a. A hierarchical information processing model for adaptation to technology change. *Systems Res.* 2(4):309–317.

ZEIGLER, B. P., and REYNOLDS, R. 1985b. Towards a theory of adaptive computer architectures. In *Proceedings, Fifth International Conference on Distributed Computer Systems,* Denver, pp. 468–475.

ZEIGLER, B. P., BELOGUS, D., and BOLSHOI, A. 1980. ESP—an interactive tool for system structuring. In *Proceedings, European Meeting on Cybernetics and Systems Research.* Hemisphere Press, Washington, D.C., pp. 439–451.

GLOSSARY

Aspect—notes a decomposition in an entity structure.
Atomic model—a model that is not further decomposable.
Behavior—that which can be directly observable about a system over time.
Bioregenerative environment—an environment that employs biological processes to provide for recycling of materials.
Biosphere—a stable, complex, evolving system containing life, essentially closed to material exchange with its surroundings.
Composition tree—contains all the information needed to synthesize a model in hierarchical fashion from components in the model base.
Coupled model—a model that is further decomposable.
Coupling—how components may be combined to construct a hierarchical system.
Declarative knowledge ("knowing that")—form of knowledge representation (compare with *procedural* knowledge).
Decomposition—how a system may be broken down into components.
Entity—conceptual components of reality for which models may reside in a model base.
Entity structure—see *system entity structure.*
Epistemological hierarchy—provides *levels of system specification* at which knowledge may be acquired, as well as conditions under which validation is possible.
Expert systems—computer software intentionally built to embody significant components of human expertise.
Facts—elements in a declarative knowledge scheme.
Formalisms—means of expressing various kinds of models.
Frame—contains knowledge about a particular object and is linked to other frames in a taxonomic relationship.
Hierarchical composition—construction of a model in a series of levels in which models constructed at one level are employed as components to form the next level.
Homomorphism—a correspondence between the states of a pair of objects that is preserved under all relevant operations.
Inference engine—a procedure for manipulating rules so as to infer new knowledge about a given situation.
Knowledge—valid representation; that is, there is a homomorphism between the representation and the reality it represents.
Model base—an organized collection of models.
Multiple entity—represents the set of currently existing model components of a given class.
Procedural knowledge ("knowing how")—representation of knowledge.
Pruning—an operation that ultimately reduces an entity structure to a *composition tree.*
Representation scheme—a means of representing reality in computerized form.
Rules—elements in a procedural knowledge scheme that can be manipulated by an inference engine.

Rule base—a collection of rules constituting procedural knowledge.
Slots—places in a representation scheme in which information of a predetermined kind may be placed.
Specialization—denotes a classification of the possible variants of an entity.
Structure—the inner constitution of a system.
System entity structure—a declarative scheme for representing the structure of a system or family of systems.
Taxonomy—hierarchical arrangement of categories.
Uniformity—in an entity structure, assures that any occurrence of an entity, aspect, or specialization carries with it the same substructure.

19

Multicommodity Ecosystem Analysis: Dealing with the Mixed Units Problem in Flow and Compartmental Analysis

ROBERT COSTANZA

Chesapeake Biological Laboratory
University of Maryland
Solomons, Maryland

BRUCE HANNON

Geography Department
and Illinois Natural History Survey
University of Illinois
Urbana, Illinois

A system is an organization that functions in a particular way. The functions of an ecosystem include transformation, circulation, and accumulation of matter and the flow of energy through and within living organisms by means of their activities and natural physical processes [Van Dyne, in *The Ecosystem Concept in Natural Resource Management*, Ch. X, p. 329 (1969), Academic Press].

Ecology has long been recognized as a multidisciplinary and integrative science. To understand the interaction between an individual organism and its environment requires an integration of knowledge from a variety of disciplines. . . . But ecology is more than the study of isolated individuals and their physical environments. There are higher levels of organization where individuals form populations, and populations interact to form communities . . . emergent properties . . . appear at these higher levels of organization . . . [Van Dyne, in *The Ecosystem Concept in Natural Resource Management*, Ch. X, p. 331 (1969), Academic Press].

INTRODUCTION

Ecology is often defined as the study of the relationships between organisms and their environment. The quantitative analysis of interconnections between species and their abiotic environment has therefore been a central issue. The mathematical analysis of interconnections is also important in several other fields. Practical quantitative analysis of interconnections in complex systems began with the economist W. W. Leontief (1941), using what has come to be called *input–output (I–O) analysis*. More recently, these concepts, sometimes called *flow analysis*, have been applied to the study of interconnections in ecosystems (Hannon, 1973, 1976, 1979, 1985a,b,c; Costanza and Neill, 1984). Related ideas were developed from a different perspective in ecology, under the heading of *compartmental analysis* (Funderlic and Heath, 1971; Hett and O'Neill, 1971; Barber, Patten, and Finn, 1979). Patten (1978, 1982) has emphasized the duality of I–O analysis in an environmentally explicit approach that defines within-system environments ("environs") of components. Retrospective analysis computes historical *input environs*, and a parallel prospective analysis, future *output environs*.

In this chapter, we (1) review the basic mathematical forms of linear, static and dynamic, flow and compartmental analyses, emphasizing their similarities and differences; (2) point out the importance of "net-input intensity factors" in combining the two approaches and the need for a hierarchical approach to their calculation; (3) discuss the necessity to deal with multiple commodities (not just, for example, carbon, energy, or nitrogen) and joint products for more realistic treatment of ecosystems; (4) develop the mathematical form for a general multicommodity system capable of dealing with joint products; and (5) present and briefly discuss examples of these applications.

We find that multicommodity, hierarchical I–O (flow) analysis can be used to calculate a set of unique net-input intensity factors that can then be used to form more comprehensive and realistic input matrices for compartmental analysis.

COMMENSURABILITY

A fundamental issue in any field that deals with interconnections among dissimilar components is *commensurability*—the ability to measure the components in the same addable (commensurable) units. Both economics and ecology must deal with this apples-and-oranges issue, although we feel its importance in ecology has not yet been fully appreciated.

Flow and compartmental analyses differ fundamentally in terms of (1) their concentration on outputs versus inputs (respectively), and (2) what things they require to add up. The fundamental reasons for these distinctions are summarized in following sections. To obtain a comprehensive understanding of ecosystem interdependence, it is necessary to understand *both* input and output relationships. We therefore seek a way to integrate these approaches.

Flow analysis requires for most of its results only that total output of each commodity in a system equal total input, without ever requiring the addition of different commodities (for example, apples and oranges or carbon and nitrogen). Compartmental analysis, on the other hand, requires a "conservative" unit of measure, such as mass, energy, or carbon content. A conservative unit is one that obeys the rule that total inputs to any process equal total outputs from it. We must therefore be able to add process inputs and outputs, and in real processes these are usually different, noncommensurable commodities. The usual simplification in ecological compartmental analysis is to arbitrarily choose one commodity (such as carbon), which is conservative, and declare that this is a valid tracer for all others. This approach can be misleading, however. Imagine using only the weight (mass) of interconnecting flows in an economic system to analyze interdependence. Bulk commodity industries like sand and gravel would be given much more weight than they deserve at the expense of service and other less mass-related industries. Likewise, choosing any one commodity in an ecosystem to analyze interdependence can be misleading.

Before compartmental analysis can yield comprehensive and realistic results for ecosystems, we must calculate and apply an appropriate set of intensity factors analogous to prices in economic systems. This chapter demonstrates several related techniques for doing this.

FLOW ACCOUNTING

In the analysis of complex dynamic systems, it is necessary to develop consistent definitions and categorizations of all the identifiable flows. We start with the diagram shown in Fig. 19.1.

A *commodity* is defined as some identifiable unit moving through the system. It can be a simple element (like carbon), a complex structure (like plant biomass), or even a service (like soil aeration). Commodities are transformed (produced, consumed, or combined into more complex commodities) in *processes*. Biomass, for example, is produced by the photosynthetic process, which combines (consumes) water, carbon dioxide, sunlight, and nutrients into plant material.

In Fig. 19.1, P is known as the production matrix. It summarizes the direct interdependencies between the system's commodities and processes as flows from rows i to columns j. Its elements are the amount of commodity i that is used by process j in the given time period. For example, P_{ij} could be the daily amount of nitrogen (i) consumed in the photosynthetic process of a particular plant species (j). This is a multicommodity system since different, noncommensurable commodities are listed along the rows. The row sums may be calculated (since they are all the same commodity), but the column sums cannot (since that would require adding, for example, nitrogen, sunlight, carbon dioxide, and water).

We begin with a system without joint products or one in which each process (column) is assumed to produce only one commodity. This ignores, for example, the oxygen produced by photosynthesis as a joint product along with the biomass. Later we relax this requirement.

The vector r is the net output of commodities from the system. It is composed of two types of commodities: (1) net export (export minus import) of each produced commodity during the given period, and (2) the amount of standing stock made during the period used in rebuilding depreciated stock (we use basal metabolism as a surrogate for these commodity flows). The Δs is a vector of net changes in stock during the time period (zero at steady state). Vector p is the total output of commodities by the system. The elements of p are the row sums of P, plus r and Δs. In other words, the total amount of each commodity produced equals the sum of the amounts sent to other internal processes, net exports, and changes in standing stock. Excluded from the row summing is the amount of the commodity i that is given off as heat in nonbasal respiration. These waste heat flows are elements of the vector w. If w and p are added, the result is a total input vector p^N. The vector e is the net input vector, that is, the amounts of commodities absorbed by the system processes, but not produced by it, during the given time period.

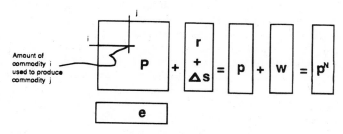

Figure 19.1 Flow accounting diagram.

RETROSPECTIVE FLOW ANALYSIS

What we shall call *flow analysis*, which is analogous to *input environ analysis* (Patten, 1982) in the original Leontief orientation looking temporally backward from outputs, proceeds in the following way given these definitions. Let $s = s_i$ be a vector of stocks; s_i is the amount of commodity i held in storage or flow delay as a stock in the system. As the exchange time interval becomes infinitesimal, the commodity balance along the row becomes

$$\dot{s}_i = p_i - \sum_j p_{ij} - r_i = p_i^N - w_i - \sum_j p_{ij} - r_i, \tag{19.1}$$

where $\dot{s}_i$ is ds_i/dt.

Note that the nonbasal waste heat vector is excluded from the balance in eq. (19.1). This exclusion indicates that w has zero *value* to the ecosystem when compared to p. In other words, w and p are different commodities. Let us assume that the flow between two components in P divided by the *receiving* component is a constant matrix $G' = (g'_{ij})$ where $g'_{ij} = p_{ij}/s_j$. Now eq. (19.1) can be rewritten as

$$\dot{s}_i = p_i - \sum_j g'_{ij} s_j - r_i, \tag{19.2}$$

or, in vector-matrix form, as

$$\dot{s} = p - G's - r. \tag{19.3}$$

Equations (19.1) or (19.2) could be used, for example, in microcosm research to determine flows from the changes in stock levels. Assuming an n component system and assuming that r can be measured, there are n^2 unknowns. The stock and net output vectors must be measured over at least n different ranges to provide sufficient data to determine G. If we assume further that the stock to flow ratio for each component is a constant diagonal matrix B, then the elements of this stock–flow matrix are $b_{ii} = s_i/p_i$, and eq. (19.3) can be further rewritten as

$$B\dot{p} = p - G'Bp - r$$
$$= (I - G'B)p - r.$$

If we combine the two assumptions, $g_{ij} = p_{ij}/p_j$ and then

$$\dot{p} = B^{-1}(I - G)p - B^{-1}r. \tag{19.4}$$

This is the standard form for the dynamic equation of the retrospective flow analysis approach. It could also be referred to as the "recipient-controlled" dynamic flow equation. The stability of future p when the system is subjected to changes in r depends entirely on the matrix which multiplies p in eq. (19.4). If the eigenvalues of that matrix are negative, p is stable. Since B and G are always positive, the eigenvalues of the matrix are always positive, and future p will always respond to changes in r in an unstable manner. Equation (19.4) can be made to respond stably by modifying r to include a feedback control in addition to the desired change in r. The feedback control can be a function of p or its derivative or integral. The control can be thought of in two ways: first, it gives the ecosystem manager appropriate coordinated steps to be taken in advance or following the desired changes in r; second, the ecosystem can be thought of as possessing certain types of these feedback controls. In this case, control theory could be coupled with ecosystem experiments to discern the mathematical form of the natural controls. Details of these control theory approaches are discussed by Hannon (1985b, c, 1986).

If $\dot{p} = 0$, eq. (19.4) becomes

$$p = (I - G)^{-1}r, \qquad (19.5)$$

which is independent of the standing stock level. Equation (19.5) is the steady-state relationship between net and total output. In an ecosystem with no material flows across its boundary, eq. (19.5) indicates the amount of total output required to make up for the structural depreciation (basal metabolism).

PROSPECTIVE COMPARTMENTAL ANALYSIS

In what we will term *compartmental analysis*, which is analogous to *output environ analysis* (Patten, 1982) in being oriented temporally forward from inputs, the inputs to a process are summed, and the total output (in this case p^N since nonmetabolic heat cannot be considered part of the stock change) is removed, leaving the change in the stock level. The inputs include the net input of nonproduced commodities from outside the system. In a form similar to eq. (19.1), the balance equation for the compartmental approach is

$$\dot{s}_j = \sum_i p_{ij} + e_j - p_j^N. \qquad (19.6)$$

We now assume that the intercompartmental flow divided by the stock of the *donor* compartment is the constant matrix G'', whose elements are $g_{ij}'' = p_{ij}/s_i$. Equation (19.6) can now be written as

$$\dot{s}_j = \sum_i s_i g_{ij}'' + e_j - p_j^N,$$

or, in vector-matrix form, as

$$\dot{s} = sG'' + e - p^N. \qquad (19.7)$$

Again, we assume that the stock to total-output flow ratio is the constant matrix B', whose elements are $b_{ij}' = s_i/p_i^N$. The two assumptions lead, as before, to the standard dynamical form of the prospective compartmental analysis approach ("donor-controlled"):

$$\dot{p}^N = -p^N(I - G^*)B'^{-1} + eB'^{-1}, \qquad (19.8)$$

where $g_{ij}^* = p_{ij}/p_i^N$.

The matrix that multiplies P^N determines the stability of the total output in response to changes in e. Since B and G are both positive matrices, the eigenvalues of the critical matrix are always negative, and the predictive solutions to eq. (19.8) are always stable. However, because the method involves summing commodities of different qualities, the value of eq. (19.8) is limited to systems measured in commensurable units. As pointed out above, commensurable units can be achieved by ignoring everything in the system except one commodity or through the use of a weighting function (see later).

The steady-state version of compartmental analysis is

$$P^N = e(I - G^*)^{-1}, \qquad (19.9)$$

which is similar to eq. (19.5) in the retrospective approach, except that here the equation is driven by net input e (rather than producing net output r), and a "donor-controlled" coefficient matrix G^*

is involved rather than the "recipient-controlled" coefficient matrix G of the backward-oriented approach. Figure 19.2 represents the mass balance specified by eq. (19.9).

Figure 19.2 clearly reveals both forms of the mixed commodity problem. The production matrix P is summed down the columns, adding together commodities of different qualities. Even though these commodities may be measured in the same physical units, they cannot be added because of their "quality" differences. For the same reasons, the quantity p^N cannot exist. It is obtained by adding physical units representing both valuable (usable) substances and waste heat.

To allow the addition required in compartmental analysis, a weighting function must be introduced. Let this be the vector ϵ. We assume that the ecosystem acts as though it is perfectly competitive in an economic sense: no producer or consumer acting alone is significant enough to affect the ϵ values; information about stocks and flows is communicated to every process in the system. Such conditions produce a unique set of weighting factors, as diagrammed in Fig. 19.3, where $\epsilon^w = 0$ means that the nonbasal metabolic heat has no value to the system or, in other words, no value relative to the commodity flows represented in P or r. The equation that represents the balance shown in Fig. 19.3 is

$$\sum_i \epsilon_i p_{ij} + e_j = \epsilon_j p_j,$$

or, in vector-matrix form,

$$\epsilon P + e = \epsilon \hat{P},$$

where $\wedge$ means a diagonalized vector. Then,

$$\epsilon = e(\hat{P} - P)^{-1},$$

and since

$$G = P\hat{P}^{-1}$$

then

$$\epsilon = e\hat{P}^{-1}(I - G)^{-1}. \tag{19.10}$$

Equation (19.10) is the solution for the needed weighting factors that allows recognition of commodity quality in the prospective compartmental analysis approach. It is derivable, however, directly from measures in retrospective flow analysis; p and G in eq. (19.10) are the total output vector and constant flow intensity matrix used in the latter approach. The analogy is extendable to the dynamic case.

In forward analysis, the weighting factors ϵ are usually implicitly assumed to be 1. This is not likely ever to be a valid assumption since, at a minimum, from the second law of thermodynamics we would expect some waste heat production. If waste heat w_j is included in p_j, the resulting ϵ values are all 1's (Hannon, 1985a). The ϵ, therefore, may be thought of as weighting factors among a diverse set of flows, with the "thermodynamic quality" of each the same along its row in $P_i + r$.

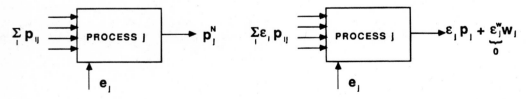

Figure 19.2 Balance diagram for prospective compartmental analysis.

Figure 19.3 Weighted balance for the compartmental analysis approach.

The quantity $\epsilon_i p_i$ is directly comparable to the quantity $\epsilon_j p_j$; the ϵ's are like economic "prices," but they arise from thermodynamic considerations.

When the compartmental approach is weighted using net input intensity factors to more closely represent reality, it equates to the flow analysis approach. Consequently, the two approaches are combined into one. If the desire is to calculate the stock and total output change due to a net output change, the past-oriented flow analysis approach is the more directly useful one. When ϵ is introduced into the future-oriented compartmental approach, the resulting output vector is identical to the one obtained from classical flow analysis. The ϵ coefficients in effect connect the dualistic perspectives on environment provided by the input and output environ concepts.

Retrospective flow analysis requires that a given commodity have the same value everywhere in the system. Otherwise, as indicated previously, summing along the rows in a matrix of physical units would not be appropriate. This assumed uniformity of commodity weighting is a result of a broader assumption of system competition. For a system to be "competitive" in the sense of these formulations, it must have many small consumers and producers each possessing all the current information about the variables that affect them. Under these conditions, no single consumer can directly affect the unit weight of a commodity.

Under certain conditions, these values can be calculated on the basis of the net inputs alone, as in eq. (19.10). If each producer has only one type of output, if all producers of the same commodity use the same inputs per unit output (common production "recipe"), and if the system has only one net input, then the ϵ elements in eq. (19.10) are the commodity unit values (Samuelson, 1966). The system produces all the desired quantities of each commodity at its unit value. We can always reduce the system to one net input (see the global ecosystem discussion and Table 19.7 later). We can also deal with systems in which each producer (process) has multiple commodity outputs or, in other words, where there are joint products (see Fig. 19.4 later).

The third requirement is more difficult to meet. In reality, each commodity producer is an averaged mix of different recipes, and the producer with the highest ϵ value is the one setting the proper system weight or unit value for that commodity. For example, suppose several types of algae exist in a particular ecosystem and that each produces biomass eaten by the same herbivore species. Suppose eq. (19.10) can be applied to this system, providing us with a *set* of ϵ for the algae. These algal ϵ are the unit *costs* for each type of biomass, but the unit *value* of all algal biomass in this ecosystem is the largest of these ϵ, the *marginal cost* of the algal mix. By the very existence of this highest-cost producer in the system, the algal consumers are indicating the maximum level of effort to which they will go to consume algae. In cases where such ϵ variation is thought to exist, a separate commodity should be defined for each variation.

COMPARING RESULTS OF COMPARTMENTAL AND FLOW ANALYSES

With the addition of two further sets of assumptions, the retrospective and prospective analyses can be compared on the same data set. The extra assumptions are consistent and symmetrical. They are based on net output for future-directed compartmental analysis and on net input for past-directed flow analysis. Since our data are for a steady-state condition, we use eqs. (19.5) and (19.9) as the basis of comparison.

Flow Analysis

For the static condition, assume that

$$e\hat{p}^{-1} = n = \text{constant.}$$

This means that net input is a linear function of total output (without nonbasal respiration). With this assumption and equation (19.5),

$$e = \hat{n}(I - G)^{-1}r. \tag{19.11}$$

Equation (19.11) is used to calculate net input e given the specified net output r.

Compartmental Analysis

In a similar way, assume that

$$(r + w)\hat{P}^{N^{-1}} = m = \text{constant}$$

and

$$w\hat{P}^{N^{-1}} = c = \text{constant}.$$

With these two assumptions and eq. (19.9), we have

$$e = r\widehat{(m - c^{-1})}(I - G^*). \tag{19.12}$$

This equation gives the forward-oriented compartmental analysis version of e if r is specified. In the applications section later, we demonstrate the strengths and weaknesses of the two approaches.

JOINT PRODUCTS IN FLOW ANALYSIS

The discussion so far has avoided reference to ecological processes that produce more than one useful output. Our analysis has recognized only that each process produces two types of output: p and w. Only one of these, p, is considered usable by the system; w is therefore a "product" not absorbed by any other process in the system. It is dropped from consideration or, equivalently, assigned an ϵ that is zero. Now we complicate the picture to better approximate reality by introducing a multiple (useful)-output format. The result is parallel to previous development: column summing and row summing become equivalent under appropriate assumptions. Figure 19.4 depicts the expanded accounting system. This diagram is interpreted as follows. The ith row sum of U plus the ith element of r plus Δs_i equals the ith row sum of V (which is equal to the ith element of p, the total output without nonbasal respiration), all for the given time period. The additional step taken in the multiple product formulation is the elaboration of p into the "make" or output matrix, V. Two assumptions are possible for the linear model. One requires summing down the columns of U, which, as argued above, is inappropriate. Under conditions of perfect competition, each commodity has a similar unit value or weight for each of its consumers. Therefore, summing commodity flows along the rows is appropriate.

A row from the accounting diagram in Fig. 19.4 can be stated in continuous time as

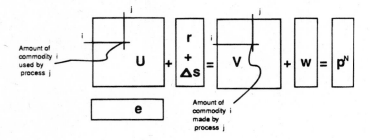

Figure 19.4 Multiproduct flow accounting diagram.

$$\sum_j V_{ij} = \sum_j U_{ij} + r_i + \dot{S}_i,$$

or, in vector-matrix form,

$$Vi = Ui + r + \dot{s}, \tag{19.13}$$

where i is a vector of 1's.

Rewriting eq. (19.13) and solving for $\dot{s}$ gives

$$\dot{s} = (V - U)i - r, \tag{19.14}$$

or

$$\dot{s} = (I - UV^{-1})Vi - r.$$

With the stock-flow assumption used above, eq. (19.14) becomes

$$\dot{p} = B^{-1}(I - UV^{-1})p - B^{-1}r. \tag{19.15}$$

By a procedure identical to eq. (19.10), the ϵ or net input intensities are (with $\dot{p} = 0$)

$$\epsilon = e(V - U)^{-1}, \tag{19.16}$$

or

$$\epsilon = e((I - UV^{-1})V)^{-1}. \tag{19.17}$$

The intensities are evaluated later for the global system. Equations (19.15) and (19.16) [or (19.17)] are the multiple input–output equivalent of the (retrospective) flow analysis approach.

It is important to note that in the joint-product flow analysis formulation it is possible to have negative entries in the $(V - U)^{-1}$ matrix (the commodities required directly and indirectly as input to processes). These negative values are a consequence of joint production. A negative value in the $(V - U)^{-1}$ matrix implies that the column process directly or indirectly produces (rather than requires) the row commodity. The less significant is joint production in the system, the fewer and smaller will be the negative elements in the $(V - U)^{-1}$ matrix.

The existence of negatives in the $(V - U)^{-1}$ matrix may also give rise to negative elements in the net input intensity vector (ϵ). Negative energy intensities would imply that production of an additional unit of the commodity in question would require less energy (directly and indirectly), rather than more. We have argued elsewhere (Hannon et al., 1986) that these negative intensities are most likely the result of poor data or poor system specification, and can be eliminated by better data or better specification of the system.

If all positive intensities are obtained for a system using eq. (19.16) or (19.17), it is possible to combine them with an additional assumption to derive a commodity by commodity flow matrix (P) and coefficient matrix (G) whose elements are guaranteed to be nonnegative and that can then be analyzed using the methods described earlier. The first step is to calculate "evaluated" U and V matrices. These matrices are defined as

$$U^* = \hat{\epsilon}U, \tag{19.18}$$

$$V^* = \hat{\epsilon}V, \tag{19.19}$$

where as before $\hat{}$ means a diagonalized vector.

A characteristic of the U^* and V^* matrices is that they are in commensurable units, and both their row and column totals are meaningful. They are analogous to dollar flow matrices in economic applications. From the definition of ϵ, the total evaluated commodity inputs to each process should equal the total evaluated commodity outputs, or

$$iU^* + e = iV^*.$$

The column-summing assumption is a way of allocating inputs to joint products in order to derive an equivalent single-product system (Hannon et al., 1985). In the context of the multiple input and output view, this assumption is related to input (prospective) compartmental analysis, except that the production matrix U and the use matrix V have been weighted first by ϵ. It assumes that input commodities are allocated to output commodities in proportion to the degree to which the output commodity represents of the *total* process output. This situation, of course, requires that we can calculate the total output of processes, which in turn requires that we measure these outputs in commensurable units. Given that V^* is known from the previous analysis, we can calculate a matrix to perform this allocation:

$$Q = V^*(\hat{iV^*})^{-1}. \qquad (19.20)$$

The elements of Q are simply the percentages each commodity flow represents of the total input to each process. A commodity by commodity flow matrix (P) can then be calculated as

$$P = UQ^T. \qquad (19.21)$$

Standard flow analysis techniques as outlined above can then be applied to the system and will yield the same ϵ as that obtained using eq. (19.16) or (19.17) (Hannon, Costanza, and Herendeen, 1986). Details of the calculations are given in Appendix A.

Similarly, we can calculate a process-by-process flow matrix (P^*) as

$$P^* = Q^{*T}U, \qquad (19.22)$$

where Q^* is a matrix of the percentages that each commodity flow represents of the total outputs of that commodity (from the evaluated V^* matrix).

APPLICATION OF THE THEORY

We have chosen an oyster reef ecosystem model (Dame and Patten, 1981) as a suitable base for comparison of compartmental and flow analyses. This model contains many cycles that overwhelm any intuition we might have about the significance of particular system interconnections. This system is shown at steady state in Fig. 19.5.

The data from Fig. 19.5 have been arranged in the proposed accounting framework (Fig. 19.1) as shown in Table 19.1. This is a unique arrangement (see Hannon, 1985a) since we have for the first time tried to estimate the basal metabolism or structural rebuilding respiration (the equivalent of depreciation in economic systems) for inclusion in the net output. There has been little interest in this important flow in the biological literature. For illustrative purposes, we have simply divided the total respiration evenly between net output and waste (Vernberg, 1975). This step, of course, reduces the difference between p and p^N and tends also to reduce any differences between the two approaches.

From the data in Table 19.1, we construct the G, G^*, $(I - G)^{-1}$, and $(I - G^*)^{-1}$ matrices and present them in Tables 19.2, 19.3, 19.4 and 19.5, respectively. In this example, the numbers in all four of the tables are unitless. In the case of mixed commodity units, G and $(I - G)^{-1}$ have a complex but identical mix of the units, while the other two matrices remain unitless.

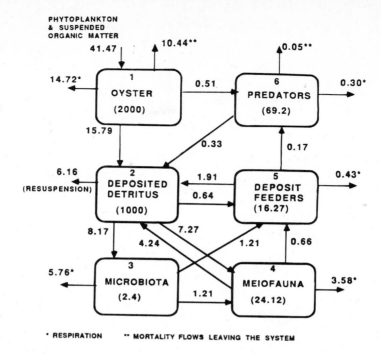

Figure 19.5 The oyster reef ecosystem model. Flow units are kcal m^{-2} day^{-1}. The stock units are kcal m^{-2}.

Table 19.6 contains the most direct comparison of the two approaches. We changed the net output vector by 1 in all categories. The response of the flow analysis approach was an increased input of phytoplankton and suspended material to the oysters of 10.95 kcal m^{-2} day^{-1}, 6.0 of these units being a response to change in net output and 4.95 being increased waste heat generated to meet the change. The response from compartmental analysis was not physically possible to obtain. Changes in phytoplankton and suspended matter absorption by *all* processes could not be determined. Therefore, the compartmental analysis approach could not be used to predict responses to net output changes irrespective of the column summing problems discussed earlier. The retrospective (flow analysis) method does give reasonable results in response to changes in net output.

When the steady-state oyster reef ecosystem is perturbed by an input increase of 1.0, flow analysis indicates that oysters would increase their net output by 0.823 and their waste heat output by 0.177, for a total P^N increase of 1.0, with none of the other processes being affected. This result is physically possible, but it seems unlikely that none of the other processes would remain unchanged by the net input increase. If the oysters consumed inputs from other parts of the system, then the flow analysis response to the net input increase would have come from every process in the system. The response of the compartmental analysis approach was more reasonable, with a positive increase in the P^N in every compartment. The largest response is from the oysters (1.0), in agreement with flow analysis. The detritus compartment responded with a total output increase larger than half that of the oysters.

It is important to note that in Table 19.6 the calculation of the ΔP^N does not require the column summing noted in eq. (19.6). Column summing would occur in static compartmental analysis only if the *e* contained net inputs of *various* commodities. Look at eq. (19.9) in another way. Because of the formation of constant G^*, the problematic summing down columns is done by vector

TABLE 19.1 Oyster reef model input–output flow matrix (*P*), along with vectors for net export plus stock replacement respiration (*r*), total output excluding waste heat (*p*), waste heat (*w*), and total output including waste heat (*p^N*)

	\multicolumn{6}{c}{*P*}						
	(1)	(2)	(3)	(4)	(5)	(6)	r	p	w	p^N
Oysters (1)	0	15.79	0	0	0	0.51	17.80	34.10	7.365	41.47
Detritus (2)	0	0	8.17	7.27	0.64	0	6.19	22.27	0	22.27
Microbiota (3)	0	0	0	1.21	1.21	0	2.875	5.295	2.875	8.17
Meiofauna (4)	0	4.24	0	0	0.66	0.17	1.75	6.65	1.75	8.4
Deposit feeders (5)	0	1.91	0	0	0	0	0.215	2.295	0.215	2.51
Predators (6)	0	0.33	0	0	0	0	0.2	0.53	0.15	0.68
Net input (*e*)	41.47	0	0	0	0	0				

TABLE 19.2 Coefficient matrix *G* for the oyster reef model; flow analysis approach

	(1)	(2)	(3)	(4)	(5)	(6)
Oysters (1)	0	0.709	0	0	0	0.962
Detritus (2)	0	0	1.543	1.093	0.279	0
Microbiota (3)	0	0	0	0.182	0.527	0
Meiofauna (4)	0	0.190	0	0	0.288	0
Deposit feeders (5)	0	0.086	0	0	0	0.321
Predators (6)	0	0.015	0	0	0	0

See eq. (19.4).

TABLE 19.3 Matrix $(I - G)^{-1}$ for the oyster reef model; flow analysis approach

	(1)	(2)	(3)	(4)	(5)	(6)
Oysters (1)	1.000	1.198	1.855	1.646	1.782	1.534
Detritus (2)	0.000	1.656	2.556	2.276	2.464	0.790
Microbiota (3)	0.000	0.144	1.223	0.380	0.794	0.255
Meiofauna (4)	0.000	0.358	0.553	1.493	0.821	0.263
Deposit feeders (5)	0.000	0.150	0.231	0.206	1.223	0.392
Predators (6)	0.000	0.025	0.038	0.034	0.037	1.012
(ϵ)	1.220	1.435	3.417	2.709	3.176	2.353

See eq. (19.5). Intensities, ϵ, calculated from eq. (19.10).

TABLE 19.4 Coefficient matrix G^* for the oyster reef model; compartmental analysis approach

	(1)	(2)	(3)	(4)	(5)	(6)
Oysters (1)	0	0.381	0	0	0	0.012
Detritus (2)	0	0	0.367	0.326	0.029	0
Microbiota (3)	0	0	0	0.148	0.148	0
Meiofauna (4)	0	0.505	0	0	0.079	0
Deposit feeders (5)	0	0.761	0	0	0	0.068
Predators (6)	0	0.485	0	0	0	0

See eq. (19.8).

TABLE 19.5 Matrix $(I - G^*)^{-1}$ for the oyster reef model; compartmental analysis approach

	(1)	(2)	(3)	(4)	(5)	(6)
Oysters (1)	1	0.537	0.197	0.204	0.061	0.016
Detritus (2)	0	1.389	0.509	0.529	0.157	0.011
Microbiota (3)	0	0.279	1.102	0.254	0.191	0.013
Meiofauna (4)	0	0.780	0.286	1.297	0.166	0.011
Deposit feeders (5)	0	1.102	0.404	0.420	1.124	0.076
Predators (6)	0	0.674	0.247	0.257	0.076	1.005

See eq. (19.9).

TABLE 19.6 Comparison of steady-state responses of compartmental versus flow analysis to changes in net input and net output

	Change in net output r or net input e:	
	$\Delta r = (1, 1, 1, 1, 1, 1)$	$\Delta e = (1, 0, 0, 0, 0, 0)$
Response from:		
Flow analysis	$\Delta e = (10.95, 0, 0, 0, 0, 0)$ from eq. (19.15)	$\Delta p^N = (1, 0, 0, 0, 0, 0)$ from eqs. (19.5) and (19.15)
Compartmental analysis	$\Delta e = (2.39, -10.25, 1.52, 3.20, 10.77, 2.58)$ from eq. (19.16)	$\Delta p^N = (1, 0.537, 0.197, 0.204, 0.061, 0.016)$ from eq. (19.9)

e. In the above example, there was only *one* entry in *e*. This single entry means there is no column summing. In fact, if this net input were used by other processes in the oyster reef system, column summing with *e* would be an acceptable procedure. But, in general in ecosystems, there may be many different types of net inputs (for example, sunlight and phosphorus). To calculate the weights (intensities) that allow these different net inputs to have the same relative value, we must apply the flow analysis approach to the largest possible ecosystem, the biosphere. At this level of analysis, there is only one net input to the ecosystem, sunlight. A commensurable *e* can be found for the analysis of subecosystems, but only with the assumption of constant G, which conflicts theoretically with the assumption of constant G^* in the compartmental analysis approach. For this reason (and because P^N contains both high-quality flows and low-quality waste flows), future-oriented compartmental analysis is not useful. Past-oriented flow analysis would not be useful either if we could not calculate the relative weights to place on the variety of net inputs that occur in the general model of a specific ecosystem. We must, therefore, demonstrate the method for calculating intensities at the biosphere or global level, since such quantities are needed to weight the multiple net inputs to subecosystems for application of the flow analysis approach to them.

HIERARCHICAL NET INPUT INTENSITIES USING A GLOBAL MULTICOMMODITY INPUT–OUTPUT MODEL

Multicommodity flow analysis is employed to calculate the net input intensity factors that can then be used to weight different commodities to make them commensurable. If there is more than one net input to the system, however, this method does not yield complete weightings. For example, consider an ecosystem with boundaries defined so that there are net inputs of both sunlight and water. In such a case we could calculate sunlight intensities *and* water intensities, but would have no way of comparing or adding the two. One approach to this problem is to define a hierarchical series of models with ever more expansive boundaries. At the level of the whole globe, there is effectively only one net input—sunlight. A global multicommodity flow analysis model can thus be used to evaluate the relative solar energy intensities of commodities at this scale. These estimates can then be used at lower levels to put different net inputs into the same units. If we can calculate the solar energy cost of fresh water via the hydrologic cycle at the global level, we can use that information to equilibrate net inputs of water and sunlight at the ecosystem level. We can speak of the net input of sunlight and the sunlight "embodied" in the water and produce a single set of more detailed net input intensities at the ecosystem level.

Tables 19.7 and 19.8 are examples of a global multicommodity flow table set up in the multiple-output, commodity-process format discussed earlier. Table 19.7 shows inputs of the listed commodities along the left to processes listed along the top. This is the U matrix. Note that the units for the commodities are different (they are not commensurable), so we cannot add down the columns of U. The outputs of each commodity from each process are listed in Table 19.8. This is the V matrix. Note that conservation holds for each commodity. The total output of each commodity from all the processes that produce it (the row sum of V) is equal to the total input of that commodity to all the processes that use it (the row sum of U), including the amount of the commodity that is "depreciated" or exported (r). Processes (columns in V) that contain more than one entry represent "joint products." At the global level, joint products are significant and unavoidable.

Using the methods described earlier, we can calculate the direct and indirect use matrix, $(V - U)^{-1}$, and the net input (in this case sunlight) intensity factors, ϵ. This matrix and vector are shown in Table 19.9. Note that Table 19.9 contains many negative values. This reflects the fact that joint products are significant. There is no guarantee that using this method will produce all positive values in ϵ. The appropriate data of U and V may have to be adjusted to yield such a vector.

TABLE 19.7 Global multicommodity input ("use") matrix (U), along with the vectors for net export plus stock replacement (r), total output excluding waste heat (p), and the net input vector (sunlight, e)

	Processes										
Commodities	Urban economy (1)	Agriculture (2)	Natural plants (3)	Animals (4)	Soil (5)	Deep ocean (6)	Surface ocean (7)	Atmosphere (8)	Deep geology (9)	Net output r	Total output p
Manufactured goods (1)	2.71	0.08	0	0	0	0	0	0	0	1.19	3.98
Agricultural products (2)	1.28	4.55	0	3.27	0	0	0	0	0	0	9.1
Natural products (3)	1.18	0	0	27.9	103.4	34.6	0	0	0.16	0.06	167.3
Nitrogen (4)	55	62.4	208	0	493.6	0	168	389.5	0	0	1376.5
Carbon (5)	0	8.2	147	0	0	15.6	37.2	110.3	0	0	318.3
Phosphorus (6)	12.6	28.5	1345.7	0	8.4	0	21	9.5	13	0	1438.7
Water vapor (7)	0	0	0	0	0	0	0	496,100	0	0	496,100
Fresh water (8)	1008	15,490	51,226	0	111,419	0	424,700	0	0	2000	605,843
Fossil fuels (9)	5	0	0	0	0	0	0	0	0	−4.93	0.07
Net input (sunlight) (e)	0	23	227	0	0	0	606	0	0		

Units are (1) manufactured goods, 10^{12} \$/y; (2) agricultural products, 10^{15} g dry wt/y; (3) natural products, 10^{15} g dry wt/y; (4) nitrogen, 10^{12} gN/y; (5) carbon, 10^{15} gC/y; (6) phosphorus, 10^{12} gP/y; (7) water vapor; km^3/y; (8) fresh water, km^3/y; (9) fossil fuel, 10^{15} gC/y; (e) sunlight, 10^{18} kcal/y. Complete references are given in Costanza and Neill (1981).

TABLE 19.8 Global multicommodity output ("make") matrix (V), along with the vector for total input

	Processes									
Commodities	Urban economy (1)	Agriculture (2)	Natural plants (3)	Animals (4)	Soil (5)	Deep ocean (6)	Surface ocean (7)	Atmosphere (8)	Deep geology (9)	Total input
Manufactured goods (1)	3.98	0	0	0	0	0	0	0	0	3.98
Agricultural products (2)	0	9.1	0	0	0	0	0	0	0	9.1
Natural products (3)	0	0	163.4	3.9	0	0	0	0	0	167.3
Nitrogen (4)	80	31	0	295	340.5	0	182	448	0	1376.5
Carbon (5)	5	6.1	73.6	14	46.5	15.6	49.5	108	0	318.3
Phosphorus (6)	14.2	0	0	0	241.3	1161.1	0	9.5	12.6	1438.7
Water vapor (7)	79	5931	50,740	0	14,650	0	424,700	0	0	496,100
Fresh water (8)	929	9829	0	0	98,985	0	0	496,100	0	605,843
Fossil fuels (9)	0	0	0	0	0	0	0	0	0.07	0.07

Units are the same as in Table 19.7.

TABLE 19.9 Matrix $(V - U)^{-1}$ for the U and V matrices shown in Tables 19.7 and 19.8

	Processes									Net input (sunlight) intensities (ϵ)
Commodities	Urban economy (1)	Agriculture (2)	Natural plants (3)	Animals (4)	Soil (5)	Deep ocean (6)	Surface ocean (7)	Atmosphere (8)	Deep geology (9)	
Manufactured goods (1)	0.79869	0.01329	−0.00029	0.00018	−0.00112	−0.00001	0.00002	0.00002	−0.00070	190.365
Agricultural products (2)	0.17925	0.21105	−0.00453	0.00281	−0.01771	−0.00013	0.00037	0.00037	−0.01112	13.997
Natural products (3)	0.07260	−0.02471	0.00916	0.00063	−0.00336	0.00027	0.00051	0.00051	0.02249	39.158
Nitrogen (4)	−0.06323	0.01735	−0.00619	0.00384	−0.02421	−0.00018	0.00050	0.00050	−0.01520	0.630
Carbon dioxide (5)	−0.00592	−0.03113	0.00256	−0.00020	0.00142	0.00008	0.00052	0.00052	0.00629	57.123
Phosphorus (6)	0.11335	−0.01646	0.01109	0.00086	−0.00289	0.00119	0.00051	0.00051	0.03216	1.167
Water vapor (7)	0.28014	0.02434	0.06136	0.00070	0.09619	0.00183	0.00070	0.00070	0.15070	0.550
Fresh water (8)	0.24934	0.01992	0.05349	0.00069	0.08183	0.00159	0.00067	0.00067	0.13136	0.550
Fossil fuels (9)	57.04948	0.94962	−0.02038	0.01265	−0.07970	−0.00061	0.00165	0.00165	14.23566	96.171

Entries in this matrix indicate the *direct* and *indirect* inputs (or outputs if negative) of the row commodities to (or from) the column processes. Net input (solar energy) intensities for the commodities based on this matrix and the direct sunlight input vector listed in Table 19.7 [using eq. (19.14)] are shown as the last column on the right.

TABLE 19.10 Evaluated U matrix in embodied solar energy units

	Processes								
Commodities	Urban economy (1)	Agriculture (2)	Natural plants (3)	Animals (4)	Soil (5)	Deep ocean (6)	Surface ocean (7)	Atmosphere (8)	Deep geology (9)
Manufactured goods (1)	515.89	15.23	0.00	0.00	0.00	0.00	0.00	0.00	0.00
Agricultural products (2)	17.92	63.69	0.00	45.77	0.00	0.00	0.00	0.00	0.00
Natural products (3)	46.21	0.00	0.00	1092.50	4048.91	0.00	0.00	0.00	6.27
Nitrogen (4)	34.65	39.31	131.03	0.00	310.95	1354.86	105.83	245.37	0.00
Carbon dioxide (5)	0.00	468.41	8397.05	0.00	0.00	0.00	2124.97	6300.64	0.00
Phosphorus (6)	14.70	33.26	1570.26	0.00	9.80	891.12	24.50	11.09	15.17
Water vapor (7)	0.00	0.00	0.00	0.00	0.00	0.00	0.00	273,068.23	0.00
Fresh water (8)	555.03	8529.11	28,206.08	0.00	61,349.57	0.00	233,848.46	0.00	0.00
Fossil fuels (9)	480.86	0.00	0.00	0.00	0.00	0.00	0.00	0.00	0.00
Net input (sun) (ϵ)	0.00	23.00	227.00	0.00	0.00	0.00	606.00	0.00	0.00
Total inputs	1665.24	9172.00	38,531.42	1138.27	65,719.23	2245.97	236,709.77	279,625.32	21.43

Each element in this matrix is calculated by multiplying the physical flow listed in Table 19.7 by the commodity's energy intensity (ϵ) listed in Table 19.9.

Tables 19.10 and 19.11 are the evaluated U and V matrices [U^* and V^* using eqs. (19.15) and (19.16)] in embodied solar energy units. Note that the total inputs to processes listed in Table 19.10 equal the total outputs from processes listed in Table 19.11. Table 19.12 is the commodity input distribution matrix (Q) calculated from V^* using eq. (19.21).

Table 19.13 is the commodity-by-commodity flow matrix (P) calculated from U and Q employing eq. (19.22). Table 19.13 shows the amount of each row commodity required to produce a unit of column commodity in the global process taken as a whole. This matrix can then be used to generate the commodity coefficient matrix (G) and the total requirements matrix $(I - G)^{-1}$ for the global system.

In Table 19.13, each commodity is measured in its own units. It is a simple matter to convert this table into a commensurable table by multiplying through each commodity by its energy intensity. The result is Table 19.14, which is a commodity by commodity flow matrix with all flows expressed in embodied solar energy units. Since column sums equal row sums for this matrix, forward compartmental analysis is directly applicable [the G^* and $(I - G^*)^{-1}$ matrices can be calculated from Table 19.14].

There is an additional way of looking at this system to make it a closer analog to the usual starting point for compartmental analysis. Table 19.14 shows the interdependence among commodities with the global process taken as a whole. We might also want to investigate the interdependence among the different global processes with the commodities aggregated and treated as a single commodity. To do this, we apply eq. (19.22) to the global U and Q^* matrices and multiply through by the energy intensities. The results are shown in Table 19.15. This matrix indicates the flow of a single hybrid commodity (again measured in embodied solar energy units) among the global processes. This hybrid commodity incorporates the interdependencies in the disaggregated commodities through the calculation of the energy intensities.

Tables 19.14 and 19.15 are interesting for what they say about the relative importance of the various commodities and processes in the global system. For example, Table 19.14 indicates the importance of fresh water as an input to the production of manufactured goods. Fresh water represents 252.24 of a total of 757.65×10^{18} solar calories/y (or 33%) of direct inputs to manufactured goods. This is slightly more than fossil fuels (29%) and turns out to be the most significant single input. Fresh water is an even more significant input to agricultural products and natural products. The total input row in Table 19.14 indicates the relative importance of the hydrologic and carbon cycles in the global system. Water is three and carbon two orders of magnitude larger than manufactured goods in terms of total inputs.

Table 19.15 yields a related but unique picture. Here the surface ocean and atmosphere (the two main processes in the hydrologic cycle) dominate the total input and output values. The urban economy is shown to be very dependent on inputs from soil, atmosphere, and deep geology (for fossil fuels).

Whether we use the commodity-by-commodity (Table 19.14) or process-by-process (Table 19.15) matrices as the starting point for further analysis depends on what questions are of interest. If the questions center on the interdependence between commodities (for example, the effect of changes in the carbon cycle on the nitrogen cycle), then the commodity-by-commodity matrix is most appropriate. If the questions center on the interdependence among processes (for example, the degree of dependence of the shallow ocean on land-based processes), then the process-by-process matrix is most appropriate. For calculating total system interdependency measures, such as Finn's (1976) cycling index or Ulanowicz's (1980) ascendancy (see Chapter 26), it is not clear which starting point is preferable, and it would be interesting to compare the two.

SUMMARY AND CONCLUSIONS

In this paper we have compared "look-forward" compartmental and "look-backward" flow analysis and have shown the latter to be more appropriate if the system is not in commensurable units. Realistic analysis of ecosystems must always include several different commodities measured in

TABLE 19.11 Evaluated V matrix in embodied solar energy units

					Processes				
Commodities	Urban economy (1)	Agriculture (2)	Natural plants (3)	Animals (4)	Soil (5)	Deep ocean (6)	Surface ocean (7)	Atmosphere (8)	Deep geology (9)
Manufactured goods (1)	757.65	0.00	0.00	0.00	0.00	0.00	0.00	0.00	0.00
Agricultural products (2)	0.00	127.37	0.00	0.00	0.00	0.00	0.00	0.00	0.00
Natural products (3)	0.00	0.00	6398.38	152.72	0.00	0.00	0.00	0.00	0.00
Nitrogen (4)	50.40	19.53	0.00	185.84	214.50	0.00	114.65	282.22	0.00
Carbon dioxide (5)	285.61	348.45	4204.24	799.72	2656.21	891.12	2827.58	6169.26	14.70
Phosphorus (6)	16.57	0.00	0.00	0.00	281.57	1354.86	0.00	11.09	0.00
Water vapor (7)	43.48	3264.60	27,928.81	0.00	8063.80	0.00	0.00	0.00	0.00
Fresh water (8)	511.53	5412.05	0.00	0.00	54,503.16	0.00	233,767.54	273,162.76	6.73
Fossil fuels (9)	0.00	0.00	0.00	0.00	0.00	0.00	0.00	0.00	21.43
Total outputs	1665.24	9172.00	38,531.42	1138.27	65,719.23	2245.97	236,709.77	279,625.32	21.43

Each element in this matrix is calculated by multiplying the physical flow listed in Table 19.8 by the commodity's energy intensity (e) listed in Table 19.9.

TABLE 19.12 Commodity input distribution matrix (Q)

					Processes				
Commodities	Urban economy (1)	Agriculture (2)	Natural plants (3)	Animals (4)	Soil (5)	Deep ocean (6)	Surface ocean (7)	Atmosphere (8)	Deep geology (9)
Manufactured goods (1)	0.454981	.000000	.000000	.000000	.000000	.000000	.000000	.000000	.000000
Agricultural products (2)	0.000000	.013887	.000000	.000000	.000000	.000000	.000000	.000000	.000000
Natural products (3)	0.000000	.000000	.166056	.134164	.000000	.000000	.000000	.000000	.000000
Nitrogen (4)	0.030264	.002129	.000000	.163263	.003264	.000000	.000484	.001009	.000000
Carbon dioxide (5)	0.171515	.037991	.109112	.702573	.040418	.396761	.011945	.022063	.000000
Phosphorus (6)	0.009950	.000000	.000000	.000000	.004284	.603239	.000000	.000040	.685929
Water vapor (7)	0.026113	.355931	.724832	.000000	.122701	.000000	.987570	.000000	.000000
Fresh water (8)	0.307178	.590062	.000000	.000000	.829333	.000000	.000000	.976888	.000000
Fossil fuels (9)	0.000000	.000000	.000000	.000000	.000000	.000000	.000000	.000000	.314071

Each element in this matrix indicates the percentage of the total inputs to the column process to be allocated to the production of the row commodity.

TABLE 19.13 Commodity-by-commodity flow matrix (P) calculated from the U matrix in Table 19.7 and the Q matrix in Table 19.12 ($P = UQ^t$)

Commodities	Manufactured goods (1)	Agricultural products (2)	Natural products (3)	Nitrogen (4)	Carbon dioxide (5)	Phosphorus (6)	Water vapor (7)	Fresh water (8)	Fossil fuels (9)
Manufactured goods (1)	1.23	0.00	0.00	0.08	0.47	0.03	0.10	0.88	0.00
Agricultural products (2)	0.58	0.06	0.44	0.58	2.69	0.01	1.65	3.08	0.00
Natural products (3)	0.54	0.00	3.74	4.93	37.71	21.44	12.72	86.12	0.05
Nitrogen (4)	25.02	0.87	34.54	3.88	65.05	2.68	400.89	843.57	0.00
Carbon dioxide (5)	0.00	0.11	24.41	0.15	25.42	9.41	146.21	112.59	0.00
Phosphorus (6)	5.73	0.40	223.46	0.49	150.88	9.08	1007.65	36.93	4.08
Water vapor (7)	0.00	0.00	0.00	500.70	10,945.25	19.67	0.00	484,634.38	0.00
Fresh water (8)	458.62	215.11	8506.39	632.85	15,927.19	487.39	475,762.24	101,853.20	0.00
Fossil fuels (9)	2.27	0.00	0.00	0.15	0.86	0.05	0.13	1.54	0.00

The elements in this matrix indicate the amounts of the row commodities required *directly* to produce the column commodities. The units are the same as in Table 19.7, as are the net output vector (r) and the total output vector (p).

TABLE 19.14 Global commodity-by-commodity flow matrix in embodied solar energy units (10^{18} embodied solar calories/year)

Commodities	Manufactured goods (1)	Agricultural products (2)	Natural products (3)	Nitrogen (4)	Carbon dioxide (5)	Phosphorus (6)	Water vapor (7)	Fresh water (8)	Fossil fuels (9)	Net output	Total output
Manufactured goods (1)	234.15	0.00	0.00	15.23	89.47	5.71	19.04	167.52	0.00	226.53	757.65
Agricultural products (2)	8.12	0.84	6.16	8.12	37.65	0.14	23.10	43.11	0.00	0.00	127.23
Natural products (3)	21.15	0.00	146.45	193.05	1476.65	839.55	498.09	3372.29	1.96	2.35	6551.52
Nitrogen (4)	15.76	0.55	21.76	2.44	40.98	1.69	252.56	531.45	0.00	0.00	867.20
Carbon dioxide (5)	0.00	6.28	1394.37	8.57	1452.07	537.53	8351.95	6431.48	0.00	0.00	18,182.25
Phosphorus (6)	6.69	0.47	260.78	0.57	176.08	10.60	1175.93	43.10	4.76	0.00	1678.96
Water vapor (7)	0.00	0.00	0.00	275.39	6019.89	10.82	0.00	266,548.91	0.00	0.00	272,855.00
Fresh water (8)	252.24	118.31	4678.51	348.07	8759.95	268.06	261,669.23	56,019.26	0.00	1191.57	333,305.21
Fossil fuels (9)	218.31	0.00	0.00	14.43	82.71	4.81	12.50	148.10	0.00	0.00	480.86
Net input (sun) (e)	1.24	0.78	43.49	1.34	46.81	0.06	852.60	0.00	474.14		
Total input	757.65	127.23	6551.52	867.20	18,182.25	1678.96	272,855.00	333,305.21	480.86		

TABLE 19.15 Global process-by-process flow matrix in embodied solar energy units (10^{18} embodied solar calories/year)

					Processes						
Processes	Urban economy (1)	Agriculture (2)	Natural plants (3)	Animals (4)	Soil (5)	Deep ocean (6)	Surface ocean (7)	Atmosphere (8)	Deep geology (9)	Net output	Total output
Urban economy (1)	518.88	37.99	197.30	0.00	110.19	13.99	390.53	167.90	0.15	229.89	1666.82
Agriculture (2)	29.71	244.43	734.68	45.77	1233.21	17.82	4721.59	2861.60	0.00	0.00	9888.81
Natural plants (3)	45.15	108.20	1939.72	1067.37	3955.79	1529.55	490.87	29,308.41	6.13	153.21	38,604.38
Animals (4)	8.20	26.99	363.40	21.85	146.28	62.74	107.22	303.55	0.13	94.63	1134.99
Soil (5)	101.50	1473.93	6119.73	0.00	10,078.43	130.10	38,457.80	9174.41	2.55	380.92	65,919.37
Deep ocean (6)	11.86	49.79	1678.66	0.00	7.91	43.66	123.90	317.68	12.24	0.00	2245.70
Surface ocean (7)	4.57	78.26	1327.24	0.00	41.05	139.01	345.46	234,761.69	0.00	823.14	237,520.43
Atmosphere (8)	465.93	7157.14	26,000.96	0.00	50,346.43	302.09	192,276.82	2215.74	0.11	345.89	279,111.10
Deep geology (9)	481.01	0.33	15.70	0.00	0.10	0.00	0.25	0.11	0.15	0.00	497.65
Net input (e)	0.00	711.75	227.00	0.00	0.00	6.73	606.00	0.00	476.20		
Total inputs	1666.82	9888.81	38,604.38	1134.99	65,919.37	2245.70	237,520.43	279,111.10	497.65		

noncommensurable units. Retrospective flow analysis can be used to derive a set of unique intensity factors that perform the same function as prices in economic systems. They allow physical flows to be put in commensurable units and the "mixed units" problem to be solved.

We have also elaborated methods to deal with hierarchical intensity factors and joint products in ecosystems and to convert commodity-by-process, joint-product systems to equivalent commodity-by-commodity, single-process systems and process-by-process, single-commodity systems. We have given a detailed description of how this works out for the global system. These methods, properly applied, can allow more realistic and comprehensive analyses of ecosystem interdependencies than are currently available.

APPENDIX A: CALCULATION DETAILS FOR THE NONNEGATIVE COMMODITY-BY-COMMODITY PRODUCTION MATRIX

From eq. (19.15), with $\dot{P} = 0$:

$$p = UV^{-1}p + r \tag{A1}$$

$$= U\hat{g}^{-1}\hat{g}V^{-1}p + r, \tag{A2}$$

where g is the vector of column sums. Thus,

$$p = U\hat{g}^{-1}(V\hat{g}^{-1})^{-1} p + r \tag{A3}$$

$$= BC^{-1}p + r. \tag{A4}$$

Matrix BC^{-1} is the definition of the "commodity-technology" assumption (Stone, 1963). With it, we assume that, regardless of where a commodity is made, its inputs per unit output are determined by the inputs per unit output of the sector where that commodity is the main output. Note that although g is contained in BC^{-1} it cancels out in UV^{-1}. Equation (A4) is similar to the flow analysis result, eq. (19.5). A somewhat symmetrical assumption is also possible. We return to a commodity balance equation that is more basic than (A1),

$$p = Ui + r, \tag{A5}$$

augment it,

$$p = U\hat{g}^{-1}g + r, \tag{A6}$$

and then realize that a matrix D^T can be defined as

$$D^T = V^T\hat{p}^{-1} \tag{A7}$$

since

$$g = V^Ti = D^Tp. \tag{A8}$$

Therefore, eq. (A6) can be written as

$$p = U\hat{g}^{-1}D^Tp + r, \tag{A9}$$

which is called the "industry-technology" assumption (Stone, 1963). This means that the inputs per unit output of a commodity are defined differently depending on the process that made the commodity. This assumption collects all such outputs together in a fictitious single-commodity output process.

But, as stated in the text, the g vector cannot be formed unless V is converted to commensurable units. If the conversion factors are the ϵ from eq. (19.16), two important things happen: first the ϵ that are produced from the "industry-technology" assumption are identical to those produced by eq. (19.16) (Hannon et al., 1986); second, eq. (A9) can be logically converted to commensurable units to produce a nonnegative commodity-by-commodity production matrix. If we completely convert eq. (A9) with the known ϵ,

$$\hat{\epsilon}p = \hat{\epsilon}U\epsilon\hat{V}^{-1}V^T\hat{\epsilon}\hat{\epsilon}^{-1}p\hat{\epsilon}p + \hat{\epsilon}r, \quad (A10)$$

which reduces to

$$p = U\epsilon\hat{V}^{-1}V^T\hat{\epsilon}p + r, \quad (A11)$$

the new commodity-by-commodity (nonnegative elements, single-commodity output per process) production matrix is

$$P = U\epsilon\hat{V}^{-1}V^T\hat{\epsilon} = UQ^T. \quad (A12)$$

This is the desired all-positive production matrix [see eq. (19.21)]. The similar matrix from the "commodity-technology" assumption is UV^{-1}, which may have negative elements due to the inversion of V.

REFERENCES

BARBER, M. C., PATTEN, B. C., and FINN, J. T. 1979. Review and evaluation of input–output flow analysis for ecological applications. In Matis, J. A., Patten, B. C., and White, G. C. (eds.), *Compartmental Analysis of Ecosystem Models*, Vol. 10, Statistical Ecology. International Cooperative Publishing House, Bertonsville, Maryland, pp. 43–72.

COSTANZA, R., and NEILL, C. 1981. The energy embodied in the products of the biosphere. In Mitsch, W. J., Bosserman, R. W., and Klopatek, J. M. (eds), *Energy and Ecological Modeling*. Elsevier, Amsterdam, pp. 745–755.

COSTANZA, R., and NEILL, C. 1984. The energy embodied in the products of ecological systems: a linear programming approach. *J. Theor. Biol.* 106:41–57.

DAME, R., and PATTEN, B. C. 1981. Analysis of energy flows in an intertidal oyster reef. *Marine Ecol. Progr. Ser.* 5:115–24.

FINN, J. T. 1976. Measures of ecosystem structure and function derived from analysis of flows. *J. Theor. Biol.* 56:363–73.

FUNDERLIC, R., and HEATH, M., 1971. Linear Compartmental Analysis of Ecosystems. Oak Ridge Nat. Lab. ORNL-IBP-71-4.

HANNON, B. 1973. The structure of ecosystems. *J. Theor. Biol.*, 41:535–546.

HANNON, B. 1976. Marginal product pricing in the ecosystem. *J. Theor. Biol.* 56:256–267.

HANNON, B. 1979. Total energy costs in ecosystems. *J. Theor. Biol.* 80:271–293.

HANNON, B. 1985a. Ecosystem flow analysis. *Can. J. Fish. Aquatic Sci.* 213:97–118.

HANNON, B. 1985b. Conditioning the ecosystem. *Math. Biol.* 75:23–42.

HANNON, B. 1985c. Linear dynamic ecosystems. *J. Theor. Biol.* 116:89–98.

HANNON, B. 1986. Ecosystem control theory. *J. Theor. Biol.* 121:417–437.

HANNON, B., COSTANZA, R., and HERENDEEN, R. 1986. Energy cost and value in ecosystems. *J. Env. Econ. Mgmt.* 13:391–401.

HETT, J., and O'NEILL, R. V. 1971. Systems Analysis of the Aleut Ecosystem. US-IBP, Deciduous Forest Biome Memo Report, 71-16. Oak Ridge National Laboratory, Oak Ridge, Tennessee.

HIPPE, P. 1983. Environ analysis of linear compartmental systems: the dynamic time invariant case. *Ecol. Mod.* 19:1–26.

KERCHER, J. 1983. Closed-form solutions to sensitivity equations in the frequency and time domains for linear models of ecosystems. *Ecol. Mod.* 18:209–221.

LEONTIEF, W. W. 1941. *The Structure of American Economy, 1919–1929: An Empirical Application of Equilibrium Analysis.* Harvard University Press, Cambridge, Massachusetts. 181 pp.

PATTEN, B. C. 1978. Systems approach to the concept of environment. *Ohio J. Sci.* 78:206–222.

PATTEN, B. C. 1982. Environs: relativistic elementary particles for ecology. *Amer. Nat.* 119:179–219.

SAMUELSON, P. A. 1966. A new theorem on nonsubstitution. In Stiglitz, J. (ed.), *The Collected Scientific Papers of Paul A. Samuelson.* Vol. 2. MIT Press, Cambridge, Massachusetts, Chap. 37, pp. 881–886.

STONE, R. 1963. *Input–Output Relationships 1954–1966. A Programme for Growth.* MIT Press, Cambridge, Massachusetts, pp. 1–41.

ULANOWICZ, R. E. 1980. An hypothesis on the development of natural communities. *J. Theor. Biol.* 85:223–245.

VERNBERG, F. J. (ed.). 1975. *Physiological Ecology of Estuarine Organisms.* University of South Carolina Press, Columbia, South Carolina, 397 pp.

Part 4

COMPLEX ECOLOGICAL ORGANIZATION: NETWORK TROPHIC DYNAMICS

The first passing in the world of a systems ecologist who was in his time a peerless advocate of complex approaches to complex ecology is cause for asking why we do research on whole ecosystems anyway, since it is so obviously difficult. If a prodigious personality like George Van Dyne could not cope with ecological holism, then who can? Is it really needed? Must science continue to pursue this exercise in futility? I (the editor) found myself asking this question in the long aftermath of George's death. I had worked for the entire month of June during that summer side by side in Zagreb with Tarzan Legović producing a qualitative algebra of indirect effects. Causality at a distance was fresh in my mind when I had to ask myself later that summer about the meaning of George's scientific life—and mine, and others'. What intuition drove us to want to know about ecology in holistic terms when most of the rest of our field was content to move along the nonsystems front of filling in the blanks of post-Darwinian mechanistic reductionism?

The life commitment of Van Dyne, and those of us who would know about holism in ecology, deserves an answer to this question: "Is 'systems study' really necessary in science, and if so, why?" The answer that gradually came to me was an unequivocal "yes," and the essential reason was "influence"—a word used by one of my students (James Hill) as the subject for a theoretical dissertation so esoteric (and fascinating) that there is no place to accommodate it yet in the ecology of today (see *Ecol. Mod.* 28:59–65 for a brief description); I became convinced there is true mystery in the connectivity of interrelated life that nonsystems science could never reveal. *Influence*—network indirect effects or cause at a distance. Every person who has ever engaged in the essential political act of saying *a*, meaning *b*, and achieving *c* knows about it. Indirect effects seem a central fact of daily life, but if they can somehow be made unimportant in general determination, then there would be no need to study systems as wholes. Science could continue to dissect complexity down to its elemental essentials and then simply fuse the whole together from a superposition of known parts. Both scientists and humanity would have it easier if causality could only be a local affair, if its ramifications to far corners and futures of systems could be truncated by processes and designs to limit spreading.

However, it is only necessary to call up the specter of radioactivity, toxic chemicals, and epidemic disease in the environment, or to take note of concepts like emergent properties or the Gaia hypothesis, to realize that causal flux in the world, although delayable, is ultimately pandemic unless or until dissipative phenomena bring it to ground. Complex ecology probably involves the whole planet as an interconnected system of hierarchically ordered subsystems at all levels of organization inside and outside the organism. Real determination is most probably the reticulated *holistic determination* (M. Bunge, *Causality*, Harvard University Press, 1959) of parts by wholes, in which indirect causes and effects acting at a "temporal" and "procedural" distance in networks of interactions enmesh and enslave components and become dominant in determination (M. Higashi and B. Patten, *Amer. Nat.* 133:288). In indirect effects is the core issue of holism versus reductionism in science—an issue requiring definitive resolution so that science can proceed with clear knowledge that "the systems approach" is or is not necessary. Bringing this issue into focus was a legacy of George Van Dyne's death. Science must determine if indirect effects are or are not significant in biodetermination in nature.

In his IBP Grasslands program, George Van Dyne established a network of research sites and scientists. An interdisciplinarian, his mind was a network of interconnected knowledge from many subjects. In scientific management, he knew the value of networking with others; without somehow interfacing a diversity of disciplines, synergism as an emergent property of systems could not emerge. Although the word "network" only infrequently appeared in his writings, he was a past master at applying the concept and its manifestation "influence," both scientifically and managerially.

Chapter 20

Trophic dynamics is probably the one traditional area of ecosystem ecology that has most given rise to a network focus. A food web is after all a network of feeding interactions. It is the archetypical ecological network through which indirect effects are expressed in ecosystems.

In Chapter 20, Gary Polis demonstrates for a sand community in a relatively "simple" desert ecosystem of the Coachella Valley, California, that food networks are highly reticulated, complex structures. His straightforward descriptive challenge to the principal tenets of contemporary food web theory underscores the requirement for theorists to consult nature often and also puts aside the vain hope of many empiricists that the world is simple enough for revelation by experiment. The biota of Polis's ecosystem includes 174 species of vascular plants, 138 of vertebrates, more than 55 species of arachnids, and an unknown (but great) number of microorganisms, insects (2000 to 3000 species estimated), Acari, and nematodes. Trophic relations are presented in a series of nested subwebs and delineations of the community. Complexity arises from the large number of interactive species, the frequency of omnivory, population age structure, looping in direct feeding relations and cycling in indirect, lack of compartmentalization, and complexity of the arthropod and soil faunas. In general, diversity and complexity in nondesert habitats tend to be greater than in deserts, so the Coachella community represents a relatively simple one in the spectrum of natural systems.

Patterns from the Coachella web are compared with theoretical predictions and "empirical generalizations" derived from catalogues of published webs. The Coachella web differs in most particulars: food chains are longer, omnivory is common, loops and cycles are not rare, connectivity is greater (species interact with many more predators and prey), top predators are rare or nonexistent, and prey–predator ratios are greater than 1. The weight of evidence argues that actual community food webs are enormously more complex than those catalogued as the basis for many theoretical analyses. Other workers have also reached similar conclusions. Polis argues, therefore, that much of extant food web theory is not very descriptive or predictive of nature. Catalogued webs are grossly incomplete in terms of both species diversity and trophic connections. That these webs depict few species, absurdly low ratios of predators on prey and prey eaten by predators, few feeding links, little omnivory, a veritable absence of looping and cycling, and a high proportion of top predators negates them as anything but oversimplified caricatures. Consequently, the practice of abstracting empirical regularities from such catalogues yields an inaccurate and artifactual view of trophic interactions. Although many of the methods developed for theoretical analyses may ultimately prove useful, right now there is a great need in the field to carry out the painstaking work of realistic description. Currently, contrary to strong literature assertions, the real food webs of most ecological communities will not support predictions developed from simplistic dynamic and graphic models.

This conclusion carries several implications. First, controversies over such issues as the causes of short chain lengths and omnivory, complexity versus stability of ecosystems, and the role of dynamics versus energetics in shaping food web patterns are based on the characteristics of unrealistically simple webs. If existing catalogues represent an inadequate data base, the controversies may collapse to the nonissues of theorists trying to explain properties and phenomena that do not exist. Second, the dominant characteristics of the Coachella Valley food web (complexity, omnivory, long chain lengths, looping, absence of compartments) are the kind that reduce stability in studies of dynamic models of simple webs. But the Coachella Valley food network of organisms, or the ecosystem for which this web is the biological basis, is not demonstrably unstable. Simplified food web models apparently say little about the structure and function of real food webs in the behavioral characteristics of ecosystems. The same conclusion might be drawn about any effort of scientists to simplify what is inherently complex; if complexity is of the essence, it cannot be abstracted away.

So, what is the usefulness of food web representations and analyses? At a minimum, properly constructed webs are empirical descriptors of the trophic-interaction backbone of community organization. Properly quantified and analyzed, they may tell us further how communities are assembled; how their functions are preserved under structural changes accompanying succession or disturbance; how, even, reticulated direct and indirect causality contribute to the evolution of species, or higher assemblages. Hypotheses and generalizations of theorists represent, always, stages in the development of knowledge—a cybernetic (error-regulated) process in which error generation is just as important and fundamental as error correction. Theory is designed to provoke concept-focused empirical work that should promote the next iteration of descriptive models and generalizations, which, in turn, encourages further observations, and so on. Many of the hypotheses and general conclusions of current food web theory are simply wrong. In being wrong, they have value, however.

For example, current theory would denote the Coachella Valley system based on its properties as being completely unstable. It is not, but to determine why not an adequate data base quantifying the described web must be assembled, the system modeled, experiments conducted on these models and validated with field observations, and new questions asked. To build even a single realistic food web, qualitatively, is an enormous task. To then quantify the flows involved is even more demanding, requiring committed laboratories and investigators, students, and technicians doing the hard work of gut analyses and related activities over long periods of time. The overwhelming complexity of natural communities makes the construction and analysis of food webs by teams of empiricists and theorists a difficult undertaking, and one wonders if the motivation to do so has yet been provided by the science. Also, there is question whether the connection of food webs to more general causal networks determining the economy of the biosphere, with implications for humanity, can be made convincing enough to compel societies to commit resources to their detailed study. At present, the author concludes, much food web theory would appear to be in critical need of reevaluation, revision, and redirection guided by a more realistic awareness of the dictates and implications of complex ecology.

Chapter 21

In Chapter 21, Robert Ulanowicz also takes aim at the widespread conception of trophic simplicity by noting that the second law of thermodynamics combined with R. L. Lindeman's "image of feeding relations as a chain of trophic transfers possesses a descriptive elegance that is hard to resist.... Yet most ten-year old school children are taught that feeding relations in an ecosystem resemble more a complicated web than a simple chain of transfers." Citing S. Cousins's observation that the trophic concept is fundamentally wrong, Ulanowicz responds that it may be necessary to abandon any direct correspondence between taxa and trophic levels. But why cannot portions of taxa be allocated to different trophic levels? This is what the author sets out to do by a scheme he calls "trophic aggregation." The methodology utilizes the basic "structure matrix" of input–output analysis, which represents the convergent sum of a power series of fractional-flow matrices. The elements of the constituent power matrices denote quantities of transferred substance over paths as long as the exponents, providing a means to assess trophic levels. If the flow network under consideration is acyclic, it becomes possible to derive a "trophic transformation matrix" whose rows and columns indicate the distribution of taxa into trophic levels and, reciprocally, of trophic levels into taxa. An example is given to demonstrate this. The method implements a superb conception, but is limited to networks without cycles; and even Lindeman's basic model, which became abstracted into the more tractable food chain, was that of a "food cycle."

To try to solve the problem, Ulanowicz suggests "that the given arbitrary network be decomposed into two constitutive networks—one containing only cycled medium and another portraying only once-through flow. . . ." He uses a "backtracking" algorithm to identify cycles and peel them away from the network, leaving an acyclic graph tree that he compresses to form a Lindeman-type chain. The procedure is illustrated by a large model (17 compartments) of a tidal marsh gut in the Crystal River of Florida (USA); the cycling in this model that is dropped out represents only ten percent of the total activity. The author notes that "the general problem of combining cycles with trophic aggregation has not been solved" and proceeds to another approach made possible by the particular character of ecosystem trophic networks. This distinguishes nontrophic transfers to detrital pools from strict-feeding transfers, and when this adjustment is made very few "food cycles" are seen to actually occur in food webs. Literature is cited to support this contention. Actual paths and cycles in "food-web" networks may be combinatorially large, but true food chains and food cycles are relatively rare among them. As Ulanowicz states, "cycling in ecosystems is overwhelmingly biogeochemical in nature." With this assumption, he devises the following procedure for trophic aggregation: (1) separate living and nonliving compartments, (2) remove cycles, (3) aggregate the remainder into discrete trophic compartments, and (4) assign all nonliving material to trophic level 1. Results are presented for the Crystal River model; it consists of a "backbone, or 'Lindeman spine,' of ever decreasing trophic transfers (the trophic pyramid . . .)" with recycling back to the detrital first level. The stepwise, monotonic decrease of serial flows captures the original Lindeman conception, and the author is gracious in his final assessment: "Lindeman was able to see beyond the immediate form of ecosystem relationships to perceive the underlying thermodynamic generator for much of organized behavior . . . it has taken almost half a century for ecologists to give concrete shape to his powerful insight."

20

Complex Food Webs

GARY A. POLIS

Department of Biology
Vanderbilt University
Nashville, Tennessee

The ecosystem as a unit is a complex level of organization. It contains both abiotic and biotic components. The order of increasing complexity is: cell < tissue < organ < population < community < ecosystem. Although the ecosystem is the most complex level, the study of a given ecosystem is less complex in many instances than the study of lower levels [Van Dyne, in Oak Ridge Nat. Lab. Tech. Rep. 3957, p. 3 (1966)].

INTRODUCTION

Biotic communities are complex natural systems. Their complexity arises from the great diversity of species in most communities and the highly reticulated trophic and other interactions among them. It is now estimated that between 5 and 50 million species exist (May, 1988; Wheeler, 1990). Most communities probably contain 10^3 to 10^5 species. Furthermore, these species interact in a myriad of ways such that, in addition to feeding relations, indirect and nontrophic interactions (for example, symbiosis and mutualism) are now recognized as also important in the maintenance of community structure and function. Trophic interactions form multiple linkages among species in a community. Trophic (interconnections between consumers and their food items are usually delineated and summarized as a "food web"—a schematic description of all the feeding linkages in a community or ecosystem. Such food webs are often extremely, and some may say intractably, complex. Yet, the prevailing paradigm of food web theory and associated empirical data argues that most webs are actually relatively simple and exhibit a number of regularities and what have been termed "empirical generalizations." It is this body of evidence that is the purpose of this chapter to challenge. The disparity between "food web theory" and normal ecological data available through field natural history observations is too great to ignore.

This chapter is divided into four main sections. I first outline the prevailing theory and current knowledge about food webs; the proposed regularities in their organization are listed. Second, I present some characteristics of biological communities that partially explain why food webs are complex. Next I sketch the food web and trophic interactions of a relatively simple but well-studied

TABLE 20.1 Partial summary of ""features observed in real food webs"

1. Chain lengths are limited to "typically three or four" trophic levels (iii): Pimm, 1982; Briand, 1983; Cohen, Briand, and Newman, 1986.
2. Omnivores are statistically "rare" (iv): Pimm, 1982; Yodzis, 1984.
3. Omnivores feed on adjacent trophic levels (v): Pimm, 1982.
4. Insects and their parasitoids are exceptions to patterns 3 and 4 (vi): Pimm, 1982; Hawkins and Lawton, 1987.
5. Webs are usually compartmentalized between but not within habitats (viii): Pimm, 1982.
6. Loops are rare or nonexistent and do not conform to "biological reality" (i): Gallopin, 1972; Cohen, 1978; Pimm, 1982; Pimm and Rice, 1987; Lawton and Warren, 1988.
7. The ratio of prey species to predator species is less than 1.0 (range 0.86 to 0.88) (x): Cohen, 1978; Briand and Cohen, 1984.
8. The proportion of top predator species to all species in a community averages 0.29: Briand and Cohen, 1984.
9. Species interact directly (as predator or prey) with only two to five other species: Cohen, 1978; Cohen, Newman, and Briand, 1985.

From Pimm, 1982; Briand and Cohen, 1984; Lawton and Warren, 1988. These "empirical generalizations" were derived by theorists from the catalogs of published food webs assembled by Briand, Cohen, and Schoenly (see text). All quoted phrases are from Chapter 10 in Pimm (1982; especially see Table 10.1). Roman numerals in parentheses refer to pattern number in this Table 10.1.

desert community; although every effort was made to shorten this description, the trophic connections are inescapably complex and require significant space to portray. I conclude by showing that the patterns and complexities observed in the described desert web are also typical of the food webs of other communities. The data presented will argue that much of the contemporary theory of food webs is, simply put, quite wrong and inadequate to describe natural communities. The data marshalled in support of most existing theory are weak, and proposed generalizations are poor descriptors of actual situations in nature.

Several approaches have been used to analyze food webs (see DeAngelis, Post, and Sugihara, 1983; May, 1986; Lawton, 1989). One approach uses models based on stability analysis (for example, Pimm, 1982; DeAngelis, Post and Sugihara, 1983; May, 1983a, 1986). Results from these models are complex and beyond the scope of this chapter to elaborate. In general, stable webs are predicted to be relatively simple and short and with few trophic levels and little omnivory or cycling (Pimm, 1982; see later). A second approach analyzes empirical webs to determine regularities in their properties (for example, Cohen, 1978, 1989; Cohen, Briand, and Newman, 1986; 1990; Briand, 1983; Pimm, 1982; Yodzis, 1984; Schoener, 1989; and Cohen, Briand and Newman, 1990). Webs used for this approach were compiled into catalogs by Cohen (1978), Briand (1983), Cohen, Brian, and Newman (1986), and Schoenly, Beaver, and Heumier (1991). Cohen, Briand, and Newman (1986; see also Cohen, Briand, and Newman, 1990) published a catalog of 113 webs, and Schoenly, Beaver, and Heumier compiled 95 insect-oriented webs. Theorists (especially Pimm and Cohen) have argued that empirical patterns derived from these catalogs are consistent with and validate predictions of their "dynamic" (Pimm, 1982; Pimm and Rice, 1987) and "cascade" (Cohen, Briand, and Newman, 1990) models. Some of these empirical patterns are summarized in Table 20.1.

GENERAL CHARACTERISTICS PROMOTING FOOD WEB COMPLEXITY

In this section I present seven characteristics of natural communities that tend to make food webs complex. These seven characteristics pose substantial problems to much previously published food web research (see also critiques by Glasser, 1983; May, 1983a; Taylor, 1984; Patten, 1985; Paine, 1988; Sprules and Bowerman, 1988; Patten, Higashi and Burns, 1990; Lawton, 1989; and Winemiller,

1990). These problems render existing catalogs of food web models totally inadequate as a data base for the types of analyses conducted. I contend that the practice of abstracting empirical regularities from these incomplete webs yields a highly inaccurate, tremendously oversimplified view of the trophic interactions within a community. In the section to follow I illustrate these problems by presenting the desert food web.

Species Diversity

Most communities contain a diverse assemblage of species. The *major* problem of catalogued food web models is that their numbers of species (or categories; see later) are far less than those in real communities. Most authors of the catalogued webs, who had other purposes for which their webs sufficed, simply ignored unfamiliar species, concentrated on taxa within their expertise, or aggregated unfamiliar species into higher categories. This practice of lumping is a severe problem. Cohen (1978) labeled lumped categories "kinds of organisms." " 'Kinds' are equivalent classes with respect to trophic relations" (Cohen, 1978, p. 7). Briand (1983, p. 253) clarifies and expands: "A 'kind of organism' (interchangeable henceforth with the term 'species') may be an individual species, or a stage in the life cycle of a size class within a single species, or it may be a collection of functionally or taxonomically related species." "Kinds" are also called "trophic species" (Briand and Cohen, 1984) and "species" (Cohen and Newman, 1985).

Examples of kinds include "basic food," "benthos," "other carnivores," "algae," "plankton," "birds," "zooplankton," "ice invertebrates," "fish," "trees and bushes," "insects," "spiders," "soil insects and mites," and "parasites" (see matrices 1, 9, 21, and, 27 in Briand, 1983). Only 28.7% of the total "kinds" in all Briand's webs are actual species; nine of his food web matrices have no real species. The "kinds" simplification has been criticized repeatedly (Glasser, 1983; May, 1983a; Taylor, 1984; Paine, 1988; Hastings, 1988; Lawton, 1989; Winemiller, 1990), even by Cohen himself (1978) in a self-critique (but see Sugihara, Schoenly, and Trombla, 1989).

The lumping problem is particularly acute for certain taxa. Plants, arthropods, parasites, trophic mutualists, and (micro- and macro-) organisms that live in the soil or benthos are rarely considered more than superficially (Pimm, 1982; Taylor, 1984). The incomplete representation of these important taxa is a serious flaw in present food web analyses. In particular, arthropods occupy a central position in terrestrial communities. The ~800,000 identified species of insects represent about 89% of all known animal species combined and 5 million to 50 million species of insects alone may exist (May, 1988; Wheeler, 1990).

Parasitoid insects are a diverse and important component of webs. Askew (1971) estimates that >10% of all animal species are parasitoids. Other parasites (Protozoa, helminths, and Acari) are a diverse and speciose group that feeds on almost all organisms and can greatly influence population dynamics and community structure (May, 1983b). Although seldom represented in food webs, parasites form another step in the flow of energy and constitute an extra trophic level (May, 1983b); some ectoparasites form yet another "gratis" level when they themselves host parasites.

Trophic mutualists (for example, endosymbiont microbes and protozoa) are totally ignored in the catalog of webs; they are not included in even one. Yet such organisms play an essential role in almost all terrestrial, marine, and aquatic webs as necessary conduits in the flow of primary productivity to either the herbivore or detritivore channel of food webs (most metazoan animals cannot digest cellulose or lignin). Fenchel (1988) outlines their key role in marine food chains. Many terrestrial vertebrate and arthropod detritivores and herbivores host cellulolytic gut symbionts that degrade plant material (see Crawford 1981, 1986, 1991) for desert detritivores and general references). These symbionts thus represent yet another "trophic level" usually not included in food web analyses.

Finally, soil and benthic infaunal organisms and their complex interactions are often ignored or greatly simplified although they are essential to understanding web structure (Cousins, 1980;

Odum and Biever, 1984; Rich, 1984). Detritivory and subterranean herbivory form one of the major pathways of energy flow in terrestrial communities (see the detritus section).

Ignoring and lumping species have produced the depauperate published webs compiled by theorists. The number of species in Cohen, Briand, and Newman's (1986) catalog ranged from 3 to 48, with an average of 16.9. All real communities have far more species, as can be illustrated by enumerating those from the sandy desert within the Coachella Valley (CV) in Riverside County, California: 174 vascular plants, 10 to >20 nonvascular plants, 138 vertebrates, >55 arachnids, and an unknown (but great) number of insects, Acari, and nematodes. The number of insect species is estimated to be 2000 to >3000. I have identified 123 *families* of insects. An ongoing and still incomplete survey of insect diversity in the adjacent University of California Deep Canyon Desert Preserve has identified 24 orders, 308 families, and >2540 species (Frommer, 1986). Intensive work reveals high diversity in some taxa. For example, 147 species of Bombyliidae (bee flies) were recorded in Deep Canyon (J. Hall, personal communication, 1986). Timberlake recorded >500 species of bees within 2 miles of Palm Springs, a city in the Coachella Valley (C. Michener, personal communication, 1988). The beetles in a nearby (85 km) sand system, Palen Dunes, include 31 families, 120 genera, and 142 species (Andrews, Hardy, and Giuliani, 1979). Published food webs are truly limited in their lists of feeding organisms compared to what are really present.

Complete Dietary Information

Published analyses of diets or lists of enemies (that is, predators, parasites, or parasitoids) suggest that most species eat and are eaten by from 10 to 10^3 other species (see later). The inadequate incorporation of these trophic links is another major weakness of catalogued food webs. Species in these webs suffer very few predators (mean 3.2 in Cohen (1978) and 2.88 in Schoenly, Beaver, and Heumier, 1991, and consumers eat very few prey (mean 2.5 and 2.35 respectively). Overall, catalogued species interact directly with only 3.2 to 4.6 other species (Cohen, Briand and Newman, 1990).

These statistics constitute gross underestimates of the enemies and foods of species in nature. This is illustrated by a yield–effort curve (Cohen, 1978) for the scorpion *Paruroctonus mesaensis* (Fig. 20.1). Such curves analyze diet as a function of the number of cumulative nights of field data collection. The number of prey items recorded is a function of the amount of time and effort devoted to observation. The number of species of scorpion prey tabulated continues to increase linearly with observation time. The 100th prey species was recorded on the 181st survey night; an asymptote was never reached in 5 years and more than 2000 hours of field collection time.

The shape of this yield–effort curve is probably typical of many other species of consumers (J. Seger, personal communication, 1990). This suggests that the amount of effort required to approach complete diets of even numerically dominant species is astronomical. It is unlikely that such effort was devoted to most species in the catalogued webs of theorists. In fact, Cohen and Newman (1988) produce a model showing that the probability of detecting a trophic link in a community of more than 30 "trophic species" is less than 0.2. Thus, a web containing all species still would inadequately describe community trophic relations unless diets were known with more confidence.

Such inadequacies are manifested in analyses of published webs. For example, catalogued webs show a high proportion (28.5%, Briand and Cohen, 1984; 46.5%, Schoenly, Beaver, and Heumier, 1991) of top predators (consumers without predators). It seems highly unlikely that one-fourth of the animals on this planet are free of predators. Probably <<1% of species do not suffer predators (see, for example, the desert food web later). The 25% figure results from grossly

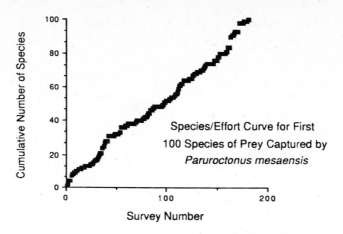

Figure 20.1 Yield-effort curve for prey capture by the scorpion *Paruroctonus mesaensis*.

incomplete data on predators and prey. For example, in the catalog of 113 webs, 57 food chains were of length 1; that is, 57 herbivores were top predators with no recorded predators (Cohen, Briand, and Newman, 1986). Top predators in this particular catalog included spiders, mites, midges, mosquitos, bees, weevils, fish larvae, blackbirds, shrews, and moles, all of which inevitably have predators of their own.

Temporal and Spatial Variation

Spatial and temporal differences in feeding ecology are undoubtedly universal. It is well known that the abundance and distribution of species, species composition, and diet all change in time and space. Such differences are often key factors shaping community patterns (see Wiens, 1976, 1991). However, it is logistically difficult to quantify spatial and temporal variation. Consequently, food webs are usually either restricted to a particular season and study site or generalized to represent some community average (Cohen, 1978; Paine, 1988). In either case, much of the dynamic richness is lost. For example, Schoenly and Cohen's (1991) analysis of 16 webs, Baird and Ulanowicz's (1989) web of the Chesapeake Bay, and Winemiller's (1990) webs of tropical aquatic ecosystems all show that many food web characteristics vary substantially through time. Each of these authors, along with Paine (1988) and Peters (1988), discuss the problems associated with the presentation of "static" food webs that ignore temporal and spatial variation.

The diets of organisms living in the CV (Coachella Valley, California) also illustrate temporal and spatial differences. Most published analyses of annual food habits of CV species note great differences through time. Several factors are contributory. Normal changes in prey composition and abundance dictate which foods are available. Prey in the CV (and elsewhere) exhibit three general phenologies: *pulsed,* short but intense population eruptions lasting a few days or weeks; *seasonal,* present for 2 to 4 months; and *annual,* available throughout the year (Polis, 1979). Feeding on prey from all three phenologies produces dietary changes over seasonal time. For example, diet similarity among adults of the same species (for example, the scorpion, *Paruroctonus mesaensis*) is a function of the time interval between months compared: diet (prey species) overlap varied from 0.73 between adjacent months to 0.0 (that is, no species eaten in common) between periods separated by five months (Polis, 1991a).

The sudden appearance of pulsed foods is often reflected in dramatic changes in diet. In most temperate and tropical habitats, mass emergence of alate termites after rains is accompanied by many predators that (temporarily) concentrate on termites. In the CV, 30 of 31 recorded predators feed on other prey when termites are not available (the exception is adult velvet mites). After heavy rains in 1976 in the CV, *Messor pergandei* ants ate 89.7% termites; normally, arthropods form 2% to 10% of their diet (Gordon, 1978).

Furthermore, most vertebrates change seasonally between plant and animal foods. Granivorous birds and rodents primarily eat seeds, but normally also feed on insects and spiders when these become abundant (Bent, 1958; Brown, Reichman, and Davidson, 1979; Brown, 1986). This switch provides protein for developing nestlings during the spring breeding season and enables permanent residency for some desert birds (Welty, 1962; Brown, 1986). In the CV, eight of 13 primarily granivorous bird species are reported to eat arthropods. Alternately, many omnivorous or primarily arthropodivorous birds and rodents in deserts consume significant quantities of seed or fruit when arthropods are unavailable (see later; Brown, Reichman, and Davidson, 1979; Brown, 1986). Many carnivorous mammals seasonally become frugivorous when fruit appears, for example, CV coyotes, kit foxes, and grasshopper mice.

The abundance, diversity, and trophic interactions among soil fauna also vary greatly as a function of time and microhabitat (Franco, Edney, and McBrayer, 1979; Whitford, 1986; Crawford, 1991; Freckman and Zak, 1991). Marked successional and seasonal changes occur. Whitford (1986) presents substantially different food webs for buried versus surface litter. Microbial populations vary with soil depth, soil moisture, and stability of substrate (often by $>200\times$). Such spatial and temporal variation produces dynamic patterns of energy and nutrient flows.

Diet is rarely the same over a species' geographical range (Welty, 1962; Wiens and Rotenberry, 1981; McCormick and Polis, 1986). Spatial differences in the composition and abundance of prey constrain the consumer to eat different foods in different places. For example, *Erodium* seeds formed 87% of the dietary biomass of the desert kangaroo rat (*Dipodomys merriami*) in Arizona (Reichman, 1975), but only 9.5% in California (Soholt, 1973). Vitt (1991) shows that dietary differences for the same species of lizards at different sites approximate the differences observed between different lizard species.

In general, similarities in diet among populations of the same species should decrease monotonically, even over relatively short distances. For example, McCormick and Polis (1986) quantified differences in the feeding ecology of *P. mesaensis* at sites 65 km apart within the CV. Predator size, prey size, feeding rate, and proportion of the same prey in the diet were each highly significantly different. Overall diet similarity was 83%.

Age Structure

Age-related changes in food and predators are not well incorporated into food web analyses. Populations are composed of age and size classes, each exhibiting significant differences in resource use, predators, and competitors (Polis, 1984a, 1988a; Werner and Gilliam, 1984). Age classes often eat different foods, expanding the species' diet (life history omnivory; Pimm and Rice, 1987). Ontogenetic diet shifts characterize species that undergo metamorphosis. The juveniles of at least 27 families of CV holometabolic insects eat radically different foods (live arthropods) from adults (plants). For species that grow slowly through a wide size range, diet changes gradually as prey size increases with predator size (for example, arachnids, reptiles; Polis, 1988a). In fact, differences in body sizes and resource uses among age classes are often equivalent to or greater than differences among most biological species (Polis, 1984a). This magnitude of change is typical of wide size-range predators (most invertebrates, hemimetabolic insects, larvae of holometabolic insects, arachnids, fish, and reptiles; Polis, 1984a).

Predators also change during growth. Juveniles are eaten by species too small to capture adults. Such developmental "escapes" are common to all communities. For example, snakes eat eggs and newborn (but not adults) of carnivorous birds and desert tortoises (see Fig. 20.7). Alternately, adults are eaten by predators that do not eat small juveniles. Thus, some predators (for example, owls and kit foxes) in the CV eat only adult scorpions (Polis, Sissom, and McCormick, 1981).

In summary, age and size differences in predators, prey, and competitors are the norm in terrestrial as well as aquatic habitats. This is contrary to Pimm and Rice's (1987) assertions and may in fact be a major determinant of population dynamics and community structure. Unfortunately, the richness that age structure contributes has largely been ignored (but see Pimm and Rice, 1987). Usually, only adults are considered or the diets of all age classes are combined. Age is recognized in only 22 of 875 "kinds" of organisms in Cohen's (1978) catalog and in only three of 422 in Briand's (1983). Age structure is often difficult to incorporate into studies; nevertheless, it is paramount to community dynamics.

Looping and Cycling

Cycling is a feeding interaction such that A eats B and B eats A (mutual predation, a cycle of length 2), or A eats B, B eats C, and C eats A (a length 3 cycle). Cannibalism is a *self-loop* (A eats A; Gallopin, 1972); a *loop* is a *cycle* of length 1. Cycles may be quite long in food webs (see Patten, 1985). Theorists dismiss loops and cycles as "unreasonable structures" and nonexistent in "real communities" (Pimm, 1982, p. 70; Gallopin, 1972, p. 266; Cohen, 1978, p. 56; May, 1983a; Cohen and Newman, 1985). Pimm (1982) summarizes from the catalog of webs: "I know of no cases in the real world with loops" (p. 67, implying no cannibalism). He has modified this view to include loops in aquatic age-structured species (Pimm and Rice, 1987), but still maintains that loops and cycles are rare in terrestrial systems, ignoring all the evidence of trophically generated energy and matter biogeochemical cycling in ecosystems.

Cannibalism (Polis, 1981) and mutual predation (Polis, Myers, and Holt, 1989) are taxonomically widespread interactions. Cannibalism was reported in >1300 species and is a key factor in the dynamics of many populations (Polis, 1981; see also Elgar and Crespi, 1992; Patten, Higashi, and Burns, 1990). Cannibalistic loops are frequent in the CV (see Figs. 20.2 through 20.7). Ontogenetic reversal of predation among age-structured species is the most common form of mutual predation: juvenile A's are eaten by B, but adult A's eat B (or juvenile B). This is observed among CV spiders, scorpions, solpugids, and predaceous insects and among lizards, snakes, and birds (see Figs. 20.5, 20.6, and 20.7). For example, gopher snakes eat eggs and young of burrowing owls, whereas adult burrowing owls eat young gopher snakes.

Mutual predation can occur independently of age structure (Polis, Myers, and Holt, 1989; Schoener, 1989; Winemiller, 1990). Two examples from the CV are as follows: (1) Black widow spiders catch three species of scorpions by using web silk to pull them up off the ground (Polis, Sissom, and McCormick, 1981); black widows traveling on the ground are captured by these same scorpions (Polis and McCormick, 1986b; Fig 20.5). (2) CV ants regularly include each other in their diets (Ryti and Case, 1988). Killing and predation of winged reproductives (after swarming) and workers (during territorial battles) are a regular interaction among ants and other social insects (see Polis, Myers, and Holt, 1989).

It should be noted that looping and cycling are dependent on the classification of "kinds." If each life cycle stage, age or size class, or individual were treated as a unique entity, cycle closure would be a rare property of food webs.

Importance of Detritus

Detritus is a broad term applied to nonliving organic matter originating from a living organism. It derives from plants [for example, wood, leaves, and roots (especially rhizodeposition), algae, and phytoplankton] and animals (feces, urine, secretions, molted skin, and dead bodies). In aquatic and marine environments, detritus ranges several orders of magnitude in size, from carcasses to particulate organic matter (POM) to dissolved organic matter (DOM). Detritus is a universal component and major energy source of all food webs simply because many organisms "leak" organic molecules, all organisms die, and all animals defecate. A large number of consumers eat dead organic material.

Most primary productivity within all communities flows directly or indirectly through the detrital component of the food web. The idea of a trophic pyramid of herbivores feeding on plants and forming the food of carnivores is an incorrect oversimplification because this part of the food chain processes from less than one-tenth to one-half of the total energy. The percentage of net primary production eaten by herbivores varies from 1% to 50%; the rest enters the detrital system (Odum and Biever, 1984). Thus, detritus removes the primacy of primary production as the bottom step in food webs in most ecosystems.

In peat-forming wetlands, detrital food chains may be the dominant type. In terrestrial systems, most primary productivity goes directly to the detrital compartment. In deserts, 12% to <33% of plant production flows via the plant→herbivore→carnivore chain; the rest goes through the soil and the detrital food chain (Wallwork, 1982; Seely and Louw, 1980; Crawford, 1991; Freckman and Zak, 1991). Macfayden (1963) suggests that 30% to 73% of the energy flow through all microarthropods is via soil detritivores.

In marine and aquatic systems, most primary production is channeled via dead organic matter to phagotrophic organisms (either directly or indirectly via microbes) and then finally up to the rest of the food chain (see Fletcher, Gray, and Jones, 1987; Patten, 1985; Salonen and Hammar, 1986; Barnes and Hughes, 1988; Fenchel, 1988; Baird and Ulanowicz, 1989; Duggins, Simenstad, and Estes, 1989; Hessen, Andersen, and Lyche, 1990; Winemiller, 1990). These authors strongly maintain that the historical and simplistic view of aquatic and marine food chains—plant to herbivore to predator—badly needs to be revised, especially with the recognition of the great importance of DOM and the "microbial loop." Large quantities of DOM leak out and are absorbed through the membranes of microbes, protozoa, algae, and even small metazoa (Fenchel, 1988). Microbes "are responsible for by far the largest part" (Fenchel, op. cit., p. 22) of the energy flow in many marine systems, especially open pelagic and soft bottom benthic communities. In the Chesapeake Bay ecosystem, detritivores process $>10\times$ more energy than herbivores (Baird and Ulanowicz, 1989). In some lake systems, most carbon in zooplankton comes from detritus, with direct links from primary consumers being low to negligible (Salonen and Hammar, 1986; Hessen, Andersen, and Lyche, 1990).

In spite of its great importance, detritus, DOM, and detritivores (especially microbes) are not well incorporated into either the descriptions or theory of food webs. Most catalogued webs concentrate on the more obvious plant→herbivore linkage. The inclusion of DOM and detritus significantly alters the way webs are now viewed. For example, Fenchel notes that inclusion of the microbial loop adds two or three more trophic levels to "classical" marine food chains. The same is true for classical desert webs (see later). "Trophic unfolding" of cyclical webs (Higashi, Burns, and Patten, 1989) leads to food chains that are potentially infinite in length. Moreover, stability models (for example, Pimm, 1982; May, 1983a, 1986) based on population dynamics are a particularly inappropriate way to understand the linkage between detritus and its consumers. Cousins (1980, 1985, 1987) is one of the few to incorporate detritus explicitly into food web analysis (also see Odum and Biever, 1984; Patten, Higashi, and Burns, 1990). Cousins disputes placing autotrophs alone at the basal position of food webs; rather, herbivory and detritivory should be considered equally important as links in a "trophic continuum." Energy, produced by autotrophs, is recycled and made available to other consumers by detritivores. Patten (in Patten, Higashi, and Burns, 1990) argues that such ignored cycling is essential to understanding food webs and the functioning of ecosystems. Inclusion of the trophic continuum and cyclic dynamics makes food chains long, increases the number of trophic levels, and decreases the importance of living primary productivity as being the only direct supplier of energy to food webs (Patten, Higashi, and Burns, 1990).

Lack of Compartmentalization within Communities

Trophic linkage among communities or within the subdivisions of communities greatly increases the complexity of food webs. Such connectivity poses great operational problems for food web analyses. To what degree are communities and their subdivisions clearly delineated in nature? Where do communities begin and end? Communities are often subdivided into "compartments" (Pimm, 1982) or different focal areas where sets of species commonly interact. Communities are often compartmentalized using vertical stratification (height in a water column or canopy, substrate depth), physical features (planktonic versus benthic, littoral versus neritic, epibenthic versus infaunal; surface versus soil biota), dominant plant species (oak, pine, kelp), or time (diurnal versus nocturnal). If sets of species exclusively or primarily interact within their compartments, food webs would be greatly simplified. If species commonly interact across compartments, webs become much more complex. The question is to what extent real communities are compartmentalized.

It appears that species frequently interact with those from other compartments (see later; Pimm, 1982; Moore, Walter, and Hunt, 1988; Winemiller, 1990). It also appears that most communities do not exhibit discrete boundaries, but are characterized by gradual changes in species composition. Even those with physical boundaries often share species and are connected trophically; for example, intertidal communities are affected by terrestrial (for example, mice and birds) and oceanic species (for example, fish) and significant energetic input from other systems (Paine, 1980; Duggins, Simenstad, and Estes, 1989). The normal distribution and dispersal of plant and animals places species across the communities recognized by investigators. Furthermore, behaviors of both consumers and their prey tend to blur the discreteness of apparent compartments (Winemiller, 1990).

Using the desert community to illustrate, species from different microhabitats and times are connected trophically to one other (see later and Polis, 1991a). At first inspection, there appear to be several distinct compartments or "energy channels," for example, diurnal versus nocturnal species, shrub versus ground species, and surface versus subsurface species. However, extensive crossover exists among compartments. Diel activity patterns of many (and perhaps most) CV species change from day to night as a function of temperature; nocturnal species eat diurnal prey (for example, scorpions on honeybees, robberflies, and mantids). Plants and detritus are eaten during all time periods by species that live both above and below the surface.

In particular, interactions within the soil are routinely studied separately from those above the soil. However, evidence strongly indicates that above- and belowground components are closely and multiply linked. Subsurface and surface herbivores feed on the same plants and may actually compete (Andersen, 1987; Seastedt, Ramundo, and Hayes, 1988; see below). Furthermore, soil detritivores feed on detritus produced by surface-dwelling plants and animals, and detritivores are a direct or indirect energy source for all secondary consumers in the CV and many other communities. As an example, the energy recycled by subsurface detritivores is exported to the surface when they become surface dwellers, for example, postmetamorphic tenebrionids.

Energy flow and strong trophic interactions among compartments characterize marine and aquatic communities. Planktonic production is a function of mixing of detritus off the bottom and subsequent nutrient input into the water column; benthic fauna depend on detrital rain from the plankton and seston above.

In general, "different channel omnivores" link communities together because they feed on prey from different microhabitats, compartments, or subwebs (Odum and Biever, 1984; Moore, Walter, and Hunt, 1988). Consumers eat available food types regardless of their compartments; they rarely specialize on particular energy channels or trophic levels. Such decompartmentalization further increases complexity.

THE COACHELLA VALLEY

The sand dune and intergrading sand flat habitat of the Coachella Valley (CV) was chosen to define the biological community. The CV, a low-elevation rain shadow desert in Riverside County, California, is well studied primarily due to the presence of the Deep Canyon Field Station. Sandy areas include 138 species of reptiles, birds, and mammals (Mayhew, 1981; Weathers, 1983; Ryan, 1968). Some species are not considered in the food web; for example, cougars and badgers occurred historically but are no longer present. Moreover, only the 56 birds (of 97 residents) that actually nest in sandy areas are considered.

CV invertebrates are less well known than vertebrates. Most data are from my own long-term research (since 1973; for example, Polis, 1979, 1991a; Polis and McCormick 1986a,b, 1987; McCormick and Polis, 1986). Information on diet and species composition was obtained from >4300 trap days for insects and from 15 years and more than 4000 hours of fieldwork. Taxonomic lists of the arthropods were also obtained from catalogs of insects (for example, Wheeler and Wheeler, 1973; Hawks, 1982; Frommer, 1986) and arachnids (Polis and McCormick, 1986b). The flora was surveyed by Zabriskie (1981).

Trophic relationships (food and predators) were determined from the literature (more than 820 papers read) and my direct work. I could not determine the diets of 18 bird and one rodent species. Interviews with scientists conducting research in the CV provided additional data. Limited space precludes the inclusion of most references and forces me to abbreviate greatly the specifics of the real food web. Interested readers can write for more information and also refer to the references cited in Polis (1991a).

A series of representative subwebs depicts how trophic relations proceed from plants and detritus to various secondary consumers. Subwebs are connected so that organisms in one web consume (or are consumed by) organisms in the next. Webs are incomplete as there are far too many species to include, and adequate dietary data are unavailable for many species. I thus concentrate on well-studied, focal species whose trophic interactions are relatively well known. Webs include only species that live in the CV and only interactions for which evidence exists. Similar complexity is expected for other, less known species. Thus, the complexity in the webs to be presented understates what actually exists.

Consumers are classified in terms of resource specialization (number of species eaten within one group, such as plants) or trophic specialization (number of different types eaten, such as plants, detritus, arthropods) (Levine, 1980). Species vary from resource specialists that eat a few species of the same resource type to trophic generalists (omnivores) eating several food types. "Parallel path omnivory" is a case of omnivory where A eats both B and C, but B also eats C. Sprules and Bowerman (1988) termed this "closed loop omnivory"; there is, however, no further cycling of energy beyond C, which only terminates two different (parallel) paths.

Results

Plant–herbivore trophic relations. Herbivory describes feeding on plant products: leaves, seeds, fruit, wood, sap, nectar, roots, and tubers. Desert herbivores include microbes, nematodes, arthropods, and vertebrates. I cannot detail the herbivores of all CV plants. Rather, I will discuss broad groups to convey the complexity in the plant–herbivore link. At least 74 families of CV arthropods are herbivorous sometime in their lives. Plant species that grow in the CV are attacked by many species of insects: more than 60 eat creosote bush, 200 consume mesquite, and 89 use ragweed (Wisdom, 1991). Both resource specialists and generalists live in the CV. For example, insects on cactus usually specialize. Damage caused by feeding on cactus facilitates several specialist and generalist fungi. Resource generalists are more common than specialists (Orians, et al., 1977; Crawford, 1981, 1986). CV generalists include the harvester ant, *Myrmex pergandei* (which uses 97 species of CV seed; Gordon, 1978) and camel crickets (16 plant species; Polis, unpublished data).

Most herbivorous arthropods are trophic specialists on plants all their lives, for example, hemimetabolic (Orthoptera, Hemiptera, Homoptera, Thysanoptera) and some holometabolic insects (curculionid, chrysomelid, scarabid, and buprestid beetles). Many holometabolic insects have larvae that are parasitic or predaceous on arthropods, but the adults feed on plants (Ferguson, 1962; Andrews, Hardy, and Giuliani, 1979; Powell and Hogue, 1979; Wasbauer and Kimsey, 1985). Trophically flexible generalists include CV harvester ants (which consume more than 40 categories of foods such as seeds, flowers, stems, spiders, and insects from at least six orders including four ant species; Ryti and Case 1988) and camel crickets (which use 15% plant detritus, 41% animal detritus, and less than 1% conspecifics).

Most CV mammals (16 of 18 species) eat plant tissue (the two bats did not). Plants (fruit) formed 0.2% to 4.1% of the diet of the largest mammal, the coyote (Johnson and Hansen, 1979). This is the smallest plant component for any of the 16 herbivorous mammals. Over 50% of the scats of the desert kit fox contain plant material. The two rabbits and gopher are the only trophic specialists; however, they are resource generalists eating many species. Antelope ground squirrels and rodents (*Dipodomys, Peromyscus, Perognathus*) feed on a seasonally changing diet of seeds and plant parts (and arthropod prey). In total, 15 mammals eat seeds (only bats and gophers do not). Nine regularly consumed more than 50% seeds in their diet. No CV mammal specializes on particular plants; for example, pocket mice fed on 27 plant species; antelope ground squirrels, 24. Except for the gopher and the two rabbits, all plant-eating mammals are trophic generalists that eat arthropods, for example, 1% to 17% for *D. merriami* and 2% to 35% for antelope ground squirrels (this species also eats vertebrates; see later).

Many (34 of 56 species) birds feed on plant parts (seeds, nectar, flowers, and fruit). Frugivorous birds are common, for example, cactus wren, phainopepla, verdin, and doves. Some birds eat fruit as a minor part of an omnivorous diet (for example, roadrunner, Scott's oriole, western tanager, western bluebird, Bewick's wren). Granivory is also common: 22 of 56 CV birds were reported to eat seeds (13 are primarily granivorous). Many insectivorous desert birds eat significant quantities of seed when insects are scarce (Brown, et al., 1979; Brown, Reichman, and Davidson, 1986; Wiens, 1991). No herbivorous birds are resource specialists. In fact, trophic specialists are rare; of 34 plant-eating birds, only five are not recorded to eat arthropods.

Two species of CV reptiles are primarily herbivorous, desert tortoise and desert iguana. Both are resource generalists eating a wide variety of plants (17 to 40 species for the tortoise). Only the tortoise is a trophic specialist. The diet of the iguana contains 1% to 5% arthropods. Five of the nine other lizards (but none of 10 snakes) consume a minor portion of plants.

Detritus and soil biota. Detritus is particularly important in desert ecosystems where the main energy flow often proceeds directly from autotrophs to detritivores (Wallwork, 1982; Seely and Louw, 1980; Crawford, 1991; Freckman and Zak, 1991). The plant–herbivore–carnivore link forms 12% to 33% of the fate of plant production in deserts; the rest goes through the soil and the detrital chain. Living plants form only 6% to 8% of the total biomass of vegetation in the Namib desert; detritus constitutes 92% to 94% (Seely and Louw, 1980).

A diverse biota lives within desert soils (Wallwork, 1982; Crawford, 1991; Freckman and Zak, 1991). Many taxa of microorganisms (fungi, yeast, bacteria, protozoa) decompose detritus in desert soils. Several families of nematodes and mites, termites, Collembola, Thysanura, burrowing cockroaches, tenebrionid larvae, some ants, millipedes, and isopods are some of the more important of the many detritivorous soil arthropods (Crawford, 1981, 1991). All are common in the CV.

Although species in these taxa degrade organic material, many include facultative or obligate herbivores on belowground plant parts (Crawford, 1981, 1986). Roots and tubers represent a rich food source in deserts. Usually, more than 50% of net primary production is allocated to belowground plant parts (Andersen, 1987). For example, in Russian deserts up to 80% to 90% (average 65%) of the plant biomass is belowground. Species from seven orders of insects, mites, nematodes, and some rodents use belowground herbivory as their major feeding mode (Andersen, 1987).

Soil organisms are quite abundant in deserts. Detritivorous arthropods represent 37% to 93% (average 73%) of all macroarthropods in four deserts (Crawford, 1991). In particular, nematodes and termites are abundant in deserts; termite biomass is often an order of magnitude greater than any other animal taxa (Whitford, 1986; Crawford, 1991; MacKay, 1991; Polis and Yamashita, 1990; Freckman and Zak, 1991).

A rich web based on detritus and underground plant parts exists within desert soils (for example, Whitford, 1986; Crawford, 1991; Freckman and Zak, 1991). Nematodes occupy several distinct trophic roles: herbivores and plant parasites, microbial feeders, fungal feeders, omnivores, omnivore-predators, and parasites. Several feeding groups of mites live in desert soils (Santos, Phillips, and Whitford, 1981). In the Mojave, Franco, Edney, and McBrayer (1979) recognized four trophic groups of soil mites: fungivores and detritivores (three families), phytophages (one), parasitic (one), and predaceous (eight). Predatory mites can be particularly abundant; Wallwork (1982) found 11 ratios of nonpredatory:predatory mites in litter in nearby Joshua Tree National Monument. These mites eat nematodes, Collembola, and other mites. The large number and diversity of predatory mites led Edney, McBrayer, and Franco (1974) to postulate the possibility of two or more predator trophic levels in the decomposer web. Wallwork (1982) and Santos, Phillips, and Whitford, (1981) suggested that decomposer pathways in the soil fauna were regulated by mites that prey on nematodes that feed on microorganisms.

Soil trophic interactions become even more complex with the inclusion of macroarthropods (36 families in the CV). Almost all larger detritivores not only eat detritus, but also feed on microorganisms (bacteria, fungi, protozoa) feeding and living in the detritus (Janzen, 1977). Furthermore, many arthropod detritivores (for example, cockroaches, tenebrionids, millipedes) are host to cellulolytic gut symbionts that degrade plant detritus (Crawford, 1991). Finally, several desert insects eat detritus directly or are predaceous on nematodes and microarthropods. In the CV these include larvae of asilid, bombylid, and therivid flies and staphylinid and clerid beetles (Edney, McBrayer, and Franco, 1974; Powell and Hogue, 1979).

Interactions within desert soils are linked to the entire community. Many surface-dwelling desert organisms (and most arthropods) either spend part of their lives in the soil (as larvae; for example, tenebrionids) or feed on macroarthropod taxa that live permanently or temporarily under the ground (see Polis, 1991a, for examples). Polis (1979) reports that 46% of the prey of *Pauroctonus mesaensis* in the CV live in the soil as larvae. Tenebrionids and termites are particularly important conduits of energy from below to above ground when they are subject to intense predation by a diverse group of arthropod and vertebrate predators (Wallwork, 1982). The 10 species of CV termites are eaten by 32 recorded predators and the 12 CV tenebrionids by 32 predators. Such predation by surface dwellers on soil insects exports much of the energy recycled by detritivores and links the soil subweb to those above the surface. Thus, even if herbivores and detritivores operate in distinct microhabitats, energy flowing further into the community merges into the bodies of predators common to both types of consumers (Odum and Biever, 1984).

Although the above studies outline trophic interactions within detritus and soil, no study explicitly analyzed these interactions in the CV. Most studies cited were conducted in deserts (the Mojave and Chihuahuan) geographically adjacent to the CV. I combined information from these studies (especially Whitford, 1986, and Freckman and Zak, 1991) with my data to construct a soil–detritus subweb (Fig. 20.2) that should describe trophic interactions within the CV. Note the complexity (even with extensive lumping), cycles, omnivory common at nonadjacent trophic levels, parallel path omnivory, and chains of four to five links. Also note the links between subsurface consumers and their surface predators.

Finally, the complex trophic interactions associated with a special class of detritus (animal feces) should be addressed. Schoenly (1983) found 30 species of arthropods in mammal dung from the Chihuahuan desert. Coprovores, predators, and omnivores (eating dung and other arthropods) were abundant. A clear succession of species occurs, with termites eventually finishing the decomposition process. Most species in all trophic classes fed on foods other than dung, thus connecting this subweb to the rest of the desert community.

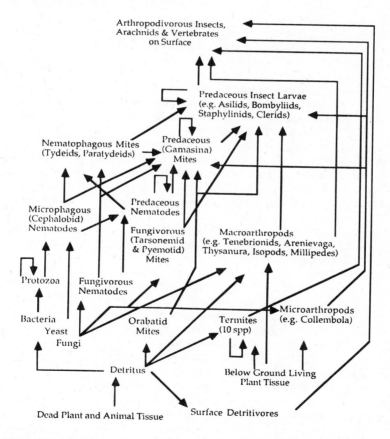

Figure 20.2 Trophic interactions within sandy soils in the Coachella Valley. See Polis (1991a) for the identity of important detritivorous species.

Carrion feeders. A rich carrion fauna occurs in deserts (Crawford, 1991) and feeding interactions are complex, involving 28 to 500 species. Some species specialize on particular tissue; others do not. Trophic specialists and generalists occur. Species composition and diversity change through time.

Studies from the Chihuahuan Desert illustrate carrion use (McKinnerney, 1978; Schoenly and Reid, 1983). In particular, McKinnerney analyzed the fauna (63 arthropod and 4 vertebrate species) associated with carrion of two rabbit species also common in the CV. Many families of Diptera, Hymenoptera, Orthoptera, Coleoptera, and Lepidoptera consume carrion directly. Coyotes and great horned owls also fed on rabbit carrion. Lizards, skunks, spiders, solpugids, assassin bugs, robber flies, and carabid and staphylinid beetles were all generalist predators on necrophagous insects. Some predators specialized; for example, larvae of clerid beetles attacked those of dermestid beetles. Ants, silphid beetles, and opiliones were omnivorous on both carrion and larvae; some larvae were those of other predators (forming cycles of the mutual predation variety).

Other interactions also occur. For example, parasitoids attack insects and spiders. Vertebrates not only ate carrion, but also eggs and larvae of necrophagous insects. Carrion feeders also eat microorganisms within the carrion (Janzen, 1977). Also, many carrion species are cannibals, for example, callophorid fly larvae.

As with other detritus, much of the energy from carrion is exported to the rest of the community. McKinnerney (1978) noted that most organisms associated with carrion were opportunistic; for example, spiders, solpugids, opiliones, ants, asilids, staphylinid beetles, reduviids, and the vertebrates not only eat insects associated with carrion (or carrion itself), but also prey on other species. In turn, all these species are eaten by other predators.

Arthropod parasitoids. Parasitoids occur in several families of flies, wasps, and beetles (20 recorded families in the CV). Parasitoids deposit eggs in or on an arthropod host; the developing larva feeds on and causes the host's death (in contrast to parasites in general). Adults usually feed on foods of plant origin. Trophic relations are generally quite complex (Askew, 1971; Price, 1975; Pimm, 1982; Hawkins and Goeden, 1984; Polis and Yamashita, 1990). Hawkins and Lawton (1987) estimate that each species of insect herbivore hosts 5 to 10 species of parasitoids.

Hawkins and Goeden (1984) studied insects that interact within galls of saltbush (*Atriplex*) in the CV. The system is complex: 67 species (40 common), at least five trophic links, and extensive omnivory (Fig. 20.3). Gall midges are trophic specialists on plants, but either resource specialists on *Atriplex* or generalists on five other plant species. Most parasitoids are primary, attacking only midge larvae or inquilines; seven of these also feed on gall tissue. Two species are facultative hyperparasitoids: *Torymus* feeds on gall tissue, gall midges, and primary parasitoids; the other, on midges, primary parasitoids, and *Torymus*. The top predator, a clerid beetle (*Phyllobaenus*), feeds on at least 17 species from all animal groups. This clerid also parasitizes 6% of *Diguetia* spider egg cases on *Atriplex* (see Fig. 20.4). Hawkins and Goeden (1984) argue that an omnivorous habit consisting of entomophagy and phytophagy is adaptive to parasitoids because it provides more potential food, growth to a larger size, and potential competitive dominance.

The trophic relations of *Photopsis* (an abundant mutillid wasp in the CV) are diagrammed in Fig. 20.5 (from Ferguson, 1962, and my own data). Females oviposit in the cells of developing larvae, and the *Photopsis* larva consumes the entire host. They are not host specific and parasitize several species of hymenopteran larvae and are hyperparasitoid on larval parasitoids of these larvae [which are other Hymenoptera, (stylopid, meloiid, and rhipiphorid) beetles, and (bombyliid) flies]. Some hyperparasitized Hymenoptera (for example, sphecid wasps) also may parasitize spiders. Up to 37% are destroyed when *Photopsis* larvae are parasitized by some of the same parasitoids (for example, sphecids) that fall host to *Photopsis*. This is an example of cycling via mutual and tertiary parasitism. Some *Photopsis* larvae also host bombyliid and stylopid (mutual) parasitoids. Furthermore, *Photopsis* larvae are cannibalistic, so there is looping. Adults are herbivores on nectar

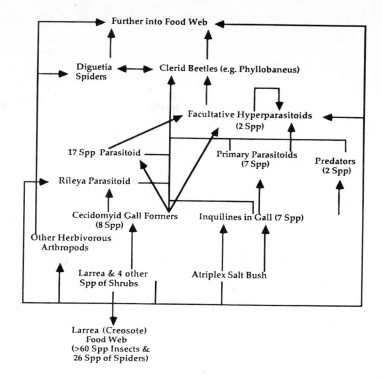

Figure 20.3 Trophic interactions within galls on saltbush, *Atriplex canescens*, within the Coachella Valley. In total, 67 species interact within galls formed by Cecidomyidae (midge) larvae. Many of these species are also involved in a subweb centering on creosote bush. Note that many species (for example, *Diguetia* and *Phyllobaenus*) occur in subsequent subwebs. An arrow returning to a taxon indicates cannibalism. Modified from Hawkins and Goeden (1984).

and pollen, but also eat ground-nesting Hymenoptera. Adults are frequent prey to many predaceous arthropods (see Figs. 20.4 and 20.6).

These parasitoid subwebs are characterized by high omnivory, parallel path omnivory, frequent looping and cycling (via cannibalism, mutual predation, and three-species cycles), and many trophic levels. Also, key species export energy from parasitoid subwebs as predators or prey in the rest of the community. Many more parasitoids common in the CV eat eggs and larvae of insects. Dipteran parasitoids (Bombyliidae, Sarcophagidae, Tachinidae) on Orthoptera, Coleoptera, and Hymenoptera; several hymenopteran parasitoids (for example, tiphiid, mutillid, sphecid, ichneumonid, and chalcid wasps) develop in Orthoptera, Neuroptera, Lepidoptera, Hymenoptera, and Diptera.

Spiders are hosts to parasitoids (pompilid, sphecid, and ichneumonid wasps; many Diptera). Adult wasps partition nests with captured spiders; developing wasp larvae eat the moribund spiders. Adults usually eat nectar. Wasps vary greatly in specificity, from resource specialists on particular families to generalists attacking several families (Wasbauer and Kimsey, 1985). Many pompilids (about 11 species) occur in the CV (Wasbauer and Kimsey, 1985): two feed trapdoor spiders to their larvae; adults drink sugar secretions from aphids and nectar from at least 10 plant species. Some CV pompilids are hyperparasitoids. Spiders are also beset by a diversity of egg parasitoids

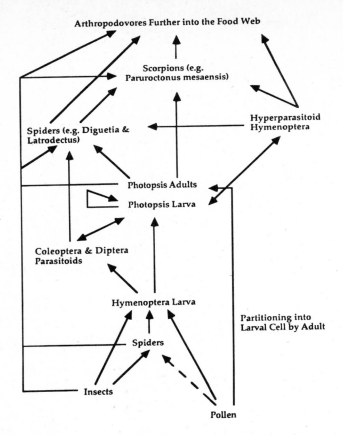

Figure 20.4 Trophic interactions involving a few of the predaceous surface arthropods within the Coachella Valley. This subweb is focused around the spiders *Diguetia mohavea* and *Latrodectus hesperus*. An arrow returning to a taxon indicates cannibalism. Note that no vertebrates are represented.

and predators (Polis and Yamashita, 1991). Kleptoparasitic insects (for example, Drosophiloidea flies) and spiders parasitize spiders by robbing captured prey.

Parasites. Parasites feed on their hosts over a long period of time; consequently, they do not cause the immediate (or, usually, the ultimate) death of their host. Although seldom represented in food webs, they are nearly ubiquitous on all metazoan animals; for example, see the Bush, Aho, and Kennedy (1990) survey of helminths on vertebrates. Thus, parasites constitute an extra trophic level usually not included in food webs. Moreover, some ectoparasites form yet another "gratis" level when they themselves host parasites. Parasites are not well studied in natural communities. Of animals living in the CV, only parasites of the coyote, lizards, and scorpions were examined in any detail. Otherwise, there are few data on the hundreds of species of endo- and ectoparasites in the sand community. The following information likely characterizes unstudied taxa.

Telford (1971) identified protozoan and helminth endoparasites in 10 CV lizards. These lizards were infected by several protozoan parasites (average 7.8 species; range 3 to 10: flagellates, ciliates, amoebas, sporozoans, and haemogregarines). Helminth parasites (1.4; 0 to 5) included

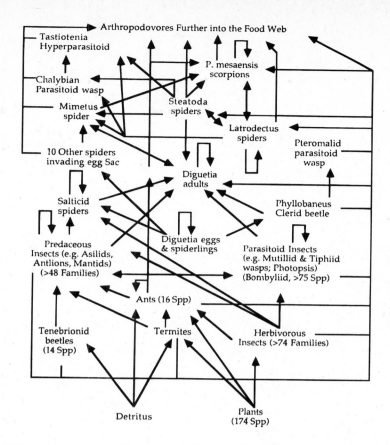

Figure 20.5 Trophic interactions involving parasitoid mutillid wasps in the genus *Photopsis*. Note that the interactions of some species are not fully represented (for example, scorpions and spiders). A double-headed arrow indicates cycling via mutual predation.

nematodes, cestodes, and Acanthocephala. In addition, each lizard hosted several mite species. Indeed, every lizard sustains its own food web, a community of parasites. Gier, Kruckenberg, and Marler (1978) discussed coyote parasites. External parasites include mange mites, ticks (seasonal), lice, and fleas. Adult fleas eat blood; flea larvae feed on organic debris, particularly adult flea feces. Internal parasites, specifically *Taenia* tapeworms, occur in 60%-95% of all stomachs. The prime intermediate host of *Taenia* is the cottontail rabbit.

It is likely that most, and probably all, of the free-living animals in the CV harbor one or more parasite species. For example, 23.4% of 1525 birds representing 112 species from deserts and other areas in the Southwestern United States had blood parasites [Woods and Herman, 1943; Welty (1962) lists the diverse parasite fauna of birds]. Inspection of CV spiders and insects almost always reveals mites. Many CV spider genera were reported with nematodes. CV scorpions host nematodes and eight species of mites (McCormick and Polis, 1990).

Arthropod predators. Predaceous arthropods are one of the most important conduits of energy flow in desert food webs. Most consumed primary productivity is utilized by herbivorous and detritivorous arthropods, rather than vertebrates (Seely and Louw, 1980; Polis, 1991a; Polis

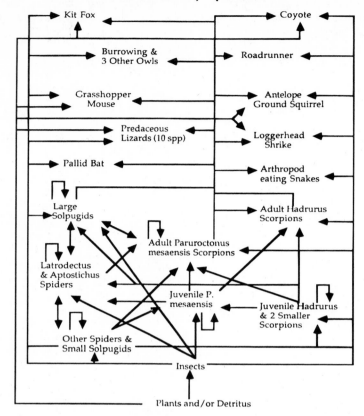

Figure 20.6 Trophic interactions centered around the prey and predators of the scorpion *Paruroctonus mesaensis*. An arrow returning to a taxon indicates cannibalism; a double-headed arrow, cycling via mutual predation. Note that the interactions of some species are not fully represented (for example, "spiders" and "insects"), and carnivory on vertebrates is not depicted (see Fig. 20.7). In total, more than 125 prey and 27 predator species have been recorded for this scorpion.

and Yamashita, 1991). Arthropods, in turn, are eaten by numerous predators, most being other arthropods (Crawford, 1981; Polis and Yamashita, 1991).

In the CV, mites (8 families), arachnids (more than 23 families), and insects (more than 21 families) are predaceous as both juveniles and adults. The complex life cycle of holometabolous insects often results in different feeding habits between life stages. At least 27 families of Coleoptera, Diptera, and Hymenoptera are trophic generalists, predaceous as larvae and herbivorous as adults. In addition, some omnivorous taxa (for example, ants, camel crickets, and gryllacridids) are occasionally but regularly predaceous.

Most arachnids are resource generalists (Polis and Yamashita, 1991). For example, *Paruroctonus mesaensis* is recorded to eat more than 125 prey species; the spider *Diguetia mohavea*, more than 70 species; and the black widow *Latrodectus hesperus*, 35 species. In fact, some scorpions and spider species scavenge dead arthropods and some spiderlings eat aerial plankton (pollen and fungal spores) trapped by their webs (Polis and Yamashita, 1991). Facultative predators (for example, the omnivores mentioned above) are trophic generalists that eat plants, plant and animal detritus, and live prey. A few predaceous arthropods tend to specialize, for example, adult velvet mites on termites and *Mimetus* spiders on other spiders.

Trophic interactions within predatory arthropods are complex. The generalized diet of most species is set by predator–prey size relationships: they catch whatever (smaller) prey they can subdue. Consequently, most smaller arthropods are potential prey, and predators commonly eat from all trophic levels, including other predators (Polis and Yamashita, 1991). For example, the diet of six CV predaceous arthropods averages 51.5% other predaceous arthropods (Table 20.2). Such predation is particularly common in deserts because predators form such a high proportion of desert arthropods (Crawford, 1991; Polis and Yamashita, 1991). Age structure also promotes predator–predator feeding interactions. Predaceous arthropods grow through a wide size range when they are potential prey to larger predators. All these factors produce extensive cannibalism (Polis, 1981), intraguild predation (Polis, Myers, and Holt, 1989), mutual predation and other cycling during ontogenetic reversals, and parallel path omnivory (see Figs. 20.4 and 20.6).

TABLE 20.2 Diet classification of some representative predators on arthropods

	Percent Arthropod Taxa in Diet		
Taxon	Predaceous and Parasitoid	Herbivorous and Detritivorous	n
Arachnids			
Hadrurus arizonensis	53	47	15
Paruroctonus luteolus	60	40	10
P. mesaensis	47	53	126
Vaejovis confusus	50	50	12
Diguetia mohavea	45	55	71
Latrodectus hesperus	54	46	35
Arachnid average ± s.d.	51.5 ± 5.4	48.5 ± 5.4	47.3
Lizards			
Callisaurus draconoides	45	55	22
Cnemidophorous tigris	46	54	15
Gambelia wislizenii	33	67	15
Phrynosoma platyrhinos	28	72	18
Uta stansburiana	39	61	28
Birds			
Blue-gray gnatcatcher	36	64	69
Burrowing owl	36	64	14
Loggerhead shrike	35	65	17
Roadrunner	35	65	23
Mammals			
Ammospermophilus leucurus	71	29	7
Antrozous pallidus	31	69	16
Onychomys torridus	40	60	15
Pipistrellus hesperis	32	68	22
Vulpes macrotus	67	33	6
Vertebrate average ± s.d.	41 ± 12.9	59 ± 12.9	20.5

A taxon is the designated unit in which the diet was classified by the author. It varies from species to families and orders. The first four arachnids are scorpions; the last two, spiders.

It is impossible to construct a food web representing all CV predaceous arthropods. Thus, subwebs are centered around three common species that I studied extensively: the scorpion *P. mesaensis* (see earlier references), and the spiders *D. mohavea* and *L. hesperus* (Polis and McCormick, 1986b; Polis and Sculteure, unpublished; Nuessly and Goeden, 1984).

Figure 20.4 diagrams trophic relations of *D. mohavea* and *L. hesperus*. Eggs and spiderlings of *D. mohavea* are eaten by species that invade the egg sac: spiders (9 families, more than 14 species), solpugids, mites, mantispids, chrysopids, and the clerid *Phyllobaenus* (from Fig. 20.3). In general, eggs and spiderlings within sacs of most spider species are attacked by these taxa, several wasp and fly families, and other spiders. Sibling cannibalism is also frequent (Polis, 1981) and occurs among *D. mohavea* spiderlings. At least three trophic levels occur in the *D. mohavea* egg sac: the clerid eats other egg predators and is itself host to a wasp parasitoid.

Adult *Diguetia* prey on more than 70 species, including 14 families of predatory insects (*Photopsis* and *Phyllobaenus* are prey) and 8 spider species, including the same species of invading salticids and the araneophagous spider *Mimetus*. Adult *D. mohavea* are fed on by *Mimetus*, *P. mesaensis*, birds, and a parasitoid pompilid wasp. At least four cases of mutual predation cycling occur.

Predators constitute 54% of *L. hesperus*'s diet (Table 20.2). It falls prey to at least seven predators, including three species that it eats (mutual predators) and other black widows. Chalybian wasps specialize on *L. hesperus* and other theridiid spiders (Wasbauer and Kimsey, 1985). A hyperparasitoid wasp preys on both cached theridiid spiders and developing chalybians.

Paruroctonus mesaensis, a large and dense scorpion, has the greatest population biomass of any CV predator (vertebrate or invertebrate). Ontogenetic shifts during growth partially explain the more than 125 species of prey, including 47% other predators (Table 20.2) and mutual predation with at least 10 species (five solpugids, three scorpions, and two spiders; young *P. mesaensis* are eaten by the same species eaten by adults). Trophic relationships of *P. mesaensis* are presented in Figs. 20.4 and 20.6.

Note the complexity of these webs: cycling and looping via mutual predation and cannibalism are frequent; parallel path omnivory is the norm; omnivorous predators feed on herbivores, detritivores, predators, and predators of predators. Consequently, chain lengths are long even without including parasitoids or cycles, for example, detritus→termites→*Messor* ants→ant lions→*Latrodectus*→*Steatoda*→*Mimetus*→*P. mesaensis*→Eremobatid solpugids→*Hadrurus* scorpion→(vertebrate subweb).

I strongly suspect that the kind of trophic interactions depicted here represent those of the hundreds of other arthropod predators in the CV. Omnivory (due to age structure, opportunism, and the generally catholic diets of these arthropods), combined with a high diversity of insect and arachnid predators, necessarily creates the observed complexity. Complexity increases even further when we consider vertebrate predators of these arthropods.

Arthropodivorous vertebrates. Arthropodivory, the consumption of arthropods, conveys that predators eat all types of arthropods (insects, arachnids, myriapods, and terrestrial Crustacea). Most vertebrates (83% of the 95 species) in the CV include arthropods in their diet (Table 20.2). Over half (58%) are primarily arthropodivorous, including 52% of the reptiles, 61% of the birds, and 50% of the mammals. Most (20 of 25) vertebrate species that are primarily carnivorous on other vertebrates also feed on arthropods. Two-thirds of the 56 primarily herbivorous or granivorous vertebrates eat arthropods, at times in large quantities (for example, 88% to 97% of the seasonal diet of the sage sparrow). In total, 71% of all reptile species, 88% of the birds, and 78% of the mammals primarily or secondarily eat insects or arachnids. Only 17 species are not reported to eat arthropods.

Arthropodivorous vertebrates are usually resource generalists that eat arthropods from all trophic categories, for example, arachnid and insect predators, detritivores, and herbivores. For example, of 36 arthropodivorous birds whose diet is detailed sufficiently, 58% eat spiders in addition to insects. Seven of the 10 lizards eat spiders and 5, scorpions. Spiders are eaten by 3 of 14 arthropod-eating mammals; scorpions, by 5. Predaceous arthropods form 28% to 71% (average 41%) of the diet of the CV vertebrate arthropodivores in Table 20.3.

Vertebrates that eat arthropods also tend to be trophic generalists. Most (28 of 55 species, or 51%) primary arthropodovores eat plants (59% of the 79 vertebrates that eat arthropods also eat plants). Of these 79 species, 32% are also carnivorous on other vertebrates.

A few arthropodivorous vertebrates tend to specialize. Ants form 89% by frequency (56% by volume) of the prey of the horned lizard, *Phrynosoma platyrhinos*. However, these lizards eat 17 other categories of prey (including spiders and solpugids), and 20% to 50% of the diet of some individuals are beetles. Worm snakes mainly eat termites and ants. No other vertebrates specialize on certain taxa of arthropods.

The trophic and resource generalization exhibited by arthropodivorous vertebrates is illustrated in the subweb focused around the predators of *P. mesaensis* (Fig. 20.6). All the birds, reptiles, and mammals that eat *P. mesaensis* also eat predaceous, detritivorous, and herbivorous insects and even plant material (44% of 16 predators). Many (81%) of the vertebrate predators on *P. mesaensis* are also carnivorous on other vertebrates (see Fig. 20.7; three of 11 arthropod predators on *P. mesaensis* also eat vertebrates).

Carnivorous vertebrates. Carnivorous vertebrates kill and eat other vertebrates. Twenty-four of the 95 CV vertebrate species are primarily carnivorous, and five others prey on vertebrates at least occasionally (Table 20.3). All carnivores are resource generalists preying on many species of vertebrates. Most (19 of 24) are trophic generalists that include arthropods (71%) or plants (33%). For example, coyotes eat other mammals (12 of 19 in the CV: rabbits, rodents, gophers, antelope ground squirrels, and even kit foxes and other coyotes), birds (including eggs and nestlings, for example, roadrunners, doves, quail), snakes (gopher and kingsnakes), lizards (horned lizards), and young tortoises, as well as arachnids (scorpions), insects, and fruit (see Fig. 20.7; Johnson and Hansen, 1979; Polis, unpublished data). Great horned owls consistently eat wolf spiders, tarantulas, centipedes, and Orthoptera in addition to rodents, squirrels, lizards, snakes, and other horned owls.

TABLE 20.3 Feeding categories of vertebrates resident in the Coachella Valley

Feeding category	Vertebrate class			
	Reptiles	Birds	Mammals	All vertebrates
Granivory	0/0/0	14/0/14	8/0/8	22/0/22
%	0/0/0	25/0/25	44/0/44	23/0/23
Herbivory	2/5/7	5/15/20	5/1/6	11/21/32
%	10/24/33	9/27/34	28/6/33	12/22/34
Arthropodivory	11/4/15	34/15/49	9/5/14	55/24/79
%	52/19/71	61/27/88	50/28/78	58/25/83
Carnivory	9/2/11	12/2/14	3/1/4	24/5/29
%	43/10/52	21/4/25	17/6/22	25/5/31
Total number of species	21	56	18	95

Species are classified as belonging primarily [food a major ($\geq$90%) component of the diet] or secondarily (food $\leq$10% of total diet) to a feeding category. Some omnivorous species (2 reptiles, 9 birds, 7 mammals) belong to two (each $\geq$33% of diet) or three (each $\geq$20% of diet) primary categories. The three numbers below each column head (for example, 11/21/32) indicate either the number (first row) or percentage (second row) of species in the primary (for example, 11 herbivorous species) or secondary (for example, 21) feeding category and the total in this category (for example, 32).

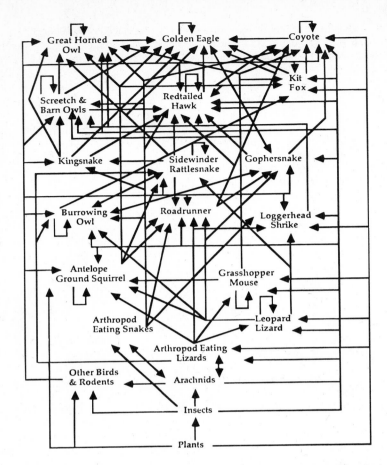

Figure 20.7 Trophic interactions involving a few of the 96 vertebrates resident in the Coachella Valley. No top predator exists in this subweb or within the Coachella Valley. An arrow returning to a taxon indicates cannibalism; a double-headed arrow, cycling via mutual predation. Note that the bottom of this subweb is simplified from the preceding five subwebs (Fig. 20.6).

Many carnivores (33%) feed on both carrion and live prey; examples include the sidewinder rattlesnake, raven, golden eagle, horned owl, red-tailed hawk, coyote, kit fox, and antelope ground squirrel. Such scavenging means that these primary carnivores also include in their diets both microorganisms (Janzen, 1977) and the rich arthropod fauna that uses decaying carcasses.

A web focusing on many of the carnivores in the CV is presented in Fig. 20.7. Note the frequency of omnivory and parallel path omnivory. Most carnivores (9 of 16 species in Fig. 20.7) are reported as cannibalistic. There are three cases of mutual predation. Each occurs because, although adult A eats B, the egg or nestling stage of A is eaten by B. For example, many snakes are nest predators (in the CV; the whipsnake, sidewinder, gophersnake, and rosy boa). These snakes eat eggs and nestlings of species (like the burrowing owl) whose adults are predators upon the same snakes (for example, gophersnakes and sidewinder rattlesnakes).

Finally, note that only the golden eagle approaches the status of top predator (a species without predators). However, even the golden eagle may not be a true top predator: at other locations, gophersnakes eat golden eagle eggs; parasitic trichomoniasis and sibling fratricide and cannibalism cause nestling mortality (30% and 8%, respectively). Thus, the Coachella Valley ecosystem represents a community and food web with potentially no top predators.

Discussion

Desert food webs.

Sandy areas in the Coachella Valley are the habitat for more than 174 species of plants, around 100 vertebrates, and thousands of arthropods, parasites, and soil organisms. These species collectively form a community connected trophically into a single web. Although only a few components of this biotic network were presented above, it is possible to summarize some general trends:

1. Each subweb is complex because of the large number of interactive species, age structure, and high omnivory. A food web representing all species would increase this complexity even more.
2. Age structure is central to food web complexity. Growth both necessitates and allows for changes in diet ("life history omnivory"), either gradually in continuous ontogenies or radically (at metamorphoses when development is discrete). Predators also change with size, and size classes also contribute to complexity.
3. Different compartments (microhabitats and times) are trophically connected (see the earlier summary of characteristics promoting complexity in food webs).

The CV ecosystem is not unique in its species diversity or properties of its food web. Other deserts are also characterized by thousands of species (Polis, 1991b). The same suite of plants, herbivores, detritivores, arthropodovores, parasitoids, parasites, and carnivores is present in all deserts. In particular, the universal existence of diverse assemblages of predaceous arthropods (Polis and Yamashita, 1991) and soil organisms (Freckman and Zak, 1991) must contribute to trophic complexities similar to that of the CV. Furthermore, omnivory is normal among desert consumers (Noy-Meir, 1974; Orians et al., 1977; Seely and Louw, 1980; Brown, 1986).

Bradley (1983) illustrated food web complexity with a source web focused on the predators of camel crickets in sandy areas of the Chihuahuan Desert. These predators (scorpions, solpugids, burrowing owls, grasshopper mice, and pallid bats) were quite omnivorous (only 2 of 27 species pairs were noninteractive, that is, not linked as predator or prey). Parallel path omnivory occurred for every predator, and cycling and looping were common: six cases of mutual predation occurred; six of eight species were cannibalistic. Finally, at least 18 three-species cycles existed.

The Namib desert dune system (Seely and Louw, 1980; Holm and Sholtz, 1980) features 136 species lumped into broad trophic groups. Detritus is the most important direct energy source and is used by many insect species (for example, 33 species of tenebrionid beetles). Herbivory is relatively unimportant. Insects are eaten by many arachnid (more than 35 species), insect (more than 12) and vertebrate (more than 25) arthropodivores (not including omnivorous ants and parasitoid wasps). Reptiles and birds eat both herbivorous and predaceous arthropods and vertebrates. The brown hyena and jackal are carnivorous on vertebrates.

The authors note several trends. (1) This web is characterized by marked temporal and spatial variation in the distribution and abundance of species. (2) Above- and belowground components are closely linked. (3) Most species (arthropods and vertebrates) are very omnivorous. Cannibalism, mutual predation, and cycling occur commonly.

TABLE 20.4 Summary food web adjacency matrix for the Coachella Valley sand community

	4	5	6	7	8	9	10	11	12	13	14	15	16	17	18	19	20	21	22	23	24	25	26	27	28	29	30
1.	X	X					X					X	X	X	X	X	X	X	X	X		X	X			X	
2.	X	X		X	X	X					X														X	X	X
3.	X	X		X	X	X					X									X				X	X	X	X
4.	X	X		X	X	X					X																
5.		X	X	X	X																						
6.			X		X																						
7.				X	X			X	X	X	X	X		X		X		X	X	X	X	X	X				
8.								X	X	X	X	X		X		X		X	X	X	X	X	X				
9.								X	X	X	X	X		X		X		X	X	X	X	X	X				
10.								X	X	X	X	X		X		X		X	X	X	X	X	X				
11.								X	X	X	X	X	X	X		X		X	X	X	X	X	X				
12.								X	X	X			X	X		X		X	X	X	X	X	X				
13.								X	X	X	X	X	X	X		X		X	X	X	X	X	X				
14.								X	X	X	X	X		X	X	X		X	X	X	X						
15.								X	X	X	X	X			X	X				X	X						
16.									X	X	X	X			X	X					X						
17.										X	X			X													
18.																											
19.																											
20.																			X	X				X	X	X	X
21.																								X	X	X	X
22.																				X				X	X	X	X
23.										X									X	X				X	X	X	X
24.										X									X	X		X	X	X	X	X	X
25.																						X	X	X	X	X	X
26.																								X	X	X	X
27.																										X	X
28.																										X	
29.																								X			X
30.																											

1. Plants and plant products
2. Detritus
3. Carrion
4. Soil microbes
5. Soil microarthropods and nematodes
6. Soil micropredators
7. Soil macroarthropods
8. Soil macroarthropod predators
9. Surface arthropod detritivores
10. Surface arthropod herbivores
11. Small arthropod predators
12. Medium arthropod predators
13. Large arthropod predators
14. Facultative arthropod predators
15. Life history arthropod omnivores
16. Spider parasitoids
17. Primary parasitoids
18. Hyperparasitoids
19. Facultative hyperparasitoids
20. Herbivorous mammals and reptiles
21. Primarily herbivorous mammals and birds
22. Small omnivorous mammals and birds
23. Predaceous mammals and birds
24. Arthropodivorous snakes
25. Primarily arthropodivorous lizards
26. Primarily carnivorous lizards
27. Primarily carnivorous snakes
28. Large, primarily predaceous birds
29. Large, primarily predaceous mammals
30. Golden eagle

Rows are prey and columns predators. The thousands of species were lumped extensively into 30 "kinds of organisms" to facilitate comparison with webs contained in the various catalogs. Entries along the diagonal indicate self-loops within that kind of organism. Entries below the diagonal denote cycles caused by mutual predation or longer path closures. Some statistics from this matrix are summarized in Table 20.5.

Comparison with published "Empirical Generalizations." As previously stated, food web theorists have produced a series of generalizations derived from catalogs of published webs (see Table 20.1). These generalizations are entering the peer-reviewed ecological literature and being accepted as truth (see May, 1986, 1988; Lawton and Warren, 1988; Lawton, 1989; Schoener, 1989). The food web of the Coachella Valley offers little support for them.

Web patterns from the CV may be compared with those from the published catalogs. An argument could be made that CV web statistics should not be calculated. Even with its demonstrated complexity, the web as described is vastly incomplete; scores of other subwebs could be presented. Any statistical treatment, therefore, represents only the level of complexity arbitrarily presented, rather than absolute web parameters. However, to facilitate comparisons, a highly simplified food web matrix of the CV community is presented in Table 20.4. The thousands of species are heavily aggregated into 30 "kinds of organisms" that form 22,220 food chains. Table 20.5 compares 19 web statistics from the CV with those from published catalogs of webs (Pimm, 1982; Briand, 1983; Cohen, Briand, and Newman, 1986; Schoener, 1989; Schoenly, Beaver, and Heumier, 1991). In the discussion that follows, I use the statistics from this simplified web and also those from the more complete web as previously described in the text.

1. The number of interactive species in the CV web is two orders of magnitude greater than the average number (17.8, Briand, 1983; 16.7, Cohen, Briand, and Newman, 1986; 24.3, Schoenly, Beaver, and Heumier, 1991) from webs analyzed by theorists. In fact, Briand and Cohen's most speciose web contains only 48 species, a diversity less than that of each CV group of plants, nematodes, mites, arachnids, bees, beetles, bombyliid flies, and birds. Winemiller (1990) similarly observed that the diversity in four (incomplete) tropical aquatic food webs is much greater (mean of 59 lumped "interactive trophic units") than those webs in the reference catalogs.

TABLE 20.5 Comparison of statistics from the Coachella Valley food web with means of the statistics from webs catalogued by Cohen and Briand

	Cohen and Briand	Schoenly et al.	Coachella Valley
Total number of "kinds" or species (S)	16.7	24.3	30
Total number of links per web (L)	31	43.1	289
Number of links per species (L/S)	1.99	2.2	9.6
Number of prey per predator	2.5	2.35	10.7
Number of predators per prey	3.2	2.88	9.6
Total prey taxa/total predator taxa	0.88	0.64	1.11
Minimum chain length	2.22	1	3
Maximum chain length	5.19	7	12* (18)
Mean chain length	2.71, 2.86	2.89	7.34*
Connectance (C = L/S[S − 1]/2)		0.25	0.49*
S × C	2–6	4.3	14.7*
% Basal species	19	16	10
% Intermediate species	52.5	38	90
% Top predators	28.5	46.5	0
% Primarily or secondarily herbivorous		14.6	60
% Primarily or secondarily saprovorous	21	35.5	37
% Omnivorous	27	22	78
% Consumers with self-loop	<1.0	<1.0	74
% Consumers with mutual predation cycle	≪1.0	≪1.0	53

Briand, 1983; Cohen, Briand, and Newman, 1986; see also Schoener, 1989 and Schoenly, Beaver, and Heumier, 1991. The Coachella statistics are derived from the highly aggregated food web matrix of Table 20.4. * Indicates that mutual links (cycles) were not used to calculate the statistic.

2. CV chain lengths average more links than 2.86 (Briand, 1983), 2.71 (Cohen, Briand, and Newman, 1986), or 2.89 (Schoenly, Beaver, and Heumier, 1991). Even excluding cycling and parasites, lengths of 6 to more than 11 are common. Both Cohen (personal communication, 1988) and I interpret short chain length as an "artifact of totally inadequate descriptions of real communities" (see also Patten, 1985). Short average lengths are derived from catalogs biased toward less complex, vertebrate-centered webs. Chains are lengthened in the CV primarily through trophic interactions among the soil biota, arthropods, and intermediate-level predators. Webs including invertebrates are typically more complex than those centering on vertebrates (see paragraph 10). Shorter chains exist (for example, plant→rabbit→eagle), but these are much less frequent simply because vertebrates form a relatively small proportion of all species when arthropods and soil biota are not ignored. For example, chains containing plants, insect herbivores, and insect parasitoids are estimated to contain over half of all known metazoans (Hawkins and Lawton, 1987). The average length of all chains in the simplified CV web is 7.3; its maximum chain length is 18 (12, with no cycles).

3. Omnivory is frequent in the CV web but statistically "rare" in catalogued webs (Pimm, 1982; Yodzis, 1984). In catalogued webs, 22% (Schoenly, Beaver, and Heumier's catalog) to 27% (Briand and Cohen's catalog) of all "kinds" are omnivorous; 78% in the simplified CV web are omnivorous. Adequate diet data, not lumping arthropods, and the inclusion of age structure partially explain the ubiquity of omnivory in the CV. It is possible that long chains in the CV exist because energy to top consumers comes from many (lower) trophic levels in addition to adjacent upper levels (see also Sprules and Bowerman, 1988).

4. Cycles are "unreasonable" to food web modelers and purported to be "very rare in terrestrial" webs (Pimm, 1982). Cycling also violates the assumption that body size relations arrange species along a cascade or hierarchy, such that a species preys on only those species below it and is preyed on only by those above it ("triangular web structure" of the cascade model: Briand and Cohen, 1987; Warren and Lawton, 1987; Lawton, 1989; Cohen, Briand, and Newman, 1990). However, cycling is neither rare nor abnormal in the CV. Most frequently, loops and cycles are a product of age structure: cannibalism and mutual predation usually result when large individuals eat smaller or younger individuals (these cycles can be integrated into the cascade model if age classes are represented separately; Cohen, personal communication, 1990). Other factors produce mutual predation (for example, group predation by ants). In the simplified community matrix, 74% of the consumers show self-loops; 44% are involved in cycles with another "kind." Analyses of other webs (Sprules and Bowerman, 1988; Winemiller, 1990) show that loops via cannibalism and mutual predation are common features.

5. CV species interact with many more species than those in catalogued webs. Cohen (1978) and Schoenly, Beaver, and Heumier (1991) calculated the number of predators on each prey (means of 3.2 and 2.88, respectively) and the number of prey per predator (2.5 and 2.35). Overall, catalogued species interact directly with 3.2 to 4.6 other species (Cohen, Newman, and Briand, 1985). Parameters from the CV are one to two orders of magnitude greater. Higher values exist first because most CV consumers eat many species (Table 20.2, diets range from 15 to more than 125 items; a few arthropod herbivores and some parasites are exceptions). Second, most species (for example mesquite) are eaten by tens to hundreds of species. Such discrepancies occur because catalogued webs lump species, and diet data are grossly inadequate. However, even the highly lumped CV web (Table 20.4) shows that each "kind" is eaten by about 10 predators and each consumer eats about 10 prey. Winemiller (1990) also observed high linkage density among species in his aquatic food webs.

6. Top predators are rare or nonexistent in the CV. Coyotes, kit foxes, horned owls, and golden eagles (the largest predators) suffer the fewest predators, but each is the reported prey of other species. This finding stands in marked contrast with catalogued webs: 28.5% (Briand and Cohen, 1984) to 46.5% (Schoenly, Beaver, and Heumier, 1991) of kinds were top predators. This great discrepancy is undoubtedly due to the inadequacy of diet information in the catalogued webs, or because these webs in general only focus on a limited subset of full trophically linked communities.

7. CV data pose great difficulty to the observation that prey–predator ratios are less than 1.0 (0.88, Briand and Cohen, 1984; 0.64, Schoenly, Beaver, and Heumier, 1991). That is, the number of organisms heading rows (denoting prey) in web matrices is less than the number heading columns (predators). It is easy to show that the ratio in the CV and other real communities should be greater than 1.0. As all heterotrophs must obtain food, every animal should head a column. Rows (prey) include plants, detritus, and all animals except those with no predators (that is, top predators). Let x be the number of animal species that are intermediate predators (both predator and prey). Then the total number of prey is x + the number of plant species (= 174 in the CV) + the number of categories of detritus and carrion; the number of predators is x + the number of top predators (= 0 or 1 in the CV). If more species of plants exist than top predators, then the prey–predator ratio will always be >1.0. Few, if any, real communities have more top predators than autotrophs. The appearance that top predators are more speciose is an artifact (see paragraph 6). Lumping obliterates the actual prey–predator ratio probably because more species of plants get lumped by food web scientists, who are zoologically oriented, than more easily recognized animals that are top predators. The CV community web exhibits a prey–predator ratio of 1.11. Prey–predator ratios less than 1.0 have been criticized by others (Glasser, 1983; Jeffries and Lawton, 1985; Paine, 1988).

8. Factors 1 to 7 make the CV web much more complex than catalogued webs. For example, the number of trophic links in the latter varies from 31 (Cohen, Briand, and Newman, 1986) to 43 (Schoenly, Beaver, and Heumier, 1991); only 2 of 133 webs had more than 100 links (Cohen, Briand, and Newman, 1986). The average CV *subweb* (Fig. 20.7) has 54.7 links; the carnivore subweb alone, 107; and the simplified community web (Table 20.4), 289. Most communities exhibit highly reticulated trophic interactions. For example, Winemiller's (1990) tropical aquatic webs averaged 514 links (range, 208 to 1243), although many species were not included.

9. The CV web questions the utility of the concept of "trophic level." A discrete trophic level is defined as a set of whole (not partial) organisms with a common number of chain links among them and primary producers. The nearly universal presence of omnivory, age structure, and detritivory and the frequency of cycling make this concept nonoperational. This is clear when the "trophic unfolding" methodology of Higashi, Burns, and Patten (1989) is employed to convert reticulated networks into isomorphic "macrochains." In these, each trophic level is a mixture of "kinds," and each "kind" is composed of a distribution of trophic levels. No single trophic level is occupied by consumers that ontogenetically, seasonally, or opportunistically eat all trophic levels of arthropods, in addition to plant material and (for carnivores) vertebrates. Cycling further blurs the concept: if A eats B and B eats A also, is B on the first, third, and (after further cycling) fifth, and so on, trophic levels (ad infinitum)? Trophic unfolding computes these relationships. The fact that detritus is itself an aggregate including "recycled" organisms that conceivably originate from many trophic levels suggests that the notion of "trophic continuum" (Cousins, 1980), as implemented by trophic unfolding, is a more accurate paradigm than that of "trophic level." In such a continuum, should detritus be called a primary producer or placed on a higher trophic level? I am not alone in criticizing the trophic level concept (see Gallopin, 1972; Cousins, 1980, 1985, 1987; Levine, 1980; Barnes and Hughes, 1988; Lawton, 1989; Winemiller, 1990; and also see Higashi, et al., 1989; Baird and Ulanowicz, 1989).

10. Patterns 4 and 5 in Table 20.1 are confirmed by the CV web. First, separate compartments did not exist within one habitat. Second, arthropod-dominated systems are more complex than those dominated by vertebrates. However, few communities are not dominated (in number of individuals or species) by arthropods (Hawkins and Lawton, 1987; May, 1988). So not lumping arthropods, the most speciose taxon on this planet, should increase the complexity of any web, including those catalogued by theorists.

Overall, a general lack of agreement exists between patterns from the CV and those from catalogs of published webs. Is the CV web unique or are catalogued webs so simplified that they have lost all realism? That catalogued webs depict so few species, such low ratios of predators on prey and prey eaten by predators, so few links, so little omnivory, a veritable absence of cycling, and such a high proportion of top predators argues strongly that they do not adequately describe real biological communities. Taylor (1984), Paine (1988), Lawton (1989), and Winemiller (1990) have reached a similar conclusion.

This conclusion carries important implications. First, controversies over such issues as the causes of short chain length and omnivory, complexity versus stability of ecosystems, and the role of dynamics versus energetics in shaping web patterns (see DeAngelis, Post, and Sugihara, 1983; Yodzis, 1984; Lawton, 1989) are based on patterns abstracted from catalogued webs. If the catalogs are an inadequate data base, these controversies may be nonissues and theorists are trying to explain phenomena that do not exist. This is a real possibility (see Lawton and Warren, 1988; Paine, 1988; Lawton, 1989). Second, characteristics of the CV web (complexity, omnivory, long chain lengths, looping and cycling, absence of compartments) considerably reduce stability in dynamic models of food webs (Pimm, 1982; Pimm and Rice, 1987). These simplified models apparently tell us little about the structure of food webs in nature (see also Hastings, 1988; and Winemiller, 1990).

SELF-CRITIQUE AND PROSPECTUS

Several issues need to be addressed in the future:

1. Are desert food webs typical of other webs? Deserts may differ in two main ways from other ecosystems. First, deserts are often considered to be relatively simple, characterized by low productivity and species richness (Noy-Meir, 1974; Seely & Louw, 1980; Wallwork, 1982; Whitford, 1986; see Polis 1991b). Should such depauperate communities translate into relatively simple food webs? If so, the complexity of the CV web can be expected to be much less than that of more typical ecosystems.

Second, are desert consumers markedly more omnivorous than other consumers? It is impossible to answer this question with rigor. I approach the issue by indicating that the features that promote omnivory in the CV food web are present in other systems as well (Table 20.6). Several authors make clear that a high level of omnivory is not restricted to deserts. Menge and Sutherland (1987) cite several studies showing that omnivory is rather frequent in some terrestrial communities and characterizes normal feeding relationships in aquatic and marine environments. Price (1975) maintains that omnivory is a normal feeding habit throughout the animal kingdom. Walter (1987, p. 228) argues that opportunistic omnivory "appears to be the common feeding behavior" in soil microarthropod populations and is common among terrestrial vertebrates and invertebrates. Moore, Walter, and Hunt (1988) provide strong evidence that omnivory is one of the most frequent and dynamically important mechanisms of trophic linkage among a diverse array of soil arthropods, protozoans, and nematodes in detrital food webs. Sprules and Bowerman (1988) concluded that omnivory is frequent and common in zooplankton assemblages. These authors also note long chains, cannibalism, and mutual predation. Winemiller (1990) found that omnivory was "extremely common in all food webs."

Regardless of whether deserts are unique, desert webs are still of general importance. Deserts occupy at least one-quarter of the earth's land surface (Seely and Lowe, 1980; Crawford, 1981), and the patterns observed in the CV and other deserts thus describe a good fraction of the terrestrial communities on this planet.

2. Should all naturally occurring trophic links be included in a community food web, or should we include only "important" links? Are some links too weak or too unusual to list (May, 1983a, 1986; Lawton, 1989; Schoener, 1989)? I included all links in the CV web. This decision

TABLE 20.6 General factors that promote omnivory in the food webs of the Coachella Valley and elsewhere

1. *"Life history omnivory"* is extremely common and widespread in aquatic, marine, and terrestrial habitats. Such dietary shifts during ontogeny may be gradual (with growth) or abrupt (with metamorphosis).
2. *Predators disregard the feeding history of prey* (equals different channel omnivory). The existence of multiple trophic levels within arthropods and within the soil biota causes consumers of these groups to feed on a diversity of trophic levels. For example, arthropodovores eat arthropods that are herbivores, detritivores, parasitoids, predators, and predators of predators.
3. *Opportunistic feeding on abundant resources* is commonly reported among consumers. Granivorous birds and rodents primarily eat seeds during most months, but normally feed on insects and spiders when arthropods become abundant in spring. Many carnivorous mammals are frugivorous when fruit appear seasonally.
4. *Foraging theory* predicts and empirical studies almost universally show that diet reflects food availability and quality. Diets expand during periods of food scarcity.
5. *Cannibalism and intraguild predation* of heterospecifics from the same trophic level regularly occur among all trophic groups (herbivores, detritivores, and predators).
6. *Arthropods, parasitoids, and gall fauna* exhibit complex feeding relations.
7. *Consumers of food in which other consumers live* regularly eat these other consumers:
   a. Scavengers not only eat carrion but the microbes and various trophic groups of arthropods that live within these foods.
   b. Frugivores and granivores commonly eat insects associated with fruit and seeds.
   c. Detrivores eat detritus, microbes, and often smaller detritivores.

See text for references.

was based on four factors. First, and most important, it would be arbitrary and impossible to evaluate which links are and are not "important." Most CV consumers include 20 to more than 50 items in their diets. Which items should be included or excluded? It is probable that at least some consumers (especially in deserts) exist or are successful because they are sufficiently flexible to utilize a number of infrequent links that sum to form an important energy source, at least periodically.

Second, it is not clear which links are important in terms of population dynamics (see also Paine, 1988). Diet and dynamics are not necessarily correlated. For example, a 1% representation of a rare species in the diet of a common species may produce much mortality to the rare species; conversely, a 100% representation of a common species in the diet of a rare species may scarcely affect the common species (see Polis, 1981, for examples). Infrequent predation events particularly should influence the dynamics of top predators, animals that are characteristically large and (consequently) relatively rare. Furthermore, a short but intensive predation event may not contribute much to the diet of a predator, but may be central to prey dynamics (Polis, 1991a).

Third, "food webs" should describe the full trophic interrelations within a community. Each link makes the description richer and more completely approaching the reality of the community. Fourth, exclusion of certain links from webs produces a systematic bias against those characteristics that foster complexity.

3. Not all diet information came from studies in the Coachella Valley. How this influences the food web is uncertain. However, the overall conclusion that food webs in nature are of great complexity would not be unduly influenced by errors arising from the use of other studies to construct the CV food web given here.

CONCLUSIONS

Food webs in the real world are much more complex than contemporary theory would have us believe. This complexity arises from the large number of species present, frequent omnivory, age structure, interconnections among compartments, and diversity of arthropod and soil or benthic faunas. It appears that much "food web theory" is not very descriptive or predictive of real nature. The catalogs of webs that have been extensively used to abstract empirical generalizations and support theory were derived from grossly incomplete representations of communities in terms of both diversity and trophic connections. Consequently, theorists have constructed an oversimplified, and, I think, invalid view of community structure. The inherent complexity of natural communities makes web construction by empiricists and analysis by theorists difficult, but this should not be cause for underrepresenting the realities of complexity that exist in nature. This whole book is dedicated to a practitioner of complex ecology as a countercurrent against a widespread tendency in the science to be simplistic and superficial in the treatment of the intricate.

What good are complex food web representations and analyses? At a minimum, they describe reality—communities and their trophic interconnections that underlie issues of assembly and system function. In an ideal world, hypotheses and generalizations made by theorists could be viewed as a stage in the evolution of understanding food webs; they are not a finished product, but in being flawed they serve the purpose of progress. Theory is designed to provoke concept-focused fieldwork that may promote the next iteration of descriptive models and generalizations, which will, in turn, encourage more empirical work, all the while approaching more accurate descriptions and predictions of reality. A strong implication of the present analysis is that many existing hypotheses and generalizations about food webs are simply wrong. Much current theory (for example, Pimm, 1982; DeAngelis, Post, and Sugihara, 1983; May, 1983a,b; Lawton and Warren, 1988), for example, would hold that the Coachella Valley food web should be completely unstable. To advance understanding on this and related issues, an adequate data base of community food webs must be assembled, experiments must be conducted, and new questions asked. This will take time, but only when the inherent complexities of food webs as networks are fully explored will a useful, heuristic theoretical framework be constructable. We should critically reevaluate where we now stand since it appears that much of existing food web theory is in clear need of revision and new direction. It is hoped that many of the avenues previously taken by the pioneers of food web study may yet be useful in the new, complex domain.

ACKNOWLEDGMENTS

Many people contributed ideas, data, and energy during the 9-year gestation of this paper. S. McCormick Carter was central to its development. S. Frommer, W. Icenogle, W. Mayhew, J. Pinto, K. Sculteure, R. Ryti, A. Muth, and F. Andrews provided data on the Coachella. The manuscript benefited greatly from suggestions by J. Brown, J. Cohen, C. Crawford, R. Holt, J. Lawton, B. Menge, J. Moore, C. Myers, E. Pianka, S. Pimm, K. Schoenly, and T. Schoener. Special thanks to Ken Schoenly for calculating many of the web statistics in Table 20.4. Fieldwork was financed partially by the U.S. National Science Foundation and the Natural Science Committee and Research Council of Vanderbilt University.

REFERENCES

ANDERSEN, D. 1987. Below ground herbivory in natural communities: a review emphasizing fossorial animals. *Quart. Rev. Biol.* 62:261–286.

ANDREWS, F. G., HARDY, A. R., and GIULIANI, D. 1979. The Coleopterous Fauna of Selected California Sand Dunes. Report, California Department of Food and Agriculture, Sacramento, California. 142 pp.

ASKEW, R. 1971. *Parasitic Insects.* American Elsevier, New York. 316 pp.

BAIRD, D. and ULANOWICZ, R. E. 1989. The seasonal dynamics of the Chesapeake Bay Ecosystem. *Ecol. Mon.* 59:329–364.

BARNES, R. S., and HUGHES, R. 1988. *Marine Ecology.* Blackwell, London. 351 pp.

BENT, A. C. 1932,1937,1938,1942,1948,1949, and 1958. *Life Histories of North American Birds.* Bulletins 162, 167, 170, 179, 195, 196, 211, U.S. National Museum, Washington, D.C.

BRADLEY, R. 1983. Complex food webs and manipulative experiments in ecology. *Oikos* 41:150–152.

BRIAND, F. 1983. Environmental control of food web structure. *Ecology* 64:253–263.

BRIAND, F., and COHEN, J. 1984. Community food webs have invariant-scale structure. *Nature* 5948:264–267.

BRIAND, F., and COHEN, J. 1987. Environmental correlates of food chain length. *Science* 238:956–960.

BROWN, J. 1986. The role of vertebrates in desert ecosystems. In W. Whitford (ed.), *Pattern and Process in Desert Ecosystems.* University of New Mexico Press, Albuquerque, New Mexico. pp. 51–71.

BROWN, J., REICHMAN, O. J., and DAVIDSON, D. 1979. Granivory in desert ecosystems. *Ann. Rev. Ecol. Systematics* 10:201–227.

BUSH, A., AHO, J., and KENNEDY, C. 1990. Ecological versus phylogenetic determinants of helminth parasite community richness. *Evol. Ecol.* 4:1–20.

COHEN, J. E. 1978. *Food Webs and Niche Space.* Monographs in Population Biology, 11. Princeton University Press, Princeton, New Jersey. 189 pp.

COHEN, J. E. 1989. Food webs and community structure. In Roughgarden, J., May, R., and Levin, S. (eds.), *Perspectives in Ecological Theory.* Princeton University Press, Princeton, New Jersey. pp. 181–202.

COHEN, J. and BRIAND, F. 1984. Trophic links of community food webs. *Proc. Nat. Acad. Sci. USA* 81:4105–4109.

COHEN, J. E., and NEWMAN, C. 1985. A stochastic theory of community food webs. I. Models and aggregated data. *Proc. Royal Soc. London B* 224:421–448.

COHEN, J. E., and NEWMAN, C. 1988. Dynamic basis of food web organization. *Ecology* 89:1655–1664.

COHEN, J. E., BRIAND, F., and NEWMAN, C. 1986. A stochastic theory of community food webs. III. Predicted and observed lengths of food chains. *Proc. Royal Soc. London B* 228:317–353.

COHEN, J. E., BRIAND, F., and NEWMAN, C. 1990. *Community Food Webs: Data and Theory.* Springer-Verlag, New York. 312 pp.

COHEN, J. E., NEWMAN, C., and BRIAND, F. 1985. A stochastic theory of community food webs. II. Individual webs. *Proc. Royal Soc. London B* 224:449–461.

COUSINS, S. 1980. A trophic continuum derived from plant structure, animal size and a detritus cascade. *J. Theor. Biol.* 82:607–618.

COUSINS, S. 1985. Ecologists build pyramids again. *New Scientist* 107:50–54.

COUSINS, S. 1987. The decline of the trophic level concept. *Trends Ecol. Evolution* 2:312–316.

CRAWFORD, C. S. 1981. *Biology of Desert Invertebrates.* Springer-Verlag, New York. 314 pp.

CRAWFORD, C. S. 1986. The role of invertebrates in desert ecosystems. In W. Whitford (ed.), *Pattern and Process in Desert Ecosystems.* University New Mexico Press, Albuquerque, New Mexico. pp. 73–92.

CRAWFORD, C. S. 1991. Macroarthropod detritivores. In Polis, G. A. (ed.), *Ecology of Desert Communities.* University of Arizona Press, Tucson, Arizona. pp. 89–112.

DEANGELIS, D. L., POST, W. M., and SUGIHARA, G. 1983. Current Trends in Food Web Theory: Report on a Food Web Workshop. Oak Ridge National Laboratory Technical Memoranda 5983, Oak Ridge, Tennessee. 137 pp.

DUGGINS, D., SIMENSTAD, C., and ESTES, J. 1989. Magnification of secondary production by kelp detritus in coastal marine ecosystems. *Science* 245:170–173.

EDNEY, E. B., MCBRAYER, J., FRANCO, P., and PHILLIPS, A. 1974. Distribution of Soil Arthropods in Rock Valley, Nevada. US/IBP Desert Biome Research Memo 74-32:53–58.

ELGAR, M. A., and CRESPI, B. J. 1992. *Cannibalism: Ecology and Evolution Among Diverse Taxa.* Oxford University Press, Oxford, New York. 361 pp.

FENCHEL, T. 1988. Marine plankton food chains. *Ann. Rev. Ecol. Systematics* 19:19–38.

FERGUSON, W. 1962. Biological characteristics of the mutellid subgenus *Photopsis* Blake and their systematic value. *Univ. Calif. Berkeley Publ. Entomology* 27:1–82.

FLETCHER, M, GRAY, T., and JONES, J. 1987. *Ecology of Microbial Communities.* Cambridge University Press, New York. 448 pp.

FRANCO, P., EDNEY, E., and MCBRAYER, J. 1979. The distribution and abundance of soil arthropods in the northern Mojave Desert. *J. Arid Environ.* 2:137–149.

FRECKMAN, D., and ZAK, J. 1991. Desert soil communities. In Polis, G. A. (ed.), *Ecology of Desert Communities.* University of Arizona Press, Tucson, Arizona. pp. 55–88.

FROMMER, S. I. 1986. A Hierarchic Listing of the Arthropods Known to Occur within the Deep Canyon Desert Transect. Deep Canyon Publications, Riverside, California. 133 pp.

GALLOPIN, G. 1972. Structural properties of food webs. In Patten, B. C. (ed.), *Systems Analysis and Simulation in Ecology,* Vol. II. Academic Press, New York, pp. 241–282.

GIER, H., KRUCKENBERG, S., and MARLER, R. 1978. Parasites and diseases of coyotes. In Bekoff, M. (ed.), *Coyotes, Biology, Behavior and Management.* Academic Press, New York, pp. 37–69.

GLASSER, J. 1983. Variation in niche breadth and trophic position: on the disparity between expected and observed species packing. *Amer. Nat.* 122:542–548.

GORDON, S. 1978. Food and Foraging Ecology of a Desert Harvester Ant, *Veromessor pergandei* (Mayr). Ph.D. dissertation, University of California at Berkeley, Berkeley, California. 158 pp.

HASTINGS, A. 1988. Food web theory and stability. *Ecology* 69:1665–1668.

HAWKINS, B., and GOEDEN, R. 1984. Organization of a parasitoid community associated with a complex of galls on *Atriplex* spp. in southern California. *Ecol. Entomol.* 9:271–292.

HAWKINS, B., and LAWTON, J. 1987. Species richness of parasitoids of British phytophagous insects. *Nature* 326:788–790.

HAWKS, D. 1982. A Checklist of the Butterflies of Deep Canyon. Deep Canyon Publications, Riverside, California. 10 pp.

HESSEN, D., ANDERSEN, T., and LYCHE, A. 1990. Carbon metabolism in a humic lake: pool sizes and cycling through zooplankton. *Limnol. Oceanogr.* 35:84–99.

HIGASHI, M., BURNS, T. P., and PATTEN, B. C. 1989. Food network unfolding: an extension of trophic dynamics for application to natural ecosystems. *J. Theor. Biol.* 140:243–261.

HOLM, E., and SCHOLTZ, C. 1980. Structure and pattern of the Namib Desert ecosystem at Gobabeb. *Madoqua* 12:5–37.

JANZEN, D. 1977. Why fruits rot, seeds mold, and meat spoils. *Amer. Nat.* 111:691–713.

JEFFRIES, M., and LAWTON, J. 1985. Predator–prey ratios in communities of freshwater invertebrates: the role of enemy free space. *Freshwater Biol.* 15:105–112.

JOHNSON, M., and HANSEN, R. 1979. Coyote food habits on the Idaho National Engineering Laboratory. *J. Wildlife Management* 43:951–956.

LAWTON, J. 1989. Food webs. In Cherrett J. (ed.), *Ecological Concepts,* Blackwell Scientific, Oxford. pp. 43–78.

LAWTON, J., and WARREN, P. 1988. Static and dynamic explanation of patterns in food webs. *Trends Ecol. Evolution* 3:242–245.

LEVINE, S. 1980. Several measures of trophic structure applicable to complex food webs. *J. Theor. Biol.* 83:195–207.

LOUW, G. N., and SEELY, M. K. 1982. *Ecology of Desert Organisms.* Longman, New York. 194 pp.

MACFAYDEN, A. 1963. The contribution of the fauna to the total soil metabolism. In Doeksen, J., and van der Drift, J. (eds.), *Soil Organisms.* North-Holland, Amsterdam. 453 pp.

MacKay, W. 1990. Ants and termites. In Polis, G. A. (ed.), *Ecology of Desert Communities.* University of Arizona Press, Tucson, Arizona, pp. 113–150.

May, R. 1983a. The structure of food webs. *Nature* 301:566–568.

May, R. 1983b. Parasitic infections as regulators of animal populations. *Amer. Scient.* 71:36–45.

May, R. 1986. The search for patterns in the balance of nature: advances and retreats. *Ecology* 67:1115–1126.

May, R. 1988. How many species are there on earth? *Science* 241:1441–1448.

Mayhew, W. W. 1981. Vertebrates and Their Habitats on the Deep Canyon Transect. Deep Canyon Publications, Riverside, California. 32 pp.

McCormick, S. J., and Polis, G. A. 1986. Comparison of the diet of *Paruroctonus mesaensis* at two sites. In Eberhard, W. G., Lubin, Y. D., and Robinson, B. (eds.). *Proc. IX International Arachnological Congress.* Smithsonian Institution Press, Washington, D.C. pp. 167–171.

McCormick, S., and Polis, G. A. 1990. Prey, predators and parasites. In Polis, G. A. (ed.), *Biology of Scorpions.* Stanford University Press, Palo Alto, California, pp. 294–320.

McKinnerney, M. 1978. Carrion communities in the northern Chihuahuan Desert. *Southwestern Naturalist* 23:563–576.

Menge, B., and Sutherland, J. 1987. Community regulation: variation in disturbance, competition, and predation in relation to environmental stress and recruitment. *Amer. Nat.* 130:730–757.

Moore, J. C., Walter, D., and Hunt, H. W. 1988. Arthropod regulation of micro- and mesobiota in below ground detrital webs. *Ann. Rev. Entomology* 33:419–439.

Noy-Meir, I. 1974. Desert ecosystems: higher trophic levels. *Ann. Rev. Ecol. Systematics* 5:195–213.

Nuessly, G., and Goeden, R. 1984. Aspects of the biology and ecology of *Diguetia mojavea* Gertsch (Araneae, Diguetidae). *J. Arachnology* 12:75–85.

Odum, E. P., and Biever, L. 1984. Resource quality, mutualism, and energy partitioning in food chains. *Amer. Nat.* 124:360–376.

Orians, G., Cates, R., Mares, M., Moldenke, A., Neff, J., Rhoades, D., Rosenzweig, M., Simpson, B., Schultz, J., and Tomoff, C. 1977. Resource utilization systems. In Orians, G., and Solbrig, O. (eds.), *Convergent Evolution in Warm Deserts.* Dowden, Hutchinson and Ross, Stroudsburg, Pennsylvania. pp. 164–224.

Paine, R. T. 1980. Food webs: linkage, interaction strength and community infrastructure. *J. Animal Ecol.* 49:667–685.

Paine, R. T. 1988. On food webs: road maps of interaction or grist for theoretical development? *Ecology* 69:1648–1654.

Patten, B. C. 1985. Energy cycling, length of food chains, and direct versus indirect effects in ecosystems. *Can. Bull. Fish. Aquatic Sci.* 213:119–138.

Patten, B. C., Higashi, M., and Burns, T. P. 1990. Trophic dynamics in ecosystem networks: significance of cycles and storage. *Ecol. Mod.* 51:1–28.

Peters, R. 1988. Some general problems for ecology illustrated by food web theory. *Ecology* 69:1673–1674.

Pimm, S. L. 1982. *Food Webs.* Chapman and Hall, New York. 219 pp.

Pimm, S. L., and Rice, J. 1987. The dynamics of multispecies, multi-life-stage models of aquatic food webs. *Theor. Population Biol.* 32:303–325.

Polis, G. A. 1979. Diet and prey phenology of the desert scorpion, *Paruroctonus mesaensis* Stahnke. *J. Zoology (London)* 188:333–346.

Polis, G. A. 1981. The evolution and dynamics of intraspecific predation. *Ann. Rev. Systematics Ecol.* 12:225–251.

Polis, G. A. 1984a. Age structure component of niche width and intraspecific resource partitioning: can age groups function as ecological species? *Amer. Nat.* 123:541–564.

POLIS, G. A. 1984b. Intraspecific predation and "infant killing" among invertebrates. In Hausfater, G., and Hrdy, S. (eds.), *Infanticide: Competitive and Evolutionary Perspectives.* Aldine Publishing, New York, pp. 87–104.

POLIS, G. A. 1988a. Exploitation competition and the evolution of interference, cannibalism and intraguild predation in age/size structured populations. In Perrson, L., and Ebenmann, B. (eds.), *Size Structured Populations: Ecology and Evolution.* Springer-Verlag, New York. pp. 185–202.

POLIS, G. A. 1988b. Trophic and behavioral responses of desert scorpions to harsh environmental periods. *J. Arid Environ.* 14:123–134.

POLIS, G. A. 1991a. Food webs in desert communities: complexity, diversity and omnivory. In Polis, G. A. (ed.), *Ecology of Desert Communities.* University of Arizona Press, Tucson, Arizona. pp. 383–438.

POLIS, G. A. 1991b. Desert communities: an overview of patterns and processes. In Polis, G. A. (ed.), *Ecology of Desert Communities.* University of Arizona Press, Tucson, Arizona. pp. 1–26.

POLIS, G. A., and MCCORMICK, S. J. 1986a. Patterns of resource use and age structure among species of desert scorpion. *J. Animal Ecol.* 55:59–73.

POLIS, G. A., and MCCORMICK, S. J. 1986b. Scorpions, spiders and solpugids: predation and competition among distantly related taxa. *Oecologia* 71:111–116.

POLIS, G. A., and MCCORMICK, S. J. 1987. Intraguild predation and competition among desert scorpions. *Ecology* 68:332–343.

POLIS, G. A., and YAMASHITA, T. 1990. Arthropod predators. In Polis, G. A. (ed.), *Ecology of Desert Communities.* University of Arizona Press, Tucson, Arizona, pp. 180–222.

POLIS, G. A., MYERS, C. A., and HOLT, R. D. 1989. The evolution and dynamics of intraguild predation between potential competitors. *Ann. Rev. Ecol. Systematics* 20:297–330.

POLIS, G. A., SISSOM, W. D., and MCCORMICK, S. J. 1981. Predators of scorpions: field data and a review. *J. Arid Environ.* 4:309–327.

POWELL, J., and HOGUE, C. 1979. *California Insects.* University of California Press, Berkeley, California. 339 pp.

PRICE, P. 1975. *Insect Ecology.* Wiley-Interscience, New York. 514 pp.

REICHMAN, O. J. 1975. Relation of desert rodent diet to available resources. *J. Mammalogy* 56:731–751.

RICH, P. 1984. Trophic-detrital interactions: vestiges of ecosystem evolution. *Amer. Nat.* 123:20–29.

RYAN, M. 1968. Mammals of Deep Canyon. The Desert Museum, Palm Springs, California. 138 pp.

RYTI, R., and CASE, T. 1988. Field experiments on desert ants: testing for competition between colonies. *Ecology* 69:1993–2003.

SALONEN, K., and HAMMAR, T. 1986. On the importance of dissolved organic matter on the nutrition of zooplankton in some lake waters. *Oecologia* 68:246–253.

SANTOS, P., PHILLIPS, J., and WHITFORD, W. 1981. The role of mites and nematodes in early stages of litter decomposition in the desert. *Ecology* 62:664–669.

SCHOENER, T. 1989. Food webs from the small to the large: probes and hypotheses. *Ecology* 70:1559–1589.

SCHOENLY, K. 1983. Arthropods associated with bovine and equine dung in an ungrazed Chihuahuan desert ecosystem. *Ann. Entomological Soc. Amer.* 76:790–796.

SCHOENLY, K., and COHEN, J. E. 1991. Temporal variation in food web structure: 16 empirical cases. *Ecol. Mon.* 61:267–298.

SCHOENLY, K., and REID, W. 1983. Community structure in carrion arthropods in the Chihuahuan desert. *J. Arid Environ.* 6:253–263.

SCHOENLY, K., BEAVER, R., and HEUMIER, T. 1991. On the trophic relations of insects: a food web approach. *Amer. Nat.* 137:597–638.

SEASTEDT, T., RAMUNDO, R., and HAYES, D. 1988. Maximization of densities of soil animals by foliage herbivory: empirical evidence, graphical and conceptual models. *Oikos* 51:243–248.

SEELY, M. K., and LOUW, G. N. 1980. First approximation of the effects of rainfall on the ecology and energetics of a Namib Desert dune ecosystem. *J. Arid Environ.* 3:25–54.

SOHOLT, L. 1973. Consumption of primary production by a population of Kangaroo rats (*Dipodomys merriami*) in the Mojave Desert. *Ecol. Mon.* 43:357–376.

SPRULES, W., and BOWERMAN, J. 1988. Omnivory and food chain lengths in zooplankton food webs. *Ecology* 69:418–426.

SUGIHARA, G., SCHOENLY, K., and TROMBLA, A. 1989. Scale invariance in food web properties. *Science* 245:48–52.

TAYLOR, J. 1984. A partial food web involving predatory gastropods on a Pacific fringing reef. *J. Exp. Marine Biol. Ecol.* 74:273–290.

TELFORD, S. 1971. A comparative study of endoparasitism among some California lizard populations. *Amer. Midl. Nat.* 83:516–554.

VITT, L. J. 1991. Reptiles. In Polis, G. A. (ed.), *Ecology of Desert Communities.* University of Arizona Press, Tucson, Arizona. pp. 249–277.

WALLWORK, J. 1982. *Desert Soil Fauna.* Praeger Publishers, New York. 296 pp.

WALTER, D. E. 1987. Trophic behavior of "mycophagous" microarthropods. *Ecology* 68:226–229.

WARREN, P. H., and LAWTON, J. H. 1987. Invertebrate predator-prey body size relationships: An explanation for upper triangular food webs and patterns in food web structure? *Oecologia* 74:231–235.

WASBAUER, M. S., and KIMSEY, L. 1985. California spider wasps of the sub-family Pompilinae. *Bull. California Insect Survey* 26:1–130.

WEATHERS, B. 1983. *Birds of Southern California's Deep Canyon.* University of California Press, Berkeley, California. 266 pp.

WELTY, J. 1962. *The Life of Birds.* W. B. Saunders, Philadelphia. 546 pp.

WERNER, E., and GILLIAM, J. 1984. The ontogenetic niche and species interactions in size-structured populations. *Ann. Rev. Ecol. Systematics* 15:393–426.

WHEELER, Q. 1990. Insect diversity and cladistic constraints. *Ann. Entomological Soc.* 83:1031–1047.

WHEELER, G. C., and WHEELER, J. 1973. *Ants of Deep Canyon.* University of California Press, Berkeley, California. 192 pp.

WHITFORD, W. G. 1986. Decomposition and nutrient cycling in deserts. In Whitford, W. G. (ed.), *Pattern and Process in Desert Ecosystems.* University of New Mexico Press, Albuquerque, New Mexico, pp. 93–118.

WHITFORD, W. G., STEINBERGER, Y., and ETTERSHANK, G. 1982. Contributions of subterranean termites to the "economy" of Chihuahuan desert ecosystems. *Oecologia* 55:298–302.

WIENS, J. A. 1976. Population responses to patchy environments. *Ann. Rev. Ecol. Systematics* 7:81–120.

WIENS, J. A. 1991. Birds. In Polis, G. A. (ed.), *Ecology of Desert Communities.* University of Arizona Press, Tucson, Arizona. pp. 278–310.

WIENS, J. A., and ROTENBERRY, J. T. 1979. Diet niche relationships among North American grassland and shrubsteppe birds. *Oecologia* 42:253–292.

WINEMILLER, K. O. 1990. Spatial and temporal variation in tropical fish trophic networks. *Ecol. Mon.* 60:331–367.

WISDOM, C. 1991. Herbivorous insects. In Polis, G. A. (ed.), *Ecology of Desert Communities.* University of Arizona Press, Tucson, Arizona. pp. 151–179.

WOODS, A., and HERMAN, C. 1943. The occurrence of blood parasites in birds from southwestern United States. *J. Parasitology* 29:187–196.

YODZIS, P. 1984. How rare is omnivory? *Ecology* 65:321–323.

ZABRISKIE, J. 1981. Plants of Deep Canyon and the Central Coachella Valley, California. University of California Philip L. Boyd Deep Canyon Desert Research Center, Riverside, California. 174 pp.

21

Ecosystem Trophic Foundations: Lindeman Exonerata

ROBERT E. ULANOWICZ

University of Maryland
Chesapeake Biological Laboratory
Solomons, Maryland

Systems analysis is basically a quantitative modeling approach to problems concerning whole complex systems. When faced with large, complex, and highly interacting systems, human judgement and intuition may lead to wrong decisions [Van Dyne, in *Ann. Rev. Ecol. Systematics* 3:348 (1972), with G. L. Swartzman)].

INTRODUCTION

> That the final statement of the structure of a biocoenosis consists of a pair of numbers, one an integer determining the level, one a fraction determining the efficiency, may even give some hint of an undiscovered type of mathematical treatment of biological communities.
>
> G. Evelyn Hutchinson
> Addendum to Lindeman (1942)

After a decade or so of relative quiesence, the field of ecology is again becoming an arena for lively debate, as new and sometimes radical concepts appear, and old, cherished ideas are vigorously challenged. One of the cornerstones of ecological thought and discussion over the past forty years has been Raymond Lindeman's (1942) concept of the trophic pyramid. Lindeman's idea is thermodynamic at its core: the total amount of energy ingested by host organisms cannot become fully available to the individuals that prey on them. That is, the amount flowing to the predators must be less than the influx to the host population. Thus, one is led to imagine a trophic pyramid of energy flow, where the amount of energy transferred during successive feeding events (as represented by the width of the pyramid) becomes progressively smaller at higher levels of feeding.

Now, there is no arguing with the second law of thermodynamics; it must prevail in the end. Also, the image of ecological feeding relations as a chain of trophic transfers possesses a descriptive elegance that is hard to resist. Perhaps these two attributes of the Lindeman scheme account for its survival as a key element in modern ecological discourse. Few ecologist are given to speaking of an ecosystem without mentioning "herbivores" or "carnivores," as though populations fit neatly into one of the links of Lindeman's trophic chain. Yet most 10-year-old schoolchildren are taught that feeding relations in an ecosystem resemble more a complicated web (e.g., Chapter 20) than a simple chain of transfers.

As Cousins (1985) remarks, "A hawk feeds at five trophic levels." In general, there appears to be no discrete mapping of real populations to integral trophic levels. While many are content to live with the conceptual ambiguities engendered by this mismatch, Cousins takes strong exception to the continued reliance on Lindeman's scheme by most ecologists: "The trophic concept is not just wrong at its edges: it is erroneous in fundamental ways that create many difficulties for ecological science."

Cousins is primarily concerned that the attention paid to the Lindeman description of ecosystems comes at the expense of that given to other concepts, most notably Elton's pyramid of numbers (Ulanowicz, 1989). To a degree, he is correct in decrying the relative lack of interest in organism size, and this writer has elsewhere encouraged the study of particle size distributions and allometric relationships in ecology in lieu of taxonomic categorization (Ulanowicz, 1981; Ulanowicz and Platt, 1985). But Cousin's insistence that Lindeman was wrong and Elton was right is ill-considered. He is perhaps too influenced by positivistic doctrine. As *descriptions* of the real world, various constructs are rarely unequivocally right or wrong. Some are simply better descriptions of events than others (Ulanowicz, 1986).

LINDEMAN'S IDEAS IN THE CONTEXT OF THE ECOSYSTEM

As a quantitative description of behavior at the level of the community, the Lindeman scheme has much to recommend it. It is necessary to abandon only the notion that the mapping from taxa to trophic levels be discrete, and then most ambiguities concerning trophic status quickly vanish. That is, a given taxon need not be assigned wholly to a single trophic level, and vice versa. It is an observed fact that the hawk apportions its activity over five different trophic levels. So what is to keep one from turning that observation into quantitative description?

There are two (interrelated) ways of making a nondiscrete trophic description. We may choose to regard dynamics at the level of the community, where trophic compartments appear discrete and the activities of each taxon are divided among the trophic groups (Higashi, Burns, and Patten, 1992). Alternatively, our focus could be on the individual taxon, and each species could be considered to feed at some noninteger trophic level that is the weighted average of the number of links in the various pathways over which it obtains sustenance. The quantification of both mappings can be achieved with use of the same data—the matrix of feeding coefficients.

TROPHIC AGGREGATION SCHEME

Suppose that species j is any member of an n-component ecological community and that its total intake of some appropriate medium (energy, carbon, nitrogen, or other) is T_j. Measurement may reveal that an amount T_{ij} of this intake comes from another member i of the community. The dimensionless ratio $g_{ij} = T_{ij}/T_j$ is called the *feeding coefficient* of j on i. It describes the extent to which j *directly* depends on i for sustenance. The coefficient g_{ij} may be considered as the entry in the ith row and jth column of an n-dimensional square matrix, $G = (g_{ij})$, the matrix of feeding coefficients. By definition, the sums down the columns of G are all less than or equal to 1. A sum down column j of less than 1 implies that j derives some of its food from outside the community.

Without loss of generality, all sustenance from outside the community may be assumed to be primary production. Imports from outside the system other than primary production can be accommodated by slight changes in the matrix algebra that follows.

A convenient property of G is that its algebraic powers provide quantitative information about *indirect* transfers in the system. For example, when we multiply the feeding coefficient matrix by itself, the result is denoted as G^2. The (i,j)th component of this product matrix represents the fraction of the whole diet of j that derives from i over all pathways of exactly two transfers. Similarly, it may be shown by mathematical induction that the (i,j)th entry in the matrix G^m represents the fraction of total input to j that left compartment i and flowed to j over all pathways of exactly m transfers.

Recalling that the components of G are all less than or equal to 1, its successive powers tend to consist of progressively smaller components and, in fact, when no recycling is present in the system, the powers of G will truncate (produce a matrix of all zeros) in at most $n-1$ steps. This opens the possibility that the sum of all the powers of G might form a convergent series, and it turns out that $I + G + G^2 + G^3 + \cdots = (I - G)^{-1}$, where I represents the multiplicative identity matrix (with 1's along the principal diagonal and zeros elsewhere).

The limit, $(I - G)^{-1}$, is called the *structure matrix* (Leontief, 1951) and may be calculated directly by matrix inversion. It contains information on all pathways of all lengths that exist in the system. In particular, Levine (1980) pointed out that the jth column of the structure matrix depicted how compartment j ultimately depended on all the other species in the community, so the sum of the jth column should yield the *average trophic position* of that species.

To apportion the given species among discrete trophic levels requires that we deal with each power of the G matrix in its turn. Let a_1 be a row vector wherein the ith component is taken to be the fraction of total input to i that enters from outside the system or, in other words, the degree to which species 1 acts as a primary producer. Multiplying a_1 on the right by G gives another row vector, call it a_2, whose ith element measures the fraction of total input to 1 which arrives after having passed through one other compartment, that is, the degree to which i acts as an "herbivore." Proceeding in a similar manner, multiplication of a_1 by G^{m-1} yields a vector a_m with elements that quantify the degree to which each compartment acts at the mth trophic level. If there are no cycles in the flow network, this series of row vectors will terminate within $n-1$ steps, and it becomes possible to form a *trophic transformation matrix, A,* whose ith row is composed of the elements of a_i.

The trophic transformation matrix A was derived by Ulanowicz and Kemp (1979) in an effort to systematize earlier efforts at trophic apportionment by Homer and Kemp (unpublished manuscript; see also Wiegert and Owen, 1971). We read the composition of the ith trophic level along the ith row of A, whereas apportionment of the jth species among the discrete trophic levels is spelled out by the corresponding elements of the jth column. The web of feeding relations can be mapped into a concatenated straight chain using A as a linear transformation (see Ulanowicz, 1986).

An example of the aggregation process can be performed on the simple, hypothetical network in Fig. 21.1. Compartment 1 receives all its input from outside the system, compartment 2 receives 50%, and 3 "produces" 25% of its throughput. The row vector a_1, therefore, looks like (1.0, 0.5, 0.25, 0), and this vector constitutes the first row of A. The feeding coefficient matrix is formed by dividing each intramural transfer by the throughput of the receiving node. For example, $g_{24} = 15/40 = 0.375$. The G matrix looks like

$$G = \begin{bmatrix} 0 & 0.5 & 0.5 & 0.125 \\ 0 & 0 & 0.25 & 0.375 \\ 0 & 0 & 0 & 0.5 \\ 0 & 0 & 0 & 0 \end{bmatrix}.$$

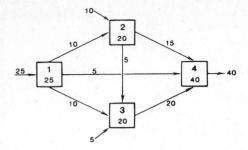

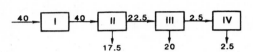

Figure 21.1 Hypothetical network of flows (arbitrary units) among four compartments. Unit 4 receives medium at three different trophic levels.

Figure 21.2 Results of trophic aggregation of the flows depicted in Fig. 21.1.

Multiplying a_1 on the right by the remaining three powers of G yields the succeeding rows of the A matrix:

$$A = \begin{bmatrix} 1.0 & 0.5 & 0.250 & 0 \\ 0 & 0.5 & 0.625 & 0.4375 \\ 0 & 0 & 0.125 & 0.5 \\ 0 & 0 & 0 & 0.0625 \end{bmatrix}.$$

Reading across the second row reveals that the "herbivore" trophic level contains 50% of compartment 2's throughput, 62.5% of 3's, and 43.8% of 4's. Conversely, looking down column 4 reveals that species 4's activity appears 43.8% at the second trophic level, 50% at the third, and 6.3% at the fourth. All columns of A sum to unity, meaning that all the activity of each taxon is accounted for in the trophic levels.

As an alternative to calculating the structure matrix, $(I - G)^{-1}$, we may compute the average trophic position of each of the four boxes in Fig. 21.1 by multiplying each member in the ith row by the value i and summing the results down the columns. This procedure yields trophic positions of 1.0, 1.5, 1.875, and 2.625 for compartments 1 through 4, respectively.

Figure 21.2 depicts the result of aggregating the network in Fig. 21.1 using the transformation A.

CYCLING CAUSES COMPLICATIONS

At this point we could conclude that the relationship between feeding webs and trophic chains is neither as impossible nor as ambiguous as many would portray it to be. We may pass readily from one depiction to the other. True, the discrete trophic compartments are mathematical constructs, but they are unique, quantifiable, and no less "real" than the results of, say, principal component analysis in mechanics or statistics.

There remains, however, one major constraint on the process: the starting network must contain no internal cycles. When cycles are present, the powers of G form an infinite sequence. It then becomes unclear how and where to artificially truncate the number of rows in the A matrix. Of course, material and energy *cycling* are critical features of any ecosystem.

To apply the trophic aggregation scheme to realistic networks with cycles, Ulanowicz (1983) has suggested that the given arbitrary network be decomposed into two constitutive networks—one containing only cycled medium and another portraying only once-through flow in the system (this having the topological structure of a graph "tree"). Removing the cycles involves (1) enumerating all the simple directed cycles, (2) weighting each cycle by an appropriate amount, and (3) subtracting each cycle from the network in such a way so that none of the residual flows becomes negative. The enumeration is accomplished using an algorithm best described as "backtracking with node ordering," and the weighting of the cycles is in proportion to both the magnitude of the smallest arc and the probability that any quantum of medium will complete the specified circuit (Ulanowicz, 1983).

For example, Fig. 21.3 schematically depicts the carbon flows (g C m^{-2} y^{-1}) among 17 ecosystem components of a tidal marsh gut in the vicinity of the Crystal River Nuclear Generating Station in Florida. This network may be decomposed into a nexus of pure cycles, as in Fig. 21.4, and a tree of transfers with dissipation (Fig. 21.5). Devoid as it is of complicating cycles, the web in Fig. 21.5 may now be aggregated into a Lindeman-type chain, as in Fig. 21.6.

We might argue that the chain in Fig. 21.6 illustrates the underlying trophic dynamics of the Crystal River system (Ulanowicz, 1986). However, there are difficulties with this portrayal. The cycles that have been excluded do participate in the trophic-dynamic process. Therefore, their exclusion from the chain in Fig. 21.6 is bound to result in distorted values of calculated trophic efficiencies. Such inaccuracy is not overwhelming in the present case, for which cycling accounts for only about 10% of total activity. However, when dealing with network flows of certain materials, cycling can constitute 80% to 90% of total activity, thereby rendering the trophic structure of the residual flows meaningless. Efforts to map the extracted cycles back onto the aggregated residual flows usually lead to a confusing jumble of feedback loops.

There is another, more subtle difficulty with aggregating the residual flows as in Fig. 21.6. In some of the trophic chains condensed from the residual flows of complicated networks, the resultant number of trophic levels turned out to be almost twice the number that had been expected. Closer inspection of the transformation revealed that the residual tree contained a few pathways wherein medium flowed up a feeding chain to a high-level predator, was transferred to the detrital compartment, and from there ascended up another feeding pathway that was independent of the starting route. By regarding the transfer from high carnivore to detritus as equivalent to a legitimate feeding relationship, we artificially increase the lengths of some pathways. By chance, the trophic aggregation in Fig. 21.6 does not appear to contain such artificially long trophic pathways.

To summarize the dilemma, trophic aggregation seems to work well in systems with only unidirectional (acyclic) flow. However, all real living systems involve recycling. Cycles inordinately complicate the trophic aggregation process, and efforts to extract them distort the quantitative results.

TROPHIC AGGREGATION WITH BIOGEOCHEMICAL CYCLING

Resolution of the problem is surprisingly easy. Those who are astonished by the simplicity of the solution are urged to meditate on the wonders of hindsight. Lest the reader be carried away by these considerations, however, it is noted that the general problem of combining cycles with trophic aggregation has not here been solved. Rather, the special structure of ecosystem networks permits a particular resolution to the problem.

Pimm (1982) and later May (1983) both remark on the rarity of cycles in the *feeding* webs of most ecosystems (however, see Patten, 1985, and Polis' results in Chapter 20). By and large, feeding relationships have been supposed to resemble topological trees. My own experience with topological analysis of many ecosystem model networks has tended to support Pimm's and May's observation. For example, of the 119 cycles identified in Fig. 21.3, only two comprised of small

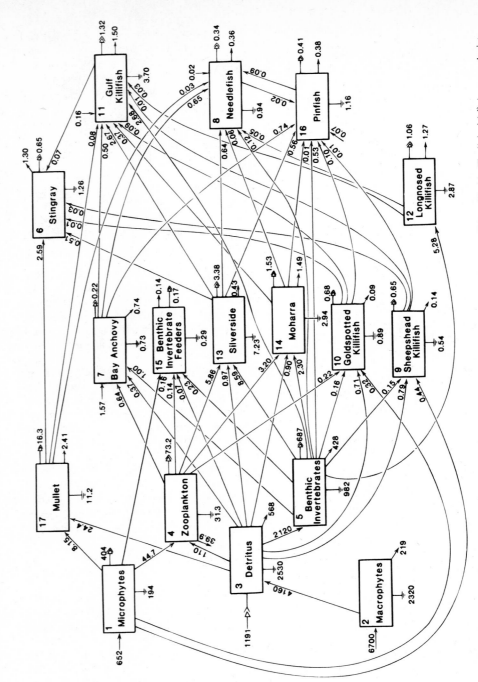

Figure 21.3 Schematic of carbon flows (mg C m^{-2} day^{-1}) among taxa of a marsh gut ecosystem, Crystal River, Florida. The linked (—▷▷) arrows depict returns to detritus. Ground symbols represent respirations (after M. Homer and W. M. Kemp, unpublished manuscript).

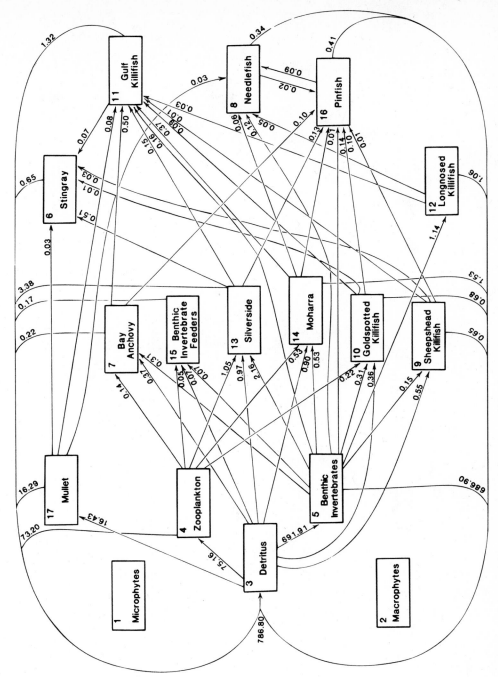

Figure 21.4 Composite nexus of all cycled flow inherent in the network in Fig. 21.3. See Fig. 21.3 for further details.

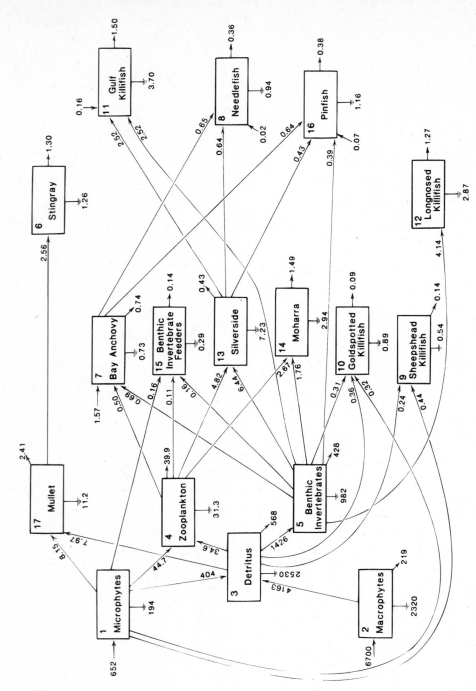

Figure 21.5 The acyclic network of carbon flows remaining after cycled flows (Fig. 21.4) have been subtracted from the original flows (Fig. 21.3).

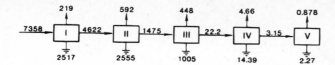

Figure 21.6 Results of trophic aggregation of the flows depicted in Fig. 21.5.

flows do not involve the detrital pool. That is, cycling in this model at least appears overwhelmingly biogeochemical in nature.

Coupling this last observation with the earlier remark about the inaccuracies occasioned by treating flows to detritus as "trophic" transfers (even though organisms of the microbial loop do "feed"), it becomes justifiable to exclude nonliving, nonfeeding compartments from the trophic aggregation. Operationally, this is easy to accomplish. We identify all the living compartments and list them first in the *n*-compartment series of taxa. Assuming that there are L taxa of feeding organisms, the trophic analysis can be carried out using only the $L \times L$ initial submatrix of the full network after any (presumably inconsequential) cycles in this submatrix have been removed by the methods described earlier. The trophic analysis performed on the submatrix of living species, free as it is from nonfeeding transfers, should prove satisfying to the intuition.

We still have the difficulty of what to do with transfers between the discrete trophic levels just created and the nonliving entities that remain. Cousins (1985) notes that during the International Biological Program it became standard practice to pool all the detrital material into a single compartment and assign this grouping to trophic level 1. For many ecosystems, this assumption is plausible. For example, in mesohaline marshes only about 5% of the primary production of emergent vegetation is consumed while the grasses are still alive, but the dead stalks of *Spartina* form the basis for a rich web of detrital feeders. It seems artificial to separate the grasses into living and detrital boxes. Of course, there is a difference in the quality of detritus issuing from organisms at different trophic levels, but, given the high degree to which the various nonliving organic elements are mixed in the environment, the assignment of detritus to any level other than the first appears hard to justify.

To summarize the various steps in the trophic aggregation algorithm:

1. Separate the compartments into living and nonliving subsets. List the living populations first.
2. Remove any cycles from the web of feeding organisms (this is usually a very small amount of the total activity).
3. Aggregate members of the feeding web into discrete trophic compartments.
4. Gather all nonliving components into a single compartment and assign this node to trophic level 1.

A schematic of the trophic aggregation matrix for this process is given in Fig. 21.7. L represents the number of living compartments and NL the number of nonliving nodes. As can be seen in the lower-right submatrix, all detrital compartments are aggregated into the *n*th position in the trophic sequence. This is merely a computational convenience, and the position in the matrix bears no relation to the trophic assignment of this aggregation, which is 1.

The results of this algorithm applied to the Crystal River network depicted in Fig. 21.3 is shown in Fig. 21.8. The detrital compartment, D, is placed directly under trophic level I to reflect its assigned level. All detritivory proceeds along the pathway $D \rightarrow$ II (2256 mg C m^{-2} d^{-1}), while direct grazing by herbivores (54.6 mg C m^{-2} d^{-1}) flows over I $\rightarrow$ II. The topology of the combination of trophic aggregation with biogeochemical cycling is remarkably similar to the way Odum (1957) originally conceived of ecosystem flows occurring in Silver Springs, Florida. This trophic aggregation is said to be canonical with respect to the algorithm that generated it because, when the output network is used as input to the program, it will remain unchanged by the operations performed on it. That is, it is irreducible according to the aggregation formula used.

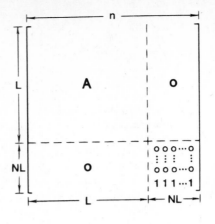

Figure 21.7 Partitioning of the trophic aggregation matrix according to two subgroups of species: *L*, living, and *NL*, nonliving.

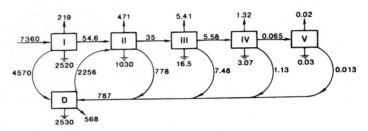

Figure 21.8 Results of trophic aggregation based on the 16 living species in Fig. 21.3. Component *D* represents the pool of all nonliving organic material.

One of the most interesting features of trophic levels is their overall efficiencies. That is, how well does each level pass on medium to the next member in the food chain? Unfortunately, there is some ambiguity concerning the efficiency of the first trophic level in Fig. 21.8. Do we divide the output of I by its input, the detritivory by the input to *D*, or some combination thereof? While individual ratios may be especially meaningful to those interested in certain topics (for example, detritivory), a more comprehensive trophic efficiency can be obtained by combining compartments I and *D* as in Fig. 21.9. Now the backbone, or *Lindeman spine*, of ever-decreasing trophic transfers (the "trophic pyramid" of earlier parlance) is readily visible and open to interpretation or comparison with the spine of other systems.

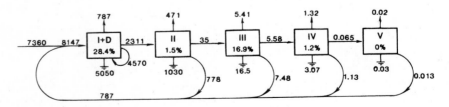

Figure 21.9 The trophic chain of Fig. 21.8 with the detrital pool and primary producers combined. The percentage figure in each box represents the trophic efficiency at that level.

CONCLUSIONS

Earlier ambiguities concerning the relationship between taxonomic category and trophic function appear to have their origins in the relatively imprecise ways in which the community of ecosystem processes was once described. More recently, the increasing number of measured networks of ecosystem processes has provided a quantitative context within which it has become possible to identify precisely the underlying trophic foundation of an ecosystem. To every complicated web of feeding relations and associated biogeochemical cycles, there corresponds a unique chain of trophic transfers in the sense of Lindeman. The ascending chain of trophic transactions is partitioned according to only the feeding processes taking place in the ecosystem. Hence, it appeals to our intuitive notion of what is meant by trophic levels. Furthermore, biogeochemical returns of medium may readily be appended to the trophic transfers to provide an accurate but uncomplicated picture of the system's underlying trophic dynamics. This ability to map arbitrarily complicated networks of ecosystems' flows into a common topological form permits the comparison of what might otherwise have appeared to be hopelessly disparate ecosystems.

Lindeman was able to see beyond the immediate form of ecosystem relationships to perceive the underlying thermodynamic generator for much of organized behavior. It is a tragedy that his genius passed so prematurely from the scene, for it has taken almost half a century for ecologists to begin to give concrete shape to his powerful insight.

ACKNOWLEDGMENTS

The author wishes to thank Steven Cousins for sending him a lucid, well-written critique of Lindeman's concepts, which engendered the present approach to the problem of how to incorporate cycling into the trophic aggregation process. Cousin's help stands as an example of how those on opposing sides of scientific issues can benefit from friendly, rational dialogue. The author is also indebted to Daniel Baird and John Field for helpful discussions concerning the trophic status of detrital material and the role of cycling in trophic dynamics. This work was supported in part by a grant from the Tidewater Administration of the State of Maryland's Department of Natural Resources.

REFERENCES

COUSINS, S. 1985. Ecologists build pyramids again. *New Scientist* 107 (1463):50–54.
HIGASHI, M., BURNS, T. P., and PATTEN, B. C. 1992. Trophic niches of species and trophic structure of ecosystems: Complementary perspectives through food network unfolding. *J. Theor. Biol.* 154:57–76.
LEONTIEF, W. W. 1951. *The Structure of the American Economy,* 2nd ed. Oxford University Press, New York. 257 pp.
LEVINE, S. 1980. Several measures of trophic structure applicable to complex food webs. *J. Theor. Biol.* 83:195–207.
LINDEMAN, R. L. 1942. The trophic-dynamic aspect of ecology. *Ecology* 23:399–418.
MAY, R. M. 1983. The structure of foodwebs. *Nature* 301:566–568.
ODUM, H. T. 1957. Trophic structure and productivity of Silver Springs, Florida. *Ecol. Mon.* 27:55–112.
PATTEN, B. C. 1985. Energy cycling in the ecosystem. *Ecol. Mod.* 28:1–71.
PIMM, S. L. 1982. *Food Webs.* Chapman and Hall, London. 219 pp.

ULANOWICZ, R. E. 1981. Models of particle-size spectra. In Platt, T., Mann, K. H., and Ulanowicz, R. E. (eds.), *Mathematical Models in Biological Oceanography.* UNESCO Press, Paris. 156 pp.

ULANOWICZ, R. E. 1983. Identifying the structure of cycling in ecosystems. *Math. Biosc.* 65:219–237.

ULANOWICZ, R. E. 1984. Community measures of marine food networks and their possible applications. In Fasham, M. J. R. (ed.), *Flows of Energy and Materials in Marine Ecosystems.* Plenum, New York. 733 pp.

ULANOWICZ, R. E. 1986. *Growth and Development, Ecosystems Phenomenology.* Springer-Verlag, New York. 203 pp.

ULANOWICZ, R. E. 1989. Energy flow and productivity in the oceans. In Grubb, P. J., and Whittaker, J. B. (eds.), *Toward a More Exact Ecology.* Blackwell Scientific Publications, Oxford, pp. 327–351.

ULANOWICZ, R. E., and KEMP, W. M. 1979. Toward canonical trophic aggregations. *Am. Nat.* 114:871–883.

ULANOWICZ, R. E., and PLATT, T. 1985. Ecosystem Theory for Biological Oceanography. Canadian Bulletin of Fisheries and Aquatic Sciences 213, Ottawa. 260 pp.

WIEGERT, R. G., and OWEN, D. F. 1971. Trophic structure, available resources and population density in terrestrial vs. aquatic ecosystems. *J. Theor. Biol.* 30:69–81.

Part 5

COMPLEX ECOLOGICAL ORGANIZATION: EXTREMAL PRINCIPLES

The theme that complex ecological systems self organize to maximize various properties of their structure and function is well represented in the theoretical literature. Not all extremal concepts advanced as optimality criteria for ecosystems are included in this section (for example, maximum biomass, energy flow, power output, emergy, and so on, or minimum entropy production), but those that are discussed indicate that the ultimate treatment of ecosystems can and probably will involve optimality considerations. George Van Dyne seemed to know this early; he was one of the first ecological practitioners to apply extremal methods from operations research and control theory to the theoretical study of ecosystems and especially to the formulation of resource management problems in optimization terms.

Chapter 22

Chapter 22 begins the closure of this book back around to the first, physical, science-oriented themes about complex ecology encountered in Part I. Coeditor Sven Jørgensen, trained originally in chemical engineering, is an avowed advocate of "hard science" approaches to the "soft science" of ecology. He introduces his chapter, championing exergy as the optimality criterion of choice for ecosystems, with the observation that "Ecosystems are soft systems in the sense that they are able to meet changes in external factors or impacts with various regulation processes on different levels. The result is that only minor changes are observed in ... function." Some things (structural) change so that others (functional) can remain invariant in a changing environment. The chapter concerns "translation of 'survival of the fittest' into thermodynamic terms."

The three main points of Darwinism are that (1) populations produce more progeny than can survive, (2) these resemble their parents more than randomly chosen individuals, and (3) they vary in heritable traits, affording a basis for natural selection. "The species are continuously tested against the prevailing conditions ... and the better they are fitted the better they are able to maintain their biomass and even grow." To interpret Darwin's theory thermodynamically requires the concept of *exergy,* the product of environmental temperature and system negentropy, and a measure of free energy relative to environment as ground. The thermodynamic reformulation of Darwin can therefore be stated in extremal terms: "an ecosystem . . . will continuously select the species that can contribute most to the maintenance or even growth of the exergy of the system." The second law of thermodynamics runs counter to this trend: entropy increase means negentropy decrease means exergy decrease also. Invoking the chemists' concept of "buffer capacity," the author observes a correspondence between exergy maximization and resistance to change.

Jørgensen repeats the holistic imperative he stated back in Chapter 1: "Ecosystems are irreducible systems in the sense that they are so complex that it is not possible to make observations and reduce these to a few simple laws." Elementary formulas for exergy are developed and illustrated with a simple ecological example. Of course, it is not possible to calculate exergy for every state variable that an ecosystem might have, but useful information can be obtained by determining relative changes in exergy for a few variables relevant to a particular situation (illustrated by phosphorus in eutrophication). Relative exergy change can be decomposed into a nonnegative exogenous component and a positive-to-negative endogenous component, and a system grows or declines in its exergy according to the balance. Models perturbed in simulation trials showed the following results: (1) exergy may increase or decrease in its transient response following perturbation, but at steady state it will always be increased; (2) buffer capacity related to the perturbed variable increases, whereas components of total buffer capacity associated with other variables may increase or decrease; (3) relative exergy may be computed as the weighted sum of system buffer capacities with respect to each experimental variable; and (4) the more state variables, processes, and feedback in a model, the higher the steady-state exergy will be. With these results, it is concluded that "it seems a workable hypothesis to use exergy and buffer capacities as a measure or indicator for the development of ecosystem structure and for changes in species composition. This does not imply that ecosystems have set up the goal to maximize exergy, but that exergy . . . is a convenient way to quantify Darwin's theory and to account for the results of the many regulating processes and feedback mechanisms that are present in ecosystems." In other words, the ecosystem may not be an actively goal seeking entity, but orientation to extrema is clearly a behavioral capability implicit in its complex ecological organization.

Chapter 23

In this chapter, Sven Jørgensen collaborates with stability theorists Dmitrii Logofet and Yuri Svirezhev (Chapter 13) to achieve more mathematical rigor in the formulation of the maximum exergy principle and to investigate the relation of the principle to general stability and special Liapunov stability. The mathematics of exergy in open system thermodynamics is reviewed, arriving at the conclusion that, at thermodynamic equilibrium, exergy change is zero: "the exergy function has an extremum at this point." Second, an exergy component can be defined with respect to every element (for example, temperature and pressure) of a physical system, and these components can be added: "the additive property of exergy with respect to different elements enables us to consider one element only [relative exergy; see Chapter 22] . . . [and t]he main results obviously remain the same for other elements too." Under this, exergy change is formulated for a mixture of ideal gases to show independence of the exergy measure from environmental characteristics and the total amount of matter and system expandability "by surrounding it with a new environment, . . . the 'hyper-environment'. . . ."

The relationship of exergy to stability derives from the "buffer capacity" inherent in exergy measure, and thus a link between exergy and Liapunov stability exists. "Heuristically, a Liapunov function is also [like exergy] a natural, consistent tool for quantifying how far the current state of the system is from the 'reference' state. . . . This idea seems very consistent with the concept of exergy, which is a measure of the free energy a system possesses relative to its environment, and an indication of how far the current state of the system is from the state of thermodynamic equilibrium with the environment." Liapunov functions for population and community models bear a close resemblance to exergy equations. A simple model of phosphorus in a pond is used to develop the relationship between exergy and Liapunov stability. An expression for exergy is derived, and it is shown that this has all the basic properties of a Liapunov function. On the other hand, it is noted that "the exergy function of the classical Lotka–Volterra predator–prey model is not a Liapunov function." This suggests a sufficient but not necessary relation between exergy and Liapunov stability.

Exergy and behavior of a system in parameter space are considered to evaluate Jørgensen's concept that "an ecosystem attempts to meet changes in external factors by developing a new structure with higher exergy under the conditions determined by the external factors." An expression for exergy change is derived that "determines the direction in which the internal parameters must change in response to given variations of the external variables. It may be regarded as quantification of Le Chatelier's principle for those systems that are *exergical at steady state*. The principle is in essence a local principle of variations." This is generalized to nonlocal conditions that, satisfied, make systems "*globally exergical at steady state*." The formulations are applied to a simple resource–consumer model for illustration, and then exergy relations in trophic chains of Volterra type closed to matter are investigated. Exergy is shown to increase with chain length, and when the network properties of cycling and storage involved in food chain lengthening are recalled, these are seen as apparent mechanisms in network organization that contribute to exergy maximization. Exergy, after all, as deviation from environmental ground is reflected in biomass: "an increase in the total amount of matter, which can be interpreted as a complication of the ecosystem structure, is always accompanied by an increase in exergy!" Network trophic dynamics (Part 4) is thus consistent with the exergy extremal principle and, through the connection to Liapunov functions, with attainment of stability. The authors conclude, "Generally, the problem of adaptational (or 'extremal') principles in biology is far from trivial. . . . [but] the evolution of closed trophic chains in parameter space proceeds in complete accordance with the principle of increased exergy."

Chapter 24

The adaptation theme just sounded, inherent in any orientation to extremal concepts, is extended in Chapter 24 by a consideration of "adaptability." The author, Michael Conrad, deals implicitly with networks in his approach; the widespread distribution of causality to all corners of well-interconnected ecological networks ensures an abundance of variability, the raw material for his adaptability phenomenon. Adaptation is the process of adjusting or conforming a specific relation of a system to its environment; adaptability is more general, concerning the power to adapt and continue function. In this chapter, Conrad images living systems as "existential computers" with active self-organizing capabilities. "The basic idea of adaptability theory is that ecological and evolutionary systems are stable only as long as they have enough problem solving power either to dissipate or control environmentally and internally generated perturbations." Short-term (successional) and long-term (evolutionary) change are his subject; he views both as problem-solving processes. There are two ways to solve problems: (1) rule based, in which actions or state changes are constrained more narrowly than by physical laws, and (2) evolutionary, unconstrained except ultimately by physics. Information processing (Chapters 25 and 26) may be equated with problem solving as long as it has survival value. Rule-based information processing, a product of evolution, is the operational mode of most biological processes. Three principles link the two modes at all levels of biological organization:

1. Information processing power increases with the number of interactions in a network, but potential for evolutionary development is maximal when only half the possible interactions occur.

2. *Self-facilitation principle:* Some features of a system make little or no specific contribution to information processing, but may facilitate the evolution of features that do and "hitchhike" along with these to increase evolutionary problem-solving power.
3. *Compensation principle:* Reduction in evolutionary problem solving is compensated for by an increase in rule-based problem solving.

Concerning the *number of interactions* in systems, the ecosystem-level problem to be solved is existential—stay in the game. This has two components, efficient use of environment and stability. Ecosystems spend more time in stable than unstable configurations, and to do this they must regenerate themselves without significant change in the pattern of their interactions. However, Jørgensen, in Chapters 22 and 23, indicates that variable structure interactions are the principal means by which ecosystems maintain constant function. Conrad would not disagree, but would only consider a system stable if it periodically regenerated a structure that was at least qualitatively similar to a preceding one. Adaptability entails all sorts of variable changes in interactive structure, but many might lead to structures that were similar as far as function is concerned, at least in any regime to be called "stable." Therefore, Conrad distinguishes strong from weak interactions, with the latter presumably more definitive of constancy of function. Jørgensen's notion of variable structure interactions may correspond to Conrad's concept that variability among weak interactions maintains the strong interactions in a system. This problem becomes more complex as the number of possible interaction patterns increases. With n components, this complexity is maximal with $n^2/2$ interactions, "just where the information processing and evolutionary power required to exploit the environment and reproduce a variety of complex patterns reach an optimal balance." If a system is strongly connected, disturbances will ramify throughout it and be destabilizing. If components are completely independent, no ramification will occur, but also there "will be no possibility of attenuating the disturbance by spreading it over the system." The role of strong versus weak interactions in finding a proper connectivity balance is discussed. Strong interactions in ecosystems are associated with "the essential predator–prey, symbiotic, and spatial affinities that define the community. The enormously high irritability of some organisms—their ability to react to a wide variety of subtle features of the milieu—suggests the importance of weak interactions." In general, in resolving the conflicting demands of interaction density, "increases in problem-solving power associated with movement in the direction of $n^2/2$ interactions will be favored if they are not offset by increase in the complexity of the regeneration problem."

Self-facilitation concerns conditions that a system must fulfill to evolve information-processing capabilities. The Darwin–Wallace postulates (having units capable of reproducing with variation but without inheritance of acquired characteristics) would seem sufficient, but not all systems satisfying these yield an evolution process. Evolution selects structure–function relations that facilitate further evolution following adaptive peaks. If a species becomes isolated at the top of such a peak, this amounts to an evolutionary bottleneck. The appearance of redundant elements, such as similar species or food-web links, increases the dimensionality of the adaptive space and produces basins of attraction into whose deepest valleys communities tend to fall. "The organization of ecosystems is a consequence of both hill climbing on an adaptive surface and community level self-facilitation of valley seeking on a potential surface. Rule-based problem solving at the level of the organism results from evolutionary hill climbing ... and at the community level [is] also influenced by successional valley seeking."

Compensation involves shifts between problem-solving modes in response to incurred costs. Thus, evolutionary versus rule-based problem solving changes in balance as the latter becomes more prominent in evolution. Problem solving may be specialization directed (closed options) and adaptability directed (open options). In an invariant environment the first is appropriate; in a changing environment, the latter. Conrad defines the *adaptability* of a system as the potential uncertainty of its environment. Hierarchical organization allows "the flow of information in a community [to pass] vertically between levels as well as horizontally.... Evolutionary adaptability is mediated by variation at the genetic level and selection at the organism level. Behavioral adaptability is manifested in terms of the actions of organisms, but is mediated by neural and subneuronal processes. Populational adaptability includes culturability of species in suitable environments ... manifested in terms of numbers of organisms [but] mediated by physiological mechanisms of reproduction. Community-level adaptability includes variability of food web structure, and this has a basis in behavioral and physiological mechanisms." Costs cause greater than necessary adaptability to tend to atrophy. In general, the adaptability of a system should no more than match the uncertainty of its environment. Hierarchical organization confers an advantage, and "the evolution of biological hierarchy is largely driven by the cyclic interplay between selection for adaptability, conversion of adaptability to specialization-directed problem solving, and consequent selection for compensating modes of adaptability.... The detritus pathway serves as a compensating mode of adaptability for the grazing pathway" in ecosystems.

Conrad recognizes four stages in the development of interaction patterns and problem-solving power in ecosystems: (1) a "preparation stage" of evolutionary problem solving, (2) the "elaboration stage" of rapid development of adaptability and specialization-directed problem solving, (3) a "compensation stage" of increased interactive complexity and reliance on rule-based mechanisms, and (4) a "destabilization stage" of adaptability atrophy, disappearance of weak interactions, and increasing vulnerability to environmental variability. Considering the relationship of complexity to stability, increasing connectivity, says the author, tends to increase problem-solving power, which tends to increase stability. However, these tendencies are not always expressed; the relationship between complexity and stability is nonsimple. Commenting on the seeming anthropocentricity of his analysis, Conrad observes that the concept of "problem solving" is no more anthropomorphic than that of "law of nature." "We need not suppose a law giver when we write down a dynamical equation, and we need not suppose that an ecosystem experiences the kinds of intentions that humans do when it goes about the business of staying in existence." A short appendix connects the developments of this chapter to the adaptability theory formalism of the author's book (*Adaptability, the Significance of Variability from Molecule to Ecosystem,* Plenum Press, New York, 1983).

Chapter 25

In Chapter 2, discussing information theory and ecology, Ramon Margalef visualized the ecosystem as a channel for communication. This conception is amplified and technically developed in Chapter 25, which provides a network-oriented treatment of information theory by Hironori Hirata. The network model of "media" (energy–matter) flows for ecosystems is a macroscopic manifestation of microphenomena such as competition, predation, and migration. Following a brief review of the history of information theory in ecology, the author begins development of Margalef's ecosystem as a communication channel: "If an ecological network is regarded as a communication channel, the information contained in the structure of the network may be theoretically defined by the concept of mutual information.... The information inherent in the network structure, that is, mutual information of the ecological channel, is a good index of self-organization of ecological networks from the macroscopic viewpoint."

Flow networks, communication channels, and the probabilistic concept of "mutual information" are formulated mathematically. Mutual information is defined as "the amount of information about input provided by the observation of channel output," and this concept is further expressed in terms of probabilities that substance is in certain compartments at certain times in network flows. The information content of an entire network is defined, and from this perspective the topic of model aggregation explored in Chapters 7 and 8 is reinvestigated. Aggregation is defined as homomorphic mapping from real to model system. "Discrete" and "weighted" aggregations are distinguished, and it is shown that information content cannot increase and generally decreases in the aggregation process. A measure of aggregation cost is developed, which it is the objective to minimize; this extremal formulation applies obviously to model making, the orientation of the two aggregation chapters (7 and 8), but more to the point here, it is applicable also to such previously considered aspects of ecological networks as optimal self-design (Chapter 19) and adaptability and problem solving (Chapter 24).

Mathematical conditions for perfect aggregation involving no information loss are developed and translated into several prescriptions for network design: No aggregation loss occurs when two elements (compartments or nodes) occur in "parallel." Two cases of this are (1) the two elements have no flow between them and no node loops (storage delays), their inflow intensities from source compartments are equal, and their outflow intensities to destination compartments are also equal; (2) the two elements exchange flow around a cycle and self-loops occur, in which case in addition to equality of their two inflow and two outflow intensities, the relative flows (intensities) in the loops and cycle must also be equal to one another. Two elements hold a perfect parallel position to one another, involving no aggregation loss, if (3) they share one inflow from a common source element and one outflow to a common destination element, with or without a direct flow from the source to the destination compartment. These elementary cases are extended to many elements in parallel rather than just two, and Hirata concludes "that parallel structures are important for perfect aggregation. . . . [and] generally tend to minimize aggregation loss." Several applications to ecosystem models are given in illustration.

The author then takes up the subject of self-organization of ecological networks in time. A *dissimilarity* measure is developed to compare two probability distributions and, formulating the probabilities in terms of network flows, the measure is applied to derive an index of self-organization (or order) for ecological networks. The index is referenced to the case of statistical independence of events and thus measures distance from the disorder state. In this, Hirata's measure of order has the same characteristic (deviation from ground) as exergy in Chapters 22 and 23, but is expressed in statistical rather than thermodynamic terms. By satisfying a Markov property, the measure can be expressed in terms of mutual information. This expression is developed in a temporal context and the conclusion drawn "that mutual information may be taken to be a natural and well-motivated measure of self-organization (succession) in the ecological network." This measure becomes the basis for ecosystem ascendency theory in the final chapter.

Chapter 26

Robert Ulanowicz's *ascendancy theory* of network organization (*Growth and Development, Ecosystems Phenomenology,* Springer-Verlag, New York, 1986) represents probably the best single example today of a coherent formal theory of complex ecology that is primary synecology and not merely a reworking of ground from physics or applied mathematics. The author would be the first to acknowledge that his theory is incomplete, and this makes it a fitting end to this book on a developing subject. George Van Dyne's life was incomplete, too, when he died; he was "rested from his labors, but his work follows after him," registered in even deeper inquiries into the nature of communal life. This is where this book remembering him should end, peering into the dusky recesses of interconnected complexity to make sense out of what is inherently obscure, but on an optimistic and ascending note, which Ulanowicz and his theory provide, quite literally.

Chapter 26 begins with natural complexity—is it organized or not? "[M]ost of the controversy is at best marginally quantitative . . . the goal here is to introduce a set of quantitative tools that can be applied to resolving the ongoing disagreement." Probabilities associated with dynamics are reviewed, leading to the concept of uncertainty, which "rises in proportion to the number of factors that serve to differentiate the various outcomes," and can be measured by the well-known Shannon–Wiener function. For application to system organization, such functions can be constructed for inputs and outputs and the uncertainty about system dynamics expressed as their sum if they are independent, or by a function based on joint probabilities of occurrence if they are not. Joint uncertainty is always less than or equal to the sum of corresponding separate uncertainties, and "the amount by which the sum of the separate uncertainties exceeds the joint uncertainty is the measure of how coherent or organized the dynamic structure is." This difference is the average mutual information as introduced by Hirata in Chapter 25; any increase in it "reflects development, and any decrease is a step towards incoherence and chaotic behavior."

After discussing difficulties in framing the mutual information measure in terms of individual organisms transferred as discrete units in network interactions, Ulanowicz takes the same approach as Hirata used and developes measures based on network flows of conservative media. The basic mathematics is all descended from input–output analysis. Growth and development are distinguished as follows. "Growth is an increase in system size," where size means total extensive activity, as in the gross national product of economics; growth is measured as the sum of all flows (total system throughput, or throughflow). An intensive measure of the articulation of a system's parts, "development is an augmentation of its organization." The product of growth and development is *ascendancy,* throughflow-scaled mutual information. The maximum ascendancy attainable is termed *development capacity,* and that unattained is *overhead.* Internally generated terms in ascendancy can be used to measure internal ascendancy, development capacity, and overhead, and internal overhead has three components: (1) *tribute,* exports of usable energy; (2) *dissipation,* exports of unusable energy; and (3) *redundancy,* the remainder, which "is related to the multiplicity of pathways connecting any two compartments in the system." It can readily be appreciated how the interplay of these various measures figures in the quantification of self-organization and how activities of individual components (like species) contribute to the system-level properties the measures represent.

Moreover, it should be possible to integrate many of the frameworks that have been on display in this book, and others like them not represented, into the general scheme of network ascendancy, broadly conceived. What would be, for example, the mutual relationships between an ascendancy regime and information, adaptability, and problem solving (Chapters 2, 24, and 25); exergy maximization (Chapters 22 and 23); stability, disturbance and stress, and feedback (Chapters 13 to 18); cybernetic concomitants of complexity (Chapter 4); the applied problems demanding attention and driving development (Chapters 9 to 12); the taxonomy of categories (Chapters 7 and 8); and the constraints of physical principles (Chapters 1, 3, 5, and 6)? The ascendancy concept, as a general network measure of organization, touches all these diverse aspects of ecological complexity. In the end, they must all come together into a unified dialogue between man and nature that finally dissolves the mystery of systemness that, in this period, eludes and defeats all ecologists.

"Nothing less than the status of ecology as a science is at stake in this dialogue," writes Ulanowicz. "If ecosystems are determined by their molecular constituents, then ecology is clearly a corollary discipline to genetics and molecular biology. If, however, autonomous elements of ecosystems can be clearly identified, then ecology takes on an importance in its own right. Even more to the point, the discovery of autonomous ecosystem behavior would stand as a significant advance in the theory of far-from-equilibrium thermodynamics and catapult ecosystem science into the very forefront of scientific inquiry—which is where it has always belonged!"

George Van Dyne saw it so a generation ago, and he would see it this way now.

22
Exergy and Ecological Systems Analysis

SVEN E. JØRGENSEN

The Royal Danish School of Pharmacy
Department of Environmental Chemistry
Copenhagen, Denmark

We have been in a period of increasing specialization in many fields of human endeavor. But now science is in the early stages of a major revolution. Many efforts are emerging toward integration, coordination, and generalization . . . [Van Dyne, in *The Ecosystem Concept in Natural Resource Management*, Ch. X, p. 364 (1969), Academic Press].

INTRODUCTION

Ecosystems are soft systems in the sense that they are able to meet changes in external factors or impacts with various regulation processes on different levels. The result is that only minor changes in function are observed; that is, the state variables are maintained almost unchanged, despite changes in external factors.

It has been widely discussed in recent years (Odum, 1983; Straskraba, 1980) how changes in ecological structure and species composition contribute to regulation at the ecosystem level.

Darwin's theory describes interspecific competition and concludes that species best fitted to prevailing conditions in an ecosystem will survive. While this theory could be used to describe changes in ecological structure and species composition, it cannot be used directly or quantitatively in ecological modeling.

This chapter is concerned with the problem of describing ecological structural changes and changes in species composition by a quantitative method developed for translation of "survival of the fittest" into thermodynamic terms. It is presented as a hypothetical theory, which attempts to unite Monod's and Prigogine's theories (Prigogine and Stengers, 1979) in the explanation of ecosystem development and evolution.

HYPOTHESIS

All species in an ecosystem are confronted with the question of how to survive or even grow under the prevailing conditions. Prevailing conditions include all the factors influencing the species, that is, all external and internal factors, including those originating with other species. These factors are dynamic; conditions are always changing and there are many species waiting to ascend if they are better adapted to emerging conditions than presently dominant species are to current conditions. The minor species in communities represent a wide spectrum of properties available for natural selection to draw on. The question is which of these species are best able to survive and grow under present conditions versus which are best able to survive and grow under conditions one time step later, two time steps later, and so on. The species must have genes whose phenotypes match these conditions enough, on average, to survive and reproduce. New environmental and genetic recombinations constantly emerge to provide material for testing of the question of which species are best fitted under the conditions prevailing just now.

These ideas are illustrated in Fig. 22.1. External factors change and select species from the genetic pool, which is slowly changed by chance. Ecological development refers to the changes caused by the dynamics of the external factors, giving sufficient time for the reactions.

Evolution is also related to the genetic pool. It is the result of the relation between external factors and gene pool dynamics. The external factors steadily change the conditions for survival, and the genetic pool steadily provides new solutions to the problem of survival.

Darwin's theory assumes that populations consist of individuals that (1) on average produce more offspring that are needed to replace them upon their death, the property of high reproduction; (2) have offspring that resemble their parents more than they resemble randomly chosen individuals in the population, the property of inheritance; and (3) vary in heritable traits influencing reproduction and survival (that is, fitness), the property of variation.

All three of these properties are inherent in Fig. 22.1. High reproduction is needed to realize a change in species composition caused by changes in external factors. Variability is represented in the short- and long-term changes in the genetic pool. Inheritance is needed to assess the effect of the fitness test. The species are continuously tested against prevailing conditions (external and internal), and the better they are fitted the better they are able to maintain their biomass and even grow. The specific rate of population growth may even be used to measure fitness. But the fitness property must be inheritable to have any effect on species composition and the ecological structure of the ecosystem.

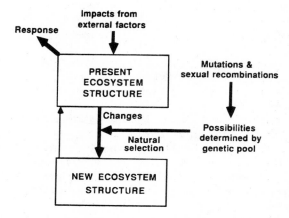

Figure 22.1 Natural selection in the ecosystem context.

Now let us turn to the translation of this presentation of Darwin's theory into thermodynamics. First, we need to define the concept of exergy (see Rant, 1956; Berman and Silverstein, 1975). *Exergy, Ex,* is defined as:

$$Ex = T_0 * \Sigma,$$

where T_0 is the temperature of the environment and Σ is the negative entropy (negentropy) of the system. This definition implies that *Ex* measures the free energy the system possesses relative to its environment. *Ex* is not a state variable in the thermodynamic sense. It depends on the state of the total system and, unlike energy, can be destroyed or consumed.

Survival will be taken to mean the maintenance of biomass, and growth is the increase of biomass. It costs free energy to construct biomass, which possesses, therefore, chemical energy, which is transferable to free energy again. Survival and growth can be measured by the thermodynamic concept of exergy, which is *the free energy relative to the environment.*

Darwin's theory can be reformulated in thermodynamic terms as follows: The prevailing conditions of an ecosystem are steadily changing, and the system will continuously select species that can contribute most to the maintenance or growth of system exergy. This does not violate the second law of thermodynamics, which states that an isolated system will always change toward increased entropy, meaning decreased negentropy or decreased exergy. Ecosystems are not isolated systems, but do receive an inflow of solar energy. This carries low entropy, whereas the radiation of heat from the ecosystem carries high entropy. If the power of solar radiation is W and the average temperature of the system is T_0, then the exergy gain per unit of time, ΔEx, is

$$\Delta Ex = T_0 * W \left(\frac{1}{T_2} - \frac{1}{T_1} \right),$$

where T_2 is the temperature of the environment and T_1 is the temperature of the sun. This exergy flow can be used to construct and maintain structure far from thermodynamic equilibrium.

Notice also that the thermodynamic translation of Darwin's theory requires that populations have the above-mentioned properties of reproduction, inheritance, and variation. The selection of species that contribute most to exergy of the system under the prevailing conditions requires a sufficient number of individuals with different properties for selection to take place. That is, reproduction and variation must be high and, once a change has occurred due to better fitness, it can be conveyed to the next generation.

Another concept has been introduced to account for the ability of the ecosystem to meet changes δ in external factors with only minor changes in state. This concept is denoted the ecological *buffer capacity*, β, and is defined as

$$\beta = \frac{\delta(\text{forcing function})}{\delta(\text{state variable})}.$$

This quantifies the ability of the ecosystem to react pliably to external factors. It also represents ecological expression of Le Chatelier's well-known equilibrium-seeking principle from chemistry. The hypothesis presented and discussed in this paper states that the ecosystem, due to its many regulation mechanisms, has an ability to maintain as high an exergy level as possible under the prevailing conditions. This also implies that essential state is maintained not necessarily in terms of the same species, but in terms of approximately the same function. There seems therefore to be a relationship between an ecosystem's buffer capacity and its ability to maintain the highest possible exergy level under the prevailing conditions.

VERIFYING HOLISTIC PRINCIPLES

The above translation of Darwin's theory into thermodynamics must be investigated in terms of both most fitted species and reactions of entire ecosystems to changes in external factors on the level of species and ecological structure.

Ecosystems are irreducible systems in the sense that they are so complex that it is not possible to make observations and reduce these to a few simple laws (Jørgensen, 1986). It is therefore necessary to support any hypothesis on ecosystems by the use of models or experimental mathematics. The idea is to construct a realistic model and observe how this model reacts. If the model behaves in accordance with the hypothesis, this will be considered support for the hypothesis. It is not sufficient to test the reactions of only one model, but many models of various ecosystems representing various situations must be investigated.

The use of models or experimental mathematics is an approach applied in nuclear physics, where observations are limited by the influence of the observer on the small atomic particles. In ecology we are in a similar situation. Here the observer is limited by the enormous complexity of the system; it is simply impossible to observe all relationships among so many components (Jørgensen, 1988). We are therefore bound to accept an uncertainty relationship also in ecology. The construction of models, including those applied for verification of holistic hypotheses, requires good knowledge of the system and the problem that we want to model in order to make the needed model simplifications correctly. It implies furthermore that the value of this verification method is highly dependent on the quality of the model.

This approach is in accordance with a change in science that has been occurring during recent decades: the acknowledgment that it is hardly possible to reduce the description of all processes in nature to a few laws. Classical physics considered only reducible systems and could therefore, come up with a few simple laws, but the sciences of more complex systems must deal with the variability and complexity that are an integral part of nature (Prigogine and Stengers, 1979).

SUPPORT FOR THE HYPOTHESIS

It is not possible to measure exergy; we can only compute it if the composition of the ecosystem is known. Mejer and Jørgensen (1979) have shown that the following equation is valid for the components of an ecosystem:

$$Ex = RT \sum_{i=1}^{n} \left[X_i^* \ln\left(\frac{X_i}{X_{eq,i}}\right) - (X_i - X_{eq,i}) \right],$$

where R is the gas constant; T is the Kelvin temperature of the environment; X_i represents the ith system component expressed in a suitable unit (for example, for phytoplankton in a lake X_i could be milligrams of phytoplankton per liter of lake water); $X_{eq,i}$ is the concentration of the ith component at thermodynamic equilibrium, which can be found in Morowitz (1968); and n is the number of components. The equation is valid for systems with an inorganic net inflow and passive outflow. It can be derived from basic thermodynamic equations (see Mejer and Jørgensen, 1979).

$X_{eq,i}$ is a very small concentration of organic components, corresponding to the probability of forming a complex organic compound in an inorganic soup (at thermodynamic equilibrium), which is also very small. Morowitz (1968) has calculated this probability and found that for proteins, carbohydrates, and fats it is about 10^{-50} mg/l. For more complex compounds, such as hormones and genes, $X_{eq,i}$ will be even smaller, and these compounds will therefore, even if they are present in very small concentrations, also contribute significantly to the exergy.

The exergy of an ecosystem can be illustrated by an example. Let us consider an algae pond that contains only one species. The phosphorus concentration of algae is 1 mg/l and in the form of orthophosphate, 0.5 mg/l. The temperature is 300 K. If we are concerned only with the contribution to the exergy from the phosphorus compounds, Ex_p, we find, by use of the above equation,

$$Ex_p = 8.31 * 300 \left[\frac{1}{31} (\ln 10^{50}) - \left(\frac{1}{31} - \frac{10^{-50}}{31} \right) \right]$$

$$+ \left[\frac{0.5}{31} \ln \frac{0.5}{1.5} - \left(\frac{0.5}{31} - \frac{1.5}{31} \right) \right]$$

$$= 9214 \text{ mJ/l} .$$

Similarly, it is possible to calculate the contribution to exergy for nitrogen and all other compounds, including the contribution originated from the more complex compounds as genes and hormones. But if we are concerned only with the changes caused by phosphorus, it is sufficient to make the computations shown to see the *relative changes* in exergy.

If zooplankton were introduced to the pond, more phosphorus would probably be bound in the organic form, and we would therefore find higher exergy. If the allocation of phosphorus were 0.9 mg/l to phytoplankton, 0.25 mg/l to zooplankton, and the remaining 0.35 mg/l as orthophosphate, it could then be calculated that the exergy would increase to 10751 mJ/l.

However, we can distinguish between two changes in exergy: a change caused directly by external factors and a change caused by the response of the living organisms to the external factors. The former is related to available resources in the ecosystem. If the phosphorus concentration is increased or decreased, the exergy will, respectively, increase or decrease. The latter change in exergy is caused by the effort of the organisms to survive and reproduce and will therefore reflect the many regulation mechanisms that an ecosystem and its organisms possess. In other words, whenever external factors are changed, we observe a change in exergy, ΔEx, that can be expressed as

$$\Delta Ex = \Delta Ex_R + \Delta Ex_E ,$$

where the subscript R refers to changes caused directly by the external factors and subscript E represents the effort of the organisms (including those in a quiescent state) to realize the best possible growth and reproduction in the given circumstances. ΔEx_R may be negative or positive, while ΔEx_E always will be ≥ 0.

In accordance with the foregoing considerations, realistic models have been developed as bases for testing the maximum exergy hypothesis. Three models were applied in the analysis: a eutrophication model, a toxic substance model, and a stream model. The eutrophication model has been used in 16 case studies, with modification from study to study. In one of these the investigations have been carried out over several years, the model being calibrated and validated, and a previously published prognosis validated (Jørgensen, 1976; Jørgensen, Mejer, and Friis, 1978); and Jørgensen et al., 1986). A conceptual diagram of one of the models is shown in Fig. 22.2.

The toxic substance model considers the effects of ionic copper on a lake ecosystem. Figure 22.2 represents the food web, and the model includes a formulation of copper uptake from water and food and the effects of copper concentration on growth and mortality. For further details, see Jørgensen et al. (1979), Jørgensen (1979), and Kamp-Nielsen (1983).

The stream model is conceptualized in Fig. 22.3. Its results are consistent with several widely used river models (Armstrong, 1977; Water Resources Engineers, 1973). The model includes how the growth of aerobic microorganisms, phytoplankton, and zooplankton is affected by oxygen concentration below a certain threshold value.

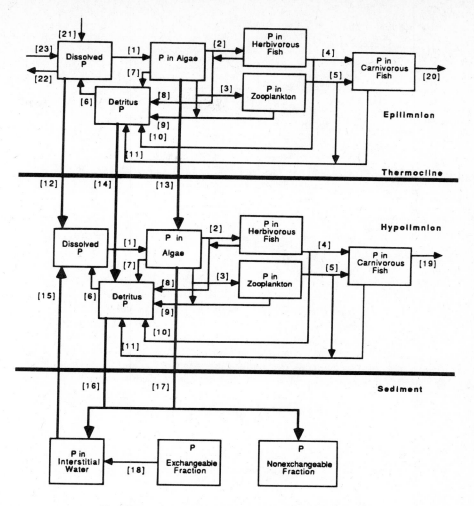

Figure 22.2 Conceptual design for a lake ecosystem model.

Perturbations were imposed on these three models, and observations on their reactions made. Changes in exergy and buffer capacities were observed and model results compared with empirical observations. These results are summarized in Table 22.1. The buffer capacities used in this table are defined in Table 22.2. Results are summarized in the following points:

1. The immediate reaction of exergy (ΔEx_R) to changes in external factors is to increase or decrease in accordance with changes in available resources. Later, when species composition has had suffcient time to react, the exergy will always increase (Table 22.1). If phosphorus input is decreased suddenly, resources are reduced and exergy initially decreases; however, changes in species composition in response to the less eutrophic water cause a subsequent rise again.
2. Buffer capacity always increases in response to changes. If phosphorus input is changed, β_P increases, while other buffer capacities may decrease.
3. Statistical analysis of results indicated a relationship between exergy and buffer capacities:

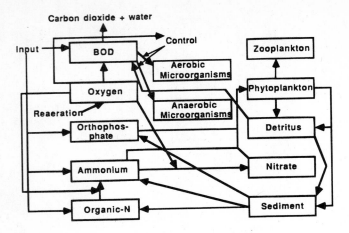

Figure 22.3 Stream model.

$$Ex = \sum a_j * \beta_j,$$

where a_j represents regression coefficients and β_j buffer capacities. Some buffer capacities may be reduced even when exergy increases, but this is more than compensated for by the increase of other buffer capacities. This explains why it has been difficult to find a relationship between ecosystem stability and species diversity. Increased phosphorus loading gives decreased diversity (Fig. 22.4), but in accordance with the model results referred to here the exergy and phosphorus buffer capacity are also increased, while other buffer capacities (see Table 22.1) decrease. Stability is, in other words, a multidimensional concept, and the relation between species diversity and stability is therefore not simple and can be revealed only by a multidimensional relation. If species diversity decreases, stability as represented by ecological buffer capacity decreases in some directions but increases in others (Jørgensen, 1990).

4. All three models are realistic and supported by good data. The responses of the models corresponded to general ecological observations.

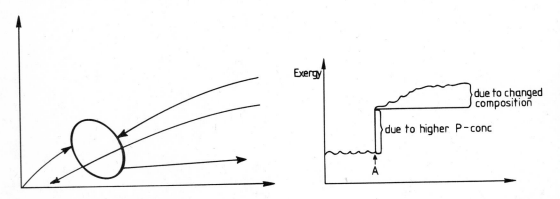

Figure 22.4 Ecological changes observed in lakes, all in accordance with the exergy principle.

Figure 22.5 Exergy = f(time) found by use of modeling as an experimental tool. The reaction to increased concentration of phosphorus at time A is shown.

TABLE 22.1 Model-generated exergy and buffer capacity changes compared to ecological observations

Number	Model used	Perturbation	Model observation	Exergy changes	Buffer capacity change	Ecological observations
1	Eutrophication	P input ↑	Smaller diversity, bigger-sized phytoplankton (slower nutrient uptake)	↑	β_P ↑, β_T and β_R ↓	See Fig. 22.7
2	Eutrophication	P input ↓	Increased diversity to a point, then decrease; smaller-sized phytoplankton (faster nutrient uptake)	↑	β_P ↑, β_T, and β_R follow diversity	See Fig. 22.7
3	Eutrophication	Nitrogen depletion (summer situation)	Nitrogen fixing species appear; decreased diversity	↑	β_N ↑	See Fig. 22.7
4	Eutrophication	Sudden temperature shift	Changes in species composition to those better adapted to the temperature	↑	β_T ↑	A trivial observation consistent with model response
5	Toxic substance	Increased input of toxic substance	Smaller diversity, but increased growth of species tolerant to toxic substance	↑	β_{TOX} ↑, β_T and β_R ↓	A trivial observation consistent with model response
6	Stream	Increased input of BOD	Smaller diversity, slower growth of aerobic microorganisms, phytoplankton and zooplankton; at high BOD concentration, oxygen is depleted and anaerobic microorganisms dominate	↑	β_{BOD} ↑, β_T and β_R ↓	See Fig. 22.8

TABLE 22.2 Definitions of buffer capacities used in Table 22.1

$$\beta_P = \frac{\partial(\text{P input})}{\partial(\text{phytoplankton concentration})}$$

$$\beta_T = \frac{\partial(\text{temperature})}{\partial(\text{phytoplankton concentration})}$$

$$\beta_R = \frac{\partial(\text{radiation})}{\partial(\text{phytoplankton concentration})}$$

$$\beta_N = \frac{\partial(\text{N input})}{\partial(\text{phytoplankton concentration})}$$

$$\beta_{TOX} = \frac{\partial(\text{toxic substance input})}{\partial(\text{reduction in phytoplankton concentration})}$$

$$\beta_{BOD} = \frac{\partial(\text{BOD})}{\partial(\text{microbiological activity})}$$

β_P, β_T, β_R, and β_N are used in the eutrophication model, while β_{TOX} is used in the copper model and β_{BOD} in the stream model.

The relation between exergy and model reactions to perturbations has been examined by another approach. It would be obvious, considering the above-mentioned results, to ask why ecosystems react to perturbations as they do. This may be explained by Darwin's theory. But it is also possible to express the result in terms of such biological and ecological properties as (1) all matter cycles, (2) there are many feedback mechanisms, and (3) the living components are able to grow. Models with such properties always react to perturbations as described in Figs. 22.5 and 22.6. In addition, the more state variables, processes, and feedback mechanisms that a model includes, the higher is its steady-state exergy. This is demonstrated in Fig. 22.7 where the exergy is shown at steady state for a series of models with increasing complexity in the form of components, processes, and feedback mechanisms.

Additional support for the maximum exergy hypothesis is taken from Prigogine and Stengers (1979). They describe biological development by use of a logistic equation:

$$\frac{dN}{dt} = rN(K - N) - mN,$$

where N is the number of individuals in a population, r is the reproduction rate, K is the carrying capacity, and m is the mortality rate. This equation corresponds to an increase in exergy up to level 1 in Fig. 22.8. However, when an ecological niche is going to be exploited by organisms, r, K, and m should not be considered as constants forever. Various species may be better adapted to the conditions and, therefore, the "constants" will develop toward more advantageous values. The

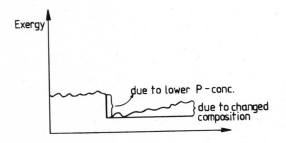

Figure 22.6 Exergy = f(time) found by use of modeling as an experimental tool. The reaction to decreased concentration of phosphorus is shown.

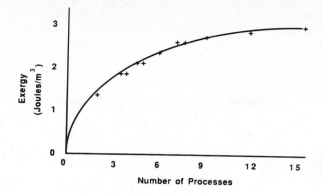

Figure 22.7 Exergy at steady state for models with increasing numbers of processes.

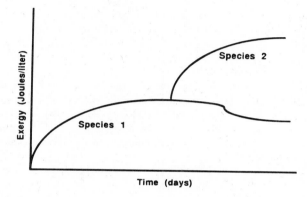

Figure 22.8 Exergy versus time for development of an ecological niche.

constants do not have the same values for all organisms of a species, but may be normally distributed around an average value. Those organisms with the best fitted values will be more and more dominant, and thereby will the constants change their values. This development could correspond to the increase in exergy shown from level 1 to level 2. In addition, many other species come to the niche and try to utilize the resources available. This may imply that the first species will be reduced in number and thereby in contribution to the system exergy, but together the two species now populating the ecological niche will contribute to a higher total exergy (Fig. 22.8).

All in all, it seems a workable hypothesis to use exergy and buffer capacities as a measure or indicator for the development of ecosystem structure and for changes in species composition. This does not imply that ecosystems have set up the goal to maximize exergy, but that exergy (and buffer capacities) is a convenient way to quantify Darwin's theory and to account for the results of the many regulating processes and feedback mechanisms that are present in ecosystems.

RELATION TO OTHER ECOSYSTEM THEORIES

Attempts have been made to describe ecosystem reactions to pertubations by other approaches. Entropy or negentropy have been applied in ecosystem theories. Entropy and exergy are related, but the latter has three clear advantages over entropy:

1. Exergy is dependent on the environment, while entropy is hard to interpret for an ecosystem which is far from equilibrium and driven by the environment.

2. Exergy is an energy concept, which makes it possible to translate it into well-known energy concepts and to such biological concepts as growth and reproduction.
3. Exergy includes the temperature of the environment in its definition, which makes it possible to compare exergy at two different temperatures, whereas comparison of entropy may be misleading if the temperatures are different.

The exergy principle is not consistent with the maximum power principle (Odum and Pinkerton, 1955, and Odum, 1983). The latter claims that the ecosystem with the quickest turnover will be selected, while the presented exergy principle claims that the system with the highest "capital" in the form of information and biomass, which can be measured by means of exergy, will win out because it is able to resist change or persist through changes caused by external or internal factors.

This is in accordance with Ulanowicz (1986), who claims that the maximum power principle may be valid at the early stages of development, but cannot describe ecosystem reactions during later stages of maturation. Ulanowicz uses the concept of *ascendancy* to explain ecosystem development. Ascendancy measures network size and organization, and Ulanowicz claims that it represents the work inherent in the creation and maintenance of order in the network flow structure. If this is correct, this concept is close to being a parallel concept to exergy; but the difference is that ascendancy results mainly from network theory, while exergy is based entirely on thermodynamics.

Finally, the relation to the theory of indirect effects should be mentioned. Indirect effects are often very important and, under certain circumstances, more important than direct effects. This is explained by system history and the storage of biomass and information in the ecological network. This storage explains the presence of indirect effects and is needed for ecosystems to store new information. Storage of biomass is exergy. So the phrase used above, "increased exergy under prevailing conditions," may be translated to "best possible improvement of indirect effects and energy cycling under the given conditions."

APPLICATION OF RESULTS IN MODELING

The result that changes in species composition are always accompanied by an increase in exergy have been applied in a eutrophication model, as presented in detail in Jørgensen, (1976) and Jørgensen, Mejer, and Friis (1978).

The maximum growth rate and respiration rate for phytoplankton were changed in the model relative to the previous value determined by calibration. Computation of the exergy was included in the model. The model was run for several values of the growth rate and respiration rate, and some of the results are illustrated in Fig. 22.9. As shown, the maximum growth rate giving maximum exergy decreases when the phosphorus input is increased. When nutrients are scarce, phytoplankton compete via their uptake rates of nutrients. Smaller species have a faster uptake due to a greater specific surface, and they grow more rapidly. On the other hand, high nutrient concentrations will not favor small species. Competition is most often not based on uptake rate, but rather on the avoidance of grazing, where a greater size is more favorable.

The eutrophication model has been used to set up a prognosis on the expected improvement of lake water quality when phosphorus input is reduced. Predicted improvements were acceptably close to observed ones. However, there is the question of whether the prognosis could be further improved. A change in species composition has taken place as a consequence of the reduction in phosphorus input. This implies that the spring bloom occurred earlier and caused a greater discrepancy between predicted and observed values. The possibilities of improving the prognosis by use of exergy as a control function were therefore tested. Results were very promising, as illustrated in Table 22.3, in which the prognosis is validated for four different cases:

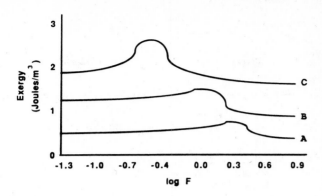

Figure 22.9 Exergy versus log F at three different levels of phosphorus in the lake water. F is the maximum growth rate for phytoplankton (which is varied) relative to the maximum growth rate selected at calibration, that is, at the present level of eutrophication (level 2). When $F = 1$, the maximum growth rate is the one found by calibration. As seen in the figure, maximum exergy occurs at different F values for the three levels of phosphorus. The three levels are (1) a eutrophication situation corresponding to twice the present level, (2) the present eutrophication level, and (3) an oligotrophic situation, corresponding to 25 times smaller input of phosphorus than at present.

TABLE 22.3 Results of prognosis on water quality improvement by a lake entrophication model

Model case	Validation	Version A	Version B
1	Y	0.79	0.72
	SD	0.18	0.08
2	Y	0.68	0.57
	SD	0.17	0.07
3	Y	0.47	0.46
	SD	0.13	0.06
4	Y	0.38	0.33
	SD	0.12	0.06

Y is a measure of predicted versus observed values over a set of water quality variables (Jørgensen et al., 1986), and SD is the standard deviation between model outputs and empirically observed values.

1. No changes in the model.
2. Exergy is introduced as the goal function and the current maximum growth rate changed in accordance with the exergy principle. The model calculated the exergy for various values of the maximum growth rate and selected the value that gives the highest exergy.
3. Silicon is introduced as the fourth nutrient to be able to include diatoms in the eutrophication model.
4. Improvements 2 and 3 are used simultaneously in the model.

The introduction of exergy as the goal function improves the model prognosis. These results indicate that the next generation of models that can account for changes in species composition can be developed by the use of exergy as a goal function. In this context the model represents biological properties including all regulation mechanisms of the system. It also represents the structure of the system and the biological components that natural selection can work on, while selection is based on the exergy principle or, in other words, a translation of Darwin's theory into thermodynamics.

The relation between exergy and the most important parameters can also be found and used in estimates of the most important parameters. This can be illustrated by a simple model of an

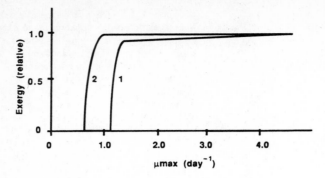

Figure 22.10 Exergy plotted versus μ_{max}. Curve 1 corresponds to $Q/V = 1.0$, and curve 2 to $Q/V = 0.5$.

algae pond, considering only the phosphorus cycle. In this simple case, the model has only two state variables: PA = phosphorus in algae and PS = soluble inorganic phosphorus. The uptake of phosphorus, the transfer of PS to PA, can be described by the use of a Michaelis–Menten expression, and the transfer from PA back to PS can be described by the use of a first-order reaction. Therefore,

$$\frac{dPA}{dt} = \frac{\mu_{max}(PS)(PA)}{K_m + PS} - (RE)(PA) - \frac{Q(PA)}{V},$$

where μ_{max}, K_m, and RE are constants. Q is the flow of water to the pond and V is its volume. The differential equation for PS is

$$\frac{dPS}{dt} = \frac{PIN(Q)}{V} + (RE)(PA) - \frac{\mu_{max}(PS)(PA)}{K_m + PS} - \frac{Q(PS)}{V},$$

where PIN is the concentration of phosphorus in the pond inflow. If the steady-state values are found from these equations, it is possible to find the corresponding exergy by use of the exergy equation presented earlier:

$$Ex = RT \ln\left(\frac{1}{PA_{eq}}\right)\left[PIN - K_m \frac{RE + Q/V}{\mu_{max} - (RE + Q/V)}\right],$$

where PA_{eq} is the concentration of PA at thermodynamic equilibrium. Figure 22.10 shows the relation

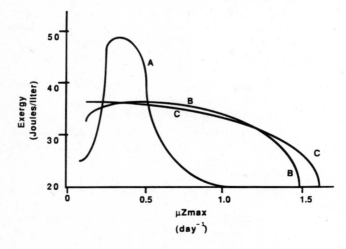

Figure 22.11 Exergy versus the maximum growth rate of zooplankton for three different Q/V values: A, $Q/V = 0.2$; B, $Q/V = 0.01$; and C, $Q/V = 0.0005$.

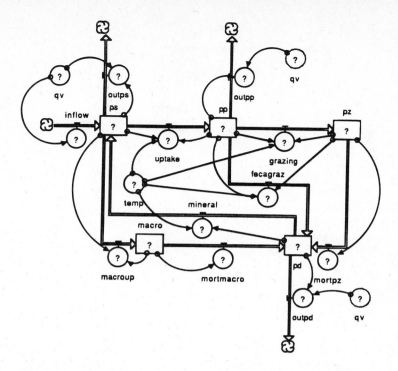

Figure 22.12 Conceptual diagram of the model developed by use of STELLA. The question marks indicate information needed to run the model.

between Ex and μ_{max} for two different Q/V values and $RE = 0.2$, $K_m = 0.02$, and $PIN = 1.0$. It is interesting from this result, which is *qualitatively* independent of RE, K_m and PIN, that:

1. μ_{max} has an absolute minimum value under which no algae can exist. These minimum values are very realistic.

2. Exergy increases very rapidly by increase of μ_{max} to just above this minimum value and then becomes almost constant. Values slightly above the minimum could therefore be expected in nature. Such values correspond to observations (Jørgensen et al., 1979). The maximum growth rate of phytoplankton is in the range of about 1 to 4 day^{-1}.

3. μ_{max} depends on Q/V. The faster the flow rate, the faster the growth rate must also be in order to maintain a certain phytoplankton concentration in the lake. This is also in accordance with observations (see Fig. 22.4).

It is not possible, or at least very difficult, to find the analytical solution for the steady state of a model if it has more than two state variables, but a relationship between the value of a crucial parameter and the exergy can be found by using a computer model. A typical result is shown in Fig. 22.11. These results are found by using a model of the phosphorus cycle in a lake. Four state variables are included: soluble phosphorus, and phosphorus in phytoplankton, zooplankton, and detritus. The model was developed by using STELLA software. The conceptual diagram is shown in Fig. 22.12 and the corresponding equations in Fig. 22.13. The model was run with fixed values for the forcing functions in 1000 days (see the equations in Fig. 22.13).

```
macro = macro + dt * ( macroup - mortmacro )
INIT(macro) = 0.005
pd = pd + dt * ( -mineral + fecagraz + mortpz - outpd + mortmacro)
INIT(pd) = 0.1
pp = pp + dt * ( uptake - grazing - outpp - fecagraz )
INIT(pp) = 0.35
ps = ps + dt * ( -uptake + inflow - outps + mineral - macroup )
INIT(ps) = 0.25
pz = pz + dt * ( grazing - mortpz )
INIT(pz) = 0.1
fecagraz = IF pp < 0.08 THEN 0.015*pp*temp ELSE 0.045*pz*temp*pp/(pp+0.05)
grazing = IF pp < 0.08 THEN 0 ELSE 0.028*pz*temp*pp/(0.1+pp)
inflow = qv*0.8
macroup = macro*ps*0.05/(0.002+ps)
mineral = 0.35*pd*temp
mortmacro = macro*0.005+ 0.001*macro^2
mortpz = 0.027*pz
outpd = qv*pd
outpp = qv*pp+0.008*pp
outps = qv*ps
qv = 0.01
temp = 1+SIN(TIME/60)
uptake = IF ps < 0.001 THEN 0 ELSE 1.0*pp*temp*ps/(0.05+ps)
```

Figure 22.13 Equations of the model shown in Fig. 22.12.

As seen from the results in Fig. 22.11, the exergy has a maximum at the maximum growth rate for zooplankton of 0.3 to 0.45 day^{-1}, depending on the flow rate Q/V. Also, these values are realistic when compared to literature values for this parameter (all other parameters were given realistic values from previous lake modeling studies or the literature), whereas the highest and lowest values are unrealistic. The same range of maximum growth rate for zooplankton is found in many eutrophication modeling studies, (for example, Jørgensen, 1986). It can also be seen from Fig. 22.12 that a high Q/V value gives a high sensitivity to the maximum growth rate, whereas a low Q/V value gives almost the same exergy for a wider range of maximum growth rate values for zooplankton.

These results lend support to the maximum exergy hypothesis presented in this chapter. But it also opens the possibility of using exergy computations to find unknown parameters or improve parameter estimation. This would be of great importance in ecological modeling.

CONCLUSIONS

The hypothesis of this chapter is that ecosystems react to perturbations by changes in species composition and structure such that (1) exergy is increased and (2) buffer capacities related to the pertubations are increased. This hypothesis is not concerned with changes in exergy due to changes in available resources.

The first point is to be considered a thermodynamic translation of Darwin's theory and the second an extension of Le Chatelier's principle applied to ecology. The two concepts, exergy and buffer capacities, are related.

The latter concept is a multidimensional stability concept. A simple stability concept will fail to describe actual ecosystem reactions to perturbations. Information on stability needs to include answers to such questions as stable to what and in which direction.

The proposed hypothesis has been supported by several modeling considerations. Ecosystems are irreducible systems, and it will not be possible to find an ecological principle by interpretation of raw observations—an ecosystem is too complex. Therefore, it is necessary to apply models or "experimental mathematics" to elucidate principles. This application of models here has clearly given support to the maximum exergy hypothesis.

The use of models with biological and ecological properties indicates that these properties are related to the ability to meet perturbations with such changes in the system that exergy is kept at its highest possible level under the given conditions and that changes caused by the perturbations are reduced as much as possible due to maintenance of a high buffer capacity.

The exergy principle has been currently used to vary parameters to account for changes in species composition. This application of the principle was rather successful as the phytoplankton growth and respiration rates were changed to reflect ecological observations of decreased eutrophication associated with smaller species with higher growth rates. These results seem promising on the point of being able to account for changes in species composition in ecological models. In the model, a computation of exergy is introduced, and the model is asked to pick the combination of the selected parameters that gives the highest exergy. Realistic ranges are given for the selected parameters; otherwise, it would be too cumbersome to find the combinations that yield the highest exergy.

Further support for the maximum exergy hypothesis, including more use of the principle in ecological modeling to account for changes in species composition, must be sought before the hypothesis can be considered a workable theory. The results are promising enough that this investment in further examinations would appear profitable.

REFERENCES

ARMSTRONG, N. E. 1977. Development and Documentation of a Mathematical Model for the Paraiba River Basin Study. Vol. 2. DOSAGM: Simulation of Water Quality in Streams and Estuaries. Technical Report CRWR-145. Center for Research in Water Resources, University of Texas, Austin, Texas. 280 pp.

BERMAN, S. M., and SILVERSTEIN, S. D. (eds). 1975. *Efficient Use of Energy. Part III. Energy Conservation and Window Systems.* American Institute of Physics, Conference Proceedings No. 25. Am. Inst. Physics, New York, pp. 245–304.

JØRGENSEN, S. E. 1976. A eutrophication model for a lake. *Ecol. Mod.* 2:147–165.

JØRGENSEN, S. E. 1979. Modelling the distribution and effects of heavy metals in aquatic ecosystems. *Ecol. Mod.* 6:199–223.

JØRGENSEN, S. E. 1986. Structural dynamic models. *Ecol. Mod.* 31:1–9.

JØRGENSEN, S. E. 1988. *Fundamentals of Ecological Modelling.* Elsevier, Amsterdam. 392 pp.

JØRGENSEN, S. E. 1990. Ecosystem theory, ecological buffer capacity, uncertainty, and complexity. *Ecol. Mod.* 52:125–133.

JØRGENSEN, S. E., MEJER, H. F., and FRIIS, M. B. 1978. Examination of a lake model. *Ecol. Mod.* 4:253–279.

JØRGENSEN, S. E., FRIIS, M. B., HENDRIKSEN, J., JØRGENSEN, L. A., and MEJER, H. F. 1979. *Handbook of Environmental Data and Ecological Parameters.* International Society for Ecological Modelling, Copenhagen, Denmark. 1180 pp.

JØRGENSEN, S. E., KAMP-NIELSEN, L., CHRISTENSEN, T., WINDOLF-NIELSEN, J., and WESTERGÅRD, B. 1986. Validation of a prognosis based upon a eutrophication model. *Ecol. Mod.* 32:165–182.

KAMP-NIELSEN, L. 1983. Sediment–water exchange models. In Jørgensen, S. E. (ed.), *Applications of Ecological Modelling in Environmental Management.* Elsevier, Amsterdam, pp. 480–540.

MEJER, H. F., and JØRGENSEN, S. E. 1979. Exergy and ecological buffer capacity. In Jørgensen, S. E., (ed.), *State of the Art in Ecological Modelling.* International Society for Ecological Modelling, Copenhagen, Denmark, pp. 829–846.

MOROWITZ, H. J. 1968. *Energy Flow in Biology.* Academic Press, New York. 280 pp.

ODUM, H. T. 1983. *Systems Ecology.* Wiley-Interscience, New York. 644 pp.

ODUM, H. T., and PINKERTON, R. C. 1955, Time's speed regulator: the optimum efficiency for maximum power output in physical and biological systems. *Amer. Scient.* 43:331–343.
PATTEN, B. C. 1982. On the quantitative dominance of indirect effects in ecosystems. In Lauenroth, W. K., Skogerboe, G. V., and Flug, M. (ed.), *State-of-the-Art in Ecological Modelling*. Elsevier, Amsterdam, pp. 27–37.
PRIGOGINE, I., and STENGERS, I. 1979. *La Nouvelle Alliance*. Gallimard, Paris. 234 pp.
RANT, Z. 1956. Exergy. *Forschung Ing. Wesens.* 22:336–344.
STRAŠKRABA, M. 1980. Cybernetic categories of ecosystem dynamics. *ISEM J.* 2:81–96.
ULANOWICZ, R. E. 1986. *Growth and Development, Ecosystems Phenomenology.* Springer-Verlag, New York. 202 pp.
Water Resources Engineers, Inc. 1973. Computer Program Documentation for the Stream Quality Model QUAL-II. Prepared for the U.S. Environmental Protection Agency, Washington, D.C. WRE, Walnut Grove, California. 322 pp.

23

Exergy Principles and Exergical Systems in Ecological Modeling

SVEN E. JØRGENSEN

The Royal Danish School of Pharmacy
Department of Environmental Chemistry
Copenhagen

DIMITRII O. LOGOFET

Institute of Atmospheric Physics
Russian Academy of Sciences
Moscow

YURI M. SVIREZHEV

Potsdam Institute for Climate Change Study
Potsdam

Most new fields rapidly gain their protagonists and antagonists. The antagonists said early, and continue to say, that it is unrealistic at present to simulate ecological systems. . . . They say such systems are far too complex and our knowledge of processes operating in these systems is far too fragmentary, both quantitatively and qualitatively. . . . Most protagonists of the modelling concept applied to ecology recognize that models are always imperfections of reality, only abstractions of the real world [Van Dyne, in *Grasslands, Systems Analysis and Man*, Int. Biol. Program 19, Chap. 14, p. 890 (1980), Cambridge U. Press].

INTRODUCTION

The notion of exergy proposed by Evans (1966) has been applied in ecological modeling by Mejer and Jørgensen (1979), Jørgensen and Mejer (1981), and Jørgensen (1982, 1990, 1992).

In a series of simulation runs of an aquatic ecosystem model, the following principle was shown to be valid: Exergy increases in the course of ecosystem evolution or, at least, it does not decrease. This principle may be violated locally in time, but on average, in a time period sufficiently large, systems evolve in accordance with this principle. It follows from this "exergy principle" that in response to a change in external forcing a system will modify its state variables and structure in such a way that exergy increases. An analogy exists between this and Le Chatelier's well-known principle from chemistry (see, for example, Sommerfeld, 1952).

The purposes of this paper are, first, to attain a formulation of this principle that is mathematically and physically rigorous, and, second, to investigate the relation the principle may have with more general concepts of stability (in particular, Liapunov stability) in mathematical ecology.

EXERGY IN THE THERMODYNAMICS OF OPEN SYSTEMS: ITS USE IN A PARTICULAR CASE OF ECOLOGICAL SYSTEMS

Let us consider a system, say A, living in its environment, A_0 (see Fig. 23.1). Each of these two entities is characterized by its temperature, T and T_0, pressure, p and p_0, chemical potentials, μ_i and μ_{i0} of the ith substance, energy, U and U_0, volume, V and V_0, entropy, S and S_0, and amounts N_i and N_{i0} of the ith substance. We consider the union, $A \cup A_0$, to be isolated (from the rest of the universe). Then, under conditions $V \ll V_0$, $U \ll U_0$, and $N_i \ll N_{i0}$, we may define a function Ex, exergy, such that its total variation dEx is

$$dEx = (T - T_0)\,dS - (p - p_0)\,dV + \sum_i (\mu_i - \mu_{i0})\,dN_i. \tag{23.1}$$

At a thermodynamic equilibrium between A and A_0 where $T = T_0$, $p = p_0$, and $\mu_i = \mu_{i0}$ (for any i), we have

$$dEx = 0;$$

that is, the exergy function has an extremum at this point. This extremum can be shown to be a minimum. It is necessary to emphasize that exergy is not a function of system state, and therefore dEx is not fully differential. Although Evans (1966) has defined exergy as

$$Ex = (T - T_0)\,S - (p - p_0)\,V + \sum_i (\mu_i - \mu_{i0})N_i, \tag{23.2}$$

with the integral of expression (23.1) as though it were fully differential, this is not quite correct. According to Evans, $Ex^{eq} = 0$ at the thermodynamic equilibrium, as follows from (23.2). Integration of (23.1) requires introduction of a constant. $Ex^{eq} = 0$ could be adopted to determine this constant.

Consider the case where pressure and temperature in the living system coincide with those in the environment, $p = p_0$ and $T = T_0$. Then

$$dEx = \sum_i (\mu_i - \mu_{i0})\,dN_i \tag{23.3}$$

or, if we define exergy for each ith component as Ex_i, we have $dEx = \sum_i dEx_i$, whereupon

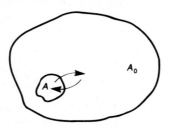

Figure 23.1 Living system, A, and its environment, A_0.

$$dEx_i = (\mu_i - \mu_{i0})\,dN_i = \mu_i\,dN_i + \mu_{i0}\,dN_{i0}$$

since $dN_i + dN_{i0} = 0$ in accordance with the mass conservation law. We will understand N_i and N_{i0} to represent concentrations of *different* elements (nitrogen, phosphorus, and so on) in the system and its environment, rather than concentrations of the same element in different ecological compartments (for example, nitrogen in phytoplankton or zooplankton).

The additive property of exergy with respect to different elements enables us to consider one element only (for example, phosphorus). Results for this will be the same for others as well. Thus, the last expression above will be written with the subscript i omitted:

$$dEx = (\mu - \mu_0)\,dN.$$

The problem now is to determine the chemical potentials μ and μ_0.

Consider a mixture of ideal gases (or dilute solutions). If N_0 is the concentration of an element in the environment and N_j its concentration in the jth (ecological) compartment of the ecosystem, then, independently of the structure of substance transitions between compartments, the following expression (Sommerfeld, 1952) obtains for the difference between chemical potentials of the element in two different compartments (the same element may be present in different chemical forms in different compartments):

$$\mu_s - \mu_k = RT \ln\left[\frac{N_s}{N_s^{eq}} - \ln\frac{N_k}{N_k^{eq}}\right], \quad s, k = 0, \ldots, n. \tag{23.4}$$

Here N_s^{eq} ($s = 0, \ldots, n$) denotes the concentration of the element in the environment or sth compartment at thermodynamic equilibrium, and n is the number of compartments.

Since μ must be the average chemical potential of the element, we have for the living system

$$\mu = \sum_{j=1}^{n} \frac{\mu_j\,dN_j}{\sum_{j=1}^{n} dN_j}, \tag{23.5}$$

where

$$\sum_{j=1}^{n} dN_j = dN \quad \text{and} \quad \sum_{j=1}^{n} dN_j = dN_0.$$

It follows from (23.5) that

$$dEx = RT \sum_{j=1}^{n} \left[\ln\frac{N_j}{N_j^{eq}} - \ln\frac{N_0}{N_0^{eq}}\right] dN_j. \tag{23.6}$$

Since the total amounts of the elements are conserved,

$$N_0 + \sum_{j=1}^{n} N_j = A \equiv \text{constant}, \tag{23.7}$$

expression (23.6) may be written in a more symmetric form as

$$dEx = RT \sum_{j=0}^{n} \ln\frac{N_j}{N_j^{eq}}\,dN_j. \tag{23.8}$$

It can also be shown that

$$dEx = RT \sum_{j=0}^{n} \frac{\partial Ex}{\partial N_j}\,dN_j,$$

where

$$Ex = RT \sum_{j=0}^{n} N_j \ln \frac{N_j}{N_j^{eq}} \qquad (23.9)$$

under condition (23.7). In other words, exergy becomes a potential function and coincides with Gibbs's free enthalpy. This is not surprising since we have used a mixture of ideal gases or dilute solutions as our thermodynamic model.

If we now recall the condition adopted above that

$$N = \sum_{j=1}^{n} N_j << N_0,$$

then the exergy function (23.9) can be represented as

$$Ex \cong RT \sum_{j=1}^{n} \left[N_j \ln \frac{N_j}{N_j^{eq}} - (N_j - N_j^{eq}) \right] \qquad (23.10)$$

to within the terms that have higher orders of smallness than ($\Sigma\ N/A$). Notice that the approximate expression (23.10) depends neither on the environmental characteristics (N_0, N_0^{eq}) nor on the total amount of matter (A_0). Hence, it can be used for any, not only closed, systems. The restriction $A \equiv$ constant (23.7) is not critical in general. Actually, let the system $A \cup A_0$ be otherwise open; that is, $N + N_0 \neq$ constants. Then we always can expand the system by surrounding it with a new environment, say A_0' (see Fig. 23.2), such that the new system $A \cup A_0 \cup A_0'$ will now be closed to matter: $N + N_0 + N_0' = A' \equiv$ constant. The exergy of system $A \cup A_0$ relative to A_0' can then be written

$$Ex(A + A_0)/A_0' = Ex(A/A_0') + Ex(A_0/A_0'),$$

or, in the case at hand,

$$\frac{Ex}{RT} = \sum_{j=0}^{n} N_j \ln \frac{N_j}{N_j^{eq}} + N_0' \ln \frac{N_0'}{N_0^{eq}}.$$

If we assume, as before, the condition

$$N_0' >> N_0 + \sum_{j=1}^{n} N_j$$

to be valid now for the "hyperenvironment," A_0', then by using (23.10) we have

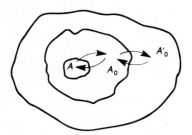

Figure 23.2 Concept of surroundings.

$$Ex \simeq RT \sum_{j=0}^{n} \left[N_j \ln \frac{N_j}{N_j^{eq}} - (N_j - N_j^{eq}) \right].$$

It is a consequence of the ideal gas or dilute solution model adopted that the exergy does not depend on a particular structure of mass flows within the ecosystem.

Extension of formula (23.9) to the case of m different elements is trivial:

$$Ex = RT \sum_{i=1}^{m} \sum_{j=0}^{n} N_{ij} \ln \frac{N_{ij}}{N_{ij}^{eq}} \qquad (23.11)$$

under the conditions

$$N_{i0} + \sum_{j=1}^{n} N_{ij} = A_i \equiv \text{constant}, \qquad i = 1, \ldots, m. \qquad (23.12)$$

This means that the mass conservation law for each particular element is valid. Sometimes the approximate formula (23.10) appears more convenient to use because there are no specific constraints imposed on the variables:

$$Ex \propto RT \sum_{s=1}^{m(n+1)} [N_s \ln \frac{N_s}{N_s^{eq}} - (N_s - N_s^{eq})]. \qquad (23.13)$$

Here, the variables N_s represent concentrations of each element of concern in every ecological compartment of the ecosystem, the elements and compartments being determined by a model description of the system.

Finally, we focus on the choice of N_s^{eq}, the concentrations at thermodynamic equilibrium. All N_s^{eq} (subscript s refers to the living system) are very small compared with equilibrium concentrations of the corresponding elements in the environment. We may identify the thermodynamic equilibrium state of an ecosystem with the state of a primordial "inorganic soup" where organic cell formations might arise by pure chance and be capable of matrix reduplication. According to estimates derived by Morowitz (1968) from the probability that life begins as a result of a fluctuation in the equilibrium ensemble, the orders of magnitudes for N_{ij}^{eq}/N_{i0}^{eq} fall in the interval between 10^{-14} and 10^{-10}.

Thus, the exergy function can be calculated practically by the above presented formulas for either a real ecosystem or its mathematical models. The next sections are devoted to what can be gained from this concept in both the theory and practice of ecological modeling.

EXERGY AND LIAPUNOV FUNCTIONS

In a number of simulation experiments on aquatic ecosystem models, Mejer and Jørgensen (1979) and Jørgensen (1982) observed an increase in exergy in response to certain changes in external conditions, that is, certain perturbations of the system. The exergy correlated positively with values of "ecological buffer capacities," defined (Mejer and Jørgensen, 1979) as the inverse sensitivity of a state variable to changes in the driving force on a model. Basically, the buffer capacity concept, representing the capability of the ecosystem to absorb perturbations, is a measure of stability with respect to perturbations of a certain kind. Therefore, the idea arises that there may be an internal link between the "increasing exergy" concept and the "Liapunov stability" concept, the most fundamental one in mathematical stability theory. Heuristically, a Liapunov function is also a natural, consistent tool for quantifying how far the current state of a system is from some reference state (see, for example, La Salle and Lefschetz, 1961; Willems, 1970). This idea seems consistent with the concept of exergy, which measures the free energy that a system possesses relative to its environment and shows how far the current state of the system is from the state of thermodynamic equilibrium with the environment.

For a number of theoretical models of population and community dynamics, Liapunov functions exist (for example, Goel, Maitra, and Montroll, 1971; Svirezhev and Logofet, 1978). These functions look similar to the exergy equations presented in the previous section, which motivates investigating the link between exergy and Liapunov stability functions. It should be emphasized that Liapunov functions, depending on their particular forms and properties, may represent either the *stability* or *instability* of a system state.

Consider a simple point model of a pond used by Mejer and Jørgensen (1979) to illustrate the exergy concept. The model has the following equations:

$$\frac{dP_s}{dt} = (P_{in} - P_s)\frac{Q}{V} - (G - M)P_a, \tag{23.14}$$

$$\frac{dP_a}{dt} = \left(G - M - \frac{Q}{V}\right)P_a, \tag{23.15}$$

$$G = G(P_s) = G'\frac{P_s}{K + P_s},$$

where the state variables P_s and P_a are concentrations of soluble and algal-bound phosphorus. P_{in} is soluble phosphorus concentration in the inflow, Q is the rate of outflow (or dilution rate), V is volume, $G = G(P_s)$ is the P uptake rate, and M is the rate of remineralization. Obvious constraints for such a model are

$$P_s > 0, \quad P_a > 0, \tag{23.16}$$

$$P_s + P_a = P_{total}, \tag{23.17}$$

where $P_{total} = P_{total}(t)$ obeys the equation

$$\frac{dP_{total}}{dt} = \left(P_{in} - P_{total}\right)\frac{Q}{V}. \tag{23.18}$$

Following Mejer and Jørgensen (1979) and considerations of the previous section, we assume the total amount of phosphorus to be conserved in the system, $P_{total}(t) \equiv$ constant. Then the system of eqs. (23.14 and 23.15) reduces to one equation:

$$\frac{d\xi}{dt} = \left[G(\xi) - M - \frac{Q}{V}\right](p_a^{eq} + \xi), \tag{23.19}$$

where the reaction coordinate is defined by any one of the conditions

$$P_a = P_a^{eq} + \xi, \tag{23.20}$$

$$P_s = P_s^{eq} - \xi,$$

while P_a^{eq} and P_s^{eq} designate concentrations at the thermodynamic equilibrium. If the inorganic soup represents the state of thermodynamic equilibrium for living nature, we have $P_a^{eq} \sim 10^{-50}$ gP/m^3 from estimates by Morowitz (Mejer and Jørgensen, 1979).

Thus, even at the beginning of organic evolution, which corresponds to the thermodynamic equilibrium state of $\xi = 0$ in (23.19), the concentration of organic matter is nonzero but a negligibly small quantity. Therefore, if we are rigorous,

$$P_a \geq P_a^{eq}. \tag{23.21}$$

This minor adjustment of the constraints must not affect the dynamic behavior of a model in other regions of the phase space. Therefore, we have to transfer the origin of our coordinate system to the *thermodynamic equilibrium* while retaining the same phase portrait as before, but within the shifted positive semiaxes.

In the general case of n living components with the thermodynamic equilibrium state $\mathbf{N}^{eq}$, the positive orthant, P^n, of the n-dimensional space must be transformed to

$$\tilde{P}^n = P^n - \mathbf{N}^{eq}.$$

Practically, this means replacing the state variables of the model by their shifted values.

Note that the above procedure does not correspond merely to the formal change of variables, $\tilde{N}_i = N_i - N_i^{eq}$ ($i = 1, \ldots, n$), since the latter keeps the phase portrait unchanged (see Figure 23.3a), whereas the purpose is to have it shifted and invariant for the transformed equations of the model (Fig. 23.3b). Since the values of $\mathbf{N}_i^{eq}$ are negligibly small, the dynamics of the model far from the equilibrium are not affected by this transformation.

Applying this general idea to the phosphorus model, eq. (23.19), we replace the state variable P_a by $P_a - P_a^{eq}$ and obtain

$$\frac{d\xi}{dt} = \left[\tilde{G}(\xi) - M - \frac{Q}{V} \right] \xi. \tag{23.22}$$

Furthermore,

$$\tilde{G}(\xi) \simeq G(\xi)$$

as

$$P_S \gg P_a^{eq}.$$

It is now to be determined, whether the exergy function is a Liapunov function, using the stationary

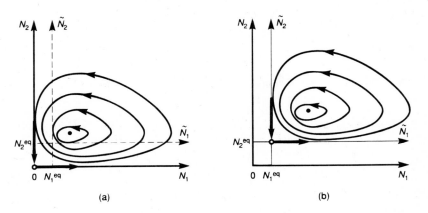

Figure 23.3 Phase portrait for a system of two population equations of prey–predator type: (a) Formal change of variables $\tilde{N}_i = N_j - N_i^{eq}$ retains the invariance of the positive orthant, $P^n = \{N_i > 0, i = 1, \ldots, n\}$; (b) "old" equations for "new" variables result in the invariance of the "shifted" orthant, $\tilde{P}^n = \{N_i > N_i^{eq}, i = 1, \ldots, n\}$.

solution of eq. (23.22), $\xi(t) \equiv 0$, corresponding to the thermodynamic equilibrium, 2 times, as the reference trajectory. We have

$$Ex(\xi) = RT\left[(P_a^{eq} + \xi) \ln \frac{P_a^{eq} + \xi}{P_a^{eq}} + (P_s^{eq} - \xi) \ln \frac{P_s^{eq} - \xi}{P_s^{eq}}\right] = 0$$

when $\xi = 0$ and, as will be shown later,

$$Ex(\xi) > 0 \quad \text{when} \quad \xi > 0.$$

Furthermore, the derivative of $Ex(\xi)$ by virtue of eqs. (23.22), that is, along the trajectories, is defined as

$$\frac{dEx(\xi)}{dt} = \frac{\partial Ex(\xi)}{\partial \xi} \cdot \frac{d\xi}{dt}.$$

By use of eqs. (23.8) and (23.20), we obtain

$$\frac{\partial Ex(\xi)}{\partial t} = RT \ln \left[\frac{(P_a^{eq} + \xi) P_s^{eq}}{P_a^{eq}(P_s^{eq} - \xi)}\right], \qquad (23.23)$$

which coincides with formula (37) from Mejer and Jørgensen (1979). As ξ is always nonnegative, expression (23.23) is always positive, except at the point $\xi = 0$ where it vanishes.

Note that

$$\frac{\partial^2 E}{\partial \xi^2} = (P_a^{eq} + \xi)^{-1} + (P_s^{eq} - \xi)^{-1} > 0$$

whenever $\xi \leq P_s^{eq}$, so function $Ex(\xi)$ has its local minimum at point $\xi = 0$: $Ex(0) = 0$ (which is also a global minimum in accordance with thermodynamic theory).

From (23.22) and (23.23) it follows that

$$\frac{dEx(\xi)}{dt} = RT \ln [\ldots] \cdot \left[G(\xi) - M - \frac{Q}{V}\right] \xi > 0 \qquad (23.24)$$

everywhere in some finite domain (excluding the point $\xi = 0$) in which the model equation gives a positive increase in the concentration of algal phosphorus, that is, whenever

$$G(\xi) > M + \frac{Q}{V}. \qquad (23.25)$$

Thus, as seen, exergy $Ex(\xi)$ possesses all the basic properties of a Liapunov function for $\xi = 0$; the derivative has the same sign as the function itself. According to Chetaev's instability theorem (Chetaev, 1955), $Ex(\xi)$ verifies *local instability* at the equilibrium state $\xi(t) \equiv 0$. This means that any initial deviation, however small, from the thermodynamic equilibrium state $\xi = 0$ will cause increase of ξ. Such behavior of model trajectories is consistent with the concept that life began by small fluctuations in the vicinity of thermodynamic equilibrium and a gradual movement away from it (Schrödinger, 1945).

Notice that we have only used (1) the invariance of the phase orthant for the model equations, and (2) the mass conservation principle of the system to test the "Liapunov" properties of exergy. We have not used (23.14) and (23.15) beyond property (23.25).

Hence, we may expect these properties to occur in a sufficiently wide class of ecological models possessing the above-mentioned characteristics. Such models could be referred to as *exergical at zero*, with the idea that model trajectories go away from thermodynamic equilibrium and exergy increases along these trajectories.

We are here dealing with a principal distinction of ecological models from those of theoretical mechanisms or chemical kinetics, where the minimum of a potential function (or the minimum of a thermodynamic potential) corresponds to a stable steady state. The dynamics of such systems are determined by use of potential functions, while for ecological models such a simple form of dynamic equations turns out practically unacceptable so that the equations have to be chosen from other considerations. These represent, in essence, a phenomenological description of the ecosystem.

From the standpoint of thermodynamics, any ecological system should be exergical at zero, but on the other hand the exergy function of the classical Lotka–Volterra prey–predator model is not a Liapunov function (see Appendix A). Its derivative along trajectories is neither positive nor negative and, furthermore, in any small vicinity of (N_1^{eq}, N_2^{eq}) there exist points $(N_1, N_2) > (N_1^{eq}, N_2^{eq})$ such that $dEx(N_1, N_2)/dt < 0$.

These observations can be explained from general Liapunov stability concepts, since the Chetaev instability theorem actually covers only the cases that, in a certain sense (that of time reversion, changing t to $-t$), are opposite to those of asymptotic stability, that is, to phase patterns of the generalized topological node type (Figs. 23.4a, b). The prey–predator case corresponds, however, to neither of these two extreme types of phase patterns (Fig. 23.4c).

We do not use the missing "exergical" property in a model as an argument against the model in general. Many theoretical models do not claim to be adequate in all regions of the phase space and, in particular, not in a critical region such as the vicinity of zero equilibrium. Most models were developed to provide a reasonable description in a region where a biological community or ecosystem was believed to function normally, for example in the vicinity of a *nontrivial* or *feasible steady state*, $\mathbf{N}^* > 0$ (with plausible values of state variables).

Liapunov stability of such steady states are frequently verified by means of a proper Liapunov function. For instance, for systems of population dynamics that are dissipative in the Volterra sense (Svirezhev and Logofet, 1978; 1992) and for their possible extensions (Pykh, 1983), the function

$$L(N) = \sum_{i=1}^{n} \left[N_i^* \ln \frac{N_i}{N_i^*} - (N_i - N_i^*) \right] \qquad (23.26)$$

and other similar functions are Liapunov functions for the steady state $\mathbf{N}^* = [N_1^*, \ldots, N_n^*] > 0$. This verifies the global stability of the steady state.

Despite all the similarity between functions (23.10) and (23.26), there are principal differences in their meanings. The expression for exergy contains values of state variables at the thermodynamic equilibrium point $\mathbf{N}^{eq}$, whereas the Liapunov function (23.26) is referred to the steady state $\mathbf{N}^*$. As accepted theoretically, $\mathbf{N}^*$ is associated with a state to which the ecosystem normally evolves or

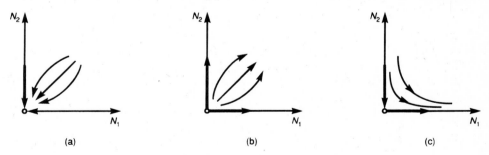

(a)　　　　　　　　　　　(b)　　　　　　　　　　　(c)

Figure 23.4 Typical phase patterns in the vicinity of the zero equilibrium state: (a) asymptotic stability (any trajectory trends to zero); (b) instability as covered by Chetaev's instability theorem (any trajectory evolves from zero); and (c) instability in the Lotka-Volterra case (where there is a trajectory that tends to zero).

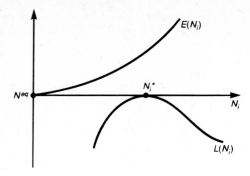

Figure 23.5 Exergy and a Liapunov function as functions of phase variables.

at which it has its normal functions. Such a state is in principle irreducible to the thermodynamic equilibrium state (Schrödinger, 1945).

In addition, there also exist mathematical reasons why the exergy function determined in (23.10) cannot be a Liapunov function for a nontrivial equilibrium $\mathbf{N}^*$. Any Liapunov function must attain its extremum [maximum in the eq. (23.26) case] at point $\mathbf{N}^*$, whereas the exergy function increases monotonically in N_i once $N_i > N_i^{eq}$, $i = 1, \ldots, n$ (Fig. 23.5).

The conclusions of this section concern the behavior of a system with the same values for its internal and external parameters (that is, coefficients in model equations). However, as stated in the introduction, the "exergy principle" considers reorganization of the system in the sense that the parameters may also change in addition to the state variables. Can the exergy principle indicate the direction in which such reorganization will evolve? Can it predict the change in system parameters in response to perturbations? These questions are discussed in the next section.

EXERGY AND THE EVOLUTION OF A SYSTEM IN PARAMETER SPACE

Let the dynamics of a system be described by the vector equation

$$\frac{d\mathbf{N}}{dt} = \mathbf{F}(\mathbf{N}; \boldsymbol{\varphi}; \boldsymbol{\mu}), \tag{23.27}$$

where $\mathbf{N}$ is a vector of state variables:

$$\mathbf{N} = [N_0, N_1, \ldots, N_n] \in P^{n+1}.$$

$\boldsymbol{\varphi} = [\varphi_1, \ldots, \varphi_l]$ represents a set of external variables, such as temperature, precipitation, or inflows of matter, that characterize the environment; $\boldsymbol{\mu} = [\mu_1, \ldots, \mu_m]$ represents a set of internal parameters inherent in the living system, such as intrinsic rates of natural increase or mortality rates for the various species. Let

$$\boldsymbol{\varphi} \in D_\varphi \subset P^l, \qquad \boldsymbol{\mu} \in D_\mu \subset P^m,$$

where the domains D_φ and D_μ are chosen from ecological constraints.

It is considered ecologically evident that there must exist relations of type $\boldsymbol{\mu} = \boldsymbol{\mu}(\boldsymbol{\varphi})$ (but not the contrary), for example, a dependence of vegetation growth rate on temperature. Therefore, at first glance it could seem unrealistic to require that the internal parameters $\boldsymbol{\mu}$ be independent of the external variables [this was expressed implicitly in (23.27)]. But, if it is recalled that such relations are usually expressed by equations with parameters no longer dependent on the variables but assumed constant, then the contradiction vanishes. It is these latter parameters that are symbolized by μ.

In the next section this point is illustrated by use of a particular example. The relation between μ and φ will evolve from the condition that the steady state must be stable when the system is evolving in the parameter space along a boundary of the stability domain.

We consider system (23.27) under the condition that

$$\langle \mathbf{N}; \mathbf{e} \rangle \equiv A \equiv \text{constant}, \qquad \mathbf{e} = [1, \ldots, 1], \qquad (23.28)$$

with symbol $\langle \cdot, \cdot \rangle$ denoting the inner product of two vectors), which expresses the mass conservation law. It is not critical since any system open to matter can be transformed into a closed system by the addition of one more equation:

$$\frac{dR}{dt} = -\langle \mathbf{F}; \mathbf{e} \rangle, \qquad (23.29)$$

such that $R + \langle \mathbf{N}; \mathbf{e} \rangle \equiv$ constant. Since constant A is in essence an external variable of the system, it will be regarded as a component of the vector φ.

Suppose now system (23.27) possesses an Ω^+ limit set (see, for example, La Salle and Lefschetz, 1961) in its phase space. The latter is assumed to be simply connected and continuously dependent on parameters φ and μ. In the simplest case, the set Ω^+ represents the unique steady state that is asymptotically stable, $\mathbf{N}^*(\varphi; \mu)$. This case will be considered next.

We shall not determine exergy throughout the total phase space, but rather on the set Ω^+ only, so that

$$Ex = RT \sum_{i=0}^{n} N_i^* \ln \frac{N_i^*}{N_i^{eq}} \qquad (23.30)$$

under the condition that

$$\sum_{i=0}^{n} N_i^* = \sum_{i=0}^{n} N_i^{eq} \equiv A.$$

In accordance with the estimates, all N_i^{eq} ($i = 1, \ldots, n$) are constant and negligibly small. Since $\mathbf{N}^* = \mathbf{N}^*(\varphi; \mu)$, $N_0^{eq} = N_0^{eq}(A)$, and N_i^{eq} can be estimated in accordance with Morowitz (1968); we have

$$Ex = Ex(\varphi; \mu), \qquad (23.31)$$

where

$$\varphi \in D_\varphi \cap \Sigma_\varphi, \qquad \mu \in D_\mu \cap \Sigma_\mu. \qquad (23.32)$$

Σ_φ and Σ_μ are the domains of $\mathbf{N}^*$'s stability in the respective parameter spaces. This implies that

$$\Sigma = \Sigma_\varphi \cdot \Sigma_\mu$$

is the stability domain in the space of all parameters.

Jørgensen (1982, p. 64) has formulated a hypothesis by which "an ecosystem attempts to meet changes in external factors by developing a new structure with higher exergy under the conditions determined by the external factors" (p. 64). How would this statement look in the framework of the formalism presented?

Suppose the system is represented by a point within the domain

$$\mathcal{B} = (D_\varphi \cap \Sigma_\varphi) \cdot (D_\mu \cap \Sigma_\mu)$$

of the total parameter space such that the variations in parameters φ and μ lead the system neither to the boundary of $\mathcal{B}$ nor out of the boundary. Let the system be at the steady state determined by the pair of parameter vectors $(\varphi_1; \mu_1)$ (see Fig. 23.6). Suppose that φ undergoes variations so slowly that there is sufficient time for new (respective) steady states to emerge. If there were no change in μ, a finite variation of φ would result in transition to a new steady state, $\mathbf{N}^*(\varphi_2; \mu_2 = \mu_1)$ (point 2 in Fig. 23.6). But if the system does change its internal parameters μ, then there is a variety of options in $\mathbf{N}^*(\varphi_2; \mu)$ restricted however, by transection of $\mathcal{B}$ with the hyperplane $\varphi = \varphi_2$ (segment $[a, b]$ in Fig. 23.6). Always there exists the maximum of $Ex(\varphi_2; \mu)$ with respect to μ in segment $[a, b]$ (in a subset of the hyperplane for a general case with certain natural constraints). Let this maximum be attained at a point $\mathbf{c} = (\varphi_2; \mu_c)$, so that

$$Ex(\varphi_2; \mu_c) > Ex(\varphi_2; \mu_2).$$

Then, according to Jørgensen's hypothesis cited above, the system has to pass from state 2 into state c and, to do so, its internal parameters μ have to change (from $\mu_1 = \mu_2$ to μ_c). If the parameters changes are considered to be sufficiently small, then there is no reason to believe that the trajectory will reach point c, but rather it should go to some intermediate point, say 3 (Fig. 23.6). Actually, the transition $1 \to 3$ will not proceed as $1 \to 2 \to 3$, but rather along a smoother curve connecting points 1 and 3.

The local variation of exergy at the passage $2 \to 3$ can be determined for sufficiently small variations $\delta\varphi$ and $\delta\mu$ as

$$\delta Ex^{(2,3)} = \left\langle \frac{\partial Ex^{(2)}}{\partial \mu}, \delta\mu \right\rangle$$

$$\simeq \left\langle \frac{\partial}{\partial \mu}\left[Ex^{(1)} + \left\langle \frac{\partial Ex^{(1)}}{\partial \varphi}, \delta\varphi \right\rangle\right], \delta\mu \right\rangle \qquad (23.33)$$

$$= \left\langle \frac{\partial Ex^{(1)}}{\partial \mu}, \delta\mu \right\rangle + \left\langle \frac{\partial}{\partial \mu} < \frac{\partial Ex^{(1)}}{\partial \varphi}, \delta\varphi >, \partial\mu \right\rangle.$$

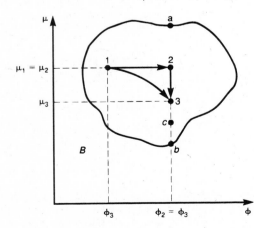

Figure 23.6 Definition of the local exergy principle. Under fixed $\varphi = \varphi_2$, exergy is maximum at point $c \in [a, b]$.

The thesis that in response to change in external variables the system modifies its internal parameters in such a way that its exergy increases can now be written in a short form as

$$\delta Ex^{(2,3)} > 0,$$

or, by using (23.33) and omitting the superscripts, in more detail as

$$\sum_{i=1}^{m} \left(\frac{\partial Ex}{\partial \mu_i} + \sum_{j=1}^{l} \frac{\partial^2 Ex}{\partial \mu_i \partial \varphi_j} \delta \varphi_j \right) \delta \mu_i > 0. \tag{23.34}$$

Expression (23.34) determines the direction in which the internal parameters must change in response to given variations of the external variables. It may be regarded as quantification of Le Chatelier's principle for those systems that are *exergical at steady state*. The principle is in essence a local variational principle.

We have introduced two different time scales: "fast" time, which is the time necessary for a steady state to be reached (the characteristic time for the transition process), and "slow" time, which is that of system reorganization (the characteristic time of evolution to the exergy maximum in the space of internal parameters). A natural generalization of the local variational principle can be formulated as follows: in a fixed environment the system evolves to a state $\boldsymbol{\mu}^*$ where

$$Ex(\boldsymbol{\varphi}; \boldsymbol{\mu}^*) = \max_{\boldsymbol{\mu}} Ex(\boldsymbol{\varphi}, \boldsymbol{\mu}) \tag{23.35}$$

under constraints

$$\boldsymbol{\varphi} = \text{constant}, \qquad \boldsymbol{\mu} \in D_\mu \cap \Sigma_\mu. \tag{23.36}$$

While the concepts of a system being exergical at zero and exergical at steady state are essentially local, the above generalization is no longer local since it refers to quite a finite domain given by the constraints (23.36). Therefore, a system that obeys (23.35) can be called *globally* [in domain (23.36)] *exergical at steady state*.

To conclude this section, notice that when deriving the local principle we presumed the variations to maintain the system inside $\mathcal{B}$, thus preserving the a priori independence among variations $\delta \boldsymbol{\varphi}$ and $\delta \boldsymbol{\mu}$ and deducing the a posteriori dependence of $\delta \boldsymbol{\mu}$ upon $\delta \boldsymbol{\varphi}$ from the condition $\delta Ex > 0$. But, if we are concerned with a case where the system is on the boundary of $\mathcal{B}$, then the variations become linked and formula (23.34) is no longer valid. In an example later, it is shown how the latter case may be treated.

APPLICATION OF THE GENERAL PRINCIPLES TO A PARTICULAR EXAMPLE

Let us consider a model of the "resource–consumer" type for a materially closed system (Svirezhev and Logofet, 1978), as given by the equations

$$\begin{aligned} \frac{dN_0}{dt} &= -g(T)N_0 N_1 + m(T)N_1, \\ \frac{dN_1}{dt} &= g(T)N_0 N_1 - m(T)N_1. \end{aligned} \tag{23.37}$$

Here, N_0 is the amount (or concentration) of the resource, N_1 is that of the consumer, T is the temperature of the system, $g(T)$ can be interpreted as a slope of the photosynthesis–temperature curve, and $m(T)$ is the consumer mortality rate (dead organic matter is supposed to be immediately converted into the resource).

It is clear that the greater is m the faster matter returns to the resource state, that is, transforms to inorganic form. Thus, parameter m can be regarded as a rate of organic matter (for example, phosphorus) remineralization or its turnover rate. The conservation law is obviously valid:

$$N_0 + N_1 \equiv A \equiv \text{constant}. \tag{23.38}$$

System (23.37) is clearly exergical at zero and has a feasible steady state,

$$N_0^* = h, \qquad N_1^* = A - h,$$

when $A > h = m/g(T)$. The steady state is continuously (and directly) dependent on parameters A and h and indirectly dependent on T since g and m are temperature dependent. The exergy function at steady state can be written as

$$Ex = RT \left\{ h \ln \frac{h}{A - N_1^{eq}} + (A - h) \ln \frac{A - h}{N_1^{eq}} \right\}. \tag{23.39}$$

Parameters A and h are regarded as the external variables, but it is more difficult to realize the internal parameters. Since both g and m depend on temperature, they are not independent parameters. Practically, we always try to represent these dependencies in the form of functions of other parameters, so that

$$\begin{aligned} g &= g(T; \rho_i), & i = 1, \ldots, L; \\ m &= m(T; \mu_j), & j = 1, \ldots, M. \end{aligned} \tag{23.40}$$

Thus, the set of independent internal parameters is given by ρ_i and μ_j.

This point is exemplified for the case where the photosynthesis curve is approximated by means of four parameters. These are the minimal and the maximal temperature bounds ($T_{\min}$ and $T_{\max}$), optimal temperature (T_{opt}), and $g_{\max} = g(T_{opt})$, the maximum rate of photosynthesis at the optimal temperature (see, for example, Racsko, 1981).

Common descriptions of the remineralization rate as a function of temperature imply either exponential dependence or its linear approximation, both being determined by two parameters, T_{cr} at which $m(T_{cr}) = 0$, and the Van't Hoff constant indicating the increase in temperature that results in a doubling of the remineralization rate (see Krapivin, Svirezhev, and Tarko, 1982).

So we have six independent internal parameters, but only some of these can vary without compromising their biological sense. The number of actual internal parameters may thus decrease somehow. Let only one of two external parameters vary, say A. Quantities g and m depend exclusively on the external parameter T, assumed now invariable. Exergy is defined in (23.39) as being dependent only on the ratio m/g. We have that the only internal parameter remaining for variations is h.

From

$$\frac{\partial Ex}{\partial h} = RT \ln \left[\frac{h}{A - h} \cdot \frac{N_1^{eq}}{A - N_1^{eq}} \right] \tag{23.41}$$

and

$$\frac{\partial^2 Ex}{\partial A \, \partial h} = -RT \left[\frac{1}{A - N_1^{eq}} + \frac{1}{A - h} \right],$$

it follows that when $A > h + N_1^{eq}$ we have

$$\frac{\partial Ex}{\partial h} < 0 \quad \text{and} \quad \frac{\partial^2 Ex}{\partial A \, \partial h} < 0.$$

The steady-state feasibility condition was formulated earlier as the condition where $A > h$. The new condition $A > h + N_1^{eq}$ is slightly more restrictive. But since, on the one hand, N_1^{eq} is very small and, on the other, $A - h = N_1^* \gg N_1^{eq}$ is a natural thermodynamic constraint, we may consider this condition to be always satisfied.

Thus, we have obtained the following constraints on A from below:

$$A > h + N_1^{eq}. \tag{23.42}$$

If we apply now the local exergy principle (23.33) to this particular case, the condition reduces to

$$\left(\frac{\partial Ex}{\partial h} + \frac{\partial^2 Ex}{\partial h\, \partial A} \delta A\right) \delta h > 0. \tag{23.43}$$

It follows immediately that, if the system obeys the local exergy principle, then h can only decrease ($\delta h < 0$) under increasing A ($\delta A > 0$). But if A decreases ($\delta A < 0$), then inequality (23.43) admits also some positive variations in h. These variations, however, depend now both on the current state of the system and the magnitude of variation in A.

To interpret these results in the ecological context of model (23.37), we might associate an increase in A with the nutrient pollution that causes eutrophication of a water body. If the aquatic ecosystem obeys the local exergy principle, then, under the eutrophication conditions, there should exist mechanisms that provide a decrease in h. This can be achieved either by increase in parameter g, the slope of the photosynthesis curve (or the maximum growth rate of algae), or by decrease in m, the remineralization rate. The latter can be interpreted either as a tendency to substitute long-living species for short-living ones in a successional process or as a tendency to retard biological turnover in the system. If we recall that $N_1^* = A - h$ can be regarded as the biomass of algae, then a decrease in h can be associated with a tendency to concentrate all matter in the organic form. Note that increasing biomass of algae in blooms is a common phenomenon under eutrophication.

Now, let temperature be the only parameter that undergoes variation. We suppose that the ratio m/g depends on the only varying parameter (in addition to the external parameter temperature), and this critical parameter might be, for instance, T_{opt}, the point of maximal $g(T)$ that can be regarded as the temperature optimum for photosynthesis. Suppose also, for simplicity, that m does not depend on T, so h depends on T and T_{opt} only via function $g(T; T_{opt})$.

Since $\partial Ex/\partial h < 0$, while $\partial g/\partial T > 0$ when $T < T_{opt}$ (Fig. 23.7), it follows that

$$\frac{\partial Ex}{\partial T} = \frac{Ex}{T} - \frac{h}{g}\frac{\partial Ex}{\partial h}\frac{\partial g}{\partial T} > 0. \tag{23.44}$$

But, if $T > T_{opt}$, then $\partial g/\partial T < 0$ and, for certain values of m, g, and T, exergy may decrease in response to increase in temperature.

As can be seen from Fig. 23.7, $\partial g/\partial T_{opt} < 0$ when $T < T_{opt}$; otherwise, $\partial g/\partial T_{opt} > 0$. Hence,

$$\frac{\partial Ex}{\partial T_{opt}} = -\frac{h}{g}\frac{\partial Ex}{\partial h}\cdot\frac{\partial g}{\partial T_{opt}} \begin{cases} < 0 \text{ when } T < T_{opt}, \\ > 0 \text{ when } T > T_{opt}. \end{cases} \tag{23.45}$$

These conditions are illustrated in Fig. 23.8. The arrows show the directions in which exergy increases. If the system obeys the exergy principle, then it clearly has to increase the temperature maximum of photosynthesis everywhere in domain I (that is, sufficiently far to the right of T_{opt}) in response to increase in temperature. But if the temperature decreases, then the increase in T_{opt} is still possible, which does not break the local principle [that is, inequality (23.34) adapted to this particular case]. If the system now has the temperature to the left of T_{opt} (domain III in Fig. 23.8), then T_{opt} must shift to the lower temperatures in response to decrease in T. Under increase in T, positive variations δT_{opt} are admissible but restricted by the inequality of the local exergy principle.

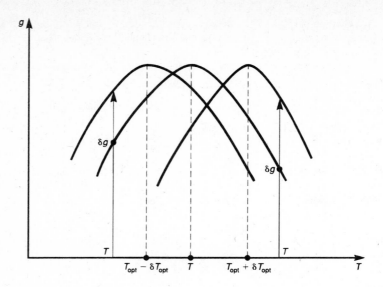

Figure 23.7 Photosynthetic curves as functions of temperature (for the determination of $\partial g/\partial T_{opt}$).

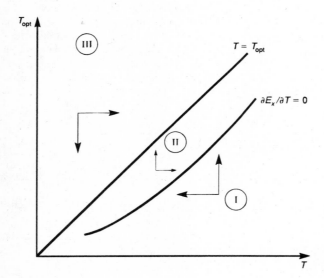

Figure 23.8 Domains of different directions for exergy to increase.

Finally, in case the system is in domain II, not far to the right of T_{opt}, and obeyed the local exergy principle, T_{opt} would shift even farther to the right in response to the decrease in temperature, the result being poorly interpreted in ecological terms.

As the calculations quickly become very cumbersome in this case, we have not verified here the local principle in the form of equation (23.34). Note, however, that it can give additional information only if the variation of external variables already brings about an exergy increase so that the variations of internal parameters that diminish exergy also become admissible; that is, the variations of internal parameters are not definite. In domain I, for instance, when temperature is increasing, both positive and negative variations, δT_{opt}, seems admissible. The local principle (23.34) then helps to restrict the variations more definitely.

Basically, the above-stated results permit an apparent ecological interpretation. Suppose a water body has a temperature that lies below the average temperature optimum of photosynthesis in its vegetation community. If the temperature decreases, then by the exergy principle the biomass has to reorganize its structure or species composition such that the average temperature optimum decreases too; that is, more cold-tolerant species become dominant. The same is true when the water body has a temperature sufficiently above the average T_{opt}. In response to further temperature increase, the system reorganizes itself in such a way that T_{opt} shifts farther to the right. This might be explained by the fact that temperature variations vary the entropy S of the system. But our main assumption that this coincides with the temperature of the environment ($T = T_0$) eliminates any explicit temperature dependence from the exergy expression [that is, $(T - T_0)dS = 0$].

EXERGY AND CLOSED TROPHIC CHAINS

While exergy is based on linear thermodynamics, most theories in mathematical ecology are principally nonlinear ones. We have chosen as a test model a materially closed trophic chain (Svirezhev and Logofet, 1978).

For a materially closed trophic chain of Volterra type, we have

$$\frac{dN_i}{dt} = N_i(-m_i + g_{i-1} N_{i-1} - g_i N_{i+1}), \qquad i = 1, \ldots, n, \tag{23.46}$$

$$N_0 = A - \sum_{i=1}^{n} N_i, \qquad N_{n+1} \equiv 0.$$

The necessary condition for a trophic chain with q trophic levels (that is, of length q) to exist has the following form (Svirezhev and Logofet, 1978):

$$A^*(q) < A < A^*(q + 1), \tag{23.47}$$

where $A^*(q)$ are some known functions of chain length, whose particular forms are determined by model parameters q_i and m_i. For instance, $A^*(1) = m_1/g_0$, so that, in order for the system to "become alive," the total amount of its matter must be higher than a certain nonzero quantity. The trophic chain of length q is associated with a steady state in system (23.46) of the form

$$\mathbf{N}_q^* = [N_0^*, N_1^*, \ldots, N_q^*, 0, \ldots, 0]$$

Under condition (23.47), this steady state can be shown to exist (that is, $N_i^* > 0$, $i = 0, \ldots, q$) and to be "ecologically" stable (Svirezhev and Logofet, 1978). According to the ideas of the section on exergy and Liapunov functions, under zeros in the steady-state pattern we mean the thermodynamic equilibrium value of the corresponding components, the main outcomes of the former theory being preserved.

To be able to interpret the results in exergy terms, we simplify system (23.46) further by assuming $g_i = g$ (for $i = 0, \ldots, n$) and $m_i = m$ ($i = 1, \ldots, n$). Hence,

$$A^*(q) = \begin{cases} \dfrac{m}{4g} q(q+2), & \text{when } q \text{ is even,} \\ \dfrac{m}{4g}(q+1)^2, & \text{when } q \text{ is odd,} \end{cases} \qquad (23.48)$$

and the quantity

$$h = \frac{m}{g}$$

becomes a key parameter for the trophic chain. It is obviously expressed in the units of biomass, so the total amount of matter can be measured in these "quanta": $A = Ch$, where C is a scalar number. It follows then from (23.48) that, when q is even,

$$N_i^* = \begin{cases} \dfrac{2A}{q+2} - \dfrac{i}{2}h, & i \text{ is even,} \\ \dfrac{h}{2}(q-i+1), & i \text{ is odd.} \end{cases} \qquad (23.49a)$$

But when q is odd,

$$N_i^* = \begin{cases} \dfrac{h}{2}(q-i+1), & i \text{ is even,} \\ \dfrac{2A}{q+1} - \dfrac{i+1}{2}, & i \text{ is odd.} \end{cases} \qquad (23.49b)$$

Suppose, for a trophic chain of length q, we choose the value of $A = A(q)$ at the middle of the interval $(A^*(q), A^*(q+1))$. Then, expressions (23.49) reduce to a fairly simple form with no dependence on whether q is odd or even:

$$N_i^* = \frac{h}{2}(q+1-i). \qquad (23.50)$$

Determined at the steady state, the exergy function then takes the form

$$Ex(q) = RT \frac{h}{2} \sum_{i=0}^{q} (q+1-i) \ln \frac{h(q+1-i)}{2N_i^{eq}}, \qquad (23.51)$$

the increment, $\Delta Ex(q)$, at the increase in chain length from $(q-1)$ to q being calculated as

$$\Delta Ex(q) = Ex(q) - Ex(q-1) \simeq RT \frac{h}{2} \ln \left\{ \frac{q(q+1)^q}{2(q+2)} \frac{h^q}{N_1^{eq} \ldots N_q^{eq}} \right\}. \qquad (23.52)$$

The values of $N_1^{eq}, \ldots, N_q^{eq}$ are naturally considered invariant at the transition from $q-1$ to q, but in order to have the conservation law preserved for the increased value of A, we also have to increase the value of N_0^{eq}:

$$N_0^{eq} = \frac{h}{4}(q+2)(q+1) - \sum_{i=1}^{n} N_i^{eq},$$

$$\frac{N_0^{eq}(q)}{N_0^{eq}}(q-1) \simeq \frac{q+2}{q}.$$

It follows immediately from (23.52) that exergy will increase with an increase in chain length if the following condition holds:

$$h > \frac{2(q + 2)}{q(q + 1)} \sqrt[q]{N_1^{eq} \ldots N_q^{eq}}. \qquad (23.53)$$

Comparing this approximate condition with the exact one derived from (23.51), we see that the two are fairly close. For example, $\Delta Ex(1) > 0$ is actually provided by the condition $h \gtrsim 2.24$ $(N_1^{eq} \ldots N_q^{eq})^{1/q}$, while the approximation (23.53) yields the condition $h > 3(N_1^{eq} \ldots N_q^{eq})^{1/q}$. The estimates become even closer when q increases.

The key role in these considerations is apparently taken by the quantity h, the quantum of matter that is necessary for life to begin in a trophic chain. Since $h = m/g$, it can be estimated, for instance, from data on phosphorus uptake or death rate of phytoplankton. This quantity can be estimated as $h = 1 gP/m^3$ for mesotrophic lakes. If we recall now the estimate for N_i^{eq} cited in the second section, then condition (23.53) can certainly be seen to hold for any trophic chain. Hence, exergy always increases with increased length of the chain (under conditions of the proper increase in A, the total matter in the ecosystem that guarantees stability of the chain with increased length). *In other words, an increase in the total amount of matter, which can be interpreted as a complication of the ecosystem structure, is always accompanied by increasing exergy.* It can be shown (Appendix B) that exergy is an increasing function of q, increasing as $q^2 \ln q$.

Now, let the number of trophic levels (chain length) be fixed (q = constant), but the total amount of matter vary, with the variations being quasistationary so that there is enough time for a new steady state to develop after a variation of A. By using (23.51), it can be shown (Appendix C) that if (for an odd q)

$$A > \frac{(q + 1)^2 h}{4} + \sum_{i=1}^{n} N_i^{eq}, \qquad (23.54)$$

then $\partial Ex/\partial A > 0$; that is, exergy is an increasing function of A. Since, by (23.48), $(q + 1)^2 h/4 = A^*(q)$ is the lower bound for a trophic chain of length q to exist, while $\mathbf{N}_i^{eq}$ are negligibly small in comparison with h, condition (23.54) can always be considered true. Thus, upon increasing the total amount of matter in the system, exergy increases too (under a fixed length of the trophic chain).

Consider, finally, the behavior of exergy when the length of a trophic chain increases [up to the values restricted by the stability constraint (23.47)] under a fixed value of A. This increase in chain length can be regarded as an evolution of the system through a series of its unstable equilibrium states with increasing number of nonzero components unless the proper steady state $\mathbf{N}_q^*$, corresponding to the given level of A, is attained. By a straightforward but fairly tedious derivation, it can be shown that exergy increases whenever the length of a trophic chain increases (under a fixed level of A sufficient to sustain the chain at the increased length).

Several observations will aid interpretation of the above results. Note that, according to the ideas presented at the end of the second section, we have assumed all the thermodynamic equilibrium values, N_i^{eq} ($i = 1, \ldots, n$), to be positive. We assume furthermore that there exist propagules of any species typical for a given system so that, once the total amount of matter has increased enough to sustain a certain number of trophic levels, those species can develop from the propagules and occupy the proper trophic level.

Another remark concerns the discreteness of q, whereby the boundary of the stability domain as determined by equalities (23.48) is a discrete set, too (see Fig. 23.9). We substitute a continuous curve for this domain to make it more visible, for example the curve passing through the middle of the segment (see Fig. 23.10). In the plane of parameters (A, q), this "stability curve" can be represented by the parabola

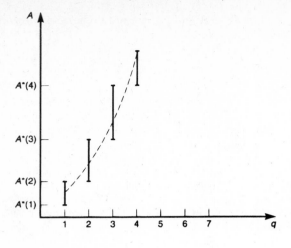

Figure 23.9 Stability domains. As an example, when $A^*(3) < A < A^*(4)$, there exist trophic chains of length 3.

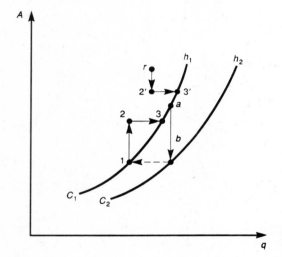

Figure 23.10 Evolution of a system on the (A, q) plane.

$$\tilde{A}^* = \frac{(q+1)(q+2)h}{4}. \qquad (23.55)$$

Let us consider now a quasi-stationary evolution of an ecosystem in the (A, q)-plane in order to determine if it conforms to the exergy principle (Fig. 23.10). Since there always exist some species propagules or seed populations, the system tends to reach the stability curve C_1. Let the system be at point 1 of the plane and parameter A be increased such that the system passes to point 2. Although this results in increased exergy, the system will tend to increase it even further by making its structure more complex, this complexity being determined by the chain length q. As a result, the system moves to point 3.

Let the system now be at point $1'$ and let parameter A decrease so that the system passes to point $2'$ with a decreased exergy value. Then, in accordance with the exergy principle, it will evolve toward point $3'$ where its exergy is increased.

Finally, let the system be at a point a on the stability curve. If parameter A now decreases, the system would appear to be bound to pass into b, but this variation is in fact forbidden because it leads the system out of the stability domain. The only path for evolution is along curve C_1 down to point 1. Structural complexity and hence exergy both decrease along this path. The system is unable to manage the decrease in exergy by any variations in its internal parameter q alone; this is forbidden by the boundary. But there is another internal parameter in the system, h. Exergy can be shown to increase when h decreases, while the stability curve shifts to the right [see eq. (23.55)]. In Fig. 23.10, the new curve is represented by curve C_2 (with $h_2 < h_1$). The variation in h can be chosen in such a way that point b falls into the stability domain and the path $a \rightarrow b$ becomes feasible, but under the condition of a decreased h and hence increased exergy. So the evolution of closed trophic chains in the parameter space proceeds in complete accordance with the principle of increased exergy.

CONCLUSION

This chapter has endeavored to provide a theoretical formulation accounting for a postulated increase in the thermodynamic concept of exergy in the development and evolution of ecological systems. For illustrative purposes we have used simple, easily interpreted models.

Generally, the problem of establishing adaptational (or "extremal") principles in biology is far from being trivial. Such principles are commonly used when information about particular biological mechanisms is scant. For example, the idea that a biological system attempts to maximize (or minimize) a certain function of its state is often an attempt to mask gaps in knowledge of the mechanisms of system function (Svirezhev and Logofet, 1978). The use of a principle like exergy, which has a thermodynamic basis, seems justified by the fact that there is no reason to believe that living systems are not obeying the laws of thermodynamics.

However, what could be some of the reasons why the exergy principle, as we have formalized it, does not always result in an adequate system response? First, the assumption that the temperature of a living system coincides with that of the environment is a very strong one. Usually, this is not true; the temperatures of living systems differ from those of their environments. This is obviously true in the case of homeothermic animals, but also the temperatures of plant communities, as shown by space monitoring, also differ from those of adjacent unvegetated areas (Vinogradov, 1984). Thus, we have to expand our description by adding the term $(T - T_0)dS$ to the exergy expression (23.3) and by adding an entropy equation to our kinetic ones.

Second, in exergy equation (23.6) we used the expression for the difference between chemical potentials that was deduced from the model of ideal gas mixtures and dilute solutions. The consequence is that the exergy of an ecosystem does not depend on ecosystem structure, and in particular the trophic structure. This contradicts one of the main paradigms of ecology, that it is the structure of matter and energy flows that determines the state of an ecosystem. Also, the basic concepts of ecosystem ecology, such as trophic levels and matter and energy cycling, are all primarily structural notions.

On the other hand, the exergy principle has shown its applicability in ecological modeling practice even in its oversimplified form as presented here. So the main conclusion is that the theory needs further development. If we recall that exergy is closely related to information, while information is, in turn, related to probabalities, we may suppose a more adequate description to be obtainable by defining exergy in terms of ecological structures. This could be considered a parallel to the definition of probabilities on algebraic structures. These are perspectives for future research.

APPENDIX A: LOTKA–VOLTERRA EXERGY

The classical Lotka–Volterra prey–predator system,

$$\left.\begin{aligned} \frac{dN_1}{dt} &= \alpha N_1 - \beta N_1 N_2 \\ \frac{dN_2}{dt} &= k\beta N_1 N_2 - mN_2 \end{aligned}\right\}, \quad \alpha, \beta, m > 0, \, 0 < k < 1, \tag{A.1}$$

is not exergical at zero in the sense of the definition given in the section on exergy and Liapunov functions.

Proof. To make use of formula (23.9) for calculating exergy, we should first expand system (A.1) with a nonliving environmental variable N_0 such that the mass conservation law holds:

$$\frac{dN_0}{dt} = -\left(\frac{dN_1}{dt} + \frac{dN_2}{dt}\right), \quad N_0 = A - N_1 - N_2. \tag{A.2}$$

Then, bearing in mind the modification to which any model should be subject in order to be considered an exergical system at zero as defined, we have by eq. (23.10) that

$$Ex = RT\left[(N_1 + N_1^{eq})\ln\frac{N_1 + N_1^{eq}}{N_1^{eq}} + (N_2 + N_2^{eq})\ln\frac{N_2 + N_2^{eq}}{N_2^{eq}} + N_0\ln\frac{N_0}{N_0^{eq}}\right]. \tag{A.3}$$

Differentiating (A.3), by virtue of system (A.1), the result is

$$\begin{aligned}\frac{1}{RT}\cdot\frac{dEx}{dt} &= \ln\frac{N_1 + N_1^{eq}}{N_1^{eq}}\cdot\frac{dN_1}{dt} + \ln\frac{N_2 + N_2^{eq}}{N_2^{eq}}\cdot\frac{dN_2}{dt} \\ &\quad - \ln\frac{A - N_1 - N_2}{A - N_1^{eq} - N_2^{eq}}\cdot\left(\frac{dN_1}{dt} + \frac{dN_2}{dt}\right) \\ &= \ln\left(1 + \frac{N_1}{N_1^{eq}}\right)\frac{dN_1}{dt} + \ln\left(1 + \frac{N_2}{N_2^{eq}}\right)\frac{dN_2}{dt} \\ &\quad + \left(\frac{dN_1}{dt} + \frac{dN_2}{dt}\right)\left[\ln\left(1 - \frac{N_1^{eq} + N_2^{eq}}{A}\right) - \ln\left(1 - \frac{N_1 + N_2}{A}\right)\right]. \end{aligned} \tag{A.4}$$

Using the approximation $\ln(1 + x) \sim x$ for a small x, we see that only the two first terms of this expression play a major role in quantitative behavior of the derivative near the zero point. If we choose N_1 and N_2 sufficiently small but with $N_1 \ll N_2$, then expression (A.4) can be shown to be negative, or positive if $N_1 \gg N_2$. Thus, the derivative dEx/dt derived from the model equations does change its sign and cannot be a Liapunov function at zero. Actually, the trajectories approach zero along one direction ($N_1 \equiv 0$) and evolve along the other ($N_2 \equiv 0$), the phase pattern having a saddle point at zero (Fig. 23.4c).

APPENDIX B: ESTIMATING THE RATE OF EXERGY INCREASE AS A FUNCTION OF INCREASING q

Assuming for simplicity that $N_i^{eq} = N^{eq} > 0$ ($i = 1, \ldots, n$; the assumption being not critical), we derive from eq. (23.51):

$$Ex(g) = \frac{RTh}{4}\sum_{i=1}^{q} i\ln\frac{ih}{2N^{eq}} + (q + 1)\ln\frac{h(q + 1)}{2N_0^{eq}}.$$

Substituting the proper integral for the sum over i, which is valid for great enough q, we have

$$Ex(g) \frac{RTh}{4} \int_0^q x \ln \frac{xh}{2N^{eq}} dx + (q+1) \ln \frac{h(q+1)}{2N_0^{eq}} = \frac{RTh}{8} q^2 \left(\ln \frac{q^h}{2N^{eq}} - \frac{1}{2} \right) + (q+1) \ln \frac{h(q+1)}{2N_0^{eq}}.$$

The first term in this sum is obviously the main one when q increases, so the order of magnitude in exergy increases as $q^2 \ln q$.

APPENDIX C: CONDITION FOR EXERGY TO INCREASE IN RESPONSE TO AN INCREMENT IN A

Let the chain length q be fixed and odd, the latter being not essential. Expression (23.51) can then be rewritten in the form

$$Ex = RT \left\{ \frac{h(q+1)}{2} \ln \left[\frac{h(q+1)}{A - \sum_{i=1}^{q} N_i^{eq}} \right] \right.$$

$$+ \sum_{i=1}^{q'} \frac{4A - h(q+1)(i+1)}{2(q+1)} \ln \frac{4A - h(q+1)(i+1)}{2(q+1)N_i^{eq}} \quad (C.1)$$

$$\left. + \sum_{i=2}^{q-1''} \frac{h(q+1-i)}{2N_i^{eq}} \ln \frac{h(q+1-i)}{2N_i^{eq}} \right\},$$

where the prime denotes summation over odd values of i and the double prime, over even values. Calculation of the partial derivative of (C.1) with respect to A results in

$$\frac{\partial Ex}{\partial A} = RT \left\{ \frac{2}{q+1} \sum_{i=1}^{q'} \ln \frac{4A - h(q+1)(i+1)}{2(q+1)N_i^{eq}} + \frac{2}{q+1} - \frac{\frac{1}{2} h(q+1)}{A - \sum_{i=1}^{q} N_i^{eq}} \right\}. \quad (C.2)$$

The derivative (C.2) is positive if the condition

$$A > \frac{(q+1)^2 h}{4} + \sum_{i=1}^{n} N_i^{eq}$$

holds, which is established in the section on exergy and closed trophic chains.

REFERENCES

CHETAEV, N. G. 1955. *The Stability of Motion*. Gostekhteoritizdat, Moscow (in Russian), 207 pp. English translation 1961. Pergamon Press, New York. 200 pp.

EVANS, R. B., CRELLIN, G. L., and TRIBUS, M. 1966. Thermoeconomic considerations of sea water demineralization. In Spiegler, K. S. (ed.), *Principles of Desalination*. Academic Press, New York, pp. 21–76.

GOEL, N. S., MAITRA, S. C., and MONTROLL, E. W. 1971. On the Volterra and other nonlinear models of interacting populations. *Rev. Modern Physics* 43:231–276.

JØRGENSEN, S. E., 1982. Exergy and buffering capacity in ecological systems. In Mitsch, W. J., Ragade, R. K., Bosserman, R. W., and Dillon, J. A. (eds.), *Energetics and Systems*. Ann Arbor Science, Ann Arbor, Michigan, pp. 61–72.

JØRGENSEN, S. E., 1990. Ecosystem theory, ecological buffer capacity, uncertainty, and complexity. *Ecol. Mod.* 52:125–133.

JØRGENSEN, S. E., 1992. Parameters, ecological constraints, and exergy. *Ecol. Mod.* 62:163–170.

JØRGENSEN, S. E., and MEJER, H. F. 1981. Application of exergy in ecological models. In Dubois, D. (ed.), *Progress in Ecological Modelling*. Editions CEBEDOC, Liege, Belgium, pp. 311–347.

JØRGENSEN, S. E., and MEJER, H. F. 1982. Exergy as a key function in ecological models. In Mitsch, W. J., Ragade, R. K., Bosserman, R. W., and Dillon, J. A. (eds.), *Energetics and Systems*. Ann Arbor Science, Ann Arbor, Michigan, pp. 587–590.

KRAPIVIN, V. F., SVIREZHEV, Y. M., and TARKO, A. M. 1982. *Mathematical Modelling of Global Biosphere Processes*. Nauka, Moscow. 271 pp. (in Russian).

LASALLE, J. P., and LEFSCHETZ, S. 1961. *Stability by Liapunov's Direct Method with Applications*. Academic Press, New York. 134 pp.

MEJER, H. F., and JØRGENSEN, S. E. 1979. Energy and ecological buffer capacity. In Jørgensen, S. E. (ed.), *State of the Art in Ecological Modelling*. International Society for Ecological Modelling, Copenhagen, pp. 829–846.

MOROWITZ, H. J. 1968. *Energy Flow in Biology*. Academic Press, New York. 179 pp.

PYKH, Y. A. 1983. *Equilibrium and Stability in Models of Population Dynamics*. Nauka, Moscow. 198 pp. (in Russian).

RACSKO, P. 1981. Simulation model of a tree as an element of forest biogeocoenoses. In *Voprosy Kibernetiki*, issue 52. Nauka, Moscow, pp. 73–110 (in Russian).

SCHRÖDINGER, E. 1945. *What Is Life?* Cambridge University Press, New York. 91 pp.

SOMMERFELD, A. J. W. 1952. *Thermodynamics and Statistical Mechanics* (Lectures on Theoretical Physics, Vol. 5). Academic Press, New York. 401 pp.

SVIREZHEV, Y. M., and LOGOFET, D. O. 1978. *Stability of Biological Communities*. Nauka, Moscow (in Russian). 352 pp. English translation 1983. Mir, Moscow. 319 pp.

SVIREZHEV, Y. M., and LOGOFET, D. O. 1992. The mathematics of community stability. In Patten, B. C., and Jørgensen, S. E. (eds.), *Complex Ecology, The Part-Whole Relation in Ecosystems*. Prentice Hall, Englewood Cliffs, New Jersey, pp. 343–371.

WILLEMS, J. L. 1970. *Stability Theory of Dynamical Systems*. Wiley-Interscience, New York. 201 pp.

VINOGRADOV, B. V. 1984. *Air–Space Monitoring of Ecosystems*. Nauka, Moscow. 318 pp. (in Russian).

24

The Ecosystem as an Existential Computer

MICHAEL CONRAD

Departments of Computer Science and Biological Sciences
Wayne State University
Detroit, Michigan

Consider the etymology of the word ecosystem. "Eco" implies environment. The term "system" implies an interacting, interdependent complex. . . . An ecosystem is an integrated complex of living and nonliving components. Each component is influenced by the others [Van Dyne, in *The Ecosystem Concept in Natural Resource Management*, Ch. X, p. 329 (1969), Academic Press].

It is necessary to study the ecosystem as a whole in order to understand [Van Dyne, in *The Ecosystem Concept in Natural Resource Management*, Ch. X, p. 331 (1969) Academic Press].

The Second Dilemma. Malthus was concerned by the fact that populations grow faster than do their means of subsistence. Malthus' dilemma, i.e., the first dilemma, is, from one point of view, an imbalance between the energy available to man and the energy he requires. Consider a second dilemma that was overlooked by Malthus when he foresaw the threat of uncontrolled population growth. The second dilemma concerns the increase in complexity, in the proliferation of the semantic environment, which accompanies the growth of population. . . . For example, when the population of an area increases, the number of contacts between members of that population increases in proportion to the square of the size of the population. The greatly increased number of contacts and interactions soon stresses our communication, transportation, and psychological systems. We may find that we may delay greatly the first dilemma, i.e., the energy imbalance, through such means as atomic energy. We have not found the solution to the second dilemma, the one caused by interactions [Van Dyne, in *The Ecosystem Concept in Natural Resource Management*, Ch. X, p. 332–3 (1969), Academic Press].

INTRODUCTION

Language and science offer two ways of describing the process of ecological succession. The first is in terms of dynamical stability. An ecological system must be stable enough to perturbation to stay in the game of existence. The second is in terms of problem solving. The ecosystem must solve the problem of staying in existence.

The dynamical point of view has a venerable history in the physical sciences. The problem-solving point of view has always been implicit in mathematics. But it is only in recent years, with the advent of computer science, that the complexity of problems and methods of solving them have been developing into a systematic body of knowledge.

Adaptability theory (Conrad, 1983) offers a framework that accommodates and cross-correlates the dynamical and problem-solving approaches. The basic idea of adaptability theory is that ecological and evolutionary systems are stable only as long as they have enough problem-solving power either to dissipate or control environmentally and internally generated perturbations.

In this chapter, I wish to look at evolution and succession as problem-solving processes and then briefly compare conclusions to those from dynamical studies. The formal systems and information theory tools of adaptability theory are essential for precise statements and argumentation. These have been thoroughly reviewed elsewhere (Conrad, 1983), although not from an explicit problem-solving point of view. The presentation in this chapter will be completely informal. A brief indication of the connection to the apparatus of adaptability theory is provided in the chapter appendix.

EVOLUTION AND INFORMATION PROCESSING

We can divide all problem solving into two modes, evolutionary and rule based. A system solves problems in a rule-based mode if its actions, or changes of state, all occur within a definite set of constraints that are narrower than the laws of physics. The actions may be determinate, or they may occur with some probability. The evolutionary mode always involves an element of randomness and structural change. The difference is that the constraints themselves change. Evolutionary problem solving is subject to no constraints apart from the laws of physics (which for the purposes of this discussion can be assumed to be an ultimate set of constraints).

For example, a typewriter such as the one I am using is a rule-based system that could be described in a simple user's manual. When I press the keys, it behaves in a predictable manner. If some of the keys were loose, an element of probability would enter. But the behavior would still be predictable in a stochastic sense. If I hit the machine with a hammer, I have gone outside the range of inputs whose effects are predictable on the basis of the manual. This is like a mutation.

I will sometimes refer to rule-based problem solving as information processing. Strictly speaking, I should use the term "rule-based information processing" since evolution is also a form of information processing. Even the distinction between rule-based and evolutionary information processing is not entirely sharp. In principle, we ought to be able to simulate evolution with a digital computer, which means that we can in large measure duplicate evolutionary problem solving with rule-based information processing (Conrad, 1981; Rizki and Conrad, 1985). Furthermore, it is not possible to predict the outcome of some rules without actually tracing them through step by step. This is the case with most computer programs. The behavior of programs can be surprising even if we know the rules they obey. Nevertheless, we can often have a pretty good intuition about whether or not a system is operating within a definite set of constraints. Finally, we should not forget that evolutionary problem solving would be impossible without self-reproduction, which is a rule-based process.

We should also consider the distinction between information processing and problem solving. It is conceivable for a system to process information extremely efficiently, but for most or all of this processing to have little use from any reasonable point of view. If the information processing has functional value, say for the survival of the system, we will call it problem solving. Problem-solving power is the same as information-processing power, but not all information processing contributes to functionally significant problem solving.

In daily life, rule-based information processing seems much more important than evolution as a mode of problem solving. Most organisms solve the problems of finding food, avoiding danger, and producing offspring by operating according to rulelike mechanisms. However, evolution is the ultimate form of problem solving in nature. This is because all the highly elaborated rule-based mechanisms are products of the evolutionary process. Some forms of learning in plastic organisms, such as human beings, may also be evolutionary in the sense of escaping predefined constraints. Immunological plasticity is a complicated rule-based mechanism spanning many levels of physiological organization, yet at the same time it utilizes an ontogenetic mechanism of antibody diversification and clonal selection that is analogous to the mechanism of natural evolution (Jerne, 1955).

Evolution and rule-based information processing are linked in three fundamental ways.

1. *Problem-solving power.* Potential information-processing power increases as the number of interactions in a system increases. But the potential for information-processing systems to develop in the course of evolution is a maximum when one-half of the possible interactions are turned on.
2. *Self-facilitation principle.* Some organizations are more suited for the evolution of information processing than others. Interactions that enhance the amenability of a system to evolution can "hitchhike" along with advances in information processing that they facilitate.
3. *Compensation principle.* The evolution of information-based mechanisms imposes constraints on evolutionary problem solving. The overall problem-solving power will decrease unless the reduction in evolutionary problem solving is compensated by the increase in information-based problem solving.

I will argue that these principles operate at all levels of biological organization and that they are decisive for the organizational changes that occur in the course of ecological succession.

PROBLEM-SOLVING POWER

Consider a system consisting of n particles. The system could be a cell, organism, ecosystem, or machine. According to presently accepted force laws, the maximum number of interactions that can occur in this system is n^2. This includes the pairwise interactions between particles and the self-interactions. In general, not all interactions are equally important. Some are frozen out entirely. As the number of constraints in the system increases, the number of frozen out interactions increases. For example, in a conventional digital computer nearly all the interactions are frozen out. This is what makes such systems effectively programmable.

Let us define *information-processing efficiency* as the number of interactions that a system actually uses for problem solving relative to the maximum number of interactions possible in that system. The maximum possible efficiency is thus $n^2/n^2 = 1$. On the basis of this definition, an ideal gas would have the highest information-processing efficiency. In fact, it would solve the problem of being itself better than any other system. But it would have no flexibility. It could not be used in alternative ways, like a conventional computer or an evolutionary system. We will define the evolutionary flexibility of a system as the number of alternative systems that can be built from the particles of which it is composed. Evolutionary flexibility reaches a maximum when the number of interactions is $n^2/2$. Thus, information-processing efficiency at the maximum evolutionary flexibility is one-half (Conrad and Hastings, 1985).

The problems solved by biological systems are not precisely defined, unlike the problem of multiplying two digits or finding the shortest route through a number of cities. Problem specification is tantamount to specification of biological function. Rarely is it possible to give a definitive statement of function. Teeth can be used for eating, for giving warnings, as tools for opening bottles, or for being elected to political office. Ambiguity of function is essential for its transformation in the course of evolution. These difficulties disappear at the level of the ecosystem as a whole, however. Here, we recall that the problem is to stay in existence. The situation is reminiscent of the existential game model proposed by Slobodkin and Rapaport (1974), except that in the latter the species are players and the payoff is existence. The existential problem is definitive at a high level of generality, without precluding the ambiguity of subordinate problems.

The existential problem has two components, efficient exploitation of the environment and stability. It is reasonable to suppose that species or communities better suited to exploiting space and other resources will displace their competitors. All else equal, an ecosystem should evolve in the direction of more efficient exploitation. If all its components were recruited to this end by the evolutionary process, it would develop in the direction of $n^2/2$ interactions, regardless of how the specific problems of exploitation are specified.

All else, however, is not equal. It is also reasonable to suppose that an ecosystem will spend more time in stable patterns of organization than in any unstable ones. To do this, it must solve the problem of regenerating itself without significant change in the pattern of interactions. The problem becomes more complex as the number of possible patterns of interactions increases, reaching a maximum complexity at $n^2/2$ interactions, just where the information-processing and evolutionary power required to exploit the environment and reproduce a variety of complex patterns reaches an optimum balance.

The second aspect of stability has to do with the ramification of perturbation. In a system with n^2 strong interactions, a disturbance to any part of the system will have drastic ramifications throughout since the components are rigidly attached. If the components are completely independent (zero interactions), no ramification will occur. But each component directly disturbed will be drastically affected, there being no possibility of attenuating the disturbance by spreading it over the system.

It is possible to compromise these conflicting demands by dividing interactions into strong and weak. To regenerate the system, it is sufficient to specify the strong interactions. The weak interactions play an important role in determining structure and function. This division reduces the contradiction in the fact that the problem of regenerating the pattern of interactions increases as problem-solving power increases. Furthermore, the weak interactions allow the effects of a disturbance on a single component to be distributed over a large number of components in a gentle way.

Strong and weak interactions play an important role at all levels of biological organization. Proteins and other biological macromolecules provide a particularly clear example. A typical protein chain contains about 300 or 400 amino acids connected by strong bonds. These fold to form a three-dimensional shape on the basis of numerous weak interactions, including interactions between amino acids at distant locations on the unfolded chain. To build a protein, it is only necessary to specify the strong bonds. The function performed by the protein is determined by weak interactions with other molecules that depend largely on shape complementarity. The situation is complicated at higher levels by the fact that signal transmission and interpretation play an increasingly important role. The internal structure of cells is determined both by specific adhesiveness of molecules and by chemical and structural networks that transmit signals over large distances. We can identify the strong interactions with specific affinities that are ultimately, although indirectly, under genetic control. All other interactions, local and global, are weak. At the level of organisms, the strong bonds correspond to obligate relationships among cells that have their origin either in specific affinities or in the ability of cells to interpret globally broadcast signals.

At the level of the ecosystem, it becomes still more difficult to provide a physical criterion for distinguishing between strong and weak interactions. For the most part a community self-assembles on the basis of weak informational interactions among organisms. This is predicated on the problem-solving activities of organisms, somewhat in the fashion that the morphogenesis of an organism is based on the information-processing activities of cells. Some of these weak interactions add up to strong interactions. Roughly speaking, we can associate strong interactions with the essential predator–prey, symbiotic, and spatial affinities that define the community. The enormously high irritability of some organisms—their ability to react to a wide variety of subtle features of the milieu—suggests the importance of weak interactions.

No universal criterion can be formulated for compromising the conflicting demands on density of interactions. Whether more or fewer interactions will be favored depends on the particularities of the system. However, we can reasonably assert that increases in problem-solving power associated with movement in the direction of $n^2/2$ interactions will be favored if they are not offset by increase in the complexity of the regeneration problem. This is most likely to be achievable if the majority of the interactions are weak, especially as weak interactions may serve to enhance stability to perturbation. Conversely, weakening of interactions should be favored whenever this can be achieved without interfering with the self-assembly of the community. An increase in the problem-solving power of the organisms that comprise the community would make this possible.

SELF-FACILITATION PRINCIPLE

Now let us look at what conditions a system must fulfill to develop its information-processing capabilities through the process of evolution. Initially, it might be thought that the postulates of the Darwin–Wallace theory are sufficient. The system must contain units that are capable of reproducing with variation (subject to the constraint that acquired characteristics are not inherited). The environment classifies these units into two groups, those that actually reproduce and those that do not. However, not all systems that satisfy these postulates yield an evolution process. For example, applying variation and selection to raw computer code does not work to produce better code. The reason is that the relationship between the structure and computational function of raw code is too delicate to be suitable as a substrate for evolution. The situation is quite different with biological organisms. Here the structure–function relations are malleable. Structural changes at the level of genes have a good chance of producing functionally viable offspring.

The gene–protein relationship again provides a good example. Due to the weak interactions responsible for folding, the relationship between amino acid sequence (which codes gene structure) and three-dimensional shape is very malleable. Single mutations or crossovers have a good chance of yielding a protein with only a slightly altered shape, and hence with only a slightly altered function. If double mutations were required to obtain an acceptable variant of the protein, the rate of evolution would scale as p^2, which is unacceptably small for any realistic values of p (p must always be a small number; otherwise, the protein would be degraded by undesirable mutations).

Suppose that a protein species reaches an evolutionary bottleneck. No single genetic event can lead to an improvement. It is always possible to add mechanistically redundant amino acids that buffer the effects of mutation on function. Such redundancy is virtually selectively neutral, so it is inevitable that it will accumulate in some organisms. Furthermore, the time required for it to accumulate is much less than the waiting time for a double mutation (since the rate of accumulation scales as p). If enough redundancy accumulates, it should be possible to bypass the bottleneck; that is, it should be possible to reach an organization that allows for an improvement in response to a single mutation. The redundancy will then hitchhike along with the advantageous traits whose appearance it facilitated. In this way the process of evolution *facilitates itself.*

Examples of the self-facilitating aspect of evolution at the level of the cell and of the organism are discussed in Chapter 10 of my book, *Adaptability* (Conrad, 1983). I referred to the principle there and in some earlier papers as the "bootstrapping" principle. This is because evolution is pictured as selecting structure–function relations that facilitate further evolution, and therefore as lifting itself up by its own bootstraps.

Let us consider how the self-facilitation principle could operate in ecological succession. For evolution of species, the useful picture is that of an adaptive peak. If a species becomes isolated atop some peak, this is equivalent to reaching an evolutionary bottleneck. If it accumulates redundancy, the dimensionality of the adaptive space increases, possibly opening up a higher-dimensional bypass to another peak. This picture is not suitable for an ecosystem taken as a whole. Here the concepts of reproduction and selection become amorphous, and the concept of adaptive peak loses meaning. However, we can replace the adaptive peak with a basin of attraction. Ecosystems should fall into wider and deeper basins of attraction during the course of succession, just as species should climb adaptive peaks in the course of evolution.

The situation is illustrated in Fig. 24.1. We picture ecological communities as falling into the lowest accessible valley, $S1$, on a potential surface. Many surrounding valleys might be deeper or wider, but not accessible. Suppose that some redundant elements appear in the community, such as redundant species, food-web links, information flows, or other redundant features at the level of the organism. From an operational point of view, elements can be considered redundant if

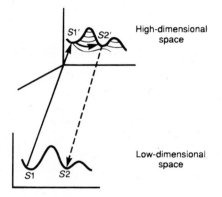

Figure 24.1 Self-facilitation of succession. The community is initially trapped in basin of attraction $S1$ in the lower-dimensional space. As a result of random accumulation of redundancy, it moves to an analogous basin $S1'$ in a higher-dimensional space. The likelihood of a bypass to the deeper or more strongly attracting basin $S2'$ is increased (since the chances of a valley occurring in a multidimensional space in general decreases with the dimensionality). $S2$ could conceivably lose redundancy due to intense selective pressures, decaying back into the originally inaccessible minimum, $S2$. But it is more likely that the redundancies will be fixed by the transition from $S1'$ to $S2'$ and by the fact that a community in $S2'$ is likely to displace neighboring communities still in $S1$. This will allow for accumulation of even further redundancy and bypass to a yet deeper basin of attraction. The process would continue until a basin is reached that is so isolated from deeper basins that no bypass can be opened up by random accumulation of redundancy. Accumulated redundancy is similar to neutral or quasi-neutral genetic variation in individual organisms. It inevitably accumulates because natural selection is imperfect as an optimizing process and much more imperfect at the community level than at the individual level. The basins are slightly less deep in the higher-dimensional space because the redundancy is a cost in terms of free energy.

removing them does not have a marked effect on other elements. It is inevitable that such redundancies would appear even if strong optimizing pressures were imposed on the community—no method of optimization would be good enough to prevent their occurrence. When they appear, they increase the dimensionality of the potential surface. The community is now located in a valley $S1'$ that differs from $S1$ only by virtue of the dimensionality of the space in which it is located.

However, the chance that it is connected to a deeper valley, $S2'$, is now increased. The greater stability of this community allows it to displace adjacent communities of the same type. The dimension-increasing redundancies will be carried along with the movement into the deeper basin of attraction that they opened up. In this way the ability of an ecological system to solve the problem of finding an organization that corresponds to a deep basin of attraction increases in the course of succession.

The organization of ecosystems is a consequence of both hill climbing on an adaptive surface and community-level self-facilitation of valley seeking on a potential surface. Rule-based problem solving at the level of the organism results from evolutionary hill climbing. Rule-based mechanisms at the community level are also influenced by successional valley seeking. It is difficult to precisely disambiguate the two processes. Successional valley seeking imposes selective constraints that act at the genetic level. Rule-based mechanisms at the community level are ultimately based on the ability of individual organisms to process information and energy. The time scales, however, are different. Time scales for genetic evolution are usually much longer than for ecological succession. A certain amount of genetic adaptation does occur on successional time scales, though, and the amount can be substantial for microbial species.

COMPENSATION AND INTERCONVERSION OF PROBLEM-SOLVING MODES

We cannot expect biological systems to support unlimited problem-solving power any more than we can expect to buy a supercomputer for the price of a minicomputer. Problem solving always involves costs. In evolutionary problem solving, the costs are associated with less efficient variants. In rule-based problem solving, the costs are associated with the construction and maintenance of physiological machinery. Thus, it is reasonable to expect that the balance between evolutionary and rule-based problem solving will change as the latter becomes more elaborated in the course of evolution.

To get a line on this, let us classify problem-solving power into four basic components.

1. *Behavioral repertoire.* The problem-solving power of a system increases as the number of possible behaviors increases.
2. *Correlation.* The problem-solving power of a system increases as its behaviors become more correlated to the environment that defines the problem.
3. *Passive problem definition.* A system may reduce the size of the problems posed by its environment by retreating into a more confined space or protecting itself with passive structural constraints.
4. *Active problem definition.* A system may alter the problems posed to it through its problem-solving activity.

It is necessary to distinguish between *specialization-directed* and *adaptability-directed* problem solving. Suppose that a biological system could be placed in a fixed, completely determinate environment. The system's ensemble of potential behaviors could, in principle, all be dedicated to functioning in this fixed environment. If we add variability to the environment, the ensemble would have to become larger. As the variety of environments in which the system can function increases, its adaptability increases. *We shall define the adaptability of a system as the potential uncertainty of its environment.* In reality, it is rare that a biological system's potential behaviors are purely specialization directed, first because retreat from all environmental variability is generally impossible, and second because the essence of evolution is the discovery of new features of the environment. The picture is also complicated by the fact that the environment of each subsystem—say each organism or species—includes other organisms and species, as well as the physical environment. As a consequence of all these internal interactions, the ensemble of behaviors in an ecosystem would become very large even if the environment could be fixed.

Biological systems have a levels-of-organization aspect, and each of the three components of problem solving, given at the beginning of the chapter, can be mediated to a greater or lesser extent by processes at any of these levels (Conrad, 1972a). We can think of the flow of information in a community as passing vertically between levels, as well as horizontally. The different levels of the hierarchy are defined by what are generally fairly natural demarcations in the spatial scale of information flow, with the top level associated with the largest spatial scale. The information may be processed at any of the levels, although more of it is probably processed at the organism level and below than at the societal and community levels. The manifestations of this information processing may, however, be much more marked at a level other than that at which the major processing occurs. For example, the development of a social organization may depend heavily on behavioral processes mediated by cellular and subcellular events, but the main phenomena are visible at the social level. Reorganization of this interaction structure in response to an environmental challenge is most naturally associated with social adaptability. It might also be accompanied by substantial behavioral changes, in which case it is associated with behavioral adaptability. But in some cases social organizations can change substantially.

We can define different modes of problem solving and adaptability in terms of the levels at which the primary manifestations occur, recognizing that supporting processes occur at many other levels. Evolutionary adaptability is mediated by variation at the genetic level and selection at the organism level. Behavioral adaptability is manifested in terms of the actions of organisms, but is mediated by neural and subneuronal processes. Populational adaptability includes culturability of species in suitable environments. This is manifested in terms of numbers of organisms, but is mediated by physiological mechanisms of reproduction. Community-level adaptability includes variability of food-web structure, and this has a basis in behavioral and physiological mechanisms.

Why is it that different modes of adaptability appear to be so closely associated with different levels of organization? Conceivably, prominent manifestations could occur at every level of processing. This, however, would be very uneconomical. Prominent manifestations occurring at different levels increase the total adaptability only to the extent that they are independent. It is best for most of the processing underlying the manifestation to be rather minimal in terms of energy requirements and to restrict all pronounced manifestations to the level at which the interaction with environment occurs. Organizations in which hierarchical structure has a fairly clear association with different spatial and temporal scales have an advantage in this respect.

Now let us return to the comment at the beginning of this section, that problem solving always entails costs. This suggests that greater than necessary adaptability will tend to atrophy. By and large, the adaptability of a system should match the uncertainty of the environment. *If constraints are imposed on one mode of adaptability, this must be compensated by expansion of other modes. It could also be compensated by enhancements of information processing that increase correlation to the environment or by retreat to a more confined environment.* The overall adaptability of a species or community can increase if it allows expansion into a wider environment. The community will inevitably be damaged if its overall adaptability falls below what turned out to be the real uncertainty of the environment. This initiates a number of adaptability-increasing processes associated with regrowth of the community. Internally directed adaptabilities, such as variabilities associated with interactions between species, may be converted to externally directed adaptabilities in this regrowth phase.

Another important process in evolution is connected with the interconvertibility of adaptability-directed and specialization-directed problem solving. Evolutionary adaptability allows species to conquer new environments or utilize a given environment in new ways. As a consequence, solving the problem of functioning in an uncertain environment may be transformed into solving problems associated with the exploitation of a given environment. There is no simple limit on the evolution of specialization-directed problem solving. An elaboration here will be sustainable as long as it brings at least enough energy into the system to offset its cost. Whether this is possible depends on the potentialities of the environment, and these can never be specified precisely. Elaboration of specialization-directed problem solving means more complicated patterns of interaction and constraint. This changes the relative costs of different modes of adaptability, leading to compensating shifts in adaptability patterns and opening up new possibilities for the conversion of adaptability-directed into specialization-directed problem solving.

The evolution of multicellular organizations provides an example. The short reproduction time and simple morphology of unicellular organisms allow for high evolutionary adaptability and culturability. Unicellular organization can also support rule-based physiological adaptability mechanisms such as inducible enzymes. Multicellular organization provides some adaptability advantages, such as better anchoring in response to mechanical disturbances and some degree of internal temperature regulation. These organizations also allow new modes of problem solving that depend on the existence of cellular networks. But multicellular organization restricts evolutionary adaptability and culturability by increasing the time required for reproduction. Higher plants compensate for these restrictions with developmental plasticity. However, the fixed growth patterns of higher animals preclude this compensation and also place strong constraints on allowable genetic variability. Here, compensations are provided by immunological and brain-based modes of adaptability. The brain-based modes open up still further modes of problem solving and new possibilities for highly elaborate social organization. These organizations in turn impose constraints on behavioral adaptability that are compensated for by the development of dedicated organs of social adaptability, such as research institutions. According to adaptability theory, the evolution of biological hierarchy is largely driven by the cyclic interplay between selection for adaptability, conversion of adaptability to specialization-directed problem solving, and consequent selection for compensating modes of adaptability.

The situation is modified at the community level since competition between organisms and communities follows different dynamics. The individual is directly subject to natural selection. If communities become unstable, strong selection pressures are brought to bear on individuals until a set of coherent relations is established. The potential for efficient exploitation increases as the number of differently specialized species increases, that is, as the number of nutritional and spatial linkages increases. Perturbations affecting the flow of energy are more gently distributed in a system with a highly ramifying food chain (Odum, 1953; MacArthur, 1955). Variability of energy flow is more important in this respect, however, since it allows for corrective rerouting (Conrad, 1972b). This means more informational linkages and less specialization. A specialized pattern of organization requires a resource-rich, constant environment to support it. As more variability is introduced, the number of available specialized niches decreases. As the environment becomes increasingly variable, the system degenerates into a patchy form of organization that prevents the ramification of disturbances. As grazing pathway organisms become more complex, the overall cycle of materials in all these systems is increasingly regulated by small organisms, especially microorganisms, of the detritus pathway. The detritus pathway serves as a compensating mode of adaptability for the grazing pathway.

TRENDS IN EVOLUTION AND SUCCESSION

Let us now look at the development of interaction patterns and problem-solving power in ecosystems in the course of evolution and succession. We have four general principles at our disposal:

1. Evolution is facilitated by redundant interactions that buffer the effects of mutation or other structural changes on function.

2. Problem-solving power increases as the number of interactions increases, with the best compromise between evolutionary and rule-based problem solving being achieved when half the possible interactions are turned on.
3. Adaptability can transform into specialization-directed problem solving, and conversely.
4. Restrictions on modes of adaptability resulting from the evolution of specialization-directed problem solving must be compensated for by expansion of other modes.

These principles operate at both the species and community levels of evolution, on overlapping time scales, and with different impacts in primitive versus highly evolved ecosystems. As a practical simplification, though, we can think in terms of the following stages:

1. *Preparation stage.* This is the period immediately following the appearance of a new form of biological organization, such as primitive self-reproducing units, primitive cells, or primitive multicellular organizations. Problem solving, including coadaptations among organizations, is mainly evolutionary at this stage. Evolution-facilitating redundancies accumulate in conjunction with the advances they facilitate, and this is accompanied by an increase in the density of interactions within the organization.

2. *Elaboration stage.* This is characterized by rapid evolution and radiation of both adaptability and specialization-directed problem solving at the species level. The increase in problem-solving power is supported by an increase in the density of interactions within organisms and societies. Rule-based problem solving plays an increasingly important role and, in the more complex organizations, evolutionary problem solving gives way to rule-based problem solving as the major mode. Physiological and behavioral adaptability mechanisms make an increasing contribution to community dynamics and stability. This means more informational transactions at the community level and hence an increase in the density of interactions. Specialized form becomes increasingly important in the system, and this means more energy-flow interactions. Self-facilitation of successional processes leads to accumulation of redundant species and to the addition of redundant interactions at the community level.

3. *Compensation stage.* Species that move in the direction of increased complexity of interactions and increased reliance on rule-based mechanisms must undergo compensating alterations in patterns of adaptability. Adaptability mechanisms underlying community stability are altered, and many species must adjust to changing ecological conditions. If compensation through existing modes is too expensive, new hierarchical levels have the opportunity to evolve. This includes special organ systems, such as the brain or immune system. In the case of the immune system, at least, this means a reemergence of an evolutionary mode of adaptability within the context of a highly coherent collection of rule-based mechanisms. The processes of self-facilitation and transformation of adaptability into specialization-directed problem solving begin to act again, leading to further increases in the density of connections. Some despecialization of species, associated with increased variability of energy flow, may occur. This means some reduction in species diversity and some replacement of strong interspecies linkages by a larger number of weak interspecies linkages. However, the resulting increased routability of energy flow may also be converted to specialization-directed exploitation of the environment through increases in nutritional specialization.

4. *Destabilization stage.* The necessary condition for evolution to stabilize is internal consistency. The adaptability pattern of a species must be consistent with environmental uncertainty and with the constraints imposed by its morphological structure. The community must solve the problem of maintaining constancy of the environment and balancing its internal birth and death rates. It is unlikely that such a stable state could last indefinitely, however. Problem-solving power sufficient to stabilize the community would also be sufficient to generate

destabilizing novelty. A regulated environment will, in effect, be a less uncertain environment. Adaptabilities in the community will atrophy. This will be accompanied by the disappearance of some of the weak linkages at the community level and some of the redundant interactions that served to open up pathways for evolution or community stabilization. The community will then be subject to destabilization by environmental variation arising from the cumulative effect of slight failures in regulation or from exogenous sources. If the perturbation is minor, the same or similar species may reestablish themselves. If it is major, opportunities will be opened up for new forms that have reached the threshold of elaboration.

The development of interaction density on successional time scales is much simpler than on evolutionary time scales. This is because species-level evolutionary processes involving self-facilitation and development of new hierarchical levels are not significant. However, new species can enter from other communities. As a consequence, the development of problem-solving power in succession to some extent recapitulates the sequence of stages in evolution. The first stage is dominated by inefficient, high-adaptability generalists or by species with a plastic genetic system and short generation time. As in phylogenesis, the evolutionary mode of problem solving becomes less important and rule-based mechanisms more important. As species with more powerful information-processing capabilities enter the system, the density of informational transactions increases. Redundancies hitchhike along with the transitions to more the stable states of organization that they facilitate. The environment becomes increasingly controlled, and the rate of community change becomes extremely slow. As in phylogenetic evolution, some adaptabilities atrophy and some of the succession-facilitating redundancies disappear. The community does not appear to have changed, but in fact it is more fragile to perturbation.

This development can be phrased in dynamical terms. A community is orbitally stable if it solves the problem of maintaining environmental constancy and of balancing its internal birth and death rates. If it learns to solve these problems in the face of stronger and more diverse perturbations, its basin of attraction becomes wider and deeper. Adaptability mechanisms that serve to dissipate disturbances provide support for this type of stability. At a finer level of description, we may be dealing with a large density of closely packed stable states, none of which is highly resistant to perturbation. High redundancy of components and interactions would provide support for this situation. We can also say that a community is structurally stable if it solves the problem of maintaining its general form in the face of major perturbations, such as the removal of a species. Structural stability should thus be highest at the earliest stages of succession, when generalist species and species with high evolutionary adaptability dominate. As succession proceeds, orbital stability becomes more important and structural stability diminishes. The density of closely packed basins of attraction at first increases, but then decreases. This corresponds to the decrease in redundant interactions and increased reliance on rule-based mechanisms that allow for basins of attraction that are deep and wide relative to perturbations the community is experiencing in its climax form of organization. The virtually invisible processes occurring during the climax stage thus render the community increasingly vulnerable to major destabilization.

COMPLEXITY AND STABILITY

Many ecosystems appear to be complex in terms of number of species and community morphology. Systems studied in the laboratory sometimes degenerate into relatively simple forms, and some communities in nature, in harsh environments, do not support a great deal of complexity or diversity. But, in general, species numbers increase in the course of succession, although often with a moderate decline in the climax stage (Margalef, 1958). The complexity of community relationships also appears to increase (Odum, 1969).

The analysis presented here accommodates this phenomenology. Potential problem-solving power increases as connectivity increases. This does not mean that any increase in connectivity will bring with it increased problem-solving power or that increases in problem-solving power are always advantageous. There is always room in a community for simple organisms and for locally simple patterns of community structure. The trend to greater complexity is due to the fact that the more complex species and more elaborate connections between and based on them take time to develop. More complex interactions may compromise stability, but this is much less likely to be the case if most interactions are weak. Large numbers of weak interactions (as exemplified in feeding relations; Chapter 20) allow for efficient information processing, high amenability to evolution, effective self-organization, and buffering against perturbation.

May (1973) presented an elegant analysis of stability and complexity that strongly suggests the nonexistence of any simple relationship between these two properties. The basic idea is that the chance of finding a valley in a state space becomes smaller as the number of conditions that must be satisfied increases. The number of conditions increases as the number of components and interconnections increases. The situation is similar to our argument that stability in terms of reproduction and perturbation decreases as the number of strong interactions increases. Interactions considered in models of the Lotka–Volterra type fall into the strong category. It is hard to represent problem-solving behavior and signal flow at the community level in purely dynamical formalisms. Thus, the number of interactions in an ecosystem could increase in the course of succession even though complexity in a dynamical abstraction decreases. This would correspond to the shift from strong to weak interactions.

Increase in complexity may also be supported by increase in adaptability. The whole point of adaptability and compensation is that highly complex suborganizations can be protected from disturbance by adaptability elsewhere in the system. This means that complexity is not properly viewed as an average property of a system. Its distribution is in general extremely inhomogeneous, in both the vertical and horizontal directions of the biological hierarchy. Furthermore, high-adaptability suborganizations may themselves be complex. This is connected to the interconvertibility of specialization- and adaptability-directed problem solving. The brain is an example.

The remaining major origin of complexity in our analysis is the self-facilitation of succession. This makes it possible for a community to escape being trapped in a valley of inefficient exploitation. Redundant connections that inevitably occur, even in the most stringent selective regimes, destabilize these inefficient valleys by opening up pathways to neighboring valleys. As a consequence, an inefficient system is unlikely to be evolutionarily or successionally stable—that is, stable to small alterations in internal structure. In final analysis, the whole evolutionary process is unstable. Species can always emerge that solve novel problems, including the problem of destabilizing the community. In this way, the emergence in evolution of unstable complex structure can catalyze both increases and decreases in community stability, and in some instances the new community structure will be both more stable and more complex.

CONCLUSION

The term "problem solving" may seem anthropomorphic. However, it is no more so than the term "law of nature." We need not suppose a lawgiver when we write down a dynamical equation, and we need not suppose that an ecosystem experiences the kinds of intentions that humans do when it goes about the business of staying in existence.

The problem-solving point of view has some advantages, though. It puts the emphasis on the elaborate hierarchy of informational processes that is pertinent to the dynamical behavior of ecosystems. Dynamical methods, to be usable, encourage the researcher to either treat these mechanisms as irrelevant or coalesce them into some easily expressible nonlinearities. It is evident that we could not describe the state-space behavior of a digital computer in this way, and information processing in such machines is a great deal simpler and less powerful than in biological systems. The problem-solving point of view is also open to novelty-generating processes such as evolution and intelligence. Here we are dealing with a system's potentialities, and it is natural to describe these externally in terms of adaptability (maximum tolerable uncertainty of the environment) and internally in terms of problem-solving power (interactions available for information processing).

APPENDIX: CONNECTION TO THE ADAPTABILITY THEORY FORMALISM

In *adaptability theory* (Conrad, 1983), the behaviors of a biological system (say, the community) and the environment are described by transition schemes, ω and ω^*. These are, in general, unknown and probabilistic. Let the entropy $H(\omega^*)$ denote the actual uncertainty of the environment and $H(\hat{\omega}^*)$ be its maximum tolerable uncertainty. We define this as the adaptability and write

$$H(\hat{\omega}) - H(\hat{\omega}|\hat{\omega}^*) + H(\hat{\omega}^*|\hat{\omega}) \equiv H(\hat{\omega}^*) \geq H(\omega^*).$$

$H(\hat{\omega})$ is the potential behavioral uncertainty, $H(\hat{\omega}|\hat{\omega}^*)$ is ability to anticipate the environment, and $H(\hat{\omega}^*|\hat{\omega})$ is indifference to environment. These measures are related to the behavioral repertoire, the correlation, and passive problem-definition components of problem solving, as discussed in the section on compensation and interconversion of modes.

We can decompose each of these entropies into a sum of effective entropies, $H(\hat{\omega}) = \Sigma H_e(\hat{\omega}_{ij})$, associated with different subcompartments of the system at different levels of organization; $\hat{\omega}_{ij}$, a time derivative, is the transition scheme of subcompartment i at level j in terms of its subcompartments at the next lower levels. The effective entropy is a normalized sum of the unconditional entropy of the subcompartment and all its entropies conditioned on the behavior of other subcompartments. Thus, it represents both the uncertain modifiability of the subcompartment and how independent this modifiability is from other compartments. The contribution to adaptability becomes greater as independence increases. The various modes of biological adaptability (evolutionary adaptability, developmental and neurobehavioral plasticity, culturability, routability of energy in food webs, and others) can be associated with specific effective entropies.

The tendency of unexercised adaptabilities to atrophy means that adaptability tends to fall to the actual uncertainty of the environment. The principle of compensation is expressed in terms of expansions and contractions in effective entropies consistent with this condition and with the costs and advantages of different forms of adaptability. Regulation of the environment can be dealt with by considering how adaptabilities in the community can serve to protect the environment from disturbances.

Adaptability increases as the difference between $H(\hat{\omega})$ and $H(\hat{\omega}|\hat{\omega}^*)$ increases, that is, as the environment provides more information about the behavior of the biological systems. In the terminology of this chapter, adaptability-directed problem solving increases. These entropies could have substantial magnitudes even in a completely definite environment. The magnitudes reflect specialization-directed problem solving, that is, the rule-based mechanisms that enable the system to exploit a given environment. They also reflect redundancy, which contributes to the reliability of information processes and to evolutionary transformability of the system.

As the number of interactions and flexible constraints in a subcompartment increases toward $n^2/2$, potential problem-solving power increases. In the case of adaptability-directed problem solving, this allows the anticipation term to decrease relative to the behavioral uncertainty term. In the case of problem solving that is purely specialization directed, the magnitudes of both terms increase equally. The cost associated with behavioral uncertainty of the subcompartment increases and, as a consequence, compensation must occur. This would be true even if the increase in problem-solving power were adaptability directed. For example, the vertebrate brain provides a great deal of adaptability-directed problem solving, but it is itself a delicate structure that must be protected from disturbances by adaptability elsewhere in the system. This means reorganizing adaptability so as to reroute disturbances originating in other subcompartments. Such buffering adaptabilities make a substantial contribution to the actual magnitudes of the entropies without increasing externally directed adaptability. They are similar in this respect to the redundancies that contribute to reliability. Adaptabilities that are internally directed can assume enormous proportions in a community due to competition among different components. Internally directed adaptabilities play an important role in evolution and may be converted to externally directed adaptability subsequent to a crisis.

The transition schemes, $\hat{\omega}$ and ω^*, are state-to-state descriptions, and this is what allows the connection to dynamics. Analogs to dynamical stability concepts, such as orbital and structural stability, are defined by imposing tolerance ("is similar to") relations on the states of the system.

REFERENCES

CONRAD, M. 1972a. Statistical and hierarchical aspects of biological organization. In Waddington, C. H. (ed.), *Towards a Theoretical Biology*, Vol. 4. Edinburgh University Press, Edinburgh, Scotland, pp. 180–221.

CONRAD, M. 1972b. Stability of foodwebs and its relation to species diversity. *J. Theor. Biol.* 34:325–335.

CONRAD, M. 1981. Algorithmic specification as a technique for computing with informal biological models. *BioSystems* 13:303–320.

CONRAD, M. 1983. *Adaptability*. Plenum Press, New York. 383 pp.

CONRAD, M., and HASTINGS, H. 1985. Scale change and the emergence of information processing primitives. *J. Theor. Biol.* 112:741–755.

JERNE, N. K. 1955. The natural selection theory of antibody formation. *Proc. Natl. Acad. Sci. USA* 41:849–857.

MACARTHUR, R. H. 1955. Fluctuations of animal populations and a measure of community stability. *Ecology* 36:533–536.

MARGALEF, R. 1958. Information theory in ecology. *Gen. Syst.* 3:36–71.

MAY, R. 1973. *Stability and Complexity in Model Ecosystems*. Princeton University Press, Princeton, New Jersey. 235 pp.

ODUM, E. P. 1953. *Fundamentals of Ecology*. Saunders, Philadelphia. 384 pp.

ODUM, E. P. 1969. The strategy of ecosystem development. *Science* 164:262–267.

RIZKI, M. M., and CONRAD, M. 1985. Evolve III: a discrete events model of an evolutionary ecosystem. *BioSystems* 18:121–133.

SLOBODKIN, L. B., and RAPAPORT, A. 1974. An optimal strategy of evolution. *Quart. Rev. Biol.* 49:181–200.

25

Information Theory and Ecological Networks

HIRONORI HIRATA

Department of Electrical and Electronics Engineering
Chiba University, Chiba

In effect, the rise of ecological thinking in science represents a more systems-oriented approach than that of the specialists, for the ecologist is concerned with interrelations, not merely with things [Van Dyne, in *The Ecosystem Concept in Natural Resource Management*, Ch. X, p. 364 (1969), Academic Press].

INTRODUCTION

One of the most useful representations of an ecosystem is its portrayal as a network. Although network representation focuses only on the direction and amount of flows, a network model is a powerful tool for the discussion of ecological systems, with both advantages and limitations.

The flow in an ecological network is a macroscopic variable. There are many "microscopic" phenomena in ecosystems, such as, predator–prey relations, competition, and migration, which aggregate to macrophenomena such as movements of mass and energy. Flows, as macrophenomena, also implicitly include micro phenomena. Network treatments cannot often refer to microscopic properties directly, but they can elaborate macroscopic properties that are collections of microproperties.

MacArthur (1955), and later Margalef (1968), introduced information theory into ecology as a possible vehicle for attempting to define macroscopic properties of ecosystems. Rutledge, Basore, and Mulholland (1976) refined earlier attempts by introducing the concept of conditional entropy as a model for choice, which would also quantify ecological stability in closed systems. Ulanowicz (1980) later conjectured that mutual information could serve as part of an index of growth and development for ecosystems. He used the term "ascendancy" for this index, which is calculated as the product of mutual information times the total amount of network throughflow (see Chapter 26). Although intuitively appealing, it is not known yet whether ascendancy will serve as a general enough measure for system growth and development.

Following these efforts, Hirata and Ulanowicz (1984) extended the Rutledge, Basore, and Mulholland (1976) model for choice to include open ecosystems (ecological networks) and estimated the amount of information in a network using the concept of the mutual information of a channel. This chapter shows how mutual information can be rigorously derived as an index of self-organization for ecological systems. Since discussion is limited to the network model, conclusions about self-organization will pertain to macroproperties of networks concerning the movement of conservative media (for example, energy or matter). The mutual information of a channel will be applied to two main topics: (1) aggregation, that is, how to frame macroscopic structure (see Chapters 7 and 8), and (2) self-organization and the measurement of succession.

Regarding an ecological network as a "communication channel" (for example, Chapter 2), the information contained in the structure of the network may be theoretically defined by the concept of mutual information. Since ecological networks are basically dynamic, this chapter implicitly concerns dynamic networks. The static case will, however, be applied to examples due to limitations of measurable data. With regard to aggregation, creating functional classifications of species means framing macroscopic structure as hierarchical orderings in the ecological network. Optimal aggregation (or grouping), as will be seen, minimizes the mutual information difference between object system and aggregated model. The information cannot increase (it generally declines) in an aggregative process. Special patterns of structure will be shown that result in no loss of information. Considering the dependency of states of the ecological network among time points, a new informational index of self-organization is proposed. If state transitions of the ecological network are simple Markovian, this index can be equivalent to the information of an ecological channel as defined by mutual information. The information inherent in network structure, that is, the mutual information of the ecological channel, is a good index of self-organization from the macroscopic viewpoint.

ECOLOGICAL NETWORKS

Consider an ecological network as shown in Fig. 25.1, consisting of n compartments each characterized by a throughflow (total flow) of some medium, such as carbon, nitrogen, mass, or energy. The detailed structure of the kth compartment is illustrated in Fig. 25.2, whose symbols are defined as follows:

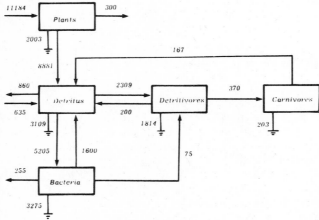

Figure 25.1 Simple example of an ecological network. Flows of energy in the Cone Spring ecosystem are schematically depicted (after Finn, 1976). Annual values for flows are given in kcal/m^2. Ground symbols ($\perp$) represent respiration.

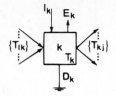 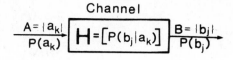

Figure 25.2 Typical kth compartment of an ecological network.

Figure 25.3 Schematic diagram of a generic channel.

T_{kj}, flow leaving the kth compartment and directly contributing to the jth compartment ($T_{kj} \geq 0$)

D_k, dissipated flow leaving the kth compartment ($D_k \geq 0$)

E_k, useful export leaving the kth compartment ($E_k \geq 0$)

I_k, input flow to the kth compartment from the system-level environment ($I_k \geq 0$)

The throughflow of the kth compartment, T_k, is defined as

$$T_k = \sum_{j=1}^{n} T_{kj} + D_k + E_k. \tag{25.1}$$

For convenience, we may use the following vector and matrix forms of these variables:

$$\begin{aligned} T &= (T_k)'_{k=1,\ldots,n} = (T_1, \ldots, T_n)' \\ D &= (D_k)'_{k=1,\ldots,n} \\ E &= (E_k)'_{k=1,\ldots,n} \\ I &= (I_k)'_{k=1,\ldots,n} \\ T^* &= [T_{kj}]_{k,j=1,\ldots,n} \end{aligned} \tag{25.2}$$

where the primes (') represent transposes. Equation (25.1) can then be expressed as

$$T = T^* \mathbf{1} + D + E, \tag{25.3}$$

where $\mathbf{1} = (1, \ldots, 1)'$. In a system at steady state, T_k also equals $\sum_{l=1}^{n} T_{lk} + I_k$; that is,

$$T = T^*\mathbf{1} + D + E = T^*\mathbf{1} + I. \tag{25.4}$$

COMMUNICATION CHANNEL AND MUTUAL INFORMATION

As in Fig. 25.3, denote the input alphabet by A containing the letters $\{a_k\}_{k=1,\ldots,n}$ and associated probabilities defined by $P(a_k)$. Denote the output alphabet by B and its letters $\{b_j\}_{j=1,\ldots,r}$ with probabilities $P(b_j)$, which need not be the same in number as the input. A memoryless channel is now completely specified by giving $P(b_j|a_k)$, that is, the probability of output b_j when the input is a_k, for $j = 1, \ldots, r$ and $k = 1, \ldots, n$. For each input there will always be an output letter, so

$$\sum_{j=1}^{r} P(b_j|a_k) = 1. \tag{25.5}$$

The probabilities $P(b_j|a_k)$, which indicate how the transition from input to output takes place, will be termed transition probabilities. $H = [P(b_j|a_k)]$ is the communication matrix of this channel.

The probability of the output letter being b_j is the sum of the probabilities of the events favorable to this occurrence and so is given by

$$P(b_j) = \sum_{k=1}^{n} P(b_j|a_k)P(a_k). \tag{25.6}$$

The entropy of A is defined by

$$H(A) = -\sum_{k=1}^{n} P(a_k) \log P(a_k), \tag{25.7}$$

where $H(A)$ represents the uncertainty about A before B is known. The entropy $H(X)$ can be thought of as a measure of the following things about X:

1. Information provided by an observation X
2. Uncertainty about X
3. Randomness of X

The conditional entropy of A given B is defined by

$$H(A|B) = -\sum_{j=1}^{r} P(b_j) \sum_{k=1}^{n} P(a_k|b_j) \log P(a_k|b_j), \tag{25.8}$$

where $H(A|B)$ represents the uncertainty about A after B is known. The mutual information of the channel, that is, the mutual information between input A and output B, is defined by

$$M(A; B) = H(A) - H(A|B) \tag{25.9}$$

$$= H(B) - H(B|A) \tag{25.10}$$

$$= \sum_{k=1}^{n} \sum_{j=1}^{r} P(b_j|a_k) P(a_k) \log \left[\frac{P(b_j|a_k)}{P(b_j)}\right]. \tag{25.11}$$

From

$$P(a_k, b_j) = P(b_j|a_k) P(a_k) = P(a_k|b_j) P(b_j), \tag{25.12}$$

the mutual information can be rewritten as

$$M(A; B) = \sum_{k=1}^{n} \sum_{j=1}^{r} P(a_k, b_j) \log \left[\frac{P(a_k, b_j)}{P(a_k)P(b_j)}\right]. \tag{25.13}$$

$M(A; B)$ represents the amount of information about the input provided by the observation of a channel output.

INFORMATION OF ECOLOGICAL NETWORKS

Theoretical Form

The model of choice for open systems is shown in Fig. 25.4, where exogenous inputs come from the 0th compartment; exports, which are used in other networks, enter the $(n + 1)$th compartment; and dissipation is collected in the $(n + 2)$th compartment. Although we could consolidate input, export, and dissipation into one global compartment, they are treated separately to facilitate extension of the theory.

Now time is introduced. The time interval for flow from one compartment to another is taken to be θ. Let a_k be the event that a single given medium (such as carbon, nitrogen, mass, or energy) passes through the kth compartment at time t_1, and b_k be the event that the medium passes through the kth compartment at time $t_1 + \theta$. We can define the following:

$P(a_k)$, probability that the medium passes through the kth compartment at time t_1
$P(b_j)$, probability that the medium passes through the jth compartment at time $t_1 + \theta$
$P(b_j|a_k)$, probability that the medium that passes through the kth compartment will be taken up by the jth compartment
$P(a_k|b_j)$, probability that the medium will pass to the kth compartment from the jth compartment, given that the medium has been taken up by the jth compartment.

From eq. (25.13), the mutual information of this ecological channel can be expressed as

$$M(N) = M(A; B) \tag{25.14}$$

$$= \sum_{k=0}^{n+2} \sum_{j=0}^{n+2} P(a_k, b_j) \log \left[\frac{P(a_k, b_j)}{P(a_k)P(b_j)} \right]. \tag{25.15}$$

$M(N)$ shows the information contained in the structure of the ecological network.

Applicable Form

We now apply the theoretical form (25.15) to a real ecological network like Fig. 25.1. Although we may theoretically think of several ways to experimentally measure and define the probabilities $P(a_k)$, $P(b_k)$, and $P(b_j|a_k)$ from the meanings given in the previous section, if measurable data are

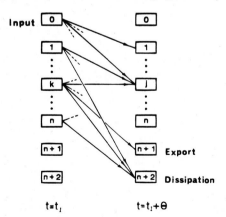

Figure 25.4 Generic model for choice in open systems.

limited as shown in Fig. 25.1, the percentage of flow distribution should be used as a surrogate for the true probabilities. Then the following definitions hold:

Q_k, percentage of the total flow that passes through the kth compartment at time t_1 ($Q_k \geq 0$, $k = 0, \ldots, n$; $Q_{n+1} = Q_{n+2} = 0$)

P_j, percentage of the total flow that passes through the jth compartment at time $t_1 + \theta$ ($P_0 = 0$; $P_j \geq 0$; $j = 1, \ldots, n + 2$)

f_{kj}, percentage of total flow through the kth compartment at time t_1 that flows into the jth compartment at time $t_1 + \theta$ ($f_{kj} \geq 0$)

e_k, percentage of flow through the kth compartment that is expected as useful flow ($e_k \geq 0$)

r_k, percentage of flow through the kth compartment that is dissipated ($r_k \geq 0$)

The relations between these variables are provided by the equations

$$P_j = \sum_{k=0}^{n} f_{kj} Q_k, \quad j = 1, \ldots, n,$$

$$P_{n+1} = \sum_{k=1}^{n} e_k Q_k, \quad (25.16)$$

$$P_{n+2} = \sum_{k=1}^{n} r_k Q_k,$$

where

$$\sum_{j=1}^{n} f_{kj} + r_k + e_k = 1, \quad k = 1, \ldots, n, \quad \sum_{j=1}^{n} f_{0j} = 1. \quad (25.17)$$

In ecological network N, we use Q_k, P_j, and f_{kj} as $P(a_k)$, $P(b_k)$, and $P(b_j|a_k)$, respectively:

$$P(a_k) = Q_k,$$
$$P(b_j) = P_j, \quad (25.18)$$
$$P(b_j|a_k) = f_{kj},$$

and we may identify variables Q_k, f_{kj}, r_k, and e_k as follows:

$$Q_k = \frac{T_k}{T}, \quad (k = 1, \ldots, n)$$
$$Q_0 = \frac{I}{T}, \quad (25.19)$$

$$f_{kj} = \frac{T_{kj}}{T_k}, \quad (k, j = 1, \ldots, n)$$
$$f_{0j} = \frac{I_j}{I}, \quad (j = 1, \ldots, n) \quad (25.20)$$
$$f_{k0} = 0, \quad (k = 0, \ldots, n)$$

$$r_k = \frac{D_k}{T_k}, \tag{25.21}$$

and

$$e_k = \frac{E_k}{T_k}, \tag{25.22}$$

where

$$T = T^* + I, \tag{25.23}$$

$$T^* = \sum_{k=1}^{n} T_k, \tag{25.24}$$

and

$$I = \sum_{k=1}^{n} I_k. \tag{25.25}$$

From equations (25.15) and (25.18) through (25.25), the mutual information of the ecological channel, that is, the information contained in the structure of a corresponding ecological network, can be expressed in terms of flows according to the following proposition:

Proposition 1. The information, $M(N)$, contained in the structure of network N is described as

$$M(N) = \frac{1}{T} \sum_{k=0}^{n} \sum_{j=1}^{n+2} T_{kj} \log\left(\frac{TT_{kj}}{T_k T_j}\right), \tag{25.26}$$

where $T_{kn+1} = D_k$, $T_{kn+2} = E_k$, and $T_{0k} = I_k$.

AGGREGATION: FRAMING MACROSCOPIC STRUCTURE

Basic Concepts of Aggregation

In ecology there is increasing interest in the fundamental importance that aggregation theory plays in constructing reasonable macroscopic models (Gardner, Cale, and O'Neill, 1982; Hirata and Ulanowicz, 1984; Sugihara et al., 1984; see also Chapters 7 and 8). The following discussion outlines some basic principles.

Definition 1. Let N and $\overline{N}$ be the sets of elements

$$N = \{\alpha_k\}_{k=1,\ldots,n} \tag{25.27}$$

$$\overline{N} = \{\beta_i\}_{i=1,\ldots,m} \quad (m \leq n) \tag{25.28}$$

such that a homomorphic mapping ϕ is made from N to $\overline{N}$ as

$$\phi : N \to \overline{N}. \tag{25.29}$$

Then N is called the original network, $\overline{N}$ the aggregated network of N, and ϕ the aggregation mapping. ϕ may be conveniently represented by a matrix as follows.

Definition 2. Define an $m \times n$ ($m \leq n$) aggregation matrix S as

$$S = [s_{ik}]_{i=1,\ldots,m,\ k=1,\ldots,n} \tag{25.30}$$

$$= \begin{bmatrix} s_{11} & s_{12} & \cdots & s_{1n} \\ s_{21} & s_{22} & \cdots & s_{2n} \\ \vdots & \vdots & & \vdots \\ s_{m1} & s_{m2} & \cdots & s_{mn} \end{bmatrix},$$

where

$$0 \leq s_{ik} \leq 1, \tag{25.31}$$

$$\sum_{i=1}^{m} s_{ik} = 1.$$

If all s_{ik} are either 0 or 1, the mapping is referred to in the following discussion as a *discrete aggregation* (Hirata, 1978). Otherwise, it is referred to as a *weighted aggregation*. We may consider the discrete aggregation as a special case of the weighted aggregation. Those positions of s_{ik} that are not zero signify which elements should be aggregated into the same group i. Because the aggregated network ($\overline{N}$) depends on the aggregation matrix (S), it may be expressed as a function of S, $\overline{N}(S)$.

Example. Consider an aggregation from N to $\overline{N}$ as follows:

$$\begin{aligned} N &= \{\alpha_1, \alpha_2, \alpha_3, \alpha_4, \alpha_5, \alpha_6\}, \\ \overline{N} &= \{\beta_1, \beta_2, \beta_3\}. \end{aligned} \tag{25.32}$$

A 6×3 matrix S describes the aggregation of N into $\overline{N}$.

Figure 25.5a illustrates the case of weighted aggregation defined by the aggregation matrix:

$$S = \begin{bmatrix} 1 & 1/2 & 0 & 0 & 0 & 0 \\ 0 & 1/2 & 1/3 & 2/3 & 0 & 0 \\ 0 & 0 & 2/3 & 1/3 & 1 & 1 \end{bmatrix}. \tag{25.33}$$

In this case,

$$\begin{aligned} \beta_1 &= \alpha_1 + (1/2)\alpha_2, \\ \beta_2 &= (1/2)\alpha_2 + (1/3)\alpha_3 + (2/3)\alpha_4, \\ \beta_3 &= (2/3)\alpha_3 + (1/3)\alpha_4 + \alpha_5 + \alpha_6. \end{aligned} \tag{25.34}$$

Figure 25.5b illustrates the case of discrete aggregation, whose aggregation matrix is

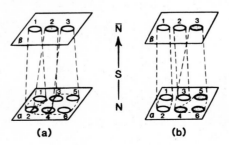

Figure 25.5 Schematized aggregation: (a) weighted aggregation; (b) discrete aggregation.

$$S = \begin{bmatrix} 1 & 1 & 0 & 0 & 0 & 0 \\ 0 & 0 & 1 & 0 & 0 & 0 \\ 0 & 0 & 0 & 1 & 1 & 1 \end{bmatrix}. \tag{25.35}$$

In this case,

$$\beta_1 = \alpha_1 + \alpha_2,$$
$$\beta_2 = \alpha_3, \tag{25.36}$$
$$\beta_3 = \alpha_4 + \alpha_5 + \alpha_6.$$

Focusing now on the process of network aggregation (both weighted and discrete), the original network and the aggregated network are characterized by the following relations.

Original Network N	Aggregated Network $\overline{N}(S)$
$A = \{a_k\}_{k=0,1,\ldots,n,n+1,n+2}$	$\overline{A} = \{\overline{a}_i\}_{i=0,1,\ldots,m,m+1,m+2}$
$P(a_k) = Q_k$	$P(\overline{a}_i) = \overline{Q}_i$
$B = \{b_j\}_{j=0,1,\ldots,n,n+1,n+2}$	$\overline{B} = \{\overline{b}_l\}_{l=0,1,\ldots,m,m+1,m+2}$
$P(b_j) = P_j$	$P(\overline{b}_l) = \overline{P}_l$
$F = [f_{kj}]$	$\overline{F} = [\overline{f}_{il}]$
$T = (T_k)'_{k=1,\ldots,n}$	$\overline{T} = (\overline{T}_i)'_{i=1,\ldots,m}$
$D = (D_k)'_{k=1,\ldots,n}$	$\overline{D} = (\overline{D}_i)'_{i=1,\ldots,m}$
$E = (E_k)'_{k=1,\ldots,n}$	$\overline{E} = (\overline{E}_i)'_{i=1,\ldots,m}$
$I = (I_k)'_{k=1,\ldots,n}$	$\overline{I} = (\overline{I}_i)'_{i=1,\ldots,m}$
$T^* = [T_{kj}]_{k,j=1,\ldots,n}$	$\overline{T}^* = [\overline{T}_{il}]_{i,l=1,\ldots,m}$

Proposition 2. The relations between the variables of the original network, N, and those of the aggregated network, $\overline{N}(S)$, are as follows:

$$\overline{T} = S\,T,$$
$$\overline{D} = S\,D,$$
$$\overline{E} = S\,E, \tag{25.37}$$
$$\overline{I} = S\,I,$$
$$\overline{T}^* = S\,T^*\,S'.$$

Proposition 3 (Hirata and Ulanowicz, 1985). During the process of network aggregation (both weighted and discrete), the following relation is maintained;

$$M(N) \geq M(\overline{N}). \tag{25.38}$$

Proposition 3 shows that information cannot increase (it is generally lost) during the process of aggregation; that is, the difference $M(\overline{N}) - M(N)$ is never negative. This loss of information may be regarded as the cost, J, of the aggregation:

$$J = M(N) - M[\overline{N}(S)], \tag{25.39}$$

and J is minimized by choosing the optimal S^* such that

$$J^* = \min_{\{S\}} J = M(N) - M[\overline{N}(S^*)]. \tag{25.40}$$

In further developments in this chapter, we shall consider only discrete aggregation.

Patterns for Perfect Aggregation

If the aggregation is done without loss, that is, $J^* = 0$, we call it *perfect aggregation*. If not ($J^* > 0$), we call it *approximate aggregation*. Perfect aggregation is the topic of this section. For simplicity, we first discuss the case of aggregating two elements and then extend these results to more general cases.

The cost J of aggregating two elements K and L is derived as

$$J_{KL} = \sum_{i=0 \, (i \neq K, L)}^{n} A_i + \sum_{j=1 \, (j \neq K, L)}^{n+2} B_j + C, \tag{25.41}$$

where

$$A_i = \frac{1}{T}\left[T_{iK} \log\left(\frac{T_{iK}}{T_K}\right) + T_{iL} \log\left(\frac{T_{iL}}{T_L}\right) - (T_{iK} + T_{iL}) \log\left(\frac{T_{iK} + T_{iL}}{T_K + T_L}\right) \right], \tag{25.42}$$

$$B_j = \frac{1}{T}\left[T_{Kj} \log\left(\frac{T_{Kj}}{T_K}\right) + T_{Lj} \log\left(\frac{T_{Lj}}{T_L}\right) - (T_{Kj} + T_{Lj}) \log\left(\frac{T_{Kj} + T_{Lj}}{T_K + T_L}\right) \right], \tag{25.43}$$

and

$$C = \frac{1}{T}\left[T_{KK} \log\left(\frac{T_{KK}}{T_K^2}\right) + T_{KL} \log\left(\frac{T_{KL}}{T_K T_L}\right) + T_{LK} \log\left(\frac{T_{LK}}{T_L T_K}\right) + T_{LL} \log\left(\frac{T_{LL}}{T_L^2}\right) \right.$$
$$\left. - (T_{KK} + T_{KL} + T_{LK} + T_{LL}) \log\left(\frac{T_{KK} + T_{KL} + T_{LK} + T_{LL}}{T_K + T_L}\right)^2 \right]. \tag{25.44}$$

Here, A_i shows the error due to aggregating the flows T_{iK} and T_{iL} (two inflows from element i) into one flow, B_j defines the error due to aggregating the flows T_{Kj} and T_{Lj} (two outflows to element j) into one flow, and C is the error due to aggregating the flows T_{KL}, T_{LK}, T_{KK}, and T_{LL} (interactional flows between element K and element L and the self-loops of elements K and L) into one flow.

Lemma 1. The quantity J_{KL} equals 0 if and only if the following three conditions hold for $i = 0, 1, \ldots, n$ ($i \neq K, L$) and $j = 1, \ldots, n+2$ ($j \neq K, L$):

$$\frac{T_{iK}}{T_K} = \frac{T_{iL}}{T_L} \tag{25.45}$$

$$\frac{T_{Kj}}{T_K} = \frac{T_{Lj}}{T_L} \tag{25.46}$$

$$\frac{T_{KK}}{T_K^2} = \frac{T_{KL}}{T_K T_L} = \frac{T_{LK}}{T_L T_K} = \frac{T_{LL}}{T_L^2}. \tag{25.47}$$

Lemma 1 follows from eqs. (25.41) through (25.44) and the conditions $A_i \geq 0$, $B_j \geq 0$, and $C \geq 0$. From Lemma 1, several special patterns for perfect aggregation are derived, as follows.

Proposition 4. There are the following two cases of perfect aggregation. In both cases, the situation of elements K and L to be aggregated may be called a parallel position.

Case a. There is no flow between elements K and L ($T_{KL} = T_{LK} = 0$) and no self-loops of elements K and L ($T_{KK} = T_{LL} = 0$), as in Fig. 25.6a, and conditions (25.45) and (25.46) in Lemma 1 should be satisfied for $i = 0, 1, \ldots, n$ ($i \neq K, L$) and $j = 1, \ldots, n + 2$ ($j \neq K, L$).

Case b. There are complete flow connections between K and L ($T_{KL} \neq 0$, $T_{LK} \neq 0$), and self-loops of elements K and L exist ($T_{KK} \neq 0$, $T_{LL} \neq 0$), as in Fig. 25.6b, and conditions (25.45) through (25.47) in Lemma 1 should be satisfied for $i = 0, 1, \ldots, n$ ($i \neq K, L$) and $j = 1, \ldots, n + 2$ ($j \neq K, L$).

The following, however, is a special and important case of Proposition 4.

Proposition 5. There is no aggregation loss, that is, $J_{KL} = 0$, when elements K and L have only one inflow from a common element and one outflow to a common element; that is, they hold a perfect parallel position as in Fig. 25.7a and b. Proposition 5 can be derived from Proposition 4. That elements K and L have only one inflow and one outflow means, respectively, that $T_{iK} = T_{Kj} = T_K$ and $T_{iL} = T_{Lj} = T_L$. Therefore, conditions (25.45) and (25.46) are satisfied. Since there is no flow between elements K and L and no self-loop of elements K and L, condition (25.47) is also satisfied.

Propositions 4 and 5 can be extended to the case of N aggregating elements as follows.

Proposition 6. There are the following two cases of perfect aggregation. In both cases, the situation of elements K_l ($l = 1, \ldots, N$) to be aggregated may be called a parallel position.

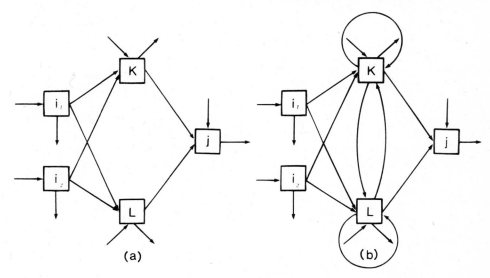

Figure 25.6 Parallel position: two types of perfect aggregation in Proposition 4.

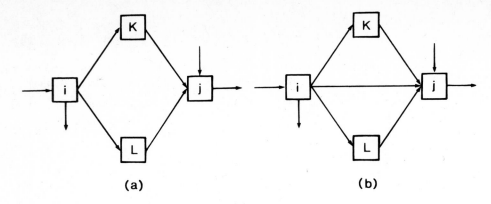

Figure 25.7 Perfect parallel position: a special type of perfect aggregation without any additional conditions in Proposition 5.

Case a. There is no flow among the N elements to be aggregated ($T_{K_i K_j} = 0$ for $i, j = 1, \ldots, N, i \neq j$) and no self-loops of elements K_l, $l = 1, \ldots, N$, ($T_{K_i K_i} = 0$ for $i = 1, \ldots, N$), as in Fig. 25.8a, and the following conditions (25.48) and (25.49) should be satisfied for $i = 0, \ldots, n$ [$i \neq K_l$ ($l = 1, \ldots, N$)] and $j = 1, \ldots, n + 2$ [$j \neq K_l$ ($l = 1, \ldots, N$)]:

$$\frac{T_{iK_1}}{T_{K_1}} = \cdots = \frac{T_{iK_N}}{T_{K_N}}, \tag{25.48}$$

$$\frac{T_{K_1 j}}{T_{K_1}} = \cdots = \frac{T_{K_N j}}{T_{K_N}}. \tag{25.49}$$

Case b. There are complete connections between K_i and K_j for $i, j = 1, \ldots, N, i \neq j$ ($T_{K_i K_j} \neq 0$ for $i, j = 1, \ldots, N, i \neq j$) and self-loops of elements K_l, $l = 1, \ldots, N$, ($T_{K_i K_i} \neq 0$ for $i = 1, \ldots, N$) as in Fig. 25.8b, and conditions (25.48) and (25.49) and the following condition (25.50) should be satisfied for $i = 0, \ldots, n$ [$i \neq K_l$ ($l = 1, \ldots, N$)] and $j = 1, \ldots, n + 2$ [$j = K_l$ ($l = 1, \ldots, N$)]:

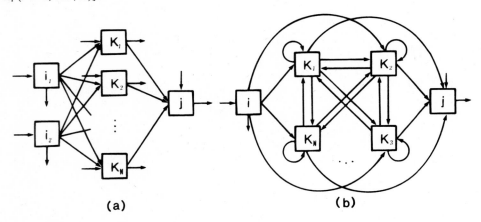

Figure 25.8 Parallel position: two types of perfect aggregation in Proposition 6.

$$\frac{T_{K_1 K_1}}{(T_{K_1})^2} = \frac{T_{K_1 K_2}}{T_{K_1} T_{K_2}} = \cdots = \frac{T_{K_m K_m}}{T_{K_m} T_{K_m}}. \tag{25.50}$$

Proposition 7. There is no aggregation loss, that is, $J_{K_1,\ldots,K_N} = 0$, when elements K_l ($l = 1, \ldots, N$) have only one inflow from a common element and one outflow to a common element; that is, they hold perfect parallel position as in Fig. 25.9a and b.

Propositions 4 through 7 show that parallel structure is important for perfect aggregation. Although these propositions are only for perfect aggregation, it is easy to demonstrate that parallel structures generally tend to minimize aggregation loss. This analytical result matches Gardner, Cale, and O'Neill's (1982) computer simulation results on network models, wherein it was found that parallel structures tend to produce less error than series structures.

Examples

Example 1. Flows among 17 compartments of a tidal marsh stream ecosystem (in mg C m^{-2} day^{-1}) were measured by Homer and Kemp (unpublished manuscript) and are shown schematically in Fig. 25.10. The ecological network of 17 elements was aggregated into seven groups, indicated by the broken perimeters in Fig. 25.10. The final macrostructure of the reduced network is illustrated in Fig. 25.11.

The organization imposed by the aggregation may be characterized as a trophic hierarchy. If we distinguish pathways of active feeding (heavy lines) from passive detrital flows (fine lines), it becomes possible to decompose the network into two acyclic subgraphs as shown in Figs. 25.12a and b. The trophic identities of the aggregated compartments in Fig. 25.12a are apparent: microphytes (I) and detritus (III) provide a food base for pelagic herbivores (IV) and benthic herbivores (V); these in turn are fed upon by carnivores (VI) and top carnivores–omnivores (VII). It is seen from the above discussion of patterns for perfect aggregation why minimizing the decrease in information of the ecological network should lump compartments in a fashion similar to trophic levels.

Example 2. As shown in Fig. 25.13, a food web based on the main groups of organisms in the North Sea has been quantitatively estimated by Steele (1974). It consists of 10 compartments.

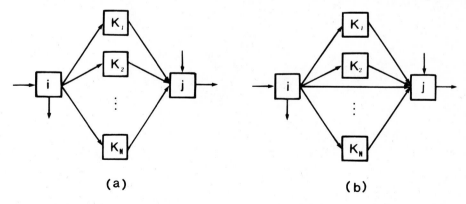

Figure 25.9 Perfect parallel position: a special type of perfect aggregation without any additional conditions in Proposition 7.

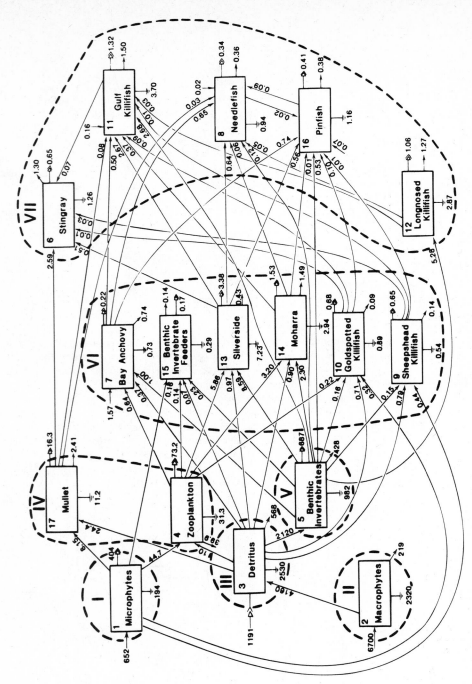

Figure 25.10 Schematic of carbon flows among 17 taxa of a marsh gut ecological network. Crystal River, Florida (after M. Homer and W. M. Kemp, unpublished). All flows are in mg C m^{-2} day^{-1}. Linked arrows (—▷) represent returns to detritus, ground symbols (⏚) are respiratory losses, simple arrows not terminating in another box depict exports from the system, and simple arrows not originating from another box are exogenous inputs. Broken lines indicate optimal aggregation into seven groups. The balance is within three-figure accuracy.

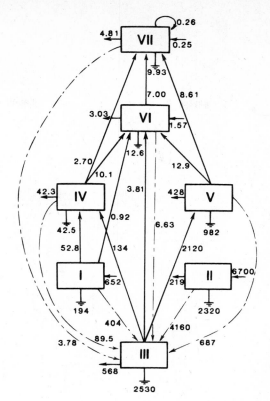

Figure 25.11 Ecological network in Fig. 25.10 after aggregation into seven groups.

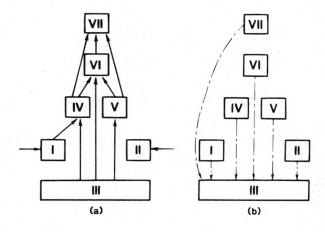

Figure 25.12 Decomposition of the macroecological network in Fig. 25.11 into acyclic subgraphs: (a) structure of the grazing chain, (b) the detrital returns.

Chap. 25   Information Theory and Ecological Networks   637

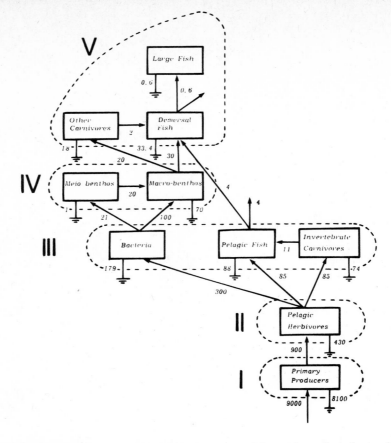

Figure 25.13 Schematic of the energy flows in a North Sea marine ecological network (after Steele, 1974, and Ulanowicz and Kemp, 1979). All flows are in kcal m^{-2} y^{-1}.

Although Steele did not provide the amounts of flows, Ulanowicz and Kemp (1979) estimated these values (in kcal m^{-2} y^{-1}). The resulting ecological network of 10 elements was aggregated into five groups as indicated by the broken perimeters in Fig. 25.13, producing the final network macroscopic structure illustrated in Fig. 25.14.

Example 3. Webster, Waide, and Patten (1975) developed models of sedimentary mineral cycles for eight ecosystem types (in kg minerals ha^{-1} y^{-1}). Figure 25.15 illustrates such a cycling network for the ocean. The original ecological network of six elements was aggregated into three groups as indicated by the broken perimeters in Fig. 25.15, producing the final macroscopic structure shown in Fig. 25.16.

INDEX OF SELF-ORGANIZATION

Self-Organization of Ecological Networks

When there are two probability distributions $P = \{p(x)\}$ and $Q = \{q(x)\}$, information $D(P, Q)$ for discrimination in favor of P against Q is defined as follows (Kullback, 1978):

$$D(P, Q) = \sum_{x \in X} p(x) \log\left[\frac{p(x)}{q(x)}\right]. \quad (25.51)$$

Since this information $D(P, Q)$ is a measure of dissimilarity, $D(P, Q)$ will tend to increase with the

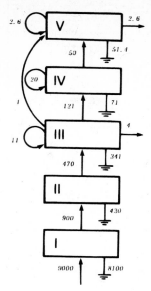

Figure 25.14 Ecological network in Fig. 25.13 after aggregation into five groups.

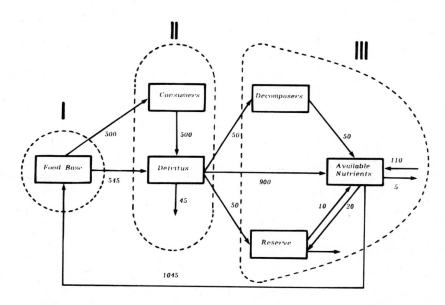

Figure 25.15 Schematic of the sedimentary mineral flows in an oceanic ecological network (after Webster, Waide, and Patten, 1975). All flows are in kg minerals ha^{-1} y^{-1}.

Chap. 25   Information Theory and Ecological Networks

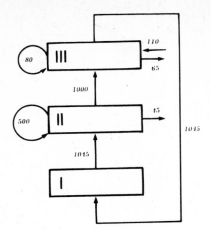

Figure 25.16 The six-compartment network of Fig. 25.15 after aggregation into three categories.

increase of dissimilarity. For example, when $P = Q$, $D(P, Q)$ has its minimum value [$D(P, Q) = 0$], and when $p(x^*) = 1$ for a particular x^* for which $q(x^*)$ is the smallest, $D(P, Q)$ is at its maximum $\{D(P, Q) = -\log[q(x^*)]\}$.

Now consider the discrete time dynamic ecological network. For this case, all variables in Fig. 25.2 may change in discrete steps according to time t_i ($i = 1, 2, \ldots, l, \ldots$) as $T_{kj}(t_i)$, $T_k(t_i)$, $D_k(t_i)$, $E_k(t_i)$, and $I_k(t_i)$.

Let a_{k_i} be the event that a single given medium (such as carbon, nitrogen, mass, or energy) passes through the kth compartment at time t_i. Setting $P = \{P(a_{k_1}, \ldots, a_{k_l})\}$ and $Q = \{P(a_{k_1}) P(a_{k_2}) \ldots P(a_{k_l})\}$ in eq. (25.51), we derive an index of self-organization in ecological networks, O_l, as follows:

$$O_l = D(P, Q) \qquad (25.52)$$
$$= \sum_{k_1, \ldots, k_l} P(a_{k_1}, \ldots, a_{k_l}) \log[P(a_{k_1}, \ldots, a_{k_l})/P(a_{k_1}) P(a_{k_2}) \ldots P(a_{k_l})],$$

where $\sum_{k_1, \ldots, k_l}$ means $\sum_{k_1=1}^{n(1)} \cdots \sum_{k_l=1}^{n(l)}$, and $n(l)$ is the number of elements at time t_l.

$P(a_{k_1}, \ldots, a_{k_l})$ is the joint probability of events $a_{k_1}, \ldots, a_{k_l}$, and it means the dependency of events $a_{k_1}, \ldots, a_{k_l}$, that is, the dependency of flow behavior (structure) of the ecological network among time points $t_1, \ldots, t_l$. As usual, all events are statistically independent if and only if

$$P(a_{k_1}, \ldots, a_{k_l}) = P(a_{k_1}) P(a_{k_2}) \ldots P(a_{k_l}). \qquad (25.53)$$

Therefore, O_l shows distance from the statistically independent case (or measure of correlation among all events). Since all events $a_{k_1}, \ldots, a_{k_l}$ being statistically independent means that the states of the ecological network are independent at all time points, O_l expresses the distance from the disorder state across time. Thus, O_l is an index of order of the ecological network. That is, since strong dependency along time is requisite for self-organization, O_l represents an index of self-organization of the ecological network in the interval from time t_1 to t_l. In eq. (25.53), it is admissible for the number of species in a network to change due to invasion, extinction, and so on, with time. When the number of species is constant, $n(l) = n$ (constant) for all l. The index of self-organization can be maximized, making it an extremal property for generalized ecological networks and thus for ecosystems.

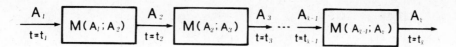

Figure 25.17 Ecological channels in cascade.

Relation between Self-organization Index and Mutual Information

If the ecological network has the simple Markov property such that

$$P\left(\frac{a_{k_l}}{a_{k_1}}, \ldots, a_{k_{l-1}}\right) = P\left(\frac{a_{k_l}}{a_{k_{l-1}}}\right), \qquad (25.54)$$

then

$$P(a_{k_1}, \ldots, a_{k_l}) = P\left(\frac{a_{k_l}}{a_{k_{l-1}}}\right) P\left(\frac{a_{k_{l-1}}}{a_{k_{l-2}}}\right) \ldots P\left(\frac{a_{k_2}}{a_{k_1}}\right) P(a_{k_1}). \qquad (25.55)$$

Substituting eq. (25.55) into (25.51) yields the following proposition.

Proposition 8. If an ecological network has the simple Markov property, the self-organization index O_l can be expressed by mutual information as follows:

$$O_l = \sum_{i=1}^{l-1} M(A_i; A_{i+1}), \qquad (25.56)$$

where

$$M(A_i; A_{i+1}) = \sum_{k_i=1}^{n(i)} \sum_{k_{i+1}=1}^{n(i+1)} P(a_{k_i}, a_{k_{i+1}}) \log\left[\frac{P(a_{k_i}, a_{k_{i+1}})}{P(a_{k_i}) P(a_{k_{i+1}})}\right]. \qquad (25.57)$$

This proposition shows that our temporal index of self-organization, O_l, of the ecological network can be expressed by the mutual information of ecological channels in cascade, as shown in Fig. 25.17. Hence, mutual information may be taken to be a natural and well-motivated measure of self-organization, to be maximized in development of the ecological network.

CONCLUSION

Although the concept of mutual information was introduced on the basis of its theoretical foundations, the emphasis of this chapter has been on real-world concerns of application. An applicable form of mutual information for ecological channels has been derived, along with examples illustrating its use with real data.

Although this chapter has been limited to considering only the aggregation problem and index of self organization, the information-theoretic perspective offers new insight into relations between information and network mechanisms operating in ecological systems, such as causality, controllability, and observability. Connection to issues of ecosystem growth and development will become more apparent in the next, and concluding, chapter.

REFERENCES

Finn, J. T. 1976. Measures of ecosystem structure and function derived from analysis of flows. *J. Theor. Biol.* 56:363–380.

Gardner, R. H., Cale, W. G., and O'Neill, R. V. 1982. Robust analysis of aggregation error. *Ecology* 63:1771–1779.

Hirata, H. 1978. Aggregation method for linear large-scale systems with random coefficients and inputs. *Int. J. Systems Sci.* 9:515–529.

Hirata, H., and Ulanowicz, R. E. 1984. Information theoretical analysis of ecological networks. *Int. J. Systems Sci.* 15:261–270.

Hirata, H., and Ulanowicz, R. E. 1985. Information theoretical analysis of the aggregation and hierarchical structure of ecological networks. *J. Theor. Biol.* 116:321–341.

Hirata, H., and Ulanowicz, R. E. 1986. Large-scale system perspectives on ecological modelling and analysis. *Ecol. Mod.* 31:79–104.

Kullback, S. 1978. *Information Theory and Statistics.* Peter Smith, Gloucester, Massachusetts. 399 pp.

MacArthur, R. H. 1955. Fluctuations of animal populations, and a measure of community stability. *Ecology* 36:533–536.

Margalef, R. 1968. *Perspectives in Ecological Theory.* University of Chicago Press, Chicago. 111 pp.

Rutledge, R. W., Basore, B. L., and Mulholland, R. J. 1976. Ecological stability: an information theory viewpoint. *J. Theor. Biol.* 57:355–371.

Steele, J. H. 1974. *The Structure of Marine Ecosystems.* Harvard University Press, Cambridge, Massachusetts. 128 pp.

Sugihara, G., Garcia, S., Gulland, J. A., Lawton, J. H., Maske, H., Paine, R. T., Platt, T., Rachor, E., Rothschild, B. J., Ursin, E. A., and Zeitzchel, B. F. K. 1984. Ecosystems dynamics: group report. In May, R. M. (ed.), *Exploitation of Marine Communities.* Springer-Verlag, New York, pp. 131–153.

Ulanowicz, R. E. 1980. An hypothesis on the development of natural communities. *J. Theor. Biol.* 85:223–245.

Ulanowicz, R. E., and Kemp, W. M. 1979. Toward canonical trophic aggregation. *Am. Nat.* 114:871–883.

Webster, J. R., Waide, J. B., and Patten, B. C. 1975. Nutrient recycling and the stability of ecosystems. In Howell, F. G., Gentry, J. B., and Smith, M. H. (eds.), Mineral Cycling in Southeastern Ecosystems. ERDA Sympos. Ser., Augusta, Georgia, May 1–3, 1974, CONF-7440513, NTIS. Springfield, Virginia, pp. 1–27.

26

Network Growth and Development: Ascendancy

ROBERT E. ULANOWICZ

University of Maryland
Chesapeake Biological Laboratory
Solomons, Maryland

We have only recently begun a revolution in ecology . . . [Van Dyne, in The Ecosystem Concept in Natural Resource Management, *Ch. X p. 333 (1969), Academic Press].*

INTRODUCTION

The ecological world, when viewed at the scale of human senses, is a myriad of seemingly random and arbitrary events. We certainly cannot dismiss out of hand the contention that this welter lacks anything akin to an overall organization (Simberloff, 1980). Yet the intuition exists in many, nurtured perhaps by centuries of reflection by artists, poets, and transcendentalist and natural philosophers, that the larger biological realm possesses some degree of coherence, order, or even organization. This issue is emotionally charged and capable of generating heated debate. However, most of the controversy is at best marginally quantitative. As in other chapters of this book, the goal here is to introduce a set of quantitative tools that can be applied to resolving the ongoing disagreement.

It is worth noting that if we existed (like Maxwell's demon) at the molecular scale, the universe would appear extremely chaotic. Uncountable numbers of molecules would streak around, rotate, and otherwise gyrate in a fashion difficult to predict. However, very large collections of these same molecules when observed at the scale of the natural senses often obey deterministic laws (for example, the ideal gas laws). An analogy to this thermodynamic situation of fine-scale chaos but order in the larger domain might be drawn in the ecological realm. But where should one begin?

PROBABILITIES ASSOCIATED WITH DYNAMICS

Probability theory, that branch of mathematics devoted to making quantitative statements on events about which we are uncertain, seems like the most natural starting point. After all, probability theory and statistics are primary tools used in quantitative biology. Similarly, they are the foundations for statistical mechanics, the attempt to reconcile atomic theory with the phenomenology of thermodynamics.

Well-defined probabilities exist only in relation to well-posed questions. A well-posed question is one for which the outcome is exclusive and the set of possible outcomes is complete. By exclusive is meant an outcome that can occur in only one of the stated options. A complete set of outcomes is simply one that spans the entire gamut of possible outcomes; that is, there is perfect accountability. For example, say that 20 types of organisms constitute the biota of a given ecosystem. An organism is chosen at random. It must belong to at most one of the chosen categories (exclusivity); otherwise, the set of categories needs to be reduced. Furthermore, it must also belong to at least one of the categories (exhaustiveness); otherwise, the set of categories should be expanded. "What is the probability that the chosen species belongs to taxon x" is a well-posed question only when the set of categories (taxa) is both exclusive and complete.

Usually, we apply probabilities to objects at or very near thermodynamic equilibrium, for example, different colored balls in a jar or the faces of a die. But living systems, and ecosystems in particular, are never such static entities. Members of the various taxa are constantly appearing and disappearing and, what is even more interesting, they undergo transformation from one category to another. These circumstances have several consequences for the application of probabilities to ecosystem dynamics.

Almost by definition, if we are concerned with ecological dynamics, the focus is not what is *in* the categories, but rather on what is *entering* or *leaving* a taxon. Thus, the expression $p(a_i)$ might be taken to mean the probability that an organism leaves taxon i at a given instant and $p(b_j)$, that an individual enters taxon j an infinitesimal interval later. Now $p(a_i)$ and $p(b_j)$ are usually not independent of each other. More often than not, a unit leaving one compartment of a set immediately enters another. Such is the case with trophic processes such as herbivory, carnivory, or detritivory. For this reason, it is useful to define a quantity known as the joint probability, $p(a_i, b_j)$, as the probability that an organism leaves i *and* enters j within some infinitesimally short interval. It should be stressed that the joint probability is not generally equal to the product of the separate probabilities [that is, $p(a_i) \cdot p(b_j)$]. In fact, the degree of this inequality is related to how well "organized" the given dynamics appear.

Like all well-defined probabilities, the joint probability should be complete; that is, any potential transition must proceed from one species in the set to another. Of course, for any given ensemble (ecosystem) of n compartments, not all exchanges originate among or transfer into one of the identified categories. Hence, any analysis must contain at least one compartment to represent the external world, the place from which and toward which these exogenous transfers proceed. For reasons later to become clear, it is useful to identify three such external categories: (1) a category 0 (zero) to serve as the source of all entities entering j, but not originating in one of the n taxa; (2) a category $n + 1$ to represent the destination of all useful things leaving the system; and (3) a category $n + 2$ to receive all units that are lost from the system and of no further use to any other similar system; that is, they are "dissipated" by the originating compartment. Examples of transitions that might occur under this categorization scheme are depicted schematically in Fig. 26.1.

Because $p(a_i)$, $p(b_j)$, and $p(a_i, b_j)$ are each complete, the former probabilities can now be written as marginal sums of the latter:

$$p(a_i) = \sum_{j=0}^{n+2} p(a_i, b_j), \qquad (26.1)$$

$$p(b_j) = \sum_{i=0}^{n+2} p(a_i, b_j). \qquad (26.2)$$

Also, by ensuring that the joint probabilities are complete, the way is paved for full accountability

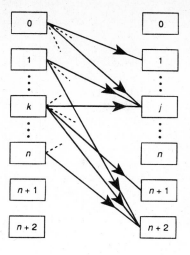

Figure 26.1 Representation of flows among the compartments of a system. Compartment 0 represents the source of all exogenous inputs, $n + 1$ the sink for all usable exports, and $n + 2$ the sink for all dissipations. Boxes on the left represent system nodes at a given time and those on the right depict the same nodes an infinitesimally short time thereafter.

of all transfers regardless of whether or not the organisms (or other appropriate units) that comprise each category are themselves conserved through time.

UNCERTAINTY

Recalling that the pivotal issue addressed in this chapter is whether ecosystems can be considered organized in any sense of the word, it is useful now to consider how to quantify the negative hypothesis—the degree to which an ecosystem is disorganized, or disordered. We are usually very uncertain about the outcome of any event occurring at the "microscopic" level of a disorganized, chaotic, or random assembly of organisms. To be more precise, uncertainty rises in proportion to the number of factors that serve to differentiate the various possible outcomes. Mathematically, this last statement is equivalent to

$$H = -K \sum_{i=1}^{m} p_i \log p_i, \qquad (26.3)$$

where H is the uncertainty attached to the distribution of probability over the m categories enumerated by the index i, and K is a scalar constant.

Equation (26.3) is not as mysterious as it is often portrayed to be, providing we focus on the phrase "number of factors that serve to differentiate" in the statement above. Suppose, for example, that we were challenged to guess an integer chosen from the range 1 to 1024. One straightforward way of targeting the choice would be to ask if the number exceeds 512. If the answer is no, the next query might be if the number exceeds 256. And so on. Using this tactic, we will determine the number after exactly 10 guesses.

Instead of guessing numbers, we might be asked to identify an organism that belongs to one of 1024 possible species that are differentiated by 10 binary choices in a taxonomic key. Again, 10 decisions will identify the organism. The 10 selections may be said to have generated the 1024 categories, each of which may be thought of as a unique combination of the constituent choices in the key. The logarithm of 1024 (the number of "combinations") to the base 2 is 10 (the number of decisions). In the case where all final categories are equally populated, the probability of any single category p_i is 1/1024, so

$$H = -K \sum_{i=1}^{1024} \frac{1}{1024} \log_2 \frac{1}{1024} \quad (26.4a)$$

$$= 10K. \quad (26.4b)$$

In the event that categories are not evenly populated, the combination (26.3) can be thought of as a probability-weighted average of $-K \log p_i$ (or, alternatively, $K \log[1/p_i]$) yielding the average uncertainty.

The numerical results of eq. (26.3) accord well with intuition. If all members are in a single category, then $H = 0$, reflecting no uncertainty. Conversely, uncertainty is maximal if units are uniformly distributed among m categories, and $H = K \log m$. Thus, the range of function (25.3) is

$$K \log m \geq H \geq 0. \quad (26.5)$$

ORGANIZATION

The extent to which a system might be disorganized is quantified by its measured uncertainty. Concerning its outputs there is the amount

$$H(a) = -K \sum_{i=0}^{n+2} p(a_i) \log p(a_i), \quad (26.6)$$

and pertaining to inputs there is the quantity

$$H(b) = -K \sum_{j=0}^{n+2} p(b_j) \log p(b_j). \quad (26.7)$$

Hence, the total uncertainty about the system's dynamics becomes $H(a) + H(b)$. If the inputs and outputs were completely independent of each other, this last sum would exactly measure the disorganization inherent in the dynamic structure.

However, inputs and outputs are not always independent, and the amount by which they are coupled quantifies the coherency of the dynamic structure. If we observe the joint behavior of inputs and outputs over a sufficient interval, it becomes possible to estimate the set of joint probabilities $p(a_i, b_j)$. The uncertainty associated with this set is

$$H(a, b) = -K \sum_{i=0}^{n+2} \sum_{j=0}^{n+2} p(a_i, b_j) \log p(a_i, b_j). \quad (26.8)$$

It can be proved that this joint uncertainty is always less than or equal to the sum of the separate uncertainties, $H(a) + H(b)$. In fact, equality pertains only to the case when inputs and outputs are independent of each other. Under the assumption of independence,

$$p(a_i, b_j) = p(a_i) \cdot p(b_j), \quad (26.9)$$

and substitution of (26.9) into eq. (26.8) along with the completeness requirements,

$$\sum_{i=0}^{n+2} p(a_i) = \sum_{i=0}^{n+2} p(b_j) = 1, \quad (26.10)$$

reveals that, when inputs and outputs are completely independent,

$$H(a, b) = H(a) + H(b). \tag{26.11}$$

More generally, however,

$$H(a) + H(b) > H(a, b), \tag{26.12}$$

and the amount by which the sum of the separate uncertainties exceeds the joint uncertainty is the measure of how coherent or organized the dynamic structure is:

$$A(a; b) = H(a) + H(b) - H(a, b), \tag{26.13}$$

where $A(a; b)$ defines the decrease in uncertainty or the degree of organization inherent in the dynamic structure. Substituting (26.8) into the right side of (26.13) and remembering (26.1) and (26.2) yields

$$A(a; b) = K \sum_{i=0}^{n+2} \sum_{j=0}^{n+2} p(a_i, b_j) \log \left[\frac{p(a_i, b_j)}{p(a_i)p(b_j)} \right]. \tag{26.14}$$

Inspection of (26.14) or (26.13) shows that the order of inputs and outputs in the argument of A is immaterial; that is, $A(a; b) = A(b; a)$.

ORGANIZATION, INFORMATION, AND DEVELOPMENT

Properly speaking, any decrease in uncertainty about or within a system can be defined as "information." As (26.13) reckons a decrease in uncertainty, it describes a quantity called the *average mutual information*. The adjective "average" comes from the averaging technique used in defining the component uncertainties in eq. (26.14), whereas the modifier "mutual" is meant to highlight the fact that A is entirely symmetric with regard to inputs and outputs.

A second and equivalent definition of *information* is given by Tribus and McIrvine (1971) as "anything which gives rise to a change in probability assignment." Thus, if nothing is known a priori about the joint behaviors of the a_i's and b_j's, then we are forced to fall back on the assumption of independence; that is, the joint probability is equated to $p(a_i)p(b_j)$. However, after actually empirically measuring the joint probabilities, they become $p(a_i, b_j)$. The information associated with this change in probability assignment is again measured by eq. (26.14).

This second definition of information emphasizes that information theory arises as a natural outgrowth of probability theory. Probability theory by itself is sufficient to analyze static or equilibrium configurations. However, when we attempt to quantify dynamic systems, the probability assignments, by definition, are subject to change. The analysis of any change in probabilities is the domain of information theory. Thus, invoking information theory to study ecosystem dynamics is not the capricious or ad hoc action many ecologists regard it to be. Rather, resorting to information theory is seen to be just as *imperative* to the study of biology as relying on probability theory! This crucial fact has been obscured by the historical accident whereby information theory was originally formulated in terms of communication theory. Once the universal nature of the information concept is more widely appreciated, resistance to the utilization of information theory in ecology should vanish.

The organization of the dynamic structure of an ecosystem has been properly quantified in terms of information variables. It is thus but a small additional step to quantify the notion of development. *Development* is an increase in organization. Any rise in $A(a; b)$ reflects development, and any decrease is a step toward incoherence and chaotic behavior.

The possibility of assigning a number to the heretofore intuitive concept of development is heartening. It allows us to recast the issue of whether ecosystems are organized as a statistical hypothesis. Kullback (1959) showed how the average mutual information, A, behaves asymptotically like the chi-square function with $(n + 2)^2$ degrees of freedom. Testing whether any observed distribution $p(a_i, b_j)$ reveals dynamical organization to any specified confidence level becomes a matter of rote substitution into tabulated formulas—a typical exercise in statistics.

ESTIMATING TRANSFER PROBABILITIES

Thus far, not much has been said about how to estimate the probabilities $p(a_i)$, $p(b_j)$, and $p(a_i, b_j)$. From eqs. (26.1) and (26.2), it can be seen that it is necessary to estimate only the joint probabilities, because the marginal probabilities are partial sums of these quantities. Usually, we estimate joint probabilities from a matrix of events. That is, the events a_i (a unit leaves compartment i) might identify the rows of the matrix, and the events b_j (a unit enters compartment j), its columns. If the events are reckoned as numbers of organisms, then the observation that one individual of i is eaten by an individual of j would constitute one entry into the (i, j)th position of the events matrix. Calling the cumulative observations x_{ij} and the total number of observations $N = \sum_{i=0}^{n+2} \sum_{j=0}^{n+2} x_{ij}$, the relative frequency of each transfer, x_{ij}/N, becomes the conventional estimator of $p(a_i, b_j)$.

The author is unaware of anyone having calculated the mutual information of ecosystem dynamics from probabilities based on numbers of individuals transferred, although the prospect is intriguing. It is possible that such an exercise might lead to significant insights into the size–frequency distributions (for example, Preston, 1948) commonly observed in nature.

The problem with using numbers of organisms to estimate probabilities is that the disparity in size and makeup of the individuals of various species brings into question the relative significance of each row in the events matrix. To circumvent this issue, ecosystem analysts have taken to choosing a common elemental currency, for example, discrete units (atoms or molecules) of carbon, nitrogen, phosphorus or their compounds, or kilocalories of chemical energy, contained in the organisms being transferred. A second, very significant convenience afforded by this choice is that these currencies are strictly conserved. Whereas individuals appear and disappear, their elemental constituents can be traced intact from compartment to compartment. On the negative side, in the shift from counting organisms to measuring transfers of material among compartments, there is a natural tendency to regard the conveyances of medium as "flows" that are continuous in time. True enough, when the organisms that embody these materials are very small and their transfers very frequent, the transformations do approximate continuous flows. However, we should never lose sight of the fact that ecosystem transfers occur mostly in discrete steps, and to measure the "flow" of, say, carbon from hares to foxes is, in essence, to observe a discontinuous process best treated by probability theory.

With this caveat in mind, attention now focuses on the flows of material from species i to species j. Call such a transfer T_{ij}. The total amount of flow occurring in the system, T, becomes simply the sum of all the individual transfers:

$$T = \sum_{i=0}^{n+2} \sum_{j=0}^{n+2} T_{ij}. \qquad (26.15)$$

Among all the activity occurring in the system, the probability of observing an atom of currency going from i to j is estimated by the fraction of the total activity that is comprised by T_{ij}:

$$p(a_i, b_j) \sim \frac{T_{ij}}{T}. \qquad (26.16)$$

By (26.1) and (26.2),

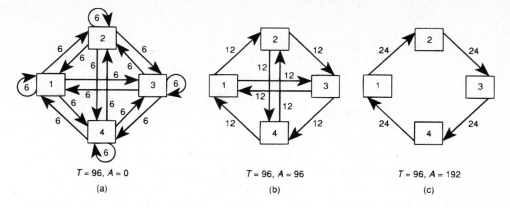

Figure 26.2 Three artificial, closed networks with differing degrees of articulation: (a) the minimally articulated configuration of 96 flow units among four nodes; (b) an intermediate level of articulation; (c) the maximally articulated configuration of flows.

$$p(a_i) \sim \sum_{j=0}^{n+2} \frac{T_{ij}}{T}, \qquad (26.17)$$

and

$$p(b_j) \sim \sum_{i=0}^{n+2} \frac{T_{ij}}{T}. \qquad (26.18)$$

Thus, substituting expressions (26.16) and (26.17) into (26.14) and simplifying,

$$A = \frac{K}{T} \sum_{j=0}^{n+2} \sum_{i=0}^{n+2} T_{ij} \log \left[\frac{T_{ij} T}{\left(\sum_{r=0}^{n+2} T_{rj} \right) \left(\sum_{s=0}^{n+2} T_{is} \right)} \right]. \qquad (26.19)$$

Equation (26.19) can be applied to any well-defined network of ecosystem "flows." It turns out that the measure A quantifies the average degree of articulation inherent in the network. For a network to be well articulated means that, if an atom of conservative substance is flowing through any particular node in the system, it is almost certain to which particular node that atom will next be transferred. By contrast, quanta at any node in a highly unarticulated network can flow almost anywhere during the next transfer. Figure 26.2 shows three networks with different degrees of articulation. In Fig. 26.2a, the network is completely unarticulated and $A = 0$. At the other extreme, the network in Fig. 26.2c is maximally articulated and A equals two units of K. Network 26.2b is intermediate between the two extremes.

SIZE AND GROWTH

When we focus on system dynamics (as opposed to a static distribution), the concept of size literally takes on a new dimension. The "size" of a dynamic system is best captured by the *amount of activity* that is occurring. If this notion sounds strange, recall that the sizes of economic communities are commonly gauged by their aggregate levels of activity, for example, their gross national products

(GNP). An analogous measure for ecosystems has already been defined as T, elsewhere called the *total system throughput* (Hannon, 1973; Finn, 1976).

Growth, in the extensive sense of the word, may be identified with any increase in the size of a system. Hence, system growth can be measured by an increase in T, just as the growth of a national economy is said to be any increase in its GNP.

UNITARY GROWTH AND DEVELOPMENT

The organization, or articulation, of a network is an intensive property of the system. That is, it is independent of system size. The dimensions of A are the same as its scalar constant, K. Usually, K is considered to be fixed by the choice of the base of logarithms used in the calculation. For example, when 2 is chosen as the logarithmic base, one unit of K is called a *bit* (binary digit). When the base e is used, a unit of K is called a *nat*; with base 10, it is termed a *hartley*; and so on.

However, Tribus (1961) suggests that the scalar constant should be used, as its name implies, to scale any intensive system measure. It has just been argued that the appropriate size measure for a flow system is its total system throughput, T. Hence, in setting K equal to T we scale the organization of a dynamic system by its size. The resultant quantity

$$A = \sum_{i=0}^{n+2} \sum_{j=0}^{n+2} T_{ij} \log \left[\frac{T_{ij} T}{\left(\sum_{r=0}^{n+2} T_{rj} \right) \left(\sum_{s=0}^{n+2} T_{is} \right)} \right] \quad (26.20)$$

becomes the product of a factor of size times a measure of organization. This resulting quantity has been termed the system *ascendancy* (Ulanowicz, 1980). It measures the ability of a system to prevail over alternative system configurations. It also measures, in a manner different from exergy (Chapters 22 and 23), the distance of a system from its most chaotic configuration, thermodynamic equilibrium.

To prevail over another configuration requires a propitious combination of size and development. A system that is big but undeveloped is a Golliath. One that is highly developed but very small is vulnerable to being extinguished by its less developed but larger neighbors. Both size and organization are required in adequate measures to produce high ascendancy.

Growth is an increase in system size; *development* is an augmentation of its organization. Thus, an increase in system ascendancy is taken to portray the *unitary* process of growth and development. That growth and development are two aspects of a single process is revealed by reference to any dictionary of the English language, where the definitions for the two characteristics can be seen to overlap significantly.

The means are now at hand to assess whether an ecosystem has undergone growth and development. Its underlying networks of carbon and the like or energy flows are estimated at two points in time. Comparing the two associated values of ascendancy will reveal whether or not growth and development have occurred during the interval.

Odum (1969) suggested 24 attributes to characterize more "mature" ecosystems. Ulanowicz (1986a) has shown how increases in most of these characteristics parallel increases in system ascendancy. This is not to imply that all increases in ascendancy are necessarily beneficial to the system or to human society interacting with it. For example, an ecosystem might react to a sudden increase in available resources by rapidly expanding in size (T), at the same time diminishing in organizational status (A/T). If the former increase more than compensates for the latter drop, the product will still rise. Such a situation has been offered as a quantitative description of the process of eutrophication (Ulanowicz, 1986b).

LIMITS TO GROWTH AND DEVELOPMENT

Growth and development, like any natural process, possesses finite limits. The lower limit has already been discussed, that is, when a_i and b_j are completely independent so that $p(a_i, b_j) = p(a_i)p(b_j)$ and $A = 0$. At the other extreme, a_i and b_j are mutually determined, so knowing a_i implies a corresponding b_j for some combination of n unique pairs (i, j). Under such conditions, $p(a_i, b_j) = p(a_i) = p(b_j)$ for each of those given pairs (i, j), and $p(a_i, b_j) = 0$ otherwise. Substituting these conditions into eq. (26.14) yields

$$C = -K \sum_{i=0}^{n+2} p(a_i) \log p(a_i), \tag{26.21a}$$

$$C = -K \sum_{j=0}^{n+2} p(b_j) \log p(b_j), \tag{26.21b}$$

where C is the maximum value that A can attain and is therefore called the *development capacity*. Notice that under these extreme circumstances $H(a, b)$, $H(a)$, and $H(b)$ all become identical in value to C. To preserve symmetry, it is best in general to identify C with $H(a, b)$ (Ulanowicz and Norden, 1990). In terms of flow estimators, the capacity (26.21a) is calculated as

$$C = -\sum_{i=0}^{n+2} \sum_{j=0}^{n+2} T_{ij} \log \frac{T_{ij}}{T}. \tag{26.22}$$

We expect that

$$C \geq A \geq 0, \tag{26.23}$$

as can be proved algebraically (McEliece, 1977). The nonnegative difference, $C - A$, is termed the system *overhead* and is generated by all real conditions that keep the system ascendancy from reaching its theoretical maximum.

Before considering in more detail the conditions that can contribute to system overhead, it is worthwhile to pause briefly and identify some of the mechanisms that contribute to an increase in ascendancy. In Chapter 4 of Ulanowicz (1986a), it was argued that the principal agent for the rise in ascendancy and the genesis of selection pressure in ecological networks is positive feedback among the members of the community (Wicken, 1984). In those systems that are alive, the ascendancy (and capacity) is dominated by the terms associated with the n members interior to the system.

To understand the nature of the overhead in more detail, we may decompose it algebraically into four separate terms:

$$C - A = I + E + S + R, \tag{26.24}$$

where

$$I = \sum_{j=1}^{n} T_{0j} \log \left[\frac{T_{0j}^2}{\left(\sum_{k=1}^{n} T_{0k}\right)\left(\sum_{m=0}^{n} T_{mj}\right)} \right] \tag{26.25}$$

$$E = -\sum_{i=1}^{n} T_{i,n+1} \log \left[\frac{T_{i,n+1}^2}{\left(\sum_{k=1}^{n+2} T_{ik}\right)\left(\sum_{m=1}^{n} T_{m,n+1}\right)} \right] \tag{26.26}$$

$$S = -\sum_{i=1}^{n} T_{i,n+2} \log \left[\frac{T_{i,n+2}^2}{\left(\sum_{k=1}^{n+2} T_{ik}\right)\left(\sum_{m=1}^{n} T_{m,n+2}\right)} \right] \qquad (26.27)$$

$$R = -\sum_{i,j=1}^{n} T_{ij} \log \left[\frac{T_{ij}^2}{\left(\sum_{k=1}^{n+2} T_{ik}\right)\left(\sum_{m=0}^{n} T_{mj}\right)} \right]. \qquad (26.28)$$

Notice that I is generated by any multiplicity in the exogenous inputs, while E, S, and R correspond to multiplicities in the exports, dissipations, and internal transfers, respectively. Elsewhere (Ulanowicz, 1980) I have called E the "tribute" to other systems, S the "dissipation," and R the internal "redundancy," that is, the multiplicity of internal pathways connecting any two compartments in the system (Rutledge, Bassore, and Mulholland, 1976).

When eq. (26.24) is written in the form

$$A = C - (I + E + S + R), \qquad (26.29)$$

a verbal narrative for the limits to increasing internal ascendancy (and thereby the overall ascendancy) becomes possible: Any increase in A is occasioned by either an increase in C, a decrease in one of the terms in the internal overhead, or some combination thereof. C, in turn, is augmented either by an increase in the scaling factor T or by an ever-finer partitioning of flow among an increasing number of nodes (Ulanowicz, 1987). The total system throughput will rise when species are maximizing their power throughput, a nonconservative strategy for survival first advocated by A. J. Lotka and later championed by H. T. Odum (Odum and Pinkerton, 1955). However, the combination of finite input flows and mandatory dissipations at each node serves ultimately to limit the rise of T. The ever-finer partitioning of compartments is likewise limited by the finite availability of resources, which implies that some of the finely partitioned nodes will inevitably become too small to persist in the face of chance environmental perturbations.

It might at first seem counterintuitive that inputs (which are required for sustenance) would contribute to the system overhead. Recall, however, that in developing systems there is a tendency for the magnitude of inputs relative to other system flows to decrease. As Odum (1969) points out, mature systems are predominated by internal activities. As for increasing ascendancy by decreasing the multiplicity of inputs, such overreliance on but few lines of sustenance jeopardizes system maintenance in the face of disruption of those remaining inputs.

Minimizing the tribute by internalizing (recycling) exports is a good way for a system to increase its ascendancy. However, that course of action, too, can have its limits. For if the exports and imports of a given system both happen to be elements in a positive cybernetic loop at some higher hierarchical level, then decreasing exports from the given system might eventually diminish its own sustenance. Such a scenario was played out on the world economic stage with the collapse of the oil cartels in the late 1970s.

Minimizing the dissipation, S, provides an analog to an illustrious principle of irreversible thermodynamics (Prigogine, 1945). However, as long as resources remain abundant, it is unlikely that the system will follow such a course, because A is more readily increased by a growing T and a widening gap between capacity and overhead. Later, however, after limitations on resources become more stringent, minimizing S becomes an appropriate route for increasing A in mature systems. Of course, S is prohibited from ever reaching zero by the second law of thermodynamics.

Finally, decreasing R reflects a more streamlined and efficient network topology. It is a natural consequence of the competition between overlapping cycles of positive feedback. Unfortunately, more efficient networks also make for more fragile structures. In systems with insufficient R, perturbations at any point in the network are likely to have disastrous consequences on downstream nodes, whereas a modicum of redundant pathways might allow for compensatory flows to downstream nodes via the less affected lines of communication (Odum, 1953). Because real environments always impose some degree of perturbation on a system, continued survival of the system will always require a nonzero level of pathway redundancy to function as "strength in reserve." We anticipate that, over the long term, the amount of redundancy retained by the network will be just sufficient to balance the rigor of the environment. In these terms, the old balance-of-nature precepts now widely rejected by many modern ecologists begin to take on a new dimension and significance.

In summary, there are hierarchical, thermodynamic, environmental, and resource-related constraints acting to retard any increase in internal ascendancy.

A SIMPLE EXAMPLE

An example of an ecological flow network found in numerous other publications is a description (Patten et al., 1976, pp. 572–574) of the energy flows among five functional components of Cone Spring, a small cold-water spring in Iowa. The compartments are depicted, along with their attendant flows (in kcal m^{-2} y^{-1}), in Fig. 26.3. Arrows not originating in a compartment portray exogenous inputs; those not terminating in a compartment denote endogenous outputs of useful energy to the system's environment; ground symbols represent dissipative flows.

These same transfers may be arrrayed in the form of an 8 × 8 matrix (Table 26.1), where energy flows from row to column designations, for example, T_{23} = 5205 kcal m^{-2} y^{-1} transferred from node 2 to 3. Inputs are arrayed along the 0 row, useful outputs down column 6, and dissipations down column 7. Substituting these T_{ij} values into eqs. (26.15), (26.20), (26.22), and (26.25) through (26.28) yields values for the total system throughput, ascendancy, capacity, and the overhead terms, respectively. These values are listed in Table 26.2 for the Cone Spring model.

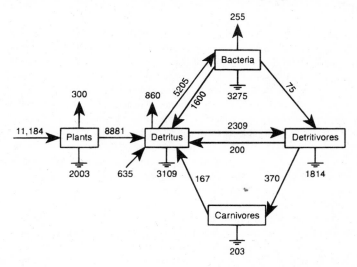

Figure 26.3 Schematic of energy flows (kcal m^{-2} y^{-1}) among functional components of the Cone Spring ecosystem. Arrows not originating from a box represent inputs from outside the system. Arrows not terminating in a compartment represent exports of usable energy out of the system. Ground symbols represent dissipations.

TABLE 26.1 Flows T_{ij} in the Cone Spring ecosystem model

	0	1	2	3	4	5	6	7
0	0	11,184	635	0	0	0	0	0
1	0	0	8,881	0	0	0	300	2,003
2	0	0	0	5,205	2,309	0	860	3,109
3	0	0	1,600	0	75	0	255	3,275
4	0	0	200	0	0	370	0	1,814
5	0	0	167	0	0	0	0	203
6	0	0	0	0	0	0	0	0
7	0	0	0	0	0	0	0	0

Row i contributes to column j. All values in kcal m^{-2} y^{-1}.

TABLE 26.2 Calculated values of ascendancy and related variables for the Cone Spring ecosystem model

Total system throughput (T)	42,445
Ascendancy (A)	56,725
Capacity (C)	93,172
Input overhead (I)	6,222
Tribute (E)	7,811
Dissipation (S)	35,274
Redundancy (R)	29,832

Units: total system throughput, kcal m^{-2} y^{-1}; all others, kcal-bits m^{-2} y^{-1}.

CONCLUDING REMARKS

The preceding development has been intended to provide a quantitative basis for investigating whether or not ecosystems undergo anything akin to growth and development. The origins and place of the described measures in the domain of probability theory have been stressed in the belief that both sides in any dialogue on the issues involved will accept definitions cast in probabilistic terms. The amount of data necessary to study the behavior of the variables introduced here for realistically modeled systems would be voluminous indeed. Hundreds of ecosystem networks and their time series would have to be examined before any outcome would become apparent. And it would require considerable effort to quantify even one such network of the scope of complexity suggested, for example, in Chapter 20. But the issue, bearing on the nature of supraorganismal organization, ranks today as one of the most philosophically intriguing in all science.

It is exciting to speculate what might happen if the evidence were to favor existence of organized behavior in ecosystems. Would such behavior be reducible to events occurring at the level of the organism or smaller, or is it conceivable that the larger scale phenomena are to a degree autonomous like their living constituent parts appear to be? At first, as embedded parts ourselves, it may seem difficult to imagine how such higher-level autonomy could arise, and even more difficult to concede that we are governed by it. But, somewhere in the feedback structure of complex reticulated relationships between subsidiary units in ecosystems may reside a cybernetic organizing principle capable of being drawn as an appropriate agent behind (semi)autonomous growth and development at the ecosystem level (Ulanowicz, 1986a; Wicken, 1984).

Nothing less than the status of ecology as a science is at stake in this dialogue. If ecosystems are determined by their molecular constituents, then ecology is clearly a corollary discipline to genetics and molecular biology. If, however, autonomous elements of ecosystems can be clearly identified, then ecology takes on an importance in its own right. Even more to the point, the discovery of autonomous ecosystem behavior would stand as a significant advance in the theory of far-from-equilibrium thermodynamics and catapult ecosystem science into the very forefront of scientific inquiry—which is where it has always belonged!

REFERENCES

FINN, J. T. 1976. Measures of ecosystem structure and function derived from analysis of flows. *J. Theor. Biol.* 56:363–380.

HANNON, B. 1973. The structure of ecosystems. *J. Theor. Biol.* 41:535–546.

KULLBACK, S. 1959. *Information Theory and Statistics.* Wiley, New York. 395 pp.

MCELIECE, R. J. 1977. *The Theory of Information and Coding.* Addison-Wesley, Reading, Massachusetts. 302 pp.

ODUM, E. P. 1953. *Fundamentals of Ecology.* Saunders, Philadelphia. 384 pp.

ODUM, E. P. 1969. The strategy of ecosystem development. *Science* 164:262–270.

ODUM, H. T., and PINKERTON, R. C. 1955. Time's speed regulator: the optimum efficiency for maximum power output in physical and biological systems. *Amer. Scient.* 43:331–343.

PATTEN, B. C., BOSSERMAN, R. W., FINN, J. T., and CALE, W. G. 1976. Propagation of cause in ecosystems. In Patten, B. C. (ed.), *Systems Analysis and Simulation in Ecology,* Vol. 4. Academic Press, New York. 593 pp.

PRESTON, F. W. 1948. The commonness and rarity of species. *Ecology* 29:254–283.

PRIGOGINE, I. 1945. Moderation et transformations irreversibles des systemes ouverts. *Bull. Class. Sci. Acad. Roy. Belg.* 31:600–606.

RUTLEDGE, R. W., BASORE, B. L., and MULHOLLAND, R. J. 1976. Ecological stability: an information theory viewpoint. *J. Theor. Biol.* 57:355–371.

SIMBERLOFF, D. 1980. A succession of paradigms in ecology: essentialism to materialism and probabilism. *Synthese* 43:3–39.

TRIBUS, M. 1961. Information theory as the basis for thermostatics and thermodynamics. *J. Appl. Mech.* 28:1–8.

TRIBUS, M., and MCIRVINE, E. C. 1971. Energy and information. *Scient. Amer.* 225(3):179–188.

ULANOWICZ, R. E. 1980. An hypothesis on the development of natural communities. *J. Theor. Biol.* 85:223–245.

ULANOWICZ, R. E. 1986a. *Growth and Development, Ecosystems Phenomenology.* Springer-Verlag, New York. 203 pp.

ULANOWICZ, R. E. 1986b. A phenomenological perspective of ecological development. In Poston, T. M., and Purdy, R. (eds.), *Aquatic Toxicology and Environmental Fate,* Vol. 9. American Society for Testing and Materials, Philadelphia, pp. 73–81.

ULANOWICZ, R. E. 1987. Growth and development: variational principles reconsidered. *European J. Operational Res.* 30:173–178.

ULANOWICZ, R. E., and NORDEN, J. S. 1990. Symmetrical overhead in flow networks. *Int. J. Systems Sci.* 21:429–437.

WICKEN, J. S. 1984. Autocatalytic cycling and self-organization in the ecology of evolution. *Nat. Syst.* 6:119–135.

Name Index

Abrams, D. A., 106, 108
Abrams, P. A., 237, 364, 464
Abrosov, N. S., 108
Acoff, R. L., 171–172
Acosta, A. J., 286
Adachi, N., 349
Adams, S. M., 264, 266
Addy, W. D., 458
Adzhabyan, N. A., 345
Aggrawala, B. D., 106
Agren, G. I., 162
Aho, J., 528
Alekseev, V.V., 346
Alexandrov, G.A., 344
Allen, J.C., 143
Allen, T.F.H., 231, 235, 239, 241, 373, 378, 380
Allison, L.J., 387
Allison, T.P., 106, 108
Almiral, H., 143
Alvarez, W., 54
Alymkulov, E.D., 231, 253, 255
Andersen, D., 521, 524
Andersen, T., 520
Anderson, W.N., 353
Ando, A., 234
Andrews, A., 228, 311–312, 317–319
Andrews, F.G., 516, 523
Aponin, J.M. 108
Applegate, V.C., 399, 405
Arrhenius, S.A., 265
Armstrong, J.S., 426, 444
Armstrong, N.E., 572
Arnold, V.I., 345
Arrow, K.J., 348
Ascher, W., 426, 444
Ashby, W.R., 106, 167, 172, 417–418, 450
Asherin, D.A., 331
Askew, R., 515, 526
Aström, M., 122
Auble, G., 228, 311–312, 317–319, 330–331
Auburn, J.S., 417, 422, 445
Austin, M.P., 106, 108
Ayala, F.J., 107

Baird, D., 517, 520, 540
Bakker, R.T., 460
Bakman, G., 185
Bakule, L., 113, 167
Banks, R.B., 285
Baranov, F.I., 196
Barber, M.C., 486

Barker, G.P., 348
Barko, J.W., 266
Barnes, R.S., 520, 540
Barrow, J.D., 460, 463
Bartell, S.M., 232
Baskerville, G.L., 312, 401, 407
Basore, B.L., 623–624, 652
Bastin, T., 90
Bazilevich, N.I., 387–388
Bazykin, A.D., 106, 108
Beaver, R., 514, 516, 538–540
Beddington, J.R., 107
Beer, A, 265
Beer, S., 417–418
Belis, M., 451
Bellman, R., 19, 122, 145, 177, 445
Belyaev, V.I., 264
Bent, A.C., 518
Berman, A. 348
Berman, S.M., 570
Bernoulli, J., 181
Berrios Loyala, R., 445
Bertalanffy, L. von, 157, 196, 221, 338, 399, 401
Beverton, R.J.H., 196
Biever, L., 516, 520–522, 525
Bird, P.M., 408
Blackburn, T.R., 378
Blau, G.E., 279, 282–283
Bobrow, D.G., 474, 476
Bohr, N., 36, 38–39
Boling, R.H., 264
Boltzmann, L., 41
Bonsdorff, E., 402, 406
Boole, G., 179–180, 182, 184
Bormann, F.H., 379, 382, 400, 403
Bosatta, E., 162
Botkin, D.B., 373, 384, 457
Boucher, D.H., 350
Boulding, K.E., 417
Bourn, D., 73–74
Bowerman, J., 514, 522, 539, 541
Bowers, C.E., 325
Boyarsky, L.L., 453
Boyd, R., 445
Boynton, W., 226, 262–263, 270, 273
Bracken, J., 304
Braudel, F., 418, 445
Brauer, F., 345
Brewer, J., 445
Briand, F., 514–516, 517, 519, 538–540
Brillouin, L., 41
Britton, N.F., 109

Brown, J., 518, 523, 535, 543
Brown, J.M., 104
Brown, J.S., 115, 123
Bruckman, G., 416
Brylinsky, M., 80, 92
Bryson, R., 445
Bumpus, D.F., 45
Bunge, M., 510
Burger, C., 86, 92
Burns, B., 56
Burns, L.A., 279–280, 282
Burns, T.P., 514, 519–521, 540, 550
Bush, A., 528
Butzer, K.W., 417–418
Byrd, R.A., 248

Cairns, J., 96, 404
Calder, W.A. III, 116
Cale, W.G., 225–226, 230–234, 237, 629, 635
Campbell, J.S., 75
Cannon, W.B., 450
Cary, J.R., 418
Casas-Vazques, J., 131
Case, T., 519, 523
Casti, J.L., 108
Cavalli–Sforza, L.L., 43
Cedarwall, M., 404, 408
Čermak, L., 77
Charley, J.L., 374
Charnes, A., 304
Chetaev, N.G., 592–593
Child, G.I., 384
Chipman, J.S., 231
Chu, C.S., 325
Clark, J.M., 420
Clark, W.C., 312
Clark, W.R., 113
Clements, F.E., 6, 52, 90, 95, 340, 453–454
Cleveland, C.J., 416
Coe, M., 73–74
Cohen, B.I., 89
Cohen, J.E. 49, 231, 237, 514–517, 519, 538–540
Cohen, Y., 122, 286–287, 289, 291–292
Colby, J.A., 264
Cole, A.H., 441, 448
Comins, H.W., 107
Confer, N.M., 263, 347
Connell, J.H., 90, 380, 454, 456–457
Conrad, M., 389, 563–565, 609–611, 614, 616–617, 621

Cook, B.G., 106, 108
Cook, E., 445
Cooper, C.F., 273
Cooper, W.W., 304
Cope, E.D., 96
Corning, P.A., 453, 461, 463
Costanza, R., 36–38, 331, 341, 485–486, 494, 499
Coughenour, M.B., 453
Cousins, S., 511, 515, 521, 540, 550, 557
Cowell, E.B., 399
Craig, P.P., 416–418, 420, 422, 445
Crawford, C.S., 515, 518, 520, 523–524, 526, 530–531, 541
Crespi, B.J., 519
Cronon, W., 417–418
Cropper, W.P., 384
Crow, M.E., 106
Cunningham, W.J., 417–418

D'Itri, F.M., 398, 404
Dame, R., 494
Darby, M.R., 426
Darwin, C., 39, 561–562, 564, 568–571, 576–577, 579, 582, 613
Dasmann, R.F., 302, 304
Datta, B.N., 348
David, G., 445
Davidson, D., 518, 523
Davidson, M., 399
Davies, N.B., 116, 122
Davies, P., 86
Davis, L.S., 297–298, 377
Dawkins, R., 115, 462
Day, J.W., 265
Day, R.W., 457
Deamer, D., 445
Dean, T.A., 457
DeAngelis, D.L., 106, 339, 450, 452, 461, 514, 541, 543
de Beer, G.R., 93
de Caprariis, P., 231, 239
de Fermat, P., 85
Demers, S., 47, 382
Dempson, B., 66
Dickinson, F.G., 420
Doman, E.R., 296
Drury, W.H., 457
Dubois, D.M., 106–107
Dubos, R., 22
Duchin, F., 248
Duggins, D., 520–521
Dunbar, M.J., 460
Dunham, S.E., 121
Durham, R.W., 280
Dwyer, R.L., 110, 384
Dyer, G., 419, 425
Dyer, M.I., 461

Easton, D., 450
Ebeling, W., 114, 141, 122
Eberhardt, L.L., 11, 14, 300
Ebling, F.J., 85
Edney, E., 518, 524
Edwards, C.J., 410
Egerton, F.N., 52, 373
Ehman, T.I., 358
Einstein, A., 168–169
Elgar, M.A., 519
Elton, C., 52, 344, 550
Emery, F.E., 171–172
Endler, J.A., 116
Engelberg, J., 453
Estes, J., 520–521
Evans, R.B., 585–586

Fay, S.B., 443
Feistel, R., 114, 122, 141
Feldman, M.W., 43
Fenchel, T., 515, 520
Ferguson, R.L., 264
Ferguson, W., 523, 526
Fey, W.R., 454–455, 457, 459
Feynman, R.P., 86–87
Fiacco, A.V., 301
Finn, J.T., 486, 501, 624, 650
Fisher, N.S., 405
Fleishman, B.S., 31–32, 166–167, 172, 174–175, 177–183, 185, 187, 191–198, 200–201, 203–206, 216, 222
Fletcher, M., 520
Flory, J., 445
Fobes, C.B., 302
Foerster, R.E., 94
Fogel, L.I., 178
Foin, T., 445
Forbes, S.A., 52
Force, D.C., 417–418
Forman, R.T.T., 398, 400–401
Forrester, J.W., 453
Fosberg, F.R., 7
Francis, G.R., 399, 409, 410
Franco, P., 518, 524
Franklin, J.F., 384
Freckman, D., 518, 520, 524–525, 535
Freidlin, M.I., 345
Friend, A.M., 398
Friis, M.B., 572, 578
Frommer, S.I., 516, 522
Funderlic, R., 486
Futuyma, D.J., 115, 123

Gaines, T.C., 420, 441
Galileo,, 169, 172
Gallopin, G., 514, 519, 540
Gantmacher, F.R., 250–251
Gard, T.C., 231
Gardner, M.R., 106, 417–418

Gardner, R.H., 231–232, 235, 238, 629, 635
Gause, G.F., 343, 362
Gaymer, R., 74
Gerritsen, J., 377
Gever, J., 416
Gibbs, J.W., 131–134, 138, 140–141, 146, 588
Gibson, C.W.D., 73
Gier, H., 529
Gilbert, E.N., 185
Gilliam, J., 518
Gini, C., 41, 48
Ginzburg, L.R., 105, 112
Giuliani, D., 516, 523
Glansdorff, F., 140
Glass, B., 11
Glass, N., 445
Glasser, J., 514–515, 540
Gleason, H.J., 95
Gleick, J., 107
Gnaiger, E., 81, 83, 88–89, 93
Gnauck, A., 105, 109, 111, 141, 143, 167
Godron, M., 398, 400–401
Goeden, R., 526–527, 532
Goel, N.S., 347, 358, 590
Goh, B.S., 106, 349, 352
Goldman, C.R., 53
Goldsborough, W.J., 265
Goldstein, J.S., 420
Golley, F.B., 86, 92
Golubic, S., 95
Goodall, D.W., 264
Goodman, D., 55, 344, 373, 417
Gopalsany, W.S.C., 106
Gordon, A.J., 406
Gordon, S., 518, 523
Gorham, E., 406
Gould, S.J., 92–93, 121
Grant, W.E., 248
Gray, T., 520
Green, J.M., 66
Green, R.H., 71–72
Greeney, W.J., 106
Grime, J.P., 456–457
Grodzinski, W., 458
Grossberg, S., 350
Grubb, P., 73–74
Gruber, B., 106
Grumm, H.R., 345
Guckenheimer, J., 107
Gunderson, L., 331
Gurney, W.S.C., 105, 107, 109
Gusen-Zadeh, S.M., 345
Gutiérrez, E., 48, 143
Gutierrez, L.T., 454–455, 457, 459

Haedrich, R.L., 69–71
Haken, H., 142–143, 179
Halfon, E., 227, 279–283, 291–292

Hall, C.A.S., 104, 416
Hall, D.J., 116
Hall, E.T., 82
Hall, J., 516
Halter, A.N., 14
Hamilton, D.B., 228, 311, 318–319, 324, 331
Hamilton, J., 73
Hamilton, W.R., 86
Hammar, T., 520
Hannon, B., 341, 485–486, 488, 490, 493–494, 506, 512, 650
Hansen, R., 523, 533
Hanski, I., 109
Hardy, A.R., 516, 523
Harleman, D.R.F., 265
Harlin, M.M., 266, 274
Harmsen, R., 248
Harner, E.J., 364
Harper, J.L., 83
Harris, L.D., 417
Harris, W.F., 387, 404, 408
Hart, J.L., 94
Hartfiel, D.J., 348
Hartig, J.H., 409–410
Hartman, W.L., 399, 401
Harvey, H.H., 403
Harwell, M.A., 106, 384
Hassel, M.P., 107
Hastings, A., 515, 541
Hastings, H., 611
Hattori, A., 266
Hawkins, B., 514, 526–527, 539–540
Hawks, D., 522
Hay, K.G., 296
Hayek, F.A., 463
Hayes, D., 521
Hayward, G.C., 458–459
Heath, M., 486
Heisenberg, W.K., 29, 32, 34, 38–39, 87, 174
Herendeen, R., 494
Herman, C., 529
Hermann, A., 226, 262
Herodotus, 52
Hessen, D., 520
Hett, J., 486
Heumier, T., 514, 516, 538–540
Heuts, M.J., 68–69
Hicks, J.R., 234, 238
Higashi, M., 510, 514, 519–521, 540, 550
Hilborn, R., 312, 316
Hill, J., 509
Himmelblau, D.W., 301, 305
Hinde, R.A., 171
Hines, W.G.S., 122
Hinson, J.M., 459
Hirata, H., 565–567, 623–624, 629–631
Hirsch, A., 312

Hirst, F.W., 420
Ho, M.W., 86
Hochachka, P.W., 116
Hoekstra, T.W., 378
Hoffman, A., 52
Hogue, C., 523–524
Holdridge, L., 375
Holling, C.S., 21, 139, 191, 228, 311–312, 331, 345, 361, 375, 380, 382, 405, 445
Holm, E., 535
Holmes, J.A., 404
Holt, R.D., 519, 531, 543
Holt, S.T., 196
Holtan, H.N., 270
Homer, M., 551, 554, 635
Hopkinson, C.S., 265
Hubbert, M.K., 445
Hughes, B.B., 416
Hughes, R., 520, 540
Hunt, H.W., 521–522, 546
Hunter, G.N., 296, 321
Hunter, P., 445
Hurd, L.E., 457
Hutchings, J.A., 69–71
Hutchinson, G.E., 106, 376, 384, 399–401, 403, 405, 417–418, 549

Iberall, A., 85, 88
Iizumi, H., 266
Ijiri, Y., 231
Iltis, H.H., 380
Innis, G., 231
Ioos, G., 141
Ivakhnenko, A.G., 178

Jacob, F., 450
Jager, Y., 374, 384, 387–388
James, G.A., 301
Janzen, D., 524, 526, 534
Jeffers, J.N.R., 273
Jeffries, C., 108, 113, 352, 354–355
Jeffries, M., 540
Jenkins, K.B., 14
Jenny, H., 5
Jerne, N.K., 611
Johnson, C.R., 348
Johnson, F.H., 266
Johnson, L., 30–33, 51, 55–56, 86
Johnson, M., 523, 533
Johnson, R.L., 319
Jones, D.D., 312, 345
Jones, J., 520
Jordan, C.F., 453, 462
Jørgensen, S.E., 29–30, 34, 39, 105, 109, 113, 132, 225, 561–564, 568, 571–572, 574, 578, 581–582, 585, 589–590, 592, 595–596
Joseph, D.D., 141
Judson, H.F., 452

Junusov, M.K., 357

Kahn, J.R., 270, 275
Kaiser, K.L.E., 279, 281
Kalff, J., 56, 116
Kaluzny, S.P., 105
Kamp-Nielsen, L., 572
Karlson, R.H., 457
Karr, R.J., 401
Kauffman, J., 410
Kauffman, R.K., 416
Kay, J.J., 410
Keenan, J.H., 91
Keith, L.B., 418
Kemp, M., 226, 262–263, 265–268, 270–275, 418, 551, 554, 635, 638
Kempf, J., 108
Kennedy, C., 528, 544
Kerner, E.H., 106
Kerr, S.R., 83
Khanin, M.A., 105, 125
Kimball, T.L., 296, 302
Kimsey, L., 523, 527, 532
Kindlmann, P., 107
King, W., 106, 116
Kitching, R.L., 361
Klee, V., 354
Kleiber, M., 83, 96
Klein, D.R., 296
Klir, G.J., 474
Kloeden, P.E., 44
Knight, B., 420
Knight, R.L., 453
Koch, A.L., 106, 112
Koestler, A., 91
Kolata, G.B., 55
Kondratieff, N.D., 416
Kornilovsky, A.N., 346
Korostyshevsky, M.A., 114
Kostitzin, V.A., 350
Kot, M., 107, 346
Kotel'nikov, V.A., 167, 172, 176
Kramer, G.R., 285
Krapivin, V.F., 598
Krebs, J.R., 116, 122
Kremer, J.N., 265
Kruckenberg, S., 529
Kuhn, T.S., 236
Kullback, S., 638, 648

Ladde, G.S., 350
Lagrange, J.L., 31, 86, 144, 203, 212–213, 336, 344–345, 347, 365–366
Lamb, H., 445
Lambert, J.H., 265
Lamprecht, Y., 122
Landau, L.D., 243
Lande, R., 115
Lane, D.A., 279
Lang, G.E., 452

Lange, O., 234, 238
Laplante, J.P., 107
La Salle, J.P., 589, 595
Lassiter, R.R., 279–280
Lawler, G. H., 75
Lawlor, L.R., 417–418
Lawton, J.H., 349, 514–515, 526, 538–541, 543
Le Chatelier, H., 563, 570, 582, 586, 597
Leffler, J.W., 384
Lefschetz, S., 589, 595
Legendre, L., 47, 382
Leigh, E.G., 92
Leontief, W.W., 248, 350, 486, 488, 551
Leopold, A.S., 297
Leopold, L.B., 315
Leppakoski, E., 406
Lerner, S., 410
Leslie, P.H., 14
Levin, B.R., 108
Levin, S.A., 104, 107–108, 113, 121–122, 345
Levine, S., 522, 540, 551
Levins, R., 263, 353
Lewontin, R.C., 53, 121, 374
Liapunov, A.M., 31, 105–106, 109, 114, 116, 336, 344–347, 349, 362, 365–366, 375, 381, 562–563, 586, 589–594, 601, 606
Lieth, M., 111
Lifshitz, E.M., 243
Likens, G.E., 379, 382, 387–388, 400
Lindeman, R.L., 453, 511–512, 549–550, 553, 558–559
Lindsey, C.C., 56
Liss, P.S., 283, 285, 291
Loehle, C., 108, 143
Logofet, D.O., 105, 335–336, 343–364, 366, 562, 585, 590, 593, 597, 601, 605
Lootsma, F.A., 432, 446
Lord Rayleigh (J.W. Strutt), 79, 81
Lotka, A.J., 44, 94, 104, 117, 157, 453, 563, 593, 606, 620, 652
Loucks, O.L., 259, 384, 408, 459
Louw, G.N., 520, 524, 529, 535, 541
Lovelock, J.E., 54, 97, 459–460, 462
Lubchenko, J., 457
Luckyanov, N.K., 226, 232, 236, 242, 244–245, 247, 249–250, 253, 255
Lundberg, P., 122
Lutz, H.J., 372
Lyche, A., 520

MacArthur, R.H., 52, 55, 83, 92–94, 344, 362–363, 373, 417, 456, 617, 623

Macfayden, A., 520
Mackay, D., 279–280, 282, 284, 290, 292, 407
MacKay, W., 524
Maelzer, P.A., 4
Maguire, H.F., 301
Maguire, R.J. 282, 292
Maitra, S.C., 347, 358, 590
Makridakis, S., 426, 444
Mann, K.H., 80, 92
Mar, B.W., 273
Margalef, R., 14, 30–31, 40, 45–46, 48, 83, 90, 405, 565, 619, 623
Margulis, L., 459
Markov, A.A., 175, 192–193, 255, 566, 624, 641
Marks, P.L., 384
Marler, R., 529
Mattson, W.J., 458
Matyas, J., 428, 432, 446
Mauersberger, P., 31, 130–131, 135, 138, 141–144, 154, 156–157, 162, 167
Maupertius, M., 86
Maxwell, J.C., 643
May, R.M., 104–108, 143, 346, 350, 353–354, 362–364, 373–374, 380, 417–418, 433, 445, 513–515, 519, 521, 538, 540–541, 543, 553, 620
Mayhew, W.W., 522
Maynard-Smith, J., 107, 122
Mayr, E., 85
McBrayer, J., 518, 524
McCall, P.J., 279, 282
McClean, M., 426, 444
McCormick, G.P., 301, 304
McCormick, S.J., 518–519, 522, 529, 532
McEliece, R.J., 651
McFarland, D., 122
McIntire, C.D., 264
McIntosh, R.P., 52, 405
McIrvine, E.C., 647
McKinnerney, M., 526
McKown, M.P., 232
McLaren, I.A., 56, 452
McManus, M., 348
McNamee, P.J., 319
McNaughton, S.J., 75, 93, 376–377, 382, 388, 417, 453
McPhail, J.D., 56
McRoy, C.P., 267
Meadows, D., 416
Medawar, P.B., 84
Mees, A.I., 44
Mejer, H.F., 132, 571–572, 585, 578, 589–590, 592
Menge, B.A., 220, 457, 541
Menten, L.M., 108, 111, 238, 267, 580

Mercer, E.H., 378
Mesarovic, M., 442
Metcalf, R., 281
Michaelis, L., 108, 111, 238, 267, 580
Michener, C., 516
Miles, L., 456
Milinaki, M., 116, 123
Miller, J.G., 416
Miller, R.B., 56
Minorsky, N., 453
Minsky, M., 474
Mitchell, B.R., 420, 430–432, 442–443
Mitchiner, J., 445
Monk, C.D., 372
Monod, J., 450, 568
Montroll, E.W., 347, 358, 590
Moon, D., 476
Moore, J.C., 521–522, 541, 543
Moore, K.A., 274
Morgan, J.J., 138
Morgenstern, O., 177
Morowitz, H.J., 81, 378, 381, 571, 589–590, 595
Morris, R.F., 417–418
Mosman, N., 445
Moulton, H.G., 420
Moylan, P.J., 350
Mueller, L.D., 107
Mulholland, R.J., 105, 345, 623–624, 652
Murdoch, W.W., 106–107
Murphy, D.A., 297
Myers, C.A., 519, 531, 543
Myrup, L., 445

Najarian, T.O., 265
Namenwirth, J.Z., 419
Navier-Stokes, C., 141
Neely, W.B., 279, 282–283
Neill, C., 486, 499
Nesmerák, I., 110
Newman, C., 514–516, 517, 519, 538–540
Newton, I., 80, 86, 89, 450, 169, 172–173, 474
Neyman, J., 14
Nicholson, A.J., 107
Nicolis, G., 81, 114, 122, 141
Nienhuis, P.H., 264
Nikaido, H., 350
Nisbet, I.C.T., 457
Nisbet, R.M., 105, 107, 109
Nitecki, M.M., 115, 123
Nixon, S.W., 265
Norden, J.S., 651
Northcote, T.G., 91
Noy-Meir, I., 535, 541
Nuessly, G., 532
Nunney, L., 106–108

O'Keefe, R., 476
O'Neill, R.V., 90, 120, 231, 234–235, 237, 248, 373, 376–382, 384, 387, 416–418, 458–460, 463, 486, 629, 635
Oaten, A., 106–107
Odell, P.L., 231, 233–234
Odum, E.P., 9, 80, 83, 90, 263, 373, 399–402, 452–453, 516, 520–522, 525, 617, 619, 650, 652–653
Odum, H.T., 83, 89, 92, 266, 270, 361, 416, 444, 453, 557, 568, 578, 652
Officer, C.B., 270
Ohm, G.S., 86, 136
Oksanen, L., 120
Oliver, B.G., 279–280, 284, 292
Ollason, J.S., 121
Olson, J.S., 21, 387, 453, 457
Onsager, L., 81, 87–89, 137, 156
Oppenheimer, J.R., 452
Orians, G., 523, 535
Orlob, G.T., 141, 143
Ortega y Gasset, J., 445
Orth, R.J., 274
Osman, R.W., 457
Oster, J.F., 107–108
Othmer, D.F., 286
Overholt, W.H., 444
Overton, W.S., 264, 378
Owen, D.F., 551

Pagels, H.R., 54–55
Paine, R.T., 269, 398, 514–515, 517, 521, 540–542
Park, R.A., 280, 286, 291, 320–321
Parry, G.D., 83, 93
Parsons, T.R., 265
Pasvolsky, L., 420
Paterson, S., 279–280, 282
Patrick, R., 406
Pattee, H.H., 378, 381
Patten, B.C., 21, 46, 49, 104, 109–110, 112–115, 123, 143, 231, 263, 270, 372–375, 377–378, 381, 383–385, 388–390, 453, 486, 488–489, 494, 510, 514, 519–521, 539–540, 550, 553, 638–639, 653
Patterson, N., 404
Pauly, D., 404
Pease, J.L., 418
Pella, J., 196
Perez, K.T., 110
Perry, R.H., 285
Pestel, E., 442
Peterman, R., 312
Peters, R.H., 116, 121, 517
Peterson, R., 106
Petraitis, P.S., 364

Petrusewicz, K., 458
Phillips, J., 524
Phillips, M., 136
Phillips,, O.M., 106, 108
Phillipson, J., 458–459
Pianka, E.R., 185, 214, 219, 221
Pichler, F., 476
Pielou, E.C., 273, 344
Pierce, J.R., 88
Pierce, M., 445
Pimm, S.L., 106, 347, 349, 361, 373–374, 376, 464, 514–515, 518, 519, 521, 526, 538–539, 541, 543, 553
Pinkerton, R.C., 89, 578, 652
Pittendrigh, C.S., 85
Planck, M., 34, 87, 91, 138
Platt, J.R., 54–55
Platt, T., 116, 550
Plemmons, R.J., 348
Polanyi, M., 87–88
Polis, G., 510, 513, 517, 518–526, 528–533, 535, 541–542, 553
Pomeroy, L.R., 385
Pontryagin, L.S., 122
Poole, R.W., 107
Post, W.M., 461, 464, 514, 541, 543
Poulson, T.L., 69–70
Powell, J., 523–524
Powell, L.E., 301
Powell, R.A., 106
Powell, T., 45
Preston, F.W., 648
Price, P., 526, 541
Prigogine, I., 30, 52, 81–82, 85, 114, 122, 140–141, 167, 179, 378, 381, 568, 571, 576, 652
Prince, H.H., 398, 404
Pykh, Y.A., 347–350, 593

Quinn, J.F., 121
Quirk, J.P., 348, 350, 353–356

Racsko, P., 598
Radtke, E., 154
Radtke, R., 70, 72
Ragsdale, H.L., 384
Ramundo, R., 521
Rant, Z., 570
Rapaport, A., 612
Rapport, D., 337, 397–401, 403, 405, 408
Rashevsky, N., 174
Rasmussen, D.I., 296
Rastetter, E.B., 231–232
Raup, D.M., 54, 86, 89
Recknagel, F., 112
Regier, H., 337, 397, 399–401, 403–410
Reichle, D.E., 387, 459–460, 463
Reichman, O.J., 518, 523

Reid, W., 526
Reiners, W.A., 54, 84, 92, 96, 388
Rejmánek, M., 107
Rexstad, E., 231
Rice, J., 514, 518, 519, 541
Rich, P., 388, 516
Richards, B.N., 374
Richardson, J., 416
Richerson, P.J., 45
Ricker, W.E., 58, 196, 209, 218, 221
Riget, F.F., 66
Rigler, F.H., 58
Riley, G.A., 45
Riley, M.J., 259
Rizki, M.M., 610
Robinson, J.V., 106
Rodin, L.E., 387–388
Roelle, J., 228, 311–312, 317–319, 330–331
Romesburg, C.H., 301
Rosen, R., 373–374
Rosenzweig, M.L., 108, 460, 462–463
Rostow, W.W., 419, 443
Rotenberry, J.T., 518
Roughgarden, J., 104, 162
Rozenblit, J.W., 480
Rubin, A.B., 141
Ruppert, R., 348, 353–356
Russell, E.S., 92
Rust, B., 231, 234
Rutledge, R.W., 623–624, 652
Ryan, M., 522
Ryder, R.A., 399, 405
Ryti, R., 519, 523

Sabersky, R.H., 286
Sadowski, D.A., 380
Sakai, A.K., 78–79
Sale, K., 407
Salonen, K., 520
Salt, G.W., 237
Salthe, S.N., 85, 120
Salvadora, L., 141
Sampson, J., 341
Samuelson, P.A., 491
Santos, P., 524
Santschi, P.H., 281
Saposnik, R., 350
Saunders, P.T., 86, 92
Schaefer, M.B., 89
Schaffer, W.M., 107, 231, 264, 460, 462
Scheer, B.T., 88
Schelske, C.L., 391
Schindler, D.W., 400, 402–403, 406
Schindler, J.E., 373, 378–379
Schlesinger, W.H., 388
Schoener, A., 457
Schoener, T., 106, 109, 457, 459, 514, 519, 538, 541

Schoenly, K., 514–517, 525–526, 538–540
Schrödinger, E., 592, 594
Schultz, A.M., 4
Schwefel, H.–P., 428, 432, 446
Schwinghamer, P., 116
Scott, E.L., 14
Sculteure, K., 532
Seastedt, T., 521
Seely, M.K., 520, 523–524, 529, 535, 541
Segel, L.A., 107
Seger, J., 516
Selye, H., 398
Semenov, S.M., 174, 196
Semevskiy, F.N., 174, 196
Semina, M.J., 116
Sengupta, J.K., 305, 307
Sepkoski, J.J., 54, 86, 89
Serrin, J., 131
Severinghaus, C.W., 301
Shaffer, W.M., 346
Shannon, C., 41, 47–48, 167, 172, 176, 224, 481, 566
Sheldon, R.W., 116
Shirk, G., 419
Sholtz, C., 535
Short, F.T., 264, 266
Shugart, H.H., 109, 384, 462
Sidorin, A.P., 345
Siljak, D.D., 112
Silverstein, S.D., 570
Silvert, W., 107
Simberloff, D., 643
Simenstad, C., 520–521
Simon, H.A., 90, 234, 264, 378, 381
Simpson, E.H., 41, 48
Simpson, G.G., 86, 89
Sinko, J.W., 107
Sissom, W.D., 519
Sjöberg, S., 108
Sklar, F.H., 36–38, 331
Slatkin, M., 107, 115, 123
Slatyer, R.O., 255, 454, 457
Slobodkin, L.B., 20, 53, 80, 612
Smart, R.M., 266
Smeed, R.J., 445
Smítalová, K., 108
Smith, E.A., 122
Smith, J.L., 107
Smith, W.B., 441
Smith, W.H., 402–403
Smith, W.R., 107
Sober, E., 115
Soholt, L., 518
Somero, G.N., 116
Sommerfeld, A.J.W., 586–587
Soodak, H., 85
Sousa, W.P., 380, 398, 457
Southern, H.N., 76
Southwood, T.R.E., 83, 93

Sparholt, H., 66
Speith, H., 445
Spengler, O., 445
Sprules, W.G., 116, 514, 522, 539, 541
Staddon, J.E.R., 459
Stallman, R., 476
Standing, J.D., 457, 466
Standridge, C.R., 477
Stanley, S.M., 95–96
Starr, T.B., 235, 373, 378
Stearns, S.C., 121
Steedman, R.J., 399, 401, 406–408
Steel, R.G.D., 305
Steele, J.H., 106, 108, 635, 638
Stefan, H.G., 259
Stefik, M.J., 476
Stein, S., 445
Stengers, I., 81–82, 568, 571, 576
Stenlund, M.H., 297
Stenseth, N.H., 106, 121–123
Sterner, R.W., 461
Stevenson, J.C., 263
Stewart, I., 375
Stinton, H., 445
Stoddart, D.M., 85
Stoddart, D.R., 73
Stokes, G.G., 141
Stommel, H., 45
Stone, R., 505
Straškraba, M., 30–31, 104, 109–111, 113, 116, 141, 143, 154, 156, 167, 568
Streeter, H.W., 286
Strefer, W., 107
Stumm, W., 138
Sugihara, G., 231, 235, 238, 514–515, 541, 543, 629
Sukachev, V.N., 4
Sulak, J.H., 78–79
Sutherland, J.P., 380, 457, 541
Sutherland, R.P., 291
Svetlosanov, V.A., 345
Svirezhev, Y.M., 105, 335–336, 343–350, 352–364, 366, 562, 585, 590, 593, 597–598, 601, 605
Swaney, D.P., 453
Swank, W.T., 372, 382, 384
Swartzman, G.L., 105, 227–228, 295, 549
Sweet, C.W., 279
Swingland, I.R., 73–74

Taber, R.D., 302, 304
Taft, J.L., 265
Takeuchi, Y., 349
Tansley, A.G., 4, 52–54, 58
Tarko, A.M., 598
Taylor, J., 514–515, 541
Tchebyshev, P., 195
Telford, S., 528

Temin, P., 419
Thaker, M.S., 286
Thom, R., 108, 160, 345
Thomann, R.V., 279
Thomas, W.R., 107
Thornton, K.W., 105, 345
Thorp, W.L., 420
Thorpe, C., 399
Thursby, G.B., 266, 274
Tinkle, D.W., 186
Tipler, F.J., 460, 463
Titus, J.E., 264, 266
Tokumara, H., 349
Torrie, J.H., 305
Tribus, M., 647, 650
Trombla, A., 515
Tsetlin, M.L., 191
Turgeon, K., 262
Turner, T., 457
Twilley, R.R., 273

Ulanowicz, R.E., 107, 113, 162, 462, 501, 511–512, 517, 520, 540, 549–551, 553, 566–567, 578, 623–624, 629, 631, 638, 643, 650–652, 654
Ulianov, N.B., 354–356
Ul'janov, N.B., 353–356
Usmanov, Z.D., 357

Valentine, W.D., 106
Van den Driessche, P., 354
Vandermond, A., 251
Van der Waerden, B.L., 353
Van Dyne, G.M., 1, 29, 34, 40, 51, 104, 130, 166, 225, 227–230, 242, 262, 279, 295–296, 335, 338–341, 343, 372, 397, 415, 446, 450, 453, 468, 485, 509–510, 513, 549, 561, 566–568, 585, 609, 623, 643
Van Montfrans, J., 264, 269
Van Valen, L., 79
Van Voris, P., 110, 384
Van't Hoff, J.H., 148, 598
Varchenko, A.N., 345
Vaughan, R.E., 77–78
Vaughn, M.R., 418
Ven, V.L., 248
Verhagen, J.H.G., 264
Vermette, S., 279
Vincent, T.L., 115, 123
Vinogradov, B.V., 605
Vitousek, P.M., 384, 462
Vogal, R.J., 398
Voinov, A.A., 258, 346
Vowles, R.H., 418
Volterra, V., 44, 104, 117, 157, 236, 344, 346–350, 352, 358–359, 361, 366, 453, 563, 593, 601, 606, 620

661

von Bertalanffy, L., 157, 196, 221, 338, 399, 401
von Forster, H., 189
von Karmen, T., 287
von Neumann, J.D., 177–178, 191
Voronkova, O.V., 245, 259

Waddington, C.H., 53, 90, 92
Waide, J.B., 336–337, 339, 372–375, 378, 380–385, 387–390, 638–639
Wallace, A.R., 564, 613
Wallwork, J., 520, 523–524, 541
Walsh, J.J., 265
Walter, D., 521–522, 541
Walters, C.J., 228, 311–312, 331
Waltz, K.N., 463
Wangersky, P.J., 109, 417–418
Ward, L.G., 263
Waring, R.H., 384
Warren, P.H., 514, 538–539, 541, 543
Wasbauer, M.S., 523, 527, 532
Waterhouse, J.C., 452
Watson, A., 86
Watson, N.H.F., 75
Watson, S., 116
Watson, V., 384, 459
Watt, K.E.F., 21, 273, 312, 338–339, 415–418, 420, 422, 425, 433, 436, 453, 467
Watts, J.A., 387
Weathers, B., 522
Weaver, W., 41
Weber, J.A., 264

Webster, J.R., 231, 235, 372, 374–375, 378, 381, 383–385, 388–390, 638–639
Weinreb, D., 476
Weiss, P.A., 378, 381
Welch, H.E., 56
Welty, J., 518, 529
Werner, E.E., 116, 518
Westra, L., 410
Wetzel, R.G., 388
Wetzel, R.L., 265
Wheeler, G.C., 513, 515, 522
Wheeler, J., 522
Whillans, T.H., 404, 407
Whitby, L.M.J., 406
White, M.L., 331, 355
Whitford, W.G., 518, 524–525, 541
Whitman, W.G., 283
Whitmore, R.C., 364
Whittaker, R.H., 52, 80, 90, 92, 375–376, 387–388, 463
Wiame, J.M., 81, 85
Wicken, J.S., 651, 654
Wickens, D.L., 428, 441
Wiegert, R.G., 265, 273, 551
Wiehe, P.O., 77–78
Wiener, N., 167, 172, 450, 566
Wiens, J.A., 52, 95–96, 517–518, 523
Wieser, W., 70–71
Wilhm, J.L., 406
Willems, J.L., 589
Wilson, D.S., 115, 461, 463
Wilson, E.O., 52, 83, 93

Winemiller, K.O., 514–515, 517, 519–521, 538–541
Wisdom, C., 523
Wissel, C., 104
Wolawer, T.G., 112
Wolf, L.L., 93, 388
Wolfram, S., 38
Wollkind, D.J., 107–108
Wood, F.A., 312
Woodhead, N., 380
Woods, A., 529
Woodwell, G.M., 92, 376, 388, 399, 406
Woolhouse, M.E., 248
Wright, C.A., 73
Wu, J.J., 285–286, 289
Wulff, F., 105

Yamashita, T., 524, 526, 528, 530–531, 535
Yaramanglou, M., 270
Yatsalo, B.I., 346
Yeager, L.E., 296
Yodzis, P., 350, 514, 539, 541
Yorke, J.A., 353
Young, J., 445

Zabriskie, J., 522
Zak, J., 518, 520, 524–525, 535
Zarull, M.A., 409–410
Zeigler, B.P., 231, 234, 340–341, 468, 470–474, 476
Zhorov, Y.N., 243
Zotin, A.I., 141
Zotin, S.J., 122

Subject Index

1,2,4 TCB, 284–286, 288–292

Abies balsamea, 78
abiotic, 5–6, 31, 112–113, 117–118, 131, 230, 239, 377, 384, 390, 399, 405, 454, 458, 486, 513
abiotic components, 113
abiotic environment, 112, 117–118, 486
abiotic fluxes, 458
abiotic state, 239
abiotic variables, 239
above-organism biological systems, 171
above-organism macrobiology, 174
aboveground production, 111
abscession, 52, 88, 90, 92, 95
abscessional law, 96
absence of cycles, 355
absence of predators, 217
absolute optimality, 120
absolute probability, 193
absolute stability, 383
absorption of incident radiation, 137, 165
absorption of light, 147
abstract systems, 14
abstraction, 12, 620
Acanthocephala, 529
Acari, 510, 515–516
accelerated dissipation, 80
access roads, 297, 302
accommodation, 398
accumulation, 4, 92, 131, 149, 214, 485, 613–614, 618
accumulation of matter, 4, 485
accuracy, 32, 35–38, 122, 169, 174, 179, 187, 219, 232, 242, 282, 301
accuracy frontier, 37
Acer, 255
acidic deposition model, 317
acidification, 402–403, 406
acidification process, 406
acquisition, 45–46, 79, 90, 96–97, 474
action, 9, 13, 18, 52, 85–89, 92–94, 96–97, 181, 220, 242, 245, 326, 397–398, 456, 473–474, 647, 652
activation energy, 147
active feeding, 635
active tissue, 83
activity, 18–19, 82, 87, 112, 167, 174, 178, 196, 220, 228, 281, 314, 321, 323, 328–329, 358, 381, 407–408, 426, 430, 512,
521, 550, 552–553, 557, 567, 576, 615, 648–649
acyclic, 358, 512, 553, 635, 637
adapt, 54, 116, 190–191, 204, 563, 565, 617
adaptability, 119, 563–567, 610, 614–622
adaptability-directed, 564, 615, 617, 620–622
adaptability-directed problem solving, 620–622
adaptability of a system, 564–565, 616
adaptability theory, 563, 565, 610, 617, 621
adaptability theory formalism, 565
adaptation, 19, 31–32, 69, 93–94, 112, 114, 116, 119–120, 133, 144, 178–179, 190–191, 193–197, 199–200, 338, 398, 404–405, 463, 563, 615
adaptation cycle, 193–194, 197
adaptation cycle extremum problems, 194
adapted system, 114, 116
adaptive capacity, 338, 399
adaptive cycle, 190–191
adaptive environmental assessment and management, 228, 311
adaptive integrity model, 192
adaptive system radiation, 31
additive decomposition, 182
adjustment, 93, 417–418, 446, 512, 591
adsorption, 280, 281
adult bulls, 321
adult females, 298, 303–307
adult males, 73, 298, 303–304, 306–307, 425
adult niche space, 64
adults, 58, 64, 66, 68–69, 73, 76, 84, 95, 213, 298, 303–307, 321, 401, 425, 518–519, 523, 526–527, 529–532, 534
AEA, 228
aerial spraying, 282
aerial surveys, 301
Aeroplane Lake, 71–72
affinities, 31, 136–138, 143–146, 154, 564, 612–613
Africa, 72, 84, 469
African savanna ecosystems, 382
age, 5, 30–31, 51, 55, 57–58, 61–62, 65–73, 76–77, 79, 84–85, 94, 106–107, 109, 112, 114, 170,
185–186, 193–194, 200, 215, 231, 236, 297–299, 301, 303–305, 307–308, 321, 365, 409, 416, 425, 431–433, 435–436, 441, 445–446, 458, 510, 518–520, 531–532, 535, 539–540, 542
age and sex class distribution, 308
age at death, 31, 67, 79, 84
age class, 58, 84, 298–299, 303, 446
age class or sex class of deer, 298
age classes, 58, 299, 303–304, 518–519, 539
age composition, 68
age distribution, 69, 425
age distribution curves, 69
age frequency distribution, 61–62, 68–70
age of maturity, 84
age-sex classes, 301, 307
age-sex groups, 305
age span, 68
age-specific dynamics, 365–366
age spread, 68, 73
age structure, 51, 66, 107, 112, 170, 297, 416, 445, 510, 518–519, 531–532, 535, 539–540
agencies, 22, 43, 229, 274–275, 282, 296, 312–316, 320–321, 323–324, 326, 330, 372, 399, 410, 462–463, 470
agency resources, 330
aggregate models, 225, 231–234, 236, 282
aggregate population equation, 237
aggregate processes, 239
aggregate search, 243
aggregated system, 226, 243, 249, 258
aggregated variables, 244–245, 247, 252
aggregating function, 248
aggregating transformations, 226, 245, 250
aggregation, 84, 195, 225–226, 230–239, 242–243, 245, 247–248, 250–251, 255, 258, 260, 264, 270, 282, 327, 358, 479–480, 512, 550–553, 557–559, 565–566, 624, 629–635, 637, 639, 641
aggregation algorithm, 258, 557
aggregation error, 226, 234–235, 237–239
aggregation laws, 243
aggregation loss, 566, 633, 635
aggregation mapping, 629

663

aggregation matrix, 557–558, 630
aggregation of linear systems, 226, 248
aggregation problem, 225, 231–232, 239, 243, 247–248, 260, 641
aggregation theory, 232, 238, 260, 629
aggregative matrix, 250
aggregative transformation, 226, 243, 249
agricultural activities, 471–472
agricultural crops, 4, 218
agricultural environmental interactions, 317
agricultural herbicides, 263
agricultural or waste treatment practices, 270
agricultural output, 437
agricultural practices, 270
agricultural products, 499–503
agricultural sector, 419, 427, 429
agriculture, 14, 75, 274, 426–427, 429, 437, 441, 468, 470, 474, 504
agriculture subsystem, 437
agro ecological problems, 13
AI derived formalisms, 477
AI frame paradigms, 476
AI representation schemes, 474
air, 5, 55, 279, 281–286, 292, 319, 401–403, 468–469
air kinematic viscosity, 286
air pollution, 401–403
air–water, 5, 283–286, 292
air–water interface, 283–286, 292
air–water interface resistance, 283
ako link, 475
Aldabra Atoll, 72–74
algae, 57, 91, 95, 106, 108, 110, 133, 149, 266, 273, 402, 469, 491, 515, 520, 572, 580–581, 599
algae interaction, 110
algae pond, 572, 580
algal blooms, 9
algal phosphorus, 592
algal species, 111
algorithm of matrix approximation, 253
allelopathy, 274
allocation, 13, 317–318, 331, 356, 421, 494, 572
allometric, 49, 550
allometric relationships, 550
alpha diversity, 41
ambient energy, 55
Amblyopsis, 69–70
Amblyopsis rosae, 69–70
Amblyopsis speleae, 69–70
American Samoa, 70, 72
amino acid, 6, 613
Ammospermophilus leucurus, 531

amoebas, 528
amorphous chaos, 183
amphipods, 69
anadromous stocks, 66
analog and digital computers, 14, 21
analog computers, 11, 14–15
analysis, 2–3, 11–14, 19, 21, 32, 41, 44, 47, 49, 91, 104, 107, 109–110, 112, 130, 134, 138, 143, 166, 179, 186–187, 199, 228, 236, 239, 242, 255, 259, 275, 279, 284, 290–292, 295–296, 302, 308–309, 311–313, 315, 324–326, 328–329, 331, 341–344, 346–347, 357–358, 360, 365–366, 372–374, 376–377, 380–381, 384, 387, 419–420, 439, 444–445, 453, 463–464, 471–473, 480, 485–498, 501, 505, 511–512, 514, 517, 521, 543, 549, 552–553, 557, 565, 567–568, 572–573, 585, 620, 644, 647
analysis of iterative procedure, 255
analytic geometry–calculus sequence, 21
analytical phase, 10, 22
anchoveta, 89
animal behavior, 122
animal detritus, 531
animal ecologists, 9, 17, 51
animal foraging, 115
animal growth rate, 111
animal–plant interactions, 111
animal populations, 7, 12, 68, 111, 308
animals, 4–5, 7, 9, 12–13, 16–17, 43, 51, 63, 67–68, 72, 74–75, 78, 91, 95, 97, 111–112, 115–116, 122, 133, 171, 189–190, 227–228, 296, 298–299, 303, 305–308, 321, 323, 328–329, 403, 469, 471, 477, 480, 499, 504, 515–516, 518, 520–521, 523–526, 528–529, 531, 540–542, 605, 617
anisotropic, 43
annual precipitation, 111
annual rainfall, 73
annual temperature, 111, 375
annual wage, 422, 435–436, 438–439
Anodonta grandis, 71
Anopheles, 116
Anse Mais, 74
ant lions, 532
Antarctic, 470
Antarctica, 53
antelope ground squirrels, 523
anthropogenic, 54, 263, 373, 377, 403

anthropogenic disturbance, 373
anthropogenic wastes, 263
Antrozous pallidus, 531
ants, 518–520, 523–524, 526, 530, 532–533, 535, 539
aphids, 527
application of technology, 430
applied ecologists, 16–17, 20, 22
approximate aggregation, 632
aquarium, 286
aquatic ecologists, 16–17
aquatic ecosystems, 23, 32, 67, 108, 113, 134, 138, 141, 147, 156–157, 291, 402, 404, 517, 585, 589
aquatic environments, 280, 282–283, 379
aquatic organisms, 135, 137
aquatic plants, 263
aquatic production, 111
aquatic systems, 5, 87, 109, 401, 406, 520
arachnids, 510, 516, 519, 522, 530–533, 538
arbitrary systems, 252
Arctic, 30–31, 51, 55–57, 59, 63–67, 72–73, 79–80, 84, 89, 91–92, 94–95, 97
Arctic Archipelago, 56
Arctic char, 30–31, 56–57, 59, 63–67, 84, 97
Arctic Circle, 56
arctic fish, 67, 72
arctic lakes, 55, 73, 79, 91, 95
arctic regions, 92
arctic systems, 89
Arctic Unit of the Fisheries Research Board of Canada, 56
Arizona, 468, 470–471, 481, 518
armaments, 420, 442–443
armaments buildup, 443
armaments industry, 420
Arrhenius relation, 265
arthropod parasitoids, 526
arthropodivorous birds, 518, 533
arthropodivorous vertebrates, 532–533
arthropodivory, 532–533
arthropodovores, 533, 535, 542
arthropods, 510, 515, 518–519, 522–535, 537, 539–542
articulation, 36–38, 567, 649–650
articulation index, 37–38
artificial barriers, 84
artificial intelligence, 341, 446, 468, 470, 473–474, 476
artificial systems, 47
ascendancy, 462, 566–567, 623, 643, 650–654
Asellus pellucidus, 69
asilids, 524, 526
aspen parkland, 23

assassin bugs, 526
assimilation, 91, 143, 146–149, 151, 236
assumptions, 36, 81, 89, 109, 112, 176–178, 214, 244, 282, 284–286, 288, 301–304, 321, 326, 364, 422, 490–491, 493–494, 498, 505–506, 512, 539, 557, 601, 605–606, 646–647
asymmetric interactions, 31, 79–80
asymmetric interdemic interactions, 31
asymmetry, 47, 80, 82, 85, 90
asymptote, 67, 187, 516
asymptotic density, 77
asymptotic estimates, 180, 184
asymptotic expressions, 179, 181
asymptotic or periodic attractor trajectories, 107
asymptotic stationary case, 193
asymptotically stable, 375, 595
Atlantic Ocean, 71
atmosphere, 4, 54, 97, 279–280, 284, 292, 342, 379, 382, 391, 406, 415, 459–460, 469, 501, 504
atmospheric gases, 459
atomic weight, 189
atomisms, 85, 87–88
Atriplex, 526–527
Atriplex canescens, 527
attainable dissipation, 51
attractors, 31, 44, 81, 88–89, 92, 94, 97, 107, 161, 336–337, 346, 375, 378, 381, 391
attribute knowledge representation, 477
autecology, 10
autocatalytic growth processes, 381
autogenic succession, 6
autonomous ecosystem, 54, 567, 655
autotroph model, 264–268, 270–271, 273–274
autotroph production, 271
autotrophic biomass, 387–389
autotrophic biomass turnover rate, 388–389
autotrophic photosynthesis, 378
autotrophic productivity, 386–387
autotrophs, 108, 227, 266, 269–270, 385, 387, 458, 521, 523, 540
available resources, 7, 74, 459–460, 572–573, 582, 650
average mutual information, 648
avian species, 325
avoidance, 116, 417, 578

bacteria, 6, 84, 95, 108, 116, 524
bacterial loop, 57
bacteriophage, 108
bacterioplankton, 258–259

bag limits, 17, 228, 296–302
Bakman's law, 185
Bakman's logarithmic law, 185
balance, 7, 30, 32, 52, 94, 131, 133, 135–138, 141, 143–144, 146, 156, 226, 231–234, 237, 265, 296, 336, 346, 352, 366, 373–374, 376, 381, 384, 426–427, 459, 471, 488–490, 505, 562, 564, 612, 615, 653
balance equation, 32, 135–137, 143, 156, 505
balance of momentum, 141
balance of nature, 30, 52, 336, 373–374, 376, 653
balance relations, 346, 352
balanced evolution, 115
balanced state, 52
balances of momentum, 135
Balanus balanoides, 220–221
balsam fir, 79, 382
Baltic Sea, 404, 407–408
Barren Grounds, 56, 58
basal area, 383
basal metabolic purposes, 87
basal metabolic rate, 69
basal metabolic requirements, 80
basal metabolism, 68, 487, 489, 494
basins of attraction, 337, 375–378, 380, 382, 384–385, 387, 390–391, 564, 614–615, 619
Batillaria minima, 71
bats, 73, 469, 523, 535
beech, 77, 255–256
Beers-Lambert relation, 265
bees, 115, 469, 516–517, 538
beetles, 516, 523–524, 526, 533, 535, 538
behavior, 20–21, 31–32, 39, 42,44–45, 52, 56, 84–85, 89, 94, 105, 107–109, 112, 114–116, 120–122, 130, 136, 142–143, 166, 170–175, 177–178, 191–192, 194–195, 200, 217, 225, 227, 230–231, 235–237, 255, 264–265, 267, 270–271, 279–283, 291, 296, 312, 316, 323, 326–328, 331, 337, 339–341, 344–346, 356, 373–375, 377–378, 380, 391, 398, 400, 403–405, 408, 410, 424, 426, 432–433, 438, 444–445, 450–453, 457, 460–464, 471–473, 476, 480, 483, 512, 541, 550, 559, 563, 567, 591–592, 594, 603, 606, 610, 620–621, 640, 646–647, 654–655
behavior and evolution of species, 337
behavior observation archiving, 480
behavior of a general system, 374

behavior of contaminants, 280, 283
behavior of ecosystems, 115, 142, 265, 391
behavior of exergy, 603
behavior of new chemicals, 282
behavior of nonlinear systems, 142
behavior of pathological ecosystems, 400
behavior of systems, 109, 170–171, 236, 374, 408, 426, 444, 453, 563, 594
behavior patterns, 471
behavioral adaptability, 565, 616–618
behavioral boundary conditions, 170
behavioral constraint, 327
behavioral ecology, 123
behavioral flexibility, 392
behavioral indeterminacy, 173
behavioral repertoire, 615, 621
behavioral segregation, 68
behavioral states, 143
behavioristic indeterminacy, 178
behavioristic interaction, 170
behaviors, 32, 174, 176–177, 196, 200, 377–381, 391–392, 462, 480, 521, 615–616, 621, 647
bell-shaped, 58, 62–63, 73, 84
bell-shaped configuration, 63
bell-shaped length frequency distribution, 58, 62
bell-shaped or dome-shaped catch curve, 58
benefit of man, 8
benefits, 37–38, 228–229, 270, 323–324, 326–327, 330, 344
benthic algae, 266
benthic and planktonic flora and fauna, 56
benthic fauna, 56, 522
benthic herbivores, 635
benthic microalgae, 265
benthic organisms, 57
benthic regions, 95
benthos, 116, 227, 242, 258, 264, 318, 515
benthos model, 264
benzene, 284–287
Bermudan tidal flats, 71
Bernoullian scheme, 181, 215
beta diversity, 41
Betula, 255
Bewick's wren, 523
bias, 66, 226, 233, 336, 416, 542
bifurcation, 32, 106, 107–108, 114, 138–139, 141–142
bifurcation points, 32, 141–142
big game, 308
bimodal configuration, 76
bimodal distribution, 78

665

bimodal length-frequency distribution, 75
bimodal size-frequency distribution, 76
bioassays, 226, 263
biochemical processes, 407
biocoenosis, 32, 144, 154, 549
bioconcentrations in fish, 281
biodegradation, 280
biodynamics, 52, 82, 96
bioenergy, 83
biogeochemical, 54, 262, 264, 336–337, 374, 378–382, 384–387, 389–391, 403, 512, 519, 553, 557, 559
biogeochemical cycles, 54, 378, 384, 391, 403, 559
biogeochemical cycling, 337, 378, 380–382, 386–387, 390–391, 519, 553, 557
biogeochemical processes, 262
biogeochemical system, 337, 378–379, 385
biogeocoenosis, 4, 358, 407
biological adaptability, 621
biological components, 133–134, 137–138, 146, 164, 579
biological control practices, 9
biological damage, 209
biological effects, 327
biological evolution, 38
biological fluctuation, 54
biological hierarchy, 565, 617, 620
biological interactions, 40–41, 357
biological laws, 133, 142
biological limitation, 95
biological organization, 114, 162, 385, 389, 452, 563, 611–612, 618
biological pest control, 357
biological populations, 364, 373, 390
biological processes, 32, 80, 91, 131–133, 134, 136–138, 141, 143–144, 146, 149, 151, 154, 156, 164–165, 231, 382, 483, 563
biological production, 47
biological reference, 54
biological systems, 14, 30, 38, 52–55, 80–82, 96, 144, 170–171, 185–186, 345, 381, 612, 615–616, 621
biological temperature, 91–93
biological time, 82, 92
biological time scales, 92
biological turnover, 385
biological variables, 131
biological, chemical, and physical interactions, 157
biologists, 16–17, 19, 21, 86, 120, 122

biology, 1, 14, 22, 32, 52, 56, 68, 90, 92, 98, 121–123, 130, 162, 167, 174, 195, 232, 339, 513, 563, 567, 605, 643, 647, 655
biomass, 7, 30, 44, 57, 67, 69, 75, 79–80, 83, 87, 89–90, 92–94, 97, 111, 134, 138, 142–145, 148–151, 153–158, 160–161, 189–190, 205–206, 267–268, 271–272, 282, 346, 352, 358–359, 361, 383–384, 387–389, 454, 458, 461, 487, 491, 518, 524, 532, 561, 563, 569–570, 578, 599, 601–602
biomass accumulation ratio, 92
biomass balance relations, 352
biomass development, 454
biomass dynamics, 358
biomass energy, 92–93
biomass fractions, 154
biomass turnover rate, 387–389
biomasses for submerged vascular plants, 272
biomatter, 54
Biome Module, 474
biome modules, 340, 471, 474, 479
biomechanics, 55
biomes, 22, 97, 375, 468–469
biomodel of trade, 220
biomodeling of commercial harvesting, 219
bioprocesses, 32, 92, 145
bioregenerative control management system, 471
bioregenerative environment, 471, 480, 483
bioregenerative environmental design, 480
bioregenerative macrocosm, 340
bioregenerative systems, 480–481
BioScience, 9
biosphere, 40, 46–47, 53–54, 81, 96, 123, 171, 173, 186, 340–341, 345, 379, 384, 391, 398, 407, 459, 462, 468–472, 474–475, 477, 479–481, 483, 498, 511
Biosphere I, 468–470
Biosphere II, 340–341, 468–471, 474, 477
biosphere model construction and validation, 480
Biosphere Module, 475
biosphereans, 469–471
biospheric boundaries, 53
biospherics, 341
biosystem, 31, 53, 80, 82, 87–88, 97, 173
biota, 54, 134, 138, 239, 338, 385, 391, 400, 404–405, 408, 459, 477, 480, 510, 521, 523–524, 539, 542, 644

biotelemetric equipment, 12
biotic, 5, 31, 46, 94, 113, 131, 137, 230, 235, 239, 336–337, 345, 374, 376, 380–382, 384–387, 390–391, 399, 401, 407, 459, 513, 535
biotic abiotic complex, 5
biotic–abiotic interaction, 113
biotic assemblages, 336, 374, 391
biotic communities, 345, 513
biotic components, 31, 113, 230, 239, 376, 513
biotic feedback, 459
biotic growth processes, 387
biotic integrity, 401
biotic mobilization, 381
biotic potential, 235
biotic processes, 382
biotic productivity, 386
biotic structure, 337, 380–381, 385
biotic turnover structure, 385, 387
bird, 95, 408, 518, 522
bird populations, 95
birds, 57, 77, 84, 92, 94–96, 318, 324, 418, 515, 518–519, 521–523, 529, 532–533, 535, 537–538, 542
birth, 201, 218, 416, 424, 433, 437, 446, 618–619
birth and death rates, 416
birth control, 446
birth probability, 218
birthrates, 424–425, 429–430, 437, 441, 443, 446
births, 200, 424, 430–434, 437, 441
bit, 459, 650
bivalve mollusks, 69, 71
black gum, 255–256
black spruce, 79
black widow spiders, 519
black widows, 519, 530, 532
blackbirds, 517
black box predictor, 328
blind African cave fish, 68
blind cave fish, 70
blind cave fishes of Zaire, 30, 68
blowdowns, 408
blue-gray gnatcatcher, 531
blue-green algae, 133
BOD concentration, 575
body forces, 135
body size relations, 539
bog, 388
Bohr's complementarity principle, 38–39
Boltzmann's entropy, 41
bombyliid flies, 524, 538
Bombyliidae, 516, 527
Boolean estimate of probability, 179
Boolean estimates, 179–180
Boolean formula, 182

bootstrapping principle, 614
boreal forest, 75, 377, 388, 405, 418
Bosnian beech forest, 77
Bothnian Sea, 408
bottom deposits, 69, 71
bottom sediments, 280
bottom-up control, 91
boundaries, 5, 8, 22, 46–47, 52–54, 83, 93, 95–97, 138, 140–141, 321, 375, 426, 498, 521
boundary, 6, 81, 96, 131, 133–136, 141, 156–157, 168, 170, 175, 283–284, 287, 320, 400, 489, 595–597, 603, 605
boundary conditions, 81, 133–134, 141, 157, 170, 175
boundary constraints, 96
boundary distinctiveness, 400
boundary layer formulation, 283
boundary linearity, 400
boundary thickness, 6
boundary values, 141, 168
boundary values of a process, 168
bounded, 5, 31, 54, 76, 105, 107, 114, 118, 142–143, 216, 238, 345, 359, 375, 377
boundedness, 118–119, 374
bounding exercise, 314–315, 324
bounding the model, 228, 314
bow hunting seasons, 302
breeding and feeding grounds, 408
breeding and spawning grounds, 405
breeding pair, 76
breeding practices, 308
breeding season, 518
breeding units, 47
Bridger Teton National Forest, 320–321
Brillouin's expression, 41
British Royal Society, 73
brucellosis infection rate, 323
bucks, 296, 305
buffer capacity, 562, 570, 573–577, 582–583, 589
buffering adaptabilities, 622
buffering capacity, 406
bulls, 321
burbot, 56
bureaucratization, 46–47
burning grasslands, 9
burrowing owls, 519, 531, 534–535
butterflies, 47

C law, 177
cactus, 523
cactus wren, 523
cadmium, 110
Caecobarbus geertsii, 68
calculus, 21, 145, 235
calf production, 323
California, 317, 339, 415, 469, 510, 516, 518, 522
Callisaurus draconoides, 531
callophorid fly larvae, 526
Calluna, 456
caloric equation, 131
calorie rations, 215
Calvaria major, 78
calves, 321
Cambrian, 95
camel crickets, 523, 530, 535
Campbell River, 408
Campbell River estuary, 408
Canada, 23, 51, 56, 98, 279, 326, 397, 410
Canada geese, 326
Canadian Department of Fisheries, 72
Canadian northwest, 30, 55, 57, 71
Canadian Northwest Territories, 55, 57
canal flow, 324
canals, 318, 324, 326
Canarium mauritianum, 78
cannibalism, 69, 519, 527–528, 530–532, 534–535, 539, 541–542
cannibalistic, 63, 519, 526, 534–535
cannibalistic char, 63
canopy, 77–78, 387, 521
canopy species, 78
capacity, 9, 14, 19, 44, 67, 72, 74, 79, 95, 108, 115, 152, 157, 176, 299, 303, 321, 324, 338, 362, 381, 386, 399, 403, 406, 426, 430, 432, 435, 442, 446, 562, 567, 570, 573–576, 583, 589, 651–654
capturing tendency, 91–93
carabid, 526
carapace length, 73–74
carbon, 4, 110, 227, 264–265, 283, 291, 342–343, 358, 462, 471, 486–487, 499–503, 520, 550, 553, 624, 627, 640, 648, 650
carbon cycle, 501
carbon dioxide, 110, 291, 471, 487, 500, 502–503
carbon flows, 553
carbon monoxide, 283
carnivores, 6, 460–461, 515, 520, 533–535, 540, 550, 635
carnivores–omnivores, 635
carnivorous vertebrates, 533
carnivory, 4, 530, 533, 644
carrion, 526, 534, 537, 540, 542
carrion feeders, 526
carrying capacity, 9, 67, 74, 152, 157, 299, 362, 430, 432
Carson Canyon, 69
cascade model, 539
catastrophe theory, 160, 345
catastrophes, 107–108, 114, 434
catastrophic disturbance, 338
catastrophic event, 46
catastrophic instability, 339, 418–419
catch, 58, 60, 62–63, 196–197, 204–209, 215, 217, 220, 519, 531
catch curve, 58, 63
catch volume, 215
catenary sequence, 406
Catostomus commersoni, 75
cats, 73
causal flux, 509
causal representations, 476
causality, 263, 341, 420, 509–511, 563, 641
cause and effect loops, 453
cause and effect relationships, 9, 21
cave adaptation, 69
cave faunas, 84
cave fishes, 30, 51, 68–69
cave fishes of the southern United States, 30, 69
cave fishes of Zaire, 30, 68
Cave Spring, 70
Cebus monkeys, 452
Cecidomyiidae, 527
celestial objects, 54
cell division, 190
cell or colony volume, 154
cell organelles, 186
cell position, 193
cells, 5, 45, 154, 186, 189–190, 192–193, 230, 450, 513, 526, 589, 611–614, 618
cellular automata, 178, 191
centipedes, 533
Central Valley of California, 317
cerithid snails, 71
Cerithium lutosum, 71
cestodes, 529
chain length, 359, 400, 511, 514, 532, 538–539, 541, 563, 601–604, 607
chalcids, 527
Chalybian wasps, 532
change, 5–7, 9, 30, 35, 41, 43–47, 52–53, 56, 58, 65–66, 83, 85–91, 93–95, 97, 115–116, 130, 140, 146, 149, 169, 172, 192, 233, 235–236, 243–245, 249–251, 259, 263, 288, 296, 299, 302, 307, 315, 325, 328, 338–340, 343, 373–374, 376–377, 382, 384, 390, 402, 405, 409, 416, 421–422, 424, 429–430, 433, 440, 452, 460, 488–489, 491, 495, 517–519, 521, 526, 535, 561–564, 569–572, 575, 577–578, 586, 591, 594, 596–597, 606, 610, 612, 615–616, 619, 640, 647

change matrix, 325
changes in concentration, 282
changes of structures, 141
channel, 30–31, 42–43, 46, 176, 515, 522, 542, 565, 624–627, 629
channel omnivores, 522
channels, 16, 46, 521–522, 565, 641
chaos, 32, 44, 52, 86, 96, 106–107, 143, 183, 346, 643
chaotic behavior, 44–45, 107, 142, 567, 647
chaotic regimes, 143
chaotic sequences, 44
Char Lake, 56, 58
characteristic time, 169–171, 597
check stations, 301
chemical, 12, 18, 32, 45, 87, 90, 97, 122, 130–134, 136–140, 142–144, 146, 152, 157, 161, 164–165, 227, 270, 279–286, 290, 292, 358, 378, 381, 386, 391, 402, 406, 415, 427, 453, 561, 570, 586–587, 593, 605, 612, 648
chemical analytical equipment, 12
chemical bond energy, 378
chemical constituents, 133–134, 137, 152, 157, 164
chemical energy, 570, 648
chemical equilibrium, 87, 90
chemical equilibrium theory, 90
chemical inputs, 427
chemical laws, 133
chemical potential, 134, 137, 165, 378, 586–587
chemical process, 146
chemical reactions, 45, 122, 132–134, 136, 157, 164–165
chemical structure, 227, 279
chemical thermodynamics, 122
chemicals, 90, 116, 133–134, 152, 189, 227, 279, 281–282, 284, 288, 290–291, 319, 407, 509
chemistry, 12, 32, 34, 130, 132, 138, 143, 162, 270, 313, 318, 383, 568, 570, 585–586
chemostats, 41, 149
Chesapeake Bay, 226, 263, 267, 270, 272–275, 517, 520
Chetaev's instability theorem, 592–593
Chihuahuan desert, 525–526, 535
chironomid larvae, 56–57
chlorophyll a, 111, 116
chlorpyrifos, 282
chromosomes, 115
cicadas, 84
ciliates, 528
Cinq Cases, 73
circadian rhythms, 46
circular or cyclic pathways, 81
circulation, 4, 382, 391, 420, 485

Civil War, 425, 433, 441
Civil War deaths, 441
Civil War debt, 425, 441
civilized man, 9
clarification of causal relationships, 328
classification, 40, 172, 179, 338, 375, 409, 417, 477–478, 484, 520, 531
clear cut, 408, 431, 454
clear cut logging operations, 408
clear cutting, 78, 382–383, 400
clerid beetles, 524, 526
climate, 73, 339, 343, 377, 404, 409, 416, 420, 585
climate change, 339, 343, 409, 416
climatic change, 6, 377
climatic control, 7
climatic models, 292
climatic variables, 377, 387
climax, 7–9, 52, 58, 66–67, 82, 86, 90, 94–95, 196, 200, 218, 336, 373–374, 456, 619
climax ecosystems, 9
climax equilibrium, 456
climax forest, 58, 66
climax stage, 619
climax state, 7, 9, 82, 196
clockwork mechanism, 451
closed, 52–54, 68, 77, 79, 81, 85, 89, 97, 108, 112, 117, 340, 345, 351, 354, 356–360, 365–367, 377, 384, 387, 404, 452, 469–470, 483, 522, 563–564, 588, 595, 597, 601, 605, 607, 623, 649
closed biogeochemical cycles, 384
closed canopies, 77, 377
closed-loop feedback system, 452
closed-loop omnivory, 522
closed loops, 357
closed systems, 52–53, 79, 81, 89, 97, 108, 112, 117, 345, 366, 404, 595, 597, 623
closed units, 53
closure, 79–80, 360, 520, 561
Cnemidophorous tigris, 531
Coachella Valley, 510–511, 516, 517–518, 522, 525, 527–528, 533–536, 538, 542–543
Coachella Valley food web, 510–511, 538, 543
coastal upwelling, 89, 388
cockroaches, 524
coconut plantations, 73
coevolution, 115, 123
coevolutionary development, 115
coevolved species, 42
cohort model, 195
cold water fishery, 324
Coleoptera, 526–527, 530
Collembola, 524

collisional cycle, 85
collisional inputs, 85
colonists, 454, 456
colonization, 113, 405, 470
color patterns, 116
Colorado, 21, 98, 227–228, 295–296, 308–309, 311
combination of deterministic, stochastic, synergetic, and cybernetic methods, 141
combustion, 88
commensurability, 486
commensurable flows, 341
commercial activity, 196
commercial and recreational fisheries, 275
commercial and recreational opportunities, 262
commercial exploitation, 220
commercial fishery, 56, 67
commercial fishing practice, 209
commercial system, 206
commodities, 341–342, 419, 421, 426, 428, 486–495, 498, 500–503, 505–506
commodity balance equation, 505
commodity-by-commodity flow matrix, 493–494, 501, 503
commodity-by-commodity production matrix, 505–506
commodity-by-commodity, single process systems, 505
commodity-by-process, joint product systems, 505
commodity flow matrix, 493–494, 501, 503
commodity flows, 487, 490, 492–494, 501, 503
commodity input distribution matrix, 501–502
commodity inputs, 341, 494
commodity outputs, 491, 494
commodity quality, 490
commodity technology assumption, 505–506
commodity units, 494
communication, 30–31, 42–43, 77, 84, 113, 176, 229, 292, 315–317, 320, 326–327, 329–330, 453, 471, 516, 539, 565, 609, 624–626, 647, 653
communication channel capacity, 176
communication channels, 30–31, 42–43, 176, 565, 624–625
communication matrix, 626
communication theory, 647
communities, 6–7, 31, 83, 90–91, 95–96, 116, 171, 173, 186, 195, 200, 209, 218, 226, 263–264, 275, 318, 324, 335–336, 344–

communities 345, 347, 349, 356, 358, 361, 364–366, 376, 402, 404, 406, 485, 511, 513–516, 519–522, 528, 539–541, 543, 549, 564, 569, 605, 612, 614–615, 617, 619, 649
communities of the biosphere, 186
community, 4–6, 9–10, 42, 83, 96, 113, 120, 123, 157, 161, 170–171, 194–198, 200–204, 206, 209, 211–214, 218–221, 230, 236, 265, 284, 320, 336, 343–352, 354, 356–358, 361–362, 364–366, 398, 400–401, 405, 407, 461, 510–511, 513–517, 519, 521–522, 524–529, 535–536, 538–543, 550–551, 559, 562, 564–565, 590, 593, 601, 613–622, 651
community adaptation, 195–196
community behavior, 200
community control, 195–196
community controlling models, 218
community dynamics, 365, 519, 590, 618
community ecology, 10
community matrix, 336, 344, 347, 349–352, 362, 365–366, 539
community matrix stability, 366
community respiration, 400
community species composition, 170–171
community stability, 9, 343–344, 365, 618, 620
community states, 197–198
community structure, 357, 400–401, 513, 515, 519, 543, 620
community suppression, 209, 211, 213
compartment model methodology, 14
compartmental analysis, 485–487, 489–492, 494–495, 497–498, 501
compartmental and flow analyses, 494
compartmental flow analysis, 341
compartmentalization, 510, 521
compartments, 13–14, 95, 280, 282, 341, 384, 454, 511–512, 521–522, 535, 540–542, 550, 552, 557–558, 565–567, 587, 589, 621, 624, 635, 644–645, 648, 652–653
compartments of an ecosystem, 13
compensation, 564–565, 611, 615, 617–618, 620–621
compensation principle, 564, 611
competing species, 107, 349, 362, 364–365, 459
competition, 6, 19, 43, 71, 93, 106, 108–109, 116, 176–177, 214, 220, 235, 237, 264–265, 269, 274, 340, 344, 349–351, 354, 356–357, 361–365, 416, 425, 440–441, 457, 459, 462, 491–492, 565, 568, 578, 617, 622–623, 652
competition coefficients, 237, 349, 362–365
competition communities, 349–350, 361, 364
competition community models, 350
competition for light, 264–265, 269
competition matrices, 351, 362–364
competition patterns, 365
competition theory, 235
competition time, 177
competitive, 49, 195, 265, 274, 361–363, 365, 403, 444, 457, 459–460, 490–491, 526
competitive exclusion principle, 361–362, 365
competitive interactions, 195, 457, 460
competitive relations, 49, 274
competitors, 115, 269, 364, 456, 518–519, 612
complex ecological organization, 30–31, 225, 335, 343, 372, 397, 415, 450, 468, 485, 509, 513, 549, 561–562, 568, 585, 609, 623, 643
complex ecosystems, 11, 104, 109, 113
complex radioelectronic systems, 175
complex systems, 13, 22, 32, 121–122, 142, 166–167, 171–172, 174–175, 190, 229, 243, 260, 438, 444–445, 486, 549, 571
complexity, 2, 4–5, 9–10, 21, 29–34, 36–40, 42, 51–52, 86–87, 89–90, 97, 104, 106, 108–110, 130, 166, 171, 178–179, 189, 225, 230–231, 239, 242, 262, 270, 275, 279, 295, 311, 314, 317, 329, 336, 339–340, 343, 373, 376–377, 398, 419, 444–445, 458, 509–511, 513–514, 521–523, 525, 532, 535, 538, 540–543, 564–567, 571, 576, 604–605, 609–610, 612–613, 618–620, 654
complexity in nature, 40
complexity of food web, 540, 543
complexity–stability debate, 336, 373
complexity–stability hypothesis, 373, 376
complexity versus stability of ecosystems, 511, 541
component mode, 37
component organism reactions, 116
component populations, 114, 345
composition, 32, 40, 68, 111, 116, 133, 170–171, 191, 306–307, 337, 363, 365, 373, 376, 380, 392, 402, 404, 406, 415, 479, 483, 517–518, 521–522, 526, 551, 562, 568–569, 571, 573, 575, 577–579, 582–583, 601
composition of species, 406
composition tree, 479, 483
computational tractability, 264
computer code, 305, 315, 613
computer gaming, 228
computer modeling, 313
computer science, 275, 341, 609–610
computer simulation, 340–341, 476, 635
computer simulation modeling, 340
computer software, 471, 483
computer technologies, 229
computerization, 301
computers, 2–3, 10–11, 13–16, 20–21, 141, 157, 225, 230, 295, 563
concentration difference, 284
concentrations of contaminants, 280
concentrations of nutrients, 138
concept of diversity, 40, 42, 48
concept of order, 90
conceptual and computational limitations, 264
conceptual and simulation models, 263
conceptual requirements, 2, 10
conditional entropy, 623, 626
conditional stability, 375, 377, 381–382, 392
conditionally stable attractors, 337, 375, 378, 381, 391
conditionally stable systems, 380, 382
Cone Spring, 624, 653–654
confined reality, 226
Congo Basin, 68
connectance, 48–49, 106, 339, 417–419, 446, 538
connectance–stability relationship, 106
connectedness, 194, 347
connective stability, 350
connectivity, 30, 41–43, 46, 48–49, 106–107, 509–510, 521, 564–565, 620
consequences of environmental change, 9
conservation laws, 169, 173, 346, 587, 589, 598, 602, 606
conservation of genetic diversity, 373
conservationists, 228

conservative and dissipative systems, 347
conservative system, 346, 353
constancy, 53, 55, 69, 84, 95, 205–206, 337, 373, 376, 564, 618–619
constant density, 63, 84
constant environmental conditions, 53, 96, 288
constant parameter systems, 112
constant states, 53
constant structure, 173
constant velocity, 89
constants, 117, 119, 150, 169, 181–182, 185, 205–206, 208, 213, 215, 219–220, 237, 251, 264, 267, 297, 378, 383, 390, 576–577, 580, 588
constraint equation, 304
construction workers, 422, 424, 435–436, 438–439
consumer–autotroph interactions, 462
consumer biomass, 80
consumer mortality rate, 597
consumer price index, 422–423
consumer purchasing power, 422–423, 436, 441–442
consumer regulation, 340, 454, 458
consumer–resource interactions 461
consumer–resource systems, 460
consumers, 6, 8, 14, 80, 237–238, 340, 356, 422–423, 425, 431, 436, 441–442, 452, 454, 458–462, 464, 469, 490–492, 513, 516, 518, 520–522, 525, 535, 538–542, 563, 597
consumers with self loop, 538
consuming populations, 237–238
consumption, 80, 137, 425–426, 429, 436, 461, 532
contaminant behavior, 280
contaminant fates, 279
contaminant loading rates, 280
contaminants, 227, 279–283, 285–288
continent, 23, 47
continental position, 391
continental shelf, 388, 436
continuity, 22, 84, 134, 141, 167, 199, 301, 398
continuity equations, 134, 141
continuity of seasons, 301
continuous control variables, 154
continuous cultivation, 108
continuous models, 107
control, 7, 9, 11, 17, 22, 31, 58, 68, 75, 79, 84, 90–91, 116, 119–122, 130–131, 142–144, 154, 156, 172–173, 177, 194–200, 202, 204–205, 229, 238, 275, 284, 297–298, 303, 308, 316–317,
319, 323–324, 326, 339–341, 357, 401, 415, 418, 424, 438, 441–442, 446, 450–451, 453–454, 456, 458, 469–472, 474, 481, 488, 561, 563, 578, 610, 612
control management systems, 471
control mechanism, 68, 424
control theory, 177, 488, 561
controllability, 32, 172, 174, 176–177, 263, 641
controlled burning, 418
controlling factors, 6–7, 143, 150, 161
controlling system, 196–197, 200, 209, 213–214
conventional ecologists, 11, 16–17
conventional extremals, 199
convergence process, 258
cooperation, 18, 20, 22, 122, 142
copepods, 56, 58, 69
copper concentration, 572
copper model, 576
coprovores, 525
coral atoll, 72
Coregonus artedii, 56
Coregonus clupeaformis, 56, 60–61, 75
Cornwallis Island, 56
Corps of Engineers, 324–325
correlation, 11, 42, 84, 308, 421, 445, 615–616, 621, 640
cosmic systems, 54
cotton wilt disease agents, 357
cottontail rabbit, 529
Cottus cognatus, 56
coupling relations, 341
covariance, 45
coves, 267–268, 274
Coweeta, 372, 375, 377, 382–383
Coweeta Basin, 375
Coweeta Hydrologic Station, 372
cows, 321
coyotes, 518, 523, 526, 528–529, 533–534, 539
Crangonyx gracilis, 69
crayfish, 402
creeks, 388
creosote bush, 523
Cretaceous, 93
crime rates, 446
criterion, 40, 47, 122, 140, 164, 171–172, 182, 234, 238, 245, 247, 252, 336, 344, 350, 354–357, 359–360, 365, 367, 473, 561, 613
criterion of separability, 247, 252
critical habitat, 404
critical nutrient, 117
critical value, 187, 199, 209, 221, 235
crop or yield prediction, 13

cropping, 197
crops, 13, 319, 387, 409, 452, 472, 478
cross processes, 145
cross products, 238
crude food exports, 438
crude oil, 418, 422–423, 431, 446
Crustacea, 532
cryptic coloration, 116
Crystal River, 512, 553, 557
crystallization, 88
cultivated systems, 209
cultivation, 108, 112, 195, 197, 200, 209–210
cultural, 46, 338, 398, 400
cultural practices, 400
currents, 85–86, 157, 280, 407, 453
cybernetic, 31–32, 42, 94, 104, 120–122, 138, 141–142, 154, 167, 172–173, 177, 340, 451–454, 458–459, 461–464, 511, 567, 652, 654
cybernetic approach, 142, 154
cybernetic aspects, 32
cybernetic behavior, 42
cybernetic concept, 340, 458
cybernetic concomitants of complexity, 567
cybernetic design, 340, 462
cybernetic development, 31
cybernetic ideas, 167, 177
cybernetic loop, 652
cybernetic mechanism, 459
cybernetic methods of satisfaction or suboptimality, 122
cybernetic optimization models, 31
cybernetic paradigm, 453–454
cybernetic properties, 32, 172
cybernetic regulation, 453
cybernetic regulators, 458
cybernetic system behavior, 452
cybernetic systems, 172–173, 340, 453–454, 462–463
cybernetics, 14, 120–121, 130, 142, 154, 167, 172–173, 339, 451–453
cybernetics development, 173
cybernetics principles, 14
cycles, 4–6, 43–45, 54, 81, 87, 97, 107, 263, 274, 336, 339, 342, 354–356, 365, 375, 378, 384, 391, 403, 416, 425, 433, 443, 445, 456, 461, 494, 501, 510, 512, 519, 525–527, 532, 535, 537–539, 551–553, 557, 559, 576, 638, 652
cycles of carbon, 4
cyclic processes, 87, 93, 96
cyclical webs, 520
cycling, 9, 19, 79–80, 84, 264, 337, 340, 378, 380–382, 386–387, 390–391, 400, 403, 445, 453,

458, 461, 501, 510, 512, 514, 519–522, 526–527, 529–532, 534–535, 539–541, 552–553, 557, 559, 563, 578, 605, 638
Cyclops scutifer, 56
Czechoslovakia, 104, 110

D stability, 348, 365
D stable, 348
damping mechanism, 67, 84
damping moment, 79, 93
darkness, 56, 68
Darwin's theory, 39, 561–562, 568–571, 576–577, 579, 582
Darwin–Wallace postulates, 564
Darwin–Wallace theory, 613
data quality and quantity, 36
Davis models, 444
daylight, 56
dead state, 112
dead system, 41
death, 31, 67, 79, 84, 175, 178, 181, 184, 186, 190, 201, 206, 215–216, 218, 221, 235, 248, 302–308, 378, 416, 509–510, 526, 528, 569, 603, 618–619
death and reproductive rates, 308
death rate estimates, 307
death rates, 235, 248, 302–307, 416, 603, 618–619
death of system, 184
deaths, 192, 194, 200, 420, 425, 433, 441, 443, 446
debt, 339, 416, 420–422, 424–425, 427–429, 433, 435–436, 438–441, 443, 445–446
debugging of programs, 15, 314, 316
decadence, 7
decay of energy, 42–43, 45
deciduous forests, 5, 377, 385, 388
decision act, 171–173
decision analysis framework, 329
decision making, 13, 47, 228, 243, 283, 297, 308, 312, 314, 326, 329, 372, 444
decision making tool, 283
decision theory, 176
decision variables, 228, 298–299, 301, 303
decompartmentalization, 522
decomposer organisms, 6
decomposers, 6, 8, 14, 358, 401, 458, 462, 524
decomposition, 4, 29, 48, 175, 182–183, 226, 338, 341, 462, 472, 476–480, 483, 637
decomposition of the system, 48, 182–183
decreased dissipation, 52, 94
deductive reasoning, 54

Deep Canyon Field Station, 522
deep geology, 342, 501, 504
deer density, 300
deer herd sizes, 298
deer herds, 296–297
deer hunting, 228, 295, 297
deer hunting pressure, 228, 295
deer kill, 228, 296, 298, 302
deer population sizes, 301, 306
deer populations, 227–228, 296, 298, 300–301, 303, 306
deer type, 298–299
defense, 47, 58, 97, 189–190, 195
defense elements, 195
defense mechanisms, 189
defoliation, 382
deforestation, 77, 114
degenerate community, 203
degradation, 20, 227, 272, 280–281, 338–339, 361, 379, 398–399, 401–402, 405–406, 409, 415
degradation processes, 227, 281
degradation sequence, 399, 402
degradative and rehabilitative processes, 338
degraded ecosystems, 397, 403
demersal fishes, 271
demographic cycles, 425
demographic–economic process, 433
demographic subsystem, 436
demographic transition, 430
demography, 428, 444–445
densities, 73, 143, 205, 215, 221, 258, 296, 363, 418
density, 63, 66–67, 69, 74, 76–77, 84, 106–107, 109, 111–112, 134–137, 144, 165, 190, 204, 218, 236–238, 296, 300, 302, 323, 327, 340, 350, 357, 362, 383, 416, 418, 452–453, 539, 564, 613, 618–619
density dependence, 106, 109, 111, 340
density dependent cannibalism, 69
density dependent (logistic) kinetics, 236
density dependent losses, 237
density dependent mortality, 323, 327
density dependent survival, 327
density independence, 340, 452
density independent, 112, 236, 238, 452
density independent growth, 112
density independent losses, 236
density self-regulation, 350, 357
dependence between dimensions, 183
dependence of recruitment, 214

dependency, 29, 72, 116, 189, 442, 624, 640
dependent state variables, 35, 374
dependent variables, 35
deposit feeders, 496–497
depressions, 53, 339, 415–417, 419–420, 424–426, 428, 433–436, 438, 440–441, 443, 445
depuration, 338
description of reality, 44
descriptive accuracy, 37–38
desert birds, 518, 523
desert community, 357, 514, 521, 525
desert detritivores, 515
desert food webs, 529, 535, 541
desert herbivores, 523
desert kangaroo rat, 518
desert kit fox, 523
desert shrubs, 388
desert soils, 524
desert tortoises, 519, 523
desert webs, 514
desertification, 9
deserts, 110, 114, 340, 357, 388, 398, 468–470, 510, 514–516, 518–526, 529, 531, 535, 541–542
desiccation, 220
desorption, 157, 282
destabilization, 107, 339, 418–420, 565, 618–619
destabilizing forces, 446
destructive inputs, 46
detection probability, 214
deterioration of the environment, 209
determinacy, 167, 177
determinate, 171
deterministic, 14, 21, 32, 107, 112, 141–143, 167–169, 171–173, 196, 221, 228, 282, 296, 302–306, 308, 346, 362, 373, 434, 456, 643
deterministic and stochastic formulations, 296
deterministic and stochastic methods, 142
deterministic and stochastic phenomena, 21
deterministic behavior, 171
deterministic description, 142
deterministic form, 303
deterministic models, 14, 196, 221
deterministic object, 167
deterministic physical systems, 172
deterministic solution, 308
deterministic systems, 107, 142, 362
deterministic theory of fisheries, 196
detrital, 57, 91, 512, 520, 522, 524, 541, 553, 557–559, 635, 637
detrital compartment, 520, 557

671

detrital feeders, 557
detrital flows, 635
detrital food chains, 520
detrital pool, 557–558
detritivores, 515, 520–521, 523–525, 532, 535, 537, 542
detritivory, 516, 521, 540, 557–558, 644
detritus, 258, 265, 271, 388, 461, 495–497, 516, 520–526, 531–532, 535, 537, 540, 542, 553, 557, 565, 581, 617, 635
detritus pathway, 565, 617
developers, 228
development, 9–10, 13, 18–21, 31–33, 38–39, 45–46, 49, 52–54, 66, 69, 80, 89–90, 92–93, 105, 109, 113, 115, 117, 120, 122, 131, 134, 136, 140, 142–145, 147, 154, 156, 161–162, 167, 173, 176, 179, 228–229, 238, 242–243, 255, 295, 311, 314–320, 323, 330–331, 338–339, 343, 356–357, 361, 365, 398, 401, 405–407, 409, 426, 430, 434, 445, 453–454, 462, 480, 492, 535, 543, 562–563, 565–569, 576–578, 605, 616–617, 619, 623, 641, 643, 647, 650–651, 654
development activities, 311
development and change, 45
development capacity, 567, 651
developmental plasticity, 617
deviation, 32, 35, 131–134, 144–145, 177, 182, 195, 233, 305, 440, 452, 563, 566, 579, 592
deviation coordinates, 131, 133
diagnostic, 338, 399
diagonal, 194, 256, 321, 348–351, 355–356, 359, 365, 367, 488, 537, 551
diagonal dominance, 349–351
diagonal stability, 348, 365
diagonally quasi-dominant, 350–351, 365
diagonally quasi-recessive, 351
diagonally recessive in columns, 351
diagonally recessive matrix, 351
diatom communities, 402
diatoms, 42, 133, 579
diets, 56, 69, 516–519, 522–523, 531–533, 534–535, 539–540, 542, 551
difference equations, 44, 107, 141–142, 157, 216, 314–316, 321, 328
differential equations, 21, 44, 105–107, 134, 139, 141, 143, 156–157, 161, 168, 243, 259, 265, 270, 282, 337, 350, 374, 473
differential equations model, 337

diffusion, 45, 81, 160, 284, 366
diffusion constant, 284
diffusive flow, 135
digital and analog computers, 11
digital computers, 2, 13–15, 21, 610–611, 621
Diguetia mohavea, 526, 528, 530–532
dilute solutions, 138, 587, 605
dimensional conversion, 327
Dipodomys, 518, 523
Diptera, 526–527, 530
direct interactions, 49, 264
directional change, 6–7, 376
directionality, 52, 86, 88, 94, 97
disclimaxes, 8
discontinuity, 44, 49
discrete, 8, 91, 95, 107, 112, 157, 167–168, 173, 175–177, 192, 199, 216, 335, 341, 473, 476, 512, 521, 535, 540, 550–552, 557, 565, 567, 603, 630–632, 640, 648
discrete aggregation, 630, 632
discrete event formalisms, 476
discrete event simulation, 341, 476
discrete models, 107
discrete population models, 335
discreteness, 167–168, 521, 603
disease, 323, 357, 400–401, 407, 418, 509
disease incidence, 400–401
disease tumors, 401
disease vector, 407
disorganized simplicity, 21
dispersion, 195, 318
dispersion values, 195
displacement, 32, 53, 94, 96, 374, 384–385
disruption, 6, 10, 22, 295, 382, 398–399, 652
disruption of ecosystems, 10, 295
disruptive or positive feedback loop, 49
dissimilarity measure, 566
dissipating toxic materials, 404
dissipation, 30–31, 51–54, 67, 69, 77, 79–91, 93–94, 96–97, 113–114, 117–118, 337, 378–381, 386–387, 390–391, 567, 553, 627, 652, 654
dissipation of heat, 54
dissipation parameters, 118
dissipation rate, 83, 89, 91, 93
dissipative, 31, 63, 77, 80–83, 85, 87, 91, 113, 117, 337, 346–352, 356, 362–364, 378–379, 381, 510, 593, 653
dissipative community matrices, 350
dissipative quasi-recessive matrices, 352

dissipative structures, 31, 81–82, 87, 337, 379, 381
dissipative systems, 117, 346, 348
dissipative units, 77, 80, 82
dissipative Volterra system, 350
dissipativeness, 348–349, 351, 362, 365
dissipativeness of a competition matrix, 362
dissolved inorganic nitrogen, 266, 271
dissolved inorganic or organic materials, 138
dissolved nutrients, 227, 270
dissolved organic matter (DOM), 520
dissolved oxygen, 138, 153, 233, 315
distress, 338, 397–398, 403
distribution, 14, 40, 42, 44–47, 58–64, 66–70, 73–78, 80, 82, 84, 91, 167–169, 179, 181, 183, 189, 209, 215–216, 235, 237, 296, 298, 301–302, 304, 308, 323, 365, 375, 382, 391, 403, 406, 408, 425, 477, 479, 501–502, 512, 517, 521, 535, 540, 563, 620, 628, 645, 648–649
distribution of age, 66
distribution of deer kill, 298
distribution range volume, 215–216
distributional patterns, 77
distributions, 14, 30, 55, 58, 65, 70–72, 74–75, 77–78, 167–169, 176, 182, 226, 236–237, 262, 267–268, 272, 302, 304, 363, 550, 566, 638, 648
disturbance, 31, 44, 46–47, 55, 67, 84, 90, 94–96, 106, 335, 337–339, 372–374, 376–377, 382, 385, 389–391, 397–399, 402, 405, 408, 456, 511, 564, 567, 612, 620
disturbance and stability, 44
disturbance rates, 456
disturbances, 9, 46, 90, 112, 116, 337–339, 372, 377–378, 384, 390–392, 398–401, 405, 408, 415, 456, 564, 617, 619, 621–622
diversification, 80, 611
diversity, 7, 9, 30, 40–44, 46–49, 51–52, 55, 80, 82–83, 86, 89–91, 94, 97, 105, 161, 171–172, 335–336, 338, 344, 373, 376, 398, 400–401, 405–406, 417–418, 445, 459–460, 510, 513, 515–516, 518, 524, 526–527, 532, 535, 538, 542–543, 574–575, 618–619
diversity and connectivity, 48
diversity and pattern, 47
diversity decline, 80, 90

diversity measurement, 41
diversity of life, 9
diversity of species, 7, 80
diversity spectra, 41, 44
diversity–stability hypothesis, 30, 55, 94, 336
divide and analyze, 225
dodo, 77–78
doe, 306
does (females), 306
dogs, 73
dollar flow matrices, 493
domain of stability, 347
dominance, 56, 67, 72, 83, 91, 93, 107, 349–352, 526
dominance hierarchy, 93, 107
dominant controlling variables, 154
dominant mode, 58
dominant species, 58, 67, 72–73, 79–80, 83, 90, 93–95, 142, 154, 401–402, 516, 569
dominant stand, 77
domination, 9
doves, 523, 533
downward hierarchical decomposition, 226
drainage basin, 406
drainage canals, 324
draining marshes, 9
drilling cost, 423
driving forces, 137–138, 143–145, 154, 156
Drosophila melanogaster, 108
Drosophiloidea, 528
drought, 405
dry season, 73
dual hierarchy, 337, 380
ducks, 317
dynamic analysis, 328
dynamic approach, 282
dynamic behavior, 31, 236, 264, 267, 331, 344, 346, 356, 377, 391, 424, 426, 438, 444, 472–473, 591
dynamic conflict, 90
dynamic equilibrium, 6–7, 31, 52, 54
dynamic fate models, 282
dynamic interactions, 118, 407
dynamic models, 192, 227, 282, 357, 364–365, 474, 511, 541
dynamic natural resource system, 329
dynamic paradigm, 328
dynamic programming, 13, 143–145, 156, 301, 308, 439
dynamic response behavior, 237
dynamic stability, 54, 107, 184, 376
dynamic structure, 567, 646–647
dynamic systems, 53, 330, 487, 647
dynamical stability, 610, 622

dynamical systems theory, 375
dynamics, 2, 6, 10, 19, 31–32, 41, 84, 96, 104–105, 107, 114, 136, 138, 140, 142, 156–157, 160–161, 169, 227, 231–232, 235, 237–238, 264, 280, 295, 306, 311, 313–315, 317–319, 323, 327, 329–331, 336, 341, 343–346, 350, 357–358, 362, 365–366, 373–385, 389–391, 426, 438–439, 450–451, 453, 458, 460–462, 464, 509–511, 515, 519, 521, 541–542, 550, 553, 559, 563, 566, 569, 590–591, 593–594, 617–618, 622, 643–644, 646–649
dynamics and functional integrity of the ecosystem, 381
dynamics of coniferous forests, 382
dynamics of continuous media, 169
dynamics of the fate of contaminants, 280
dynamics versus energetics, 511, 541

earth, 4, 35, 38, 53–55, 81, 227, 382, 459–460, 468, 470–471, 474, 481
earth functions, 53
earthworms, 461
ecesis, 6
ecodeme level, 83, 93
ecodemes, 30–31, 79–80, 82–85, 93, 96
ecodevelopment, 338, 398, 409
ecological aggregation theory, 232
ecological channels, 641
ecological community, 4, 550
ecological competition, 237
ecological complexity, 29, 31, 225, 340
ecological concept of stability, 366
ecological consequences of the reliability limit law, 184
ecological degradation, 398
ecological development, 569
ecological efficiency, 80
ecological flow network, 653
ecological holism, 509
ecological interactions, 116, 238–239, 263, 273
ecological interpretation, 166, 187, 194, 213, 601
ecological modeling, 31, 34, 38, 227, 248, 262–263, 273–275, 356, 568, 582–583, 585, 589, 605
ecological models, 38, 109, 113, 120, 142, 243, 245, 291, 336, 344, 346, 444, 583, 593
ecological networks, 366, 510, 563, 565–566, 578, 623–625, 627–629, 635, 637–641, 651

ecological niche overlap, 336, 361, 365
ecological niches, 43, 336, 361, 365, 576–577
ecological or global modeling, 434
ecological organization, 30–31, 42, 122, 225, 335, 343, 372, 397, 415, 450, 468, 485, 509, 513, 549, 561–562, 568, 585, 609, 623, 643
ecological pyramids, 339
ecological research reserves, 76
ecological segregation, 46
ecological stability, 337, 339, 345–347, 362, 364–366, 378, 417, 446, 460, 623
ecological stability analysis, 366, 378
ecological stability theory, 365, 417
ecological succession, 43, 46, 90, 399, 454, 457, 462, 610–611, 614–615
ecological systems analysts, 308
ecological theory, 21, 95, 104–105, 112–113, 231, 237, 373, 430, 452
ecological time, 31, 52, 83, 90, 92
ecological transfer efficiency, 80
ecology, 1–5, 9–16, 18–23, 29–34, 36, 38–42, 44, 52, 84, 97–98, 104–105, 108, 112, 121, 123, 162, 166–167, 173–174, 226, 228–229, 230–232, 234, 236–239, 242, 247, 274–275, 283, 295, 311, 313, 331, 335–341, 345, 353, 364–366, 373, 380, 415–416, 418, 431, 445–446, 451, 453–454, 469, 481, 485–486, 509–511, 517–518, 543, 549–550, 561, 565–567, 571, 582, 585–586, 601, 605, 623, 629, 643, 647, 654–655
econometric, 444
economic, 8, 42, 120, 122, 142, 177, 231, 234, 243, 275, 308, 312, 315–316, 318–319, 397, 409–410, 415–416, 420, 424–425, 427, 430, 433, 438–439, 441–443, 445–446, 453, 486–487, 490–491, 493–494, 505, 649, 652
economic activity, 430
economic assistance, 446
economic changes, 439
economic cycles, 425
economic development, 409
economic downturn, 442
economic fluctuations, 439
economic growth, 425, 446
economic growth rates in industrial societies, 446
economic models, 142, 243

economic perturbation, 446
economic return, 308
economic stagnation, 420
economic stress, 442–443
economic systems, 120, 243, 453, 494, 505
economic systems modeling, 243
economic theory, 234, 445
economics, 248, 313, 340–341, 344, 348, 350, 354, 416, 424, 444, 463, 486, 567
economy, 9, 19, 22, 45, 56, 76, 83, 93, 122, 339, 342, 397, 416, 420–421, 423, 425, 443, 445, 499–502, 504, 511, 650
ecosystem adaptation, 398
ecosystem analysis, 130, 134, 372, 376–377, 381, 384, 485
ecosystem ascendency theory, 566
ecosystem behavior, 20, 114–115, 230, 567, 655
ecosystem biogeochemistry, 384
ecosystem breakdown, collapse, and devolution, 405
ecosystem change, 6
ecosystem channel, 46
ecosystem component reactions, 110
ecosystem components, 2, 5, 13, 114, 134, 357, 553
ecosystem composition, 392
ecosystem concept, 2–5, 34, 40, 51, 104, 113, 130, 166, 242, 295, 415, 450, 485, 568, 609, 623
ecosystem constraints, 87, 373
ecosystem degradation, 402
ecosystem development, 80, 117, 131, 462, 578
ecosystem disintegration, 401
ecosystem distress syndrome, 338, 397
ecosystem diversity, 80, 89
ecosystem dynamics, 114, 136, 138, 142, 264, 373–378, 380, 384, 644, 647
ecosystem dynamics and stability, 373–375
ecosystem ecology, 10, 84, 239, 510, 605
ecosystem existence, 114–115
ecosystem fate models, 280
ecosystem functional integrity, 376–377
ecosystem growth and development, 641
ecosystem hierarchy, 83, 94
ecosystem integrity, 376, 378, 410
ecosystem interdependence, 342, 486
ecosystem level and population level regulating processes, 433
ecosystem level feedback loops, 457

ecosystem manipulation, 9
ecosystem metabolism, 384, 386–387, 389
ecosystem modeling, 115, 141–142, 145, 154, 232, 234, 236, 264, 269
ecosystem networks, 553, 654
ecosystem nonlinearity, 109
ecosystem optimality, 115
ecosystem organization, 113, 375, 377, 380, 384, 386–387, 391
ecosystem organization and dynamics, 380
ecosystem pathology, 397–398, 405–406
ecosystem property, 5
ecosystem reality, 109
ecosystem recovery, 382–383, 391, 402, 409
ecosystem relative stability, 372, 383–384, 391
ecosystem resilience, 391
ecosystem resistance, 390–391
ecosystem responses, 110, 373, 377
ecosystem simulation models, 113
ecosystem stability, 336–337, 372–378, 380, 383, 392, 418
ecosystem states, 409
ecosystem stress, 338
ecosystem stress adaptation, 338
ecosystem structures, 108, 119, 378, 380–382, 384, 387, 562–563, 577, 603, 605
ecosystem terminology, 4
ecosystem theory, 31, 104, 121, 232, 416
ecosystem therapy, 410
ecosystem transformations, 398
ecosystems, 1–20, 22–23, 29–35, 39–47, 49, 51, 53–58, 67–68, 72–73, 77, 79–80, 82–85, 87–96, 104–105, 108–123, 130–136, 138–143, 145, 147, 153–154, 156–157, 161, 165–166, 171, 173, 225–228, 230–232, 234, 236, 239, 242–243, 258, 260, 262–270, 280–281, 291, 295, 309, 311, 335–338, 340, 340–346, 352–353, 357–358, 360–361, 364–365, 372–392, 397–410, 415–416, 418, 430, 433, 450–454, 456–464, 468–471, 480, 485–491, 494–495, 498, 501, 505, 509–513, 517, 519–521, 523, 535, 541, 549–550, 552–553, 557, 559, 561–574, 577–578, 582, 585, 587, 589, 593, 595, 599, 603–605, 609–620, 623–624, 635, 638, 640–641, 643–645, 647–650, 653–655
ecotoxicological models, 281

educational taxes, 446
effect of overcatch, 209
effective roughness height, 286
effectiveness, 13, 20, 36–38, 63, 85, 297
effectiveness frontier, 38
effectiveness index, 38
effectivity, 32, 167, 172, 174–175, 178–179, 182–184, 186–187, 192
effectivity parameter, 186
efficiency, 13, 80, 84, 89, 93, 214, 308, 358, 400, 406, 427, 446, 461–462, 549, 558, 611
effluents from pulp mills, 404
eggs, 69, 76, 417, 469, 519, 526–527, 532–535
elasticity, 191
electric fields, 136
electrical conduction, 81
electrodynamics, 86
electromagnetic energy, 135–136, 141
electronic equipment, 12
electronic nets, 48–49
electronic surveillance of herd size, 301
element cycling, 381, 391
element duplication, 175
element mobilization, 378, 391
element recycling, 378
elementary interaction processes, 170
elementary versus emergent processes, 32
elements, 4, 6, 13–14, 17, 31–32, 38, 48, 52, 55, 114–117, 122, 167, 170–172, 175, 177–178, 181, 183–185, 187–195, 199–201, 204, 215, 217, 222, 253–254, 256, 315, 317, 321, 331, 337, 341, 352–353, 358, 374, 378–379, 381–382, 385, 387, 390–391, 408, 410, 415–416, 418, 453, 470, 472–473, 483, 487–489, 491, 493–494, 503, 506, 512, 551, 557, 562, 564, 566–567, 587, 589, 614–615, 630, 632–635, 638, 640, 652, 655
elevation, 318, 324–325, 375, 377, 428, 522
elimination, 90, 166, 176, 178, 184, 281, 346, 361, 364
elk, 9, 229, 317, 319–321, 323, 327
elk migration, 321
elk movements, 319, 321
elk population dynamics submodel, 317
elk populations, 229, 319
elk submodel, 321, 323, 327
embayments, 267–268, 271–272
emergent laws, 462

emergent processes, 32, 170
emergent properties, 84, 225, 485, 509
emigration, 303, 418
empirical (or process) relationships, 235
empirical experiments, 274
empirical representation, 476
encounter principle, 346
encounter probability, 214
endosymbiont microbes, 515
energetic input, 521
energetics, 96, 400, 511, 541
energy, 1, 4–9, 13–15, 19, 30–31, 35, 42–47, 51–55, 67, 69, 75, 79–97, 108, 113, 117, 131–136, 138, 141, 143, 146–147, 149–151, 157, 164–165, 167, 169, 178, 186–187, 229, 232, 238, 265, 320, 337, 339, 341, 343, 345, 358, 360–361, 373–374, 378–382, 386–387, 390–392, 400, 416, 420–424, 427–429, 431, 434, 436, 439–441, 443–446, 451, 453, 458, 460, 462–464, 468, 470–471, 485–486, 493, 498, 500–504, 515–516, 518–527, 529, 535, 539, 542–543, 549–550, 552, 561–562, 565, 567, 570, 578, 586, 590, 605, 609, 614–618, 621, 623–624, 627, 638, 640, 648, 650, 653
energy ambience, 81
energy and entropy, 52–53, 135
energy and entropy exchange, 52
energy and nutrient flows, 463–464, 518
energy balance, 131, 135
energy balance equations, 135
energy base, 460
energy channels, 521–522
energy consumer, 431
energy crises, 422, 446
energy currents, 85
energy cycling, 552, 578, 605
energy deficit, 146, 150–151, 165
energy demand, 149, 444
energy depletion, 440
energy development, 229
energy discovery and production, 422
energy dissipation, 53, 69, 79, 93–94, 113, 337, 378, 380–381, 387, 390–391
energy distribution, 82, 91
energy fixation, 381, 386
energy flow, 5–6, 19, 30, 51–53, 75, 79–84, 86–87, 91–92, 96, 238, 345, 361, 453, 516, 520, 522–523, 529, 549, 561, 605, 617–618, 638, 650, 653

energy flow model, 238
energy flux, 30–31, 52, 80–83, 86–87, 93, 143
energy inflow, 361
energy input, 8, 30, 52–55, 80–83, 87, 89, 117, 361, 379, 427
energy input (leaf fall), 361
energy input costs, 427
energy input cycle, 53–54, 81, 83
energy input regime, 55
energy intensity, 341, 500–502
energy intensity coefficient, 341
energy–matter, 31, 341, 565
energy–matter and signal flows, 341
energy–matter open, 31
energy pathways, 55
energy potentials, 85
energy prices, 429, 434, 440
energy processing, 373, 392, 400
energy production, 420, 422, 424, 431, 436, 439, 441, 458
energy production system, 436
energy quality, 42
energy requirements, 69, 616
energy reserves, 53, 422, 434, 441
energy resources, 420, 422, 431
energy sources, 416, 520–521, 535, 542
energy storage, 146, 149
energy storages and flows, 416
energy subsystem, 428, 434
energy supply shortage, 421
energy transfer, 92
energy uptake, 80
energy use, 434
energy values, 91
enforcement, 296, 301
England, 76, 220, 418, 422
Engraulis ringens, 89
enthalpy, 132, 135, 138, 146, 148, 588
entity–aspect relation, 478
entity–specialization relation, 478
entity-structure axiom, 480
entrophication model, 579
entropy, 30–31, 30–32, 41–43, 52–54, 80–83, 86–87, 86–94, 96–97, 130–132, 134–141, 140–146, 156, 164–165, 168, 344, 381, 561–562, 570, 577–578, 586, 601, 605, 621, 623, 626
entropy balance, 32, 136–138, 141, 143, 146
entropy balance equations, 32, 136–137, 143
entropy control, 31, 130–131
entropy export, 32, 140
entropy flow density, 137
entropy flux density, 136
entropy increase, 43, 81–82, 562
entropy macroparameter, 168

entropy principle, 32, 131, 139, 156
entropy production, 30, 32, 81–83, 86–92, 94, 96–97, 131, 136, 139–142, 144–145, 156, 165, 381, 561
entropy reducing processes, 131
environment, 4–10, 13, 17, 32, 47, 54, 69, 72, 83, 88, 94–96, 112, 114–118, 170, 173–176, 178, 190–192, 194–198, 200, 206, 209, 212, 227, 279–282, 291, 318, 336–337, 340, 343, 363, 373, 376–378, 380–383, 385–387, 391, 397, 401, 405, 415, 417–418, 445, 450–452, 454–456, 459–460, 470–472, 480, 483, 485–486, 491, 509, 557, 561–565, 570–571, 577–578, 586–590, 594, 597, 601, 605, 609, 612, 615–619, 621, 625, 653
environmental behavior, 173, 178, 227, 279, 281
environmental behavior of a contaminant, 281
environmental carrying capacity, 364
environmental change, 9, 83, 97
environmental community, 320
environmental concentrations of pollutants, 280
environmental conditions, 31–32, 53, 79, 96, 105, 112, 115–116, 178, 191, 206, 214, 219–220, 264–265, 274, 283, 288–289, 291, 459, 472
environmental conditions at the air–water interface, 283
environmental constancy, 53, 619
environmental constraints, 67, 138, 140, 373, 391, 462
environmental contaminants, 317
environmental control systems, 471
environmental deterioration, 206, 208
environmental effects, 173, 198
environmental factors, 53, 280, 377, 407
environmental fluctuations, 53, 55, 67, 161, 337
environmental health, 410
environmental heterogeneity, 107
environmental influences, 104
environmental loss, 232–233
environmental noise, 175
environmental oscillations, 377–378, 380–382, 384
environmental pressure, 179, 191, 196–197
environmental problems, 18, 39, 329, 445
environmental quality, 452, 454

675

environmental resources, 204–206, 208–209, 214
environmental signals, 87–88, 381
environmental standards, 280
environmental systems, 187
environmental temperature, 92, 561
environmental variability, 85, 565, 616
environmental variables, 112, 116, 235, 374, 382, 606
environmental variation, 376, 619
environs, 486
epibiota, 227, 264
epibiota model, 264
epicoenology, 338, 398, 407–408, 410
epidemic disease, 509
epifauna, 265, 271
epiflora, 265–266, 268–269, 271–272, 274–275
epifloral biomass, 268
epigeal organisms, 68
epiphytic sediments, 266
equidurability, 182
equilibrium, 6–8, 30–32, 44, 51–54, 80–82, 87–90, 94, 96–98, 107, 112–114, 131–134, 139, 164, 196, 227, 232, 234, 237, 280–282, 284, 344–347, 349–353, 359, 361–367, 378, 381, 387, 391, 452, 456, 459, 461, 470, 562, 567, 570–571, 577, 580, 586–587, 589–594, 601, 603, 644, 650, 655
equilibrium and dynamic models, 227, 282
equilibrium catch, 196
equilibrium conditions, 227, 280
equilibrium level, 452
equilibrium models, 281–282
equilibrium-seeking principle, 570
equilibrium solution, 346
equilibrium stability, 351
equilibrium state, 112, 344–347, 350, 359, 362–363, 590–594
equilibrium system, 81–82
equilibrium thermodynamics, 82, 567, 655
equipartitioning energy, 85
equipment, 12, 20, 204, 213–214, 227, 427
equipollence, 88
ergoclines, 46–47
ergodic, 169, 173
ergodic processes, 173
ergodic systems, 173
ergodicity, 166, 169
ergodynamic equilibrium, 31–32, 81, 83, 88–90, 94, 96–97
Erodium seeds, 518
erosion, 6, 31, 272, 402

erosion cycles, 6
erosion of stability, 31
error measures, 226
error statistic, 233
errors, 3, 8, 226, 237, 239, 267–268, 281, 307, 316, 542
Esox lucius, 56, 75
essential elements, 374, 381–382, 385, 387, 391
establishment, 52, 66, 69, 76–77, 88, 95, 173, 280
establishment population, 95
estimates of deviation, 195
estimation algorithm, 432
estimation of parameters, 300
estuaries, 262–264, 268, 270, 319, 388, 404–408
estuarine ecology, 275
estuarine ecosystem, 226
estuarine littoral, 388
estuarine macrophyte communities, 226
estuarine macrophyte ecosystems, 269
estuarine resources, 263
estuarine sediment loading, 272
estuarine systems, 263–264
estuarine waters, 267, 269
estuarine watershed, 274
Euclidean distance, 169
Europe, 407, 433–434, 446
European virgin forests, 30, 64, 77
eustress, 398
eutropic lakes, 239
eutrophication, 274, 391, 399, 401–402, 404, 562, 572, 575–576, 578–579, 582, 599, 650
eutrophication model, 572, 576, 578
eutrophication modeling, 582
evenness, 48–49
evolution, 31, 38, 43–46, 52–53, 69, 85–86, 88–89, 92, 94, 97, 113–115, 121, 131, 133, 140, 143, 164, 170–171, 178, 337, 377, 380–381, 385, 410, 445, 451, 459–462, 511, 543, 563–565, 568–569, 585, 590, 594, 597, 603–605, 610–622
evolution of a system, 594, 604
evolution of closed trophic chains, 563
evolutional modeling, 178
evolutionary, 6, 31, 52–53, 80, 82–84, 86, 89–90, 93–94, 96–97, 114–116, 120–123, 141, 170, 337, 374, 377, 379–381, 385, 391–392, 405, 445, 451, 459–460, 462, 563–565, 610–621
evolutionary adaptability, 565, 616–617, 619
evolutionary biology, 121–123

evolutionary convergence, 380, 392
evolutionary ecology, 121
evolutionary equilibrium, 52
evolutionary flexibility, 611
evolutionary history, 377, 392, 405
evolutionary homeostasis, 53
evolutionary incorporation, 380
evolutionary models, 121
evolutionary processes, 90, 94, 97, 141, 445, 611–612, 619–620
evolutionary tendencies, 52
evolutionary theory, 121
evolutionary time, 6, 31, 52, 80, 82–83, 86, 89–90, 97, 122, 170, 385, 391, 460, 619
exchange, 52–54, 56, 123, 131, 134, 136, 170–171, 173, 175–179, 181–183, 196, 267–268, 274, 281, 285, 291, 337–338, 374, 391, 419, 426, 470–471, 483, 488, 566
exchange constant, 285, 291
exchange of biomatter, 54
exchange of matter, 131
executive, 341, 471–472
exergical at steady state, 563, 597
exergical at zero, 593, 597–598, 606
exergical systems, 585
exergy, 32, 132, 561–563, 566–568, 570–583, 585–607, 650
exergy extremal principle, 563
exergy function, 562–563, 586, 588–589, 591, 593–594, 598, 602
exergy maximization, 562–563, 567
exergy maximum, 597
exergy measure, 562
exergy principles, 574, 578–579, 583, 585–586, 594, 596, 599–601, 604–605
exergy value, 605
existence, 11, 30, 32, 41, 53–55, 68, 84–85, 90, 94–96, 114–115, 120, 134, 142, 161, 167, 172, 175, 179, 234, 244, 254, 340, 345–347, 350, 353–354, 360–364, 416, 419, 459, 462–463, 469, 471, 481, 491, 493, 535, 542, 565, 610, 612, 617, 620, 654
existential computers, 563
existential game model, 612
exogenous disturbances, 391
exotic species, 400–401, 408
expansion of trade, 213
expected value, 195, 213, 216–217, 219
expenditure, 69, 80, 427
experience, 19–20, 22–23, 34, 47, 85, 94, 174, 230, 235–236, 296–297, 306, 311, 314, 327, 372, 444, 553
experiment stations, 3, 22–23

experimental control, 11
experimental design, 11, 237
experimental disturbance, 382–383
experimental manipulation, 382–383
experimental mathematics, 38–39, 571, 582
experimental phase, 10, 22
experiments, 10–12, 20, 111, 178, 226, 263, 265, 267–268, 272, 274, 283, 288–292, 338, 372, 382–383, 438, 457, 488, 511, 543, 589
expert systems, 229, 331, 471–472, 483
explanatory power, 29, 340, 464
exploitation, 7, 9, 89, 108, 220, 339, 417, 612, 617–618, 620
exploited populations, 58, 196, 221
exploited resource, 208
exploiting, 213, 422, 612
exploratory drilling, 424
exponential, 14, 58, 78, 84, 222, 235, 272, 427, 431, 433, 435, 439–440, 598
export of entropy, 139
exports, 419, 426, 429, 438, 441, 443, 487, 524, 567, 627, 645, 652–653
external disturbances, 112, 401
external driving, 31, 108, 111, 117, 399
external energy, 44–46
external flux, 5
external flux potentials, 5
external forces, 135
external impacts, 131
external influences, 338, 399, 405
external nutrients, 149–152
external resource inflow, 359–360, 365
external resources, 212
extinction, 42, 78, 90, 136, 157, 366, 460, 640
extrapolations from the laboratory, 281
extreme desert, 388
extreme properties of ecological systems, 336
extremum expressions, 203
extremum problem, 187, 204
eye development, 69
eyes, 69, 97

F optimized expression, 210
facilitating homeostasis, 339, 417–418
facilitation, 330, 340, 454–457, 564, 611, 613–615, 618–620
facilitation model, 454, 456–457
facilitation succession, 454–455
Fagus, 255

far-from-equilibrium, 31, 51, 53, 81–82, 96, 113–114, 378, 381, 387, 567, 577, 655
far-from-equilibrium states, 378, 381, 387
far-from-equilibrium thermodynamics, 567, 655
farm debt, 427, 438, 441
farm income, 428, 438, 441, 443
farm interest payments, 441
farm management, 13
farm mortgage debt, 428
farm operators, 427
farm output, 437
farm overhead, 441
farm sector, 427–428
farmers, 426
farmland, 76, 275
farms, 13, 427–429, 437–438, 441–443, 469
fast homeostasis, 339, 417
fate models, 279–280, 282–284, 291
fate of toxic contaminants, 279–280, 291
fate predictions, 282
fathead minnow, 402
fauna, 7, 56, 73, 468, 518, 522, 524, 526, 529, 534, 542
fawns, 298, 303–307
feasibility, 324, 364–366
feasible equilibrium, 344, 346–347, 349–353, 366
fecundity, 68, 83, 96
federal experiment stations, 3, 22
federal government, 421, 425, 428–429, 435–436, 438, 440–441
feedback, 19, 43, 49, 53, 108, 119, 142, 177, 270, 275, 328, 335, 339–340, 345, 350, 415, 417–420, 428, 438, 441–442, 450–457, 459–462, 488, 553, 562, 567, 576–577, 651–652, 654
feedback control, 339, 415, 418, 441–442, 488
feedback control loops, 418
feedback cybernetic system, 451
feedback effects, 108, 275, 328, 452
feedback effects (recycling of nutrients), 108
feedback in ecosystems, 340, 450
feedback interactions, 461
feedback loops, 43, 49, 454–457, 553
feedback mechanisms, 119, 142, 339, 345, 419, 562, 576–577
feedback models, 339
feedback networks, 460
feedback oscillation, 453
feedback relationship, 451
feedback systems, 339–340, 452–453
feedbacks, 114, 172, 264, 270, 340, 450–454, 457, 459, 462–463

feeding, 21, 66, 80, 106, 108–109, 111, 226, 263, 321, 358, 408, 452, 460, 464, 510–513, 516–520, 523–524, 526, 530–531, 533, 541–542, 549–553, 557, 559, 620, 635
feeding areas, 226, 263
feeding coefficient, 550–551
feeding coefficient matrix, 551
feeding ecology, 517–518
feeding linkages, 513
feeding rates, 111, 518
feeding relations, 511, 513, 542, 549–551, 559, 620
feeding webs, 552–553, 557
felling forests, 9
female harvest, 307
fenitrothion, 282
feral populations, 73
fertility, 17, 68
fertilization, 3, 9, 267–268, 274, 400
fertilization rates, 267
fibrosity, 46
Fick's law, 284
field conditions, 12, 268, 281
field observation tagging procedures, 301
field research, 238
field studies, 226, 235, 263
filter feeders, 110
filtering, 46, 110, 226, 263, 378, 380
fin damage, 401
fine-grained landscape, 339, 417
finite difference relation, 217
Finn's cycling index, 501
fire, 9, 398, 405
firewood, 422
first-order kinetics, 227, 280
fish, 17, 22, 30, 51, 55–59, 62–70, 72, 84–85, 91, 94, 194, 200, 205, 226–228, 242, 263, 270–272, 275, 281–282, 308–309, 311, 315, 318–320, 324, 328, 330–331, 401–402, 405, 408, 469–470, 515, 517, 519, 521
fish abundance, 271
fish and wildlife resources, 228, 311, 319
fish biomass, 57, 271, 282
fish larvae, 517
fish pond, 282
fish populations, 30, 51, 55, 57–58, 64, 67, 72, 85, 318–319, 401
fish production, 55, 270
fish schools, 85, 200
fish species, 56–58, 91, 328, 402
fish stocks, 57–58
fisheries, 4, 33, 51, 56, 72, 75, 89, 97–98, 196, 238, 262–264, 270, 275, 296, 302, 405
fisheries data, 75

677

fisheries harvest, 264
fisheries management, 33, 238, 275
fisheries management policies, 275
fisheries managers, 405
fisheries production, 262
fisheries science, 72
fishery, 56, 66–67, 196, 227, 313, 324, 405
fishery potential, 56
fishery science, 313
fishery theory, 196
fishes, 30, 51, 68–69, 258, 271, 308
fishing, 57–58, 66, 75, 84, 134, 196, 206, 208–209, 296, 328, 402
fishing birds, 57
fishing effort, 196, 206, 208
fishing gear, 296
fishing pressure, 75, 328
fishing regime, 66
fishing theory, 58
fitness, 79, 122, 235, 237, 464, 569–570
fittest, 39, 568
fixed or time varying parameters, 114
fixed state structures, 114
fixed structure, 192
flagellates, 528
flattened pyramid, 339, 417
fleas, 529
fledglings, 76
fleet characteristics, 296
flexibility, 18, 296, 392, 611
FLEXIPLEX, 305
flies, 516, 524, 526, 528, 538
Fliessgleichgewichte, 139, 144
flood control alternatives, 323–324, 326
flooding, 324, 326
floods, 68, 398, 405
flora, 7, 56, 73, 77, 468, 522
Florida, 242, 361, 469, 512, 553, 557
flow accounting, 341, 487, 492
flow analysis, 341, 486, 488, 490–498, 501, 505
flow behavior, 640
flow culture, 41
flow intensity matrix, 490
flow networks, 565
flow of energy, 4, 9, 91–92, 343, 379, 485, 515, 617
flow of matter, 136
flow of nutrients, 8, 401
flow pattern, 81
flow structures, 81
flow through system, 345
flowers, 91, 523
flows, 53–54, 91–92, 97, 131, 137, 140, 142, 265, 318, 339, 341–342, 361, 416, 429, 453, 463–464, 472, 475, 486–490, 492, 498, 501, 505, 511–512, 518, 520, 552–553, 557, 559, 565–567, 589, 605, 614, 623–624, 628–629, 632, 635, 638–639, 645, 648–650, 652–654
flows of energy, 54, 97, 131, 453, 624
flows of energy and matter, 54, 131
fluctuating (unstable), 95
fluctuation, 7, 52–54, 58, 67, 84, 430, 433, 439, 589
fluctuations, 6, 53–55, 67, 80, 88, 94–95, 140, 142, 161, 239, 337, 345, 385, 401, 419–421, 429–430, 432, 439, 446, 452, 459, 592
fluctuations in element availability, 385
flushing, 267–268, 270, 274, 404
flushing rate, 274
flux, 5–8, 14, 30–31, 52, 80–84, 86–87, 93, 95, 136, 264, 284–285, 339, 405, 418, 446, 509
flux of nutrients, 5
food, 7, 9, 13, 23, 54, 56–58, 67–69, 75–76, 80, 83, 91–93, 95, 97, 106, 114, 116, 214–217, 222, 270–271, 275, 321, 323, 328, 347, 361, 376, 400, 407, 418, 428, 438, 443, 453, 464, 469, 510–522, 524, 526, 528–529, 532–533, 535–536, 538–543, 550, 558, 563–565, 572, 611, 614, 616–617, 621, 635
food chain hierarchy, 67
food chain length, 400
food chains, 9, 54, 56–58, 67, 83, 91–92, 95, 270, 361, 400, 510, 512, 515–516, 520–521, 538, 558, 563, 617
food concentration, 106
food cycle, 512
food habits, 518
food limitation, 323
food object, 222
food organisms, 76
food particles, 215–216
food products, 9
food requirement, 321, 323
food resources, 75–76
food scarcity, 68, 542
food spawning areas, 214
food supply, 69, 321, 323, 328
food web analyses, 515, 518
food web community matrices, 347
food web complexity, 535
food web dynamics, 453
food web links, 564, 614
food web matrix, 538
food web models, 515
food web networks, 512
food web structure, 565, 616
food web theory, 453, 510–511, 513, 543
food webs, 7, 114, 347, 407, 453, 464, 510–518, 520–522, 529, 532, 535–536, 538–543, 564–565, 572, 614, 616, 635
foods, 17, 401, 418, 516, 518–519, 523, 525–526, 542
forage, 75, 298–299, 307, 321, 323, 405
forage quantity and quality, 298
forced and dynamic process, 44
forced process, 44
forced response simulations, 237
forcing functions, 38, 266, 268, 270, 581
forcing inputs, 46
forest and grassland ecosystems, 398
forest clear-cutting experiments, 382
forest development, 255
forest dominants, 95
forest ecosystem recovery, 383
forest ecosystems, 336, 402–403
forest floor, 302
forest removal, 382
forest simulation models, 382
forest stands, 387, 454
forest succession, 255, 457
forest systems, 58, 95
forests, 4–5, 9, 30, 64, 77–79, 92, 110, 318, 336, 361, 373, 376–377, 382–383, 385, 387–388, 422, 470
formalism, 32, 113, 131, 225, 231–232, 358, 378, 383, 475, 565, 595, 621
formalisms, 473–474, 476–477, 483, 620
FORTRAN, 14, 316–317
forward-oriented compartmental analysis, 492
fossil energy, 422, 424, 427, 431, 434, 439
fossil energy production, 422, 424, 439
fossil energy reserves, 434
fossil fuel, 342, 422, 424, 429–432, 434–438, 442, 499–503
fossil fuel production, 424, 430, 435, 437
fossil fuel reserves, 422, 442
four trophic-level systems, 108
fractional flow matrices, 512
fragmentation, 338
frame hierarchy, 474
France, 430–432, 434, 442
free area, 218
free energy, 53–54, 85, 96, 378, 562, 570, 590, 614

free energy gradient, 378
free enthalpy, 132, 138, 146, 588
free trade, 195, 197, 200, 206, 209, 214
French births, 431, 433–434
frequency of oscillation, 387
frequency signals, 378, 380, 391
frequency structure, 385, 387, 392
freshwater, 30, 51, 56, 71, 92, 97, 116–117, 227, 324, 328, 342, 387–388, 399, 407, 498–503
freshwater marsh, 324
freshwater mollusks, 71
freshwater phytoplankton, 116
freshwater systems, 92
friction, 136, 169, 285
friction velocity, 285
frozen lakes, 53
fuel consumption, 425
fuel price index, 434, 436
fuel prices, 425, 435–436
fuel production, 424, 430, 435, 437, 440
fumigation, 405
function, 6, 10, 13–14, 18, 23, 35, 44, 54, 58, 63, 67, 77–79, 82, 84–86, 94, 97, 117, 119–122, 132, 140, 142, 145, 147–148, 150–155, 157, 159, 169, 176–177, 182, 184, 186–187, 192–193, 196, 208, 211, 215, 220, 222, 226, 228, 232, 234–236, 243–249, 251–253, 257–258, 262, 265, 269, 285–286, 297–300, 302–303, 305, 307–309, 321, 335, 337–338, 340, 347, 349, 360, 364–365, 367, 376, 379–382, 384, 391, 404, 425, 427, 431–432, 434–436, 439–440, 453, 458, 477, 488–491, 505, 511, 513, 516, 518, 521–522, 543, 559, 561–564, 566–568, 570, 578–579, 586, 588–589, 591–594, 598–599, 602–603, 605–606, 612–614, 616–617, 630, 646, 648, 653
function Q, 120, 122, 360
functional complexity, 110
functional integrity, 376–377, 381–382, 390, 392
functional interaction, 48
functioning, 4, 8, 54–55, 80–81, 96, 114, 130, 139–141, 173, 175, 183, 192, 339, 344, 416, 469, 521, 615, 617
functions, 4, 14–15, 38, 53–54, 76, 81–83, 85, 89–90, 97, 120, 130, 132–133, 137–138, 141, 143–147, 149, 154, 156, 161, 166, 182, 184, 187, 193, 196–197, 199–200, 202–203, 213, 226,
232–234, 243–249, 253, 262, 266, 268, 270, 297–298, 302, 313, 317, 338, 359, 361–364, 366, 397, 401, 409, 424, 454, 468, 471–472, 485, 511, 562–563, 566, 581, 589–590, 593–594, 598, 600–601, 606
fundamental biology, 90
fundamental L property (self-organizability), 191
funding, 22, 410
fungal feeders, 524
fungi, 6, 402–403, 456, 523–524
fungivores, 524
future-directed flow analysis, 491
future states, 191

Gaia, 97, 123, 462, 509
Gaia hypothesis, 97, 462, 509
Gall midges, 526
Gambelia wislizenii, 531
game and fish management, 308
game herd age and sex structure, 308
game management, 296–298, 302
game management units, 296, 302
game managers, 296–298, 300–301, 308
game theoretic formulations, 122
game theory, 13, 115, 120, 174, 176–177, 458
gas, 12, 20, 85, 95, 138, 164, 169, 189, 201–202, 218, 284–287, 289, 323, 424, 471–472, 477, 480, 571, 589, 605, 611, 643
gas balance, 471
gas chromatographs, 20
gas constant, 138, 164, 284, 571
gas laws, 643
gas mixtures, 605
gas molecules, 85
gas phase, 284, 287, 289
gas resistance, 285
gaseous exchange, 53
gases, 8, 459, 471, 477, 562, 587–588
gasoline tax, 446
Gavia, 57
Gavia Lake, 58, 62–63, 65, 84
gear, 58–59, 62, 204, 296
gene–protein relationship, 613
general ecology, 313
general physiology, 214
general stability, 346, 562
general stress syndrome, 399, 402–403
general systems theory, 340, 453
generalist predators, 526
generalists, 16, 522–523, 526–527, 530–533, 619
generation time, 83, 619
genetic, 6, 46, 67, 83–84, 94, 112, 114–116, 121–122, 191, 235, 373, 450, 565, 569, 612–617, 619
genetic codes, 6
genetic control, 84
genetic drift, 112
genetic evolution, 114, 615
genetic identity, 122
genetic makeup, 67, 83, 94
genetic material, 115
genetic mechanisms, 121
genetic memory, 191
genetic systems, 450
genetic variability, 235, 617
genomes, 30
genotypes, 461
geochemical matrix, 378
geographic area, 314, 323
geographic boundary, 320
geographic gradient, 92
geographic information systems, 229
geologic boundaries, 93
geological time, 54
geometric progression, 258
geophysical and climatological changes, 90
German economy, 443
Germany, 343, 420, 442–443, 585
giant clam, 70, 72
giant land tortoise, 74
giant tortoises, 30, 72–73
giant tortoises of the Seychelles, 30
Gibbs relation, 131–134, 138, 140–141, 146
gill net, 58–59, 63, 75
Gini–Simpson index, 41
glacial moraine, 89
glaciation, 377
Gleasonian individualism, 95
global biosphere, 341, 462
global model, 426, 429, 431
global multicommodity input, 499
global multicommodity output, 499
global patterns of circulation, 391
global response, 110
global stability, 109, 349–350, 364, 593
global succession, 47
goal, 9, 13, 18, 31–32, 40, 85, 104–105, 112, 120, 122, 142, 167, 171–172, 174–175, 178–179, 182–184, 190–194, 197, 200, 202, 233, 410, 452–454, 457, 459, 463, 470–471, 479, 562, 566, 577, 579, 643
goal function, 120, 122, 142, 193, 579
goal functionals, 179, 182
goats, 73, 469
golden eagles, 534–535, 537, 539

679

gopher, 519, 523, 533
gophersnakes, 519, 534–535
government, 16, 274–275, 296, 320, 410, 418, 420–421, 424–425, 428–429, 433, 435–436, 438, 440–442, 445–446
government agencies, 275, 296
government debt, 425, 429
Grand Teton National Park, 320–321
Grande Terre, 72–74
granivory, 523, 533
grass, 95, 319, 382, 398
grass prairie, 398
grass production, 382
grasses, 387, 557
grasshopper mice, 518, 535
grassland ecosystems, 23, 309, 398
grasslands, 5, 9, 22–23, 75, 95, 104, 166, 227, 262, 279, 295, 309, 343, 376–377, 388, 398, 405, 456, 510, 585
gray birch, 255–256
grazed, 67, 398, 452
grazing, 17, 75, 79, 91, 95, 138, 144, 151, 159–160, 216, 269, 274–275, 377, 382, 452, 557, 578, 637, 565, 617
grazing animals, 75, 95
grazing chain, 637
grazing lawn, 75
grazing pathway, 617
grazing pressure, 269, 274
great age, 51, 67, 79, 94
Great Bear Lake, 56
Great Depression, 440–441
great horned owls, 526, 533
Great Lakes, 284, 288, 292, 391, 408–410
Great Lakes community, 284
Great Plains, 23
Great Smokey Mountains, 375
great tit, 85
greenhouse, 54
Greenland, 66
Gros Ventre, 321
Gros Ventre River drainage, 321
ground flora, 77
ground state, 87, 96
groundwater, 6, 318–319, 324
groundwater inflows, 324
groundwater response, 324
group dynamics, 311, 313, 327
group dynamics procedures, 311
growing season, 268, 274, 325, 478
growth, 51, 58, 66–67, 70–73, 76, 83–84, 93, 110–112, 116, 149, 156–159, 167, 175, 188–189, 206, 208, 217, 222, 231, 235, 263, 265, 268–270, 272–274, 302, 319–320, 350, 381, 387, 400, 405, 407, 416, 422, 425–426, 431, 433, 436, 441–442, 446, 451, 454, 458, 472, 519, 526, 532, 535, 542, 562, 566–567, 569–570, 572, 575, 578–579, 580–583, 594, 599, 609, 617, 623, 641, 643, 649–651, 654
growth and interaction parameters, 110
growth and production to maintenance, 400
growth curve, 70–71, 302
growth cycles, 274
growth pattern, 51, 67, 73, 84
growth periods, 269
growth rate parameter, 231
growth rates, 58, 66, 110–111, 116, 149, 231, 235, 302, 416, 426, 433, 446, 578–583, 594, 599
growth stages, 84
Gulf of Bothnia, 408
Gulf Stream, 407
guppie, 116
Gustavia trees, 452

h-age migratory population, 194
habitat, 9, 66–67, 69, 95, 226, 264, 270–271, 275, 313, 317–320, 324–325, 328, 331, 340, 362, 402, 404, 453, 522, 535, 540
habitat for wintering waterfowl, 317
habitat requirements, 313, 320
habitat structure, 404
habitat suitability, 317, 324–325
habitat suitability indexes, 317
Habitation Module, 474
habitats, 69, 97, 235, 263–264, 269, 275, 311, 318, 328, 468–470, 510, 514, 518–519, 542
Hadrurus arizonensis, 531
haemogregarines, 528
half life, 288, 292
half life of a contaminant, 288
Hamilton's first principle of function, 86
harmful community, 209
harmonic analysis, 110
Harney Lake, 324
hartley, 650
harvest, 93, 219, 228, 264, 296, 301, 303–308, 316, 319, 321, 323, 328, 338
harvester ant, 523
harvesters, 296
harvesting, 204, 206, 208–209, 211, 214, 306, 405, 409, 470–471, 480
harvesting efforts, 206, 208–209, 211
harvesting equipment, 204, 214
hatching, 69
hatchlings, 74, 76
hawks, 534, 550

hazard level, 282
HCB, 284–286, 288–292
heat, 53–54, 56, 81, 86, 136, 378, 469, 472, 487–490, 495–496, 499, 570
heat conduction, 81
heat flow, 136
heavy industrial markets, 442
heavy industry, 427
HEC 5 model, 325
Heisenberg principle, 29
Heisenberg's principle of indeterminacy, 174
Heisenberg's uncertainty principle, 32, 39, 87
Heisenberg's uncertainty relation, 34
helminths, 515, 528
Heming Lake, 75–76, 85, 96, 98
Hemiptera, 523
Henry's law constant, 281, 284, 286, 288, 290
herbicide loading, 271–272
herbicide runoff, 274
herbicides, 227, 263, 270–273, 274–275
herbivore trophic level, 552
herbivores, 6, 238, 462, 491, 515–516, 520–521, 523–526, 532, 535, 537, 539, 542, 550–552, 557, 635
herbivorous invertebrates, 227, 270
herbivory, 4, 227, 269–270, 516, 521, 523–524, 529–533, 535, 537–538, 644
herbs, 403
herd, 296–299, 301–303, 305–308, 317, 320–321, 323
herd and forage protection, 299
herd composition, 306–307
herd–forage relationships, 298
herd management, 296
herd ranges, 317
herd segments, 320, 323
herd size, 301, 303, 305–307
heredity, 52, 92, 115
heterochrony, 52, 92–93, 97
heterogeneity, 41, 80, 82, 107, 231
heterotrophs, 108, 458, 540
hexachlorobenzene, 284–285
Hexagenia, 402
hierarchical approach to biological systems, 381
hierarchical approaches to ecosystem ecology, 239
hierarchical array, 263
hierarchical biogeochemical systems, 337, 374, 380, 389
hierarchical level, 75, 171, 235, 379, 652
hierarchical levels of organization, 339

hierarchical net input intensities, 498
hierarchical or scale dependent systems, 379, 381
hierarchical organization, 31, 85, 230, 235, 378, 381, 564–565
hierarchically organized model, 434
hierarchy, 29, 31, 49, 67, 79, 83, 87, 90–94, 107–108, 142–143, 171–172, 174, 186, 337, 344, 347–349, 351–352, 356, 365, 380–381, 392, 471, 474–476, 483, 539, 565, 616–617, 620, 635
hierarchy concepts, 337, 474
hierarchy formation, 31, 90
hierarchy of biological systems, 186
hierarchy of functional processes, 392
hierarchy of stability notions, 348
hierarchy of stability properties, 352
hierarchy of stable community matrix, 365
High Arctic, 94
high densities, 73
high diversity systems, 41
high forest, 77
high-frequency phenomena, 132
high-resolution process models, 239
high technology, 435
high turnover systems, 373
higher level behavior, 225
highly complex models, 426
historical development of systems, 45–46
holism, 227–228, 509–510
holistic approaches, 39
holistic determination, 510
holistic ecosystem responses, 110
holistic hypotheses, 571
holistic principles, 121, 571
holons, 91
home ranges, 116
homeokinesis, 31, 82, 84–85, 87
homeorhesis, 53, 340, 452, 454, 457
homeostasis, 30–31, 53, 85–86, 172, 177, 339–340, 417–418, 452, 454, 460
homeostatic controls, 6
homeostatic mechanism, 52–53
Homo sapiens, 93
homogeneity, 63, 80, 82, 85
homogenization, 80
homomorphic (many-to-one) mappings, 29
homomorphism, 231–232, 473–474, 483
Homoptera, 523
honeybees, 403, 521
horizontal plane, 46
horizontal transport, 400

horizontally developed communities, 336
horizontally structured communities, 336, 358, 361
horned lizard, 533
horned owls, 526, 533, 539
host–parasite interactions, 107, 356
hostility, 442–443
human, 8–9, 16, 35, 42, 55–56, 84, 93–95, 110, 112, 114, 116, 122, 168, 171, 189, 191, 227–228, 232, 262–264, 270, 296, 319, 321, 328, 336–339, 343, 373, 376, 397–399, 401–403, 405, 407–410, 415–419, 433–434, 450, 452, 459, 468–472, 477, 479–480, 483, 549, 565, 568, 611, 620, 643, 650
human abuses, 409
human activity, 112, 228, 263, 270, 321, 408
human development, 228
human engineered ecosystem, 469
human habitation, 470
human hunting, 328
human impact, 56, 110, 227
human interaction, 410
human life span, 336, 373
human population densities, 296
human populations, 9, 296, 319, 321, 418, 433
human presence, 338
human pressures, 398
human recreational activities, 321
human representational system, 122
human resources, 35
human settlement, 338, 407–408
human societies, 84, 262, 339, 416–419, 650
human stresses, 398–399, 401–402
human stresses on ecosystems, 398
human systems, 264, 270, 434
human uses, 264
hunter–deer contacts, 300
hunter distribution, 296
hunter management, 296
hunter success, 300–302, 321
hunter types, 298
hunters, 227–228, 296–302, 321, 445
hunters' bag limits, 296
hunters' choice, 296
hunting, 228–229, 295–302, 304–305, 308, 321, 328
hunting access, 302
hunting effectiveness, 297
hunting load, 296
hunting management, 296–297, 302
hunting mortality, 321
hunting pressure, 228, 295–300, 302, 308

hunting region, 298
hunting season, 228, 296–298, 301, 321
hunting success, 302, 308
husbandry, 338, 397–398, 410
hydrocarbons, 133, 284
hydrodynamic controls, 387
hydrodynamic models, 292
hydrodynamic processes, 160, 379, 382
hydrodynamic turbulence, 385
hydrodynamics, 107
hydrogen, 189, 236
hydrologic analysis, 325
hydrologic chemical runoff model, 270
hydrologic conditions, 326
hydrologic cycles, 4, 97, 498, 501
hydrologic record, 324
hydrologic regime, 325
hydrologic system, 329
hydrologic versus carbon cycles, 342
hydrological and atmospheric currents, 407
hydrology, 130, 313, 315, 319, 324–325, 372
hydrology submodel, 315, 324–325
hydrolysis, 280–282
hydrosphere, 4
Hymenoptera, 526–527, 530
hyperbolic conditions, 186
hyperbolic growth, 159
hyperbolic law, 167
hyperbolic predetermination, 187
hyperbolic reliability law, 185
hyperenvironment, 588
hypereutrophication, 114
hyperparasitoids, 526–527, 532, 537
hypotheses, 10–12, 21, 121, 198–199, 227, 273–274, 312, 344, 416, 468, 511, 543, 571

I law, 176
IBP (International Biological Programs), 3, 9, 22, 111, 309, 444
ice cover, 56
ice invertebrates, 588
ichneumonid, 527
ichthyology, 30
Idaho, 296
identification (experimental verification), 174
identification of parameters, 218, 221
iguana, 523
image formation, 116
immigration, 112, 303, 404
immortality, 175
impact assessment, 296, 326
impact predictions, 326

681

impingement, 404
indefinite persistence, 340
indeterminacy, 29, 87, 166, 173–174, 178
indeterminacy relation, 29
indeterminate, 171
index, 36–38, 41, 47, 167, 192, 251, 265, 383, 386–387, 419, 421–425, 427–428, 430, 434–438, 459, 501, 565–566, 623–624, 638, 640–641, 645
index of accuracy, 37
index of effectiveness, 37
index of abundance, 324
index of habitat suitability, 324
index of species diversity, 344
Indian Ocean, 72, 77
indifference, 44, 621
indifferent pattern, 44
indirect effects, 30, 49, 112, 509–510, 578
indirect interactions, 49
indirect transfers, 551
inductive approach, 130
industrial automation, 177
industrial discharges, 284
industrial–environmental interactions, 317
industry technology assumption, 505
infauna, 271
inflation rates, 339, 426
inflow, 131, 135, 158, 160, 358–361, 365, 566, 570–571, 580, 590, 633, 635
inflow intensities, 566
inflow of energy, 135
inflow/outflow of phytoplankton biomass, 158
influence, 4, 6, 8, 19, 36, 42, 58, 88, 132, 144, 147–148, 196, 198, 214, 263–264, 270, 281, 296, 299, 307, 353, 373, 376, 397, 399, 415, 452, 458, 509–510, 515, 542, 571
information, 6, 14, 16, 21, 29–31, 35, 38–47, 67, 85, 87–88, 113, 123, 130–131, 142, 144, 166–167, 172, 176, 179, 194, 205, 225, 227–229, 242–243, 262, 266–267, 270, 273, 280–282, 297–298, 300–301, 307–308, 312, 314–317, 323, 328, 331, 335, 340–341, 344, 372, 399, 418, 450, 453, 462, 468, 470–472, 474, 477, 479, 483–484, 490–491, 498, 516, 522, 525, 528, 540, 542, 551, 562–567, 578, 581–582, 600, 605, 610–616, 619–621, 623–627, 629, 631, 635, 638, 641, 647–648
information entropy, 344

information flow, 31, 616
information increase, 43
information processing, 341, 470, 563–564, 610–613, 616, 619–621
information processing efficiency, 611
information theory, 14, 30, 40, 43, 172, 176, 179, 205, 565, 623, 647
information transfers, 314
ingestion, 236
inheritance, 475, 564, 569–570
inhibition, 111, 248, 340, 456–457
inhibition models, 340, 456–457
initial value of a process, 168
initial variables, 226, 245
input (forcing) variables, 112
input and output flows, 53
input and output variables, 314
input cycle, 53–54, 81, 83
input environ analysis, 488
input environs, 486
input–output, 232, 316–317, 341, 486, 493, 496, 498, 512, 567
input–output flow, 341, 496
input–output functions, 317
input–output relationships), 232
input overhead, 654
input variability, 31, 110, 114
inputs, 8, 30–31, 44–47, 52–55, 80–83, 85, 87, 89, 94, 109–114, 117, 133, 227, 232–234, 263, 269–270, 272–274, 280, 314–317, 319, 324, 341–342, 361, 374, 379, 399, 427, 437, 442, 470, 474, 486–489, 491–505, 512, 521–522, 551, 557–558, 565–567, 573, 575–576, 578–579, 610, 625–627, 645–647, 652–654
inputs of energy, 44, 47
inquilines, 526
insect grazers, 458
insect herbivores, 539
insect or fungus attacks, 403
insect parasitoids, 539
insect pests, 357
insecticides, 116
insects, 17, 357, 403, 416, 418, 458, 460, 469, 510, 514–516, 518–520, 522–524, 526–530, 532–533, 535, 539, 542
in situ, 162, 274
in situ observations, 162, 263
instability, 8, 30, 55, 94, 114–115, 140, 143, 158, 160, 164, 338–339, 377, 397, 415, 417–420, 424–426, 429, 436, 446, 460, 590, 592–593
instantaneous dissipation, 89
instantaneous probabilities of births and deaths, 200

instantaneous probability, 181, 206, 215–216
integral macroparameters, 186
integrated research, 22
integrity, 53, 168, 184, 192, 194, 230, 376–378, 381–382, 390, 392, 398, 401–402, 410
intensities, 148, 399, 493, 497–498, 500–501, 566
intensity factors, 341–342, 486–487, 491, 498, 505
Intensive Agriculture Module, 474
interacting molecules, 172
interaction density, 564, 619
interaction law, 170
interaction of separable variables, 258
interaction patterns, 564–565, 617
interactions, 4, 6, 10–11, 20, 31–32, 35, 40–41, 44, 46–49, 51–52, 66, 79–80, 82–85, 88, 95, 104–105, 107, 110–118, 120, 122–123, 157, 161, 167, 169–170, 172–173, 175, 177–178, 187, 192, 194–197, 213, 228–229, 231–232, 234, 238–239, 243, 250, 252, 258, 263–267, 269–271, 273–275, 283, 300, 312–313, 315–319, 321, 328, 330, 338, 340, 345–348, 350–352, 354, 357–358, 361, 365, 379–380, 400–401, 403–404, 407, 410, 416, 418, 420, 425, 433, 442, 444, 452, 454, 457, 460–464, 471, 485, 510–511, 513, 515, 518–522, 524–532, 534, 539–540, 549, 563–565, 567, 609, 611–613, 616–621, 650
interactions between trophic levels, 358
interactive gaming, 314, 316
interactive species, 510, 535, 538
interactive testing and debugging, 316
interactive trophic units, 538
interconnections, 13, 341, 486, 494, 513, 542–543, 620
interdisciplinary communication, 315–316
interdisciplinary team, 16, 18–19
interest, 9, 33, 41–42, 54, 78, 97, 107, 206, 231–232, 237, 239, 242–243, 255, 285–286, 296, 312–314, 317, 326, 328, 339, 372–373, 376–377, 399, 415–416, 420, 422, 424, 426–427, 429, 438, 441, 444, 458, 460, 494, 501, 550, 629
interest and inflation rates, 339, 426
interface, 47, 270, 283–286, 292, 312, 331, 403, 472

interfaces, 5, 8, 46–47, 404
intermittent reproduction, 69
internal and external constraints, 80, 89
internal and physiological resources, 212
internal chemical reactions, 133
internal control mechanisms, 131
internal damping, 94, 96
internal energy, 131–132, 134–135, 146, 149–150, 164
internal equilibration, 32
internal feedbacks, 450
internal interactions, 264, 616
internal overhead, 567, 652
internal self-regulation, 58, 67
internalization, 46, 79
International Biological Program (IBP), 3, 9, 22, 111, 309, 444
international migration, 437
interplanetary travel, 8
interpretation, 11–12, 55, 166, 171, 173, 187, 190, 194, 202, 212–213, 218, 297, 313, 336, 341, 350, 362, 372, 383–384, 390, 435, 453, 463, 479, 558, 582, 601, 603, 612
interprocess relationships, 170
interspecific interactions, 350, 454
intertidal community, 220
interval flow, 324
interviews with hunters, 301
intraguild predation, 531, 542
intraspecific competition, 351
intraspecific regulation, 350–351
intrinsic noise, 143
introduction of error, 239
introduction of new species, 77
intuitive aggregation, 250
intuitive reality, 234
Inuit family, 66
Inuvik, 71–72
invasion, 364, 640
invasion of species, 364
inventory, 5, 13, 301, 416
inverse size metabolic law, 83
invertebrate population, 57
invertebrates, 57, 91, 227, 264, 269–270, 468, 515, 519, 522, 539, 541
investments, 446
ion exchange, 281
ion transport, 149
ionic copper, 572
ionizing irradiation, 13
Iowa, 653
irreducibility, 30, 225
irreducible systems, 38–39, 562, 571, 582
irreversibility, 82, 90, 94
irreversible thermodynamics, 138, 380

irrigation, 324, 471
irrigation diversions, 324
isa link, 474–475
island biogeography, 52
islands, 47, 51–52, 56, 72–73, 77, 319
isolated system, 30, 53–54, 89, 570
isolation, 21, 30, 53–55, 70, 72–73, 77, 79, 89, 274, 314, 335, 379, 408, 440, 443, 462, 469, 485, 564, 570, 586, 614
isomerization, 281
isopods, 69, 524

Jackson Hole elk submodel, 327
Jackson Hole, Wyoming, 229
Jacobian, 245–246, 249, 359, 367
Jacobian matrix, 245–246, 249
jobs, 435
joint arrival, 179
joint interaction, 170
joint probability, 48, 183, 567, 640, 644, 647–648
joint production, 493
joint products, 341, 486–487, 491–492, 494, 498, 505
joint uncertainty, 567, 646–647
joint withdrawal, 214
Joshua Tree National Monument, 524
Jungle Module, 474
juvenile cave forms, 68
juvenile cohorts, 215, 219
juvenile fish, 58, 66
juvenile replacement stock, 51, 84, 94
juvenile segment, 68
juvenile stages, 195
juvenile stocks, 31
juveniles, 31, 51, 55, 58, 63–64, 66–69, 74, 84, 94–95, 195, 213, 215, 219, 221, 519, 530

K-selected species, 83
Keller Lake, 56, 60
Kelvin scale, 91
Kelvin temperature, 571
Kent Peninsula, 56
Keretella cochlearis, 56
Keratella hiemalis, 56
keystone predators, 269
kill, 228, 296, 298–303, 307, 533
kill fractions, 303
kill rate, 303
kill success, 298, 300
kinetic, 85–87, 135, 141, 267, 274, 381, 605
kinetic, internal, and electromagnetic energy, 141
kinetic energy, 85, 135
kinetics of petrochemical reactions, 243

kit foxes, 518–519, 523, 533–534, 539
kleptoparasites, 528
knowledge, 5, 9, 11, 16, 20, 29, 34, 36, 38, 51, 130–131, 142–143, 166, 171, 218, 225, 227, 234, 236, 242–243, 279, 292, 301, 308, 313, 327, 330, 341, 445, 462, 468, 470–481, 483–485, 510–511, 513, 571, 585, 605, 610
knowledge representation, 476, 479–480
Kondratieff waves, 416
Kronecker's delta, 254
Kuroshio Current, 407

L–V (see also Lotka-Volterra), 104–107, 109–110, 112–113, 117
L–V equations, 104, 109–110, 112–113
L–V models, 106–107, 112, 117
L–V systems, 106
Latrodectus hesperus, 530, 532
labor, 423–425, 435, 440–442, 444, 446
labor force, 424–425, 441, 446
labor market, 435, 441
laboratories, 3, 22–23, 154, 162, 262, 511
laboratory data, 147, 280–281, 292
laboratory microcosms, 226, 263, 469
laboratory studies, 130
laboratory toxicity, 327
Labrador, 66
Lac La Martre, 61–62
lack of integrity, 184
lacustrine limnetic, 388
lacustrine littoral, 388
lagged feedback control, 339, 442
lagoon, 70
Lagopus lagopus scoticus, 85
Lagrange stability, 105, 114, 344–345, 347
Lagrangian multipliers, 144
lake char (lake trout), 56
lake ecosystems, 55, 239, 258
lake elevation, 324–325
Lake Erie, 399, 401
Lake Hazen, 56, 58–59
Lake Michigan sand dunes, 457
Lake Ontario, 280, 287–288, 292
Lake St. Clair, 287–288
lake surface area, 324
lake trout, 56, 60, 62, 402
lake volume, 324
lake whitefish, 56, 60–61, 75
lakes, 30, 45, 51–67, 71–73, 75–76, 79, 84–85, 91, 95–96. 98, 110–112, 114, 116, 133, 154, 229, 233, 239, 248, 258–259, 280,

lakes (cont'd), 282, 284, 287–288, 292, 318, 323–326, 345, 391, 398–399, 401–404, 406–410, 457, 520, 571–574, 578–579, 581–582, 603
land, 56, 74, 89, 117, 229, 270, 317–320, 427, 445, 459, 501, 541
land ownership, 320
land use patterns, 229
land use practices, 317
landscape, 5, 255, 338–339, 344, 377, 400–401, 406–408, 417–418, 450, 461, 470
landscape productivity, 400
landscape units, 470
landscapes, 97, 400, 407
Lange-Hicks condition, 234
Lange-Hicks criterion, 238
language, 16, 30, 41, 43–45, 49, 419, 452, 610, 650
larval stages, 69
Latrodectus, 528, 531
Latrodectus hesperus, 528, 531
law of thermodynamics, 32, 88, 131, 136–137, 141, 162, 345, 490, 511, 549, 570, 652
laws of matter, 169
laws of potential effectivity, 174
Le Chatelier's principle, 563, 582, 597
leaf area, 265, 383
leaf area index, 383
leaf biomass, 383
least action, 52, 85–89, 94, 96–97
least action attractor, 88
least attainable dissipation, 51
least cost rations for livestock, 13
least dissipation, 30–31, 52, 67, 77, 79–88, 94, 96
least dissipation of energy, 81
least entropy production, 83, 86–89, 94, 96
leaves, 20, 91, 105, 226, 232, 265–266, 268, 520, 523, 644, 648
lemma, 217, 632–633
length, 44, 51, 58–66, 68–78, 84, 144, 156, 164, 176, 268, 301, 327, 354–355, 359–360, 365–367, 400, 416, 418–419, 430, 445, 452, 516, 519–520, 538–539, 541, 563, 601–604, 607
length and age frequency distribution, 61–62
length class, 62–63, 66, 68
length frequency, 58–60, 62–63, 66, 70, 72–75, 84
length frequency distribution, 58–60, 62–63, 66, 73–75
length modes, 66
Leontief orientation, 488
Lepidoptera, 526–527

levels of organization, 5, 10, 17, 29, 90–91, 230, 235–237, 339, 380, 402, 485, 513, 616, 621
Liapunov approach, 346
Liapunov asymptotic stability, 109
Liapunov concept, 344–345, 365
Liapunov functions, 140, 347, 349, 381, 562–563, 589–594,, 601, 606
Liapunov stability, 105–106, 109, 114, 336, 344–345, 347, 362, 365–366, 562–563, 586, 589–590, 593
Liapunov stability analysis, 109
lice, 529
licenses, 228, 296, 299, 301
life cycles, 327, 403, 426, 456, 515, 520, 530
life expectancy, 236, 437
life history, 77, 264, 377–378, 380, 385, 391, 519, 535, 537, 542
life history adaptations, 380
life history characteristics, 377
life history evolution, 385
life history omnivory, 519, 535, 542
life space, 218
life span, 44, 51, 57, 64, 79, 84, 92, 94, 206, 218, 336, 373, 376
life support system, 470
life zones, 97
light, 39, 42, 45, 54–56, 68, 73, 78, 81, 85, 110–111, 116, 119, 136–138, 143, 145–149, 152–155, 157, 159–161, 263–270, 273–274, 456–457, 460
light attenuation, 265, 269, 274
light conditions, 116, 263
light depletion, 456
light energy, 81, 136
light intensity, 111, 137–138, 143, 145, 148–149, 152–155, 159–160
light levels, 265
light resources, 270
limes convergens, 47
limes divergens, 47
limestone, 68
limit cycle behavior, 375
limit cycles, 107, 375, 390, 433, 445
limit law of potential effectivity, 179
limit laws, 173
limitations, 3, 9, 12, 21, 29, 34, 39, 196, 200, 205, 209, 211–214, 227–229, 264, 274, 279, 298–299, 301, 330, 344, 362, 381, 415, 623–624, 652
limited food supply, 69
limited populations, 114
limiting resources, 214, 422, 430, 446
limits of precision, 259
Limnocalanus macrurus, 58
limnology, 382

Lindeman chain, 553
Lindeman scheme, 550
Lindeman spine, 512, 558
Lindeman's trophic pyramid concept, 549
linear, 13, 21, 31, 35, 57, 105–107, 109–111, 114, 123, 137, 142, 145, 156–157, 193, 202, 214, 226–228, 231–235, 237–238, 243–244, 248–253, 265, 268, 270, 280–283, 291, 295, 297, 303–308, 346–347, 349–350, 353, 356, 362–363, 367, 406, 431, 486, 491–492, 551, 598, 601
linear, nonlinear, and dynamic programming, 13
linear aggregation, 251
linear aggregative transformation, 249
linear algebra, 21
linear and nonlinear formulations, 307
linear and nonlinear mass balance models, 231
linear and nonlinear models, 107
linear and nonlinear programming problems, 305
linear approximation, 349–350, 367, 598
linear combination, 249–251
linear constraint inequalities, 297
linear cross effects, 156
linear deterministic case, 305
linear deterministic problem, 306
linear donor control, 238
linear dynamics, 31
linear equations, 244
linear formulations, 281–282, 307
linear independent lines, 250
linear inputs, 111
linear L–V equations, 109
linear models, 106, 109, 226–227, 237, 248, 251, 305, 492
linear objective function, 297, 305
linear programming, 228, 295, 297, 303–305, 308
linear recipient control, 238
linear regression models, 281
linear relation, 291
linear separability, 226, 251, 253
linear sequence, 406
linear structure, 193
linear system aggregation, 248
linear systems, 105–106, 109, 226, 231, 243, 248–252, 362
linear thermodynamics, 601
linear transformation, 362, 551
linear trophic function, 248
linearization, 106, 109–111, 123, 360, 367
linearization hypothesis, 110

linearization matrix, 360, 367
linearized ecosystem response, 110
linkage density, 539
links, 7, 42, 48–49, 218, 347, 353–357, 475, 510, 516, 520–521, 524–526, 538–542, 550, 564, 614
liquid and gas phase boundary, 287
liquid fuel, 446
liquid layer, 286
liquid phase, 284, 291
lithosphere, 4
litter, 462, 518, 524
Little Nauyuk Lake, 62–64
littoral region, 263
live births, 437, 441
livestock, 13
living organisms, 4, 8, 82, 172, 391, 459, 468, 485, 572
lizards, 186, 469, 518–519, 523, 526, 528, 533, 537
loading, 227, 263, 271–274, 280, 399, 406, 574
loading rates, 271, 273, 280, 406
loans, 446
local disturbances, 338
local entropy production, 32, 131, 136, 140, 165
local equilibrium, 131–133
local equilibrium hypothesis, 131
local exergy principle, 599
local fishery, 66
local instability, 592
local level, 53, 96
local principle, 563, 597, 599–601
local stability, 109, 349, 364, 381
locality, 54, 111
log-normal, 14
logarithmic character, 47
logarithmic law of growth, 167
loggerhead shrike, 531
logging, 229, 320, 323, 408
logic, 10, 45, 459–460
logical tree, 54
long lagged feedback control, 442
long lived, 73
long-term ecosystem research, 22
long-term forest responses, 372
long-term impact on man, 22
long-wave theory, 339
longevity, 96
look-backward flow analysis, 501
look-forward compartmental analysis, 501
looking-outward matrix, 228, 314–315, 321–322, 327–328
looking-outward matrix exercise, 314–315, 321–322, 328
loons (*Gavia* spp.), 57
looping, 510–511, 519–520, 526–527, 532, 535, 541
loss functions, 233–234

loss of habitat, 271
Lota lota, 56
Lotka–Volterra equations, 453
Lotka–Volterra exergy, 606
Lotka–Volterra model, 157
Lotka–Volterra models of population dynamics, 104
Lotka–Volterra predator-prey model, 563
Lotka–Volterra prey-predator model, 593
Lotka–Volterra prey-predator system, 606
Lotka–Volterra type, 620
low-diversity systems, 41
low energy, 44, 46–47, 378–379
low kill, 302
low nutrient input, 114
Lower Great Lakes, 408
lower-level properties, 225
lowland Michigan, 79
lynx–hare, 45

M-matrix, 249, 350
m-separability, 245
m-separable, 245
machinery, 427–428, 615
macroarthropod predators, 537
macroarthropods, 524, 537
macrobenthic communities, 404
macroclimate, 4, 6, 14–15, 377
macroclimatic influences, 14
macroeconomics, 420
macroecosystems, 6, 20
macroinvertebrates, 272
macrolevel, 32, 174, 183, 195, 200, 213, 338, 415
macroparameters, 168–170, 173–175, 183–184, 186
macrophyte abundance, 270, 274
macrophyte bed, 404
macrophyte biomass, 271
macrophyte communities, 226, 263, 275
macrophyte decline, 263, 274
macrophyte dominated community, 265
macrophyte ecology and physiology, 274
macrophyte ecosystem dynamics, 264
macrophyte ecosystem model, 264
macrophyte ecosystems, 226, 262, 264, 267, 269–270
macrophyte growth, 263, 265, 268, 273–274
macrophyte growth equation, 265
macrophyte leaf surfaces, 264–265
macrophyte leaves, 266, 268
macrophyte management model, 270–272, 274–275

macrophyte management modeling, 269
macrophyte nitrogen uptake, 266
macrophyte physiology, 273
macrophyte plants, 266
macrophyte populations, 270
macrophyte production, 269–270, 272
macrophyte shoot biomass, 268
macrophyte simulation models, 264
macrophyte species, 265–266
macrophyte subsystem, 265
macrophytes, 226–227, 262–275, 404
macroscopic (phenomenological) theory, 147
macroscopic ecosystem properties, 376, 388
macroscopic model parameters, 162
macroscopic state, 142, 390
macrosystem, 231
Madagascar, 72, 469
Magicicada spp., 84
Maine Department of Inland Fisheries and Game, 302
mainframe computer, 316
maintenance, 7, 42, 52–53, 66–67, 69, 73, 75, 80–81, 83, 85, 88, 94, 97, 138, 263, 384, 400, 433, 470–471, 513, 562, 570, 578, 583, 615, 652
Malabar, 72–73
male : female ratios, 307
Malheur Lake, 229, 323–326
Malheur Lake hydrology submodel, 325
Malheur Lake model, 326
Malheur Lake National Wildlife Refuge, 229
Malheur model, 317
Malheur National Wildlife Refuge, 324
malnutrition, 215, 218
MAM Model, 200–202, 204–206, 209–214, 218–219
MAM or MEM models, 204–205, 209, 211
mammal populations, 51, 55, 67, 95
mammals, 30, 73, 84, 94, 458, 468, 518, 522–523, 532–533, 537, 542
man, 4–9, 16, 18, 22, 69, 104, 166, 279, 295, 297, 308, 338–339, 343, 373, 397, 415, 458, 516, 567, 585, 609
man-created environmental problems, 18
man-made systems, 339
man–nature relationship, 338
managed wildlife population, 317
management, 3–4, 8, 13, 16, 19–21, 33, 40, 51, 108, 112, 130, 141–142,

685

management (cont'd), 144, 166, 177, 226–229, 238, 242, 263, 266, 269–275, 295–298, 301–304, 308, 311–315, 317–321, 324, 327–331, 340, 372, 390–391, 415, 418, 420, 444, 450, 470–472, 480, 485, 510, 561, 568, 609, 623, 643
management actions, 226, 313, 330–331
management activities, 144, 313
management agencies, 321
management alternatives, 227, 302, 315
management and control actions, 340
management and decision making, 312
management areas, 301
management decision process, 270
management decisions, 13, 228, 312, 372
management intervention, 390
management issues, 317, 330–331
management modeling, 269–270
management modeling framework, 269–270
management models, 227, 266, 270–272, 274–275
management of a natural resource system, 328
management of natural resources, 228
management of renewable resources, 4, 13
management of water resources, 141
management or development scenarios, 315
management regions, 296–298, 302–303
management science, 418
management studies, 308
management units, 228, 296, 308, 324
managers, 16, 228, 296–297, 301, 308, 313, 331, 405, 418
mangrove swamps, 388
manipulation of ecosystems, 2, 8
manipulators, 8, 296, 397, 415
Manitoba, 30, 51, 75
mantids, 521
mantispids, 532
manufactured goods, 342, 499–503
marine and atmospheric fronts, 47
marine fouling organisms, 209
marine microcosm, 110
marine mollusks, 69–70
marine pelagic systems, 108
marine reef algal bed, 388
market supersaturation, 428, 435, 439, 441, 443
Markov model of forest succession, 255

Markov property, 566, 641
Markovian transition matrix, 192
Marsh Module, 474
marshes, 9, 388, 557
marshlands, 408
mass, 13, 21, 32, 131–138, 141, 144, 146, 149–152, 154, 156–157, 164–165, 231–234, 237, 267, 287, 385, 389, 391, 406, 443, 486–487, 490, 518, 587, 589, 592, 595, 606, 623–624, 627, 640
mass balance, 134, 141
mass balance aggregation model, 234
mass balance characteristics, 232
mass balance equations, 32, 133, 137–138, 144, 156
mass balance flow model, 237
mass conservation law, 587, 589, 606
mass conservation principle, 592
mass density, 134, 165
mass fraction, 134, 149–151, 164–165
mass of nutrients, 13
mass or energy transfers, 232
mass transfer, 137, 146, 287
mass transfer rates, 287
material (energetic) stability, 32, 191
material and energy cycling, 552
material and energy feedbacks, 462
material composition, 32, 191
material cycles, 81, 87
material flows, 489
material volume, 135
materialistic independence, 470
mathematical abstractions, 11
mathematical analysis, 13, 199, 486
mathematical approximation, 122
mathematical argument, 11–12
mathematical conclusions, 11–12
mathematical economics, 248, 344, 354
mathematical expectation, 192, 200–201, 204, 214, 217
mathematical experimentation, 11
mathematical flow accounting procedure, 341
mathematical formalism, 113, 232, 383
mathematical formalization, 191
mathematical model of volatilization, 291
mathematical modeling, 11, 243
mathematical models, 11, 36, 45, 114, 178, 196, 260, 279–281, 344, 347, 358, 365, 453, 462, 589
mathematical models of ecosystems, 260

mathematical models of wetlands, 36
mathematical programming, 13, 297–299, 302
mathematical stability, 337, 366, 374–376, 589
mathematical system, 11
mathematical techniques, 10
mathematical theory of ecological aggregation, 231
mathematical theory of trophic chains, 361
mathematical tools, 11, 13, 15–16
matrices, 6, 14–15, 117, 175–176, 192–194, 199, 245–247, 249–256, 258–259, 288, 314–315, 321, 325, 327–328, 336, 344, 347–356, 359–360, 362–367, 378, 487–494, 496–506, 512, 536–539, 550–552, 557–558, 589, 625626, 629–630, 648, 653
matrix notation, 117
matrix of feeding coefficients, 550
matrix representation, 14–15
matrix sign stability, 336, 353
matrix space, 349
matrix stability properties, 336, 344
matrix stability subsets, 356, 365
matrix theory, 348, 366
matter, 4, 7, 10, 13–15, 18, 22, 31, 41–42, 47, 53–54, 86–87, 108, 113, 122, 131, 136, 138, 169, 187, 228–229, 268, 270–271, 313, 341, 358, 373,, 378–379, 385, 387–388, 392, 417, 450, 459–460, 468, 485, 495, 501, 519–520, 562–563, 565, 576, 588, 590, 594–595, 597–599, 601–603, 605, 624, 648
matter and energy balances, 358
matter and energy cycling, 605
matter and energy flows, 605
matter and energy processing, 373, 392
maturation sequences, 337, 399
mature ecosystem, 89, 402
mature forest, 78
maturity, 84, 90
Mauritius, 30, 77
maximal metabolic viability, 185
maximal survivability, 188
maximization of yields, 209
maximum age, 70
maximum entropy, 52, 80–81, 89, 96
maximum exergy hypothesis, 572, 576, 582–583
maximum power principle, 578
mayflies, 402
meadow microcosm, 110
mean age, 31, 55, 68, 79, 84

mean length, 84
mean life span, 64, 79, 84, 92, 218
measure of diversity, 40, 48
mechanism, 29, 225
mechanism–reductionism, 225
mechanism response stimulus, 191
mechanistic detail, 226
mechanistic relations, 226
meiofauna, 496–497
meloiid, 526
MEM model, 200–202, 204–205, 209–212, 214, 221
memory effects, 132
mesh selection, 63
mesh selectivity, 58
mesh sizes, 58–60, 75
mesolevel, 32, 166, 174, 195, 200, 213
mesomodel, 212
mesotrophic lakes, 239, 603
mesquite, 523
message, 46, 452, 476
Messor pergandei, 518
metabolic demands, 83
metabolic effects, 185
metabolic elements, 184
metabolic rates, 67, 69, 83, 92, 96, 385, 389
metabolism, 52, 68, 92, 345, 384, 386–387, 389, 391, 405, 487, 489, 494
metastability, 400–401
metazoa, 520, 539
meteorites, 54
meteorological inputs, 117
metropolitan districts, 296
Mexico, 23, 95, 469
mice, 73, 518, 521, 523, 535
Michaelis-Menten relation, 108, 111, 238, 267, 580
Michigan, 30, 78–79, 403, 457, 609
microalgae, 265, 274
microarthropods, 520, 524, 537
microbes, 385, 469, 477, 515, 520, 523, 537, 542
microbial ecologists, 51
microbial feeders, 524
microbial loop, 520, 557
microbiota, 496–497
micro-bomb calorimeters, 12
microchemical buffering, 403
microclimate, 280
microcomputers, 316
microcosms, 8, 52, 110, 226, 263, 287–288, 384, 469, 488
microecosystems, 8
microlevels, 32, 174, 195, 200, 239
microorganisms, 4, 510, 524, 526, 534, 572, 575, 617
microphytes, 635
micropredators, 537

microscopic thermodynamics, 31
microstates, 169, 183
microsystem, 231
Middle East, 9, 418
midges, 516, 526
migration, 6, 92, 94–95, 193, 320–321, 323–324, 329, 339, 366, 417, 437, 446, 565, 623
migration patterns, 92, 320
migration principles, 339, 417, 446
migratory tendency, 96
military buildup, 420
millipedes, 524
MIM model, 200, 204–205, 212, 215, 221
Mimetus, 531–532
Mimusops maxima, 78
mineral resources, 418, 422
minerals, 456–457, 638–639
minimax condition, 94
minimax principle, 173, 177
minimum entropy production, 30, 81–82, 561
minimum tillage agriculture, 274
Missouri, 297
mites, 515–516, 518, 524, 529–532, 538
mitigation plans, 317
mixed conifers, 79
mixed hardwoods, 390
mixed stock ocean harvest, 328
mixing, 46, 111, 157, 268, 274, 385, 522
mixing processes, 385
mixing rates, 268
mobile invertebrates, 227, 264
modal configuration, 66
modal length, 58, 66, 68–69
modal length class, 66, 68
modal size, 57, 66, 73, 79, 84
modal size class, 73, 84
modal size groups, 79
modal value, 62–63, 66, 76
model accuracy, 37
model aggregation, 225–226, 230, 242, 260, 282, 565
model behavior, 108, 316, 326, 344
model calculations, 314
model complexity, 36, 339, 444
model conceptualization, 315
model construction, 226, 314, 327, 480
model debugging, 15, 314, 316
model descriptions, 274
model development, 316–317
model dimensions, 242
model discrimination, 291
model equations, 130, 142, 232, 353, 366, 592, 594, 606
model equilibrium, 365
model exams, 280

model formulation, 288
model instability, 115
model invalidation, 316
model of reliability, 192
model of survivability, 189
model of the regenerative system, 193
model output, 231, 268, 272, 313, 316, 444
model populations, 107
model separability, 245
model simplicity, 426
model simulations, 268
model stability, 357
model uncertainty, 283
modeling, 11, 14, 29, 31, 34, 38–39, 44, 115, 120, 130, 133, 138, 141–143, 145, 150–151, 154, 162, 178, 225–234, 236–237, 239, 242–243, 248, 260, 262–265, 269–270, 273–275, 280, 291, 311–313, 320, 323–324, 326–331, 336, 339–341, 356–357, 416, 424, 426, 434, 444, 470, 472–474, 476, 481, 549, 568, 576, 578, 582–583, 585, 589, 605
modeling and simulation, 339, 416
modeling complex systems, 260
modeling ecosystems, 44, 232
modeling guided workshops, 228
modeling inadequacy, 115
modeling limitations, 39
modeling objectives, 233
modeling workshops, 311, 320, 323–324, 329–331
models of density-dependent populations, 107
models of ecosystems, 39, 44, 260
models of human society, 416
models of mutualism, 350
models of natural systems, 231, 450
models of succession, 374, 462
modern ecosystem theory, 416
modes, 37, 58, 66, 75–76, 78, 563–565, 610, 615–618, 621
module decomposition, 477
moisture, 382, 385, 391, 471, 518
molal volume, 285–286
molar mass, 134, 137–138, 151–152, 164
mole numbers of the ecosystem, 132
molecular biology processes, 167
molecular weight, 283, 285–286, 290
molecular weight of water, 285
molecules, 54, 85, 87, 95, 172, 186, 520, 612, 643, 648
moles, 517
molluscan predator, Thais, 220
mollusks, 30, 69–71, 94

momentum, 29, 34, 131, 133–136, 141, 157, 328
money supply growth rates, 426
monitoring data, 282
monitoring, 75, 282, 317, 402
monoprocess, 169, 171
Monte Carlo methods, 14
morphological structure, 618
morphology, 617, 619
mortality, 58, 69, 93, 112, 151, 157–158, 160, 215, 303–304, 307, 319, 321, 323, 325, 327–328, 358, 365, 535, 542, 572, 576, 594, 597
mortality rates, 93, 307, 321, 323, 576, 594, 597
mortgage debt, 428
mosquito control, 116, 319
mosquitos, 116, 517
most action, 52, 87–89, 92, 94
most action attractor, 88–89
most dissipation, 87
motile, 218
Mougeotia, 402
movement and degradation of pollutants, 20
movement of the system, 122
movement of water, 329
moving average, 421, 430–431
mud, 70, 324
Mud Lake, 324
mule deer (*Odocoileus hemionus*), 9, 297
multicommodity, hierarchical I–O (flow) analysis, 486
multicommodity approach, 341
multicommodity ecosystem analysis, 485
multicommodity flow analysis, 498
multicommodity flows, 341
multidimensional stability concept, 582
multidisciplinary research, 29
multimeshed nets, 58
multiple commodity inputs and outputs, 341
multiple entity ES, 478
multiple input–output, 493
multiple interacting species, 117
multiple stationary states, 158, 160–161
multiple stresses, 398, 407
multiplicative decomposition, 183
multiproduct ecosystem flow accounting, 341
multiproduct flow, 492
multispecies lakes, 57
multispecies models, 106, 109, 349
multispecific populations, 345
multistage decision processes, 145
mussels, 71–72, 220

mutation ability, 115
mutation mechanism, 179
mutations, 613
mutillid wasps, 526–527, 529
mutual information, 565–567, 623–627, 629, 641, 648
mutual predation, 519, 526–527, 529–532, 534–535, 538–539, 541
mutual predation cycle, 538
mutual proportionality, 253
mutualism, 338, 350, 354, 356, 400–401, 513
myriapods, 532
Myrmex pergandei, 523
Mysis relicta, 402

N-cell space, 192
n-dimensional function, 245, 247
N-dimensional phase space, 168
n-species communities, 335
n-species models, 106
Namaycush Lake, 58
Namib Desert, 535
nat, 650
natality rates, 307
nation states, 446
National Elk Refuge, 320–321
national energy models, 444
national laboratories, 3, 22–23
national parks, 9
nations, 339, 419–421, 425–426, 434, 439–440, 442–444, 446
natural, 3–5, 9–10, 16, 19, 33, 38, 40, 43, 47, 51, 58, 69, 94–95, 104, 109, 115–117, 123, 130, 166–167, 169, 172, 174, 182–183, 189, 197, 199–200, 214, 217, 220, 226–228, 230–232, 234, 237, 239, 242, 248, 262–263, 272–275, 281, 295, 302–303, 308, 311–313, 318–319, 321, 323, 325, 327–331, 335, 337–338, 340, 346, 348, 351–352, 358, 362, 365, 372–374, 376–382, 390, 397–399, 401–402, 404–405, 408–410, 415, 433, 445, 450, 452, 454, 456, 461, 463, 469–471, 485, 488, 499–504, 510–511, 513–514, 528, 543, 559, 561–562, 566, 568–569, 579, 589, 594, 596–597, 599, 609, 611, 614, 616–617, 621, 623, 641, 643, 647–648, 651–652
natural adaptation, 197, 199–200
natural complexity, 33, 566
natural disturbance, 337–338, 397–399, 401–402, 408
natural ecosystems, 94, 117, 340, 373–374, 376, 378, 382
natural growth, 433

natural herd units, 303
natural history, 380–381
natural mortality, 58, 69, 319
natural observations, 109, 116
natural population, 237
natural products, 499–503
natural resource conflicts, 263
natural resource management, 3, 40, 51, 130, 273, 295, 302, 308, 331, 415, 450, 485, 568, 609, 643
natural resource systems, 228, 311–313, 327–330
natural resources, 227–228, 275, 295, 318, 372, 559
natural selection, 43, 47, 115–116, 123, 454, 461, 463, 561, 569, 579, 614, 617
natural self-organization, 408
natural systems, 38, 47, 230–231, 239, 405, 450, 452, 510, 513
naturalistic appreciation, 47
naturalistic contemplation, 40
nature of disturbances, 377
Nauyuk Lake, 56, 62–64, 66
Navier–Stokes equation, 141
near equilibrium states, 82
near symmetry, 89
necessity, 20, 55, 94, 166, 169, 173, 190, 229, 244–245, 247, 303, 355, 460, 486
necessity of condition, 355
necrophagous insects, 526
negative balance, 7
negative entropy, 31, 42, 570
negative exponential curves, 58, 78, 84
negative exponential distribution, 78
negative feedback, 49, 339, 418, 452
negative feedback loops, 49
negative interaction, 400
negentropy, 31, 96, 561–562, 570, 577
negentropy flux, 31
nekton, 227, 264
nekton model, 264
nematodes, 510, 516, 524, 529, 537–538, 541
nervous system, 450
nestlings, 518, 533–534
net export, 487, 496, 499
net inflow, 571
net input, 341, 486–487, 489, 491–493, 495–500, 503–504
net input intensity factors, 341, 486, 491, 498
net landscape production, 400
net output, 7, 487–489, 491–492, 494–495, 497, 503
net primary production rate, 155
net production, 273, 385
nets, 48–49, 58, 60, 75, 473

network aggregation, 631
network closure, 80
network flow, 14, 553, 565–567, 578
network growth and development, 643
network interactions, 567
network models, 341, 565, 623–624, 635
network of feedbacks, 452, 459
network organization, 563, 566
network size, 578
network structure, 119, 565, 624
network theory, 578
network throughflow, 623
network trophic dynamics, 563
neural nets, 49
New Mexico, 95
Newton's laws, 80, 86, 89, 450
Newtonian laws of mechanics, 169, 173
Newtonian mechanical systems, 172
niche boundaries, 83
niche overlap, 336, 361–362, 364
niche periphery, 66
niche space, 64, 83, 95, 460
niches, 21, 362–365, 463, 617
nine-spined stickleback, 56
nitrate, 149
nitrogen, 4, 117, 227, 264–267, 271, 273–274, 358, 384, 462, 486–487, 499–503, 550, 572, 575, 587, 624, 627, 640, 648
nitrogen (water column and sediment pore water), 265
nitrogen cycle, 501
nitrogen depletion, 575
nitrogen fixing, 575
nitrogen to carbon conversion, 265
nitrogen uptake relation, 265
noise, 11, 14, 32, 55, 80, 89, 110, 143, 169, 172, 174–176, 302, 418
noise immunity, 172, 174–176, 175–176
noise/signal ratio, 418
nondecomposable matrix, 355
nondestructive sampling, 12
nondiagnostic stress symptoms, 338
nonequilibrium, 32, 52–53, 81–82, 89, 131–132, 136, 139, 167, 179, 282, 378, 381, 451
nonequilibrium systems, 53, 89, 132
nonequilibrium thermodynamics, 81, 131, 378, 381, 451
nonfeeding transfers, 557
nonindigenous species, 77
nonlinear, 13, 31, 35, 106–111, 114, 119, 123, 133–134, 137, 139, 154 143 141–145, 147, 154, 156–157, 204, 228, 231–234, 259, 265, 282, 295–297, 300–301, 304–308, 432, 601

nonlinear character, 119
nonlinear dependence, 204
nonlinear deterministic constraint, 304
nonlinear difference, 141, 143
nonlinear differential equation, 232
nonlinear feeding, 106, 109
nonlinear formulations, 109, 307
nonlinear functions, 137, 144–145, 154, 156
nonlinear inequality constraints, 228
nonlinear interactions, 111, 234
nonlinear irreversible processes, 133
nonlinear mass balance models, 231, 233
nonlinear methods, 123
nonlinear programming, 228, 295–297, 301, 305, 308
nonlinear programming algorithm, 301
nonlinear programming formulation, 305, 308
nonlinear programming problems, 297, 301, 305, 308
nonlinear relations, 143–144, 154, 300
nonlinear system, 143, 259
nonlinear system of differential equations, 259
nonlinear thermodynamics, 133–134
nonlinearities, 106, 109–110, 114, 300, 620
nonlinearity of ecosystem structure, 119
nonliving organic matter, 387–388, 520
nonmechanical energy, 135–136
nonnegative commodity-by-commodity production matrix, 505–506
nonrenewable resources, 45
nonrepeating main values, 194
nonreplenished generation, 185
nonstochastic coefficients, 304
nontrivial equilibrium, 594
nontrivial solution, 244
nontrophic interactions, 513
nonviability, 48, 181, 186
normal, 14, 30, 62, 68, 71, 225, 235, 237, 302, 304–305, 307, 324, 331, 337, 339, 349, 351, 363–364, 398–399, 408, 416, 456, 513, 518, 521, 535, 541, 594
normalized resources, 202
normalized withdrawals, 202
North America, 433–434
North American Indians, 418
North Carolina, 372
North Sea, 635, 638
northern cavefish, 69
northern lower Michigan, 30, 78

northern Manitoba, 75
northern pike, 56, 75
northern tundra, 95, 110
nourishment, 216
nucleotide sequences, 6
Nuculana pernula, 69–71
number of divisions, 37
number of species, 41, 43, 47–49, 56, 93, 107, 344, 464, 516, 522, 533, 542, 619, 640
nutrient–algae–predator model, 108
nutrient and sediment losses from farmlands, 275
nutrient and seston concentrations, 269
nutrient and toxic loadings, 408
nutrient availability, 456
nutrient balance, 7
nutrient competition, 265
nutrient concentrations, 138, 143, 145, 149–150, 152, 155, 159–160, 268, 383, 578
nutrient cycles, 5
nutrient cycling, 9, 19, 400, 403, 453, 458, 461
nutrient enrichment, 269, 274
nutrient exchange, 274
nutrient export, 383
nutrient flow, 454
nutrient flux, 264, 270, 405
nutrient inputs, 269
nutrient loading, 263, 272
nutrient pollution, 599
nutrient–prey system, 108
nutrient processing, 406
nutrient recycling, 338, 400, 402–403, 405, 453–454, 461
nutrient regeneration, 275
nutrient storage, 150–151
nutrient uptake, 149, 154, 273, 575
nutrient uptake kinetics, 273
nutrients, 5, 7–9, 13, 19, 45, 97, 106, 108, 110–111, 114, 117, 133, 138, 143, 145–146, 149–155, 159–161, 164, 226–227, 258, 263–265, 267–275, 338, 340–341, 356, 383, 400–408, 453–454, 456, 458–464, 468, 487, 518, 522, 575, 578–579, 599
Nyssa, 255

oak–hickory forest, 377
object realization, 167
objective function, 228, 297–300, 302–303, 305, 307–308
objective of model construction, 327
objective teleology, 32, 174, 198
observations, 29, 34–35, 38–39, 45, 55–56, 58, 105, 109–110, 116, 123, 162, 174, 263, 274, 281–282, 361, 380–382, 402,

observations (*cont'd*), 462–463, 511, 513, 562, 571, 573–575, 581–583, 603, 648
ocean water, 283
oceanic environment, 112
oceanic plankton, 116
oceanography, 262, 382
oceans, 5, 45, 51, 54, 71–73, 77, 95–96, 283, 328, 340, 382, 385, 388, 391, 407, 469–470, 499–502, 504, 638
octanol–water partition coefficient, 281
Odocoileus hemionus, 297
offspring, 69, 451–452, 569, 611, 613
Ohm's law, 86
oil, 323, 416, 418–419, 422–424, 431, 434, 446, 652
oil discovery cost, 423
oil imports, 419, 446
oil production, 422
oil resources, 422, 434
oligotrophic lakes, 239, 248, 258
omnivore predators, 524
omnivores, 514, 522, 524–525, 531, 537, 635
omnivorous, 347, 518, 523, 526, 530, 532–533, 535, 537–539, 541
omnivorous links, 347
omnivory, 339, 417, 419, 446, 510–511, 514, 519, 522, 525–527, 531–532, 534–535, 539, 540–542
omnivory principle, 417
Oncorhynchus spp., 84, 94
one-level model, 175
one-step projection, 328
one-way processes, 146
Onega Lake, 259
Onsager coefficients, 156
Onsager's reciprocity relations, 87
Onsager's theorem, 89
ontogeny, 93, 542
Onychomys torridus, 531
open, 7, 31, 53, 66, 77, 94–96, 108, 112–113, 117, 134, 138, 231, 267–268, 271–272, 291, 297, 335, 337–338, 348, 351, 356, 358–359, 361, 365–367, 378–379, 385, 387–388, 404, 407, 452, 469–470, 520, 558, 562, 564, 586, 588, 595, 617, 619, 621, 624, 627
open canopy, 77, 387
open canopy woody species, 387
open chain, 358–359, 366–367
open ecosystems, 338, 624
open embayment, 267–268, 271
open estuary, 388
open loop, 452
open ocean, 95, 385, 388
open shrub steppe, 95

open system, 53, 96, 108, 117, 134, 138, 562, 586, 595, 627
open trophic chain, 361
open water, 387
open water ecosystems, 387
openness, 53, 400, 404
operational space, 114, 116
operations research, 2, 13–14, 19, 295, 297, 308, 418, 561
operative characteristic, 191, 193
opiliones, 526
opportunism, 532
optimal, 31–32, 69, 115, 120, 138, 171, 174–179, 183, 187–188, 190, 194, 196, 200, 202–203, 205, 208–209, 211–214, 279, 288, 301–302, 306–307, 439, 458–459, 461, 564, 566, 598, 624, 632
optimal catch, 196
optimal control, 177, 196, 200, 202
optimal control problem, 202
optimal environmental conditions, 288
optimal kill, 302, 307
optimal selection, 120
optimal values, 187, 200, 203, 212, 214
optimal withdrawal, 205
optimality, 30, 32, 114–115, 120–123, 459, 561
optimality principle, 30, 32, 138, 174
optimality theory, 115, 459
optimally preadapted, 191
optimization, 13, 19, 31–32, 121–122, 142–145, 154, 177, 186–187, 194, 205, 212, 228, 297, 301, 305, 308, 329, 454, 459–462, 464, 561, 615
optimization formulations, 121
optimization framework, 228, 297
optimization methodology, 122
optimization of control, 205
optimization principle, 142, 144–145, 154
optimization problem, 305
optimum foraging models, 122
optimum foraging theory, 116
optimum hunting, 297
optimum principle, 174
Orconectes, 402
order, 3–5, 9, 19–20, 35, 43, 52, 83, 86, 88, 90, 104, 130, 139, 142, 162, 169–172, 174, 182–183, 213, 218, 222, 227–228, 230, 232, 244, 246–247, 249–250, 258, 265–266, 280–281, 291, 315, 317, 329, 346, 349, 352, 359, 378, 384, 387, 415, 420, 463, 494, 513, 524, 566, 571,

578, 580–581, 601–602, 604, 606–607, 609, 640, 643, 647
Oregon, 229, 324
organ cells, 186
organelle molecules, 186
organic matter, 387–388, 459, 520, 590, 597–598
organic matter standing crop, 387
organic matter storage, 384, 459
organic structure, 378–379, 381–382, 387–390
organic wastes, 9
organic world, 52
organism complex, 53
organism interactions, 113, 116
organism–population–community paradigm, 236
organisms, 4–6, 8, 10, 40, 43, 53, 57, 67–68, 76, 80, 82, 84–85, 87, 93, 95–96, 104, 106, 108–109, 113–116, 118–123, 135, 137–138, 161, 167, 171–174, 190, 209, 230, 232, 235–236, 242, 264, 385, 391, 398, 401–402, 406–407, 450–451, 459–460, 462–463, 468, 471, 485–486, 510–511, 515–516, 518–520, 522, 524, 526, 535, 537–538, 540, 549–550, 557, 564–565, 567, 572, 576–577, 611–618, 620, 635, 644–645, 648, 654
organization, 3–5, 10, 17, 22, 29–31, 41–42, 46, 80, 82, 84–85, 90–91, 106, 113–115, 122, 133, 138–140, 142, 162, 166, 179, 190, 225, 230, 235–237, 239, 263, 335, 339–340, 343, 372, 375–382, 384–387, 389–391, 397, 402, 408, 415, 450–452, 462, 468, 474, 480–481, 485, 509–510, 513, 549, 561–568, 578, 585, 609, 611–613, 615–621, 623–624, 635, 638, 640–641, 643, 646–648, 650, 654
organization of ecosystems, 239, 391, 462, 615
organizational integrity, 381
organizational state, 382, 390–391
organizational status, 650
organized complexity, 21
organized simplicity, 21
organizing centers, 406, 408
organizing principles, 329–330
organs, 5, 47, 69, 97, 122, 186, 230, 405, 513, 617–618
ORNL Health Physics Division, 3
orthophosphate, 572
Orthoptera, 523, 526–527, 533
oscillations, 76, 107, 345, 375, 377–382, 385, 387, 416, 419, 424, 430, 432–434, 441, 445, 461, 464

oscillatory damping, 55, 76, 445
outdoor mesocosms, 263
outfitting and ranching industries, 320
outflow, 131, 158, 160, 566, 571, 590, 633, 635
outflow intensities, 566
output environ analysis, 489
output matrix, 492
output rates, 238
output signals, 110
outputs, 7, 20, 45, 53, 81–82, 89, 93, 97, 109–110, 133, 231–232, 238, 267–268, 270, 272, 284, 313–317, 329, 331, 341–342, 372, 415, 420, 427, 437, 439, 444, 450, 458, 470, 486–506, 512, 557–558, 561, 565–567, 579, 625–626, 646–647, 653
overcatch effect, 166, 196, 207–208
overcatching, 195
overfishing, 296, 399
overgrazing, 9, 299
overharvesting, 408
overpopulation, 296, 339, 415
oversimplification, 339
overwintering death rates, 302
owls, 30, 76, 85, 519, 526, 533, 535, 539
Oxford, 76, 372
Oxford University, 76
oxidation, 280–281
oxygen, 54, 113, 138, 147–149, 152–153, 155, 233, 236, 283–286, 291, 315, 406, 487, 572, 575
oxygen concentration, 147, 152–153, 155, 315, 406
oxygen loading rates, 406
oxygen organisms, 113
oyster reef, 494–498
oysters, 495–497
Ozark cavefish, 69

Pacific flyway, 317, 324
Pacific salmon, 84, 94
PAHs, 284, 292
pallid bats, 535
parabolic predetermination, 187
paradox of enrichment, 108
parallel and hierarchical species organizations, 108
parallel case, 238
parallel path omnivory, 525, 527, 531–532, 534–535
parallel position, 566, 633–635
parameter effectivities, 186
parameter values, 107–108, 122, 254, 280, 282–283, 291, 307, 429–430, 432, 435–439
parameters, 12, 22, 36, 44–45, 94, 109–110, 112, 114–122, 138,
141–142, 157, 162, 174, 192–196, 200, 205, 208–209, 211–214, 218–222, 226, 237, 243, 253–254, 280–284, 288, 290–291, 297, 299–301, 304, 315, 324, 329, 344, 347, 359, 361, 365–367, 374, 418, 432, 437–438, 453, 473–475, 477, 538–539, 563, 579, 582–583, 594–598, 600–601, 603
parameters of relation, 213
parasites, 75, 107, 209, 356–357, 401–402, 452, 515–516, 524, 526, 528–529, 535, 539
parasitism, 4, 338, 401, 526
parasitoids, 514–516, 526–527, 529, 531–532, 535, 537, 539, 542
parent material, 6–7
parental investment, 69
part–whole relation, 31, 337
partial pressure, 132
partial specific entropy, 131
partial specific internal energy, 131
partially aromatic hydrocarbons, 284
partially closed trophic chain, 358–360, 366–367
particle size distributions, 550
particulate matter, 131, 268, 270–271, 385
partitioning, 327, 558, 652
Paruroctonus mesaensis, 516–518, 530, 532
Parus major, 85
passive problem definition, 615, 621
past-directed flow analysis, 491
pasture, 112
patches, 30, 41, 461
patchiness, 46–47, 107, 161
pathology, 338, 397–398, 405–408
pattern, 44–47, 51, 66–67, 72–74, 77, 81–82, 84, 88, 90–91, 96, 106, 176, 179, 196, 296, 323, 327, 329, 331, 345, 347–348, 350–353, 355, 357, 364–365, 382, 391, 403, 408, 420–421, 425, 428, 430, 440, 442, 454, 473, 514, 564, 601, 606, 612, 617–618
pattern formation, 46
pattern matching, 47
pattern of species interactions, 347–348
pattern of symmetry, 90
Paruroctonus luteolus, 531
peace treaties, 420
pelagic herbivores, 635
perfect aggregation, 566, 632–635
perfect parallel position, 566, 633–635
performance of the system, 450
periodic system, 193

permeability, 52
permit limitation, 301
permits, 11, 84, 205, 296, 298–299, 301, 445, 553, 559
Perognathus, 523
Peromyscus, 523
persistence, 43–44, 337, 340, 361, 373–374, 376, 380, 382, 384, 392, 470
persistence and functional integrity of the ecosystem, 382
persistence of local differences, 44
persistence of species assemblages, 380, 392
perturbations, 22, 79, 107, 109, 112, 336–337, 339, 343, 345, 357, 408, 418–420, 430, 433, 437, 445–446, 458, 562–563, 573, 576, 582–583, 589, 594, 610, 612–613, 617, 619–620, 652–653
Peru, 89
pest control, 357, 471
PEST model, 280
pesticides, 8, 469
pests, 116, 209, 357
pH, 18, 53, 143–144, 147, 149, 281, 402, 406
phainopepla, 523
phase space, 168, 345, 347, 352, 591, 593, 595
phase vector, 168
phenol toxins, 462
phenomenological, 32, 121, 138, 147, 173, 593
phenotypes, 43, 569
phlogiston, 42
phosphorus, 4, 111, 117, 274, 358, 498–500, 502–503, 562–563, 572–574, 576, 578–581, 587, 590–592, 598, 603, 648
photochemical reactions, 144
photolysis, 280
photons, 42, 87–88, 144, 265
Photopsis, 526, 532
photosynthesis, 4, 133, 146–149, 151, 154, 164, 265, 378, 403, 487, 597–599, 601
photosynthesis curve, 598–599
photosynthesis–irradiance function, 265
photosynthetic curves, 600
photosynthetic groups, 264, 270
Phrynosoma platyrhinos, 531, 533
Phyllobaenus, 526–527, 532
physical, 4, 10–13, 16, 18, 20, 31–32, 42–43, 47, 52–54, 85, 92, 94, 97, 114, 118, 120, 122, 130–131, 133, 138–140, 142–144, 157, 161, 166–167, 169, 171–177, 179, 190–191, 227, 243, 267,

691

physical (cont'd), 281–283, 291–292, 315–317, 321, 342, 345, 377, 400, 404–405, 415, 451–453, 468, 470, 472, 477, 485, 490–491, 500, 502, 505, 521, 561–563, 567, 610, 613, 616
physical, biological, and economic interactions, 316
physical, chemical, and biological laws, 142
physical, chemical, and biological processes, 138
physical analytical equipment, 12
physical and cybernetic theories, 167
physical environment, 13, 47, 616
physical ergodicity, 166
physical laws, 31, 114, 563
physical sciences, 18, 451, 610
physical systems, 54, 120, 172–173, 562
physicochemical environments, 336
physics, 1, 3, 21, 29–30, 32, 34, 38–39, 86, 92, 122, 130, 132–133, 138, 143, 162, 166–169, 171–175, 243, 291, 343, 563, 566, 571, 585, 610
physics laws, 172
physiological ecology, 236
physiological life span, 57
physiology, 5, 214, 273–274
phytoplankton, 42, 111, 116, 142, 149, 151, 153–154, 156–161, 235, 258, 265–266, 268–269, 271, 274, 402, 495, 520, 571–572, 575–576, 578–579, 581, 583, 587, 603
phytoplankton biomass, 157–158, 160
phytoplankton communities, 116
phytoplankton development, 161
phytoplankton dynamics, 160
phytoplankton growth, 157, 159
phytoplankton populations, 235
phytoplankton species, 151, 153–154, 156, 158
Picea, 78–79, 456
Picea mariana, 79
picture analysis, 47
pike population, 75
Pimephales promelas, 402
Pipistrellus hesperis, 531
Planck's constant, 34, 87
Planck's theorem, 91
planetary biosphere, 123
plankton, 45, 56, 116, 157, 227, 317–318, 336, 376, 385, 387, 515, 522, 530
plankton populations, 116, 317
planktonic, 56, 58, 106, 273, 373, 385, 387, 391, 521–522

planktonic autotrophs, 387
planktonic ecosystems, 385, 373, 391
planktonic production, 522
plant, root, and shoot uptake, 274
plant and animal populations, 7, 12
plant communities, 218, 264, 605
plant–consumer relation, 458
plant disease, 418
plant ecologists, 9, 17, 51
plant–herbivore-carnivore link, 524
plant–herbivore link, 523
plant–herbivore trophic relations, 523
plant nutrient, 264
plant nutrients, 263
plant populations, 6
plant production, 273, 458, 520, 524
plant resources, 238
plant species, 95, 463, 487, 521, 523, 526–527, 540
plant strategies, 456–457
plant succession, 453
plant tissue, 523
plants, 6–7, 9, 12–13, 16–17, 30, 42, 51, 63, 67, 77–78, 91, 95, 111–112, 115–116, 189, 218, 238, 263–266, 270, 272–274, 340, 387, 403, 405, 410, 418, 436, 453, 456–458, 461–463, 468–469, 471, 477–478, 480, 487, 499–500, 502, 504, 510, 515–516, 518–524, 526–527, 531, 533, 535, 537–540, 605, 617
plasticity, 116, 611, 617, 621
Pleistocene, 56
plots, 5, 37–38, 78–79, 456
plume dynamics, 317, 329
pocket mice, 523
Poecilia reticulata, 116
point of recruitment, 77
point source pollution, 406
Poisson, 14
policy analysis, 313
political, 8, 121, 297, 313, 397, 415, 419, 444, 450, 509, 612
political decisions, 297
pollen and fungal spores, 530
pollination, 403
pollutant, 406–407
pollutant or nutrient stress, 407
pollution, 5, 11, 18, 112, 226, 280, 282, 338–339, 401–403, 406, 408–409, 415, 599
pollution stress, 402–403, 406–407
pollution trends, 280
polyprocess, 167, 169, 171
ponds, 267–268, 282, 292, 401, 563, 572, 580, 590
population age structure, 170, 445, 510

population and community dynamics, 590
population and community models, 562
population behavior, 107
population biology, 232
population cycles, 339
population data, 55, 430
population density, 63, 66, 143, 296, 350, 416, 452
population dynamics, 32, 84, 104–105, 107, 138, 156–157, 161, 317, 319, 330, 336, 345–346, 350, 357, 462, 515, 519, 521, 542, 593
population dynamics equations, 346
population dynamics models, 105, 336, 346
population dynamics theory, 345
population ecology, 10, 21, 105
population estimates, 299, 307
population fitness, 235, 237
population fluctuations, 94, 452
population genetics, 232, 237
population growth, 58, 157, 235, 319–320, 350, 416, 436, 441, 569, 609
population growth rates, 416
population increase, 115
population interactions, 6
population levels, 67, 452
population management, 227
population models, 31, 104–105, 115, 335–336, 345, 352
population numbers, 248, 328
population phenomena, 430
population self-regulation, 400, 455
population size, 228, 236, 301, 306–308, 328, 346, 424, 430, 432, 452
population stability, 107
population structure, 63, 67, 76
population time constants, 264
population trajectories, 425
populations, 5–7, 9–12, 14, 17, 21, 23, 30–32, 51–53, 55–58, 60, 63–64, 66–77, 79–80, 83–85, 92, 94–96, 104–105, 107–109, 111–112, 114–116, 120, 122, 133, 138, 143, 145, 147, 153, 156–157, 161, 170–173, 186, 189, 194–196, 200–202, 204, 212–215, 218, 220–221, 226–233, 235–238, 248, 256, 263–264, 270–271, 296, 298–304, 306–308, 313, 317–321, 323, 325, 327–328, 330, 335–337, 339–340, 343, 345–346, 350, 352, 357, 364–366, 373–376, 380, 390–392, 400–404, 406, 416, 418, 424–425, 429–434, 435–436, 437, 440–441, 443, 445–446,

451–453, 455, 457, 460–463, 485, 510, 513, 515, 518–519, 521, 532, 541–542, 549–550, 557, 561–562, 569–570, 576, 590–591, 593, 604, 609
pore water nitrogen, 267
position, 29, 34, 52, 67, 82, 94–95, 97, 116, 119, 132, 192–193, 313, 365, 391, 515, 521, 551–552, 557, 566, 633–635, 648
positive balance, 7
positive D dissipativeness, 348
positive dominating diagonal, 350
positive feedback loops, 49, 452
postdisturbance dynamics, 390
postharvest survival, 321
Potamogeton perfoliatus, 265–267
potential effectivity, 167, 172, 174–175, 179, 192
potential energy, 86–87
potential functions, 593
potential noise immunity, 172, 176
power output, 89, 561
preadaptation, 32, 178–179, 187, 190–191, 405
preadaptive cycle, 191
prebiotic state, 81
Precambrian, 95
precipitation, 17, 110–112, 117, 314, 324, 375, 379, 594
precybernetics, 340
predaceous, 519, 523–524, 527–533, 535, 537
predation, 21, 57, 67, 73, 79, 116, 221, 235, 336, 352, 355–356, 365, 460, 464, 519, 524, 526–527, 529–532, 534–535, 537–539, 541–542, 565
predation communities, 336, 356
predation nonsymmetry, 352
predation pressure, 235
predation regimes, 116
predation subgraph, 355–356
predator avoidance, 116
predator–predator feeding interactions, 531
predator–prey, 43, 49, 105, 107, 116, 226, 247, 335, 345, 460, 463, 531, 563–564, 613, 623
predator–prey models, 105, 226, 563
predator–prey relations, 43, 116, 623
predator–prey size relationships, 531
predator–prey systems, 345, 463
predator–prey type models, 335
predator–prey type systems, 247
predators, 9, 31, 43, 47, 49, 57, 83, 105–110, 115–117, 123, 195, 215–221, 226, 247–248, 269, 271, 275, 328, 335, 345, 349, 352–357, 361, 365, 452, 459–460, 463–464, 469, 496–497,

510, 514, 516–520, 522, 524–535, 537–542, 549, 553, 563–564, 591, 593, 606, 613, 623
predators' numbers, 222
predators of pests, 357
predetermination, 183, 187
predictability, 32, 116, 121, 142–143, 383
predictions, 38, 44, 107, 116, 120–121, 227–228, 279, 281–282, 288, 315, 326, 460, 471, 474, 480, 510–511, 514, 543
predictive ability, 329, 336, 339
predictive calculations, 11
predictive capacity, 72
predictive power, 122–123, 328
predictive tool, 329
predictive utility, 444
preferences, 131, 142, 167
pregnancy rate, 323
pre-reproductive size classes, 74
preseason, 298
pressure, 53, 75, 82, 132–134, 147, 179, 191–192, 195–198, 200, 212–214, 219, 228, 235, 269, 274, 281, 283–284, 290–291, 295–300, 302, 320, 328, 461, 562, 586, 651
pressure for development, 320
pressure on the community, 195, 212
prevention of change, 53
prey abundance, 328
prey composition, 518
prey numbers, 219
prey organisms, 106, 109
prey patchiness, 107
prey populations, 105, 157, 248
prey–predator chain, 352
prey–predator interactions, 105, 116, 195, 352
prey–predator models, 31, 105
prey–predator pairs, 353
prey–predator population studies, 109
prey–predator ratios, 510, 540
prey–predator relations 356
prey quantity, 222
prey size, 116, 518, 531
prey species, 106–107, 116, 236, 514, 516, 518, 530
prices, 341, 416, 422, 425, 428–431, 435–436, 440, 446, 487, 491, 505
Prigogine conclusion, 30
Prigogine thermodynamics, 114
primary consumers, 8, 520
primary optimal structuring, 183
primary parasitoids, 526, 537
primary producers, 6, 9, 43, 57–58, 80, 83, 91, 95, 133, 238, 265,

270–271, 360, 458, 460, 540, 551, 558
primary production, 56, 143–146, 151–155, 157, 160, 164, 265, 267, 338, 361, 377, 400, 406, 454, 460–461, 520, 524, 551, 557
primary productivity, 235, 388, 400, 402–403, 405, 520–521, 529
primary succession, 6–7, 454
principle of equifinality, 401
principle of increased exergy, 563
principle of least action, 85–88, 94, 96–97
principle of section equidurability, 182
principle of species individuality, 463
probabilistic distribution, 46
probability, 11, 13, 21, 32, 41, 48, 90, 94, 96, 106, 167–168, 171, 173, 175–179, 181, 183–186, 192–195, 200–201, 204, 206, 214–221, 255, 282, 300–302, 304–305, 404, 420, 434, 442–443, 516, 553, 565–567, 571, 589, 605, 610, 625–628, 638, 640, 643–648, 654
probability distributions, 301, 304
probability of survival, 181, 201
probability statements, 11
probability theory, 643, 647–648, 654
problem solving, 331, 563–567, 610–613, 615–622
problem-solving power, 563–565, 610–613, 615, 617–621
process-by-process flow matrix, 494, 504
process-by-process, single commodity systems, 505
process macrostate, 168
process microstate, 168–169
process rates, 143, 145, 147–148, 235
process stability, 105
processes, 4, 6–14, 18, 29–32, 38–42, 44–47, 52–53, 55, 72, 80–85, 87–91, 93–97, 104–105, 111, 113–115, 117–118, 120–122, 130–134, 136–149, 151–152, 154, 156–157, 160, 164–165, 167–171, 173, 175–177, 215, 225, 227–228, 230–232, 235–239, 254–255, 258, 262–263, 270, 272, 274, 279–284, 291–292, 308, 312–317, 325, 327–331, 335, 337–338, 340, 366, 372, 374, 377–382, 385, 387, 391–392, 397–398, 400–401, 403–409, 416, 426, 430–434, 438, 441, 445, 450, 453–454,

processes (cont'd), 456–457, 460, 471, 483, 485–487, 489–495, 498, 500–502, 504–506, 509, 511, 520, 525, 551–553, 557, 559, 561–566, 568, 571, 576–577, 585, 597, 599, 610–621, 624, 631, 644, 648, 650–651
processes of repair and maintenance, 83
producers, 6, 8–9, 14, 43, 58, 80, 83, 91, 95, 133, 214, 227, 238, 265, 270–271, 360, 426, 458, 460, 490–491, 540, 551, 558
production, 18, 30–32, 41, 45, 47, 54–56, 75, 81–83, 86–94, 96–97, 111, 131, 136–146, 151–157, 160, 164–165, 262, 264–265, 267–274, 318–319, 323–324, 328, 338, 361, 377, 382, 385, 400, 406, 417, 420, 422–424, 426–427, 429–432, 434–441, 450, 454, 458–462, 469, 487, 490–491, 493–494, 501–502, 505–506, 520, 522, 524, 551, 557, 561
production matrix, 487, 494, 505–506
production of fossil fuel, 422, 430, 432, 435–438
productive capacity, 386, 435
productivity, 5, 7, 235, 338, 382, 385–388, 398, 400, 402–403, 405, 409, 428, 461, 515, 520–521, 529, 541
progeny, 213–214, 561
progressive succession, 7
property, 5, 7, 30, 32, 44, 79, 85, 90–91, 97, 105, 109–110, 139, 171–174, 178, 189, 191, 226, 234, 239, 337, 344–345, 347–348, 350–353, 377, 385–386, 400, 460, 510, 520, 551, 562, 566, 569, 587, 592–593, 620, 640–641, 650
propositions, 31, 113, 115, 344, 633, 635
protein denaturation, 266
protoorganisms, 88
protozoa, 515, 520, 524
pseudoalgebraic languages, 14
pseudononlinearity, 109–110, 114
public debt, 416, 421, 424–425, 428, 435–436, 438, 440–441, 445
public debt retirement, 441
public lands, 441
pulp mills, 404, 408
Pungitius pungitius, 56
purchases, 446
purchasing power, 422–423, 425–426, 436, 441–442
pure control phenomenon, 177

purposeful systems, 171–173
purposefulness, 171–172, 174

quail, 533
qualitative analysis, 357, 365
quality of soil, 461
quantitative analysis, 32, 110, 486
quantitative behavior, 606
quantitative resource assessment, 263
quantum mechanical considerations, 169
quantum mechanics, 34–35
quantum of energy, 360
quantum phenomena, 87
quasi-dominant interaction matrix, 351
quasi-limit cycling, 445
quasi-recessive interaction matrix, 351
queuing theory, 14

R aggregatable, 226, 250
R law, 175, 189–190
r selected species, 83, 400
rabbits, 523, 526, 529, 533
radiation, 1, 3, 13, 31, 110–112, 117, 131, 135–137, 165, 378, 382, 470, 570, 618
Radiation Ecology Section, 1, 3
radioactive wastes, 5
radioactivity, 509
radioisotopes, 13
radionuclide loading, 399
ragweed, 523
rain forest, 30, 77, 388, 462, 468–470
rain forest of Mauritius, 77
rain shadow desert, 522
rainfall, 73, 377, 382
Rainforest Module, 474
rainy season, 68
random, 6, 13–14, 41, 46–47, 177, 180, 216–217, 235, 304, 345, 347, 366, 408, 452, 614, 643–645
random function generators, 14
random impacts, 46
random processes, 13
random sampling, 235
randomness, 610, 626
range, 7, 31, 49, 55–58, 75, 77, 81, 93, 110–111, 114–115, 130, 137, 167, 204, 215–218, 222, 227, 235, 243, 263–264, 282, 285, 289, 292, 296–299, 301, 303, 307–308, 321, 323, 326–327, 338, 340, 360, 362, 376, 406, 408, 474, 514, 518–519, 528, 531, 539–540, 581–582, 610, 645–646
range condition, 297, 308
range condition survey, 297

range conditions, 307
range quality, 299
ranges, 4, 9, 14, 107, 116, 282, 317, 319, 321, 323, 328, 381, 488, 520, 583
Raphus cucullatus, 77
rapid atmospheric change, 339
rapid resource exploitation, 339, 417
rapid return, 87
rapid turnover, 92, 384
rate equations, 133
rate–force relations, 138
rate kinetics, 266
rate limiting, 264
rate of accumulation, 613
rate of dissipation, 31, 80–81, 83, 91, 94
rate of ecosystem metabolism, 389
rate of ecosystem response, 374
rate of energy flow, 6, 83–84
rate of eutrophication, 404
rate of exergy increase, 606
rate of recovery, 384–385, 390–391
rate of replacement, 93
rate of retirement, 436
rate of uptake, 358
rate processes, 377
rates, 14, 17, 51, 82–83, 87, 90, 92–93, 96, 116, 136–138, 141, 143–149, 154, 156, 227–228, 231, 235, 238, 248, 267–269, 271, 273, 280, 286–287, 289, 291–292, 302–308, 323, 325, 339, 348, 360, 365, 374, 376, 382–385, 387, 391, 406, 416, 422, 424, 426, 429, 433, 436, 441, 446, 454, 456, 460, 578, 583, 594, 618–619
rates of biological processes, 138, 141, 149, 154
rates of change, 325, 376
rates of ecosystem recovery, 382–383
rates of natural increase or decrease, 348
rates of recycling and uptake, 391
ration, 215, 217, 220, 222
RC law, 178, 190
reaction, 6, 90–92, 110–111, 116, 131, 134, 148, 164–165, 417, 446, 573–574, 576, 580, 590
reaction enthalpy, 148
reaction times, 90
real estate, 427
real matrices, 250
real parameters, 45
real predictive capability, 228
real problems, 229
real square matrix, 253
real systems, 225, 231, 260, 464

real world, 11, 21, 97, 105, 229, 262, 273, 331, 444, 473–474, 519, 542, 550, 585, 641
real world ecosystem, 262
real world stability, 105
real world system, 11
realism in ecological models, 113
recession, 428, 434, 436
recipient-controlled dynamic flow equation, 488
reciprocal effects, 452
reciprocal relations, 81, 88
recolonization, 408
recovery, 7, 77, 326, 338, 382–385, 390–391, 398–399, 402, 404, 406, 408–409, 441
recovery characteristics, 338
recovery phase, 408
recreation, 23, 227, 229, 320–321
recreational activities, 264, 321
recreational purposes, 410
recruitment, 67, 69, 76–77, 95, 196–197, 209, 212–214, 328, 402
recruitment failures, 402
recruitment mechanisms, 69
recruitment size, 95
recruits, 63–64, 69, 76–77, 84
recycling of nutrients, 108, 462
recycling rates, 384
red grouse, 85
red maple, 255–256
reduced complexity, 230, 239
reducible systems, 38, 571
reducing processes, 131, 146
reduction, 63, 69, 89, 96, 112, 264, 271, 280–281, 401, 403, 409, 433, 441, 564, 578, 611, 618
reductionism, 225, 338, 509–510
reductionist–mechanistic mode, 29
redundancy, 42, 264, 376, 392, 404, 567, 613–614, 619, 621, 652–654
reduviids, 526
reference system, 53–54, 95, 226, 232
reflection, 167, 172, 643
refractive indexes, 86
refuge from predators, 271, 275
regeneration, 32, 77–78, 178, 190–191, 194–195, 275, 385, 564, 613
regenerative system control, 195
regenerative system model, 192
regenerative systems, 179, 190, 192–193, 195
regional and global cycles, 263
regional managers, 308
regression, 6, 11, 47, 281, 326, 443, 574
regularity, 7, 41, 81
regularity of populations, 7
regulated feedback system, 340

regulation, 58, 67, 76–77, 85, 138, 295–296, 340, 350–352, 355, 357, 400, 452–455, 458, 460, 561, 568, 570, 572, 579, 617, 619, 621
regulation mechanisms, 570, 572, 579
regulation of biogeochemical cycles, 403
regulation of exchange, transformation, and storage, 337, 374
regulation processes, 561, 568
regulation scheme, 85
regulatory agencies, 282
regulatory processes, 85
regulatory purposes, 282
rehabilitation, 338, 409
rehabilitative therapy, 409
rejuvenation, 397–399, 401
rejuvenation processes, 398
relationship, 33, 35–36, 43, 84, 87, 106, 110–111, 148, 167, 170, 175, 179, 190–191, 195, 227, 231, 274, 299–300, 307, 337–339, 347, 377, 387–389, 429, 444, 451, 473–475, 483, 489, 552–553, 559, 562–563, 565, 570–571, 573–574, 581, 613, 620
relative accuracy, 35
relative connectivity, 48–49
relative error, 226, 233, 238
relative removal rate, 287
relative stability, 372, 383–384, 391
relative survivability model, 173
relay floristics, 454
reliability, 32, 167, 172–175, 181, 183–185, 189–190, 192, 621–622
reliability element, 184
reliability R law, 189–190
reliability section, 184
reliability subsystem, 183
relief, 6–7
remediation, 409–410
remineralization, 590, 598–599
removable product, 9
removal time, 282
renewable resource management, 3, 8, 20, 311
renewable resources, 4, 13–14, 228
renewal, 151, 338
reorganization, 346, 594, 597, 616
replacement populations, 307
replacements, 69, 84, 402
replenishing yearlings, 305, 307
replenishment, 197–198, 201, 204, 206, 209
replicable subsystems (species), 42
representability of functions, 244
representable in aggregated variables, 244
representation capability, 331, 475

reproduction, 68–69, 83–84, 115, 187, 212, 214, 303–304, 319, 565, 569–570, 572, 576, 578, 610, 614, 616–617
reproduction and protection of progeny, 214
reproduction rate, 303, 576
reproductive capacity, 403
reproductive condition, 68
reproductive females, 307
reproductive organs (intercalary meristems), 405
reproductive rates, 83, 307–308
reproductive resource, 214
reptiles, 51, 94, 519, 522–523, 532–533, 535, 537
reptilian populations, 73
research design, 263
research needs, 316, 330
research priorities, 312
reservoirs, 110, 132, 154, 318, 324
reset succession, 46
residence time, 282
resilience, 30, 53, 139, 191, 336–338, 345, 372, 374–375, 381, 383–385, 388–391
resistance, 30, 53, 79, 82, 220, 272, 283, 285–286, 289–292, 336–338, 372, 374, 381, 383–385, 388–391, 398, 404–405, 562, 647
resistance–resilience model, 336, 372, 381, 383–385
resistance stability, 79
resonance, 87–88, 93
resource amount, 204
resource B, 197, 209
resource base, 418, 430
resource competition, 237, 264
resource consumer, 356, 464, 563, 597
resource cycle, 358
resource decision making, 228
resource dependence, 211
resource depletion, 339, 415, 424
resource development and management, 320
resource evaluation models, 270
resource flux, 339, 418, 446
resource harvesters, 296
resource inflow, 359–360, 365
resource K, 186
resource management, 3, 8, 16, 20, 40, 51, 130, 229, 242, 263, 269–270, 273–274, 295–296, 302, 308, 311, 331, 372, 415, 450, 485, 561, 568, 609, 623, 643
resource management models, 269
resource managers, 16, 228, 308
resource quantity, 200–201, 208, 218
resource space, 364–365

resource specialists, 522–523, 526–527
resource specialization, 522
resource state, 237, 598
resource supplies, 339, 419
resource systems, 228–229, 311–314, 316, 327–330, 460
resource two consumer, 238
resource use efficiency, 400
resources, 3–4, 6–8, 13–14, 16, 20, 22–23, 31, 34–35, 40, 45, 51, 68, 74–76, 79, 91, 93, 108, 130, 141, 166, 170, 173–174, 176–177, 186, 196–197, 200–206, 208–209, 211–215, 218, 227–229, 237–238, 242, 263–264, 269–270, 273–275, 292, 295–296, 302, 308, 311–314, 316, 318–320, 324, 327–331, 339–340, 356, 358–362, 364–365, 372, 398, 400, 415, 417–420, 422, 424, 430–431, 434–435, 444–446, 450, 452, 459–461, 463–464, 485, 511, 518–519, 522–523, 526–527, 530, 532–533, 542, 559, 561, 563, 568, 572–573, 577, 582, 597–598, 609, 612, 617, 623, 643, 650, 652–653
resources of producers, 214
respiration, 116, 133, 138, 143–149, 151, 154–155, 157, 164, 235–236, 400, 487, 491–492, 494, 496, 578, 583, 624
respiration rate, 116, 235, 578, 583
response nonlinearity, 110
response–stimulus, 174, 191, 198–199, 203
restoration, 226, 263, 338, 409, 456
restricting relationship, 195
retail outlets, 427
retirement of war debt, 420, 425, 440, 446
retrogression, 6–7
retrospective flow analysis, 488, 490–491, 505
revolution, 46–47, 295, 568, 643
Reynolds number, 286, 291
rhipiphorid, 526
rhizodeposition, 520
rhizome, 267
rigidity, 40, 48, 418
risk analysis, 296
river basin, 317–318, 324
river delta, 89
river reaches, 317–318
rivers, 56, 398, 404, 407–408
road construction, 323
roadrunner, 523, 531, 533
robber flies, 526
robust property, 234
rodents, 518, 523–524, 533, 542
role of feedback, 340, 450
root mycorrhizal associations, 385
root versus shoot uptake, 274
roots, 272, 456, 520, 523–524
Rose Atoll, 70, 72
rosy boa, 534
rotifer, 56
roughness Reynolds number, 286
round whitefish, 56
Rouse–Hurvitz criterion, 354
row commodity, 493, 501–502
ruderals, 456
rule-based, 563–565, 610–611, 615, 617–619, 621
rule-based information processing, 563, 610–611
rule-based mechanisms, 565, 615, 618–619, 621
rule-based problems, 564, 610, 615, 618
rule-based problem solving, 564, 610, 615, 618
runoff, 270, 274, 324, 326
runoff scenarios, 324

s-dimensional vector, 254
saddle point, 199
sage sparrow, 532
Sahara, 114
salamanders, 69
salt tidal marsh, 388
saltbush, 526–527
Salvelinus alpinus, 56, 59, 63–64, 66
Salvelinus namaycush, 56, 402
samples, 42, 55, 58–59, 63, 69, 74
sampling, 11–12, 58–59, 63, 66, 68–71, 74, 95, 227, 235, 268, 378, 380, 392, 416
sampling error, 95
sampling gear, 58
sampling procedure, 68, 74
sampling program, 63
sampling strategies, 378
sand community, 510, 528, 536
sand dunes, 522
sanitary sewer, 406
saplings, 66, 78, 95
saprovory, 538
Sarcophagidae, 527
satellite monitoring, 301
satiation, 106, 109
Saudi Arabian crude oil, 431
savanna, 340, 377, 382, 388, 468–470, 474
Savanna Module, 474
scale, 7, 11, 17–18, 32, 91, 115, 120, 149, 151, 162, 226, 231, 237, 239, 252, 263, 317–318, 327, 329–331, 335–337, 373, 377–383, 385, 387, 390–391, 403, 408, 420, 443, 459, 468–471, 498, 613, 616, 643, 650, 654
scale-dependent analyses of ecosystem dynamics, 378
scale-dependent ecosystem organization, 391
scale-dependent filters, 380
scale-dependent mean values, 385
scale-dependent properties, 337, 377
scale-dependent view of ecosystems, 382
scale-dependent windows, 380
scale of observation, 377
scale-specific versions of theory, 162
scale of space and time, 375–376, 378–380, 382–383, 387, 389–391
scaling factor, 37, 652
scenario development, 314–315
school, 21, 34, 194, 205, 222, 242, 511, 568, 585
schools, 85, 200, 205, 215–216
science of complex systems, 229
scientific and statistical methods, 10
scientific decision making, 13
scorpions, 516–519, 521, 528–533, 535
Scott's oriole, 523
sculpins, 56, 402
SDG, 353–357, 359–360, 365, 367
sea, 66, 116–117, 319, 338, 398, 404, 407–408, 635, 638
sea palm community, 398
seas, 95, 398
seascapes, 5, 97
season, 17, 68–69, 73, 200, 202, 212–213, 228, 268, 274, 296–299, 301–303, 306, 319, 321, 323, 325, 405, 478, 517–518
seasonal bag limit, 299
seasonal calculations, 321
seasonal changes, 47, 112, 518
seasonal conditions, 214
seasonal or annual scale, 239
seasonality, 108
seawater, 459
second law of thermodynamics, 32, 88, 131, 136–137, 141, 162, 345, 490, 511, 549, 570, 652
second-order damped oscillator, 384
secondary consumers, 521–522
secondary optimization, 186–187
secondary production, 271
secondary succession, 7, 399, 408, 454, 456
sedentary, 218
sediment, 263–266, 271–272, 274–275, 282, 318, 403
sediment inputs, 272
sediment loading, 272, 274

sediment pore water, 265–266
sediment–water interface, 403
sedimentary mineral flows, 639
sedimentation, 45, 270, 404
sediments, 226–227, 263–264, 266, 270–274, 280–281, 404
seeding, 471
seedlings, 66, 78, 95
selection, 11, 43–44, 46–47, 63, 93, 105, 115–116, 120, 122–123, 313, 344, 417, 454, 461–463, 561, 565, 569–570, 579, 611, 613–614, 616–617, 651
selection criterion, 122
selection pressure, 617, 651
selective poisons, 13
self-adaptation, 133
self-competition, 106, 109
self-design, 114, 566
self-facilitating aspect of evolution, 614
self-facilitation, 564, 611, 613–615, 618–620
self-facilitation principle, 564, 611, 613–614
self-inhibition function, 248
self-limitation, 350
self-limiting species, 356, 365
self-loops, 519, 537–539, 566, 632–634
self-organizability, 32, 172, 174, 178, 191
self-organization, 82, 122, 133, 142, 179, 408, 565–567, 620, 624, 638, 640–641
self-organization effect, 179
self-organization processes, 142
self-organize, 561
self-organizing patterns, 453
self-organizing processes, 130, 401
self-organizing system, 116
self-perfection, 178
self-regulating mechanisms, 67
self-regulation, 58, 67, 350–352, 355, 357, 400, 455
self-regulation dominance, 352
self-regulation loop, 355
self-shading effects, 120
self-sufficient environments, 341
self-sustaining ecosystems, 470
semiarid, 73, 324
semidesert grasslands, 23
senescence, 7, 69
sense organs, 69
sensitive species, 112, 402–403, 406
sensitivity, 112, 268, 270–271, 274–275, 582, 589
sensitivity analyses, 268, 270–271, 274–275
separability, 226, 243, 245, 247–248, 251–253, 255

separability coefficients, 253
separability conditions, 248, 251
separability of linear models, 226, 251
separability of linear systems, 243
separability of variables, 252
sequence, 21, 43, 45, 59, 83, 90, 140, 168, 171, 174, 191, 258, 264, 270, 308, 330, 368, 399, 402–403, 406, 409, 428, 441, 552, 557, 613, 619
series aggregations, 238
service activities, 435
service personnel, 437
seston, 142, 265–266, 268–269, 522
set theory, 353
sex, 236, 296, 298–299, 301, 303–308, 321
sex and age cohorts, 321
sex classes, 298–299, 301, 303, 307–308
sexual selection, 116
Seychelles, 30, 72–73
shading effects, 120, 271
Shannon's diversity, 48
Shannon's index, 41, 47
Shannon–Wiener function, 566
shape of a plane, 35
shear stress, 285
shifting mosaic steady-state model, 382
shoot biomass, 268
shore erosion, 272
shore lines, 5
short-lived species, 400
shrews, 517
shrub-steppe biome, 96
shrubs, 77, 403
side effects, 225
sidewinder rattlesnake, 534
sifting, 71
sign equivalence matrices, 355
sign pattern, 353, 357
sign stability, 336, 344, 352–357, 359–360, 365, 367
sign-stability criterion, 336, 355–357, 359–360, 365, 367
sign-stability theory, 357
sign-stable communities, 356
sign-stable patterns, 344, 354, 356
sign structure, 350, 352–353
signal coding and decoding, 176
signal space, 88
signal system, 81, 93
signal-to-noise ratio, 89
signed directed graph (SDG), 353, 365
silica depletion, 391
silicon, 579
silphid beetles, 526
siltation, 399

Silver Springs, 361, 557
SIMCON, 316
Simplex method, 446
simulation analysis, 292
simulation control package, 317
simulation controller, 316
simulation devices, 11
simulation dynamic model, 357
simulation experiments, 272, 274, 589
simulation modeling, 225–226, 228–229, 237, 260, 263–264, 311–312, 320, 323, 326–331, 340–341
simulation modeling theory, 260
simulation modeling workshops, 311, 320, 323, 329–330
simulation models, 14, 37, 113, 228, 231, 242, 258, 262–264, 271, 274, 312–314, 316, 324, 326–328, 330–331, 340, 343, 382, 416, 434, 439, 442, 468, 471–472, 476
simulation of nets, 49
simulation programs, 434
simulation runs, 288
simulation studies, 109, 121, 238, 269, 271
simulations, 36, 237, 264, 268, 270–271, 288–289, 438, 443, 480
simultaneous survival, 178
single-commodity output, 505–506
single-commodity systems, 505
single state-variable model, 234
sinks, 54, 134, 226, 263, 284, 292, 403, 645
size, 2, 4–5, 8, 20, 30–31, 34, 37–38, 40, 44–45, 51, 55–59, 62–63, 66–67, 70, 72–74, 76, 79, 83–85, 94–96, 112, 116, 140, 205, 228, 235, 264, 297, 300–301, 303, 305–308, 317, 328, 350, 390–391, 403, 416, 423–424, 430, 432–433, 436, 440–441, 452, 477, 515, 518–520, 526, 531, 535, 539, 550, 567, 578, 609, 615, 648–650
size–age relationship, 84
size class, 55, 70, 73, 79, 84, 515, 520
size distribution, 70, 403
size-frequency distributions, 70
size structure, 66–67, 84
size of schools, 205
skew-symmetric matrices, 349
skunks, 526
slime molds, 84
small biota, 400
snakes, 519, 523, 533–534, 537
snowfall factor, 321, 323
snowshoe hare, 418
social data, 416

social disturbances, 9, 415
social living, 85
social organization, 106, 616–617
social structure, 107, 109, 114
social systems, 49, 170, 311–312, 416–417
societal instability, 415
society, 9, 73, 84, 97–98, 262, 270, 339, 415–420, 422–428, 430, 435–438, 441–442, 445–446, 463, 511, 618, 650
socioeconomic consequences, 227
socioeconomic considerations, 264
socioeconomic environmental systems, 445
socioeconomic sphere, 227
socioeconomic systems, 269
sociopolitical sphere, 338
soil, 5–7, 17–18, 280–282, 313, 342, 398, 400, 402, 407, 454, 456–457, 461, 487, 501, 504, 510, 515, 518, 520–521, 523–525, 535, 537, 539, 541–542
soil acidity, 457
soil arthropods, 524, 541
soil biota, 521, 523, 539, 542
soil depth, 454, 518
soil detritivores, 520
soil detritus subweb, 525
soil fauna, 518, 524
soil integrity, 398
soil microbes, 537
soil moisture, 518
soil organisms, 524, 535
soil properties, 7
soil quality, 461
soil water, 5, 282
soil–water interfaces, 5
solar energy, 87, 341, 391, 498, 500–504, 570
solar energy cost, 498
solar energy inputs, 87
solar energy units, 341, 500–504
solar radiation, 110, 112, 378, 382, 570
solar radiation inputs, 110
solitons, 45
solpugids, 519, 526, 532–533, 535
solute export, 383
solution caves, 68
Sonoran Desert, 468
sorption, 157, 272
sources, 22, 95, 134, 231, 265–266, 274, 301, 387–388, 416–420, 428, 445–446, 619
South Park, 321
South Yellowstone, 321
Southern Green Bay, 403–404
southern United States, 30, 69–70
Soviet Union, 32, 470
SO_x fumigation, 405

space, 6, 20, 37, 41, 43–47, 64, 66, 83, 88, 92, 95–96, 105–107, 109, 114, 116, 119, 122, 140, 143–145, 161, 167–169, 174, 192–193, 200–201, 204, 218, 229, 258, 319, 336–337, 345, 347, 349, 352–353, 356, 364–365, 373–383, 385–387, 389–391, 407–408, 434, 454, 460, 468, 470–471, 514, 517, 522, 563–564, 591, 593–597, 605, 612, 614–615, 620–621
space and time, 6, 41, 43–44, 144, 167, 373, 375–376, 378–383, 387, 389–391, 407–408
space mode, 37
space scales, 6, 229
space stations, 470
space–time dynamics of ecosystem structure, 382
space–time levels, 200
space–time patterns, 377, 382
space–time scale, 377–378, 382, 387
space–time scale of analysis, 387
Spartina, 557
spatial, 6, 37, 41, 47–48, 92, 114, 142, 161, 168, 189, 231–232, 264, 269, 280, 282, 292, 314, 317, 320, 323, 325, 327–328, 331, 338, 365, 377, 398, 406–407, 454, 456, 463, 517–518, 535, 564, 613, 616–617
spatial, age, and social structures of populations, 114
spatial and temporal resolution, 314
spatial distribution, 189, 365, 406
spatial heterogeneity, 41, 231
spatial patterns, 47, 463
spatial proximity, 41, 48, 463
spatial separation of habitats, 269
spatial variability, 280, 292
spatially local, 30
spatiotemporal, 138, 337, 373, 377, 380, 391
spawning locations, 402
spawning season, 405
specialization directed, 564–565, 615–618, 621
speciation, 114
species, 6–7, 9–10, 14–15, 19, 30, 38, 40–49, 51, 56–58, 67–69, 71–73, 75, 77–80, 83–85, 89–97, 104, 106–112, 114–118, 120, 122–123, 133, 137, 142, 144–146, 149–154, 156, 158, 161, 164–165, 170–171, 189–190, 194, 209, 231–232, 235–236, 248, 256, 263, 265–266, 297, 308, 320, 324–325, 328, 331, 335, 337–338, 344–354, 356–

359, 361–365, 373, 376–378, 380, 385, 387, 391–392, 398, 400–406, 408–410, 418, 451–452, 454–464, 468–469, 471, 474, 486–487, 491, 510–511, 513–519, 521–535, 537–542, 550–552, 557–558, 561–562, 564–565, 567–579, 582–583, 594, 599, 601, 603–604, 612–620, 624, 640, 644–645, 648, 652
species assemblages, 337, 380, 392
species community, 194
species complement, 89, 93–94
species complexity, 189
species composition, 116, 170–171, 337, 365, 373, 376, 380, 517, 521–522, 526, 562, 568–569, 573, 575, 577–579, 583, 601
species diversity, 90, 161, 338, 344, 376, 398, 400–401, 405–406, 510, 515, 535, 574, 618
species formation, 170–171
species interactions, 112, 346–348, 354, 365
species optimization, 464
species persistence, 337, 373, 376
species populations, 51, 79–80, 83, 92, 104, 337, 373, 380, 391–392, 451, 460, 463
species richness, 56, 89, 406, 541
species strategies, 456–457, 462
species turnover, 376
specific enthalpy, 135
specific volume, 131, 134, 165
spectra of diversity, 41, 46
spectral analysis, 107
spectral analytical methods, 110
spectral composition of light, 111
speed, 20, 40, 43, 53, 222, 283, 285–286, 288–289, 291, 404, 445
sphecids, 526–527
spider parasitoids, 537
spiders, 469, 515–516, 518–519, 523, 526–533, 542
spontaneous behavior, 173
spontaneous systems, 172
sporozoans, 528
spread of age, 79
spreading risk, 339
springs, 388
spruce budworm populations, 418
stability, 7, 9, 30–32, 44, 47, 52–55, 58, 79–80, 83, 94–95, 105–110, 112–115, 138–140, 142, 184, 189–191, 195, 235, 237, 335–339, 343–368, 372–378, 380–384, 391–392, 401, 416–419, 445–446, 459–460, 470, 488–489, 511, 514, 518, 520, 541, 562–565, 567, 574, 582, 586, 589–590, 593, 595, 603–605,

610, 612–613, 615, 618–620, 622–623
stability analysis, 109, 112, 344, 346–347, 366, 514
stability and complexity of ecosystem models, 336
stability behavior, 107, 109
stability conditions, 140, 354–355, 359–360, 364–365, 367–368
stability curve, 603–605
stability domain, 347, 595, 603–605
stability investigation, 113
stability measure for competition structures, 362
stability of complex systems, 190
stability of differential equations, 337
stability of ecosystems, 352, 374, 376, 384, 511, 541
stability of horizontally structured communities, 336
stability of isolated populations, 335
stability of spatially distributed ecosystems, 336
stability of trophic chains, 358
stability principles, 55, 339, 345, 417, 446
stability principles of biomechanics, 55
stability studies, 31, 105, 108, 113, 349
stability subsets, 347, 356–357, 365
stability theory, 335, 339, 345, 357, 365, 373, 375, 377, 416–417, 445, 589
stability versus diversity, 105
stabilization, 6, 53, 107, 112, 459–460, 619
stabilizing mechanisms, 114, 406
stable, 6–9, 30, 32, 52–53, 55, 58, 90, 105, 107–109, 114, 139–141, 144–145, 160–161, 191, 196, 206, 217, 237, 303, 336–337, 343–346, 348–349, 351–356, 359–360, 362–368, 373–378, 380–383, 391–392, 433, 459–461, 469–470, 483, 488–489, 514, 563–564, 582, 593, 595, 601, 610, 612, 618–620
stable attractors, 337, 375, 378, 381
stable behavior, 345
stable climax, 52, 58, 90
stable coexistence combinations of species, 364
stable community, 354, 362, 365
stable competition models, 237
stable dynamic regime, 139
stable equilibria, 9, 109
stable limit cycles, 433
stable regime, 141
stable species composition, 365

stable stationary reference state, 144
stable stationary state, 140, 145, 160–161
stable steady state, 382, 593
stand density, 383
standard deviation, 35, 579
standard of living, 437, 439
standing modes, 76
standing stock, 487, 489
staphylinid beetles, 524, 526
staple food, 57
starvation, 77, 296, 298
stasis, 30, 52, 79, 94, 340
state, 3, 5, 7, 9, 16, 21–23, 30–32, 34–37, 42, 51–53, 55, 58, 67, 69, 76–77, 79–91, 94–96, 98, 105, 111–112, 114–115, 117–120, 122, 131–134, 137–138, 140–147, 153–154, 157, 160–161, 166, 169–170, 173, 178, 181, 190–191, 194, 196–198, 218, 227–228, 232–234, 237, 239, 255, 258–259, 264–266, 270, 274, 280, 282–283, 289, 291, 295–299, 302, 304, 308–309, 313, 323–324, 337–339, 344–347, 350–351, 359, 362–364, 373–378, 381–384, 386–387, 390–391, 397–399, 401–405, 408–410, 415, 420, 422, 425–428, 435, 439, 442, 444–445, 455–456, 460–461, 469, 472–474, 476, 487, 489, 491, 494–495, 497, 559, 562–563, 566, 568, 570, 572, 576–577, 580–581, 586, 589–599, 601–603, 605, 609–610, 618, 620–622, 624–625, 640
state and federal experiment stations, 3
state displacement, 384
state equations, 119, 233
state of climax, 94, 218
state space, 105, 114, 337, 374–377, 381–382, 386–387, 391, 620–621
state-space behavior, 621
state-space diagrams, 375
state transitions, 624
state variables, 34–37, 111, 117–119, 133, 137–138, 140–147, 157, 161, 227, 232–234, 237, 258, 264–266, 270, 280, 282–283, 374, 422, 428, 439, 444, 562, 568, 570, 576, 580–581, 586, 589–591, 593–594
statewide harvest, 228
stationary climax state, 196, 200
stationary state, 30, 32, 52, 55, 67, 81, 139–140, 144–145, 158, 160–161, 164, 170, 173, 190–191

stationary value, 194
statistical distributions, 14, 262
statistical fluctuations, 142
statistical inference, 11
statistical mechanics, 95, 106, 644
statistical methods, 10–11, 171
statistical physics, 172
statistics, 21, 40, 176–177, 233, 313, 397, 420–421, 424–425, 427–428, 430–432, 445, 516, 537–538, 543, 552, 643, 648
steady state, 7, 52–53, 67, 81–84, 87, 90–91, 94, 105, 112, 194, 259, 270, 304, 308, 345, 382, 487, 489, 491, 494–495, 497, 562–563, 576–577, 580–581, 593, 595–598, 601–603, 625
steady-state box models, 270
steady-state ecosystem, 84
steady-state exergy, 562, 576
steady-state feasibility condition, 598
steady-state population, 304
steady-state relationship, 489
steady-state value, 105
STELLA, 581
stem-diameter frequency, 77–78
steppe grasslands, 5
stewardship, 398, 410
stimulus–response, 174, 191, 198–199, 203
stimulus–response hypothesis, 198–199, 203
stochastic, 14, 21, 31–32, 39, 44, 110, 114–115, 138, 141–142, 166–168, 171–173, 191, 193–196, 200, 228, 231, 282, 296–297, 301–308, 350, 362, 434, 452, 610
stochastic analysis, 308
stochastic automata, 191
stochastic character, 166
stochastic coefficients, 228
stochastic constraint, 304
stochastic death rates, 305
stochastic fate models, 282
stochastic interactions, 31
stochastic methods, 142
stochastic migration of elements, 193
stochastic models, 14, 21, 296
stochastic noise, 110
stochastic object, 167
stochastic parameters, 304
stochastic phenomena, 21
stochastic system, 44
stochastic theory of community adaptation, 195
stochastic vector, 167–168
stochasticity, 106, 167

stock, 4, 51, 58, 66–67, 84, 94, 196–197, 316, 328, 422, 437, 487–489, 491, 493, 495–496, 499
stock configuration, 67
stock–flow assumption, 493
stock–flow matrix, 488
stock replacement respiration, 496
stocking, 17, 308, 470
stocks, 31, 56–58, 197, 268, 296, 488, 490
stoichiometric coefficients, 137, 165
storage, 79–80, 87, 133, 146, 149–151, 226, 263, 316, 323–324, 337, 374, 379, 384, 391, 416, 458–459, 488, 563, 566, 578
storage delay, 79, 566
storage or flow delay, 488
storages and sinks, 226, 263
stored energy, 87, 146–147, 149, 151
storms, 302, 398
strange attractors, 346
strategic goals, 32, 171, 174–175
stratification, 45, 288, 521
stratified stability, 381
stratum, 77
stream, 233, 372, 383, 401, 407, 469, 572, 574–576, 635, 653
stream chemistry, 383
stream ecosystems, 372, 401, 635
stream model, 572, 574, 576
stream nutrient concentrations, 383
streamflow, 383
streams, 17, 319, 388, 398, 401, 407
stress, 71, 113, 120, 135, 174, 266, 274, 285, 335, 338–339, 397–408, 417, 442–443, 451, 456, 567
stress-adapted, 397, 405
stress buffering, 405
stress effects, 397–398, 406
stress gradients, 406–407
stress intervention, 400, 405
stress tolerators, 456
stress-compensating capability, 405
stress-induced transformations, 404
stressed degraded, 338
stressed state, 399
stresses, 9, 121, 135, 263, 314, 328, 338, 397–408, 444, 456, 609
stressor, 399
striving, 97, 167
Strix aluco, 76
stromatolites, 95
structural change, 141, 170, 511, 568, 610, 613, 617
structural characteristic time, 170
structural configuration, 401
structural deformation, 190
structural degeneration, 184
structural effect, 170, 185
structural evolution, 178
structural mutation, 170

structural optimization, 194
structural organization, 190
structural stability, 108, 459, 619
structural state, 178
structural viability, 185
structure, 9–10, 22–23, 30, 36, 43, 45–47, 51, 57–58, 63, 66–67, 72–73, 76–77, 83–85, 94, 107–109, 112, 114–121, 131, 138–142, 166, 170, 173, 175, 177, 183–184, 190–196, 200, 227, 230, 232, 246, 262, 265, 270, 279–284, 291–292, 297, 308–309, 312, 315, 320, 323–324, 327, 329, 335, 337, 340–341, 344, 348–353, 357–358, 363, 365, 377–382, 384–385, 387–392, 398, 400–401, 404, 416, 418, 420, 426, 429–430, 439, 442, 444–445, 463, 469–470, 472, 474, 476–477, 479–481, 483–484, 487, 510–513, 515, 518–519, 531–532, 535, 539–543, 549, 551–553, 561–565, 567–571, 577–579, 582, 586–587, 589, 595, 601, 603–605, 612–614, 616, 618, 620, 622, 624, 627, 629, 635, 637–638, 640, 646–647, 654
structure and dynamics of ecosystems, 142
structure matrix, 512, 551–552
structures, 30–32, 43, 47, 51, 69, 81–82, 87, 107–109, 114–115, 120, 130, 133, 141–142, 166, 173–174, 176, 178, 183, 258, 284, 313, 330, 336–337, 339, 347, 355–358, 361–362, 365, 378–379, 381–382, 385, 390–391, 399, 408, 418–419, 470, 474, 510, 519, 564, 566, 605, 635, 653
subgroup reports, 314–315
subject matter expertise, 313
submodel definition, 314
submodel development, 315
submodel linking, 314, 316
submodel programming, 314
submodels, 227–228, 270, 280, 283, 314–317, 321, 323–325, 327, 329
suboptimality, 114–115, 122
substance, 172, 287, 456, 512, 565, 572, 575–576, 586–587, 649
substantial, 32, 95, 169–170, 173, 263, 269, 314, 330, 392, 514, 621 615–616, 622
substitution, 243, 252, 281, 344, 646, 648
substrate, 108, 111, 187, 454, 518, 521, 613

substrate concentration, 111
substrate splitting, 187
substrates, 113, 116
subsystem models, 226, 264
subsystem simulations, 264
subsystems, 42–43, 46, 97, 120, 142–143, 174, 183–184, 186, 227, 264, 314, 339, 373, 379, 416, 423, 428, 434, 510
succession, 6–7, 43, 46–47, 52, 80, 83, 88–90, 92, 95, 176, 196, 200, 232, 255, 318, 325, 330, 336, 340, 373–374, 399–400, 408, 419, 453–458, 462, 511, 525, 566, 610–611, 614–615, 617, 619–620, 624
succession models, 454, 457
succession reversal, 400
successional change, 45
successional or ecological time, 92
successional process, 454, 457, 599
successional states, 218
sufficiency of conditions, 247
sufficiency proof, 355
suitable applications, 326, 329
sulfur, 459
summary predator, 220
sunlight, 54, 266, 271, 487, 498–500
superadditive law of system effectivity, 187
supplementary energy, 132, 134, 164
supply instability, 446
suppression, 66, 95, 195, 197, 209–210, 382, 446
surface, 4, 8, 47, 54, 56, 75, 81, 87, 135, 190, 265, 282–283, 287, 319, 324, 365, 377, 400, 470, 501, 504, 518, 521, 524–525, 528, 537, 541, 564, 578, 614–615
surface inflow, 324
surface layer, 54
surface litter, 518
surface temperature, 56
surface tension, 47
survivability, 173–174, 177–178, 182, 184, 186–192
survivability model, 173, 187, 191–192
survivability RC law, 190
survival, 30, 39, 43, 53, 83–84, 115, 171, 175, 178, 181, 190, 195, 201, 215, 217, 219, 221, 321, 327, 452, 550, 561, 563, 568–570, 611, 652–653
survival potential, 53
survival probability, 215, 221
survival rate, 83
suspended particulate matter, 271
suspended sediments, 226–227, 263, 274, 281
suspended solids, 404

swan mussel, 71–72
Swedish coast, 404
switching of predators, 106, 109
syllogism, 54
symbionts, 515, 524
symbiosis, 4, 513
symmetric (reciprocal) interactions, 31
symmetric interactions, 79, 82
symmetric state, 87
symmetry, 31, 80, 82, 87–90, 651
synergetic, 32, 138, 141–142, 470
synergetics, 142, 179
synthesis of research results, 317
synthetic ecosystems, 470
synthetic theory of the ecosystem, 378
system, 4–14, 20–23, 30–32, 35–39, 41–45, 48–49, 52–55, 58, 67–68, 79–83, 86–95, 97, 105–106, 108–110, 112–114, 116–120, 122, 130–133, 135, 137–143, 157, 164, 166–167, 170–178, 181–198, 200, 206, 209, 213–214, 216–217, 225–226, 228–229, 231–235, 237–238, 242–244, 247–253, 255, 258–259, 262, 264, 269, 279–281, 283–284, 291, 312–316, 318, 326–331, 336–341, 343–348, 350, 352–353, 356–364, 366–367, 373–381, 383, 385–386, 398–401, 403–408, 415–422, 424–429, 431, 433–434, 436, 438–446, 450–454, 456–464, 468–472, 474–477, 479–481, 483–495, 498, 501, 505, 510–511, 516, 520, 526, 535, 543, 551, 553, 561–567, 569–571, 577–579, 583, 586–597, 599–601, 603–606, 609–613, 615–625, 644–654
system ascendancy, 650–651
system autonomy, 55, 80
system behavior, 39, 122, 170–171, 174, 255, 377–378, 452, 476, 480
system boundaries, 52–53
system configuration, 94
system destabilization, 339, 419
system dynamics, 32, 237, 373, 566, 646, 649
system entity structure, 341, 472, 476–477, 480–481, 483–484
system entity structure model base, 476, 480–481
system exergy, 570, 577
system expandability, 562
system flow chart, 428
system function, 605
system growth and development, 623
system hierarchy, 171
system history, 578
system-level environment, 625
system linearization, 106
system macrostate, 170–171
system maintenance, 53
system microstate, 170
system model, 166, 192, 228
system negentropy, 561
system open to matter, 595
system organization, 80, 166, 378, 381, 462, 566
system overhead, 651–652
system performance, 117, 119, 427, 471
system perturbation, 112
system reliability, 189
system reorganization, 597
system stability, 235, 338, 345, 419
system state, 181, 233–234, 374, 427, 586, 590
system stationary state, 173
system structure, 170, 312, 337, 476
system-theoretic representation, 476
system theory, 339, 416, 476
system throughput, 650, 652–654
system trajectories, 255, 375
system versus environment, 32
systemological development, 173
systemology, 32, 166–167, 171–174, 177
systems analysis, 2–3, 13, 21, 104, 166, 228, 279, 295, 308, 311–312, 331, 343, 453, 549, 568, 585
systems analysis applications, 2, 13
systems ecologists, 1–3, 10, 12–21, 51, 113, 339, 416, 458, 509
systems ecology, 1–5, 9–16, 18, 20–23, 33, 104, 229, 283, 295, 331, 340–341, 365–366, 446, 453, 469, 481
systems methodology, 120
systems theory, 32, 49, 340, 375, 451, 453, 473–474, 481
systems view, 228

Tachinidae, 527
tactical, 32, 171–172, 174–176, 178, 183, 194
tactical goals, 171–172, 174, 194
Taenia tapeworms, 529
Tajikestan SSR, 357
Takamaka Grove, 74
Tansley's axiom, 54
tarantulas, 533
target species, 75
tawny owls, 30, 76, 85
taxonomic, 42, 242, 341, 472, 474, 476, 478, 522, 550, 559, 645
taxonomic categorization, 550
taxonomic groups, 42
taxonomic relation, 341
taxonomic units, 242
Taylor series, 258
Tchebyshev inequality, 195
telemetric devices, 20
temperate deciduous forest, 385, 388
temperate evergreen forest, 388
temperate regions, 57
temperature, 56, 91–93, 110–112, 116–117, 119, 132–134, 137–138, 143–145, 147–149, 150–155, 157, 159–161, 164, 235, 259, 265–267, 283–286, 288–292, 375, 379, 404, 406, 459, 471, 474, 521, 561–562, 570–572, 575–576, 578, 586, 594, 597–601, 605, 617
temperature acclimation, 116
temperature conditions, 288–289, 291
temperature correction factor, 286
temperature inputs, 112
temperature interactions, 267
temperature kinetics, 265
temperature–precipitation axis, 110
temporal distributions, 268
temporal dynamics, 231, 344, 377
temporal scale, 377
tenebrionid beetles, 535
tenebrionid larvae, 524
terminal buds, 452
terminal predator, 83
terminology, 4, 10, 340–341, 355, 358, 470
termites, 469, 518, 524–525, 531–533
ternary systems, 49
terrestrial, 9, 12, 23, 63, 67, 73, 95, 109–111, 375, 387–388, 401, 405–406, 515–516, 519–521, 532, 539, 541–542
terrestrial production, 111
terrestrial vegetation, 110
territories, 55, 57, 71, 76–77
tertiary periods, 93
tertiary production, 271
test hypotheses, 10
Testudo gigantea, 73
Teton Park, 321
Thais, 220–221
theorems, 11, 226, 233–234, 348, 417–418
theoretical analyses, 511
theoretical behavior, 291
theoretical conceptualization, 380
theoretical dependence, 189
theoretical ecology, 31–32, 162, 230, 234, 237–238, 353
theoretical ecosystem models, 104–105, 112
theoretical phase, 10

theoretical physics, 168, 173
theoretical predictions, 288, 510
theories of physical fields, 169
theory of aggregation, 231, 243
theory of aquatic ecosystems, 138, 141
theory of automata and control, 177
theory of complementarity, 36
theory of fish school formation, 205
theory of indirect effects, 578
theory of internal variables, 131
theory of multispecies systems, 347
theory of nonconflict games, 177
theory of nonequilibrium states, 167
theory of optimal pattern classification, 179
theory of pattern identification, 176
theory of potential effectivity, 167, 175, 179, 192
theory of potential noise immunity, 176
theory of qualitative stability, 353
theory of reliability, 175
theory of resource competition, 237
theory of self-organizability, 178
theory of stability, 344, 365
theridiid spiders, 532
therivid flies, 524
thermoclines, 47
thermodynamic approach, 167
thermodynamic concept, 605
thermodynamic considerations, 32, 107, 491
thermodynamic dissipation, 30
thermodynamic engines, 85
thermodynamic equations, 158, 571
thermodynamic equilibrium, 30, 52–53, 81–82, 88–89, 94, 96–97, 132, 562, 570–571, 580, 586–587, 589–592, 603, 644
thermodynamic equilibrium state, 591–592
thermodynamic formalism, 32
thermodynamic foundations, 30
thermodynamic laws, 133
thermodynamic method, 142
thermodynamic model, 588
thermodynamic principles, 52, 143
thermodynamic processes, 168
thermodynamic situation, 643
thermodynamic stability, 140, 345
thermodynamic state, 85
thermodynamic systems, 53, 81
thermodynamic terms, 39, 91, 561, 566, 568, 570
thermodynamic theory, 138–139, 141, 143, 156, 592
thermodynamic theory of ecosystems, 138–139, 141
thermodynamic threshold, 141

thermodynamics, 30–32, 39–40, 44–45, 52–54, 80–82, 85, 87–89, 91, 94, 96–97, 107, 113–114, 122, 131–134, 136–144, 150, 154, 156–158, 162, 167–168, 336–337, 345, 378, 380–381, 386, 451, 490–491, 511–512, 549, 559, 561–562, 566–568, 570–571, 578–580, 582, 586–594, 599, 601, 603, 605, 643–644, 652–653, 655
thermodynamics of ecosystems, 132–133
thermostatic equilibrium, 32, 131, 133, 139, 164
three-dimensional matrix, 15
three trophic-level systems, 108
threshold (step) function, 235
threshold and nonlinear feeding, 109
trichomoniasis, 535
Thuja occidentalis, 78–79
Thysanoptera, 523
Thysanura, 524
ticks, 529
tidal exchange, 274
tidal flushing, 267
tidal marsh, 388, 512, 553, 635
tidal marsh stream, 635
tidal mixing, 268
Tigrovaya Balka Reserve, 357
timber and range management, 296
timber products, 470
time, 5–8, 13–14, 16, 19–20, 22, 30–33, 35, 37–38, 41–47, 52–55, 58, 62–63, 67, 69, 73, 75, 77, 79–97, 105–107, 109–110, 114, 117, 119–122, 130–136, 139, 142–145, 148–149, 151, 154, 156–157, 160–161, 164, 166–172, 174–177, 179, 183–184, 191–194, 200, 206, 214–218, 220–222, 225, 227, 229, 231–234, 237, 242, 251, 253, 256, 264–265, 282, 301, 308, 313–317, 321, 324–329, 331, 336–337, 339, 345, 358, 366, 373–385, 387, 389–391, 400, 405–408, 416, 418–420, 422, 424–430, 432–434, 436, 438–445, 452–454, 458, 460, 464, 468–469, 472, 474, 476, 483, 487–488, 492, 494, 509, 511, 516–518, 521, 526, 528, 543, 564, 566, 569–570, 573–574, 576–577, 586, 593, 596–597, 603, 611–613, 615, 617–621, 624, 627–628, 640, 645, 648, 650, 654
time and space scales, 6, 229
time delay, 30, 53, 79, 81, 85–87, 96
time dependence, 117, 119, 148
time-dependent solutions, 161

time-dependent variable, 119
time derivatives, 117, 374
time frame, 90, 313–314, 316, 339
time-independent controlling factors, 161
time-independent states, 143
time lags, 106, 109, 149, 151, 418, 425, 433, 452
time mode, 37
time of adaptation, 191
time scale, 7, 149, 151, 237, 329, 331, 377–378, 382, 387
time scales, 90, 92–93, 200, 327, 376, 380, 385, 453, 460, 597, 615, 618–619
time-sequential impact, 218
time-series analysis, 110
time series of data, 231
time variation, 135–136
timeless order, 90
tiphiids, 527
tissue, 5, 83, 230, 385, 450, 513, 523, 526
tolerance, 111, 225, 235, 239, 340, 454, 456–457, 622
tolerance models, 456–457
toluene, 287
tools, 2–3, 5, 10–13, 15–17, 20–22, 142, 166, 213–214, 295, 566, 610, 612, 643
top predator, 514, 526, 534–535
topography, 44, 375, 391
tortoise population, 72–73, 75
tortoise turf, 75
tortoises, 30, 72–75, 469, 519, 533
tortoiseshell, 73
Torymus, 526
total community biomass, 346
total detritus, 388
total ecosystem research, 22–23
total effectivity, 175, 183
total error, 226, 233
total herd, 305–306
total inputs, 342, 486, 500–502, 504
total kill, 296, 298, 303
total mass, 135
total output, 486–487, 489–492, 494–496, 498, 501–503
total-output flow ratio, 489
total stability, 348, 365
total system state, 234
total system throughput, 650, 652–654
totally stable matrices, 348–349
tourism, 321
TOXFATE model, 280, 292
toxic chemicals, 509
toxic concentrations, 403
toxic contaminants, 279–281, 283–284, 291–292
toxic loading, 399

toxic substances, 572, 575–576
toxic substance model, 572
toxicants, 391
toxicity, 271–272, 327
toxics, 406
trade, 38, 166, 195–197, 200, 206, 209, 213–214, 220, 416, 419, 426
trade balances, 426
trade conditions, 196, 220
trade effects, 206, 209
trade efforts, 206, 209, 220
trade in oil, 419
trade populations, 166
trade system, 196
trajectory in phase space, 168
transfer, 13–15, 80, 87, 92, 113, 136–137, 146, 280–285, 287, 291, 331, 345, 358, 379, 426, 458, 551, 553, 580, 591, 644, 648–649
transfer and degradation parameters, 281
transfer functions, 14–15
transfer of energy, 345
transfer of mass, 136
transfer of matter and energy, 13
transfer probabilities, 648
transfer rates, 283, 285
transformations, 4, 80, 110, 177, 190, 204, 226, 236–237, 243, 245–246, 249–250, 253, 255, 270, 337–338, 362–363, 374, 398, 404–406, 409, 479, 485, 512, 551, 553, 591, 612, 618, 644, 648
transient response, 231, 233, 238, 562
transition, 44, 63, 66, 142, 192, 200, 255, 284, 292, 326, 428, 430, 469, 596–597, 602, 614, 621–622, 626, 644
transition (succession) regime, 200
transition pathways, 326
transition probabilities, 192, 255, 626
transition process, 597
transition zone, 142, 284, 292, 469
transitional and stationary processes, 169
transitions, 140, 143, 325, 587, 619, 624, 644
transport and removal processes, 282
transport processes, 81
transport, flushing, and transformation, 270
transportation, 13, 444–445, 609
transportation demand, 444
transposition, 249, 253, 351
trawls, 69–70
tree density, 383

tree species, 77–78, 256, 377
trees, 43, 64, 77–78, 91, 94–95, 361, 403, 452, 456, 460, 462, 515, 553
trend, 16, 21, 47, 49, 52, 67, 69, 79–80, 82–86, 89–90, 94–97, 136, 308, 416, 421, 427–428, 430, 432, 435, 437, 439–442, 562, 620
Triaenophorus crassus, 75
Triaenophorus nodulosus, 75
triangular web structure, 539
tribute, 339, 567, 652, 654
trichlorobenzene, 284–285
Tridacna maxima, 70, 72
Trinidad, 116
trivial predation subgraph, 355
trophic aggregation, 512, 550, 552–553, 557–559
trophic aggregation scheme, 550, 553
trophic and resource generalization, 533
trophic apportionment, 551
trophic chain length, 359, 602–603
trophic chain stability, 361
trophic chains, 335–336, 344, 358–361, 365, 550, 552–553, 558, 563, 601–605, 607
trophic chains and communities, 335
trophic coefficients, 248
trophic compartments, 512, 550, 552, 557
trophic complexity, 376–377
trophic connections, 510, 514, 543
trophic continuum, 521
trophic-dynamic process, 553
trophic dynamics, 341, 509–510, 553, 559, 563
trophic efficiency, 558
trophic function, 247–248, 359, 361, 365–366, 559
trophic gradient, 116
trophic graph, 353
trophic interactions, 231, 264, 345, 358, 511, 513, 515, 518, 522, 524–525, 527–532, 534, 539–540
trophic interconnections, 543
trophic interrelations, 542
trophic-level population models, 105
trophic levels, 14–15, 40, 75, 105, 108, 114, 226, 238–239, 264, 347, 356, 358, 360–361, 365, 453, 462, 512, 514–515, 520–522, 524–525, 531–532, 539–540, 542, 550–553, 557–559, 601, 603, 605, 635
trophic linkage, 521, 541
trophic links, 516, 526, 540–541
trophic mutualists, 515
trophic organization, 138
trophic pathways, 553

trophic position, 551–552
trophic processes, 644
trophic pyramid, 108, 512, 520, 549, 558
trophic pyramid of herbivores, 520
trophic relations, 264, 510, 515–516, 522–523, 526, 532
trophic roles, 524
trophic simplicity, 511
trophic specialists, 523, 526
trophic specialization, 522
trophic species, 40, 515–516
trophic structure, 553, 605
trophic transfers, 511–512, 549, 557–559
trophic transformation matrix, 512, 551
trophic unfolding, 520, 540
trophic webs, 275, 349
tropical, 51, 73, 79–80, 89, 92, 112, 373, 377, 388, 469, 517–518, 538, 540
tropical ecosystems, 112
tropical island, 51
tropical rain forests, 79
tropical seasonal forest, 388
tropical systems, 80, 92
tropics, 73, 86, 89, 92, 94, 112
trouble-search algorithms, 175
troughs, 66, 76
true (reference) state, 233
trunk diameter, 77–79
trunk-diameter distribution, 77–78
trunk-diameter frequency, 78
Tucson, Arizona, 340, 468, 470
tundra, 30, 55, 94–96, 110, 388
turbidity, 402
turbulence, 286, 385, 406
turbulent diffusion, 45
turbulent energy, 45
turbulent flow, 286
turbulent mixing, 111
turbulent transport, 157
turnover, 46, 384
turnover of nutrients, 356
turnover of structures, 43
turnover rate, 22, 83, 92, 383–385, 387–389, 598
turnover structure, 385, 387
turnover time, 83–84, 91–92, 384
two-layer boundary system, 284
two-layer volatilization model, 284
two-level model, 175, 177, 181
two-species models, 106–107, 109, 458
types, 10, 13–14, 16, 18, 21–22, 43, 53, 109, 116, 134, 138, 143, 200, 204, 226, 232, 238, 242, 298–299, 301, 313, 319, 324–330, 336, 343–344, 347, 354, 358, 361, 365, 376–377, 382,

types (*cont'd*), 384–385, 387–388, 398–399, 401, 409, 416–419, 422, 433, 445, 450, 454, 456, 461, 463, 477, 479, 487–488, 491–492, 498, 515, 522, 525, 532, 593, 633–634, 638, 644

U.S. Army Corps of Engineers, 324
U.S. births, 434
U.S. Department of Commerce, Bureau of the Census, 421
U.S. dollars, 435
U.S. Environmental Protection Agency, 282
U.S. Fish and Wildlife Service, 311, 318, 324, 331
U.S. fossil fuel resource, 431
U.S. Geological Survey gaging stations, 324
U.S. oil discovery costs, 446
U.S. population, 437
U.S. wholesale price index, 419, 438
Ulanowicz's ascendancy, 501
uncertainty, 29, 32, 34–36, 38–39, 44, 49, 87, 96, 225, 227, 231, 280–281, 283, 291–292, 297, 307, 564–567, 571, 616, 618, 621, 626, 645–647
uncertainty in time, 35
uncertainty limits, 36
uncertainty of model structure, 291
uncertainty principle, 29, 32, 39, 87, 225
uncertainty relation, 34–35, 38
understory plants, 91
unemployment, 420, 443, 446
unemployment rates, 446
ungulate populations, 9
ungulates, 461
uniform age, 77
uniform distribution, 42, 80, 237
uniform size, 31, 51, 55, 79, 84
uniformity, 31, 42, 51, 55, 57–58, 60, 67, 75, 77, 79–80, 82–85, 87, 89–90, 94, 122, 192, 235, 237, 457, 480, 484, 491
uniformity in size, 79, 85, 94
unimodal distribution, 76
United Kingdom, 30, 442
United States, 30, 69–70, 84, 297, 324, 338, 408, 418–419, 421–428, 430–433, 437, 441, 443, 470, 529
universities, 21–23
unlimited existence (immortality), 175
unnatural relation, 352
unstable regime, 141
unstable stationary states, 161
unstressed ecosystems, 337
upland rain forest, 30, 77–78

Upper Twin Cave, 70
uptake, 80, 133, 138, 143–144, 146–147, 149–151, 154, 164, 265–267, 273–274, 282, 358, 391, 572, 575, 578, 580, 590, 603
uptake rate, 149–151, 164, 267, 578, 590
urban economy, 342, 499–500, 502, 504
urban watersheds, 401
urbanization, 338, 398, 405, 409
user interface, 331
Uta stansburiana, 531
Uta strasburiana, 224
utilizable resources, 93
utilization, 7, 69, 84, 93, 166, 216, 309, 327, 362–364, 474, 647

vacancy rate, 428
Vaejovis confusus, 531
Van Dyne philosophy, 229
Van Dyne systems, 338
Van't Hoff's reaction isobar, 148
vapor pressure, 281, 283, 290
variability, 31, 36, 39, 85, 89, 94, 110, 114, 119–120, 172, 226, 231, 235, 280–282, 292, 307, 338, 361, 399, 401, 563–565, 569, 571, 615–618
variable age, 51, 57
variable aggregation methods, 260
variable life span, 51
variable naming, 314
variable structure, 114, 564
variable structure interactions, 564
variation, 37, 58, 84, 135–136, 231, 337, 361, 376, 380, 421, 430, 439–440, 491, 517–518, 535, 564–565, 569–570, 586, 596, 599–600, 603, 605, 613–614, 616, 619
variational principle, 81, 597
vascular plants, 263, 265, 270, 272, 387, 510, 516
vector space, 106
vegetation, 4–7, 9, 77, 110–111, 313, 317–318, 324–326, 375, 388, 398, 404, 406, 418, 454, 456, 458, 462, 524, 557, 594, 601
vegetation distribution, 375
vegetation dynamics, 313
vegetation strategies, 456
vegetation submodel, 325
vegetation types, 326
vehicle propulsion, 446
velocity, 38, 54, 89, 164, 285–286, 289, 291
verdin, 523
vertebral scute, 73
vertebrate grazing, 377, 382

vertebrates, 73, 76, 84, 377, 382, 510, 515–516, 518, 522–524, 526, 528–535, 539–541, 622
vertical movements, 45
vertically structured communities, 336, 344, 358
viability of biological systems, 185
virtual memory capability, 316
viruses, 43, 108
volatile chemicals, 290–291
volatile compounds, 279, 290
volatility, 279, 284, 288, 290–292
volatilization, 227, 279–280, 282–284, 286–289, 291–292
volatilization models, 279, 284, 291
volatilization of chlorobenzenes, 292
volatilization process, 291–292
volatilization rates, 283, 286–289, 291–292
volatilization submodel, 283
volcanic eruptions, 433, 445
Volterra dissipativeness, 348
Volterra equations, 236, 453
Volterra form of trophic functions, 366
Volterra models, 104, 344, 346, 358
Volterra prey–predator systems, 349
Volterra systems, 347, 350
Volterra type models, 346, 563, 601, 620
volume, 11, 131–132, 134–136, 140–141, 154, 165, 190, 215–216, 231, 237, 258, 282–283, 285–287, 324, 341, 343, 416, 423, 426, 432, 437, 444, 476, 479, 481, 533, 580, 586, 590
von Bertalanffy's formula, 221
von Bertalanffy's formula for juvenile individuals, 221
von Karman constant, 287
Vulpes macrotus, 531

wage index, 438
wages, 422–425, 429–430, 434–436, 439–441, 443, 446
war, 11, 19, 339, 415–421, 425–428, 432–435, 437, 439–446
war cycles, 339
war deaths, 420, 425, 433, 441, 443, 446
war debt, 339, 420, 425, 433, 435, 439–441, 443, 445–446
Washington, 262, 295, 398
wasps, 469, 526–527, 529, 532, 535
waste disposal, 264
waste flows, 498
waste heat, 487–488, 490, 495–496, 499
waste heat production, 490

waste inputs, 263
wastes, 5, 9, 263–264, 270, 469, 487–488, 490, 494–496, 498–499
water, 5–6, 8, 23, 42, 55, 57, 70, 82, 134, 138, 141, 143, 150, 152, 155, 227, 236, 242, 258, 263–268, 270–271, 273–274, 279, 281–289, 291–292, 313, 317–319, 324–327, 329, 331, 342–343, 361, 387, 401, 403–404, 406–407, 460, 468–469, 471, 487, 498–503, 521–522, 571–573, 578–580, 599, 601, 653
water birds, 57
water body, 138, 242, 279, 282, 287–288, 404, 599, 601
water boundary, 284
water budgets, 324
water chemistry, 313
water column, 265–267, 274, 318, 521–522
water column nutrients, 274
water exchange, 268
water intakes, 404
water management, 317–318, 331
water molecule, 236
water organisms, 138
water quality, 141, 143, 150, 263, 270–271, 273, 318–319, 331, 578–579
water quality modeling, 143, 150
water quality models, 141
water solubility, 281, 283
water surface, 265
water temperature, 292
water vapor, 285, 499–500, 502–503
water viscosity, 286
water volume, 258, 287
waterfowl, 226, 263, 317–319, 324, 331, 408
waterfowl production habitat, 324
watershed, 17, 263, 270, 272, 274, 297, 317–319, 382–383
watershed basins, 297
watershed manipulation experiments, 382
watershed-scale clear-cutting experiments, 382
watersheds, 4, 47, 297, 383, 401, 410
wave action, 220, 398
wave breaking, 289
wave generating mechanism, 420, 428
wavelength, 42, 419, 421, 425, 433, 444

weakly stable, 374–375
weapons, 97, 298, 443
weather, 115, 169, 171, 296, 302
weather conditions, 115, 302
web matrix, 538
web of trophic interactions, 358
web patterns, 511, 538, 541
web structure, 515, 539, 565, 616
webs, 7, 275, 347, 358, 453, 464, 510–522, 524, 529, 532, 534–536, 538–543, 550–551, 553, 557, 559, 564–565, 572, 614, 616, 635
weed control, 275
weeds, 209, 218
weevils, 517
weighted, 196, 235, 490–491, 494, 550, 562, 565, 630–631, 646
weighted aggregation, 630
western bluebird, 523
western tanager, 523
wet meadow, 325
wet season, 73
wet tropics, 73, 86, 89
wetland ecology, 313
wetland vegetation, 326, 398
wetlands, 36, 319, 398, 404–405, 520
whales, 116
white blood corpuscles, 189
white cedar, 79
white suckers, 30, 75–76, 96
white-tailed deer, 301
whole organism, 450
whole-plant uptake, 274
whole-scale condition, 252
wholesale price index, 419, 421, 430, 435–436, 438
wildfire, 337, 377, 380, 382, 387
wildfire suppression, 382
wildland resources, 14
wildlife, 4, 228–229, 311, 313, 317–321, 324–326, 330–331, 418
wildlife calculations, 325
wildlife cycle, 418
wildlife management, 313, 320
wildlife populations, 325
wildlife refuges, 331
wildlife resources, 228, 311, 319–320
wildlife species, 320
wind, 82, 274, 280, 283–286, 288–292, 453
wind conditions, 291–292
wind currents, 453
wind-induced turbulence, 286

wind mixing, 274
wind speed, 283, 284–286, 288–289, 291
wind tunnel, 286, 291
wind velocity, 285–286, 288–289, 291
windstorms, 380, 387
Winnipeg, 51, 75, 97
winter, 56, 76, 259, 296, 298, 303, 304–306, 319–321, 323, 325, 327
winter death loss, 305
winter feeding grounds, 321
winter mortality, 303, 323, 325
winter starvation, 296
winter temperatures, 56, 259
withdrawal regions, 214
withdrawals, 202–203, 206, 209, 212, 214, 218
wolf spiders, 533
wood, 312, 422, 520, 523
woodland–shrubland, 388
woodlands, 76
woody species, 387, 398
workshop modeling, 311–312, 326–328, 330–331
workshop simulation models, 228–229, 311–312, 316, 326–327, 329
world depressions, 434
world economy, 339, 416
world energy market, 431, 434, 445
world society, 419, 430
World War I, 428, 432–433, 440, 443
world wars, 419, 428
Wyoming, 229, 320
Wytham Woods, 76, 85

yearling bulls, 321
yearling females, 298, 303–306
yearling males, 298, 303–304, 306
yearlings, 303, 305–308
yeast, 524
Yellowstone National Park, 9, 320–321
yield losses, 211
yields and energy flows, 361
Yoldia thraciaeformis, 69–70
Yosemite National Park, 9

Zaire, 30, 51, 68–69
Zaire caves, 69
zero equilibrium, 593
zero error, 233–234, 238
zooplankton, 108, 156–157, 161, 165, 242, 258, 271, 515, 520, 541, 572, 575, 580–582, 587